SOUTHEAST ASIA

SOUTHEAST ASIA
Diversity and Development

Editors:

Thomas R. Leinbach
Richard Ulack

University of Kentucky

Cartographic Design and Production:

Richard Gilbreath & Donna Gilbreath

Gyula Pauer Cartographic Information Laboratory
University of Kentucky

Prentice Hall
Upper Saddle River, New Jersey 07458

Library of Congress Cataloging-in-Publication Data

Leinbach, Thomas R.
Southeast Asia: diversity and development / Thomas R. Leinbach, Richard Ulack
p. cm.
Includes bibliographical references and index.
ISBN 0-13-825126-0
1. Regionalism—Asia, Southeastern. 2. Asia, Southeastern—Economic integration. 3. Asia, Southeastern—Economic policy. I. Ulack, Richard– II. Title.

HT395.A85 L45 2000 99-047022
338.959—dc21 CIP

Executive Editor: Daniel Kaveney
Executive Managing Editor: Kathleen Schiaparelli
Art Director: Jayne Conte
Cover Designer: Donna Gilbreath
Cover Photos: Thomas R. Leinbach and Richard Ulack
Manufacturing Manager: Trudy Piscotti
Buyer: Michael Bell
Editorial Assistant: Margaret Ziegler
Production Supervision/Composition: WestWords, Inc.

Printed in the United States of America

10 9 8 7 6 5 4 3 2 1

ISBN 0-13-825126-0

Prentice-Hall International (UK) Limited, *London*
Prentice-Hall of Australia Pty. Limited, *Sydney*
Prentice-Hall Canada Inc., *Toronto*
Prentice-Hall Hispanoamericana, S.A., *Mexico*
Prentice-Hall of India Private Limited, *New Delhi*
Prentice-Hall of Japan, Inc., *Tokyo*
Prentice-Hall (*Singapore*) Pte. Ltd.
Editora Prentice-Hall do Brasil, Ltda., *Rio de Janeiro*

To our spouses, Marge and Karen, and our children, Amy, Jessica, and Christopher, without whose love, support, encouragement, and presence during our field "excursions" would have made this effort much more difficult.

Contents

Preface and Acknowledgments

In 1994, after a number of conversations about the need to provide a book that would convey our excitement about Southeast Asia to a broad audience, we finally took the plunge. Our initial plan was to author this work ourselves but we quickly became aware of the massive task this would entail given the wide range of issues and depth of material, ideas, and concepts which needed to be covered. Moreover through our careers we have worked with numerous geographers who have developed specific interests in Southeast Asia and who would be able to write more competently on particular topics or countries. Consequently we decided to produce an edited work. A number of active and expert researchers and teachers, particularly in Southeast Asia, could have written essays either on specific countries or systematic themes but are not represented as authors. We trust nonetheless that they will read and use this work and offer constructive comments to us.

As we began to conceive and map out this project, the countries of Southeast Asia with only several exceptions were among the most vibrant and dynamic in the developing world and appeared to be models of development promising a bright future. This picture maintained for much of the writing period. Then in mid-1997 the Asian "economic flu" appeared—first in Thailand and subsequently diffused through the region. We obviously were obliged to change our generally positive script to account for this unfortunate series of events. However, we felt it inappropriate, and indeed almost impossible, to rewrite the materials, which had been completed to that point. In fact both of us remained confident that it was simply a matter of time before the economies would rebound. In August 1999 while still a bit early to say that the corner has been turned, there is abundant evidence to suggest that most of the economies are on the mend. Having noted this, it is clear that the dramatic rates of growth and progress witnessed during the previous two decades will be much more difficult to attain in the future. This is so for a variety of reasons ranging from political circumstances to the simple fact that there is as a result of enabling mechanisms (transport and communications), the reach of the multi-national firm, and the spread of liberalization policies much keener competition and a more even playing field among the more progressive nations than previously.

Our appetites for Southeast Asia were whetted by the writings of Charles Fisher and Donald Fryer among other scholars. We hope this book captures some of the flavor, insights and scholarship exemplified in the writings of both of these pioneer scholars in the region. As our interests emerged Peter Gould and Fred Wernstedt influenced us in different ways at the Pennsylvania State University. We owe them an immense debt of gratitude for their teaching over the years. We were bitten by the Southeast Asian "bug" early and have never regretted investing virtually the whole of our careers in this piece of geography. We greatly appreciate the experiences we have had while carrying out research and teaching in the region over the past thirty years. Especially important to us are the Southeast Asian colleagues with whom we have worked and continue to collaborate. Needless to say added bonuses have been the opportunity to meet, interact, and debate with fascinating people from all walks of life, travel widely throughout the region, and sample the spectacular cuisines. Regrettably or perhaps fortunately we have not dwelt on the latter although they clearly do deserve special attention.

As always it is impossible to thank everyone who has assisted, befriended, housed, and aided us over the years in connection with our learning experiences but we would like to mention several individuals and institutions. In Malaysia, the University of Malaya and the Science University of Malaysia (Universiti Sains Malaysia) provided bases for one of us (TRL) in initial forays for dissertation and follow-up research. Subsequently in Indonesia, the University of Indonesia, University of North Sumatra, Gadjah Mada University, Sriwijaya University in Palembang and University of Pattimura in Ambon, Moluccas have been helpful in a variety of ways. The latter two especially served as bases for research on Indonesia's transmigration program. In addition the National University of Singapore, especially through the Department of Geography and the Centre for Advanced Studies, has been generous in offering support and a venue for discussions of scholarship. Also the Institute for Southeast Asian Studies in Singapore has been a rich mine of information and we are grateful for the opportunities to use this facility. In the Philippines, three institutions deserve special mention. Affiliation with the Department of Geography at the University of the Philippines-Diliman through a Fulbright award opened up many doors and has proved to be a long-lasting friendship. More recently, affiliation and collaboration with scholars at Xavier University in Cagayan de Oro City and the University of San Carlos in Cebu City has afforded research and learning opportunities that would not have otherwise been possible.

Lastly we wish to gratefully acknowledge the financial assistance provided over the years by the National Science Foundation, Fulbright-Hays, the Ford Foundation, the International Development Research Centre (Canada), the National Geographic Society, and especially the University of Kentucky.

We wish to give special praise and thanks to Richard and Donna Gilbreath in the Department of Geography's Cartographic Laboratory for their spectacular graphics, maps, and editorial assistance in producing this volume. A major portion of its worth is due to their talents. We also want to acknowledge those geography students who, under the guidance of University of Kentucky Cartography Lab director Dick Gilbreath, assisted in producing the graphics that appear in this volume. Those individuals include former graduate student Stephen Hanna and undergraduate majors Ryan Adcock, Joseph David, Brandon Jett, David Lafferty, James Martin, Taylor Vasek, Matthew Wagoner, and Reid Webb.

We are grateful to the six reviewers who read and commented on this work. Excellent and extensive comments, especially from geographers Ralph Lenz and Jack Williams, were very helpful in improving the final product.

Finally we truly hope that in some small way this work will repay the confidence and investment that institutions and individuals have made in our scholarship. It is also our hope that this work will truly excite more research by both students and faculty and expose this spectacular region to a much larger audience.

Thomas R. Leinbach
Richard Ulack
Lexington, Kentucky

Contributors

EDITORS

Thomas R. Leinbach, Professor, Department of Geography, University of Kentucky, Lexington, KY 40506-0027
email: *leinbach@pop.uky.edu*

Richard Ulack, Professor, Department of Geography, University of Kentucky, Lexington, KY 40506-0027
email: *ulack@pop.uky.edu*

CONTRIBUTORS

Christopher A. Airriess, Associate Professor, Department of Geography, Ball State University, Muncie, IN 47306-0470
email: *cairries@wp.bsu.edu*

John T. Bowen, Jr., Assistant Professor, Department of Geography, University of Wisconsin, Oshkosh, WI 54901-8642
email: *BowenJ@vaxa.cis.uwosh.edu*

Jessica Rothenberg-Aalami, Ph.D. candidate, Department of Geography, University of Oregon, Eugene, OR 97403

Carolyn L. Cartier, Assistant Professor, Department of Geography, University of Southern California, Los Angeles, CA 90089-0255
email: *ccartier@usc.edu*

Cecile Cutler, Department of Geography, Flinders University of South Australia, GPO Box 2100, Adelaide 5001, Australia

Vincent J. Del Casino, Jr., Ph.D. candidate, Department of Geography, University of Kentucky, Lexington, KY 40506-0027
email: *vjdelc1@pop.uky.edu*

Dean Forbes, Professor, Department of Geography, Flinders University of South Australia, GPO Box 2100, Adelaide 5001, Australia
email: *dean.forbes@flinders.edu.au*

Jon Goss, Associate Professor, Department of Geography, University of Hawaii at Manoa, Honolulu, HI 96822
email: *jgoss@hawaii.edu*

James A. Hafner, Professor, Department of Geosciences, University of Massachusetts, Amherst, MA 01003-0026
email: *hafner@geo.umass.edu*

Graeme Hugo, Professor, Department of Geography, University of Adelaide, The University of Adelaide, Box 498, GPO, Adelaide 5001, Australia
email: *ghugo@arts.adelaide.edu.au*

Robert E. Huke, Professor Emeritus, Department of Geography, Dartmouth College, Hanover, NH 03755-3571
email: *Robert.E.Huke@dartmouth.edu*

David M. Kummer, Assistant Professor, Department of Social Sciences, SUNY Westchester Community College, Valhalla, NY 10595-1698
email: *david.kummer@sunywcc.edu*

Thomas R. Leinbach, Professor, Department of Geography, University of Kentucky, Lexington, KY 40506-0027
email: *leinbach@pop.uky.edu*

Robert R. Reed, Associate Professor, Department of Geography, University of California, Berkeley, CA 94720-4740
email: *rrreed@garnet.berkeley.edu*

Richard Ulack, Professor, Department of Geography, University of Kentucky, Lexington, KY 40506-0027
email: *ulack@pop.uky.edu*

Acronyms

ADB	Asian Development Bank
AFTA	ASEAN Free Trade Area
AIDS	acquired immuno-deficiency syndrome
ALS	Area Licensing Scheme (Singapore)
APEC	Asia-Pacific Economic Cooperation
ASEAN	Association of Southeast Asian Nations
BKKBN	National Family Planning Co-ordinating Board (Malaysia)
BMR	Bangkok Metropolitan Region
BN	*Barisan Nasional*; English translation= National Front (Malaysia)
BOI	Board of Investment (Thailand)
BPS	*Biro Pusat Statistik* (Indonesia)
CARP	Comprehensive Agrarian Reform Program (Philippines)
CAS	Country Assistance Strategies (World Bank)
CAW	Committee for Asian Women
CBD	central business district
CBR	crude birth rate
CDR	crude death rate
CEDAW	Convention on the Elimination of All Forms of Discrimination Against Women
CIA	Central Intelligence Agency (U.S.)
CMEA	Council for Mutual Economic Assistance
CPF	Central Provident Fund (Singapore)
CPM	Communisty Party of Malaysia
CPP	Cambodian People's Party
CPP	Communist Party of the Philippines
DAWN	Development Alternatives with Women for a New Era
EAEC	East Asia Economic Caucus
EAEG	East Asia Economic Grouping
EBMR	Extented Bangkok Metropolitan Region
EEZ	Exclusive Economic Zone
EIU	Economic Intelligence Unit
EMR	extended metropolitan region
ENGENDER	Center for Environment, Gender, and Development
EOI	export-oriented industrialization
EPZ	export-processing zone
ERP	Electronic Road Pricing (Singapore)
ESCAP	*see* UNESCAP
EU	European Union
FAO	*see* UNFAO
FDI	foreign direct investment
FELDA	Federal Land Development Authority (Malaysia)
Fretilin	in Portuguese, *Frente Revolutionaria de Timor-Leste Independete*; English translation=Revolutionary Front for an Independent East Timor
Funcinpec	in French, *Front Uni National pour un Cambodge Indépendant Nutre, Pacifique, et Coopératif*; English translation= National United Front for an Independent, Neutral, and Cooperative Cambodia
FTZ	free trade zone
FVP	Forest Village Program (Thailand)
GAD	gender and development
GATT	General Agreement on Trade and Tariffs
GDI	gender-related development index
GDL	gendered division of labor
GDP	Gross Domestic Product
GEM	gender empowerment measure
GLD	Guided Land Development (Indonesia)
GNP	Gross National Product
GRDP	Gross Regional Domestic Product
GRC	Group Representation Constituencies (Singapore)
GT	growth triangle
HDB	Housing Development Board (Singapore)
HDI	human development index
HIV	human immuno-deficiency virus
HYV	high-yielding varieties (of rice)
IBRD	International Bank for Reconstruction and Development
ILO	International Labor Office
IMF	International Monetary Fund
IPM	integrated pest management

IPTN	*Industri Pesawat Terbang Nusantara*; English translation=National Aircraft Industry (Indonesia)
IRRI	International Rice Research Institute (located in Los Baños, the Philippines)
ISI	import-substitution industrialization
IVDU	intravenous drug user
KIP	Kampung Improvement Program (Indonesia)
KL	Kuala Lumpur
KLIA	Kuala Lumpur International Airport
KTM	*Keratapi Tanah Melayu* (Malaysia's government-operated railway)
KVC	Klang Valley conurbation
LDC	less-developed country
LICADHO	Cambodian League for the Promotion and Defence of Human Rights
LKAAM	*Lembaga Kerpatan Adat Alam Minangkabau*; English translation= Association of Adat Councils of the Minangkabau World
LMW	licensed manufacturing warehouse
LNG	liquified natural gas
LPDR	Lao People's Democratic Republic
LPRP	Lao People's Revolutionary Party
LRT	light rail transit
MCA	Malayan Chinese Association
MDC	more-developed country
MIC	Malaysian Indian Congress
MILF	Moro Islamic Liberation Front (Philippines)
MISC	Malaysian International Shipping Corporation
MNC	multinational corporation
MNLF	Moro National Liberation Front (Philippines)
MP	member of parliament
MRT	Mass Rapid Transit
MV	modern variety (of rice)
NAFTA	North American Free Trade Agreement
NDP	National Development Policy (Malaysia)
NEP	New Economic Policy (Malaysia)
NEZ	New Economic Zone
NGO	non-governmental organization
NIDL	new international division of labor
NIE	newly-industrializing economy
NLD	National League for Democracy (Burma)
NPA	New People's Army (Philippines)
NPP	New Population Policy (Malaysia)
OCW	overseas contract worker
ODA	Official Development Assistance
OECD	Organization for Economic Cooperation and Development
OPEC	Organization of Petroleum Exporting Countries
OPM	*Organisasi Papua Merdeka*; English translation=Free Papua Movement
PAL	Philippine Airlines
PAP	People's Action Party (Singapore)
PAP	Poverty Alleviation Program (Thailand)
PAS	*Partai Islam Sa-Malaysia*; English translation=Islamic Party of Malaysia
PE	public enterprise (Malaysia)
PKI	Communist Party of Indonesia
PNG	Papua New Guinea
PSR	Port of Singapore Authority
PVO	private voluntary organization
R & R	rest and recreation
RDA	Regional Development Authority (Malaysia)
RJCP	Rural Job Creation Program (Thailand)
SAe	Singapore Aerospace
SAP	structural adjustment policy
SDU	Social Development Unit (Singapore)
SEATO	Southeast Asia Treaty Organization
SIA	Singapore Airlines
SLORC	State Law and Order Restoration Council (Burma)
SMI	small- and medium-sized industries
SOE	state owned and operated enterprise
SPDC	State Peace and Development Council (Burma)
SRT	State Railway of Thailand
STD	sexually-transmitted disease
TAT	Tourist Authority of Thailand
TFR	total fertility rate
UMNO	United Malays National Organization (Malaysia)
UNDP	United Nations Development Program
UNESCAP	United Nations Economic and Social Commission for Asia and the Pacific
UNFAO	United Nations Food and Agricultural Organization
UNHCR	United Nations High Commission for Refugees
UNIDO	United Nations International Development Organization
UNTAC	United Nations Transitional Authority in Cambodia
VCP	Vietnam Communist Party
VOC	*Vereenigde Oost-Indisch Compagnie*; English translation=United East India Company
VWU	Vietnam Women's Union
WHO	World Health Organization
WID	Women in Development
WTO	World Tourism Organization

1 An Opening View

RICHARD ULACK THOMAS R. LEINBACH

As we approach the twenty-first century, there is a heightened interest in the Third World where traditional modes of production and economic, cultural, social, and political relations have been experiencing dramatic change. Especially as a result of communications and information technology, external economic linkages, political alliances, and the imposition of institutional reform, new approaches to development have begun to filter through the global structure. This is true in Africa and Latin America but is most intense in Asia. The driving forces of the need to improve livelihoods and to develop competitive advantage is being felt at all spatial scales in a variety of economic, social, and political ways. This has come about in part by a continuing round of crises related to political processes, cultural conflicts, and agricultural capabilities in various parts of the world. Also important is the increasing global interdependence brought about by the uneven distribution of resources and the emergence of an international division of labor. Furthermore, there is increased recognition and concern with problems related to uneven development at the global, national, and regional scales. It is clear that before we can appreciate and understand the differing responses to these forces of change and to understand more completely the reasons for uneven development, we need to thoroughly appreciate the physical, historical, cultural, social, and economic bases that give rise to such problems (Dixon and Smith, 1997; Rigg, 1997). We propose to accomplish this in the context of one of the world's major regions: Southeast Asia. Through examination of the character and great diversities in the region and its individual countries, we assess the past and present patterns of development in the region. Emphasis is on the key forces affecting and producing current developmental change as the individual states seek to gain footholds in the international economy. In short, the key (and closely related) themes to be used throughout this text are: "Diversity" and "Development."

The Southeast Asian region represents an ideal regional laboratory in which to examine development and its human and physical bases. The region, which contains the 10 countries of Brunei Darussalam, Burma, Cambodia, Indonesia, Malaysia, Laos, the Philippines, Thailand, Singapore, and Vietnam, displays a wide range of development settings and conditions both between and within states. These range from the chaos and isolation prompted by political forces in Burma, Laos, and Cambodia, to the economic renewal (*doi moi*) in Vietnam, to the high level of technological and trade-driven economic development in Singapore, a city-state that has been able to capitalize uniquely on its physical situation and human resources. In short, although there are clearly areas of stagnation and backwardness, the development picture and prospects for the future in most of the region, notwithstanding the current economic downturn, is quite bright. The countries within ASEAN (Association of Southeast Asian Nations) including Brunei Darussalam, Indonesia, Malaysia, the Philippines, Thailand, Singapore, Vietnam, and most recently Burma, Cambodia, and Laos have recorded, with the exception of the newest members, strong recent economic performances. Indeed, this Asian "common market" may serve as a model for other developing regions.

Yet even in those countries where economic progress has been recorded and ostensibly the equality of income distribution has increased, there exist many problems related to poverty and spatial inequality. Underlying the conspicuous symbols of "economic boom" is a clearly perceived wealth gap (Robison and Goodman, 1996). Along with this is a democratization movement where new political and humanitarian concerns are being expressed (Vatikiotis, 1996b; Dixon and Smith, 1997). These gaps in wealth and political expression have produced significant unrest in Burma, Indonesia, the Philippines, Thailand, and recently Malaysia. The protesters are farmers who have been displaced to make way for dams, upland minority groups affected by forestry concessions, Chinese who have been targets of resentment, and others affected by development forces. There are slum dwellers caught in construction frenzies and those whose health has been adversely affected by the race for economic gains in textile factories and elsewhere. Increasingly, groups and individuals are speaking out with views contrary to those of the well-entrenched, often authoritarian political leaders and their parties, as well as local

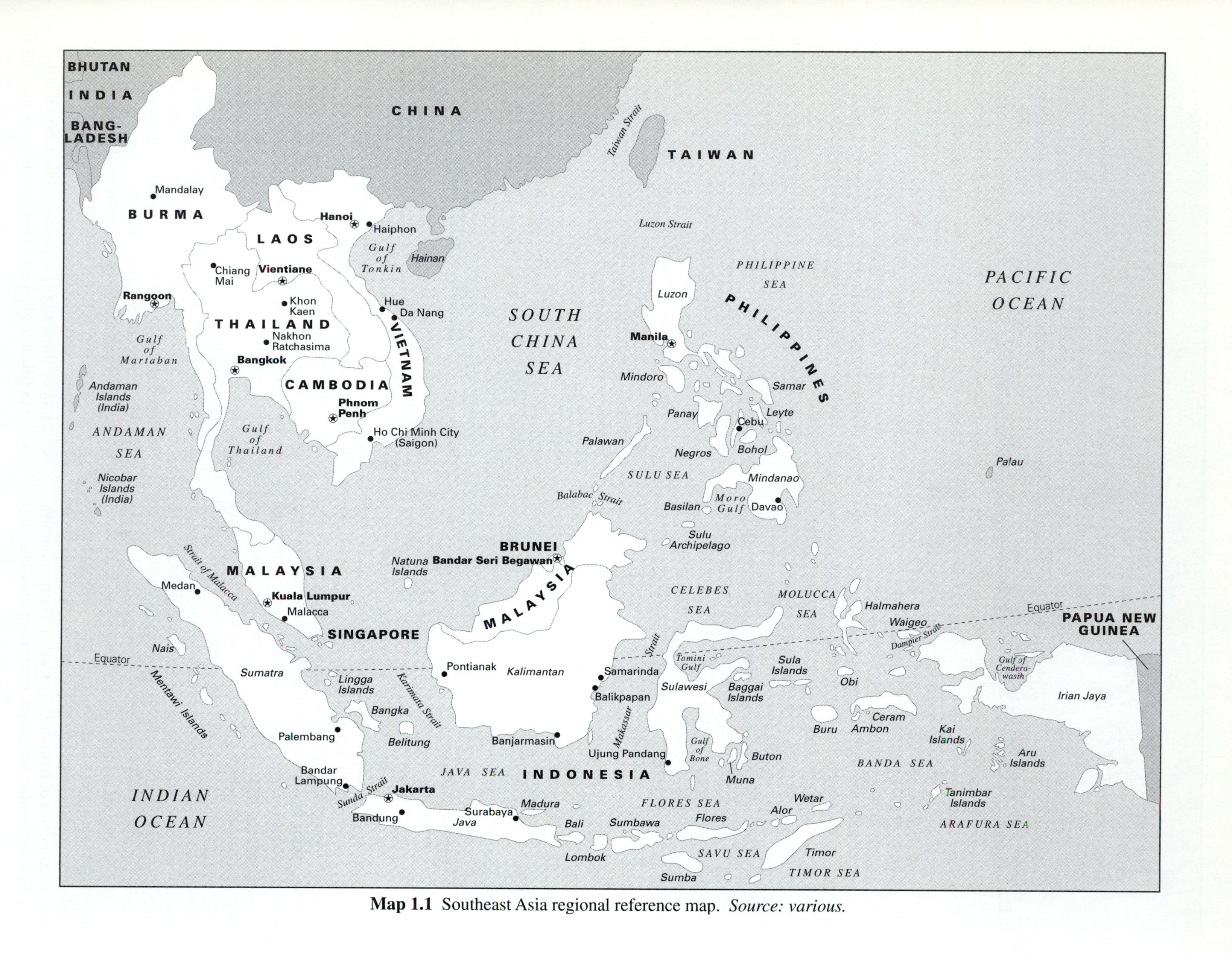

Map 1.1 Southeast Asia regional reference map. *Source: various.*

elites, and are demanding a greater role in governance. The depth of these struggles has in fact become enlarged. A major purpose of this volume will be to identify the diversity of problems that exist, at the transnational, national, subnational, and local scales. Thus, for example, one aspect of this work will be to examine issues related to the differences between the regional and national cores and the periphery to help us unravel problems. We also explore the concept of liberalization and associated policies as well as social and cultural implications of development. An aim will be to examine development as a product of, and its impact on, human capital and physical resource diversity and capability.

Our intent is to portray the very diverse physical and human landscapes that are present in the region. This involves to some extent a historical analysis of the human patterns we currently observe. Dominant elements in this portrayal will be the indigenous cultures, external forces, population and settlement characteristics, and economic patterns. Simultaneously we wish to convey the roots and nature of present-day development problems, progress, and issues that face Southeast Asia. Our emphasis will be on the contemporary period, but, clearly, to understand these current patterns we must spend some time examining the past.

THE CONCEPT OF THE REGION

The meaning behind and understanding of the term *the Southeast Asian region* is intriguing and somewhat elusive for it connotes many things to many people and rightfully reflects one's personal experiences and interests. For example, for many—including the editors—Southeast Asia's meaning lies in the appreciation and diversity of specific landscapes, cultures, peoples, and cuisines that are part of our knowledge of Indonesia and the Philippines. To others, the defining characteristic may be the recent dynamics of her numerous economies that capture the identity of the region. Prior to World War II, these countries were best known through their colonial affiliations and were, first and foremost, sources of agricultural goods, resources, and minerals that were extracted from mines and plantations. In fact, all of the countries, save one—Thailand—experienced this relationship.

Quite apart from individual and collective perceptions, it is fairly clear that the concept of Southeast Asia as a geographical region is relatively recent and in fact came into usage first during World War II as the Western powers were concerned with military operations given the Japanese invasion of the area (Fisher, 1971). The expression *Southeast Asia* was adopted and gradually came to prevail as a term to define the peninsulas between India and China, as well as the Indonesian and Philippine archipelagoes. In the historical context the region was distinctive because of its location lying between India and China, two areas with huge populations and quite distinctive cultural characteristics. Thus Southeast Asia was positioned to serve as a crossroads over and through which trade and the movement of peoples, ideas, and innovations passed. In many respects this is still true in the contemporary period. Part of Southeast Asia's allure and identity lies in the cultural attributes that have been acquired as a result of this unique spatial situation. Indeed it is this cultural richness that continues to draw tourists into the region in ever increasing numbers—a fact of considerable consequence for development. Yet notwithstanding this aspect, today Southeast Asia has developed and matured so that the region has become an attractive destination for capital investment and, increasingly, labor opportunities that lure migrants from within and outside the region. As a result of the rapid maturation in development, the countries have gained a new political status and are increasingly major actors in international politics and the global economy. This is remarkable, in fact, for a region with a mere 1.5 million square miles (4 million square kilometers) or less than 3 percent of the landmass of the globe, less than 9 percent (more than 500 million) of its total population, and where most countries have only recently gained independence.

Traditional Perspectives

Much of the discussion of Southeast Asia has emphasized its uniqueness on ethnic–linguistic, cultural, and even environmental grounds. The region has been portrayed as a complex human and physical mosaic that makes it distinctive. Moreover, it has been suggested that in the past, individual countries' connections beyond the region have often been of greater importance than those to neighbors within the region. Furthermore, independence brought the realization that the basic cohesion, thought so desirable, was missing. The diversity of ethnic groups, the dominance and political power of particular groups—for example the Javanese in Indonesia—and the extraordinary accumulation of wealth by still others (the Chinese) continues to be felt in a variety of disruptive ways (Brown, 1994). Outside ties in respect to particular countries do remain quite strong. Indonesia, Malaysia, the Philippines, and Vietnam, for example, all maintain well-defined trade and tourism linkages that reflect their former colonial associations. Yet over time there has been a strengthening of intranational and intraregional complementarities which bind the countries and the region together. The growth of ASEAN reflects this in part. In many ways the elements of centrifugality, the notion of the region as a shatter zone and the views of uniqueness and diversity as a divisive force have abated considerably, although they have not disappeared. Rather, attention has come to focus on new forces and concerns as the pace of development has quickened and the region has become more entwined in the global economy. The characterization of national development patterns into a dichotomy of "progress" and "stagnation" is much less accurate than it was 20 years ago (Fryer, 1979). Certainly Burma, Laos, and Cambodia are trailing the development pace seen elsewhere in the region, but the Indonesian and Vietnamese situations reflect dramatic change. Still another development anomaly has been the Philippines, where after a recent period of stagnation in the 1980s, positive aspects have emerged and growth appears to be on the ascent once again. Although clearly some old attitudes and development complexities remain, contemporary Southeast Asia reflects a much different situation and presence than that of even 20 years ago. Environmental, economic, political, cultural, and societal factors as they are reflected in and impinge on development in local-, regional-, and global-scale perspectives, have come to take center stage. The intent and thrust of this new volume is to capture these concerns and forces that have critical impacts on the countries of the region.

Contemporary Forces

Twenty years ago the countries of Southeast Asia were hailed, along with the "newly industrializing countries" of Asia and Latin America, as the bright lights of the international economy. As we will see this has, in part, come true, but it has not occurred without setbacks and the need for some critical policy decisions that have had strong impacts on people, institutions, and environment-resources. As a result of economic downturns, including mounting indebtedness in the 1980s, increased pressure was applied to the states of the region to engage in major structural adjustments of their economies. Essentially this has meant redrafting priorities and improving efficiency in governance and the production of national revenues. Since the 1970s there has been a gradual shift in industrial emphasis from import- to export-oriented industrialization (both to be discussed fully in Chapter 7). Simultaneously there has been a greater internationalization of capital reflected in the growth of multinational corporations and the emergence of borderless economies. Basically a rise in wages in major industrial countries, together with increased productivity and technological advancements have allowed the disaggregation of some important production processes. Yet the cheap-labor advantage that the Southeast Asian countries once had has disappeared (with a current exception perhaps of Indonesia, given the effects of the recent financial crisis in that country), and now new strategies are being sought by the countries of the region to remain competitive and attractive investments. Clearly the "fourth wave" of global restructuring, where the consequence of advances in technology are negating formerly inherent advantages in low-wage labor, allows international capital to be much more selective about the production processes it locates in Southeast Asia and elsewhere.

The growing interconnectedness and complexity of the global economy has had tremendous ramifications not only on industry and other economic activities per se but also within environmental, political, cultural, and social realms as well. As we set the stage for the reader's journey into our volume on Southeast Asia, we would like to call attention to several themes that are both defining and an integral part of developmental change as we approach the twenty-first century. These are:

- the relationship between environment and development;
- declining population growth, gender issues, aging, the changing role of the family, and the economic and political situations of cultural minority groups;
- interrelationships between urbanization, population mobility, and employment;
- borderless economies, transnational production and restructuring development efforts;
- institutional changes—laws and regulations affecting development and in particular foreign investment and transnational corporations. Especially conspicuous is the pattern and pace of deregulation and privatization—collectively "liberalization;"
- and finally, the regionally varying traditional and changing roles of the state in these matters.

Part I

The Developmental Context

2 The Physical Environment

DAVID M. KUMMER

There is a remarkable diversity of physical environments in Southeast Asia, ranging from the volcanic peaks of Indonesia and mangrove shorelines of Vietnam to the virgin forests of the highlands of Burma. This natural diversity has had an enormous impact on the relationship between society and nature in the region. The intent of this overview is to provide a broad outline of the physical environment of the region and to demonstrate why it is of crucial importance to an understanding of contemporary patterns of diversity and development. Knowledge about the physical environment is essential to a better understanding of both patterns and processes of social, political, economic, and environmental change in Southeast Asia. The first half of this chapter will provide an overview of the region's physical geography and its natural resources, and the second half will utilize this information to inform discussions regarding present-day environmental issues in the region. A true understanding of current environmental change is possible only with knowledge of the human societies and physical environment in which they exist. The relationship between society and the environment is key.

LANDFORMS

The region is conventionally divided into two major zones: peninsular and insular, or island, Southeast Asia. Peninsular Southeast Asia consists of the mainland countries of Burma, Cambodia, Laos, Thailand, and Vietnam, and insular Southeast Asia is comprised of Brunei, Indonesia, Malaysian Borneo (Sarawak and Sabah), the Philippines, and Singapore. Although peninsular Malaysia is physically part of the mainland, it is usually considered part of the insular region because culturally and historically it is more closely related to this area. The total land area of peninsular and insular Southeast Asia is approximately equal; however, they represent very different geological histories.

Peninsular Southeast Asia, the entire island of Borneo, and the floor of the intervening South China Sea (the Sunda shelf) are all part of the Eurasian tectonic plate. The Philippine and Indonesian archipelagoes, on the other hand, are on the margin where the northeast-moving Australian plate and the west-moving Philippine and Pacific plates are colliding with the Eurasian plate. Because the Philippine, Pacific, and Australian plates are oceanic plates and therefore heavier than the continental Eurasian plate, they are being forced down into the Earth's crust. These areas where continental plates are overriding oceanic plates are called subduction zones and are characterized by a series of deep-sea trenches offshore that coincide approximately with the boundaries of the various plates. In fact, they almost completely encircle Southeast Asia (Map 2.1). It is in the vicinity of such subduction zones that mountain-building activities including earthquakes and volcanic activity takes place. Indeed, the Indonesian and Philippine archipelagoes have been created as a result of the collision of these tectonic plates and are, respectively, the first- and second-largest archipelagoes in the world in terms of the number of islands they contain; Indonesia is made up of about 13,000 islands and the Philippines has more than 7,000 islands.

Southeast Asia is one of the most conspicuous areas of seismic (earthquake) activity in the world. In fact, "Because of the high population density, more people are subjected to some degree of earthquake risk in Southeast Asia than anywhere else in the world. Over 200 million people live in an area in which there is at least one great earthquake every decade, a large earthquake every year, and perhaps a thousand small earthquakes, some of which cause damage, every year" (Arnold, 1986, p.15). Indonesia alone claims approximately 22 percent of the world's active volcanoes.

The onshore counterpart to the oceanic trenches is a series of mountain ranges that parallel the subduction zones. As a result of this tectonic activity, the Philippines and Indonesia are home to a large number of dormant and active volcanoes as well as a great deal of earthquake activity. Both nations are part of the "Rim of Fire" that surrounds the Pacific Ocean. Two of the largest volcanic eruptions in the nineteenth century occurred in Indonesia, Tambora in 1815 and on tiny Krakatau Island in 1883. Effects are often devastating; in the case of Krakatau the 2,640-ft (800-m) peak of the volcano collapsed to 1,000 ft (300 m) below sea level, leaving only a small portion of

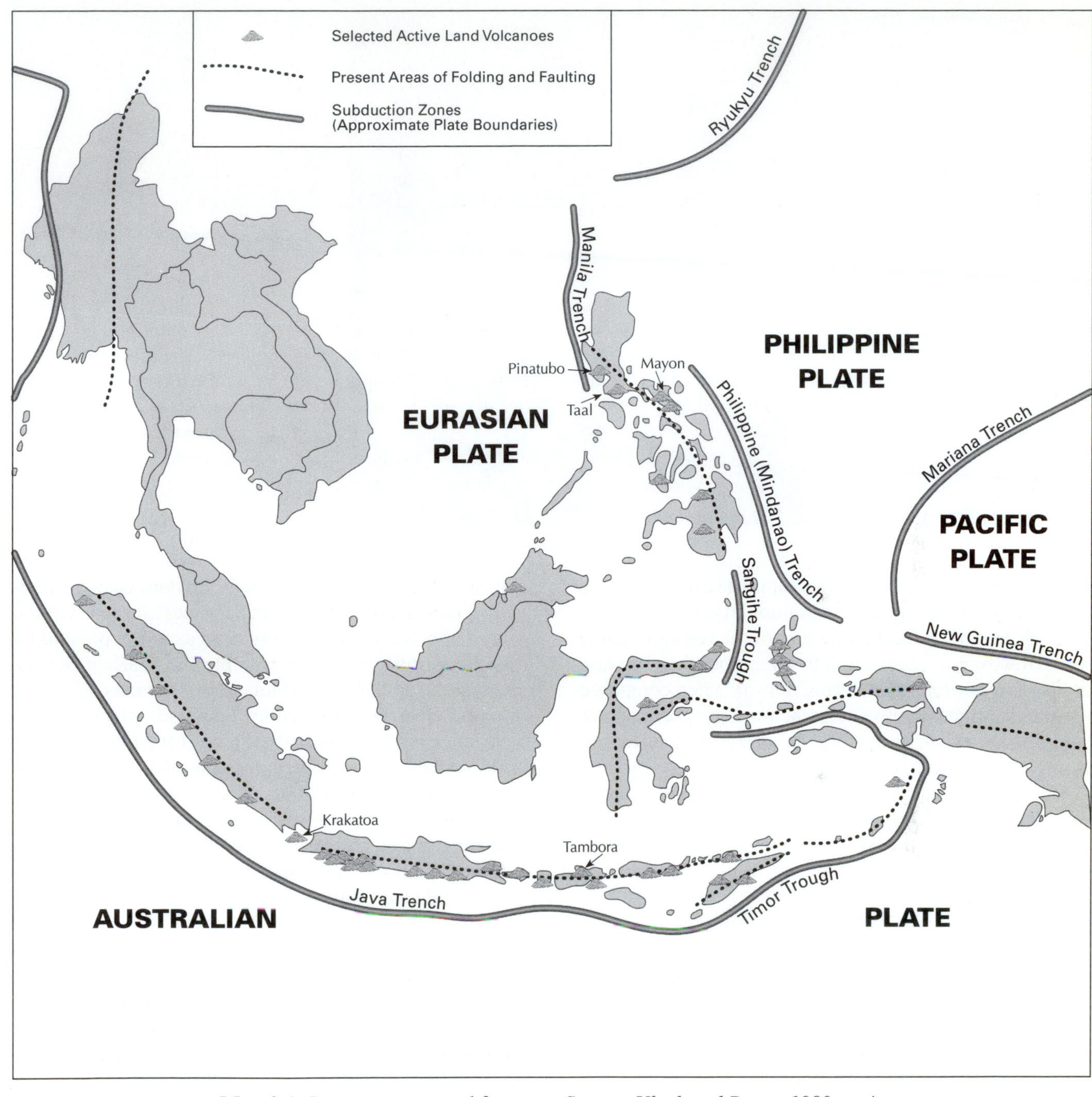

Map 2.1 Structure zones and features. *Source: Ulack and Pauer, 1989, p. 4.*

the island standing above sea level. Ash from the eruption colored sunsets around the world for two years and gave rise to the "Chelsea sunset" paintings in England. In addition, a tsunami (tidal wave) was generated that killed 36,000 people in nearby Java and Sumatra. The largest volcanic eruption in the twentieth century occurred on the island of Luzon in the Philippines at Mt. Pinatubo in 1991. While the explosion of Mt. Pinatubo resulted in fewer than 200 deaths, it displaced well over a 100,000 people and destroyed tens of thousands of hectares of productive agricultural land. Moreover, the lahars (a rapid flow of water, mud, and volcanic debris) that occur during the rainy season will continue for years to come.

At the same time, the long-term effect of volcanic activity in both Indonesia and the Philippines has generally been to produce soils that, by tropical standards, are relatively fertile. This results from the fact that most of the volcanic material in the area is basic rather than acidic in composition. The soils that have evolved from this

Photo 2.1 A volcanic landscape in west Java. (Leinbach)

nonacidic volcanic material can support sustained levels of agricultural activity. The high population densities on Java are, in part, at least a function of the fertile volcanic soils. However, it is important to remember that, because of its relatively high population density when compared to other seismically active regions, more people in Southeast Asia are at risk from volcanoes and earthquakes than in any other major world region.

Because the insular portion of the region is geologically younger, in general, than the peninsular region, the relief of island Southeast Asia tends to be more severe; that is, it is characterized by steeper slopes and higher elevations. Generally, there are very few large areas of little or no relief in the entire region. The largest areas of relatively flat land are found in river valleys and the larger deltas of the Irrawaddy, Chao Phraya, Mekong, and Red Rivers; the only exceptions to this are the larger inland basins of Burma, Thailand, and Cambodia. As a topographic map reveals, much of Southeast Asia may be characterized as having mountainous or hilly terrain.

The settlement pattern of Southeast Asia has been dominated by densely populated river valleys or deltas separated by hilly, forested areas with very low population densities. These intervening upland forest areas have acted as barriers between the more heavily populated settlement clusters. Over time, successive immigrant groups have settled in the lowland coastal, valley, and delta areas of the region. However, as more recent and more-populous groups have come into the region, they have "pushed" earlier, less-sophisticated groups into upland areas, thereby bringing about divisions and conflicts between upland, or "hill," peoples, and lowland groups. Today, the lowland areas are populated by the dominant ethnic group of each country including the Burmese in Burma, Thais in Thailand, Khmers in Cambodia, and Javanese in Java. On the other hand, the upland, forested areas are populated by ethnic minority groups, for example, the Shans in Burma, and the Yao and Miao in Thailand. As a result of the very different environments that lowland and upland peoples have inhabited for thousands of years, very different livelihood systems have evolved. In general, uplanders have engaged in some form of dry field or shifting cultivation known as *swidden*, and lowlanders have practiced intensive wet rice cultivation, or *sawah*. These general patterns continue today and are important for understanding several critical environmental issues in the region.

Most of the coastal plains in Southeast Asia are relatively narrow. There are only two significant exceptions to this statement: the deltas of the major rivers, such as the Irrawaddy and the Mekong, and the eastern coast of Sumatra and the southern coast of Borneo. Both of these coastal areas, however, are occupied by swamps and are not suitable for large-scale agriculture and dense settlement. In short, due to the narrowness of its coastal plains and its generally hilly to mountainous terrain, Southeast Asia overall has had a relative shortage of cultivable land. For example, land under crops or in pasture comprises only 7 percent of the total land area of Laos, 15 percent in Malaysia, 16 percent in Burma, and 21 percent in Cambodia and Vietnam. This is to be contrasted with the United States where almost 50 percent of all land is under some form of agriculture. The long-term adaptation to this situation involved the evolution of numerous forms of land-extensive shifting cultivation, and labor-intensive agriculture (particularly irrigated rice but also including backyard gardens and the growing of crops on terraces).

The geology of Southeast Asia is remarkably complex and is the result of hundreds of millions of years of rifting and subsequent colliding of landmasses. This process continues today. In a very general sense, Southeast Asia is composed of three geologic regions, and insular Southeast Asia is the result of the collision of the various oceanic plates with the Eurasian plate. Peninsular Southeast Asia, on the other hand, is composed of two geologic units: a central core located primarily in Cambodia, eastern Thailand, and southern Vietnam, and a mountainous region that surrounds this low lying core. This mountainous area includes northern Burma and Thailand, Laos, and most of Vietnam and is the result of more-recent uplift.

The mountainous and semimountainous terrain and the island nature of many of the countries of the region combine to create one of the distinctive characteristics of the area as a whole—its fragmented nature. This in turn has led to considerable diversity with regard to adaptation to the natural environment. It has also contributed to great socioeconomic and cultural diversity in the region. Indeed, it is quite appropriate that this fragmented region was labeled by geographers a "shatterbelt" and the "Balkans of the Orient" (Broek, 1944; Fisher, 1962).

By way of summary, an appreciation of the geomorphology of the region is important because it highlights features of the area that are helpful in trying to understand past and ongoing events. First, the remarkable physical geographic variability has resulted in an equally remarkable variation in land use and cultural practices. This diversity is one of the distinguishing characteristics of Southeast Asia. Second, the relative lack of flat, easily tilled agricultural land continues to have an effect on the types of agriculture that can be practiced. Lowland agriculture, particularly irrigated rice has, in general, been very labor intensive. Terraced agriculture (primarily for irrigated rice) is also found throughout the region. Classic examples are the rice terraces of northern Luzon, the Philippines, and the island of Bali in Indonesia. It remains to be seen whether or not these adaptations to the natural environment can provide useful examples of sustainability for contemporary society. In particular, as populations continue to increase and expand into hilly and mountainous areas, a major question is whether or not more-intensive agricultural systems can be devised to work in upland areas without, at the same time, causing excessive environmental damage.

CLIMATE

Nearly the entire Southeast Asian region can be said to be part of the humid tropics; that is, the region is warm all year round and receives abundant but temporally and spatially uneven rainfall (Map 2.2). Overall, the annual average temperature is approximately 80°F (27°C), and seasonal variation in temperature is usually less than the diurnal variation; that is, the difference between daytime and nighttime temperatures is greater than the difference in temperatures between the winter and summer months.

In addition to latitudinal position, temperature is also a function of altitude as it decreases with increasing elevation. This decrease, referred to as the environmental lapse rate, is equal to about 3°F per 1,000 ft increase in elevation. The result is that because the region as a whole is so hilly or mountainous, almost all nations in Southeast Asia have areas of higher elevations where cooler temperatures prevail. It is at higher elevations that hill stations emerged during the colonial period; today, these hill stations are often referred to as upland resorts. In the Philippines, for example, one of the first things the U.S. personnel did following the defeat of the Filipinos after the Spanish-American War was to develop the city of Baguio in northern Luzon as a retreat to escape the heat and humidity of summertime Manila. The area around Baguio is now a major producer of temperate fruits and vegetables for the Philippine domestic market and is a popular tourist destination. Other examples of such hill stations include Dalat in southern Vietnam, the Cameron Highlands in peninsular Malaysia, and the Karo Highlands of North Sumatra. Some of these hill stations—Bandung on Java is the best example—have become major metropolitan areas (Photo 2.2).

Although it is difficult to provide an average rainfall figure because there is so much variation within the region, a figure of 80 in. (200 cm) per year would be considered normal in many areas. At the same time, however, a distinction must be made between equatorial and tropical, or monsoon, rainfall regimes. Equatorial areas are on or near the equator, from about 10 degrees north to 10 degrees south of the equator. Equatorial regions receive rainfall throughout the year; that is, they do not have a dry period. Singapore is a good example of an area in which rainfall is evenly distributed throughout the year (Figure 2.1). Tropical monsoon climate regions generally affect areas more than 10 degrees north and south of the equator, and such areas are characterized as having distinct wet and dry seasons. Manila and Rangoon are good examples of weather stations that are characterized by monsoon conditions (Figure 2.1). In turn, this seasonal variation in rainfall is largely a function of the monsoons, winds (and their accompanying rains) that alternate direction on a regular basis (Map 2.2). The term *monsoon* is derived from the Arabic word *mausin,* which refers to the time of the year the wind occurs and which made possible the seasonal trips across the Indian Ocean. These winds are caused by the seasonal migration of high- and low-pressure areas, which, in turn, are caused by the differential heating of land and water, movement of the subtropical jet stream, and other factors. During

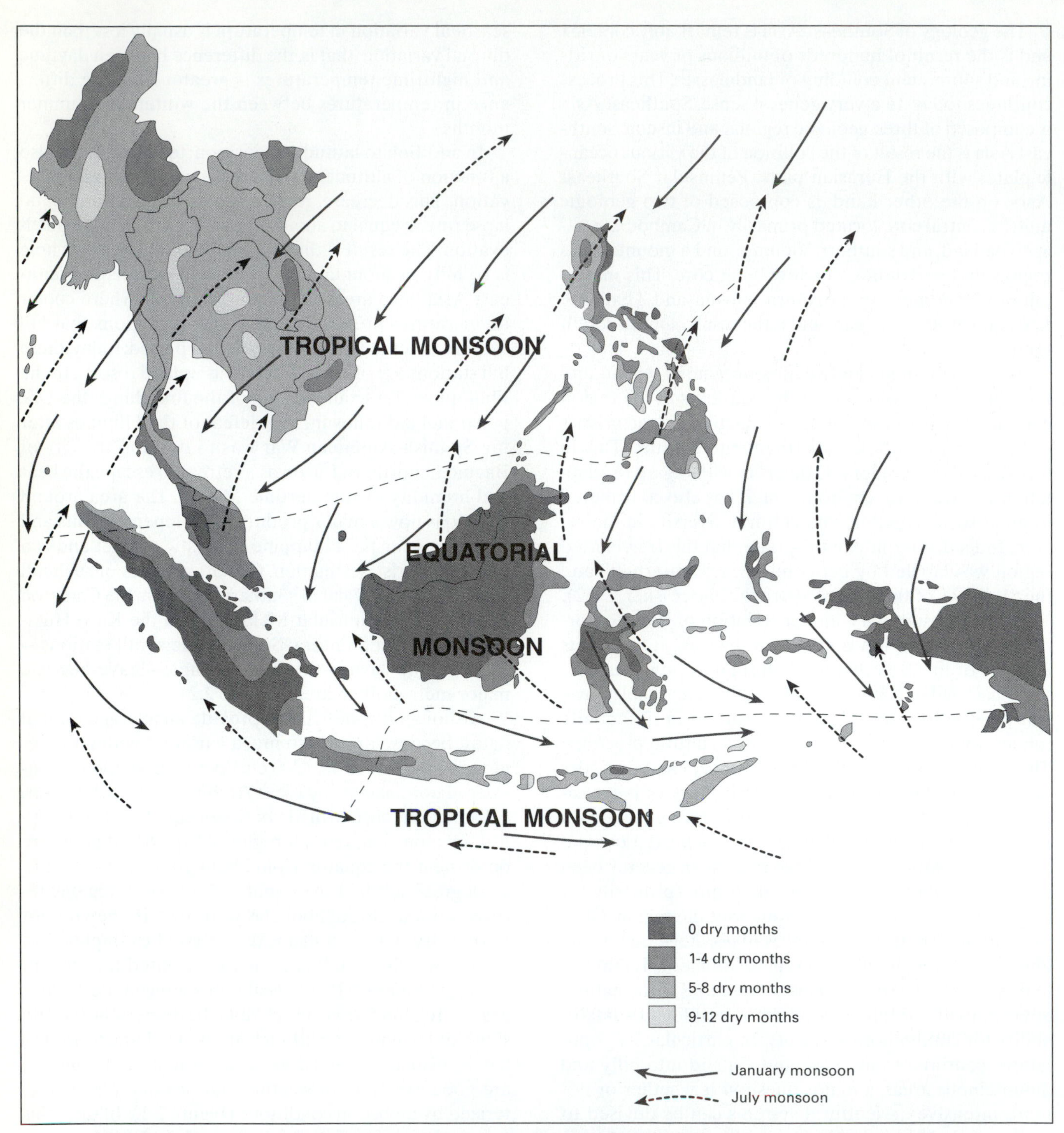

Map 2.2 Climate Patterns: Prevailing winds and length of dry season. *Source: Ulack and Pauer, 1989, p. 6.*

the winter months, the monsoon winds blow from the northeast off the Asian landmass, and during the summer months they originate from the southwest off the Indian Ocean. Equatorial regions receive both monsoons and, hence, are wet all year; tropical areas, on the other hand, usually receive only one monsoon a year and, hence, have a dry period at some point during the year, sometimes as long as six months. It is important to recognize that although monsoons are the dominant factor in seasonal variations of rainfall, the rainfall pattern in Southeast Asia is the result of a variety of complex and interlinked forces. Other factors that are important include cyclonic disturbances, localized land–sea breezes, and the presence of mountains.

The western Pacific Ocean is the origin of a severe tropical weather disturbance called a typhoon (called a

Photo 2.2 Guest house in Fraser's Hill, Pahang, Malaysia, a hill station. (Leinbach)

hurricane in North America and a cyclone in south Asia). These weather systems primarily effect the northern two-thirds of the Philippines but may also reach parts of peninsular Southeast Asia, particularly Vietnam. They are accompanied by high winds and heavy rainfall. In fact, until recently, the world record for precipitation in a 24-hour period was held by the city of Baguio in the Philippines, where in 1911 46 in. (117 cm) of rain fell during a typhoon, or *baguio*. On occasion, tropical thunderstorms can match the intensity of typhoons. In 1991, the Ormoc region of Leyte, the Philippines, was hit with such a storm; it is estimated that almost 16 in. (40 cm) of rainfall occurred in a 4- to 6-hour period. In the massive flooding that followed, approximately 6,000 people were killed.

Tropical rains tend to be more intense than temperate rains; that is, the amount that falls in any given time period is greater. Because erosion is, to a certain extent, a function of the amount and intensity of rainfall, this necessarily means that tropical regions like Southeast Asia are more vulnerable to erosion than temperate regions. This is particularly the case in steeply sloped upland areas. The problem is compounded if planting results in the removal of much of the original ground cover. In short, the hilly topography and intense rainfall mean that many upland areas in the region are environmentally fragile. This point is of direct relevance to any discussion regarding future uses of the upland areas in Southeast Asia, especially for agriculture.

Another important factor that has an effect on rainfall is the presence of mountains. Air masses are forced to rise over mountain barriers because they cool and their ability to hold moisture declines, resulting in rainfall called orographic precipitation. This means that the windward side of a mountain range usually has a significantly greater amount of rainfall than the leeward, or "rain-shadow," side. The effect of mountain ranges on local precipitation is a function of the length and height of the mountains, the prevailing winds, and the absence or presence of monsoons. Examples of areas that experience orographic precipitation include almost the entire western coast of Sumatra, the northeast and northwest coasts of Luzon, and coastal Burma. Good examples of the rain-shadow effect include the island of Cebu in the central Philippines and the "dry zone" of central Burma. Both of these areas receive less rainfall because of higher elevations in surrounding areas. On a smaller geographic scale, the presence of mountains means that there can be much local variation in rainfall, which may influence the type of agriculture practiced.

The length of the dry period is a crucial factor controlling the natural vegetation in an area and the type of agriculture that is possible. If we define a *dry month* as one that receives less than 4 in. (10 cm), then most of peninsular Southeast Asia has an average dry season of approximately 5 months. Many islands in eastern Indonesia, such as Flores and Timor, also have dry periods of at least 5 months. In other situations, the dry period can last as long as 8 or 9 months such as is the case in central Burma. Under these circumstances *sawah* agriculture can only occur with irrigation. In short, it is not just the total amount of rainfall that is important but its seasonality. Even though the entire region is part of the humid tropics, quite a few areas, particularly further from the equator and effected by the presence of mountain ranges, have a lengthy dry period that restricts agriculture

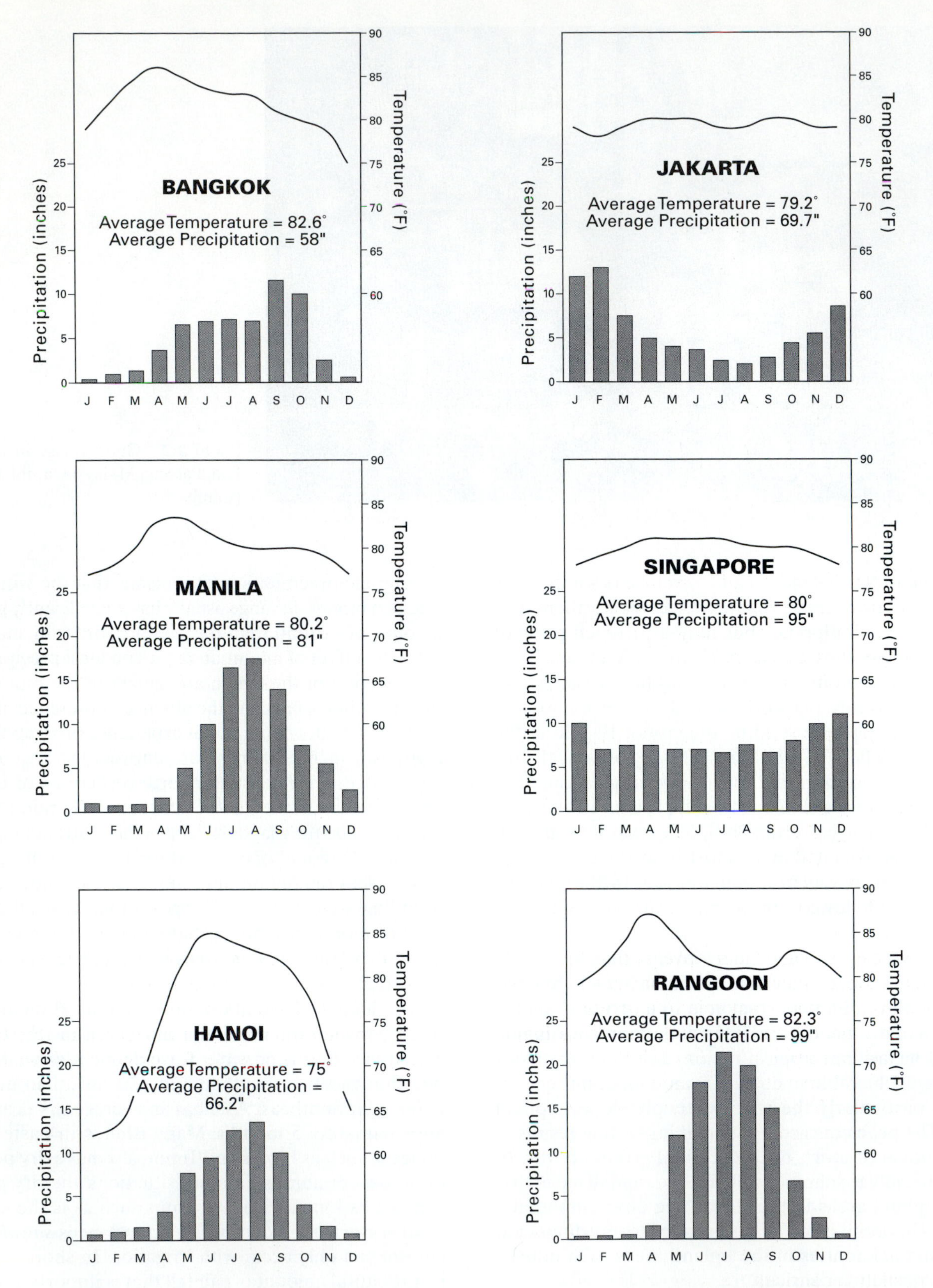

Figure 2.1 Climographs for selected cities in Southeast Asia. *Source: Ulack and Pauer, 1989.*

Photo 2.3 Flooded street, Samarinda, Kalimantan. (Wood)

and has a major effect on the vegetation that can grow there. Lastly, the timing of the rains is crucial because, in the absence of irrigation, the agricultural season cannot begin until the rains arrive. Overall, however, the region does not suffer from a shortage of water.

DRAINAGE PATTERNS AND WATER RESOURCES

An outstanding characteristic of Southeast Asia is the relative ease of water transportation throughout much of the area. For the region as a whole, the territory of oceanic water (Andaman Sea, South China Sea, Gulf of Thailand, and the waters surrounding the Indonesian and Philippine archipelagoes) is approximately four times the combined land area of the ten countries. The island nature of Southeast Asia and the relatively protected nature of its ocean waters means that maritime activity has historically played an important role in unifying the area. At the same time, on the mainland, numerous major rivers have facilitated transportation and communication between the coast and inland communities (see Map 1.1). Moving from west to east these major, generally north-south flowing rivers are as follows: the Irrawaddy, Salween, Chao Phraya, Mekong, and the Red. With the exception of the Salween, which is primarily confined to narrow gorges and flows through rugged terrain with low population densities, all of the above river valleys represent areas of intensive agricultural activity and relatively high population densities. Most important, these rivers have large deltas that are also intensively cultivated. Such areas have some of the highest population densities in the region.

The Mekong deserves special mention because it is the largest and most important river and because it flows through all five mainland Southeast Asian countries (Map 2.3). Its drainage basin is approximately 300,000 sq mi (800,000 sq km), making it by far the largest river in Southeast Asia. Moreover, the Mekong originates in the Tibetan highlands of China and thus it could become the gateway to southern China. China, Cambodia, Laos, and Thailand each have approximately 20–25 percent of the entire area of the Mekong Basin while Burma and Vietnam have considerably smaller areas at 3 percent and 8 percent, respectively.

An interesting hydrological feature of the Mekong Basin is the Tonle Sap (or "Great Lake") in Cambodia. Cambodia can be likened to a saucer, with the Great Lake occupying the shallow center of the country, which is surrounded by areas of higher elevation. It was on the shores of Tonle Sap that the Angkorian civilization flourished between the ninth and thirteenth centuries. The lake was a source of water for irrigation, as well as an area with high fishery yields. The Tonle Sap is connected to the Mekong by the Tonle Sap River, but during the annual summer floods along the lower Mekong, the flow actually reverses and water flows from the Mekong to the Tonle Sap. During this time, the surface area of the Tonle Sap is from two to four times its dry season area of approximately 1,000 sq mi (2,600 sq km). In effect, the Tonle Sap acts as a giant overflow basin, or "safety valve," and moderates the effects of floods along the lower Mekong.

At the present time, the Mekong is much underutilized. Despite the size of its drainage area and its 2,600 mile

BOX 2.1 Water Shortages in Metro Cebu

Although there is an abundance of water in Southeast Asia as a whole, local shortages can occur, particularly in areas of relatively low precipitation. A case in point is metropolitan Cebu on the island of Cebu in the central Philippines.

The population of Cebu City and the surrounding metropolitan area is approximately one million; it is the second-largest urban area in the Philippines, and it is growing rapidly in terms of population, jobs created, and building construction. However, the number-one constraint on future growth that Cebu City must deal with soon is the lack of freshwater. Cebu City is on a very narrow coastal plain, and so many deep wells have been sunk in the urban area that saltwater intrusion is now a problem in some downtown areas. Plans to help alleviate the prospect of water shortages in the future include: (1) better regulation of drilling in the urban area; (2) reforestation of denuded watersheds to increase infiltration of rainwater; (3) the building of a large dam and reservoir in the urban watershed; and (4) the building of dams and reservoirs in nearby watersheds and pumping the water to Cebu City. While all four of these proposals have been discussed at length, they have not resulted in any concrete activity.

An even more ambitious plan which is now under investigation would involve the pumping of freshwater from the nearby island of Bohol to Cebu City. This would necessitate building a 30-km-long pipeline several hundred feet under the surface of the Bohol Strait, which separates the two islands. Such an engineering project has never been attempted in the world before, and it is doubtful that it would succeed; however, the fact that municipal officials are considering such a proposal is an indication of how serious the water problem in Cebu has become.

(4,200 km) length, there is relatively little river traffic. Although there are numerous reasons for this lack of traffic, the major factor is most certainly related to the nearly continuous state of warfare in most parts of the region in the post-World War II era. The conflicts have included the various wars of Vietnamese independence, the revolutions in Laos and Cambodia, the Sino-Vietnamese border war of 1979, and the unrest in large parts of northern and eastern Burma. This political instability has resulted in a lack of investment in infrastructure, especially roads, bridges, and dams. The Asian Development Bank (ADB) is in the process of preparing feasibility studies on various aspects of potential projects in the area. In addition to bridges over the Mekong and the clearing of rapids along stretches of the river, these include the creation of extensive road and rail linkages among the various countries of the region. Three forms of development involving the river are being targeted: tourism, trade, and hydroelectric potential. Included in the plans is to dam the river in as many as five different places, which would effect the stream flow and animal life. New bridges and road and rail connections would drastically increase trade and make isolated communities more accessible to outside influences. Production of electricity would increase dramatically, and production and consumption of goods would also increase, as would the processing of agricultural products and tropical woods. Migration into the area would most likely increase, adding to a population of 50 million already there (Hori, 1993). In short, planners see the development of the Mekong River valley as a way to increase economic activity in the region, create a more integrated regional economy among the six countries, generate goodwill among the member countries, and provide the southern Chinese province of Yunnan with an alternate route to the sea. In fact, this process of economic integration has already begun with the completion in 1994 of the first bridge across the Mekong connecting Laos to Thailand (Murray, 1994).

There are of course significant potential environmental consequences involved in the large-scale development of the Mekong. On the assumption that development of the Mekong Valley as presently envisioned does occur and increases incomes and the standard of living for those affected, the question that is appropriate to ask is: "Does this justify the environmental changes that will necessarily occur?" Changes would most likely include increased deforestation, increased soil erosion as agriculture expands and forests decline, and the extinction of some species leading to a decline in biodiversity. If development of the Mekong River does indeed raise living standards and these gains are widely shared by the local inhabitants, then many people would argue that the environmental changes brought about by this development are justified. The situation is much more problematic, however, if the benefits of development are not widely shared and, instead, accrue to only a small group of elites who are most likely the wealthier members of society to begin with. In short, it is not a question of whether or not environmental change will occur; rather, it is a question of how much environmental change there will be and who will benefit.

With the exception of several larger rivers on the island of Borneo, the Malay Peninsula and insular Southeast Asia do not contain any major navigable rivers. The vast majority of the rivers in insular Southeast Asia are short and have relatively steep gradients, meaning that their inland navigability is generally quite limited. At the

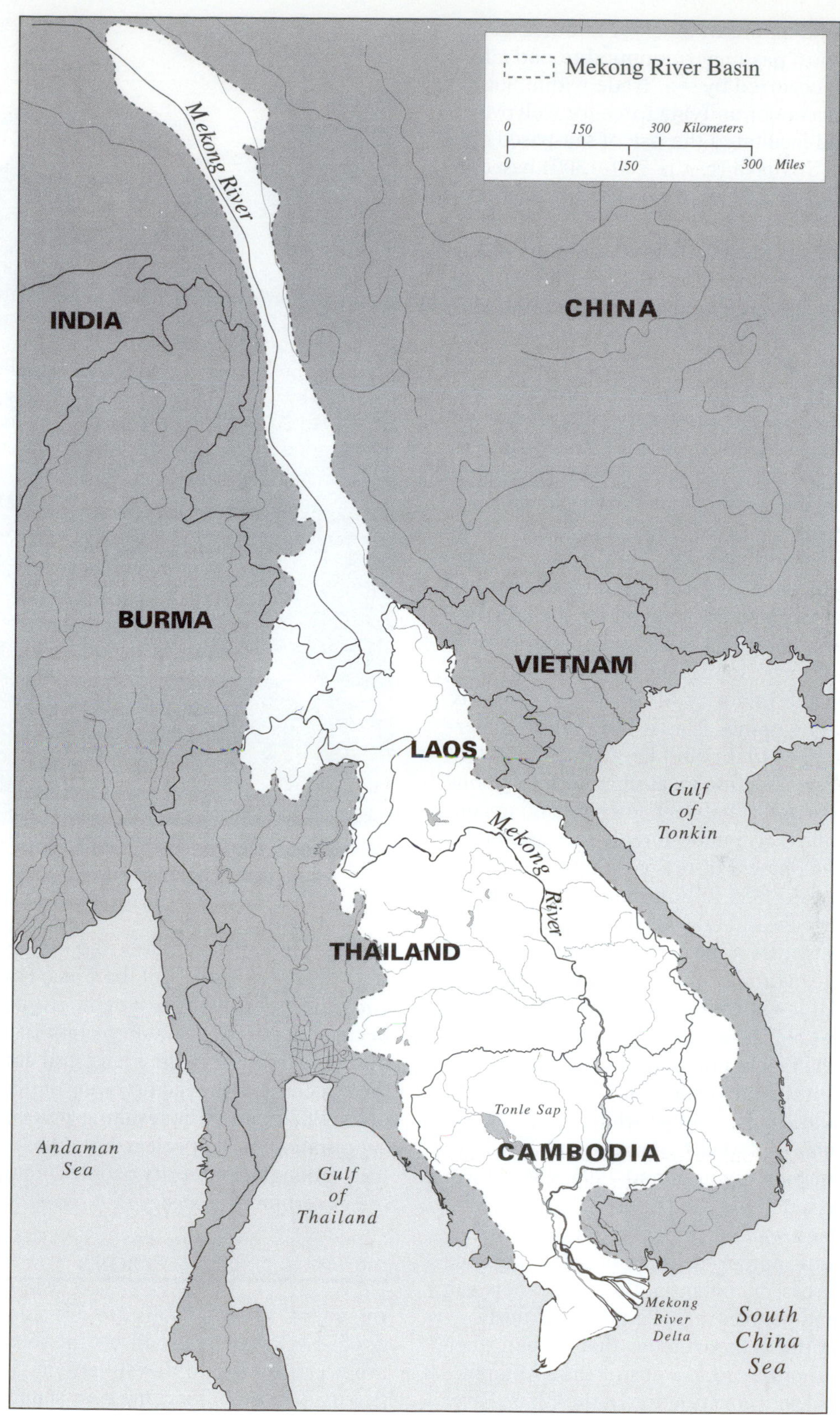

Map 2.3 Mekong River Basin. *Sources: Various.*

same time, the island nature of much of Southeast Asia has dictated that most passenger, commodity, and information flows have occurred by sea. Trade within Southeast Asia has been a major unifying force for well over a thousand years and facilitated the rise of sea-based empires, for example, Srivijaya (c. A.D. 700–1300) based in southeast Sumatra, and Majapahit focused in east Java (c. AD 1200-1500).

Given the relatively steep gradients of many of the rivers and the large amounts of rainfall, Southeast Asia would appear to have a fairly large potential for the generation of hydroelectric power. One of the more successful projects has been at Maria Cristina Falls on the Agus River on the northern coast of Mindanao, the Philippines (Photo 2.4). The electricity generated from this project has helped lead to the industrialization of nearby Iligan City. At the same time, however, to maintain a relatively constant flow of water in the Agus River, Lake Lanao, the source of the Agus, has often been allowed to drop below normal levels. This has had a negative impact on the ecology of the lake which, in turn, has led to reduced fish yields for those living on the lake. Because most of the inhabitants of Lake Lanao are Muslim, the ecological damage to the lake has led to increased tensions between the minority Muslims and majority Christians in the Philippines. This is but one of numerous regional examples that could be given where "development" projects affecting local upland populations tend to exacerbate conflict between lowland and upland peoples. Such conflicts will undoubtedly increase in the future as such development efforts continue.

Photo 2.4 Maria Cristina Falls, Iligan City, Mindanao. (Ulack)

The controversial US$6 billion dam project on the Bakun River in Sarawak, Malaysia, is an example of a project that demonstrates some of the possible trade-offs involved with large hydroelectric projects. Opponents of the project say that it will flood an area about the size of Singapore, destroy 172,000 acres (70,000 ha) of rain forest, and involve the displacement of about 9,000 native Dayak peoples. On the other hand, it would produce a huge amount of electricity, most of which would be exported to peninsular Malaysia. Construction has not yet started on the dam itself, but as of mid-1995, a Malaysian contractor started to fell trees in the area that will be inundated by the dam waters ("Logging On," 1995). Because almost all sites appropriate for dams are going to be found in upland areas, the potential for conflict between lowland and upland peoples must be taken seriously.

An example of this is the violence that ensued when former President Ferdinand Marcos of the Philippines tried to push ahead with construction of the Chico River dam project in northern Luzon in the 1970s and 1980s. The goal of the Philippine government was to increase the output of hydroelectric power to lessen dependence on imported oil and to increase the availability of electricity to lowland areas to further the process of industrialization. Unfortunately, the indigenous people, collectively called Igorots, of the Chico River basin were not consulted about the government plans, and when it became apparent that completion of the project would mean the flooding of their ancestral domains, they sought the assistance of the military wing of the Communist Party of the Philippines. The resulting violence finally led to the cancellation of the project, but it has left a legacy of distrust among the minority peoples of northern Luzon and the government, a legacy that remains to the present.

SOILS

In general, tropical soils are not as agriculturally productive as temperate soils, and the soils of Southeast Asia are no exception to this statement. This is the case for two main reasons. First, the year-round high temperature and availability of water means that bacterial activity is high and thus organic material is rapidly broken down. The end result is that many tropical soils contain few organic nutrients and that the organic material, if present at all, is found only in the top layer of soil. This is not a problem for the natural vegetation found in tropical areas because root systems have adapted to cope with this situation. In short, most soils in the region are deficient in the organic material that agricultural crops need. Second, the amount and intensity of rainfall means that nutrients are rapidly leached from the soil. The high amount of rainfall in the tropics as compared to temperate environments means that some basic minerals in the soil such as calcium are dissolved and leached from the soil rapidly. Soils, produced in humid tropical conditions, that are lacking in nutrients and minerals are called laterites or latosols, and the process that produces them is called laterization. The end product of this process is a hard bricklike structure that has been used in construction, including that at the ancient temple complex of Angkor. Lateritic soils are widespread throughout Southeast Asia. In summary, the luxuriant vegetation that was observed by Europeans when they first arrived in the early sixteenth century and that still occurs in certain parts of the region does not demonstrate that the soils have a high agricultural potential. Instead, it is a reflection of an ecosystem based on a rapid recycling of nutrients, which are stored primarily in living vegetation. This of course has very important implications for ongoing and planned efforts to expand the area in the region devoted to agriculture.

When the original forest cover is removed, soils in the region tend to be degraded rapidly for several reasons. First, the process of removal of the forest can damage the soils itself. For example, commercial logging using heavy equipment can result in soil compaction, or poorly constructed logging roads may directly cause considerable erosion. Second, with the forest cover removed, soils are now directly exposed to the elements. This means that inevitably rates of erosion will increase rapidly. Third, forest removal followed by agriculture will exacerbate the erosion problem because the planting of annual crops means that the soil will be harrowed and plowed several times a year. In addition, because most of these soils are deficient in organic material and many important minerals, agriculture, to succeed, must increasingly rely on heavy applications of fertilizer. Although fertilizer can increase crop yields, there are several drawbacks. First, the ability to buy sufficient fertilizer is often beyond the financial means of farmers, particularly poor ones, and, thus, fertilizer may not be used in the proper amount. Second, long-term use of fertilizer can damage the structure of the soils themselves and therefore make it necessary to increase the amount applied over time. Third, due to the amount and intensity of rainfall, fertilizer leaches out of the system quickly and ends up in nearby waterways where damage to wildlife and water quality may be severe.

In general, relatively little is known about the long-term viability of many agricultural systems in the tropics. This is especially so in the areas of poor soils that are found throughout the region, particularly in upland areas. Considerable research effort is presently being undertaken in the Philippines, Thailand, and Indonesia to devise agricultural or agroforestry systems that can perform better in upland environments with poor soils. As we have seen, the soils of Southeast Asia are relatively impoverished. There are, however, two major exceptions to this general statement.

The high levels of rainfall on peninsular Southeast Asia mean that the rivers on the mainland transport large volumes of water, especially during the rainy season. This, combined with their relatively steep gradients, means that they have the capacity to carry large amounts of sediment. Over time, these rivers have developed large valleys and deltas upon which these sediments have been deposited. The resulting alluvial soils are well known for their high fertility. In addition, the annual flooding in these areas means that new alluvial soil is deposited annually and, thus, high soil fertility is maintained. Such areas are of course well known for intensive wet rice cultivation, or *sawah*. The upland areas between these river systems are, in general, characterized by poor soils, which preclude intensive agriculture. The major form of agriculture practiced in these areas has usually been swidden.

The second major exception is found principally in insular Southeast Asia, an area as we have seen that lacks the large river systems found on the mainland. Recall that it is the insular portion of the region wherein much tectonic activity is taking place. Volcanic activity in this area, especially in parts of Indonesia and the Philippines, has produced fertile volcanic soils that can support intensive agriculture. The premier example of this phenomenon is the island of Java in Indonesia where rural population densities commonly exceed 2,500 people per sq mi (965 people per sq km). As noted earlier, some volcanic soils are too acidic to be of much agricultural use but, in general, many of the volcanic soils in Indonesia and the Philippines are basic soils and therefore quite suitable for intensive agriculture.

NATURAL VEGETATION

The natural vegetation in Southeast Asia is a result of a series of related influences, primarily differences in climate, landforms, and soils. In addition, Southeast Asia is at the crossroads of two great floral (and faunal) realms, the Asian or Oriental in the west and the Australian in the east, and as such it has been influenced by the plant and animal life of both of these realms. The boundary, or transition zone, between the Oriental and Australian realms is in the eastern part of Indonesia. To the west of what is known as Wallace's line (modified later by Weber and Huxley) is the Oriental realm and to the east is the Australian realm (Map 2.4). Here are found the rather peculiar animals known as marsupials, which include

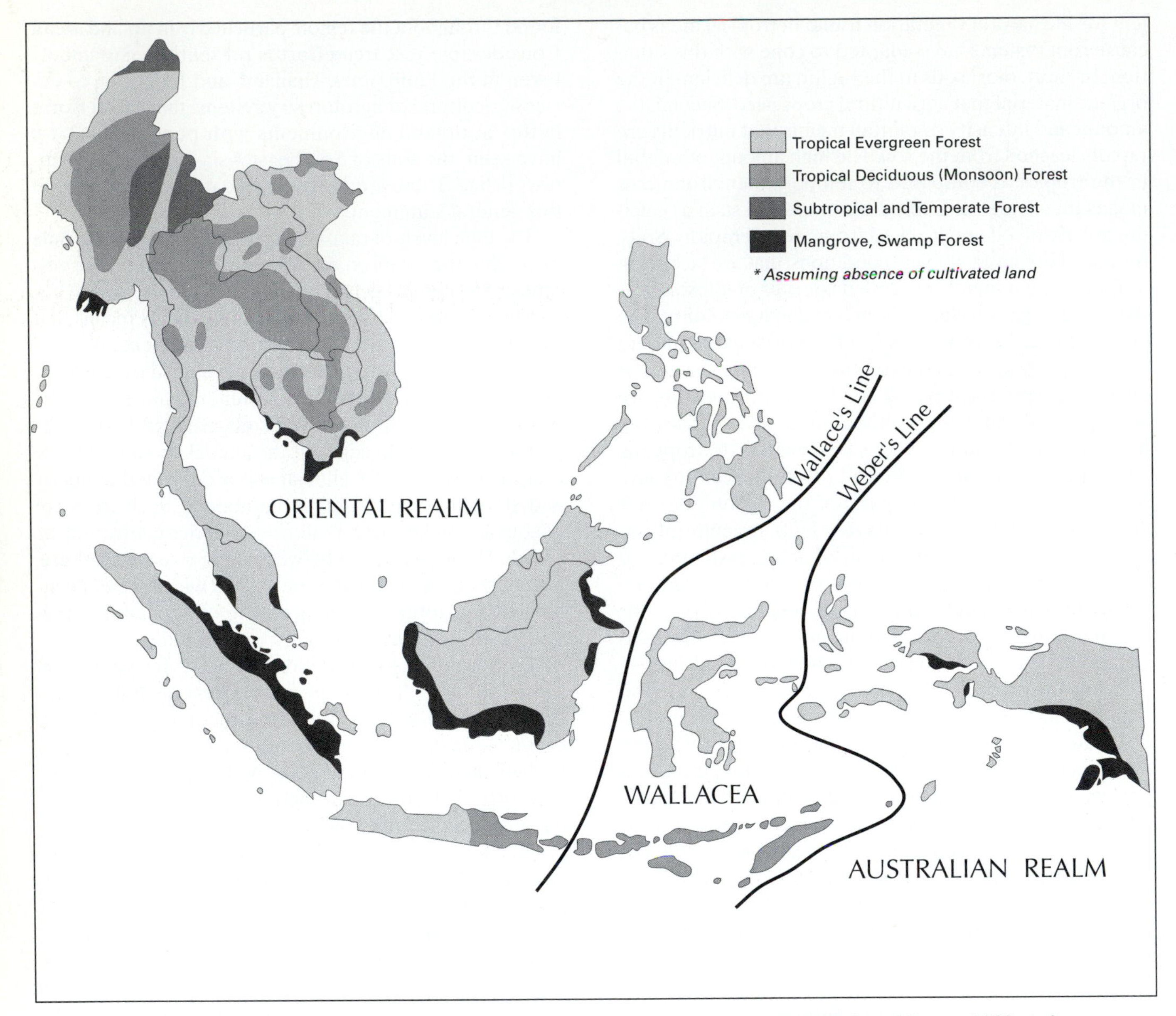

Map 2.4 Natural vegetation regions* and biogeographical realms. *Source: Ulack and Pauer, 1989, p. 9.*

kangaroos, wombats, and koala bears. During the Pleistocene epoch, or Ice Age, that began about 2 million years ago and ended only 10,000 years ago, most of insular Southeast Asia was attached to the Asian landmass because sea levels were much lower than they are presently. Land areas then exposed included the continental shelf areas, ocean areas that are less than 650 ft (200 m) deep. In Southeast Asia during the Pleistocene the existence of the Sunda Shelf, or Sundaland, meant that Borneo, Java, Sumatra, and Palawan were connected to the mainland and the Sahul Shelf, also known as Sahulland, connected the island of New Guinea and other, smaller islands to the Australian continent. In the area between the lines drawn by Huxley, Wallace, and Weber is an area called Wallacea, where greater ocean depths kept the areas physically separated by at least 40 mi (65 km). Thus, the two great biogeographical and vegetative realms were separated by a water barrier that precluded many types of life-forms from diffusing across the barrier (Map 2.5). It is estimated that the first humans, people of the Australoid racial group, did not reach those areas to the east of the barrier (the easternmost islands of Indonesia, New Guinea, and Australia) until perhaps 40,000 years ago. The end result is that the flora (and fauna) of Southeast Asia is remarkably diverse. For nearly the entire area, the natural climax vegetation (i.e., the terminal plant community that results from long-term adaptation to its environment) is one of three major forest types found in the

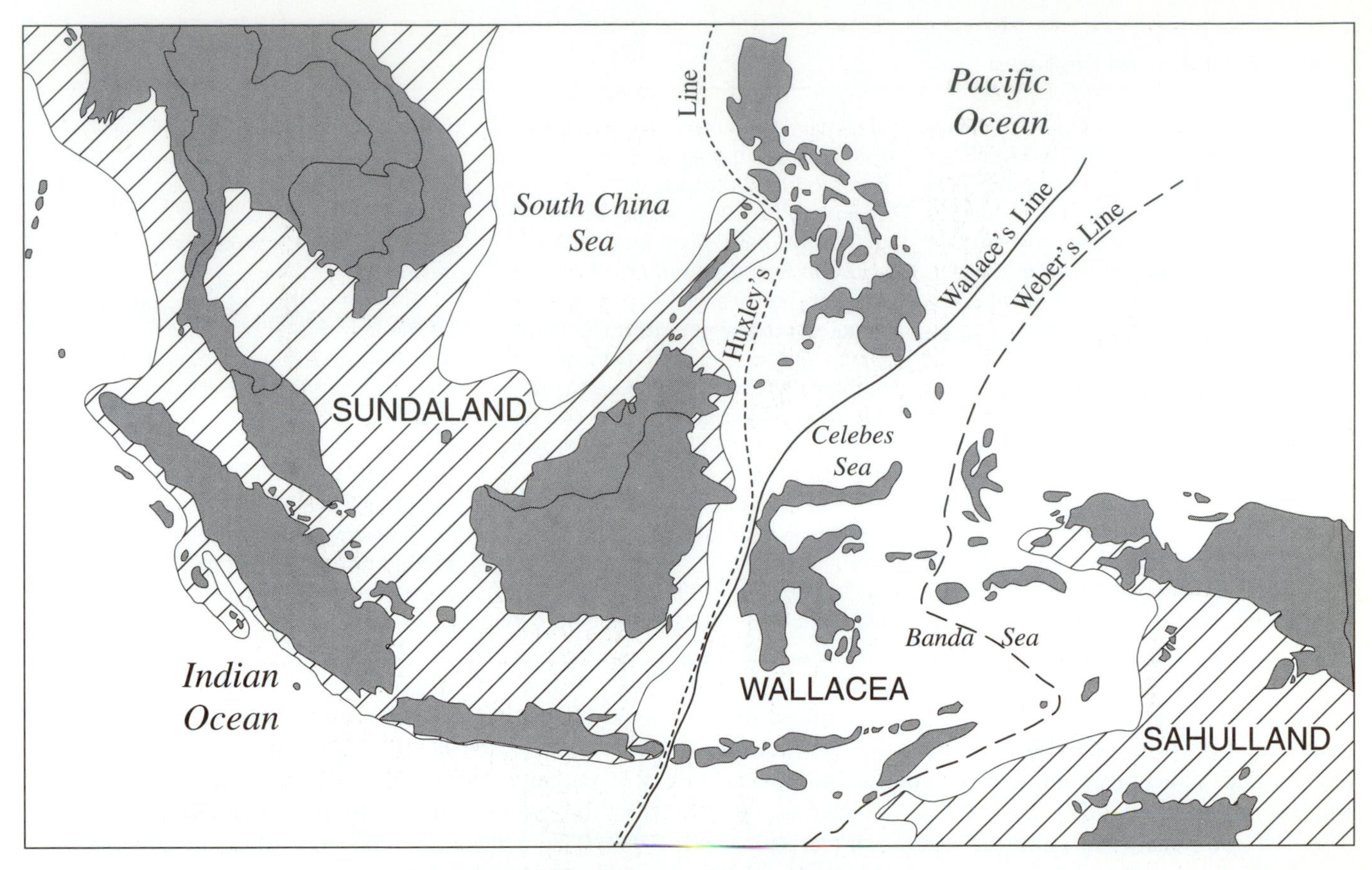

Map 2.5 Sunda and Sahul. *Source: Various.*

region. These three major natural forest types are evergreen, deciduous or monsoon, and mangrove (Map 2.4).

The tropical evergreen forest occurs in areas of high rainfall all year round; as such, the forest retains its green foliage throughout the year. This type is found in all countries throughout both peninsular and insular Southeast Asia with the exception of Singapore which, for all practical purposes, has no original forest cover left. The deciduous forest (often referred to as a monsoon forest) occurs where there is a distinct dry season; trees lose some or all of their leaves when the stress caused by lack of water is greatest. This forest type occurs primarily in south-central Indonesia and in large parts of inland Burma and Thailand. These forests are particularly vulnerable to fires caused by human or natural disturbances because the long dry season makes them more susceptible to burning than the evergreen tropical forest.

The third major forest type is mangrove. Tropical tree species in these forests have adapted themselves to live on the saline muds of tidal zones and occur along protected coastlines. In Southeast Asia they grow mostly in Indonesia and on the deltas of the major rivers on the mainland. In Indonesia, mangrove forests are located on or within a few miles of coastlines and are especially important on the east coast of Sumatra and on Kalimantan. Throughout the region an estimated 13.3 million acres (5.4 million hectares) of mangrove forest are left; they contain 20–40 different mangrove species. Indeed, the region accounts for a large share of the world's remaining mangrove forests. Of the more than 100 countries that have some mangrove area, Indonesia ranks first in the world with 10 million acres (4 million hectares); Malaysia is fifth, and Burma ranks eighth. The Philippines has only 350,000 acres (140,000 ha) left; it is estimated that 70 percent of its total mangrove area was lost between 1920 and 1990 (World Resources Institute, 1996–97). Mangrove forests play a vital role in protecting coasts from erosion and in providing a suitable habitat for many animals important to the marine food chain. Mangroves have come under intense pressure in the post-World War II period for several reasons. First, they have been sought as a source of high quality fuelwood, used for domestic consumption by the majority of the population, especially in rural areas, of most of the region's countries. Second and equally important, they have suffered from the expansion of urban areas, most of which have of course developed along coastal areas. Other reasons include pollution from mining and agriculture; the damming of rivers, which alters salinity; and, especially in the past 10 years or so, conversion of mangroves to commercial

BOX 2.2 Fires and the Forest

Fire is one of the major threats to forests all over the world. In 1997, the Southeast Asian region was beset by an unusually dry season and this, combined with fires set by people, caused extensive damage and haze, particularly in Indonesia. Satellite imagery led analysts to the conclusion that fires were being deliberately set in Kalimantan, Sumatra, Sulawesi, and Irian Jaya to clear areas for oil palm plantations and, to a lesser extent, for rubber estates and tree plantations. In Kalimantan and Sumatra it was estimated that nearly 2 million acres (750,000 ha) of forests burned (M. Hiebert, 1997). This has occurred in spite of a 1995 Indonesian government ban on burning forests to clear land. Certainly a growing global demand for palm oil, a demand which has increased Indonesian exports of the commodity by 32 percent in the five-year period to 1997, has boosted the demand for new plantation lands. In addition to Indonesia, smog from the fires has been carried by winds to Singapore and Malaysia, threatening the health and livelihood of the population in these two countries. In Kuala Lumpur and other Malaysian cities, air pollution levels reached "unhealthy" levels causing many to wear surgical masks. In Sarawak pollution levels attained "extremely hazardous" levels, causing government offices, business, and schools to close (Cohen, 1997). In Indonesia, it was concluded that the smog from the fires was the principal cause of the crash of one commercial Singapore Airline flight in Medan, northern Sumatra, and the sinking of one commercial passenger ship in which scores of passengers perished.

This is not the first time, nor will it be the last, that smoke from fires set in Indonesia have caused devastation. Fires also occurred in 1982–83 when 9 million acres (3.6 million hectares) were destroyed, and in 1991 and 1994. Following the 1994 fires, ASEAN officials established an ASEAN Cooperation Plan on Transboundary Pollution which included an agreement to work toward preventing forest fires.

fish and shrimp ponds (Photo 2.5). High Japanese demand for shrimp, for example, has led to the destruction of thousands of hectares of mangroves in the Philippines, Indonesia, and Thailand.

All of Southeast Asia's natural vegetation has undergone extensive and rapid transformation (Maps 2.6a and 2.6b). These changes will be discussed in more detail in the section on contemporary environmental problems, but at this point it is important to note the considerable and critical commercial value of many of the forests in Southeast Asia. In the post-World War II period the majority of tropical wood products have come from Southeast Asia; in the 1980s it is estimated that approximately 85 percent of all tropical wood product exports came from the region, primarily Indonesia and Malaysia (Gillis, 1988). The high commercial value of the forests in Southeast Asia is in contrast to tropical forests in other regions which, due to the large number of different species of trees per hectare, have been of limited commercial value. With the exception of some of the forests in Western Africa, the forests of Southeast Asia are unique among tropical forests in having high-valued tree species in large

Photo 2.5 Commercial fishponds in central Java. (Leinbach)

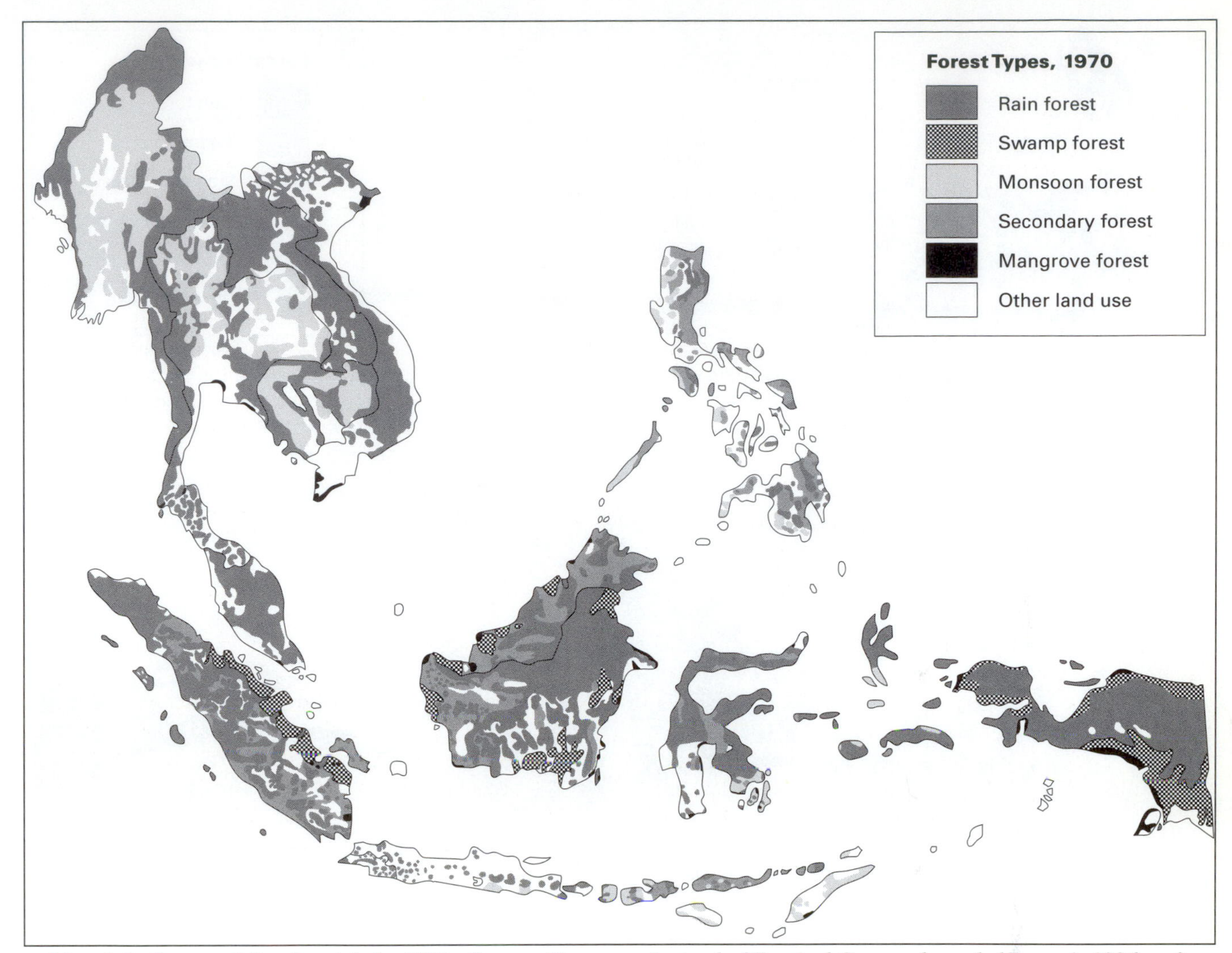

Map 2.6a Forests of Southeast Asia, 1970. *Source: Singapore Journal of Tropical Geography, vol. 17, no. 1, 1996, p. 3.*

numbers per unit area. While these forests still contain a great deal of tree species diversity, their commercial yields as measured by cubic meters of wood per hectare can be very high. Two forest types stand out in this regard, the teak and dipterocarp forests of the region.

Burma and Thailand are the world's primary producers of teak, a hardwood used in shipbuilding and furniture making, which is found in both natural forests and on plantations. However, the forests of Southeast Asia, which have undergone the most rapid transformation recently, are those composed of tree species from the family *Dipterocarpaceae*. Dipterocarps, as they are commonly called, occur in fairly even stands that contain large amounts of harvestable wood per hectare. In addition, there is strong overseas demand for these logs and the wood products, primarily plywood and veneer, derived from them. Although Japan has been by far the most important destination of wood products from Southeast Asia, other important consumers include the United States, Taiwan, Hong Kong, and South Korea. The result has been massive logging of dipterocarp forests in the past 50 years, particularly in the Philippines, Indonesia, and Malaysia.

Finally, mention must be made of changes in vegetation brought about by deliberate replanting programs. As a result of the obvious effects of deforestation and domestic and international pressure, several of the governments in the region, notably the Philippines, Indonesia, Malaysia, and Thailand, have implemented large-scale reforestation programs. In addition, some logging companies that have pulp and paper mills have also initiated reforestation programs to ensure themselves of a steady supply of softwood in the future. Of importance for this section is that virtually all of these programs have involved about a half-dozen tree species, all of them exotic, that is, not native to the region. Because reforestation efforts in the region will most likely increase in the future, the long-term effect of these programs on biodiversity is an important concern.

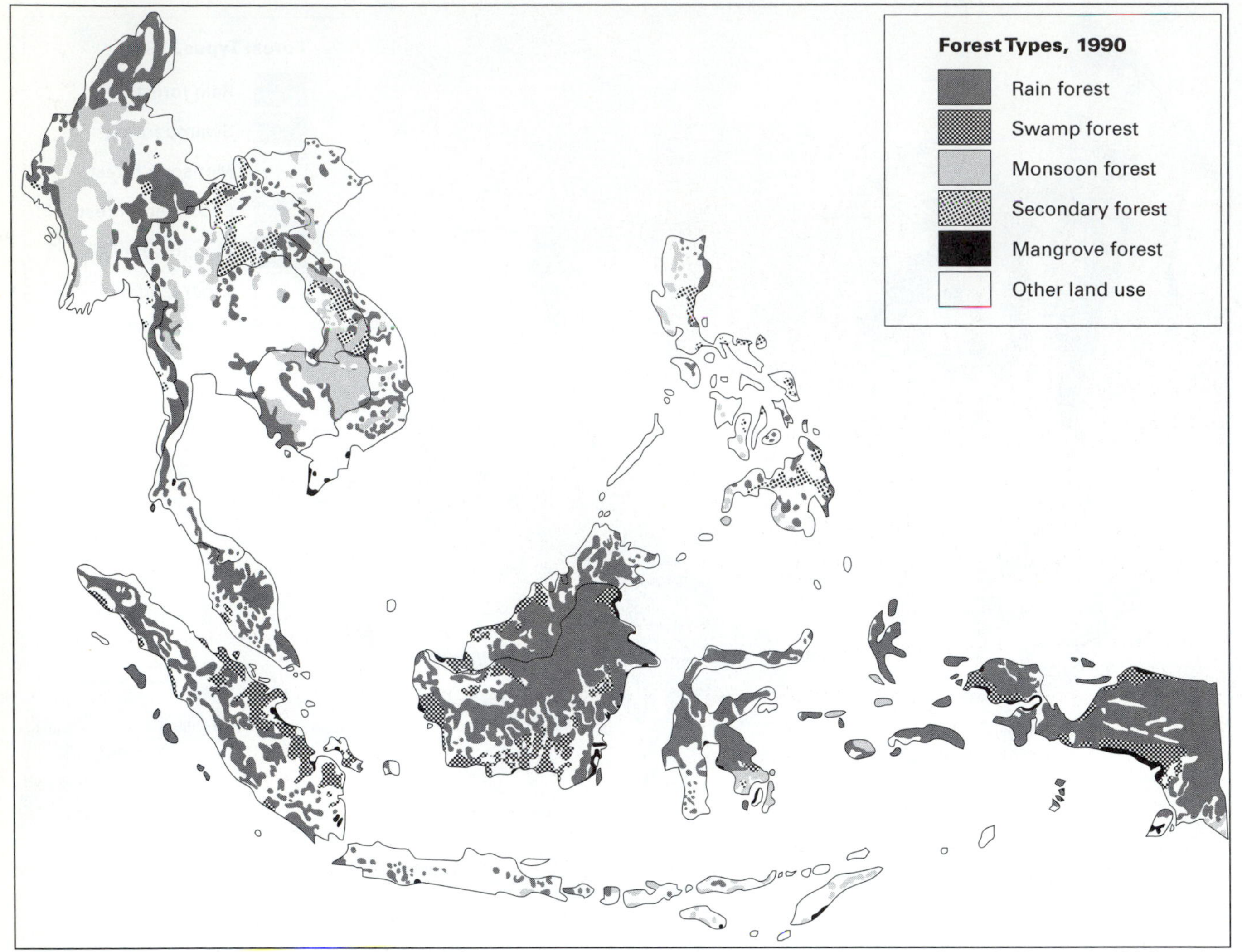

Map 2.6b Forests of Southeast Asia, 1990. *Source: Singapore Journal of Tropical Geography, vol. 17, no. 1, 1996, p. 4.*

MINERALS AND ENERGY RESOURCES

In general, Southeast Asia is relatively poor in terms of its mineral endowment. Two countries in the region, tiny Brunei and Singapore, have virtually no minerals, and, as a result of political instability, four other countries have no significant mineral exports (Burma, Laos, Cambodia, and Vietnam) at the present time. Thus, the only four countries with substantial mineral production and export are Indonesia, Malaysia, the Philippines, and Thailand.

Tin is the most significant mineral ore from Southeast Asia—30 percent of all tin produced in the world comes from the region. This, however, represents a sharp drop from the late 1970s when the region produced approximately 50 percent of global output (World Resources Institute, 1994). Indonesia, Thailand, and Malaysia ranked second, sixth, and tenth, respectively, in terms of world production in 1994 (World Resources Institute, 1996, p. 291). Of special interest is the fact that Malaysia, which was formerly the world's largest producer of tin, has been eclipsed. This is the result of two factors: competition from lower-cost producers both inside and outside (China, Peru, Brazil, and Bolivia) the region, and because the most easily worked tin deposits in Malaysia have already been mined.

Another important mineral is nickel, Indonesia is the world's third-largest producer with the bulk of production coming from Irian Jaya. In the future, Irian Jaya will continue to play a major role in Indonesian mineral production as several large mines are opened. These mines will produce primarily copper and gold. As with dam projects, the environmental consequences of developing large mining projects in what is essentially virgin forests are the negative effects on both the physical environment and the indigenous peoples of the area. The Philippines used to be a major producer of nickel, chromite, and copper, but output has declined considerably in recent years. The Philippines also has substantial gold deposits; in fact,

Photo 2.6 Tin mine in Perak, Peninsular Malaysia. (Leinbach)

after South Africa, the Philippines has more gold per unit area than any other country in the world.

Of more importance for the region as a whole are resources of petroleum and natural gas. Approximately 5 percent of world production occurs in Southeast Asia. The three major producers are Indonesia (a member of the Organization of Petroleum Exporting Countries, or OPEC), Malaysia, and Brunei. In addition, Vietnam is rapidly increasing production and exports of petroleum. In all four countries, the majority of the petroleum and natural gas production is exported, with Japan the major market. Such exports generate between 10 and 15 billion dollars a year in export earnings. In the case of Brunei, its gross national product (GNP) per capita is one of the highest in the world due to these exports. In the cases of Malaysia and Indonesia, these export earnings have consistently been one of their largest sources of foreign exchange. Approximately 20 percent of Indonesia's foreign exchange earnings comes from oil and natural gas, and 10 percent of Malaysia'a earnings are from the export of oil and natural gas. Almost all of Brunei's exports are derived from energy; even Vietnam, which had minimal exports of oil until recently, now derives approximately 25 percent of its foreign exchange earnings from oil. The region has significant reserves of oil and natural gas, but they account for only slightly more than 1 percent and 3 percent of world reserves, respectively (Table 2.1).

Although oil and natural gas dominates the commercial energy sector of the region, there are other sources of power. Coal is important in Thailand, Vietnam, and Indonesia, and geothermal energy is important in the Philippines. In fact, the Philippines is the world's second-largest producer of geothermal power after the United States. In addition, hydroelectric power is found throughout the

Country	Crude Oil (million metric tons)	Natural Gas (billion cubic meters)	Hydroelectricity: Known Exploitable Potential (megawatts)	Hydroelectricity: Installed Capacity, 1993 (megawatts)
Burma	7	278	160,000	288
Cambodia	n.a.	n.a.	83,000	0
Indonesia	759	2,000	709,000	2,169
Laos	n.a.	n.a.	22,638	235
Malaysia	585	2,150	59,229	1,439
Philippines	33	98	31,951	2,055
Thailand	27	175	8,169	2,459
Vietnam	68	105	6,490	1,864

Table 2.1 Proved recoverable reserves of crude oil and natural gas, and hydroelectric potential and capacity, 1993. *Source: World Resources Institute, 1996, p. 289.*

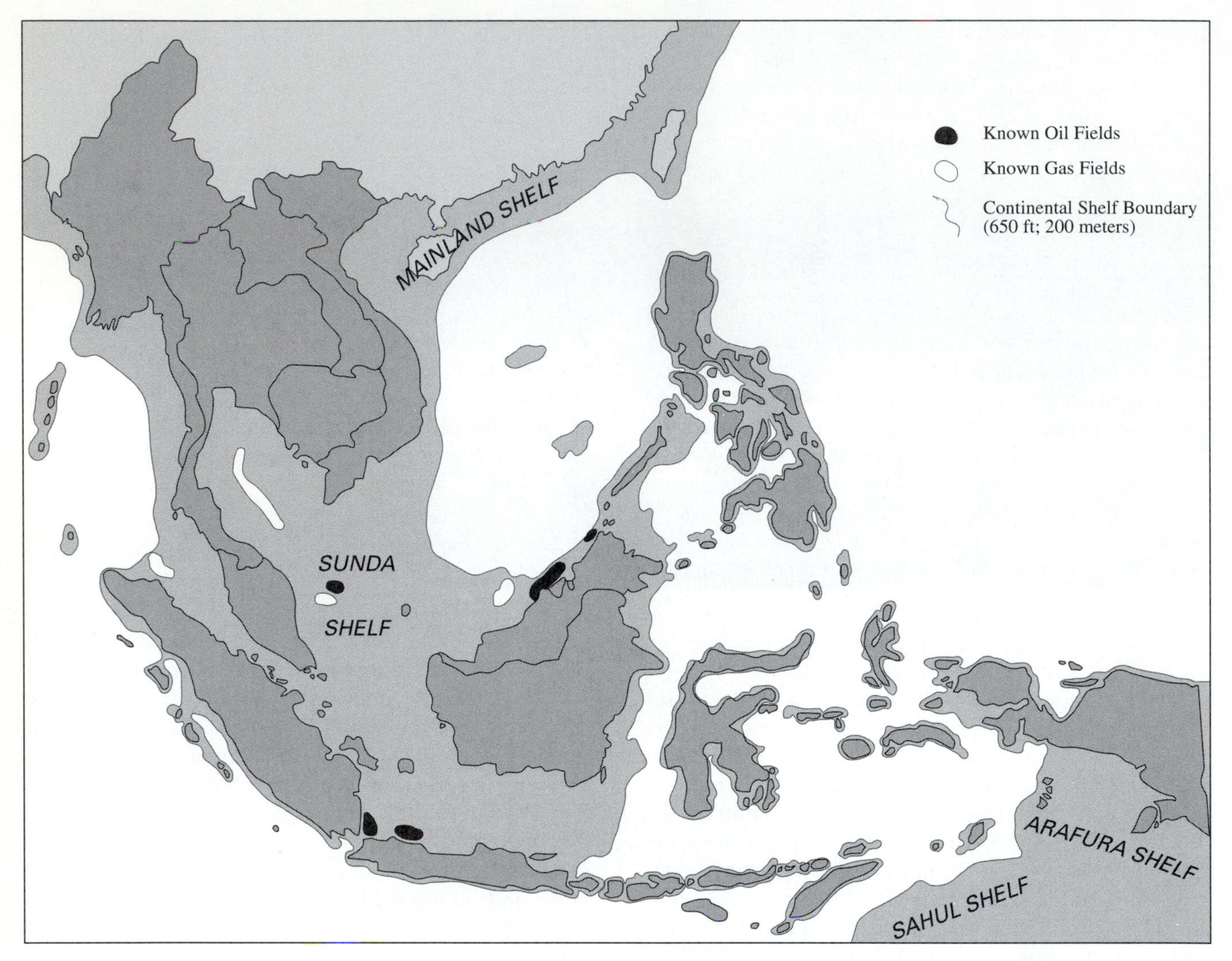

Map 2.7 Continental shelf and known offshore oil and gas fields. *Source: Ulack and Pauer, 1989, p. 8.*

region but, in general, presently plays a relatively small role in terms of its overall contribution to total electricity generation. Hydroelectricity constitutes a significant future source of energy for the region (Table 2.1). Finally, mention must be made of the fact that biomass (vegetative matter, primarily in the form of fuelwood and charcoal) is an important source of energy, particularly in the rural areas.

Consumption of energy is increasing rapidly in the region, particularly in those countries that have industrialized and urbanized the most. Singapore's per capita consumption of commercial energy is equal to that of most Western European countries. It is expected that the region as a whole will continue to experience strong increases in per capita energy demand well into the twenty-first century and this raises several major questions. First, because most energy production is based on fossil fuels and these are a major contributor to greenhouse gases, the long-term effect of increased energy consumption on global climate change cannot be ignored. If per capita consumption of energy in the region increased to the level of Singapore, that would mean that Southeast Asia would produce as much, or more, greenhouse gas than either North America or all of Western Europe. In other words, the economic development of the region could have global environmental effects. Second, because most of the production of oil and natural gas in the region occurs in offshore areas, the potential negative environmental effects of increased production, for example, oil spills, should also be considered. An offshore area that may achieve notoriety in the future is the South China Sea where the Spratly Islands are located (see Map 10.3). This disputed archipelago is claimed in part or in its entirety by China, Taiwan, Vietnam, Malaysia, the Philippines, and Brunei. All but Brunei have had troops on at least one of the tiny islands in the Spratly group even though not one island in the archipelago has freshwater. While some of the claims to the Spratly Islands are based on historical or geopolitical considerations, much of the tension over the islands stems from the fact that experts believe they are located over substantial deposits of oil and natural gas. In fact, both Vietnam and China have recently awarded exploration contracts to Western oil companies in regions which are close to the Spratlys. In short, the Spratly Islands are an example of overlapping territorial claims exacerbated by the probable presence of fossil fuels.

MAJOR ENVIRONMENTAL ISSUES

Before addressing some of the specific environmental issues of regional importance, it is important to keep in mind that for almost all Southeast Asian countries a major goal is rapid economic growth through development. Singapore has for decades been considered one of the four "Little Dragons" (along with Hong Kong, South Korea, and Taiwan) and now has one of the highest standards of living in the world. Brunei, because of its energy exports, has one of the highest per capita incomes in the world. As of the late 1990s, many observers consider Malaysia and Thailand to be the next candidates for newly industrializing economy (NIE) status. Both Vietnam and Indonesia have recently liberalized their economies and are actively courting foreign investment. Of the remaining four countries, Burma and Cambodia continue to be preoccupied with domestic strife, Laos is starting to make its first tentative moves to expand its interaction with the outside world, and the Philippines has made economic growth a priority.

In short, with the exception of those countries that are experiencing internal turmoil, all the countries of Southeast Asia are committed to rapid, long-term economic growth based on the model of the industrialized West. In fact, the area as a whole has been one of the most rapidly growing areas in the Third World during the post-World War II era. For example, some areas of Malaysia and Thailand have gone from primarily agricultural to industrial-based communities in as little as one or two generations. In many cases, the rates of socioeconomic change observed in Southeast Asia are occurring much more rapidly than anything that has happened in the industrialized West. It is equally important to note that the major reason economic growth is being pursued so vigorously by most Southeast Asian countries is that it is viewed as the best means to eliminate poverty. Although there are other reasons to advocate economic growth, most countries in the region still have large proportions of their populations living in poverty. Any discussion of Southeast Asia's present and future environment must recognize that rapid economic growth will most likely be the norm in the region for many decades to come; as an example, the stated goal of Malaysia's "Vision 2020" is to achieve developed country status by the year 2020.

Two other points are appropriate here. First, environmental change brought about by human activity has a very early origin in the region. Human induced environmental change started with settled agriculture that may have occurred as long as 6,000 or more years ago. Moreover, small-scale deforestation, particularly in accessible lowland areas, was most likely simultaneous with sedentary agriculture. What is unique about the contemporary period is that change has been extremely rapid and of wide geographic extent. The second point is that rapid rates of economic, social, and environmental change mean that our ability to predict future events and anticipate future problems is necessarily limited. It is important to keep in mind that no observer of Southeast Asia in the 1940s and 1950s came close to predicting the incredibly rapid economic and demographic changes that have occurred since then. As late as the 1960s and 1970s, influential agencies such as the United Nations and the Asian Development Bank were advocating more rapid exploitation of natural resources with an almost complete disregard for the environment. The prominence given to environmental issues, particularly in a Third World context, is relatively new. In short, not only do environments and economies change rapidly but so do our perceptions of these changes and their relative importance.

Deforestation

Most knowledgeable observers and researchers would agree that the most severe environmental problem in Southeast Asia has been rapid and large-scale removal of forest cover and the loss of soil that has accompanied this removal. Three reasons are commonly given for this state of affairs: shifting cultivation, the expansion of settled agriculture, and logging.

Traditional shifting cultivation, or swidden, is an agricultural system that is land extensive and suitable for areas with low population densities. It involves cutting down the existing forest, drying and burning the woody material, growing crops for one or two years, and then abandoning the field as weeds increase and yields decline due to the loss of organic material. If population density is low and land is available, the original cultivators will let the abandoned land lie fallow for anywhere from 10 to 100 years before returning to repeat the cycle. Shifting cultivation has been practiced all over the world and, in many cases, can be considered to be a form of sustainable agriculture; however, the system can break down when there is not enough land to maintain long fallows or when subjected to outside forces such as in-migration or logging. If fallow periods are too short, then the forest will not have sufficient time to return and the creation of grassland areas will occur. In fact, there were fairly large areas of grassland in the region before the Europeans arrived around 1500. Today, swidden is still practiced in most Southeast Asian countries, particularly in relatively isolated upland areas such as parts of Mindanao,

Kalimantan, Irian Jaya, Laos, and northern Burma and Thailand. However, these systems are coming under stress as the area available to them decreases and, thus, it is most likely that traditional shifting cultivation is causing some deforestation in the region today. The availability of chain saws also means that larger areas of forest can be cut down more quickly (Photo 2.7).

In addition to traditional shifting cultivation, deforestation is also caused by both spontaneous and planned migration to forested or formerly forested areas. In most cases, it is primarily undertaken by lowland families who are seeking to engage in subsistence agriculture. In general, these families have no choice due to the absence of economic opportunities in lowland agriculture or urban areas. On a large scale, this process has occurred throughout the Philippines in the postwar period, in parts of Indonesia (Sumatra, Kalimantan), and in northern Thailand. In short, poverty is forcing poor families to engage in frontier migration and, hence, these people are sometimes referred to as "shifted cultivators." An outstanding characteristic of many, if not most, of these households is that they are not familiar with the environment they are migrating to and, therefore, are not as knowledgeable regarding appropriate agricultural techniques as are the local inhabitants. In addition, many of these frontier areas are environmentally fragile because they have hilly or mountainous topographies and, therefore, agriculture is often being practiced on lands that are steep (e.g., parts of Sumatra, northern Thailand, and parts of Mindanao). The shifting cultivation practiced by migrants in these environments is usually much more destructive than that practiced by long-term residents.

The last major agent of deforestation is large-scale commercial logging. The forests of Southeast Asia have a high commercial value. Recall that they contain tree species like teak and dipterocarps that are valued on the international market and that these species occur in large numbers per unit area. Because stocking densities are high, yields in terms of cubic meters of wood are also high per hectare. These forests have been heavily logged in the post-World War II period with many of the wood products exported to Japan and the United States. As we have noted, most of today's exported tropical wood products originate in Southeast Asia; indeed, more than 50 percent of all tropical wood exports come from the island of Borneo alone (Brookfield et al., 1990) (Photo 2.8). An important factor in explaining the rate and scale of logging in Southeast Asia has been that it has been very lucrative for a small number of individuals, primarily owners of logging companies and their political allies. The military has also benefited handsomely in countries such Thailand, the Philippines, and Indonesia. The entire process has been accompanied by considerable corruption, particularly noticeable in Indonesia, Malaysia, the Philippines, and Thailand (Callaham and Buckman, 1981; Vitug, 1993; Sricharatchanya, 1987; Broad and Cavanagh, 1994).

There have been three major forms of corruption in the forestry sector. The first is illegal logging, which is logging for commercial purposes without a permit. The second is underinvoicing timber exports and the smuggling of logs overseas. In the case of the Philippines, it is estimated that the vast majority of logs exported are done so illegally ("A Brazilian Tale," 1989). The third is the granting of concessions either for political or monetary favors.

Photo 2.7 Shifting cultivation clearing done with chain saws in central Borneo, East Kalimantan Province. (Kummer)

Photo 2.8 Logs from central Borneo floated down Mahakam River arriving at the port of Samarinda, East Kalimantan Province. (Kummer)

All three forms of corruption have been immensely profitable to those with the right connections because of the high price that timber exports fetch on the world market. Despite the attention increasingly being devoted to the environment in general and to the tropical rain forest in particular, corruption in the forestry sector is still common. Although not the focus of this chapter, the issue of corruption is important because it indicates that some people may personally gain from and, thus, have a personal interest in the continuation of deforestation. Because some benefit from environmental destruction, an understanding of who gains from the rapid deforestation that is occurring in Southeast Asia is important in attempting to halt or reverse the process. In other words, the political economy of forest resources, that is, who has access to and control of the forests, is among the most important features of their use today.

There are numerous positive effects of deforestation. First, the process provides employment. Loggers, haulers, and those who use tropical wood products, such as furniture manufacturers, all gain a livelihood from the harvesting of timber. In an area where poverty and unemployment may be high, this can be an important benefit. Second, the export of tropical wood products can provide substantial foreign exchange earnings. This has certainly been the case for Indonesia, Malaysia, and the Philippines and, to a lesser extent, for Burma, Cambodia, and Laos. Lastly, the elimination of forest cover allows for the expansion of agriculture whether it be of the subsistence or commercial type. In other words, the natural forest is replaced with an agricultural system that can produce subsistence and cash crops. Malaysia in particular and Thailand to a lesser extent have both generated substantial export earnings from the expansion of commercial agriculture onto previously forested lands. In short, from the viewpoint of many individuals, deforestation represents development because the process of deforestation itself provides jobs and foreign exchange. In the eyes of some, the replacement of the natural forest with agriculture represents a "better," that is, a more "productive" use of the land.

However, there are also clearly several negative aspects to deforestation. The process of removing forest cover, particularly if done on a large-scale basis such as capital intensive commercial logging, causes considerable damage to the remaining forest and the soil. If logging is followed by short-term agriculture and the land is abandoned after a year or two, the resulting landscape is almost invariably severely eroded. In short, rather than the natural forest being replaced by a productive system of agriculture, it is more often the case that the productive forest is replaced by a badly degraded land surface that is not capable of sustaining intensive agriculture. The extensive areas of almost completely unproductive grasslands throughout Southeast Asia are eloquent testimony to a destructive process of recent deforestation (Photo 2.9).

A key element in the creation of these grasslands is the process of soil erosion. Due to the elimination of many large trees and the damage to the soils caused by heavy equipment, logging itself can often result in a substantial increase in erosion. The process is completed when agriculturists remove the remainder of the vegetative cover, plow the land, grow crops, and then abandon their farms.

Photo 2.9 Grassland area on Cebu (central Philippines) that was under forest cover as recently as 15 years ago. (Kummer)

The resulting soil erosion can often be so severe that the land, for all practical purposes, is of little value for agriculture. In addition to these on-site effects of soil erosion, there are several off-site effects as well. Most eroded soil is deposited in water bodies. This siltation can harm animal life and raise riverbed levels, which can increase the chance and severity of flooding. In addition, siltation can accumulate behind dams and reduce the economic life of hydroelectric projects. It has also been found that eroded soil is reaching the coast, where it is being deposited on coral reefs, effectively killing them.

Another major negative aspect of deforestation in the region is its effect on the biodiversity of the forest. Deforestation has led to the extinction of numerous forms of life. In the Philippines, where most of the original forest cover has been removed, the Philippine Plant Inventory Project has estimated that as many as 80 percent of all species have already become extinct as a result of deforestation. While it is difficult to quantify the precise effect of this loss of biodiversity, many observers feel that, in the long run, this may be the most serious aspect of tropical deforestation. The tropical rain forest contains more biodiversity than any other ecosystem in the world and the impoverished ecosystems in Southeast Asia today may be one of the long-term legacies of the rapid and widespread deforestation that has occurred since 1945.

Land degradation and the loss of biodiversity are the two most serious negative consequences of deforestation, but there are others. One is the impact on the numerous ethnic minorities who live in the forests. One well-known group in this regard is the Penans of Sarawak, Malaysia. Only a minority of all Penans are still nomadic hunters and gatherers; however, they have established roadblocks in Sarawak to prevent further logging on their traditional lands and have achieved considerable international notoriety (Colchester, 1989). In this case, deforestation necessarily means the destruction of their homeland and, quite likely, their culture. It has been claimed by some that cultural diversity is akin to biological diversity, and if this equivalence is accepted, then loss of distinct cultures is an irretrievable loss of part of the world's cultural heritage. More immediately, the case of the nomadic Penan demonstrates that in the process of development some "lose" and some "win." Whether or not this is inevitable will be discussed in the concluding section. In general, the people who live in forest or upland areas tend to be ethnic minorities in their own countries; as such, deforestation almost invariably means the destruction of the homelands of people who are not lowlanders and are not from the dominant ethnic group. This process is occurring in the uplands of virtually every country in the region.

Lastly, deforestation necessarily involves the destruction of the amenity resources of the forest. Given the concerted effort by many of the nations of Southeast Asia to increase tourism as part of the process of development, large-scale removal of the forests will make future efforts regarding ecotourism, for example, more difficult. In short, deforestation assumes that the amenity value of the forest is zero or, at best, very low, a proposition that is patently false.

Because deforestation involves costs and benefits to society and to individuals, an evaluation of its overall effect is difficult. This is particularly the case because many of the costs of deforestation cannot be expressed in monetary terms; for example; "How much is a species worth?"

In addition, the winners and the losers are almost always different groups of people, and it is difficult to make comparisons across groups; for example, "If poor minority people in the uplands benefit from cutting down the forest for fuelwood but this causes erosion that negatively effects the yields of middle-class farmers in the lowlands, who is to say that society as a whole is better or worse off?"

The above questions notwithstanding, almost all observers are agreed that land degradation in Southeast Asia as a result of deforestation and poor agricultural practices is the most serious environmental problem confronting the region today. This is so for several reasons. First, land degradation covers a large area. Every country in the region suffers from this problem. Second, large numbers of people are effected, not only in the rural areas where the land degradation is occurring but also in lowland and urban areas that are impacted by the off-site effects of siltation and flooding. Because deforestation is still occurring at a rapid rate and agriculture is spreading into more marginal hilly or mountainous areas, it is safe to assume that land degradation is an ongoing problem and, in fact, may be worsening (Photo 2.10). Eventually, agricultural yields may start to decline if the process continues much longer. While there is some evidence to indicate that agricultural yields per hectare have declined in some areas, there is no evidence to prove that agricultural productivity over larger areas is declining. If such evidence became available, it would be a dramatic indication of just how far the process of land degradation had reached.

What has happened to the commercial forests of Southeast Asia since 1945 is not a new phenomenon and it is certainly not unique. It is very similar to what has happened to other valuable resources as they were incorporated into the global capitalistic system for the first time, for example, the southern and midwestern forests of the United States in the nineteenth century, fur-bearing animals of the northern Pacific Ocean in the eighteenth and nineteenth centuries, or gold deposits in Australia, Canada, and the United States in the nineteenth century. These "first-generation staples" are never exploited in an environmentally sound manner; instead, rapid booms are followed by rapid depletion and, as the resource becomes more and more depleted and degraded, a reaction designed to conserve the resource develops. In the case of the forests in Southeast Asia, it is obvious that the region as a whole is still in the rapid-depletion phase; at the same time, more and more voices critical of this path are being heard. Whether or not the voices favoring conservation and preservation will have a significant impact is unclear at this point. In this regard, two points are significant: First, there has been a remarkable increase in nongovernmental organizations (NGOs) that deal with environmental issues in most countries in the region; second, several countries, most notably Malaysia and Indonesia, have developed a series of national parks to protect unique and representative environments. Although some of these parks exist on paper only, there is a rising demand from within the region for conservation and preservation of natural resources.

Resettlement Projects and Their Impacts

As noted above, deforestation can result from spontaneous migration or if planned settlement occurs in previously forested regions. In terms of the latter type,

Photo 2.10 Smallholder agriculture moving up the slopes of Mt. Maquiling, Laguna Province, the Philippines. (Kummer)

planned settlement, there are several examples of such deliberate expansion of agriculture through large-scale government programs in the post-World War II period.

Governments as different as Thailand, Vietnam, and the Philippines have all had land development programs since 1945; the prominent example of this is the Indonesian Transmigration Program, which has moved several million families from the densely populated Inner Islands of Java, Madura, Bali, and Lombok to the Outer Islands of Sumatra, Kalimantan, Sulawesi, and Irian Jaya where population densities are much lower and agricultural land is available (Photo 2.11). Malaysia has also had an extensive land development program that has concentrated on the conversion of tropical forests to rubber and oil palm plantations. This is the principal reason that peninsular Malaysia has been one of the largest producers of rubber and oil palm for the past 20 years. In the case of Malaysia, it is important to keep in mind that the replacement of natural forest by commercial agricultural plantations has been deliberately undertaken by the national government. The conversion process has, to a large extent, been directed by government planners through the Federal Land Development Authority (FELDA). It is reasonable to argue that the Malaysian people today enjoy a standard of living higher than most of their neighbors, in part because of the success of developments in plantation agriculture (Sutton, 1989).

Both the FELDA and Transmigration programs have demonstrated that large-scale resettlement and land-conversion projects are possible in a tropical environment. Of the two, however, FELDA has been the more economically successful in terms of raising standards of living and producing tropical products for export. At the same time, both programs have been criticized for their perceived negative impacts on the environment. The conversion of tropical forest to smallholder or plantation agriculture has necessarily meant the nearly complete destruction of the original vegetation in the receiving area (Secrett, 1986). The negative environmental effects include: species extinction, soil erosion which has resulted in land degradation and siltation of waterways and the displacement of indigenous peoples. In addition, there is some question as to the long-term agricultural viability of some of the projects, particularly the Transmigration Program. If the projects end up being abandoned due to declining agricultural yields and demand, then a valuable natural resource (the rain forest) will have been converted to an asset of very little value (abandoned agricultural land) with no long-term benefit to society as a whole.

Urbanization

Economic growth, at least in the Western sense, has meant urbanization as agriculture declines in relative importance to the more rapid growth of the manufacturing and service sectors. Indeed, urbanization has been proceeding steadily in many of the countries in the region, and with it has come the attendant problems of congestion and air and water pollution. Although this chapter will not cover these problems in depth, it is appropriate to point out that with the increased concentration of economic activity in urban areas, these issues have become more pressing. Because economic activity is certain to increase rapidly in the future (the urban population of Southeast Asia may increase by three times in the next 30 years), it seems rea-

Photo 2.11 A transmigrant settler's home in South Sumatra, Indonesia depicting cleared forest. (Leinbach)

sonable to expect that the geographic extent and magnitude of these problems will also increase. In fact, much of the expansion of metropolitan regions is taking place on formerly intensively farmed agricultural land (Photo 2.12). In such "extended metropolitan regions" (EMR) as Jakarta, Manila, and Bangkok, these problems are already well advanced; in all three cities air pollution is already a serious problem. In Jakarta, approximately 1 percent of the income of its inhabitants is spent on boiling water because the untreated water is not safe to drink. Commuting times from place of residence to place of work have increased dramatically in the region's largest cities. In short, there is a rapidly growing body of evidence to indicate that, on the basis of different indicators, some aspects of the quality of life in several major urban centers of Southeast Asia are decreasing (Brookfield and Byron, 1993; Ooi, 1987).

In addition to the environmental problems concomitant with deforestation, agricultural settlement, and urbanization, there are of course other environmental issues that affect Southeast Asia. For example, mining has caused localized land and water pollution; excessive fishing and inappropriate marine fishing practices such as dynamiting have caused the destruction of coral reefs, particularly in the Philippines—in fact, the degradation of coral reefs is the marine equivalent of deforestation. Chemical pollution as a result of Green Revolution technologies is also widespread. Lastly, depletion of groundwater reserves is occurring near some major urban areas; for example, Bangkok is literally sinking a little every year. In short, there is a broad range of environmental problems in Southeast Asia, as there are in all regions of the world.

It would be unfair and incorrect to give the impression that the environmental picture in the region is completely negative. It is necessary to keep in mind that the environmental change that has occurred in the region has, in many instances, been accompanied by genuine economic growth, which has resulted in improved standards of living. This will continue in the future. In addition, as countries in the region become wealthier, the demand for an improved environment will hopefully increase and the economic ability to provide for it will also increase. This is what has occurred in Japan, South Korea, and Taiwan and is very likely to occur in Southeast Asia.

SOLUTIONS

It has been argued that the major environmental problems faced by Southeast Asia are land degradation in rural areas as a result of improper land use, and pollution in urban areas as a result of rapid economic development. Given the present rapid rates of economic growth and the commitment of the region's governments to further this growth, it is to be expected that urban and rural stresses will increase. This section will examine several solutions that have been proposed.

Putting a Price on the Environment

Economists would argue that one of the main reasons environmental damage occurs is because use of the environment is free. For example, there is no economic cost involved in polluting; hence, the solution is to "get the prices right," that is, to put a value on using the environment. This could be done, for instance, by charging logging

Photo 2.12 The construction of subdivisions near Manila on former agricultural land. (Kummer)

companies for each cubic meter of wood harvested or by charging companies and households for each kilogram of waste discharged into a river or stream. The idea is that if users (whether they be firms, governments, or individuals) are required to pay the cost of the negative aspects of their behavior, they will reduce or eliminate these activities. Putting a price on using the environment would raise the price of resources and encourage increased efficiency and conservation. A good example of the effects of pricing occurs in irrigation: If irrigation water is free or is available at very low cost, it encourages overuse; for irrigation water to be used properly, however, it must be priced properly. Notice that this argument does not just apply to Third World countries; the remarkable misuse of water resources in the southwestern United States is primarily the result of the national government providing water at a greatly subsidized rate. In general, economists argue that the only way to ensure that natural resources are used properly is to guarantee that their pricing take into account both private and social costs. They would further argue that the best way for this to happen is through the market. In short, society must change conventional approaches to provide financial incentives for environmentally sound use of natural resources. In the case of commercial logging in Southeast Asia, for example, the above line of reasoning would require that loggers pay substantially higher fees than they do now for each cubic meter of timber removed. If "excessive profits" were eliminated, the rate of commercial logging would slow considerably.

Sustainable Development

Sustainable development is a vague term even though it is widely used (National Research Council, 1993; Wilbanks, 1994). In a very general sense, sustainable development refers to a type of economic growth that is not environmentally destructive. Such a process would mean that the present generation would pass onto the succeeding generation a natural resource base equivalent to what it had inherited. But sustainable development may also include local participation, empowerment, an equitable sharing of society's resources, limits on consumption and energy use, and a new ethical relationship to the environment. While the goals of sustainability are easy enough to state in a general sense, it is much more difficult to devise a means to attain them. In terms of specific policies or goals, what does sustainable development mean? Is continued economic growth compatible with sustainability, or is there a contradiction between the two? If there is a contradiction between the two, how does one reconcile reducing poverty in the region and at the same time preserving the environment? This trade-off is particularly acute in Southeast Asia where, as previously mentioned, almost all governments agree that economic growth is the best way to solve the region's widespread poverty (Aiken, et al., 1982).

At the same time, the case is increasingly being made that economic growth and an active concern for the environment are not incompatible. Three major arguments are put forth in support of this viewpoint. First, the present path of development is destroying numerous productive natural assets such as forests, coral reefs, and the soil; that is, these natural resources are being used in an unsustainable manner. If these resources were managed on a sustained yield basis, they would be able to provide for economic growth for the long term. Second, the economic activity based on these natural resources results in significant externalities (effects on third parties) such as destruction of animal and plant life, soil erosion, and harmful effects to human beings caused by, for instance, polluted water and air. In this regard, one of the key factors to be aware of is agricultural productivity. If crop yields per hectare start to decline at the national level because of soil degradation, then the question of environmental damage will have entered a dangerous new phase. Third, it is becoming clear that the economic costs to undo the effects of past environmental neglect may be quite high. Several countries of East Asia, including Japan, South Korea, and Taiwan, have been forced to spend billions of dollars to improve their air and water, particularly in urban areas. Significant sums of money could have been saved if these negative environmental effects had been avoided in the first place. Overall, the case is being made that respect for the environment makes economic sense.

At the same time, tens of millions of people in Southeast Asia still live in poverty, and there is a strong feeling that the best way to eliminate poverty quickly is through economic growth. Even though conservation/preservation of the natural environment is often suggested as a desirable goal in and of itself, many Third World countries are quick to point out that there may be significant opportunity costs involved. The cost of preserving a virgin forest is the foregone profit from logging that will not be earned or the people who remain unemployed. In some cases, these will be substantial as the export earnings that Indonesia and Malaysia earn from timber indicate. Two logical questions are: Who should pay the costs of conservation? and Why should Third World countries conserve their biodiversity if they do not benefit financially? One of the reasons that the First World is economically dominant today is because of profligate exploitation of natural resources in the past. In fact, the vast majority of the earth's resources are still being consumed by the First World. To some Third World countries, the First World emphasis on conservation and biodiversity is a deliberate attempt to prevent Third World countries from developing economically. At the same time, it deflects attention from the fact that the First World has engaged in more environmentally damaging behavior than the Third World.

Including the Environment in the National Income Accounts

Another approach advocated by economists is to change the way that economic activity is measured. The stock of natural resources is presently not counted in the national income accounts of nations; as a result, destruction of a valuable natural asset such as forests is counted as a contribution to GNP. Logging increases GNP because it produces a product that has value on the market; however, the decreasing value of the forest asset is nowhere recorded. If the national income accounts included forests (and coral reefs, biodiversity, and other natural resources) as assets, then their destruction would show up as a negative entry, and decision makers could more clearly see the effect that economic activity was having on the national heritage.

Local Control of Natural Resources

Another approach to environmental degradation and, indeed, development as a whole, is the increased emphasis given to local control of natural resources. The idea is that local communities, if given actual control of local resources, are in a better position to manage them for the benefit of the entire locality than agents from outside the community, whether they be firms or government agencies from the capital city. Local control would necessarily mean the decentralization of many government activities and, almost by definition, an emphasis on the small-scale. A key element of community management is indigenous/traditional knowledge. The notion here is that over a period of time, rural communities have successfully adapted to their local environment and, in fact, have become expert in appropriate local land use. Local control would give more credence to this knowledge than is presently done. Local control also implies local participation by citizens and NGOs. In fact, NGOs in Southeast Asia are now much more accepted as part of the development process than they were just five years ago, and empowerment of local communities (and indigenous peoples) is even recognized by some governments as an integral part of sustainable development. In short, the issue of local control is now being presented as an alternative to the present development path with its emphasis on largescale industrialization and urbanization. There is an increasing emphasis on small-scale, participation, and empowerment in development discussions.

While many of these discussions have not led to action, the fact that such ideas are being openly debated is an indication that they are being seriously considered. The Philippines has already created a Muslim Autonomous Area in parts of Mindanao and the Sulu Archipelago, and the idea of increased regional or provincial autonomy on a nationwide basis is being considered. It is too early to tell how successful these efforts will be.

Social Forestry

In the forestry sector, a great deal of emphasis in the past 15 years has been on what is called social or community forestry. In its most basic formulation, social forestry is the simple recognition that forests provide more than just commercial timber. They provide a gamut of economic and environmental goods and services such as wildlife, oils, resins, medicinal herbs, rattan, bamboo, fuelwood, and watershed protection. In addition, many of these products are of benefit primarily to the poorer members of society. From the simple recognition that forests provide noncommercial timber resources have come attempts to legitimate the claims of people other than loggers to the products of the forests. Another aspect of social forestry has been the emphasis on local control of community-based projects such as reforestation. Lastly, social forestry has encouraged the active participation of community groups and NGOs in project design and implementation. Overall, this social view of the forests and what they should be used for is diametrically opposed to the view of the traditional forester who sees the forest primarily as a source of commercial timber. The change of viewpoint is refreshing; unfortunately, in reality, the rhetoric of social forestry has been greater than its actual accomplishments. Throughout Southeast Asia, it is difficult to point to many large-scale social forestry projects that have been successful.

Political Economy

Environmental change occurs in a complex natural and social framework. One of the critical variables of the social framework has to do with the distribution of wealth and power and the rules of the game that regulate access to natural assets. It should come as no surprise that it is usually those with the wealth and power who determine who will have access to valuable natural resources, and it would appear that they usually appropriate the most valuable resources for themselves. This has particularly been the case with prime agricultural land and commercial forests. In the case of commercial logging in Southeast Asia, the governments of Burma, Indonesia, Malaysia, the Philippines, and Thailand have all supported rapid destruction of their forests in the name of development. What is not so well known is that the governments in these countries primarily represent the interests of the military and of political and economic elites. In this context, a fair question to ask is, "How can these governments be made to be more democratic?" or, to follow up on our discussion above, "What right do we have to put pressure on these governments to change to meet our standards?" Once the political economy of resource use is made a central point of the analysis, the issue of how to devise more-representative and environmentally benign socioeconomic systems becomes a lot more difficult because people in positions of power virtually never give them up voluntarily. The discussion has now expanded to include development, the environment, and representative government.

CONCLUDING COMMENTS

Environmental change is the result of the interaction of the physical environment with natural and social processes. An underlying premise of this chapter is that a true understanding of contemporary environmental issues in Southeast Asia requires knowledge of both the physical geography of the region and the forces leading to change in society. This does not mean that one has to be an expert in geology, climatology, and hydrology before meaningful statements about environmental issues can be made. Rather, it means that one should be aware of the relationship, for example, of monsoon rains and agriculture or of soil erosion and slope of the land if developments in Southeast Asian agriculture are to be understood.

The environment of Southeast Asia has undergone widespread and profound change since 1945 and this process of change will continue in the foreseeable future. By the year 2025 there will be fewer virgin forests, more grasslands, fewer species of flora and fauna, more land devoted to agriculture, more and larger urban areas with their attendant environmental problems, and more people (the population of the region is projected to grow from 500 million in 1996 to 718 million in the year 2025, an increase of approximately 50 percent). At the same time, average standards of living will have increased. In short, the people of Southeast Asia will use their environment in a manner similar to the Japanese, Europeans, and North Americans. However, the major difference will be that the rate of change in Southeast Asia will be far greater.

This conventional "vision" of development is increasingly being challenged within Southeast Asia by a diverse group of people who are motivated by a concern for the environment, sustainable development, the rights of indigenous people to preserve their way of life, and the rights of those who have been marginalized by development, for example, the unemployed, landless and, all too often, women. In short, the "environment" and the overall path of "development" are becoming an issue of contention between different classes, regions, and, in some cases, ethnic groups. There is a risk that the winners and losers in the process of growth will become increasingly polarized.

On a more optimistic note, although the discussion regarding the political economy of resource use in Southeast Asia was not particularly hopeful, it does point out one feature of the present situation that is often neglected: In many cases, the governments of the region do have control over the natural resources at their disposal. The best example of this is government regulation of logging because virtually all large-scale logging in Southeast Asia is taking place on government forestland and under government regulations. Although the process has been remarkably destructive, it has not occurred in a vacuum; theoretically speaking, the governments of Southeast Asia could have exercised more control over this process if they had wanted to. This has not been the case for two major reasons. First, the Western model of development sees industrialization as the key to economic growth. This necessarily entails the conversion of natural resources into industrial inputs. Second, and as already discussed, the process has been primarily designed to benefit a small elite. Once again, the political economy of resource use in the region is of paramount importance.

Finally, while it is difficult to generalize about sociopolitical developments in each of the 10 countries of the region, it would appear that there is now more democratic, public space available for different environmental viewpoints to be expressed. This movement promises to bring more groups into the discussion about the nature of development and its effects on the environment. It is part of a wider democratization process occurring in these countries, but it is also a reflection of the fact that increasing numbers of Southeast Asians are no longer content to allow environmental destruction to continue as it has in the recent past.

3 Historical and Cultural Patterns

ROBERT R. REED

This chapter will highlight a common path of cultural history that binds people throughout Southeast Asia. After reflecting on the hazy advent of humans in the region, our attention focuses on the evolution of prehistoric hunting and gathering societies into semimigratory agriculturalists around six millennia ago. We then explore the pivotal protohistoric transition—beginning about 500 B.C.—when Southeast Asian farmers became increasingly sedentary while also developing long-distance trade with South and East Asia. These commercial relationships in turn facilitated cross-cultural linkages with Indian and Chinese civilizations that fostered critical sociopolitical transformations leading to the emergence of complex societies, the beginnings of urbanism, and the rise of territorial states in Southeast Asia. Following a summary survey of the greatest precolonial kingdoms, we describe the forward movement of European powers into the region beginning in 1511 and lasting for 450 years. The chapter concludes with reflections on the material, social, and institutional legacies of Western rule, which set the stage for independent nation-building after World War II and a more autonomous trajectory of Southeast Asian history.

HISTORICAL IMAGES OF SOUTHEAST ASIA

Before World War II, almost all literature on the history and culture of Southeast Asia was framed geographically in terms of individual Western colonies rather than the region as a whole. These narratives—artificially structured by imperial boundaries that transected indigenous states and ethnic territories—generated the mental map of a compartmentalized area devoid of binding cultural traditions. Their authors proved quite adept in portraying the European presence in tropical Asia by focusing on Western military conquests and administrative structures along with colonial development schemes and systems of social change. Such topically and spatially biased historical construction promoted the idea of an always fragmented area comprised of functionally isolated cultures that were derivative of India and China or defined by powerful metropoles in Europe or North America (Legge, 1994, p. 1–15).

While promoting the image of a balkanized Southeast Asia, most historians before World War II ignored the precolonial past or discounted ancient yet persisting sociocultural processes that facilitated regional integration. Nor did they study parallel forms of politico-economic evolution in the Westernized colonies or encourage comprehensive research on such issues as the maldistribution of wealth, rural-urban inequities, and class conflict in Southeast Asia's emerging nations. Consequently they failed to nurture the concept of a unitary region in which a substantial sector of humankind shares a singular environmental domain, similar subsistence strategies, an interlinked cultural and institutional heritage, and certain converging historical experiences antedating the arrival of European colonizers (Emmerson, 1984).

Even as a large corpus of "country studies" and paternalistic chronicles accumulated late in the colonial era, the intellectual groundwork for an integrated perspective on Southeast Asia was being laid by pioneering historians and social scientists. The foremost of these, J. C. van Leur, argued in the mid-1930s that a substratum of indigenous culture remained intact throughout the region despite the introduction of Indian and Chinese politico-religious concepts, the impact of Islam, and the transforming power of European colonialism. In his opinion these recent forces were of superficial influence. Challenging the dogmas of Eurocentrism, he advocated an autonomous area history unfettered by Western prejudices and sentiments as well as open to new avenues of inquiry and innovative modes of interpretation (Leur, 1955). A decade later, Georges Cœdès (1944) in his classic work on the Indianization of Southeast Asia evaluated the complex processes of religious and social change that produced the region's earliest kingdoms. Others soon followed the path opened by van Leur and Cœdès. In

A History of South-East Asia, D.G.E. Hall (1955) demonstrated the utility of comparative studies built on the works of historians and social scientists with mutual interests in bygone cultures and polities. He thus set the stage for modern researchers in many disciplines who envision Southeast Asia as an integral region inhabited by diverse peoples sharing similar historical experiences and a common cultural matrix.

After World War II, a new generation of scholars began to conceive of national identities that seemingly reflected an ancestral map of localized societies and ethnic groupings. As they articulated the ingredients of this precolonial heritage, their collective writings began to project the vision of enduring cultural patterns and an abiding regional identity. These historians also questioned conventional periodizations based on European historiography; scrutinized the development of indigenous polities; verified cultural continuities at the grassroots level over recent millennia; and expanded the scope of inquiry from the colonial era to both earlier and later periods (Reynolds, 1995). Such shifts in topics, emphases, and paradigms heralded a refreshingly original and intellectually challenging treatment of Southeast Asian history and cultural evolution.

Acknowledging the significance of prewar debates about the processes and consequences of external influences on Southeast Asians, some scholars are now pursuing investigations that confirm the enduring integrity of a distinctive indigenous heritage even amidst later evolving civilizations born of Indian, Chinese, and Muslim cultural intrusion. Their writings have effectively authenticated an autonomous area history derived from basal beliefs and traditions. A second yet associated subject—the much examined but still intriguing process of Indianization—also continues to stimulate provocative historical work. The importance of research concerning the cross-cultural transfer of ideas, which undergirds all studies on the protohistoric origins of regional polities, remains undeniable. It was through precolonial connections between South and Southeast Asia that religious, social, and institutional forms diffused eastward to foster new social orders anchored by the innovation of kingship, the implantation of three major belief systems (Hinduism, Buddhism, and Islam), and a derivative blossoming of cities and territorial states in deltaic, coastal, and riverine situations. Finally, a resolute contingent of historians and social scientists continues to analyze the multifaceted colonial experience of Southeast Asians. Freed of Eurocentricism, they now offer balanced commentaries on imperialism as seen through the eyes of both rulers and ruled. Additionally they stress the resilience of ancient cultures despite the onslaught of Western invaders, the intrusiveness of their institutions, and the potency of some foreign ideas.

INDIGENOUS HERITAGE

In recent decades a number of cultural geographers—notably Carl O. Sauer (1952) and Paul Wheatley (1983)—have studied the prehistoric and protohistoric past of Southeast Asia. Like co-workers in archaeology and history, they reject the former image of this region as a "cultural backwater" and testify to its role as a key arena of human innovation. Even in the nineteenth century a few prescient scholars were portraying Southeast Asia as a home of hominids who left Africa more than a million years ago and ventured eastward. Always a realm of great peninsulas, lengthy coastal plains punctuated by deltaic marshlands, and myriad islands, Southeast Asia later became a springboard for the world's first great maritime undertakings. Some 40,000 to 45,000 years ago Southeast Asia was the hearth of Australo-Melanesian peoples who migrated to Australia and islands of the Western Pacific. Many millennia later, in late prehistoric times, Malay voyagers from its archipelagoes journeyed across the Indian Ocean to East Africa. An area of incredible environmental diversity, Southeast Asia may have been an early center of plant and animal domestication besides being the botanical source of staple foods (sago, rice, certain yams, and taro) and other cultigens (bananas, some citrus fruits, mango, breadfruit, sugarcane, rambutan, and betel). During the last 4,000 years it has been a scene of multidirectional migrations and a base for Austronesian (Malay–Polynesian) seafarers, who sailed and settled along a 9,500-mi (15,288-km) oceanic expanse from Easter Island to Madagascar. Only 2,000 years ago Southeast Asia became one of the world's last areas of independent urban genesis. Indeed for millennia this region has been an important stage in the intriguing saga of human diaspora, indigenous creativity, cultural evolution, and environmental change.

Human Presence before History

For more than a century archaeologists have charted the prehistoric existence of archaic hominid species in Southeast Asia and speculated about their modes of ecological adaptation. With his discovery in 1891 of a humanoid skull-cap in central Java (Solo River valley), Eugène Dubois confirmed the ancient presence of *Homo erectus* ("Java Man," or *Pithecanthropus erectus*) in Indonesia. Additional sites in Indonesia, Vietnam, and elsewhere have yielded significant fossils of primordial humans together with a large inventory of stone tools and other artifacts. These discoveries are contributing to an ever more-detailed hominid family tree, especially for the period beginning about 40 millennia ago. Current progress in archaeology is also stimulating revised interpretations of settlement patterns, material culture, environmental knowledge, and impacts on habitat.

Despite extensive finds, the ancestral record of Southeast Asia's first humanlike inhabitants seems woefully incomplete. This reflects species demographics and environmental circumstances. First, the population of *Homo erectus* was minuscule throughout its long regional presence. Second, fossils weathered rapidly or disintegrated in the hot and humid climate. Bodies were likewise scavenged by animals or scattered through cumulative geomorphologic processes. As a result, the meager physical remains of the earliest hominids appear in situations of secondary deposition and are not easily linked to stone tools or primary living sites. Third, global climatic changes during the Pleistocene often jeopardized the already scant regional vestiges of humankind. With the waxing and waning of continental glaciers, Southeast Asia witnessed repeated sea-level adjustments (between 150 and 400 ft, or 46 and 122 m) accompanied by parallel expansions or contractions of land areas. During epochs of water capture by massive ice caps and corollary oceanic retreats, seabeds in the region's western sector were exposed and fused with Borneo, Java, Sumatra, and the Malay Peninsula to form the subcontinent of Sundaland (Map 2.1). Periodic drownings of the Sunda shelf came in turn as warmer conditions triggered glacial melting. On some islands, the rising waters entombed *erectus* under tropical seas and destroyed countless habitation locales. Fragile fossil remains were also lost along the western and southern peripheries of Sundaland, where tectonic and volcanic forces periodically reshaped the landscape. Despite all, enough is known of prehistoric humans in Southeast Asia to permit summary commentary on the presence of *Homo erectus* and the more recent arrival of *Homo sapiens.*

Archaeological evidence suggests that early hominids occupied an environmental niche in Southeast Asia similar to that favored by their distant East African ancestors. Comprised of grasslands broken by brushy forest, the savannas of Sundaland expanded dramatically during each glacial advance. In these periods temperatures and precipitation were marginally lower, vegetation gradually changed, and sea levels dropped to expose plains on once submerged ocean floors. Dense equatorial forests on Borneo, Sumatra, and Sulawesi apparently withstood the ice ages, but in an attenuated form because of cooler temperatures and less rainfall. These same conditions favored an evolution of grassy corridors within emerging forests and woodlands on the exposed seabeds, thus allowing human settlement. Many species of mammals (archaic forms of pig, panther, elephant, bear, deer, and other range animals) thrived on such parklands.

Because the carrying capacity of rainforest is quite low for humans, it is likely that *erectus* foragers preferred the savannas of Sundaland during times of peak glaciation in middle and late Pleistocene times (750,000 to 10,000 years ago). With their high mammalian biomass, diversity of birds, and variety of edible plants, the parklands offered food security. Hominid populations thus fluctuated in harmony with global climatic changes, which triggered major sequential ecological shifts from dense forests to grasslands. With each expansion of the polar ice sheets, a secondarily growing Sundaland accommodated larger numbers of animals and humans. The return of warmer global conditions, glacial melting, and rising seas fostered equally dramatic environmental changes. These included widespread flooding of lowlands, wholesale plant and animal extinctions, declining human populations, and isolation of some hominid subgroups, which were decoupled from the mainland Asian gene pool because they lacked maritime skills.

The fate of *Homo erectus* in Southeast Asia still puzzles archaeologists. Until recently most scholars detected both a distinctive morphological lineage among the original hominids of Sundaland and strong ancestral linkages to early Chinese populations. They further perceived a smooth pattern of evolutionary change with some gene transfer to later humans in Australia and Southeast Asia. Today an alternative view—constructed on a model of evolution that postulates episodic origins of species followed by stasis (the "Noah's Ark" or punctuated equilibrium theory) as proposed by Stephen Jay Gould and Niles Eldredge—assumes that *erectus* represented an evolutionary dead end beyond Africa and left no genetic issue in tropical Asia. This perspective suggests that *Homo sapiens* derived exclusively from an African fountainhead of hominid stock. From there, beginning some 50 millennia ago, modern humans apparently spread throughout the Old World. While debate continues, a few prehistorians are cautiously suggesting that *erectus* might in fact have achieved some transfer of genes to later Southeast Asians through transitional hominid forms analogous to the likely offspring of Neanderthals and *sapiens* in Europe. Even without archaeological proof, they note morphological similarities in certain late *erectus* and early *sapiens* fossils as well as persuasive cultural clues, namely, circumstantial evidence in key regional sites for stone-tool associations linking these possibly related hominids (Bellwood, 1985, p. 1–68; 1994, p. 55–73).

Following the appearance of *sapiens* in Southeast Asia in about 40,000 B.C., the cultural inheritance of prehistoric humans began to take on real clarity. Based mainly on stratified cave deposits subject to reliable radiocarbon dating, the emerging archaeological record displays verifiable links with historical societies and living peoples in terms of skeletal remains, weapons, tools, and fossilized plants substantiating the region's agricultural transformation.

Although the majority of Southeast Asians today are part of the Southern Mongoloid biological sector of humankind and so share a common origin in China's subtropical mountains, they are actually relative newcomers to the region. Their immediate and extant antecedents (Negritos and Melanesians) seem to share a common *sapiens* ancestry with earlier Australo-Melanesians who ultimately settled lands to the east and south. The earliest Mongoloid people began to drift into Southeast Asia at the end of the Pleistocene with a steady increase in numbers during the past 7,000 years. Enjoying a crucial demographic advantage resulting from a potent combination of agricultural and maritime skills, they gradually replaced indigenous Australo-Melanesian foragers whose only regional descendants now dwell in or near New Guinea. The process of gene flow continues even now and is expressed geographically in the clinal zone of eastern Indonesia where Southern Mongoloid peoples are supplanting native Melanesians. In this area the intermarriage of markedly different peoples has produced a gradual biological gradient of human phenotypic differences that are apparent even to untrained observers.

The other survivors of Southeast Asia's indigenous stock are scattered Negritos, who once occupied vast island territories. Now surviving only in mountainous sectors of the Philippines, the Malay Peninsula, and the Andaman Islands, these short-statured and dark-skinned people were archetypal hunter-gatherers. Never numerous, theirs was a semimigratory existence in forest habitats. Only recently did Negritic peoples become farmers under the influence of nearby agriculturists. Still they readily adopted the languages of dominant neighbors and expanded their collecting activities to sustain petty trade with outsiders. Yet the Negritos always resisted irreversible assimilation by retreating to remote areas. Today they can rarely implement this survival strategy. Sadly, the Negritos are threatened by deforestation of essential habitats, agricultural settlement by lowlanders, tourism in once isolated mountain homelands, and the corrosive influences of modernity. Only in the Andaman Islands have Negritic groups preserved some semblance of their ancient lifeways.

HOABINHIAN HUNTERS AND GATHERERS

It was long assumed that the foraging peoples of Southeast Asia lived a nearly effortless existence in intrinsically bountiful habitats. They supposedly obtained food, shelter, and clothing with minimal expenditures of time and energy because fertile soils and gentle climates sustained magnificent forests that served as inexhaustible storehouses of domestic provisions. Such assumptions were incorrect. While adaptive strategies in tropical milieus differ from those of the colder midlatitudes, where labor inputs and harvests are highly seasonal, neither the equatorial rain forest nor the monsoon forest can be envisioned as a cornucopia. The binding ingredient of both environments is a broad yet thin resource base embracing a diversity of fauna and flora. Hence the subsistence forms fashioned to utilize these ecosystems include a variety of dietary items rather than a few staples. Such settings favor small and mobile social units. The resulting hunting-and-gathering system includes two key elements: a generalized tool kit and a broad-spectrum pattern of foraging over extensive territories.

Crystallizing at the interface of the Pleistocene and Holocene periods, the Hoabinhian cultural adaptation loomed large on mainland Southeast Asia and Sumatra until the advent of agricultural societies nearly 6,000 years ago (Map 3.1). Ancestral to the Negritos and perhaps other peoples, the Hoabinhians (named for a type-site near Hanoi) were foragers whose livelihood combined gathering, hunting, scavenging, and fishing. They made distinctive flaked tools and domestic artifacts (grinding and pounding stones, bone points, spatulate tools, and pottery) that testify to the use of assorted plant and animal resources in forest, riverside, and coastal habitats. This adaptation is verified by fish, snake, reptile, and animal bones, and turtle and mollusk shells associated with ancient camps. The less abundant plant remains show no convincing marks of domestication; yet some may have been protected in the wild before collection. In short, their economy rested on a broad base of resources but remained opportunistic. These prehistoric folk evidently utilized any available foodstuff that could be killed, trapped, or harvested.

Although the Hoabinhians ranged widely to amass provisions for their tiny bands, they found seasonal shelter in caverns and rock crevices of karstic limestone hills. It may be inferred from the archaeological record that they sometimes stayed in these secure and comfortable places for weeks or even months. The floors of some Southeast Asian caves provide clear evidence of their long presence in the form of tool assemblages, weapons, and potsherds. The production of pottery by the Hoabinhians represents a rare adaptation among mobile hunter-gatherer societies. In addition, many habitation areas contain heaps of consolidated seashells and freshwater snails. Some shell mounds stood 20 ft (6 m) high and measured 100 ft (30 m) in diameter before being destroyed for the modern manufacture of lime (Bellwood, 1979, p. 53–82; 1985, p. 159–203).

Even while exploiting an ecologically complex base of forest, riverine, and maritime resources, the Hoabinhians also displayed a telling religious impulse by contemplating an afterlife and designing rites to communicate with spirits. They not only buried their dead adjacent to encampments but also placed corpses in flexed positions before the onset of rigor mortis or during secondary burial. Possibly for ritual purposes, some bodies were dusted with reddening hematite. A few writers believe that such entombment indicates forethought about super-

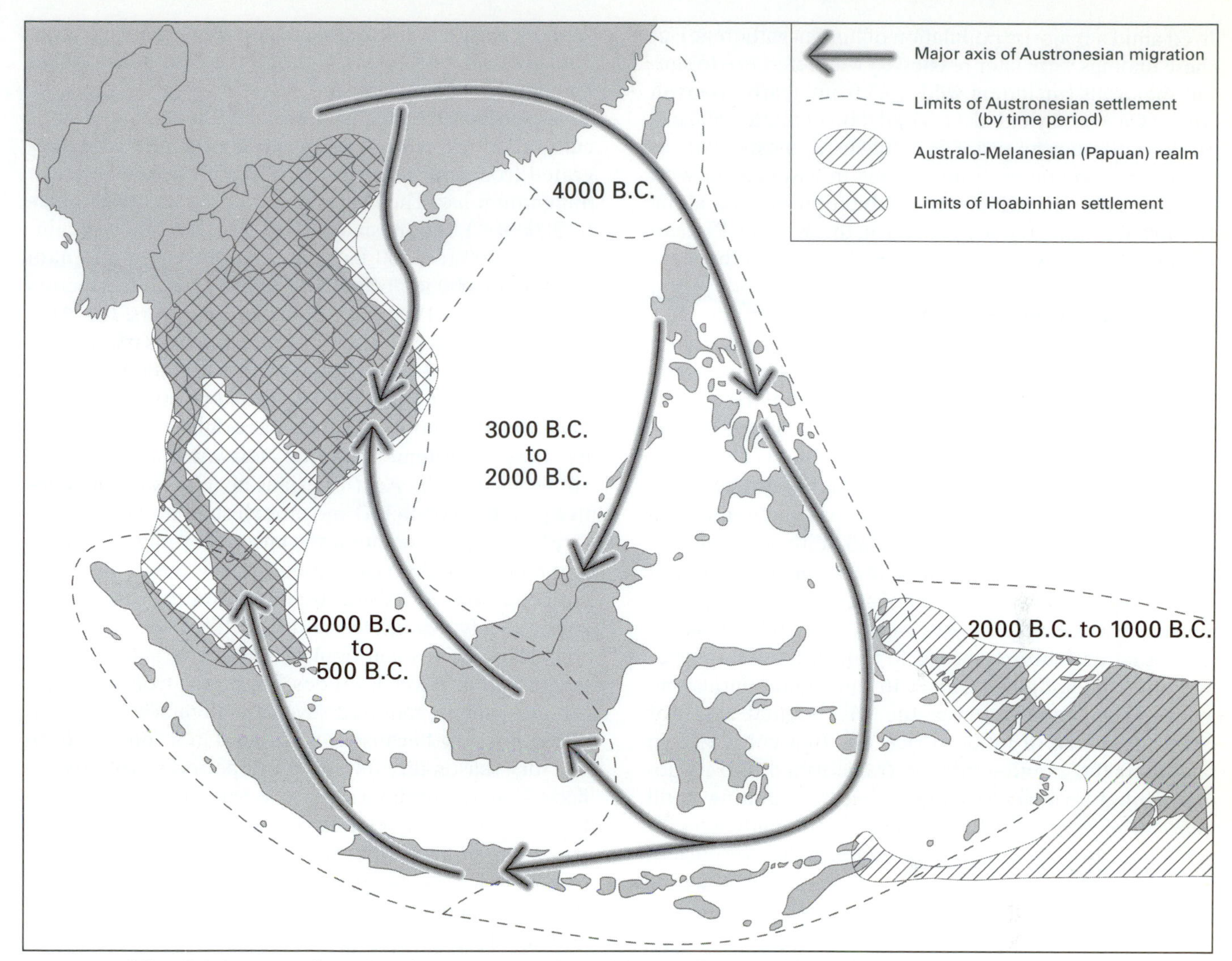

Map 3.1 Austronesian migration. *Source: adapted from Bellwood, 1985, p. 121; Pluvier, 1995, plate 2.*

natural forces and routinized human relationships with denizens of the spirit world (Wheatley, 1965, p. 14–17).

Origins of Agriculture

Substantially influenced by Sauer, who in the 1950s postulated an independent center of domestication in monsoon lands flanking the Bay of Bengal, some scholars ascribe to the Hoabinhians a key role in the origins of agriculture. In his quest for the Old World hearth of cultivation, Sauer (1952) envisioned a human and environmental matrix where the following factors intertwined: (1) a tropical realm marked by an enormous diversity of fauna and flora that provided genetic grist for cultigens and domesticated animals; (2) aboriginal peoples with time to manage "jungle gardens" in a leisurely fashion and "woodland skills" (use of digging sticks for foraging, expertise in fire management for conversion of forests to meadows, and a keen sense of habitat) that were transferable to farming; (3) nascent cultivators who remained semisedentary and so could protect fields from pests; and (4) agriculture based on vegetative plantings of tubers, root, and tree crops. A convergence of these conditions seemingly commenced around 10,000 B.C., or during the florescence of Haobinhian culture. Though plausible, the Sauer model is difficult to test because of a sparse archaeological record. Yet his intriguing ideas were endorsed by respected scholars until recent research fostered a reappraisal of Southeast Asia's role as a center of domestication.

Though failing to sustain Sauer's scenario, modern investigators have greatly refined our understanding of the human experience in Southeast Asia about six millennia ago when farming societies first appeared. Exaggerated claims of domesticated plant remains dating from before 7000 B.C. and nearly as aged pottery in Hoabinhian deposits at Spirit Cave (northwest Thailand) once suggested the presence of ancient agriculturists who supposedly lived amid a majority population of hunter-gatherers. But these findings were later reassessed with reference to similar materials (including wild rice) from nearby Banyan Valley Cave and yielded a revised date of about 3500 B.C. In short, there is no proof of cultigens, domesticated animals, or permanent communities in association with these or other early Holocene habitation sites, let alone the bronze industry proposed for the Khorat Plateau about 3000 B.C. Archaeological evidence from throughout the region now confirms an exclusive dependence on wild plants and animals until the fourth millennium B.C. Even then the domestication process transpired not on the mainland but in the island realm. Scholars presently believe that an agricultural revolution within the bounds of Southeast Asia occurred only once: among Melanesian peoples in the highlands of New Guinea beginning shortly before 4000 B.C. But the products of this hearth in terms of cultigens (sugarcane, bananas, pandanus, and taro) and farming techniques initially diffused eastward to Oceania and did not immediately impact Southeast Asians to the West. As a result of archaeological and linguistic research over the past two decades, it has become clear that agriculture spread to Southeast Asia proper mainly from southern China. In this mountainous and ecologically complex area the wild annual ancestor (*Oryza nivara*) of modern rice was first cultivated in about 5000 B.C. and gradually transformed into its domesticated form (*Oryza sativa*) before radiating to all parts of monsoon Asia. Although not the only agricultural indicator, rice tillage does confirm the presence of more settled peoples who necessarily remain in place for five or six months each year to prepare, plant, weed, and harvest fields.

On the Southeast Asian mainland the dissemination of farming almost surely involved the southerly drift of Mongoloid peoples who spoke Austro-Asiatic languages and gradually replaced earlier Hoabinhian foragers. Though their avenues of expansion await future clarification, there is no doubt about the concurrent crystallization of a Neolithic mode of technology anchored in farming and animal husbandry. It is certain that agricultural societies were securely established by 3000 B.C. in the once fertile flood plains of Thailand's Khorat Plateau, where the archaeological sites of Non Nok Tha and Ban Chiang are yielding rice, other cultigens, domesticated animals (cattle, dogs, chickens, and pigs), wild mammals, pottery, and burials testifying to semipermanent communities; by the latter centuries of the third millennium B.C. in the Red River valley of Vietnam, where the Phung Nguyen inheritance is considered ancestral to the rich Dong-S'on culture that impacted all of Southeast Asia; and by the middle of the second millennium B.C. along a corridor embracing peninsular Thailand and Malaysia, where the Ban Kao culture marked an abrupt arrival of new Mongoloid farmers with their distinctive tool kits. In short, early agricultural societies proliferated in coastal and lowland niches throughout the Southeast Asian mainland between 3500 and 1000 B.C.

In Southeast Asia's island realm, the agents of agricultural diffusion were Austronesian peoples who emigrated from southern China to Taiwan late in the fifth millennium B.C. They then migrated to the Philippines (c. 3000 B.C.), Indonesia (c. 2500 B.C.), the Malay Peninsula (c. 2000 to 1500 B.C.), and southeastern Vietnam (c. 1500 to 1000 B.C.), as well as into uninhabited sectors of Oceania (c. 1500 B.C.). Peerless seafarers, the Austronesians followed a strategy of rapid coastal movement, but only gradual penetration of the mountain interiors on larger islands (Map 3.1). At the same time they readily interacted and melded with aboriginal groups along frontiers of settlement, thereby spawning a diversity of ethnic groups. The Austronesians proved to be an adaptive people who altered and embellished traditional pottery forms, tools, and domestic kits as they migrated. They also adjusted their basic agricultural system to fit different conditions of habitat by incorporating local plants (tubers, leafy vegetables, and fruit trees) into cropping complexes while abandoning or modifying cultigens that languished in lower latitudes. Even rice, tropical Asia's foremost staple, required several millennia of equatorial conditioning because it originated in a zone of sharp wet–dry seasonality and its photoperiod sensitivity reflected a midlatitudinal climate. When introduced to continuously hot, rainy, cloudy, and humid lands, rice plants produced prolific leafy growth but immature grains. Yet following long-term environmental tempering, it now thrives throughout Southeast Asia.

The advent of agriculture was pivotal in the interlinked social and environmental history of Southeast Asia. Resonating a growing complexity of technology and communal arrangements from the forth millennium B.C., the archaeological record confirms the arrival of Austronesians and other southern Mongoloid peoples who transformed forests into fields and so heralded increasing populations and densities of settlement. Unlike hunters and gatherers who frequently changed campsites, these immigrants were pioneering cultivators who only supplemented their primary livelihood with foraging and fishing. They necessarily remained rooted for much of the year to tend farms and protect harvests. As the newcomers' settlements proliferated, their material culture began to reflect local lifeways and distinctive character. By the same token, systems of exchange became more intricate and embraced growing communities of farmers in concert with hunter-gatherers who traded utility products collected from remote forests. The dawn of agriculture in Southeast Asia thus inaugurated a general retreat of indigenous foraging peoples into interior mountain havens and introduction of the most powerful instrument of ecological change ever devised by humankind.

PROTOHISTORIC TRANSITION

During the dim epoch of Southeast Asian protohistory from the middle of the first millennium B.C. to about the third century A.D., the Neolithic tradition that had prevailed in the lowlands for 3000 years was waning. In its place some coastal and riparian peoples were combining metallurgy and fixed-field agriculture, which provided a solid base for sedentary settlement. But even this rich material endowment failed to foster a spontaneous emergence of complex societies as evinced by cities and states.

Dong-S'on Cultural Inheritance

The new and distinctive preurban culture—generically identified as the Dong-S'on heritage—materialized in northern Vietnam about 700 B.C. and spread southward through migration and trade. A composite of Chinese innovations veneered on indigenous environmental wisdom and societal forms, this cultural assemblage was named after a village in Thanh Hoa province where archaeologists have unearthed abundant metal artifacts (utensils, weapons, and ornaments) marking its classic configuration. The most remarkable finds are large bronze drums. Decorated with intricate motifs (birds, animals, houses, and people) and side-by-side running geometric designs, these magnificent objects attest to sophisticated metallurgical skills derived from China. A few archaeologists once claimed that this technology arose about 3500 B.C. in a Thai hearth of bronze manufacture. But scholars now believe it is of Chinese origin and impacted Southeast Asia in the second millennium B.C. An iron industry appeared a thousand years later. Both metals diffused widely after 500 B.C. Still the advent of metallurgy did not signal an abrupt termination of the Neolithic. Instead the protohistoric transfer of technology and ideas was a slow process. Some Dong S'on communities coexisted for centuries alongside more conservative groups who continued to use stone weapons and tools.

Like their Neolithic forebears, who practiced either transient jungle gardening or swidden (shifting cultivation), the Dong-S'on peoples were farmers. However they did not depend exclusively on shifting agriculture; instead they procured food through the practice of *sawah* (wet rice) cultivation complemented by swidden, foraging, and fishing. Their mixed economy illustrated a successful shift from mainly migratory to increasingly sedentary farming. Forsaking the time-honored instruments of fire, ax, and dibble used by shifting cultivators for field tillage, some Dong-S'on folk employed domesticated oxen or water buffalo to prepare flooded farm plots. While freeing them from a wandering existence, wet-rice cultivation also fostered a clustering of population in localities favorable to irrigation and the evolution of social structures that ensured communal cooperation. During the early stages of transition from swidden to *sawah,* settlements were probably temporary affairs. But they gradually stabilized as ever more sedentary farmers improved terraced fields and embryonic hydraulic works and as they recognized the extent of their collective labor investments in a permanent agricultural system (Bellwood, 1985, p. 271–317; 1994, p. 115–136).

Hamlets and villages of the Dong-S'on folk had only small populations and displayed a morphology akin to countless settlements in Southeast Asia today. Each community consisted of a loosely nucleated cluster of bamboo-with-thatch houses situated among kitchen gardens. Influential leaders built larger structures to validate their power and prestige, but few settlements displayed significant variations in house size, architecture, or construction materials. Wet-rice fields framed this residential core and yielded the staple food of the Dong-S'on farmers, who also raised vegetables, fruits, and animals. Their diet was supplemented with numerous wild products. Given the difficulties of overland travel during the rainy season, it is likely that barter among these dispersed settlements remained limited and seasonal. Most communities functioned as closed ecosystems nestled within marshlands or woodlands. In coastal situations such isolation was partly ameliorated by the changing monsoon, for some Dong-S'on peoples were expert mariners who not only plied Southeast Asian seas but also sailed to distant ports on the Indian Ocean.

The social life and religious beliefs of people in protohistoric Southeast Asia continue to be subjects of speculation. Researchers have gleaned some data on these critical aspects of culture from the archaeological record and by studying contemporary relict groups. Such evidence suggests that Dong-S' on folk were animists who revered nature and venerated their ancestors. They often buried the dead in huge jars or caves and graced hills or coastal promontories with shrines, thereby testifying to a tradition of cosmological dualism. Women enjoyed high status in this society and played an important role in village affairs. In fact some communities may have recognized descent through the maternal line (Cœdès, 1968, p. 3–13).

Sedentarization

Some 2,000 years ago, Southeast Asia was already inhabited by diverse peoples whose cultures and economies reflected kaleidoscopic adaptations. Mobile Negritic foragers retained their lifeways in the forested interiors of larger islands. In riverine and deltaic habitats, semisedentary folk lived by hunting and fishing with some gathering and tending of wild plants. Along coasts and in lowlands of the archipelago, Austronesian migrants were developing systems of root tillage, fruit growing, and swidden. Some metal-using societies of

Dong-S'on derivation—whose heritage embraced agriculture and seafaring, a belief system involving animism and cosmological dualism, impressive art, and animal domestication for food, draft, and sacrificial purposes—had made a crucial transition to wet-rice and permanent settlement. Despite obvious differences, the subsistence strategies of these coexisting peoples were all geared for the exploitation of generalized forest or marine ecosystems. Even Dong-S'on *sawah* farmers did some gardening, shifting cultivation, hunting, and fishing.

Although the process of sedentarization transformed cultural landscapes in certain lowland districts of protohistoric Southeast Asia, associated processes of social change remained slow and village communities small. The usual mode of organization involved kin clustering and ascriptive personal status tempered by egalitarian customs. Social control and political influence did not derive from material wealth or land ownership. Instead sociopolitical position reflected a person's ability to exploit kinship alliances, forge advantageous marriages, and impress fellow villagers through courage and good judgment. Aspiring leaders who earned trust and respect were also responsible for managing local systems of defense, trade, and magico-religious practice. Chieftains confirmed their personal prestige and strength by acquiring economic dependents and enslaving enemies. Yet such earned power seldom received the endorsement of customary law and was rarely heritable. Hence we may conclude that chiefly status and derivative authority relationships within Dong-S'on communities were highly particularized and often ephemeral (Wheatley, 1983, p. 43–117).

Most chieftains in preurban Southeast Asia presumably conformed without question to traditions defining local authority relationships. But not all. Occasionally ambitious leaders tried to subjugate their neighbors through such methods as marriages of convenience, the purchase of labor rights, and petty warfare. In the Philippines, and probably elsewhere in the region, this process resulted in the growth of scattered supravillage chiefdoms. Each was ruled by one individual supported by a group of loyal followers (Hutterer, 1977; Reed, 1978, p. 1–10). But without any means of legitimation other than their own organizational abilities and charisma, the leaders of confederations were unable to consolidate and preserve their territorial gains for future generations. Few such enterprising men could forge a durable system of governance. All grappled with the perplexing problem of ruling restive client communities. They also faced rivals who hoped to expand their own power. Likewise ambitious leaders were encumbered by traditional law that applied only to local communities and did not sanction chiefdoms. Consequently these precarious alliances depended on the ingenuity and political dexterity of exceptional individuals—described variously as "men of prowess" or "big men" by Southeast Asian scholars—who through sheer force of will and skill extended their power into nearby settlements. Still the resulting confederations were fragile societal arrangements because each depended on one charismatic person (Wheatley, 1983, p. 273–303; Wolters, 1979). Without common principles of legitimation and a durable bureaucratic framework, such embryonic polities inevitably remained transitory in time and space.

The ever more sedentary coastal peoples of protohistoric Southeast Asia clearly possessed a rich material culture. Nevertheless they failed to create integrative institutions that might have fostered a consolidation of villages leading to the development of complex societies. In fact the instruments that finally triggered the twin processes of state formation and urban generation evolved not from an indigenous cultural matrix but were borrowed from the older civilizations of India and China.

DAWN OF HISTORY IN SOUTHEAST ASIA

Even as debate continues on the dynamics of Indianization in Southeast Asia, scholars concur on key elements of the diffusion process. They agree that the eastward spread of Indian culture advanced slowly in time and space with transmission seldom requiring force. Likewise it involved countless individuals acting independently with sailors and merchants constructing the first maritime bonds between South Asia and the mystical *Suvarnadvipa* (Peninsula or Island of Gold). By the same token historians identify this exploratory period as the last centuries of the protohistoric era and suggest a considerable increase in seaborne commerce about 2000 years ago (Map 3.2).

Indianization as Prelude

Until recently it was assumed that Indians were the foremost if not the only agents of culture transfer between South and Southeast Asia. Few scholars considered the idea of a two-way exchange in which Austronesians might have carried trade westward to India. Instead writers mythologized the Indian mariner as an heroic pioneer who sometimes settled in the "realms of gold" after acquiring power and prominence through commerce. Inspired by visions of past glory, a few South Asian historians even envisaged a process of territorial conquest followed by the migration of Hindu settlers and the founding of colonies. Predictably such commentators usually depicted Southeast Asians as simple or passive folk who never ventured beyond their home environs and lived without polity (Majumdar, 1963). This once convincing model of Indianization is now rejected by serious area specialists. Contemporary scholars admit that some Indian merchants

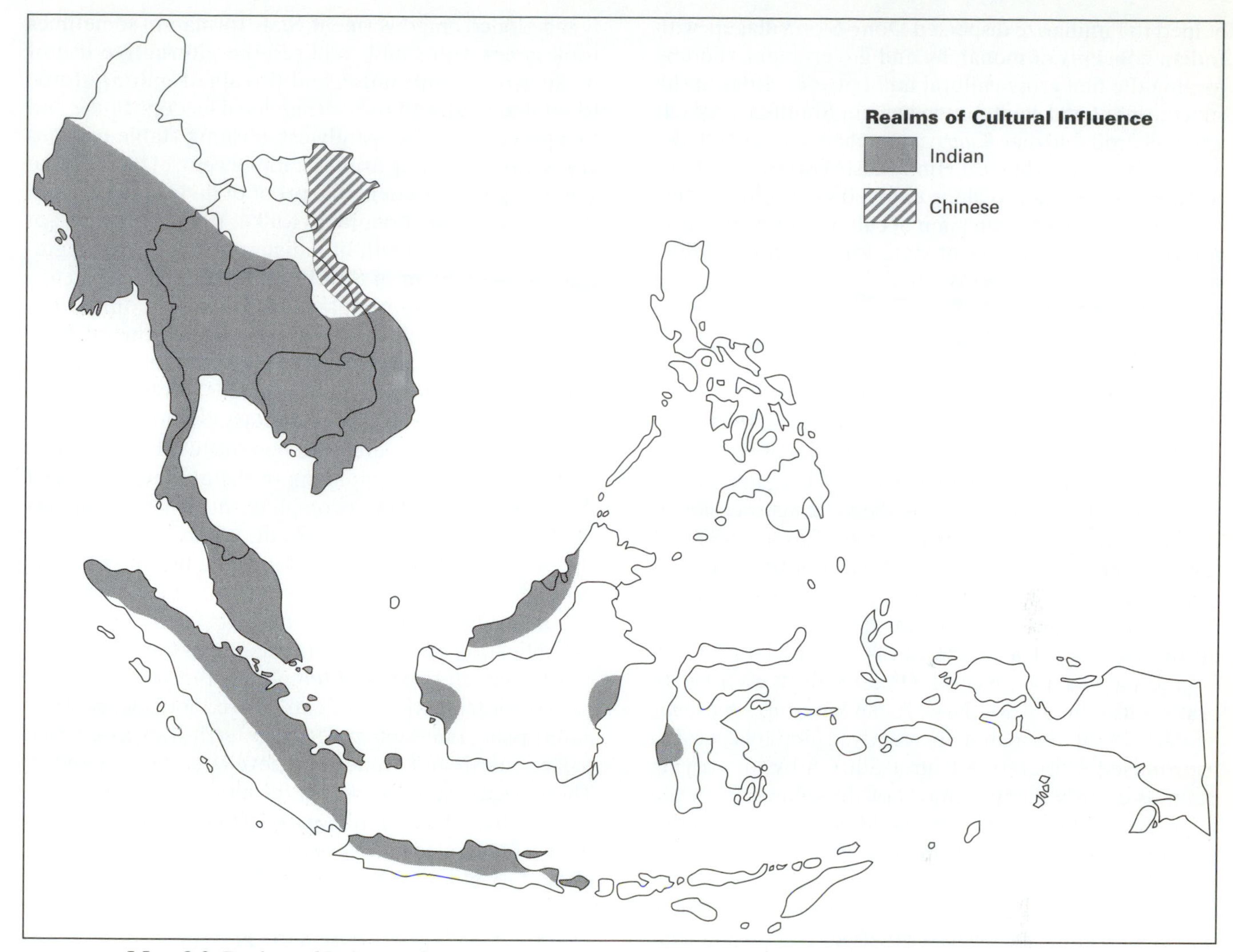

Map 3.2 Realms of Indian and Chinese cultural influence circa A.D. 500. *Source: Pluvier, 1995, plate 3.*

likely played a proactive role in the maritime exchange across the Bay of Bengal and so inaugurated the eastward spread of South Asian civilization. While peddling merchandise, they may have introduced coastal dwellers to certain material elements of Indian culture (weaponry, clothing, jewelry, and tools) and told inquisitive chieftains about fabled lands to the west. Yet it is doubtful that such traders and common sailors could have executed the cross-cultural transfer of elaborate systems of state administration, Hindu religious belief and rite, military organization, royal coronation, and court ritual. Most were presumably lower-caste adventurers who knew little about the abstruse aspects of their tradition. They apparently had few contacts abroad beyond the market place, were limited in dealings with local chieftains, and had little impact as agents of change.

Like their Indian counterparts, Southeast Asian seafarers bound for India's coastal emporia could not have been the cardinal brokers of a complex cultural assemblage. Such visitors were probably confined to the commercial quarters of ports and restricted in their communications with high-caste Hindus because of prohibitions against social intercourse with *Mlecch'a* (potentially polluting foreigners). Such ordinary mariners played an important role in defining the international exchange between South and Southeast Asia. However, they had only a minor part in conveying the critical ingredients of Hindu civilization to their homelands. Because most sailors and peddlers were illiterate folk, it is doubtful that they grasped the subtleties of Indian court life or the ritualized temple culture, let alone disseminated such complex information to ranking chieftains in *Suvarnadvipa.*

Some scholars believe that one group may have proved the exception: the aristocratic *Ksatriya* (warrior caste in Hinduism) merchants. Given their high status, refinement, and wealth, these traders garnered respect wherever they went and maintained a cultivated lifestyle even while living among less-sophisticated people. They helped to familiarize dispersed Dong-S'on villagers with Indian concepts of monarchy and government, thereby forging the first cross-cultural link between elites of the subcontinent and chieftains in ancient Southeast Asia. It appears probable that *Ksatriya* merchants pioneered the way for Brahman (highest Hindu caste) priests and their retainers. These literati alone enjoyed sufficient prestige to orchestrate the transmission of cultural forms that triggered the twin processes of state formation and urban genesis in *Suvarnadvipa* (Mabbett, 1977; Wheatley, 1975).

Politico-Religious Components of Complex Society

By the early centuries A.D., certain ambitious chieftains in Southeast Asia were beginning to comprehend the Indian model of royal governance. Inspired by tales of Hindu and Buddhist princes who commanded vast kingdoms, these supravillage leaders doubtless wished to construct similar polities with a guaranteed right-to-rule. We may infer that they recognized the utility of Indian politico-religious forms in circumventing customary laws, controlling obstinate client villagers, and legitimizing new territorial acquisitions. The idea of the *deva-raja* (god–king)—an organizing principle which later became the basis for elaborate cults defining Southeast Asian kingship—was particularly attractive to more perceptive chieftains because it promised a liberation from tradition by identifying them as earthly manifestations of the divine. But even the most outstanding leaders could not claim this indispensable mark of authority through a simple affirmation of intent. Their apotheosis required religious consecration. Such ceremonies could be performed only by Hindu priests who knew the magical rituals and sacral language of royal investiture, understood court protocol, and could affirm the divine lineage of aspiring chieftains. Sometime between the first and third centuries A.D., therefore, certain Southeast Asian leaders began to derive sociopolitical advantage from their knowledge of Indian civilization by convincing either itinerant or immigrant Brahmans to sanction their deification (Heine-Geldern, 1942; Kulke, 1978; Mabbett, 1969).

A critical institutional catalyst to the process of state formation, the god–king concept became the keystone for successive indigenous kingdoms in Southeast Asia. But in the earliest centuries of Indianization, elevation to kingship did not always guarantee unchallenged supremacy. Even after acquiring formal legitimacy through Hindu or Buddhist rites, new divine rulers and their immediate successors still had to win the loyalty of village leaders who had lost customary privileges; neutralize localized belief systems; build a unitary society with guidance from Indian advisors; and routinize the bureaucratic operations of their kingdoms. In other words, it was not sufficient for self-proclaimed *deva-raja* to simply repudiate tribal traditions and claim heavenly sanctioned empowerment. State formation sometimes took generations and required the alternative use of persuasion, compromise, and threats of military force to subdue finally all recalcitrant local leaders. In the end this process of successfully establishing stable bureaucracies and standing armies under a central officialdom was accepted as undeniable proof of divine authority in the eyes of subject peoples (Kulke, 1986). It is thus apparent that the growth of indigenous states and beginnings of urbanism in Southeast Asia involved a convergence of many people with mutually compatible interests: native chieftains anxious to legitimate their right-to-rule, Brahmans with religious standing to certify such authority, the new monarchs' kinfolk or allies who garnered prime civil or military postings, and a citizenry open to the perceived benefits of civilization.

Though our understanding of the earliest Southeast Asian polities remains incomplete, the general elements of state form are quite apparent. While functioning as the political fulcrum in each kingdom, the ruler clearly depended on priestly literati to organize the administrative and religious infrastructures. Obviously the principal obligation of resident Brahmans was to preside over ceremonies of court and temple. At the same time the priests' organizational talents proved indispensable as enterprising kings expanded their territories, assembled palace guards and armies, and developed bureaucracies. Though subject to the will of the monarch and his royal council, the civil and military staffs in larger states ultimately functioned as separate branches of government or hereditary elites (Wheatley, 1983).

URBANISM AND POLITY

When Portuguese explorers first glimpsed Southeast Asia as an integral region early in the sixteenth century, only the Philippines remained without cities and states. Elsewhere, Europeans discovered flourishing kingdoms and a rich urban heritage that had yielded two distinctive settlement forms. Along the coasts they found many autonomous city-states that were part of a loosely knit maritime trading network stretching from the western rim of the Indian Ocean to the Sea of Japan. The influence and affluence of these ports mirrored their comparative breadth of commercial outreach, quality of leadership, commercial relations with China, and control of commodities extracted from mines, fields, and forests in nearby riverine or deltaic hinterlands. Naturally the sea, not the land, provided the economic lifeblood for such city-states. Indeed these communities often deployed fleets of trading and war vessels that controlled the seaways far beyond their immediate environs.

In the great river valleys of mainland Southeast Asia and on Java, the Europeans encountered a second settlement form—the sacred city. Each served as the ceremonial capital, military stronghold, marketing center, and administrative hub of an agrarian kingdom. Although the inland cities were sometimes places of lively international trade and even had segregated quarters for foreigners, their essential role was politico-religious in nature and their prosperity ultimately derived from levies on peasant farmers. Commerce remained a lesser source of wealth. Despite differences in function, population, and morphology, there can be no doubt that the city-states and sacred cities originated long before the arrival of Westerners and so offered undeniable proof of indigenous urbanism and state formation (McGee, 1967, p. 29–41).

Urban Genesis and Elaboration

Although scholars have identified the religious and political components of statecraft and urban genesis in Southeast Asia, questions concerning the spatial and morphological details of early cities will be resolved only through archaeological research. Still it is now clear that the western sector of Southeast Asia was a discrete realm of nuclear urbanism and state formation stimulated by Indianization. Admittedly Brahmans played a critical role by disseminating institutional archetypes that nurtured urban genesis and the development of regional polities. But the resultant cities and kingdoms of *Suvarnadvipa* were unquestionably indigenous cultural entities and remained enduring reservoirs of domestic culture. None can be described as products of Indian imperialism; none were founded by outsiders who stayed on to dominate trade and society; and none endured governance by alien officialdoms from India. Supreme authority reposed in the hands of creative native rulers who were served by supportive local elites. While informed and tutored by Brahmans (and later Buddhist priests), the resultant Southeast Asian leadership ultimately supervised the civil and military officialdoms. Kings and their advisors were also charged with preserving and protecting religious institutions that provided the essential foundation for urbanism and statehood.

For centuries after the advent of kingship and urbanism in Southeast Asia, the great majority of plains people and all highlanders continued to dwell in rural situations even as a markedly different way-of-life crystallized in urban centers. Through the ministrations of Brahmans, chieftains were transformed into god–kings, shamans joined the ecclesiastic professions of Hinduism or Buddhism, prominent villagers acquired training as literati, and tribal warriors became disciplined soldiers. In concert with the general eastward transfer of Indian ideas and institutions—such as Hindu mythology, court protocol, written laws, art and architecture, and town-planning theory—successive chiefdoms were transmuted from folk cultures into civilized societies. The urban pivots of such metamorphosis became locales of sophisticated religious, political, and social activities that remained mysterious to surrounding and once independent rural populations that had been gradually transformed into subject peasantries. The two poles of this fundamental cultural bifurcation were expressed on the landscape as the agricultural village, where people retained a semblance of their Dong-S'on heritage, and the *nagara* (capital city), which served as the repository of civilization under the aegis of a divine monarch.

Even in the earliest cities, temples served as focal points of societal interaction and settlement. They were probably first built under the direction of Hindu priests who wished to discharge their ceremonial duties in a fitting milieu. Sacred groves, spirit houses, and ancestral shrines plainly failed to provide an appropriate setting for such impressive rituals as royal investiture and clerical consecration; for sheltering the palladium of state (the lingam, or revered phallic symbol of Shiva with its essence of divine power); and for offering sanctuary to priestly and managerial literati. Consequently enormous resources were devoted to temple construction. The temple complex was indeed an ubiquitous element of morphology in indigenous Southeast Asian cities and was sometimes built on a monumental scale in the inland capitals.

In the archetypal urban plan of traditional Southeast Asia, the king's residence adjoined the national sanctuary to form a palace precinct or occupied a site nearby. Although the size and decoration of the royal compound reflected comparative national resources, even in the wealthiest kingdoms these buildings were seldom made of permanent materials. Until influenced by Western ideas, in fact, most rulers observed sumptuary laws that confined humans to wooden or bamboo houses because only the gods enjoyed an absolute right to reside in structures built of stone or brick. The dwellings of ranking bureaucrats and servants, national treasury, rice granaries, monasteries, and court garrison were usually situated near or within the temple–palace complex and often secured by palisades, walls, and moats. Merchants, craft specialists serving the royal household and state officialdom, and the general citizenry established their communities in the suburban area under the shadow of the walled sacred city (Wheatley, 1983).

Sacred Capitals

In spite of a common origin within a milieu of Indianized culture, the sacred cities and coastal city-states of Southeast Asia were quite different in their functional roles and morphology. Far more impressive in size, symbolic design, and architectural opulence, the inland ceremonial centers could be ranked by their breadth of territorial outreach, number of client villages, and stockpiles of food. Though the ruling elite extracted resources from tributary villages through a variety of methods (military duty, service as craftsworkers, drudgery on public projects, and agricultural levies), the wealth of court and capital was ultimately equated with traditional rights to labor. Everywhere the enduring obligation of the peasant to the ruler was calculated in terms of corvée or its equivalents. Indeed mandatory work was recognized as the king's due, and his rightful share of local labor pools was apportioned to high-ranking villagers and the state according to inherited status and acquired influence. By appropriating the skills or toil of the peasantry, the urban elite verified its hegemony while at the same time validating a primary role of the sacred city as an effective instrument of redistributive activities.

The sacred capitals of Southeast Asia were preeminently fountainheads of indigenous civilization. Like the archetypal cities of orthogenetic transformation envisaged by Robert Redfield and Milton Singer (1954), their institutions evolved during the misty period of "primary urbanization;" their societies engendered distinctive "national cultures;" and their literati supervised the urban communities. These priestly advisors and ranking bureaucrats mediated all social and technological changes in terms of prevailing religious and moral standards, thereby assessing the usefulness of innovations and the potential impacts of outsiders. Even in sacred cities with major markets or pilgrimage centers, the literate elite monitored the spread of revolutionary ideas or politico-religious heterodoxies through segregation of merchants and pilgrims in designated quarters. The result was a uniform set of cultural norms that integrated the urban community and linked residents of the capital with the peasantry in a shared civilization.

The urban landscape was also designed to foster a distinctive national identity and promote territorial integration. It is true that the morphology of the typical sacred city is often represented as a solidly constructed temple-and-palace nucleus, embraced by an administrative district and military garrison and embedded in a matrix of markets, craft districts, ethnic and occupational quarters, and scattered houses of common laborers. But this portrait remains only a sketch. Every ceremonial center was in truth fashioned to ensure an enduring sacred experience for city residents and rural visitors alike, thereby increasing their dependence on the *deva-raja* and his council. Variations in topography obviously necessitated adjustments in the placement of buildings and fortifications, and likewise local cultural peculiarities and disparities in royal wealth produced differences in architecture and urban plans. But the archaeological record testifies to the fact that all interior capitals shared a fundamental purpose as the symbolic platform for integrative ceremonial activities through which royal and priestly elites cultivated a common identity and binding loyalty that infused the citizenry of each kingdom.

In the classic sacred city of Southeast Asia, the spatial design clearly reproduced a cosmological master plan configured to guarantee consonance between heaven and earth and thereby promote harmony and prosperity for both city and state. The capital—conceived by believers as the primary point of intersection between human and divine planes of existence—was constructed according to a widely recognized celestial archetype patterned after the cosmic designs of either Hinduism or Buddhism. Although marked by differences in detail, both models of the universe embraced a notion of heavenly concentricity (continental, orographic, and oceanic) with a singular hub (the majestic Mount *Meru*) that was inscribed on every urban landscape. In realms where people followed Hinduism, sacred cities often centered on a hillock (natural or artificial) or a temple that represented the cardinal mountain of the gods, provided an arena for the *deva-raja* cult, and symbolized the axis of the state. In kingdoms adhering to Theravada Buddhism, a belief system that remains without a pantheon of gods or cultic cliques, the royal palace symbolized Mount *Meru*. Among the other design elements of the inland capitals were foursquare walls or palisades around the palace precinct that duplicated the mountain chains of Indian cosmic diagrams; moats that depicted the encircling oceans of the universe; and avenues, gates, and bridges aligned with the cardinal directions of the compass to highlight the transfer of divine energy from the temple-palace complex to distant villages in the kingdom (Heine-Geldern, 1942). Even today the ruins of Pagan, Angkor, Ayutthaya, and other sacred cities vividly attest to their layout as earthly replicas of the cosmos designed to orchestrate an integration of religious, political, and social life within a unitary civilization.

To ordinary people and leaders alike, Southeast Asia's inland capitals were not only major population centers, primary consumers of regional surpluses, and key administrative headquarters but also the actual mechanisms of universalizing authority. As urban thrones of divine kings who occupied awesome temple-palace compounds, dominated urban life, and ruled vast territories, their essential role as *axis mundi* or pivots of all creation appeared beyond challenge. It was natural for contemporary populations to believe that these monumental assemblages actually served as celestial conduits through which divine power entered the profane world of humankind and from there radiated throughout the kingdom. Accordingly each sacred city functioned as an earthly microcosmos where elites and peasantry alike attained a sublime sense of spiritual security in built environs that seemed a virtual image of the heavenly macrocosmos. Through the mundane activities of daily worship, religious merit-making, and periodic ceremonial activities they experienced personal contentment and a collective faith in the politico-religious integrity of the kingdom (Wheatley, 1969, 1975).

Port Cities

For more than a millennium before the onset of Western rule, the smaller coastal city-state also proliferated in Southeast Asia. The prosperity of such ports was a direct function of the breadth of their maritime trade. Unlike the inland capitals, which depended on regular harvests of sprawling agricultural areas, the coastal cities required little more than a secure harbor, an open market, a sizable merchant fleet, a supportive host of piratical privateers who offered protection during crises, and a capable leadership. Clearly centers of heterogenetic transformation (Redfield and Singer, 1954), the city-states embraced a diverse community of traders and sailors from East and South Asia and beyond. Their cosmopolitan character was renewed biannually as foreign traders arrived and departed with the changing monsoons. Nearly all visitors came for commerce, yet also became agents of cultural and technological change. The ensuing diversity of religious and moral norms was tolerated as long as everyone respected the sanctity of an unencumbered market.

The morphology of port cities deviated sharply from the clearly symbolic and durably built sacred cities of Southeast Asia. In contrast to the monumental buildings, solid walls, stately gates, and grand avenues of inland capitals, the city-states were constructed of bamboo, wood, and thatch. Sometimes a few stone or brick fortlets flanked the coastal cities, but their encircling fortifications usually consisted of log stockades and shallow moats. Early Western explorers often testified to the flimsy and chaotic appearance of city-states, further noting they were plagued by fires because of their highly flammable building materials. In some ports the palace precinct was vaguely imprinted by an Indian cosmic design, but additional evidence of planning could be discerned only in the segregated quarters for foreign traders. Overall the morphology of the classic port city appeared localized in design, disorderly in aspect, and focused on the marketplace.

Though distinctly different in functions and morphology, the sacred cities and ports had several features in common. None prospered for more than a few centuries without decline; some proved ephemeral in place; and many were quite populous by demographic standards of the time (Reid, 1993, p. 67–77, 302–303). As nodes in a far-flung Asian network of coastal emporia, the city-states of Southeast Asia flourished or suffered recession in accord with either regional or international shifts in trade. When Portuguese forces assaulted Malacca (1509), for example, it was experiencing an economic and cultural florescence as the critical fulcrum of commerce between India and China and a springboard of Muslim outreach to countries further east. Yet the aptly described "city that was made for merchandise" was barely a century old and was prospering even as other places declined. Such was the developmental scenario of Brunei, Tumasik (now Singapore), Macassar (now Ujung Pandang), Indragiri, Banten, and scores of other ports that waxed and waned between the fifth and eighteenth centuries A.D. At the same time Southeast Asia's sacred cities sometimes underwent even more dramatic changes that echoed the fortunes of war, dynastic decline, and the whims of monarchs. In Burma—even as recently as the middle of the nineteenth century—the opulent city of Amarapura with its population of more than 100,000 people was abruptly abandoned by King Mindon in favor of a costly new urban foundation at Mandalay (1857), which supposedly possessed a more auspicious location than the old capital. The significance of such a relocational process cannot be overemphasized, as scores of kings proved willing to leave established inland capitals simply because they associated certain misfortunes of court or kingdom with particular urban centers (Reed, 1976a, p. 22–23). Such monarchs were firmly convinced that the success of their reigns was dependent on the reestablishment of sacred cities elsewhere and in accordance with propitious astrological signs.

A KALEIDOSCOPE OF STATES

Before the crystallization of complex societies in Southeast Asia, patterns of territoriality reflected the subsistence requirements and usufruct rights of families and localized kin groups. The resulting checkerboard of minuscule and overlapping home territories incorporated the broad ranges of hunters and gatherers, loosely restricted forests of swidden cultivators, and lowland ecosystems comprised of woodlands and fields claimed by *sawah* farmers. Because of continuing labor investments in irrigation systems, wet-rice agriculture may have engendered a heightened sense of ownership. But conflicts born of competing territorial claims were rare as long as population densities remained low and resources proved sufficient. It may thus be inferred that the principle of individual or communal tenurial rights was still unfamiliar to most people. Accordingly the boundaries separating home territories of foragers, farming communities, and even chiefdoms proved flexible, indefinite, or nonexistent.

The beginnings of Indianization heralded a new sense of territoriality grounded in the notion of exclusive royal rights to the labor and produce of all subjects within the domain of each *deva-raja*. Through religious conversion, military threats, and claims to village resources, new urban elites inaugurated an adaptive process by which rural communities were merged into larger spatial systems—the precolonial kingdoms of Southeast Asia. With the successful sociopolitical fusion of hamlets, villages, and chiefdoms, god–kings began to demarcate dependent territories or hinterlands. This is not to suggest that boundaries separating the proliferating states were firmly fixed. Because more isolated communities proved better able to reject imperial demands, tension always endured between urbanized centers and rural peripheries. In the resulting "galactic polities" posited by Tambiah (1977), with their clearly defined capitals and hazy frontiers, kings exercised a tight control over satellite villages in the immediate environs of larger cities. There the royal authorities could easily discipline stubborn populations still longing for lost independence. But in outlying areas, the rulers' influence diminished through distance decay, allowing quasi-independent or factious communities to break free periodically from the orbit of state dominion. Even loyal villages along unstable frontiers occasionally challenged tribute obligations imposed by distant urban officialdoms. Consequently the boundaries separating kingdoms remained indistinct and might be best described as broad or fluctuating border zones. As regional polities prospered and declined, therefore, Southeast Asia became a veritable patchwork quilt of kaleidoscopically changing territorial states (Bentley, 1986).

Chinese Influences in Vietnam

Notwithstanding the pervasive impact of Indianization, the first urban and political foundations in Southeast Asia were born of imperialism emanating from China. The earliest Chinese garrisons appeared in the agricultural zone of the Red River valley in northern Vietnam following armed conquest in 111 B.C. Divided into three commanderies, this area was ruled by China until A.D. 939 and was transformed through relentless programs of religious, social, and administrative Sinification. While bending to Chinese rule, the Vietnamese steadfastly preserved their own cultural integrity. At the same time, they selectively borrowed Chinese ideas and institutions that blended with their own beliefs and practices.

After nearly 10 centuries of Chinese sovereignty, the Vietnamese identity represented an amalgam of cultural ingredients drawn from the traditions of rulers and ruled alike. Predictably resident officials from China and Sinified city elites both espoused the imperial culture of Confucianism, Chinese classical education, Mahayana Buddhism, and an intrusive bureaucracy. Working together, an influential native mandarinate and wealthy gentry conveyed much Chinese wisdom to the urban people of Vietnam. But the transmission of an integral civilization remained incomplete. In a myriad villages traditional folk culture survived the onslaught of foreign ideas and institutions. At the same time the vernacular language continued as the conversational vehicle of the masses. While incorporating useful loan words and a Chinese writing system, this non-Sinitic tongue highlighted and validated the ethnic and cultural distinctiveness of the Vietnamese (Taylor, 1983).

When the Vietnamese finally expelled their oppressors and proclaimed their own national identity, they naturally adopted the bureaucratic and societal model of China. It served well for a millennium, providing a solid organizational structure for territorial expansion and state governance. In spite of periodic Chinese threats and incursions, peasant rebellions, palace intrigues, and dynastic conflicts, the Vietnamese increased in numbers and their nation's strength grew steadily. From the Red River lowlands, they advanced southward through the tandem efforts of armies and peasant migrants who settled newly won lands. Their cumulative conquests of resident Cham and Khmer populations continued until halted by the French in the nineteenth century.

Early Indianized Kingdoms

As China orchestrated its conquest of northern Vietnam and created the first documented Southeast Asian polity, the Isthmus of Kra and the Malay Peninsula became the setting of near contemporary Indianization and state formation (Map 3.3). Materializing in the second century A.D. and located near modern Pattani, Langkasuka was the foremost of these Hindu kingdoms. At its zenith this strategically situated state commanded a safe overland route between the Gulf of Thailand and the Andaman Sea, as well as scattered agricultural niches on both coasts. About the same time, a number of small city-states appeared along the Siam-Malayan corridor at key transshipment points in a trading system stretching from Arabia to East Asia. Each port prospered as long as it commanded a shortcut across the isthmus for traders to portage their merchandise without fear of shoreline pirates.

Sometime in the first or second centuries A.D., Funan—the first of Southeast Asia's great Indianized states—crystallized in southern Vietnam and Cambodia. Its sacred capital of Vyadhapura was situated near the apex of the Mekong River. Comprised of ceremonial centers, administrative towns, and protourban chiefdoms, Funan prospered for 500 years and reached a fifth-century zenith when it dominated client states, absorbed vassal ethnic chiefdoms, and conquered lands extending from coastal Vietnam to southern Burma. The key to its territorial reach, long prosperity, and reputation lay in a geographical situation that permitted a balanced development of agriculture and international trade. In an era when sea traffic between China and ports on the Isthmus of Kra was conveyed by shallow-draft coastal vessels, Funan's emporium at present-day Oc Eo became the natural break point for mariners awaiting biannual shifts in mon-

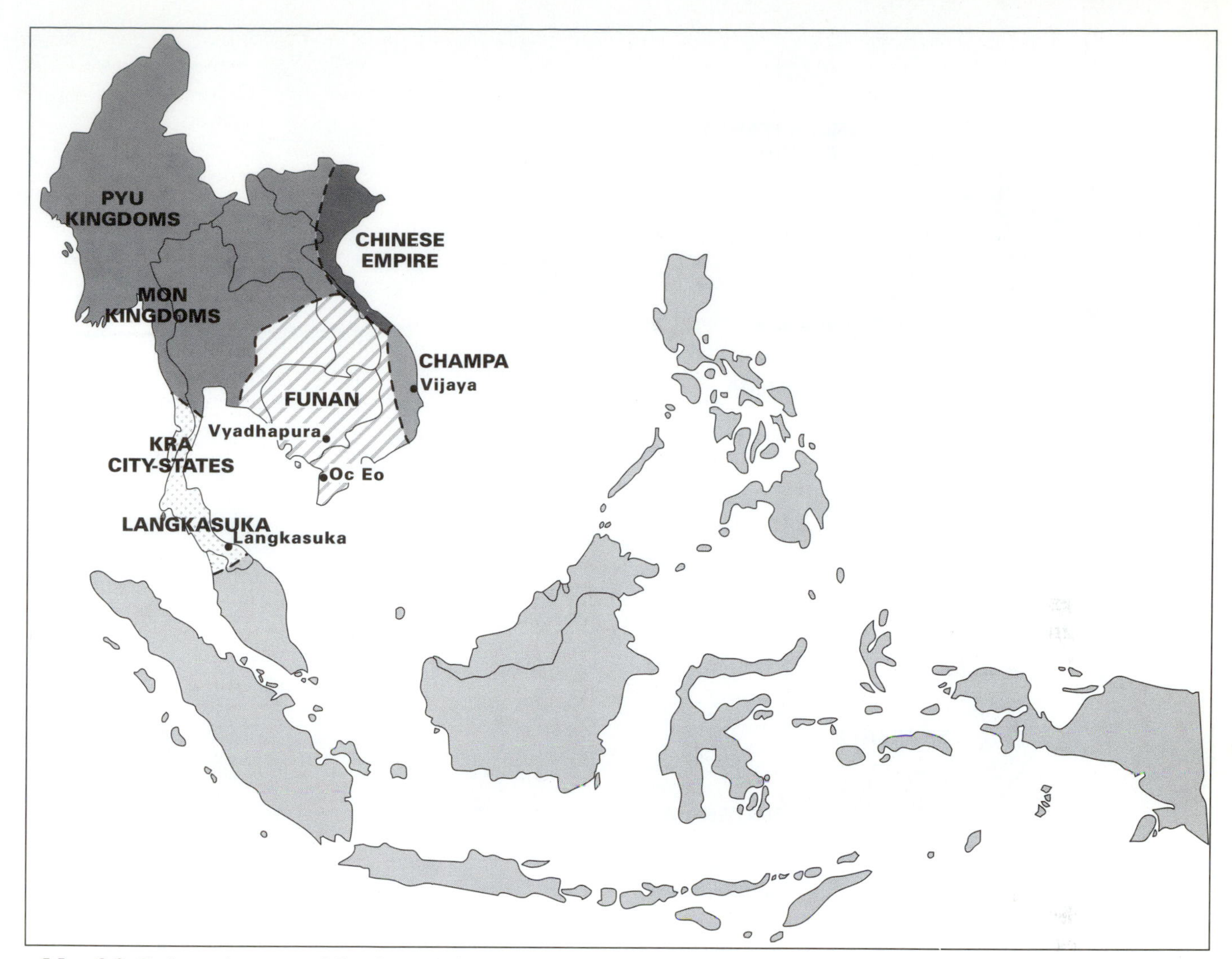

Map 3.3 Embryonic states of Southeast Asia, second to sixth centuries A.D. *Source: adapted from Pluvier, 1995, plate 4.*

soon winds. During their lengthy layovers, this famed city provided traders with a safe harbor and ample provisions (Map 3.4). As shipbuilding techniques and navigational skills improved, and seafarers opened direct routes across the South China Sea, Oc Eo declined as a nexus of commerce. Despite efforts to recast its economy in an exclusively agrarian framework, Funan succumbed to the rising Khmer state of Chenla in the sixth century A.D. (Hall, 1982). Inheriting Funan's cultural legacy, this former dependency soon split into two parts ("Land Chenla" in the north and "Water Chenla" in marshlands of the south) and dissolved at the end of the eighth century under pressure by raiders from peripheral territories and Sailendra invaders from Java.

Champa—a state in central Vietnam dating from the early period of Indianization—emerged not as a unitary Hindu kingdom but as an amalgam of small coastal polities demarcated by the formidable Truong Son mountains and the South China Sea. Ruled by kinglets and occupied by Austronesians who shared the cultural legacy of Southeast Asia's island peoples, the Cham kingdoms were sustained by a diverse economy of wet-rice agriculture, fishing, maritime commerce, and petty trade with highlanders. Occasionally these essentially autonomous political entities coalesced under powerful monarchs to repel Vietnamese invaders on their northern border, spar with the Chinese, battle sea raiders, invade Khmer territories on their southwestern flank (even as far as Angkor), or gather strength for overseas expeditions reaching from Java to China. Such episodes were consummated through the corollary construction of splendid ceremonial centers such as Indrapura (ninth century) and later Vijaya. But these regional confederations under all-powerful monarchs remained fragile and soon fractured into smaller polities commanded by insurgent local leaders. Decisively defeated by Vietnam in 1471, the Chams surrendered most of their territory north of Da Nang. In the seventeenth century Champa's last territories fell to the Vietnamese, though a puppet Cham king ruled in modern Thuan Hai province until 1832.

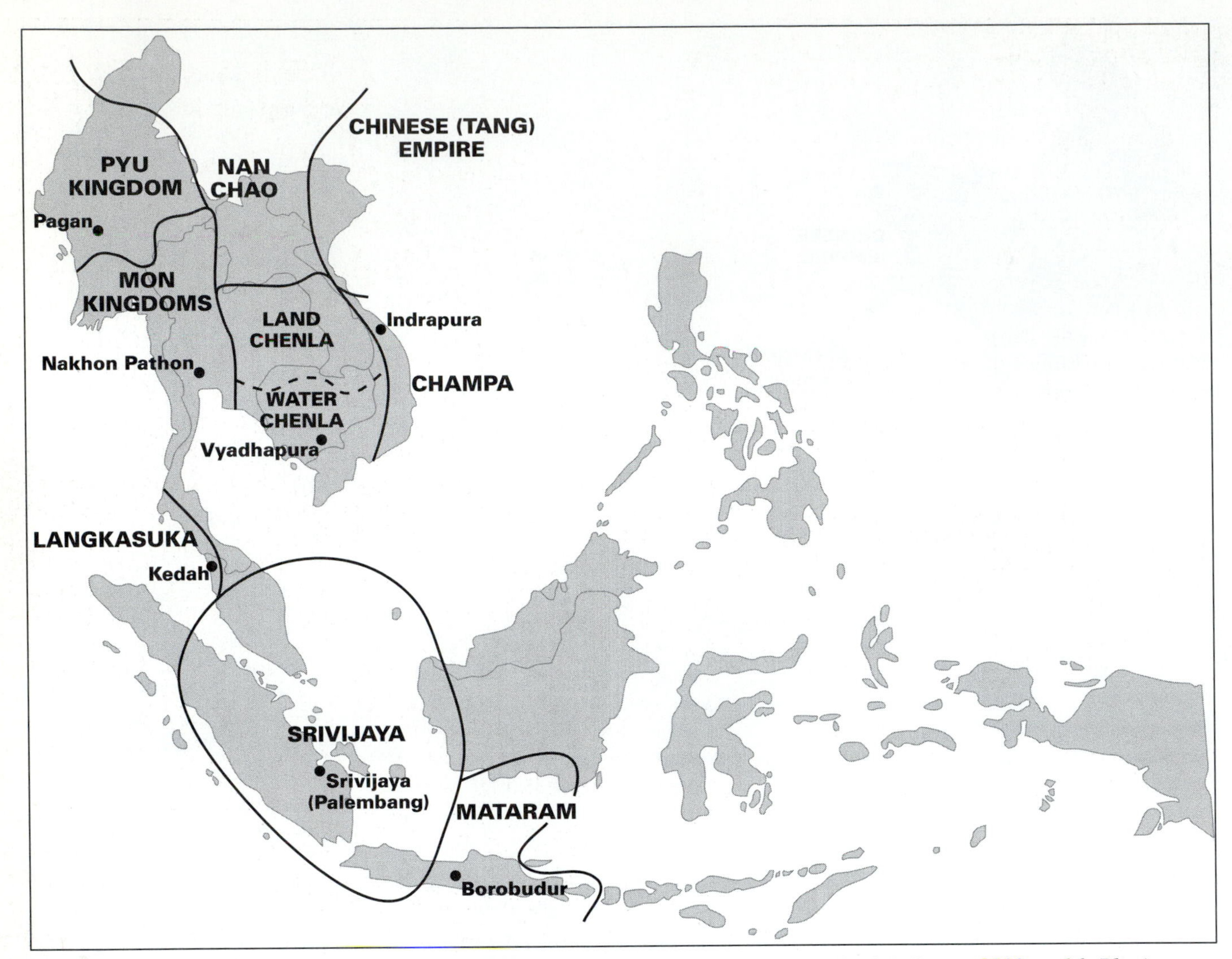

Map 3.4 Major empires of Southeast Asia, eighth and ninth centuries A.D. *Source: Ulack & Pauer, 1989, p. 16; Pluvier, 1995, plate 5.*

Champa is remembered today for its remarkable longevity and its episodic glory as revealed in monumental ruins. Indeed, the latter-day Chams—a minority Malay people in Vietnam and Cambodia—still cherish their rich cultural heritage.

Two other important states emerged on the mainland of ancient Southeast Asia. First, the Mon kingdom of Dvaravati (seventh and eighth centuries A.D.) occupied Thailand's lowlands and straddled the Burmese border. Sustained by *sawah,* its people also did portage across the northern sector of the Isthmus of Kra and traded with South Asia. Inspired by the Singhalese belief system, Dvaravati's traders and sailors assisted in the eastward diffusion of Theravāda Buddhism. Second, the Pyu peoples of Burma between the third and ninth centuries fashioned a series of urban principalities in the Irrawaddy valley that were supported by dense agrarian populations. Primarily administrative and ceremonial centers, these cities also maintained overland trade linkages with India and China. In later centuries, the Pyu were influenced by their fervently Buddhist Mon neighbors, whose kingdoms centered on the sacred cities of Pegu and Thaton (Wheatley, 1983).

Archipelagic Empires

With the decline of Funan's port of Oc Eo and the shift of maritime traffic from coastal waters to the high seas, Southeast Asia's archipelagic realm took on new geopolitical importance. From the sixth century until the time of Western hegemony, it was the scene of a vital international exchange involving seafarers from East and South Asia, the Arab world, and Southeast Asia. This broad trading complex involved a dynamic and variegated system of supravillage confederations, city-states, pirate bases, shoreline empires, and Javanese kingdoms. Against a backdrop of evolving commercial networks, changing regional rivalries, and shifting imperial fortunes, three empires reigned supreme (Srivijaya, the Mataram states, and Majapahit) and a fourth (Malacca) achieved prominence before falling to Portugal (Map 3.5).

Srivijaya, which thrived on a monopoly control of trade funneling through the Strait of Malacca, replaced Funan late in the seventh century A.D. as Southeast Asia's foremost commercial power. A powerful sea-state, Srivijaya enjoyed sweeping dominion over coastal territories in the southwestern quarter of the region. It remained a major power until the mid-eleventh century A.D. and endured for two more centuries in a much attenuated form. The glory of this maritime empire is echoed in Chinese dynastic records, Buddhist pilgrim chronicles, Arab accounts, and epigraphic inscriptions on scattered memorial stones commemorating military conquests and royal deeds. Despite limited archaeological evidence, scholars believe that the capital of Srivijaya was a sprawling settlement built on pilings and rafts near contemporary Palembang in Sumatra. This city functioned at once as the paramount regional entrepôt, a center of religious pilgrimage, and the seat of powerful monarchs. Its rulers preserved their authority by employing pirates as traders or naval forces, building alliances with upriver chieftains who controlled resources in riverine hinterlands, and forging a strong politico-economic relationship with China. There is no doubt that the Chinese appreciated Srivijaya's role as gatekeeper of the East–West exchange and so nurtured mutually beneficial diplomatic relationships that guaranteed preferential trade for both parties. Each opened its ports to the other for commerce, ship repairs, and residential quartering of seafarers. Thus bound to a distant but powerful empire, Srivijaya prospered or declined in consonance with the fortunes of China.

While Srivijaya's fame was secured by trade, her cosmopolitan capital functioned as a key station in the circular Buddhist pilgrimage between China and India. Converted to Mahayana Buddhism in the seventh century A.D., the urban elite clearly derived political legitimacy from this universalizing religion. At the same time,

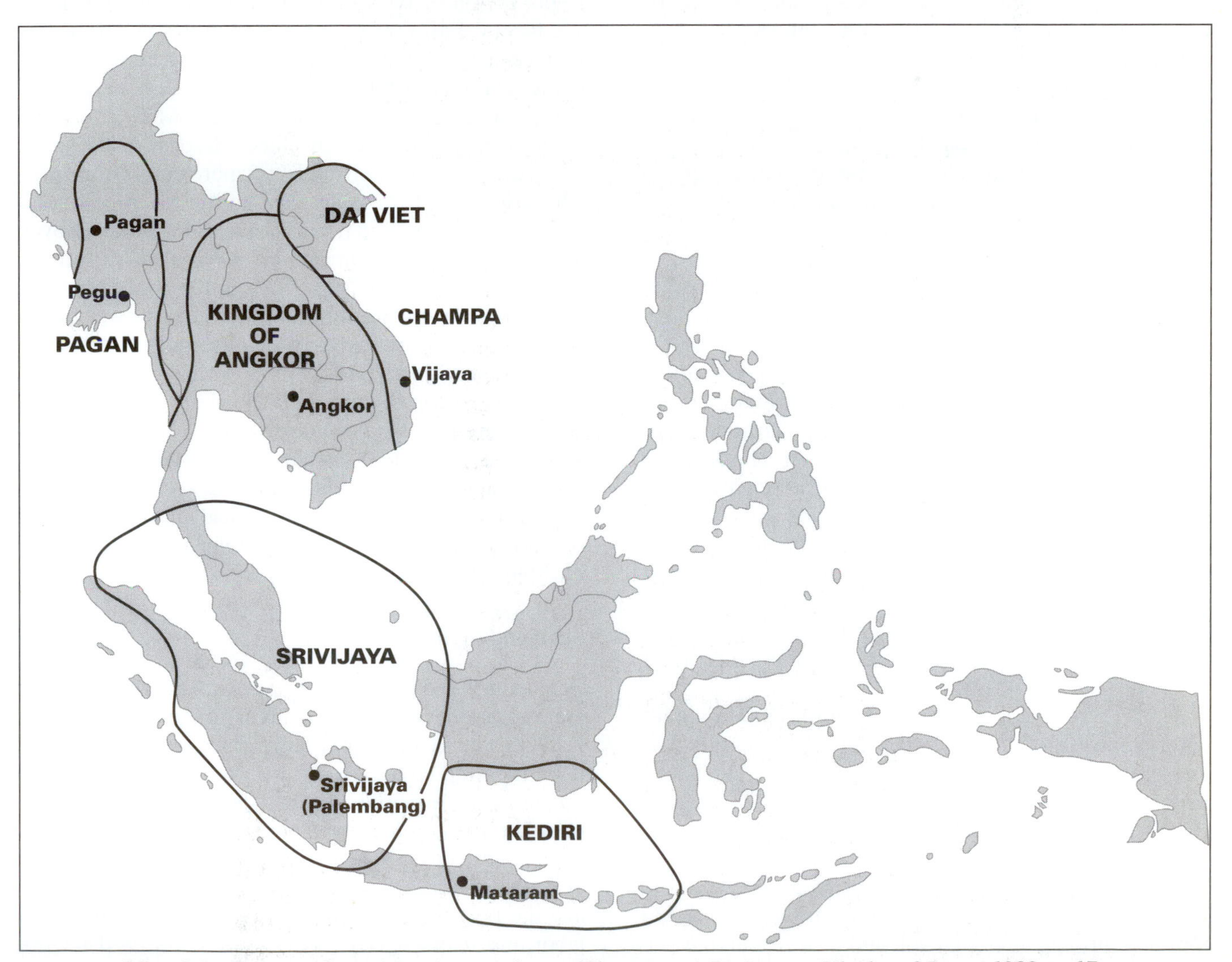

Map 3.5 Major empires of Southeast Asia, twelfth century A.D. *Source: Ulack and Pauer, 1989, p. 17.*

it provided a vehicle for the Srivijayan literati to participate in a sophisticated dialogue with scholars and religious leaders in monastic communities of India, Sri Lanka, and China.

Repeatedly surviving insurrections and the ninth-century loss of its kingship to Sailendra interlopers, Srivijaya endured with strength until challenged by a South Asian imperial power in 1025 A.D. Provoked by excessive port taxes and import duties, Chola corsairs from Tanjore in that year sacked many ports along the Strait of Malacca and shattered forever this maritime trade monopoly. In the subsequent two-century restructuring of regional commerce, the Sumatran city-state of Jambi assumed Srivijaya's organizational role but reigned over a much weakened state. Competing ports soon proliferated on Java's north coast and the Malay Peninsula (Wolters, 1967, 1970). Though remembered in regional legends and foreign chronicles, Srivijaya never recovered its former glory.

Opponents of Srivijaya, the successive Mataram polities of central Java flourished under different dynastic leaderships (Sailendra, Mataram, Kadiri, and Singhasari) between the seventh and thirteenth centuries A.D. Sustained by a dualistic economy of agriculture and maritime trade, they comprised Java's first land-based empires with dense populations of *sawah* farmers, integrated marketing networks, and a capacity for large-scale civil (hydraulic systems) and religious (temple construction) infrastructural projects. Their agrarian economies were extremely durable, remaining productive even during times of fluctuating commercial conditions, wrenching dynastic warfare, religious change and costly temple-building projects, military expeditions against mainland states (Chenla, Champa, and city-states on the Malay Peninsula), and exhausting conflict with Srivijaya. Lasting for centuries in one form or another, the Mataram states are noteworthy for their classic form as agrarian kingdoms and a dramatic transformation from Hinduism to Mahayana Buddhism under the Sailendras (eighth–ninth centuries). The associated spiritual fervor and collective expressions of faith produced some of the most magnificent architectural monuments in Southeast Asia (e.g., the Hindu temples of Prambanan and the Buddhist shrine of Borobudur), testifying to Java's rank as an international center of religious learning and intellectual ferment (K. Hall, 1994, p. 202–218).

Heir to the ancient Mataram empires, Majapahit emerged in East Java late in the thirteenth century A.D. and survived for nearly 200 years. Following tremendous expansion under Singhasāri's warrior-king Kertanagara (1268–1292), who projected Javanese power into the domain of Srivijaya, conquered the Sumatran kingdom of Malayu, and acquired feudal authority elsewhere in the Indonesian archipelago, the Mataram epoch ended abruptly through internal revolt and an invasion by Mongol forces from China. After expelling the Chinese early in the fourteenth century, Majapahit arose as a potent regional empire through the statesmanlike guidance of Gajah Mada. Though plagued by dynastic struggles, civil war, and external threats, Majapahit filled the political vacuum produced by a declining Srivijaya and remained secure for generations because of a solid agricultural base (Hall, 1985). Its imperial power began to fade only with the disruptive infiltration of Islam, the emergence of upstart Malacca as a regional force, and the rise of independent ports on the north coast of Java (Map 3.6).

As Majapahit declined, a new maritime power assumed a commanding position on the Strait of Malacca and developed in the mold of the Srivijayan empire. It centered on the city-state of Malacca, which was established late in the fourteenth century A.D. by the Malay ruler Iskandar Shah (also called Paramesvara). His earlier base at Tumasik (Singapore) had proved vulnerable to Siamese raiders, thus prompting the move to a safer site. Malacca soon became renowned as the foremost entrepôt in Southeast Asia. Its security was guaranteed by sea forces that dominated nearby ports of the Malay Peninsula and Sumatra. Always nurturing friendly diplomatic relations with China, Malacca's rulers periodically dispatched tributary missions to the Ming court and pledged an open market for Chinese traders. Additionally they courted Tamil and Gujerati merchants from India who offered direct connections to the Middle East, as well as merchants from north Javanese ports that enjoyed access to the spice islands of Indonesia (the Moluccas). At the same time, the Malay leadership adopted Islam. This belief system accentuated Malacca's power by solidifying the linkages of its royal house and citizenry with co-religionist Indian and Arab traders, while simultaneously converting the city into the chief conduit of the Muslim faith to other Malay peoples (Thomaz, 1993; Wolters, 1970). The dizzying rise of Malacca demonstrated the continuing utility of a "sea-state model" in precolonial Southeast Asia, the critical importance of cooperation with China, and the significance of Islam as a linchpin of geopolitical power.

Mainland Agrarian Domains

Beginning in the ninth century A.D. with the rise of its first great agrarian civilization (Angkor), mainland Southeast Asia became a changing tapestry of kingdoms with broad territories and large populations whose loyalty reflected proximity to the sacred city. Such societies required strong rulership and surpluses from *sawah* agriculture. The power and expanse of these Indianized states—whether Hindu or Buddhist—resonated the ability of urban elites to manage the peasantry through military conscription, draft labor, and an efficient marketing system for food and do-

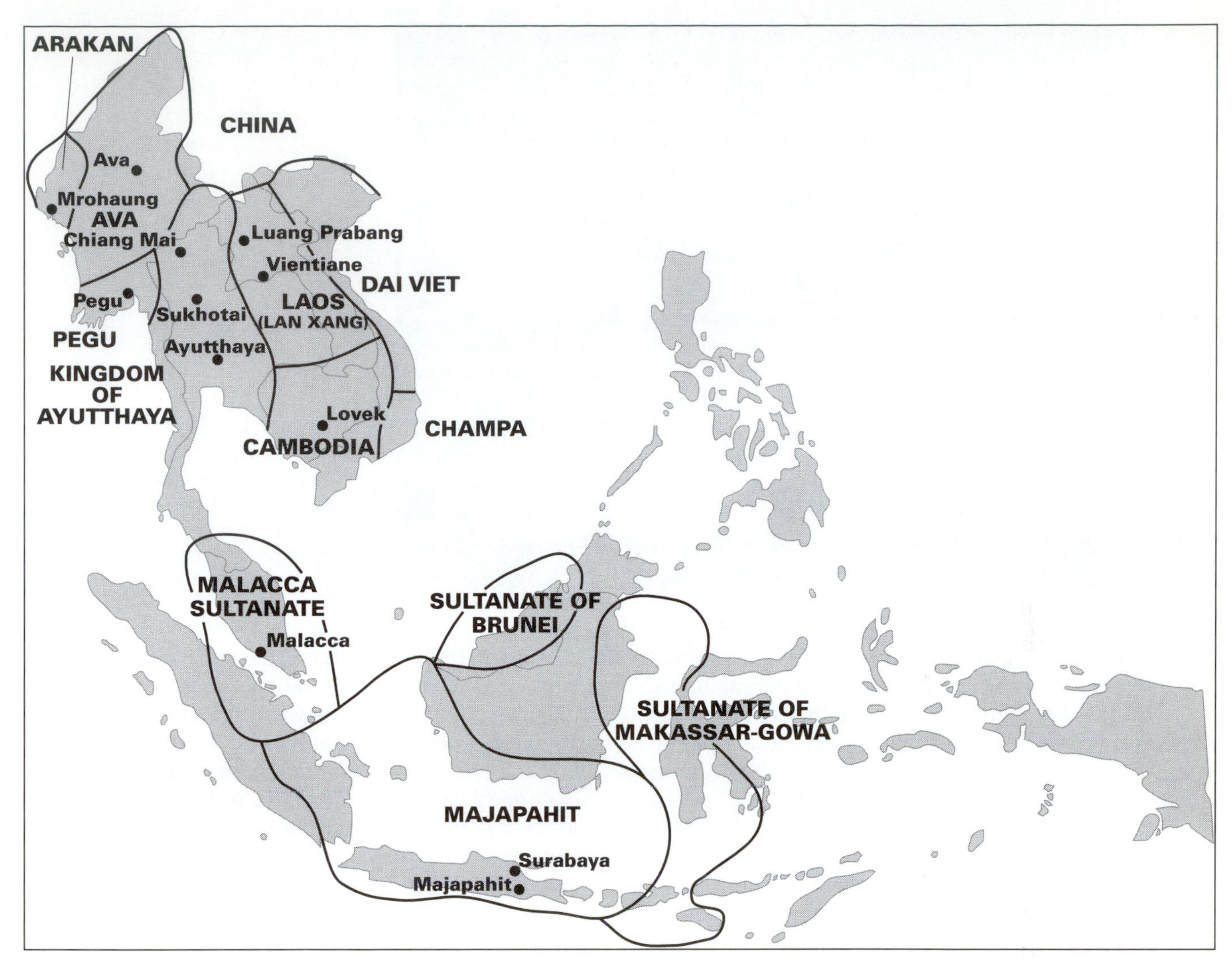

Map 3.6 Major empires of Southeast Asia, fifteenth century A.D. *Source: Ulack & Pauer, 1989, p. 18; Pluvier, 1995, plates 13, 17.*

mestic goods. Their decline resulted from costly wars, dynastic conflicts, environmental degradation corollary to the construction of resplendent capitals, and expansion of peoples from southern China (especially Tai groups and Burmans) who displaced settled populations and destroyed their polities. Though space does not permit detailed commentary on the classical mainland states, several of the great empires warrant summary narratives.

Born of a territorial reunification of the two Chenlas, which had earlier supplanted Funan, the renowned Khmer kingdom of Angkor emerged in the ninth century A.D. at the northern end of the Tonle Sap basin. From this extraordinary reservoir—produced by seasonal backflows from the Mekong River during its monsoonal spate—many generations of Khmer have harvested abundant supplies of fish while growing *sawah* on flooded lakeside lowlands. The natural system of irrigation, which produces a single annual crop of rice, was expanded by various kings through massive engineering projects resulting in an intricate hydraulic network. Angkor's incomparable fluvial resource not only guaranteed regular food supplies but also indirectly fostered development of a large bureaucracy to supervise waterworks and manage surpluses.

From the time of Angkor's foundation under King Jayavarman II (802) to its demise early in the fifteenth century A.D., successive rulers embraced the idea of the *deva-raja* to validate their claims of sovereignty. Even after Mahayana Buddhism superseded Hinduism in the twelfth century, the god–king cult persisted as an anchor of statehood. To provide earthly manifestation of a divine right-to-rule, monarchs mobilized vast human and material resources for the creation of immense temple complexes reproducing Indian cosmic designs and confirming Brahmanical ideas of kingship. Each religious compound was endowed with villages subject to corvée,

Photo 3.1 The Angkorian Kingdom flourished from the ninth century until its demise in the fifteenth century. Angkor Wat, built in the twelfth century, ranks among the largest and most elaborate religious structures ever constructed. (Duiker)

shrine servants or slaves, and elite donors who earned merit by underwriting construction and maintenance. Such religious foundations testified to the importance of temples not only as ceremonial arenas but also as instruments of national integration. Even as the Khmer empire expanded to embrace not only Cambodia but also the delta of the Mekong River, much of its watershed in Laos and Thailand, and a swath of territory reaching to the Gulf of Martaban, kings granted certain newly conquered lands to the priestly establishments and thereby ensured support for the proliferating temples.

If architectural splendor, artistic creativity, and monumental construction are measures of a great civilization, Angkor sets the standard for Southeast Asia. Indeed many impressive ruins dating from the Angkor period still mark territories once ruled by the Khmer kings. But the greatest clustering of religious structures, and site of several capitals, is near modern Siem Reap. Conceived as a monument to and sepulcher for King Suryavarman II (1113–1150), who reunified the kingdom after bitter internal strife and campaigned successfully against Cham, Mon, and Vietnamese enemies, Angkor Wat ranks among the largest and most elaborate religious structures ever built. There is no doubt that this vast temple compound (5,500 by 4,700 ft, or 1,676 by 1,433 m), huge moat (550 ft, or 168 m wide)—with its incredible sculptural detail and low-relief ornamentation, stupendous entrance gate, great causeway bordered by giant Naga (divine serpents) balustrades, and sprawling sanctum sanctorum graced by a remarkable tower symbolizing Mount *Meru* and four secondary towers depicting mountains of the Hindu cosmos—represents one of the most spectacular architectural creations of the human mind. No less spectacular, Angkor Thom was later fashioned by King Jayavarman VII (1181–c. 1215) as a national capital. Foursquared (2.2 mi, or 3.5 km per side) according to the Indian cosmic model with an encircling wall (25 ft, or 1.6 m high) and moat (300 ft, or 91 m wide), grand entrances and avenues, fine palaces, and impressive government buildings, the jewel of this ancient city was the Bayon temple. The forest of shrines surrounding the Bayon's magnificent central tower, which features four massive sculpted faces (the city's patron deity Bodhisattva Avalokiteshvara or perhaps the king), comprise a truly marvelous religious structure (Cœdès, 1963, 1968; Groslier and Arthaud, 1966). Even today, it is not difficult to believe that Angkor Wat and Angkor Thom with their heavenly design and impeccable detail offered vivid and undeniable evidence of the legitimacy of god–kingship to ancient worshipers.

Scholars still dispute the cause of Angkor's dissolution. Some argue that the extravagance of temple construction simply exhausted the Khmers. Others suggest that the irrigation system deteriorated and collapsed; that disease ravaged the kingdom because its waterworks provided a fine habitat for the vectors of malaria (anopheline mosquitoes); that the late advent of Theravāda Buddhism destabilized the rigid social order associated with earlier religious tradition; or that attacks by Thai armies finally broke the national spirit and treasury of Southeast Asia's archetypal mainland state.

While Angkor prospered, declined, and dissolved between the ninth and fifteenth centuries A.D., other kingdoms appeared and disappeared on its flanks. To the East, Champa withered under the constant military and demographic pressure of southwardly moving Vietnamese. To the North, various Tai peoples (Shan, Lao, Thai, and oth-

BOX 3.1 Angkor Thom: Abode of the Gods

While Angkor Wat is a free-standing temple designed exclusively as a sanctified stage for imperial funerary rites, group worship tied to the ceremonial calendar, and spontaneous personal prayers, the neighboring complex of Angkor Thom was conceived and constructed as a replica of the cosmos, axis of the universe, and national capital. It was thus an archetypal sacred city with integrated administrative, economic, religious, and social functions.

Almost 5 sq mi (13 sq km) in extent, Angkor Thom's heavenly inspired morphology featured grand avenues that marked the cardinal directions of the compass, divided urban space into four quarters, and centered on the Bayon with its spectacular towers, gargantuan statuary, and splendid bas-reliefs depicting Hindu myths, conflict with Champa, and aspects of Khmer culture. This extraordinary religious shrine along with nearby Baphuon, a lesser pyramidal temple representing Mt. *Meru*, served as the hub of public religious life. The northwest sector of the capital was dominated by a royal compound consisting of a palace, gardens, pools, chapels, and numerous wooden structures occupied by advisors, guards, and servants. Angkor Thom's other three districts embraced the houses and workplaces of essential government personnel and notable urban residents, including the ranking priesthood and literati, civil functionaries, court artisans and entertainers, military troops, and major merchants. Suburban areas immediately outside the city's enclosing walls and moats were densely settled by domestic craftsworkers, petty traders, market gardeners, and common laborers who lived in bamboo and thatch buildings situated in a matrix of temples, great artificial lakes (Baray) and other irrigation works, dooryard gardens, and rice fields. Together they sustained the urban lifestyle of elite classes living within the walled capital. After conquest by Thai invaders and the shift of Khmer power to Phnom Penh in the fifteenth century, Angkor Thom was abandoned by all but a few caretaker Buddhist monks and was soon reclaimed by tropical forest. Rediscovered and partly restored during the French era, it became a focus of archaeological investigation and nascent tourism that endured until the bloody Khmer Rouge victory in 1975. Ravaged by war and looted for priceless religious statues and decorative art for the past two decades, the ancient ruins of Angkor Thom and adjacent temples are now being salvaged and protected for growing numbers of international tourists.

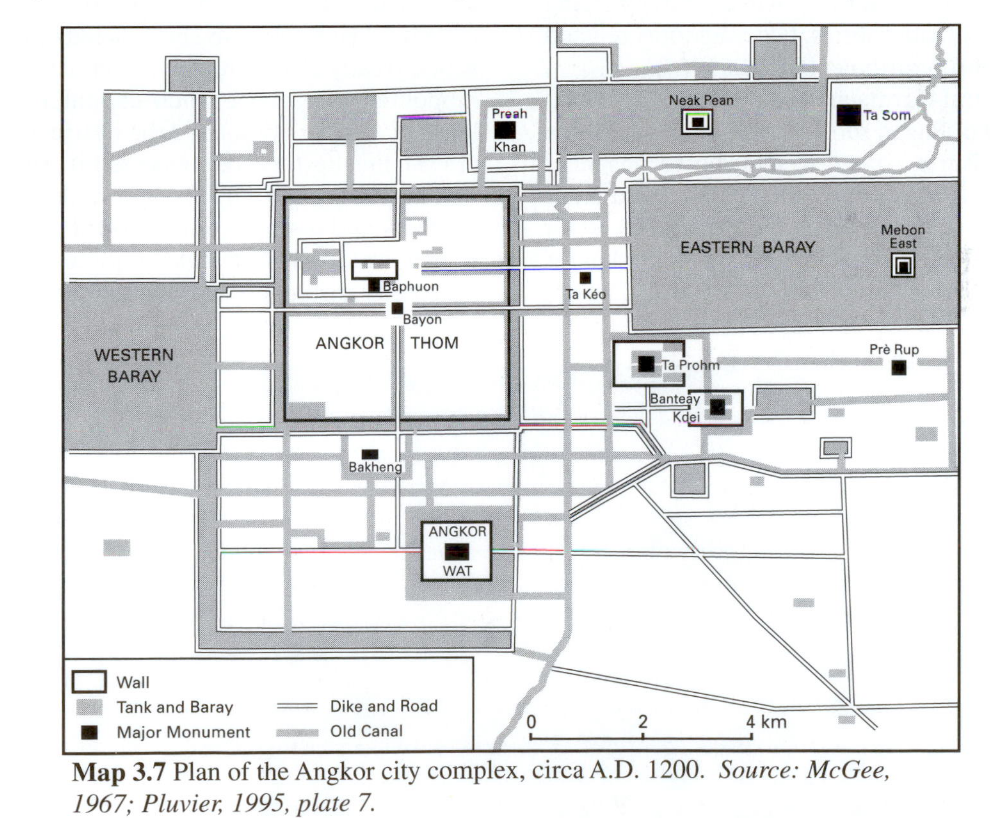

Map 3.7 Plan of the Angkor city complex, circa A.D. 1200. *Source: McGee, 1967; Pluvier, 1995, plate 7.*

ers), who were then leaving western Yunnan, began to develop ethnopolitical identities, gather military strength, and organize chiefdoms. One such became the long-lasting but unstable kingdom of Lan Na in the borderland of Burma and Thailand. It centered on the sacred city of Chiang Mai (founded in 1296) and functioned as a hub of Tai religious and cultural life. Slightly earlier (1240s), several Tai chieftains joined to expel Angkor's regional governors in Thailand's central plain and establish the first Siamese state. For nearly 200 years, this kingdom of Sukhothai flourished and at one point reached from Laos to the Salween River and southward to Ligor (Nakhon Si Thammarat) on the Isthmus of Kra. To the West, the small kingdoms and urban principalities of the Pyu and Mon peoples were on the defensive. After 700 years of enduring polity, the last of the Pyu states fell to Tai invaders in the ninth century A.D. Emerging in the lower stretches of the Sittang, a series of Mon kingdoms (Dvaravati, Thaton, Pegu, Canasa, and other short-lived states) existed between the sixth and the sixteenth centuries A.D. in an area straddling the Burma–Thai border. Among the earliest Theravada Buddhists in Southeast Asia, they fervently proselytized their neighbors. Harassed on three fronts by Burman and Tai invaders, the Mon territories shrank steadily and their last kingdom of Pegu (or Hanthawaddy) finally disappeared in 1540. Further West, the Arakanese forged a unique identity in a series of coastal kingdoms that first crystallized in the fifth century A.D. Despite sea raids by the Cholas and overland sorties by various Burmese peoples, they not only survived but in 1433 founded a substantial state centered on the splendid city of Mrohaung (Mrauk-U). Amidst this changing assemblage of mainland polities, two kingdoms reigned supreme: Pagan and Ayutthaya (Cœdés, 1968).

Situated in central Burma and founded in 849, the immense sacred city of Pagan served for about 350 years as capital of the first Burman state. Readily borrowing ideas and institutions from resident Pyu people, the immigrant Burmans soon controlled most of the Irrawaddy River valley and fashioned a classical agrarian state. Under King Anawrahta, who reigned from 1044 to 1077, Pagan's territories expanded southward to Mergui and northward to Bhamo. He, along with King Kyanzittha, also played a critical role in making Theravāda Buddhism the religion of state and the catalyst to a blossoming Burmese civilization. Pagan evolved into an archetypal ceremonial complex: the urban throne of rulers, seat of a powerful officialdom, home of a large monastic community, and cosmicized axis of empire. The cultural landscape of this sprawling capital (26 sq mi, or 67 sq km) was riddled with thousands of well-endowed temples (Aung-Thwin, 1985). In spite of internal and external threats, Pagan endured until late in the thirteenth century, when a Mongol invasion and Shan incursions disrupted the rural economy and destroyed the government.

Taking advantage of Angkor's increasing weakness, the obscure adventurer Prince U Thong in 1351 established the kingdom of Ayutthaya in the riverine lowlands of Thailand's Chao Phraya. His power derived from an odd aggregation of Tai migrants who had recently professed Buddhism, their Mon co-religionists, and a supportive Chinese trading community. In a familiar pattern, this mainland state was secured by an agrarian economy, organized according to ancient ideas of divine kingship, governed by literati, and seated in a sacred city. But the rulers of Ayutthaya also encouraged international trade, which greatly contributed to imperial wealth and the cosmopolitan character of their capital. Following the collapse of Angkor (1431) and the destruction of Sukhothai (1438), this upstart state asserted its regional authority on land and sea. Ayutthaya certainly suffered dynastic conflicts and disruptive invasions by the Burmese and French. Yet its skillful leadership nurtured a long-lasting prosperity through effective administration, continuing

Photo 3.2 Founded in the mid ninth century, Pagan was the capital of the first Burman state for about 350 years. Photo is of Thatbyinnyu Temple in Pagan, built in the twelfth century. (Reed)

politico-military vigilance and occasional aggressive moves against other kingdoms, cautious dealings with Western colonial powers, and systematic efforts to develop overseas commercial links. Ayutthaya remained a viable state for more than four centuries until conquered by the Burmese in 1767.

WESTERN INTERVENTION

While acknowledging the multifaceted impact of European imperialism in Southeast Asia, it should be emphasized that many social and ecological elements of modern society are of precolonial origin and so survived the seemingly overpowering onslaught of ideas and institutions emanating from the West. Such continuities of indigenous origin, which broadly frame this culture region, are often disregarded or dismissed by popular and scholarly writers. Like an earlier generation of South Asian historians, who were inclined to portray Southeast Asia as a product of ancient Indian imperial intervention, some Western historians have overlooked the binding ingredients of local traditions while overemphasizing the legacy of Europe.

Indigenous Counterpoint

Certainly in the realm of Southeast Asian religions, an enduring continuity of regional distribution from the sixteenth century to the present is apparent even to casual observers. At the time of European intervention, mountain and forest dwellers everywhere subscribed to animism; most lowland peoples on the mainland had accepted Theravāda Buddhism (except for the Vietnamese, who held to Mahayana Buddhism and various Chinese cults); coastal peoples throughout the archipelago were adopting Islam; and the more cosmopolitan port cities seem to have been a confusion of belief systems. Apart from Catholic sectors of the Philippines, scattered highlands with Protestant minorities, and larger cities embracing people of diverse faiths, the contemporary map of religions would readily correspond with that of the early 1500s (Map 3.8).

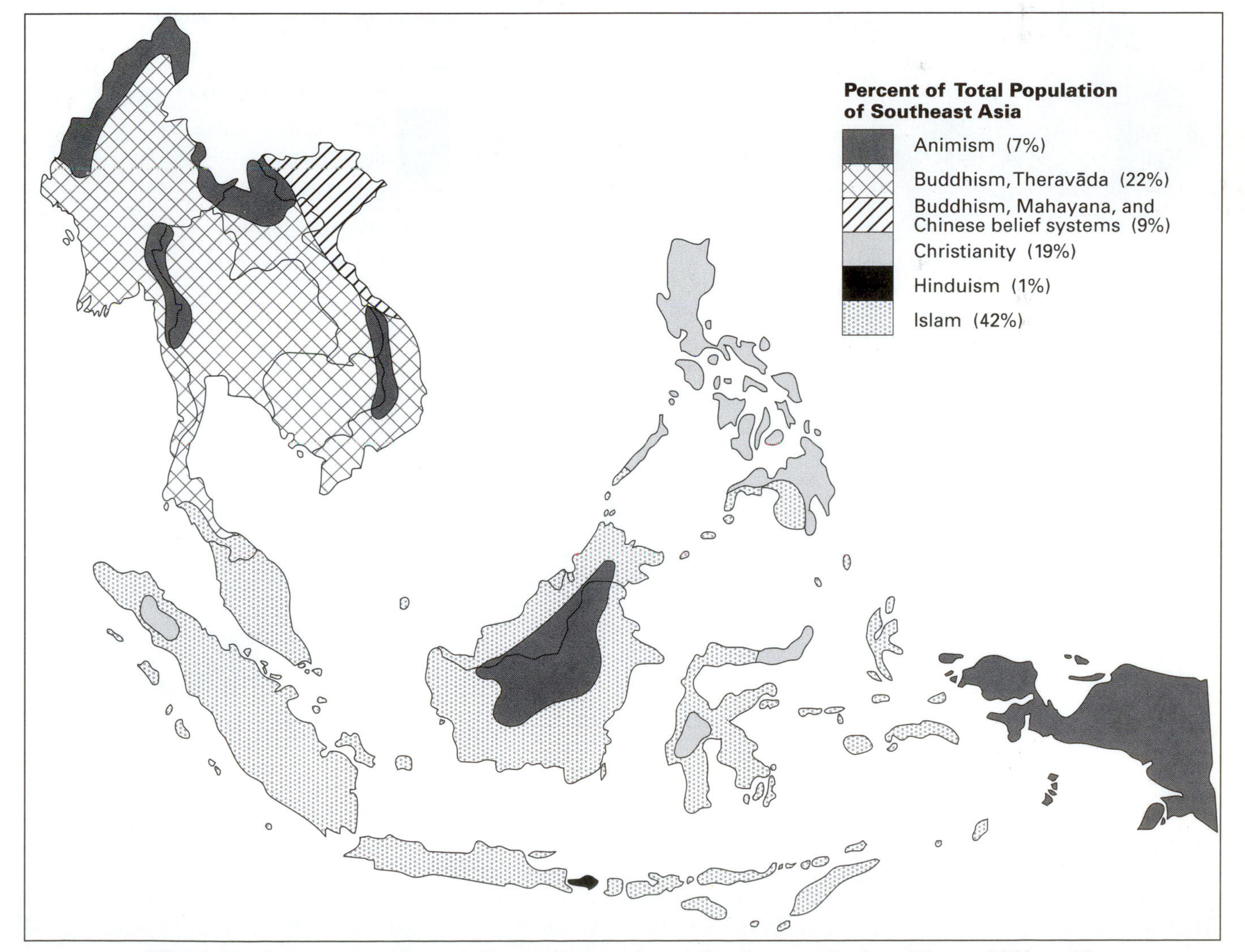

Map 3.8 Religions of Southeast Asia, twentieth century A.D. *Source: Ulack and Pauer, 1989, p. 29.*

Though spatial details have changed, the ethnic map of Southeast Asia today reflects an overall pattern that has been little modified during the past five centuries. Malay peoples continue to dominate the archipelago, while the Burmans, Thai, Khmer, Laotians, and Vietnamese still occupy the mainland realms held by their forebears in the sixteenth century. At the same time the uplands everywhere remain a shatterbelt of minorities. Only the cities and towns of Southeast Asia offer significant exception with their large Chinese populations (and Indians in Malaysia, Burma, and Singapore) resulting from immigration during the European colonial era. Some overseas Chinese still nurture the heritage of their forebears, but the majority have been integrated into societies of their chosen homelands and have often intermarried with indigenous peoples to produce communities of mixed cultural inheritance (Map 3.9).

At the same time there is a fundamental constancy in both the material and ethereal expressions of rural tradition. Today, as in the precolonial past, the great majority of farmers are *sawah* or swidden cultivators who also tend kitchen gardens and forage (where appropriate) from the forest and the sea. Even in the present era of unitary government and orchestrated allegiance to the state, Southeast Asian villagers remain under the strong influence of local leaders, nurture regional art, prefer vernacular to national languages, tend to marry within their own ethnic group, retain a wealth of regional environmental wisdom, and are custodians of socially instructive folklore. Although the Europeans imposed new political forms, transformed economies, fostered modern educational systems, proselytized Christianity, and promoted an "idea of progress" (Bury, 1920) proclaiming the superiority of Western culture and technology, they did not, nor could they, sever the taproot of tradition. Indeed one of the hallmarks of our time is a reawakening of interest in the indigenous past to strengthen present-day nationalism.

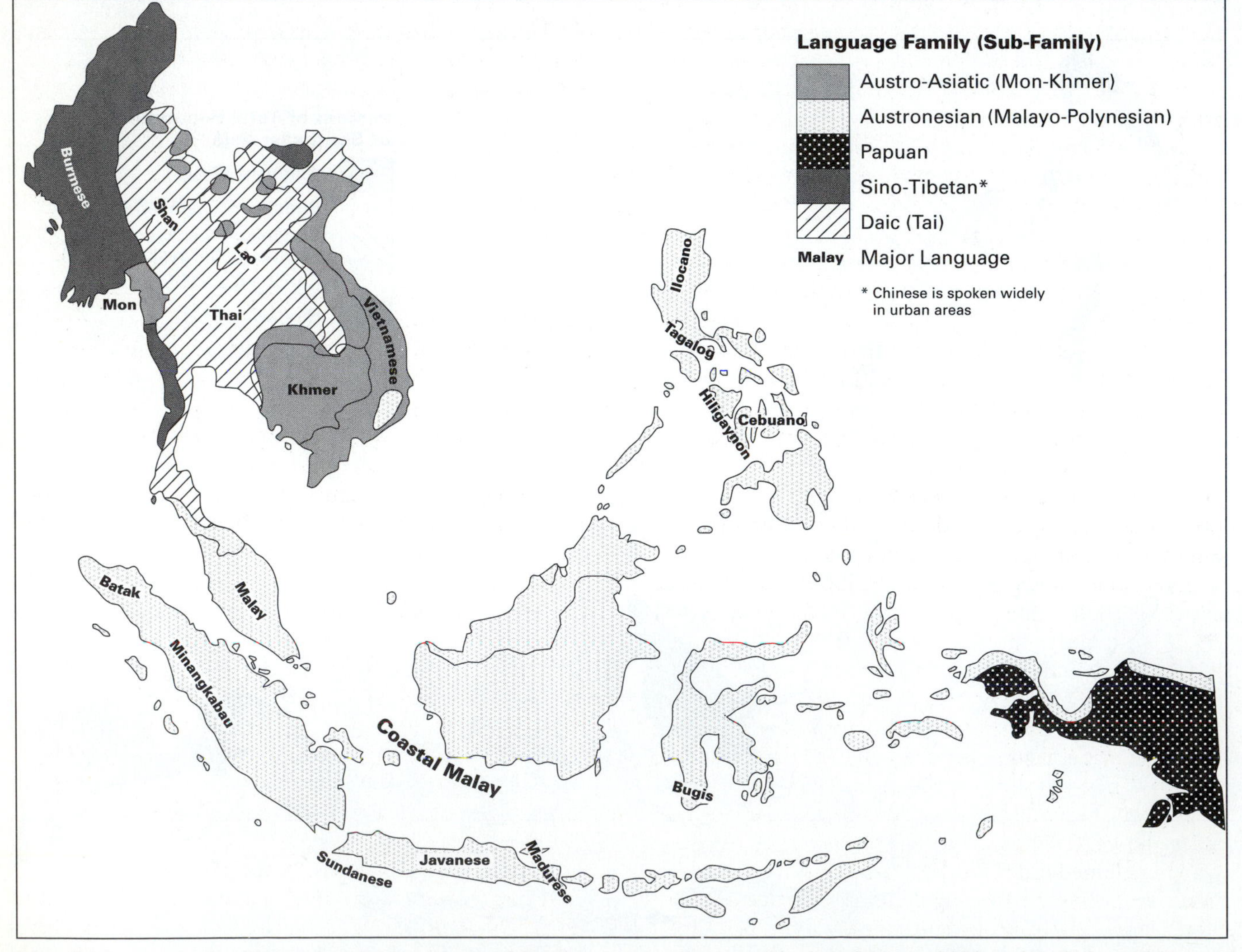

Map 3.9 Language families and major languages of Southeast Asia. *Source: Ulack and Pauer, 1989, p. 27; Fisher, 1964, p. 99.*

Early Imperial Strategy

An abiding Western presence in Southeast Asia commenced with the fall of Malacca to the Portuguese in 1511, but did not immediately inaugurate a period of systematic regional conquest. For almost three centuries, most governments and chartered companies confined their imperial activities to matters of commerce and avoided serious involvement in the internal affairs of indigenous states. Mutually satisfactory trading relationships between Europeans and Asians could often be arranged without direct control over production systems, so few Westerners found it necessary to advocate colonial rule.

The pioneering Portuguese, whose human resources were extremely limited because of a sparse national population in Iberia, forged a model of overseas outreach that required little investment beyond the foundation of one well-fortified urban garrison and a far-flung network of so-called factory (trading station) settlements where luxury goods were collected for transshipment to distant markets. Their single colonial capital functioned additionally as an international commercial emporium, the seat of religious institutions, an administrative headquarters, and a cultural nerve center where metropolitan values could best be perpetuated. By supervising only one multifunctional city and a scattering of small trading posts served by a few merchants, a small detachment of soldiers, and native servants, the Portuguese together with their Dutch and English successors avoided excessive operational expenditures and markedly increased profits. To enhance mercantile receipts early in the colonial era further, the Westerners shied away from prolonged and exhausting territorial conflicts. Force was indeed applied against recalcitrant local rulers who attempted to sabotage productive trading relationships, but there is no evidence that the early European colonizers relished the use of their armies or navies. They fully realized that sustained imperial campaigns ultimately proved very costly and seldom resulted in improved or long-lasting commercial arrangements.

Even when drawn into war to facilitate trade and successful in their military operations, the Europeans hesitated to shoulder the tasks of governance. The Dutch United East India Company (*Vereenigde Oost-Indisch Compagnie,* or VOC), for example, was periodically involved in major land wars on Java and punitive naval operations throughout the Indonesian archipelago, but generally proved unwilling to assume day-to-day administrative responsibility over newly won territories. Instead the VOC chose to bolster tractable native leaders who maintained order through traditional laws and sanctions. Accordingly a growing number of Javanese princes were compelled to acknowledge Dutch suzerainty. Under the resultant system of "indirect rule," indigenous potentates enjoyed considerable latitude in managing their domains as long as they did not break agreements with colonial authorities in Batavia (Jakarta). The territorial holdings of the VOC thus grew slowly and without benefit of a master plan for expansion. While peace reigned and trade remained unfettered, most Europeans refrained from interference in local political affairs. Prior to 1800, therefore, Westerners in Southeast Asia only rarely attempted the overthrow of indigenous kingdoms, destruction of sacred cities, or acquisition of lands. Only in the Hispanic Philippines did sweeping political, economic, and cultural changes occur during the early period of European imperial intervention.

Portuguese Pathfinders

Driven by a desire to neutralize the near absolute control of Venice and its Muslim partners over European trade with Asia, the Portuguese during the fifteenth century probed southward along the western coast of Africa in search of a direct route to India. In 1487 Bartolomeu Dias finally rounded the Cape of Good Hope. A decade later Vasco da Gama reached Calicut in India and returned safely to Lisbon with spices and other expensive luxuries that garnered fabulous profits. Not content to participate in an open competition with Arab and Indian merchants and anxious to arrest the continuing diffusion of Islam, Portugal's field commanders acted quickly to fashion a monopoly that would guarantee profitable commerce.

Under the bold and inspiring leadership of the viceroy of the East, Afonso de Albuquerque (c. 1459–1515), the Portuguese in 1510 established their heavily fortified Asian headquarters in Goa on India's western coast together with other strategically situated ports and factories along the littoral of Africa and Asia. Capitalizing on a decisive naval superiority, they attempted to construct a system of exchange almost like institutionalized piracy by neutralizing competing communities of Asian merchants, controlling commodities flows, selling letters of passage, and destroying Muslim maritime forces on the Indian Ocean. While Goa functioned as the axis of Portugal's empire, two great garrisoned cities secured its flanks. Covering one of the oceanic entrances to the heartland of Islam (the Persian Gulf), Hormuz kept Arab navies and trading fleets at bay. With command of the narrows between Sumatra and the Malay Peninsula, Malacca effectively opened the door to the riches of Japan, China, and the Spice Islands (the Moluccas) of Southeast Asia.

After learning of Malacca, the Portuguese Crown in 1509 dispatched an exploratory expedition of four ships under Diego López de Sequeira to locate this famed emporium, assess its geopolitical potential, and negotiate a treaty of friendship with the local ruler that would allow the permanent residence of a Portuguese commercial agent. At first well received by Sultan Mahmud Shah, Sequeira soon provoked suspicion among the international traders of Malacca by breaking ground for a heavily fortified factory. Convinced of the mounting threat to his sovereignty, the Sultan took several hostages and expelled the Portuguese with some loss of life. After assembling a strike force of 1,200 men and 18 ships, Albuquerque retaliated by laying siege to Malacca in April of 1511 and took the city in four months.

Though successful in war, the Portuguese soon discovered that the conquest of Malacca and other military victories did not automatically afford absolute dominance over Asian trade. In some years their imperious mercantile scheme yielded impressive profits. But it could never translate into a stable monopoly as long as the Portuguese failed to subjugate the primary producers of spices (cloves, pepper, cinnamon, nutmeg, and mace) and to control all exports from Southeast Asia. Certainly the strategic location of Malacca provided considerable commercial leverage. Yet it remained a city-state without a significant rural hinterland, and the Portuguese ultimately depended on distant and dispersed indigenous suppliers for spices, other luxuries, and food supplies. Indeed the flow of all colonial trade through Malacca required the goodwill and continuing involvement of Asian merchants in an intricate trading network. But Portuguese participation in regional commerce was often complicated by their monopolistic practices, the arrogance of many colonists, and sporadic proselytizing by Catholic missionaries. This resulted in repeated and terribly costly assaults on Malacca by rival polities along the strait as well as disruptive local wars that angered Muslims everywhere in monsoon Asia. Consequently many merchants shifted their operations to friendlier ports in Aceh, Johor, or northern Java, where comparatively open markets remained unencumbered by high duties, anchorage fees, and intrusive colonial regulations.

Portugal was also hampered by other problems. Almost from the onset of colonization, the officialdom in Malacca seemed plagued by corruption and bureaucratic inefficiency. While publicly endorsing the Crown monopoly, almost all Portuguese governors and other ranking administrators continued to trade for personal gain in direct violation of royal command. At the same time, Portugal's initial superiority in weaponry, ship design, and military organization rapidly dissipated. Most of its commercial competitors on the Strait of Malacca, together with the rulers of many other kingdoms and chiefdoms elsewhere in Southeast Asia, actually employed disaffected Portuguese as mercenaries and advisors to modernize their navies and armies. By the end of the sixteenth century, the early Portuguese technological and organizational advantages were effectively neutralized. Finally, Portugal's Asian enterprise never attracted sufficient Western personnel to operate a trading monopoly embracing all of monsoon Asia. In fact its colonial settlements were not even capable of renewing their own European populations because very few white women immigrated to the East. Albuquerque and his successors attempted to remedy this problem by encouraging intermarriage between Portuguese men of lower social rank and Asian women. This policy did produce scattered communities of Christianized and westernized Eurasians who remained loyal to Portugal over the centuries, but they were never numerous and their collective presence did not resolve a persisting and chronic shortage of colonists. Although the Portuguese established heavily fortified factory towns in the heart of the Moluccas (Ternate, Tidore, and Ambon) to better manage spice production, these outposts proved short-lived in the face of growing indigenous (and later Dutch) opposition.

Portugal failed to monopolize regional trade, but its thin line of imperial forces clearly impacted traditional systems of commerce and left lasting cultural imprints. During a brief tenure in the Moluccas, the Portuguese converted more than 50,000 Indonesians to Catholicism and their descendants retained a strong cultural orientation toward Europe. Other westernized and Christianized communities of Eurasians materialized and survived in the colonial cities of Malacca, Dili (Timor), and Macao (southern China). Even in areas outside their direct control, the Portuguese influenced Southeast Asians through loan words relating to technology, religion, and abstractions of Western origin (Boxer, 1969; Meilink-Roelofsz, 1962).

Spanish Conquistadores

Although European imperialism in Asia was driven mainly by an appetite for wealth, the Hispanic-Filipino experience in the Philippines proved a partial exception. From the onset, Spanish conquistadores were motivated by a mixture of economic, political, and religious goals. Unlike other Europeans, who before the nineteenth century amplified their commercial profits by avoiding political entanglements, debilitating wars, and administrative responsibilities, the Spaniards pursued a multifaceted program of direct involvement with conquered peoples through acquisition of territory, religious conversion, and cultural transformation in addition to economic exploitation. Even a cursory reading of Spanish laws governing overseas possessions, the writings of Catholic philosophers concerned with human rights, or letters between administrators on the colonial frontier and officials in Iberia suggests that most Spaniards shared several motives: a predictable craving for easy riches, a fervent commitment to advancing Christianity, and a desire to expand Spain's

BOX 3.2 Malacca: From Traders to Tourists

Following its meteoric rise during the fifteenth century as a pivot of commerce with maritime connections from China to Arabia, the Muslim entrepôt of Malacca was abruptly transformed into a vehicle of European imperialism after conquest by Portugal in 1511. Wisely maintaining the established structure of international and intraregional trade forged long before the advent of colonialism, the Portuguese proclaimed their new imperial order by gradually introducing a monopolistic politico-economic system designed to penetrate localized markets throughout Southeast Asia. Interestingly this mercantilist strategy did not necessitate or even promote fundamental alterations in the functional form or physical layout of Malacca.

Unlike the westernized capitals of Manila and Batavia, which were platted as morphological replicas of contemporary Spanish and Dutch cities, Malacca underwent significant spatial reconfiguration only in the European quarter (Fort A'Famosa) situated on St. Paul's hill and its flanks. The Portuguese immediately enclosed this highly defensible site, which was framed on two sides by the sea and on another by the Malacca River, with massive stone walls and imposing bastions. Within their seemingly impregnable fortress, the Portuguese constructed administrative buildings, residences for the governor and bishop, a jail, Catholic churches and monasteries, a Jesuit college, two hospitals, military barracks, and private houses. Outside the walled city, Malacca remained essentially unchanged. Situated amidst dooryard gardens, rice fields, fruit orchards, coconut groves, and swamp lands, its clearly demarcated ethnic quarters retained their pre-colonial sites (Chinese in Kampung China, Javanese in Kampung Java, Indians in Kampung Kling, and various Malay communities). Only one new communal assemblage emerged—a group of Eurasians living in the area of Medan Portugis.

Colonial rule by the Dutch between 1641 and 1824, which coincided with a steady decline in the commercial fortunes of Malacca, brought only superficial changes to the urban landscape. They erected a substantial town hall (Stadhuys) on the main square, refurbished some municipal buildings, reconstructed sections of the fortress, and built the outlying Fort St. John to guard the southern approach to the city. But the design of the port and its environs remained intact. Likewise the British, who governed from 1824 to 1957, preserved Malacca's basic plan. While dismantling the city walls to prevent subsequent use by either invaders or local insurgents, their other morphological contributions consisted only of modern infrastructural improvements, several public buildings, a few commercial establishments, and private residences.

At the time of Malaysian independence, Malacca functioned as a minor local port, capital of a small state, and an embryonic center of tourism. Recognizing the significance of its well preserved urban core, which embraces dozens of western architectural treasures as well as many significant Chinese, Indian, and Malay structures dating from the sixteenth to the nineteenth centuries, local and state authorities have instituted progressive policies of historic preservation designed to protect the city's cultural landscape and promote both domestic and international tourism. Today Malacca attracts thousands of visitors daily and is widely recognized as a virtual museum of the colonial era.

Photo 3.3 Malacca's main square: the Dutch influence. (Reed)

imperial boundaries. In the course of three failed maritime expeditions during the sixteenth century and Miguel López de Legazpi's successful enterprise (1565–1572), authorities in Spain and Mexico portrayed Philippine colonization as a coordinated effort of church and state involving a military vanguard, a stable civil officialdom, merchant investors, and a host of missionaries. Together they were charged with implementing a system of governance and exploitation based on tested practices in Hispanic America.

Like other Europeans, most Spaniards embarked on their long and hazardous voyages to Southeast Asia with visions of fame and fortune. Indeed it is unlikely that the Spanish Crown could have retained its distant Asian dependency for more than three centuries without alluring promises of easy wealth. Unless motivated by greed, few colonists would have risked their lives on the 7,500 mi (12,070 km) outbound passage on the galleons from Acapulco along the easterly trade winds or the especially hazardous return run from Manila sailing stormy westerly winds for five or six months at higher latitudes on the North Pacific. During the sixteenth and early seventeenth centuries, the Spanish Crown fostered this continuing "quest for gold and spices" through a series of imperial ventures in both East and Southeast Asia. Even while converting Manila into an opulent entrepôt linking Asia and Mexico, Spain launched a dozen exploratory probes to southern China, Cambodia, Siam, Borneo, and the Moluccas in an ongoing endeavor to expand the *conquista.* Some Spaniards seriously dreamed of conquering Japan and China, but most simply wished to participate in the profitable spice trade. Following four unsuccessful efforts, they gained a foothold in the Spice Islands through the foundation of a settlement on Ternate in 1606. Spain held this costly and hard-won outpost until 1662, when it was abandoned to strengthen garrisons on Luzon in the face of a threatened invasion by the Chinese corsair Koxinga (Cheng Ch'eng-kung). Nevertheless, Spanish adventurers continued to settle in the Philippines because of royal assurances that all colonists could purchase shares in the lucrative galleon trade between Acapulco and Manila.

Although driven in part by an unquenchable hunger for riches, Hispanic imperialism in the Philippines embodied a second and unique dimension in the context of Southeast Asian colonial history—a formative religious component. In the Philippines, as in the New World, the Spanish *conquista* was envisaged as a joint undertaking of soldiers charged with neutralizing Filipino resistance and missionaries responsible for consolidating the military victories through religious conversion. Unlike other Westerners in the region, most Spaniards endorsed the Christian mission in Southeast Asia. Their individual and collective allegiance to the Catholic Church—reinforced during centuries of conflict and periodic wars with Iberian Muslims (Moors)—provided a strong motive for religious war with the Islamicized peoples (Moros) of Sulu, Palawan, and Mindanao. This seemingly endless struggle, which was punctuated by bitter battles on land and sea, represented the concluding phase of a grand design to convert all Filipinos to Christianity and resettle them in stable lowland communities. Clearly the Catholic missionary enterprise in Spain's Philippine colony was a pillar of imperial policy.

The mixed commercial and politico-religious character of Spanish imperialism translated into a spectrum of opportunities for colonists of various social and economic backgrounds. Over the centuries, a small but steady stream of European friars sought spiritual fulfillment as missionaries in *poblaciónes* (towns) and barrios (villages) throughout the lowlands of Luzon, the Visayas, and parts of Mindanao. The great majority of these priests labored in isolated parishes until death. Other Spaniards, especially individuals of humble birth, sought adventure as soldiers and sailors in the colony's armed forces, which repeatedly challenged the hated Moros and warred on occasion with the Dutch navy, Anglo-Indian invaders, and Chinese pirates. A smaller contingent of Spanish aristocrats found rewarding niches in the administrative and judicial bureaucracies of Manila. Lastly, almost all Western colonists participated in the galleon trade and some prospered (Lyon, 1990; Phelan, 1959).

Although the Hispanic community of the Philippines was fragmented along lines of wealth, social standing, and profession, as well as rank in the civil, military, and religious officialdoms, all Spaniards stood solidly with the Crown in a never-ending effort to foster the Christianization and Hispanization of Filipinos. Accordingly they endorsed imperial policies designed to institute direct colonial rule everywhere in the Philippines; to subdue the Muslims of Sulu and Mindanao and mountain peoples of northern Luzon; to stimulate production of food surpluses through imposition of the *encomienda* (a dualistic system of tribute and labor organization); and to consolidate the *conquista* by resettling (*reducción*) the formerly dispersed Filipino population in permanent colonial cities, towns, and villages where their religious and cultural transformation could be supervised by European priests (Doeppers, 1972; Reed, 1978).

In addition to introducing Christianity, a Hispanicized cultural milieu, a national government, and urbanism in an archipelago devoid of indigenous cities, the Spaniards fostered several other profound changes in the Philippines. To guarantee the food supplies of growing urban communities, swidden cultivators were encouraged to become *sawah* farmers. This transformation also produced a heightened sense of private ownership, leading by the eighteenth century to formal tenurial rights in provincial areas as well as cities. Additionally it provided the basis for increasing class differentiation and land tenure conflicts as more sophisticated Filipinos secured their wealth through the land acquisition. At the same time, the Spaniards facilitated a significant addition to the ethnic mix of the archipelago. Because the eastern leg of the galleon trade was monopolized by Chinese traders, each year for more than two centuries Manila witnessed the arrival and departure of 30 to 40 junks carrying between 6,000 and 12,000 crewmen and merchants. The majority returned to China, but thousands remained. Despite lingering racism and occasional massacres, these settlers provided the essential economic sinew of Manila, Cebu, and later other cities, where they monopolized almost all crafts and trades, worked as ordinary laborers, and played a critical role as moneylenders. Most Chinese immigrants were men, so they generally took Filipina wives. Their Sino-Filipino descendants, who often acquired the economic skills of their fathers while deriving an Hispanicized cultural inheritance from their mothers, have contributed significantly to Philippine economic life as petty traders, skilled craftsmen, plantation owners, and urban investors. In disproportionate numbers, Chinese mestizos have also joined the professions and entered politics (Wickberg, 1965).

Although the Western population remained small throughout Spain's tenure in the Philippines, colonial officials did not promote intermarriage to anchor Hispanic suzerainty through a community of loyal Eurasians. Instead Filipinos were granted limited responsibility in various military, civil, and religious institutions. From the onset of Hispanic rule, the Spaniards recruited native warriors into the imperial military and later the provincial constabulary. Filipinos also comprised the bulk of the crews on eastbound galleons. By the same token, local *datu* (chiefs) became the *gobernadorcillos* (petty governors, or municipal magistrates) in most towns. As native representatives of the Crown and agents of Hispanization, they derived prestige and politico-economic leverage for themselves and their families. Finally, despite opposition among Spanish colonists, the Catholic Church permitted the ordination of Filipinos as priests. By end of the Hispanic period, there were more than 600 Filipino members of the clergy. Along with the Chinese mestizos, these groups provided the intellectual and organizational foundation for the first credible nationalist movement in Southeast Asia during the second half of the nineteenth century (Map 3.10).

Dutch Company Traders

Late in the sixteenth century, as the second generation of Spaniards consolidated their position in the Philippines by institutionalizing the galleon trade, inaugurating the *reducción* program, and fortifying Manila with a moat and massive stone walls, the Dutch and British arrived on the Southeast Asian scene. Their presence testified to a shift in European military and economic power from Iberia to several nations on the North Sea. During this period of geopolitical struggle, the port at Lisbon was periodically closed to shippers who had been distributing spices and other exotic Asian goods to northern European markets. Consequently many Dutch merchants began to contemplate direct trade with the East.

The Netherlands was well prepared for the ensuing maritime and economic contest with other Western nations and various indigenous states in Southeast Asia. Its ship commanders had already honed their technical and organizational expertise in the Baltic trade and coastal commerce elsewhere in Europe, while Dutch seamen had been schooled on fishing fleets that plied the northern seas of Europe and during exploratory expeditions in the New World. By the same token, its naval and military forces (along with the national citizenry) were hardened in their bitter struggle for independence from Spain and religious freedom as Protestants. Equally important, the merchant society of the Netherlands proved quite sophisticated in banking, investment, and commerce. Although constrained by a small national population, the Dutch effectively countered this problem by hiring the British, Germans, Scandinavians, and others for overseas service. Likewise, a poverty of certain resources for the building and outfitting of ships (especially high quality lumber, tars, and a metallurgical industry) was neutralized by the immense financial resources of Dutch entrepreneurs, who could purchase required materials from far-flung suppliers.

Guided by sailors from the Netherlands who had served as seamen on Portuguese and Spanish ships operating in the waters of tropical Asia, the first Dutch vessels reached Southeast Asia in 1596. These inaugural expeditions were funded by individual mercantile houses in different Dutch cities and so represented extremely risky undertakings. The loss of a single vessel could bankrupt an entire urban community. Accordingly in 1602, the governing States-General of the Netherlands chartered the United East India Company (VOC) to bring investors throughout the nation under a single corporate umbrella with broad economic, political, and military authority. Its charter guaranteed the VOC a monopoly of all Dutch commerce in Asia in addition to comprehensive rights of political sovereignty in territories under its command. In short it was a commercial institution with the powers of an independent polity. As a joint-stock company, the VOC provided insurance against disasters and facilitated an equitable distribution of profits. And as a surrogate Dutch government, it could make treaties with native leaders, administer colonial territories, and employ the ultimate sanction of war against native armies as well as naval forces of the Portuguese, Spanish, and British.

When they first gathered cargoes in the Moluccas for shipment to Europe, the Dutch seemed content to follow

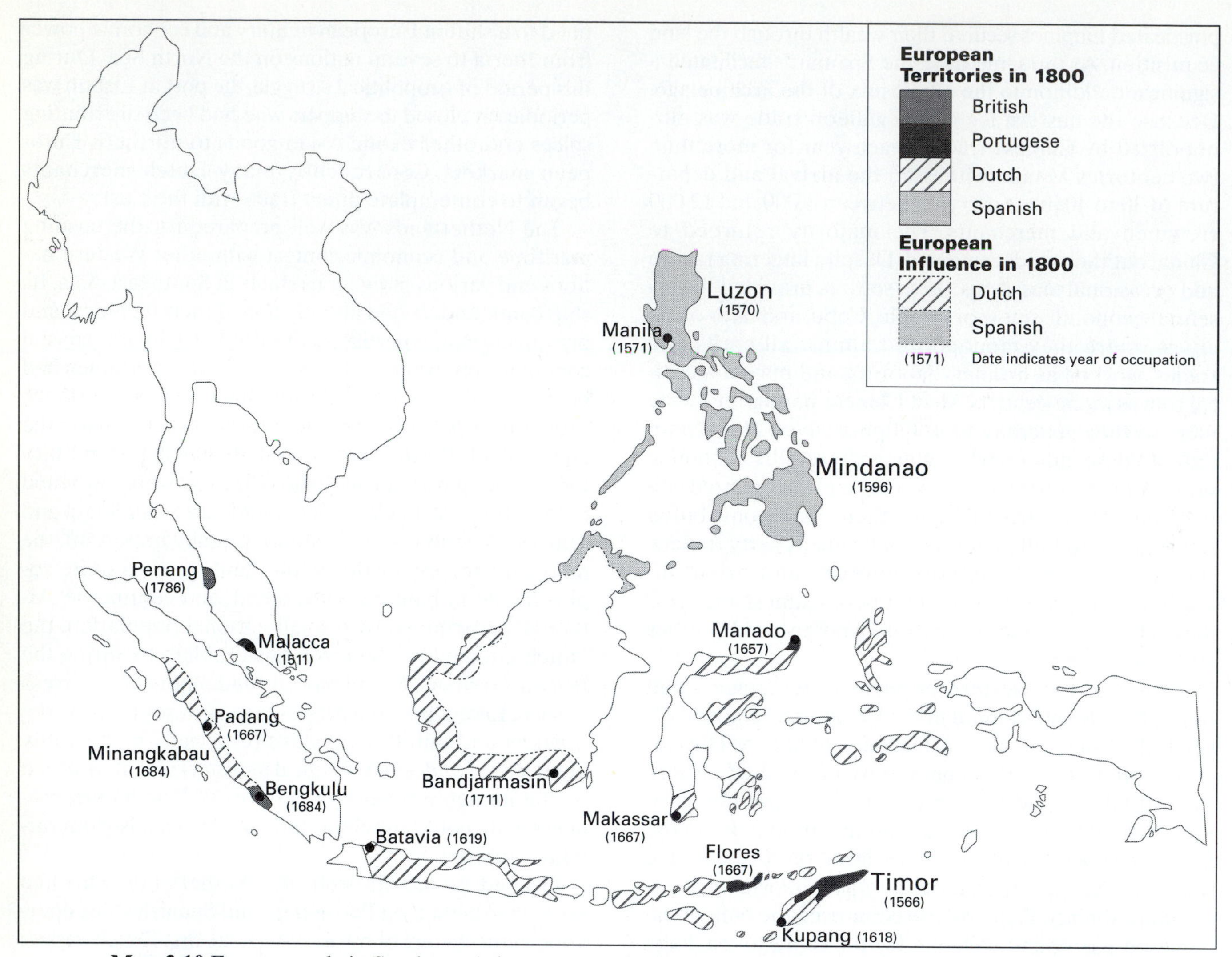

Map 3.10 European rule in Southeast Asia, seventeenth and eighteenth centuries A.D. *Source: Hall, 1964.*

the locally defined and comparatively open trade practices of the Southeast Asians. They respected customary marketing rules, used indigenous leaders as intermediaries, and avoided costly conflicts. Their commercial sense and timing proved auspicious, for the Portuguese and Spaniards were already hated because of arrogance in both trade and personal behavior. Through seemingly equitable treaties with the rulers of Ambon and Ternate, along with punishing attacks on their Iberian antagonists, the Dutch carefully sowed goodwill among the Spice Islanders. Even before inauguration of the VOC, they were thus trading successfully in the Moluccas and had been given access to the port of Banten in western Java. But this prelude of unfettered commerce lasted for only two decades.

Under the aggressive leadership of Governor-General Jan Pieterzoon Coen (1618–1629), the VOC rejected the traditional system of commerce where overproduction remained a problem and smuggling always jeopardized profits. Instead the Dutch instituted strict curbs on harvests of pepper and spices to gain monopolistic control over supplies. This policy involved frequent military expeditions in crop-producing regions and accretive territorial conquests to better monitor the cultivation of cloves, mace, and nutmeg. To expand and implement this crude form of harvest management effectively, the Dutch frequently made forays into areas outside their imperial jurisdiction to obliterate groves of spice-producing trees, kill uncooperative farmers, and destroy villages that resisted their production bans. Additionally they conducted lightning-strike naval raids against coastal communities involved in smuggling and imposed their dominion on the flourishing emporium of Makassar late in the 1660s. The VOC also endeavored to control intra-Asian trade flowing through the Indonesian archipelago by establishing an extensive network of well-fortified cities and factories in South Asia (Surat, Colombo, and other places), East Asia (particularly Formosa, Canton, and Deshima), and mainland Southeast Asia (Malacca, Ayutthaya, and smaller ports). In short, the company fashioned an empire not unlike those of Srivijaya or precolonial Malacca and ruled mainly through control of maritime traffic. This two-pronged Dutch strategy involving limitations on the supply of key commodities and regional trade regulation was secured for almost 200 years through expulsion of the British from the Moluccas (1623) and elimination of the Portuguese as an effective regional force following the Dutch capture of Malacca (1641).

Headquartered in the colonial capital of Batavia (Jakarta), the VOC projected its power throughout the Indonesian archipelago by means of garrisoned cities and towns that were sustained and protected by maritime forces that enjoyed a sweeping command of the seas. While the Dutch often extended their political influence by exploiting local disputes among indigenous polities, they were also ready to accede to war and conquer new territories to gain incremental commercial advantage. This was demonstrated in dealings with the sultanate of Mataram, which laid siege to Batavia in 1629. The Javanese were soundly defeated through the combined efforts of Dutch naval forces that destroyed their supply ships and colonial land forces that stubbornly held Batavia against overwhelming odds. Humbled but not destroyed, the huge native army retreated home to Central Java to reorganize and rebuild its resources. After a nearly 50-year standoff, the VOC gathered sufficient strength to overcome Mataram in the 1670s and so imposed Dutch territorial authority on Batavia's eastern flank. During the 1680s, territories in West Java were likewise pacified through the conquest of Banten. Afterward the VOC could pursue monopoly commerce in Indonesia from its secure Batavian headquarters.

While the VOC initially posed as just another actor in Southeast Asian commerce, the monopolistic program introduced by Coen ultimately destroyed a number of indigenous states, transformed regional economies, and impacted societies in unpredictable ways. After pushing aside the Portuguese and the British, the Dutch not only rearranged trade and modified export-crop distributions but also deepened their penetration of Indonesian social life. Gradually the native rulers of Java and key polities in the outer islands were stripped of their independence and made beholden to the Company. They became in effect the administrative agents and commercial guarantors of the VOC, losing ancient rights of governance within their own territories and legitimacy on the international stage (see Map 3.10).

But even while their political power eroded, the traditional leadership of Indonesia along with countless ordinary people found freedom in an area beyond easy Dutch influence: the practice of religion. The growth of VOC control was almost synonymous with a general diffusion of the Muslim belief system outside of already Islamicized port cities and into rural lowlands. Indeed it may have been partly fueled by the Dutch because their expanding commerce provided a vehicle for itinerant Arab (and some Indian) merchants who proclaimed their faith everywhere they went. In corollary fashion a small number of wealthy and privileged Muslim converts undertook the pilgrimage to Mecca. The return of these respected hajjis often triggered religious awakenings and fostered a shared sense of international fellowship with people in Africa, Asia, and Europe.

A final and profound social result of VOC control was the rise of a powerful Chinese community, especially in cities and towns. Most came to Indonesia as itinerant traders, and many stayed. As in the Philippines, these hard-working sojourners dwelt in designated urban quarters, faced racism, and in 1740 endured a frightful massacre in and around Batavia. Still the Chinese prospered under the Dutch, filling the entire spectrum of crafts, trades, and professions. Some immigrants who began as ordinary laborers later became entrepreneurs, and others grew powerful through the lending of money to Indonesians and Westerners alike. Predictably their commercial influence increased steadily throughout the colonial era.

As a result of changing conditions in Asia and Europe, the once supreme Dutch VOC slowly lost its economic strength and institutional integrity late in the eighteenth century. Though still significant, profits from spices decreased steadily as their production increased elsewhere in the world. Yet the VOC failed to promote alternative commercial crops other than coffee. Even as profits declined, its operating costs escalated because of increased territorial responsibilities, a like expansion of bureaucracy, and widespread corruption. At the same time the financial and human resources of the Netherlands were sorely depleted by all-encompassing European wars in the several decades before and after 1800, so the metropolitan economy had little elasticity. On December 31, 1799, the bankrupt VOC ceased to exist as its charter expired and the government of the Netherlands assumed its tremendous burden of debt and territorial possessions (Boxer, 1965; Meilink-Roelofsz, 1962).

Early in the nineteenth century, while the Napoleonic wars raged in Europe, Indonesia mirrored power relations in Europe. Under Herman W. Daendels (1808–1811), a supporter and appointee of Napoleon Bonaparte, the VOC's administration was reformed and the military infrastructure improved in anticipation of confrontation with Great Britain. Later, during a five-year occupation of Java by the British (1811–1816), Sir Thomas Stamford Raffles overhauled the colonial bureaucracy, promoted infrastructural development, and advocated economic and social reform. When the Dutch returned, they spent the next 15 years reimposing their suzerainty over increasingly rebellious Indonesians. This difficult process depleted the colonial treasury and sapped the energy of a disheartened officialdom whose prestige had been compromised during the European wars and the interlude of British rule. Their task was further complicated by the bloody Java War, which was one of the longest (1825–1830) and most destructive of Southeast Asia's colonial conflicts. It pitted a mobile imperial army equipped with modern arms against a guerrilla force of Javanese peasants led by Prince Dipo Negoro of the Jogjakarta sultanate. Determined to expel the Europeans from Indonesia, he led a prolonged uprising that devastated Central Java and resulted in more than 200,000 deaths.

After reclaiming their authority and exiling Dipo Negoro, the Dutch in 1830 introduced a new scheme for the governance and fiscal salvation of the renamed Netherlands East Indies. The *cultuurstelsel,* or "Culture System," was designed to regularize the production and delivery of agricultural exports. Building on a program of coffee quotas used earlier by the VOC in the Preanger region, this system required that the peasantry set aside 20 percent of village lands for growing specified commercial crops or that all male farmers spend 66 days annually working on government lands. Marketed exclusively by a national trading institution, the harvests in only three years began to produce enormous profits. But these funds were never used to aid and uplift the Indonesians; instead they were redirected to the metropole to pay off the Dutch national debt, reduce annual tax burdens on citizens of the Netherlands, and fund massive infrastructural and military defense projects. Sadly the Culture System converted the Javanese into part-time serfs and even fostered famines. After 40 years of oppressive operation—vividly portrayed in the stirring novel *Max Havelaar* on the plight of Indonesian farmers—this crushing agricultural program was abolished.

With implementation of the Agrarian Law of 1870, which offered some protection to small farmers, the rural realm of Indonesia was finally opened to private and foreign investment (agricultural and industrial). But even while relinquishing their monopoly control, the Dutch imposed heavy taxes, utilized forced labor for local projects, and failed to provide credit for ordinary farmers. The Indonesians thus remained vulnerable to exploitation by economically skilled Chinese and European entrepreneurs. Even though the government introduced educational, economic, and welfare reforms at the turn of the twentieth century, the several centuries of exploitation, social disruption, and political disorganization had already produced sufficient anti-Dutch sentiment to fuel the beginnings of a nationalist movement that could not be extinguished by even the most powerful of imperial armies (Ricklefs, 1981).

British Imperial Tenacity

The British Crown actually chartered the English East India Company for the development of trade in tropical Asia on December 31, 1600, or two years before the Dutch founded the VOC. But England did not become a major power in Southeast Asia until the nineteenth century. At first cooperating with the Dutch to challenge the Iberians, the British in 1623 abandoned all but a few factory stations in the region following the unexpected arrest, summary trial, and beheading of English traders by VOC officials in the Moluccas (the long-remembered "Massacre of Amboina"). The East India Company's regional headquarters in Bengkulu (Benkoelen) was not even founded until 1684. Situated in a peripheral location on the southwestern coast of Sumatra, it failed to prosper as a center of the pepper and spice trade. Recognizing the strong commitment and superior strength of the Dutch in Southeast Asia during the seventeenth and eighteenth centuries, the British simply chose to focus their resources in South Asia.

It was not until late in the eighteenth century that the British even constructed a sufficiently secure base in India to challenge Dutch supremacy in Southeast Asia. By this time the maritime dominance of Britain on the Indian Ocean seemed assured. Commerce with South Asia was substantial and growing, and the English East India Company was orchestrating a comparatively free system of trade that would ultimately embrace most of monsoon Asia. The island of Penang, Britain's first significant territorial claim in Southeast Asia (1786), functioned for a short time as a prominent port in this commercial network. But it was soon surpassed by Singapore. Founded in 1819 by Raffles (then the company's lieutenant-governor in Bengkulu), who clearly understood the strategic disadvantages of Penang and fully appreciated the geographic position of the small island at the tip the Malay Peninsula, Singapore immediately prospered and emerged as the regional nerve center of free trade. Along with Malacca, it was formally ceded by the Dutch in 1824 while the British simultaneously relinquished Bengkulu.

As decades wore on, Britain continued to expand its colonial holdings when circumstances warranted. First came control of the coastal regions of Burma. In the First Anglo-Burmese War (1824–1825), the Burmese badly underestimated the company's strength and were soundly defeated. However the company's campaign in Burma also represented a heavy drain on its financial resources and was plagued by deadly diseases. Yet it resulted in the acquisition of Arakan and Tenasserim. The Second Anglo-Burmese War (1852) was a very short affair, for the well-equipped British army quickly routed Burmese forces and took control of Lower Burma with its vast teak forests and strategic port at Rangoon. In a third confrontation (November 1885) with the still sovereign but much smaller kingdom of Burma, the British conquered its capital and sacred city of Mandalay in only 15 days with a loss of 10 soldiers. In the following year the remaining territories of Upper Burma were annexed by the company, thus creating a territorial buffer between India and a rapidly crystallizing French colonial realm in Indochina.

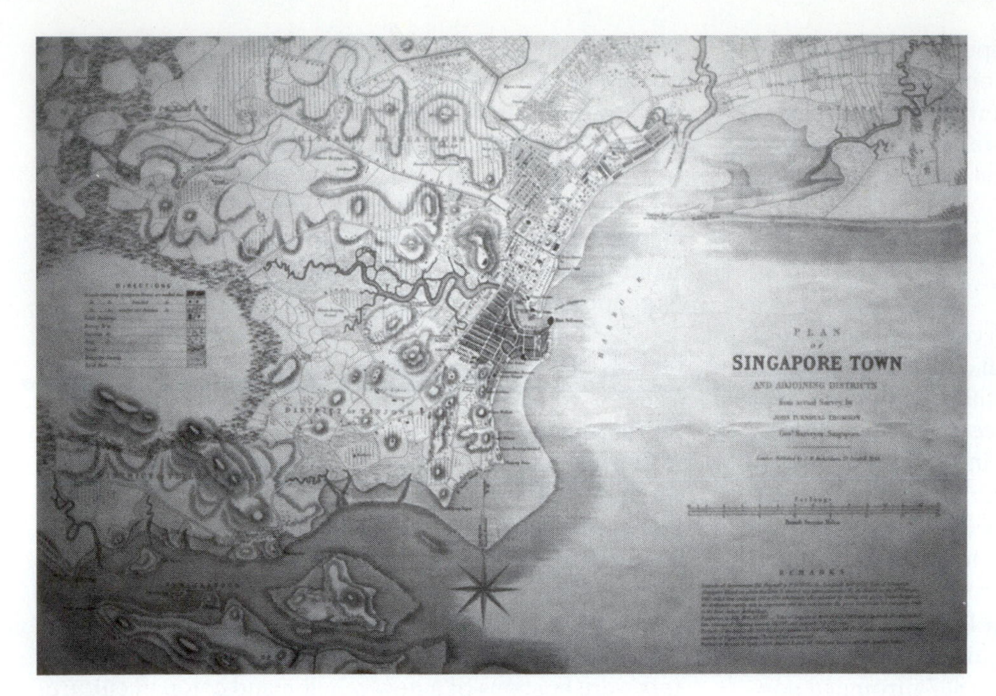

Photo 3.4 Early Map of Singapore. (Ulack)

As part of a grand design to outflank France while simultaneously securing the already colonized Straits Settlements (Singapore, Penang, and Malacca) and protecting newly discovered tin resources, the British late in the nineteenth century began to formalize their growing presence in the Malay Peninsula. They first established a protectorate system (1874) embracing the four critical tin-mining states—Larut, Selangor, Sungei Ujong, and Perak. Several years later like agreements were signed with Pahang (1888), Negri Sembilan (1888), and Johor (1895). Early in the twentieth century, Britain acquired four other states (Kelantan, Perlis, Terengganu, and Kedah) from Thailand through gentle political coercion confirmed by treaty arrangement (1909). Finally, the British sphere of influence was further extended through several curious private ventures in northwestern Borneo. There Britain intervened by means of a chartered company in North Borneo and a personal kingdom developed by the English adventurer James Brooke. After assisting the Sultan of Brunei during a time of rebellion, Brooke had been proclaimed raja of Sarawak (1846). His heirs continued to rule until 1946, when Great Britain assumed sovereignty (Andaya and Andaya, 1982; Chew and Lee, 1991; Cowan, 1961; Hall, 1950; Htin Aung, 1967; Runciman, 1960; Tregonning, 1964).

A French Mission

French colonial intrusion was rather drawn out before culminating in the acquisition of Vietnam, Cambodia, and Laos. Driven initially by religious zeal, a few Catholic missionaries late in the 1600s labored to secure a beachhead in Ayutthaya by gaining the confidence of King Narai and then assumed this would lead to political linkages between Siam and France. But when Narai died (1688), so did the earliest French imperial plan. Missionary activities continued during the next century, but with less intensity and mainly in Vietnam. There a small corps of priests made some conversions. Additionally they became involved in Vietnamese politics following the stormy Tayson "Rebellion" against corrupt Trinh rule in Tonkin. Bishop Pigneau de Béhaine gave support and counsel to a young prince of the Nguyen family later known as Gia Long—the emperor of Annam. Thus the French were well situated when Gia Long captured in succession Saigon (1788), Hue (1801), and Hanoi (1802). While establishing a new dynasty that unified Cochin-China (Mekong delta), Annam (Central Coast), and Tonkin (Red River delta and nearby lowlands), the emperor with the assistance of advisors from France also modernized his military, created a navy, and fortified many urban centers. Still the Vietnamese leadership did not trust the French, and their apprehension ultimately proved justified.

Under the pretext of religious persecution against Christians and harassment of Catholic missionaries, the French during the 1840s began their imperial intervention in Vietnam with occasional displays of naval force. Decisive action followed in 1858 with the capture of Danang. Saigon fell in the following year, triggering serious Vietnamese resistance that was countered through the systematic taking of all Cochin-China by 1866. Then France formally claimed Cambodia from Thailand (1867), converted Annam and Tonkin (1885) into helpless protectorates, and finally secured Laos (1893) to complete the so-called Indo-Chinese Union. Despite the apparent regularity of French imperial expansion in Southeast Asia, it was in fact punctuated by frontal warfare and siege, widespread guerrilla actions, unpredictable brigandry, and border clashes, as well as naval actions with China (Buttinger, 1968; Osborne, 1969).

American Latecomers

Long before the United States even contemplated a colonial adventure in the Philippines, an elite class of educated and often wealthy Filipinos (the *illustrados*) were beginning to dream of independence. Many had imbibed of liberal thought while studying in the universities of Spain and elsewhere in Europe. Their general restiveness, which was fueled by the haughtiness of Spaniards, perceived discrimination against native priests in the Catholic hierarchy, and severe Spanish repression following the Cavite Mutiny of Filipino troops in 1872, intensified late in the nineteenth century. The Filipino hope for freedom finally found popular expression in 1887 through a powerful novel that both echoed and fashioned history: *Noli Me Tangere.* In this commentary on the Hispanic-Filipino experience, José Rizal exposed the growing corruption of colonial life in the Philippines and called for reform. With his second reformist novel, *El Filibusterismo,* Rizal incurred the wrath of the Spanish officialdom, was banished for a time to Mindanao, and in 1896 was executed. Meanwhile, the spirit of protest had united the *illustrados* with ordinary Filipinos, ignited insurrection, and produced a short-lived Philippine Republic.

From the U.S. perspective, conquest of the Philippines was simply part of a broad assault on the decrepit Spanish empire that yielded a number of strategic possessions: Puerto Rico, the Philippines, Guam, and other Pacific islands. But the cost proved considerable. After an easy naval victory over the Spaniards in May of 1898, the U.S. hesitated until August before assaulting Manila. By that time Filipino revolutionaries, who at first viewed Admiral George A. Dewey and his forces as a liberators, controlled many Philippine provinces and encircled Manila. Following their landing, the U.S. forces seemed less like allies and more like conquerors. Consequently the Filipinos had little choice but to challenge the new invaders. The resulting war lasted for three bloody years in Luzon and the Visayas, and a simmering guerrilla conflict continued in some Muslim areas of Mindanao until 1915.

In the annals of Southeast Asian history, the U.S. episode in the Philippines was rather unique. It lasted only 48 years; was marked by early implementation of public education for the masses and widespread use of English; featured a rapid Filipinization of the bureaucracy and military; witnessed the first elective legislature (the Philippine Assembly) in the entire region (1907); ushered in a long-term and two-way cultural exchange; and seemed to be sealed by an uncommon degree of cooperation and friendship between rulers and ruled. Finally, the United Sates departed as promised under peaceful conditions in 1946. While thus championing nation-building through an attenuated form of imperialism, the United States continued to pursue policies that facilitated its entrepreneurial activities and guaranteed a military presence in the Philippines for decades after independence (Stanley, 1974).

LEGACY OF THE WEST

By the nineteenth century, the merchant fleets and navies of Europe had been sailing the waters of Southeast Asia for almost 300 years. After discovering a flourishing regional commerce, they founded several heavily fortified colonial capitals and a number of garrisoned cities. Each anchored a thin network of factory settlements from which Westerners conducted localized trade. In the always exceptional Philippines, the Hispanicized urban centers were beacons of a new religion and colonial culture. To the indigenous peoples, the Portuguese, Dutch, and English at first appeared little different from Chinese, Indian, and Arab merchants who preceded them. They were few in number and seemed obsessed with mercantile affairs. Their technical knowledge of ship design, navigation, and armaments provided significant leverage in negotiating initial trading agreements, but the Europeans were unable to convert this advantage into airtight commercial monopolies. It is true that Westerners ultimately dominated the international maritime exchange, but many smaller regional systems of "country trade" continued to flourish on seas and in riverine realms outside their easy control.

Nor were Westerners in command of much Southeast Asian territory until the nineteenth century. Only in the Philippines did the *conquista* imply a systematic taking of land and direct governance of both urban and rural peoples. Admittedly the Dutch imposed territorial rule on various islands in the Moluccas to monitor spice production and on Java to construct a secure hinterland for Batavia. But most native rulers continued to enjoy their ancient right-to-rule, while occasionally cooperating with ephemeral European allies or employing individual Westerners as mercenaries, weapons makers, and advisors. Accordingly a majority of indigenous people in the lowlands and all mountain dwellers lived beyond the political reach of the colonizers and their traditional ways of life continued intact.

The European impact on the ordinary people of mainland Southeast Asia during the early centuries of imperial intervention remained quite superficial. This partly reflected the low volume of Western trade with the great agrarian kingdoms, because the key commodities (spices and pepper) then exported from Southeast Asia came mainly from the archipelago. It also testified to the relative strength of the inland states, where power was measured by rice surpluses and labor equivalents derived mainly from large agricultural populations. Although Western weapons experts and adventurers occasionally provided advice to the rulerships of Burma, Arakan, Siam, and other kingdoms that subsequently served to intensify regional wars, they seldom challenged the prevailing political order. Admittedly, the ancient enmities and complex rivalries of the mainland states clearly obstructed trade and limited the Western impact. But it would have been difficult for pioneering colonial merchants with their sparse human resources to overthrow the still-strong mainland polities. Though European traders, mercenaries, and ambassadors served as information brokers between the mainland agrarian kingdoms and the West, their presence offered little political benefit to any parties before the nineteenth century.

Shaping the Future during the Nineteenth Century

After three centuries of "pin-prick imperialism," Southeast Asia in the nineteenth century became the regional stage of broad territorial acquisitions by the European powers and fundamental transformations of society, economy, and culture. This all-pervasive form of colonialism was a direct offspring of the Industrial Revolution in the West. As vanguards of modern science and pioneers in industrialization, the nations of Europe enjoyed a decisive advantage over their Asian competitors. Consequently they proved able to unilaterally chart the path of development and modernization in Southeast Asia.

Revolutionary changes in transportation and communications during and after the 1840s greatly facilitated the management of Europe's Asian colonies and permitted massive increases in the volume of East–West trade. For one thing, steamships effectively replaced clippers and other long-haul vessels on the world's oceans. They thus rendered obsolete both general haulage sailing ships and Chinese junks in the multidirectional trade on Southeast Asia's regional seas. Concurrently the opening of the Suez Canal in 1869 both shortened the distance and time of transit between Asian and European destinations. This increased the profitability of trade while also reducing the risks and human discomfort associated with long voyages. By 1865 Europe was linked to India by telegraph, and in succeeding years the lines reached eastward to the nations of Southeast Asia. It is difficult indeed to overstate the importance of this development, for metropolitan governments and Western companies could now respond immediately to crises as well as supervise all daily public and private operations in the distant colonies.

While Europeans tightened their grip in colonial Southeast Asia, a slowly modernizing Thailand retained its freedom as a self-governing polity. Though situated between the domains of British Burma and French Indochina, the Thai state survived by ceding territory as necessary on each flank. Indeed Thailand's kings and their counselors proved astute in appraising the economic and military power of various Western nations and avoided confrontations that might have provoked intervention. Likewise Britain and France understood the utility of a buffer zone between their spheres of influence and so respected the integrity of Thailand. To keep the Europeans at bay, the Thai leadership modernized the nation's bureaucracy, military, educational system, and economy at a pace equaling or exceeding that of nearby colonies. To facilitate this process, the government employed many Western advisors while also dispatching hundreds of Thai students abroad for education and technical training. Trade flourished in Bangkok after 1800, but the Thai prevented any nation from gaining commercial supremacy by welcoming merchants from all countries. By matching or exceeding the development and modernization of its colonized neighbors, Thailand alone remained free (Map 3.11).

As Westerners expanded and consolidated their power in Southeast Asia, they introduced revolutionary ideas and inaugurated processes that transformed economies, societies, and states. They also displayed an uncommon concern about the frontiers and boundaries of their colonies. Whereas the urban officialdoms of Southeast Asia's indigenous "galactic polities" exercised absolute authority in the environs of sacred cities, considerable power over settled agrarian peoples, and diminishing control over outlying populations of swidden farmers and foragers (Tambiah, 1977), the European rulers insisted on a uniformity of political discipline to the very limits of their territorial dominions. Accordingly great national resources, both financial and human, were expended in the second half of the nineteenth century and the early decades of the twentieth as colonial armies warred against indigenous peoples to secure frontier zones and political boundaries. The Dutch frontal and guerrilla campaigns in Aceh over several generations are legendary for their length, bloodshed, and wanton destruction on both sides. Almost the same can be said for Spanish and U.S. struggles against the Muslims of Mindanao, British forays against the Burmese, and French battles with the Vietnamese. It is also noteworthy that the Western militaries and police faced hundreds of smaller and shorter insurrections, peasant protests, and localized demonstrations during the heyday of imperialism. When negotiations failed, these confrontations were usually resolved to European advantage through armed force. Accordingly the Westerners capitalized on their military and technological superiority to fix boundaries which generally remain intact today (Steinberg, 1971; Tarling, 1994a).

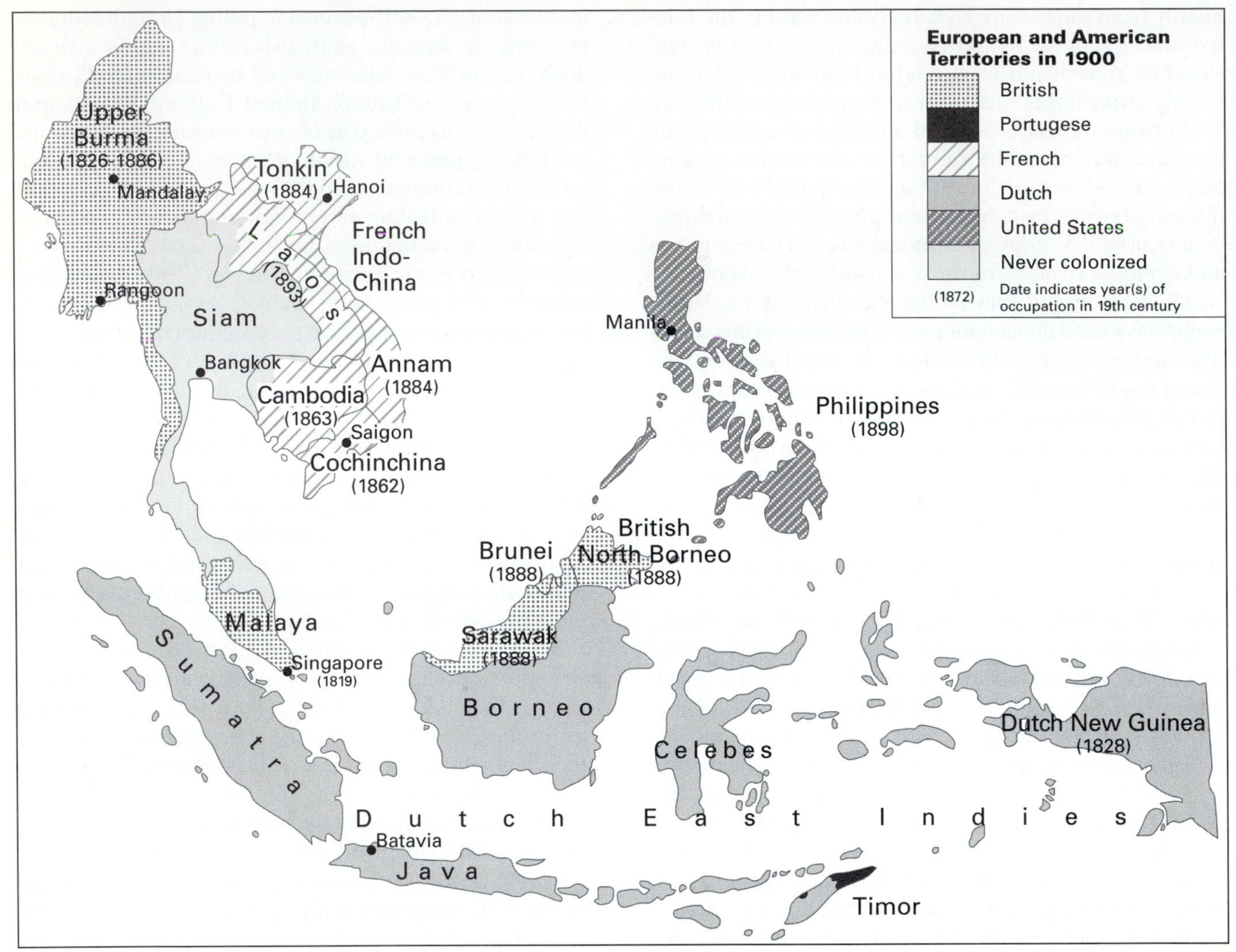

Map 3.11 Western rule in Southeast Asia in 1900. *Source: Hall, 1964.*

Enduring Impacts of Colonialism

Predictably the Western umbrella of imposed peace and order fostered an expanding commerce geared to the pressing need of the industrial metropoles for raw materials and markets. Unlike the earlier East–West trade in highly profitable luxury goods, the newly evolving exchange involved bulk cargoes of eastward bound manufactured products from Europe and westward-bound rice, lumber, minerals, and plantation produce from Southeast Asia. A sufficient output of such commodities was guaranteed in each colony through an amalgam of European capital, organizational experience, and technological expertise; Chinese, Indian, and native Southeast Asian labor and entrepreneurial activities; and the enterprise of indigenous people working as individuals, families, or communities bound by common residential, ethnic, and religious interests. In the resultant process of development, Southeast Asian economies and landscapes were fundamentally and forever changed.

From about 1850 to 1940, *sawah* cultivation expanded dramatically in the deltaic lowlands of mainland Southeast Asia, on Java, and in Central Luzon with an increasing shift from subsistence farming to commercial production. This was accompanied by massive migrations of farming peoples and increasing densities of population as homesteaders cleared lowland forests, opened mangroves, dug drainage canals, and built irrigation systems. The mounting food surpluses of Burma, Thailand, and Indochina flowed to the West and to growing populations on plantations, in mines, and in urban centers of Southeast Asia. Rice was also exported to the cities of South Asia, industrializing Japan, and treaty ports of China.

It may seem a contradiction that the dramatic expansion of *sawah* in Southeast Asia was paralleled by a decreasing well-being of farming families and a diminishing security of rural populations. The explanation is simple and straightforward. Because the new colonial agriculture that produced massive rice surpluses was defined by an external market and was not regulated by subsistence requirements, cultivators sold most of their produce and were so enveloped by a monetary economy with its burden of homesteading expenses, labor costs, taxes, and unpredictable fluctuations in the value of harvests. Because few farmers fully understood such monetized agriculture or had capital for start-up and maintenance expenditures, they often depended upon alien moneylenders (Indian Chettyars in Burma, Chinese mestizos in the Philippines, and the Chinese in Indonesia, Malaya, and Indochina) or indigenous but absentee landlords. In retrospect the results appear predictable. Rates of tenancy increased in realms of commercial rice cultivation; traditional forms of mutual assistance were replaced by low-paid occasional labor; poverty became commonplace; and good farming practices were sacrificed for profits.

Even as *sawah* clothed the landscapes of deltas, coastal zones, and riverine lowlands, the plantation appeared in certain parts of Southeast Asia during the nineteenth century. Combining farm work and industrial processing activities, and born of a wedlock between Western finance and Asian labor, plantation agriculture required substantial capital for forest clearage, planting, infrastructural investment (factory and warehousing facilities, irrigation, roads and narrow-gauge railways, and workers' quarters), and ongoing operations (cultivation, harvesting, processing, and shipping). Moreover, most plantations did not yield any produce until the third to the seventh year, depending on the type of crop. Still the promise of large profits in growing international markets guaranteed a steady flow of capital and labor. Plantations proliferated in the Philippines (sugar, abaca, pineapples, copra); Malaya, Sarawak, and North Borneo (rubber, oil palm, tea, copra); Indonesia (sugar, oil palm, tea, coffee, copra, spices, cinchona, cinnamon); Thailand (rubber, oil palm); Burma (teak); Cambodia (rubber); and Vietnam (rubber, sericulture, tea). During the late colonial era, the common handmaidens of such plantation development were the serious exploitation of indigenous and alien labor (Chinese and Indian workers in Malaya), episodic protests and rebellions, and periods of prosperity followed by depression.

While orchestrating agricultural transformation, the Europeans also promoted mineral extraction and forest industries. Though mining and lumbering remained secondary to commercial agriculture, such development was extremely important in local niches throughout the region, attracting considerable Western capital and some immigrant labor. British Malaya, for example, owed its relative affluence in part to placer tin-mining operations established by the Chinese late in the nineteenth century that set the standard for late-coming European companies. Fueled by U.S. capital, the Philippines had major gold and copper mining operations early in the twentieth century. Under British supervision, the mines of Burma were significant sources of lead, silver, tungsten, petroleum, and precious stones (especially rubies and jade). By the same token Indonesia had risen to importance as an oil producer before World War II. Coal deposits are scattered throughout Southeast Asia and locally important, fostering a significant extraction industry in Vietnam. Elsewhere small-to-medium size deposits of the foregoing minerals and others (iron ore, nickel, chromium, and bauxite) had already been identified or were being measured for development by colonial entrepreneurs. Lumbering operations, many funded by small Chinese capitalists, were ubiquitous in the region and provided building materials for local construction and some export. Burma, Thailand, and Indonesia enjoyed a significant international trade in forest products such as teak, other hardwoods, and rattan.

To integrate the evolving agricultural and extractive economies, all European powers built modern infrastructures and elaborated the existing urban systems. Before 1800 most trade in the coastal and riverine sectors of Southeast Asia had been waterborne. But the development of far-flung plantations, mines, and lumbering operations soon necessitated the construction of modern railway systems, all-weather roads, strategically situated ports, irrigation projects, and communications networks. At the same time the Europeans secured their colonial acquisitions through comparatively dense networks of towns and cities, which at once facilitated the process of economic change and assured the effective supervision of indigenous peoples. With citizenries of 10,000 to 100,000 people by the 1930s, most district or provincial capitals emerged as functionally diverse places that combined commercial, administrative, military, communications, educational, and light manufacturing activities. In marked contrast to these multifunctional urban centers, the Westerners also founded numerous highly specialized settlements: mining camps, army cantonments and naval bases, plantation towns, railway communities, and hill stations (costly mountain resorts for Europeans and Westernized elites). Colonial capitals occupied the apex of each urban system.

Space precludes comparative discussion of either the institutional or the morphological forms of Southeast Asia's major colonial cities, for each was deeply imprinted by distinctive imperial policies emanating from the metropole and the prevailing design theory at various stages of urban development. Nevertheless, several common characteristics warrant mention. First, all of the capitals and some of the major secondary cities of colonial Southeast Asia (Rangoon, Singapore, Batavia, Saigon, Hanoi, Manila, Bangkok, Surabaya, Cebu, Makassar (now Ujung Pandang), Medan, Penang, and others) shared tidewater or deltaic sites where transportation networks that served agricultural hinterlands were easily linked to international systems of exchange. To utilize the inherent locational advantages of these ports effectively, the Europeans in the last decades of the nineteenth century invested heavily in a modern infrastructure (harbor improvements, maintenance shops, fuel storage facilities, warehousing complexes, and so on) to serve both exporting and importing activities. Second, the colonial capitals were preeminently administrative nerve centers from which Westerners ruled over their respective dependencies. Almost without exception, the ranking civil and military officials with their large supportive bureaucracies dwelt in the major cities. Third, and undoubtedly the most critical functional characteristic of the larger cities, was their pronounced diversity of economic activities. All larger cities contained an integrated complex of banks, insurance companies, shipping firms, and other institutions through which Europeans manipulated colonial economies. Additionally the main urban centers became arenas of industrialization and processing platforms for goods leaving and entering Southeast Asia. Fourth, the colonial capitals together with most secondary cities displayed conspicuous ethnic diversity and a cosmopolitan atmosphere. Not only did the coastal cities accommodate sometimes complex indigenous citizenries and small groups of Europeans, but they often sheltered large Chinese communities as well as other Asians (especially Indians in Burma, Malaya, and Singapore; Arabs in Indonesia). Although some alien Asians amassed immense wealth through real estate holdings, commercial dealings, and investment in industrialization, most remained part of a vast force of skilled and common tertiary laborers who filled all employment niches. Finally, the colonial capitals served as the foremost beacons of Western education and culture. By 1900 the large cities boasted of public and sectarian institutions providing clerical, technical, and higher educational training for aspiring students of various ethnicities. Corollary to such progress in education was the florescence of nationalism in this century among the indigenous urban elites of Southeast Asia.

As the colonial capitals of Southeast Asia elaborated their functions, they also expanded dramatically in population. Like the older garrisoned cities of Manila and Batavia, most administrative centers (Rangoon, Singapore, Hanoi, Saigon, and Bangkok) of later foundation embraced at least 200,000 people by the dawn of the twentieth century. During the five ensuing decades of imperialism, the pace of urbanization quickened as major cities doubled or even tripled their citizenries. This growth in the capitals was paralleled by their significant relative gain over cities and towns of lower rank in the urban hierarchy. By 1900 the foremost urban centers in the various colonies of Southeast Asia proved to be at least two times the size of the second-ranking cities. In the decades that followed, the great coastal emporia continued to exceed the rate of population growth achieved in smaller urban centers. By all measures, these were classic "primate cities" (Jefferson, 1939). In terms of population, as well as diversity of functions, the colonial capitals of Southeast Asia effectively dominated their respective national hinterlands long before the advent of independence following World War II.

Japanese Heralds of Change

During the heyday of Western imperialism, the ancient rivalries that once divided the polities, ethnic groups, and religious communities of Southeast Asia seemed of secondary importance as people attempted to improve their individual fortunes and increasingly identified with colonies soon destined for nationhood. In retrospect, this was a time of profound change when the regional economy became firmly integrated into the emergent "world system;" when new technologies and occupations completely transformed the employment mix for ordinary urban and rural populations; when even the poorest households began to utilize ordinary products of modern industrialization (milled cloth, needles and safety pins, electric items, packaged foods and cleaners, bicycles, buses and trains, medicines, and scores of other goods); when the masses gained access to modern education; and when increasingly sophisticated people of the region began to utilize Western ideas and doctrines in challenging imperialism. In a fitting paradox, the gospel of nationalism—which had earlier given moral sanction to European expansionists—inspired successive generations of Southeast Asian patriots to protest, resist, and revolt against their Western oppressors. After Japan's dramatic and unexpected victories at the onset of World War II proved that the technologies and techniques of modern warfare could be turned against their European inventors, the future of colonialism in tropical Asia became a dead issue.

Early in 1941 few people in Southeast Asia imagined that the seemingly indestructible power structures of European imperialism would be swept away in less than two decades. But the Japanese advance in December gave new meaning to already seething anti-Western sentiments. The clarion call from Tokyo—"Asia for the Asiatics"—soon stimulated nationalist feelings everywhere in Southeast Asia that translated into the development of revolutionary political parties and revolts against the external oppressors. Initially welcomed as liberators, the Japanese in time revealed their own imperial goals. As World War II wore on, the hardships of war and occupation gradually cooled the ardor of Southeast Asians for Japan's leadership against the West. Indeed the Japanese seemed as harsh as their predecessors.

Predictably as the military strength of the allies increased, the Japanese intensified their already brutal policies of forced labor, resource exploitation, persecution of local Chinese (especially in Malaya, Singapore, and

Indonesia), and oppression of Western sympathizers. Still these hard realities did not quench the fires of nationalism. With the defeat of Japan, the overwhelming majority of Southeast Asians took strength, accepted guidance, and sought protection not from erstwhile colonizers, but from their own political leaders and armies of national liberation. Despite its brevity, the Japanese interlude was critical because it proved that the West had lost its aura of invincibility and that political legitimacy is ultimately the child of indigenous sentiment. The peoples of Southeast Asia were thus posed at the end of World War II to pursue the difficult process of modern nation-building without the overlordship of Europeans, Americans, or Japanese.

4 Demographic and Social Patterns

GRAEME HUGO

The last two decades have witnessed unprecedented economic transformation in Southeast Asia and, as both a cause and consequence, demographic and social change has been equally profound. The population of the Southeast Asian region as defined in this book in 1997 numbered 501 million or 8.6 percent of the estimated world total (Population Reference Bureau, 1997). Hence in any consideration of the world, this region must loom large. Although Southeast Asia contains some of the world's fastest growing economies and most rapidly changing societies, it also includes some of its poorest nations. Indeed, Southeast Asia is nothing if not a very diverse region and this is reflected in its demographic and social patterns. Yet the region does have distinctiveness, which Charles Fisher (1964, p. 9) perceptively referred to as residing in its common diversity rather than in any close knit unity: tropical and maritime, focal but fragmented, ethnically and culturally diverse, plural alike in economy and society. Southeast Asia clearly possesses a distinctive personality of its own and is more than an indeterminate borderland between India and China. The regions geographical and ethnocultural diversity is matched with a wide range of national politico-economic systems, demographic patterns, and social systems. Both the demography and social systems are undergoing rapid change and the present chapter seeks to show these main patterns of change.

The chapter begins with an examination of some aspects of social change in Southeast Asia. It is argued that this change has not attracted as much attention as economic development but has been no less important. Most important have been changes in levels of education. There are however, inequalities in the availability of education both within and between countries in the region. Another important social change relates to the shifts occurring in the structure and functioning of the family—the basic social unit in the region. Other changes in labor force, the role and status of women, and service provision are also discussed. However, there are vast differences between the countries in the region in relation to social development and these relate closely to economic differences.

The second part of the chapter addresses the change in the population of the region. This has been no less dramatic and as important as the economic change taking place. Again, great diversity and rapidity of change is in evidence. First of all some of the differences in the region with relation to population growth, distribution, and density are discussed. Then each of the major demographic processes shaping that growth are taken in turn—mortality, fertility, and migration. Each of these processes has changed profoundly in recent years with major implications for the countries involved. Some aspects of changing age distribution of the population in the region are discussed and finally some aspects of population policies and programs in the region are discussed. No other region in the world has seen so many ambitious attempts by governments to intervene to change the patterns of population growth and distribution.

SOCIAL PATTERNS

Regional Inequalities

One of the most salient features impinging upon social and demographic change in the last two decades in Southeast Asia has been the fast growth of the economies of most countries in the region up until the economic crisis of 1997–1998. This is evident in Table 4.1, which shows that in the countries for which data are available, rates of economic growth have generally been more than double the global average during the period since 1980. While there are no equivalent data for Vietnam, Cambodia, Laos, and Burma, there are indications of the beginnings of significant economic growth in those nations as well. This is especially true of Vietnam, whose economy is estimated to be growing at about 8 percent per annum in the mid-1990s. Southeast Asia contains

Indicator	Country									All Low Income Economies	All Middle Income Economies	All High Income Economies
	Indonesia	Philippines	Thailand	Malaysia	Singapore	Vietnam	Cambodia	Laos	Burma			
GNP Per Capita												
•Dollars 1995	980	1050	2,740	3,890	26,730	240	270	350	*	430	2,390	24,930
•Average Annual Growth (%), 1985--95	6.0	1.5	8.4	5.7	6.2	--	--	2.7	--	3.8	-0.7	1.9
Average Annual Growth (%) in GDP												
•1970--80	7.2	6.0	7.1	7.9	8.3	--	--	--	--	--	--	3.2
•1980--90	6.1	1.2	7.6	5.2	6.7	--	--	--	--	6.0	1.9	3.2
•1990--95	7.6	2.3	8.4	8.7	10.5	8.3	6.4	6.5	--	6.8	0.1	2.0
Percent of Labor Force in Agriculture												
•1960	75	61	84	63	8	--	82	83	--	--	--	--
•1980	59	52	71	41	2	73	76	80	67	73	38	9
•1990	57	45	64	27	0	72	74	78	--	69	32	5
Average Annual Growth in Labor Force (%)												
•1980--1990	2.9	2.7	2.6	2.8	2.3	2.7	2.8	2.3	1.9	2.2	2.1	1.2
•1990--1995	2.5	2.7	1.3	2.7	1.7	1.9	2.5	2.7	1.7	1.7	1.8	0.9
Population per Physician, 1990	7,030	8,120	4,360	2,590	820	--	--	4,380	12,900	11,190**	2,020	420
Number Enrolled in Primary Education as % of Age Group, 1991	116	110	113	93	108	--	--	98	102	101	104	104
Number Enrolled in Secondary Education as % of Age Group, 1991	45	74	33	58	70	--	--	22	20	41	55	93
Number Enrolled in Tertiary Education as % of Population Aged 20--24, 1991	10	28	16	7	--	--	--	1	--	3	18	50
Adult Illiteracy Rate (%), 1995	16	5	6	17	9	6	35	43	19	34	18	--
Central Government Expenditure, 1995 (% of total expenditure)												
•Defense	6.2	10.6	--	12.7	37.4	--	--	--	--	--	--	--
•Social Services	70.4	26.3	57.7	48.0	48.5	--	--	--	--	--	--	--

Note: In some cases indicator #6 exceeds 100 percent because enrollment ratios are calculated as the numbers enrolled in primary schools divided by those aged 6-12 whereas many children aged over12 are enrolled in primary school.

** Estimated as low income ($765 or less).*
*** Excludes China and India.*

Table 4.1 Some basic social and economic indicators in the mid-1990s. *Source: World Bank, World Development Reports 1994, 1997.*

some of the world's most dynamic economies in Singapore, Malaysia, and Thailand but also some countries that have generally experienced slow growth (e.g., the Philippines) and others whose lack of development is reflected in their dearth of basic economic data. However, although economic growth has been substantial in most countries, there is greater variation between countries in the base on which this growth is occurring and the extent to which the benefits of growth are shared by groups and regions within those countries.

There is enormous variation between and within countries in the region with respect to economic and social development. A detailed examination of this complex topic is not possible here but a few points need to be made. The World Bank (1997, p. 207) divides developing countries with more than one million inhabitants into low- and middle-income economies (with 1995 GNP per person of US$765 being the threshold). By their classification, four Southeast Asian countries (Cambodia, Vietnam, Laos, and Burma) can be classified as low-income countries. The Philippines, Indonesia, and Thailand are rated as lower–middle-income countries and Malaysia is considered an upper–middle-income country. Countries with per capita incomes of US$9386 or more were considered high income and Singapore and Brunei were in this category in 1995. Hence there is considerable intraregional variation in average levels of income. Moreover, within countries there are considerable inequalities between groups and regions with respect to income. Measurement of income is extremely difficult and unreliable

Country	Year	Percentage of Share of Income or Expenditure by Percentile Groups of Households						
		Lowest 20%	2nd Quintile	3rd Quintile	4th Quintile	Highest 20%	Highest 10%	Gini Index
Laos	1992	9.6	12.9	16.3	21.0	40.2	26.4	30.4
Indonesia	1993	8.7	12.3	16.3	22.1	40.7	25.8	31.7
Thailand	1992	5.6	8.7	13.0	20.0	52.7	37.1	46.2
Philippines	1988	6.5	10.1	14.4	21.2	47.8	32.1	40.7
Malaysia	1989	4.6	8.3	13.0	20.4	53.7	37.9	48.4
Singapore	1982	5.1	9.9	14.6	21.4	48.9	33.5	n.a.
Vietnam	1993	7.8	11.4	15.4	21.4	44.0	29.0	35.7

Table 4.2 Distribution of income or consumption in the late 1980s or early 1990s. *Source: World Bank, World Development Report 1997, p. 222-3.*

in some countries of the region because many workers are employed in semisubsistence and informal sector occupations. Nevertheless, Table 4.2 shows that in the few countries for which income distribution data are available, the distribution of income is quite unequal.

In most countries in the region, poverty is one of the most fundamental problems impinging significantly on social and demographic change. Again there are substantial measurement problems. There is considerable controversy about some national estimates of poverty, and poverty standards vary greatly between countries. Table 4.3 presents some official estimates for five large nations in the region of the population officially considered to be in poverty. Certainly Indonesia, Thailand, and Malaysia have experienced substantial declines in the proportion of their population living in poverty while more than one-half of the population in the former Indochinese countries, Burma, and the Philippines still live in poverty. Moreover, there are significant group, regional and rural–urban differences within nations in the incidence of poverty.

To illustrate such trends in poverty it is useful to examine a single country in the region. Indonesia, the largest nation in the region, despite recording two decades of rapid economic growth, is still squarely within the ranks of the world's LDCs (less-developed countries). In 1992 it was estimated that GNP per capita was still only US$670—the lowest of the ASEAN nations (World Bank, 1994, p. 162). The government's official estimates of Indonesians living below the poverty line indicate that the number fell from 54.2 million in 1976 to 27.2 million in 1990 (Indonesia, 1993, p. 1). There is considerable debate about the level of poverty in Indonesia, although not about the fact that its incidence has declined over the last three decades. Substantial variations exist between different parts of Indonesia in the incidence of poverty. Map 4.1 shows that in one set of estimates the proportion of people living in poverty in 1990 varied between 1.3 percent in the capital, Jakarta to 45.6 percent in East Nusa Tenggara (Bidani and Ravallion, 1993, p. 53). The highest incidence of poverty tends to be in Southern Sumatra, Central and East Java, Nusa Tenggara, West Kalimantan, Sulawesi, and Moluccas.

Although the last two decades have seen substantial economic and political change in Southeast Asia, most countries in the region remain squarely within the group of nations designated as developing. The social indicators listed in Table 4.1 reflect this and, although much data are missing for the poorest countries in the region and one must be cautious in interpreting the data that are available, they do establish the broad context of social and economic change in the region against which social and demographic shifts need to be considered. The

Country	Urban (%)		Rural (%)		Poverty Line 1990 US$/Year
	1970s	1980s	1970s	1980s	
Indonesia	39	20	40	16	124
Malaysia	19	8	54	23	314
Philippines	51	40	66	54	260
Thailand	13	7	36	29	n.a.
Burma	n.a.	n.a.	40	n.a.	n.a.

Table 4.3 Population considered to be living in poverty, 1970s and 1980s. *Source: ILO 1993, p. 85; UNESCAP, 1993c, p. 4.*

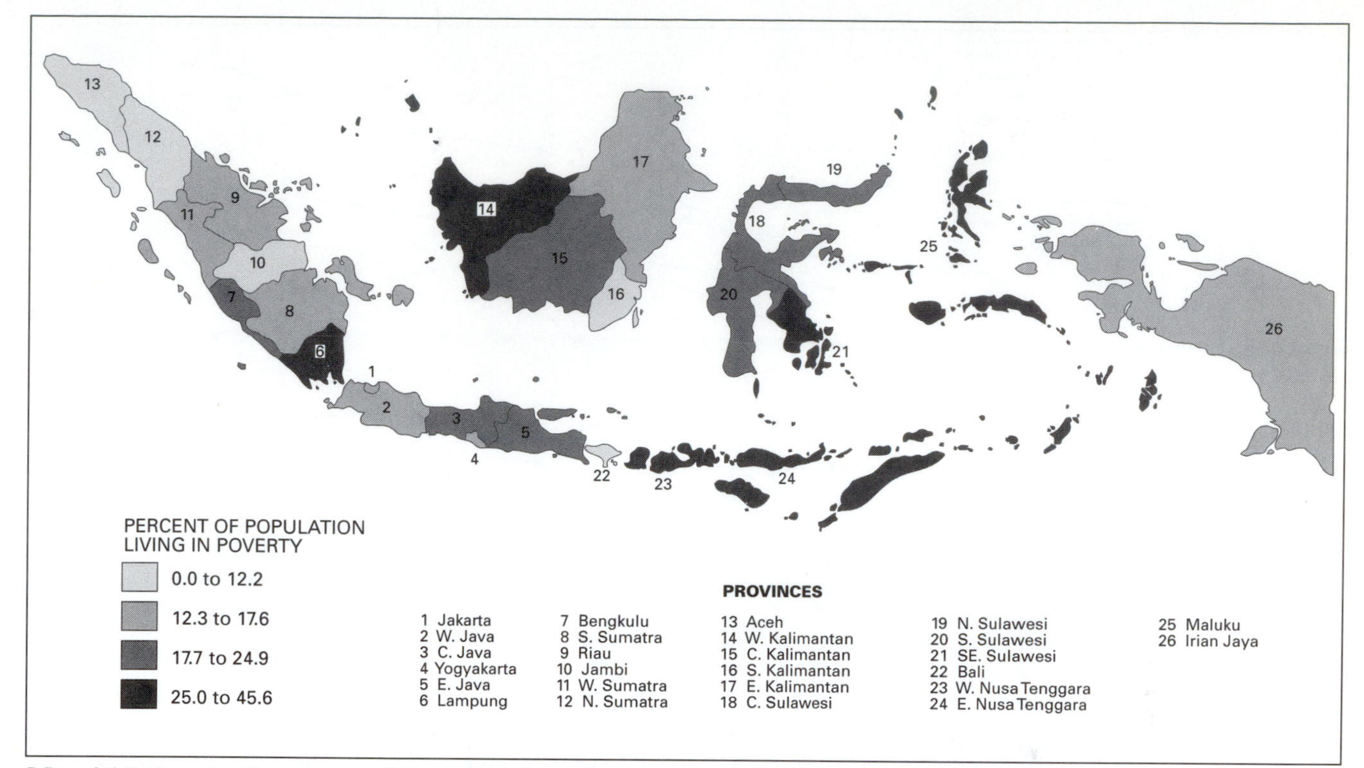

Map 4.1 Indonesia: Percentage of population living in poverty by province, 1990. *Source: Bidani and Ravallion, 1993, p. 53.*

table indicates the substantial structural changes occurring in national economies whereby, despite considerable labor force growth, there has been a significant decline in the proportions of workers being employed in agriculture. This has been due not only to employment creation in other sectors but also to the substantial labor displacement in agriculture despite expansion of agricultural production. The latter has been associated with increased commercialization and mechanization within the agricultural sector.

Education

Few changes have been of greater significance in Southeast Asia than those in education. Primary education has become compulsory in most countries of the region. However, there are large intercountry differences in the quality of the mass primary education provided, the extent to which primary children drop out, the extent to which students are able to proceed to secondary and tertiary education, and the degrees in equity in access to education. Nevertheless, one of the most massive changes in the region has been that, whereas at the close of the colonial period a small minority of the children in the region were receiving a modern education and the majority of the population were illiterate, the bulk of children in the region now receive at least some primary school (see Table 4.1). Modern education has had a profound effect economically and socially in Southeast Asia. Education opens up to children a wider horizon than the local village; it enables them to read and hence become more productive in the work they take up; it enhances the use of healthier lifestyles, and the acceptance of family planning, and the acceptance of new ideas; and it helps prepare children for nonagricultural occupations. Socially it leads to a challenging of tradition, erosion of traditional authority, and an influx of new ideas.

Although the majority of children in the region obtain some primary schooling, there is vast variation in and between countries in the quality and training of teachers, the schooling facilities and equipment, class size, and drop-out rates. Some of the differences between countries are reflected in literacy data (Table 4.1) and trends in secondary school attendance (Figure 4.1). Clearly education levels are highest in Singapore and the Philippines although Malaysia is catching up with the majority of children gaining a secondary education. In Thailand, Indonesia, and Malaysia about one-third of children reach secondary school although the government of Indonesia has set the objective of all children gaining three years of secondary education by early next century. Data on the other countries are not available but undoubtedly the penetration of secondary school is lower.

Several countries of the region have indicated that the development of their human resources is one of their most

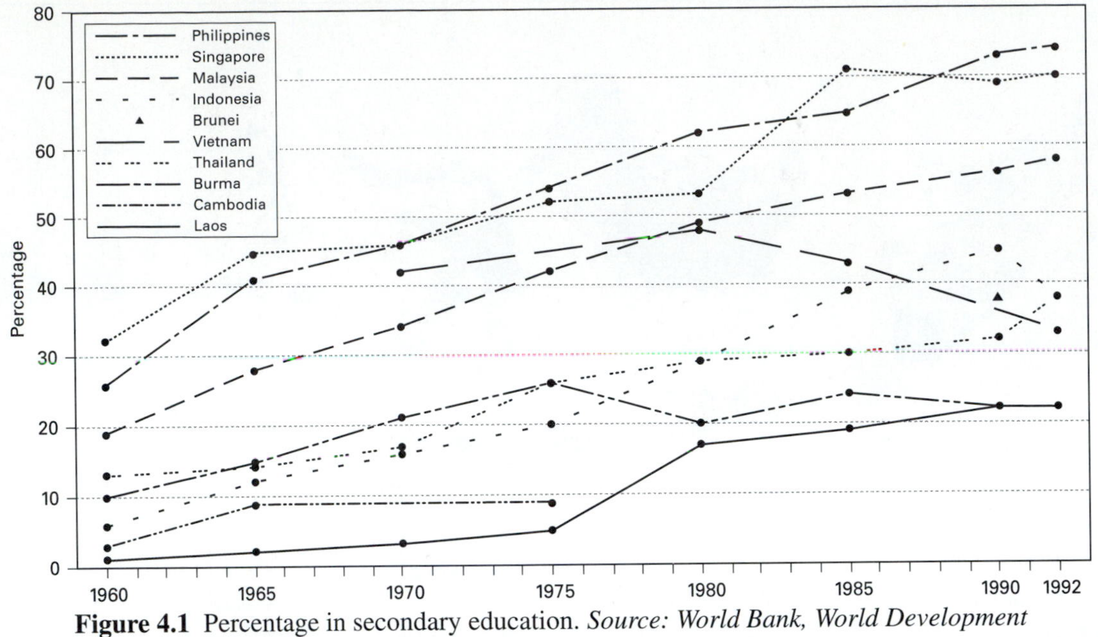

Figure 4.1 Percentage in secondary education. *Source: World Bank, World Development Reports, various years.*

urgent priorities and the proportions of national budgets spent on children are quite high, although often the amounts per 1,000 children are low. There is a great deal of debate as to whether limited funding should be put predominantly into providing all children with a high quality basic education or whether expensive tertiary education should be developed for a minority of the population. Nevertheless, vast strides have been made in education in the region. It is interesting for example that the Singapore mathematics curriculum has been adopted in California schools. There can be no doubt that despite problems in quality of education, mismatches between training outputs and labor market needs, and regional, gender, and class inequalities in access to education, the advent of mass education has been one of the greatest forces for social change in the region. Human resource development remains one of the major challenges facing governments in the region (Ogawa, Jones, and Williamson, 1993) and a significant priority in their budgets (Table 4.1).

Access to Social Services

The inequalities which exist in most national societies in the region are all too starkly evident in the data in Table 4.1 showing the generally limited access to, and availability of, basic social services. Hence, the population per physician in most of the countries is very high. In Singapore there are 820 persons per physician; in Burma it exceeds 10,000; it is also very high in Indonesia and the Philippines. Government expenditure on health is very low in absolute and relative terms. The minuscule expenditure on services for people can be illustrated by examining the amount spent per capita on health and education in some of the largest countries in the region. In Indonesia in 1993 this amounted to US$19, in the Philippines $36, in Thailand $103, and Malaysia $251 (*ASEAN*, 1997, p. 81). This compares to well over US$1,000 in developed nations. Only Brunei and Singapore have adequate levels of health and education provision; however, expenditures in Vietnam, Burma, Cambodia, and Laos are lower than those for the countries given above.

One of the major agents of social change in Southeast Asia in recent years has been the mass media, especially radio, television, and cinema. The latter two are dominated by western productions and have many direct and indirect influences on the population in Southeast Asia. Television especially has become all pervasive in Southeast Asia. While data are not available for all countries, in general there has been massive growth in the number of television sets in the region (Figure 4.2). Moreover, owners of television sets in the region tend to share them readily with less-fortunate neighbors.

Radio, television, cinemas, books, magazines, and newspapers have been critical in the social changes sweeping across the region. The largely western orientation of such media (and in education) have been important modes of spreading modern ideas and different ways of doing things as well as Western ideas of the family, intergenerational relations, peer relations, and gender relations. No observer of Southeast Asia during the last two decades can have failed to notice the ubiquitous presence and influence of the mass media. In Indonesia, for example, the proportion of households with a television set increased from 10 in 1980 to 26 in 1990 (Hugo, 1993b).

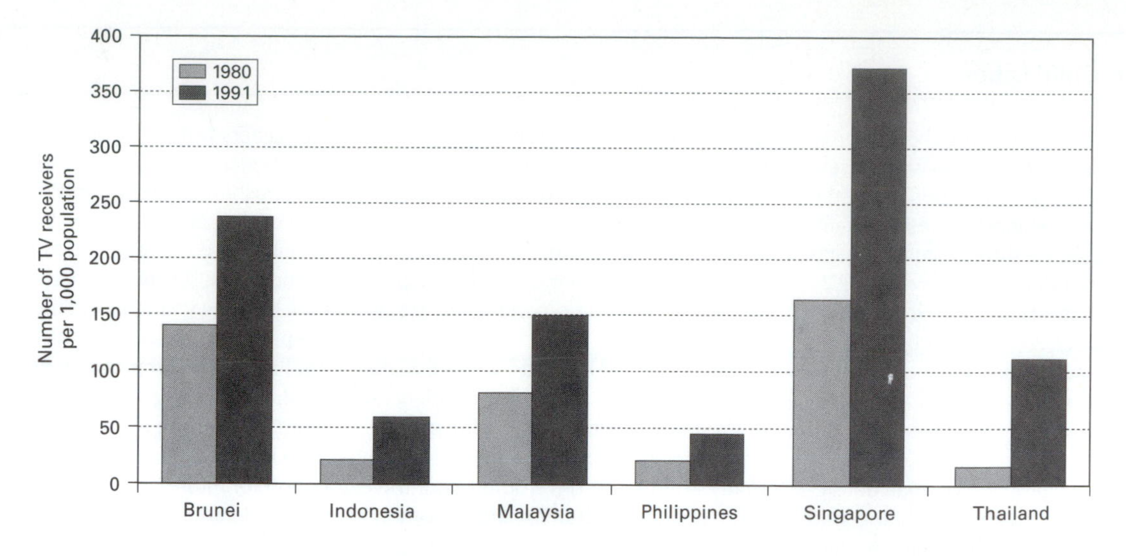

Figure 4.2 Number of TV receivers per 1,000 inhabitants, selected countries, 1980 and 1991. *Source: ASEAN 1997, p. 125*

In Vietnam more than four out of five households in Hanoi and Ho Chi Minh City have television sets. The development of satellite television has meant that it now reaches into the most remote parts of the region.

The Family

The family is the most fundamental unit of social organization in the Southeast Asian region. However, the family has been changing greatly in its structure and function and this is having a profound effect on many aspects of life in the region. Mason (1992) has pointed out there is no such thing as a single type of traditional or modern family in Southeast Asia. Nevertheless, it is possible to make some meaningful generalizations, especially about the way families are changing with respect to their size, structure, and functioning.

The most fundamental change that is occurring relates to a shift away from the extended family toward a predominance of the nuclear family. The definitions of these families relate to the attitudes and obligations between generations, not so much to their patterns of residence. In a traditional family the predominant sense of obligation and major *net* flows of wealth over a lifetime are

Photo 4.1 An Indonesian family. (Leinbach)

BOX 4.1 Child Labor

Child labor is illegal throughout the region but children continue to work in several nations. Because it is outlawed, the numbers involved tend to be uncertain. However, Table 4.4 shows some estimates of the numbers of children working in three of the largest countries. It is difficult to obtain estimates of the number of child workers because many children are unpaid family workers in fields and factories and work in the informal sector of the economy to assist the family's income. There is also the problem of small but significant numbers of children being sold into virtual slavery, sometimes by their parents, and of children being trafficked within or between countries. This is one of the results of poverty and cannot be stopped by simple legislation or boycotts which can be counterproductive. In countries like Thailand children work, sometimes alongside parents, in sweatshops.

Sanctions and boycotts, say critics, do little more than assuage the consciences of Western consumers—grinding poverty and a lack of affordable schooling leave many children in developing countries with no choice but to work. Forcing youngsters out of factories before they have better alternatives, the critics argue, risks pushing them into more dangerous jobs or deeper into penury. So, how best to help these children? International aid agencies are starting to focus on education. Keeping youngsters in the classroom keeps them off the assembly lines. In Thailand, for instance, as compulsory public schooling has been expanded, more children are studying and fewer are working. Education also lays the foundation for future economic growth (Fairclough 1996, p. 54).

Country	Child Laborers (mil)		Population Aged Under 16 Years (mil)
	Government Estimate	Unofficial Estimate	
Indonesia	2.2	3.3	69
Philippines	0.8	5.0	29
Thailand	1.5	4.0	20

Table 4.4 Estimates of child labor in some Southeast Asian countries in the mid-1990s. *Source: Far Eastern Economic Review, March 7, 1996, p. 56.*

toward one's parents. Hence, in such families the elders decide what work each family member will do, who and when they will marry, who receives income earned by family members, and who controls the economy of the family. In a nuclear family system there is a shift whereby the major sense of obligation is felt toward one's spouse and children rather than parents and the net flow of wealth is downward from parents to children. This doesn't mean that the older generation is abandoned; it simply means that people have greater control over where they work, in what they do with their earnings, and who and when they will marry. The older generation does not have as great economic and social control over younger generations.

It is clear that in Southeast Asia this movement from a predominance of traditional families toward a predominance of nuclear families has been rapid in the last quarter-century. The reasons for this change are associated with a number of factors. These include the spread of universal modern education, which has not only changed peoples ideas but also has to a large extent taken children out of the workforce and reduced their economic value, although child labor remains an important problem in the region. It also has been influenced by more Southeast Asians working in nonagricultural, nonfamily jobs, which means that the older members of the family lose power over their children and grandchildren. Mass media has also been important in changing people's attitudes.

Some of the indicators of this shift from an emotionally extended to a nuclear family in Southeast Asia are as follows:

- Substantial changes in *marriage* patterns involving increases in first age at marriage of women and, whereas two decades ago most marriages in the region were arranged by parents, now the majority are love marriages (Jones, 1994; Leete and Alam, 1993).
- A decline in the average number of children borne by women and in the size of the family.
- An increase in population mobility, which has meant that the extended family is much more spread out geographically than anytime in the past.
- A decline in *patriarchal* (and in a few areas, *matriarchal*) authority whereby decision making, power, and control of the earnings of all family members were concentrated in the oldest members.
- A decline in *polygmy*, that is, husbands having several wives.

- A decline in the family as a *unit of production* as more family members are employed on a commercial or wage basis in formal sector activities.
- Increased female participation in education and workforce outside of the home.

Although some of these changes are difficult to quantify, there can be no doubt that the patterns are widespread and of great significance in influencing demographic and social changes in the region.

The Changing Role of Women

In most traditional societies in Southeast Asia men and women were assigned quite specific roles, and the status of women tended to be concentrated on reproduction and home-maintenance roles although their productive role in agriculture is frequently underestimated (see also Chapter 11). Although the status of women in Southeast Asia was not as low as in East or South Asia, it was generally lower than that of men. This situation is changing in the region with considerable implications although there is considerable variation within the region. As Mason, (1995, p. 18) states: "In most parts of Asia, except the most backward, women today appear to be better off than their counterparts of 20–30 years ago and also better off in relation to men." Table 4.5 presents some important dimensions of this change in Southeast Asia.

Although there are some considerable problems in the underestimation of women's labor-force participation in the region, it is apparent that a greater proportion of women are working outside the home. In some cases economic change is creating more opportunities specifically for women. In particular, the development of labor-intensive factories in electronics, clothing, toys, footwear, and cosmetics has mainly employed young, unmarried, educated women because of their alleged greater docility, nimbleness, and willingness to work for lower wages than males. However, this has opened up opportunities for exploitation of such women who often experience low wages, poor conditions, excessive working hours, and dismissal upon marriage or pregnancy. It is apparent also that female participation in primary and secondary education has improved greatly in the region in recent decades in relation to men (Table 4.5). Both of these developments, along with the spread of mass media, have led many women to question traditional forms of organization, which in turn frequently has led to their having lower status and experiencing discrimination in the labor market, in the family, and in society generally.

In her review of the situation of women in Asia, Mason (1995, p. 3) concludes that, although Southeast Asian women have not traditionally enjoyed equality with men, their situation has been on the whole less oppressed than those in East and South Asia. Her analysis of education levels, employment, age at marriage, health and longevity, and sex preferences for children generally shows an improvement in the situation of women in Southeast Asia during the last two decades. These changes have been influenced by, and have impinged

	Female Workforce					Females per 1,000 Males in Education			
	Participation Rate			Percent Female		Primary		Secondary	
Country	15-20 Years Ago	Most Recent	Percent Difference	1980	1995	1970	Latest	1970	Latest
Indonesia	37	43	+5.3	35	40	84	93	59	62
Laos	51	55	+4.0	45	47	59	77	36	66
Malaysia	36	44	+7.6	47	47	88	95	69	104
Philippines	37	36	-1.2	33	37	n.a.	94	n.a.	99
Singapore	40	49	+8.4	35	38	88	90	103	100
Thailand	50	53	+3.0	47	46	88	95	69	97
Vietnam	47	49	+2.1	48	49	n.a.	n.a.	n.a.	n.a.
Burma	n.a.	n.a.	n.a.	n.a.	n.a.	89	n.a.	65	n.a.
Cambodia	n.a.	n.a.	n.a.	56	53	n.a.	n.a.	n.a.	n.a.

Table 4.5 Indicators of change in women's roles. *Source: UNESCAP, 1993c, p. 4-39; World Bank, World Development Report 1997, p. 220.*

upon, the rapid demographic transitions occurring in some of the Southeast Asian countries.

Labor-Force Issues

The economies of Southeast Asian countries have undergone massive change in recent decades, experiencing almost uninterrupted economic growth between the 1970s and 1997, although in Vietnam, Cambodia, Laos, Burma, and the Philippines there have been significant interruptions due to political factors and warfare. Moreover, the economies have undergone significant structural change. Generally agriculture has declined in employment significance, although it still is the largest single employer in most states in the region. The second structural change is in the growth of the formal sector at the expense of the informal sector. Nevertheless, the informal sector remains the largest sector of employment in most of the countries. Although the lines between the formal and informal sectors in Southeast Asia are not sharply drawn it is possible to generalize about the major differences between the formal and informal sectors. The informal sector employs more workers than the formal sector in most Southeast Asian countries; hence an understanding of how it works is crucial to understanding the region.

The workforces of the region are being rapidly transformed. In most countries the young age structure and increasing female participation are leading to their workforces growing substantially more rapidly than their total populations. Workforce age groups are growing faster than the entire population in all countries except Singapore. This presents the countries of the region with very substantial challenges in creating sufficient job opportunities to absorb the rapid increases in the labor force, especially given the 1997–98 economic crisis. It must be realized that it will be some time before reductions in fertility result in reductions in the growth rates of the labor force. Only in Singapore has the age structure changed to a degree such that the workforce age group will decline in the early twenty-first century. As was indicated in the previous section, females are making up an increasing proportion of the workforce; this is creating another series of policy issues because women workers are frequently exposed to exploitation, especially in the sex industry, domestic service, and factory work—important areas of absorption of female labor.

However, it is not only with respect to size that the workforce is changing rapidly; it is also experiencing substantial structural change (Figure 4.3). In most countries agriculture is declining in its share of the workforce and industry and other areas have become more significant. In some regions the agricultural workforce has begun to decline as a result of commercialization and mechanization in agriculture. In most countries the proportion working in industry has begun to increase. In Indonesia for example, the agricultural workforce proportion declined from 60 percent in 1980 to 44 percent in 1995 while the proportion working in manufacturing increased from 8 percent to 13 percent. There was an abrupt decline in the agricultural workforce in the Indonesian provinces of West Java and Central Sulawesi. This represents a huge change—for centuries agriculture has been the major sector in which increases in the workforce have been absorbed (Geertz, 1963a). Singapore represents an interesting exception, with the economy now so advanced that the proportion of workers in industry declined from 44 percent to 36 percent between 1980 and 1990 and the proportion in the tertiary and quaternary sectors increased from 54 percent to 64 percent. The major differences between nations in the levels of living standards are demonstrated in Figure 4.4.

Unemployment remains low, being under 5 percent in most countries of the region, less than one-half the levels prevailing in European and North American countries for much of the 1980s and 1990s. This is very largely a function of the virtual lack of unemployment benefits throughout the region, which means that the poor are forced to take any employment regardless of whether the employment is consistent with their aspirations, has low productivity and status, and produces low income. Hence, unemployment is largely restricted to children from relatives of well-to-do families who can afford to support their children through a period of unemployment. Most of the unemployed are concentrated in the highly educated group. This partly reflects an education system that is not meeting the needs of the labor market in several of the countries. The new economy needs managers, technicians, computer technicians, accountants, and engineers, but still the emphasis in the education system are on such areas as administration, the arts, and law. This results to some extent in a mismatch between the skills of the graduates of the tertiary training system and the needs of the labor market.

On the other hand, there are high levels of *underemployment* in several of the Southeast Asian countries including the largest nations. In these nations the surplus labor means that many workers are not working as many hours as they wish, are highly unproductive in the output per hour they produce, or have their skills severely underutilized. It is not unusual for almost one-third of workers to be classified as underemployed. Clearly many of the workers who in a developed country would be unemployed became underemployed in Southeast Asia because they cannot afford to be unemployed.

POPULATION, SIZE, GROWTH, AND DISTRIBUTION

Population Size and Growth

Most Southeast Asian countries participated in the 1990 round of censuses but the quality of population data varies greatly across the region. The last decade has cer-

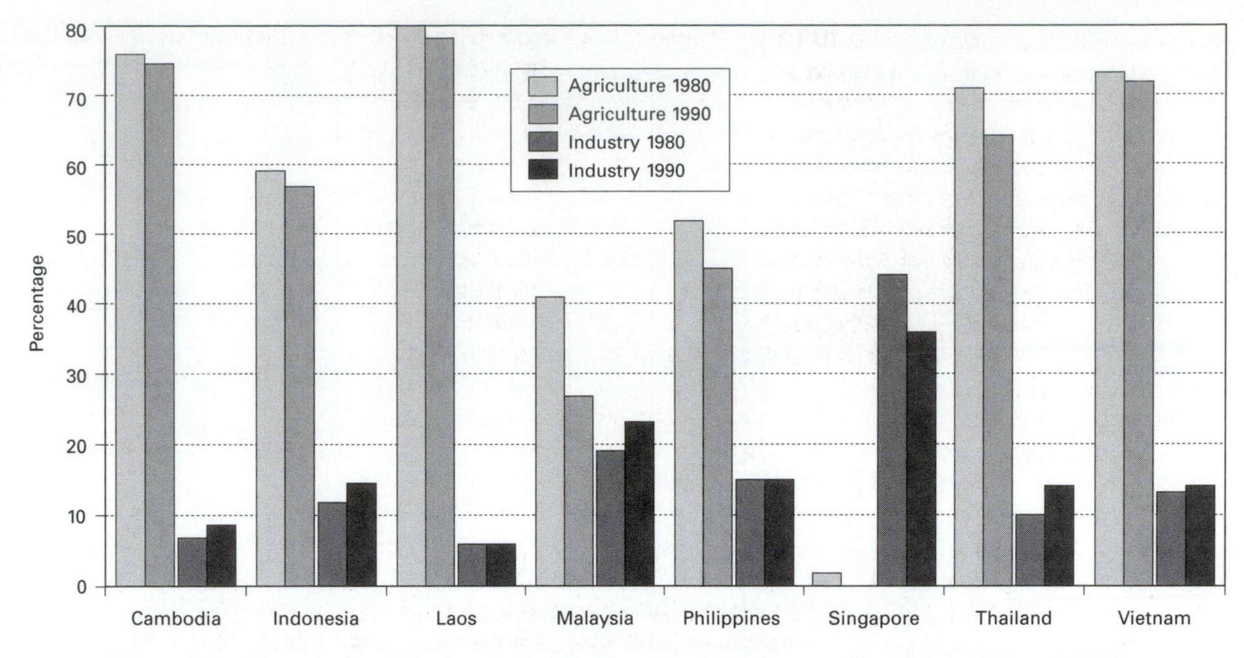

Figure 4.3 Percentage of the population employed in agriculture and industry, 1980 and 1990.
Source: World Bank, World Development Report 1997.

tainly witnessed a substantial improvement in the availability of demographic information in the region. Several countries have undertaken rounds of the Demographic and Health Survey while most have increased the numbers of national population surveys undertaken. Nevertheless, the profound disruption of earlier times in Vietnam, Cambodia, Laos, and Burma has meant that the knowledge of the demography of those countries is limited compared with that of the rapidly developing countries such as Thailand, Malaysia, Singapore, and Indonesia. Table 4.7 shows some of the main demographic features of contemporary Southeast Asian countries and reflects the great variation between nations in the size, dynamics, and growth of their populations. These variations will be examined in more detail in this section.

There is wide variation in the size of the national populations in the Southeast Asian region. Indonesia is by far the largest nation having more than 40 percent of Southeast Asian residents. On the other hand, four nations in the region had less than 10 million residents. Table 4.6 shows however, that Indonesia has one of the slowest annual rates of natural population increase, with only Singapore, Thailand, and Vietnam growing at a slower rate in 1997. On the other hand, the Philippines is growing at well above the region's average and indeed in excess of the average for all developing countries. Time series data shows that most countries have experienced reductions in their population growth rates since the 1970s, although some very substantial differences are evident (Figure 4.5). Singapore was the first country in the region to complete the demographic transition to stable low rates of population growth. Most spectacular, however, is the change in Thailand, where growth rates have almost halved since 1980 to 1.1 percent per annum. The Philippines and Malaysia, on the other hand, while having similar rates of growth to Thailand in the 1970s continue to experience population increases in excess of 2 percent per annum as do Burma and Brunei. The former Indochinese states show an interesting trajectory with population growth rates increasing after peace was restored in the mid-1970s and 1980s after decades of conflict. Accordingly, they had the highest rates of growth in the region in the early 1990s.

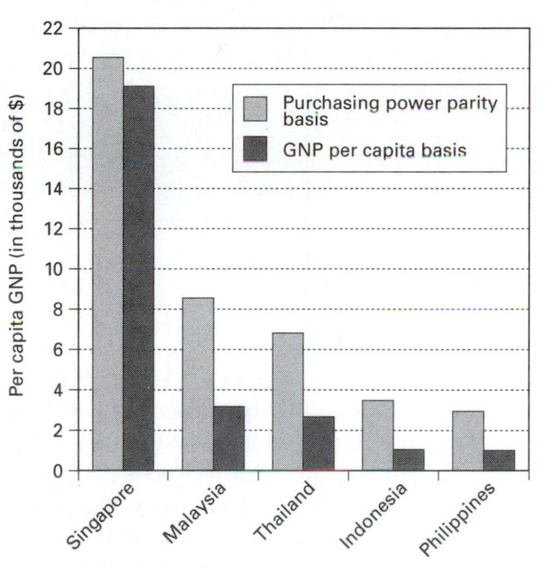

Figure 4.4 Two measures of living standards for selected Southeast Asian countries.
Source: Far Eastern Economic Review, January 12, 1995, p. 81.

Indicator	Country										All Southeast Asia	All Less Developed Countries	All More Developed Countries
	Indonesia	Philippines	Thailand	Malaysia	Singapore	Vietnam	Cambodia	Laos	Burma	Brunei			
Total Population (millions)	204.3	73.4	60.1	21.0	3.5	75.1	11.2	5.1	46.8	0.3	501	4,666	1,175
Natural Increase (annual %)	1.7	2.3	1.1	2.2	1.1	1.6	2.9	2.8	1.9	2.3	1.8	1.8	0.1
Doubling Time in Years (at current rate)	40	30	63	31	64	43	24	25	36	30	39	38	564
Projected Population to 2025 (millions)	276.4	113.5	71.1	32.8	4.5	103.9	22.8	9.8	72.2	0.5	707	6,810	1,226
Infant Mortality Rate	66	34	32	11	4.0	38	111	102	49	11.2	52	64	9
Life Expectancy at Birth (years)	62	66	69	72	76	67	49	52	61	71	64	63	75
Total Fertility Rate	2.9	4.1	1.9	3.3	1.7	3.1	5.8	6.1	4.0	3.4	3.2	3.4	1.6
Percent Using Family Planning (all methods)	55	40	66	48	65	65	–	–	17	–	52	54	66
Percent Population Aged Under 15 Years	34	38	30	36	23	40	46	45	36	35	36	35	20
Percent Population Aged 65 or More	4	4	4	4	7	5	3	3	4	3	4	5	14
Urban Population (%)	31	47	19	51	100	20	13	19	25	67	30	36	74

Table 4.6 Some basic demographic indicators, 1997. *Source: Population Reference Bureau, 1997.*

It is estimated that Southeast Asia's population is currently growing at about 1.8 percent per annum (Population Reference Bureau, 1997). If this is the case, then the region's growth rate has gone back to levels obtained in the pre-1950s period (United Nations, 1979, p. 176). Southeast Asia's population was growing at about 2.03 percent per annum in the early 1950s; this increased gradually to about 2.7 percent in the early 1970s and fell to 2 percent in the 1980s. However, Figure 4.5 shows that there was considerable variation between the countries in the changing patterns of population growth over time. A clear pattern is evident in the time series data. In the 1950s comparatively well-off countries such as Singapore and Brunei had the highest rates of growth, while in other countries with similar high levels of fertility, higher levels of mortality kept population growth rates at about or below 2 percent. In the 1960s several of these latter countries experienced mortality declines that saw their growth rates rise. The subsequent return to growth levels at or near those of the early 1950s has been fostered in many cases by fertility declines, and immigration is becoming more significant in the low fertility countries. Nevertheless, within this broad pattern there is considerable variation from country to country.

Despite recent reductions in population expansion, Southeast Asia's present population-growth rate of 1.8 percent is well above the average for the entire Asian region which is currently at 1.6 percent per annum (Population Reference Bureau, 1997) because of the low rate in China but, is close to the average for developing countries as a whole. Behind the shifts in population growth shown in Figure 4.5 lie some major changes in fertility, mortality, and migration, and it is to a consideration of these processes we now turn.

Mortality Trends

As Pressat (1970) has pointed out, the ultimate inequality is inequality in the face of death. In the mid-1990s the average expectation of life at birth for a baby born in

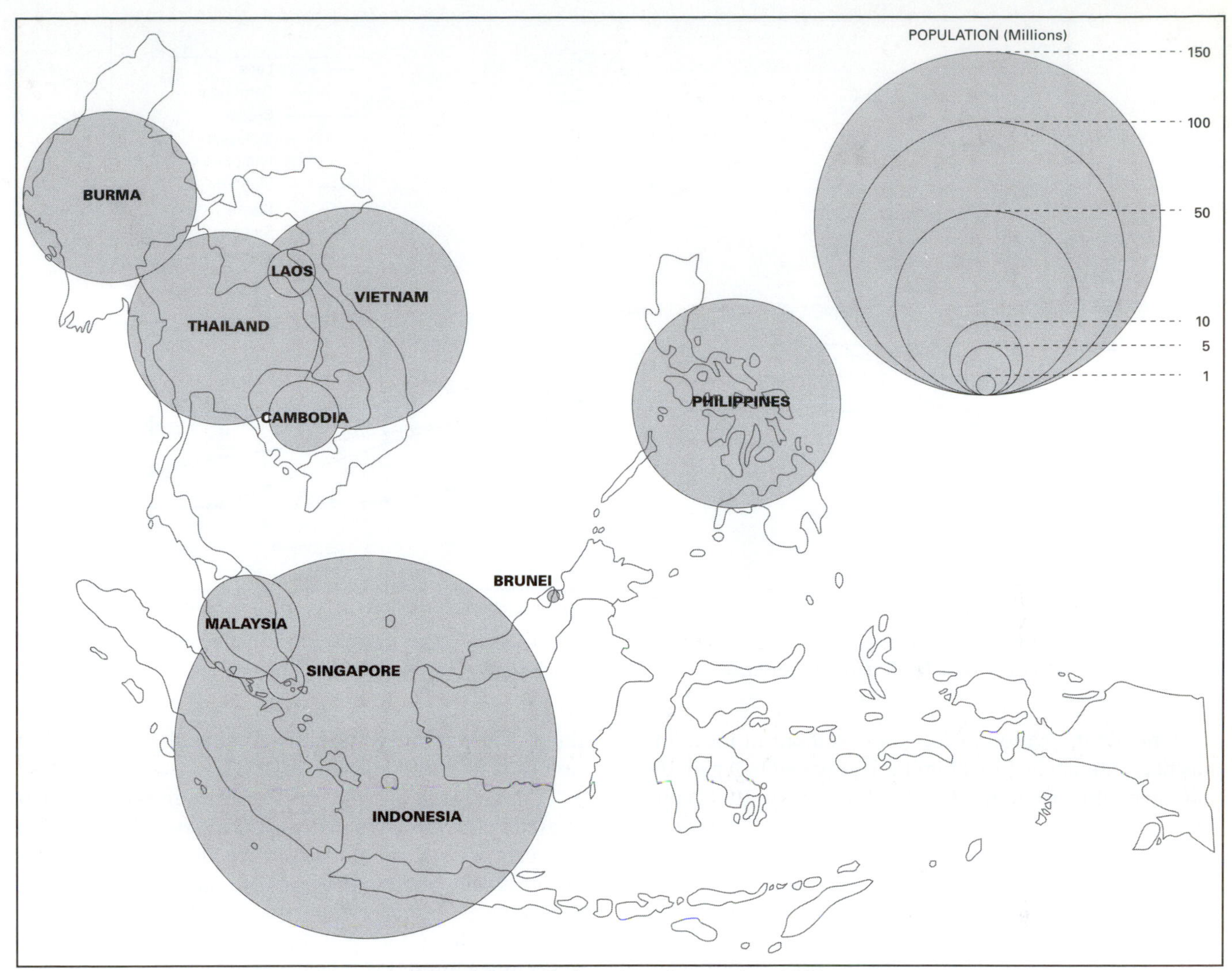

Map 4.2 Population size of countries, 1993.

Southeast Asia was 62 years if it was male and 66 if it was female. This is still well below levels in Western Europe (73 and 80) and North America (72 and 79) but it represents improvement over the last four decades. Although in many cases the data are poor and, as Ruzicka (1983, p. 4) points out, should be interpreted as representing the central point of a range of possible values rather than an exact actual level of mortality, the life expectancies shown in Figure 4.6 indicate that several countries have recorded an increase in life expectancy during the post-World War II period.

There are still, however, differences between Southeast Asian countries in mortality ranging from life expectancies of 76 years in Singapore to 52 years in Laos. The overall improvements in mortality have been due to "the introduction of comparatively cheap and effective technologies for the control of disease vectors, immunization and effective disease treatment . . . outbreaks of epidemics became rare . . . large scale famines also disappeared" (Ruzicka, 1983, p. 4). In Malaysia, Singapore, and Brunei where life expectancies are now in excess of 70 years, it is clear that standards of living and provision of health services have advanced to the levels typical of Euro-American societies. In the countries with life expectancies in the 60s (Thailand, Philippines, Burma, Indonesia, and Vietnam) there are indications that despite growth in prosperity, government spending on health has been limited and that there is a need for major improvements in provision and access to health services to substantially reduce mortality further. The legacy of long years of conflict and dislocation is still seen in very high levels of mortality in Cambodia and Laos. Nevertheless, Figure4.6 shows that Southeast Asians in the 1990s are living between 13 percent and 27 percent longer than their compatriots of the 1950s. This represents one of the greatest achievements of the countries of the region since World War II.

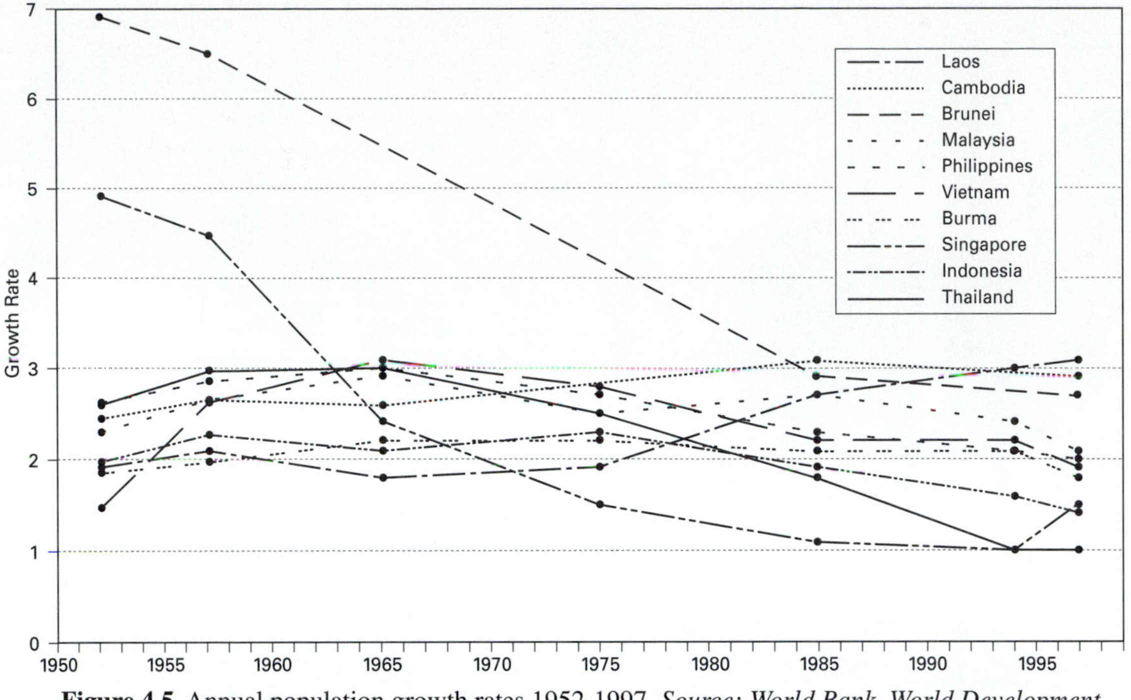

Figure 4.5 Annual population growth rates 1952-1997. *Source: World Bank, World Development Report 1983; UNESCAP Population Data Sheets 1983, 1997; United Nations 1979, 1994a.*

A major indication of the improvement in mortality has been the decline in infant mortality levels. Again the data available are limited and often of poor quality, but the data suggests that there have been some massive reductions in infant mortality during the last decade, especially in Singapore, Malaysia, Philippines, and Thailand (Figure 4.6). A number of factors have been involved in this decline. These include reduction in fertility, which has led to a reduction in high risk deaths, expanded immunization, improved health provision, improved diets of infants and mothers, and reduction in poverty (Leinbach, 1988). Nevertheless, in several countries in the region 50 or more babies out of every 1,000 born alive die before they reach their first birthday—more than five times the levels in Euro-American nations. Reducing mortality remains a major priority in most of Southeast Asia, especially in light of some evidence of a slowdown in mortality improvement during the last decade. In 1962 a United Nations report on mortality suggested that: "It may not be too much to hope that, within a decade or two, the vast majority of the world's peoples will have an expectation of life at birth of 65 years or more" (Gwatkin, 1980, p. 617). Figure 4.6 shows that by 1970–1980, only two Southeast Asian countries met that goal and, three decades later, four of the ten countries still have yet to reach that target. It has been suggested that the rapid mortality declines of the 1950s and 1960s were achieved by large scale technological initiatives and that such improvements could not be sustained without major improvements in the socioeconomic conditions of the bulk of the population (Gwatkin, 1980; Ruzicka, 1983).

A major area of concern that arose in the late-1980s related to the spread of HIV/AIDS in several countries in the region. Data are of course very limited but Table 4.7 presents some information on four nations in the region indicating a significant degree of prevalence, especially in Thailand. Chin (1995, p. 9) has argued that by the end of the century the AIDS-specific mortality rate in Thailand will surpass the current mortality rate for all causes of death among young and middle-aged adults and concludes that the HIV infection will continue to spread, although its distribution will vary greatly among countries in the region.

In a recent assessment of the potential for spread of the HIV/AIDS epidemic in Asia, three Southeast Asian countries were in the rapidly increasing category (Cambodia, Burma, and Thailand) and five were in the potential for rapid increase category (Indonesia, Laos, Malaysia, Philippines, and Vietnam) (Brown and Xenos, 1994).

Fertility Trends

No aspect of the demography of the Southeast Asian region attracted more attention in the 1970s and 1980s than the decline in fertility that began in several countries during this period. Between 1965–1970 and 1985–1990, the total fertility rate (TFR) in Southeast Asia declined from 5.81 to 3.73. This represented a remarkable change

Country	Adults 20-49 Years (Million)	Estimated Number HIV Cases (Thousands)	Prevalence Rate (Percent Infected)
Burma	15	250	1.700
Indonesia	75	20	0.025
Philippines	25	17.5	0.070
Thailand	23	500	2.200

Table 4.7 Estimated HIV prevalence and prevalence rates among adults aged 20-49 years. *Source: Chin, 1995, p.5.*

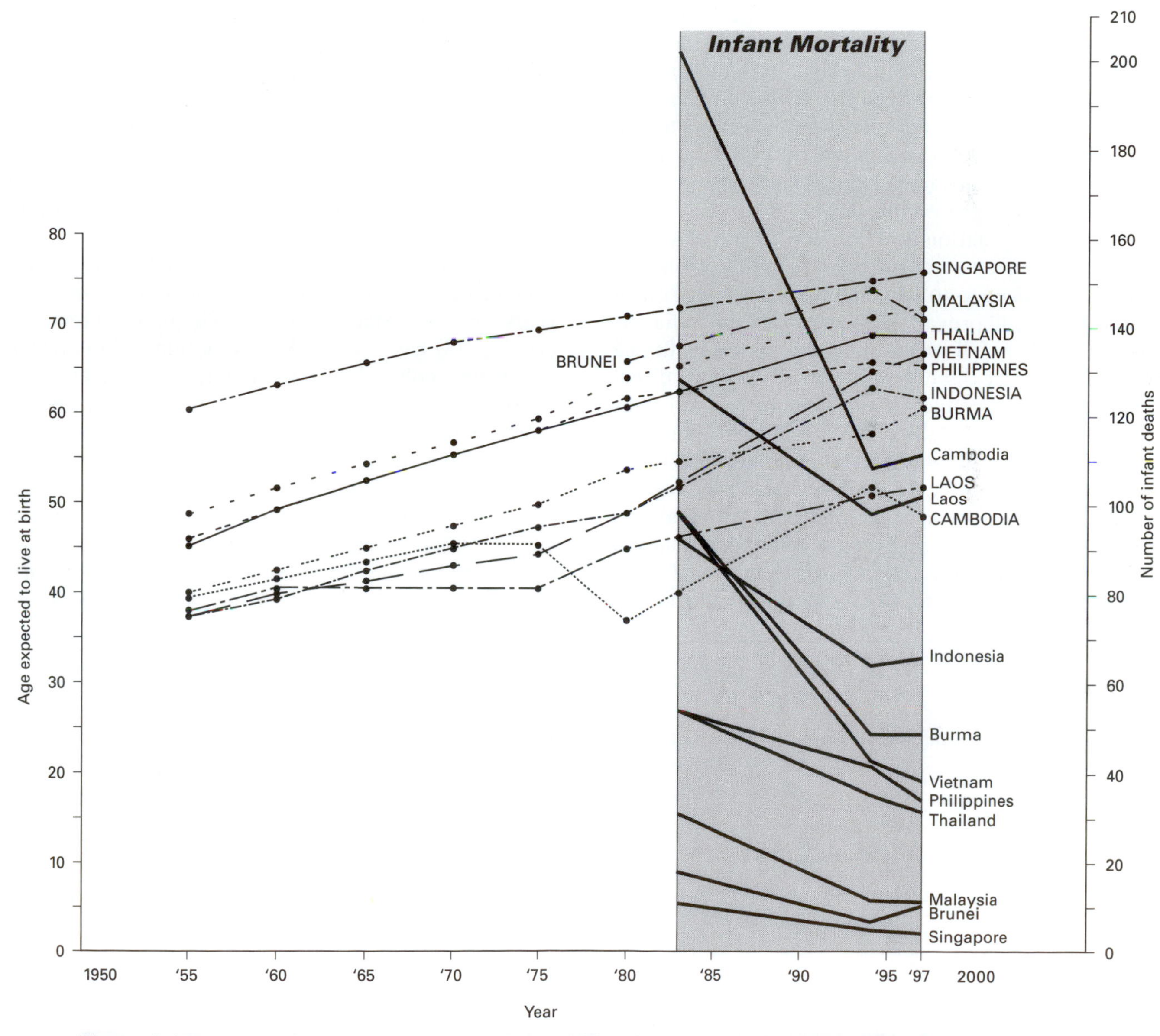

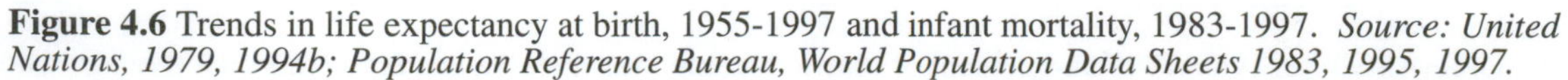

Figure 4.6 Trends in life expectancy at birth, 1955-1997 and infant mortality, 1983-1997. *Source: United Nations, 1979, 1994b; Population Reference Bureau, World Population Data Sheets 1983, 1995, 1997.*

whereby within two decades the average number of children born declined by more than two. There is, however, a substantial variation between Southeast Asian countries in the extent to which they have passed through the transition from high fertility (and high mortality) to low fertility and low mortality (Leete and Alam, 1993).

This is evident in Table 4.8, which shows that there is a huge range in the region not only in contemporary TFRs (from 6.7 in Laos to 1.7 in Singapore) but also in the nature and pace of change in fertility during the last three decades (from a decline of 67 percent in Thailand to a gain of 18 percent in Laos). Nevertheless, the dominant pattern is one of rapidly declining fertility, and this represents one of the most significant, striking, and unanticipated trends in the region in the last quarter-century. As a World Bank (1993, p. 15) report on the Asian Economic Miracle pointed out, this demographic transition may be the cause as well as the effect of important elements of economic growth in the region. It has been shown (Ahlberg, 1994) that reductions in fertility have a positive impact on economic growth in LDCs. Clearly if economic growth is absorbed by population increases per capita income will grow more slowly.

The fertility situation in the region ranges from Singapore, which completed its transition to low fertility almost two decades ago, to Laos and Cambodia, where high fertility levels still prevail. In the case of Cambodia it is clear that fertility fell dramatically during the 1970s as a result of the tragic disruption of war, invasion, and civil strife that was associated with major outflow of refugees and horrific loss of life (Meng, 1981). It is usual in such circumstances for low fertility to rebound (Hugo, 1984) and this is apparent in Table 4.8. Laos has by far the highest fertility in the region, although there are strong indications that national carrying capacity has been reached; the government remains pronatalist although birth spacing has been endorsed to protect mothers and children whose mortality stands at high levels (Fauveau et al., 1993). Hence both Laos and Cambodia are essentially pretransitional societies that have yet to experience any significant fertility decline. Fertility levels are also high in Burma, but it appears that it has entered the early stages of the transition with the TFR declining from 5.8 in the early 1970s to 4.2 in the 1990s.

At the other extreme is Singapore, which has one of the lowest levels of fertility outside Europe. In Singapore, however, official concern about below replacement fertility goes back to 1983 when then-Prime Minister Lee Kwan Yew expressed "concern about the trend of many university educated women to remain unmarried and the possible impact this might have on the future of Singapore's population ... Lee had hinted that incentives would be needed to coax single, educated women to marry, and those married to have more than one child" (Kulkarni, 1983, p. 23). This has subsequently developed into a comprehensive pronatalist policy aimed relatively at better-educated women in the Singapore population (Yap, 1995). Nevertheless, the slowing down of the growth of Singapore's workforce in an era of rapid economic growth has seen it turn to international migration to meet its burgeoning workforce needs.

Other nations in the region that have followed somewhat pronatalist paths in their fertility policies in recent years have been Malaysia and the Philippines. Freedman (1995, p. 15) has described the situation in Malaysia as one where "political power, building on pronatalist religious and cultural values, can have pronatalist behavioral effects despite considerable economic and social development." The decline in Malay fertility stalled and in fact rose between 1978 and 1986 while that of the Chinese and Indian populations continued to fall (Leete and Tan, 1993). This was in response to a strong movement backed by Muslim religious forces and the ruling political party to include a pronatalist policy that would see Malaysia

	Total Fertility Rate						Percent Change 1960-95	Contraceptive Use	
Country	1960-65	1965-70	1970-75	1975-80	1985-90	1990-95		Year	Percent
Burma	5.8	5.6	5.8	5.5	4.5	4.2	-28	--	--
Cambodia	6.2	6.2	5.5	--	4.6	5.3	-15	--	--
Indonesia	5.4	5.6	5.5	4.8	3.5	2.9	-46	1991	50
Laos	5.7	6.1	6.1	6.1	6.7	6.7	18	1992	15
Malaysia	6.7	5.9	4.7	4.2	4.0	3.6	-46	1988	48
Philippines	--	6.0	--	--	4.3	3.9	-54*	1993	40
Singapore	4.9	3.4	2.7	1.9	1.7	1.7	-65	1982	74
Thailand	6.4	6.1	5.0	4.3	2.6	2.1	-67	1987	66
Vietnam	6.9	6.9	6.4	5.6	4.2	3.9	-43	1988	53

**percent change 1965-1995*

Table 4.8 Total fertility rate, 1960-1995 and percent of married women (or their husbands) of reproductive age who are using a contraceptive method. *Sources: UNESCAP, 1984, p. 4; United Nations, 1994a and 1994b; Fauveau, et al., 1993.*

have a population of 70 million by the year 2100. In the Philippines the decline in fertility has been slower than its ASEAN neighbors with comparable socioeconomic situations because the "Philippine family planning program is relatively ineffective, possibly in part because the Roman Catholic Church, unique to the Philippines in Asia, has strongly opposed the program" (Freedman, 1995, p. 16).

Perhaps the most striking pattern in Table 4.8 is the fact that Thailand's TFR in the 1990s was only one-third that of the 1960s. Indeed Thailand has now joined Singapore as the only countries in Southeast Asia with a below-replacement-level of fertility. While Thailand has experienced significant social and economic change and has had a highly effective noncoercive family planning program, cultural factors including the dominance of the Buddhist religion, the predominance of nuclear families, and high levels of female autonomy have played an important role (Freedman, 1995, p. 20). The achievement of Indonesia in reducing its fertility by one-half between the late 1960s and mid-1990s has been almost as impressive. In Indonesia the constellation of factors associated with the fertility decline are somewhat different, involving not only substantial social and economic change and a highly effective family planning program but also an administrative structure that facilitates action at the grassroots level for such programs (Freedman, 1995, p. 10) (Photo 4.2).

Photo 4.2 Family planning billboard in Palembang, Sumatra announcing "Dua Anak Cukup"—"Two Children are Enough." (Leinbach)

In Vietnam the long period of war prior to 1975 influenced fertility and population growth and indeed is still being felt in that there is a shortage of men in the 20–39 age group where women make up 52.8 percent of the population (Hiebert, 1994a). The results of the 1989 census demonstrated that fertility levels in Vietnam had in fact been declining for several decades (Feeney and Xenos, 1992; Hull, 1990). However, the rapid growth of the population in the 1980s convinced the government to develop an active family planning policy (Hiebert, 1993b). Although this has had a degree of success, it has been suggested that sustaining change may be difficult due to poverty (Kaufman and Gunawan, 1994).

The broad trend of fertility decline shown in Table 4.8 is of enormous significance for the future development of Southeast Asia. It is important to note that this decline has occurred in some contexts in which broad-based major economic development has not occurred. Indeed it is the experience of such countries as Thailand, Vietnam, and Indonesia that have helped break down support for purely economic explanations of fertility decline. Table 4.8 shows that in those countries more than one-half of eligible women were using modern contraceptive methods in the early 1980s. With the exception of Singapore, these prevalence rates would have been almost negligible during the 1960s, so these data reflect a social change of major proportions in Indonesia, Malaysia, the Philippines, Vietnam, and Thailand. Box 4.2 shows how the decline in fertility in Southeast Asia in the last quarter-century has altered the consensus about why fertility has declined.

As Freedman (1995, p. 5) has pointed out, "The important question is not whether across-the-board socioeconomic development was often instrumental to the Asian fertility decline but rather what were the circumstances under which fertility declined with little or with less than broad-based socioeconomic development." He identifies four such areas:

- Strong government commitment to family planning together with active, efficient, well-organized family planning programs.
- Broad-based economic development is not always necessary for major fertility declines—limited, specific impacts such as improved health, education (especially of women), and policies to improve the status of women can be influential.
- Ideas are a potential agent of change. Hence, mass media, and education have an important impact.
- Culture makes a difference in whether or not groups are receptive to family planning.

***BOX 4.2* Southeast Asia's Demographic Transition**

Since the early years of this century, demographers have found that as societies experience economic development and social change there tends to be a consistency in the sequence of changes that occur in vital rates (that is, crude birthrates and crude death rates). This led to the formation of the Demographic Transition Model depicted in Figure 4.7. This model identifies a sequence of connected stages as development proceeds. The first or pretransitional phase is of a traditional society in which there is a high level of fertility and high but unstable mortality rates.

The instability of the latter results from periods of particularly high mortality caused by famines, wars, and infectious-disease epidemics. During this phase overall population growth is slow because the gains in population during a period when death rates are lower are canceled out by the losses when the death rates are high. Conventional versions of the model suggest that the transition is initiated when in response to the onset of modernization, improved living conditions and control of disease bring about a decline in death rates. At first fertility rates remain high so that the early stages of the transition are ones of high natural increase rates and of substantial population growth. In the third or late-transitional phase there is a stabilization of the death rate at low levels and a reduction in the birth rate so that population growth rates level off as the birthrates and the death rates begin to converge. The fall in fertility is traditionally ascribed to a perceived increase in the burden of rearing and educating children due to the forces associated with the growth of an urban industrial society (including greater emphasis on individualism and less on the traditional family) and the greatly improved probability of survival of infants and children supported by improved methods and more widespread knowledge and practice of contraception. The final stage, like the initial situation, is one of relative equilibrium or slight growth but with birth rates and death rates both being at *low* levels, with some fluctuations in fertility due to changes in economic and, to a lesser extent, social trends.

This well-known generalized descriptive model is inductive in origin, which means that it was developed from observation and description of the experience of Western Europe, North America, and Australia by demographers working in the early part of this century. This de-

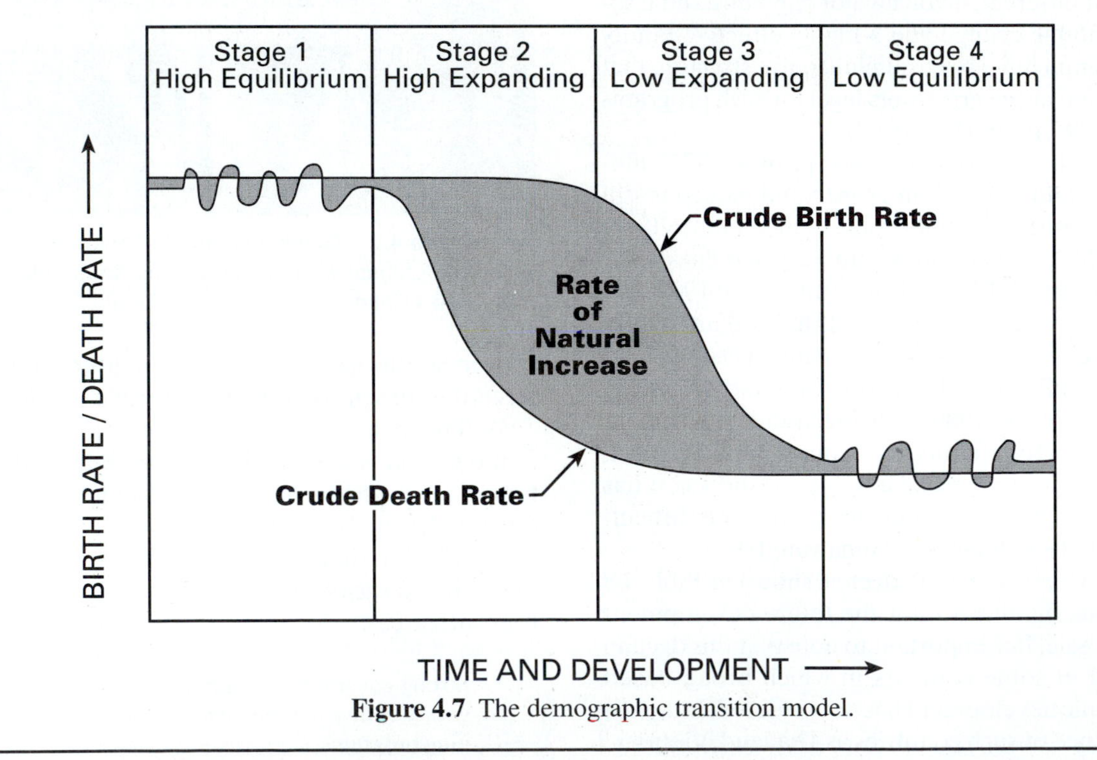

Figure 4.7 The demographic transition model.

Clearly there are multiple pathways (Freedman, 1995, p. 24) to lower fertility, and in Southeast Asia varying mixes of economic development, family planning programs, and education that impinges on different cultures has produced a wide range of fertility outcomes but with most now having achieved significant fertility declines.

Changing Age Structure

The major shifts in fertility (and to a lesser extent in mortality) described above have profound implications for the age structure of Southeast Asian countries. The inevitable corollary of the widespread pattern of sustained fertility decline is aging. All Southeast Asian nations are

scriptive part of the Demographic Transition Model, however, has not been accepted by demographers as being universally applicable, even in Europe. For example, whereas the traditional description of the Transition attributes the rapid growth of European populations during the nineteenth and early twentieth century to a decline in mortality that preceded a decline in fertility, recent studies based on newly developed analytical methods indicate that in much of Europe declines in fertility and mortality occurred at roughly the same time (e.g., France), while in some cases the fertility decline *preceded* mortality declines (e.g., Germany and the Netherlands). Clearly, there were and are many deviations from the pattern described in the classical demographic transition. Much of the debate in demography and a major focus of research relates to *what actually initiates the decline in fertility.* This knowledge is of course not only important from a theoretical perspective but also for policy. Until the early 1970s it was generally accepted that the fertility transition was initiated largely by economic forces and that it depended upon major breakthroughs in improving average levels of income, industrialization, and urbanization. More recently it has been suggested that the key process is not economic but social. The social change may be speeded up by economic change, but the latter is a sufficient rather than a necessary condition. For the fertility transition to begin there needs to be a change in the family structure and functioning that involves a shift in intergenerational relationships. It is a move from an emotionally extended family to a nuclear family, and although while economic changes can be associated with this shift, it is considered that the more powerful underlying forces are associated with achievement of universal education and the penetration of mass media.

Figure 4.8 shows the trajectory of crude birthrates and death rates in Indonesia for the last 40 years and projected trends for the next 30 years and this is typical of patterns in Southeast Asia. Quite clearly in terms of the Demographic Transition Model, the fertility transition has well and truly begun in Indonesia, and it is well into the third phase of the Demographic Transition in which natural increase is falling as the nation moves toward a low equilibrium situation.

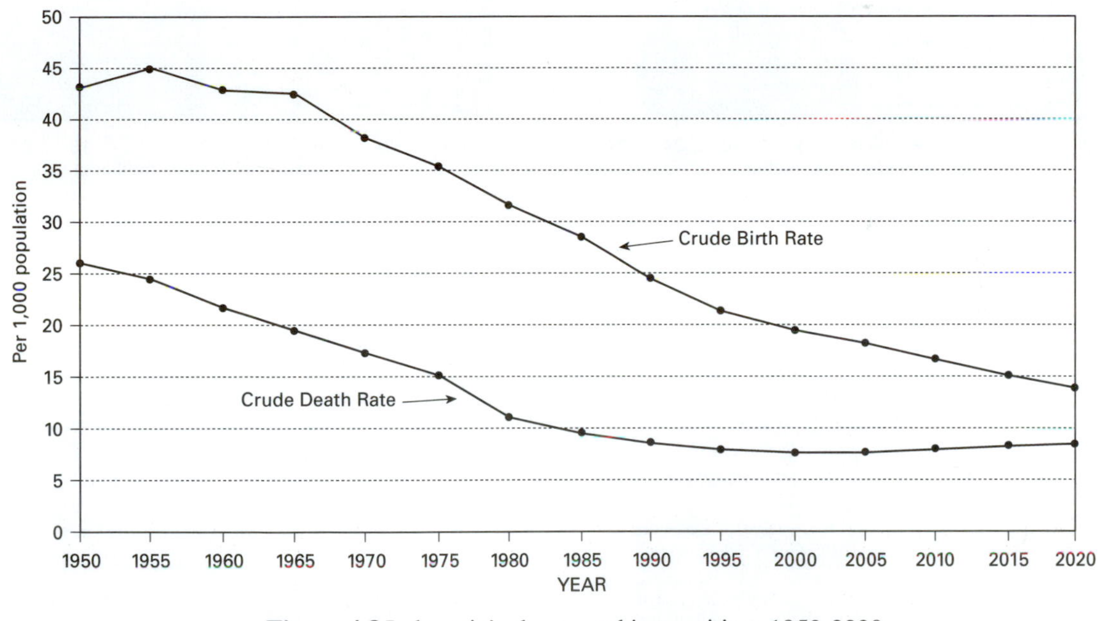

Figure 4.8 Indonesia's demographic transition, 1950-2020.

currently experiencing a growth in the proportions of their populations aged 65 years and older. Indeed aging is becoming an issue of increasing significance in the region with the certainty of rapid absolute and relative growth in the elderly in prospect for the next 25 years (Hermalin, 1995). In the mid-1990s Southeast Asia's population remained an emphatically young one with 37 percent of the population being less than 15 years of age. This is not only almost twice the proportion in this group in MDCs (more-developed countries) (20 percent) but above the average for all LDCs (35 percent). On the other hand, there was only 4 percent of age 65 years and over in Southeast Asia while in MDCs 13 percent were in this group. Hence, the region's population must be

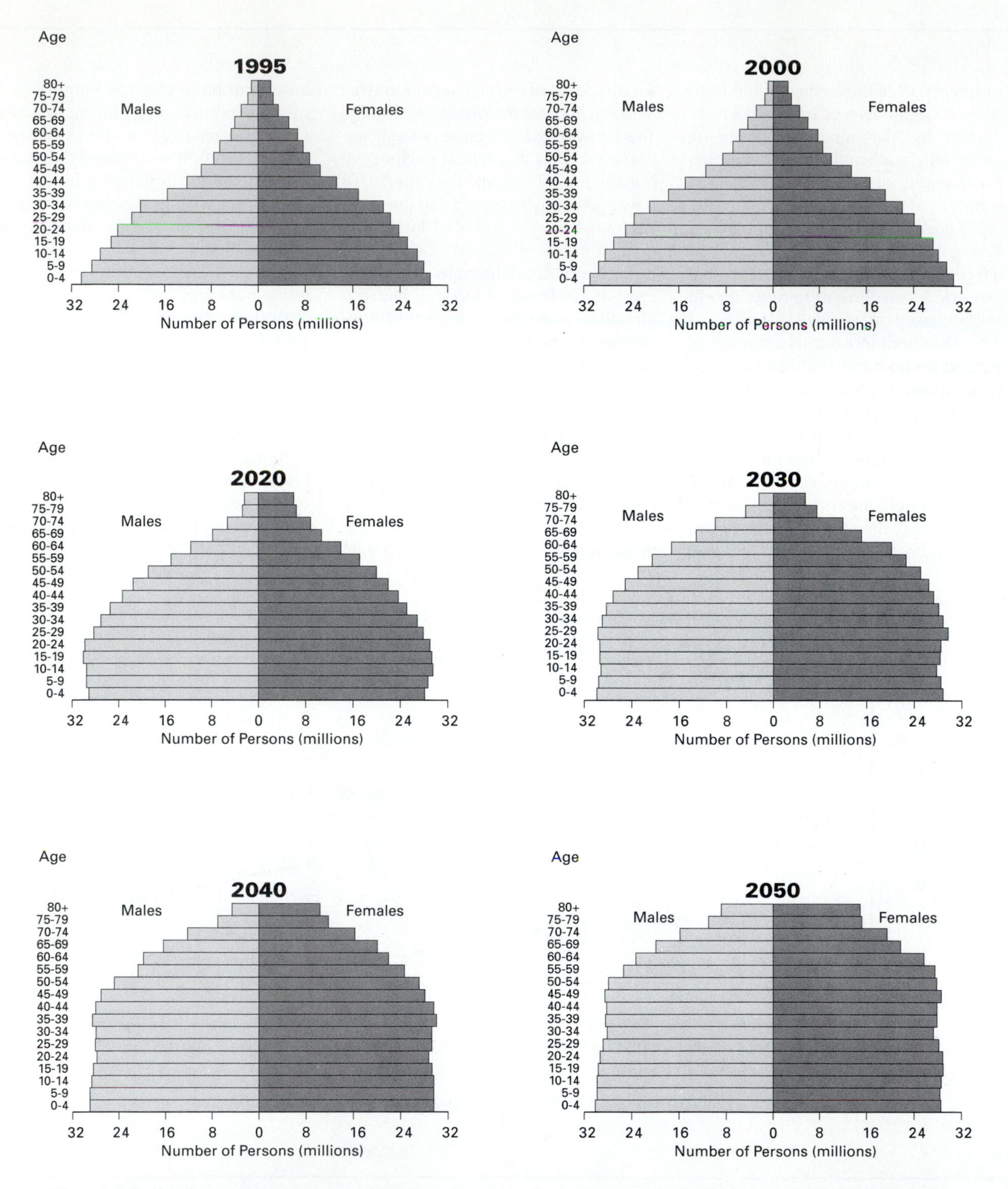

Figure 4.9 Southeast Asia: Age-sex structure of projected population, 1995, 2000, 2020, 2030, 2040, 2050.
Source: United Nations 1994a.

characterized as a very young one, albeit one wherein a rapidly growing elderly community is emerging.

The impending fundamental shift that will occur in the age composition of the population of Southeast Asia is apparent in Figure 4.9. At present the age structure of the region follows a more or less even pyramidal shape of a relatively fast growing population with comparatively high mortality and fertility. However, over the next half-century reductions in fertility and increases in life expectancy will see the age structure move toward a more pillar type of shape in which dependent children make up a smaller proportion of the population.

In the 1990s the number of elderly people in Southeast Asia will increase by more than one-third while that of the working-age population will increase by 25 percent and the dependent-child population by 7 percent. However, it is important to note that although the balance in the dependent population is shifting away from children toward the elderly, the numerical size of youth dependents is approximately eight times greater than that of elderly dependents in the region. Hence while there is undoubtedly a need to focus more on the needs of the elderly in Southeast Asia, there will continue for the next few decades to be very large numbers of children entering the school-age groups and entering the labor markets of the region as was described earlier in Table 4.5. Moreover, from the perspective of population growth it is apparent from Figure 4.9 that despite the impressive fertility declines of the last two decades in Southeast Asia, the high fertility of the 1950s, 1960s and 1970s coupled with improvements in mortality have built into the age structure of the region tremendous momentum for growth. Hence even if the cohorts of women born in the 1960s and 1970s only have half as many children (or less) than their mothers', the fact that there are several times more of them in the child-bearing ages than was the case in their mothers generation means that the total numbers of babies born will be greater than in the earlier generation. Hence it takes a considerable time for heavily reduced fertility to be reflected in major declines in population growth.

The patterns presented in Figure 4.9 for all of Southeast Asia tend to mask considerable intercountry variation in age structure due to the quite different recent demographic histories of the 10 countries. This is apparent in the large variation in age structures of the individual countries reflected in Table 4.6 and in the age pyramids presented in Figure 4.10. The age structures of Burma, Laos, and the Philippines represent broad-based pyramids reflecting sustained high fertility and high mortality. Brunei represents a relatively high fertility structure but one in which mortality also is relatively low. The impact of war and tragic events of the 1970s is evident in the Cambodian age structure while the effects of war explain the very small numbers in the older adult ages of Vietnam.

The effects of substantial fertility declines have caused a steepening in the younger ages in the Indonesia pyramid and an undercutting of that for Thailand. Malaysia's age structure shows that the upswing in fertility in the 1980s caused a particularly broad base of children under 10 in 1990 compared with the size of the cohort born in the 1970s. Singapore has one of the most distinctive age structures in the region with substantial undercutting evident below age 25 in 1990. Hence the numbers coming into the labor force in Singapore are currently declining. There is a wide variation between countries in the region in the proportions of national populations aged below 15 years but only Singapore (23 percent) and Thailand (31) were below the average for LDCs (35 percent) and proportions were especially high in Cambodia (46 percent), Laos (45 percent), the Philippines (40 percent), and Vietnam (39 percent). On the other hand, proportions aged 65 years and older are low ranging from 3 percent in Brunei, Cambodia, Laos, and the Philippines, to 7 percent in Singapore.

The immediate outlook for most of the countries of the region is for continued rapid growth, albeit in many cases at declining rates, of school-age children. Countries like Singapore, Thailand, and Indonesia can anticipate decreasing numbers in the younger schooling years. In most countries labor forces will continue to grow faster than the population as a whole due to the continued passage of high fertility cohorts in the working ages (and increased female workforce participation). However, in Singapore there is a rapidly declining growth of the labor force and in Indonesia and Thailand the rate of growth has begun to decline in the final years of this century.

Hence the demand for education services and job opportunities will continue to increase in most countries in the region, although the pressure will begin to ease in some nations like Indonesia and Thailand so that there will be the chance to improve the quality of education and reduce rates of youth underemployment. On the other end of the age spectrum, growth of the elderly population, from a relatively small base, will be very rapid in all countries. The long-term outlook suggests that during the next half-century the aged populations will grow between three times in Singapore and eight times in Brunei (Table 4.9). This presents significant challenges to the nations in the region since few have social security systems in place and there is evidence that the traditional family-based support systems are not necessarily going to be able to maintain the levels of well-being that they have in the past (Hugo, 1997a).

Population Density

We will now turn to population distribution and the processes that have affected this over recent years. There are huge contrasts in population density within Southeast Asia. By and large these variations reflect the resource endowment of areas, but they are also indicative of the sophistication of the agricultural systems practiced. Although the highest population densities recorded are in large cities, it is clear from Map 4.3 that there are wide contrasts in population density in primarily rural areas. The highest population densities are

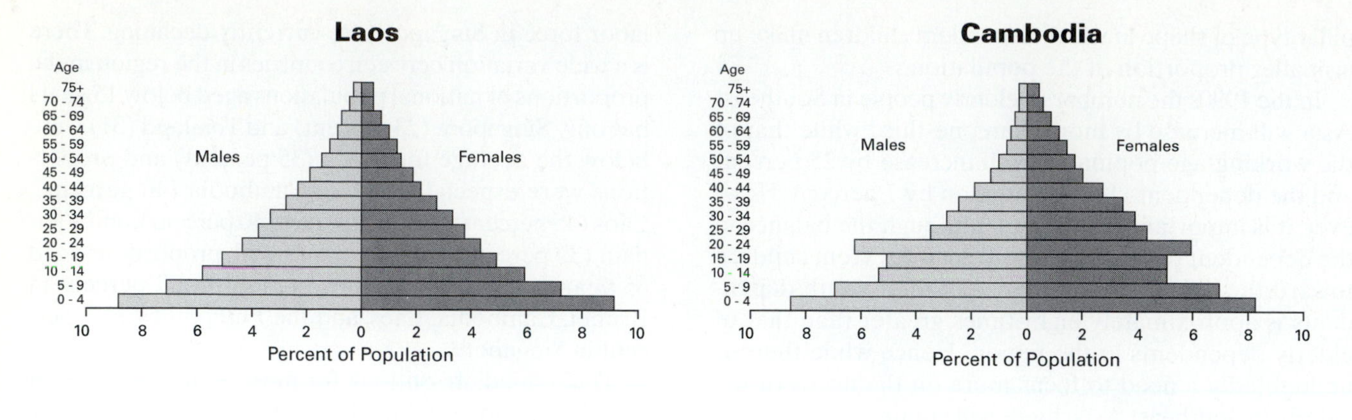

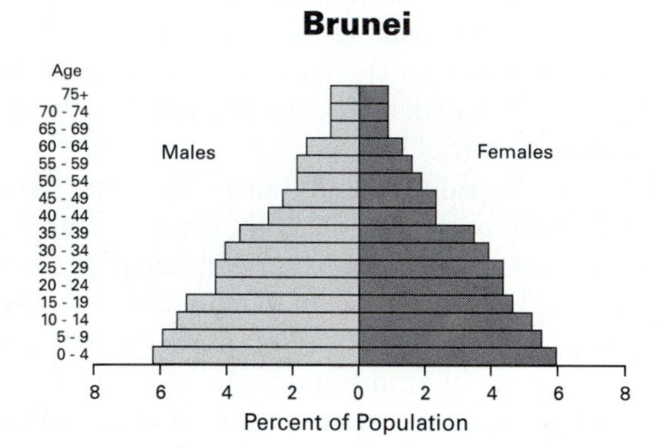

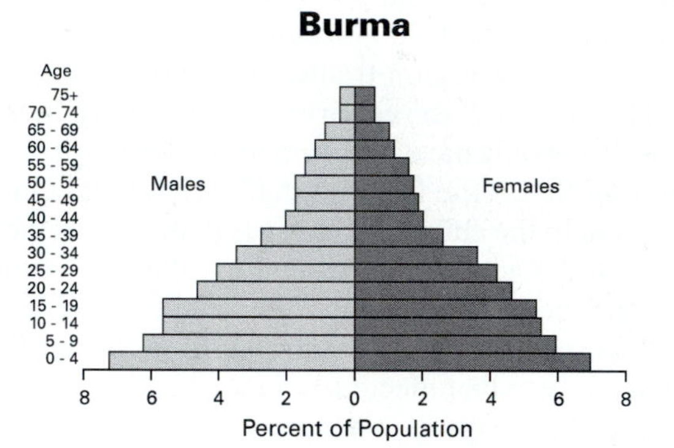

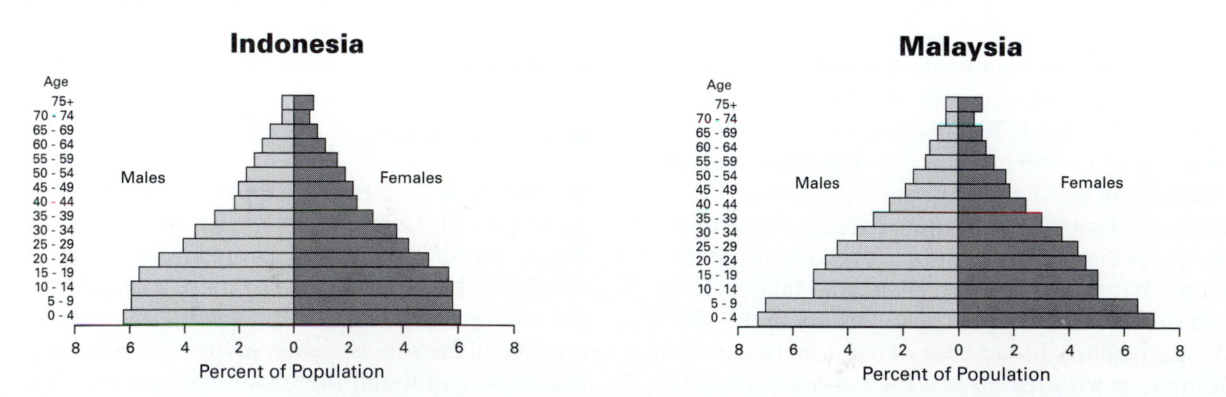

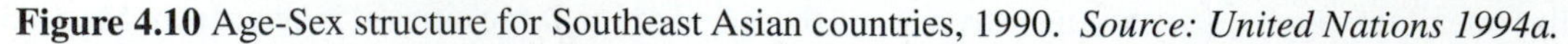

Figure 4.10 Age-Sex structure for Southeast Asian countries, 1990. *Source: United Nations 1994a.*

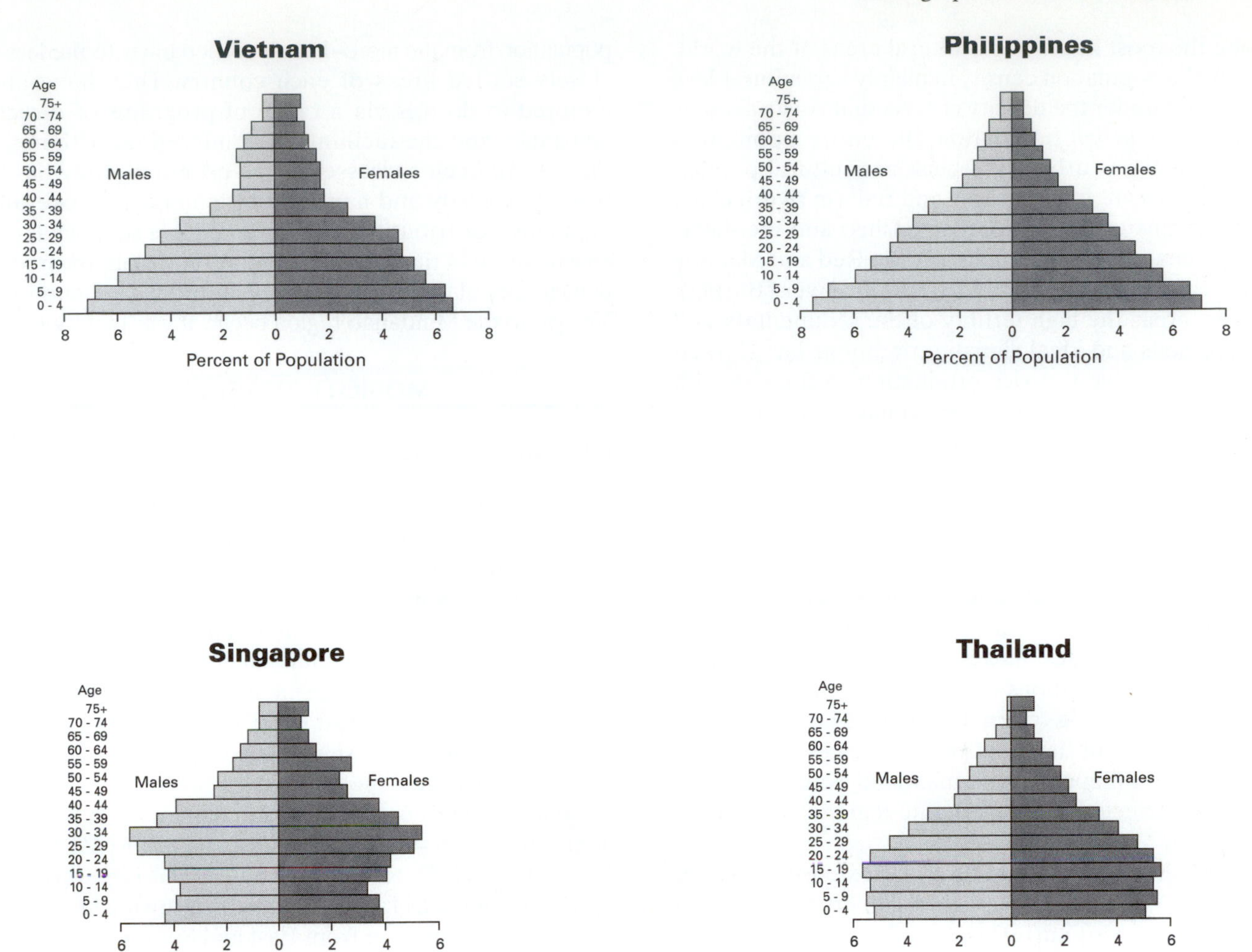

Figure 4.10 (cont.) Age-Sex structure for Southeast Asian countries, 1990. *Source: United Nations 1994a.*

Country	Population 65 Years and Over					Total Population	
	1995 (thousands)	2050 (thousands)	Percent Change	Percent 65+ 1995	Percent 65+ 2050	Median Age 1995	Median Age 2050
Brunei	10	88	780	3.5	17.9	23.6	38.9
Burma	1,894	10,975	479	4.1	11.6	21.3	35.5
Cambodia	269	2,053	663	2.6	7.8	18.0	32.2
Indonesia	8,573	50,183	485	4.3	15.7	23.1	37.7
Laos	146	979	568	3.0	7.5	17.6	32.7
Malaysia	788	5,697	623	3.9	15.0	21.7	37.8
Philippines	2,293	17,270	653	3.4	13.3	20.8	36.7
Singapore	192	781	307	6.7	23.6	32.2	42.9
Thailand	2,932	15,596	432	5.0	19.0	25.5	39.3
Vietnam	3,630	20,946	477	4.9	14.6	21.1	37.7

Table 4.9 Projected growth of population aged 65 years and over, 1995-2050. *Source: United Nations, 1994a and 1994b.*

among the most heavily agricultural areas of the world. Indeed, the population density in mainly agricultural Java is such that under the density criteria that Australia uses to define its *urban* population, the entire population would classify as urban. The peak agricultural population densities are found in Java and Bali (in riverine and coastal plains and intermontane basins) and the deltas of the Menam River (in Thailand), the Red and Mekong Rivers (in Vietnam), and the Irrawaddy River (Burma). In these areas the high fertility of the sedimentary and volcanic soils and ideal climatic conditions has allowed highly productive wet-rice production to flourish with 2–3 crops per year. Geertz (1963a) has argued that wet rice cultivation (*sawah* in Indonesia and Malaysia) has the capacity to respond to increased labor inputs with greater productivity. Hence, it has an absorptive capacity that has allowed high concentrations of agricultural population to build up over time. In other parts of Southeast Asia there are smaller areas of high-quality soil and ideal climatic conditions; these Buchanan (1967) refers to as favorable ecological niches that have allowed similar local high concentrations of population, but not on the scale as in those areas referred to above. Many such areas are to be found in the river valleys, basins, and coastal plains of the Philippines and mainland Southeast Asia. These are reflected in the moderate regional population densities in Map 4.3.

It will also be noted that there are extensive areas of Southeast Asia that have less than 20 persons per sq mi (50 persons per sq km). These predominantly occur in inner and northern mainland Southeast Asia and the outer islands of Indonesia. In many of these areas there is a predominance of poor conditions for agriculture—hilly slopes, leached soils, poor drainage, poor availability of water, and unfavorable climatic conditions. Many of these areas are in the outer islands of Indonesia—Kalimantan, Irian Jaya, Sulawesi, and Sumatra—and northern Thailand and central Burma. Vietnam and Malaysia were traditionally heavily forested. However, they have been subject to massive clearing for cash crops, forestry, and agriculture, and there have been considerable environmental results with massive loss of forest, loss of diversity, soil erosion, and forest fires. These areas are only able to maintain a lighter agricultural population density because the cash crops feature palm oil, rubber, tea, cocoa, and coffee. The eastern part of Indonesia is generally poor environmentally and dryland agriculture is dominant.

It is apparent from Map 4.3 that there are major population density differences between various parts of individual countries. In each of the main countries of the region there have been attempts by governments in both colonial and independence times to even out population distribution within countries. Programs such as Transmigration in Indonesia, FELDA in Malaysia, and government-sponsored settlement and colonization in the Philippines have attempted to shift population from the more-densely settled parts to the less-closely settled areas of each country. They have attempted to do this via a range of programs of direct assistance and the facilitation of migration. Although there have been successes in the programs, they have been very costly and have had little impact on overall population distribution. The Philippines is somewhat of an exception in that once the migration stream from the densely populated rural areas of central Luzon and the Visayas to the Mindanao region began, it was maintained.

MOBILITY PATTERNS

International Migration

One of the most significant developments in Southeast Asia during the last two decades has been an increase in not only the scale of international population movement but also its economic and social significance. Moreover, movements involving permanent and temporary, legal and illegal, forced and voluntary migrations have become more varied and also increasingly complex in their spatial patterning with almost every country in the region becoming involved as a significant origin and/or destination of international movements.

First, there has been a substantial increase in permanent settler migration from Southeast Asia to MDCs in North America, Oceania, and Europe. This is readily apparent in Figure 4.11, which shows the growing tempo of south–north migration from East and Southeast Asia in recent years. This has been facilitated by the official removal of racist restrictions on immigration to these areas in the 1970s and the acceptance of family reunion as a basis for immigration during the last quarter-century. The Philippines has become a major source of such migrants as have Vietnam, Malaysia, and Thailand, but Indonesian migrants have tended to go to intra-Asian destinations. Asian communities including Southeast Asian groups are now significant minorities in most major Euro-American and Japanese cities.

A second important component of Southeast Asian emigration to MDCs has been the flow of refugees, predominantly those from the former Indochinese nations. The total refugee picture in Southeast Asia has been dominated since the reunification of Vietnam in 1975 by outflows from these countries. Between reunification and the early 1990s, some 2 million had landed in neighboring countries (see Map 18.8). However, as Billard (1983, p. 24) points out, "it is possible to count those who finally arrived . . . no one will ever know how many were lost at sea. The wrecks and human remnants washed ashore on the beaches of Southeast Asian countries give only a faint idea of the extent of the tragedy." In any case at least 2 percent of Vietnam's total population has left the country since 1975. During that period the outflow from

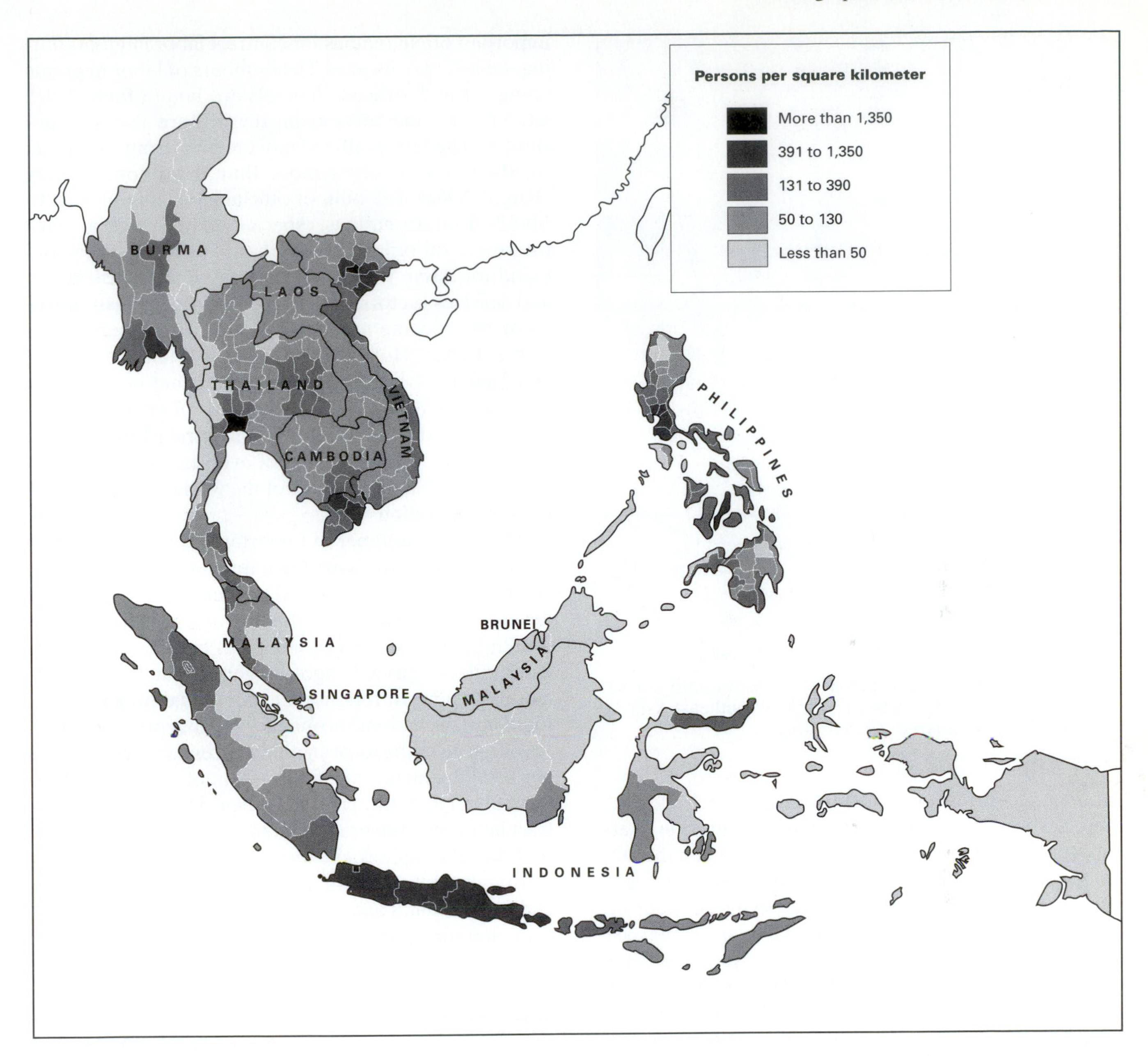

Map 4.3 Southeast Asia population density, 1991. *Source: calculated from recent national censuses.*

Laos and Cambodia has made up more than 5 percent of their total combined national populations. Although almost all of these refugees landed in an Asian country of first asylum, the majority has been resettled in a third country outside of Southeast Asia. With the political change and the opening up of countries such as Vietnam, the flows are being reduced. Nevertheless, the movement out of Vietnam, Cambodia, and Laos were still very much in evidence in the early 1990s with large numbers in China (287,000 from Vietnam; 4,200 from Laos), Hong Kong (60,000 Vietnamese), Indonesia (18,700), Malaysia (12,500), Philippines (18,000), and Thailand (370,000 Cambodians; 59,000 Lao; 15,700 Vietnamese). In recent years there has been a great deal of controversy with many people claiming and being refused refugee status and substantial numbers being repatriated. There are also smaller but locally significant flows that have been important in recent years. Perhaps the most substantial of these has been the refugee outflow from Burma. There have been longstanding flows of Rohinga Muslims into Bangladesh due to persecution in predominantly Buddhist Burma. In 1992, some 300,000 entered Bangladesh (Rogge, 1993, p. 4). There has been movement across the Burma-Thailand border for many years, but crackdowns by the military government in Burma caused large-scale movements into Thailand in 1991–92, involving especially

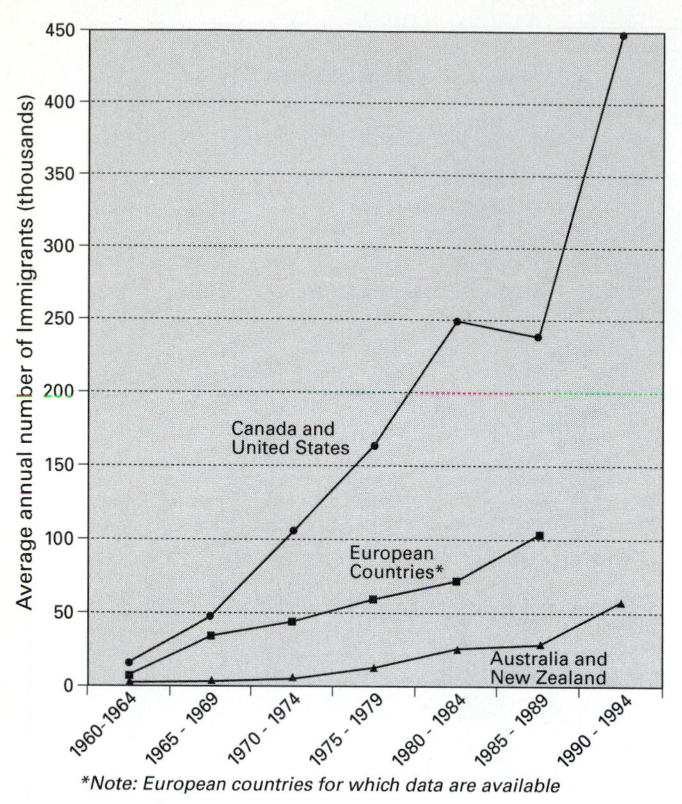

Figure 4.11 Average annual number of immigrants to developed countries from East and Southeast Asia, 1960-1994. *Source: Chapter author.*

students and intellectuals and focusing on Bangkok. There are now more than a half-million Burmese in Thailand, many of them refugees, and they have been the target of government attempts to repatriate them in light of the financial crisis of 1997–98. Other significant refugee flows have included the movement of as many as 200,000 Muslim Filipinos from Mindanao into Sabah in East Malaysia and the much smaller flow of Irianese refugees from Irian Jaya in Indonesia to Papua New Guinea.

In addition to the permanent international migrations, Southeast Asia has been one of the world's most important origin regions for contract labor migrants during the last two decades. The numbers of labor migrants going through official channels are large (Table 4.10), although the numbers moving illegally are also very substantial. The largest illegal movement—from Indonesia to Malaysia—involves more than a million workers (Hugo, 1996a). The bulk of official labor migrants go to Middle Eastern nations, especially Saudi Arabia. Since the rise in oil prices in 1973 there has been a huge demand for Asian workers, especially in the construction and services sectors. More recently, intra-Asian movement has become more significant, with Japan, South Korea, Taiwan, Hong Kong, Singapore, Malaysia, Brunei, and Thailand being particularly important destinations. The growing diversity of destinations of contract workers from Southeast Asia is well illustrated in Map 4.4, which shows the distribution of overseas contract workers from the Philippines, one of the world's leading countries of emigration.

There are a number of important issues surrounding the deployment of workers overseas from Southeast Asian countries. The general consensus is that the net developmental impact of the movement is positive. In the Philippines, export of labor has become the largest earner of foreign exchange—larger than any single commodity export. In Indonesia export of labor is now factored into the national economic planning process (Hugo, 1996a). One of the most significant trends in recent years has been the increasing involvement of women in all international migration flows but especially in that of contract labor. Women predominate in the legal flows out of both the Philippines and Indonesia. Much controversy surrounds this as many of the women are engaged in work as domestic maids and entertainers and have been subject to exploitation (Heyzer, Nijeholt, and Weerakoon, 1994). Other significant trends include the increasing level of undocumented migration (Bilsborrow et al., 1997) and the development of an immigration industry in which gatekeepers of various kinds are crucial to the whole labor migration process (Goss and Lindquist, 1995). The bulk of

Country	Year	Total Overseas	Period	Total Deployed
Burma	1995	415,000	1989-92	35,248
Indonesia	1997	1,000,000	1969-97*	2,072,304
Philippines	1997	6,100,000	1984-95	6,299,556
Thailand	1995	445,000	1973-95	1,529,694
Vietnam	1995	195,000	1997	178,000

*To August 1997

Table 4.10 Official deployment of workers, 1971-1997. *Source: Chapter author, various.*

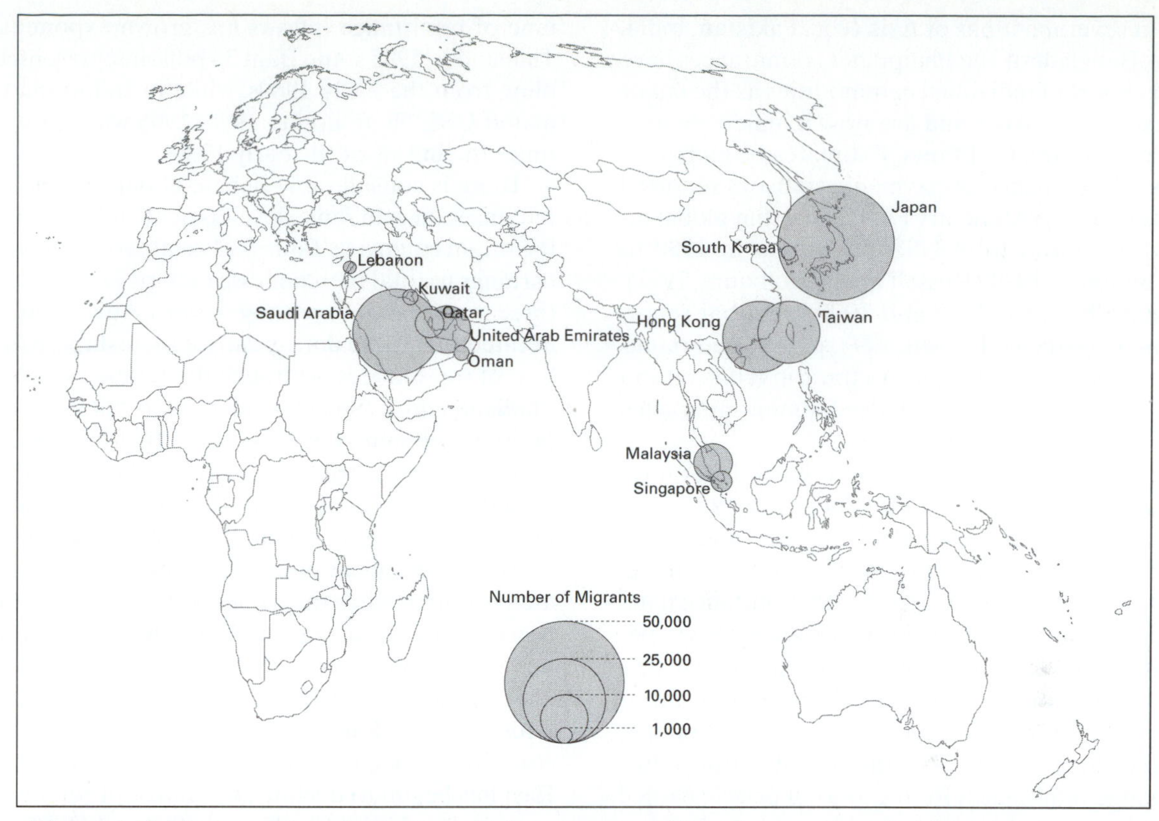

Map 4.4 Philippines: Main destination of labor migrants, 1994. *Source: Compiled from data in Rimban, 1995.*

labor migration in Asia involves unskilled workers, although in the Philippines there are some complaints that the movement robs the country of skilled workers who otherwise could assist in the development of the country (Hugo, 1995a).

Another significant flow of international migration is from MDCs *into* Southeast Asian countries. The rapid economic growth of the region has in some cases outpaced the capacity of countries to supply the skilled human resources needed to sustain and maintain that growth. As a result there has been a substantial influx of skilled workers, many from Europe, North America, and Australasia, to Singapore, Malaysia, Thailand, and Indonesia (Vatikiotis, Clifford, and Macbeth, 1994, p. 32). In Singapore the numbers are so large that official figures are not released, but it is estimated that there are more than 200,000 expatriate workers. In Indonesia, a country with more than 30 percent underemployment (Hugo, 1993a), the officially registered international workers doubled from 20,761 to 41,422 in 1994, and they are only the tip of the iceberg of the foreign workers in the country. It is interesting also that return migration of highly skilled Asians to their home countries is increasing. Indeed several nations in the region have special policies and programs to encourage them to return (Hugo, 1995a). One of the major areas relates to remittances that countries attempt to capture as a source of foreign exchange.

Remittances

It is sometimes overlooked that migration from one country to another initiates (and often is initiated by) a number of other flows of information, ideas, people, goods, and capital. From the perspective of countries sending migrants overseas, the associated capital flows can be outward as when permanent emigrants take their accumulated assets with them or they can be inward when permanent or temporary absentees remit earnings back to families remaining in the country of origin. In Southeast Asia the former does occur as was the case, for example, in the late 1980s when there was some fear in Malaysia of "capital flight" associated with the permanent emigration from Malaysia (Seaward, 1988). However, the major impacts of migration-related capital flows have been due to receipt of remittances from nationals and former nationals in foreign countries.

Remittances were once discounted as being of limited scale and impact, but detailed measurement of money sent back by migrants to their place of origin and of their impacts has led to a significant revision of conventional wisdom. In several nations of Asia (e.g., Pakistan, India, Sri Lanka, Bangladesh, the Philippines) remittances have displaced export of individual commodities as the major foreign exchange earner and are now a major element in international financial flows. Estimates of global remittances using balance-of-payments statistics reported by the International Monetary Fund show that global remittances increased from US$43.3 billion in 1980 to US$71.1 billion in 1990 (Russell and Teitelbaum, 1992). The increase between 1989 and 1990 in global remittances was 17 percent (Russell, 1992, p. 269) and remittances were second in value only to trade in crude oil and were larger than the total world development-assistance commitment.

In Southeast Asian countries the official data on remittances are substantial underestimates of the actual flows because they do not capture money and goods brought home by migrants themselves and sent home with friends. Nevertheless, it is clear that remittance flows are substantial and increasing in significance. This is evident in Figure 4.12, which shows trends in official estimates of remittances into the three major labor exporting nations in Southeast Asia. In the Philippines, for example, Figure 4.12 shows that in 1994 official foreign exchange remittances from export of people earned US$3.28 billion—a third the size of *total* merchandise exports from the country. The rapid increase of remittances in recent years is evident in the diagram, with an increase of 32 percent between 1993 and 1994. When it is considered that the official data do not capture money and goods sent and brought home by Filipinos overseas through unofficial channels, the central significance of international migration in the Philippines economy can be appreciated. The role of unofficial channels can be seen from the fact that in 1990, when the Philippine National Bank extended its services to Italy, Amsterdam, Germany, and Madrid, official per capita labor income from Europe more than doubled above 1989 levels (Russell, 1991, p. 20). Similarly, in Thailand and Indonesia the volume of remittance inflows has grown exponentially. In Thailand in 1995 some Baht 34 billion represented a doubling from the early 1980s, while in Indonesia the estimated US$356 million in 1994–1995 was more than ten times the inflow of the early 1980s.

There is considerable debate about the impact of remittances on economic development and social change in the origin countries. Table 4.11 relates official remittance earnings to total merchandise exports and imports in the three major labor exporting nations. Again it must be reiterated that these data greatly underestimate the actual flow of remittances. Although the relative significance in Thailand and, especially, Indonesia is not as great as that in the Philippines, it will be noted that in both countries there has been a significant increase above the 1980–1992 period in the ratio of remittances to exports and imports in terms of foreign exchange officially recognized.

The origins of remittances to Southeast Asian countries are becoming more diversified as the number of destination countries is increasing. Again this is well exemplified in the Philippines where Map 4.5 shows the main origins of remittances officially recognized in 1994. Clearly the United States dominates the pattern of remittances accounting for some 57 percent of the total. The Middle East has begun to decline as a source of remittances, accounting for US$130 million in 1994 and US$173 million in 1993. In 1990 the Middle East accounted for 9 percent of remittances but in 1994 this had declined to 4 percent. On the other hand, Asian origins provided US$75.4 million (6 percent) of remittances in 1990 but US$381.4 (13 percent) in 1994.

We still know very little about the impact of remittances on development in the sending areas of Southeast Asia. In the past there has been a tendency to trivialize their impacts and to write them off as being spent only on consumption that was often to the benefit of regions or countries other than the origin areas where the goods purchased were manufactured. However, these views are being strongly challenged in em-

Country	Year	Workers Remittances (r)	Total Merchandise (Millions of Dollars)		(r/x)100	(r/m)100
			Exports (x)	Imports (m)		
Indonesia	1980	33	21,908	10,834	0.2	0.3
	1992	264	33,815	27,280	0.8	1.0
Philippines	1980	421	5,744	8,295	7.3	5.0
	1992	2,222	9,790	15,465	22.7	14.4
Thailand	1979	1991	5,240	7,158	3.6	2.7
	1992	1,500	32,473	40,466	4.6	3.7

Table 4.11 Workers' remittances relative to exports and imports, 1980-1992. *Source: Chapter author, various.*

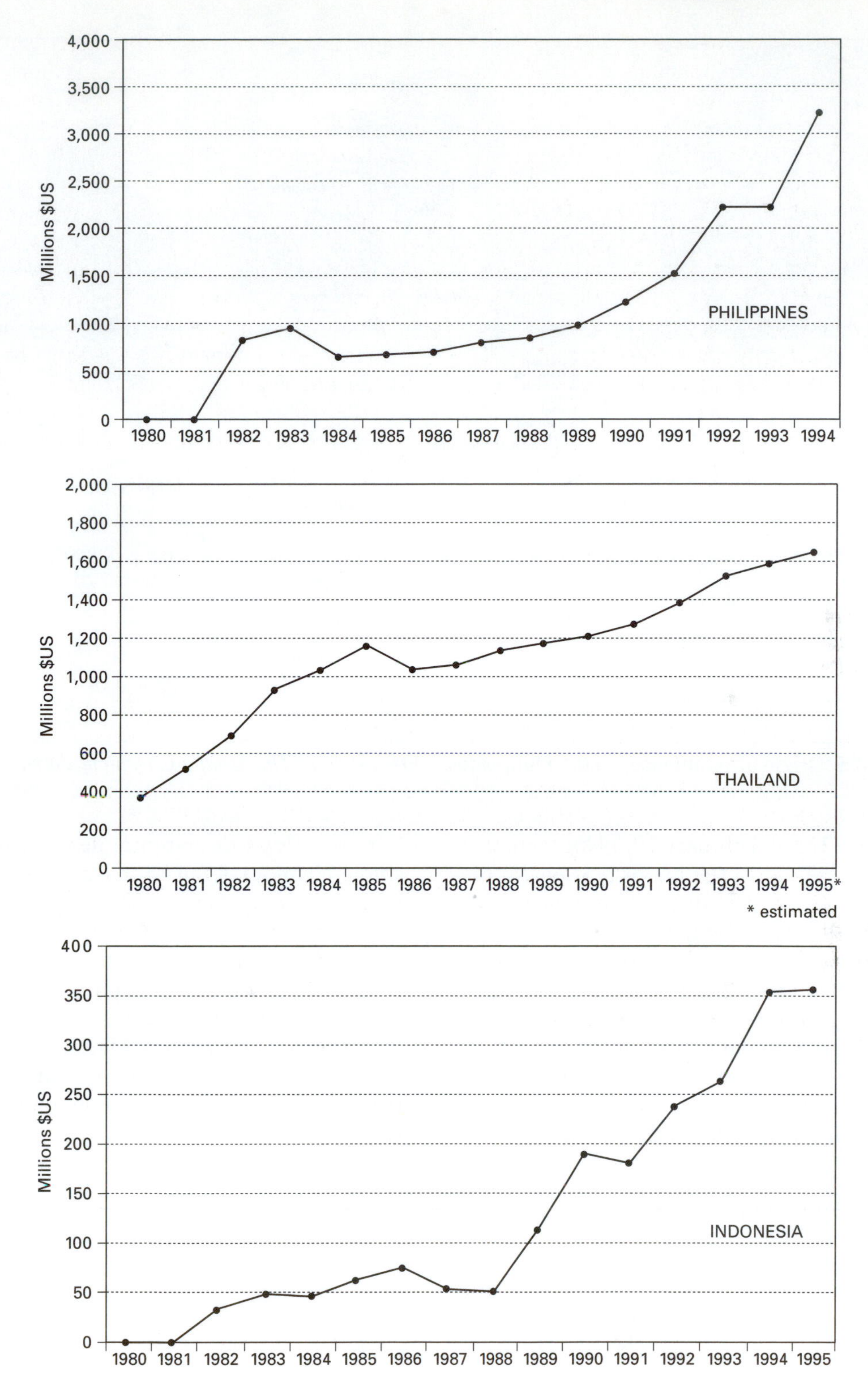

Figure 4.12 Growth of remittances to the Philippines, Thailand, and Indonesia 1980-1995. *Source: Chapter author, various.*

pirical studies. Certainly a high proportion of remittances are expended on consumption, especially upon purchase and improvement of housing. However, it is clear that these expenditures have substantial second- and third-round multiplier effects in the areas to which migrants return. In the Philippines it is estimated that some 8 million jobs are supported by remittance income (*Manila Chronicle*, July 18, 1995). Indeed the *Far Eastern*

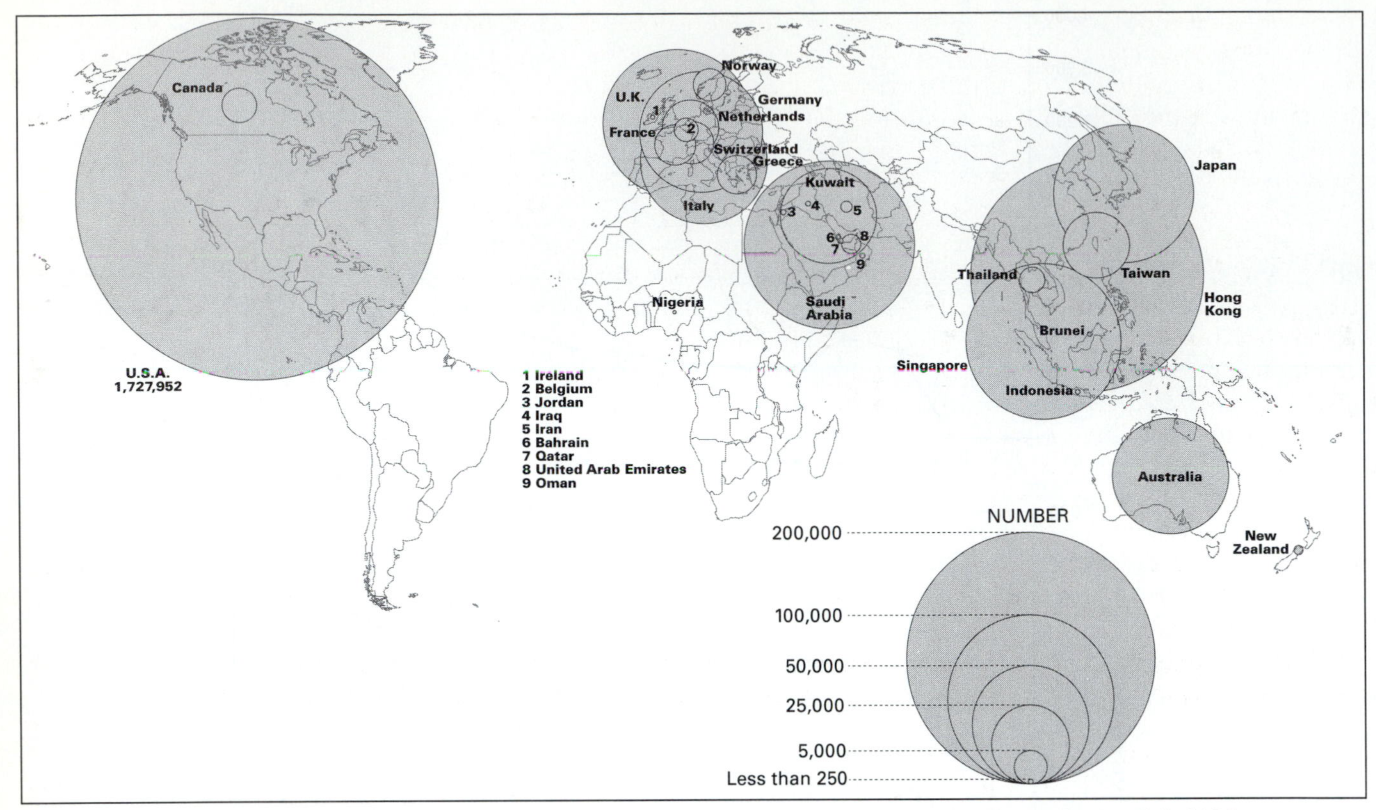

Map 4.5 Origin of remittances to the Philippines, 1994. *Source: The Manila Chronicle, April 5, 1995.*

Economic Review (Tiglao, February 29, 1996c) reported that remittances were one of the major reasons for the turnaround in the growth rate of the Philippines gross domestic product (GDP) as Table 4.12 shows.

Moreover, in most cases the expenditure and education of siblings and children, of purchase of agricultural land, and in setting up small enterprises is significant. Myths regarding the minimal development impacts of remittances have recently been debunked by Brown and Foster (1995). Nevertheless, our knowledge in this area remains limited (Russell, 1992) and the impact of remittance flows on national development needs to be studied in more detail. Moreover the regional impact of remittances also need to be studied. In Indonesia, for example, West Nusa Tenggara is one of the nation's poorest provinces, and its 10,000 migrant workers in Malaysia send back Rupiah 120 billion each year—substantially greater than the total provincial budget of Rupiah 80.4 billion (*Indonesian Observer*, March 18, 1995). Of course remittances from urban based migrants to their rural based families *within* Southeast Asian nations can be treated similarly.

		Billions of Dollars					
Administration	Average annual GDP growth rate	(a) Capital flight (-)/ return (+)	(b) Remittances from overseas workers	(c) Net direct foreign investment	(d) Portfolio investment inflow	(e) Foreign loan disburse-ments	TOTAL (a) to (e)
Marcos 1971-80	5.8%	-5.8	1.9	0.5	*	12.9	9.5
Marcos 1981-85	-1.1%	-2.1	3.7	0.3	*	11.6	13.5
Aquino 1986-91	3.8%	+4.1	6.0	0.5	0.9	10.9	22.4
Ramos 1992-95	3.0%	n.a.	10.7**	2.2	9.4**	18.8**	41.1

** Negligible, included in (c)*
***Figures to September, 1995.*

Table 4.12 Philippines: Sources of foreign exchange and GDP growth, 1971-1995. *Source: Tiglao, Feb. 29, 1996, p. 26.*

There has also been an increase in other forms of movement, especially that of short duration for business, tourism, and study. In Australia alone the numbers of full fee-paying tertiary students (the bulk of them from Asia) has increased from 2,168 in 1986 to 63,013 in 1993.

Circulation

To even the most casual observer of Southeast Asia during the last two decades, one of the most dramatic changes has been the increase in the levels and complexity of personal spatial mobility. The friction of distance on people's ability to range over a large area in search of work and to participate in other activities has been greatly reduced with considerable implications for social and economic change. Unfortunately, data collection relating to population movement remains poor within the region. Nevertheless, we will attempt to illustrate this important change (Table 4.13) by showing the rapid increase in the number of passenger vehicles per 1,000 residents in several Southeast Asian countries, except in the Philippines where the 1980s were a period of economic decline and political disruption (Table 4.13). Indeed, there has been an even more rapid increase in the numbers using public transportation (especially buses and minibuses) and motorcycles (Hugo, 1997a). This has produced a pattern whereby more and more Southeast Asians have developed a range of mobility strategies that allow them to take advantage of job opportunities distant from their homes without having to make a permanent relocation of their home and family (Leinbach, 1983a). The time and money costs of travel have reduced so much that people are able to, for example, live in rural areas where they can minimize risk by having rural work, take advantage of cheaper costs in the village for housing and food, and bring up their children in more traditional rural surroundings, while still earning in urban areas.

These strategies can be divided into two categories. The first is commuting, which involves people traveling long distances on a daily basis to obtain work at a distant location. This is particularly common in the rural areas surrounding cities. Second, circular migrants (or circulators) from more distant areas can move on a regular basis, ranging from a week to a number of years, from one area to work in another area. Both types of movement are especially common around major urban centers, but also there are rural-to-rural commuters and circular migrants (Leinbach and Suwarno, 1985).

Country	Passenger Cars per 1,000 Residents 1980	1991
Brunei	220	436
Indonesia	4.3	8.1
Malaysia	50.7	113.2
Philippines	9.7	7.3
Singapore	92.5	110.2
Thailand	8.8	22.9

Table 4.13 Number of passenger cars per 1,000 population, 1980 and 1991. *Source: ASEAN, 1997, p. 124.*

Long-distance commuting is usually associated with MDCs, but it is of enormous importance in contemporary Southeast Asia and is increasing in significance with massive improvements in transport, especially public transport. Nevertheless, commuting long distances using private vehicles (especially bicycles and motorcycles) is of major importance around major cities. To take one example, commuting has become a massive issue around Bangkok where in the year 2000, 3.7 million of the Bangkok metropolitan area's anticipated population of 10.8 million will be living outside Bangkok itself. Similarly in Indonesia, 11 million of a total of more than 20 million in 1995 in the Greater Jakarta area (Jabotabek) lived outside the capital city district of Jakarta. Clearly with the bulk of job opportunities still in the main capital cities, there is massive daily commuting by bus, train, bicycle, motorcycle, minibus, and car from the peripheries. Commuting scale and distances have increased greatly. These have produced enormous increases in pollution with substantial environmental and health consequences (e.g., see Punpuing, 1997). Moreover, these workers spend a large proportion of their wages and working days on commuting. Also, rural dwellers are ranging much more widely not only to commute to urban areas but also to other rural areas to work.

Circular migration to urban areas and to other rural areas is increasingly being adopted by rural dwellers in Southeast Asia. In such areas off-farm income has been a critical element of rural based households (Leinbach and Watkins, 1998). In the rural areas of Java, Thailand, and the Philippines it would appear that almost one-half of rural households have some off-farm income. This comes either from remittances from within or outside the country, commuting or circular migration of family members. Why has this form of mobility become so significant in rural Southeast Asia?

- This type of mobility strategy is highly compatible with work participation in the urban informal sector because the flexible time commitments allow time to circulate between urban area and village. Similarly, the ease of entry to the urban formal sector is a factor.
- Participation of circular migrants in work in both the urban and rural sectors spreads the risk of the family's income by diversifying the families portfolio of income-earning opportunities. This was recently demonstrated

with the onset of the 1997–98 economic crisis when many rural migrants in cities in Indonesia and Thailand returned to their villages.

- The cost-of-living in urban areas (especially the megacities) is considerably higher than in rural areas so that keeping the family in the village and earning in the city while spending in the village allows earnings to go further.
- Many urban jobs, especially those in the informal sector, can be readily combined with regular visits to the home village.
- The transport system in Southeast Asia has been greatly improved in cost, diversity, and speed, making shuttling between village and city without massive costs or loss of time physically and financially possible (Leinbach, 1983a).
- Job options in the village, especially during the seasonal increases in demand for labor (such as harvesting time) are able to be kept open.
- Many urban-based employers in the informal and formal sectors provide barracks-type accommodation for circular migrant workers, especially in the megacities.
- Often the movement is part of a family-based labor allocation strategy in which some family members are sent out of the village to contribute to the village-based family's income.
- In many cases there is a social preference for living and bringing up children in the village of origin where there are perceived to be fewer negative nontraditional influences.
- Social networks are crucial in the development of this form of migration. Most circular migrants make their initial movements with other experienced migrants from their village or join family and friends who are already established at the destination. They often share accommodations and/or jobs with other circular migrants from their village of origin.
- As is the case with international migration, recruiters, and middlemen of various types have played a significant role in the increase in circular migration. They are especially active in the recruitment of rural workers to work in the urban-based construction sector.
- In recent years females have become heavily involved in circular migration. Up until the 1970s males predominated in this movement but with increased education, social change, development of social networks to protect women at the destination, and improved transport women have become more heavily involved. A large amount of internal (and international) circular migration in the region involves young women moving to cities to work as domestic workers.

Circular migration was identified at an early stage as being an important mobility strategy in Southeast Asia (Goldstein, 1978) but it has increased exponentially during the last two decades with the increasing significance of the forces identified above. As with international labor migration in Southeast Asia, we are unsure as to whether these temporary migrants eventually settle permanently at the destination.

Internal Migration

A major part of the mobility transition in the region has been an increase in the extent to which people are moving their place of permanent residence. This is difficult to quantify because migration questions included in censuses in the region generally only capture a selective subset of longer distance more or less permanent internal migration (Hugo, 1982). However, as limited as those data are, they do show increased rates of migration over the last few decades. To demonstrate this, Table 4.14 shows the increasing proportion of Malaysians who have moved

Year	Definition of Migration	Number	Percent of Population
1957	Living outside state of birth	515,410	10.0
1970	Living outside state of birth	953,652	12.0
1980	Living outside state of birth	1.9 million	13.8
1991	Living outside state of birth	3.2 million	18.2
1970-80	In 1980 living outside state of residence in 1970	1.1 million	10.0
1975-80	In 1980 living outside state of residence in 1975	783,000	6.8
1986-91	In 1991 living outside state of residence in 1986	1,145,600	7.5
1985-86	In 1986 living outside state of residence in 1985	312,903	2.4
1970	Ever moved between districts	n.a.	27.9
1980	Ever moved between districts	n.a.	45.4

Table 4.14 Malaysia measures of migration, 1957-1990. *Source: Hugo, Lim, and Narayan, 1989, p. 47; Malaysia, Department of Statistics, 1995, p. 90.*

Year	Country	Type of Migration	Males % of Total	Females % of Total
1971	Indonesia	Percent ever lived in another province	6.29	5.06
1985	Indonesia	Percent ever lived in another province	8.37	7.29
1990	Indonesia	Percent ever lived in another province	10.62	9.03
1995	Indonesia	Percent ever lived in another province	11.19	10.00
1975	Philippines	Percent lived in other province in 1970	3.80	4.07
1990	Philippines	Percent lived in other province in 1985	4.15	4.64

Table 4.15 Indonesia and Philippines: Changes in long distance migration, 1971-1995. *Source: Hugo, 1997a; Jackson, forthcoming.*

within the country during the 1957–1990 period. There is a clear pattern of increase in the rate of longer distance permanent migration in the country and this undoubtedly is the case also for shorter distance permanent relocations within states. Similar patterns were observed in other countries in the region as Table 4.15 shows. Table 4.15 also shows that female migration has increased in significance in the region's internal migration in recent years.

In terms of the spatial patterning of internal migration, one of the dominant trends is redistribution of population from rural-to-urban areas, although detecting this is made difficult by migration data in several nations not identifying whether migrant origins are urban or rural. Unraveling the pattern of rural-to-urban migration and its role in urbanization in Southeast Asia is daunting if not impossible. Huge variations in the definition of what constitutes an urban area not only between countries but within countries at different points in time frustrate any attempt at precision. Nevertheless, the rapidity and enormity of urban growth in the region cannot be doubted. It is one of the world's fastest urbanizing regions. The tempo of urbanization is evident in Table 4.16 that shows that whereas in 1980 less than a quarter of Southeast Asians lived in urban areas, by 1994 this had risen to a third and by the end of the second decade of the twenty-first century more than one-half of all Southeast Asians will be city dwellers. However, there is substantial variation between the nations of the region in the degree of urbanization, ranging from the city-state of Singapore to Cambodia, Laos, Thailand, and Vietnam, where one-fifth of the population are classified as urban (Table 4.16).

The higher rates of growth of urban compared to rural populations that have been experienced during the last quarter-century are potentially the result of any or all of the following:

- Areas formerly classified as rural being reclassified as urban.
- Fertility of urban-based women being higher than their rural based counterparts.

Country	1950	1960	1970	1975	1980	1994	Projected 2025
Brunei	26	43	67	70	76	58	73
Burma	16	19	23	25	29	26	n/a
Cambodia	10	11	12	13	15	20	44
Indonesia	12	15	17	18	21	34	61
Laos	7	8	10	11	15	21	45
Malaysia	21	25	27	28	30	53	73
Philippines	27	30	33	34	39	53	74
Singapore	80	78	75	74	100	100	100
Thailand	10	13	13	14	17	20	39
Vietnam	12	15	18	20	19	21	39
Southeast Asia	15	18	20	21	24	33	55

Table 4.16 Percent of population living in urban areas for select years, 1950-1994 and projected 2025. *Source: United Nations, 1980, p. 161; Population Reference Bureau, 1983; United Nations, 1994b.*

- Mortality of urban-based people being lower than their rural-based counterparts.
- A net redistribution of people from rural to urban areas through migration.
- A net gain of migrants from overseas is greater than in rural areas.

Mortality differentials in Southeast Asia are favoring a more rapid growth of urban population than rural populations. However, this differential is probably only sufficient to counterbalance the effect in the opposite direction caused by lower fertility levels prevailing in urban than in rural areas. Hence, overall natural increase (i.e., births minus deaths) is not contributing significantly to *urbanization*, that is, the increasing *proportions* of national populations living in urban areas. On the other hand, natural increase is contributing significantly to *urban growth* in Southeast Asia because fertility levels are still high relative to MDCs and the populations of such cities are quite young, with large proportions in the childbearing age groups. Hence Table 4.17 shows that over the entire Southeast Asian region natural increase contributed

	1980-1985		1990-1995		2000-2005 Projected	
Region/ Country	Natural Increase	Migration and Reclassification	Natural Increase	Migration and Reclassification	Natural Increase	Migration and Reclassification
Asia	**37.8**	**62.2**	**40.7**	**59.3**	**42.6**	**57.4**
East Asia	**21.9**	**78.1**	**28.0**	**72.0**	**30.3**	**69.7**
China	15.9	84.1	23.7	76.3	27.2	72.8
Hong Kong	61.0	39.0	58.2	41.8	75.3	24.7
Japan	81.5	18.5	78.5	21.5	70.3	29.7
Mongolia	93.6	6.4	85.6	14.4	73.4	26.6
Rep. of Korea	35.5	64.5	38.8	61.2	50.7	49.3
Southeast Asia	**49.1**	**50.9**	**44.9**	**55.1**	**41.7**	**58.3**
Burma	110.0	-10.0	63.2	36.8	44.5	55.5
Cambodia	70.9	29.1	49.5	50.5	30.6	69.4
Indonesia	35.2	64.8	37.0	63.0	36.7	63.3
Laos	43.8	56.2	44.7	55.3	43.8	56.2
Malaysia*	22.0	78.0	38.0	62.0	40.0	60.0
Philippines	66.0	34.0	62.4	37.6	57.0	43.0
Singapore	100.1	-0.1	100.1	-.01	98.9	1.1
Thailand	39.6	60.4	31.4	68.6	31.2	68.8
Vietnam	1.7	28.3	50.5	49.5	38.1	61.9
South Asia	**56.0**	**44.0**	**54.4**	**45.6**	**47.0**	**53.0**
Afghanistan	-589.1	689.1	30.4	69.6	46.1	53.9
Bangladesh	37.6	62.4	39.9	60.1	41.9	58.1
Bhutan	41.5	58.5	34.8	65.2	34.1	65.9
India	55.6	44.4	52.1	47.9	44.8	55.2
Iran	56.0	44.0	81.4	18.6	71.0	29.0
Nepal	31.9	68.1	32.5	67.5	32.6	67.4
Pakistan	69.0	31.0	67.8	32.2	54.9	45.1
Sri Lanka	170.8	-70.8	66.8	33.2	33.6	66.4

Note: The contribution of natural increase is calculated by assuming that the urban population has the same rate of natural increase as the national or regional population. The category of migration and reclassification is calculated as a residual.
**Government data.*

Table 4.17 Components of urban growth by country or region (percentage of urban growth). *Source: UNESCAP, 1993b.*

49.1 percent of urban growth in the 1980s and 44.9 percent in the 1990s. There is considerable variation between countries in the extent to which natural increase is contributing to urban population growth and reclassification of rural areas.

It is apparent from Table 4.17 that net migration gains explain virtually all of the increases in levels of urbanization and provide more than one-half of urban population growth. Unfortunately, it is very difficult to separate these two components in the data that are available for many Asian countries. It is apparent that reclassification has been significant in many countries. For example, it was reported in Indonesia that the number of *desa* (the basic building block area units for urban and rural classification) that were classified as urban doubled between the 1980 and 1990 Censuses from 3,500 to around 6,700. While in such cases many people will have been classified as rural in 1980 and urban in 1990 without having changed their place of residence, it must be remembered that the reclassification has almost certainly been triggered by a substantial in-migration and increase in population density. Hence, reclassification itself often involves significant rural-to-urban migration.

Although the data we have available are very limited, it can be concluded that rural-to-urban migration has played a major role in both the rapid urbanization and urban growth in Asian countries over the last two decades. Again, to take Indonesia as an example, Table 4.18 shows that urban Indonesians are close to their rural migrant origins. For example, in 1995 the 17.7 percent of urban dwellers in Indonesian cities were migrants from another province. Because intraprovincial migrants outnumber their interprovincial counterparts by up to five to one (Hugo, 1975), it is apparent that migrants or their offspring make up the majority of Indonesian urban dwellers. This is apparent in the data for Jakarta, which is a totally urban area.

There can be no doubt that migration within Southeast Asian countries has become increasingly focused on the major metropolitan regions within the various countries. To take Peninsular Malaysia as an example, Map 4.6 shows the patterns of interstate migration in the 1970s, which include:

- The megacity of Kuala Lumpur and its surrounding state of Selangor was a major focus of migration. Clearly much of this movement was rural-to-urban in nature.
- There was a significant movement into the growing agricultural state of Pahang. Much of this was associated with the FELDA schemes whereby the government paid for settlers in the more densely settled agricultural areas such as those on the west coast to open up new, largely cash-crop settlements in newly cleared areas of Pahang.

However, by 1986–91 the pattern had changed significantly with the overwhelming trend toward a dominance to the megaurban region of Kuala Lumpur and Selangor and Pahang no longer a significant destination of interstate migrants. A similar pattern is repeated in Indonesia (Hugo, 1997a) and the Philippines (Jackson, forthcoming).

Southeast Asia has seen some of the world's most ambitious attempts to resettle rural populations from densely settled areas in agricultural colonies located in less-densely settled parts of the country. Indonesia, Vietnam, Thailand, the Philippines, and Malaysia have all had government-sponsored resettlement programs and they remain active in most of these countries, albeit at a lesser scale than earlier in the postwar years (Leinbach, 1989). The large countries in the region are characterized by substantial interregional differences in population density, although those often reflect just as large differences in resource endowments. Nevertheless, there does remain some limited potential in some areas for agricultural settlement expansion, although it is unlikely to be at levels that prevailed in earlier years. Government-sponsored settlement programs such as the Transmigration Program in Indonesia have been criticized for their high costs and environmental and social impact in destination areas, but they have resulted in a substantial redistribution of population in many instances. In Vietnam the government initiated an ambitious program to redistribute people from north

Year	Percent Born in Another Province		
	Indonesia	Indonesia Urban	Jakarta
1961	n.a.	n.a.	48.4
1971	6.1	n.a.	41.1
1980	7.3	17.2	40.7
1990	9.1	17.3	39.0
1995	10.6	17.7	37.8

Table 4.18 Indonesia, Indonesia Urban, and Jakarta: Percent of population ever lived in another province. *Source: Indonesian censuses, chapter author.*

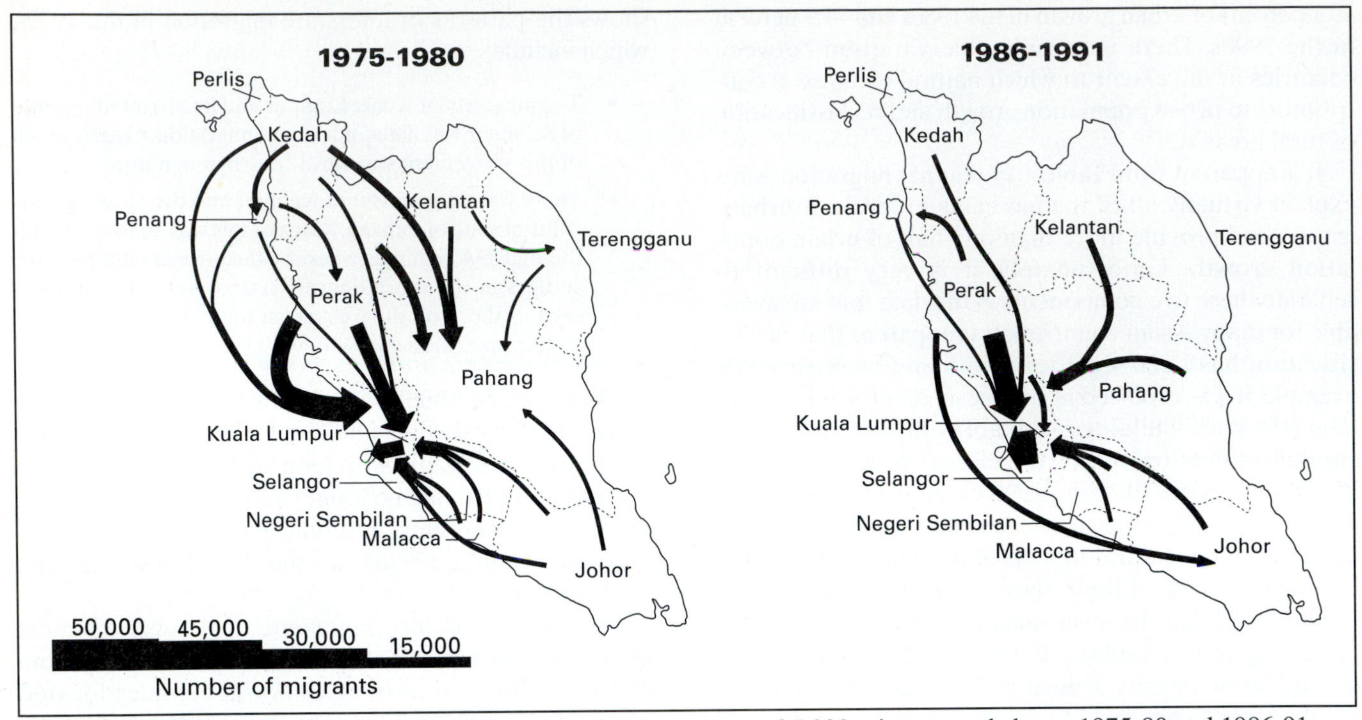

Map 4.6 Peninsular Malaysia: Net migration streams between states of 5,000 migrants and above, 1975-80 and 1986-91.
Source: Department of Statistics, Malaysia, 1995.

to south and to decongest southern cities (Desbarats, 1987).

The increase of international, internal, and circular migration has a number of important implications for development in Southeast Asia. It not only results in increasing the population at the place of destination and decreasing it at the origin, but because migration is a highly selective process the impact of the migration on political, social, and economic change can be out of proportion to the sheer numbers involved. For example all migration in Asia is selective of young adults, which results in a younging of the population in destination areas and an aging of that in the origin. Moreover, migration, especially internal migration, is selective of the most educated and the most entrepreneurial and those with leadership ability. Hence, major cities in Southeast Asia tend to have younger, more educated populations than rural areas. This may lead to a diminution of social and economic potential in rural areas. Moreover, the accumulation of young, educated people in cities is not only of benefit to urban economies, but when these people are subject to unemployment or are unable to fulfill their aspirations the potential for questioning of the status quo and for conflict appears. Nevertheless, the role of migration in influencing regional development and creating and exacerbating differences between areas is much debated in the region.

CONCLUSION

The last two decades have witnessed enormous demographic and social change in the Southeast Asian region. Moreover, there is little indication that the pace of change will slacken in the near future. Table 4.19 presents the United Nations' projections for the countries of the region into the early decades of twenty-first century. The overall prospect for the region is a reduction of population growth rates, but the table shows that very wide variations in the tempo of growth between countries are anticipated. Indeed, Southeast Asia not only has countries that have completed or almost completed the demographic transition but also several that are still in the earliest stages of that transition. That pattern of diversity and contrast is the defining characteristic of the region. Although it contains some of the fastest growing economies in the world, it also has some of the most lagging nations, and this is reflected in their demographies.

In looking to the future influences on population change in Southeast Asia, there is much to be optimistic about, despite the downturn of 1997–98. The conflicts that have hampered development in several countries have largely been resolved. However, significant poverty remains in many parts of the region and overcoming this must be a major priority. Population developments can both contribute to, and will be substantially influenced by, poverty. Reduction of population growth must remain a priority in several countries in the region. The experience of the last two decades has been that fertility decline can be considerably facilitated through provision of efficient, well-run family planning and maternal child health services and through programs to lift both education levels and the status of women generally. Although many of the population issues that have challenged Southeast Asian policy makers over the last two decades remain—the need to reduce mortality, especially child, infant and maternal mortality, fertility reduction, pressure of growing youth population on education services and increasing job availability, the need for improving the status of women, and rapid growth of urban areas—the coming decades will bring new challenges and opportunities. These include the increasing scale and significance of international migration, the rapid growth of elderly populations, the growth of megacities, pressures of population on fragile environments, a shift from the extended to the nuclear family, and in some areas the spread of the HIV/AIDS epidemic. It is difficult to anticipate how successful the policy makers will be in addressing and overcoming these substantial challenges. However, it is encouraging to consider that very few, if any, commentators writing about and researching Southeast Asian population issues in the early 1970s anticipated or predicted the degree of success that several nations in the region achieved in their attempts to address the population problems of that time. Those efforts have contributed significantly to the region having some of the world's most dynamic economies and most rapidly improving average levels of living.

Country	1950 (Thousands)	Percent per Year	1990 (Thousands)	Percent per Year	Projected 2000 (Thousands)	Percent per Year	Projected 2025 (Thousands)
Brunei	48	4.28	257	2.15	318	1.17	425
Burma	17,832	2.15	41,825	2.12	51,567	1.54	75,604
Cambodia	4,346	1.64	8,336	2.41	10,580	1.85	16,716
Indonesia	79,538	2.12	184,283	1.45	212,731	1.04	275,598
Laos	1,755	2.21	4,202	2.89	5,592	2.10	9,411
Malaysia	6,110	2.72	17,891	2.21	22,263	1.37	31,274
Philippines	20,988	2.76	62,437	2.00	76,091	1.30	105,147
Singapore	1,022	2.47	2,710	0.94	2,976	0.43	3,309
Thailand	20,010	2.54	54,677	1.13	61,202	0.67	72,264
Vietnam	29,954	2.02	66,688	2.03	81,516	1.45	116,958
Southeast Asia	181,603	2.26	443,306	1.70	524,836	1.20	706,706

Table 4.19 Recent and projected population change, 1950-2025. *Source: United Nations, 1994a.*

5 Urbanization

JON GOSS

With the exception of Singapore and Brunei Darussalam, Southeast Asian societies are still predominantly rural, exhibiting relatively low levels of urbanization, that is, with a small proportion of the total population living in urban settlements (Figure 5.1). Until recently they also experienced relatively slow rates of urbanization compared with other parts of the developing world (Ogawa, 1985), but these have increased quite dramatically in the last two decades due to the combined effect of rural to urban migration, relatively high natural birthrates in urban centers, and, to a lesser extent, changes in administrative boundaries that define urban settlements (Figure 5.2).

Urban centers have always exerted a "pull" on rural populations in Southeast Asia, as parents have sent their children to be educated in modern ways, and peasant farmers have sought fortunes in urban markets or temporary employment in urban activities (O'Connor, 1983). The attraction of large cities has increased dramatically, however, with economic development, and particularly the introduction of modern communications and transport technologies, together with a rapid expansion of income opportunities and a rising demand for consumer goods. Successful migrants impress their rural friends and relatives with their urban lifestyle and provide contacts and resources in the city, both reinforcing the desire and providing the necessary means for others to follow. The presence of relatives and friends in the city is vital to the success of many migrants, and established migrants even actively recruit new migrants for employment in their own household or in petty enterprises. Such chain migration is a cumulative process, initially selective of the most informed and ambitious, but expanding to incorporate elaborate migrant networks reaching down the urban hierarchy into remote rural areas, drawing the populations of towns and villages to the city and into the urban economy.

At the same time, people are "pushed" from the rural areas by deteriorating conditions and restricted opportu-

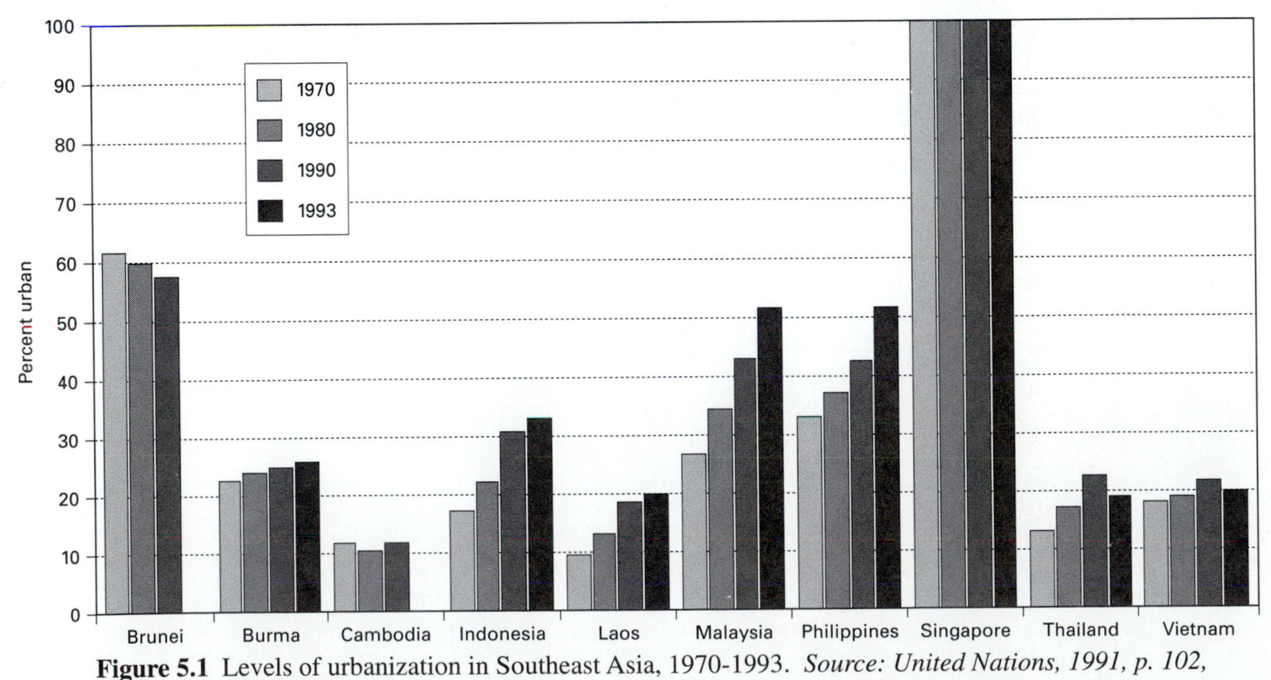

Figure 5.1 Levels of urbanization in Southeast Asia, 1970-1993. *Source: United Nations, 1991, p. 102, 106; World Bank, 1995, p. 222-3.*

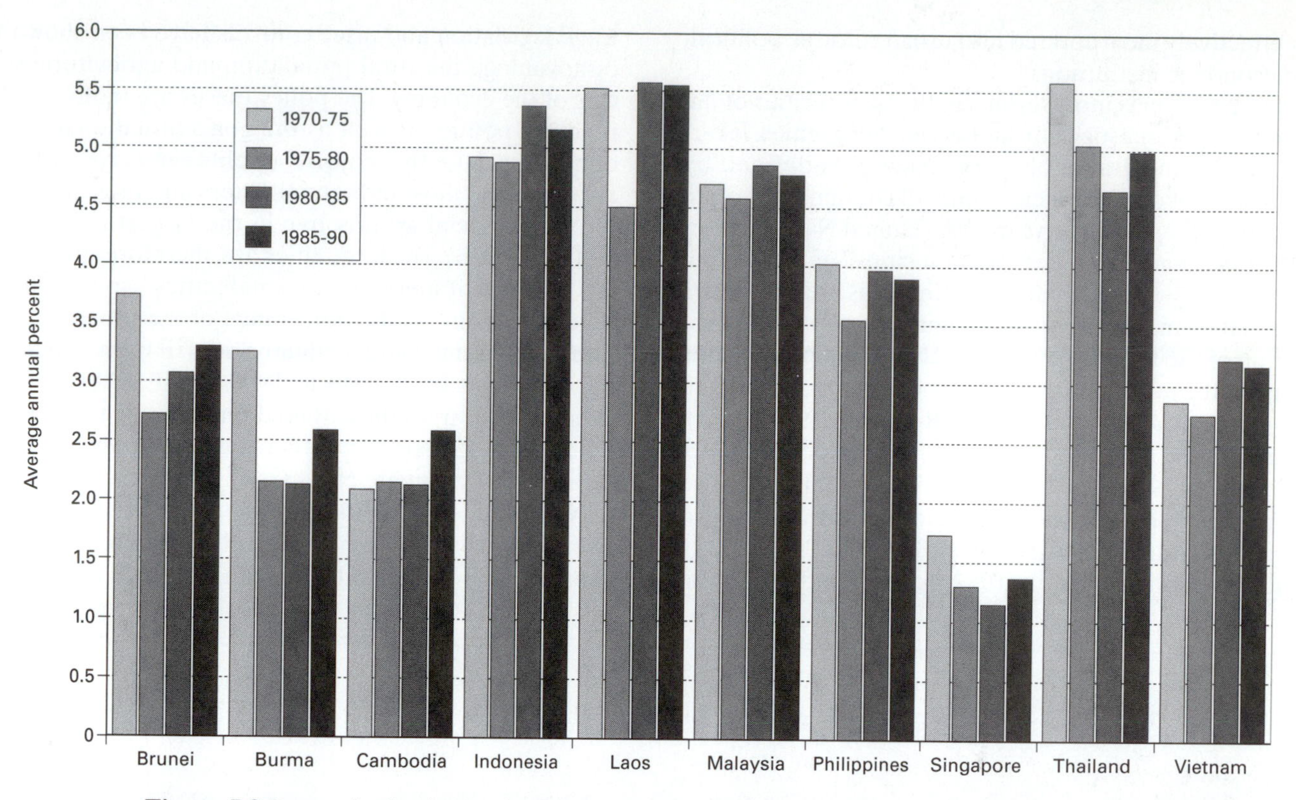

Figure 5.2 Rates of urbanization in Southeast Asia, 1970-1990. *Source: United Nations, 1991, p. 166.*

nity associated with technological changes in agricultural production, landlessness, poverty, and in some cases, political instability (for example, during the communist insurgency or "Emergency" in Malaya, 1948–60; during the nationalist insurgencies in Burma in the 1950s; during the Hukbalahap insurgency in the Philippines in the 1940s and 1950s; during the war in South Vietnam in the 1960s; and during Khmer Rouge guerrilla activity in Cambodia during the 1980s). These conditions are exacerbated by high rates of natural population growth in rural areas.

Partly as a result of the high proportion of first generation migrants in urban areas, population fertility rates in cities are still relatively high, even though they are declining as the industrial economy expands and women in the workforce delay childbearing (Ogawa, 1985, p. 92). Rates vary, but generally natural increase accounts for about one-half of the urban population growth in Southeast Asian countries—during the 1970s, for example, natural growth accounted for 60 percent of the population increase in the Philippines, 45 percent in Indonesia, 49 percent in Malaysia, and 39 percent in Thailand (Ogawa, 1985).

The rapid growth of some cities is also in part the result of changing definitions of the urban area as metropolitan boundaries are extended to incorporate more of the surrounding area. For example, part of the 5.2 percent per year urban growth rate in Indonesia between 1971 and 1980 is an artifact of boundary changes, and the real rate is likely to be nearer 4 percent (Gardiner, 1992, p. 267); up to 10 percent of the growth of Bangkok in the 1970s may have been due to annexation of surrounding areas (Changrien and Stimson, 1992, p. 2); and Kuala Lumpur literally doubled in size in 1974 with the creation of the Federal Territory (Yeung, 1988).

With the obvious exception of Singapore, a modern city-state, the rapid rate of urbanization in most Southeast Asian countries in recent years has generally not been matched by a sustained development of the national market economy or expanded administrative capacity of local governments sufficient to provide secure employment, shelter, and basic services such as sanitation, water, and electricity for the majority of the urban population. The city therefore is unable to fully incorporate its population into a modern urban way of life, leading to the suggestion that Southeast Asian cities are overurbanized (McGee, 1967), or even that they exhibit pseudourbanization (Dwyer, 1968). Moreover, many urban residents are not committed to permanent residence in the city and so return regularly or periodically to their place of origin, often during the periods of peak agricultural labor demand, when they are ill, and for "vacation" from city life. Thus they retain a base in their village to which they desire ultimately to return and live out their lives. Due to the prevalence of circular migration, many migrants effectively live in both rural and urban worlds and in both subsistence and commercial economies, which only reinforces the appearance that the majority of urban residents are not effectively incorporated into urban cultural, political, and economic institutions.

This problem is most serious in the largest cities of the region, and most especially in the three megacities, Jakarta, Manila, and Bangkok—a megacity, as defined by the United Nations, is a city that will contain more than 10 million people by the year 2000 (United Nations, 1986, 1987, 1989)—and in other "million cities" of the region. Map 5.1 shows the location of these as well as other major cities in the region, and Figure 5.3 shows the proportion of the total urban and total national populations living in these cities. Due to their centrality in national and regional culture and politics (Robinson, 1987, p. 172), their integration into the global capitalist economy (Friedmann, 1986), and biases in government policies, the large cities dominate the urban hierarchy. National and regional development strategies (Simmons, 1988, p. 88), education policies (Salih, 1982, p. 155), and minimum wage legislation and price controls have been shown to disadvantage the rural population and agricultural sectors of the economy. The policy biases are motivated by practical politics as well as the goals of national development because the majority of elites and civil servants, as well as the most influential voters and most politically volatile social groups, live in the largest cities (UNESCAP, 1993b). The large cities are, therefore, growing at the expense of medium and small cities—in the Philippines, for example, the percentage of the urban population living in small and medium cities fell from 86 percent to 53 percent from 1948 to 1980 (UNESCAP 1993b). Where these large cities attain dominance over the urban hierarchies of their respective countries, they are often referred to as primate cities (see Box 5.1).

While these large cities are the product of continued polarization within the national urban hierarchy, they are spontaneously decentralizing at the regional level by ex-

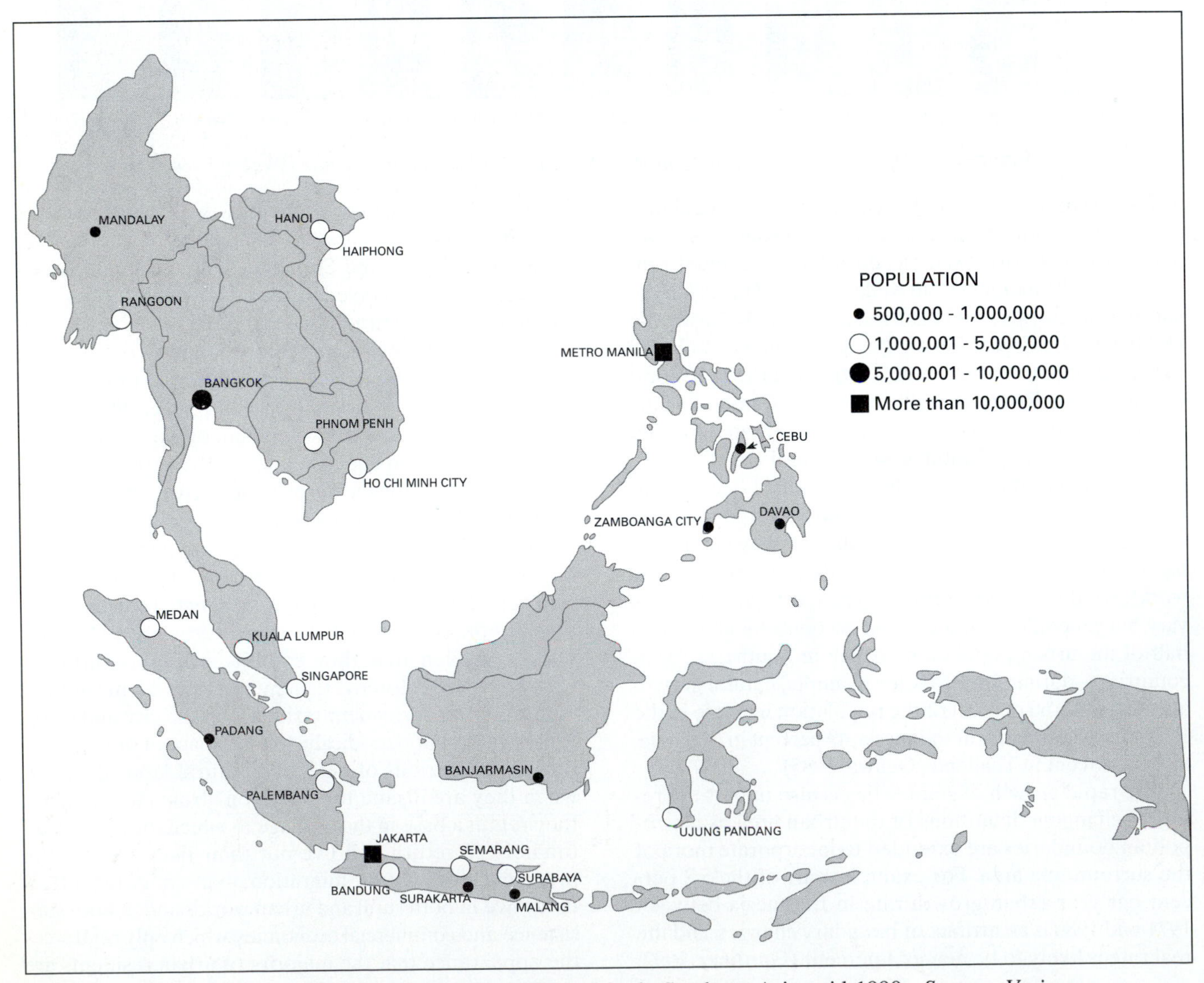

Map 5.1 Distribution of megacities and million cities in Southeast Asia, mid-1990s. *Source: Various.*

BOX 5.1 The Primate Cities of Southeast Asia

Where a single city dominates the country in terms of population and function—such as is particularly the case of Bangkok, Manila, Phnom Penh, Vientiane, and Rangoon—it is said to be a primate city (Table 5.1). Bangkok is the most extreme case in the region and it has, in fact, been called the world's pre-eminent primate city (Sternstein, 1984). Bangkok has grown particularly rapidly in the modern era, due to the rapid capitalization of agricultural production that has created shortages of cultivable land. In addition, the upgrading and expansion of Thailand's transportation infrastructure improved the means for displaced workers to migrate to the city. Bangkok is 34 times the size of Nonthaburi, Thailand's second city, and 51 times the size of Chiang Mai (Ruland, 1992, p. 23); it is responsible for 95 percent of Thai foreign commodity trade and 49 percent of GDP (Rigg, 1991, p. 186); and the Bangkok Metropolitan Area (BMA) generates 77 percent of manufactured goods and 90 percent of all industrial employment in Thailand (Pakkasem, 1987).

Primacy is argued to cause inefficient and inequitable distortions to the national political economy, as well as diseconomies of scale in the primate city itself. In Thailand, Malaysia, Indonesia, the Philippines, and Vietnam, policies have been developed to encourage more balanced growth (Robinson, 1987; Drakakis-Smith and Rimmer, 1982), although the evidence for this inefficiency and the wisdom of such policies are increasingly disputed (Richardson, 1989).

Country	Percent Urban Population in Largest City	Primacy Index*
Singapore	100	n.a.
Vietnam	24	1.50 (1997)
Malaysia	19	1.64 (1991)
Brunei	39	2.22 (1988)
Indonesia	17	3.33 (1990)
Burma	39	4.71 (1983)
Laos	52	5.84 (1992)
Philippines	29	8.83 (1990)
Cambodia	44	12.00 (1991)
Thailand	69	34.43 (1990)

Table 5.1 Indices of primacy. *Source: Various.*
*The Primacy Index is derived by dividing the population of the largest metropolitan area by that of the second largest metropolitan area.

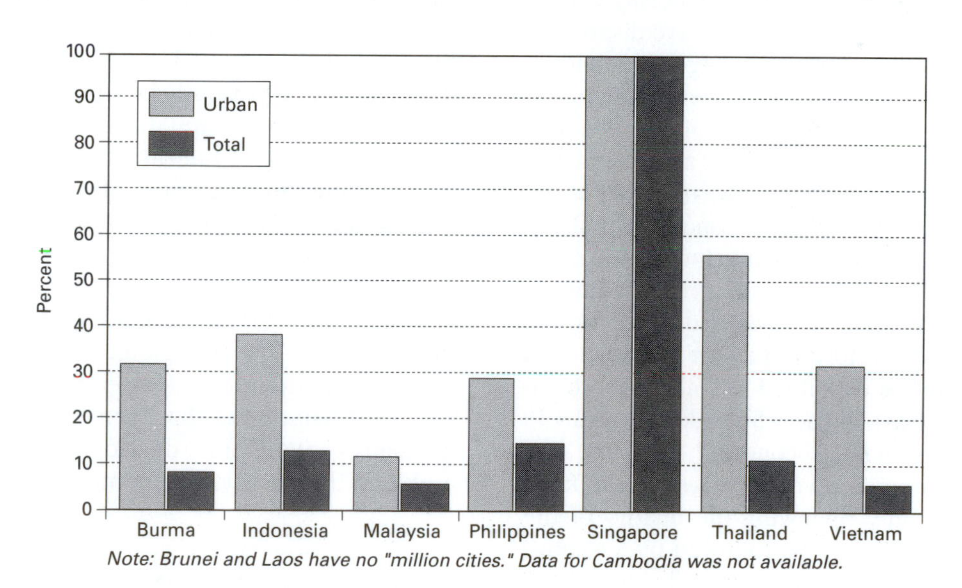

Figure 5.3 Proportion of total urban and national populations living in "million cities." *Source: World Bank, 1995, p. 222-3.*

pansion of industry and residential subdivisions into the surrounding countryside, producing what have been called extended metropolitan regions, or EMRs (Ginsburg, Koppel, and McGee, 1991; McGee and Robinson, 1995). This applies particularly in the case of Bangkok Metropolitan Area (BMA), Metropolitan Manila, and the Jabotabek region (Jakarta, Bogor, Tanggerang, Bekasi). McGee (1989, p. 93-95) has suggested the compound verb *kotadesasi*—derived from the words for town/village in Bahasa Indonesia—to describe a uniquely Asian form of urban development, whereby nonagricultural activities develop in rural areas on the urban periphery and in corridors between cities. Such areas contrast markedly with the conventional suburb (also, however, an increasingly regular feature of urbanization in Southeast Asia) in that they are characterized by diverse economic activity, intense mixed land uses, high mobility of people and circulation of goods, high levels of participation of females in the nonagricultural labor force, and an absence of governmental regulation.

The characteristics of chain migration, circular migration, overurbanization, primacy, and metropolitan extension described above combine to distinguish the contemporary Southeast Asian city, although the model with which it is implicitly being contrasted is, of course, Euro-American. In fact, as we examine the history of the Southeast Asian city we will see that urban development has long been influenced by external models and ideologies imposed upon these societies. As we look at the contemporary Southeast Asian city, we will realize that our understanding of the situation, the way we see the city, is informed by existing models of "normal" urbanization and by western political ideologies. It is difficult for western-educated academics and planners to avoid seeing the city from this perspective. Jurgen Ruland (1992, p. 1), for example, has argued that the concept of *primacy* has led not only to inappropriate policies but has also distorted the perceptions of academic observers, who have disproportionately researched the conditions in large cities and ignored smaller urban centers. As we review some of the contemporary problems of Southeast Asian cities, we shall see that what is really at stake is precisely a conception of the proper form of urbanization, or the meaning of the city. The question ultimately becomes one of "urbanization for whom?" or who has the "right to the city," to make a living and build a living place in the urban centers of Southeast Asia? Is this right to be extended only to the modern citizen who can afford to pay for access to land and services, or can it be expanded to include those engaged in traditional ways of life who lack the means to compete in a capitalist market economy?

We will begin by providing a context for these difficult questions, first by briefly reviewing the history of the Southeast Asian city. This will be followed by a sojourn into the heart of the city as seen through the eyes of different academic observers. We will then discuss urban policies that emerge with three contrasting views of the city. Finally, we will consider some of the problems of Southeast Asian cities, including employment, housing, service provision, and environmental conditions. Before that, however, there are two notes of caution to be sounded.

First, it is conventional practice in comparative accounts of urbanization to issue the reader a warning about the validity and reliability of data on urbanization due to inconsistencies in definition and measurement. The quantitative data presented on Southeast Asian urbanization are derived from diverse sources and use conflicting definitions. These data should be viewed as crude approximations and are only generally reliable—especially on sensitive problems such as employment, housing, and poverty. For example, data on urbanization for Vietnam, Cambodia, Laos, and Burma are almost nonexistent, and those that are available should be taken as gross approximations only. This said, however, as Kamil Salih (1982, p. 147) has argued, data quality is not really the important issue, and the regular reports of statistics tend to become tiresome and irrelevant. What is more important is to develop an understanding of urban processes and ways of life in cities and to examine concrete proposals for the alleviation of some obvious problems.

Second, there is, of course, considerable difference in the ways of life of Southeast Asian cities, between countries with various levels of socioeconomic development and contrasting cultures, and even within countries according to the relative size, the function, and the particular local contexts of cities. Each Southeast Asian city has a character unique to its own national and regional context, or its own "sense of place:" from the "sanitized serenity" of the orderly modern city-state of Singapore ("A Tale of Two Systems," 1994); to the sprawling, chaotic cities of Jakarta, Surabaya, Bangkok, and Manila, with their extreme contrasts of wealth and poverty and their increasingly chronic environmental problems; to the large "postsocialist" cities of Ho Chi Minh City, Hanoi, and Rangoon, which presently lack the extremity of the contrasts of the largest capitalist cities but seem bound to follow their lead; to the rapidly growing and economically dynamic "middle" cities of Cebu, Medan, Bandung, Penang, and Chiang Mai; to the slowly awakening but still sleepy "market" cities of Phnom Penh, Mandalay, Vientiane, and Savannakhet; and to the dozens of smaller cities and towns that serve as administrative, market, and cultural centers in their regions. To speak of the Southeast Asian city is therefore to employ an abstraction. Still, Southeast Asian cities share some common historical antecedents, contemporary development policies, and urban problems that justify some limited generalizations.

THE SOUTHEAST ASIAN CITY—A BRIEF HISTORY

The Precolonial City

The concentration of population in any urban center can only be sustained by resources obtained from beyond the city walls through trade or tribute. In the historic city, favorable terms of trade and the right to tribute were maintained by military force, civil regulations, and/or religious law, and such authority necessarily depended for its moral legitimacy upon a monopoly over goods, technologies, and ideologies that are imported from outside of the particular urban society. The urban center is therefore always a material and moral institution that integrates its hinterland into a local social system and is in turn integrated into a "global" urban system (O'Connor, 1983, p. 33). Nowhere, perhaps, is this "two-faced" nature of cities more evident than in Southeast Asia.

With the exception of the Philippine archipelago and New Guinea, which had no precolonial urban tradition (Reed, 1976a), urban centers developed throughout the region as early as the first century A.D. (McGee, 1967). The distribution of these historic cities is shown in Map 5.2 and they were of two principal types: the sacred city and the trading city. The sacred cities were agrarian capitals and centers of Indianized spiritual authority based on great temple complexes, such as those of Angkor Wat (Angkor, Cambodia), Mataram (central Java), Majapahit (East Java), Pagan (Burma), Srikshetra (Pyu, lower Burma), Vyadhapura (Chenla, southern Vietnam), and Indrapura (Champa, Vietnam). The sacred city was located and designed according to cosmological principles and was at once an ideological focus for the realm and a concrete symbol of the moral order (Reed, 1976a, p. 20–22; O'Connor, 1983, p. 34; Keyes, 1977). The sacred city was generally located inland and organized productive activity in its agricultural hinterlands. The modest urban population depended upon the ritual right of the ruler to appropriate part of the surplus produced by the peasantry, a right that depended on royalty exemplifying the ideals of the religious order, preserving sacred features of the city, and maintaining military rule. The sacred cities rose and fell with the military fortunes of the kings and agricultural productivity and were frequently relocated upon the advice of court astrologers.

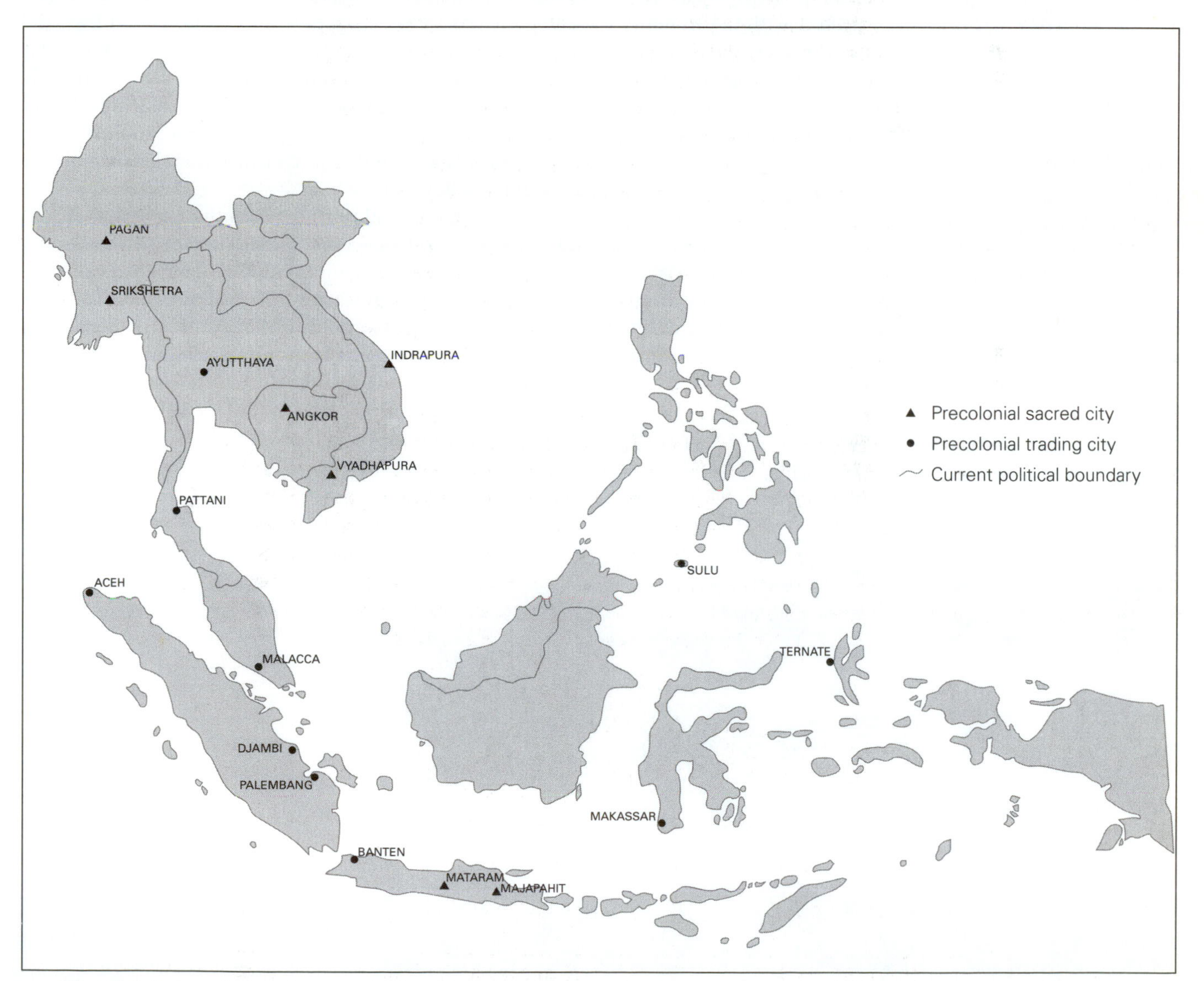

Map 5.2 Selected historic, precolonial cities of Southeast Asia. *Source: McGee, 1967, p. 44; Sardesai, 1989, p. 46.*

The trading cities, on the other hand, were bustling emporia that supplied commodities such as forest and marine products (camphor, sea cucumber, and tortoise shell), spices (especially pepper, cloves, and nutmeg) and slaves to the world economies of the Mediterranean, India, and China via complex international trading networks. They were located on coasts or rivers and included cities such as Palembang (capital of Srivijaya), Ayutthaya, Malacca, Atjeh (now Aceh), Bantam (now Banten), and Makassar (now Ujung Panjang). Such cities depended on monopoly control over regional and maritime trade, and they derived their incomes principally from commerce, taxes levied in their markets and harbors, piracy, and slavery (McGee, 1967, p. 32; Keyfitz, 1961, p. 349). This monopoly depended in part on the military strength of their rulers and in part on maintaining favorable relationships with Indian, Chinese, and Arab traders, which was necessary to gain access to prestige goods from overseas (Hutterer, 1977, p. 181). These cosmopolitan urban centers were fortified, with the Indianized elites living inside the walls and native and foreign traders outside, and during periods of prosperity they may have reached 50,000 to 100,000 people, comparable in size with the contemporary European city-states (Reid, 1983, p. 239).

The Colonial City

Colonial cities combined the functions of the two types of indigenous urbanism in Southeast Asia in that they dominated their agricultural hinterlands, extracting the surplus produced by the peasantry through taxation and forced labor, and exercised monopolies over trade, acting as transfer points in the export of commodities to world markets (O'Connor, 1983, p. 1–3; Keyfitz, 1961, p. 350). Some precolonial trading cities retained their functions under colonialism—for example Malacca and Makassar became important garrisons and ports—and some indigenous inland capitals such as Phnom Penh, Jogjakarta, Mandalay, and Hanoi, as well as many smaller regional centers, were "adopted" as administrative centers in the colonial political economy which organized the collection of surplus from the rural hinterland and exported it to European and colonial markets (Smith and Nemeth, 1986, 130). In other cases, however, new cities were founded for these purposes. In the Philippines, for example, first Cebu in 1565 and then Manila in 1571 were founded as administrative centers of the archipelago, and Manila became a vital entrepôt in the Spanish galleon trade between China and Mexico. Batavia (Jakarta) was founded as a port and colonial capital in 1619, and it became the most important entrepôt in the Dutch East Indies trade (Photo 5.1). Typically, in the early colonial city, the European officials lived in a fortified area surrounded by the semifortified quarters of foreign merchants and mestizos, while the indigenous population lived in the ethnically segregated and occupationally specialized settlements beyond.

In the late colonial period, when the Industrial Revolution expanded European demand for the raw materials of Southeast Asia, the general model of urbanization that emerged was of a hierarchy of settlement focused on ports at the head of agricultural and mining hinterlands. Some major port cities were founded on natural harbors

Photo 5.1 Historic city core—Sunda Kelapa, Jakarta. (Leinbach)

or at break-of-bulk transportation points for the export of plantation products and minerals, such as Singapore in 1819 and Kuala Lumpur in 1858, while other existing towns were redesigned for new port and administrative functions, such as Rangoon and Saigon, both in the 1850s. Even the established colonial cities of Manila and Batavia reoriented from maritime entrepôt trade to become headlinks and administrative capitals for their hinterlands (Reed, 1976a). These new cities became the nerve centers of economic exploitation and political administration of the colony: here the institutions of capitalism—the banks, trading houses, and shipping companies—organized the production of agricultural and mineral commodities to be exported overseas and the distribution of industrial commodities imported from Europe; and here also the institutions of colonial governance imposed an administrative order upon the colony, partly for the purposes of economic exploitation but also to "civilize" its society. The colonial capitals, in particular, became primate cities, dominating demographically, politically, economically, and culturally the rural hinterlands of the colonial society—Rangoon in British Burma, Singapore in Malaya, Jakarta in the Dutch East Indies, Manila in the Philippines, and Saigon in Indochina. The case of Bangkok, whose primacy originates in centralist ideologies and security concerns of the absolute monarchy (Ruland, 1992, p. 24), and which has maintained its centrality within the Kingdom of Siam (now Thailand) to the present (Withington, 1985, p. 97), suggests that it was not colonialism per se, but the conditions of integration of the regional economy into the world system that produced primacy. Meanwhile, the precolonial agrarian capitals and trading cities that were not incorporated into the mercantile economy and colonial administrative hierarchies tended to stagnate or decline, some disappearing entirely from the map of urbanization until they recently became archaeological sites or tourist destinations.

In a sense, the colonial period saw the internal structure of the Southeast Asian city turned inside out (Steinberg et al., 1985, p. 247). First, in the precolonial city the religious, military, and court functions were located in the center, while the market and the merchants were confined to the outskirts. In the colonial city, on the other hand, commerce was located in the heart of the city with the administrative functions. Second, the new colonial elite, following patterns being established in Europe, retreated to exclusive residential districts outside the congested center of the city. Urban society and urban space nevertheless remained rigidly divided by ethnicity and to a lesser extent by occupation as the new social structure was built into the physical landscape. Also, the colonial administrations introduced private property and the formal titling of urban land, and land use was publicly planned by a professional class of bureaucrats to maintain a rational segregation of functions, resulting in a general urban form that was characteristic of colonial Southeast Asian cities. In this model, physical planning was employed to prevent contamination of the civilized elite by contact with the masses; for reasons of hygiene, sanitation, and political stability, the spacious and well-ordered European residential and commercial districts were isolated from those of the local populations. The indigenous population, who provided labor for essential urban services lived precariously on untitled and unserviced land in "urban villages," or artisanal quarters, often identified by place of origin of the migrants as well as by occupation. The nonindigenous ethnic minority, typically the Chinese, lived in the crowded shop-houses of the non-Western commercial district adjacent to the European commercial district and the port. Although despised and periodically persecuted by both the Europeans and indigenous people, the Chinese traders were vital to the circulation and distribution of commodities because of the lack of an indigenous commercial class in Southeast Asian society and the reluctance of Europeans to engage in retail trade. The Chinese traders, transporters, and creditors acted as vital intermediaries between the rural peasant and the European wholesalers, and Chinatown was an essential feature of the urban landscape, one that typically remains today.

The precolonial and early colonial city had drawn upon external symbolic systems made physically manifest in the segregated temple complex and the fortified garrison to legitimate authority over the rural periphery. Similarly, the foreign economic and civic institutions of the late colonial era were housed in imposing buildings of exotic architecture in the commercial and administrative districts of the city. Unlike the precolonial city, however, and again with the partial exception of the Philippines, authority originated in the perceived rationality and cultural superiority of the Europeans in governance and commerce rather than religion or military might. The "classic colonial" cities were centers of modernization and were carefully planned according to European ideas of urbanism (King, 1990, p. 47–48). The grand European architecture and landscaping of governors' mansions, railway stations, parks, banks, trading houses, libraries, and gentlemen's clubs typically made few concessions to local environmental conditions or local culture.

THEORIZING THE SOUTHEAST ASIAN CITY

Two key points emerge from this discussion of the history of Southeast Asian urbanization. First, urban centers have always functioned simultaneously to mobilize and concentrate the material surplus produced by rural hinterlands for export to metropolitan countries, and to disseminate

Photo 5.2 Chinese shophouses in Petaling Jaya, Malaysia. (Leinbach)

foreign ideologies and ways of life from the metropolitan countries. Although there clearly is a historic urban tradition in Southeast Asia, for most of its history the city has been dependent on integration with external political authority and economic structures. In the case of the contemporary Third World, Armstrong and McGee (1985, p. 41) have theorized the city as at once both a "theater of accumulation [for] modern commerce, finance, and industrial activity" and a "center of diffusion [of] lifestyles, customs, tastes, fashions, and consumer habits of modern industrial society." Their model is a very useful means to understand and depict the Southeast Asia city.

Armstrong and McGee (1985) consider that the precise function of the city in the global economy leads to differentiation between cities, depending historically on the nature of the colonial relationship, and contemporaneously on the role of the national economy in the international division of labor. Focusing on the contemporary city, for example, it can be argued that foreign investment in Indonesia, particularly compared with Thailand and Malaysia, has been in resource extraction rather than manufacturing, leading to lower rates of urbanization and higher proportions of service and trade employment in its cities (Forbes and Thrift, 1987). The relatively high rates of urbanization in the Philippines and Malaysia, on the other hand, are partly due to the growth of manufacturing under colonial administrations in the early twentieth century and to growth in intermediate cities such Cebu and Penang following high levels of foreign investment in recent years (Ruland, 1992, p. 286; Tiglao, 1991, p. 60). Similarly, the lack of foreign investment and the small middle class of Cambodia, Burma, and Laos have restricted the development of urban manufacturing, finance, and real estate, leading to the dominance of petty trade in the cities of Phnom Penh, Rangoon, and Vientiane. The various conditions of incorporation into the global economy produce a divergence in patterns of urbanization in the region.

On the other hand, Armstrong and McGee (1985) also argue that the global media, educational and cultural institutions, and the imitative lifestyles of urban elites promote a convergence of cultures and experiences of urbanization. Even the cities of previously communist or socialist regimes evince the influence of the forces of westernization as planners dictate efficient forms for the city; private capital is used to construct office buildings, condominiums, shopping centers and suburban homes modeled on those of the West; and middle-classes pursue materialist and individualist urban lifestyles. The spatial order of the contemporary city, just like the precolonial temple complex and the colonial administrative center, functions to legitimate symbolically the dominance of urban life by foreign-influenced economic elites and their values. Similarly, even though the simplicity and honesty of indigenous rural society was romanticized in nationalist ideology during the struggle for independence and may still be evoked for political purposes, the urban elites generally show contempt for peasant culture and lifestyle. To Southeast Asian urbanites the terms *peasant*, *villager*, and *provincial* are used to imply a lack of education, ambition, and sophistication, while from the village or province the city dweller appears to be cosmopolitan and knowledgeable of modern ways.

The problem with the model proposed by Armstrong and McGee, however, is that the forces of convergence and divergence are not so simply cultural and economic, respectively. For example, the economic functions of the larger national and regional cities are becoming increasingly similar due to the internationalization of producer services—banking, insurance, commercial real estate, and business services (Forbes and Thrift, 1987)—the globalization of consumerism, suburbanization, property speculation, and the restructuring of manufacturing under the New International Division of Labor (NIDL). The ideologies of economic growth have remarkable currency throughout Southeast Asia at present, and foreign capital, in alliance with the urban economic elites (Armstrong and McGee, 1985, p. 11), plays a critical role in the accumulation process, providing capital and expertise for the development of national capitalism and funneling commodities and profits into the global economy. The result of the penetration of consumer, financial and property markets into Southeast Asian economies is a convergence of patterns and functions of urbanization, illustrated symbolically, for example, by the dominance of the corporate headquarters of the Central Business District (CBD) over the urban skyline, and practically by the increasing effect of decisions made in those headquarters over the everyday lives of the population.

On the other hand, Southeast Asian political elites are concerned with nation-building, as well as with economic growth, and they seek to maintain their authority in terms of indigenous conceptions of power, as well as through modern Western-style institutions. The governments of Indonesia, Malaysia, and even cosmopolitan Singapore, for example, have expressed concern about the "Westernization" of the new urban middle-classes and the influence of global media on their lifestyles and tastes (particularly their conspicuous consumption and support for free press and representative democracy). In the interests of nationalism, political elites may therefore explicitly resist cultural convergence around modern, western values. Thus cultural identity is expressed in the urban landscape as planners and architects attempt to represent nationalist values in architecture and urban space, such as the monumental forms of Jakarta that mix modernism with symbols of Indonesian national identity (Ford, 1993) and the neotraditionalism of Phnom Penh (Karan and Bladen, 1985, p. 28). As the forces of modernization remake the cities of Vietnam, national and international planning agencies are concerned to preserve the heritage landscapes of the Old Sector of central Hanoi (Logan, 1995), and again, even in Singapore, the modernist city par excellence, attempts are being made to preserve what remains of its Chinatown after decades of redevelopment (Photo 5.3). In these latter cases, of course, planners are also motivated by the demands of tourism for which the built environment has become a valuable cultural and commercial resource. A report of Singapore's Ministry of Trade and Tourism argued, for example, that "in our effort to build up a modern metropolis, we have removed aspects of our Oriental mystique and charm which are best symbolized in old buildings, traditional activities and bustling roadside activities" (Wong, 1984).

This suggests that the distinction between the economic process of divergence and cultural process of convergence is not clear and that we might better conceive of both the economic and cultural spheres of urbanization, and more generally of development, in Southeast Asia as subject to tensions between the forces of globalization and localization, and modern and traditional ways of life.

These examples also illustrate the second point that emerges from the foregoing discussion of the historical city in Southeast Asia—that is, that urban centers are the locus of articulation between global and local processes

Photo 5.3 Refurbished urban dwellings in Emerald Hill, Singapore. (Leinbach)

and between modern and traditional values. Although the financial district, shopping malls and supermarkets, condominiums, residential suburbs, and highways are evidence of the global nature of contemporary urbanization (Harvey, 1973, p. 278), the peculiar characteristics of Chinatown, produce markets, squatter and slum settlements, and modes of transportation such as trishaws, tricycles, bemos (three-wheeled passenger vehicles), and jeepneys, are a testament to the adaptability of local urbanism. Thus, while some features of indigenous urbanism have been lost in the transition from precolonialism to colonial and to postcolonial regimes, others have been conserved, even as their meaning and function have been partly transformed. This is apparent, for example, in the form of the contemporary Southeast Asian city, which clearly conserves elements of the colonial city—the old colonial government district has been taken over by the modern government of the independent state, and Chinatown now includes modern Asian banks as well as traditional shop-houses—while new forms such as ultramodern CBD and suburbs have emerged (Photo 5.2). Figure 5.4 illustrates a model of the spatial structure of the contemporary city, and Photo 5.4 provides a graphic contrast of both the appearance and production of squatter housing and the ultramodern business district on the skyline. In the contemporary Southeast Asian city we are constantly reminded of the presence of the past in the urban landscape and the adaptive capacity of traditional technologies, social relationships, and ways of life. How we see this contrast is critical to our experience of the Southeast Asian city, our theorization of urban processes, our identification of urban problems, and the proposals we make for their solution.

EXPERIENCING THE SOUTHEAST ASIAN CITY

At late afternoon in a Southeast Asian city, the main streets are jammed with luxury private cars, crowded public buses, and various forms of nonmotorized transportation such as bicycles and trishaws. Workers are returning from offices, construction sites, and factories; students are returning from afternoon school; and housekeepers are going to market to buy ingredients for the evening meal. The sidewalks, curbs, and many of the side streets are choked with vendors plying their trade—from young boys and girls to old men and women. Some have laid out mats on which they have spread their wares—fruit, vegetables, and spices in small piles suitable for a single meal and costing only a few cents. Others sell prepared foods and drinks from tables or carts, and the air is thick with smoke and the smell of barbecuing meats and corn. People stop to exchange information about market prices and community news, while vendors and their customers haggle at length over prices in various languages and dialects.

Most of the pedestrians and the vendors live in a dense residential community nearby. The settlement is perhaps illegal, built by squatters on government land or private-

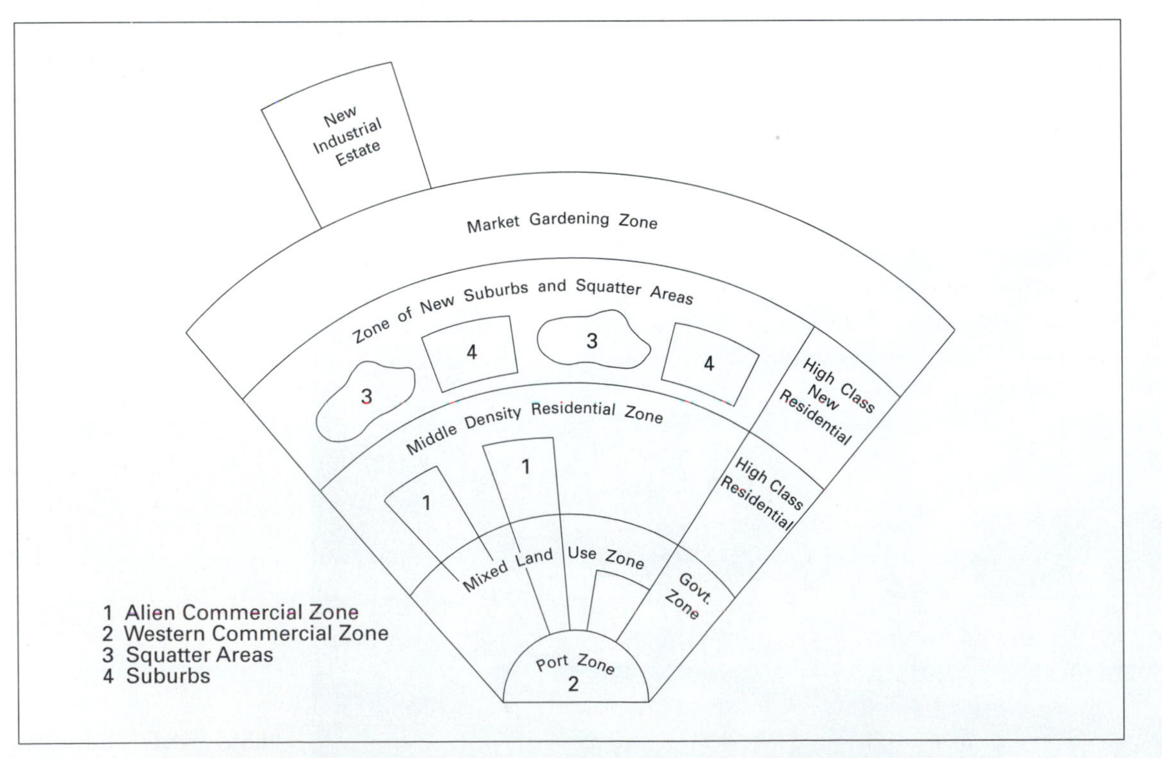

Figure 5.4 Model of the urban form of the contemporary Southeast Asian city.
Source: McGee, 1967, p. 128.

Photo 5.4 Squatter housing—Central Business District contrast, Manila. (Goss)

ly owned lots. The makeshift shelters flaunt regulatory standards and house unlicensed stores and workshops, as well as chickens and pigs. In some of the more substantial houses, built incrementally over time, rooms or bed spaces are shared with extended family or rented out to tenants. Residents buy water from peddlers or fetch it from a public standpipe (Photo 5.5); electricity is ingeniously pirated from utility lines; and sewage is carried by open drains to the nearby canal or river. The area regularly floods in the rainy season and periodically, perhaps every 10 years or so, the wooden houses are consumed by fire. The community is probably also threatened with eviction to make way for urban redevelopment. Despite all this, and partly to deal with their hardship, everyone seems to be a member of neighborhood, religious, regional, political, social, and charitable associations: there are revolving credit associations; committees for electricity and water services; neighborhood beautification and fire protection; and there are all kinds of informal reciprocal networks among neighbors.

These scenes and this sense of community contrast markedly with the orderly financial, commercial, and middle-class residential districts of the city. Here the production of urban space is organized for specialized functions by real estate developers and is subject to strict controls imposed by government. This urban space is rationally planned, different land uses are segregated, and the architecture is modern with careful attention to aesthetics as well as functionality. It consciously resembles the modern ideal of urbanism and presents itself as the model for the future. The people who work, shop and live in these districts are predominantly members of the growing urban middle class, whose individualist and materialist values and lifestyles are more closely related to

Photo 5.5 Water vendor in Jakarta. (Leinbach)

the middle classes of Western cities than the urban poor in the Southeast Asian city. They live in suburbs in nuclear families (although usually also with live-in servants); they work at secure salaried jobs in modern office buildings; they shop in supermarkets and department stores; and they travel in air-conditioned cars. They seek to isolate themselves from the reality that is the life of the majority of urban residents.

It is not surprising, given this contrast in the built environment and ways of life, that observers have tended to conceive of the Southeast Asian city as a dual city, composed of separate modern and traditional components. One part of the city seems to be oriented toward the rural areas and traditions, while the other embraces modernity and Western urban values. If the dualistic conception of the city is common, however, there is disagreement about the nature of the relationship between the two components and about the role that traditional urbanism might play in socioeconomic development. There are, in fact, three distinct "ways of seeing" the colorful scenes of traditional commerce and community described above and of conceiving the relationship between the "two cities" within a contemporary Southeast Asian city.

Some observers, for example, will focus on the apparent disorder, the obstruction of vehicular circulation, the long hours of labor, the poor quality of goods, the lack of skills, and the obvious inefficiencies of the traditional marketing system. They may also be shocked by the crowded rooms, the inadequate sanitation, the lack of privacy, and the insecurity of residential tenure. The low productivity of labor and low standards of living seem more reminiscent of backward rural areas than a modern city, and the majority of the population appear to be *in* the city, but not *of* the city (Mangin, 1970); that is, they are perceived to be marginal to modern urban society—economically marginal because they do not seem to be involved in the modern urban economy, socially marginal because they seem to be excluded from the formal organizations and associations of urban life, culturally marginal because they seem to maintain rural values and customs, politically marginal because they seem to lack standing with government institutions, and geographically marginal because they live in unserviced slums on the periphery of the city, on river banks and railroad tracks, or in other dangerous or undesirable locations (Nelson, 1969, p. 5).

In adapting to conditions of marginality, the institutions of the traditional city may actually frustrate progress. For example, it has been noted that the markets of Southeast Asian cities evince a peculiar economic logic in that the detailed division of labor and the rapid circulation of credit and commodities absorb labor but frustrate innovation that would increase productivity, a phenomenon called urban involution (Armstrong and McGee, 1968). Similarly, in response to a lack of opportunity the poor may adopt fatalistic attitudes, marked by a lack of a work ethic and inability to save and by exploitation of their children rather than provision of education for their futures. The effect is a vicious cycle from which poor households cannot easily escape—the so-called culture of poverty (Jellinek, 1991; Jocano, 1984).

To observers taking this perspective, it would seem that the urban poor, having sought escape from rural poverty by migrating to the city, only become trapped into urban poverty, to remain parasitic on the urban economy, obstructing efficient development, and threatening political stability. This pessimistic view is perfectly expressed by a Filipino development expert (Laquian, 1972, p. 17):

> In place of subsistence farming or fishing in remote villages, marginal living in the big city is a more likely and diverting springboard towards a better life. At the same time, however, the consequences of substandard living conditions in an overcrowded slum are of a different order and magnitude compared to its effects in a small village. Squalor, disease, crime, and dissidence set in and tend to overwhelm the basic faith, humanity, and decency of the residents. They become ripe for political agitators who play on their weaknesses to transform them from simple dignified individuals making the most of their lot to improve themselves, into whining malcontents who act as though the world would stop and lay itself at their feet if they simply shouted loud enough.

Viewing the city from this perspective, we might recommend strategies to eliminate vendors and nonmotorized transport, to evict squatters, to prevent or restrict migration, and to incorporate as much of the existing urban population as possible into the modern and formal city institutions through the provision of public markets, the construction of public housing, and the promotion of industrialization.

Alternatively, some observers are impressed by the capacity of traditional urban institutions to provide remunerative employment for the poor, while also providing for the daily needs of consumers, and to provide cheap housing despite rapidly increasing prices of urban land. These observers see a logical order hidden behind the apparent disorder of the market and identify an informal system that ensures that capital, credit, and goods circulate extremely rapidly, that monopolies do not develop, and thus that almost everyone has access to the means to make a living. Competition is intense, but the market is a training ground for entrepreneurial skills that successful vendors may employ in the modern economy. Similarly, squatter communities produce living places for the majority of the population who are excluded from private or public housing development. Over time, as the residents accumulate savings and increase their sweat equity investment, houses are expanded and improved, and the communities stabilize and gain access to services, gradually becoming part of the fabric of the city (Ulack, 1978). Some households may even be able to accumulate sufficient wealth in their "illegal" property that they are able to sell and move on to private subdivisions. Moreover, income-generating activities such as outworking, petty retailing, cooked foods, and renting of rooms that take place in these settlements are particularly important to women who tend to be restricted by the gender division of labor to activities that do not conflict with their child-rearing and housekeeping roles.

From this perspective, it seems that government restrictions on the activities of vendors and evictions of squatter communities inadvertently frustrate the potential for "bottom up" development and that governments would therefore be better advised to provide positive incentives in the form of training and credit to the petty entrepreneurs of what is often called the informal sector. Contrasted with the formal sector, or the industrial economy, the informal sector is characterized by the small scale of operations, low levels of capitalization, appropriateness of technologies, competitiveness, ease of entry, and lack of government regulation. It is argued that the promotion of this sector—through lifting of restrictions, the provision of skills training and credit, and the development of forward and backward linkages with the formal sector—would benefit national development not only by providing employment and housing for the poor but also by enhancing the competitiveness of national industries because the cheap goods, services, and housing produced by petty enterprises would reduce wages and the overall costs of the production and distribution of commodities.

Finally, a third group of observers, while impressed by the adaptability of the traditional economy, might instead focus on the low incomes, long hours, poor working conditions, lack of security, and the exploitation of child labor in petty enterprises, as well as the crowding, poor sanitation, and the exploitation of family labor in "self-help" housing. They might also note that while the informal sector is a useful description of the appearances of the everyday activities of many of the poor, it is a catch-all term for a variety of production relations. A closer examination of the situation would reveal that far from being independent "bootstraps" capitalists, many of those identified with the "informal sector" are petty commodity producers who are dependent on owners of capital and land or on permission from government officials and that they inevitably pay interest, rent, and protection money (Goss, 1990). Even as some of the poor manage to obtain social mobility through their petty enterprises and houses—and their success is vital as a model to others—the majority unfortunately do not. Similarly, the high rates of participation of women in the informal sector reflects the low return and insecurity of typical informal sector activities and are thus a symptom of the exploitation of female labor rather than its relative emancipation.

From this perspective, exploitation of the poor not only occurs through particular social relations within the enterprise but must also be seen as a general characteristic of dependent urbanization whereby urban development is not an autonomous national process but is determined by the position of the national economy and city in the international division of labor (Smith and Nemeth, 1986; Friedmann, 1986). The urban scenes described above exhibit structural inequalities and exploitative relations between the rural and urban areas, the urban poor and the elites, and between Southeast Asian economies and the developed world. Thus, although the poor are not marginal, in that they are thoroughly integrated into urban relations of exploitation, neither are they fully incorporated into modern urban institutions, as they do retain some material and cultural attachment to the rural areas and the subsistence economy. The social relationships under which they earn their livings and obtain their shelter in the city often depend upon the preservation of traditional values and social relationships as well as economic ties to rural origins. As we have seen, many periodically return to their villages, and many more will do so when they are old or too ill to continue working and living in the city. It is partly this connection that explains the low returns to labor that the majority of the population tolerates, which in turn reduces the cost of urban goods and services and lowers average urban wages which the formal sector would have to pay its workers. From this perspective it is therefore argued that petty commodity producers of goods and services subsidize the profits of national and foreign capital invested in the city and that national development can only be effectively furthered by a radical change in these relations, including, for example, rural and urban land reform, elimination of class bias in development policies, and restrictions on foreign investment.

URBAN POLICY IMPLICATIONS

Although our description is undoubtedly guilty of caricature, these "ways of seeing" are broadly consistent with the main approaches taken by academics, development agencies, and national government planners to Southeast Asian urbanization. The observers appear to agree on the symptoms—a marked contrast between the city and the countryside, between the traditional and modern sectors, and between rich and poor—but they disagree on the diagnosis and potential treatments.

The first, and still dominant view places priority on modernization and economic growth under the assumption that this will ultimately provide jobs and services for the marginal population. Government policies are focused on the development of national industry, and infrastructure and services are concentrated in the larger cities to take advantage of market demand and scale economies. The inevitable growth of population in the larger cities, however, reduces efficiency (introducing diseconomies of scale), threatens political instability, and is also an embarrassing reminder to elites of the relative underdevelopment of the economy. Policies have been introduced to limit migration to large cities, such as the use of identification cards to "close" Jakarta to migrants in the 1960s and the restriction of urban services to established residents in Manila in the 1970s (Simmons, 1988, p. 92–94). In some extreme cases governments attempted to return the "surplus" urban population to their rural origins or to relocate them to the agricultural frontier. In Jakarta, for example, during the 1970s vendors and becak (trishaw) riders operating illegally were regularly rounded up and offered their fare back to their village or transmigration to the outer islands in exchange for leniency (Photo 5.6). Due to suspicion of the counterrevolutionary tendencies of the urban middle classes and the ethnic Chinese, Vietnam forcibly relocated urban populations during the late 1970s (Thrift and Forbes, 1986; Hiebert, 1989, p. 42; Dwyer, 1990, p. 305). The Vietnamese government still attempts to restrict migration to cities and operates a massive population-redistribution program such that present levels of urbanization remain lower than at the time of reunification (Forbes, 1995, p. 796). In the most extreme case, the population of Phnom Penh was forcibly evacuated upon the victory of the Khmer Rouge in 1975 and sent to reeducation camps where large numbers died. For the most part, however, and beyond these extreme examples, neither the policies to restrict migration nor relocate urban populations have met with any long-term success. Even Phnom Penh is now growing extremely rapidly—by as much as 50 percent since 1989—as refugees from the border camps and peasants return to the city.

The urban policies of Southeast Asian governments changed during the 1970s, however, partly due to the considerable influence of international agencies such as the World Bank, as they adopted the second view outlined above and came to favor the decentralization of growth and promotion of the "informal sector." Planned decentralization of growth in regional growth poles and industrial estates became the standard planning ideology (Rondinelli, 1990). Thailand's Fourth National Economic and Social Development Plan (1977–81), for example, called for development of the eastern seaboard and five regional growth centers (Ruland, 1992); the Philippines Development Plan (1977–82) included, among other provisions, a ban on location of new industries within 31 mi (50 km) of Metro Manila, the development of new industrial estates throughout the provinces, and the development of regional cities; and the Third Malaysia Plan (1978–82) included rural industrialization programs, new towns, and Regional Development Authorities in regional cities. These policies aimed to provide "countermagnets" to the large cities and to provide alternative opportunities for rural migrants to improve their socioeconomic conditions.

Recognizing the threat that the burgeoning urban population posed to political stability and economic development, attention was focused on providing for the basic needs of the poor urban majority and to enhance their integration into urban institutions. Development experts advised the deregulation of the economy and particularly the reduction of costly bureaucratic controls on productive activities and the elimination of the subsidies and other privileges enjoyed by formal sector industries. With the financial and technical assistance of international agencies, governments sought to provide credit, infrastructure, land, training, and services to encourage informal-sector solutions to urban problems. Rhetoric shifted from centralized planning concerned with overall efficiency in the allocation of urban resources to community and participatory planning.

The new policies and rhetoric did not, however, stem the flow of migrants nor limit the growth of the urban population, and informal sector-promotion policies and self-help housing did little to address structural inequalities because benefits often accrued to owners of capital and to the middle classes. More recently, rising interest rates have limited the credit available for such policies, and the concern of development agencies has shifted to the rationalization of urban management and indirect intervention in urban processes through planning, land-use controls, and tax administration, rather than direct intervention in job creation, infrastructure development, and welfare provision (Gould, 1992, p. v; World Bank, 1991b).

This policy shift highlights the influence of external forces on urban development and adds further evidence for the validity of the third approach, although we realize that prospects for radical change have not been good and are perhaps diminishing. As the national economy develops and is increasingly incorporated into global capitalism, as the middle class expands and urban land markets develop, and as international aid agencies refocus their priorities, so governments have again become less sympathetic to the poorer and politically weaker majority of the urban population. In Indonesia, for example, self-help institutions have recently been restricted (Cohen, 1994, p. 32), the becak has recently been entirely outlawed, and thousands of street vendors were banned in the lead up to the 1992 elections (Vatikiotis, 1990). Even in Vietnam the new middle classes show increasing indifference and contempt toward the poor (Hiebert, 1994b). But poverty cannot be effectively treated by displacing it or ignoring it, and it remains the root of long-term and growing problems of Southeast Asian urbanization.

THE PROBLEMS OF SOUTHEAST ASIAN CITIES

The principal problem of the Southeast Asian city, again with the exceptions of Singapore and Bandar Seri Begawan, is the relative and absolute poverty of the majority of the population. Poverty must be viewed, however, as a multidimensional phenomenon with economic, social, physical, political, legal, and cultural components and as a process in which these components interact to prevent low-income households from improving their conditions.

Employment and income. Most obviously perhaps, poverty is defined by a lack of income sufficient to provide for the minimum needs of the household. Although the figures are always somewhat unreliable due to problems in definition and measurement, it is estimated, for example, that 32 percent of the population of Metro Manila live below the established poverty line, while 10 percent live in ultrapoverty on the equivalent of US$35–50 per month (McBeth and Goertzen, 1991, p. 30). Similarly, 29 percent of Indonesia's urban population and 13 percent of urban Malaysians live below the basic-needs poverty line (UNESCAP, 1993b, p. 4.5).

In Southeast Asian cities, due to the low productivity of industry, lack of job security and employment benefits, and limited representation in labor organizations, many workers even in the formal sector receive insufficient wages to support their households. Similarly, the majority of petty commodity producers earn low and variable incomes due to fluctuations in demand for their products and services and unforeseen expenses (such as bribes paid to officials). To maximize income and minimize risk, poor households therefore typically adopt multiple income-generating strategies in various types of remunerative activity: In the community of predominantly poor households surveyed in Manila in 1987, for example, there was an average of 2.48 workers per household and 26 percent of these workers had more than one form of employment (Goss, 1990, p. 184–189). Perhaps an example, or case study, is the best way to illustrate the diverse individual and household strategies that lower income households might employ to allocate labor and capital between various formal-sector and informal-sector activities to maximize income (Box 5.2).

Early dualistic approaches to Southeast Asian urbanization presumed that unskilled rural migrants would work as self-employed traders or as contract laborers, eventually becoming permanent wage workers in the city. We now know, however, that not only is obtaining access to wage labor extremely difficult, it is not necessarily more remunerative or more stable. First, although formal sector employment is expanding with economic development in most Southeast Asian countries, the informal sector provides a large proportion of jobs in urban centers. For example, 65 percent of urban employment in Jakarta, 50 percent in Metro Manila, and 49 percent in Bangkok is estimated to work in firms with less than 10 employees (UNESCAP, 1993b, p. 3.22). The informal sector dominates the trade and services sector, which accounts for 70 percent of Bangkok employment and 72 percent of Metro Manila's labor force (UNESCAP 1993a, p. 14). Second, formal sector employers reduce costs, maintain discipline, and increase flexibility by hiring large numbers of casual or probationary workers and limiting permanent hires to those with the right credentials and personal connections with management. Again, research in Manila, for example, reveals that large factories selectively employ in full-time positions those who are related to established and trusted workers and that supervisors, for example, will often directly recruit workers from their home villagers (Goss, 1990). Other temporary workers are often terminated before the end of their probationary period. Labor markets are highly segmented and there is limited mobility between types of employment. This also applies even to wage-labor employment in small businesses and parts of the "informal sector," which are undoubtedly more competitive and exclusionary than previously thought.

Photo 5.6 Jakarta becak driver. (Goss)

BOX 5.2 A Case Study: the Garavillas

The Garavillas are not as poor as many of their neighbors because they are long-term residents in Metro Manila and have managed to acquire property, but they are nevertheless low–income squatters and must struggle daily to make ends meet by an ingenious combination of income-generating strategies and sharing of domestic duties.

Edilberto, a police officer now deceased, originally built this wooden house for his second wife in 1963. Its original four rooms have been extended on several occasions, and there are now 11 rooms, 5 of which are rented to tenants. The tenants are all long-term residents and are all related to the Garavillas through marriage or godparenthood or are workmates, so rents are very low. The regular income nevertheless is a vital supplement to pooled household incomes.

The family divides into two units—one upstairs and one downstairs. Two of Edilberto's sons live with their families in the two rooms downstairs, and five of his daughters, three of whom are married, live in the three rooms upstairs. The two groups share the rental income and remittance income from a third son who has recently obtained a contract to work as a construction laborer in the Middle East (he provided most of the skilled labor employed in expanding the house). The management of the various incomes and the sharing of household labor within and between the two groups is extraordinarily complex.

Living downstairs, Edward is a school-bus driver and his wife Lilia, who sends money every six months, is a nursing aid in Abu Dhabi. They have three children of school age. Apolinario is a stable hand and his wife Gina, with the help of Edward's oldest daughter, does the cooking and most of the cleaning for both of these families. They have a baby son. An unfixed portion of Lilia's remittance is put towards general expenses, and the two families form a household because they effectively share a budget and eat together. It is doubtful whether Apolinario's family could subsist on his irregular and very low income without this subsidy to their living costs and, on the other hand, Gina's labor is vital to supporting Edward and his family in the absence of its own (female) homemaker.

Turning to the upstairs household, Angelina is the landlord. She keeps part of the rent for repairs to the house, gives part to the downstairs household, and uses the remainder for marketing, as well as unfixed but regular contributions from the wage earners: Angelina's husband is a municipal clerk (and who moonlights as a handyman), Nina is a local messenger, and Rosa is a clerk in a private company. Cooking duties for the household are shared by Angelina and her two unmarried sisters: Violeta, an unemployed nursing graduate, and Teresita, who is slightly retarded.

Household income is supplemented by selling both the pigs and the ducks that Nina and Teresita raise in and behind the house and the meat they occasionally barbecue in the alley in front of the house. The household obtains income in kind and various benefits through Rosa's husband, who has been unemployed for almost a year but is active in the squatter's organization and is an acknowledged "street leader." Angelina and her husband have three school-age children, Rosa and her husband have an adult daughter, and the household includes three children of another of Edilberto's sons (he is separated from his wife and works as a jeepney driver in another part of the city, paying small amounts of money for his children when he is able to save).

Finally, laundry of the two households is often done collectively by whatever female labor is available, although some is also done within the five family groups, and cleaning of the shared living rooms, baby-sitting, and sometimes cooking is also shared.

Petty commodity production, on the other hand, is generally more open, provided an individual has access to the necessary capital. The success of vendors, hawkers, trishaw riders, and scrap collectors depends on establishing such a territory, developing competitive skills, obtaining protection from patrons, and soliciting regular clientele. Failure rates are high because the market is extremely competitive, but even among the most territorial of trades, established participants will usually accommodate newcomers. This relative openness of such activities is the result of a notion of a right to make a living, a notion that involves recognition, however reluctant, of the need of another to earn an income. This is consistent with the notion of urban involution but without the negative connotations because the concern is not overall productivity but provision of access to resources necessary to sustain a household. Moreover, although it may be true that individual incomes decrease as more people exercise this right to make a living in a closed market, it is not primarily competition from other petty producers that prevents accumulation of capital and social mobility. Researchers adopting the third perspective outlined above have found that petty commodity producers are often effectively working for members of the middle-classes, government employees and even some large corporations. Street vendors and trishaw riders, for example, rent equipment from owners or purchase goods on consignment from dealers at marked-up prices, and some previously independent producers are now dependent distributors of factory products. This applies to many ice cream vendors in Jakarta, for example, who, as a result of the "taste transfer"—the preference of consumers for modern and especially Western-style products (Drakakis-Smith, 1991)—distribute factory produced ice-cream for a commission rather than making and selling their own (Douglass, 1992, p. 49). In almost any squatter community one is also likely to find households producing handicrafts and garments for export to international markets, and these workers include children and elderly who cannot be legally employed in a factory, as well as women whose domestic role confines them to the home. Detailed research on such relationships reveals that many of the working poor in the so-called informal sector are exploited indirectly by the formal sector through usury (high interest rates), rents, and unequal exchange such that the returns they receive as profits, or piece rates paid by subcontractors, are much lower than would be made if they were legally employed wage workers. It is argued, therefore, that such workers are dependent on and are exploited by the formal sector and that they are effectively "disguised" wage workers. Many of what are conventionally seen as the informal sector are then effectively workers for national and even international capital, and despite their "traditional" appearance they are a part of the world capitalist system. Even if they are not directly exploited by the formal sector, some would argue that the cheap goods and services that petty commodity producers provide to urban populations indirectly reduce the wage costs to national and international capital.

Land and housing. Partly due to lack of income, the poor live in substandard built environments that are vulnerable to flood, fire, and pollution—often without formal tenure rights. They are excluded by their low incomes and lack of legal tenure from basic services, including health, education, and utilities, which further reduce their quality of life. In most Southeast Asian cities, land—a necessary condition for shelter—is simply too expensive for the urban poor to buy or even to rent in the formal market. This is due to a complex combination of historical factors. In the case of the Philippines, for example, the skewed distribution of land ownership inherited from the colonial era has created a situation in which prominent families own huge tracts of prime land within the boundaries of contemporary cities, and they release it only very slowly for development (Solon, 1987, p. 5–6). Throughout the region, even now in the former socialist countries, land speculation has become a vital pursuit of the political elites. In what Yoshihara (1988) has described as the "ersatz capitalism" of Southeast Asia's economies, the manufacturing sector is underdeveloped, and often dominated by foreign or Chinese capital, and indigenous owners of capital seek "easy" profits in rent and interest, rather than risking productive investment. Real estate is particularly attractive because it is also a "hedge" against inflation and, for the political and military elites, an insurance against the instabilities of office.

Also, more recently, overseas Chinese capital from Taiwan and Hong Kong and from other Southeast Asian countries has allied with indigenous Chinese capital and has come to dominate the national construction sector. The result has been rampant land speculation and construction booms that produced rapid inflation in housing costs in the major Southeast Asian cities during the late 1980s (Friedland, 1990, p. 54). International Chinese investment in real estate results from networks of business connections, the cultural importance of providing for subsequent generations, and the cultural value attached to land and property (Goldberg, 1985), and in the case of Southeast Asian Chinese, from a persistent fear of discrimination in the host country. Exploiting ethnic and business connections with naturalized Southeast Asian Chinese, capital from Hong Kong, Singapore, and Taiwan (seeking new outlets for investment due to the long-term political risk in Hong Kong, strict regulation in Singapore, and competitive markets in Taiwan), has been invested heavily in real estate in Southeast Asia. Similarly, although Vietnamese ethnic Chinese are only cautiously reinvesting in the post-*doi moi* economy, they are important connections in the investment of overseas Chinese capital from Hong Kong, Taiwan, and the ASEAN countries in the real estate sector of Vietnam's major cities. There are also signs that the recent development of export industries in Cambodia and of the construction industry in Phnom Penh is dominated by Khmer Chinese capital (Hiebert, 1991, p. 84), and even the phenomenal growth of Cebu is partly due to investment by overseas Chinese capital in connection with the large Chinese-Filipino business community in that city (Tiglao, 1991, p. 62).

Throughout the region the various interests in property development have consolidated into powerful corporations that combine all stages of land development from finance and land assembly to construction and sale. These oligopolistic firms exert considerable control over the nature, direction, and rate of urban development, accelerating the absorption of land into the urban real estate market for office development, condominiums, and middle-class subdivisions, while the land available for low-income tenants and home builders is increasingly restricted. Unfortunately, in most cases national and urban governments lack the necessary political will, financial capacity, and bureaucratic efficiency to regulate the land market effectively, whether directly through the planning mechanism and exercise of eminent domain or indirectly through taxation. For example, in the Philippines, a vicious circle has developed whereby the political influence of the landowners (disproportionately represented in the legislature) and the middle classes prevents urban land reform and even enforcement of zoning regulations. Such influences also mandate market-price compensation for any land appropriated by government and frustrate the efficient assessment and collection of property taxes. This in turn encourages speculation and increases property prices. Thus with limited revenues local governments lack the means to regulate the market effectively, let alone purchase land for low-income housing (Goss, forthcoming). Effective land-use planning is also compromised by changes in personnel and the hopeless jurisdictional fragmentation of most metropolitan administrations (Richardson, 1989, p. 369; Tasker, 1990, p. 53), to say nothing of infamous widespread (although not universal) corruption in local governments.

As a result of the increasing price of land, the ratio of shelter costs to household income is much higher and is increasing more rapidly in Southeast Asian countries than in most of the developed world—in Malaysia and the Philippines, for example, the ratio is 2.1 and 1.6 times the ratio in the United States (UNESCAP 1993b: 3.16). Consequently, a significant proportion of the urban population in most Southeast Asian cities, and in the megacities in particular, is priced out of the market and forced to occupy empty lands illegally. In Metro Manila, for example, squatters constitute 40 percent of the population living in at least 415 communities (McBeth and Goertzen, 1991, p. 30; Zablan, 1990, p. 11), and 45 percent of the population lives in slum conditions (UNESCAP, 1993b, p. 5.3); in Bangkok about 23 percent of the population lives in 1,020 slum and squatter areas, probably just under half of which are under threat of eviction (Angel and Pornchokchai, 1989, p. 137; Boonyabancha, 1983, p. 257); and in both Kuala Lumpur and Jakarta squatters constitute about 25 percent of the population (Dwyer, 1990, p. 302; Hoffman, 1992), although it should be noted that the proportion of land that is colonized illegally by squatters is lower in Jakarta than other megacities due to the persistence of traditional land rights in urban villages, or *kampung* (Hoffman, 1992, p. 331–32).

Government agencies in Southeast Asia are generally hostile to the spontaneous disorder of slum and squatter communities, even if they provide shelter for a significant proportion of the urban population. Squatters obstruct development (where, in a capitalist market, land is allocated to the highest, most capital-intensive use), contravene numerous laws, complicate the provision of urban services, and are an affront to the modern sensibilities of government representatives very much concerned with the image of their city. Squatters are also often well-organized; thus, they petition for services, invade unused urban lands, and resist evictions in an attempt to preserve their right to a living place in the city. In the name of progress, squatters are evicted and slums redeveloped or, at the very least, barricaded, and hidden from view. In the Philippines, for example, even as recently as November 1996, the embarrassment of elites at the extent of poverty in the city led to the temporary boarding up of squatter settlements and the dispersal of vendors along routes between the airport, luxury hotels, and the conference centers so that those attending the APEC summit meeting would not have to witness inequality and underdevelopment.

Only the government of Singapore has successfully addressed the problem of squatters and slums, which it inherited, along with other national problems, on secession from the Federation of Malaysia in 1965. Singapore's leadership, concerned about the tiny city-state's vulnerability to internal ethnic and class tensions, external aggression from its larger neighbors, and fluctuations in the world economy, has used strict control over the economy, society, and the built environment to create an efficient global city. The state, for example, through a thoroughly professional bureaucracy, administers a master plan for land use, employs compulsory acquisition of land for development, levies charges on private development, and is the owner of nearly all land that it leases for development (Leung, 1987). It solved the housing problem by accommodating the majority of its population in government-built flats. The Housing and Urban Development Board (HDB) built 500,000 housing units between 1960 and 1985, providing shelter for 85 percent of the population, the majority (62 percent) of which is owner occupied. This massive housing program successfully incorporated minorities, reduced ethnic tensions, and provided a safety net for petty entrepreneurialism. It also forced high rates of domestic saving and national investment, exerted a multiplier effect on the national economy, and provided a minimum standard of living for workers even while wages and labor organizations were controlled to attract foreign capital (Castells, Goh, and Kwok, 1990; Yeung, 1985). Although residents have had to tolerate both dense high-rise living, with little flexibility as to location, size, and style, and all workers have had to participate in a compulsory payroll savings scheme to finance housing construction (Corey, Fletcher, and Moscove, 1992), this must be acknowledged as a remarkable success.

Other Southeast Asian countries possess unified agencies charged with developing public housing. But they lack the necessary economic resources and political will, and they face a problem of far greater magnitude—as a city-state Singapore has not had to cope with massive rural-to-urban migration nor conflicts between national and provincial governments (Dwyer, 1990, p. 302). As a consequence, they have provided little housing for the poor and public housing has usually benefited only government workers, such as teachers, police or military, or, in the case of Malaysia, particular ethnic groups (Ruland, 1992, p. 267). Without the secure, formal employment that economic growth has guaranteed to virtually all residents in Singapore, the urban poor are unlikely to be able to make regular payments even for subsidized housing and may even find that their adapted survival strategies are compromised by the spatial structure of public housing (Goss, forthcoming).

Somewhat paradoxically, perhaps, the income-generating strategies of petty commodity producers may depend on the spatial concentration of urban poverty in the slum and squatter settlements of the Southeast Asian city and on the particular ecology of poor communities. First, partly due to land assembly costs, public housing is usually located on the urban periphery where the journey-to-work in the congested city strains household budgets and consumes precious productive time. Second, due to this location and the low relative densities of living, there are fewer opportunities for individuals to sell goods and services, such as cheap foods, petty manufactures, and repairs, to other poor households. Third, due to notions of the ideal separation of workplace and living place and of the ideal living densities in formal housing developments, there are usually regulations restricting entrepreneurial activities such as trading and petty manufacturing in the home and also the modification of living space to accommodate subtenants. Research in Manila found that 59 percent of squatter residences housed some kind of petty enterprise—from the manufacturing of soy sauce to pig raising, and from small stores to "gambling dens"—and that 67 percent of houses generated rental incomes for squatter households (Goss, 1990). Fourth, both the physical design of public-housing units and regulations as well, prevent livestock raising and subsistence production that are vital to supplement household income—for example, in Jakarta and Bangkok, respectively, up to 15 percent and 19 percent of food for the urban poor is self-produced (Evers, 1981; Evers and Korff, 1986). Fifth, public housing imposes a particular order on social life based on the norm of a single nuclear family and privacy, while the poor rely on extremely adaptable household forms and social networks maintained by intense everyday contacts in the crowded informal public spaces of the slum—for example, in alleys, at water pumps, and in front of small stores. Sixth, in public-housing units, services such as electricity, water, and sanitation are provided by the state or commercial sources, and residents have little control over the quality and pricing of these services. Finally, much of the productive and reproductive work within the poor community is done by women who are able to tend to their children and younger siblings, keep house, raise livestock, produce handicrafts, sell cooked foods and general goods, or manage tenants, without leaving their home (Goss, forthcoming; Jellinek, 1991; Thorbek, 1987). The design and regulations of public housing do not provide sufficient flexibility for these diverse household economies and so reduce income opportunities of the household.

As a result of these constraints on their survival strategies and the fact that the poor do not have savings to meet sudden medical expenses or to provide entrepreneurial capital, beneficiaries of public housing notoriously sell their rights, even if this is technically illegal, and the real beneficiaries become the middle-class households or even speculative investors that buy them (Lindauer, 1981). Partly as a result of these failures, in recent years governments have all but abandoned direct construction of public housing in favor of sites and services and upgrading schemes in which the government provides tenure security and minimal services, while the poor construct and improve their own housing. Self-help solutions have been promoted and subsidized by international agencies such as the World Bank and have had some local successes, but they have generally been plagued by the same problem of "downward raiding" in which immediate beneficiaries, who have qualified by length of residence and low household income, sell out to wealthier individuals who benefit from long-term investment.

Photo 5.7 Polluted canal used for waste disposal, Jakarta. (Leinbach)

In self-help housing schemes poorer residents provide their labor to improve the physical infrastructure in upgraded communities or to construct their individual homes in sites and services schemes—where house lots and minimal facilities such as outhouses, sewage, and drainage are provided. In upgraded squatter and slum communities the beneficiaries have often been involved in many years of organizational efforts to protect their community from eviction and development, or to obtain relocation sites, but as property values increase, the properties become attractive for investment by middle classes. The poorer households, who are often unable to keep up with payments—most schemes require the poor to pay to recover costs—often sell out. On the other hand, where the authorities have relocated squatters, sites and services have often been provided on the periphery of the city where land is cheapest. In Metro Manila, for example, during the antisquatter drives of the martial-law years in the 1970s, the urban poor were evicted and relocated far from employment opportunities, markets, and relatives, and the beneficiaries have returned to squatter settlements or slums, abandoning the lots and outhouses with which they were provided (Keyes, 1983; Westfall, 1990; see also Boonyabancha, 1983).

Lessons have been learned, however, and some novel schemes have been developed to cope with these failings. Under Jakarta's Guided Land Development (GLD), for example, sites and services are prepared by the government in consultation with local residents, and under Bangkok's land sharing schemes, squatters are provided with a portion of the land for their residential needs, and the remainder is developed by the government or private landowner. Such schemes have received limited application, however, again due to the lack of financial resources and political commitment from urban administrations.

Urban service provision. As we have seen, in most Southeast Asian cities, the poor, by virtue of their low incomes and often illegal occupation of land, are excluded from basic services such as piped water, sewerage and electricity and live under threat of flooding, fire, and contagious disease. Table 5.2 provides some figures on the limited service provision in Southeast Asian cities. Through personal experience of squatter living in Manila, for example, it was found that human waste flowed in open drains or was wrapped in newspaper and thrown into the canal (the so-called wrap-and-throw method), water was obtained from standpipes, electricity was pirated from nearby utility lines, and garbage was burned after it was picked over by scavengers (Goss, 1990). It should be noted that scavenging provides a vital means of income and a service as it efficiently recycled waste material (Goss, 1990; Sicular, 1992). A recent study in Hanoi, for example, has found that scavenging provides a livelihood for about 6,000 people and is responsible for collecting one-third of the city's refuse

Country or City	Piped Water	Piped Sewerage	Solid Waste Collection
Manila	58	15	75
Jakarta	30	40	25
Burma	7	n/a	n/a
Bangkok	50	n/a	75
Vietnam	57	33	n/a

Table 5.2 Percent of urban population receiving services in selected countries or cities. *Source: UNESCAP, 1993a, 1993b.*

(Hiebert, 1993, p. 36). Although it might be similarly argued that the private provision of water and electricity within settlements of the poor provides livelihoods, for the most part, profits accrue to wealthier households that can afford to pay for infrastructure and that then sell services to their neighbors at inflated prices.

While poor communities organize to provide their own services, they lack sufficient capital to purchase equipment and supplies, and this is often provided by politicians and government agencies in exchange for promises of political support. This widespread process is a means by which the poor are integrated into the political system, but it also reproduces a politics of patronage and particularism. Some attempts have been made to provide services on an urban scale, again with support of the World Bank and other international agencies, under the philosophy of self-help. Indonesia's *Kampung* Improvement Program (KIP) is often cited as model of self-help service provision. Begun in 1969, the program upgrades physical infrastructure of the "urban villages" of Indonesian cities—providing walkways, drainage, water supply, community bathroom and laundry facilities, and garbage disposal—by combining local labor with government capital and expertise, supported by loans and technical assistance from the World Bank Asia Development Bank, and the Government of the Netherlands (United Nations, 1989). About 90 percent of Jakarta's *kampungs* have been upgraded, with obvious visible success. Research shows, however, that the improvements have not benefited all households equally—the program was often co-opted by the wealthiest residents, and subsequent increases in land prices and rents have sometimes forced the poorest *kampung* residents to leave (Jellinek, 1991; Karamoy and Dias, 1986). This is thus a general problem of self-help solutions to urban housing and service provision, and it seems to challenge any attempt to address the problems of urban poverty at the level of the community or even across a city without more radical and widespread social reforms (Photo 5.7).

The urban environment. The poor are most vulnerable to environmental pollution and its consequences, which include poor health and sociopsychological stress that may reinforce poverty. Singapore is famous for its cleanliness and order, but most other cities in Southeast Asia have serious environmental problems, and some of the largest cities have reached or even exceeded their ecological limits (Douglass, 1992, p. 13). The most obvious problem perhaps is air pollution that results from the combustion of fossil fuels in power generation, motor vehicles, and industrial processes, from the incineration of solid waste, and occasionally from forest fires in surrounding areas during the dry season. The largest cities, especially Bangkok, Manila, and Jakarta, have high levels of sulfur dioxide, nitrogen oxide, carbon monoxide, ozone, lead, and suspended particulate matter that can lead to respiratory and other diseases. In Bangkok, for example, particulate matter and carbon monoxide are, respectively, two to three times and 50 percent higher than international standards (Westlake, 1990, p. 56).

As much as 70 percent of all air pollution is produced by motor vehicles (Handley, 1990, p. 44), and this source is increasing dramatically in the large cities as the middle class expands and as governments promote motorization in the name of economic efficiency—this despite the obvious advantages of nonmotorized vehicles in these contexts, which include their low cost and absorption of labor, as well as freedom from pollution (Replogle, 1992). In Bangkok, for example, car ownership doubled from 1977 to 1987 and increased again by almost 80 percent in 1989 to 1991 alone—in a city that has a very low proportion of road surface to total area: 9 percent compared with London at 17 percent and New York at 23 percent (UNESCAP, 1993a: p. 105). As a result more than 2 million vehicles crawl at an average speed of less than 6 mi (10 km) per hour (Westlake, 1990, p. 56). It is the urban poor, many of whom work in the streets and many of whom cannot afford the increasing time and money costs of commuting, who pay the highest costs of pollution and congestion. Of the largest cities only Singapore has effective traffic controls; since 1975 Singapore has attempted to reduce congestion by a system of area licensing in which low-occupancy vehicles without permits are not allowed to enter designated areas during peak hours without displaying special licenses. Very recently Singapore has begun to implement electronic road pricing in its core area. Mass transit offers a partial solution, and monorails are under construction in Manila and Kuala Lumpur, while trains provide vital relief in Jakarta, but without the strict discipline of Singapore's society, it will be hard to limit the number of private cars in the urban area, particularly as the suburbs expand inexorably to accommodate the new middle classes.

A second obvious problem is the pollution of surface water by upstream industries, leaking septic tanks, and the dumping of solid wastes by residential communities. Rivers and canals in the big cities are often biologically dead and are a potentially dangerous source of bacteria and poisons, especially the Ciliwung in Jakarta, the Pasig in Manila, and the lower Chao Phraya in Bangkok. Environmental pollution is associated with the two main causes of infant mortality—diarrhea and respiratory diseases—and the children of the poor, due to their residential location and lack of access to health care, are particularly vulnerable. In the poor *kampungs* of Indonesia, for example, infant mortality is five times the Jakarta average (UNESCAP, 1993b, p. 4.54), and industrial pollution of rivers and canals discharging into the Java Sea has caused mercury poisoning and brain damage in children (Hadiwinoto and Leitmann, 1994).

A third, less obvious environmental problem is caused by the increased demand for water by expanding cities. For example, urban sprawl and residential development in the uplands to the south of Jakarta has reduced water recharge, even as the extraction of groundwater has increased, and as a result saltwater intrusion occurs on the coast and the city is subsiding. Similarly, in Bangkok, overdraft and seawater intrusion are contaminating water supplies, and this city, which is built on marine clays, is also sinking, exacerbating the effects of seasonal flooding (Sharma, 1986). Such problems occur even in smaller coastal cities such as Cebu, and again it is the poor who do not have access to piped water and who live on flood-prone lands that are most vulnerable to these effects. We might speculate on the combined effects of rapid population growth, groundwater depletion, poor drainage, and sea-level rise on Southeast Asia's cities—where the majority of the urban populations live (Figure 5.5). An

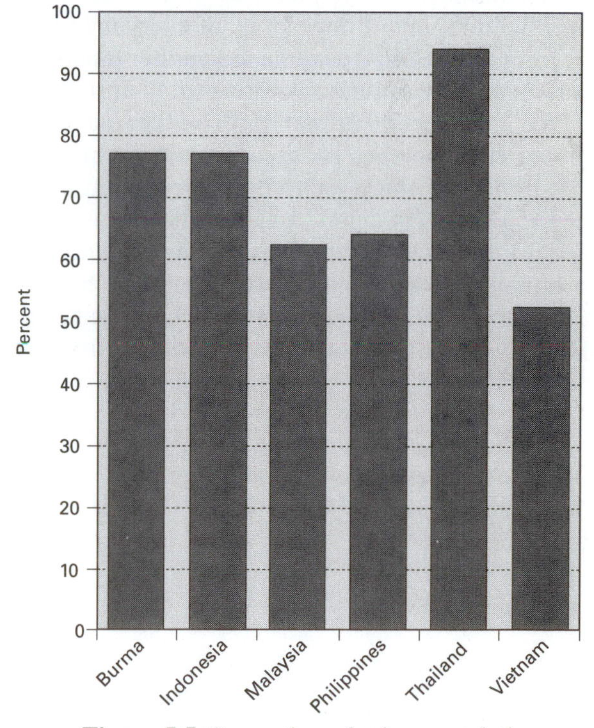

Figure 5.5 Proportion of urban population living in large coastal cities. *Source: Social Indicators, 1990, p. 50.*

environmental crisis seems almost inevitable (Douglass, 1989).

CONCLUSION

In this discussion we have taken a problem-oriented approach to Southeast Asian urbanization, following the convention of western observers who have used medical analogies in "diagnosing" urban maladies and suggesting possible "treatments" for over urbanized conditions. In a sense, this is the perfect model for academics and international agencies because like Western medical practice it puts the expert in the active role vis-à-vis a passive patient. The international consultant, backed up by academic theory and the considerable economic clout of multinational development agencies, is seen to possess a critical, objective perspective and to be in a position to inform Southeast Asian governments of the most appropriate policies. This assumes, of course, that the governments are receptive to such advice and are equally desirous to improve the conditions of the urban poor. In concluding this chapter we would like to challenge this view.

A Right to the City

First, our personal experience of living in Southeast Asian cities, even in Jakarta and Manila, the two largest and by all accounts most problem-ridden cities in the region, is by no means negative, despite the poverty and insecurity and the undoubted difficulties of everyday life. The poor communities on the whole are neither fatalistic nor depressed but are actively struggling for their right to the city. One must remember that the drive that makes farmers leave their villages for the city is the same as that which motivates the poor to earn their livings and to build their homes under extremely difficult conditions—the desire to obtain better lives for their children. This desire may have its origins in rural cultures, but in the urban communities it has led to the development of some quite modern urban institutions, including active associations and organizations of workers and squatters. In other words, the very people who we want to help are busy helping themselves.

However, the manner in which they are helping themselves is perceived as threatening by some Southeast Asia governments—particularly perhaps after the well-organized urban poor in Manila progressed from providing numbers for Marcos's political rallies to mobilizing for his downfall in February 1986 in the so-called People's Power Revolution. Vendors and squatters are seen by authoritarian or weakly democratic governments to be a threat to the efficiency of the urban economy and the political stability of the state, and so their activities are limited by discriminatory laws and direct repression. We might ask why this is the case, because surely the activities of such organizations are essential to a truly self-help solution to urban problems.

We believe this is because the struggle of the vendors and squatters is literally over the meaning of urbanization and over the right to the city. We have suggested that workers in what is conventionally called the informal sector are concerned to establish a right to a living and that squatters are concerned to establish a right to a living place. Together these rights provide a means for the household to survive in the city and for it to maintain secure access to urban space. Organizations that articulate and pursue this right challenge the very notions of private property and an orderly and modern urban development central to the self-identity of the middle classes and the urban elites that make up Southeast Asian governments (see Westfall, 1990). The poor are organizing for a radically different notion of citizenship, based on need rather than on access to the economic and political resources of the modern capitalist economy and bureaucratic state. In the Southeast Asian city, this struggle for territory and over the meaning of the urban development process in general is perhaps of equal, or greater, significance than the perennial struggle over wages and working conditions in the formal sector (also Evers, 1983). It is this that we think the "geography" of the Southeast Asian city should really be all about.

6 Perspectives on Agriculture and Rural Development

JAMES A HAFNER

The history of Southeast Asia has been intimately associated with agriculture and the rural sector throughout most of the past two millennia. Many of the early states prospered through trade and commerce; yet the region's unique sense of place was significantly shaped by the natural environment. Marked by a warm climate, abundant rainfall, and ubiquitous waterways, lifestyles developed that were dominated by the forest, padi cultivation in river valleys, and fishing. The people grew common crops, used similar technology, ate food in similar manners, and lived in houses sharing common materials and structural features. By the fifteenth century, surpluses from these traditional agricultural systems supported the development of emerging urban core areas while spices and rice were being traded as far away as China and Europe.

During the 400 years of the colonial period, large areas of subsistence cultivation gave way to both smallholder and plantation agriculture. Spices, rice, rubber, sugar, and timber were produced for export, new crops and production technologies were introduced, and the rural economy became increasingly incorporated into the global market. In the contemporary era, natural resource-based development and investments in infrastructure and technology to increase agricultural production have helped to underwrite rapid economic growth in many of the ASEAN nations. While these policies have contributed to rapid industrialization, they have also exacted a high price in environmental degradation, rising urban-rural and regional income gaps, and persistent rural poverty. The broad themes of regional diversity and development, then, are especially appropriate to agriculture and the rural sector in Southeast Asia.

This chapter begins with several related perspectives on agriculture and rural development in Southeast Asia. The first focuses on development and change in agricultural production and land use during the past 20 years. This leads to consideration of some of the general factors affecting agrarian change and agricultural modernization. Because agricultural activities and the rural sector are so closely related, these elements are then considered in terms of the larger issue of policies and strategies for rural development. Our attention then turns to the region's major agricultural systems, patterns of land ownership and tenure, and land reform as a policy tool intended to redress social and economic inequities in rural areas. Arguably, one of the more important influences on rural agrarian change has come from the commercialization of agriculture, particularly the dissemination of the Green Revolution technology. This 'package of technology' is considered in terms of the factors affecting its uneven pattern of adoption, the arguments about its adverse social and economic impacts on the rural sector, and the problems surrounding its sustainability in the future. Finally, we consider the larger issue of rural development strategies, how the focus of rural development has changed, and whether the goals of growth and equity have contributed to an alleviation of rural poverty. We conclude with a consideration of key issues and trends which will influence the region's rural agrarian future.

AGRICULTURAL PRODUCTION AND LAND USE

Despite the rapid growth of urban areas and the attention given to expanding manufacturing and industrial activities, the majority of Southeast Asia's population still lives in rural areas and depends on agriculture as their main form of economic livelihood. However, the performance of the agricultural sector during the past several decades offers a striking contrast to most other areas of the developing world. Agricultural output has risen by more than 60 percent since the early 1970s, production of padi has more than doubled, and exports of food and industrial crops have risen significantly. This encouraging record has taken place in the context of considerable diversity in agricultural systems. Pockets of intensive market gardening on the urban periphery contrast sharply with the densely settled and intensively cultivated padi landscapes of central Java and with areas of sparsely settled and still relatively undisturbed tropical forest. Although these production environments reflect the agricultural diversity found in Southeast Asia, the strength and continued importance of agriculture throughout all countries in the region cannot be questioned. This can be illustrated with selected agricultural data for the past 20 years.

Agriculture continues to employ the most people in all of the countries in Southeast Asia except Singapore and will continue to do so into the twenty-first century. This is readily apparent by the fact that more than one-half the total regional labor force is engaged in agriculture, although its share of the labor force has been declining in most countries for some time (Table 6.1). The agricultural share of gross domestic product (GDP) has also been declining in all countries for almost 30 years but at rather different rates. This change has been most significant in Thailand, a decrease from 40 percent in 1960 to only 13 percent in 1991, where rapid gains in the industrial, mining, and service sectors have taken place in the last decade. Despite this shift, agricultural exports such as rice, rubber, and tapioca are among Thailand's top exports. Similar sectoral shifts have also occurred in Malaysia where manufacturing's share of GDP showed a 20 percent gain between 1960 and 1988, and it now ranks ahead of agriculture in its share of GDP and in total exports. Nevertheless, like Thailand and Indonesia, a major share of Malaysian manufacturing continues to be based in agro-industries, wood products, and petroleum production. Primary resource extraction, minerals, timber, and agriculture remain important components of the region's economies even though the role of industry has grown significantly.

The regional changes in land use however, have not been as dramatic as the growth in the nonagricultural sectors of the economy. Arguably, the most significant change regionwide has been in forested area. Since 1973 the proportion of the total area identified as forest has shrunk from 61 percent to less than 48 percent, and although this trend has slowed it is expected to continue into the next century (Figure 6.1). Due to differences in accounting methods and definitions of forest, this situation is perhaps more serious than is reflected in official figures (Kummer and Turner, 1994). Since the late 1960s satellite imagery has provided more-precise measurements of forested area, but these data are too often at a scale that does not allow differentiation between undisturbed primary forest, logged or "managed forest" and secondary regrowth forest. A reasonable estimate is that less than one-third of the region's land area is forested and remains largely undisturbed by human interference (Box 6.1).

Across the region, agricultural land has increased by only 15 percent in the last 20 years. The notable exception to this pattern has been Thailand where agricultural land area increased by almost one-third, primarily as the result of forest clearance and the spread of cash export crops such as cassava. Most of the statistically significant change in land use has taken place in the catch-all category of *other* which accounts for more than 28 percent of regional land use. Here however, comparisons between countries are difficult because of apparent differences in how land is classified in some countries. For example, predominantly urban Singapore has more than 90 percent of its land in the *Other* category; yet the decidedly rural agrarian countries of Vietnam, Burma, and Laos have strikingly large percentages of their land also in this category. Despite these problems, it is not unreasonable to assume that the expansion of arable land throughout the region has been small. This is of some significance when one considers the rapid growth of population during the past 20 years and that the rising demand for food has been satisfied primarily through the intensified use of existing arable land. Increased yields and production of food staples such as rice, which have resulted from the adoption of Green Revolution technologies, have been instrumental in this regard. This is even more significant because of the slow growth in irrigated cropland.

Irrigation is a major factor in increasing agricultural productivity, as it allows for multiple cropping and the use of modern padi varieties while reducing drought losses. However, since 1977–79 when international lending and assistance for irrigation in Southeast Asia peaked at US$630 million, overall investment had fallen to US$202 million in 1986–87, or one-half the level of the 1970s (Svendsen and Rosegrant, 1994). This has also been accompanied by a sharp decline in the annual rate of growth in irrigated area; 4 percent between 1980 and 1985 to 1.5 percent from 1985 to 1988. There are a number of factors accounting for this change, but the main reasons have been declining international padi prices and the rising per-hectare costs of new irrigation facilities and production inputs. The existing irrigation facilities are also confronted with problems of inefficient water use, siltation, and long-deferred maintenance that will need to be addressed if levels of productivity are to be maintained (Box 6.2).

Despite the slow growth in irrigation facilities, Southeast Asia has seen some major improvements in agricultural crop yields and total production (Table 6.2). While these changes are most evident with respect to padi production, export crop production has also grown as well, especially in Indonesia, the Philippines, and Thailand. Since 1980 maize production in Indonesia and the Philippines has increased by more than 65 percent; the output of cassava in Thailand and Indonesia by almost one-third; and smaller but significant gains in coconut production have also taken place in all three countries. Indonesia has also emerged as the world's second-largest oil palm

Country	Agricultural Share of GDP (Percent)			Agricultural Share of Labor Force (Percent)			Land Use as Percent of Total Area								Irrigated Land as Percent of Cropland	
							Cropland		Pasture		Forest		Other			
	1960	1980	1991	1965	1980	1993	1975	1993	1975	1993	1975	1993	1975	1993	1975	1993
Burma	33	46	48	-	53	45	15.2	15.3	0.6	0.5	48.9	49.3	35.4	34.9	3.0	3.8
Cambodia	-	-	49	82	75	69	17.3	13.6	3.3	11.3	75.8	65.7	3.7	9.4	28	37
Indonesia	54	26	21	75	58	46	10.9	17.1	6.8	6.5	67.5	61.7	14.9	14.6	12	13
Laos	-	60	60	80	76	70	3.7	3.5	3.5	3.5	61.5	55.2	31.3	37.8	7	7
Malaysia	36	24	23	63	42	30	14.2	15.0	0.1	0.1	68.4	67.8	17.3	17.2	10	10
Philippines	26	23	22	61	52	45	24.6	30.8	2.8	4.2	45.2	45.6	27.3	19.3	16	20
Singapore	8	0.3	0.3	8	2	1	13.1	1.6	-	-	4.9	4.9	82.0	93.4	-	-
Thailand	40	25	13	84	71	62	32.6	40.7	1.1	1.5	36.2	26.4	30.1	31.4	16	19
Vietnam	81	-	37	81	68	58	18.9	20.6	0.3	1.0	41.6	29.6	38.6	48.8	24	29

Table 6.1 Selected agricultural data for Southeast Asia *Source: Asian Development Bank, 1993: World Resources Institute, 1994; UN FAO Production Yearbooks.*

BOX 6.1 Deforestation

Calculating rates of deforestation in Southeast Asia is difficult because of different definitions of *forest* and methods used to categorize vegetation regrowth. For example, some definitions of *forest* are based on the types of tree species that predominate in particular areas. The prevalence of dipterocarps in the ever-moist lowlands of Sumatra, Peninsular Malaysia, and Kalimantan are associated with tropical rain forests. The FAO however, classifies forests without regard to tree species but in terms of their structure, the openness of the forest canopy, and how extensively they have been cleared for agriculture. Some national forest agencies have even included grasslands and pasture in their calculations of forest.

Putting this aside for the moment, consider the estimates of annual deforestation rates for the two subregions of Southeast Asia shown below (Figure 6.1). How do these annual rates compare with more familiar areas in the United States? Between 1976–80 and 1986–90, the rates of annual deforestation for both the insular and continental subregions more than doubled or, in more familiar terms, increased annually by an area roughly two-thirds the size of Connecticut to one equivalent to the entire state of New Jersey. Between 1986–90, an area equal to the size of Massachusetts and Connecticut was being deforested each year in all of Southeast Asia (see also Map 2.6a, 2.6b: Forests of Southeast Asia).

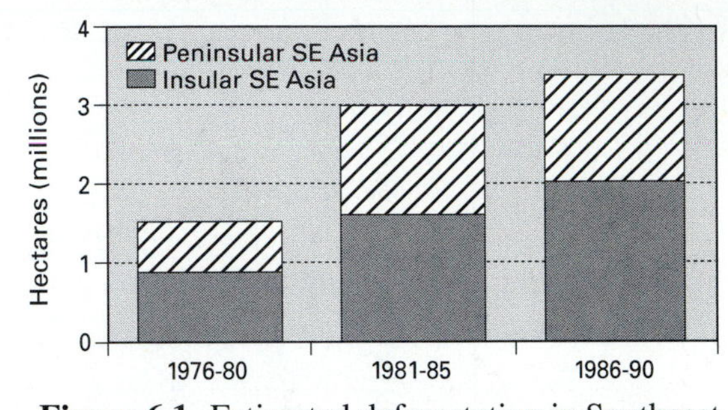

Figure 6.1 Estimated deforestation in Southeast Asia. *Source: UNFAO Production Yearbooks.*

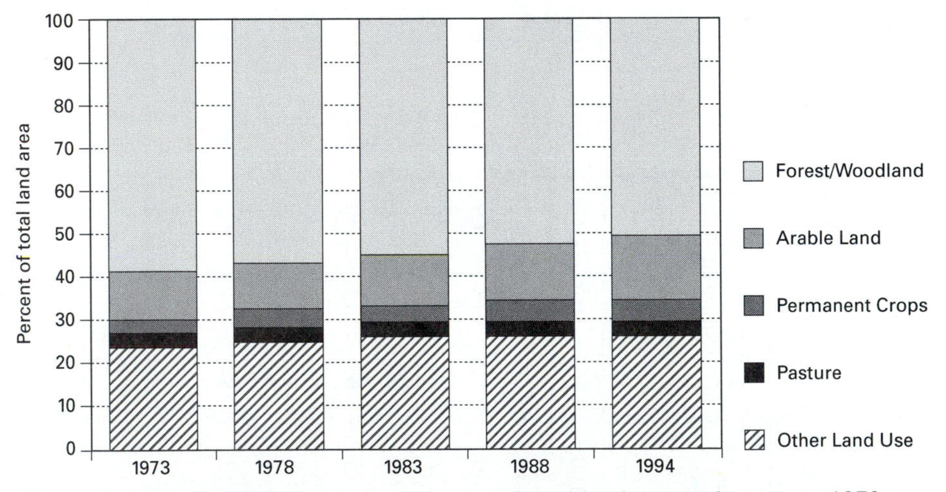

Figure 6.2 Regional land use as percentage of total land area, select years, 1973-1994. *Source: UNFAO Production Yearbooks.*

producer after Malaysia with 1995 output at 4.56 million tons and an increase in planted area since 1969 of more than 450,000 ha. Stable or improved world market prices, new export markets, improved production techniques, and the expansion of cultivated area are among the factors contributing to this growth. Large areas of forest in Indonesia have been burned to clear areas for oil palm plantations, and the smoke from these fires has been a major cause of the haze that has covered much of the region in the late 1990s (see Box 2.2). In contrast, the effect of market forces—low demand and prices—have contributed to a decline in natural rubber production in Malaysia. Between 1980 and 1992, rubber production declined by some 330,000 tons (300,000 metric tons), and almost 5 million acres (2 million hectares) of rubber land were converted to more-profitable oil palm cultivation in 1990. This shift has enabled Thailand to emerge as the world's leading natural-rubber producer—satisfying 25 percent of global demand. Although these trends indicate an increased diversity in agricultural production throughout the region, the success of this diversification remains tied to global market demand and prices.

In contrast to earlier periods, increased padi production has come primarily from the higher yields and mul-

BOX 6.2 Water and Irrigation: The Paradox

Water for crop irrigation is among the most critical natural resources for agriculture in central Thailand, the primary region producing rice for export. Although many areas are flooded in the rainy season, dry season water demands must be met by importing water from the North. This paradox has resulted from the lack of new investments in dam and irrigation facilities since 1957, intensification of rice production, and increased water demand from nonagricultural users. The existing irrigation system has also become increasingly inefficient due to water loss through porous canal linings, siltation, incursion of aquatic weeds, and damage to canal structures. These problems have been aggravated by increasing water demands from urban and industrial growth, residential development, recreational needs, and dry-season cultivation. In 1994 Thailand had 25 golf courses in operation and another 30 under construction. With an average size of 533 acres (213 ha), when all these courses are completed they will consume more than 476 million cubic meters of water per year; an amount equal to one-half of the surface water used by urban centers in 1991. This may only aggravate the growing dry-season water shortages for agriculture that has seen more farmers shift from rice to field crops, which require much less water. This problem may only worsen with population growth and the increasing demands from nonagricultural water users.

Country	Padi		Maize		Coconut		Cassava		Rubber		Sugar Cane	
	1980	1992	1980	1992	1980	1992	1980	1992	1980	1992	1980	1992
Burma	13.3	12.6	0.16	0.19	0.1	0.18	-	-	-	-	2.0	2.4
Cambodia	1.7	2.2	0.1	0.05	.03	0.05	-	-	-	0.03	-	-
Indonesia	29.6	45.8	3.9	6.7	10.9	13.0	13.7	16.3	-	-	-	-
Laos	1.0	1.6	0.03	0.7	1.2	1.0	0.07	0.22	-	-	-	-
Malaysia	2.0	1.8	0.01	0.04	1.2	1.0	-	-	1.5	1.2	-	-
Philippines	7.7	9.1	3.1	5.1	8.6	9.0	-	-	0.10	0.17	22.3	21.5
Thailand	17.4	20.0	3.0	3.5	0.74	1.4	17.7	22.0	0.50	1.4	18.6	48.4
Vietnam	11.7	18.5	0.42	0.66	0.3	1.0	-	-	0.04	0.07	-	-

Table 6.2 Crop production (million tons) by country, 1980 & 1992. *Source: UN FAO Production Yearbooks.*

tiple cropping made possible in many areas by the adoption of Green Revolution technology. Throughout the region, multiple cropping has also increased, contributing to a 13 percent increase in harvested area and a 53 percent gain in production during the past decade. Indonesia has shown the greatest gains in both padi yields and production, and in 1989 mean yields were more than double those of Thailand, the other major regional padi producer. Padi productivity in Vietnam and the Philippines has also risen significantly, but with only a modest growth in harvested areas of 7 and 13 percent, respectively. Malaysia, by contrast, has expanded production by only 13 percent, and the padi farming sector has remained an area of relative poverty in an otherwise industrializing country. These changes for both padi production and other agricultural crops can be briefly summarized by using the FAO index of per capita agricultural productivity (Figure 6.3). The most surprising growth has come in Cambodia, although this is due less to gains in productivity than other factors. Within Southeast Asia, Thailand, Indonesia, and Vietnam have the largest proportion of their areas in arable land. Yet, the pattern of change in agricultural production is quite different, rising faster in Indonesia and Vietnam than in Thailand where overall productivity has been rather stagnant for a decade. The reasons for this lag in Thailand have been inadequate wet-and-dry-season control of irrigation and rainwater, less intensive farming on larger holdings, and the availability of land that has meant less pressure to intensify production. A similar pattern has also taken place in the Philippines where inadequate development of irrigation, the priority given to export crops, and rising fertilizer prices have contributed to stagnant or declining production.

These conditions lead to the inevitable question: How can the agricultural sector be expected to perform in the future? The area of agricultural land will undoubtedly continue to expand, but the limits to this expansion may not be very far in the future. In many countries under existing production conditions, the limits of arable land are rapidly being approached. This land "frontier" has already been reached in Thailand and Malaysia. The intensification of production can also be improved

Photo 6.1 Green Revolution inputs (fertilizer, insecticide) in Bandung Store Window. (Leinbach)

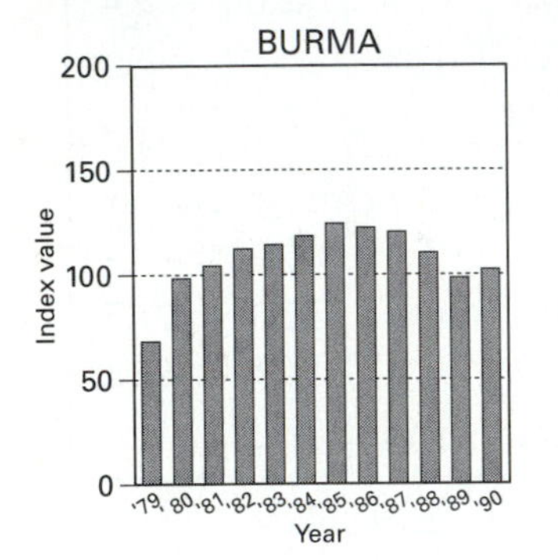

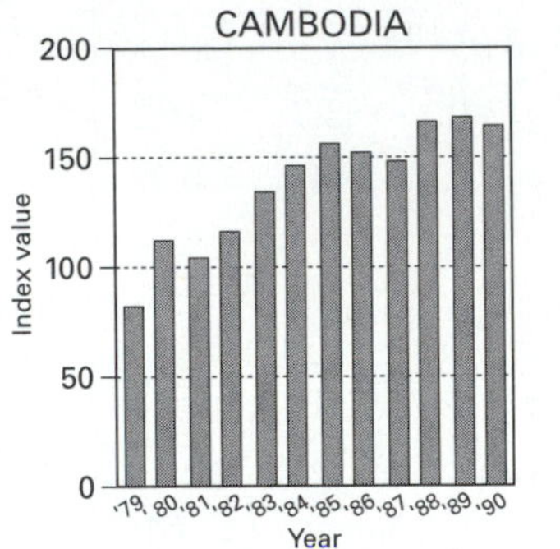

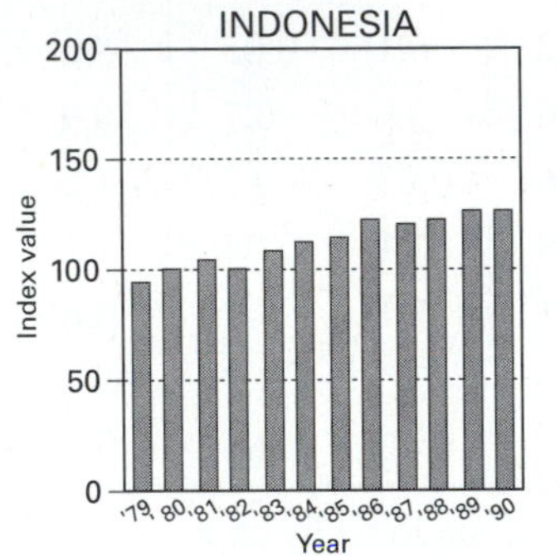

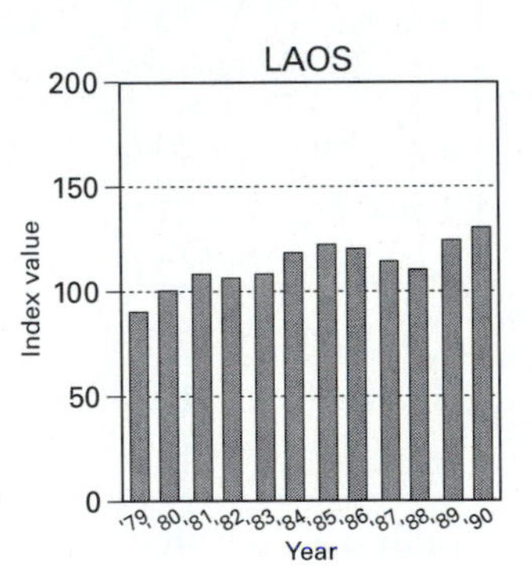

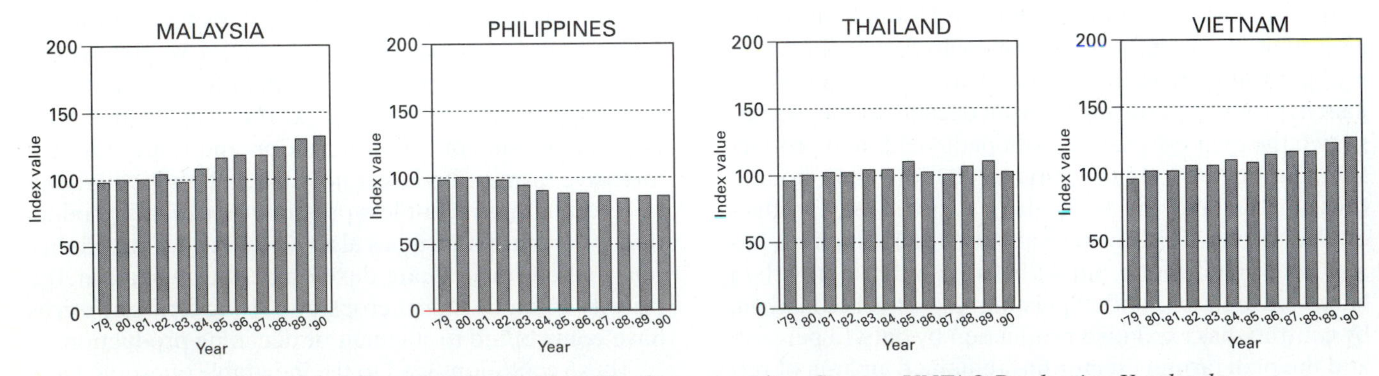

Figure 6.3 Index of agricultural production per capita. *Source: UNFAO Production Yearbooks.*

throughout the region, primarily through the development of new production methods to convert traditional land-extensive practices in the uplands to ones that are more land intensive. The prospects for this are arguably uncertain but may well be determined by current research to create more sustainable agricultural systems in Southeast Asia. The historical view of Southeast Asia as a resource-abundant region has begun to change rapidly in the past decade. This suggests a future in which the constraints on resources will become increasingly important and more efficient and in which sustainable forms of agricultural production will need to be developed.

AGRARIAN CHANGE IN SOUTHEAST ASIA

Many of the changes that have emerged in rural Southeast Asia in recent decades have been shaped by three major forces: population growth, commercialization resulting from penetration of market economies and new technologies, and the growing influence of state-led development. While high levels of population growth have been associated with natural resource degradation, they have also heightened demand for food, land, and water and have contributed to the intensification of land use and increased agricultural outputs. Explanations of what happens to peasant agrarian economies under conditions of high population pressure and agricultural modernization have been the subject of considerable discussion. Perhaps the most influential but debatable of these, Esther Boserup's (1965) theoretical statement on the relationship between population growth and agricultural change and Clifford Geertz's (1963a) "agricultural involution" thesis, are especially relevant to this brief consideration of agrarian change in Southeast Asia.

Boserup has argued that population density explains geographical variations in farming practices and increases in population tend to produce changes in those practices. Thus, when population growth begins to exceed the productive potential of a particular agricultural system, that society will find innovative ways of accommodating this growth rather than experience famine and starvation. From this will emerge a new or adapted agricultural system that is more productive and in the long-run will result in the "advancement" of agriculture. In modern-day Java, for example, the structure and productivity of the wet-padi agricultural system has evolved to the extent that it supports agricultural population densities in excess of 386 people per square mile (1,000 people per square kilometers) while contributing almost two-thirds of Indonesia's total padi production (Fox, 1993). This has led to self-sufficiency in padi production and the general absence of famine, despite a population increase of more than 75 million people since 1900 without opening new land to cultivation. This has been accomplished by a continual improvement in the wet-padi agricultural system so that padi yields have more than doubled in this century.

Geertz's study of the nineteenth and early twentieth century Dutch colonial economy in Java and his concept of *agricultural involution* offers a somewhat different interpretation (Geertz, 1963a). Colonial Java was seen as a relatively undifferentiated egalitarian rural society consisting of millions of small, labor-intensive subsistence farms. In response to colonial demands to increase sugar production, and in contrast to Boserup's general argument, the rural peasant economy absorbed the growth in population by more carefully managing their agricultural resources and increasing the per hectare productivity of their *sawah* or wet-padi holdings. This enabled padi and other food-crop production to increase steadily even though total output failed to keep up with population growth. For example, per-capita padi output declined from about 0.20 tons in 1880 to 0.13 tons in 1965, an all-time low, when increased double cropping and adoption of new production technology resulted in growth in yields and total production. Geertz also argued that this capability to absorb population despite rather stable yields, was made possible by various leveling mechanisms such as labor sharing, tenancy relations, and public feasts, which acted to redistribute wealth and minimize social and economic differences in village society. This not only spread life risks so that even the poorest were assured of survival but assured access to a limited resource base for all members of rural society. It also had the effect of inhibiting economic differentiation because the notion of "shared poverty" enabled rural society to accommodate a growing population on an ever more crowded land base.

Elsewhere in Southeast Asia the increased demands for food and other natural resources related to population growth have been met by expanding the amount of land under cultivation. Population pressures have pushed cultivation from the high-yielding lowland plains and valleys into the more fragile and less productive hill and upland environments where crop yields are lower, watersheds have become eroded, and forestlands have been denuded. For example, in the Philippines cultivated land more than doubled between 1960 and 1990, and most of this new cropland was in uplands which comprise more than one-half the land area of the country. The fact that rates of upland population growth have been 15 percent above the national average in the last decade suggests that population has contributed to these changes. Population growth and pressure on land resources have also contributed to Thailand's rapid expansion in cultivated area since the 1960s.

But, it would be an oversimplification to accept population growth as the only factor contributing to these processes. According to Blaikie and Brookfield (1987), growing numbers of people account for part of these changes, but they do not in themselves provide a sufficient explanation. Poor land management, poverty, inappropriate technology, unequal access to land, market forces, and other factors may affect conditions differently across the region. The importance of looking beyond population as an explanatory factor is suggested by the fact that rural populations in the region are generally growing at rates much below those for urban areas. In some countries such as Indonesia and Malaysia, rural areas are expected to have negative population growth early in the next century (Concepcion, 1993). Future growth in production may therefore depend more on the social and economic conditions of production than any growth in land under cultivation.

Commercialization has also become an increasingly pervasive force affecting the rural economy in recent decades. The traditional subsistence focus of rural households producing primarily to meet domestic needs has become undermined by the demands of a cash economy, various forms of state taxation, and the adoption of more capital intensive production systems. More capital intensive crops, such as padi and field crops produced for external markets, increase the farmers need for fertilizers and machinery, make them more dependent on commercial markets, and weaken their self-sufficiency associated with more traditional cultivation systems. These conditions are most present in areas where the commercialization of padi production under the Green Revolution has taken place. It has been argued that this has led to economic and social differentiation of rural communities, polarization into opposing groups of commercial farmers and landless agricultural laborers, and declines in the number of more traditional small-scale farmers.

This trend also has increasingly important implications for employment, unemployment, and how household labor is allocated between on-farm and off-farm work. Until a decade ago, rather little attention was paid to the rural nonfarm economy and rural industrialization. A key factor explaining this lack of interest has been industrialization-oriented development strategies that were expected to create new jobs in both the urban and rural nonfarm sectors. This was expected to help alleviate problems of rural poverty, reduce unemployment, and absorb surplus rural labor resulting from agricultural commercialization. The slow pace of rural industrialization and growth in the nonfarm economy in most countries suggests this has not occurred. Among the reasons for this are the limited capacity of agricultural intensification to absorb labor; the trend toward capital- rather than labor-intensive industrial growth; a bias in new manufacturing investment toward the larger urban centers; and insufficient credit, technology, and infrastructure in rural areas. More recently—in the past decade—rural industrialization and the development of nonfarm employment activities have become important policy and development priorities in many countries. Among the arguments supporting these strategies are: (1) the slowing of rural-to-urban migration and lessening of pressures on limited urban services and infrastructure; (2) easing the loss of skilled labor from rural areas while providing more modern workforce skills; (3) expanded use of underemployed rural labor; and (4) the benefits of rural industrialization for diminishing sectoral and regional inequalities. Although these benefits are seen as a potential panacea for the social and economic inequities in the rural sector, finding viable strategies to stimulate off-farm employment remains a major challenge in many countries. Some of these issues are discussed in the context of Industrial Processes and Development in Chapter 7 and for Indonesia and Thailand in Chapters 12 and 16.

In the last 30 years of development focused upon economic growth, increased productivity and improved income, the state has become an important agent of change in rural areas (Lea and Chandhri, 1983). Through state-led development planning, the setting of investment priorities, and deciding how resources will be allocated, the state has exercised considerable power in determining the pattern of agrarian change. Land reform and settlement programs, agricultural price controls, the selective provision of credit and creation of infrastructure, and even the promotion of family planning programs reflect government efforts to focus the pattern and structure of change in rural areas. In most ASEAN nations this has involved adopting a broadly "capitalist" or "market" approach to development through which traditional rural values, institutions, and control over resources have become increasingly incorporated into the mainstream economic and political policies of the nation. However, this trend has not gone unchallenged in some countries. Popular opposition to power abuse by the state and elites, mismanagement of natural resources, and social and economic polarization have become more common in Thailand and the Philippines, and "people's movements" have gained increasing political influence. The work of James Scott, and particularly his *Weapons of the Weak* (1985) as well as Philip Hirsch's *Development Dilemmas in Rural Thailand* (1990) offer effective examinations of this theme. How this conflict between policies of state-led industrial growth and public pressures for greater political and economic equity is addressed may affect the the strong economic growth that has been enjoyed over the past several decades.

RURAL DEVELOPMENT: GROWTH AND EQUITY

The approach of national governments toward development of the rural sector has changed in recent decades. During the immediate post-World War II development era, governments thought in terms of economic growth and increased productivity, particularly in agriculture. The policies designed to pursue those goals focused on individual economic sectors, such as agriculture, transport, or industry. By the 1970s continued evidence of lagging rural incomes, persistent poverty, and unemployment suggested that the benefits from this approach were not trickling-down to a majority of the rural population nor were they being shared equitably. Since then, more governments have begun to design and implement policies and strategies which specifically target the rural sector and ensure that the benefits from that growth would be shared equitably (Johnston and Clark, 1982). This more balanced approach has been typical of rural development planning in many of the ASEAN nations.

In 1961 the United Nations declared the 1960s as a "development decade" with an emphasis on providing the aid necessary for improving economic and infrastructure development: roads, power plants, irrigation facilities. These priorities were consistent with the goals of improving agricultural production, increasing savings rates, and accumulating capital for industrialization. The development and promotion of improved agricultural technology, particularly the Green Revolution, is the most prominent example of this approach. Although agricultural productivity did increase in some countries, the impacts of this improved technology were limited geographically and a large majority of rural farmers did not benefit. This led to a reevaluation of rural development strategies in light of the failure of the benefits of growth-centered policies to "trickle-down" to a large majority of the rural population. By the 1970s a second development decade emerged that emphasized what has been termed basic needs or efforts intended to improve the economic conditions of the small farmer with credit programs, agricultural assistance, dissemination of family planning technology, improved public health and nutrition, and disease prevention programs. The success of Thailand's internationally recognized family planning program, land reform begun during the New Society era of the Marcos Administration (1972–1986) in the Philippines, and the Provincial Development Program in East Java of the 1970s can be generally associated with this approach. In some cases, these strategies were packaged as "integrated" rural development programs because they focused on simultaneously addressing a range of social, economic, and agricultural needs for a targeted group of communities or a region. The Integrated Bicol River Basin Program in the Philippines is one notable example of this approach.

Contemporary rural development strategies in many countries in Southeast Asia contain a mixture of programs and objectives. These often include the goals of growth in incomes and productivity, meeting basic needs, and creating greater equity in rural areas in terms of access to resources and opportunity. There is nevertheless, a disturbing lack of commitment of resources, innovative program development, and sustained support for rural development efforts in most of the export-oriented industrializing countries in the region.

AGRICULTURAL SYSTEMS

The bulk of agricultural production in Southeast Asia takes place in small-scale, traditional, subsistence agricultural systems that have prevailed in the region for millennia. To describe these systems as *traditional* in the sense of remaining unchanged over time misrepresents the extent of environmental, technological, and economic change and adaptation that is typical of Southeast Asian agriculture. For centuries farmers have modified their production environments, experimented with new production techniques, and grown an expanding array of new commercial crops. Although pockets of traditional subsistence agriculture remain in more isolated areas of central Mindanao, Kalimantan, and Irian Jaya, these are increasingly exceptions. Commercialization and the spread of a cash economy have also modified the social structures that underpin agriculture. Land is increasingly less a common property resource than a commodity to be owned, bought, or sold; integration of the rural village into the national and international economies has affected how village and household labor is used; and landlessness and the rise of a wage labor class are more common today than at any time in the past. While visual impressions of the agricultural landscape may suggest it has changed little over time, the social, economic, and even biophysical frameworks behind these images have been significantly altered. This point emphasizes the importance of understanding agricultural systems in terms of both their natural and human components and how their interactions influence these systems.

The dominant agricultural systems in the region are swidden or shifting cultivation, *sawah* or wet-padi (rice) cultivation, and commercial plantation. Each of these systems takes on a variety of forms in different parts of the region, and most landscapes are a patchwork of a number of these systems. Their character also ranges from traditional subsistence to those heavily influenced by Western technology. Other forms of agricultural production such as home gardens and intensive commercial vegetable or market gardening are also important but are not discussed here. The related issues of land ownership, land tenure, and land reform that are central to understanding the problems of rural development and poverty in the region are also discussed.

Swidden, *Sawah*, and Plantation Agriculture

Swidden agriculture or shifting cultivation, is perhaps the most common form of cultivation for crops other than padi in Southeast Asia, especially in upland areas and lowlands with sparse populations. This type of agriculture is characterized by a pattern of clearing and then alternative cultivation of small plots of land for two or three years before abandoning them to fallow for periods of years or even decades. The related term *slash and burn*, also used to identify this form of cultivation, refers to the practice of preparing fields for cultivation by first cutting or "slashing" the vegetation and allowing it to dry before it is to be burned. With the ash providing a natural fertilizer for the soil, the cleared area is then planted with a diversity of crops such as dry padi, beans, tuber and root crops, chili peppers, squash, herbs, and maize. Crop yields are typically high for the first several years, but they will decline rapidly as soil nutrients become exhausted, weeds and insects deplete yields, and soils increase in acidity. Because only small areas are cultivated in any one year, but a large land area is required to accommodate the frequent opening of new swidden, this is defined as a land-extensive agricultural system.

While technological change has been typical of the more commercialized forms of agriculture, especially wet-padi cultivation, the technology of swidden cultivation has remained simple. Knives and axes are used for clearing, cutting, and harvesting; fire has long been a basic tool to reduce cleared vegetation; and sharpened poles or dibble sticks are used to poke holes in the soil for planting seeds. Agricultural inputs such as commercial fertilizers and pesticides are seldom used by swidden cultivators, so the simplicity of this technology contrasts sharply with the diversity of crops that may be cultivated on a single plot. In northern Thailand, the Lua tribe plant dry padi as the primary crop with dozens of other crops mixed together in the same swidden (Srimongkol and Marten, 1986; Kunstadter et al., 1978). It is this diversity of cultivated crops and the approach to planting that shape the ways farmers perceive the available natural resources, structure their agricultural decisions, and influence how they organize their labor. For example, the cultivation of different species of dry padi allows harvest periods to be staggered so that labor demands for harvesting can be distributed over longer time periods. Moreover this crop diversity also has a profound impact on how swidden systems perform, serving as a form of insurance so that potential risk is spread across the full range of cultivated crops rather than on a single crop. The diversity of this "field in the forest" then mirrors the natural ecological diversity of the tropical forest.

However, the practice of swidden agriculture has become the focus of a number of legal, social, and environmental conflicts. Land ownership among many indigenous minority populations practicing swidden agricultural systems is frequently communal or tribal, but some form of private property usually exists among other cultivators. The fact that these rights may not be legitimized with formal land titles has become an increasingly important issue where swidden agriculture has been used by local farmers to clear land in state-controlled forest (Poffenberger, 1990; Hafner and Apichatvullop, 1990a). Examples of this are found in state-local conflicts over access and user rights to forestlands and over movements to map the traditional resource domains of indigenous peoples in the Philippines, Thailand, and Indonesia. Scientists, environmentalists, and local governments have also increasingly debated whether swidden cultivation is a harmful and inefficient use of resources or is a rational and environmentally benign form of agriculture. Where population growth and migration to uplands have placed increasing pressure on upland ecosystems, there has been a marked shortening of fallow periods, the conversion of swidden to permanent cultivation, and deforestation and soil erosion. Official government views claim that upland shifting cultivation has been a primary cause of environmental degradation, especially deforestation. In Thailand, the state has used this argument to justify periodic forced resettlement of upland minority communities and of other populations illegally residing in national reserved forests. The problems of illicit opium cultivation among upland swidden communities in the infamous "Golden Triangle" of Burma, Thailand, and Laos have received frequent attention for more than four decades (See Map 17.11). Despite the efforts of the U.S. Drug Enforcement Agency in encouraging upland communities to replace opium with other cash crops, opium's high economic value, the mobility of swidden cultivators, and the cross-border movement of opium has been relatively little affected. Finally, one of the top priorities set out in guidelines being developed for achieving sustainable agriculture in Southeast Asia focuses on upland areas subject to extensive swidden cultivation (Asian Development Bank, 1993b). While these systems represent remnants of the historical past in many parts of Southeast Asia, contemporary conditions indicate that significant changes in the distribution and practice of swidden agriculture will be essential if sustainable agricultural development is to be realized.

Rice-based agriculture is arguably the most important agricultural production system in Southeast Asia. Rice is clearly the preferred dietary staple throughout the region and *padi*, a general term for rice cultivation, is cultivated on 50 percent or more of all land. Next to wheat, it is the most widely grown and consumed grain in the world and is believed to have originally been domesticated in monsoon Asia. The long history of padi culture in Southeast Asia has also seen the crop interwoven into the social and cultural fabric of most societies, forming a symbolic part of their daily and seasonal activities, rituals, ceremonies, religion, and folklore. In Thailand, the King still presides over the annual plowing ceremony, casting rice into a plowed furrow as a symbolic ritual to seek good favor for the upcoming agricultural season. Perhaps nowhere else in the world is a single crop so intimately related to the people and cultures of a region than is rice in Southeast Asia.

Padi cultivation as a form of permanent agriculture is practiced under both rain-fed and irrigated conditions. It is the dominant cropping system in most of lowland Southeast Asia, and irrigated padi accounts for one-third to three-quarters of all padi crop area in Thailand, the Philippines, and Indonesia (David and Otsuka, 1994, p. 20). Padi is also grown in more densely populated upland

Photo 6.2 Planting rice seedlings, Thailand. (Hafner)

Photo 6.3 Rice terraces, Banaue, northern Luzon, the Philippines. (Hafner)

areas where terracing of steeper slopes has been developed to facilitate permanent agriculture. The padi terraces of Banaue in the mountains of northern Luzon illustrate one of the most remarkable examples of how irrigated padi cultivation on terraces has enabled marginal land to be brought under cultivation.

Padi under flooded conditions, either on terraces or through natural flooding of the landscape, can be cultivated year-round if adequate irrigation is available. All of these production conditions are referred to as *sawah* agriculture, after the Indonesian word for flooded padi fields. Where irrigation is not possible, padi is grown under rain-fed conditions only during the rainy season, a condition most common in upland areas where land is scarce. Among the main padi-producing nations in Southeast Asia, rain-fed padi still accounts for a significant amount of the total area under cultivation: 62 percent in Thailand, 37 percent in the Philippines, and 16 percent in Indonesia (David and Otsuka, 1994).

In contrast to swidden agriculture, field preparation and cultivation of padi in lowland areas is both a labor- and capital-intensive activity. Planting and harvesting have traditionally involved a high degree of cooperative labor exchange, especially in villages where padi is the main crop. Where there is a distinct dry season, primarily in mainland Southeast Asia, field preparation typically begins in late April and May before the seasonal monsoon rains begin. Planting is done with either the broadcast method or by transplanting individual seedlings from seed beds, although transplanted padi generally provides higher yields. Weeding of the newly planted fields is also important to minimize competition from weeds during the early stages of plant growth. Crop harvesting generally takes place from three to five months after planting, depending on the growth period or photoperiod sensitivity of the variety of padi planted. Farmers may also plant other crops, such as bananas, cassava, yams, and beans, around the edges of their padi fields and use house-lot gardens to raise vegetables and fruit for domestic consumption. Fish are also common in padi fields and provide an important source of protein for the household diet. Rain-fed padi cultivation in upland areas also generally involves some use of chemical fertilizers and pesticides, although usually only a single crop is grown during the rainy season.

The impacts of modern technology on padi production are evident in the increased use of mechanized plowing and planting, mechanical padi threshers, and applications of fertilizers and chemical pesticides. These types of inputs are widespread in wet-padi agriculture and have contributed to increased production, especially where year-round availability of irrigation water allows for almost continuous cultivation. Where modern varieties (MVs) are planted, it is common for all padis in a large area to be planted to the same variety, although dozens of different padi varieties may be found in more-traditional production areas. Mechanization has also increased labor efficiency on larger holdings, and some of the most highly mechanized areas of padi production in Southeast Asia are found in the Mudah Scheme of Kedah state, northwest Malaysia. Increased mechanization has also resulted in new or unanticipated environmental and social costs. In Bulacan and Batangas Provinces in the Philippines, for example, the use of mechanical threshers has displaced women who traditionally did this work, forcing them into the lower forms of wage labor as simple field-workers in rice agriculture (Heyzer, 1987). These technological changes in padi cultivation and some of their social and biological implications will be examined in the context of the Green Revolution later in this chapter.

Among all the agricultural systems in Southeast Asia, plantation agriculture is arguably the best example of how Western economic forces have incorporated land and labor into the world market during the past 200 years. By the early nineteenth century, European colonial powers had introduced a range of tropical and subtropical crops as the basis for large-scale plantation agriculture. These early plantations took the form of both large-scale monocrop systems that were owned and managed by European companies and smallholder farmers that produced for international markets. The rubber estates established in West Malaysia around the turn of this century by British interests are most often associated with the plantation sector in Southeast Asia. But a century earlier Chinese and Europeans were already engaged in commercial plantation agriculture producing pepper, tapioca, sugarcane, and spices. In recent years Malaysian government policies have transferred control of large

Photo 6.4 Mechanized plow, "the iron buffalo". (Hafner)

segments of the European estate sector to local interests. Many of the early plantation companies have formed the core of contemporary multinational agribusiness corporations such as Castle and Cook, Dole, and Del Monte. Although plantation agriculture has undergone significant change, it remains an important commercial agricultural system in many countries in the region.

The vast majority of plantation agriculture has remained concentrated near the equator where high humidity and temperatures are best suited to year-round production and the uniform use of labor and equipment. The cultivated area devoted to plantation agriculture is very limited, however, occupying less than one-sixth of the land under padi and with almost three-quarters of this area under smallholding production (Fryer, 1990, p. 187). Rubber has been the most important plantation crop and still occupies almost one-half of the large-scale plantation land in Southeast Asia, although three-quarters of it is cultivated by smallholder farmers. In Malaysia and Indonesia for example, 80 percent of the total rubber land, or 9 million acres (3.6 million hectares), is operated as smallholdings while only about 1 million acres (410,000 ha) remain in estates. The growing importance of smallholders in the production of rubber is suggested by the fact that Thailand surpassed Malaysia in 1991 as the world's leading producer and 91 percent of the country's 4.2 million acres (1.7 million hectares) of rubber land are operated by more than 800,000 smallholder producers (Figure 6.4).

In areas further from the equator having distinct wet and dry seasons, the more dominant examples of plantation agriculture are found in the sugar and coconut industries of the central Philippines, and the export fruit industry in Mindanao. In contrast to the trend toward smallholder producers in the plantation crop sector elsewhere in the region, an opposite trend has developed in the export fruit industry in Mindanao. Since the late 1970s, the policies of the former Marcos administration, proximity to the growing Japanese market, and an abundance of cheap labor have enabled corporate and multinational agribusiness to gain access to large areas of land for plantation use. In 1990, 16 multinationals, their domestic affiliates, and local agribusiness corporations controlled almost 400,000 acres (160,000 ha) of banana, pineapple, and oil palm plantation land in Mindanao. This represents an increase of 43 percent in the area of plantation holdings since the rapid expansion of the export fruit plantation sector began in 1978–79. The multinational agribusiness firms of Del Monte and Dole control half of Mindanao's arable land, including land rented from the government, large landowners, or through contract arrangements with large- and medium-size landowners. It is perhaps not surprising that the labor involved in these plantations is predominantly hired workers, working conditions and wages are poor, and little of this plantation land has been affected by the national land-reform program.

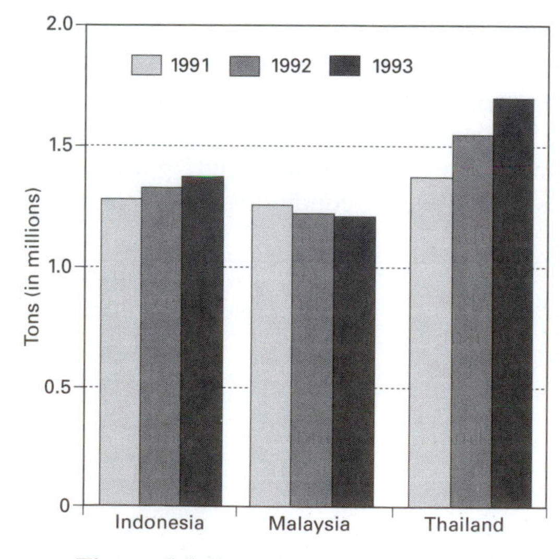

Figure 6.4 Natural rubber production.
Source: UNFAO Production Yearbooks.

Photo 6.5 Rubber-plantation worker, Peninsular Malaysia. (Leinbach)

In the context of growing concerns about the sustainability of Southeast Asian agricultural systems, the future of the plantation sector will depend upon several issues. Political and social opposition to the control of large areas of productive land by foreign corporations and to large landholdings under any form of ownership has grown throughout the region. There will therefore be a need to shift ownership to the private sector if these systems are to remain competitive. At the same time the size of such private land holdings must be limited. The social, economic, and health conditions of plantation workers must also be addressed because they represent some of the poorest and most disadvantaged segments of the agricultural population.

LAND OWNERSHIP AND TENURE

The pattern of land ownership in Southeast Asia was traditionally determined by how the land was used, although in fact it belonged to the state or the king, sultan, or datus in whom authority was vested. While land was an abundant resource used primarily to satisfy household subsistence needs, it was labor that was often the limiting factor in production. Property rights over land were therefore only loosely defined and the concept of *formal ownership* was not introduced until the colonial era. Two processes have changed this situation: land scarcity linked to population growth and commercialization of the rural economy. Land has become a commodity like rice that can be bought, sold, and rented, and various rules now control or limit access to it. Land acquisition has become an important means of accumulating capital and wealth. Among the outcomes of these processes has been fragmentation of land holdings, increased tenancy, and the consolidation of larger amounts of land owned by a smaller share of the population. These trends are illustrated below with data for Indonesia, the Philippines, and Thailand.

As is evident in Table 6.3, the average farm size for three of the larger agricultural countries in the region is generally small, roughly less than 3 ha. This pattern is especially prominent in Indonesia where almost two-thirds of the farms are less than 1 ha. In the Philippines and Thailand, average farm size is somewhat larger, and medium to large farms above 3 ha account for more than one-third of all farms and more than one-half of the total farmland. The general pattern of small farm size has been the subject of much debate, especially over the efficiency of smaller holdings, the links between farm size and rural poverty and the inability of small farmers to benefit from the adoption of modern production technology such as that offered by the Green Revolution. However, recent studies have shown that neither farm size nor tenure significantly affect the adoption of modern padi varieties (David and Otsuka, 1994). At the same time however, the amount of land cultivated by each household has begun to decline, and this process is reflected in the higher measures indicating land concentration in Table 6.3. Although this has been true in many areas of Southeast Asia, it has been more dramatic in areas of higher population density. For example, average farm size in Central Thailand in 1953 was 13 acres (5.47 ha) but had shrunk to 9 acres (3.6 ha) by 1983, and rice land per capita in Java has declined from 0.070 ha in 1940 to 0.038 ha in 1980. Despite efforts to implement the redistribution of land in some countries, large farms and plantations are still common. In central Luzon, the large padi haciendas of several hundred hectares that once existed have been abolished by land reform programs, but others in sugarcane and the plantation sector still persist. In fact, farms of more than 61 acres (25 ha) still account for more than 14 percent of all farms in the Philippines. Ironically, the most prominent example of this is the 14,800-acre (6,000-ha) Hacienda Luisita, a large sugar plantation owned by the family of former Philippine President Corazon Aquino, whose administration came to power on a land reform platform.

Determining the land tenure or ownership status of Southeast Asian farmers is a somewhat more problematic task. This is due to the absence of comprehensive, accurate, and comparable data. For example, in 1985 only 12 percent of the more than 24 million hectares of farmland in Thailand had formal title deeds and 40 percent of that land remained "undocumented," placing precise estimates of land-tenure conditions in the realm of educat-

Country	Average Farm Size (ha)	Percentage of Farms and Farmland				Index of Land Concentration (Gini coefficient)	Ownership (percent total farmland)		
		<1 ha	Area	>3 ha	Area		Owner	Part Owner/ Share Tenant	Tenant
Indonesia	1.05	65.8	26.1	6.7	33.4	0.56	63.6	18.2	18.2
Philippines	2.83	13.6	1.9	38.9	75.9	0.60	29.0	71.0	
Thailand	3.57	15.4	2.6	34.3	55.9	0.45	83.2	5.0	11.8

Table 6.3 Farm size and ownership in Indonesia, Philippines, and Thailand. *Source: after David and Otsuka, 1994; Table 5.1, p. 263.*

ed guesswork. Overall however, the owner-cultivator is the dominant form of land tenure in the region and includes about 70 percent of the farmland. Part owner-operated farms include those where some land is owned and some is rented. The third category of tenant involves farmers who operate land as share tenants, paying part or all of the land rent with crop harvests or with some combination of cash and crop harvests. Share tenancy is especially common in Indonesia and the Philippines due to the imperfect implementation of tenancy reform programs, while fixed-rent tenancy with cash rental payments is much more common in Thailand. There are marked regional variations in tenure patterns within countries. In central Thailand where agricultural commercialization and population densities are high, 50–60 percent of the farmland in some provinces is operated by tenants, while 85 percent of the farm holdings in the poorer northeast are owner operated. Similarly in Java, where 50–80 percent of all *sawah* holdings are owner operated, the proportion of tenant-operated *sawah* holdings in villages varies from 17 percent in West Java to 50 percent in East Java.

The increase in tenant farmers and fragmentation of land holdings has been accompanied by a rise in the number of landless agricultural laborers. Precise measures of this population are unavailable, but in Thailand and the Philippines their numbers have been conservatively estimated at 2.4 million and at least 5 million, respectively (Putzel, 1992). Although there are many causes for this situation including loss of land through indebtedness and canceled lease agreements, their numbers are expected to rise as available land resources dwindle, population growth continues, and redistributive land-reform programs remain ineffective.

The Land-Reform Experience

Most of the world's poor live in rural areas of Asia and a significant majority of them farm smallholdings or are share tenants and landless agricultural laborers. Indonesia's Transmigration program, the FELDA land settlement scheme in Malaysia, and land reform in the Philippines represent programs that have attempted to redistribute land to the landless and rural poor, reduce rents, and increase the security of tenure for rural farmers (Photo 6.6). Because the distribution of farm land and the incidence of tenancy vary greatly in these countries, the impact of redistributive land reform to alleviate rural poverty is likely to be different. The experience with these policies in Southeast Asia presents at best very mixed results and the details of various programs cannot be adequately examined here. However, a brief outline of the land-reform program in the Philippines will serve to define some of the major problems with these policies.

In the Philippines, land-reform programs based on the 1963 Agricultural Land Reform Code were implemented vigorously under martial law, which was proclaimed in 1972. This policy applied only to tenant-operated padi and corn land which was to be gradually transferred to the tenants, but excluded owner-cultivated land and allowed landlords to keep 7 ha of land. Any land holdings above this threshold were to be taken for redistribution, and the "new" tenant owners would then receive a Certificate of Land Transfer (CLT). After making annual payments to the Government Land Bank for 15 years, the former tenants would receive rights to the land. However, because of various loopholes in the law, landowners were able to evade the law by switching from corn and

Photo 6.6 FELDA land settlement scheme, Perak, Peninsular Malaysia. (Leinbach)

padi to other crops, by changing the operational status of the land from tenants to wage laborers, and by subdividing land among family members and friends to evade the 7 ha limit. Many landlords openly discouraged tenants from asserting their rights under the law by denying them access to irrigation water, filing civil and criminal suits against them, and forcing tenants to move their houses to new locations. Tenants were often evicted under the pretext that they had voluntarily surrendered the land for owners to resume cultivation, forcing the tenants to join the ranks of landless agricultural laborers. Studies in some padi-growing villages in Luzon have shown that nearly one-half of the landless laborers had once been tenants in villages where land reform was "effectively" implemented. By 1986 new land rights or CLTs had been issued for only 800,000 of the 6 million hectares under corn and padi in the entire country, or less than 13 percent of these holdings (Boyce, 1993, p. 136).

While Philippine and U.S. government sources portrayed the program as a resounding success, more-objective observers consider these claims as nonsense. But it would be an exaggeration to conclude that land reform has accomplished nothing. Some of the larger tenant holdings (2–3 ha) did benefit because many of the large padi haciendas in Luzon were broken up and the incidence of share tenancy farms in padi areas that adopted the modern padi varieties declined from 70 percent in 1970 to 10 percent by the mid-1980s. Although the available evidence is not conclusive, land reform in the Philippines appears to have aggravated rural poverty by causing increased tenant eviction, forcing thousands of tenants to become landless agricultural laborers, and adversely affecting agricultural productivity. Given the shrinking land resources in much of Southeast Asia, the landless population will continue to increase and with it political and social pressures for greater equity in land distribution and ownership will also increase.

THE GREEN REVOLUTION AND SUSTAINABLE AGRICULTURE

The technological revolution in padi production, commonly called the Green Revolution, developed in the 1960s at a time when population growth and lagging agricultural production had raised concerns about regionwide food shortages. That this has not occurred is in great measure due to the gains in padi production that have resulted from this Green Revolution. The package of production technology was based on genetically engineered high-yielding varieties (HYVs) of padi developed at the International Rice Research Institute (IRRI) at Los Baños in the Philippines. Since the mid-1960s the adoption of these modern padi varieties and the agrotechnology required to make best use of them have transformed padi production in many parts of the region. Between 1969 and 1989 regional padi production more than doubled and countries such as Indonesia and the Philippines have reached the threshold of national food grain self-sufficiency. Much of this growth has come from the increased yield per hectare made possible by the adoption of the fertilizer-responsive, high-yielding or modern varieties (MVs) of padi introduced in 1966.

These unprecedented gains in grain yields however, have not taken place everywhere in the region. Padi production and yields vary widely within and between different production environments that are influenced by the availability of irrigation, and debate lingers over the adverse social, economic, and ecological impacts of this new technology. New biological and economic concerns also challenge the continued productivity and future sustainability of this technology. The limits of the new technology, problems that have emerged in its wake, and future requirements needed to sustain the momentum of the Green Revolution are considered in this section.

Adoption of the New Technology

The Green Revolution in padi was based on a package of technology. That package included MVs of padi; the use of irrigation, chemical fertilizers, and pesticides; improved methods of farm management; and various forms of government support and economic incentives. The term *modern variety* refers to the short-statured, stiff-strawed, fertilizer responsive, nonphotoperiod-sensitive Indica padi varieties, typified by the IR8 variety, which were developed at IRRI between 1962 and 1966 and first released for distribution throughout the region. MVs that followed IR8 incorporated new traits—improved grain quality, greater pest resistance, shorter growth periods—that further encouraged its adoption. By the late 1960s, it is estimated that almost 25 percent of Asia's padi land was planted to IR8 or similar semidwarf varieties and almost 40 percent by 1984 (Fox, 1993, p. 215). In 1982, IRRI estimated that one variety, IR36, was grown on 27 million acres (11 million hectares) of padi land, making it the most widely grown strain of any crop at any time in world history. These varieties were also very responsive to fertilizer and performed best under irrigated conditions where chemical fertilizers were liberally used. In addition, pests and disease must be controlled by chemicals and the use of varietal resistance seeds. Finally, economic incentives and subsidies ideally must be provided. National programs to introduce this new technology and provide credit so that farmers could adopt the new strains were also implemented in many countries.

The major source of increased padi output has come from improved yields. Trends in padi yields for countries in Southeast Asia are shown in Figure 6.4 while gains in harvested area, production, and irrigated area are recorded in Table 6.4. Between 1974–76 and 1993–94 total padi

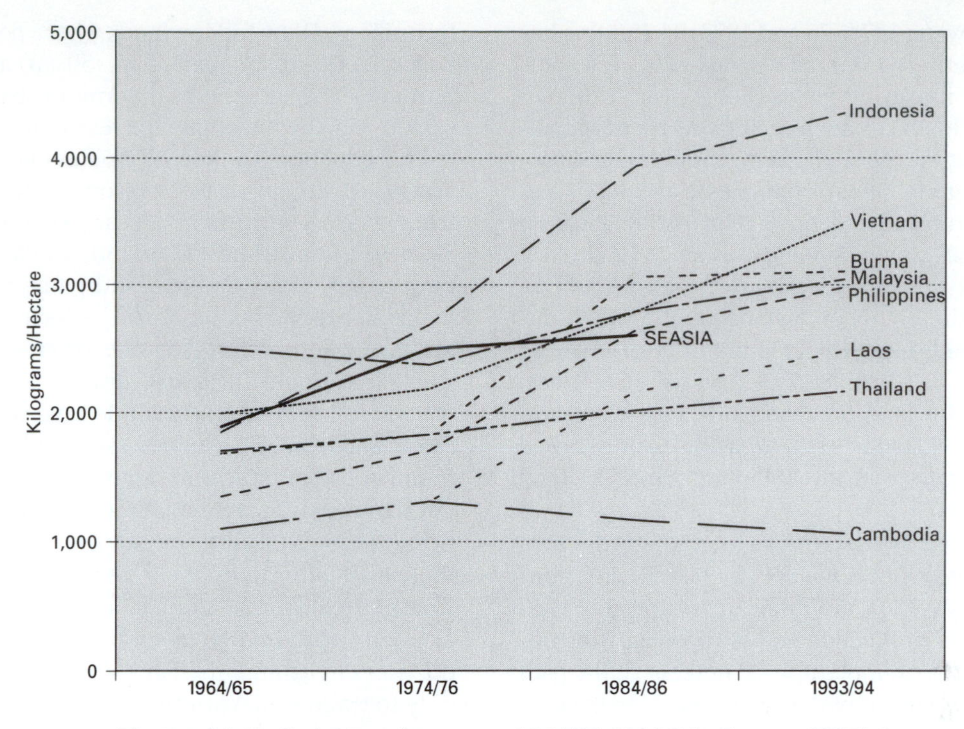

Figure 6.5 Padi yields, select years, 1964/65-1993/94. *Source: UNFAO Production Yearbooks.*

Country	Harvested Area (1000 ha)			Production (1000 metric tons)			Irrigated Area (1000 ha)		
	1974/76	1984/86	1993/94	1974/76	1984/86	1993/94	1976	1986	1993
Brunei	2	1	1	3	2	1	0	1	1
Burma	4942	4744	6477	9037	14825	19057	985	1059	1068
Cambodia	1002	1717	1700	1312	1933	1800	89	90	92
Indonesia	8458	9846	10646	22705	38815	48245	4900	5418	8215
Laos	682	616	639	891	1427	1853	50	120	125
Malaysia	741	640	685	2029	1795	2040	312	336	340
Philippines	3555	3365	3350	6092	8882	10150	1070	1460	1580
Thailand	7952	9811	8482	14858	19756	18447	2448	3912	4400
Vietnam	5134	5682	6500	11213	15859	22500	1200	1790	1850

Table 6.4 Trends in padi harvested area, production, and irrigated area. *Source: UN FAO Production Yearbooks.*

production in the region rose from 68.1 million metric tons to 124.1 million metric tons. Of this expansion the most dramatic gains in both production and yields have taken place in Indonesia. Since 1964–65, padi yields in Indonesia more than doubled from 1,850 kg per hectare to almost 4,360 kg per hectare by 1993–94, and total padi production has risen from 22.7 to 48.2 tons per hectare between 1974–76 and 1993–94. This has been the highest yield increase for any of the Southeast Asian countries and is, to a large extent, the result of the adoption of the new padi technology. The Philippines and Vietnam also experienced yield increases between 1964–65 and 1993–94 of 120 percent and 74 percent, respectively, and significant growth in total production. Elsewhere in Southeast Asia this new technology has had a much more limited impact. Grain yields showed a small increase in Malaysia by the early 1980s but have changed little since then as a result of low returns, shortages of labor, poor management, and occasional drought. Cambodia's situation is even less satisfactory as yields have declined due to continued internal instability. Burma also had a significant increase in padi yields, 85 percent between 1964–65 and 1993–94, although this has not been due to the adoption of modern varieties (Win and Win, 1990). Thailand's padi production also grew by 19 percent between 1974–76 and 1993–94, but this has resulted almost entirely from the expansion in area cultivated rather than improved yields (Table 6.4). Although these gains in yields have been facilitated by public investments in irrigation, especially in the more favored production environments, most countries have less than one-fifth of their cropland under irrigation and the increase in irrigated area has slowed significantly since 1985.

This mixed pattern of modern variety adoption is related, in part, to the highly diverse production environments in which padi is grown. Five major padi-growing environments can be identified based on water regimes: irrigated, rain-fed lowland, tidal wetland, deep water, and upland. The distribution of padi area for each of these environments in Indonesia, the Philippines and Thailand is shown in Figure 6.6.

Irrigated areas are most favorable for padi production, produce the highest yields, and allow for a second or third crop during the dry season, but these areas represent relatively small areas of production and are unevenly distributed. In Thailand for example, only 30 percent of the padi area is irrigated; 85 percent of that share occurs in the central region, which produces two-thirds of the country's total padi output. By contrast, 77 percent and 47 percent of the Indonesian and Philippine padi areas are irrigated, respectively. Yet, even in these environments, yields have been affected by a wide range of factors. Climatic conditions, including drought and flooding, have often had adverse impacts on padi yields over large areas. This was especially true in 1972, when widespread flooding occurred, and in the drought years of 1982 and 1987. Disease and insect outbreaks have been important, and variations in the supply and costs of fertilizers have also acted to dampen padi production, especially during the 1970s oil crisis.

Nonirrigated production environments present more diverse conditions affecting the performance of the new technology. Rain-fed lowlands where padi is grown with water depths of less than 20 in. (50 cm) are at least as important as irrigated areas in terms of size throughout the region; yet they account for less than one-half of total padi production. In the Philippines and Indonesia, these areas have adequate wet-season rainfall and water control, and MVs perform almost as well as in irrigated areas. However, in northeast Thailand, which contains most of the country's rain-fed lowland padi area, drought and highly limited irrigation facilities have meant that farmers have continued to grow the lower-yielding but more-stable traditional glutinous varieties of padi. Other areas of rainfed lowland padi include tidal wetlands near seacoasts and in inland estuaries, especially the tidal swamps found in Kalimantan and Sumatra, but because these environments suffer from salinity and adverse soil conditions, their suitability for MVs is very limited. Finally, there are deepwater padi areas where crops are grown in water reaching depths of 10 ft (3 m) and upland environments that involve a smaller proportion of all production environments. The latter are typically planted only to traditional varieties of padi.

Southeast Asian padi output has experienced remarkable growth since the mid-1960s due to the Green Revolution. Yet the suitability of this technology to irrigated and rain-fed lowland environments has limited its benefits to these areas. Therefore, the availability of irrigation has been a critical factor in both levels of adoption and growth in yields between countries and across different environments. Countries with higher rates of modern variety adoption, such as Indonesia and the Philippines, have a higher ratio of irrigated to total padi area. Similarly, fertilizer use and padi yield per hectare are highest in Indonesia and lowest in Thailand, reflecting the differences in modern variety adoption rates and irrigation ratios (Figure 6.7). At the same time, the gaps in productivity between different production environments have grown wider. This has increased disparities in yields and total production between irrigated and nonirrigated areas, adding to persistent debate over the adverse impact of the Green Revolution on incomes, rural social differentiation, landlessness, and poverty. There is also growing evidence that a new range of problems may threaten the capability of this technology to meet the region's future demands for rice.

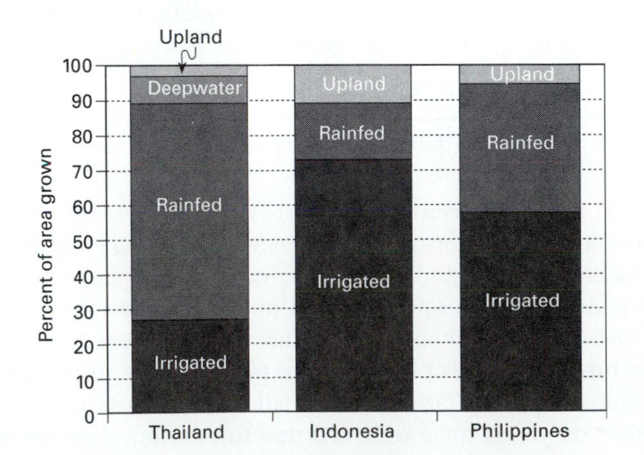

Figure 6.6 Distribution of rice crop area by production environment. *Source: David and Otsuka, 1994, p. 20.*

Sustainability of the Green Revolution

Despite the enthusiasm that accompanied the initial promotion of the Green Revolution, this technology is being confronted by a new range of problems. Irrigation has been an important ingredient for the success of the Green Revolution, but the high costs, limited distribution of inputs, and inefficiency due to poor management have in-

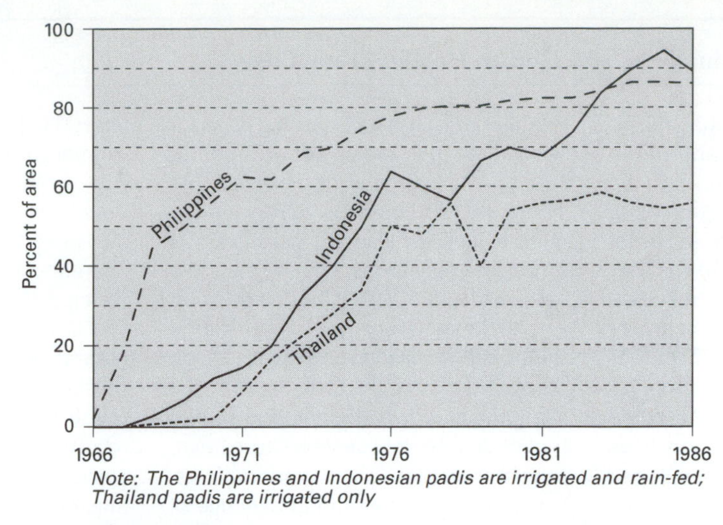

Figure 6.7 Adoption rates of modern rice varieties.

creasingly become problems in many areas. The annual rate of increase in irrigated area in Southeast Asia has declined from an already low rate of 3–4 percent between 1970 and 1985 to only 1.5 percent in the last half of the 1980s. Prospects for any future expansion in irrigated area appear limited. Rising construction costs, the lack of suitable new sites for irrigation, and low returns from past projects have led to this situation. Irrigation facilities in the Philippines have been adequate for adoption of the new padi technology on a wide scale; yet they have not enabled yields to reach their full potential because of the scaling back of planned irrigation projects, problems with organizing and sustaining communal irrigation systems, and the high costs of government water-control development (Boyce, 1993). In Central Thailand, most existing irrigation systems are old and poorly maintained, and irrigation water is used inefficiently leading to shortages for dry season cropping.

The genetic vulnerability of the new HYV monocrop padi systems has emerged as a potential challenge to the sustainability of this technology. In traditional padi agriculture, the diversity of cultivated species and their different levels of susceptibility to pests has buffered them from the adverse impacts of insects and pest infestations. But, as MVs have spread over large areas, this genetic diversity and the protection it provides against pest and plant diseases has been greatly reduced. With the loss of varietal resistance, new or previously minor pests such as the brown plant hopper, which attacks padi by feeding on fluids in the padi stalks, have adapted to these environments and caused widespread crop damage. In 1971 a tungro-virus outbreak in central Luzon resulted in the loss of one-third of the padi crop, and the brown plant hopper has been a recurring problem in Indonesia and Vietnam over the last two decades. This has led to a perpetual race between the pest and the plant breeder, which is typical of much modern agriculture. New varieties with genes resistant to these pests have been released, but only buying time until the pests adapted to and overcame the protection offered by these new varieties. This strategy enabled Indonesia to increase its total padi production by 127 percent between 1968 and 1984, despite recurrent problems with the brown plant hopper in the mid 1980s.

Greater economic and health problems have also arisen with the indiscriminate and repeated use of the wide spectrum of insecticides that reduce the natural enemies of padi pests. Pesticide use and their residues have posed increasing hazards to human health, harm the larger padi ecosystem, contaminate aquatic food resources, and have led to a "pesticide treadmill" where ever larger doses of more potent pesticides have been required to counteract pesticide-resistant insects and plant diseases (Box 6.3). Potential problems also occur with the increasing costs of fertilizers, especially nitrogen, because most nitrogen fertilizer is produced from fossil fuels. Its price is therefore susceptible to instability in the global energy market and the possibility of a long-run rise in prices as fuel stocks are depleted. In light of these conditions, a strong case can be made for an integrated-pest-management (IPM) strategy that stresses the development of host plant resistance, biological controls, and pest-reducing agronomic practices such as crop rotation. In 1986 Indonesia became the first country in the world to establish a nationwide IPM strategy for its padi crop. This program has involved training 2.5 million farmers in the techniques of IPM, removal of all government subsidies for pesticides, and replacing pesticide spraying with biological and cultural controls. The continued growth in Indonesian padi production since 1986 demonstrates both the success of this program and its value as an effective pest-management strategy for other padi producing areas throughout the region.

BOX 6.3 Pesticides, Productivity, and Human Health: Rising Costs

During the past several decades pesticides have played a significant role in increasing agricultural production in the developing world. During the early years of the Green Revolution in Southeast Asia, pesticides and herbicides were heavily promoted to control pests, minimize crop losses, and maximize yields. Many of these padi pesticides such as methyl parathion and monocrotophos, defined as extremely hazardous by the World Health Organization, have been banned for use in the developed world but are still sold to developing nations whose growing market accounts for 31 percent of world pesticide exports. Although the creation of improved padi varieties with broader pest resistant traits has reduced the need for pesticides, their use in the Philippines rose by 10 percent per year in the 1980s. In 1990 the most lethal of these chemicals were also the most popular among Filipino farmers due to their low cost, wide availability, and reputation for wide-spectrum toxicity—despite their extreme hazardousness.

In Nueva Ecija Province of the Philippines recent studies have shown that pesticide exposure can cause heart, lung, nerve, blood, and skin problems. Farmers often have little knowledge of safe pesticide use, handling, and storage that has caused unnecessary exposure, poisoning and death. More than 4,000 cases of acute pesticide poisoning and 603 deaths were recorded in government hospitals from 1980-87, perhaps a small fraction of the many unreported cases. Many pesticides may also persist in the environment for up to 15 years, contaminating soils, groundwater, and a wide range of wildlife, insects, and fish. Ironically however, natural pest control which focuses on preserving natural pest predators was found to be the most economical and resulted in significantly lower health costs. However, natural pest control and integrated pest management (IPM) is still the exception throughout Southeast Asia despite the notable success reported in Indonesia (Rola and Pingali, 1993).

Photo 6.7 Pest control, Kedah, Peninsular Malaysia. (Leinbach)

Since the first release of IR8 in 1966, there have been persistent criticisms that the Green Revolution would create economic and social inequality. It has been argued that income gaps would grow between the larger and more commercialized farmers who are better able economically to adopt the new technology than the small farmers. Further it has been argued that landlessness would increase among small and tenant farmers as large farmers consolidated land holdings through tenant eviction and land purchases. Increased farm sizes and incomes among larger farmers would promote mechanization, which would further reduce rural employment opportunities and wage rates for landless households. The resultant economic and social differentiation would appear as rising rural poverty, growing income gaps, expanded rural to urban migration, and the marginalization of a larger segment of the rural population. The fears surrounding this "equity issue" have for the most part proven to be largely unfounded. Recent empirical studies have shown that farm size, land-ownership status, and other social and institutional factors did not significantly affect the diffusion rate of MVs (David and Otsuka, 1994).

On the other hand, there is little disagreement that the income gap between farmers in the more-favored production environments and subsistence farmers in less-favored rain-fed uplands has widened. This suggests that further research, aimed at breeding a continuing stream of pest-resistant padi varieties, and expanded integrated pest management can reduce risks of crop loss by fostering greater genetic diversity. This will ultimately contribute to a more sustainable agriculture in Southeast Asia. Bringing the benefits of the new technology to these more marginal padi-production environments remains one of the unmet challenges of the Green Revolution.

Priorities for Sustainable Agriculture

In the last decade the new religion of development has become *sustainability*, although its precise meaning and how it should be accomplished are still debatable. While the Green Revolution has been successful in increasing padi production, the site-specific nature of the new technology has limited its impact. Its benefits have not been extended to areas of poorer soils, traditional mixed-cropping systems under rain-fed conditions, and places where weak agricultural infrastructure exists. Because past and present agricultural efforts have focused only on one or two types of production, the development of cropping systems for upland environments, for example, has been seriously neglected. These conditions have created a new set of imperatives embedded in the matters of population growth, rising needs for food and fiber, and the persistence of rural poverty and unemployment. Population growth has continued to increase exponentially, creating needs for more food, employment, health care, housing, and education. Seventy percent more rice will be needed in the year 2025 than was produced in 1993 if malnutrition and poverty are to be overcome. Therefore, other production systems must also evolve to meet local, national, and international needs; each of these systems has different requirements and serve a somewhat different clientele. This suggests that a new vision of sustainable agriculture will be needed for Southeast Asia, one which will include a wider spectrum of social, environmental, and equity considerations.

Sustainable agriculture can be defined as, "that [production system] which can evolve indefinitely toward greater productivity and human utility, enhance protection and conservation of the natural resource base, and ensure a favorable balance with the environment" (Havener, 1992, p. 5). This means increased and more stable productivity, using natural resources to conserve and enhance the quality of the environment, improving the quality of life, and creating equity across generations. For these criteria to be met, the deficiencies in existing agricultural systems must be understood, solutions to the current patterns of degradation found, and difficult policy choices made between different needs and uses of resources. For example, choices will have to be made between increased food production and more export products, short-term economic development and long-term environmental quality, and whether to maximize production on existing arable land or give priority to rehabilitating degraded and unproductive land. What then are some of the priorities for achieving sustainable agriculture in Southeast Asia?

The priorities for sustainable agriculture can be framed in terms of three broad categories of agricultural systems; upland, lowland, and industrial-plantation systems. Upland areas present the most difficult and problematic environments for achieving sustainability. Their development lags far behind areas of more favorable soils and water resources, and social and environmental pressures such as migration and deforestation are creating a shift in agricultural development emphasis. In these areas we need to better understand which agricultural systems can achieve optimum productivity and how traditional knowledge systems can provide insights into why such systems work or do not work. Conditions must also be created so that farmers can make ecologically sound long-term investments in the land to preserve its production potential. This must also involve accumulating the biological capital—trees, perennial crops, soil carbon—needed to offset poor soil and water resources. Critical to this transition will be the provision of transport, market, institutional, and technological infrastructure needed to meet these development needs.

Lowland environments present a different set of priorities because of their higher levels of development and better soil, water, and infrastructure. Here the major priorities must be in developing better community-based water resource management, improving crop diversity to control pests, introducing integrated pest management to limit the use of pesticides and reduce environmental and human risk, and promoting better nutrient management to replace the current dependence on genetically engineered crops. Incentives for private sector investment to develop new varieties of cereal or root crops will also be important in the future.

In contrast to the smaller-scale and higher labor inputs of upland and lowland agricultural systems are the capital-intensive industrial-plantation systems. These large-scale operations produce crops for export. Examples are the banana and pineapple plantations typical of southern Mindanao and oil palm in Malaysia. Among the key priorities for sustainability in these systems are the need to shift toward more private sector ownership, adopt integrated-pest-management programs and better worker-safety programs, and achieve better social equity in wages, working conditions, and access to land (Asian Development Bank and Winrock International, 1992).

Although many nations in Southeast Asia are moving rapidly toward becoming industrialized, achieving that goal is unlikely to come at the expense of the agricultural sector and the environmental resources that sustain a majority of the region's population. Their future development will depend on efforts to achieve a more sustainable agriculture that increases productivity, conserves natural resources, and alleviates poverty and hunger. While these are not problems unique to Southeast Asia, the region is perhaps better positioned to address these problems successfully than any other area of the developing world.

RURAL DEVELOPMENT: STRATEGY AND PROGRAMS

The need to transform rural areas has been noted in national development plans throughout the region for over four decades. Since the early 1950s, the dominant development paradigm in most ASEAN nations has emphasized increasing production and economic growth. These priorities have been pursued through sectoral-specific investments in agricultural technology and infrastructure focused on selected geographic areas and segments of the rural population where the most immediate and rapid returns on these investments were expected. In the 1950s for example, 30 percent of the Malaysian budget went to commerce, industry, rubber replanting, and infrastructure; more than 70 percent of these allocations were for improving the export sector in urban locations; and little emphasis was placed on improving rural living conditions. The benefits of this rapid growth were expected to trickle down to other social and economic segments of society, extending "development" throughout the rural landscape. Despite the general gains in agricultural productivity that resulted from this approach, there was little evidence of a developmental transformation of the rural sector. By the early 1970s, the condition of the majority of the rural population showed little improvement and in many cases had deteriorated. Large segments of the rural population had slipped further into poverty, urban-rural and regional income disparities had increased, and rural underemployment and unemployment had grown.

Since then, many governments have begun to acknowledge that rural development is both a complex process and one that can not be realized simply by addressing technical problems. The priorities in this "new" development paradigm were captured by the frequent use of such slogans as "redistribution with growth," "growth with equity," "participation," and "grassroots development." These slogans have accompanied new government programs to meet basic rural needs in health, education, and employment, and targeted the poor in isolated and "by-passed" rural areas. These types of policies, which seek to address a wider variety of rural needs and basic social and economic inequality, have also been included more frequently as elements of national development plans. Such policies also have been recognized by international development organizations. Although it is debatable whether this shift in emphasis has involved a fundamental abandonment of the earlier production and growth-centered paradigm for development, it does acknowledge that rural development is a process that will require both new strategies and more varied approaches than was the case in the past.

Rural Development Strategies

The strategies associated with rural development activities in Southeast Asia can be grouped into four broad categories: technocratic, reformist, radical, and alternative. The general objectives, methods, programs, and impacts of each strategy are indicated in Table 6.5. The first three categories represent the formal policies and programs of national governments and their development agencies in the original ASEAN countries and the socialist states of Laos and Vietnam. Alternative strategies include those that are typical of the activities of nongovernmental organizations (NGOs) and private voluntary organizations (PVOs) that operate independently of the national development bureaucracy.

Most of the ASEAN states have used rural development policies that have a strong technocratic element designed to increase productivity, especially in agriculture. Technocratic strategies are often characterized as "development from above" because they are based in policies and programs defined and implemented by the state. The types of activities that are generally associated with this approach have emphasized improving infrastructure, modernizing agriculture, and providing larger amounts of production inputs such as fertilizers. As discussed earlier, the Green Revolution with its improved production technology, irrigation, and rural credit is most widely associated with this technocratic strategy. By the mid-1970s the general failure of technocratic strategies to bring about trickle-down growth contributed to a more egalitarian reformist approach to rural development.

Reformist strategies have continued to emphasize increased productivity and income, although these goals were often stated as secondary to the priority of meeting basic needs, decentralizing the development process, increasing local participation, and alleviating rural poverty. This emphasis on equity-based development has included efforts to redistribute power and income, widen access to resources—especially through land reform—and strengthen the planning and management capabilities of government organizations at regional and local levels. Local participation has frequently been a central slogan of reformist rural development programs, although how this has been defined and the actual extent

	Technocratic	Reformist	Radical	Alternative
Objective(s)	•Increase productivity	•Increase productivity •Redistribute income and access to resources	•Social change to create socialism •Egalitarian society	•Empower rural poor •Change unequal social and political structures
Method(s)	•Technical change in agriculture •Expand infrastructure and input facilities	•Decentralize development decision-making •Improve institutional capabilities •Mobilize local resources	•Collectivize rural land and labor •Centralized state planning	•Organize rural poor •Promote participation in decision-making •build local peoples organizations
Programs	•Green Revolution •Infrastructure and production inputs	•Land reform •Cooperative formation	•Cooperatives and producer collectives	•Delivery of welfare services in health, nutrition •Improving self-reliance in food and income
Impacts and Problems	•Increased production •Expanding road, water, and irrigation services •Increased tenancy and landlessness	•Improved local decision-making and management capability •Limited job-creation and income distribution •Neglects problems of rural poor	•Greater rural-urban income equality	•Increased self-reliance and self-worth •Community organizing •Limited sustainability •Hard to replicate activities at larger scale •Limited technical resources

Table 6.5 Main types of rural development strategy. *Source: adapted from Cheema, 1985; Holloway, 1989.*

of participation has varied greatly and with limited success. Land-reform programs such as those implemented under the "land to the tiller" program of the early Marcos administration in the Philippines, the 1975 Agricultural Land Reform Program in Thailand, and various elements of the New Economic Policy implemented in Malaysia since 1971 are examples of efforts to widen access to land resources, redistribute income, and reduce levels of rural poverty. Other reformist rural development strategies have focused on decentralizing development decision making, improving the management and implementation capabilities of provincial and local organizations, and increasing local participation in these activities. Thailand's Poverty Alleviation Program discussed below had many of these elements, although its priority in upgrading and expanding infrastructure and public works facilities to improve agricultural production is reminiscent of earlier technocratic strategies.

Radical strategies are associated with the socialist states of Burma, Vietnam, Cambodia, and Laos. Their adoption of policies for rural social transformation rely on fundamental social change, a rejection of capitalism, and the goals of creating an egalitarian and just rural society. They also rest on the removal of the structural causes for rural poverty by introducing radical land reform, reducing income disparities, reorganizing land and labor through agricultural collectivization, and using local government and party institutions to identify and implement programs for rural areas. In principle, the main beneficiaries of this approach have been the rural poor and disadvantaged groups in society. The general experiences of both Vietnam and Laos after 1975 are illustrative of this approach. Their policies were patterned after the Soviet-style system of centralized state planning that created strong central government control in the allocation of resources to both agriculture and industry, and collectivization of agriculture through "production cooperatives" as a means to achieve a social and technological revolution of the countryside (Long, 1988; White, 1988). In both Vietnam and Laos this strategy was neither as coercive or ruthless as it was in the case of Cambodia under the Pol Pot regime.

Alternative strategies are associated with the activities of NEOs and PVOs, which have emerged as active participants in promoting rural development in Southeast Asia in the past 15 years. Although the size of the populations affected and scope of their activities are limited, they provide an alternative to the more conventional programs of government. Their approach is based on a fundamental commitment to empowering the rural poor, increasing local participation in making the critical decisions that affect people's lives, and overcoming the social and economic structures that dominate rural society and are seen as the main causes of rural poverty. An Indonesian nongovernmental organization, the Institute for Socio-Economic Research, Education, and Information, has used the traditional Islamic charitable tax system in Central Java to form a fund to help the very poor finance income-generation and productive enterprise (Holloway, 1989). As this suggests, NGO and PVO programs are typically small-scale, community-based efforts focused on such specific needs as community forestry, sustainable agriculture, appropriate technology, and self-help programs in health and nutrition. Their activities are also not widely acknowledged by many governments and are often seen as threats to the existing authority and "expertise" of the mainline development bureaucracy. However, there is growing support for the lessons they can provide on how to work with local people to achieve meaningful and sustainable development, especially in addressing rural needs that are beyond the capability or scope of most conventional rural development programs.

Growth and Equity in Thailand

Since the early 1980s, Thailand's national development plans have placed a higher priority on investing in the development of the rural sector. This marks a departure from the four previous Five-Year National Development Plans and mirrors similar trends in other ASEAN countries. This shift toward equity-based rural development policies has been a response to the failure of earlier plans to meet the diverse social and economic needs of a rural sector where incomes have declined relative to urban areas, poverty has persisted, and productivity has improved only slowly. Despite the government's slowness in acknowledging these and other problems, its recognition of the importance of addressing these problems is an encouraging sign. Several examples from these efforts will help to illustrate some of these strategies and the problems which are being confronted.

Thailand's Fifth National Social and Economic Development Plan (1982–86) gave top priority to tackling the problems of the rural sector. This plan outlined a new equity model for rural development under the title of the Poverty Alleviation Program (PAP). The PAP was a set of programs focused on tackling a wide variety of rural problems, especially in areas of high and persistent rural poverty. A major goal was to tailor development assistance so that it met the different conditions and needs of rural areas. Some of those needs included improved crop marketing services; providing basic services in health care, nutrition, potable water supplies; and livestock and fishery projects to meet dietary protein deficiencies. Because poverty alleviation has been its top priority, these basic services and production programs were provided to 216 districts in 37 provinces in the northeastern, north, and southern regions of the country under the related Poverty-Stricken Area Development Plan. The population in these areas totaled 7.5 million and accounted for about 75 percent of all those classified as poor in these regions. Also, unlike most previous rural development strategies, the PAP has stressed cooperation and coordination among the nine separate state, provincial, and local organizations, ministries, and agencies responsible for implementing different aspects of these programs.

Also incorporated into the PAP was the older Rural Job Creation Program (RJCP) initiated in 1980 to increase rural incomes and employment opportunities. The RJCP had several objectives that were similar to those of the PAP; these were to decentralize control over budgets to community-level organizations, to strengthen local capabilities to design and implement projects, and to increase local incomes by providing wage labor employment in building new projects. Between 1980 and 1985 approximately 13 percent of the national budget was directed to this program. In keeping with the priority to target poor rural areas, funds were allocated on a regional basis depending on the level of rural poverty and income. Therefore, the poorer Northeast region for example received three times the funds provided to the more developed Central Region and at least twice the funds allocated to the North. Although the RJCPs policies allowed a wide range of projects to be supported such as irrigation canals, health centers, and new schools, the governments top priority was the building of new irrigation canals to increase agricultural production (Surarerks, 1986).

These examples of recent trends in rural development policy and programs in Thailand contain some useful lessons. Decentralizing the development process and shifting decision making and resources to the local level has helped to overcome a long-established "crisis of confidence" among rural people about government commitment to their needs. In some cases this has also improved local capability to create and manage development activities. However, the governments' priorities for expanding agricultural infrastructure such as small-scale irrigation dams frequently replaced the villager's requests for new schools, day-care centers, and public toilets. Corruption and local mismanagement of funds were also common problems with many district councils. Orders for materials and building contracts were sublet to favored contractors whose use of inferior design and construction materials resulted in a high rate of dam failure, especially in the northeast. Also, increasing local incomes by providing dry-season wage labor employment added only 5 percent to the average annual household income, and those opportunities declined rapidly as wage laborers were replaced by mechanized construction techniques. Finally, if the goal of these programs has been to increase incomes and reduce poverty, existing statistics for the Northeast region paint a rather dismal picture. Between 1981 and 1988, the incidence of rural poverty increased by more than 5 percent and per capita household income declined by almost 6 percent.

Despite the rhetoric of decentralization, participation, and equity in its policy documents, a significant gap has remained between stated policy and actual project implementation. Local participation and control over decision making has been a distinct weakness because the choices of development activities remain defined by central planning authorities. The failure of the states' centralized development organizations to decentralize decision making is reflected in how both programs have used rather uniform solutions for the diverse needs of heterogeneous rural areas rather than allowing target areas to define activities most appropriate to their own needs. Another weakness arises from the problems encountered in coordinating the activities of multiple agencies involved in implementing rural development programs (Suthasupa, 1987). These issues have been supported by a growing body of opinion that suggests that not all types of rural development activities can be effectively implemented by centralized state planning agencies and that more effective forms of rural development may be realized through government collaboration with NGOs and the private sector.

Poverty and Rural Development

Rural development in Southeast Asia has aimed at more than improving physical infrastructure, raising agricultural productivity, and increasing rural income levels. Governments have also made social and institutional development, the improvement of the quality of rural life, poverty reduction, and more-equitable distribution of resources high priorities of rural development policy, particularly in the last several decades. In assessing the progress of these programs one must confront the basic question of how progress is to be measured, especially because there are a large number of interrelated factors affecting these efforts. However, if one of the primary goals of rural development is poverty reduction, as it has been in a number of countries, changes in poverty levels may provide one indication of the success of rural development programs.

It is clear from the data provided in Table 6.6 that poverty is primarily a phenomenon of rural areas, that two-thirds or more of the poor are in rural areas, and that it remains at unacceptably high levels. This situation has been relieved only slightly by the increases in urban poverty related to the growth of urban populations and has probably been aggravated by the robust growth and rising incomes of the urban-based manufacturing sector of the economies of most of these countries. However, the proportion of the poor in rural areas was significantly less in the late 1980s than it was a decade or more earlier. In Indonesia the proportion of the rural population that was poor has declined from two-fifths to less than one-fifth since 1976, although in absolute terms this still represented more than 20 million people. In Malaysia, poverty levels have remained almost unchanged, while in the Philippines and Thailand they have also declined. But the absolute size of this population in the Philippines, 10.5 million families, is in itself a staggering statistic. Also disguised in these data are the important questions of the age and gender composition of the rural poor, and especially the position of rural women whose contributions to the household and national economy have long been undervalued. Valuable insights into these issues, especially with respect to women and poverty, can be found in Noeleen Heyzer's *Women Farmers and Rural Change in Asia* (1987). Data comparable to that shown in Table 6.6 for the noncapitalist nations of Burma, Cambodia, Laos, and Vietnam however, is less readily available. Nonetheless, if we use the less reliable criterion of GNP per capita as one gauge of relative poverty, all of these nations fall below the US$250 level set by the World Bank, making them among the poorest countries in the world. In Vietnam for example, the per capita poverty line of US$100 set in 1995 by the World Bank translates into more than one-half of that country's population being below this level; 76 percent of that number are rural farmers. A similar situation also exists in Laos where at least 85 percent of the population lives in rural areas and average GNP per capita is less than US$250 per year.

The explanations for these patterns are complex, but a number of factors continue to be conspicuously associated with rural poverty. The evidence reveals a higher incidence of poverty among those dependent solely on agriculture, among those with larger families, among household heads with little or no formal education, among cultivators of small-farm holdings, among the landless and wage laborers, and among those who have limited access to safe drinking water, sanitary facilities, health care, and education. Other characteristics related to rural poverty are limited access to land and credit, unequal distribution of land, and the lack of comprehensive land reforms. Fluctuating commodity prices and weakening terms of trade have also worked against the small farmer producing for the export market. Policies for economic growth and industrialization have compounded these conditions by generally neglecting the rural sector and omitting any actions to address depressed rural wages. In addition there has been too little emphasis placed on improving local nonfarm employment as a way to increase incomes and provide wider choices for the use of rural labor (Saith 1992; Rigg 1997, p. 163–65). The fact that both national and rural development programs have not effectively met these needs is suggested by data on poverty. Most of Southeast Asia's poor are in rural areas where levels of poverty remain high. For the industrializing countries of the region, this situation poses some serious obstacles to their progress toward completing the transition to a new development stage.

RURAL AGRARIAN FUTURES

In many respects, the fabric of Southeast Asian society is and remains dominated by agriculture and the rural sector, and this will not change significantly in the immediate future. On the other hand, the growth-centered development policies of the ASEAN states, new economic reforms in some of the socialist states, and the factors promoting agrarian change are unlikely to diminish in the future. This will mean that rural social, economic, and environmental systems will continue to be altered by these forces. What then does this suggest about the region's agrarian future?

While the initial phase of rapid economic growth in many countries was fueled by the exploitation of natural resources, the dwindling land frontier and degraded natural resource base will impose increasing pressures on how resources are used and allocated. Farmers in many areas will need to intensify their use of land and adopt more-productive farm-management technologies. This can only be accomplished with improved production technologies, an expansion of rural credit, more eq-

Country	Rural Poor* (Percent of Total Poor) 1991	Rural Population		Rural Population	
		% Poor	Year	% Poor	Year
Indonesia	67.2	40	1976	16	1987
Malaysia	68.9	36	1970	38	1989
Philippines	61.9	70	1971	54	1988
Thailand	83.4	34	1969	26	1988

** Poverty levels are based on individual country definitions.*

Table 6.6 Rural poverty in selected Southeast Asian countries. *Source: UNDP, 1993; Quibria, 1993, Table 1.3, p. 34*

uitable land distribution, and the use of new farming systems that will better fit the environmental conditions and agricultural needs of farmers in the more marginal upland environments. The growing population of landless agricultural workers and even larger numbers of rural-urban migrants means that new strategies are needed to meet social and economic needs in both rural and urban areas. High priority will have to be given to rural industrialization, improving rural infrastructure, investing in sustainable and productive nonfarm employment, and intensifying agricultural production. These changes will be necessary if the rural landless and agricultural laborers are to participate more fully in and benefit from the development process. Although productive employment in urban centers must also be found for rural to urban migrants, the cyclical or short-term nature of some of this migration may see increasing numbers returning to rural areas, especially when economic recessions depress opportunities in urban economies. These priorities are especially important because pressures to control access to forest land and improve the management of these resources will increasingly limit their capability to absorb the large numbers of the rural poor and landless which occurred in the past.

Responding to these needs has traditionally been the domain of national governments and the state development bureaucracy. The general failure of state-led development to address the needs of all segments of rural society in an equitable and sustainable manner suggests that changes are required in this process. The experiences of the many small-scale rural development activities of NGOs and PVOs can provide useful lessons to better inform this process. More collaboration between the government and private sector could also lead to greater efficiency and economies of scale if their separate strengths are focused on those tasks and needs that each can accomplish best. Equitable growth and development in rural areas will also depend on increasing the involvement of rural people. If rural populations are provided greater opportunity to participate in defining and contributing to activities that meet their needs, rural development can become the sustainable process which has been a goal of national development policy for decades.

7 Industrialization and Trade

THOMAS R. LEINBACH JOHN T. BOWEN, JR.

EMERGENCE OF SOUTHEAST ASIA IN THE GLOBAL ECONOMY

The emergence of Southeast Asia as a critical arena in the world economy continues to capture the attention of development specialists and business decision makers (C. Dixon, 1991). Although the region's output is overshadowed by that of its more industrialized Northeast Asian neighbors, the area is increasingly important. Its emergence as a vital component of the world economy was evident by World War II as the region's markets and resources became a focus of rivalry between the United States and Japan. Indeed, Japanese dependence on oil from the East Indies became its most dangerous vulnerability. U.S. submarine attacks on Japanese tankers culminated in severe oil shortages by 1944, leaving Japan powerless to wage war at sea or in the air. A half-century later, Southeast Asia is again a focal point for global competition—but now the competition is over consumer markets, international investments, and high paying jobs.

The area is far from uniform in terms of development, but there are features in common. Most striking since the early 1970s have been the rapid development of industry, massive inflows of foreign direct investment, the emergence of authoritative regimes, and complex economic bureaucracies. The critical role played by foreign investment was highlighted by the financial crisis that swept the region in 1997 and 1998. Weaknesses in several key export industries, especially semiconductors, combined with fiercer competition from other major export regions, especially China, forced a devaluation of Southeast Asian currencies. As the currencies fell in value relative to the U.S. dollar and other important currencies, the region's exports became more competitive on world markets. But the devaluations also provoked some foreign investors to take their money out of the region. That outflow exacerbated the region's financial crisis and led to projections of slower economic growth.

Despite the 1997–98 setback, most observers expect an eventual return to the robust economic growth that has characterized the region since the 1960s. The region's strengths, including a large market potential (especially in the growing middle class), a high savings rate, low taxation, governments committed to reform, and an openness to trade and foreign investment by multinational corporations, provide a solid foundation for further accelerated economic development.

Our intent in this chapter is to discuss the nature of industrial development and trade within the region focusing upon the period since 1980. The success of economic policies emphasizing manufacturing has been striking. As a result, much of our attention is devoted to this aspect of economic development and the factors that characterize and affect it. Although export-oriented industrialization has been a leading factor in the region's growth since the 1960s, the globalization of Southeast Asia's service sector is a more recent phenomenon. Telecommunications, transportation, tourism, and international business services (like banking and insurance) have played an prominent role in the 1990s. As a result, the region's huge trade flows with the rest of the world include a wide range of manufactured goods from tennis shoes to software, as well as a variety of higher-order services.

CHANGING ECONOMIC STRUCTURES

Looking at the region overall, annual average growth in gross domestic product (GDP) between 1980 and 1993 reveals how dynamic, with only a few exceptions, the various economies have been (Table 7.1). The strongest growth has been recorded in Thailand (8.2 percent per year) followed by Singapore (6.9 percent), Malaysia (6.2 percent), and Indonesia (5.8 percent). Within the original ASEAN states, the Philippines performance was most disappointing (1.4 percent), although more recently that economy has exhibited much stronger growth. Surprisingly the Laotian economy has recorded good progress, albeit from a relatively small base, despite the disadvantages of political upheaval. Given the lack of a 1980 base figure, growth measures for Cambodia and Vietnam are unavailable. However, in the mid-1990s Vietnam's economy was expanding at an annual rate of more than 8 percent, while persistent political turmoil continued to stymie

Country	1993 GDP (Million $)	Distribution of GDP–1993 (Percent)				Aggregated GDP Growth 1980–93 (Percentage)
		Agriculture	Industry	Manufacturing	Services	
Burma	-	63	9	7	28	0.8
Cambodia	-	-	-	-	-	-
Indonesia	144,707	19	39	22	42	5.8
Laos	1,334	51	18	13	31	4.8
Malaysia	64,450	21	35	19	44	6.2
Philippines	54,068	22	33	24	45	1.4
Singapore	55,153	0	37	28	63	6.9
Thailand	124,682	10	39	28	51	8.2
Vietnam	12,834	29	28	22	42	-

Table 7.1 Distribution and growth of gross domestic product (GDP), 1980 and 1993. *Source: World Bank Development Reports, 1980 and 1993.*

Cambodia's economy. The salient features of growth in the numerous economies since are captured below.

Singapore

Singapore is a special case in ASEAN with a GDP per capita five times higher than that of its nearest rival, Malaysia. Singapore's strategic location and harbor have contributed to its emergence as one of the world's largest oil refining centers. More important has been the creation of robust manufacturing and financial service sectors. Although the government has promoted private enterprise, it maintains a pervasive role in overall planning and development. Although the state owns or indirectly controls a variety of enterprises, such as Singapore Airlines and the Development Bank of Singapore, the government's philosophy is that economic growth must come from the private sector. Since independence, a major priority has been industrialization. The success of this effort is reflected in the fact that the industrial sector's (which includes mining, construction, electricity, and gas) contribution to GDP doubled between 1960 and 1980. In 1993, industry accounted for 37 percent of GDP. Since the mid-1980s, the government has also given great emphasis to revitalizing the service sector, which accounts for the other 63 percent of GDP. As with other export-oriented economies in the region, the Singapore economy is highly sensitive to developments in major foreign markets, especially the United States, Europe, and Japan. High technology manufacturing and international services, including tourism, are the keys to future development.

The Newly Industrializing Economies (NIEs)

Thailand's agricultural sector accounted for 75 percent of total employment and 23 percent of GDP in the mid-1980s. A decade later, agriculture had dipped to 10 percent of GDP, and industry had grown from 27 percent to 39 percent. This shift in the balance of the two sectors reflects, above all, an aggressive drive toward manufacturing. Although the price terms of trade have deteriorated and the country is dependent upon imported oil for energy, export growth has been strong.

By the early-1990s, the Thai economy was booming, fueled by strong domestic demand and surging exports. Garments were the top export but computers and components, as well as agricultural goods such as rice, rubber, and shrimp were also important. The country enjoyed strong private investment as reflected by General Motors' 1996 decision to open a $750 million assembly plant south of Bangkok on the eastern seaboard with the intent of doubling its share of the Asian auto market to 10 percent by 2005. The decision to locate here, rather than in the Philippines, was based upon a strong financial incentives package, the strength of the domestic vehicle market, and a well-established supplier base. In return for GM's investment, Thailand will finance a $15 million automotive training institute, highlighting a major constraint on future Thai growth: the shortage of skilled workers. This deficiency is also slowing the shift from labor-intensive manufacturing to higher value-added production. In addition, necessary infrastructure projects such as Bangkok's mass rapid transit system, a new port to replace Klong Toey, and a new Bangkok international airport have been delayed and mismanaged by the government. The 1997 crisis in the Thai economy that led to similar crises in Malaysia, Indonesia, and the Philippines, was blamed on corruption and government incompetence.

The Philippine economy grew rapidly between 1960 and 1982 (on average nearly 6 percent per year). During this period, agriculture gradually declined, mining growth increased, and manufacturing grew strongly. But the limited size of the domestic market and poor export competitiveness resulted in only sluggish growth in the 1980s. The share of manufacturing has actually declined since the early 1970s. As elsewhere, agriculture accounts for a large share of employment but the service sector has absorbed most of the new entrants to the labor force.

Corruption, entrepreneurial inefficiencies, heavy protective barriers, and a foreign exchange crisis adversely affected the Philippine economy during the 1980s. Although the period was devastating, the 1990s brought clear signs that a turnaround had been achieved. GDP growth in the early 1990s averaged 5.5 percent and industrial output recorded significant growth too. How did the economy expand so quickly? The answer is by attracting massive net inward investment (more than $2.4 billion per year). In addition, Philippine nationals working overseas remitted $4 billion per year and continue to be the major source of foreign exchange earnings for the country. Nonetheless, too little of the investment reached the farm sector and real agricultural output growth was the lowest in Southeast Asia. Low government revenues prevented the necessary investment in irrigation and transport infrastructure that were so badly needed to improve productivity.

Malaysia has had a strong manufacturing performance. From 1970 to 1980 manufacturing grew by 7.6 percent annually, increasing to 10.3 percent per year from 1980 to 1993 (Table 7.1). Since 1960, industry and agriculture reversed positions in their contribution to the economy. Although Malaysia's resources (especially rubber, oil palm, tin, and timber) were hurt by depressed commodity prices, diversification into technologically sophisticated manufactured exports enabled the economy to thrive. At the same time, the government aggressively sought to shift the economic strength from foreign and Malaysian Chinese enterprises to the indigenous Malaysians—the *bumiputras*.

Despite the dramatic 6.2 percent annual average growth of the economy between 1980 and 1993, several concerns persisted. Income distribution among the country's main racial groups remained a problem as the Malay and Indian communities still controlled less wealth than the Chinese. There was also a concern about "overheating" in the economy that had been fueled by large scale construction projects including the Petronas Towers (the tallest buildings in the world) (Photo 7.1), a new international airport for Kuala Lumpur, and a new satellite city (Putrajaya) which will serve as Malaysia's new capital. The Malaysian economy has been largely investment driven and a key question is whether this will continue. Foreign investment has been attracted by the country's political stability, low cost labor (including a relatively inexpensive science and engineering workforce), effective infrastructure, and an openness to investment. In the 1990s, investment in labor-intensive industries has been deemphasized and the key has been to obtain the right mix of knowledge- and technology-intensive industries.

Photo 7.1 Petronas Towers, Kuala Lumpur. (Ulack)

A major concern has been the shortage of labor caused by the country's success. Some companies, such as Mattel, abandoned the production of some products in Penang—Barbie dolls for example—because of the scarcity of workers. To overcome this shortage, labor has been imported from Indonesia and elsewhere. A key aspect of future success is whether Malaysia will be able to remain competitive with her neighbors, especially Singapore, Thailand, Vietnam, and Indonesia. Another concern is the damage to Malaysia's environment caused by the country's stunning industrial growth.

Indonesia's economy grew rapidly under the Suharto government. Since 1974, growth has been largely export-led in the sense that exports grew more rapidly than the total economy. Although Indonesia benefited from rising world prices for its vast oil exports in the 1970s and early 1980s, there was a strong push toward manufacturing from the mid-1980s, reducing the reliance on oil and liquefied natural gas (LNG). To attract foreign investment and boost exports, liberalization has been a major emphasis of policy makers. These reforms have included the relaxation of regulations, reductions in subsidies, devaluation of the rupiah, rephasing of the public sector investment program, and reforms in the financial sector. Foreign direct investment was strong, especially in large industrial projects. In particular, Indonesia gained an advantage over its neighbors in competing for labor-intensive manufacturing investment, especially in the shoe industry.

Yet Indonesia is still viewed skeptically by many potential investors. Corruption remains a significant problem and the rules of business favor the well connected. A more "transparent" economy is needed where procedures and rules are well defined and understood by all. Indeed the deregulation "packages" must be complemented by a more effective competition policy to break monopolies, cartels, and other constraints on free competition. Tourism, which is to be the major foreign exchange earner by the turn of the century, has suffered from underdeveloped infrastructure and poorly trained labor. Another major concern is the looming oil crunch. With oil production stagnating and domestic consumption rising at 7 percent annually, Indonesia will soon become a net oil importer to cover industrial needs. New discoveries are running well below the level of production and demand is growing. Liberalized prices may have to be instituted and would come as a blow to an economy where fuel has always been heavily subsidized. In 1997, a liter of diesel fuel in Indonesia cost 16 cents compared with 68 cents in Japan and 24 cents in Vietnam. More than the price issue, rising domestic demand may reduce Indonesia's oil export earnings that still account for $10 billion annually. Finally, a persistent challenge has been the need to channel funds through banks to small-scale businesses in the less developed parts of Indonesia.

Three decades of growth in Indonesia have come to a halt as the financial crisis has spread across Asia. Despite a balanced budget, low inflation and a healthy trade account, the Indonesian currency came under attack by speculators. In late 1997 aggressive banks and businesses found themselves overextended with huge debts in dollars that could not be paid with a severely depreciated rupiah. As prices began to escalate, the Chinese community and political leadership came under attack. The action to remove subsidies on fuel and rice as part of an International Monetary Fund plan for restructuring helped to trigger the resignation of Suharto on May 21, 1998. As in other countries in the region, repair of the economy will require huge financial restructuring.

The Less-Developed Economies

An emerging trend in the region is the spread of economic growth from the Newly Industrializing Economies (NIEs) to the Less-Developed Economies (LDEs) of the region. Vietnam, as a new member of ASEAN, has a small GDP compared to the larger members but it is clear that a strategy for growth is being marshaled. A series of major, market-oriented reforms within the framework of socialist ideology have been implemented since the mid-1980s. The 10-year-old *doi moi* economic reforms included changes in pricing policy, reform and privatization of state enterprises, new foreign investment regulations, and adjustment to the removal of Russian aid. In 1993, industry accounted for 28 percent of GDP while agriculture and services accounted for 29 and 42 percent, respectively. In the 1990s, the economy's growth averaged nearly 10 percent per year. Exports, led by oil, rice, coffee, and some light manufactured goods grew by 25 percent, to $4.5 billion. Foreign investment is also being attracted to Vietnam both from ASEAN and beyond. Economic growth, however, has not occurred without some significant obstacles. In the mid-1990s, Vietnam's legal system was still inadequate for a modern market economy, corruption was widespread, and infrastructure dilapidated. More serious was the lack of political will for desperately needed reforms beyond those that were forced by *doi moi*. One reflection of the slow pace of change was the persistence of regulatory restrictions that disadvantaged private enterprises in favor of state firms.

Burma, once one of the world's leading rice exporters and a country with a solid resource base, has clearly lagged in economic progress. This is due in large part to political disruption and the poor implementation of economic reform measures that enjoyed some brief success in the 1970s. By the mid-1990s, agriculture, fisheries, and forestry still accounted for more than 63 percent of output and 68 percent of employment. GDP growth of nearly 7 percent was almost totally related to agriculture. Some observers feel that such growth will help liberalize state trading monopolies while promoting the private sector. Others say that state-controlled, outdated labor practices continue to exploit the rural farmers. Foreign investment has grown (within ASEAN, Singapore is important) despite a grossly overvalued kyat, but has focused on energy and services, not agriculture, the real economic force. Ending Burma's isolation and ethnic conflicts are prerequisites for more rapid economic growth. A more democratic form of government is essential for progressive change to take place.

Likewise, Cambodia and Laos have been scenes of political upheaval and their economic potential has been unrealized. Economic development, however, is beginning to move forward in Laos in the shape of some foreign investment and industrial growth. Laos adopted far-reaching economic reforms in 1986 to give greater scope for private sector development, but the impact of these reforms has been muted by inadequate infrastructure. Poor roads, for instance, limit farmer's access to markets, new higher-yielding seed varieties, pesticides, and fertilizers (Rigg, 1995). Thailand has shown interest in investing in both Cambodia and Laos, but the two smaller countries view their former adversary and powerful neighbor warily. Both countries have pinned their development strategies on integration into the greater Mekong subregion, which will link the economies of Laos, Cambodia, Burma, Vietnam, Thailand, and Southwest China.

THE GROWTH OF MANUFACTURING

The industrialization of Southeast Asia has progressed rapidly during the last three decades. In 1960, the share of manufacturing in GDP was less than 10 percent in most countries in the region. Thirty-five years later, the share of manufacturing exceeds 20 percent in all but Brunei and the LDEs (Table 7.2). In the region's most advanced economies, Singapore, Thailand, and Malaysia, manufacturing's share is about 30 percent.

Much of the success of the manufacturing sector has been achieved by following a strategy of *export-oriented industrialization* beginning in the late 1960s. By exporting to the vast global market, the manufacturing sector has been able to achieve higher rates of growth than the other sectors of the region's economies (Table 7.3). In the more industrialized economies of the region, manufacturing is a *leading sector*; its robust growth raises the rate of growth for the whole economy. The emergence of a vibrant manufacturing sector in Southeast Asia is one factor giving rise to the increasingly large and affluent middle class in the region. The growth of the middle class has led to sharply higher consumption of manufactured goods like automobiles and VCRs. As a result, domestic demand within the region has become a key force driving industrialization in the 1990s.

Country	1960	1993
Brunei	n/a	2*
Burma	8	7
Cambodia	n/a	5
Indonesia	8	22
Laos	n/a	13
Malaysia	9	30
Philippines	20	23
Singapore	12	28
Thailand	13	28
Vietnam	n/a	22

** 1990 data.*

Table 7.2 Contribution of manufacturing to GDP (percent). *Source: Dicken, 1998, Table 8.1;* World Bank, *1995; EIU, 1995c and 1995d.*

Country	1985–1990 (percent)	1990–1994 (percent)
Indonesia		
Manufacturing	10.7	9.5
GDP	6.1	6.5
Malaysia		
Manufacturing	14.1	12.3
GDP	7.1	8.3
Philippines		
Manufacturing	5.7	0.6
GDP	5.0	1.5
Singapore		
Manufacturing	11.8	7.5
GDP	8.2	8.4
Thailand		
Manufacturing	15.6	11.6
GDP	10.8	8.2

Table 7.3 Comparison of manufacturing and GDP growth. *Source: World Bank (1995) and various issues of Singapore "Monthly Digest of Statistics" (1986–1996).*

Manufacturing in the Colonial Era

The development of industry in Southeast Asia during the colonial period was modest. Two types of manufacturing prevailed. First, there was limited processing of raw material in preparation for export to major markets in Europe, the United States, and elsewhere in Asia. For example, a handful of tin-smelting factories were established in Malaya in the nineteenth century as the colony emerged as the world's most important source of tin. Similarly, there was some semiprocessing of natural resources in the Philippines such as coconut oil, copra products, and copper. But despite the abundance of natural resources throughout the region, industry was slow to develop under colonization. Instead, the region's exports (whether in raw or semiprocessed form) were used to feed factories located in the home countries of the colonial powers (Kaur, 1985).

The other type of manufacturing that developed in the nineteenth and early twentieth centuries was light industry that produced rudimentary goods such as cigars, beer, and medicated oils (to relieve headaches) for domestic consumption. Some of the small firms that emerged a century ago serving local markets have since become giant conglomerates. The San Miguel Corporation, for example, began as a beer brewery in Manila in 1890 (Photo 7.2); today it controls 90 percent of the Philippine beer market and is the dominant foreign beer in Hong Kong and China. San Miguel has also become the Philippines' top producer of soft drinks, ice cream, fruit juices, milk, and processed shrimp. San Miguel and a handful of other Southeast Asian manufacturers grew rapidly during the colonial era, but they were exceptions. The majority of manufactured goods were imported, and the proportion of the workforce employed in industry was small.

Photo 7.2 San Miguel Brewery, Cebu City, the Philippines. (Ulack)

Early Import-Substitution Industrialization

From the 1960s, most Southeast Asian governments gave high priority to industrialization. In part, industrialization was intended to reduce dependence on imported goods and to create jobs for the large numbers of urban unemployed. But the importance of industrialization went beyond merely economic considerations. It was equated with the modernization of societies that were still largely rural and agrarian (Krongkaew, 1992). In newly independent countries, industrialization was a potent political symbol as well. The development of indigenous industries meant achieving *economic* as well as political independence from the West. In Indonesia, this kind of economic nationalism was an especially strong force and remains a powerful influence on industrial policy today (Schwarz, 1995).

From the late 1950s to the 1960s, the most common strategy for achieving rapid industrial growth in these developing countries was through *import-substitution industrialization* (ISI). Under ISI, steep tariffs were levied against selected manufactured imports. In Malaysia, for instance, import tariffs introduced in the 1960s increased the prices of some imported goods to several times what they would have been otherwise (Table 7.4). These tariffs gave domestically manufactured products a competitive advantage. Initially, the products protected under ISI were *consumer* goods (e.g., processed foods, clothing), but it was expected that as these industries grew, the demand for indigenously manufactured *producer* goods (e.g., steel, industrial chemicals) would also expand, ultimately giving rise to an integrated domestic industrial structure.

Type of import	1962	1969
Clothing	25	400
Food Processing	5	65
Paper Products	40	140
Plastic Products	15	265
Textiles	55	110
Tobacco Products	60	125
Transport Equipment	n/a	135

Table 7.4 Effective rate of protection on selected Malaysian imports (percent). *Source: Jomo and Edwards, 1993, Table 1.8.*

In pursuit of ISI, the state dominated the private sector by imposing tariffs, import quotas, and financial incentives for domestic production. The state's strong role was consistent with the prevailing view in the 1960s that the state could not rely on the private sector to generate economic development but should instead take the lead (Kraiyudh, 1995).

ISI worked well, at first. New industries, such as processed foods, pharmaceuticals, and leather goods, emerged in the region. Manufacturing became a substantially more important part of the economy. In Malaysia, for instance, the proportion of GDP accounted for by manufacturing rose from 6.3 percent in 1957 to 10.4 percent in 1965. Other countries which adopted ISI experienced similarly rapid growth at first. This period has been labeled the easy or exuberant phase of ISI.

By the late 1960s, however, Malaysia and the other countries of Southeast Asia began to confront the severe limitations of ISI. Most importantly, the domestic markets of these relatively poor developing countries were too small to sustain industrialization. Firms producing goods under ISI were unable to achieve economies of scale and therefore costs were high. The high cost of producing goods under ISI further compounded the problem of small domestic markets because the high prices hindered market growth (Pranee, 1995). Moreover, the growing middle class had developed a preference for foreign imports for their prestige value or for their perceived superiority in quality.

Domestic markets were soon saturated and growth under ISI slowed. The easy stage had run its course, and it became inevitable that this strategy could not be used to achieve longer-term industrialization. In most countries, the result of ISI was a small, inefficient industrial sector confined to a narrow range of consumer goods. ISI had exacerbated trade deficit woes (due to dwindling exports and expensive capital imports such as machinery for ISI factories) and raised costs to consumers. Although jobs had been created, the employment impact was muted by the tendency of ISI factories to be capital-intensive rather than labor-intensive (Falkus, 1995). In relation to the very large numbers of unemployed and underemployed in the region, ISI was clearly inadequate.

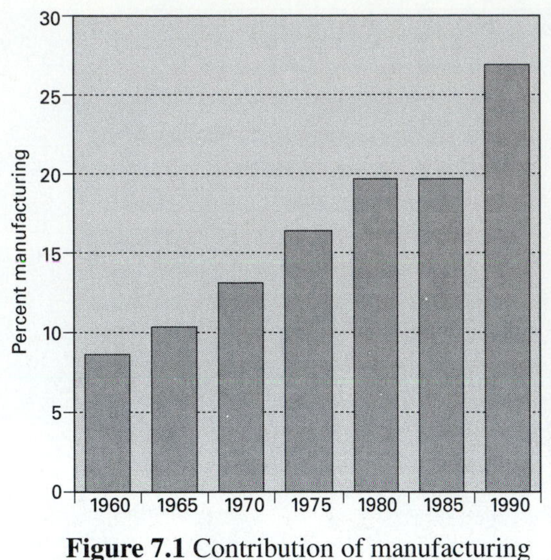

Figure 7.1 Contribution of manufacturing to Malaysia's GDP. *Source: Jomo and Edwards. 1993.*

The Turn to Export-Oriented Industrialization

Beginning in the mid-1960s, countries in Southeast Asia which had relied on ISI began to give greater emphasis to *export-oriented industrialization* (EOI) as a more viable development strategy. Although the two strategies shared a common objective, namely to industrialize as rapidly as possible and create thousands of new jobs, EOI implied a fundamentally different relationship between developing countries and the global economy. Under ISI, developing countries had sought to isolate selected industries from global competition. In contrast, EOI called for a much closer integration with the global economy. Rather than avoiding international competition, developing countries aggressively took advantage of their significant competitive advantages in the global economy, especially low labor costs, to accelerate industrialization. In the case of Malaysia, for instance, the pace of industrialization accelerated sharply during the 1970s as the government implemented an aggressive strategy of export-oriented growth (Figure 7.1).

For Singapore, the first Southeast Asian nation to emphasize export-oriented industrialization, the adoption of this strategy was almost unavoidable. Singapore's expulsion from the Malaysian Federation left the city-state with a domestic market far too small to support import substitution (Grice and Drakakis-Smith, 1985). For the regions' other economies, however, the adoption of an export-oriented strategy has been more contentious and is still incomplete. Favored individuals and firms benefited from the protection they enjoyed under ISI and have resisted its dismantling.

A common theme in the region is that overcoming these entrenched interests has required a crisis that allows each economy to take a new direction. In Malaysia, the violent race riots of 1969 and the subsequent adoption of the New Economic Policy (NEP) gave added impetus to the government's industrialization efforts. Rapid industrialization was considered essential to reduce poverty while creating thousands of modern jobs for Malays migrating from rural areas to the country's cities (Jomo and Edwards, 1993). In Indonesia, the collapse of oil prices in the mid-1980s sparked a renewal of that country's industrialization drive, and in the Philippines, where protectionist policies were still very strong in the early 1980s (Figure 7.2), the demise of the Ferdinand Marcos administration led to a stronger emphasis on export promotion.

External Forces Behind EOI

The turn to EOI occurred at a time when conditions in the global economy were ripe for an export push by developing countries. After World War II, international trade expanded rapidly as the major world economies experienced a surge in economic growth and adopted more liberal trade regulations. In the 1950s, these conditions had helped Japan recover from its wartime devastation. Later, in the 1960s, developing countries in Southeast Asia also began to take advantage of the favorable world trading environment to achieve rapid export growth in developed country markets.

The boom in world trade meant that firms in countries with higher production costs experienced increas-

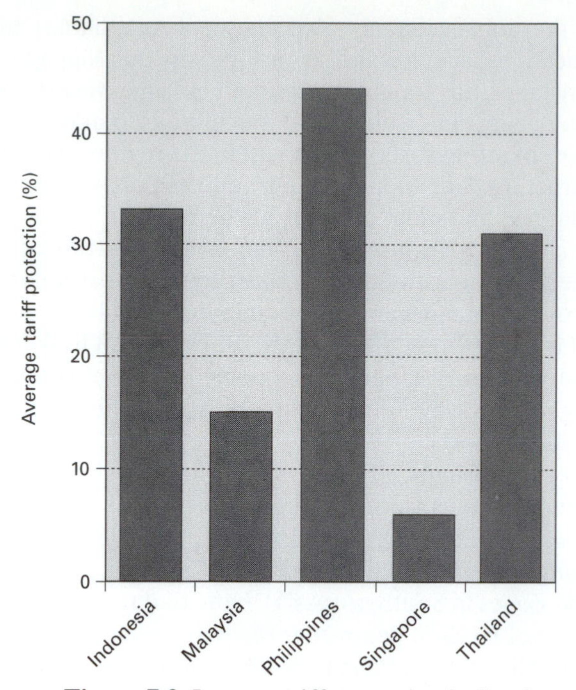

Figure 7.2 Import tariff protection in Southeast Asia, early 1980s. *Source: Suphachalasai, 1995. Table 3.5.*

ingly fierce competition from cheaper imports, especially from Japan. To deal with this threat, firms began to look for ways to reduce their production costs. One means of doing so was to shift part of their operations to lower-cost countries, a process referred to as the *internationalization of production* (Dicken, 1998).

The internationalization of production was greatly facilitated by advances in communications and transportation technology. In particular, the development of containerization (in sea freight) and larger jet aircraft reduced transport costs and travel time, making a globally dispersed production process economically feasible. Also, the extension of more reliable communications facilities in developing countries and the introduction of satellite communications allowed better coordination and integration among manufacturing facilities separated by thousands of miles. At the same time, the integration of computers into business management has helped firms to control complex, global operations.

These conditions—a liberal world trading regime, increasing pressure on firms in the advanced industrialized countries to reduce production costs, faster and more reliable means of transportation and communication—set the stage for a new international division of labor (NIDL) in the 1960s. The NIDL marked a departure from the traditional bisection of the world into a few industrialized countries (the core) and a much larger number of developing countries (the periphery) that were integrated into the global economy mainly as raw material exporters (Hill, 1987). From the 1960s the relocation of factories to developing countries gave these economies a new role as exporters of manufactured goods.

Although the NIDL has presented developing countries with new opportunities, the periphery remains dependent on the core of advanced industrialized countries. Most multinational corporations (MNCs) have shifted mainly labor-intensive functions to the periphery and have been quick to move their operations when host countries become too expensive or when new, less expensive sites are opened up (e.g., in Vietnam). Conversely, the more highly skilled positions, management, control of capital investments and technology, and profits remain in the MNCs headquarters in the core countries.

Moreover, developing countries face increased dependence on fluctuations in global markets. A century ago, Southeast Asia was vulnerable to shifts in the world prices for tin, rubber, and other natural resources. Today, the region is still vulnerable to global demand, but today the commodities at stake are semiconductors and software packages. Adverse swings in global demand for electronics were a contributing factor in the region's economic crisis which began in 1997 (Salih and Young, 1987). Other industries in the periphery are also vulnerable to this threat.

Although the NIDL has not liberated developing countries from their dependence on the developed world, there is little doubt that changes in global patterns of industrialization have resulted in millions of new jobs in Southeast Asia. The challenge for the periphery is not to escape dependency, which is an inevitable feature of an ever more-integrated global economy. Instead, the challenge is to move up rapidly in the NIDL, to progress from specializing in simple, labor-intensive manufacturing toward capital-intensive, knowledge-intensive manufacturing—from exporting radios to exporting tiny, delicate silicon wafers worth thousands of dollars apiece. Several Southeast Asian countries have managed this transition effectively.

The Role of the State in EOI

The state has played a decisive role in Southeast Asia's accelerated, export-oriented industrialization. The speed of industrial development in the region is vital for two reasons. First, in the long run, higher living standards are more easily achieved as a country's industrial sector becomes more sophisticated. As a country makes the transition from specializing in simple, labor-intensive operations (assembling toys) to more knowledge-intensive operations (assembling aircraft), wage levels generally go up in accordance with higher skill levels and the greater value that each worker adds to the final product. For the state, it is imperative that this transition toward a more sophisticated industrial structure occur as quickly as possible because stagnation would undermine the state's legitimacy. Southeast Asian governments have based their right to rule on their ability to deliver rapid improvements in incomes and living conditions. In Indonesia, for instance, President Suharto justified his government's harsh response to the Jakarta riots of 1996 by pointing out the tremendous economic advances Indonesia had made under his autocratic rule.

Second, in today's dynamic and highly competitive global economy, no nation can afford to stand still. Indonesia's advantage in the shoe industry, for instance, is being eroded constantly as countries with lower costs (e.g., Vietnam) develop the infrastructure, financial incentives, and other inducements necessary to attract investment by shoe manufacturers. Industrialization in the developing world usually involves the use of existing technology—often technology that was pioneered and is owned by MNCs. The spread of existing technology makes it possible for whole new industries to emerge quickly in developing countries. The resulting speed of industrial development in the developing world puts great pressure on firms and governments to adapt quickly.

Much of the recent industrialization in the region has been fueled by footloose capital. Unlike manufacturing that needs a large nearby supply of a natural resource, such as tin or coal, most of the manufacturing operations set up by MNCs in Southeast Asia require only one major local resource: labor. MNCs therefore have greater latitude in deciding where to locate such plants; they are *footloose* investments. Even manufacturing plants set up by local firms rather than foreign MNCs can be footloose. Increasingly, manufacturing firms from such countries as Singapore and Malaysia have begun to relocate some operations to such lower-cost neighboring countries as Indonesia and the Philippines. These firms have become home-grown MNCs and, like their foreign-based counterparts, are prone to invest overseas to expand, reduce operating costs, or gain better access to markets.

Foreign investors have been offered a battery of financial incentives such as special tax rebates and low import and export duties. In several Southeast Asian countries, even more generous incentives have been granted to firms that set up manufacturing operations in designated Free Trade Zones (FTZs). Investors in Malaysian FTZs, for instance, are eligible for pioneer status providing complete exemption from taxes for 5–10 years, and goods that are exported from or imported into an FTZ are exempt from customs duties (Leinbach, 1982; Rajah, 1993). Restrictions on foreign ownership and remittance of dividends to foreign countries are also more relaxed in FTZs.

To complement these financial incentives, governments have spent billions on infrastructure, especially those kinds of infrastructure necessary for integration with the global economy: airports, seaports, reliable communications facilities, and so on. One way that Singapore, for instance, has sought to maintain an edge over neighboring countries (which have much larger markets and lower operating costs) is to invest in the region's best infrastructure. Singapore's seaport and airport are among the largest and most efficient in the world. Conversely, Thailand's economic growth has been constrained by inadequate investment in infrastructure, most notably seaports and land transportation in the Bangkok metropolitan area (Wong, 1993). In the early 1990s, it took four to seven times longer for goods to move through Bangkok's Klong Toey port than through Singapore's world class port.

Not surprisingly, FTZs have been located close to key international transportation facilities; for example, the largest FTZ in Malaysia, at Bayan Lepas, is adjacent to Penang's international airport (Map 7.1: Clusters of Free Trade Zones in Southeast Asia). Note that there are four clusters of FTZs in Peninsular Malaysia, each near major metropolitan areas (Penang, Kuala Lumpur, Johor Bharu, and Malacca). These clusters have contributed to the unequal pattern of development in the country because they reinforce the advantage of the western coast that had its roots in the colonial era. That area is also home to Malaysia's ambitious new infrastructure project, the Multimedia Supercorridor (MSC) (Box 7.1). This project is intended to attract new investment from manufacturers producing software and other kinds of multimedia goods.

The Philippines has also used FTZs to promote industrialization. Several were set up by the state in the early 1970s. Sites in Cebu, Baguio, and other outlying areas were intended to spread industrial growth. Since the early 1990s, nine privately owned FTZs have contributed to the rejuvenation of the Philippines' manufacturing sector. The two largest are the Laguna International & Industrial Park (near Manila) and the Subic Special Economic & Freeport Zone. The latter was established at the former U.S. naval base on Subic Bay. The excellent infrastructure left behind by the U.S. personel enticed the U.S. air express delivery firm, Federal Express, to establish its Asian hub at Subic Bay. The rapid air connections to points in the region and in the United States that FedEx now offers have been an additional factor attracting manufacturing investments from such companies as the Taiwanese computer manufacturer, Acer, that require fast, worldwide distribution of their output.

Conversely, for Burma and the countries of Indochina, the infrastructure necessary to integrate into the global economy is much more basic. The most common obstacle faced by manufacturers in these countries is the lack of electricity. In the early 1990s, for instance, foreign investors in the Ho Chi Minh City region had no power for at least one working day each week (Nash, 1993).

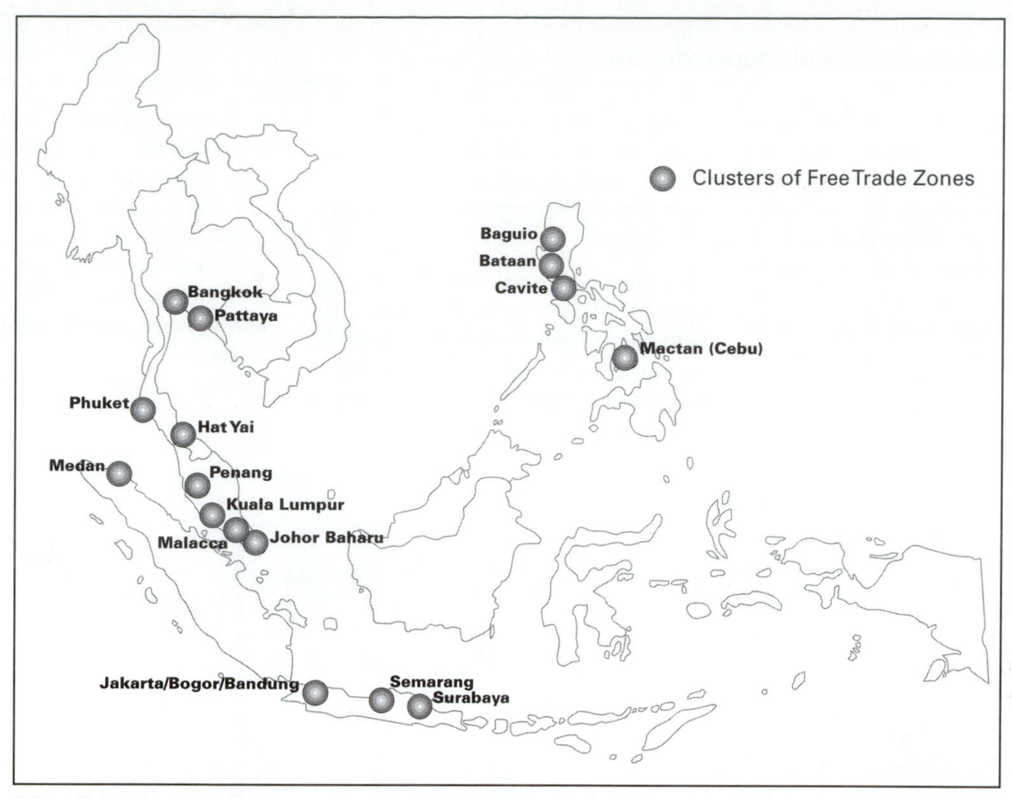

Map 7.1 Clusters of free trade zones in Southeast Asia. *Sources: Rimmer, 1994a; Sopiee, 1991; UNIDO, 1991.*

Labor Costs. Government investments in infrastructure and financial incentives have been complemented by efforts to manage the industrial labor force to limit wage growth and eradicate labor unrest. In the late 1960s, for instance, the government of Singapore passed legislation curtailing the power of organized labor. As a result of these initiatives, the amount of work time lost to strikes fell drastically (Rodan, 1989). Similar efforts to create a pliant workforce were undertaken in Malaysia and Thailand (Rajah, 1993; Phongpaichit and Baker, 1995).

Today, labor costs in these countries have risen far above the levels when rapid industrialization began. It is important to note that governments in these countries *want* wages to go up. Indeed, rising wages are a critical factor in the continued popularity of the governments in the region. Yet it is equally important that wage growth does not outstrip the growth in labor productivity (output per worker per hour). The government in Singapore implemented policies that led to sharply higher labor costs in the early 1980s. Productivity could not keep pace, and many labor-intensive manufacturers abandoned the city-state. Their departure was one factor contributing to a recession in the mid-1980s; to facilitate economic recovery, the government scaled back some of the wage increases introduced earlier in the decade (Rodan, 1989). Figure 7.3 highlights the imbalance between labor productivity growth and wage growth in Singapore in the early 1980s.

Labor Quality. Worker skill levels also have a decisive influence on patterns of industrialization in the region. Basic literacy and numeracy are often prerequisites even for the most simple operations. Singapore's relatively well-educated population gave the city-state an early advantage in attracting MNCs in the 1960s (Rodan, 1989). Today, the emphasis has gone beyond basic numeracy and literacy (though these remain crucial); instead, greater attention is given to technical and science education. For example, the availability of inexpensive, highly trained professionals such as engineers has helped Singapore and Malaysia attract billion-dollar investments from high tech MNCs.

In neither country did this pool of engineering talent emerge by accident. Rather, both countries have pursued very specific human resource development policies to manage the number of people trained in science, engineering, and technical fields to meet the growing needs of industry (Ali, 1992). Tertiary institutions, for instance, have given great emphasis to programs in chemical, electrical, and mechanical engineering; and student enrollments in economically less critical fields like the humanities are limited by government policy.

Political Stability. Finally, investors have been drawn by the region's relatively stable governments because these governments maintain predictable policies toward

BOX 7.1 Malaysia's Multimedia Supercorridor

The Multimedia Supercorridor (MSC) is a bold attempt to leapfrog Malaysia's economy into the information age. The MSC will occupy an area 9.3 mi (15 km) wide and 24.8 mi (40 km) long, stretching from Kuala Lumpur in the north to Malaysia's new Kuala Lumpur International Airport in the south (Map 7.2). In addition to the region's newest airport, the MSC will also contain a new north-south highway tying into an expressway system that extends from the Thai border to Singapore and a backbone network of very high-volume optical-fiber links ("MSC Plan Gets High Priority," 1996). The zone's physical infrastructure is to be complemented by a host of financial incentives, including up to 20 years of tax-free status. In addition, the Malaysian government intends to implement a host of cyberlaws covering such areas as digital signatures and intellectual property rights to make the MSC more enticing to multimedia corporations.

Malaysia has targeted such major international players as Microsoft, Netscape, Sony, and Sun Microsystems with this venture. The idea of MSC is to get these corporations to view Malaysia as much more than a relatively low-wage export platform. Instead, the country would become home to a whole gamut of research, development, design, manufacturing, marketing, and distribution operations for the multimedia industry. The types of products and services that the government expects to thrive in the MSC include:

- application-specific integrated circuits (which control the operation of such things as car engines and home appliances)
- wafer fabrication
- electronic publishing
- animation services
- distant learning software and hardware

Malaysia planned for the MSC to be open for business by 1998, but it is likely that the financial crisis that swept the region in 1997 will delay full implementation. In particular, the construction of a new city, called Putrajaya, in the middle of the MSC will be slowed. Putrajaya is envisioned as an "intelligent" city serving as Malaysia's new administrative capital to ease the burden on overcrowded Kuala Lumpur. In the longer term, the success of the MSC concept will depend on how adeptly Malaysia can match the requirements of the fast-moving information technology industry.

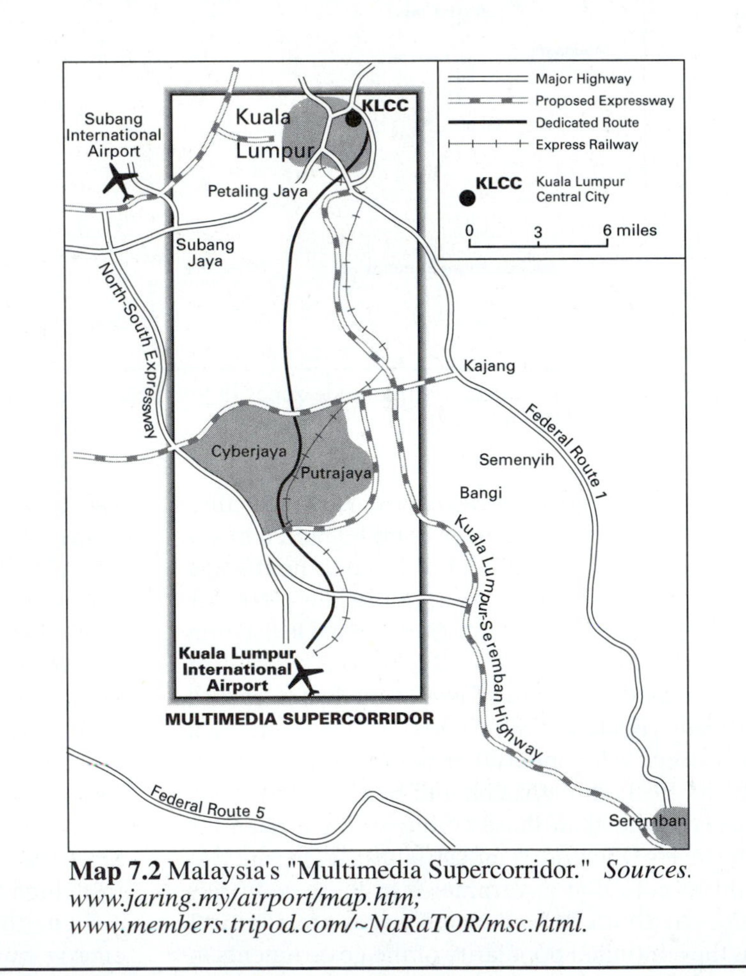

Map 7.2 Malaysia's "Multimedia Supercorridor." *Sources. www.jaring.my/airport/map.htm; www.members.tripod.com/~NaRaTOR/msc.html.*

international investment. Accordingly, MNCs are more likely to make billion-dollar investments in Singapore or Malaysia than in Burma. There were reports in the mid-1990s that potential investors in Indonesia were deterred by uncertainties about the succession of President Suharto ("Indonesia: Okay, They Say," 1996).

Yet the importance of government stability should not be taken too far. Although it is true that three of the region's four most successful economies—Singapore, Malaysia, and Indonesia—were each ruled by a single party for more than 30 years, the other strong performer in Southeast Asia has had a much different political history. Thailand has not only had frequent changes in government since the 1960s, but it has also experienced military coups, communist insurgency, and mass protests in Bangkok (Samudavanija, 1995). In the past Thailand was able to overcome this instability because behind the frequent changes in government there has been an elite group of top bureaucrats who have consistently pursued economic development. The economic bureaucracies maintained tight relationships with the corporate business sector, and they worked togeth-

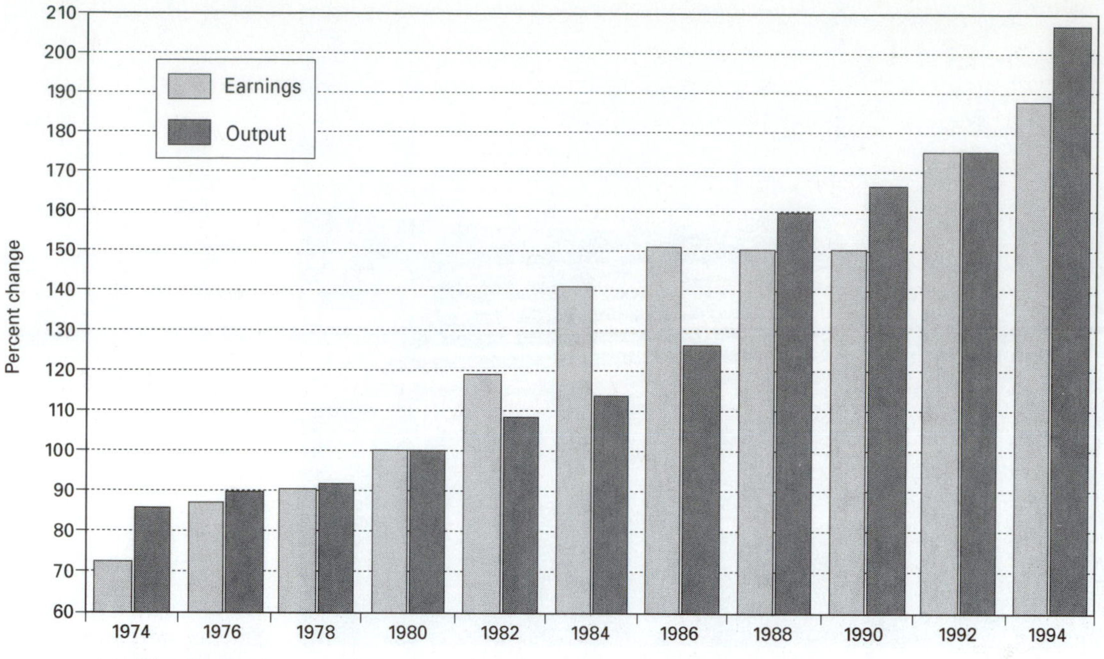

Note: The data for the comparison of manufacturing wages and productivity in Singapore are based on an index level of 100 for 1980.

Figure 7.3 Change in manufacturing earnings and output per employee, Singapore, 1974–1994 (even years). *Source: World Bank, 1989; UNESCAP, 1996.*

er to promote Thailand's rapid industrialization (Samudavanija, 1995).

An ugly backdrop to the region's reputation for efficient, competent, stable governments has been the climate of corruption that pervades much of Southeast Asia. Some long-serving regimes have established far-reaching patronage networks that reward government cronies with state contracts or jobs or policy favoritism. In Indonesia, President Suharto's children and other relatives amassed spectacular fortunes through their privileged positions in everything from toll-highway construction to plastic imports. In Thailand, corruption has played out in the form of widespread vote-buying during the country's frequent elections. It was estimated that $800 million was spent buying votes in the 1996 election (Mydans, 1997a). The result was a parliament that was slow to adopt needed financial remedies because many of its leaders or their patrons had personal interests in banks that would have been hurt by the austerity measures. This ineffective government was not considered much of a problem during the country's boom years but crippled the ability of the economic bureaucracies to deal with Thailand's first major financial crisis.

Patterns of MNC Investment in Southeast Asia

Different kinds of MNCs are drawn to Southeast Asia for different reasons. Some set up manufacturing plants in Southeast Asia to tap the local market. For example, global beverage companies Coca Cola and Pepsico have been among the first to set up manufacturing operations in newly opened economies like Vietnam. Other MNCs come to Southeast Asia to tap natural resources in the region. Examples include the Japanese-funded Asahan aluminum smelter in Sumatra (Photo 7.3), which uses alumina from Australia and locally generated hydroelectricity; the Thai owned shrimp processing plants in coastal Burma; and oil refineries in Singapore set up by foreign MNCs to tap output from Indonesia and Brunei. But the most important type of MNC activity in the region is manufacturing for export. For these firms, Southeast Asia is an *export platform*: Goods are manufactured in the region, often using imported components, and then exported for sales worldwide.

Although a huge variety of goods are manufactured for export by MNCs in Southeast Asia, two broad industrial sectors are particularly important—textiles (including clothing and shoes) and electronics. It was in these two industries that much of the early MNC investment took place in the 1960s and they remain critical today. But in the 1990s, a rough dichotomy may be said to characterize the region: For the least industrialized economies (Indonesia, Vietnam, Laos, and even Brunei), manufactured exports are dominated by textiles, clothing, and leather goods (especially shoes); for the more industrialized economies (Singapore, Malaysia, and to a lesser extent Thailand and the Philippines), electronics dominate.

The textile, clothing, and shoe sector has often served as a kind of entry-level industry for the integration of developing countries into the global economy; for example,

Photo 7.3 Asahan Aluminum Smelter, northern Sumatra. (Airriess)

Japanese textile firms began operations in Thailand in the late 1950s (Phongpaichit and Baker, 1995), and textiles are now one of the main industries in which MNCs are active in Vietnam. Even Laos has managed to gain a foothold in the industry with export sales of the Laotian garment industry swelling from US $2 million in 1989 to US$25 million in 1992 (EIU, 1995a).

The textile industry has proven relatively easy to transplant to developing countries because the technology is simple. For MNCs, there are two strong incentives to relocate production abroad. One, of course, is to escape high production costs in the home country (Table 7.5). The second is to gain additional quotas for markets in the United States and the European community. These countries have imposed quotas on textile, clothing, and shoe imports from lower-cost developing countries, especially Hong Kong, South Korea, and Taiwan (Dicken, 1998). Accordingly, much of the foreign direct investment in this industry in Southeast Asia has been by MNCs from these three countries (O'Connor, 1993a). More recently, some countries in the region have become major sources of foreign direct investment (FDI) in the textile industry as well. Thai firms, for example, have been among the major investors in the Laotian textile industry

Textiles have been described as a sunset industry in the more advanced economies of Southeast Asia and data on the contribution of the sector to total exports do show lower figures for Malaysia and Singapore (Figure 7.4). As labor costs have risen in the latter two countries, their attraction to textile multinationals has waned. It is possible, however, to compensate for higher labor costs by increasing productivity through such new technologies as numerically controlled cutters, laser cutting systems, and computer-aided design; but these technological advances have been slow to diffuse to Southeast Asia (Wiboonchutikula, 1990). Instead, they have been applied in the home countries of the MNCs while the more labor-intensive assembly operations are carried out at shifting locations in the developing world (O'Connor, 1993a).

Country	Average Total Hourly Labor Costs (U.S.$)	Ratio to U.S. Cost (%)
U.S.	10.02	100
Japan	13.92	139
Germany	16.46	164
Taiwan	4.56	46
South Korea	3.22	32
Hong Kong	3.05	30
Mexico	2.21	22
Turkey	1.82	18
Malaysia	0.86	9
Philippines	0.67	7
China	0.37	4
Indonesia	0.25	2
Sri Lanka	0.24	2

Table 7.5 Labor cost comparison for spinning and weaving, 1990. *Source: O'Connor, 1993a, Table 9.4.*

In contrast to textiles, the electronics industry has continued to offer new opportunities for the development of ever more sophisticated operations. As a result, electronics dominates the manufactured exports of countries

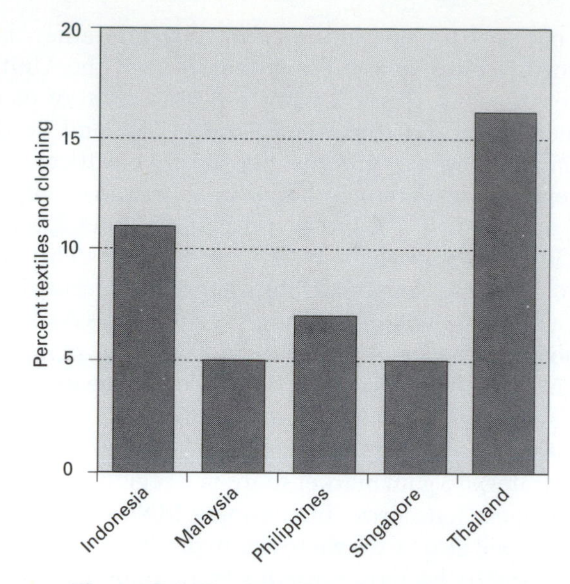

Figure 7.4 Contribution of textiles and clothing to merchandise exports, 1990.
Source: O'Connor, 1993a, Table 9.1.

as disparate as Singapore, Malaysia, and the Philippines, but what these countries produce for the global electronics market varies considerably. The pattern of development of the industry in Southeast Asia has been likened to flying geese, with Singapore remaining at the technological frontier in the region while the other countries trail some distance behind (Chalmers, 1991). Malaysia's Multimedia Supercorridor is a bold attempt to challenge this pattern by leapfrogging ahead of the city-state.

The electronics industry is grouped into three categories: (1) consumer electronics such as radios and TVs, (2) electronic components such as memory chips that are used in a wide variety of electronic goods, and (3) industrial electronics such as telecommunications equipment. Initially, much of the investment by MNCs in the region was in plants that produce consumer goods for the local market. In the 1960s, factories to assemble TVs, radios, and phonographs were established across the region (Chalmers, 1991).

Beginning in the late 1960s, MNC investments in manufacturing for export rather than for local markets began to accelerate. The new investments were in electronic components rather than consumer goods. The first semiconductor plant in the region was established in Singapore in 1967 by Texas Instruments. Silicon wafers that had been manufactured under exacting standards in the developed world were flown to Southeast Asia for the labor-intensive process of cutting the wafers into chips and then assembling integrated and discrete circuits. Singapore was the first country to emerge as an export-processing center and, by the mid-1970s, ranked first in the world in semiconductor exports (Dicken, 1998). As costs in Singapore rose, new investment shifted to Malaysia and the Philippines so that by the early 1980s, these two countries ranked first and second, respectively, in the world in semiconductor exports (Chalmers, 1991). In the mid-1990s, Southeast Asia remained the world's top region for semiconductor exports.

Although an increasingly diverse range of other electronic components are manufactured in the region (e.g., resistors, diodes, and transformers), industrial policy in Southeast Asia continues to give particular emphasis to the semiconductor industry. In the mid-1990s, both Singapore and Malaysia gave high priority to the development of wafer fabrication plants. Wafer fabrication is the next level up in the division of labor in the semiconductor industry, and for national governments wafer fabrication offers the lure of high-paying, professional jobs. Initially, wafer fabrication took place only in the United States and other advanced economies because the process requires not only a pool of skilled semiconductor engineers who can fine-tune each plant to achieve acceptable yields of good wafers, but also ample supplies of electricity, pure water, specialty gases, and electronics-grade chemicals.

Yet relentless price competition among chip manufacturers has forced MNCs to look for ways to shift wafer fabrication to developing countries. Three other factors have facilitated the relocation of wafer fabs to Southeast Asia (O'Connor, 1993b):

- The process technology is increasingly mature and is therefore easier to transfer to a developing country without experience in wafer fabrication;
- Several Southeast Asian countries have decades of experience in the electronics industry and have built up the managerial and technological prowess necessary to operate the most sophisticated manufacturing plants;
- The consumer electronics industry, which is the main market for chips, is increasingly centered in Asia. Firms can cut transport costs and production time by locating wafer fabs closer to these users.

With these factors at work, wafer fabrication plants were established in both Singapore and Malaysia by the early 1990s. In keeping with the flying geese pattern, Singapore moved a further step ahead in 1996 when it became the first place in Southeast Asia to manufacture photo masks ("S'pore Chip Industry Receives $21M Boost," 1996). Masks, a vital element in semiconductor manufacturing, are high-precision quartz plates embedded with microscopic images of electronic circuits. The Singapore facility is intended to supply the semiconductor industry across Asia.

Yet some have pointed out the risks of specializing in electronics. Both Singapore and Malaysia depend on electronics, including everything from Pentium chips to home computers to electric toasters, for more than 50 percent of their manufactured exports. This dependence, combined with the openness and dynamism of the industry, make for a dangerous situation during cyclical downturns in the global electronics market.

Manufacturing for the Asian Market

MNCs have long been attracted to the large domestic markets in the region. But historically most MNC manufacturing for domestic consumers has been in simple, low-cost commodities like processed food and beverages, tobacco products, pharmaceuticals, and health and beauty products. MNCs in these industries use their global brand names (e.g., Pepsi, Nescafé) to gain a significant share of these young markets. More recently, in the more developed economies of the region, the growth of the middle class has given a new slant to MNC investment. Now MNCs are moving into local production of more complex, higher-priced consumer durables or big ticket items.

This transition is illustrated by the automobile industry. By the early 1990s, Southeast Asia was the fastest-growing automobile market in the world (Figure 7.5 and Box 7.2). Most automobiles sold in the region are assembled locally from imported kits in which most of the components are manufactured overseas. Increasingly, however, the volume of sales in the region's more lucrative markets has reached levels which justify development of full-scale vehicle production rather than just assembly of imported kits.

Thailand, with one of the fastest-growing vehicle markets in Asia and the world's largest market for trucks outside the United States, has been the main beneficiary of this trend (Gibney, 1996). Ford and Mazda are building a truck plant in Thailand that could produce 150,000 pick-ups per year for sale in Asia and Europe. Rather than importing most of the components from the United States or Japan, the new plant will draw most of its intermediate inputs from other plants in Thailand. Ford is setting up plants to produce plastic and electrical components as well as air-conditioning systems to supply the new auto plant ("Ford to Invest United States $53 Million in Two Components Plants in Thailand," 1996). And as noted above, General Motors announced its own ambitious plan to make Thailand the center of its Asian automotive strategy. A new US $750 million plant will produce 100,000 cars a year for markets in Japan, Southeast Asia, and Australia (Dunne, 1996).

For Ford and GM, investment in Thailand is one key to a strategy to gain market share in a region dominated by Japanese auto firms. Increasingly, MNCs in other industries will also be drawn to set up production in Southeast Asia to be closer to the front lines in what is becoming one of the world's most lucrative consumer markets. Figure 7.6, for instance, illustrates the explosive growth in sales of color TVs during the early 1990s. Southeast Asian consumers were purchasing goods and services at a feverish pace, and this phenomenon is one of the trends that will continue to shape the pattern of industrialization in the region as the economies recover.

As MNCs enter the Southeast Asian market to produce goods for local competition, they confront increasingly fierce competition from local firms. In the auto industry, for example, both Malaysia and Indonesia have set up national car manufacturers. These firms have enjoyed varying degrees of success, but across the region MNC brands continue to have overall dominance because of their technological, marketing, and distribution advantages as well as greater economies of scale.

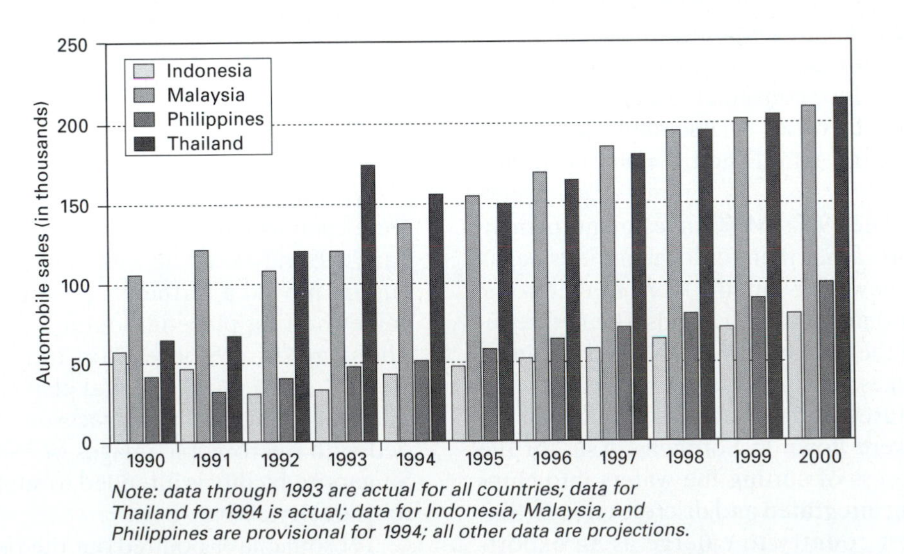

Figure 7.5 Car sales in selected Southeast Asian countries, 1990-2000. *Source: Maxton, 1995.*

BOX 7.2 The Development of the Auto Industry in Southeast Asia

Until the 1980s, the auto industry in Southeast Asia comprised dozens of small plants where cars were assembled for sale in each national market. The small scale of operation kept costs at these labor-intensive operations high, but they persisted because most governments in the region had import substitution policies that forbade the import of fully assembled cars. In Malaysia, for instance, most of the world's car manufacturers set up plants in the country, and by 1980 there were 11 assemblers producing 122 models. The local content of cars assembled in these plants was low; generally, only the tires, batteries, and paint were Malaysian-made (Jayasankaran, 1993).

Following the example of South Korea and other East Asian countries, Malaysia became the first Southeast Asian country to try to develop a national auto industry. In the early 1980s, Malaysia had a higher level of car ownership than South Korea (one car for every 21 people in Malaysia versus one car for every 146 people in South Korea); yet while the Korean auto industry was already becoming a factor in the global economy, the auto industry's contribution to Malaysia's GDP was less than one percent in 1983.

Malaysia turned to another East Asian country, Japan, to get its national car project started. In 1982, the new company, called Proton, was established as a joint venture between the Malaysian government and the Japanese car manufacturer Mitsubishi (Machado, 1992). The first Proton rolled out of the factory in April 1986. The car was met by a wave of national enthusiasm, but sales were poor as Malaysia was then in the midst of a severe recession (Jayasankaran, 1993). Although the project had initially been aimed at the domestic market only, Proton was forced to begin to exploit overseas markets early. The Malaysian product proved moderately successful in foreign markets. In Britain, Proton received favorable quality ratings and sold 80,000 units between 1989 and 1995 (Jayasankaran, 1993). In Singapore, the Proton Wira was the best-selling model in 1995 ("Malaysia's Proton Wira is Best-Selling Model Here," 1996).

Yet Malaysia remains Proton's most important market by far. Heavy import duties on other cars sold in Malaysia make Proton about 40 percent less expensive than competing cars in the same class. Through 1995, Proton had manufactured 500,000 cars with roughly 400,000 sold in the domestic market (Figure 7.6). This dependence on Malaysian sales leaves Proton with a significant problem: market saturation. Accordingly, Proton is giving greater emphasis to overseas expansion. In 1996, Proton began production in Vietnam and announced a plan to manufacture cars in the Philippines for the domestic market there ("Dr M Launches Sale of Proton Wira in Vietnam," 1996; Labita, 1996).

Buoyed by Proton's success, Proton established a second national car company, called Perodua, in conjunction with the Japanese company Daihatsu in 1991. Perodua produces smaller, lighter vehicles than Proton, and the two companies do not compete directly with one another. Perodua targets families who already own a Proton but need a second car.

More recently, Indonesia tried to develop its own national car industry. In 1996, a joint venture between a company controlled by one of President Suharto's sons and the Korean car manufacturer Kia was established to produce a national car called Timor. The Timor was initially manufactured in Korea but production was scheduled to be transferred to an Indonesian factory in 1998. Like Proton, the Timor enjoyed exemption from certain taxes, making it much less expensive than competitors. In contrast to the Malaysian experience, however, Timor failed to win much support. Indeed, the project was surrounded by controversy as critics objected to the favoritism shown to one of Suharto's sons.

The national car projects in Malaysia and Indonesia are exceptions to the broader trend in the region towards greater openness to investment by foreign automakers. With the region's largest auto market in the mid-1990s, Thailand attracted large investments by General Motors and Ford. Honda and Toyota also intend to build their largest Asian factories, outside of Japan, in Thailand. But even Vietnam, with a much smaller market, is attracting new investment. In the mid-1990s, Ford began construction of an assembly plant south of Ho Chi Minh City that will produce cars and vans (Gibney, 1996).

The Southeast Asian auto industry was adversely affected by the 1997 financial crisis in the region. The downturn was especially severe in Thailand where sales fell as much as 73 percent versus 1996 levels. As a result, there were 29,000 lay-offs in the automobile and spare-parts industries in the first nine months of 1997. The devaluation of regional currencies forced car prices higher because most cars assembled in the region have a high foreign content and those parts cost more to import after the devaluation. With many banks in crisis, new car buyers in 1997 had to pay higher interest rates, dampening an already weak market. In the Philippines, banks required a 50-percent down payment for new car purchases. Malaysia's two national car companies, Proton and Perodua, weathered the financial crisis with fewer ill effects, in part because they were not as adversely affected by the plunge in regional currencies. In fact, the weaker ringgit was expected to bolster Proton's export sales.

Foreign Direct Investment (FDI) Source Countries. FDI in the manufacturing sector of Southeast Asia has four main sources: United States, the European community (EC), Japan, and The Four Tigers (Hong Kong, Taiwan, South Korea, and Singapore). Each of these source areas has different investment characteristics, and their relative importance has changed over time. In particular, intra-Asian investment has become increasingly important as

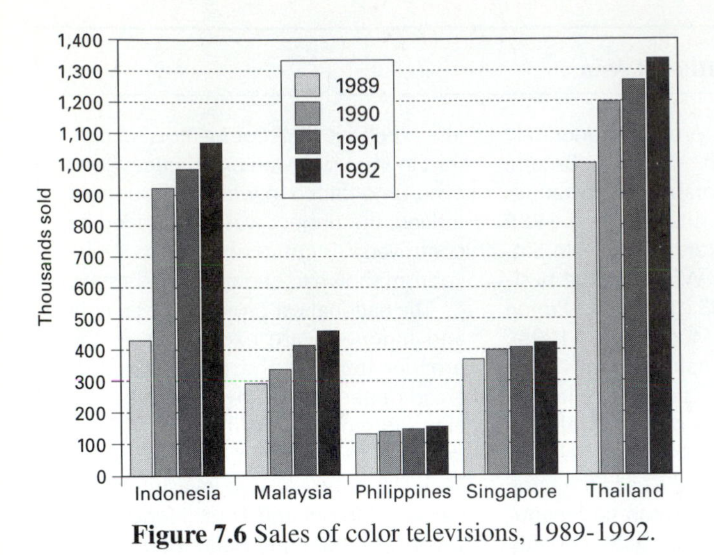

Figure 7.6 Sales of color televisions, 1989-1992.

Japan and the Four Tigers have looked to investment in Southeast Asia as a means of reducing costs and gaining access to rapidly growing markets in the region's emerging economies.

Figure 7.7 compares the importance of these four main sources of FDI for several host countries in Southeast Asia. Investment by the United States is most important in the Philippines, which has traditionally had strong economic ties to the United States, and in Singapore, a major export platform for U.S. electronics MNCs. Conversely, U.S. investment was negligible in Vietnam in the late 1980s as diplomatic links had not yet been normalized, but by the mid-1990s this had begun to change. Meanwhile, Japanese firms were major investors in every country in the region. Their share of total FDI was highest in Thailand and Singapore. In Thailand, Japanese firms invested heavily to tap the country's abundant natural resources, especially timber. Conversely, in Singapore the electronics industry was the most important arena for Japanese investment. Finally, investment by the Four Tigers was most significant in Malaysia, where neighboring Singapore was the single largest source of FDI. In contrast, Singapore itself attracted little investment from the other Tigers due the lack of complementarity in their economies.

In terms of total FDI worldwide by U.S. MNCs, Southeast Asia has not been a major region for investment. Far more money has been invested in nearby Canada and Europe. Even within Southeast Asia, manufacturing has not been the main activity for U.S. MNCs; rather, they have given greater emphasis to the oil industry, especially in Indonesia. Nevertheless, U.S. firms have been critical to the industrialization process in Southeast Asia because of their dominant position at the technological frontier of the electronics industry. U.S. firms have helped to raise the technological sophistication of the economies in the region, particularly Singapore, Malaysia, and the Philippines; for example, the largest manufacturer in Singapore is Seagate, a U.S. manufacturer of disk drives. In addition to the plant in Singapore another is due to open in Cebu, Philippines. Apart from electronics, the other main areas for manufacturing investments by U.S. MNCs

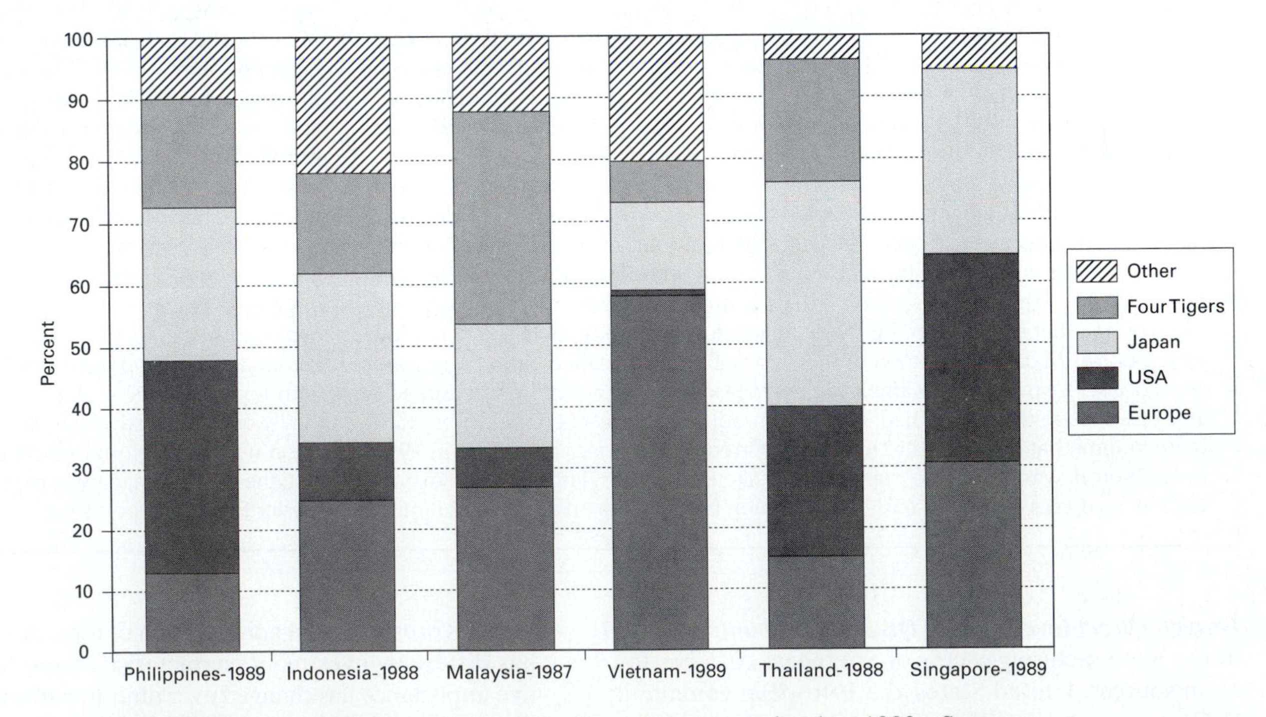

Figure 7.7 Sources of FDI in selected Southeast Asian countries, late 1980s. *Source: United Nations, 1992.*

have been textiles and leather products, such as shoes (Alburo, Bautista, and Gochoco, 1992). Here, too, U.S. firms have not been the largest investors, but their presence is important because their global marketing strength (e.g., Nike, Reebok) gives locally manufactured goods access to worldwide markets.

The main manufacturing sectors for European MNCs include electronics, chemicals, and food production. With respect to the latter for instance, the Swiss giant Nestle has used global recognition of its brand names to gain a significant market share across the region and has five plants in Malaysia alone. Nestle is representative of the increasing number of MNCs that are drawn to the region by its consumers. As wage levels in the region's economies continue to rise, some MNCs that currently use Southeast Asia as a low-cost export platform will reinvest elsewhere, but other MNCs will be drawn by the region's growing spending power.

Electronics firms from Europe with substantial export-oriented manufacturing facilities in Southeast Asia include conglomerates such Philips and Siemens, both of which produce semiconductors in the region. Finally, chemicals and oil refining comprise a third major area of activity for European MNCs. The world's major oil companies are drawn to Southeast Asia to be close to the rich fields of Indonesia and Brunei. For chemical firms such as the British giant Imperial Chemical Industries, the motivation for internationalization is different. Many of their products (e.g., explosives, paints, fertilizers) have high transport costs. Locating in Southeast Asia brings them closer to their market.

Japanese firms were among the first to set up manufacturing subsidiaries in Southeast Asia, beginning as early as the 1950s, though the number was small (Blomqvist, 1995). By the 1970s, the flow of investments from Japan had accelerated as the rising yen made it more expensive to produce at home. Most of these early investments were in light industry such as textiles. As the yen continued to appreciate, especially after 1985, more industries have been compelled to expand overseas to reduce costs. Consequently, recent Japanese investment in Southeast Asia has tended toward more technologically sophisticated and capital-intensive industries.

Japanese firms have displayed a strong Asian bias in their overseas investments and are among the largest investors in Southeast Asia. In Thailand, for instance, Japanese firms accounted for more than one-third of all FDI in 1988 (Blomqvist, 1995).

Although Japanese economic activity and investment influence were of paramount importance in Southeast Asia from the 1970s onward, this picture has now changed. As a result of a severe recession and bank failures in Japan, Japanese economic activity in the region has declined dramatically even as post-crisis growth in Southeast Asia resumes. Japanese foreign direct investment in Southeast Asia fell by 71 percent in 1998. Trade figures also reflect Japan's declining importance. While Japan accounted for 25 percent of ASEAN countries' imports and exports in 1989; in 1997 it was 16 percent. The exodus of Japanese capital is apparent in bank lending, investment and trade statistics. In place of new investment the Japanese government, largely through the Miyazawa Plan, contributed $48 billion to both Asian governments and Japanese firms with operations in crisis countries. The Export-Import Bank of Japan (Jexim) alone contributed $8.2 billion in 1998 to Japanese companies and their affiliates in Asia. Roughly 60 percent of this amount has gone to operations in Thailand. Indonesia accounts for about 24 percent of Jexim's spending in the region. The intent is to rescue hard hit Japanese firms by allowing them to keep their overseas operations running and even buying out the weak local partners in order to maintain a presence in the region. (Sender, 1999)

The Four Tigers comprise one of the fastest-growing sources of FDI in the region (Blomqvist, 1995). Before the 1980s, this source was insignificant. But as in Japan, rapid economic growth has led to conditions that make foreign investment imperative for many firms. These conditions included labor shortages, higher labor costs, and appreciation of the home currency. As a result, the Four Tigers were among the top investors in the emerging economies of Southeast Asia. In Thailand, for instance, the rank of investor nations in 1988 was (1) Japan, (2) the U.S., (3) Singapore, and (4) Taiwan. In the late 1980s, Singapore was the only Southeast Asian country where investment by MNCs from the Tigers was unimportant. As noted above, the similarity of Singapore's economic structure and the fact that it too is a high-cost production center have limited its attraction to firms from the other Tigers.

In the rest of the region there are patterns distinguishing investment by MNCs from Korea and Taiwan, on the one hand, and Singapore and Hong Kong, on the other. Manufacturing MNCs from Korea and Taiwan tend to use Southeast Asia as a low-cost export platform to increase the competitiveness of their products (textiles, shoes, electronics) on the global marketplace (Wie, 1991). Investments from Singapore and Hong Kong, conversely, are less export oriented; rather, firms from these countries target the host countries' domestic markets and specialize in the production of consumer items such as food and beverages. For example, Singapore has been a leading foreign investor in the Malaysia food industry (Ali and Wong, 1993).

It is likely that investment flows *within* Southeast Asia will increase. Already Singapore is a major investor throughout the region, including the emerging economies of Burma, Laos, Cambodia, and Vietnam. Malaysian firms are also increasingly active in these frontier economies; for instance, the Malaysian auto firm Proton began to assemble cars in Vietnam in the mid-1990s ("Dr M Launches Sale of Proton Wira in Vietnam," 1996). Thailand, too, is drawing on its historical ties and proximity to these countries to promote economic integration through FDI by Thai firms.

Evaluating the Role of MNCs in Southeast Asia

Governments across Southeast Asia continue to step up their efforts to attract investments by foreign MNCs. Clearly, MNCs have a critical role to play in these countries' development strategies. In particular, the region's developing economies hope to duplicate the success of Singapore in harnessing MNCs to achieve rapid, sustained economic development. Twenty years ago, MNCs in the city-state were mainly low-wage, labor-intensive employers. Since then, they have expanded the size, scope, and sophistication of their operations in Singapore. Today, the jobs they create in Singapore include a wide-range of skill-intensive and knowledge-intensive positions, including highly trained production operators in computer disk drive plants, software engineers, and regional marketing managers.

Through careful and effective government management, Singapore, and more recently Malaysia, clearly have moved up in the NIDL. Elsewhere in the region, however, a job with an MNC still usually means working in an unskilled position, under difficult conditions, for wages that are extremely low by international standards. For example, in the early 1990s, workers at the Nike factory in Tangerang, Indonesia, earned 30 times less than athletic shoe factory workers in North Carolina (Barff and Austen, 1993). The low wages are, in fact, the chief inducement for Nike and other MNCs to shift shoe production away from the United States (their main market) to Southeast Asia.

For most workers in MNC manufacturing plants, conditions are difficult. In a typical plant, the work is fast paced, repetitive, and closely supervised. The standard workweek is long and overtime is often mandatory. Quotas and competitions are used to pressure workers to raise productivity. Eyesight deterioration is common in the many electronics factories where workers spend their days looking though microscopes at conveyor belts laden with delicate electronic components. Rotating shiftwork and exposure to workplace chemicals also take a toll on workers' health (Lin, 1987).

Nike again came under fire in 1997 when an audit of a factory near Ho Chi Minh City found that workers were exposed to 177 times the legal limit of a carcinogen that harms the liver, kidneys, and nervous system. The factory, which opened in 1995, is operated by a subcontractor and is one of the largest in the region, producing 400,000 pairs of athletic shoes per month. Nearly 10,000 workers labor 65 hours a week in hot, noisy conditions for weekly wages of $10 (Greenhouse, 1997).

The difficult conditions have led to high turnover in many MNC plants. In all countries of the region, MNC factories often recruit young, unmarried women (who will work for less because they are not the chief income source for their families). Young women migrate from rural areas to take up such jobs, work for several years before they become "burnt-out" and quit to return to their rural homes (Lin, 1987). This sort of pattern detracts from the positive employment creation impact of the MNCs because relatively few long-term positions are created for male heads of households. Chapter 11, Gender and Development, contains other case studies and discusses more deeply the role of women in development.

In broad terms, it is fair to say that MNCs have played a commendable role in the development of Singapore and Malaysia, while their role in the region's other economies is more dubious. However, it should be stressed that the most important feature of the MNCs' role in Southeast Asia is its dynamism. The current emphasis of MNC operations in, say, Indonesia, on low-wage, labor-intensive operations is not permanent. By creating favorable conditions (including good transportation and communication infrastructure, a well-trained workforce, transparent regulations that do not encumber business transactions), Indonesia can attract the kind of increasingly knowledge-intensive MNC operations that now typify Singapore and Malaysia.

It is the dynamism of MNC investments and operations in the region that create both opportunities and threats for Southeast Asian countries. A country with minimal export-oriented manufacturing can industrialize rapidly by creating the right conditions for MNCs. Conversely, Singapore, which now seems far ahead of its neighbors in terms of industrial development, can lose its edge if the MNCs that now dominate the city-state's economy find that other, less-expensive countries offer a competitive business climate.

State-Owned Enterprises

Simultaneous with efforts to attract MNCs, governments in Southeast Asia have sought to cultivate specific industries through state-owned enterprises (SOEs). SOEs have been set up in technologically demanding industries (e.g., the aerospace industry in Singapore and Indonesia) and in industries with huge capital requirements. In these industries, the state has taken the lead because the private sector was deemed either unwilling or incapable of making massive investments that would show only very slow returns.

State-owned enterprises have been an important element of industrial policy, particularly in Singapore, Indonesia, and Malaysia. They have been relatively less prominent in Thailand and the Philippines. In Burma and the countries of Indochina, industrialization has just begun and industrial policy is in its infancy, but it is likely that state enterprises will be significant in the development of manufacturing in these countries too, given their weak domestic private sectors and traditions of strong state involvement in the economy.

Indonesia

In Indonesia, state-owned manufacturing enterprises had their genesis in the oil boom of the 1970s when the windfall profits from oil sales were poured into glamour projects in high profile, heavy industries including steel, weapons, telecommunications, and shipbuilding (Schwarz, 1995a). The most ambitious project of all was IPTN (National Aircraft Industry), Indonesia's aircraft manufacturer. IPTN illustrates the pivotal role of these enterprises in Indonesia's industrialization effort. They have been set up not just for the sake of import substitution; rather, these enterprises have been the key element in a strategy to leapfrog Indonesia to the forefront of industrial development. The future of both the N-250 and the N-2130 has been jeopardized by Indonesia's financial crisis.

IPTN merits close consideration for it exemplifies both the stunning achievements of Indonesian industrialization and the disappointing failures of SOEs. IPTN was established in 1976 and must be regarded as an extraordinarily audacious move by what was then one of the world's poorest countries. Part of the reason for Indonesia's early entry into the aircraft industry lies in the close relationship between President Suharto and Dr. B.J. Habibie, IPTNs founder (Hill and Pang, 1988). Habibie had returned to Indonesia in 1974 from a very senior position in the West German aircraft industry and was reportedly invited personally by Suharto to lead Indonesia's drive to develop and use modern technology.

IPTN differs significantly from the aerospace industry in other newly industrializing countries (e.g., Singapore, South Korea, Brazil). In these other countries, the industry developed in phases such as maintenance and repair followed by component manufacturing followed by aircraft assembly. In contrast, IPTN chose to go directly to assembling complete aircraft in Indonesia (Hill and Pang, 1988). Initially, IPTN assembled aircraft components designed and manufactured by foreign producers. Later, in the 1980s, IPTN entered into an arrangement with the Spanish firm CASA to jointly produce an aircraft called the CN-235. ("IPTN Targets Design Recognition in Second Decade of Operations," 1989). Finally, in the early 1990s, IPTN took the next step by completely designing its own aircraft, a 70-seat turboprop passenger plane called the N-250, which took its maiden flight on August 10, 1995—the fiftieth anniversary of Indonesian independence.

The N-250 is a remarkable achievement, but many have questioned the wisdom of Indonesia's massive investment in IPTN and other large state enterprises. It has been estimated that at least $1.6 billion has been invested in IPTN since 1976, including $400 million for the N-250 program alone. The commercial success of IPTN has been very limited: Only 260 aircraft were assembled and produced between 1976 and 1990, virtually all of them sold in Indonesia. Indonesian airlines have been under heavy pressure to purchase IPTN products. The head of one state-owned domestic airline, Merpati, was dismissed when he refused to buy 16 CN-235s.

Moreover, IPTN's development impact has been quite small. IPTN has indeed gained significant expertise in high tech areas such as advanced metallurgy, but so far there has been relatively few spillover effects on the rest of the economy. IPTN is still heavily dependent on imported components and employed as many as 350 foreign engineers at the peak of the N-250 program. Geographically, IPTNs impact is limited to Bandung, the technological hub southeast of Jakarta in West Java province.

Yet IPTN's defenders point out that the European consortium, Airbus Industrie, became the world's second-most-important commercial aircraft manufacturer only after decades of massive, state-funded losses. IPTN, they argue, will have a significant advantage in the industry because its costs (especially labor costs) are much lower. From this perspective, IPTN will only be a viable player in the industry when it gains more experience and develops a wider product line. In 1995, shortly after the maiden flight of the N-250, IPTN announced that it would develop a medium range passenger jet called the N-2130.

Malaysia. State-owned manufacturing enterprises have also had a mixed record in Malaysia. Whereas Indonesia's state enterprises reflected a nationalist ideology that had its roots in the early post-World War II period, in Malaysia they were the product of more immediate circumstances in the 1970s. Having achieved considerable success though export-oriented growth, Malaysia's government set a new course for industrialization based on powerful new SOEs. Instead of relying mainly on MNCs to create jobs in light, export-oriented industries, the state made heavy industrial enterprises the driving force in industrialization (Jomo and Edwards, 1993).

The change in Malaysia's industrialization strategy reflected several goals. First, the establishment of heavy industries (e.g., cement, steel, etc.) was intended to lead to a more integrated economy. MNC investment had led to *enclave industrialization* in which some areas of the country, especially the FTZs near Penang, were industrialized but were not well integrated into the rest of the Malaysian economy. Second, heavy industrialization was meant to help achieve the country's racial equality goals by creating more jobs for Malays. Finally, heavy industrialization was a key to Malaysia's vision of becoming a developed country. Malaysian Prime Minister Mahathir had encouraged Malays to look east toward the example set by

Photo 7.4 Proton Sega, Malaysia's national automobile. (Ulack)

Japan, South Korea, and Taiwan, all of which had achieved remarkable levels of economic development within a generation (Jomo and Edwards, 1993).

To replicate the success of more-developed Asian countries, the Malaysian government set up a number of new state-owned enterprises beginning in the late 1970s in industries such as steel, cement, petrochemicals, and shipbuilding (Jomo and Edwards, 1993). Some of these have been spectacular failures: Most notably, the national steel manufacturer Perwaja suffered billions of dollars in accumulated losses ("KL Confident in Turning Round Perwaja Steel," 1996). Set up in 1982 with Japanese technology from Nippon Steel, Perwaja was undermined by financial mismanagement and overcapacity in the global steel industry.

As in the case of Indonesia, however, Malaysia has one industrial state enterprise that has become the pride of the nation: the national car manufacturer, Proton (a Malay acronym for National Automobile Enterprise Company), has enjoyed considerable success since it was established in 1982 (Jayasankaran, 1993) (Photo 7.4). Yet, like IPTN, Proton has drawn the ire of many critics as well. They argue that the government's investment in Proton could have been more wisely spent to deliver a bigger development impact.

With cumulative sales of more than 500,000 cars by 1995, Proton has had a more impressive record of commercial success than IPTN. It is also fair to say that Proton has had a bigger development impact. In particular, Proton has spawned a network of more than 100 suppliers based in Malaysia. Initially, the *local content*, by value, of Proton cars was just 36 percent (Jayasankaran, 1993)

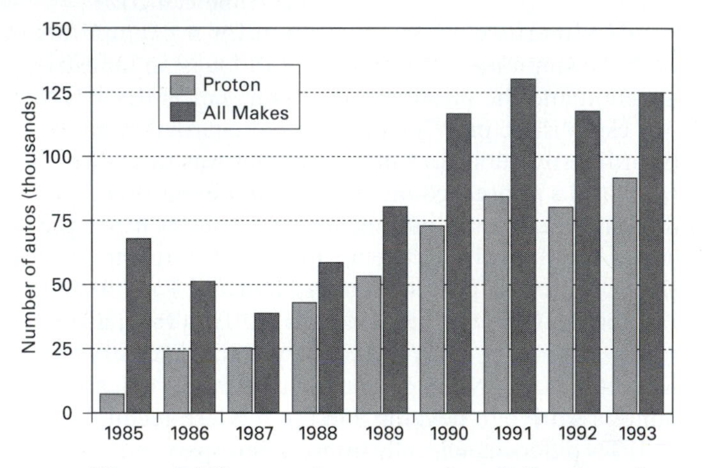

Figure 7.8 Proton sales versus sales of all makes of automobiles, Malaysia, 1985-1993. *Source: "The Dream for a Malaysian Car," published by Proton in English and Bahasa Malaya, 1994.*

and only 13 components (e.g., tires, batteries, seat belts, glass, etc.) were produced locally; almost everything else was imported from Japan. But by 1995, local content had reached 80 percent ("Car Industry Told to Cut Reliance on Japanese Parts," 1996). As the local content has risen, Malaysian firms have gained expertise in a wider range of industries related to the auto industry. One reason that the Proton project has had this bigger spillover effect on the rest of the economy, in contrast to IPTN, is because the technology in the auto industry is at a lower level and technical standards are less demanding. For the same reason, it has been easier for Proton to generate jobs for the target group, indigenous Malays (*bumiputras*), because much of the workforce requires only minimal entry skill levels. More than 90 percent of Proton's employees are Malays (Jayasankaran, 1993). Like IPTN, however, Proton has had a limited geographical employment impact. The car maker's factories are concentrated in the already industrialized Klang Valley west of Kuala Lumpur.

Singapore. Compared to its neighbors, Singapore has pursued a less flamboyant approach to use of state enterprises to promote industrial development. Rather, the city-state has used this type of firm to further its standing in key *industrial clusters*. While market forces determine within which industrial clusters Singapore has an advantage, the state takes the lead in exceptionally large projects that might be beyond the scope of the private sector ("Government Will Remain Involved in Fostering Economic Growth," 1996). One cluster where Singapore clearly has an advantage is sophisticated electronics. To promote the sector, Singapore's government formed a joint venture called Tech Semiconductor with three foreign MNCs: Texas Instruments, Hewlett-Packard, and Canon. Tech Semiconductor operates a $330 million wafer fabrication plant manufacturing state-of-the-art four-kilobyte memory chips.

Another industry that Singapore has tried to cultivate is aerospace manufacturing, and here too the state has taken the lead. In particular, Singapore Aerospace, a state-owned firm, has developed expertise in a range of activities extending from repair and refurbishment to components manufacture and assembly (Hill and Pang, 1988). It is important to note however that Tech Semiconductor and Singapore Aerospace, unlike Proton and IPTN, have not grown up behind walls of government protection. On the contrary, both compete directly against firms in the same industry in Singapore. In the semiconductor industry, for instance, there are already several wafer fabrication plants operating in the city-state and the government is keen to attract more. It is crucial, therefore, that the government not be perceived to be extending any special favors to Tech Semiconductor.

Private Manufacturing Enterprises

Although the technological advancement of Southeast Asia's industrial sector has been led mainly by state-owned enterprises and foreign MNCs, there are a large number of privately owned manufacturing firms throughout the region. In fact, these firms account for a very substantial portion of total manufacturing employment, especially in the less developed economies. In the Philippines, for instance, private firms with five or fewer employees account for 30 percent of manufacturing value-added but employ more than 60 percent of manufacturing workers (EIU, 1995b).

Throughout the region, from the most developed to the least developed economies, new emphasis is being placed on the development of private enterprises. In the newly emerging economies such as Vietnam, for instance, private enterprises are stepping into the void left by the closure of many inefficient state-owned enterprises. At the other end of the spectrum, in Singapore and Malaysia, small and medium-size enterprises are being promoted as a way of overcoming reliance on footloose international capital.

The scope of private industry in Southeast Asia is truly vast, ranging from tiny backyard brickmakers to giant agro-industrial conglomerates. Small, privately owned firms often manufacture various types of handicrafts, ranging from decorative batik cloth in Indonesia (Photo 7.5) to shellcrafts in the Philippines. Handicrafts can be a significant contributor to local economies. Cebu, for instance, is the world's leading rattan-furniture manufacturing capital, and all of the firms in this thriving local industry are Philippine owned.

Conversely, many of the largest firms are in the food, tobacco, and beverage industries. In Indonesia, for instance, several of the most powerful manufacturing conglomerates are centered on production of clove cigarettes for the domestic market (Schwarz, 1995). In Thailand, huge firms have emerged out of the rice-milling, chicken-processing, and feed-mill industries (Phongpaichit and Baker, 1995). For example, in the 1960s, the Thai company Charoen Pokphands (CP) main business was in feed mills and integrated production of poultry products. By the 1990s, CP had become a massive conglomerate that was active in industries ranging from oil refining, chemicals, and beer brewing to transportation and finance, and it had invested heavily overseas with operations in China, Indonesia, Taiwan, Singapore, and Malaysia (Phongpaichit and Baker, 1995).

CP was established in the 1920s by a Chinese immigrant to Thailand, and its spectacular success since then is just one example of the critical role played by the overseas Chinese in Southeast Asian economies. The Chinese comprise a majority only in Singapore, but they dominate the private sector of almost every country in the region. In the Philippines, with a population that is just 1 percent Chinese, two-thirds of the sales of the country's top 67 companies were accounted for by Chinese-owned firms in the late 1980s. In Thailand, at least 90 of the 100 wealthiest business families are Chinese despite the fact that only 10 percent of the country's population is Chinese (Rohwer, 1996).

Across Southeast Asia, Chinese businesspeople carry on a tradition of family-centered entrepreneurship. The overseas Chinese typically started out as traders and then spread to other major sectors such as property, commodities, and shipping. Most Chinese firms continue to be small enterprises in which the family patriarch marshals the labor resources of the entire family in what are often very labor-intensive operations. These businesses are

Photo 7.5 Commercial batik production, Jogjakarta, Java. (Leinbach)

linked in informal networks defined by kinship, a common dialect, or a shared origin in a Chinese village, and such resources as capital for new expansion are exchanged through these networks.

The prosperity and cultural distinctiveness of the Chinese have engendered a variety of state policies to limit their economic influence (Rigg, 1997). Malaysia's NEP was a prominent example. A more extreme case was the persecution of Chinese in Vietnam during the period leading up to and following China's incursion into northern Vietnam in 1979. Hundreds of thousands were forced to move to new economic zones along the country's frontier between 1975 and 1978. By 1979, a massive wave of Chinese boat people had begun to flee the country. Elsewhere in the region, Chinese businesspeople have sought to escape the brunt of discriminatory policies by forming mutually beneficial relationships with key government leaders. For example, Liem Sioe Leong, one of the richest men in Southeast Asia, has been described as closely allied with the Suharto family.

As the region's governments give greater emphasis to private-sector initiatives and small and medium-size firms, the scope and degree of discrimination against the Chinese will ease. In Vietnam, the Chinese minority has enjoyed greater economic freedom since the beginning of reforms there in the 1980s. In Malaysia, the NEP has been replaced by a somewhat more moderate program since 1991.

Southeast Asian governments, especially in the more advanced economies, would like to encourage more privately owned enterprises to expand to the rest of the region. In Singapore and Malaysia, high labor costs and comparatively small local markets limit the growth opportunities for small and medium-size enterprises. Far greater opportunities are offered by the large, rapidly growing markets in nearby countries. So governments have sought to help firms to go regional with their expertise in, say, food processing.

Ironically, the countries in the region to which Singaporean, Malaysian, and Thai firms are now expanding have themselves sought to stimulate the growth of locally owned private manufacturing enterprises. In Vietnam, for instance, government attitudes toward the private sector have undergone a transformation in the last decade. The government officially abolished all private enterprises in the country by 1979, although many continued to function informally. By 1988, however, the state's position had changed to one of creating favorable conditions for the private sector (Truong and Gates, 1992). To achieve these goals, reforms were introduced to secure the rights of private property owners. Nevertheless, many obstacles remain: limited numbers of skilled workers, capital shortages, antiquated technology, and the lingering advantages that state-owned enterprises still enjoy. As a result, private enterprises account for only a tiny, albeit growing, share of production.

Throughout the region, development of private manufacturing enterprises will continue to receive high priority in state industrialization strategies. On the one hand, governments want to limit their economic dependence on foreign MNCs that often have minimal commitment to their host countries. On the other hand, the role of state-owned enterprises is necessarily limited by budget constraints. Moreover, as the economies of the region develop, there is a larger pool of talent, expertise, and entrepreneurship that can be harnessed by local firms.

THE SERVICE SECTOR

The rapid growth of service industries within the global economy is one of the most striking developments of the past few decades. Although emphasis in economic development is often placed primarily upon manufacturing, it is clear that the services sector also plays a central role. Services account for the largest share of gross domestic product in many countries, and the services sector is increasingly the major source of employment in many developing countries. A major theme in the global economy is the increased internationalization of services for there is a stronger emphasis on both trade in services and also foreign direct investment in services. A major debate continues over the matter of liberalization of trade in services, which is more difficult to reform than trade in goods because of the special features of services.

Services, in contrast to manufactured goods, are intangible, perishable, and must be consumed at the same time and place as their production. The services sector is very diverse and ranges from the provision of sophisticated financial services and medical treatments to more-common courier service, waste disposal, and food and beverage outlets, as well as a variety of personal services. Although we have noted that services are quite distinct from manufactured goods, the distinction is not really sharp. This is so because most goods that are purchased do provide a service and most services require products in their delivery process. Services can be categorized into consumer or producer (business) services. Such a distinction is based on the final use of a particular service. Travel agents are an example of the former, while management consulting and air cargo are examples of the latter. Yet in the last analysis, while some services may be firmly distinguished by their final-use allocation, still other services may be used by both individual consumers and firms. Thus most services may be viewed as mixed.

It is important, in light of the above, to view services within a production chain, that is, a chain of linked functions. Transactions occur as a commodity moves from an initial stage to the final disposition. For example, we may view materials being procured, then transformed, sold, distributed, and finally enhanced through services. The key here is that services have become a major source of value added and a critical ingredient in competitive strength. The production of personal computers (PCs) and the marketing of software and maintenance contracts is an example of this production chain concept.

As noted above, the internationalization of services has been conspicuous in the global economy, and in Southeast Asia more particularly. The internationalization of services takes place in two ways; first, a services firm establishes an outlet in a foreign country, or second, a service is actually traded across international boundaries. Examples of the former include the Hard Rock Cafes that have sprung up in the region's major cities (Photo 7.6). Within the traded services category we can draw a distinction between transportation and such private services as management, finance, advertising, and professional/technical services. These have enjoyed especially strong growth recently.

Trade statistics do not accurately reflect the extent to which services have become internationalized because services are bound into the products that are traded, and indeed many services are internalized within the operations

Photo 7.6 Hard Rock Cafe, Kuala Lumpur. (Ulack)

of multinational corporations. Examples of such hidden services include the shipment by air of clothing from factories in Indonesia to shopping malls in the United States and the legal advice given by an MNCs headquarters in London to its subsidiary in Singapore. Information technology has allowed, in some cases at least, for trade in services to occur where indeed the consumer and producer are separated by space (Daniels, 1995). The debate on the liberalization of trade in services is focused not on the matter of trade but rather the conditions under which firms and individuals will be permitted a presence in a particular market. Thus, services become closely linked to FDI. Two basic issues involved in this are information technology and government regulation.

Although agriculture and manufacturing are more frequently discussed as the driving forces in development, services are critically important for several reasons. First, services are required as new products are developed. Second, service activities themselves stimulate the growth of additional services. Finally, services provide, in part, a solution to many governments' continuing need to create employment. As Southeast Asian economies have begun to expand and mature, services' output has grown more rapidly than agriculture. Certainly this is true over the region for the period 1970–1993. This is normal because the share of services generally rises in line with the growth of GDP. During the period from 1970 to 1980 in the ASEAN countries, services' growth was strong but did not exceed industry or manufacturing growth in any of the nations. In the more recent period from 1980 to 1993, the growth in services exceeded that of industry and manufacturing only in the Philippines and Singapore. In Indonesia for the same period, services' growth exceeded industry but not manufacturing growth (Table 7.6). This fact reflects the strong push to develop an industrial base in the economies of Thailand, Malaysia, and Indonesia.

Services contribute significantly to total employment throughout the region. In 1993, the sector's share ranged from a low figure of 17 percent in the struggling economy of Laos to 60 percent in Singapore where the development strategy has specifically emphasized services' exports. On average, services account for 31 percent of employment in Southeast Asia.

There are several interesting features of the service sector's development in these countries. Most striking is the dramatic transformation that has occurred since the 1970s in technology. This technological advance has had significant impact upon transport, financial services, telecommunications, and a wide range of personal services. Examples include the diffusion of cellular phones, personal pagers, automated teller machines (ATMs), and information technology in the logistics revolution (through which firms precisely control the movements of raw materials and finished goods in their production chains). The resultant change, measured by productivity increases and new products and processes, has been just as important as the developments of the Green Revolution in agriculture and the shift to higher value-added goods in manufacturing. The transformation has been an essential prerequisite to the region's rapid growth (Hill, 1993).

There is tremendous diversity in the service sectors of the region's countries, reflecting their different stages of economic development, resource endowments and utilization, and economic orientation. As noted above, Singapore is most conspicuous because services are dominant and provide most of the country's output and employment. After a recession in the mid-1980s highlighted the dangers of relying solely on manufactured exports, Singapore's government has sought to strengthen the service sector's contribution to economic growth. The city-state has become a center for transportation and financial services in the region and indeed the world. In countries such as Indonesia, Malaysia, the Philippines, and especially Thailand (where employment in services contributes only 19 percent to the total), there remains a shift of resources

Country	Growth of GDP: 1980–93				Employment Percent: 1993		
	Agriculture	Industry	Manufacturing	Services	Agriculture	Industry	Services
Burma	0.6	1.4	0.1	0.9	53	19	28
Cambodia	--	--	--	--	--	--	--
Indonesia	3.2	6.3	11.8	6.9	57	13	30
Laos	--	--	--	--	76	7	17
Malaysia	3.5	8.2	10.3	5.5	42	19	39
Philippines	1.2	-0.1	0.8	2.9	52	16	33
Singapore	-6.4	6.2	7.2	7.4	2	38	60
Thailand	3.8	11	10.8	7.7	71	10	19
Vietnam	--	--	--	--	68	12	20

Note: Industry includes mining, construction, electricity, and gas.

Table 7.6 Growth of GDP by sector and GDP's sectoral employment share, 1993. *Source: World Bank, 1995.*

from agriculture into industry, and the services sector is now begining to develop more fully.

A significant feature of the region's service activities is the considerable variation within sectors. In each country there is a fascinating range of activities from the very sophisticated lending and funds management of large banks to small self-employed vendors selling a variety of goods and services. Within the transport sector alone services range from those provided by the national airlines to very local, often irregularly scheduled minivan operations that provide access for local people between villages and service centers. The urban informal sector illustrates the huge diversity of simple service occupations carried out by individuals. Despite highly unstable earnings, the absence of employment security, and harassment by municipal governments, such employment provides vital interim or even long-term employment as rural migrants settle in the city. Further discussion of the urban informal sector is contained in Chapter 5, Urbanization.

Services are strongly affected by government operations and investment behavior. The retention of a large civil service, such as in Indonesia, and especially the investment in physical and social infrastructure are major sources of employment. The expansion of public sector activities is, of course, most conspicuous in periods of strong revenue growth such as the oil boom of the 1970s. In Indonesia employment opportunities resulted from the need to refurbish and create new infrastructure, which had deteriorated. A massive highway and rural-roads program, including "self-help and cash labor" programs to create simple roads and irrigation works, especially were quite important. The strong thrust in resettlement efforts in both Indonesia and Malaysia where land in the form of small farms for poor Javanese and *bumiputras* was also significant in this respect.

As a prime example of the growth of services in the region, perhaps no example is more instructive than financial operations. As economies have grown and employment has expanded, the increased monetization and circulation of capital has generated a response from the financial services industry. At a very basic level the emergence of international banking firms on top of the layer of publicly and privately owned local banks as well as the expansion of branches into rural areas has continued to diffuse over time. But more important has been the evolution of a much more sophisticated financial system that is linked to international markets. In this development the governments became involved through the establishment of the macroeconomic and regulatory environment, especially through the setting of interest rates and controls on loans and investment. As liberalization has gained favor these regulatory strictures have been relaxed to stimulate financial development. A sign of this relaxation is also evident in the development of nonbanking services such as insurance, the stock market, and pension funds.

Although the liberalization of the financial services sector has enabled it to play a more important role in the region's economic development, many observers blamed poorly supervised banks for causing the 1997 financial crisis in the region [see Box 7.3: The Financial Crisis: Causes and Consequences]. After deregulation in 1988, Indonesia, for example, had one of the most open banking systems in the world, and more than 200 new banks were established (Mydans, 1997b). Many of these banks borrowed money from overseas (in U.S. dollars for instance) and then aggressively loaned the money in Indonesia. By 1996, more than 25 percent of loans in Indonesia had gone bad and, when the rupiah fell in 1997, dozens of banks and the businesses that depended on them were pushed into financial distress. Similar problems were experienced in Thailand and, to a lesser extent, in Malaysia and the Philippines.

TRADE: ENGINE FOR GROWTH?

It is clear that the Southeast Asian region is trade oriented. The majority of the countries have a large external sector and a generally high trade-output ratio (Table 7.7). The reasons for this are numerous. First, the important institutional preconditions are present. Among the most important are political stability and a well-developed financial sector. But in addition, a high rate of capital formation must be recognized. In all of these countries whenever investment takes place, much of the machinery and other required materials are imported. Thus rising investment and increasing imports are interrelated. Because of this the real mechanism for growth in the Southeast Asian economies must be export expansion, which simultaneously increases their ability to import critical requirements. The ratio of exports to total production for the majority of the Southeast Asian countries is higher than the average ratio for all developing countries. From this it is important to recognize that the outputs are therefore more sensitive to the impact of external changes than in other countries. Because the export-generated growth ties the region to the international system of dependency dominated by the United States, Europe, and Japan, it is true that trade is both an engine for growth and a mechanism for dependency (Wong, 1980). But, in general, the growth benefits of trade exceed the backwash effects of trade dependency.

Patterns of Trade

To say that the trade of Southeast Asia is considerable is perhaps an understatement. One measure of this is that the region's trade is roughly one-third that of the United States. The importance of trade for the various countries of the region is depicted in Table 7.7. As we would expect, the five original ASEAN states dominate the trade pattern. Singapore, reflecting its traditional status as entrepôt

BOX 7.3 The 1997 Financial Crisis: Causes and Consequences

By the mid-1990s, Southeast Asia's largest economies had been expanding almost without interruption for more than a generation, and the consensus in the region was that this prosperity would continue. Buoyed by confidence in their future, millions of Southeast Asians adopted more-comfortable middle-class lifestyles and imports surged.

Then in 1996, signs of weakness began to appear on the horizon. Exports, the engine of the region's economic growth, grew at a much slower pace. The slowdown had several causes. Rapid wage growth in such labor-tight manufacturing centers as Penang and Singapore made their output less competitive on the world market. Meanwhile, business was lost to other thriving export economies, especially China. Finally, 1996 was a poor year for the region's principal export—electronics.

As imports boomed and exports sagged, banks in troubled economies such as Thailand and Indonesia continued to borrow heavily from overseas. Much of this money was invested in speculative property deals and other nonproductive assets. Interest rates in the United States and Japan were lower than in Southeast Asia, so borrowing overseas could yield handsome profits for banks in the region; the stable exchange rates between the Southeast Asian currencies and the yen and the dollar made this a seemingly safe practice.

But by mid-1997, currency traders around the world were beginning to speculate that Southeast Asian governments would be forced to devalue their currencies to correct the adverse trends in exports and imports. As the traders forced the currencies down, investors across the region began to sell their baht, ringgit, pesos, and rupiah to buy U.S. dollars and other safe haven currencies. That exodus of funds accelerated the drop in the currencies' value.

The wave of currency devaluations left many banks in the region unable to repay their foreign debts. Some banks were forced to close, with devastating consequences for the businesses that depended on them. Consumers and firms suddenly found it much more difficult to borrow money and economic growth slowed.

The financial crisis had stark consequences, particularly in Thailand where it was expected that as many as 2 million Thais could lose their jobs by the end of 1998. An early casualty was the government of Thai Prime Minister Chavalit Yongchaiyudh. More generally, the quick reversal of the region's fortunes presented a political challenge across Southeast Asia for governments that had staked their legitimacy on achieving steady economic growth.

To end the crisis, Thailand, Indonesia, and the Philippines each turned to the International Monetary Fund (IMF) for billions of dollars worth of aid. The package for Indonesia was initially worth $17.2 billion, the third-largest bailout in the IMF's history. The IMF assistance would be used to restore the strength of each country's financial sector. The IMF's managing director accurately described the crisis and the subsequent recovery process as a culture shock of the greatest violence for nations that had grown used to robust, steady growth.

Country	Exports (million $)	Imports (million $)	Exports as % GNP		Export Growth 1980-1993 (percent)
			1986	1993	
Brunei	2,370	1,176	75.6	64	--
Burma	583	814	3.6	2	-2.5
Cambodia	--	--	--	--	--
Indonesia	33,612	28,086	18.0	23	6.7
Laos	80	353	2.1	5	--
Malaysia	47,122	45,657	47.0	73	12.6
Philippines	11,089	18,757	15.2	21	3.4
Singapore	74,012	85,234	117.4	134	12.7
Thailand	36,800	46,058	20.6	30	15.5
Vietnam	--	--	1.7	--	--

Table 7.7 Imports and exports by value. *Source: World Bank, 1995.*

and a country in which trade is essential to growth, is dominant. Given this entrepôt status, it is also worth noting that re-exports accounted for as much as 37 percent of total exports in 1988. Clearly the dynamic economies of Thailand, Malaysia, and Indonesia stand out in these figures. The latter country's position in these data is striking given the fact that industrialization in Indonesia was retarded by nearly three decades of political turmoil. The Philippines, in contrast, exhibits almost the opposite pattern. Starting early and then hitting the limits of import substitution, the government made the shift to an export orientation. However, as noted above, for a variety of reasons the manufacturing growth collapsed. Malaysia and especially Thailand have experienced no swings in the pol-

Country	Fuels and Minerals		Other Primary Commodities		Machinery and Transport Equipment		Other Manufacturers		Textiles and Clothing	
	1970	1993	1970	1993	1970	1993	1970	1993	1970	1993
Brunei	--	--	--	--	--	--	--	--	--	--
Burma	7	7	92	82	0	2	2	9	1	0
Cambodia	--	--	--	--	--	--	--	--	--	--
Indonesia	44	32	54	15	0	5	1	48	0	17
Laos	36	--	33	--	30	--	1	--	3	--
Malaysia	30	14	63	21	2	41	6	24	1	6
Philippines	23	7	70	17	0	19	8	58	2	9
Singapore	25	14	45	6	11	55	20	25	6	4
Thailand	15	2	77	26	0	28	8	45	8	15
Vietnam	--	--	--	--	--	--	--	--	--	--

Table 7.8 Structure of merchandise exports (percentage share). *Source: World Bank, 1995.*

icy environment, apart from Malaysia's state-led, heavy industrialization drive of the mid-1980s. Direct state involvement has been low and the trade regime has become gradually more open, leading to export-oriented growth. Trade policies have been one of the successes in the reform movement that has swept the region since the 1980s. Most critical protection regimes have been simplified and, especially for manufactures, reduced.

Singapore was clearly the leader in the pursuit of an export strategy followed by Thailand, Malaysia, the Philippines, and Indonesia a decade later. Export orientation has meant different things in the various economies. In Singapore it has meant literally free trade and in the relatively open economies of Malaysia and Thailand, trade has been encouraged by the establishment of special free zones to attract investors, as we have discussed above, as well as other incentive packages. In Indonesia and the Philippines different circumstances apply. In the latter country macroeconomic policies did not enhance competitiveness as cronyism, rent seeking, and political instability negated gains. In Indonesia, the oil boom rendered most other exports uncompetitive and meant that reform efforts were weakened. The deflated oil market of the 1980s was the real trigger for reform in Indonesia, and that country has now begun to reap some of the benefits of these changes (Leinbach, 1995).

Given the structural changes noted above in the various economies, it comes as no surprise that a dramatic transformation has occurred in the region's exports (Table 7.8). The dominance of rubber and tin in Malaysia, oil and rubber in Indonesia, and rice in Thailand have been altered dramatically. The column in Table 7.8 headed "Other Primary Commodities" reveals the striking turnaround in the structure of exports between 1970 and 1993, a relatively short period of time in terms of development. In keeping with the implications of the NIDL, manufactures are now the major export item in all countries of ASEAN (except Vietnam and Brunei) and account for more than 40 percent of total merchandise exports. Electronics, textiles, and garments, footwear, and furniture have replaced the agricultural exports of previous years. In Cambodia, Burma, and Laos, vital exports remain timber, rubber, and rice.

The widespread and steadily increasing export emphasis among countries in the region has, as we might suspect, significantly affected the region's trade patterns. The colonial era was dominated by trade flows that were oriented largely to the home markets of the colonizing nations. These ties are still observable but to varying degrees today. It is especially strong in the case of the Philippines with trade ties to the United States, somewhat less in the case of Malaysia to the United Kingdom, and dramatically less in the case of Indonesia to the Netherlands. Although the United States and Europe are important trading partners in the region, the most significant change that has come about is the growth of Japanese influence within Southeast Asia. Indeed, trade within Asia in general has deepened considerably as economic structures have changed and competition has become more intense. This trade has especially taken on significant proportions where complementarity, proximity, and a variety of initiatives have reduced barriers to commerce. Japan developed as a major economic partner for most of the region's countries in the 1970s and 1980s through aid, investments, and trade. Because of the admiration of Japan and its achievements by the region's leaders and the high degree of complementarity amongst the economies, it is still the dominant partner for many countries in the region. In the case of imports, Japan is the principal partner for all the ASEAN nations. On the other hand, relatively closed economies that have not adopted trade reforms have different orientations: For example, Burma and Laos maintain their strongest import links with China and Thailand, respectively; Vietnam's major trading partner is Singapore, which accounts for 27 percent of imports and 35 percent of exports; the United States, Europe, and Japan are the major destinations for exports from the NIEs although Malaysia and Singapore, as a result of their geographical and historical ties, maintain primary export relations with each other.

Although the major industrial nations are critical trading partners for the Southeast Asian countries, the dynamics of global economic change have led to new relationships. In particular, the most conspicuous is the emerging force of several Tigers (South Korea, Taiwan, and Hong Kong). Since the mid-1980s these nations have gradually developed a relationship with ASEAN not unlike that which occurred several decades earlier with Japan. Reflecting their own structural change, these three economies have begun to shed labor-intensive activities and as a consequence have developed stronger ties to the economies of the Philippines, Thailand, Malaysia, Indonesia, and even Singapore, with which they share a similar level of development. More recently, these nations have begun to look to investment and labor supplies in Vietnam as well.

Trade Policies. Trade policy is an important instrument by which the pace and pattern of industrial development can be guided, but the exact form and sequencing of a trade strategy to carry out such development remains in debate. Some countries seek to promote growth through domestic industrialization and import substitution, that is, where trade and industrial incentives are biased toward production for the domestic market. Such a strategy relies on discretionary interventions that usually are characterized by controls as well as high and variable tariff protection and restrictions. Other countries aim to promote economic development through the expansion of foreign trade and by maintaining an "open door" policy toward private foreign investment and foreign aid. This approach emphasizes linkages to the global economy through exports and enhanced import capacity. Further, it is critical that this strategy does not bias incentives toward the domestic market. Thus neutrality of incentives is desirable with respect to production for home and export markets. Yet this should not be construed to mean that government does not intervene. Indeed, many governments, especially among the East Asian NIEs and as we shall see the ASEAN group, rely heavily on government intervention to achieve market orientation.

The terms *inward-* and *outward-looking trade strategies* provide a convenient way to contrast these two different attitudes and policies (Myint, 1984; Bhattacharya and Linn, 1988). Thus pure outward-looking, in contrast to inward-looking, policies pursue free trade and export expansion goals instead of protectionist and import-substitution policies. Open type domestic and private foreign-investment policies characterize the former, while the latter's policies are more insulating and restrictive in this regard.

As suggested earlier, Singapore has been a free-trade-oriented nation, recognizing that that posture was essential to growth. Although the government has adopted a free-trade position, it has also intervened in maintaining a large state-enterprise sector and fiscal incentives. In Thailand, direct state involvement has traditionally been low, and the trade regime has gradually become more open. In Malaysia also, growth has been rapid and policy firm. A brief venture in state-dominated industrialization in the 1980s was quickly abandoned.

Trade policies, indeed, have been one of the most successful areas of reform in the region during the 1980s. In both Indonesia and the Philippines, inward-oriented regimes have been revamped. Protection regimes were simplified, and the rate of protection for manufactures has been reduced. Both countries recognized that protection was harmful to agricultural development and export promotion and counterproductive in regard to protecting their export interests in international trade negotiations. The outcome in Indonesia has been superior to that in the Philippines. This is so because in Indonesia the cohesive polity and high-level commitment were critical to producing and installing changes. In the Philippines, reform was hampered by political instability and fragmented governance, as well as infrastructural bottlenecks such as the delivery of electricity.

In Indonesia trade reform has attacked inefficiencies in the import environment. Especially critical has been the reduction of nontariff barriers (such as quotas, licensing, and local content requirements) and the overall reduction of tariffs. Hand in hand with these changes in specific trade measures has been a gradual liberalization of regulations governing foreign private capital. For example, foreign investors may now operate in more industries in joint ventures with domestic enterprises and equity limits have been eased. The Indonesian government intends to continue liberalizing with regards to the import environment and to harmonize her tariff structure with GATT (General Agreement on Tariffs and Trade) and AFTA (ASEAN Free-Trade Agreement) (Leinbach, 1995).

ASEAN Free-Trade Agreement (AFTA). As competition for export markets has increased, there has been rising interest among the ASEAN countries in creating stronger trade linkages within the region, but this has been difficult because of uneven tariff regulations. Consequently in 1993, AFTA was launched. The agreement aimed to create an integrated market of 430 million people with a combined GDP of more than $293 billion. The plan was to reduce tariffs (in some cases as high as 40 percent) to 5 percent by 2008. The gradual reduction of tariffs by all countries in the group is in the process of being implemented. Just as important, however, are the removal of nontariff barriers such as import quotas. A key goal is to have ASEAN move from a political group-

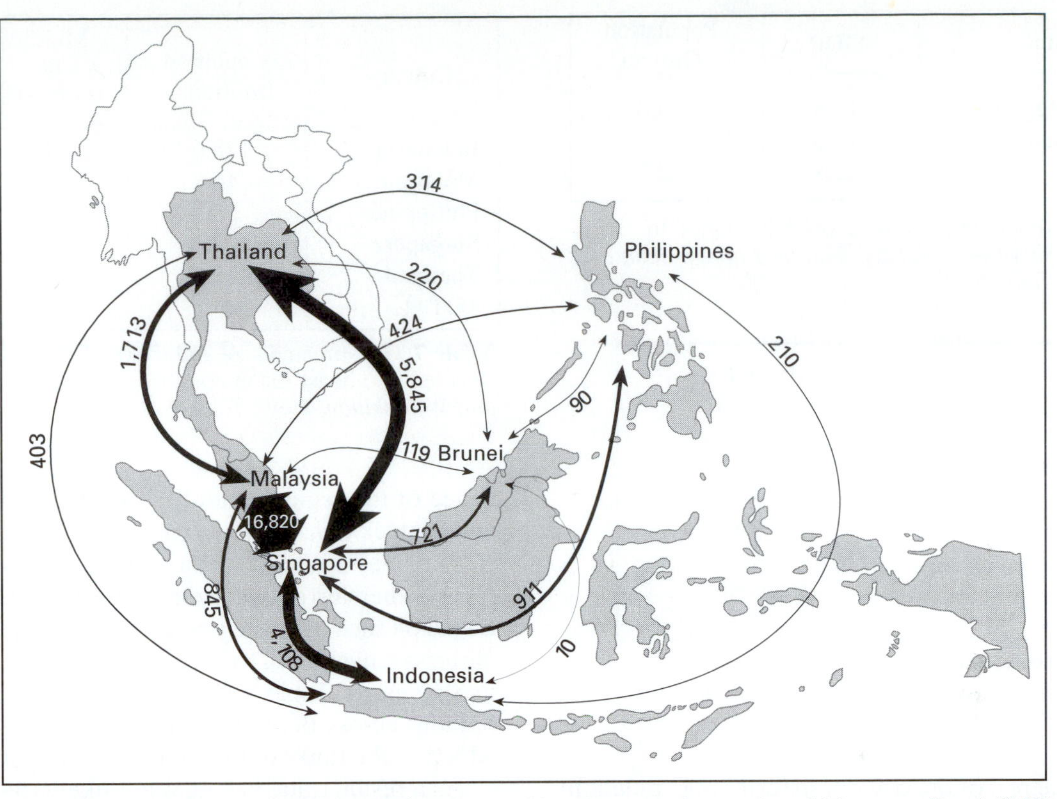

Map 7.3 Trade flows in ASEAN, US$ million, 1991. *Source: International Monetary Fund (IMF)*

ing to an economic one. Of the seven members, Singapore has the most to gain because its trade flows with other ASEAN members are the largest and because its port is the region's largest transshipment center. A clear problem is that those countries that are expanding rapidly with far-flung export markets feel less need to implement tariff reductions with partners within the region. Thailand is a case in point. Time will tell whether this venture, which has attracted worldwide attention, will be successful. One measure of this will be the changes within the region in trade flows that are now clearly dominated by Singapore. The weak trade links between the Philippines and other members of ASEAN are clear (Map 7.3).

INDUSTRIALIZATION AND TRADE IN THE REGION

Export-oriented industrialization has been at the forefront of Southeast Asia's remarkable economic development. Rapid expansion of manufacturing and exports have sustained economic growth rates that are among the highest in the world. Yet there is still vast untapped potential for expansion. The region's northern frontier (Burma, Laos, Cambodia, Vietnam) has only recently opened up to trade and remains overwhelmingly agrarian. Even in the manufacturing powerhouses of the region, such as Malaysia and Thailand, there is tremendous scope for further development—by moving toward more-sophisticated, specialized, and higher value-added manufactured exports.

Yet there are also constraints upon the further development of industry. These include: the global trade regime, changes in world demand for Southeast Asia's exports, and increasing competition from other developing areas, especially China and India. Coping with these international pressures will be a critical challenge for the national governments in the region.

Opportunities for Growth

Though vitally important to the economies of most Southeast Asian countries, industrialization has been largely limited to a handful of areas within the region. Most manufacturing enterprises, and especially those in higher-tech industries, have been set up either in urban areas or in FTZs (which are themselves often located within urban areas). As a result of this concentration of activity a wide gap has opened up between the productivity of manufacturing centers, on the one hand, and less-industrialized areas, on the other.

In Thailand, for instance, the GDP per capita of the Bangkok metropolitan region has grown much faster than the rest of the country (Table 7.9). As a result, GDP per capita for Bangkok was three times higher than the average for the whole country in 1989 (Table 7.10). This disparity is one reason why Bangkok has also grown

Thailand	GDP	Population Growth
Bangkok Region	15.8	5.0
Rest of Country	11.3	2.3
TOTAL	12.8	2.3

Table 7.9 GDP and population growth by region in Thailand, 1960-1989 (percent). *Source: Phongpaichit and Baker, 1995, Table 5.11.*

Region	GDP per Capita (Baht)
Bangkok Region	96,239
Central	30,587
North	18,833
North-East	11,981
South	21,955
TOTAL	30,028

Table 7.10 GDP per capita by region in Thailand, 1989. *Source: Phongpaichit and Baker, 1995, Table 5.10.*

Country	Population (millions)	Middle Class (millions)	Percent Middle Class
Indonesia	175.6	14.0	8
Malaysia	16.9	2.5	15
Philippines	58.7	7.0	12
Singapore	2.7	1.4	50
Thailand	51.6	6.2	12
TOTAL	305.5	31.1	10

Table 7.11 Estimated size and share of the middle class in selected Southeast Asian countries, 1988. *Source: Hughes and Woldekidan, 1994, Table 6.*

faster than the rest of the country in terms of population: the city's economic success means higher wages and better job prospects for rural–urban migrants.

But in the long run, such concentration of activity is undesirable for economic, social, and political reasons. As an example of the economic cost, one might consider the potential investors who have been deterred from investing in Thailand because of the severe congestion in Bangkok. To overcome this problem, Thailand's government has been promoting development of the eastern seaboard since the early 1970s (Phongpaichit and Baker, 1995). Most significantly, the new international seaport for the country was built at Laem Chabang, far to the south of Bangkok. (see Map 16.1).

Other industrializing countries in the region have faced the same challenge and have also sought to spread development away from primate cities. The Philippines, for instance (as noted earlier), tried to encourage industrialization away from Manila in the 1970s by establishing FTZs at Baguio in the north and Cebu in the south. Both areas have attracted some investments from global electronics MNCs. Similarly, Sabah and Sarawak have lagged far behind Peninsular Malaysia in terms of industrialization, but in the early 1990s there were some developments (such as the establishment of a U.S.-owned disk-drive factory near Kuching in Sarawak) that promised a brighter future for industry in East Malaysia.

The spread of industrialization will be one of the major trends affecting the economic landscape of Southeast Asia in the future. Another key trend will be the emergence of the region's middle class. One study (Hughes and Woldekidan, 1994) estimated the size of the middle class in five Southeast Asian nations based on such criteria as ownership of televisions and access to a telephone. Based on such data, the percentage of people in the middle class ranged from 8 percent in Indonesia to 50 percent in Singapore (Table 7.11). With sustained, rapid development across the region, millions will continue to be added to the ranks of the middle class each year.

As a result, trade balances are likely to shift as eager consumers purchase imported goods. But at the same time, the bigger local market is likely to encourage both foreign and domestic firms to produce more consumer goods within the region. That development will in turn make it viable to manufacture an increasingly broad variety of producer goods (steel, chemicals) in Southeast Asia. Ironically, by fostering the emergence of a huge middle class, export-oriented development will indirectly help to achieve the main goal of import-substitution industrialization: integrated, domestic manufacturing of a wide variety of consumer and producer goods.

Challenges

The emergence of the middle class in Southeast Asia reflects another trend: rising labor costs. One of the central challenges faced by governments in the region will be how to increase wages without undermining national competitiveness. The region remains dependent on MNCs for much of its investment; yet today MNCs have an unprecedented choice of emerging markets in which to invest. Countries from East Europe, Latin America, Africa, and elsewhere in Asia are liberalizing their economies and opening up for investment. The competition for limited international investments puts considerable pressure on national governments. In particular, the tension between the demands of domestic constituents, on the one hand, and the prerogatives of international investors, on the other, requires a government that can effectively channel resources to education and infrastructure. These have been the two keys to the successful development of higher value-added manufacturing in Malaysia and especially Singapore.

The importance of manufactured exports to the economies of Southeast Asia means that their future growth depends not only on their competitiveness versus other developing countries but also on the health of the global economy and the world-trading regime. In particular, several Southeast Asian nations are heavily reliant on just one type of export: electronics. If global demand for electronics tapers off or if new trade barriers are imposed, these economies will be hard hit.

One alternative for economies confronting the limits of the global marketplace for merchandise exports is to follow Singapore's lead and target the service sector for rejuvenation. Outside of the city-state, none of the governments of the region have given much attention to the development of financial services, with the exception of Malaysia's offshore financial center on the island of Labuan near Brunei. Development of financial services elsewhere in the region will require an improvement in the effectiveness of regulatory authorities.

Another alternative is to develop stronger trading ties with countries outside the developed world. Singapore, Malaysia, and Thailand have all shown a keen desire to form links with the emerging economies of Burma, Laos, Cambodia, and Vietnam. These new linkages reduce the dependence of Southeast Asia's industrializing economies on Western markets. In the 1990s, Singapore and Malaysia both have looked to developing countries beyond Southeast Asia for new markets and investment opportunities. Malaysia, for instance, has sought to bolster its economic relations with other Muslim countries including Iran, where the Malaysian oil company Petronas is a major investor in new exploration.

The Role of the State

Certainly, such external variables as competition from other developing countries and the global trading environment will have a substantial impact on the industrialization of Southeast Asia, but internal variables will also play a pivotal role. One of these variables is the role of the state. Historically, the state has actively intervened in the development of industry and trade. Indeed, considerable credit has been given to national governments for the pace of the region's industrialization. Yet some have suggested that the state should now give freer rein to private enterprise, not only in the former command economies but even in such successful economies as Malaysia.

The automobile industry has been cited as an example of how state intervention can hinder the development of industry ("Bumper to Bumper," 1996). Thailand and Malaysia, for instance, have adopted different strategies toward the auto industry. As noted above, Malaysia aggressively cultivated its own national car through a state-owned enterprise. Heavy import duties were levied on foreign imports to ensure a large market share for Proton. In contrast, Thailand has had the least-restrictive policy of any Southeast Asian country toward the auto industry. How have these two approaches fared? In Malaysia, Proton is beginning to confront the limits of a small home market. But Thailand has attracted investments from virtually every major auto MNC, many of which are using the country as an export base for markets as far afield as Australia, North Asia, and Europe.

The example of the automobile industry indicates the difficult choices that national governments face in forming an industrialization strategy. Is it better to have a national car that is designed, manufactured, distributed, and sold domestically, or is it better to create more jobs by inviting MNCs to set up export-oriented assembly operations? The answer depends on development priorities and how development itself is defined by national decision makers. Ultimately, the choices they make will have a far-reaching impact on the changing patterns of trade and industrialization in Southeast Asia.

8 Tourism

RICHARD ULACK VINCENT J. DEL CASINO, JR.

> If I had to nominate one region of the world as my favorite for traveling, the one area I'd choose if I had to quit traveling everywhere else, I would have no second thoughts—South-east Asia. There is simply more variety here than almost any other region of the world. In food, religion, culture, Southeast Asia has everything you could possibly ask for. (Wheeler, 1985, p. 7)

Tourism has become the driving force of many economies throughout the developing world. This being the case, particular developing countries have adopted tourism as *the* policy of development in an attempt to bring in foreign currency, counteract balance-of-payments deficits, and stimulate "positive" economic development. Moreover, as Stephen Britton argues, tourism ". . . is one of the quintessential features of mass consumer culture and modern life" (1991, p. 451). In regional texts, however, the topic of tourism has traditionally been a part of chapters on economic development or the economy, if addressed at all. But tourism has become big business in the rapidly developing economies of the Third World—very big business indeed—and we believe it warrants separate coverage here.

At the global level, tourism is the largest business in the world, with annual receipts for both domestic and international tourism exceeding US$2 trillion, which represents about 12 percent of the world economy. This staggering figure is 5 percent of all global sales of goods and services (Rafferty, 1992, p. 2). Tourism has also become the largest source of employment in the world, in both the formal and informal sectors of the economy. According to estimates, Third World employment in travel and tourism alone numbers more than 50 million workers, and tourism is the single largest employer in almost every country, accounting for 1 out of every 16 jobs worldwide.

In addition to the economic impacts, tourism has played a critical role in constructing and reconstructing the image of Southeast Asia as a region. Tourism, similar to other forms of capitalistic production, is subject to commodification. Tourism places are therefore never "authentic" (MacCannell, 1989) but are instead social constructions. For such constructions to be maintained, tourists must visit and interact with the tourist site (Urry, 1990). In response to the ebbs and flows of tourists, therefore, governments and entrepreneurs continually redesign tourism infrastructures and place "identities" to meet tourist demands. Through this process of creating and re-creating place identities, Southeast Asia has become tied to specific identity markers—such as exotic, unique, and sometimes erotic for example—which are intended to maintain its position as an exciting tourist destination. As Shields (1991, p. 60) argues, "[i]n more sociological terms, they [tourism sites] are labeled. Partly through ongoing interaction, a site acquires its own history; partly through its relation with other sites, it acquires connotations and symbolic meanings." We should therefore, in any study of tourism, recognize that the meanings and images assigned to tourist destinations are not necessarily natural or representative; but instead often function to appease the tourist. It is thus critical to any overview of tourism to examine not only the economic patterns of supply and demand but also the social construction of tourism spaces. Moreover, we need to recognize that the practices tied to the consumption of tourism may be interpreted in terms of race, class, and gender. As Swain (1995, p. 247) notes: "Tourism, as leisured travel . . . and the industry that supports it, is built on human relations, and thus impacts and is impacted by global and local gendered [class, and race] relations."

The increasing importance of tourism and its impacts in Southeast Asia and elsewhere has meant an increase in tourism research by geographers and other social scientists. While very few geographers of tourism have examined issues related to the social construction of tourism places (see Shaw and Williams, 1994 and Del Casino, 1995, 1996 as exceptions), there are other areas that have been broadly covered. Within geography, for example, the broad topic areas of tourism research include the spatial patterns of supply and demand; the geography of resorts; the analysis of tourist movements and flows; the social, economic, and environmental impacts of tourism; and models of tourist space (Pearce, 1989, p. 4–5). There

Photo 8.1 Sign of "Authentic Thai Restaurant" in Thailand. (Del Casino)

are a number of examples of work by geographers in Southeast Asia, and most recent studies address some aspect of tourism and national development and planning (Din, 1993), tourist flows (Oppermann, 1992), or the impact of tourism on some of the region's favorite tourist destinations including Bali (Hussey, 1989) and northern and southern Thailand (Dearden, 1991; Williamson, 1992). Generally, research on tourism and travel has been criticized on the grounds that "geographers working in the field have been reluctant to recognize explicitly the capitalistic nature of the phenomenon they are researching. This problem is of fundamental importance as it has meant an absence of an adequate theoretical foundation for our understanding of the dynamics of the industry and the social activities it involves" (Britton, 1991, p. 451). It is through such a lens that we intend to examine tourism in the Southeast Asian region. In this chapter we will examine both the capitalist practices of tourism in Southeast Asia and how these practices have functioned to construct Southeast Asia for the tourist.

THE CHARACTERISTICS OF TOURISM IN SOUTHEAST ASIA

The Demand for Tourism

In 1993 there were more than 500 million international tourist arrivals worldwide and by the year 2000 it is estimated there will be 637 million tourist arrivals (United Nations Statistical Yearbook, 1993; World Tourism Organization, 1994). The industry ranked third in revenues among all export industries and accounted for 7.5 percent of all world exports and 30 percent of international trade in services (World Tourism Organization, 1994). The Southeast Asian region is among the fastest-growing areas of the world in terms of tourism. The 10 nations of the region received 5 percent of the world's tourist arrivals in 1993 (and more than 95 percent of all arrivals were in the five ASEAN member states), but the number of tourists to visit the region increased from 17.4 million in 1989 to more than 24 million in 1993, or by 38.8 percent. This figure is nearly twice that of the growth of world tourism during the same period, 19.6 percent (United Nations Statistical Yearbook, 1993).

As is so of the number of visitor arrivals, receipts from international tourism also increased in all Southeast Asian nations by nearly 75 percent compared to the world average of 42 percent. Indeed, tourism has become a leading source of earnings from foreign exchange in Thailand (where it ranks first), the Philippines (second), Singapore (third), Malaysia (fourth), and Indonesia (fourth). Even Vietnam, Burma, Laos, and Cambodia have witnessed increasing numbers of international tourists and tourist receipts, but political and economic instability, coupled with poor infrastructure to support tourism, have meant that growth in tourism has been slower in these nations. But for the Southeast Asian region as well as most of East Asia and the Pacific, it can be said a tourism boom is underway. By one estimate, tourist arrivals in the (East) Asia-Pacific region are projected to increase at 7 percent annually until the end of the century, a much higher rate than the global average of about 4.5 percent per annum (Hitchcock et al., 1993, p. 1). By the year 2000 it is estimated the entire East Asia-Pacific region will account for

17.9 percent of all international tourists, up from 11.5 percent in 1991 (C. Hall, 1994, p. 198).

It is also important to note that rising incomes in the ASEAN member states mean that a considerable number of *outbound* tourists are traveling *from* Southeast Asian nations to international destinations. By 1990 nearly one-quarter of all tourists arriving in Japan and the newly industrializing economies (NIEs) of Hong Kong, Taiwan, and South Korea were from ASEAN, and nearly two-fifths (37 percent) of international travelers in ASEAN were from member nations (Hitchcock, 1993, p. 2). As such travel increases, so too will the net effect on foreign exchange derived from tourism earnings. One Southeast Asian nation, Malaysia, is already experiencing a net deficit in part because many Malaysians visit Singapore (US$1.96 billion spent by outbound visitors versus US$1.876 billion in tourist receipts; United Nations Statistical Yearbook, 1993).

The nearly 25 million tourist arrivals in Southeast Asia in 1993 were of course not evenly distributed throughout the region but rather reflected those nations, cities, and sites that have long been important attractions, those places that have the amenities or the infrastructure to support large numbers of visitors, and those places that have been politically stable in recent years. The latter factor has meant that Vietnam, Laos, Cambodia, and Burma have not been major attractions in recent decades, although there is evidence, especially with regard to Vietnam, that these places will become significant destinations for tourists again in the near future (e.g., see Lenz, 1993).

Since the mid-1980s the major regional attractions have been the region's largest metropolitan areas (Jakarta, Manila, Bangkok, Saigon, Singapore, and Kuala Lumpur); the precolonial historic places of central and eastern Java including Yogyakarta and nearby Borobudur; and the beaches and coral reefs of Bali, Thailand, and Malaysia, including Denpasar, Pattaya and Phuket, and Penang, respectively (Map 8.1). Upland resorts, or hill stations, are also popular destinations for

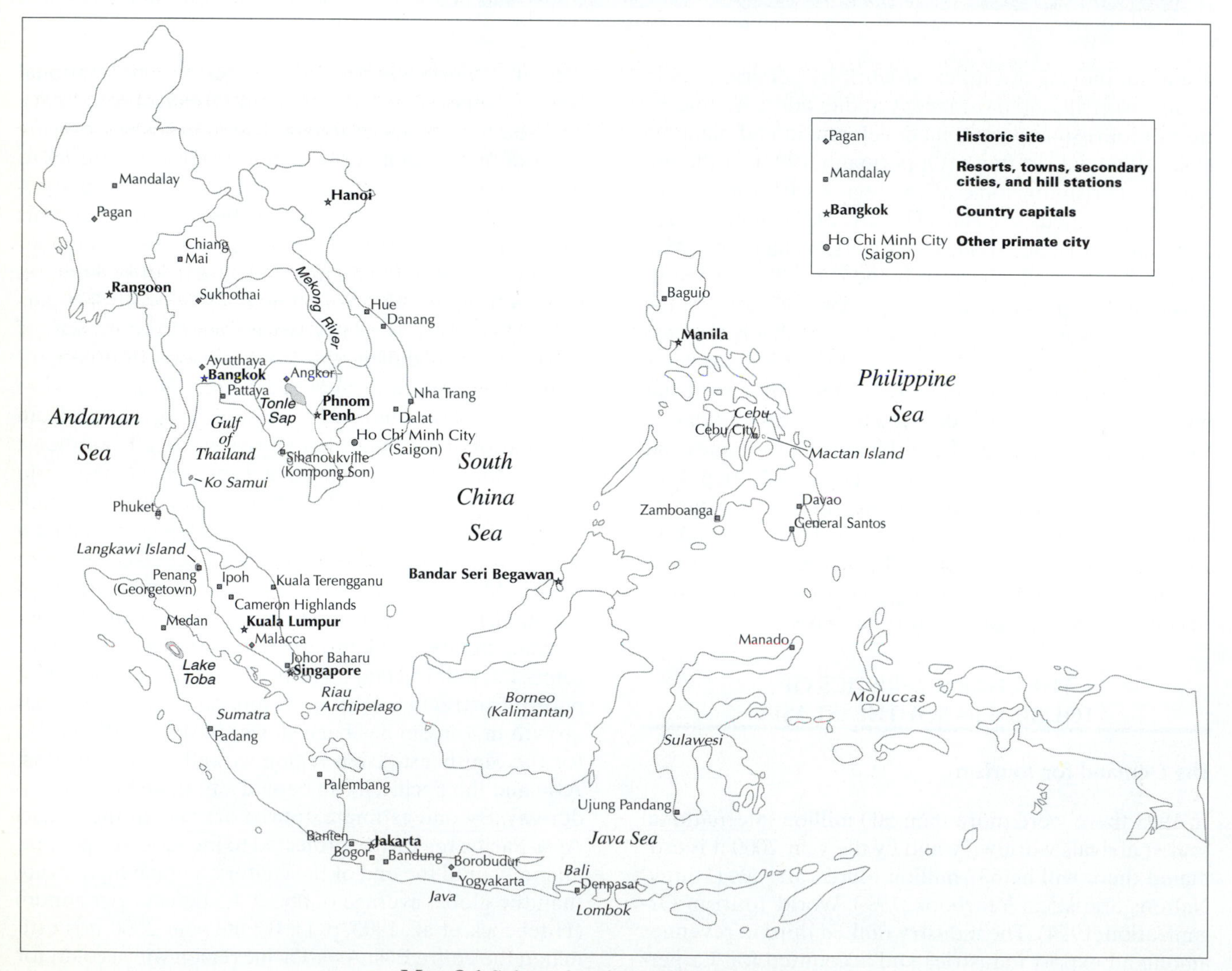

Map 8.1 Selected major tourist destinations.

Photo 8.2 Beach at Phuket, Thailand. (Leinbach)

local elites and the middle class and for international visitors (e.g. Aiken, 1987; Crossette, 1998; Reed, 1976b, 1979). Thus, Baguio (the Philippines), Cameron Highlands and Genting Highlands (Malaysia), Bandung and Bogor (Indonesia), and Dalat (Vietnam) are all examples of such sites. Increasingly, such new types of tourism as environmental tourism (ecotourism) and historic tourism are opening additional destinations for the tourist. Such places include northern Thailand (centered on Chiang Mai), the national parks of Malaysia and Indonesia, and Angkor Wat, Phnom Penh, Hue, and Hanoi, which are slowly being redeveloped after two decades of war. Finally, some of the region's medium-size cities, including Cebu, General Santos, and Davao in the Philippines, Medan in northern Sumatra, and Georgetown and Ipoh in Malaysia, are destinations where tourist arrivals have increased.

It has been the five original ASEAN member states (Indonesia, the Philippines, Malaysia, Thailand, and Singapore) that have witnessed the greatest number of tourists, accounting for more than 95 percent of regional visitors in 1993 (United Nations, 1995). Clearly, Malaysia, Singapore, and Thailand were the leading Southeast Asian destinations in 1993 (Figure 8.1). But the nations that have

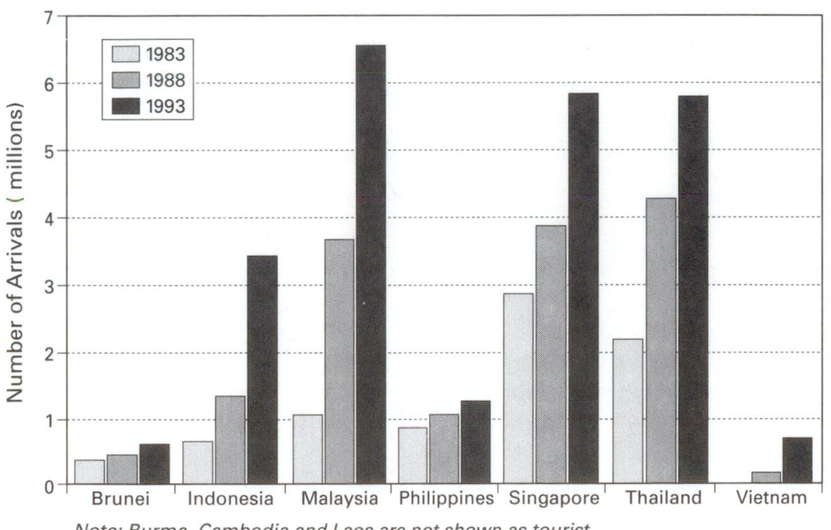

Figure 8.1 Tourist arrivals: 1983, 1988, and 1993. *Source: World Tourism Organization, various years.*

realized the most rapid rise in tourist arrivals between 1988 and 1993 have been Vietnam and Indonesia, where tourist numbers have increased 353 percent and 162 percent, respectively.

In 1993, the leading destination in the region, and the third-leading destination in Asia after China and Hong Kong, was Malaysia with more than 6.5 million visitors, an increase of nearly 80 percent since 1988 and 519 percent since 1983. One reason for the rapid increase in tourist visits is related to the success of the Malaysian government-sponsored Visit Malaysia Year 1990, which was an international promotion and advertising campaign to encourage visitors to vacation in Malaysia. Such a tactic has been used by several of the ASEAN governments. Estimates are that visitor arrivals to Malaysia will expand by 7–8 percent annually and reach 12.5 million by the year 2000. The large number of arrivals in Malaysia is in part explained by the fact that the single largest group are Singaporeans, most of whom arrive via road across the Johor Causeway. The next largest visitor arrival origins are Thailand and Japan (Table 8.1). Together, East and Southeast Asia and the Pacific regions accounted for 84 percent of all visitor arrivals to Malaysia in 1993.

In line with Singapore's policy of economic diversification, tourism has been actively promoted. An indication of this is illustrated by evaluating the increase in the number of visitor arrivals since the mid-1980s (Figure 8.1). Between 1983 and 1993, arrivals increased by 103 percent to total 5.8 million, and during the period 1988 to 1993, arrivals increased by 51 percent; in 1994, nearly 7 million visitors arrived. The number of hotel bookings has also improved, from an occupancy rate of 70 percent in 1987 to 83 percent in 1994, one of the highest such rates in the region. The tourist industry's expansion is likely to be affected by a shortage of hotels as well as by other internal factors regarding the competitiveness and quality of tourist services. External factors, such as the growth of the world economy, air liberalization measures, and the availability of seats from airlines, could also influence its development. In any case, one set of estimates states that whereas Singapore will be able to accommodate a maximum of 10.3 million tourists by the year 2000, 9 million would be a more appropriate and manageable number (Khan, et al., 1996, p. 223).

Thailand was the third-most-important country destination in 1993 with 5.76 million visitor arrivals, and tourism has been Thailand's principal source of foreign exchange since 1982. The megacity of Bangkok and its myriad attractions including shopping, architecture, and commercialized sex remains Thailand's major destination for tourists, but other tourist destinations include the beach resorts of Pattaya and Phuket in the south; more recently, ecotourism and trekking in the north that centers on Chiang Mai as well as heritage tourism to ancient cities, including Sukhothai and Ayutthaya (both on the United Nations World Heritage List), are becoming in-

Nation	Total Number of Visitors (Thousands)	Country of Origin (Percent)										
		ASEAN	Japan	Taiwan	Hong Kong	Australia	USA	UK	Germany	France	Netherlands	TOTAL
Brunei												
1984	309,080	84.2	0.5	0.0	0.2	0.9	1.2	3.0	0.3	0.1	0.6	91.0
1988	457,410	93.2	0.4	0.3	0.0	0.5	0.7	2.0	0.2	0.2	0.4	98.0
1992	500,259	90.9	0.9	0.2	0.2	0.9	0.5	3.1	0.3	0.3	0.7	97.8
Indonesia												
1984	700,910	29.8	12.6	1.8	2.0	15.3	7.2	5.1	4.5	3.4	4.5	86.2
1988	1,301,049	36.6	12.1	2.6	2.2	11.4	5.2	4.8	4.5	2.9	4.6	86.9
1992	3,064,161	40.0	12.9	7.2	2.6	7.7	4.1	3.8	3.9	2.0	2.8	86.9
Malaysia												
1984	2,779,081	78.2	4.0	0.6	1.2	2.7	1.5	2.1	0.9	0.4	0.3	91.9
1988	4,846,320	78.5	4.3	1.8	0.9	1.6	1.6	2.3	1.0	0.5	0.4	92.9
1992	6,016,209	75.1	4.3	3.4	1.6	2.0	1.3	2.4	0.8	0.4	0.4	91.7
Philippines												
1984	816,712	7.1	19.7	4.9	7.8	6.2	24.2	2.7	3.0	1.0	0.6	77.1
1988	1,043,114	5.7	17.4	5.4	12.8	4.3	20.3	2.4	2.6	0.8	0.6	72.3
1992	1,152,952	4.8	19.2	10.6	5.7	4.4	19.2	3.4	3.1	1.1	0.6	72.3
Singapore												
1984	2,991,430	35.3	12.4	2.7	3.3	9.7	5.9	5.0	2.4	1.4	1.3	79.5
1988	4,186,091	27.7	16.3	3.6	3.0	8.6	5.4	6.1	2.8	1.5	1.5	76.5
1992	5,989,940	30.2	16.7	6.4	3.9	6.4	4.8	5.1	2.7	1.2	1.3	78.8
Thailand												
1984	2,346,709	33.8	9.5	4.8	2.0	3.3	6.6	5.7	5.0	3.4	1.2	75.3
1988	4,230,737	28.2	10.6	4.5	3.6	3.3	6.1	6.6	4.5	3.7	1.2	72.3
1992	5,136,443	22.5	11.1	7.9	5.7	4.0	5.3	4.6	5.4	3.8	1.6	71.9

Table 8.1 Tourist arrivals by country of origin, 1984, 1988, and 1992. *Source: World Tourism Organization, 1994.*

creasingly important (Peleggi, 1996). Whereas the number of tourist arrivals continues to rise, the increase is not as rapid as other ASEAN members, and there have been years recently where the number of arrivals in Thailand has actually declined. There are several factors that account for a recent slowing in the number of visitor arrivals to Thailand. One reason for the decline in tourism in the early 1990s was the 1991 Gulf War, which negatively affected travel and tourism all over the world. Other reasons included increased competition from other, less expensive destinations in Asia and the Pacific; recession in the developed nations; the military coup d'état of 1991 and the international coverage of the military crackdown on student uprising against the military dictatorship in 1992, which happened to coincide with the hosting of the Miss Universe Pageant (see Van Esterick, 1994); the impact of the AIDS epidemic; and the negative publicity of Thailand's role in the international drug trade and in the sex industry. In the latter case, the Thai government through the Tourist Authority of Thailand (TAT), is taking steps to discourage sex tours and also has developed cash-crop alternatives to growing the opium poppy.

Photo 8.3 Tourist site for the Golden Triangle, northern Thailand. (Sirijintakarn)

One other Southeast Asian destination noted for sex tourism, the Philippines, through its Department of Tourism (DOT) is also letting it be known that such tours are no longer welcome. The negative publicity associated with sex tourism, especially as it relates to children, and the emergence of growing international political organization against pedophilia through, for example, End Child Prostitution in Asian Tourism (EPCAT), will continue to work toward decreasing sex tourism (Richter, 1994, p. 49). One of the major origins for sex tourists has been Japan, but the development of greater women's rights in Japan has brought about social and political changes that are bound to curtail the number of Japanese men who travel to sex destinations (Leheny, 1995). The scale of the industry is, however, huge: for example, it has been estimated there are 800,000 child prostitutes in Thailand and 60,000 in the Philippines; some 20,000 young girls and boys are brought from Burma to Thailand each year for use in the sex industry (Asia Watch Women's Rights Project, 1993; Mirkinson, 1994).[1] In preparing for a continued decline in tourism in Thailand, TAT has begun to establish links with Vietnam and Laos, with a view to joint development of tourism and Thai investments in the area. Thai investments in international tourism may become significant in the former Indochinese states, as it has elsewhere in the world. One example is the 1994 acquisition of the German hotel chain Kempinski (and its 18 European hotels) by a consortium of Thai investors headed by the Dusit Thani group, owners of the well-known 26-year-old Dusit Thani Hotel in Bangkok (Vatikiotis, 1996a, p. 77).

Of the five leading regional tourist destinations, visitor arrivals to Indonesia have increased the most rapidly between 1988 and 1993, by more than 160 percent (Figure 8.1). In 1994 tourism continued its rapid growth as there were estimated to be 4 million tourists, an increase of 18 percent more than 1993 arrivals. The vast majority of tourists to Indonesia visit Java and Bali, the latter well known for its scenery, cultural uniqueness, and Hindu temples and festivals. As ecotourism continues to expand so, too, will travel to the Outer Islands, especially Sumatra, Kalimantan, Sulawesi, and Lombok. As is currently so for most other Southeast Asian nations, visitor arrivals from other Asian and Pacific nations were most significant, as was demonstrated by the fact that Singapore, Japan, Taiwan, Malaysia, and Australia were the leading tourist origins with 23 percent, 13 percent, 11 percent, 7.5 percent, and 7 percent of the total, respectively. Favorable economic conditions and measures taken to increase nonoil exports in the 1980s, were

[1] The number for Thailand is highly contested by the Thai government, which places the figure at less than 100,000. The numbers are tenuous at best and both sets of numbers can be called into question. Suffice to say that child prostitution, however, is present and should be treated as a serious issue in not only Thailand but in the Philippines and Bali as well.

important impetuses to increases in the tourism sector. Specifically, two major currency devaluations (in 1983 and 1986) and changes designed to simplify government regulation of private sector activities (e.g., reducing the number of licenses required to build a new hotel to one) and visitor arrivals [e.g., granting of additional landing rights to foreign airlines in Jakarta, Medan, and Bali; abolition of visa requirements for ASEAN and OECD (Organization for Economic Cooperation and Development) nationals staying less than two months] have facilitated tourist visits. As a relative latecomer among ASEAN nations, the potential for tourism in Indonesia is great as the nation can attract many people who have developed a taste for Southeast Asian travel but who have not yet visited Indonesia (Booth, 1990, p. 69). This is so despite the U.S. State Department's traveler advice that warns tourists to avoid Indonesia, given recent political and economic turbulence.

In the Philippines, tourist numbers have fluctuated in response to the political instability of the country since the early 1980s. In 1973–78 arrivals were rising by an average of 33 percent a year. There was a slackening in growth in the following four years, mainly reflecting international trends, and a sharp decline thereafter, with arrivals at 773,000 in 1985, compared with just more than 1 million in 1980 (Europa, 1991, p. 879). Among ASEAN states, the Philippines has experienced the slowest growth in visitor arrivals between 1988 and 1993, from just more than 1 million visitors in 1988 to about 1.25 million, or an increase of 21.8 percent (United Nations, 1993 and 1995). Tourism accounted for more than one-fifth of total merchandise export earnings (1993) and was the second leading foreign exchange earner for the country. The largest groups of tourist visitors each year are from the United States and Japan, which each accounted for about one-fifth of all visitors in 1993. *Balikbayans*, or overseas Filipinos, accounted for about one-tenth of visitors. Until the Aquino presidency in 1986, tourism was hampered by insurgency movements associated with the difficult economic times of the late Marcos era. Aquino used the excesses of the Marcos era to spur tourism (e.g., opening Malacanañg Palace so that tourists could witness the excesses) and to legitimize her own presidency. But the seven attempted coups during Aquino's six years as president was a factor that kept the number of arrivals lower than anticipated. More recently, the AIDS epidemic and occasional natural disasters (e.g., the eruption of Mt. Pinatubo in 1991 and various typhoons) have affected the number of arrivals. In 1994, it was estimated that 1.57 million tourists arrived, an increase of 26 percent more than in 1993 and one of the largest increases since the 1970s. Both Aquino and her successor, Fidel Ramos, have attempted to improve the image of the Philippines and tourism by divesting many of the tourist hotels and lodges that were holdovers from the Marcos era and by attempting to curb sex tourism; unfortunately, the record for these recent administrations has been at best mixed.

The other countries of the Southeast Asian region are well aware of the potential economic benefits from development of the tourist sector and all, most notably Vietnam, Cambodia, and Burma, are developing, or redeveloping, tourism. In the late 1960s, for example, about 60,000 tourists visited Cambodia each year. For all practical purposes, tourism ended during the war years; since 1986 tours of Phnom Penh and Angkor Wat were again available, but as recently as 1987 Cambodia received fewer than 1,000 tourists, most of whom were from socialist countries and Japan. Beginning in 1989 Cambodia began to encourage foreign capital investment because of its potential as a source of foreign exchange. By 1989, the number of tourists visiting Cambodia increased to about 2,000 but decreased to 1,542 in 1990 because of the political situation. In 1990 the number increased to 3,000; by this time improvements had been made in hotel and restaurants for tourists. In 1994 more than 175,000 tourists arrived (Europa, 1995, p. 191). The largest project to be built with investment from Malaysia is the resort and casino complex on Naga Island near Sihanoukville, which also involves the renovation of the airport at Sihanoukville. Malaysian capital is also responsible for plans to build a racetrack and sports complex near Phnom Penh, and to expand the capital city's international airport.

As noted above, Vietnam has witnessed a veritable tourism explosion in the 1990s in large part because the government has facilitated tourism by providing visas more easily and removing travel restrictions to interior areas. Indeed, it was estimated that 1 million tourists visited the country in 1994, a two-thirds increase over the 600,000 that visited in 1993 and more than 450 percent more than the 180,000 who traveled to Vietnam in 1992. Revenues have likewise increased from about US$80 million in 1992 to US$363 million in 1994 (Europa, 1995, p. 1064). Tourism in Vietnam is still in an early stage of development and, as has been the case elsewhere in the Third World (e.g., the Caribbean), it has been the wealth-package tourists who have been the first to arrive (Lenz, 1993, p. 2). Hotels have been constructed to accommodate the relatively small numbers in this and the second group of tourist arrivals (the upper middle class), but infrastructure for the third and potentially largest group, the average middle-class traveler, is still inadequate. As is so for most Southeast Asian nations, it is the primate city (Saigon in this case) that is the primary tourist destination and the place with the most (and best) hotels, restaurants, organized tours, and, arguably, climate. The other major destinations are Hue, Danang, Nha Trang, and Hanoi (Lenz, 1993, p. 6). Vietnam's most important upland resort, Dalat, is another important destination

(Reed, 1995). Given recent historic events, visitors to Vietnam from France and the United States will no doubt continue to increase as adequate infrastructure for tourists becomes available; already increasing numbers of veterans of the war in Vietnam are returning to visit.

In Burma, too, the exploitation of tourism potential has been regarded in recent years as one way of increasing foreign exchange earnings. However, during the political instability and unrest in 1988 the government stopped issuing tourist visas. The new government again began to issue visas in mid-1989, allowing visitors who were on package tours to remain for as long as two weeks in the country. It was not until 1993, when 50,000 tourists arrived, that the number of visitors reached the levels attained prior to 1988. The government hoped to attract one-half million visitors in 1996–97, which had been declared "Visit Myanmar Year," but the problems of freedom in Burma that have been emphasized by Nobel Peace Prize-winner Aung San Suu Kyi and her supporters may inhibit not only the hoped-for increase in tourism but also development more generally.

As is so for assistance and trade, the geographical origins of tourists to the region reflects in part the colonial past (Table 8.1). Thus, the United States, Great Britain, France, and, in the case of Indonesia, the Netherlands generally had disproportionate shares of tourist arrivals in the nations previously colonized. In more recent years German visitors have been most conspicuous among the "Euro-tourists." Since the 1980s, the Japanese have become the largest single group of extraregional visitors. This is not surprising given the high income levels of the average Japanese, and Japan's economic interests in and close proximity to Southeast Asia. Certainly Japanese travel to the region as well as financial investment in the region can be expected to continue to increase in the future. In many ways no other single nation outside the region has had a greater influence than Japan during the 1990s.

Supplying the Demands for Tourism: The Commodification of Culture, History, and Leisure

Southeast Asia is a "vast and variegated region" with many historical, environmental, and cultural sites (Savage, 1984). From the ethnic and cultural tours of Bali and Northern Thailand to the adventure tours of Malaysia to the historical tours of Angkor Wat, tourism has become a critical component of development plans throughout Southeast Asia. The various forms of tourism will necessarily have a differential impact across the region. To differentiate impact, therefore, it is necessary to create a typology for tourism and to define each type in the context of Southeast Asia. For our purposes here we refer to four major categories of tourism: ethnic and cultural tourism; historical (or heritage) tourism; environmental tourism (ecotourism); and recreational tourism. Before examining the various forms of tourism, however, we need to examine the physical infrastructures that are produced for tourism—the airports, roads, and lodging facilities, for example. This is necessary because tourist visits necessitate alterations to the physical infrastructure to accommodate those visits. Following a brief examination of the physical infrastructures of tourism, we will investigate how the social systems of tourism are constructed for a tourist's consumption. In this section, then, we address the infrastructures of tourism and the way in which places have been commodified to meet the demands of particular types of tourism.

Building a Physical Infrastructure for Tourism

The infrastructure for tourism refers to the hotels and resorts, restaurants, shopping, recreational facilities such as golf courses, and beaches that are constructed for tourists and that are directly used by tourists. In addition, an adequate-support infrastructure is needed that includes transportation (roads, harbors, airports, parking), communications (telephones and increasingly advanced access to new systems such as the Internet), and medical and public utilities (electricity, sewage disposal, water). Some support infrastructure is often already in place, but in the development of a new beach area, upland resort, or national park, for example, completely new transportation, communications, and utilities systems may need to be constructed, sometimes entirely for the tourist. Whereas such development may serve the resident population, the infrastructure provided does not generate direct revenues; in short, tourism development can be a costly or even money-losing proposition.

There is of course great variation throughout the region with regard to the distribution of such infrastructure. One such direct indicator of infrastructure is hotel capacity. Overall, there were more than 400,000 hotel rooms in the five ASEAN member countries in 1991 that accounted for nearly one-third of all hotel rooms in the East Asian and Pacific region (World Tourism Organization, 1994, vol. 1, p. 74, 81). Within the region, nearly 90 percent of all hotel rooms were located in just three nations in 1992: Thailand, Indonesia, and Malaysia, which ranked tenth, fourteenth, and twenty-fifth, respectively, among the world's nations in terms of hotel capacity. Between 1988 and 1992 the hotel capacity for the ASEAN countries increased by nearly one-third, an increase that will no doubt continue as the number of tourists increase and tourism grows in Vietnam (which joined ASEAN in 1995).

Clearly, continuing to expand the infrastructure for tourism to permit more tourists to visit has some direct impact on local and national economies. Building hotels, for example, means employment for construction workers and revenues for local construction firms. Unfortunately for the host nation, the construction firms and hotels are often foreign owned, and the employment at upper managerial levels is usually reserved for foreign nationals from those countries that own the businesses. The locals, who comprise the vast majority of workers in tourism, are employed in unskilled or semiskilled jobs that are low paying, and often the employment is seasonal and erratic. Economic impact studies of tourism have often been set in the context of dependency theory (or its more recent "revisions") whereby the discussion emphasizes the external domination of the industry and the broad historical context and political-economic structures that were established during the colonial period. In the terms of such theorists the hotels, restaurants, and leisure activities developed in the world's periphery (i.e., the Third World) are often Western owned and managed; very little of the tourism revenues remain in the host country but rather find their way back (as "backward leakages" that result from the repatriation of profits) to the global core nations of the United States, Japan, Germany, Great Britain, and so forth. As Britton notes, "... while all participants in the industry hierarchy profit to a degree, the overall direction of capital accumulation is *up* the hierarchy" (1982, p. 346 as quoted in Pearce, 1989, p. 94).

The design of the physical infrastructure for tourism is strongly connected to the commodification of places and their constructions as particular types of tourism sites. Here we want to examine the varying types of tourism found in Southeast Asia to understand better how each developed, what resources are used, and how the image of particular places are constructed to support and maintain each tourist destination's uniqueness. Again, the four major categories of tourism used here are: ethnic and cultural tourism; historical (or heritage) tourism; environmental tourism (ecotourism); and recreational tourism. We rely on this typology to detail some of the debates and discussions that have been highlighted by researchers of Southeast Asian tourism.

Tourism Types and the Social Systems of Tourism

Regardless of the type of tourism, all forms are linked in some way or another to the cultural and developmental policies of a particular nation. Perhaps that is why some of the most controversial literature in tourism studies revolves around the development of cultural and ethnic tourism. More than any other form of tourism, cultural and ethnic tourism relies directly on specific definitions of a people's icons, images, and identities (e.g., the link of specific ethnic identities with particular characteristics and places). The construction of a place as a tourist destination is not an overnight phenomena, however, but is a *process* that develops over centuries, is continually changing to meet the new expectations of tourists, and is generally guided by government officials and tourism entrepreneurs. The formation of an image of place is, for example, often linked to colonial occupation and to the travel literature that accompanied the period of conquest and later "touring" by both European and Asian elites. For some, Bali is an interesting site in part because of its "history" as, "... an island which had long been reputed in the West for its plunderous salvage of shipwrecks and barbarous sacrifice of widows on the funeral pyre. . . . " (Picard, 1993, p. 74). These images, and many others, guide travelers' geographic imagination and in part drive government policy in ethnic and cultural tourism development.

The development of such culture industries has had consequences beyond the complications caused by the interactions between local hosts and foreign tourists. Given the rewards from cultural and ethnic tourism, the government has a direct stake in the development and definition of what is a local culture. As Shaw and Williams (1994) argue, "... private capital cannot guarantee the sustained production of those tourism and leisure activities that a society values ..." (p. 11) and therefore government intervention is critical. In the case of ethnic tourism, the government sets the standards of cultural and ethnic norms. From this, a debate often emerges between local leaders and state officials as to what defines both ethnicity and culture in particular places. Thus, the question of what is *the* "culture" of a particular place and people is always suspect.

Despite these complications, there are some broad conceptual differences between ethnic and cultural tourism. For Wood (1984), ethnic tourism is primarily linked to those activities which focus on cultural practices. On the other hand, cultural tourism is focused more closely on the physical artifacts and markers of everyday life in a particular place. Thus, for example, ethnic tourism would involve, "special performances supplementing other forms of tourism such as a visit to a historical site whereas cultural tourism provides, a physical setting for other forms of tourism which imprints them with a sense of uniqueness" (Wood, 1984, p. 361). For the most part both ethnic and cultural tourism have been packaged components of a particular tour: specific performances organized at designated places and times or stops on a tour of specific cultural icons.

In recent years, however, there has been a distinct shift for some tourists from organized tours to the travel of trekkers—people who want to leave the well-worn path of the average tourist and experience the more authentic experiences of nonurban village life. In the context of ethnic tourism this has led to a proliferation of trekking outfits specifically designed around excursions into the "jungles" of Southeast Asia to witness local village life. But even this form of tourism has a series of different levels from the one-day trek to an "organized" village to a more extensive week-long trip that includes elephant

Photo 8.4 Entrance to "Thailand in Miniature" (TIMland) tourist attraction, near Bangkok. (Ulack)

rides; a series of different villages, each representative of another ethnic group; and perhaps, as in the case of Northern Thailand, a rafting trip down the Mekong River.

What has developed in many cases is a proliferation of tourist amenities in close proximity to ethnic and cultural sites. Various forms of cultural capital alter the urban and rural landscapes of tourist destinations thus increasing, or in some cases diminishing, their value as "unique." Changes to the built environment, such as resorts like Club Med, are perhaps the most obvious, while others such as organized spectacles and festivals appear on the surface to be authentic and original but are in reality "staged" (MacCannell, 1989). Night bazaars in Chiang Mai and cultural tours of Bali are but two examples. On the other hand, this form of tourism also functions to protect and preserve ethnic and cultural life that might otherwise have fallen prey to the goals of ethnic and cultural nationalists. In the case of Bali, the importance of Balinese culture to tourism and development has helped preserve the different cultural forms found there, distancing the peoples of that island from the process of Indonesianization. Similarly, in Northern Thailand schools have developed programs that teach about various ethnic traditions to maintain the richness of peoples' cultures for the purposes of ethnic-tourism promotion.

Historical tourism is also linked to specific experiences that hold particular meaning for tourists. Similar to the questions raised about culture and ethnicity, we can ask questions of histories and their representations. Who, for example, is doing the representing of what? Indeed, places such as Angkor Wat in Cambodia and Borobudur on Java have become important sites for tourists. Large majestic monuments such as these signal that Southeast Asia was very early the site of some of the world's largest cities. Images of a "magnificent" and unique past is represented in such places, but such places do not reveal the whole "story." Class, race, and gender relations, for example, are often absent from such historical monuments (Lerner, 1986). In short, the discourses represented in these places have been specifically selected and tell only part of the story.

Although some tourists and trekkers are interested in cultural, ethnic, and historic sites of Southeast Asia, others travel to the region to explore the tropical forests of Malaysia, Thailand, and Indonesia, the coral reefs of the archipelagoes, "exotic" rivers such as the Mekong, or the mountain ranges of northern Thailand and Sarawak. Governments have reacted to this new wave of environmental tourism, or ecotourism, by creating national parks and wildlife refuges such as the orangutan sanctuaries in Kalimantan and Sumatra which preserve "pristine" spaces for the tourist. Indeed, environmental tourism has become an intricate part of many local economies although the number of ecotourists relative to all tourists remains relatively small. Trekkers seeking adventure and views of reportedly uncharted spaces travel to Southeast Asia to tour these regions. Others, including avid climbers, hikers, and divers, come to the region to challenge the steep rock face along the Gulf of Thailand or the great oceanic depths of the Philippines. In the wake of the recent global environmental crises, some see the new wave of ecotourism as a viable and sustainable method for both increasing tourism flows and preserving valuable resources such as rain forests and oceans and their incredibly rich biodiversity. Indeed, environmental tourism is often discussed in the same breath with sustainable development. As Parnwell argues, ". . . the growth of special interest tourism, such as ecotourism (safaris, bird-watching, wildlife photography, landscape painting, even organized hunting trips . . .), has in many instances helped to generate a great awareness of the aesthetic value of natural ecosystems amongst both the promoters and consumers of tourism resources" (1993, p. 287).

Linked in part to environmental tourism are many forms of recreational tourism. Tourists often travel to enjoy the scenic seasides and mountain resorts available throughout Southeast Asia, relaxing and enjoying the favorable climate and pristine waters. The region has thus developed a plethora of resort areas that cater to scuba divers, golfers, hunters, animal watchers, and others who seek a brief escape from the everyday. Even Vietnam has climbed onto the recreational tourism bandwagon. The Vietnamese government has begun construction of several golf courses outside Ho Chi Minh City to satisfy the recreation needs of the new wave of businesspeople who continue to take advantage of Vietnam's new economic trade policies. Resort corporations such as Club Med, for example, have been established along seacoasts in Southeast Asia, and large hotel chains have situated their hotels in close proximity to both downtown areas and golf resorts to cater to the enormous influx of business tourists who want also to play golf.

The most controversial form of recreational tourism has little to do with physical infrastructure and natural resources, however. Instead many men travel to Thailand and the Philippines from Japan and other Asian nations, Australia, the United States, Europe, and elsewhere to visit one of many sex tourism sites. Such travel is linked to the construction of particular sexualities—particularly female sexualities—as passive, available, and exotic (Del Casino, 1995). This has led to the international trafficking of women from Burma and China to accommodate the demand (Asia Watch Women's Rights Project, 1993). There are major districts in Bangkok and Manila, as well as smaller towns and resorts such as Pattaya in Thailand, that cater to the sex trade. Whereas some tourism researchers place sex tourism under the umbrella of recreational tourism, it can and should be separated from the forms of recreational tourism, at least empirically, and perhaps typologically as well. Although all are controversial, sex tourism is unique in that in some cases it involves the sexual exploitation of men and women, boys and girls. In the post-World War II period, Bangkok has developed a reputation as a center for the sex industry, despite the fact that prostitution has been illegal in Thailand since 1960. Tourism has been an important and rapidly growing part of Thailand's economy, and this rapid growth owes much to the sex industry (see Meyer, 1988; Thanh-Dam, 1990; Del Casino, 1995). In some cases, those working in the sex industry in Thailand and the Philippines have been exposed to HIV; the explosive spread of the human immunodeficiency virus (HIV, which causes acquired immune deficiency syndrome—AIDS), is taking a huge toll in Thailand. The World Health Organization (WHO) reported in 1993 that the number of people infected with HIV in Thailand had grown from 50,000 to 450,000 in three years. Most of these cases, however, are among those working and participating in the local sex tourism industry, as "high-class" commercial sex workers in Thailand have a much lower incidence of HIV than do those catering to the indigenous population. This may be, in part, due to the awareness of sex tourists in regard to HIV transmission and the ability of women working in larger and higher paying sex tourism sites to demand that customers use condoms. All sex workers, however, are always at risk for physical violence, despite the cautions they might take.

In this section we have briefly examined the four major types of tourism and have suggested that in Southeast Asia, at least, there is a fifth type, sex tourism. It is important to remember the following two notions: first, the production of tourism is like any other commodity; and, second, whereas we have separated out the various forms

Photo 8.5 Cordoba Reef Village Resort, Mactan Island, the Philippines, opened in 1997. (Ulack)

Photo 8.6 Advertisements in Patpong, Bangkok's entertainment district. (Del Casino)

of tourism into distinct categories, the types we have identified are fluid and overlapping. It is quite likely that any one tourist could explore one or all the types on a single visit. What follows in the next section builds on our discussion here—that is, an examination of the economic, social, cultural, and environmental impacts that tourism has had on Southeast Asia.

THE IMPACTS OF TOURISM

Because it is an industry that can be developed quickly, tourism is increasingly the choice of government officials and planners whose economies are suffering a severe balance-of-payments problem; it is seen as a quick fix to the economic and employment problems that are common. Furthermore, loans from international financial agencies for infrastructure projects in support of tourism (e.g., roads, ports) have been accessible to developing countries for some time. Unfortunately, it must also be remembered that there are negative socioeconomic, cultural, and environmental impacts associated with tourism. As we have noted, the transmission of HIV through sex tourism, the alteration and even loss of cultural traditions and ethnic identities (although sometimes the impact is preservation of such traditions), coastal pollution, and the destruction of coral reefs are but a few Southeast Asian examples. On the other hand, there are benefits to tourism besides the economic impacts suggested by the numbers noted earlier. These include the increased travel and contact among peoples of different cultures that arguably promotes a better understanding of and thus appreciation for cultural difference. Whereas many tourists only see the hotels, major attractions, and carefully orchestrated cultural performances that are part of the typical package tour, increasing numbers of trekkers and others are finding their way to many parts of the region. Clearly, tourism has played an important role in the establishment of national parks and other similar areas designed to preserve natural habitats (e.g., the areas set aside for the orangutan and the Komodo dragon in Indonesia), cultures, or historical sites.

Economic Impacts

Calculating the economic costs and benefits of tourism or its impact on balance of payments for a nation can be a simple or difficult task, depending upon one's definition of costs and benefits. Most simply, one can for example analyze annual data from the World Tourism Organization on tourism receipts and tourism expenditures. Receipts are the revenues generated by international visitors, and expenditures are those made by nationals who go abroad as visitors. Subtracting one from the other yields a measure of net earnings from tourism. Thus, for example, in 1992 Singapore had the highest receipts among the region's nations (and the twelfth highest in the world), US$5.2 billion. It also had the highest regional tourism expenditures (and ranked twenty-first in the world), US$2.34 billion (these and other data in this section are from World Tourism Organization, 1994). Thus, the difference, US$2.86 billion, yielded the net earnings from tourism for Singapore. Similarly, net earnings for Thailand, Indonesia, and Malaysia in 1992 were US$3.24 billion, US$1.56 billion, and US$28 million, respectively. In the latter case, Malaysia, tourism receipts and expenditures were almost identical (US$1.77 billion vs. US$1.74 billion) and thus earnings were small. Such figures do not tell us very much about the net contribution of tourism because the two figures, receipts and expenditures, are largely independent of one another (see Pearce, 1989, p. 194–199). What is not included in such calculations, of course, are the backward leakages mentioned previously. Such leakages are among the most important costs in tourism and include costs of imported goods and services that are used by tourists (e.g., whiskey, wine, processed foodstuffs such as canned and jarred goods, recreational equipment), imported construction materials, promotion and publicity abroad by national tourist boards such as TAT, remittances by foreign tourist companies and expatriate workers, interest payments on foreign investment loans, and other forms of repatriation of profits where there is significant involvement of multinational corporations (MNCs) or international development (i.e., lending) agencies.

The Southeast Asian destination where social scientists have probably studied tourism more than anywhere else is Bali (not surprising because such social science research usually "necessitates" research in the field and because Bali's beaches and scenic beauty certainly makes it an excellent field site!). This is partly because Bali is a good example of one of the most successful tourist destinations in the region, at least from the standpoint of such measures as tourism receipts and number of arrivals. It is also because the island province's tourist development has been guided by national development policy. The Indonesian government established tourism as a mechanism for economic development during its fourth five-year development plan (*Repelita IV*, 1984–1989) and its goal was for tourism to become the nation's second-largest foreign exchange earner, after oil and gas, by the end of *Repelita V* in 1994 (C. Hall, 1994, p. 69). Although Bali became an important tourist destination during the colonial period, the island was singled out for very active promotion as an international tourism center beginning in the 1970s. One indication of this is that during the 1970s and 1980s the annual number of foreign visitors to Bali increased more than twentyfold to 700,000 (Hitchcock, 1993, p. 18). The economic impact of tourism on Bali has been quite significant as is evidenced by the fact that in the early 1990s the industry, including its internationally renowned handicrafts, accounted for one-fifth of Bali's total gross provincial product (Wall, 1993; as cited in C. Hall, 1994, p. 77). Tourism, however, has grown too rapidly on the island; for example, in the three-year period from 1988 to 1991, the number of star-rated hotel rooms on Bali increased by more than 150 percent, from 1,745 to 4,500. As hotels and support infrastructure have been added so that more and more tourists can visit the island, it has become clear that, Bali cannot continue to grow at such a rapid rate without placing enormous stress on the physical infrastructure and social fabric of the island (C. Hall, 1994, p. 77). Beach erosion, demands on the water supply, and disposal of wastes are but a few of the problems already evident on Bali. Even given such negative impacts, one field-based study found that most residents of the island view tourism positively and hope to have more tourists visit, although those residents who were nearest the resorts had more reservations about the positive impacts of tourism (Wall, 1996). One way in which the national government hopes to take some of the pressure off Bali and other major destinations is by developing new sites and resorts in other parts of the vast Indonesian archipelago. Environmental tourism, especially that which is marine based, and historical tourism, are two types of tourism that are especially well suited to the country. Examples of tourism projects that are being developed are those in two of Southeast Asia's newly designated "growth triangles:" the Riau archipelago in the Johor–Singapore–Riau growth triangle and the Gulf of Tomini area in north Sulawesi, part of the Brunei–Indonesia–Malaysia–Philippines—East ASEAN Growth Area (BIMP—EAGA), established in 1994 to promote regional development and integration (Rimmer, 1994; also see Chapter 13). The Moluccas region (which is near Australia) and the Anyer–Banten region of western Java are also being developed based on marine resources and, in the case of Banten, wildlife (this is the home of the rare Java rhinoceros).

An example from ecotourism in Southeast Asia, trekking, demonstrates how a type of tourism is more locally developed and economically impacts the area's residents. Still quite small in numbers, trekking in northern Thailand is growing and brings in an estimated US$2 million annually from 100,000 visitors; most of this is "profit" for the hill people and thus has a major impact on the villages that are affected (Dearden, 1991).

Social and Cultural Impacts

The question of the social and cultural impacts of tourism has been debated since tourism studies began to appear in the late 1950s and early 1960s. Supported by United Nations development policies, tourism has been pursued by many developing nations in the past 40 years as *the* approach to development. Almost immediately, questions of the social and cultural impact of tourism on a host country have been addressed by economists, sociologists, anthropologists, and in some cases geographers. The early predominant school, the modernization theorists, argued that traditional society was an impediment to development, and therefore a society's social system must be sacrificed for development to occur. Tourism for these theorists was seen as a means to promote Western, and so-called modern, values in the developing world. Thus, tourism was an intricate component of the rise of capitalism, the nuclear family, and other Western ideals (Wood, 1993, p. 50). Other social scientists, particularly sociologists and anthropologists, have challenged the basic assumptions of modernization theorists, arguing instead that the loss of cultural identities and "traditional values" in a given host society is a detriment and not a benefit of tourism. In particular, feminist theorists have approached tourism through an examination of the changes in gender roles and identities resulting from changes in the economies and cultures of host countries (Enloe, 1989; Kinnaird et al., 1994; Swain, 1995; Wilkinson

Photo 8.7 Igorots on roadside in Baguio, the Philippines. Tourists can pay for a "photo opportunity". (Ulack)

and Pratiwi, 1995). Other authors analyze the impacts of tourism and conclude that tourism results in the commodification of culture and facilitates the destruction of a society's "authenticity." In other words, tourism is "staged authenticity." As Wood notes (1993, p. 52), however, "what modernization theorists welcomed, most early students of tourism's impact lamented, but both made similar assumptions about the nature of the process of cultural change"—modernity and tradition were opposites and incompatible.

More recent studies on impacts have taken a more critical perspective. It is argued that tourism and culture, like modernity and traditionalism, are not two opposites waiting to collide; instead, they are dialectically related (see Wood, 1993, p. 57). We should therefore look at how local culture informs and is informed by tourism and vice versa. In examining this interaction for example, researchers have concluded that it is more than a simple one-way impact of the tourist's culture on the host's; rather, the relationship, albeit uneven, is two way. Indeed, the tourist takes back cultural constructs of the place visited.

In the context of Southeast Asia the impact of tourism on the cultural landscape initially appears obvious. Large hotels, cultural demonstrations as well as sophisticated infrastructural changes and alterations to the social structure of the family are all present. Jobs have been created not just in the formal sector—hotel clerks, baggage handlers, official tour guides—but in the informal sector—commercial sex work, and food and craft vendors (which may or may not be sanctioned by the government). Rural-to-urban migration has increased to meet the needs of the growing service and manufacturing economies of the countries' largest cities. Southeast Asian cultures are packaged to meet the place-myths created by tourist representations. The region is oftentimes constructed as a tropical and exotic paradise, with images of sultry women, long white-sand beaches, and mystical and ancient relics drawing the tourist to the uniqueness of region. Cultural demonstrations and ethnic villages in the hills of Northern Thailand or on the island of Bali are commercialized and constructed to fit the images that the tourist has anticipated. The impacts have of course been profound. In some cases, ethnic and cultural rituals have been resurrected specifically for the tourism market. In other cases, ritual and performances are enhanced or altered to satisfy a particular niche market; for example, to satisfy the imagination of someone traveling to Thailand for sex or for someone touring the mountains of Sarawak. Pico Iyer's anecdotal account of his trip to Thailand articulates the chaos and change that has surrounded the rapid development of tourism:

> "Welcome, my friend! Welcome to Bangkok!" cried a small man in sunglasses hurrying toward me, hand outstretched and smile well-rehearsed. Outside the dumpy single block of Don Muang Airport, the tropical dusk was thick with sultriness. On every side, lank-haired, open-shirted cabbies were whispering solicitations. Smooth-skinned soldiers were fingering $100 bills. Girls in loose shirts slouched past, insouciance in their smiling eyes. (Iyer, 1989, p. 287)

Tourism and culture are inextricably linked—one cannot find the point at which tourism impacted culture or at which point culture affected the form tourism takes in a particular place. But that does not mean that we cannot recognize that change has occurred and that local social structures have been altered by tourism practices. For example, the obvious demands for female labor in all sectors of the service economy—both informal and formal—have altered the social structure of the family. In Thailand, for example, young women, and girls have in many cases, become the breadwinners for their family (Phongpaichit, 1978), and in the Javanese village of Pangandaran, "more women have become employed in the informal sector, particularly in informal trading, and have more control over their lives as they can be at least partially economically independent" (Wilkinson and Pratiwi, 1995, p. 297). Of course, local consumption patterns in host countries have also changed, often dramatically, as Western and Japanese marketing strategies have meant that major cities have begun to resemble the modern Western urban landscape—Baskin Robbins, Pizza Hut, and, McDonald's, to name just a few, are commonplace in most of the large Southeast Asian cities.

The result of such rapid urbanization and changes to the social structure of Southeast Asian countries is that tourism policies and advertising must constantly reinvent the particular country to maintain its uniqueness and attractiveness to the tourist. If a place does not meet the expectations of the tourist, it could and often does lose its share of the tourist market (Butler, 1980). What has become apparent however, is that the state has taken more direct control of the construction and maintenance of a country's particular cultural identity in the wake of recent development trends. Docile workforces, satisfactory recreational facilities with available and amiable service sector labor, and exotic images of ethnic minorities are all critical components of state tourism policies.

In some cases, like Bali, tourism was utilized by the regional government in opposition to the central government to assert an individual identity within the context of the state. Balinese leaders and elites believe, for example, that tourism, and in particular cultural tourism, would "foster Balinese culture" and "should be used by the Balinese as a means to further their own ends" (Picard, 1993, p. 84). The Balinese have calculated the economic and political benefits and weighed these against what they perceived to be the cultural consequences. In the context of these political, economic, and social changes that began in the 1970s, the notion of what is and is not culture has shifted. Since the influx of tourism and the commercialization of Balinese cultural and physical artifacts, the emphasis of culture has been placed on "the arts" as opposed to cultural "values" (Picard, 1993, p. 89–90). Interestingly, cultural tourism in Bali has become an attraction to many Balinese as well as tourists. Thus tourism is not simply a product of satisfying the outsider but is often a well-calculated product formulated to reassert from the top down what is meant by "culture." It is critical to recognize this and be aware that many studies of tourism have often been one sided, either looking at the tourist experience or the host reaction. Instead, what should be understood is that tourism is a complex social system in itself, not separate from, but an intricate part of both local and global cultural and social systems.

It is much easier to criticize the tourism practices of the state and the capital than it is to recognize the intricacies and complexities of tourism practices related to cultural and social change. In Southeast Asia, tourism is not simply a product of consumption for the tourist; it is simultaneously a process of commodification of local culture and a means of organization and control for the state. Thus the organization of tourism in Southeast Asia has had profound impacts, allowing for the consolidation of cultural capital among an elite few. But it also allows for moments of resistance to the state and the capital by local groups that utilize their valued positions as tourism commodities to reassert and reconnect to their own cultural experiences and traditions. We must recognize that "culture" and "tourism" broadly conceived do not address the complex interaction of these two. Context within both local and global social systems, therefore, becomes the key component of any understanding of how tourism will impact a particular society. (See Box 8.2 for an example of Thailand's tourism policies during the past three decades.)

Environmental Impacts

With regard to the natural environment it would seem that governments, planners, developers, *and* the tourists themselves should all have one thing in common: clearly, it is in the best interest of all parties to protect and preserve the natural environment that is the attraction for the tourist in the first place. In other words, if the specific environmental attraction (mountain landscape, rain forest, rare species, coral reef, beach) that draws tourists is not kept in an "unspoiled" state, the tourist will simply choose an alternative destination. Unfortunately, tourism often poses a severe threat to the very environment and natural resources that are the magnets for tourists:

> "Tourism is responsible for fundamentally transforming, and in some cases severely damaging, the natural resources upon which the industry has been built over the last two decades or so. Even areas which are ostensibly being protected because of their natural beauty or ecological vulnerability, such as the Pulau Redang marine park off the coast of Terengganu in Malaysia, and the marine parks of Phang Nga Province off Thailand's southern An-

daman coastline, are currently being developed for their tourism potential." (Parnwell, 1993, p. 287)

Whereas the spatial extent of environmental attractions is great in Southeast Asia, there are a relatively small number of key centers for environmental tourism, and it is these key centers, which are spatially concentrated, that are being intensively exploited. Research on environmental impacts has demonstrated that the growth in number of tourists has necessitated expansion of the physical infrastructure in support of more tourists and that such development can negatively impact such tourism centers. Where such development is controlled, and where for example adequate sewage and trash-disposal systems are built, the impact on the environment can be minimized. Where such is not the case, the environment can of course be radically and negatively altered. Again, the classic case in the region is Bali, and as we have seen, it is an island where the number of tourist arrivals continues to increase very rapidly and construction of hotels and related facilities also continues to explode. But even a form of environmental tourism that is spatially extensive (rather than concentrated), such as trekking, has some potential negative environmental impacts. Dearden (1991), for example, in his study of trekking in northern Thailand found that such impacts included soil erosion, human waste disposal (disease), and further destruction of forests (for fuelwood and bamboo for making rafts for trekkers to use on rivers).

There are many other less well-known "key" tourism centers in the region that have become major tourist attractions because of, originally at least, the natural environmental amenities. One such place is the relatively small, 96.5 sq-mi (250 sq-km) island of Ko Samui, about 10 mi (16 km) off the coast of the Isthmus of Kra in the Gulf of Thailand. The beaches and coral reefs that attracted the first tourists have been changed irreversibly by the 220 hotels and bungalows that had been constructed by the end of the 1980s, by the regular ferry service, and frequent flights to the airstrip constructed for the tourist boom, and "by the bars, restaurants, shops, entertainment, vehicle rental facilities, prostitution, and so on which are a familiar feature of many coastal resorts in Thailand today" (Parnwell, 1993, p. 289). Furthermore, very few of the local people are benefiting from this boom; rather it is the national and international companies that are reaping the lion's share of the profits. Some one-quarter-million tourists arrived in Ko Samui in 1987 and it was estimated that more than 1 million would arrive annually by the end of the century.

In Thailand, the beach resorts of Phuket and Pattaya are other major tourist destinations, and both are suffering from major environmental problems. Pattaya, located about 50 mi (80 km) south of Bangkok, was visited by nearly 3 million tourists in 1989. A place that became important as a tourist destination in the 1960s when many U.S. GIs traveled to it for R&R (rest and recreation), Pattaya is today a place where the natural environment (beaches, water, and coral reefs) has been severely affected; for example, by the end of the 1980s it was dangerous to swim in the waters around Pattaya because of the raw sewage that was being dumped into the sea. Perhaps one positive outcome from the potential environmental disaster at Pattaya is that the government decided in 1989 to suspend large scale tourism-development proposals in

Photo 8.8 Bridge connecting Cebu City to Mactan Island, the Philippines. (Ulack)

BOX 8.1 Tourism Development on Macan Island, Cebu

Mactan Island, a small 24 sq-mi (62 sq-km) island, is connected to nearby Cebu Island and metropolitan Cebu by the Mactan Bridge. This connection and proximity to the Philippines' second-largest metropolitan area, as well as the location of Mactan International Airport (serving Cebu City but located on Mactan), has meant that population growth has spilled over so that today much of the island (most notably Lapu-Lapu City) is part of the larger Cebu metropolitan area. These factors have also been critical to the island's industrial and tourist developments. In 1979 the Mactan Export Processing Zone (MEPZ) opened and was located adjacent to the international airport (MEPZ II opened in the mid-1990s). By the mid-1990s, more than 100 factories were in operation in MEPZ and these industries employed more than 30,000 workers. With regard to tourism, there were only two resorts that catered to international tourists in the early 1980s; by the late 1990s there were more than 15 such resorts open for business along the island's 10-mi (16-km) long east coast, with others being added each year (Photo 8.5). The industrial and tourism developments on the island have translated into rapid population growth as well, as in-migrants from other parts of the Visayan Islands have moved to Mactan in search of jobs. During the 1980s decade, Mactan Island's population increased by more than 46 percent, more rapidly than that of the country (26 percent), Cebu Province (26.5 percent), the National Capital Region (i.e., Metro Manila, 34 percent), or the rest of the Cebu metropolitan area (33 percent). With the exception of the National Capital Region (NCR), Mactan's population again grew more rapidly (19 percent in both cases) than any of the other places noted during the 1990–95 intercensal period. Mactan's total population in 1995 was nearly 200,000, a high density for an island so resource poor.

Such rapid tourism development may be surprising because Mactan has very little freshwater and practically no natural beaches; rather, it is a coralline island with poor soils. The beaches that do exist are located at resorts and have been constructed with sand brought in from other parts of the Visayas. Furthermore, the roads serving the resorts on the eastern portion of the island are generally of very poor quality; in short, the infrastructure to support Mactan's rapid tourism development is not keeping pace with the construction of resorts. The vast majority of foreign tourists who visit this rapidly growing tourist mecca fly directly from their origins to the Mactan International Airport; by far, most come from the East Asian nations of Japan (more than one-third of foreign tourists come from Japan), Taiwan, and Hong Kong, respectively. The two major western origins are the United States and Germany.

Why is Mactan experiencing such rapid development of its tourist industry? Clearly, the location of the international airport (which has six weekly flights from Hong Kong, five from Japan, three from Singapore, and two from Taipei, as well as several from Australia, Malaysia, and Korea; additionally, there are a dozen flights from Manila daily and more than 15 others daily from other domestic destinations) and near-by Cebu City, the Philippines' second largest metropolitan area, have much to offer in the way of attractions. There is golf available in Cebu City (at very low cost, which for the Japanese golfer almost makes the journey worth it, given the exorbitant golf fees in Japan), shopping, historic sites, and casino gambling. Mactan also has gambling casinos (there are two, one at the airport hotel and one on a stationary cruise ship), as well as numerous opportunities for scuba diving, snorkeling, and other water sports. Furthermore, the hospitality industry (i.e., sex) continues to attract male visitors, although as we have noted the Philippine government is working toward ending the sex tours and other such activities that were well known during the Marcos years. The Cebu office of the Department of Tourism summed up its attractions as follows: "Mention Cebu and warm vivid images instantly fill one's mind. Fine sandy beaches, fantastic diving sites, five-star resorts, beautiful Cebuanas [i.e., women from Cebu], friendly hospitable people. . . sunny days and sultry nights" (as quoted in Chant and McIlwaine, 1995, p. 65). This statement is a good example of how a local agency "constructs" a tourist destination for its visitors.

19 of the country's national parks, particularly those with such fragile marine environments as coral reefs (C. Hall, 1994, p. 102).

In addition to the negative impacts that can be directly associated with the growth of tourism in key centers such as Bali or Pattaya, there are also negative environmental impacts caused by growing local populations. Tourism in the Lake Toba region of northern Sumatra, for example, is being negatively affected because the local people are burning forests near the lake and polluting the lake. In coastal areas of Thailand, the Philippines, and elsewhere, local fishermen use explosives to catch fish, thereby destroying coral reef systems. Indeed, coral reefs have been affected throughout the world; up to 40 percent of East Asia's reefs have already been lost. In Thailand, reefs are threatened by cyanide and dynamite fishing, tourist flotsam, industrial jetsam, and human sewage. A common strategy of the ASEAN governments is to promote environmental education among such populations. The message is that unless the environment is preserved, the tourism industry will not be sustainable. Additionally, the major tourism-generating nations have all passed national laws protecting the environment. Thailand, for example, passed a National Environmental Act in 1992, the

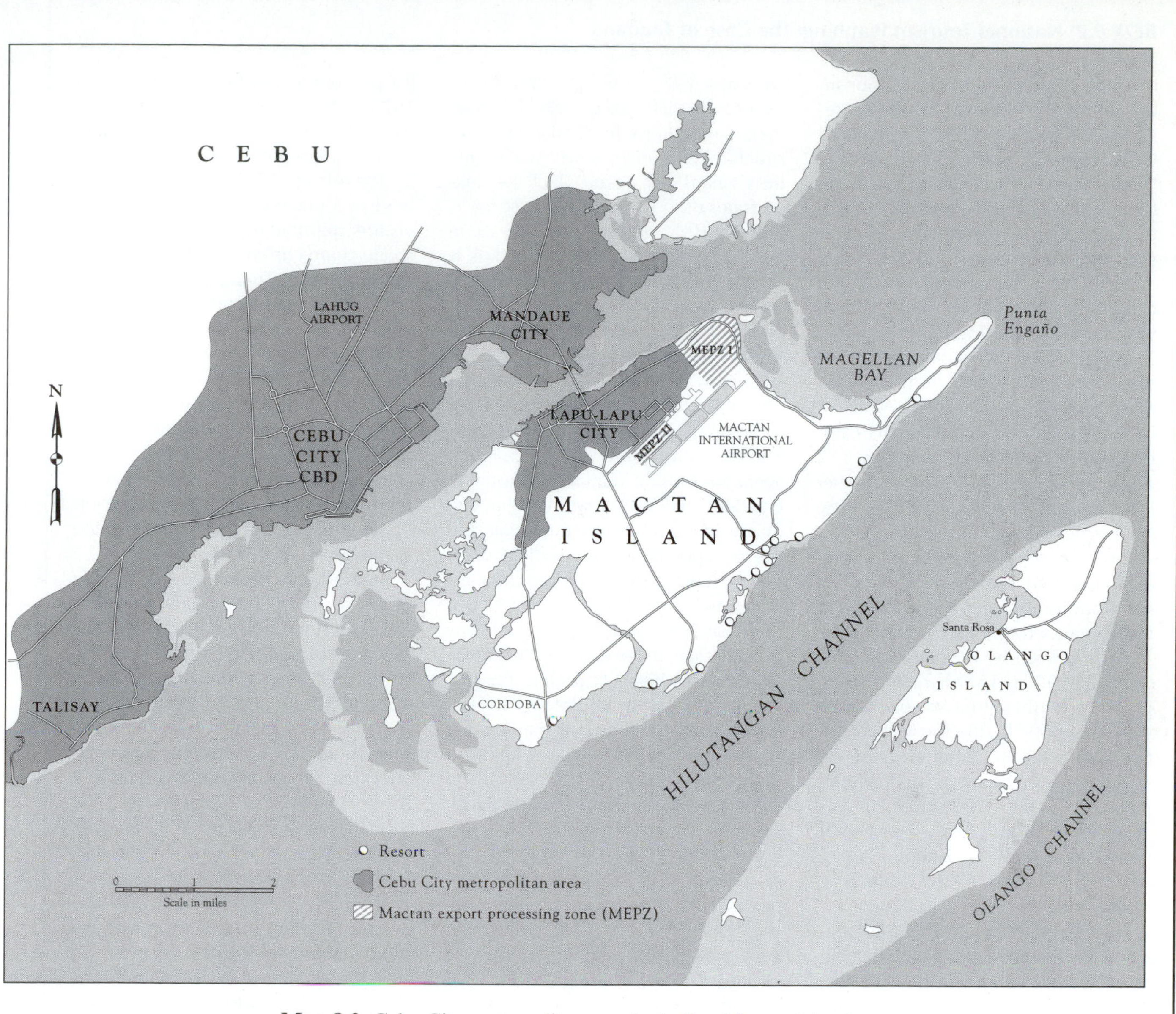

Map 8.2 Cebu City metropolitan area including Mactan Island.

Indonesian government requires tourism developers to conduct an environmental impact assessment to demonstrate that the project will have minimal negative environmental impacts; Malaysia, similarly, requires that an impact assessment be approved by the Department of Environment before a project can begin. The success of such education and legislation, however, is limited by severe constraints in implementation and enforcement. As C. Hall notes with respect to Indonesia:

> ... the problem facing the protection of the Indonesian environment is not really lack of awareness or ability to prevent pollution, it is enforcement.... In the longer term, as with many other countries in the [Pacific Rim] region, it will be the implementation of laws and regulations that will be critical in determining the environmental well-being of tourism regions (1994, p. 78).

CONCLUSION

In this chapter we have examined the complex processes associated with the development of tourism in Southeast Asia. This has been done by featuring particular case

BOX 8.2 National Tourism Planning: The Case of Thailand

In 1960, Thailand formed the Tourism Organization of Thailand (TOT), later renamed the Tourism Agency of Thailand (TAT). In 1967 Thailand held its first "Visit Thailand Year," an international campaign to increase travel to the country. Since then Thailand has held other such annual events in 1980, 1987, and in 1996, the latest coinciding with the celebration of King Pumibon's fiftieth year on the throne. Initially, the goal of TAT was to formulate a policy of development to lure tourists from the mass markets of Europe and North America; in the 1970s, TAT expanded its campaigns to include Asian countries. TAT, supported by the World Bank and other international development agencies, "recommended more public sector investment in infrastructure, i.e., airport expansion and maintenance, [and the] development of provincial sites and resorts" (quoting the *Bangkok Bank Monthly Review*, October 1972, pp. 489–90 in Thanh-Dam, 1990, p. 163). Further, tourism developers were supported by the Tourism Promotion Law that introduced a "guarantee against nationalization, against state monopoly of the sale of products similar to those produced by a promoted person, and against price control" (quoting United Nations, ESCAP, 1980, p. 99 in Thanh-Dam, 1990, p. 163).

The emergence of Thailand as a major world tourist destination coincided with U.S. involvement in the Vietnam War. Although Thailand had been a tourist destination before the mid-1960s (one of the first guide books of Thailand written in English was produced in 1904 by J.J. Antonio), the numbers did not begin to rise rapidly until the late 1960s and early 1970s (Meyer, 1989). To coincide with this rapid growth in visitors, major alterations in Thailand's infrastructure for tourism were made during the Vietnam War, a time which the United States used Thailand as a base for its growing R&R industry. The increase in visiting soldiers to areas in Bangkok & Pattaya (originally a fishing village) and around U.S. air bases in Northeastern Thailand prompted an increase not only in basic recreational tourism, such as swimming, sailing, hiking, and so forth, but helped foster a large commercial sex industry. In terms of economic impact, 1969 was the year in which U.S. forces spent the most in Thailand, an estimated US$22 million (Meyer, 1988, p. 69). This was viewed by both officials and the local population as a vehicle for securing hard currency within the cash economy.

But as U.S. troops began to withdraw from Vietnam in the early 1970s and the R&R industry began to decline, the government moved to fill the tourist void. By 1972, the state-owned Thai Airways had started its first regular round-trip flight to Western Europe, specifically to Copenhagen. The infrastructure developed for the R&R industry was rapidly converted to support the growing flow of European tourists. Asian tourists were also sought after the U.S. withdrawal, and Japan became Thailand's "main economic partner and tourist-generating country" (Meyer, 1988, p. 73). By 1972 overseas visitors to Thailand reached 594,000 per year, spending an average of US$166 per tourist. "The soldiers cleared the way, the tourists eagerly followed. Tourist arrivals increased by an astonishing and unprecedented 20 percent a year between 1960 and 1975" (Meyer, 1988, p. 73). It was estimated that Thailand was visited by more than 7 million tourists in 1996.

The role of TAT was to establish Thailand as a place of choice for overseas visitors, particularly those who would spend "hard currencies" (Figure 8.2). In the early years, Thailand relied heavily not on its environmental attractions but on its human attractions, particularly female commercial sex workers; indeed, one of the by-products of the Vietnam War was a well-established commercial sex industry. Throughout the 1970s and early 1980s Thai entrepreneurs, with the support of government officials, avidly promoted tourism as a means of development and with it tacit support of the commercial sex industry as well.

Interestingly, prostitution, or the sale of sex for money, had been illegal in Thailand since 1960. Regardless of this fact and supported by other legislation such as the Service Establishment Act of 1966 which made the "entertainment" industry a legal enterprise, the commercial sex work industry and particularly female sexual labor became a major commodity for Thailand. By 1982 tourism had surpassed rice as the number-one foreign exchange earner in Thailand, and the government had already begun to diversify its tourism industry. For example, trekking in the North into the Golden Triangle, a place infamous for its large opium, trade to examine the flora and fauna while riding atop an elephant became quite popular. Many trekkers also sought out the opportunity to meet and take photographs of local ethnic groups, known euphemistically as the "hill tribes."

studies and by the development of a framework for examining tourism geographies. Tourism places are not viewed as mere sites of supply or objects of demand but as social constructions. Place images and "myths" (Shields 1991) are produced and reproduced by capitalists and government officials in an attempt to maintain tourist flows. In addition, we have examined the ways in which tourist places are produced and reproduced through their use by tourists. No place can be a tourist place unless the tourist chooses to examine that space. When they do, they not only change what that place means but they also take with them what that place has to offer. Thus touring is not simply a ritual process of seeking and collecting images of another place; instead we also learn about ourselves and others when we travel to various sites around the world.

More specifically, tourism in Southeast Asia has become an increasingly important industry and has been

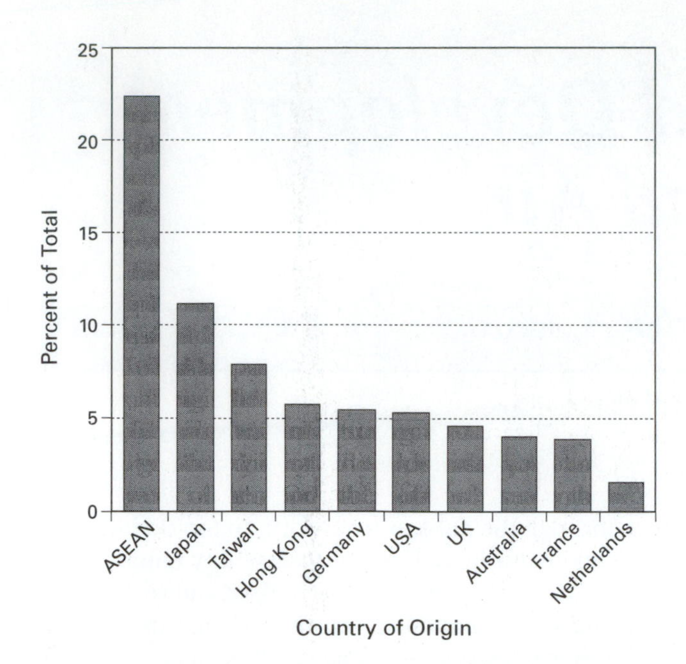

Figure 8.2 Thailand tourist arrivals by country of origin, 1992. *Source: World Trade Organization, various years.*

seen by many within and outside the region as a viable development alternative in the future. We must be careful, however, not to assume that because hotel chains and fancy resorts are being built, "development" is occurring. We must ask who is building those places and for whom. Few profits may actually "return" to the host country, and those profits that stay in-country may benefit a select few, even increasing class differences. These impacts, along with others related to cultural values, social structures, and the natural environment, make tourism studies a vital component of future geographic research. Geographers must continue to interrogate the ways in which tourism development impacts and is impacted by both local culture and global forces. Our examination of tourism in Southeast Asia demonstrates that the question of the significance of tourism to national or local development still begs for answers; what does seem clear is that diverse environmental and cultural systems still thrive and are maintained within a complex and often contradictory system of globalization.

Photo 8.9 Sign at Jerantut, Pahang advertising the wildlife of Malaysia's Taman Negra National Park, the country's oldest (established as a preservation area in 1937) and largest (1,676 sq. mi. or 4,434 sq. km.) national park. Located in central peninsular Malaysia, the dense rainforest of the park harbors many of the 59 protected species in the country including tigers and Asian elephants. (Leinbach)

9 Transport and Development: Land, Sea, and Air

THOMAS R. LEINBACH   CHRISTOPHER A. AIRRIESS   JOHN T. BOWEN

It is clear that a number of basic ingredients have been key to the process of rapid economic change in Southeast Asia. Among the most important factors have been the development of a transportation networks and related services that have more effectively moved people and goods. An underlying theme throughout this chapter, one that often functions as a hidden or poorly understood component of the development process, is the role of road and rail transport in regional and national development and the way in which sea and air transportation assume a critical function in promoting economic integration with an emerging global economy.

The important and pervasive role of transport in national and regional development has long been recognized. Economic growth potential requires a well-developed transport network and the need to maintain and upgrade transport capacity to keep pace with economic growth (Hilling, 1996; Owen, 1987). The function normally attributed to transport is mainly a supportive role, that is a "necessary but not sufficient" interactor with development. Although under some special circumstances transport may in fact act as a primary catalytic agent, it normally is no longer perceived this way. Transport becomes only one of the many, albeit critical, services and elements required for development to move forward.

Inefficient or nonexistent transport perpetuates subsistence agriculture and stalls the transformation and integration of the rural sector into the larger national economy. At the regional level, too, the poor quality or absence of transport infrastructure or services can severely constrain not only economic growth but political and social maturation as well. The exploitation of resources and the linking of markets and products underscore the fundamental economic impact that results from the provision of transport. Too often, however, transport is viewed only from a narrow economic perspective. Yet as transport geographically expands, there are far-reaching implications for the cultural, social, and political spheres of national life because just as commodities flow through the links of a network, so too do ideas and innovations. Improved spatial accessibility fundamentally alters the relative location of a place and restructures the pattern of communication, which itself is the basis for social change. Access may also result in an erosion of resistance to change so that transport may indeed be viewed as a critical factor in the spread and acceptance of innovations. Expanded mobility clearly also has an impact on the search for employment and basic needs, including the delivery of health and other social services. Finally, improved transport can bind countries within specific regions closer together so that economic opportunities are created to promote greater regional cooperation with resulting mutual benefits. In an economically dynamic region such as Southeast Asia, the potential for improved regional economic integration is substantial.

EARLY ROAD AND RAIL PATTERNS

In many ways, the early growth of the road and rail networks in Peninsular Malaysia (then Malaya) illustrates a more general explanation that applies to the other Southeast Asian countries perhaps with only slight modification (Map 9.1). With the objective of expanding trade between Southeast Asia and Great Britain in the early decades of the nineteenth century, the British influence in Malaya had spread from Penang to other Straits Settlements, especially Malacca and Singapore. The first roads in these settlements supported the important functions of the residential administrators in revenue collection and the maintenance of peace and order. These early primitive links thus aided in impressing political control but also provided access to rural agricultural areas and small towns.

By 1850, tin mining became well established in other regions of the western half of the peninsula, which were controlled by Malay rulers. Although tin lodes closest to the sea were exploited first, the mining activity gradually diffused inland as the early deposits became depleted.

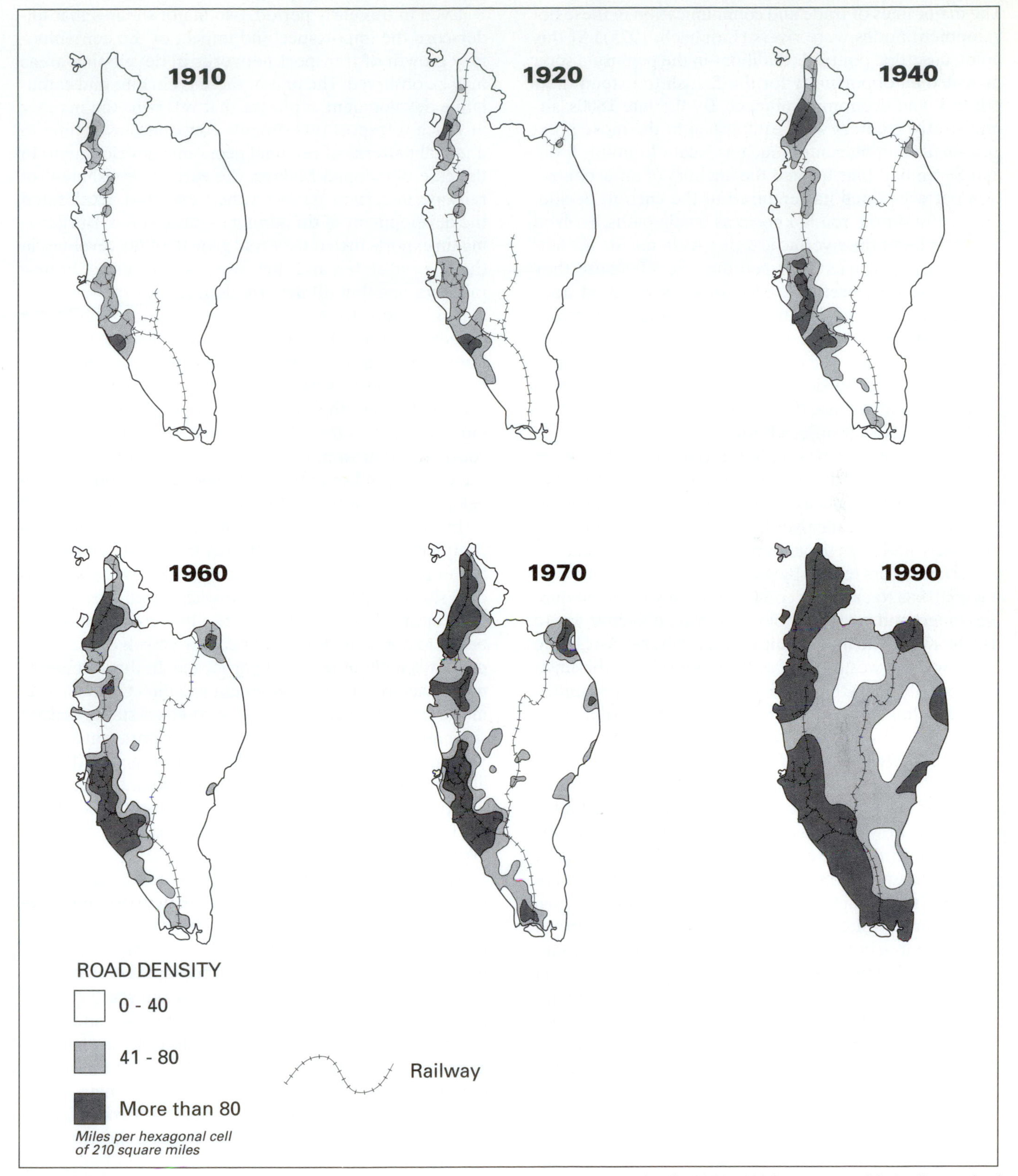

Map 9.1 Growth of road and rail network in peninsular Malaysia. *Source: Leinbach, 1975; 1990 map (unpublished) supplied by author.*

The major lines of trade and communication in these development nodes were rivers (Leinbach, 1975). At this point, unsettled political conditions in the peninsula soon provided an opportunity for the British to extend their political and economic influence. By the late 1800s, administrative centers were established in the more prosperous tin-mining centers such as Kuala Lumpur, Ipoh, and Seremban that formed the anchors of an urbanization pattern which has endured to the current period. Gradually simple roads, known as bridle paths, evolved to supplement the riverine arteries. As in the Straits Settlements, cart tracks displaced the rivers because they provided a more useful base to support settlement, agriculture, and mining. Yet the British administrators felt that a more dependable and substantial means of transport was required. The elephants that served as the primary beasts of burden on these cart tracks proved unsatisfactory because they were too slow, had a limited payload, and often suffered from tender feet.

Thus the railway was added to this simple transport system and provided a more reliable and cheaper form of haulage. The first railways were intended to complement rather than replace the river-based transport system. Short lines were laid from the uppermost navigable stretches of important rivers to nearby mines. Although the immediate goal was to establish connections only between mining centers and ports, the bolder, ultimate objective was to use the railway to link mining and agricultural districts up and down the western half of the peninsula. In this environment, the growth of the road network also responded rapidly as links emerged from railway terminals that complemented railway construction.

Thus, the twin needs of political-administrative control and mining access provided the initial stimuli for the growth of the Malayan road and rail networks. Coupled with these, however, was the strong growth of commercial agriculture in the early 1900s. The sector provided a major demand for transport services. The early crops of gambier (an astringent extract used in medicines, dyeing, and tannin), sugarcane, pepper, and cassava were grown on the estates in the Straits Settlements and elsewhere by Chinese immigrants, often in an extensive rather than an intensive manner. There is clear evidence that planting activity spread concurrently with the road system and that cultivation often declined when the distance from a road or market proved limiting. When rubber became the major cash crop, accessibility was as critical for the growth of this industrial crop as it had been for tin. Consequently, the rubber industry spread into areas already served with a transport system and, in particular, near areas leading to a major port where rubber could be exported and capital, consumer goods, and labor could be brought into the estates. By the 1920s, the railways spanned the entire length of the western peninsula and penetrated the central interior as well.

Even in this early period, two major themes that underscore the importance and impact of the contemporary growth of transport networks in developing areas may be observed. The first of these is circular and cumulative development, a phrase that refers to the manner in which transport investments contribute over time to unequal patterns of regional economic development. In the case of colonial Malaya, the early establishment of railways in certain states on the west coast accelerated the development of tin mining in those states. Burgeoning tin exports fueled the rapid growth of tax revenues in the favored states, and they were able to invest in new railways, opening up new tin deposits for exploitation. Again, revenue from taxes on tin exports grew, and more money could be devoted to additional infrastructure. Later, when rubber supplanted tin as the most important Malayan export, these states continued to have an advantage because their denser railway networks enabled more land to be planted with rubber. Consequently, they continued to grow more rapidly than most other parts of the country and over time established a cumulative development advantage, which persists to this day.

Initially road and rail transport in Malaya were complementary—as rail and river transport had been—but in the early twentieth century, the railways lost not only short-haul but long-haul traffic as well. The former was captured by road transport while coastal shipping began to dominate the latter because vessels could make port calls without paying a harbor fee. In the absence of mining and plantation investment and direct colonial rule in the more traditional Malay east-coast states, coastal shipping dominated long-distance transport centered on Singapore. Only as World War II approached did the emerging east-coast urban center of Kuantan begin to function as a collection node for east-coast products as the transpeninsular road connected east and west coasts (Ward, 1964).

As in Malaya, the development of a road and rail network in Indonesia was dictated by the need to exploit resources and strengthen the administrative reach of Dutch rule. Government investment in long-distance transport was very much restricted to the economic core island of Java. The main link in the Javanese system was the Great Post Road built by the Dutch between 1808 and 1810. The road ran from the Sunda Strait to Batavia (Jakarta) and then across the north coast via the agricultural heart of Central Java to East Java. In this case, the main link in the highway system provided essentially the route for the first State Railway (Fisher, 1964). Later, a shorter route via Cirebon, Semarang, and the lower Solo Valley was used for both the main highway and railway lines. For the most part, government transport investment in the outer islands was left to private capital and the native rulers from which private interests leased land. The resulting rail and a good part of the road network in the outer islands was a series of fragmented segments serving as transport corridors connecting plantations and mines with the various ports through which lightly processed primary products were exported (Airriess, 1995). Except for far northern Aceh where a narrow-gauge track was laid for military purposes, the three unconnected rail lines serving colonial investment enclaves on Sumatra were centered on the urban-port nodes of Medan, Padang, and Palembang. The absence of a spatially integrated railway in Sumatra and the other Outer Islands reflects a decision in the 1930s to concentrate on developing a basic road system in these islands. Nevertheless, despite substantial road building in the Outer Islands, the return on government investment was perceived as being greatest in Java, and this single island received 70 percent of first-class asphalt roads in the archipelago.

The even more fragmented nature of the archipelagic Philippines explains the decision to concentrate the limited resources for transport investment, both road and rail, on the colonial core island of Luzon. The growth of ports in the rest of the archipelago, however, stimulated the road network as well; inland populations grew and thus required connections with the small ports which serviced interisland shipping. In developing a rudimentary road and rail network, the Spanish worked to modify a strong indigenous maritime tradition (Wernstedt and Spencer, 1967), but the stability and inherent importance of this maritime emphasis is evident in the contemporary patterns of interaction.

Although Thailand did not experience colonial rule, there existed policies encouraging exposure to Western influences under King Chulalongkorn in the late nineteenth and early twentieth centuries. Prior to 1900, land transport to aid regional development did not exist because medium-distance commerce relied on an extensive network of waterways in the Chao Phraya delta. Railway construction only commenced at the turn of the century for essentially political rather than economic purposes; the railway to the Khorat Plateau in 1900, for example, was designed to bring remote provinces into easy communication with Bangkok. Similarly, the northern extension of the railway, which reached Chiang Mai only in 1921, was pushed through to prevent the north from drifting into the British Burmese orbit, as over the course of many years Chinese merchants had generated considerable trade between northern Thai towns and Lower Burma (Fisher, 1964).

The development of the railways, however, did become a catalyst for economic growth in Thailand during the 1900–1940 period. This development, along with the growth of external demand for rice, were the two most important factors in the early development drive of the country. Once the railways were built, they stimulated the production of rice and other commodities for shipment to Bangkok, thus expanding, in the course of time, the dominance of the national capital as the central transport hub. In Thailand, as in Malaysia, the early growth of roads occurred as feeders to the railway system and its main nodes. Similarly, postal and telegraph communications followed the railways, utilizing the rights of way and access that had been created (World Bank, 1959; Leinbach, 1982a).

Until the late 1920s, feeder links were the major elements in the crude, fledgling road system of Thailand. In the 1930s, the road network paralleled portions of the inland waterway and then gradually expanded to link up inland provincial centers. The several and sometimes contrasting impacts of investment in the road system are well illustrated by the Thai case. As the road network expanded over the Bangkok-centered central region between 1917 and 1967, rice yields declined steadily and new acreage expanded at only a modest rate. This negative impact on yields and productivity is perhaps explained by the spread of oppressive tenancy and credit systems that were brought by entrepreneurs into new areas along expanding transport routes. In contrast, road transport expansion in northern and northeastern Thailand had a more stimulating effect. In the late 1940s and early 1950s, highways expanded in these provinces significantly, and the result increased interconnection of market towns. This combined with better marketing systems, improved demand, and hybrid seeds did produce dramatic productivity increases. These "change clusters" illustrate how transport improvements combined with other favorable development attributes may spark faster development and improved living standards. In the case of Thailand, the contrasting impact of road network expansion in the drier uplands reflects the absence of institutional restraints found in the wet-rice environment of the lower plain (Hafner, 1972).

In French Indochina, road and rail transport was poorly developed throughout much of the colonial period. Until 1913, the only north–south road link consisted of the "Mandarin Road" constructed by the Annamite emperors to link the centrally located capital of Hue with its northern and southern domains. A single integrated railway between Hanoi and Saigon was only completed in 1936 because of the relative absence of colonial investment in the central regions of this elongated possession, rivalry between the two colonial administrative centers, and the north's preoccupation with developing economic ties to southern China. Both Cambodia and Laos lacked rail connections to Vietnam, and landlocked Laos supported only three poorly surfaced roads to the Vietnam coast by the end of the colonial period. Throughout the same period, Laos depended on the navigationally hazardous Mekong River route for connections to the Saigon region.

Although British Burma possessed a reasonably good system of north–south oriented roads and railways, inland river navigation which centered on the Irrawaddy

River provided the lion's share of internal transport. Passengers and cargo in British-owned vessels reached to 30 mi (48 km) from the Chinese border, some 900 mi (1,450 km) from the delta (Fisher, 1964). The Rangoon and Irrawaddy Valley State Railway did play an important role, however, in helping Burma to emerge as a major rice exporter. Many branch lines were built from this railway for the exclusive use of large rice mills.

TRANSPORT IN REGIONAL DEVELOPMENT

The road and rail networks suffered greatly from the devastation of World War II and the lack of maintenance in the immediate postwar period. But with the basic networks in place by the 1960s, transport began to receive more attention from government planners. The role of roads in promoting economic growth and nation building has been especially evident in the national economic plans of Malaysia, Indonesia, and Thailand. Unlike previous national economic plans, these policy pronouncements explicitly recognize the important role of roads as a development tool.

Rural Development in Malaysia

In 1970s Malaysia, transport was given a seminal role in the New Economic Policy which sought to expand Malay or *bumiputra* participation in the economy by expansion of the road network in rural Malay regions as well as the promotion of Malay participation in the road transport industry. For example, through Federal Land Development Authority (FELDA) and other agriculturally based regional development schemes in Johor, Pahang, Negeri Sembilan, Kedah, Perak, and Terengganu, more than 1,490 mi (2,400 km) of development and feeder roads were built between 1971 and 1980. These roads provided internal village access and also linked settlement clusters to the main road network. Roads were also extended within the Muda and Kemubu rice irrigation schemes of Kedah and Kelantan. The improvement of rural access in specific development schemes has been supplemented by other efforts. Most noteworthy is the effort in Malaysia to accelerate in situ agricultural schemes through the improvement of farm-to-market roads. The rural road development programs form an important part of the overall integrated rural-development package for raising the productivity and incomes of rural people.

Overall, the implementation of road expansion programs resulted in nearly a 30 percent gain in the total road network across Malaysia by the early 1980s. Although all states benefited, low-income Kedah and Kelantan received the greatest attention. With a relatively low population density, compact scale, abundant land for development, and strong revenues generated from oil palm, tin, and rubber, the improvement of rural access in peninsular Malaysia has been the most dramatic in the region save for Thailand (Barwell, et al., 1985, p. 5-20).

Indonesian Rural Road Improvements

Indonesia possesses a very low road density serving a very high population ratio when compared to other countries in Southeast Asia: for example, the Philippines and Thailand each have over four times the road density of Indonesia. It is estimated that in the early 1990s, approximately 20 percent of Indonesian villages had no vehicle access to main highways; outside Java, 30 percent of the villages had no such access. It has been recognized for some time that the lack of a serviceable feeder-road network remains a major obstacle to national cohesion, civil administration, and economic development (Leinbach, 1983a). On the last point, an acute need in Indonesia has been to improve rural agriculture through extension and support services and more efficient marketing. A related problem has been the inadequate delivery of basic social services (Leinbach, 1981, 1983b).

The need for expanded road investment was recognized during the Third National Plan (1979–84) as an explicit component of the country's broader rural-development strategy. The greater attention given to rural roads builds on earlier efforts to accelerate their development. For example, a Food-for-Work Program implemented during the 1969–1974 period provided food to villagers in return for labor on a variety of rural infrastructural projects that included improvement of some 2,670 mi (4,300 km) of rural roads. Hampered by the inherent problems of in-kind compensation, the revised 1974 Rural Works Program was implemented based on supplemental cash incomes for underemployed labor in poor subdistricts. The construction of rural roads and other infrastructural improvements was part of the government's long-term goal of improving food production, employment opportunities, and income distribution. Although the program achieved some measurable successes, a number of weaknesses in project implementation were encountered as well (Box 9.1).

Rural Roads in Thailand

Thailand possesses an extensive rural road system, comprising nearly 66,447 mi (107,000 km). More than 80 percent of this total fall under the jurisdiction of the Ministry of the Interior through its various agencies. As in Malaysia and Indonesia, some of the rural roads have been built by special agencies in support of specific regional development projects. An example of this is the office of Accelerated Rural Development (ARD), which constructs rural roads intended to promote economic development and aid in the counterinsurgency activities of the Thai gov-

BOX 9.1 Road Improvements and Rural Development in Indonesia

There exists a persistent concern among transport planners that the economic impact of road projects in rural areas largely favors and indeed increases the wealth of both urban-based business interests and landholding rural elite at the expense of the rural poor. Greater spatial access is the essential issue in road development, and the question is: Who primarily benefits from the economic opportunity afforded by road development? In 1978, a survey of 36 rural feeder roads was conducted to assess the economic and social impact of these projects completed under the 1974 Rural Works Program. Based on 2,500 structured interviews of individuals living within 3 mi (5 km) from road projects, this postconstruction survey yielded some important results critical to understanding the role of transport as a catalytic agent of rural development (Leinbach, 1983b).

The most basic economic impact was felt by the employment and income generation among the families engaged in road-improvement labor. As expected, basic food and other consumables accounted for the major purchases earned from these wages. A secondary but significant amount of expenditures was dedicated to purchases directly related to development. In six villages in West Java, for example, 7 percent of labor earnings were spent on purchasing fertilizer and 6 percent of earnings on repaying family debt. Equally as important is the fact that in 20 of the 36 project areas, more than 60 percent of project wages were spent locally.

In almost every project area, incomes increased as the ability of farmers to market their crops in a once-distant market center improved. Newly established and regularly scheduled transport services, usually in the form of minivans, were utilized by middlemen marketers who bought local produce to be sold at market. The impact of road access, then, was to transfer the point of sale closer to the farmer. It is not definitely known, however, whether the middleman or the farmer benefited most from the increased spatial access resulting from road construction or improvement. The spatial mobility of farmers increased dramatically once the feeder road was in place. According to 35 percent of survey respondents, trips to market and visits to relatives increased, and some 65 percent of respondents indicated that the frequency of visits to health centers increased as well. The flow of information as a result of improved accessibility also increased. Information regarding credit sources, medical care, off-season employment opportunities, agricultural innovations, and crop prices, for example, increased as roads allowed greater face-to-face, interpersonal exchange. In virtually all project areas, respondents indicated that outside sources were most important in gaining beneficial information.

Despite the seeming success of many rural road improvement projects as an important catalyst of development, a handful of problems were identified. One is the inadequate selection criteria in that, too often, roads serve too few people. Some projects were less than cost effective because roads were too close to parallel linkages that provided greater access for the bulk of the local population. In addition, roads were sometimes positioned to benefit influential individuals or government enterprises and thus did not directly serve the rural poor. Perhaps the most serious shortcoming is the absence of maintenance, primarily because monies were not included in the initial project budget. Although these various problems are inherent to the planning process, success is furthered as long as rural road projects are designed in conjunction with other development plans. If so, roads assume an important ingredient in pushing rural development forward.

ernment (Jones, 1984). Normally, in the case of resettled areas, such roads would gradually come under the control of the provincial office of the Department of Highways.

The regional distribution of rural roads (northeast 45 percent, north 22 percent, central 21 percent, and south 13 percent) does fit the population distribution reasonably well. Greater-than-expected road densities are accounted for by the political significance of the northeast and rice production in the central region. The Thai concern with inequalities in rural incomes, as in the other Southeast Asian countries, is manifested in efforts to improve the distribution of rural roads. Much of the inequality is the consequence (and cause) of economic differences among regions. The northeast is one of the poorest regions, where more than 40 percent of the population are below the poverty line. A major poverty-reduction strategy has been aimed at the northeast and in this effort, land and road development have played major roles.

Although the rural road system in Thailand is one of the most extensive and well placed in the region, many of these roads are seasonal and become virtually impassable in the wet season. While some improvements and upkeep have been undertaken through employment generation and village and regional development programs (especially in the northeast), much more must be accomplished to achieve the full use of this significant communications network. The bulk of the roads receive little maintenance; this deficiency is critical in any attempt to attract agroprocessing and other small industries to start up in rural areas. Moreover, the absence of an improved road network means that health and education services are underutilized. Coupled with this, it should be noted that there is also a great need to upgrade provincial roads. This is especially true of the nearly 7,452 mi (12,000 km) of unimproved roads that provide rural access.

Developmental Impact

Rural road construction has come about as a result of the need to improve incomes and expand agricultural productivity in rural areas. In Indonesia, Malaysia, the Philippines, and Thailand there is also interest in providing transport access to the poorest groups and areas. A major question, then, concerns the impact of road network expansion and upgrading in light of the considerable expenditures that have been dedicated to the transport sector. It is clear that without access provision in rural areas, development progress will falter and stall. The primary issue therefore is not *whether* road networks should be provided but, rather, *where* and *how*. Evidence from numerous surveys over the past decade clearly shows that in many cases, improved transport has significant and positive impacts, particularly with reference to improved crop marketing and production, as well as greater access to educational and health-care facilities (Hughes, 1969; Leinbach, 1983b).

Negative impacts of rural road development, however, also exist. Too often, for example, there remains a strong spatial and distance bias in rural development where needs are seen only near urban areas and along main roads (Chambers, 1983). Road expansion may also induce rural–urban migration through which talented individuals migrate to urban areas so that the brightest hopes in village development take their skills to urban areas. In some cases, only the wealthiest individuals benefit from improved access; thus the income gaps in rural pockets actually widen instead of narrow (McCall, 1977). Entrepreneurs with enough capital to buy a light truck or bus are able to achieve rapid income growth and often join a small middle class.

In the last analysis, the income-distribution effects of rural road provision and upgrading are dependent on agricultural potential in the area, the bargaining power of rural labor, and the provision of personal and goods transport services. Road investment will be most successful when it is accompanied by complementary investment in rural electrification, irrigation, and credit for farmers and extension services (Leinbach, 1995). Critical too, but often overlooked, is the long-term or periodic maintenance program allowing for the year-round use of roads (Howe and Richards, 1984). Southeast Asian governments have often used funds for road expansion rather than for maintaining the road network.

Appropriate Transport Facilities

A widely held concern is that improvements in transport infrastructure, services, and facilities in the past have been inappropriate to, and ineffective in matching, the transport needs of the poorest people. There exist more appropriate transport technologies that can better meet the needs, resources, and situations of the vast rural populations (Hilling, 1996; Barwell and Howe, 1979). The transport needs of the rural poor may be viewed in terms of on-farm and off-farm activities. In the former case, tasks related to the movement of seeds and plants, the application of fertilizers and insecticides as well as harvesting dominate. Nearly all of these needs consist of relatively small loads (under 330 lb, or 150 kg) moving over short distances (under 6.2 mi, or 10 km). Travel for water and firewood is also important, and their collection efforts alone often consume three to six hours per day. On-farm transport requirements are a clear burden and are often major constraints on productivity and income generation. Off-farm activities refer to those movements between the farm and the market or town. Farmers need reliable and cheap services to acquire inputs and to market agricultural commodities.

Unfortunately, most efforts at improving rural transport are synonymous with road access by conventional motor vehicles. To improve transport delivery, a wide variety of vehicles must be encouraged that are more suitable, given the wide variations in incomes, road surfaces, social and economic systems, and topography. Examples of these basic vehicles range from improvements in human carrying devices to pedal-driven and simple motorized vehicles. A major advantage of these simple vehicles is that they may be used on less costly roads that are ideal for low-volume and remote situations that are now poorly served.

Higher-Order Road Development

While the rural road network is currently a major concern of transport planners, higher-order or longer-distance linkages are being forged to provide a new thrust to basic connectivity in the highway network (Map 9.2). At the same time, these major development or penetration links are intended to stimulate growth in areas where poor transport was a binding constraint on regional development.

In Indonesia, the Trans-Sumatra Highway is a prime example of such a linkage. The concept of a cross-insular highway developed in the 1970s with the intent that it would be a catalytic force and stimulate development along its length (Map 9.3). The highway, which stretches from Banda Aceh to Tanjungkarang in Lampung, is actually a linkage of roads built at various times, mostly during the Dutch period. Many sections of the highway have recently been upgraded and rehabilitated using funds from the World Bank. Government funds have been used on completing gaps in the 1,675-mi (2,700-km) highway, especially in support of the nearby transmigration and plantation areas. The final 192-mi (309-km) link in South Sumatra was completed in 1985. Although the catalytic development impact potential of the highway was overestimated, the goal of a through highway linking

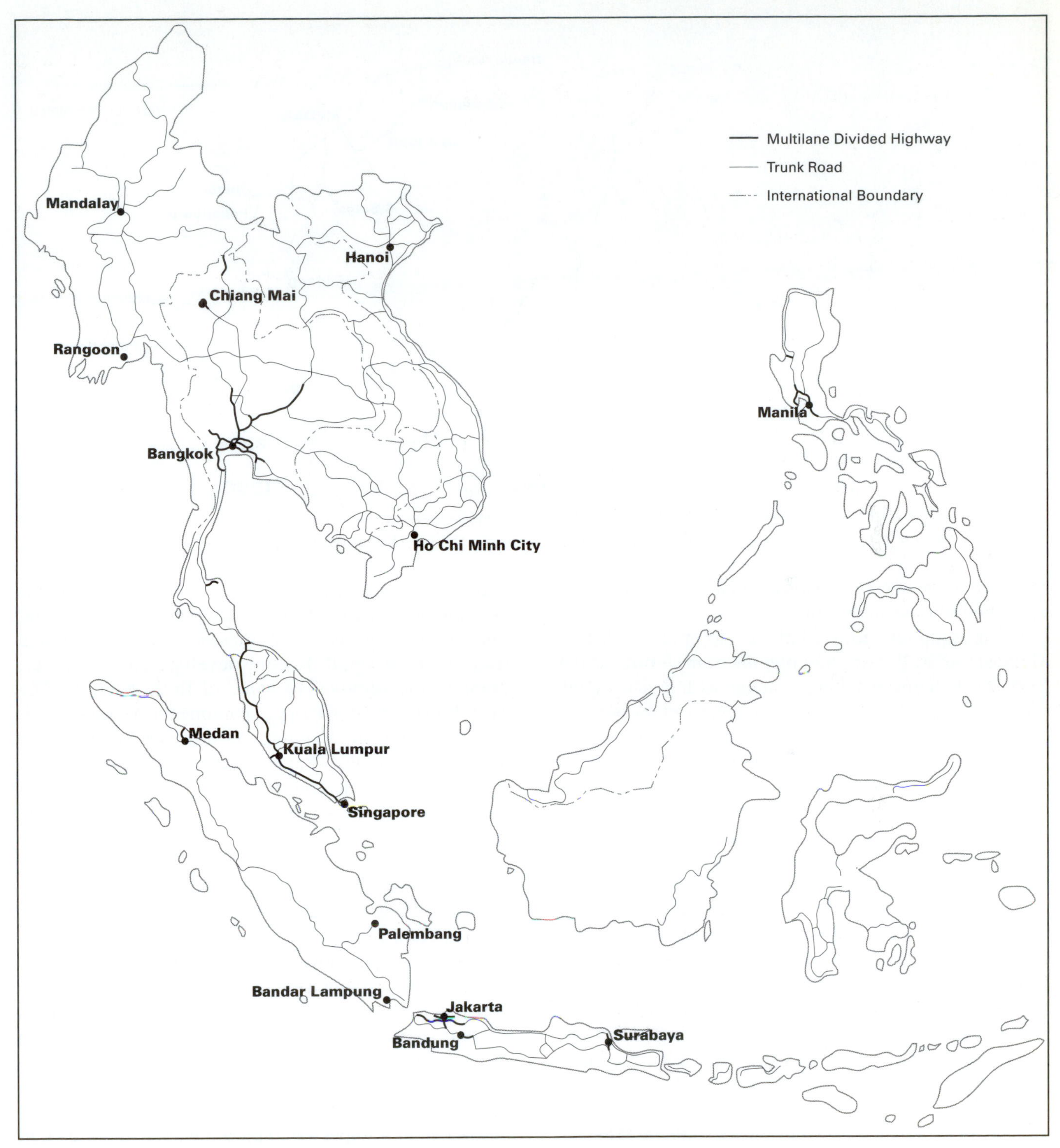

Map 9.2 Regional highways in Southeast Asia.

the major towns of Sumatra has merit. There is little doubt that the improved access will lower costs, generate some new development, and improve trading activities within the island and the major market areas of Java. Indeed with an efficient vehicular ferry link between Sumatra and Java, some interisland shipping trade has suffered. The planned Trans-Sulawesi Highway, which will link Ujung Pandang with Menado on the north coast, is another higher-order road project promoting long-distance access. A similar highway is planned for Kalimantan and, when completed, will dramatically improve interior road access in this forested and road-poor region. The Pan Philippine Highway, an 88,606-mi (1,430-km) route that serves as the primary north–south artery in the country

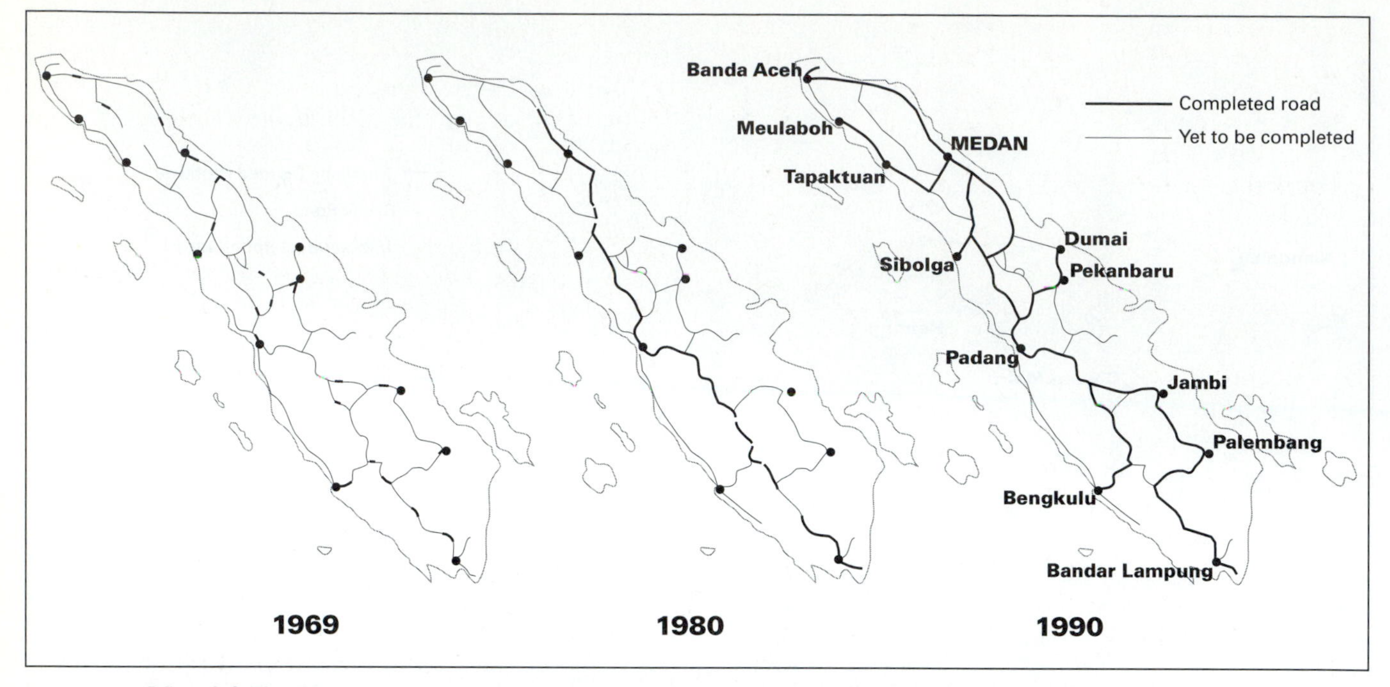

Map 9.3 The Trans-Sumatra Highway development stages. *Source: Indonesia Development News, vol. 5.*

with ferry links between islands, is analogous to these developmental links in Indonesia.

Similar higher-order road developments exist in Malaysia as well. For example, the completion of the East–West Highway linking Kelantan with Penang in the north of peninsular Malaysia was carried out for both security and development purposes. In this respect, commercial traffic from the Kelantan region has been provided with a much shorter route to the major port of Penang. Another major development road in Malaysia is the Kuala Krai–Gua Musang–Kuala Lipis road in Kelantan and Pahang. The construction of this road has provided access to southern Ulu Kelantan which had been accessible only by rail and seasonally by small river craft. The link is intended to stimulate development of timber resources and integrated timber-based industries in both southern Kelantan and the Kuala Lipis region of Pahang. Connections to port facilities on both coasts have existed since the road's completion in 1983. A similar link, the Jerangau–Jabor Highway, has also been constructed in Terengganu. Finally, a series of new development road projects were completed by the mid-1980s in both Sabah and Sarawak. Most conspicuous were segments that allowed a land transport link between Kuching and Miri and the improvement of various links of the Pan-Borneo Highway.

The developmental impact potential of transport in mainland Southeast Asia is perhaps most conspicuous in the Mekong Basin. Although the Mekong River is an important artery for short distance commerce, many navigational barriers such as falls and seasonal low water levels presently preclude the river from serving as an avenue for long-distance trade. Cross-boundary trade is also hampered by the lack of interconnecting roads and railways, and it is because of these barriers that regional trade remains small. Recent developments, however, point to the removal of some of these barriers. The Thai–Lao Friendship Bridge spanning the Mekong at Vientiane opened in 1994 and now provides isolated Laos limited access to the Thai highway network and beyond. Although economic opportunities for Laos can only increase, some negative impacts most likely will include monopolization of trade by Thais, increased deforestation in Laos, and, from the Lao perspective, cultural pollution originating from more-developed and "worldly" Thailand. Plans also call for a handful of bridges linking Thailand and Burma, but the resulting trade volume is very much dependent on a more open economy in Burma. With a population of more than 93 million, the potential for transport to act as a catalyst in development is great, as this area (including the southern Chinese province of Yunnan) is one of the few remaining marginal areas in Southeast Asia. Despite the currently immature markets in Laos, Cambodia, and Burma, forging physical links is expected to promote deeper, more open and regulated trade (as opposed to the illegal commerce that is now rife across the river).

In the lowlands, the countries have agreed to conduct feasibility studies for upgrading the transport infrastructure that already links Thailand, Laos, Burma, and Vietnam. Among the highest-priority projects is the upgrading of a primary road linking Bangkok, Phnom Penh, and Ho Chi Minh City and another route linking the Vietnamese coast at Danang through Laos to northeast Thailand. This latter route allows Laos more direct access to the sea and offers economic promise for regional development in central Vietnam. A 1996 agreement between the Thai and Laotian governments to construct a bridge across the Mekong River at Savannakhet, Laos, provides greater assurances that this route will function as a catalyst for increased trade volume. Indeed, this is one of several bridges being funded by Thailand to promote cross-border trade.

The primary north–south highway in Vietnam is only now receiving funds for upgrading; much of domestic north–south trade continues to be carried out by coastal trading vessels. With few exceptions, these new or upgraded long-distance high-priority linkages throughout Southeast Asia are simply two-lane, unlimited access highways rather than the multilane transport corridors characteristic of developed countries.

To better accommodate dramatic increases of traffic volume and eliminate transport bottlenecks associated with rapid economic growth, a handful of Southeast Asian states have constructed a network of interurban toll roads (Johansen, 1989). Unlike their counterparts in industrialized countries, however, Southeast Asian governments have chosen to operate toll-roads through the private rather than the public sector. In the late 1970s, Indonesia was the first country to adopt the toll-road strategy; the Jagorawi Highway linking Jakarta with Bogor was the earliest interurban facility. Toll roads also extend to the west of Jakarta and south of Surabaya. By the late 1980s, 175 mi (280 km) of roads were tolled and half of those served the greater Jakarta region. Outside Java there is only one toll-road bypass and this links the port of Belawan with Medan in North Sumatra. By the mid-1990s, an additional 466 mi (750 km) of toll roads comprising 18 separate projects were in various stages of planning or construction throughout Java. Consideration is also being given to the construction of toll bridges between Java and Bali and Java and Madura.

A similar strategy has been used in Malaysia to construct highways. Under the auspices of the Malaysian Highway Authority, approximately 578 mi (930 km) of roads have been upgraded and function as toll segments throughout peninsular Malaysia. Most noteworthy is the 526-mi (847-km) North–South Highway completed in 1993 that reduces driving time between Kuala Lumpur and Singapore from six to four-and-one-half hours. The North–South Highway illustrates both the attraction and dangers of private highway development. Spurred by monetary incentives, the developer finished the highway way ahead of schedule. Yet the project has been surrounded by allegations of scandals, specifically that the developing company, which is indirectly owned by Malaysia's governing party, was given unfair advantages in the bidding process and now enjoys a contract that virtually guarantees its profitability. Connected to the North–South Highway are a number of toll roads around the Kuala Lumpur urban region and a bridge segment that links Penang Island with the peninsula.

In the Philippines, toll-road construction has been less aggressive, with only two toll roads in place by the late 1980s. The first was the South Expressway stretching between Manila and Batangas, but the tolls recently have been removed. The second is the North Diversion toll road linking Manila and Papampanga; this 62-mi (100-km) link reaches in the direction of the northern resort and trade center of Baguio. Although the same needs occur

Photo 9.1 Toll road, Jagorawi Highway, Java. (Leinbach)

in Thailand and Vietnam, the strategy of toll roads has not received as much attention in these countries.

RAILWAY DEVELOPMENT

In general, most railways in Southeast Asia were built prior to road transport and enjoyed almost a monopoly position. The function was to provide basic transport for goods and people where few, if any, reliable alternatives were available. Under these conditions, some unjustified or misinformed investments took place. As the road network expanded and competition increased, the financial position of the railways gradually deteriorated. The response to this development in some cases was to regulate the transport sector, but, more commonly, direct subsidies to the railways were provided. Thus government officials and not commercially oriented managers were given control of pricing, investments, hiring, and marketing. This situation resulted in the further deterioration of the financial positions and quality of operations. Generally, the period from 1970 to the mid-1980s revealed an overall decline (or at best stagnation in some systems) in the rail share of freight and passenger traffic. Nonetheless, the railways may still play an important role in the transport sector of the various countries under review. To be successful, railways must provide efficient services when compared to road transport. This is dependent on both institutional and market factors, which include the cost-effectiveness of operations, the structure of the network, the ability to respond to new needs, and the elimination of unprofitable links and services. As illustrations of the development role of railways in Southeast Asia, we have selected the examples of Indonesia, Malaysia, and Thailand for further discussion (Map 9.4).

Java-Centered Development in Indonesia

The first railway line in Indonesia was a 16-mi (26-km) section from Semarang to Tanggung in Central Java. This link, built by the Dutch colonial administration in 1864, initiated a rail construction boom across Java and Sumatra that involved 12 private railway companies along with the colonial government-owned lines. Between 1880 and 1900, rail lines were constructed between Jakarta and Solo via Bogor, Bandung, and Yogyakarta, as well as between Cirebon, Surabaya, and Probolinggo on the north coast of Java. At the turn of the century, systematic rail construction of port-centered lines commenced in south, west and northeast Sumatra servicing plantation interests. With government energies focused on Java, rail construction in Sumatra was almost exclusively through private capital.

In 1950, a consolidation among the several companies was arranged whereby the Indonesian government assumed ownership of the entire system. The Indonesian State Railway now encompasses about 4,220 mi (6,800 km). Seventy percent of this network is found in Java with the remainder consisting of three disconnected lines in Sumatra. As a result of inadequate maintenance and equipment, the railways have gradually deteriorated with the passage of time. Although a rehabilitation program has been applied to the railways, it was started later than that in the road sector and took much longer to carry out.

The railways carry more than 40 percent of all passengers over distances exceeding 310 mi (500 km). Java rail carries most (82 percent) of the passengers, and of this, 27 percent is accounted for by the larger Jakarta-centered urban region of Jabotabek. Merak, Jakarta, Bogor, Bandung, Semarang, Yogyakarta, and Surabaya are the busiest rail-passenger nodes in Java. Passenger traffic for Java was 44 million in the early 1990s and 3 million for Sumatra in the same period. In part reflecting the colonial bias for freight haulage, more than half of the freight total is carried on the three systems in Sumatra. The North Sumatra system hauls largely palm oil and rubber from plantations in the surrounding region to Belawan. Coal and cement in connection with the Indarung cement plant in Padang are the main commodities carried in West Sumatra. The South Sumatra system distributes oil from Palembang, coal from Baturaja, and cement and other goods in the Palembang hinterland. In Java, the distribution of bulk liquid fuel and fertilizer from the major cities on the north coast account for more than one-third of the total freight traffic.

Some significant structural changes have occurred in the pattern of railway traffic between 1970 and the early 1990s. In general, average passenger journey lengths have increased. In addition, the railway has increasingly become a carrier of bulk commodities and long-distance passengers. Agricultural-commodity hauls have declined in volume, while a few major bulk goods such as petroleum, cement, and fertilizer have assumed a larger proportion of the total. In spite of the growth of traffic, the system's financial performance has declined sharply. As an illustration, petroleum products, the biggest single traffic item, have been hauled at rates below the average tariff because of government subsides for state-owned petroleum products.

It is clear that the long-term future of the railway lies in its competitive position for bulk commodities in particular, though nonbulk commodities are important as well. Under a proposed development plan, bulk commodities would be carried by contract, and the rates charged would cover long-run variable costs. The government has also decided to improve its nonbulk services to compete better with road transport and that tariffs here must also cover the long-run variable costs. Trans-

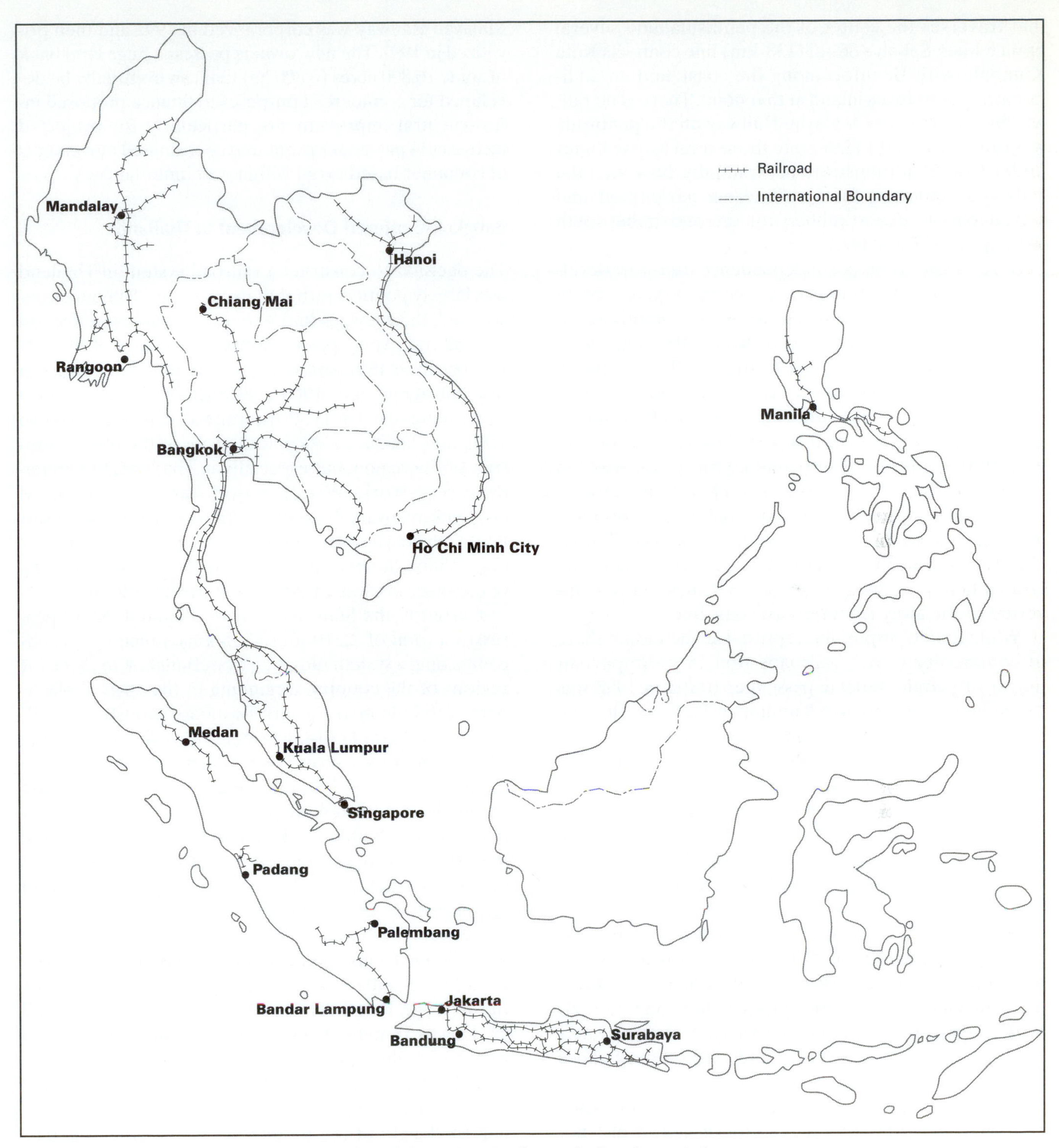

Map 9.4 Regional railways in Southeast Asia.

port planners also believe that the railway can be an economically efficient carrier of medium- and long-range passengers. Higher-class and long-distance passenger services are those that exhibit the strongest growth, but these users are not among the poorest elements who need the subsidy.

The West Coast of Peninsular Malaysia

The railroad network of Peninsular Malaysia measures 1,294 mi (2,084 km) of track and consists of a west coast line that connects to the Thai state-owned railways and Singapore's Mass Rapid Transit System, an east coast line that traverses the center of the peninsula, and several branch lines. Sabah's 86-mi (138-km) line connects Kota Kinabalu with Beaufort along the coast, and an additional segment turns inland at that point. There is no railway in Sarawak. The Malayan Railway on the peninsula was originally built to integrate the several Malay States under British administration. Gradually, however, the railway's economic function became paramount and served to move export rubber, iron ore, and timber south to the port of Singapore.

Several important postindependence transport developments have had an impact on the railway system. In the late 1960s, the Malaysian government commissioned expansion at the major ports of Klang, Penang, Johor, and Kuantan to attract port traffic that had formerly moved via rail to Singapore. Lengths of haul were lowered significantly, and the railway's freight tonnage growth stagnated during this period of agricultural and industrial expansion. The strong highway construction program in the peninsula clearly has had an adverse effect on the railway and has allowed highway operators to increase their share of the land transport market. Particularly significant in this context was the opening of the Ulu Kelantan highway in the early 1980s that has diverted traffic away from the east coast line.

While road transport has captured an increasing share of commodity traffic, railroads remain an important mover of people. Indeed, passenger traffic in 1992 was 7.4 million in contrast to 5.9 million in 1966, and the passenger component of operating revenues increased from 28 percent in 1967 to more than 40 percent in the early 1980s. Evidence shows that passenger traffic will continue to account for a major part of operations, particularly in light of planned improvements such as wider-gauge tracks accommodating faster trains, double tracking, and new hardware to be utilized between west coast urban centers. The railway goods traffic reveals much slower-paced growth in relation to the country's sustained export boom during the past 20 years; in 1966, 3.4 million tons moved by rail, and that figure only increased slightly to 3.6 million tons in 1980 and 4.3 million tones in 1991. With freight traffic revenues declining by about 3.3 percent per annum between 1980 and 1985, railway goods traffic is clearly stagnating.

These revenue declines are attributed to increased competition from road transport operators in a short-distance transport environment, the capacity constraint of the railway, and obligations such as operating unprofitable passenger services and branch lines for social and political reasons. As elsewhere in the region, the Malayan Railway is attempting to operate on a commercial basis within a competitive environment but is inhibited by too many financial constraints and nonfinancial obligations. The near-term future, however, may witness a dramatic improvement in the railway's operating environment. Indeed, the Malayan Railway was corporatized in 1992 and then privatized in 1997. The new owners possess a huge land bank of some 16,800 acres (6,800 ha) that can eventually be developed for commercial purposes to finance proposed infrastructural improvements, particularly for projected increases in passenger traffic and coordinated movements of container-based cargo within port hinterlands.

Bangkok-Centered Development in Thailand

The decision to construct a railroad system in Thailand was largely political rather than economic. But once constructed, the railways had a strong stimulative effect on agricultural export production. Construction of the first line began in 1892 and reached Nakhon Ratchasima in the central region by 1900, a distance of 165 mi (265 km) from Bangkok. By 1930 the railway had increased to 1,777 mi (2,862 km). As was the case in the other countries of the region, but especially in Thailand, the immediate post-World War II period was a time of rapid economic recovery. Total rail traffic soon exceeded prewar levels, causing a severe strain on the system. A considerable rehabilitation of the rail network and the beginning of dieselization was carried out in the late 1950s.

Currently, the State Railway of Thailand (SRT) possesses a total of 2,319 mi (3,735 km), much of this encompassing a system radiating from Bangkok to all major regions of the country. Beginning in the early 1980s as part of the planned Eastern Seaboard industrial development, a 81-mi (130-km) single track line connecting Chachoengsao, a station 37 mi (60 km) from Bangkok, to the deep-sea port of Sattahip was constructed. Because of the SRT's persistent inability to service Thailand's export boom, a double-track system linking Bangkok and the new Eastern Seaboard port of Laem Chabang was under construction but has been delayed as a result of the general economic situation.

Between 1975 and 1993, passenger traffic grew from 62 million to 87 million, an annual growth of only 2.2 percent. In the same period passenger-km grew faster for the average length of journey increased from 36 mi (92 km) to 57 mi (169 km). Passenger-kilometers in 1993 were 14.7 million having grown from 11.6 million in 1990. Clearly, rail-passenger traffic is the greatest of all the countries in the region. With a rail system just over one-half the length of the Indonesian system, the Thai passenger traffic is nearly double that in Indonesia.

Freight traffic between 1975 and 1993 grew at an annual rate of 3.1 percent (from 4.8 to 7.5 million tons), in contrast to an increase of 5.6 percent between the years of 1975 and 1980. An average increase in freight tariffs of 73 percent in 1981 coupled with the international recession led to a decline in freight traffic to 5.3 million tons in 1983. The current situation in the Thai economy will certainly have a similar negative effect on both passenger and freight traffic. The main commodities carried are petroleum and industrial products, building materials, cement, rice, lumber and logs. The length of haul, on average, is about 298 mi (480 km) (*Thailand, Statistical Yearbook,* 1994).

A considerable problem in Thailand, and in fact elsewhere in the region, is the continued subsidy of uneconomic lines and services. Although the three main lines of the State Railway of Thailand could be profitable, there are six to eight branch lines in addition to the Bangkok metro commuter services where profitability is very unlikely. These branch lines have been operated for security reasons and also to encourage regional development. This issue, of course, is not exclusively a Southeast Asian problem for it applies to other developing countries and indeed to several Western nations as well.

Coupled with this problem, there is a broader financial concern affecting the SRT. Presently, revenues are not sufficient to meet operating expenses, debt interest, and depreciation. The problem developed in the late 1970s and is rooted in the dilemma of the political and social impacts of upward tariff changes. In light of these budget difficulties, the government has, like Indonesia, formulated a policy of reducing subsidies by permitting price increases for certain goods and services. However, in the case of the SRT, such increments must be accompanied by other measures. Improved efficiency in train operations, maintenance, and fuel consumption as well as the possible leasing of railway-owned land to private concerns in Bangkok are possible solutions. Continued modernization and maintenance are badly needed at the same time to cope with the growing traffic demand. Thus, the current and near-term period for the SRT is a watershed in which revenues must grow faster through a variety of measures if the railway is not to develop a very deep and irreversible dependency on government subsidies. The government continues, however, to exhibit direct control over rail transport as the previously mentioned high-speed rail project to Laem Chabang is 80 percent financed by public funds.

URBAN TRANSPORT

Before leaving our discussion of land transport and development, devoting attention to land transport within Southeast Asian cities is certainly warranted. Although the countries of Southeast Asia are not as highly urbanized when compared to other developing world regions, urban transport provision is a central component of development policies and urban planning budgets because it functions as a critical means of circulating people and goods as well as a significant generator of employment and income. To service the wide array of urban transport needs, a diverse and often uncoordinated system of urban transport modes include water taxis, human and motor-powered bicycles, motorcycles, buses, taxis, trains, private autos, jeepneys, and high technology mass transit systems.

Public Sector Transport

Because of their riverine environments, water transport has traditionally been an important form of passenger movement in cities such as Saigon, Bangkok, Rangoon, and Bandar Seri Begawan. The importance of water transport has declined, however, even in Bangkok with its many canals. Here land transport has gradually supplanted the traditional form of transport. The lone exception perhaps is Bandar Seri Begawan where some 53 percent of the city's inhabitants reside in the stilt dwelling suburb of Kampung Ayer which requires a short-distance water-taxi shuttle service.

Another form of more traditional urban transport that is also declining in popularity is human-powered, three-wheeled bicycle seating two persons in a covered seat to the rear or side. In use in many cities by World War II, this mode of transport is mostly used by lower-income patrons for short trips such as conveying school children or housewives making trips to the market based on negotiated fares. Collectively known in English as pedicabs and trishaws, or becak in Indonesia and cyclo in Vietnam, Laos, and Cambodia, these vehicles are often owner operated or a single owner of many vehicles leases to operators. In Phnom Penh, where some 9,000 cyclos comprise a large slice of the public transport system, the majority of operators are poorer rural migrants who often dedicate more than 12 hours per day to this activity (Etherington and Simon, 1996). In high-income and Chinese-dominated Penang and Singapore, pedicabs comprise part of the tourist landscape representing the Asian experience. More common since the 1950s are motorized pedicabs called the *samlor* or *tuk-tuk* in Thailand, *saamlaw* in Laos, and *bajaj* in Indonesia (Photo 9.2). Unlike the nonmotorized version, motors allow drivers to cruise for riders but share the characteristics of negotiated fares and access to the narrow streets of low-income areas.

An even more common form of flexible passenger transport is a small trucklike vehicle generically called jitneys with lengthwise seats accommodating an average of a dozen customers. Known as *bemo*, *opelet*, *mikrolet,* and *kolt* in Indonesia, *songthau* or *silor* in Thailand, and the elaborately decorated jeepney in the Philippines, these vehicles operate on licensed and fixed routes, picking up and setting down passengers on demand as close as possible to the customer's final destination. Strong directional flows to universities, central business districts, major bus terminals, and into fringe rural areas are common. Oftentimes, however, jitney drivers operate unlicensed vehicles and on routes that can be viewed as the operator's response to changing passenger demand. In the early 1980s, for example, 30 percent of jeepneys in

Photo 9.2 Jakarta bajaj. (Leinbach)

Manila were operating without a license and some 11 percent of routes were illegal. The great demand for jitney services in the cities of Indonesia, the Philippines, and Thailand has spawned a thriving domestic small-scale industry in jitney manufacturing and repair and thus is an important employment generator.

These various modes of urban passenger services are commonly referred to as forms of informal-sector transport or *paratransit* in that they are a cross between conventional public transport such as buses, taxis, and rapid transit and the various forms of private transport such as motorcycles and autos (Rimmer, 1980; Rimmer and Dick, 1980). Unlike more formal forms of passenger transport which are highly regulated and inflexible, paratransit is highly competitive, demand responsive, flexible, small scale, and characterized by the use of adapted rather than purpose-built technology. In an environment where personal ownership of transport is cost prohibitive and more formal types of public transport are often too costly and geographically inflexible, paratransit is seen as a viable alternative.

Despite the attractions of paratransit, a major trend in Southeast Asian transport has been toward greater reliance on conventional transport modes. This trend has been encapsulated in the phrase modernization and incorporation. *Modernization* refers to the increasing size, complexity, and technological sophistication of vehicles used in these cities; *incorporation* refers to the increasing scale of public transport providers—from small-scale entrepreneurs to large state or private-sector bureaucracies. Singapore clearly illustrates the process of incorporation. Between 1971 and 1973, the city-state's government forced 11 public bus companies to merge into a single company, Singapore Bus Service (Photo 9.3). In Bangkok, the Bangkok Mass Transit Authority was created as an amalgamation of 24 companies in 1976. Another example is Jakarta where the state-owned public bus company acquired most of its competitors by 1979.

Kuala Lumpur was considered an exception to the pattern of incorporation because many small companies operated a combined fleet of more than 400 minibuses, a form of paratransit. But in 1996, the government forced all minibuses off Kuala Lumpur's roads in favor of Intrakota, a private company that has been given the contract to provide public bus service in the city. Southeast Asian governments favor larger bus companies because they are believed to reduce congestion and offer more integrated metropolitanwide services. Yet from another perspective the bus companies are often a poor substitute for the paratransit systems they replace. After Intrakota took over in Kuala Lumpur, many commuters were stranded because they did not live near designated bus stops.

Except in Singapore, bus fleets are usually old, poorly maintained, and overcrowded; in the mid-1980s, half or more of Bangkok's and Jakarta's fleets were nonoperational. The ability to modernize fleets is hampered by government policies to keep fares deliberately low to appease urban dwellers who oftentimes dedicate up to 10 percent of income to bus transport. Because high import tariffs make obtaining new vehicles expensive, it is not surprising that paratransit has eroded bus-service market share. Only with various forms of government protection have many large city bus systems survived.

One of the common solutions to the perceived inefficiency of many public transport providers is the development of a rail or subway system. Until recently, only

Photo 9.3 Singapore double decker bus. (Ulack)

Jakarta and Manila possessed a traditional commuter-train system, but a number of cities now support or are planning to construct modern rail-based transport systems. Singapore was the first with the ultramodern Mass Rapid Transit (MRT), which began construction in 1982 and now covers 51 mi (83 km) with 48 fully automated stations and much of the track lying underground to minimize the impact on Singapore's very limited supply of land (Photo 9.3; see also Maps 14.7 and 14.8). Half of Singapore's population lives within .62 mi (1 km) of the routes. Plans call for the addition of a light rail system (LRT) to connect satellite towns in the northeast and northwest regions of the island to the MRT. Kuala Lumpur is in the throes of providing an MRT in a monumental and costly effort to make rail the infrastructural centerpiece of the congested 31-mi long (50-km) Klang Valley. In 1995 the corporatized Keratapi Tanah Melayu Berhad (KTMB) launched its first double-track electric commuter-train service between downtown Kuala Lumpur and the northwestern suburban satellite town of Rawang. The line now extends 30 mi (50 km) south to the town of Seremban in Negri Sembilan state. Although first recommended in 1981, Kuala Lumpur is just now moving forward with an LRT system, based on two privately owned companies providing the east–west and north–south links, respectively. The first stage of an east–west line was completed in 1996, and the entire system will probably be complete by 1998. Supported by a bus-feeder system as well as being integrated with KTMB's electric commuter trains, Kuala Lumpur's LRT is being touted as the showpiece of Asia (Menon 1995).

Further modernization of urban rail transport has taken the form of elevated or subway systems in selected primate capitals. In Manila, a 9-mi (15-km) LRT with 16 stations transects the western side of the city from north to south. Construction of a second LRT system is underway, and the financing of a third stage was completed in late 1996. Jakarta possesses a single, downtown-centered 6-mi (9-km) elevated rail line completed in 1992. A 9-mi (14.5-km) north–south subway, however, is expected to be completed by 2001. The projected average single-ticket cost, however, is approximately six times the cost of the average bus fare.

Bangkok, meanwhile, is the world's largest city outside of China without a mass rapid-transit rail system. By the mid-1990s three separate projects were underway. The most important of these was a 37-mi (60-km) elevated line under the direction of Hopewell, a Hong Kong-based company. The project was delayed for years by political infighting and then by controversy over whether the line should be elevated or underground. Finally, construction began in 1993, but then, after just 10 percent of the project had been completed, Hopewell pulled out in 1997 due to mounting financial difficulties. As a result, the prospect of a major rail system alleviating Bangkok's horrific traffic woes has receded further into the future.

Private-Sector Transport

The use of private transport correlates negatively with city size and positively with level of economic development. In Indonesia for example, the use of private transport is lower in the primate city of Jakarta when compared to the intermediate-size city of Bandung, but as a cross-country comparison, private transport use in more affluent Kuala Lumpur is higher than in Jakarta. In the least-developed countries and among the poorest of

Photo 9.4 Mass Rapid Transit (MRT) infrastructure, Singapore. (Leinbach)

the urban population where residence and workplace are within close proximity, fully half choose walking as the primary mode of travel. Bicycles are an important mode of travel in many medium and small-size cities, as are motorcycles and scooters, which are often symbols of economic mobility in the least-developed countries. Although almost exclusively based on the import of parts from Japan, domestic motorcycle assembly plants make the smaller-powered vehicles more affordable.

Rapid economic growth in ASEAN countries has witnessed a dramatic increase in discretionary family income, and this has resulted in greater levels of car ownership. Passenger-car ownership is positively correlated with level of economic development; in 1990, for example, passenger-car ownership per 1,000 persons is shown in Figure 9.1. In the regional aggregate, passenger-car ownership per 1,000 persons increased 41 percent over the 1980–1990 period.

This rapid increase of car ownership (despite strong disincentives such as high purchase and use taxes), coupled with the inability of urban governments to provide adequate road facilities, has led to severe congestion, particularly in the largest cities. With greater personal mobility, urban expansion has not been accompanied by a coordinated and hierarchical system of main arteries and feeder roads. In rapidly growing Jabotabek, for example, approximately 400,000 new cars are on the road each year, but the annual road network expanded by only 4 percent during the mid-1990s. As a result, the density of traffic on Jakarta's roads is more than three times the level of Singapore's (Figure 9.2). The traffic problems of Bangkok are notorious. Average travel speeds dip to 4 mi

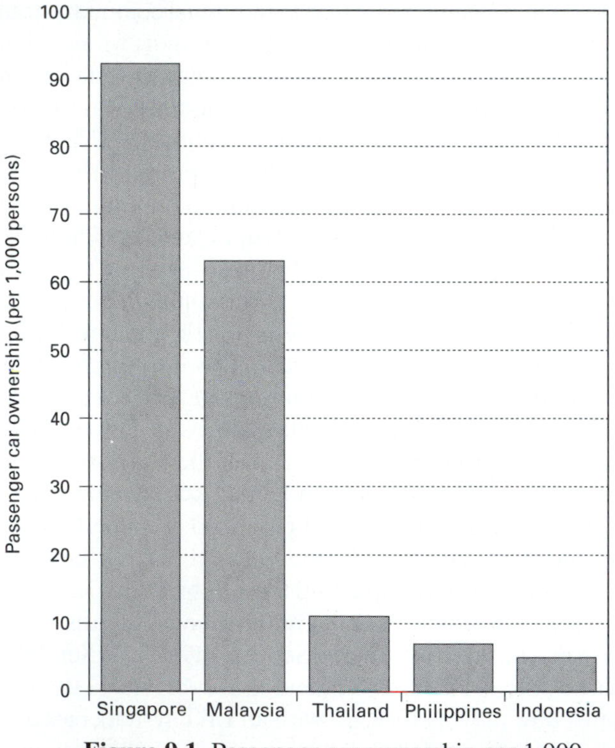

Figure 9.1 Passenger car ownership per 1,000 persons in selected Southeast Asian countries, 1990.

(6 km) per hour at the peak. In Bangkok roads make up only about 7 percent of its land surface in comparison to 15 to 20 percent in most cities. Congestion induced delays cost the city an estimated $4 million per day.

Adding to the congestion problems are street vendors conducting business on the road edge, slower moving

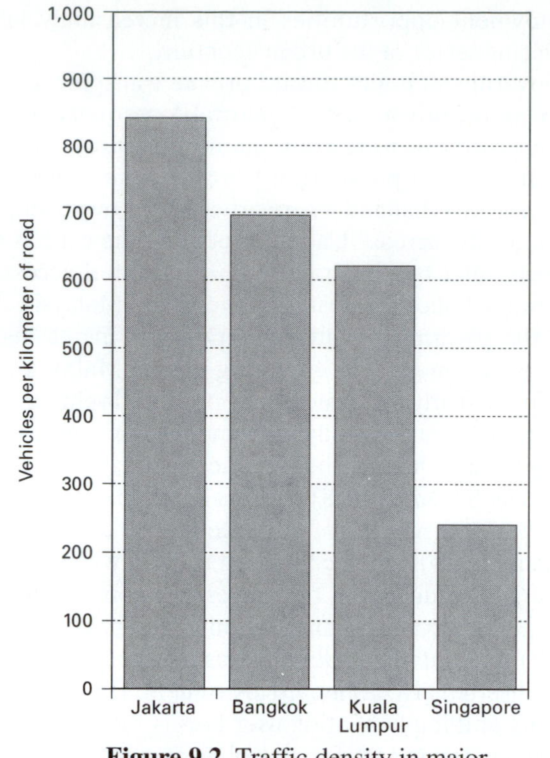

Figure 9.2 Traffic density in major Southeast Asian cities. *Source: Land Transport Authority, 1996.*

paratransit, pedestrians sharing roads with cars in the absence of adequate sidewalk infrastructure; because of inadequate parking facilities, parked cars often consume more road space than moving cars. The result is that rush hour traffic actually lasts throughout the daylight hours, and it is not uncommon for commuters to spend three to four hours a day on the road. If the trip is only a short distance, it is often quicker to walk. When compared with industrialized countries, accident rates are high, particularly with motorcycles. When coupled with the widespread use of cheaper diesel fuel, congestion substantially increases atmospheric pollution. Levels of carbon monoxide, lead, and dust are dangerously high in cities such as Manila, Jakarta, and especially Bangkok. Handkerchiefs over pedestrian mouths and noses are an increasingly common sight. In Bangkok it is estimated that water and air pollution cost $2 billion a year.

Perhaps most critical, no effective instruments have been adopted for restraining traffic in most urban regions. For example in Jakarta, experimentation with a high-occupancy-vehicle (HOV) strategy (where automobiles must carry at least two passengers at peak periods) was a distinct failure. The major exception to this is Singapore where an effective areal licensing scheme that taxes vehicles for entry to certain urban zones at peak times has been in operation for more than a decade. This scheme has recently been improved on by the implementation of electronic road pricing (ERP). Under this system, automobiles are fitted with sensors and can be tracked as they enter zones of heavy traffic at peak congestion periods (see Urban Transport Policy section below).

In a very active land market, rapidly escalating prices have made it difficult for the governments to acquire land for local distributor roads and have encouraged both high-density central development and periurban sprawl. Yet in the attempt to alleviate urban congestion, in part caused by the surge of private auto use, governments are frantically constructing limited access expressways to more efficiently improve intraurban travel. Singapore possesses an expressway system that has been in place for the past 10 years, and the Shah Alam and Klang Valley Expressways in Kuala Lumpur have assisted in alleviating east–west congestion. A 2005 master plan proposes expressways joining downtown with fringe suburban development to the north, south, and east of downtown Kuala Lumpur. Other primate cities are only now just beginning to bring both publicly and privately financed projects on line. Thailand possesses one major intraurban toll road; the very profitable 18-mi (27-km) elevated Bangkok Expressway allows for smoother traffic flow in this, the most congested capital city in the region. Currently under construction is a 26-mi (42-km) expressway that when coupled with the Bangkok Expressway will form a ring around the capital city. In Indonesia, present intraurban projects include the Jakarta Inner Ring Road and a Bandung Bypass. Yet recent large-scale infrastructure projects such as mass transit systems (e.g., in Bangkok and Jakarta) and limited access roads have not been very successful. Government and private-sector conflicts over investment levels by participants and fare structures (which must be sufficiently high for the private sector to make profits on their investment) have in part been at fault. Currently, however, the financial crisis where several of the economies are in recession make such huge projects prohibitively expensive and an unwise use of scarce resources under financial restructuring plans.

Urban Transport Policy

The role of national governments in planning urban transport is pervasive and is anchored in the belief that transport provision is too complex and important to the nation's welfare to be left to the private sector. This basic philosophy is based on a handful of economic assumptions. First and foremost is that private interests do not possess the needed financial requirements for infrastructural development and maintenance. Second, because transport directly affects urban land use, only the government is capable of coordinating both transport and land-use planning to achieve the goal of a better organized city. Third, if transport provision were left totally to the private sector, monopolies or oligopolies would emerge resulting in reduced competition and higher costs to the transport consumer. Finally, if no government control existed, many private operators would not feel obliged to serve the poorest segments of the urban population because greater profits are realized elsewhere. Although government policy directly determines the relative mix of lower versus higher technology urban-transport modes as well as the strength of public versus private interests, in reality, policies strive to promote a compromise between the two. A simple, but conceptually rewarding scheme to better understand the role of government policy in determining the profile of public urban transport is to classify public transport into "unconventional" or "conventional" policy philosophies. The former refers to the informal or noncorporate sector of low-technology and personally controlled pedicabs and jitney modes, while the latter refers to the formal or corporate sector of larger-scale modern technologies comprising buses and trains (Spencer, 1989).

As a general rule, governments have adopted this conventional philosophy. Some examples illustrate this trend. In Manila, the government restricted the number of jeepney franchises along primary bus routes, and by the early 1970s, the Singapore government through economic disincentives, drove the 8,000 "pirate" taxis (taxis that operated much like jitneys) out of business so that the bus system could survive. Many municipal authorities have banned many pedal and motorized trishaws from the entire city or city center because of the congestion resulting from slow speeds and frequent stops. Pedicabs are thus forced to operate in less-trafficked areas, or at night out of view of authorities. The decades-long consolidation of bus lines in Singapore, Bangkok, and Manila point to greater economies of scale service.

The unconventional philosophy, which supports the role of the small-scale operator in transport provision, is not without supporters in the government, in part because of World Bank support. Creating more one-way streets and the segregation of modes by specialized road lanes would, for example, keep traffic more efficiently flowing. But at the heart of the unconventional approach is the belief that tight regulation of transport is not economically efficient in that it reduces the role of the market in meeting transport demand and consumes scarce government finances. Highly subsidized government-operated bus lines, for example, operate in the red and are still too expensive for the large pool of low-income transport consumers. Financing high-technology mass transit and expressways is simply a subsidy for the higher income population who are the primary users of this new infrastructure and does not increase spatial access to areas where the poor reside. In addition, high-cost transport provision reduces employment opportunities in this increasingly labor-displacing sector of the urban economy.

Government policy toward private transport is either to accept the private car as a natural consequence of economic growth or to restrain the use of private automobiles in favor of public transport. In countries such as Malaysia and Thailand where private auto ownership has dramatically increased, accommodating the car has simply been met by constructing more limited access expressways. Policies to restrain auto use in Malaysia clash with the government's unstated policy of increasing car ownership among the emerging ethnic-Malay middle class, particularly for domestically produced vehicles. Only Singapore has successfully implemented an auto restraint policy through high license fees and user taxes. The Area Licensing Scheme (ALS) charges autos to enter the city. As a result the number of cars entering the city was lower in the mid-1980s when compared to 1975 when the ALS was first introduced. In the newest version, the government has implemented an electronic road pricing scheme (ERP) that automatically deducts user fees from smart cards inserted in a dashboard instrument. This form of extracting and implementing user fees is the harbinger of further automated traffic-control measures. Promotion of costly mass transit in other primate cities, however, will no doubt involve similar auto restraint measures.

PORT AND MARITIME TRANSPORT DEVELOPMENT

In few other major global regions have ports and maritime transport development assumed such a critical role in economic change as in Southeast Asia. Dependence on water transport is especially characteristic of island Southeast Asia where the archipelagic environment necessitates the movement of goods and people by sea. This rich maritime tradition emerged long before the colonial period as numerous port-based polities controlled vast stretches of maritime space and the trade associated with long-distance Indian, Chinese, and later Arab merchant activity. During the colonial period, the volume of maritime trade dramatically increased as European vessels loaded with primary exports and imported manufactures filled regional shipping lanes as capitalist production spread and incorporated peripheral regions such as Southeast Asia into a single global production system. As nodes linking land and maritime transport space, colonial ports were often a prominent part of the cityscape of the primary as well as the secondary colonial administrative centers.

Because transport is a critical component of production, the current phase of industrialization in Southeast Asia, anchored largely by investment from western and East Asian multinational corporations (also referred to as offshore investment), has dramatically transformed the nature of port operations and maritime transport. The widespread use of container technology has necessitated the morphological transformation of ports, and much of the container transport has been captured by shipping firms from the industrialized world. The dependence of the industrialized world, particularly Japan, on mineral and fossil fuel-rich Southeast Asia has necessitated the construction of numerous resource-based ports as well. The focus on international transport linkages has engendered the problem of transport dualism as the international maritime sector is favored for faster development than its domestic counterpart. The role of national economic and transport policy and the impact of international lending institutions are critical to understanding the rapidly changing configuration of port and maritime transport development in Southeast Asia.

Port Development

The ports of Southeast Asia were ill prepared to act as catalysts for the growth of manufacturing and the emergence of a robust trade regime during the 1960s. With port morphologies and operations rooted in the colonial period, ports were unable to service their emerging export-based national economies. Problems common to most primate ports included poor maintenance of storage sheds and loading equipment causing serious dockside congestion and low productivity; failure to dredge shallow approaches to allow access to larger vessels; poorly developed management and labor skills; and, most important, a highly regulated transport sector that hampered efficiency and kept costs high. Although governments were confronted with more-pressing development needs, the lack of attention given to improving port infrastructure and operations was a result of the inability of government decision makers to recognize the serious costs associated with inefficient port operation.

With increased exports and pressure from foreign shipping firms that utilized container technology to modernize port infrastructure and operations, governments began to remove the numerous institutional barriers that prevented the smooth flow of trade. This meant dissolving the inflexible public port authorities and creating a more efficient port sector through leasing or selling port interests to the private sector. Deregulation of port operations was required if governments were to receive from international lending institutions the necessary loans to construct capital intensive container terminals (Airriess, 1989). Financial inducements to deregulate ports were substantial because between 1968 and 1988, the Asian Development Bank funded approximately 41 percent of total ASEAN port project costs, and between 1980 and 1986, the World Bank funded approximately 48 percent of port-related projects in ASEAN. Identifying the catalytic role of international lending institutions in influencing the direction of port development, Robinson (1989, p. 146) concludes that "The ASEAN port development process has, in no small measure, been an aid-induced process." Only Thailand, among the mainland Southeast Asian states participated in this aid-enhanced port development process.

Containerization and Regional Traffic Patterns

Because Southeast Asia is the most rapidly industrializing major global region, it is not surprising that this region was also the major global focus for container traffic growth during the 1980s and 1990s. The rapid growth of export-oriented manufacturing has necessitated the use of space-collapsing container technology to reduce high transport costs associated with traditional break-of-bulk shipping (Airriess, 1993). Greater integration of Southeast Asia's economy with the industrialized world required the adoption of industrial world transport technology in Southeast Asia. The regional penetration of container technology has been so successful that fully 60 percent of cargoes once shipped break-of-bulk are now shipped utilizing containers and, in Singapore, containerized cargo accounts for 93 percent of traffic through the port. No other global developing region supports a more efficient container transport system than that of Southeast Asia.

The spatial distribution of container traffic in Southeast Asia illustrates regional levels of export-induced economic growth through time (Map 9.5). By 1980, the export-oriented ASEAN states adopted container transport not only at the primate cityports where export-oriented manufacturing linked to the global economy is concentrated but also at secondary cityports such as Penang and Cebu. These cities attracted offshore manufacturing facilities because of the increasingly high cost of land and urban congestion in primate cities during the 1980s. In contrast, the export trade of Burma, Cambodia, and Vietnam was not significant enough to warrant the construction of dedicated container port facilities.

By 1993–1994, the diffusion of container transport penetrated the secondary ASEAN ports of Tanjung Perak (Surabaya), Belawan (Medan) in Indonesia, a handful of tertiary ports in both Peninsular and East Malaysia, and Davao, Iloilo, and Zamboanga in the Philippines as lower-value primary product exports began to move mainly in containers. The early 1990s also witnessed the increased penetration of container traffic in mainland Southeast Asia as well. Especially significant is the emergence of Laem Chabang in Thailand's industrializing eastern seaboard region. This privately operated port competes with the shallow and congested riverine port of Bangkok to service the country's export boom. Significant also is the port of Ho Chi Minh City (Saigon) in Vietnam: Increased foreign investment, particularly since the admission of Vietnam into ASEAN in 1995, has dramatically increased container traffic. The government plans to construct new ports to replace the older break-of-bulk ports

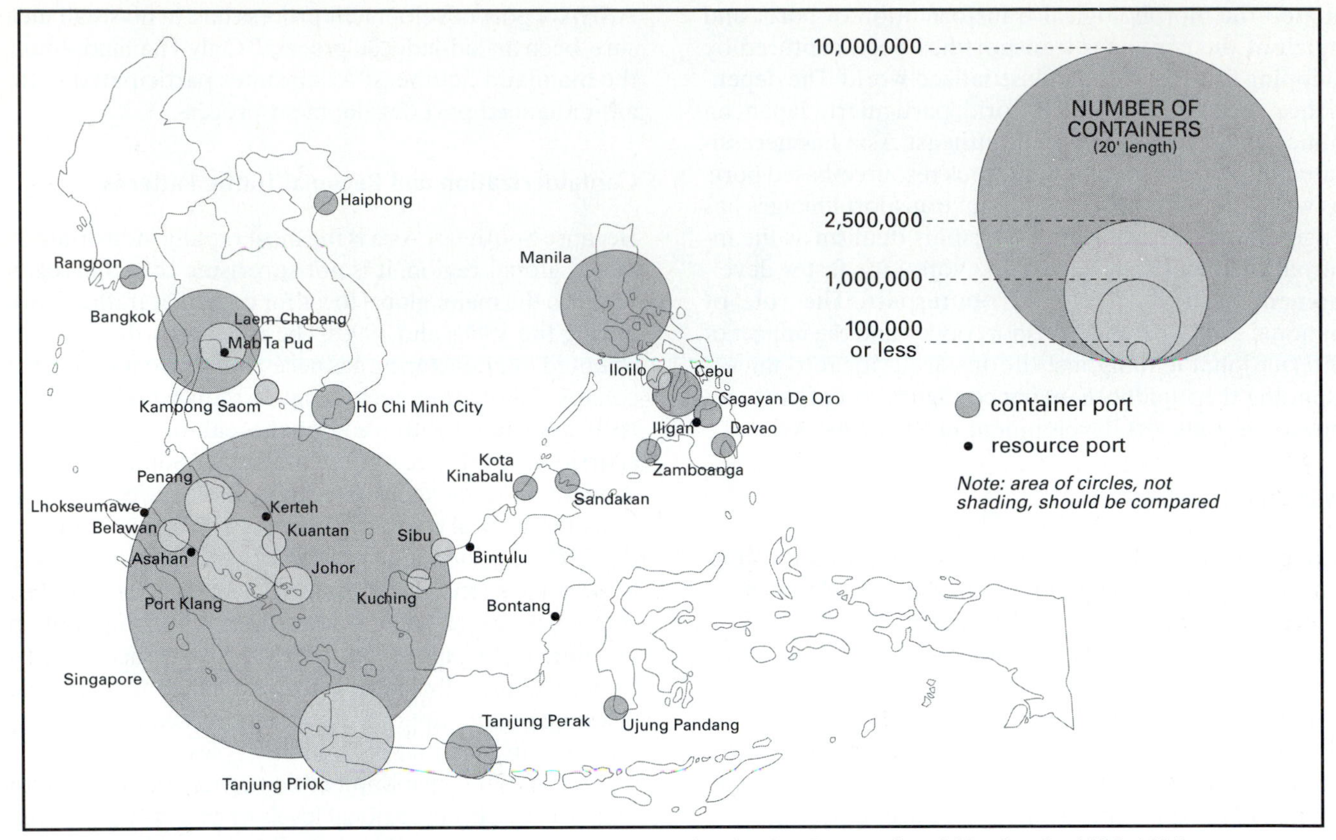

Map 9.5 Container traffic, 1993-94. *Source:* Containerisation International Yearbook, 1990-1995.

in Ho Chi Minh City and Haiphong. In the south, the small open-water port of Vung Tau, some 47 mi (75 km) from Ho Chi Minh City, has been chosen as the location of a dedicated container port to replace the three shallow riverine/delta ports that currently serve the south.

Accounting for approximately 58 percent of regional container traffic in 1994, Singapore is Southeast Asia's primate container port and annually competes with Hong Kong as the busiest container port in the world (Photo 9.5). As an early location of offshore manufacturing facilities, Singapore was the first to construct container terminal facilities in 1972; it was not until the late 1970s that other ASEAN primate ports constructed similar facilities. The interdependent factors of Singapore's geographic situation and the economic logic of container transport also explain its dominant maritime transport position. Strategically situated along primary transoceanic routes as well as centrally located within ASEAN, Singapore functions as a regional transshipment hub or "load center" for containers in both the export and import trade of neighboring countries. In 1995, approximately 80 percent of the containers passing across Singapore's three container terminal facilities were transshipped containers from neighboring countries. Some 20 percent of Singapore's container throughput are comprised of Malaysian imports and exports, and approximately 80 percent of Indonesian container traffic is transshipped through the city states container terminals. The resulting regional network of container ports resembles a hierarchy of traffic nodes centered on Singapore (Table 9.1).

Because the capital-intensive nature of transoceanic container transport dictates calling only at Singapore, smaller container feeder vessels shuttle cargoes to and from Singapore and are then loaded or unloaded onto much larger oceanic vessels. Singapore-centered feeder routes include Singapore–Tanjung Priok–Surabaya, Singapore–Bangkok, and Laem Chabang, Singapore–Belawan–Penang–Port

Port Class	Traffic Node	TEU* (millions)
Regional Load Center	Singapore	11.85
Primate Port	Port Klang	1.13
Secondary Port	Penang	0.43
Tertiary Port	Kuantan	0.02

**TEU=Twenty foot equivalent unit*

Table 9.1 The Singapore-Peninsular Malaysia container port hierarchy, 1995. *Source: Various 1996 issues of* Lloyd's List Maritime Asia.

Photo 9.5 Tanjong Pagar terminal, the oldest of Singapore's four dedicated container terminals. (Airriess)

Klang, and Singapore–Saigon. Singapore's feeder reach also stretches to the Bay of Bengal ports and even to a handful of ports in the Persian Gulf. Historical inertia is important as this rational feeder system strongly resembles the regional maritime-transport network structure during the colonial period when smaller steam and diesel vessels fed cargoes to larger oceanic vessels in Singapore. Because of greater proximity, container cargoes in the Philippines are transshipped to the load centers of Hong Kong and Kaohsiung, Taiwan. Although Singapore's status as a premier container load center is a by-product of fortunate geographical location, the government's evolving seaport policy plays a critical role in the ports success as well (Box 9.2).

Container Fleet Development

Although the benefits of a more deregulated maritime transport industry have served the export-oriented national economies of Southeast Asia well, regionally based shipping firms have not been able to capture the lucrative economic linkages associated with industrialization. Despite pronounced cargo reservation schemes (that reserve some or all of a nation's sea freight for national shipping lines) and indirect financial incentives from national governments to promote fleet development during the 1970s and early 1980s, the capital intensive nature of container shipping has forced governments to focus on promoting trade rather than utilizing scarce financial resources to promote a national fleet. Indeed, recognizing that market-driven operations are economically more efficient, governments have pushed for greater privatization of their respective national shipping industries. Reliance on container fleets of the industrialized world to function as "handmaidens" for the economic development of Southeast Asian countries has resulted in substantial foreign-currency outflows for shipping services. In 1987 for example, ASEAN countries excluding Singapore experienced an outflow of approximately US$3.5 billion, which accounts for 7 percent of the total debit in national current accounts. Malaysian and Thai-flagged vessels, for example, carried less than 20 percent of their respective country's foreign cargo in 1997.

In 1988, the total container capacity of ASEAN vessels serving the high-volume routes between ASEAN, and North America and Western Europe did not equal the container capacity of the single largest Western shipping company serving those same routes. Within ASEAN itself, Singapore's Neptune Orient Line (NOL), which is becoming a major global concern, accounted for 52 percent of ASEAN's total container-vessel capacity. Foreign control of shipping is similarly strong on the increasingly important ASEAN–East Asia trade; East Asian firms, and Western firms functioning as cross-traders, possessed 65 percent of container-vessel capacity. ASEAN container fleets fared better on the short-distance feeder routes centered on Singapore; possessing approximately 62 percent of vessel capacity share, Singapore dominated with 60 percent of feeder-vessel capacity share (Airriess, 1993).

In the medium-term future at least, greater participation by Southeast Asian shipping firms in their own foreign trade is precluded by the very nature of multinational container shipping. First, offshore manufacturing facilities commonly sign long-term contracts with foreign shipping firms to transport goods to consuming markets. These

BOX 9.2 Development Policy and Singapore as a Global Maritime Center

During the 1990s, Singapore competed with Hong Kong as the world's busiest container port and was also selected by the global transport industry as the world's most efficient port. These superlatives are not surprising because the government has purposefully dovetailed its development policies with the city-state's emerging role as a "global maritime center" that possesses the "one-stop"entrepôt functions of trading, storing, packaging, banking, insurance, communications, and marketing (Chia, 1989; Ho, 1996). Unlike most free-market economies elsewhere in the world, the government early on adopted a syncretic economic policy that combined both government monopoly power and laissez-faire philosophies. In 1964 the Port of Singapore Authority (PSA) was established as a government statutory board to construct and operate port infrastructure, but it simultaneously promoted a highly competitive and free-market port environment.

The PSA, for example, has continually rationalized port labor and systematically adopts efficient dockside container-handling equipment, allowing for the most productive container terminals in the region. Port tariffs are regionally competitive, and the port was the first to guarantee berthing space for mainline and feeder-container shipping companies as well. PSA developed and maintains a state-of-the-art integrated telecommunications network that is linked to banks, warehouses, and shipping companies and is crucial to its status as a load center because efficient information transfer lies at the heart of the logistics-dependent global container-transport industry.

The port of Singapore provides an ideal example of transport that functions as a catalytic agent to economic development because of its "growth pole" (Richardson, 1989) multipliers of functionally related transport industries. As of 1995, PSA possessed some 11 subsidiaries and associated companies that provided producer services. Some of these "propulsive" producer service industries have reached beyond Singapore and have sold their port planning, engineering and operational expertise to ports in countries such as Italy, Ghana, Brazil, China, South Korea, Indonesia, Vietnam, and India. It is because of these multiplier effects that the PSA is Singapore's most profitable government enterprise; in 1995, for example, the PSA possessed an operating surplus of approximately US$550 million, an increase of some 20 percent more than 1994.

These producer services are critical to PSA's continued success as a "global port company" because its role as a regional hub will gradually lessen as neighboring primate container ports improve their ability to handle mainline vessels. Port Klang and Laem Chabang, for example, are already attracting a handful of transoceanic container vessels. Singapore's ability, however, to anticipate demand systematically will, for the medium-term future at least, be able to keep one step ahead of the regional competition. One example is the new Pasir Panjang container terminal, which is the largest and most technologically sophisticated facility in the world. Perhaps more importantly, Singapore's systematic and aggressive adoption of increasingly sophisticated information technologies will also ensure its present position as a regional container load center. Despite its great success and intimate synergies with the national economy, the PSA was corporatized in 1997 with sights on eventual privatization so that the port might become even more competitive.

global logistics conglomerates possess the capital to provide the value-added services of inventory control and distribution made possible by a computer-based information network. Second, foreign shipping firms operating in the liner trades operate consortia to provide greater economies of scale. As of the early 1990s only Singapore's NOL and Malaysia's MISC (Malaysian International Shipping Corporation) have joined these oceanic liner consortia. The potential control of port-hinterland container transport by foreign and a handful of domestic transport concerns will also mean the liquidation of small-scale indigenous firms traditionally engaged in the short-distance carriage of break-of-bulk traffic.

Resource-Based Ports

Paralleling the dramatic transformation of ports associated with export-based manufacturing has been the proliferation of large-scale port construction linked to resource-based development (Robinson, 1985). Whether for export or domestic markets, these resource-oriented port complexes offering break-of-bulk and reduced transport cost advantages have been viewed as mechanisms for national and regional economic development strategies. Because resource-based port complexes are often located in economically peripheral regions, industries promote greater spatial equity in economic development through hoped for growth-pole multiplier effects, meaning that the expansion of activity at a port propels the growth of related industries in the surrounding area. Resource-based industries also assist in structurally broadening the national economy and are a natural replacement for an inefficient import-substitution sector that economically benefited the core and discriminated against peripheral regions. The globalization of markets for raw materials and energy fuels, in which Southeast Asia is well endowed, is an additional catalyst for the emergence of port-based resource processing industries. The role of Japan is significant; this resource-poor country has become increasingly reliant on resource-rich

Southeast Asia to sustain economic growth, and in turn these same countries depend on Japan for capital and technology to promote industrial development.

One important category of port industrial complexes consists of a joint venture, in which foreign capital and domestic government investment are used to construct and operate resource-processing industries for export as well as domestic uses. Perhaps the largest of these are the ports of Laem Chabang and Mab Ta Pud along the southeastern seaboard of Thailand. Oil and natural gas from the Gulf of Thailand are piped onshore to fuel energy-intensive industries, many being petrochemical related. Another Japanese-funded resource project is the Asahan aluminum-smelter complex in North Sumatra, Indonesia. Utilizing local hydroelectric power and alumina feedstock from Australia to produce aluminum ingots, the Asahan project is the only aluminum smelter and the largest individual Japanese investment in resource processing in Southeast Asia (Airriess and Kohno, 1993).

Additional resource-based, port-industrial complexes, often with foreign equity participation dot Southeast Asian shores. Off the east coast of Peninsular Malaysia, oil and natural gas are piped ashore to a terminal and processing complex at Kerteh, Terengganu. Similar port facilities exist at the LNG plant at Bintulu, Sarawak. Large-scale agroprocessing port facilities also exist in Johor Bharu. In Indonesia, port facilities have been developed for LNG exports in Bontang, East Kalimantan and Lhokseumawe, Aceh. Downstream agrochemical industries have emerged with urea fertilizer being the most important. The Arun fertilizer port facility is the first ASEAN industrial joint venture, with Malaysia, Thailand, the Philippines, and Singapore holding a combined 40 percent equity share. Although presently a small producer at the global scale, five coal-loading terminals in East and South Kalimantan serve Indonesia's export trade. With increasing foreign investment and substantial interior reserves to qualify Indonesia as a major exporter, more-efficient loading terminals lining the strategic Makassar Straits are being planned. Providing virtually no oil and natural gas for downstream processing, the ports of hydrocarbon-rich Brunei possess few industrial-based growth pole functions. While many public and private ports serve mining and agroprocessing operations in the Philippines, the only large-scale port-industry facilities serving peripheral resource regions are at Iligan where a steel mill is located.

Much like international container shipping, resource-based exports are monopolized by foreign bulk carriers. The export of crude oil and natural gas from Indonesia and Brunei, for example, move almost exclusively in the hulls of foreign tankers. In 1995, only 2 percent of Indonesian exports were carried by domestic registered vessels. While the high capital costs of purchasing vessels is a problem, it must also be recognized, particularly in the case of exports to Japan, that investors in joint venture, resource-processing facilities also operate shipping services through well-known *keiretsu*, which are the large industrial and financial cliques of Japan. Perhaps the notable exception to foreign control is Malaysia's exports of Bintulu LNG to Japan and Taiwan. Malaysian International Shipping Corporation has shipped LNG to Japan since the early 1980s, and now Petronas, the national oil corporation, is diversifying into the capital-intensive LNG shipping business as well. Although this Petronas subsidiary is partly owned by Shell and Mitsubishi, Malaysia is now able to control freight costs, which constitute one-third of the total cost of the LNG business.

Domestic Port and Fleet Development

Until the 1960s, domestic maritime transport flourished as road networks were still not adequately developed to displace multipurpose vessels engaged in coastal trade. The completion of long-distance highways during the 1970s and 1980s that parallel the coasts of Southeast Asia's many elongated coasts, coupled with roads crossing large islands, have captured much of this coastal general-cargo trade. Perhaps only in Indonesian Kalimantan, Sulawesi, Irian Jaya, and some medium-size islands in the Philippines has the absence of road development prolonged the traditional importance of coastal shipping. By virtue of their fragmented archipelagic morphology, Indonesia and the Philippines continue to rely heavily on domestic maritime transport to integrate national space economically and politically. A comparison of domestic port and maritime transport development in these territorially large Southeast Asian states provides an opportunity to examine the interdependent linkages between transport and economic development.

Indonesia The challenge of transport provision in a vast archipelagic state where much of the eastern half of the country is comprised of many small islands, low population densities, and low levels of development are monumental. Although the Dutch company Koninklijke Paketvaart Maatschappij (KPM) provided regular transport services to the most isolated regions of the archipelago, the company focused on connecting individual islands with foreign markets. This resulted in a dendritic shipping network tied to large ports with few horizontal links between islands. The KPM was dissolved on the country's independence, and the Indonesian government then constructed a four-tier and highly regulated interisland transport system of both government-owned and subsidized private services, which in the early 1980s comprised almost 8,000 vessels serving hundreds of routes and ports. The primary objectives of government intervention were to protect the national shipping fleet from external competition, to establish captive markets for the domestic shipbuilding industry, and to ensure that regular services would be provided at a reasonable cost to all regions of the archipelago. Essentially, government-determined freight rates on high-volume routes were artificially high to subsidize low-volume routes. This created, however, a high-cost economy that functioned as a barrier to development in these lagging eastern islands (Dick, 1985).

Under pressures of financial austerity caused by depressed global oil prices in the early 1980s and the desire to promote nontraditional exports, this high-cost system was dissolved to promote a more competitive, market-driven domestic maritime transport system. Several consequences of this dramatic policy shift have now materialized. First, the number of interisland operators and vessels has expanded and freight rates have declined, but only on those most profitable routes focused on western ports with international links such as Jakarta (Tanjung Priok), Surabaya (Tanjung Perak), and Medan (Belawan). These same ports were corporatized in 1985. Smaller ports, however, particularly in far-eastern Indonesia suffered as private firms have withdrawn services to chase cargo along more profitable routes. Although small-scale shipping such as motorized and sail *perahu rakyat* have entered this less-profitable trade because of resulting demand, government transport subsidies in less-developed eastern Indonesia have actually increased since the late 1980s (Photo 9.6).

Second, because these new policies allowed the chartering of less-expensive foreign vessels, national shipping firms which invested heavily in domestically built vessels are now saddled with high-priced vessels. Third, containerization is not well developed in the trading patterns of eastern Indonesia because of the absence of port facilities and the insufficient generation of cargo on the return trip to Java. In the short term then, the deregulation of maritime transport has heightened economic stagnation in the less-developed regions of eastern Indonesia.

The Philippines Compared to Indonesia, domestic maritime transport in the Philippines is far more developed due to both physical and human factors. First, the archipelago is far more compact and anchored by numerous large islands allowing for the development of distinct regional economies. Second, government policy has promoted industrial dispersion, and this coupled with a more spatially even distribution of urban centers has created greater demand for interisland transport than in Indonesia. Third, the development of interisland shipping was accomplished through private investment and control with little direct government assistance.

The interisland fleet is composed of a diverse array of vessels. The use of fishing vessels and smaller watercraft such as the outriggered *banca* is extensive; they are crucial to the transport of passengers and cargo on short-distance island-hopping routes. Somewhat unique to Southeast Asia and traditionally not used in Indonesia are barges, lighters, and tugboats which in the early 1980s comprised approximately 40 percent of interisland vessel capacity. Unlike Indonesia, where wide stretches of water separate islands, the flexible nature of barge service is suited to the shallow and protected waters of the southern islands and the simple pier infrastructure characteristic of many small, municipally operated and specialized private ports that numbered 542 in the early 1980s.

Photo 9.6 Traditional commercial sailing craft, Surabaya, Java. (Leinbach)

The efficiency of transporting interisland cargoes has been greatly enhanced by the systematic modification of break-of-bulk vessels into container carriers. As a result of government provided financial incentives to modernize the aging break-of-bulk fleet, along with loans from international lending institutions to improve port facilities, containerization has rapidly penetrated the traditional break-of-bulk trade. The share of containerized cargo grew from only 3.6 percent in 1978 to 29 percent in 1980 and 47 percent in 1986. No other Southeast Asian state supports a more developed domestic container transport network than the Philippines.

Centered on the North Harbor in Manila and secondarily in Cebu, the dedicated domestic container transport network in the mid 1980s included the six Mindanao ports of Cagayan de Oro, Iligan, Zamboanga, General Santos, Davao, and Surigao, the three Visayan ports of Cebu, Iloilo, and Dumaguete, and Puerto Princesa on Palawan. The north-south, Manila-centered domestic containerized trade is far more balanced than Indonesia's; in 1985 for example, fewer than 10 percent of inward containers from the southern ports were returned empty. Although a substantial amount of agricultural-related products such as animal feed, cereals, fertilizers, and livestock moved southward from Manila, the flow is dominated by higher-value manufactured goods while the return trade northward is comprised primarily of agricultural products. The 1990s have witnessed reduced government regulation which traditionally has meant lower profits in this already marginally profitable industry. Coupled with higher rates of economic growth during the early 1990s, the number of domestic shipping firms almost doubled between 1991 and 1995. Many of these firms are products of joint ventures between domestic and foreign shipping firms. Despite such progress, however, persistent institutional inefficiencies make it more costly to transport goods on domestic routes than on longer-distance international routes.

Although government and private interests provide for a relatively efficient domestic cargo trade, such is not the case for domestic passenger flows. The Philippines has one of the highest frequencies of marine accidents and fatalities in the world because inspection of vessels is poor. In addition, the possibilities for accidents is heightened because, with an even spread of population, interisland mobility is substantial. In 1986, 14 million passenger trips were undertaken in a country with a total 1990 population of 62 million. The tragic sinking of the heavily overloaded ferry *Doña Paz* off Mindoro in 1987 with more than 2,000 fatalities, is symptomatic of the hazardous nature of interisland passenger service. Although marine safety is poor in Indonesia as well, the concentration of population on western islands connected by short-distance ferry service between Sumatra, Java, Madura, and Bali reduces the potential for serious marine accidents. In 1982, only 4 million passenger trips were taken, and this in a state supporting triple the population of the Philippines. Marine safety is also a serious concern in the Strait of Malacca where some 600 ships transit this busiest strait in the world every day. In 1992 alone, large-scale collisions involved a U.S. naval vessel, a cruise ship, an ocean trawler, a container vessel, and three supertankers.

AIR TRANSPORT

Although a much more recent element of Southeast Asia's transport system, air transportation has rapidly gained a critical role in the region's economic development. Domestic air networks have served to integrate isolated communities into national economies, especially in peripheral regions of Indonesia, Malaysia, and the Philippines where alternative forms of domestic transportation, such as roads and railways, are poorly developed or impossible. Yet as important as national integration is to development, the level of integration sought by the countries of Southeast Asia today is global. International air links integrate these countries into the global economy. Indeed, for the region's export-oriented economies, good international airports and strongly competitive national airlines are crucial as an increasing proportion of the region's exports move by air—including microelectronics, running shoes, and orchids. Moreover, air transport infrastructure is a prerequisite for successful tourism development. No wonder, then, that Vietnam, Laos, Cambodia, and Burma have each placed great emphasis on improving their airports and airlines in the 1990s. Much like the region's more industrialized countries, these four nations intend to accelerate their development through better air transport linkages with the rest of the world.

Domestic Services

Though the emphasis of Southeast Asian airlines is increasingly on international traffic, domestic services continue to grow rapidly as development spreads from primate cities to other urban centers. Indeed, some of the domestic trunk routes in the region are as heavily traveled as important routes in the United States, Europe, or Japan. Bangkok–Phuket in Thailand, Jakarta–Surabaya in Indonesia, Kuala Lumpur–Penang in Malaysia, and Manila–Cebu in the Philippines are each served by more than 100 flights a week. These key cities are hubs for domestic networks stretching to more than 200 cities and towns in Southeast Asia, including more than 50 in Indonesia alone.

National governments in Malaysia, Indonesia, and the Philippines have used the airline industry as a tool of development in remote, economically underdeveloped areas. State-owned airlines have been required to operate "missionary routes" from urban centers to small communities that are often cut off from other forms of transportation. A small turboprop aircraft might fly a missionary route as seldom as once a week, landing at a roughshod airport hacked out of a forest.

Missionary routes are typically unprofitable. National governments mandate fares that are too low to recover an airline's operating costs. The low fares are designed to make air travel more affordable. This principle has been applied not just to missionary routes but also to domestic routes more generally. For example, Philippine Airlines claimed that 60 percent of its domestic routes were money losers. In Thailand, Thai International noted that the government-imposed fares it was allowed to charge in the early 1990s averaged only 1.93 baht (US$0.08) per kilometer, or one-fifth the cost of riding in one of Bangkok's famous *tuk-tuk* taxis. Raising domestic fares is difficult because the governments that set fares are reluctant to antagonize travelers who depend on cheap transportation. Thus, fares tend to stagnate for years without regard for the rising costs an airline may face. For example, domestic fares had not risen for 10 years in Malaysia before a small increase in 1992.

Nor is it easy for airlines to shed loss-making domestic routes. Malaysia Airlines has been criticized in the 1990s for neglecting its domestic network in favor of more-lucrative international routes. Critics charge that the airline is reneging on its historic mission to contribute to Malaysia's social and economic integration. Ironically, even as Malaysia Airlines struggles with a largely unprofitable domestic network, it is losing traffic on some of its most heavily traveled and most profitable routes in Peninsular Malaysia. The completion of Malaysia's North–South Highway has cut air traffic on routes from Kuala Lumpur to Penang, Ipoh, and Johor Baharu. The same phenomenon could affect air traffic on short-haul routes in other Southeast Asian nations as car ownership becomes more common and highway infrastructure is improved.

Conversely, in Burma, Vietnam, Laos, and Cambodia, domestic air transportation is still in an early phase of development. Safety is a serious issue in these countries. Burma Airways has had an especially poor safety record. Between 1974 and 1989, nine of the airline's Dutch-built Fokker turboprop aircraft crashed. Fleet renewal is a high priority for airlines in these countries. Lao Aviation, for example, is replacing some of its old Soviet-built airplanes with new French-built turboprop aircraft. Another priority is development of new airports and upgrading of existing ones. Vietnam, for instance, plans to double the number of airports in the country to 32 by the year 2010 while greatly expanding the airports of Ho Chi Minh City, Hanoi, and Danang—each of which is expected to experience up to a twelvefold increase in traffic by that date.

Southeast Asia in the Global Airline Industry

Until the regional financial crisis, which began in 1997, international air traffic to and from Southeast Asia grew more rapidly than in nearly any other world region. The growth of the region's economies has been accompanied by increased international travel by managers, financiers, engineers, salespersons, and technical personnel. In addition, the affluence that economic growth has brought to many Southeast Asians has given rise to more international tourists from the region. Likewise, the development of new tourist attractions in the region continues to draw increased numbers of visitors from around the world.

International air freight in Southeast Asia, meanwhile, is growing even more rapidly than passenger traffic. Many of the region's exports, whether fresh tuna from Bali or computer chips from the Philippines, are well suited to shipment by air. Seaborne shipping is less expensive but, for goods that are of great value, lightweight, small, or perishable, air freight is often more economical. Most of the world's air-freight airlines, including Federal Express and United Parcel Service from the United States and Cargolux from Europe, serve the region. FedEx has made an especially large commitment to the region with its Asia One network centered on a new hub at the former U.S. airbase at Subic Bay, the Philippines. Freighters operate around the clock from Subic Bay linking major points in Asia to one another and the U.S. market. In addition to foreign competitors like FedEx, the airlines of Southeast Asia also carry a significant amount of cargo in the bellies of passenger aircraft and in dedicated cargo aircraft. Singapore Airlines fleet, for instance, includes seven Boeing 747s built to carry freight exclusively.

Much of the air traffic, both passenger and freight, flowing between Southeast Asia and the rest of the world is funneled through either Bangkok or Singapore. In fact, the routes between Singapore and other large Southeast Asian capitals are among the most heavily traveled in the world. Singapore and Bangkok are also major transit hubs for long-haul interregional traffic. Singapore, for instance, is an important transit stop on the "kangaroo route" between Europe and Australia. In the 1990s, Singapore received more international air traffic than its Thai rival, but in the future, Singapore may lose its dominance to Bangkok. The Thai capital is better positioned to serve as a gateway to Burma, Laos, Cambodia, and Vietnam. Bangkok also enjoys a superior location with respect to Asia's future economic giants, China and India.

Another contender is Kuala Lumpur. The Malaysian government has built a sprawling new airport in Sepang, 43 mi (70 km) south of the capital. The first phase of the project, to build a two-runway airport, was completed in 1998 at an approximate cost of US$3.9 billion. The new facility has an annual capacity of 25 million passengers, about twice the capacity of Kuala Lumpur's previous airport that was squeezed on one side by a rubber plantation and on the other by a military base. Jakarta and Manila are less important as regional hubs due to their locations away from the region's core and their proximity to much larger hubs. Jakarta is firmly in the shadow of Singapore, while Manila is dominated by nearby Hong Kong and Taipei. Consequently, Manila and Jakarta receive significantly fewer international flights per week than Singapore, Bangkok, and Kuala Lumpur (Map 9.6).

Manila is important as an air-freight hub, however. Two major international air-freight carriers, DHL and TNT, operate regional hubs at Manila, although these are smaller operations than Federal Express's facility 50 mi (80 km) to the north at Subic Bay. The Philippines has been favored by these air-freight carriers because it is well positioned to serve markets in both Northeast and Southeast Asia, offers an abundant supply of English-speaking workers, and is relatively liberal in granting landing rights to foreign carriers.

International Airline Hubs and Economic Development

The stakes are very high in this battle of international airline hubs. National governments invest billions in new airports because the potential rewards are just as great. Certainly, Singapore's importance as a global hub has contributed to its economic development. Multinational corporations (MNCs) searching for investment sites appreciate Singapore's excellent connections to destinations around the world. In 1997, the 58-passenger airlines serving Changi Airport offered nonstop flights to nearly 100 cities in 42 countries. For export-oriented manufacturers in Singapore, these links mean easy access to global markets. For the banks and other firms in Singapore's business services industry, the thousands of flights that arrive and depart each week mean immediate access to international clients and colleagues. As a result, Singapore's importance as a hub has helped the city-state attract billions of dollars in foreign direct investment each year.

A second benefit of the hub in Singapore has been its impact on the growth of Singapore Airlines (SIA). The hub in Singapore has helped SIA in two ways. First, SIA has been able to tap into the whole Southeast Asian market. The many flights between Singapore and other cities in the region, even if they are operated by other airlines, feed traffic onto SIA's flights to and from destinations outside the region. For example, a traveler between Bandar Seri Begawan and Amsterdam might fly Royal Brunei Airlines to Singapore and then connect to an SIA flight to the Netherlands. Second, the popularity of Changi Airport with foreign airlines has helped SIA acquire landing rights around the world. Airlines from outside the region often use Singapore as a gateway to Southeast Asia. The government of Singapore has welcomed any foreign airline seeking landing rights at Changi Airport, asking in return that SIA be given landing rights in

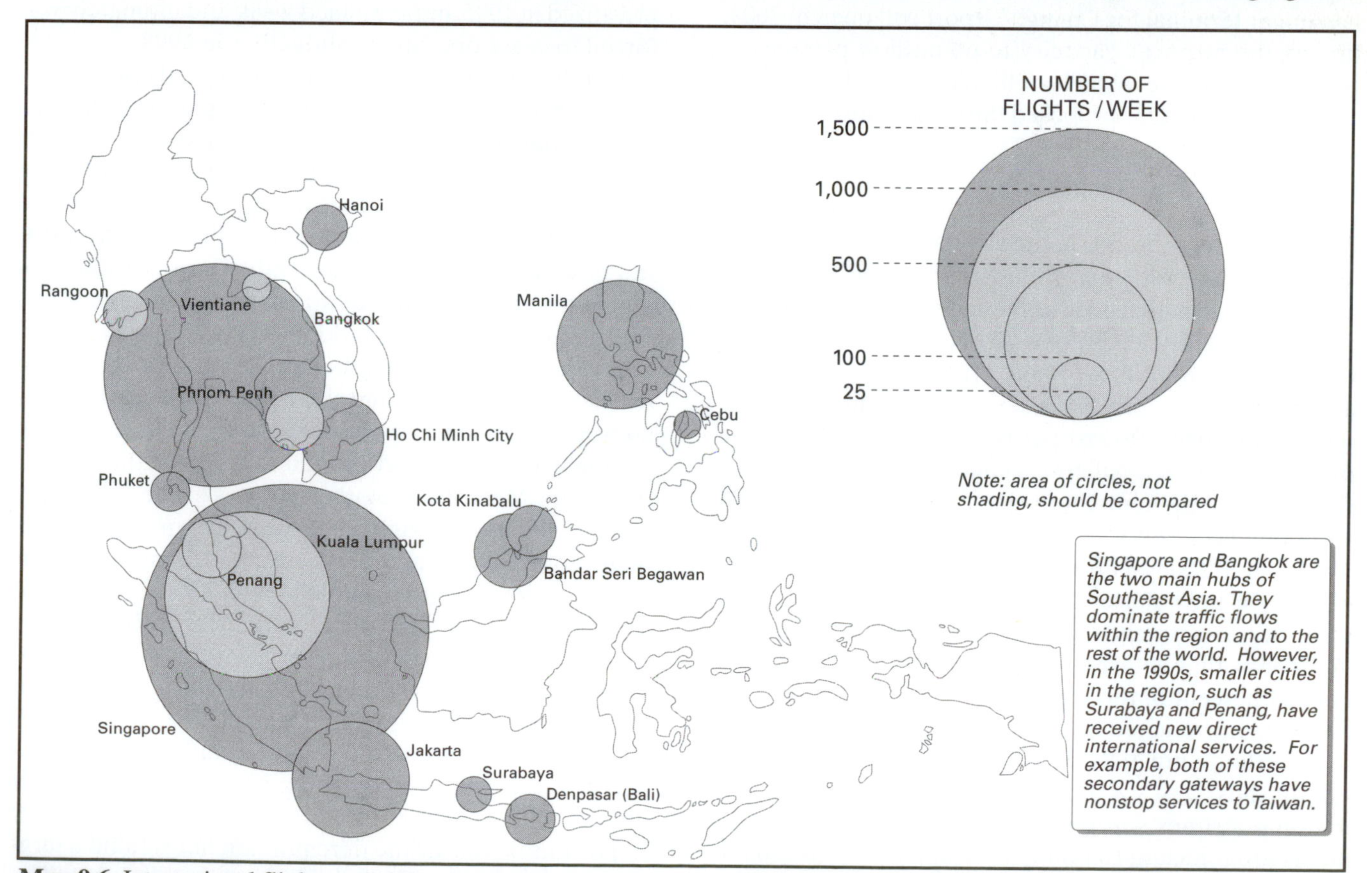

Map 9.6 International flights to Southeast Asian gateways. *Source:* OAG Desktop Guide: Worldwide, *vol. 22, no. 9, November 1997.*

the home country of the foreign airline. This strategy has won SIA landing rights around the world. These two advantages, regional traffic feed and abundant landing rights, helped make SIA the single most-profitable airline in the world in the early 1990s.

Airline hubs also bring tourists and contribute to the emergence of airport-related activities. A significant portion of Singapore's thriving tourist trade is made up of visitors who stop over in the city-state for a few days before boarding a flight for their final destination. A potentially much more lucrative type of tourism is the international conference business. Singapore is Asia's most important conference city, a distinction its global air connections have helped it win. Meanwhile, the area around Changi Airport is home to several firms specializing in air-freight repair and refurbishment as well as many aircraft logistics centers. Much like Singapore's port, the city-state's airport functions as a multiplier producing growth pole as well.

Clearly, the potential economic benefits of an international hub are substantial. Malaysia's development of a new international airport is intended to capture more of those benefits for its own gateway city, while Thailand and the Philippines are planning major new airports near Bangkok and Manila for much the same reasons. But Singapore will not concede its advantage willingly. A third passenger terminal for Changi Airport will open in 2004, raising the airport's capacity to 64 million passengers from the current level of 44 million per year. Land reclamation has begun for proposed third and fourth runways.

Liberalization and Competition

One of the major trends in the airline industry of Southeast Asia is liberalization. That is, national governments are playing a smaller role in the industry, giving airlines greater flexibility to respond to market forces. Liberalization is manifest in two ways: privatization and deregulation. Since the mid-1980s, several of the airlines in the region have been at least partly privatized. At the same time, government regulations concerning airline routes, fares, and equipment choice have been eased (Bowen and Leinbach, 1995).

SIA and Malaysian Airlines were the first to be privatized. In 1985, shares in both carriers were sold to private investors. In each case, however, the state held on to a controlling interest. The government of Singapore kept 54 percent of the shares in SIA; and the Malaysian government retained a similar percentage of the shares in its airline until 1994. The partial nature of privatization in these cases is a reflection of the vital roles that national airlines still play in Southeast Asia. Governments are reluctant to relinquish ultimate control to the private sector, given the importance of air services to economic development and even national defense. Moreover, these same national functions may make an airline less attractive to private investors. Malaysian Airlines' ambitious fleet and route expansion during the 1980s and early 1990s helped Malaysia forge links with markets around the world, but it also cut into profits that otherwise would have been distributed to shareholders. In February 1994, the privatization of Malaysian Airlines entered a new phase. Frustrated by the airline's poor financial performance, Malaysia's federal government sold most of its remaining stake to a wealthy Malay entrepreneur. Private control helped to reinvigorate the airline, spark an overhaul of its strategy, and make it more competitive in the global airline industry.

Frustration was also one of the primary reasons behind the privatization of Philippine Airlines (PAL). In the early post- World War II period, PAL was one of Asia's most prestigious airlines, but by the 1980s, it was the weakest of the five large Southeast Asian carriers (SIA, Malaysian Airlines, PAL, Thai International, and Garuda Indonesia). Some of its problems were far beyond its control. Political instability and lackluster economic growth in the Philippines hurt the carrier. In addition, its once prized routes to the United States became unprofitable as airline competition over the Pacific became increasingly fierce in the 1980s. The airline was privatized in 1992, but remained weak and ultimately was forced to cease operations altogether in 1998.

In 1992, Thai International was also partly privatized. The government sold 8 percent of its shares in the carrier to financial institutions, private investors, and airline employees. That means that among the region's big airlines, only Garuda Indonesia has not been exposed to privatization. In Indonesia, liberalization has a different emphasis. For decades, only Garuda Indonesia was allowed to operate jet aircraft, and, with very few exceptions, only Garuda Indonesia was allowed to fly on international routes. The other airlines in Indonesia were restricted to serving domestic routes with smaller turboprop aircraft. In 1989, however, Garuda gave most of its domestic routes and the jets it used on them to its subsidiary Merpati. The other airlines argued that they too should have the right to use jets. Beginning in 1990, their request was granted. By 1994, three more airlines, Sempati, Bouraq, and Mandala, had also begun to use jets on domestic routes. The other hurdle for Indonesia's small airlines was the right to serve international routes. That hurdle was crossed in 1991 when the Indonesian government granted the right to smaller carriers to compete on the route from Jakarta to Singapore. Later, Sempati also received the right to add Taipei and Perth to its network.

For Indonesia, airline deregulation helps fulfill a larger goal: to promote the nation's aerospace industry. The government hopes that by loosening some restrictions, it can accelerate the growth of Indonesia's airlines, which will in turn become major customers for IPTN, the country's aircraft manufacturer. IPTN is currently developing a new 60–70 passenger turboprop aircraft called the N-250. Indonesian airlines, including Merpati, Bouraq, and Sempati, have already made commitments to buy this new aircraft. Even when Indonesian airlines buy from foreign manufacturers, however, IPTN can benefit. In 1993, Sempati and another Indonesian airline became the first customers for a new aircraft, the Fokker 70, produced by the Dutch company Fokker; to win the contract, Fokker agreed to subcontract the manufacture of some Fokker 70 components to IPTN. IPTN already manufactures components for the Fokker 100 jet which Sempati started using in 1991. Such deals give IPTN experience with state-of-the-art Western aerospace technology that can later be incorporated in aircraft that IPTN designs and builds independently. It is important, therefore, to IPTN that the Indonesian airline industry grow now and in the future. Deregulation is one means of achieving that objective.

Secondary Gateways and New Carriers

Sempati was one of a number of new Southeast Asian carriers which were launched in the 1980s and early 1990s. These carriers generally operate smaller aircraft than the better-established flag carriers and serve the heavily traveled domestic routes that can sustain competition. A few of these carriers have entered international markets as well, offering direct services to such secondary gateways as Medan, Indonesia; Kota Kinabalu, Malaysia; and Chiang Mai, Thailand. Yet while a few new carriers have prospered, many have failed. Indeed, even Sempati Airlines, which had enjoyed so much success since its founding in 1989, went out of business during the political and economic upheavals in Indonesia.

Sempati illustrated the importance of political clout for new airlines that want to challenge the flag carriers. Many observers contended that Sempati was favored because it was partly owned by one of President Suharto's sons and partly by the military, a powerful combination in Indonesia. Conversely, tiny Bangkok Airways has languished for years because the Thai government refused to let it serve any route already served by Thai Airways International. To get around this restriction, Bangkok Airways has build its own airfields at tourist destinations, especially on the country's coast. More recently, a new Thai airline has emerged. Orient Thai Airline was set up by former executives from Thai International Airways and they have been able to secure permission to operate to most of Thailand's major cities as well as several international destinations in Southeast Asia.

The Philippines has witnessed the sharpest increase in airline competition in the 1990s. As recently as 1994, PAL had a complete monopoly on domestic routes; but during the next three years, four new airlines, some funded by foreign investors, invaded PAL's stronghold. These new carriers have all focused on the trunk routes to Manila, resulting in fierce competition to such big cities as Cebu and Davao.

Even Vietnam and Burma, though still at a very early stage in aviation development, have spawned new competitors. In Burma, a Singapore-owned company called Mandalay Airways serves major points alongside the flag carrier, Myanmar Airways International, and in Vietnam, Pacific Airlines was set up with French foreign investment in 1992 and operates to the country's four largest cities as well as Taiwan and Macau.

The emergence of secondary carriers has been paralleled by the rapid growth of international services to secondary gateways. Pacific Airlines, for instance, has launched nonstop international services from Danang to Macau while GrandAir operates from the new civilian airport at the former Clark Air Force Base in the Philippines to Hong Kong. The region's largest secondary carrier is Silkair, a subsidiary of SIA. Silkair serves 20 secondary points in the region, including many tourist destinations such as Phuket, Thailand, and Palau Tioman, Malaysia. But Silkair's network also includes a number of emerging business centers including Pekanbaru, a Sumatran city that occupies an important place in Indonesia's oil industry. The new services by such airlines as Silkair give international travelers easier access to secondary destinations and help to accelerate the spread of economic development away from the region's primate cities (Map 9.7).

Aviation Opportunities in Burma, Laos, Cambodia, and Vietnam

The potential impact of air transportation on development has inspired the governments of Southeast Asia's four less-developed countries to call on outside help to jump-start their airline industries. For example, to boost tourism the government of Burma formed Myanmar Airways International, replacing accident-prone Burma Airways in a joint venture with a consortium controlled by the sultan of Brunei. In 1993, the new airline began to fly from Rangoon to Singapore, Bangkok, and Hong Kong with a Boeing 757 acquired from Royal Brunei Airlines. It also relied on pilots, maintenance staff, and foreign sales offices borrowed from the Brunei carrier.

Thai firms, meanwhile, have been heavily involved in the development of aviation in Cambodia and, to a lesser extent, Laos. In fact, one of the rapidly growing routes in the region, Bangkok to Phnom Penh, was pioneered by Cambodian carriers that were partly owned by Thai

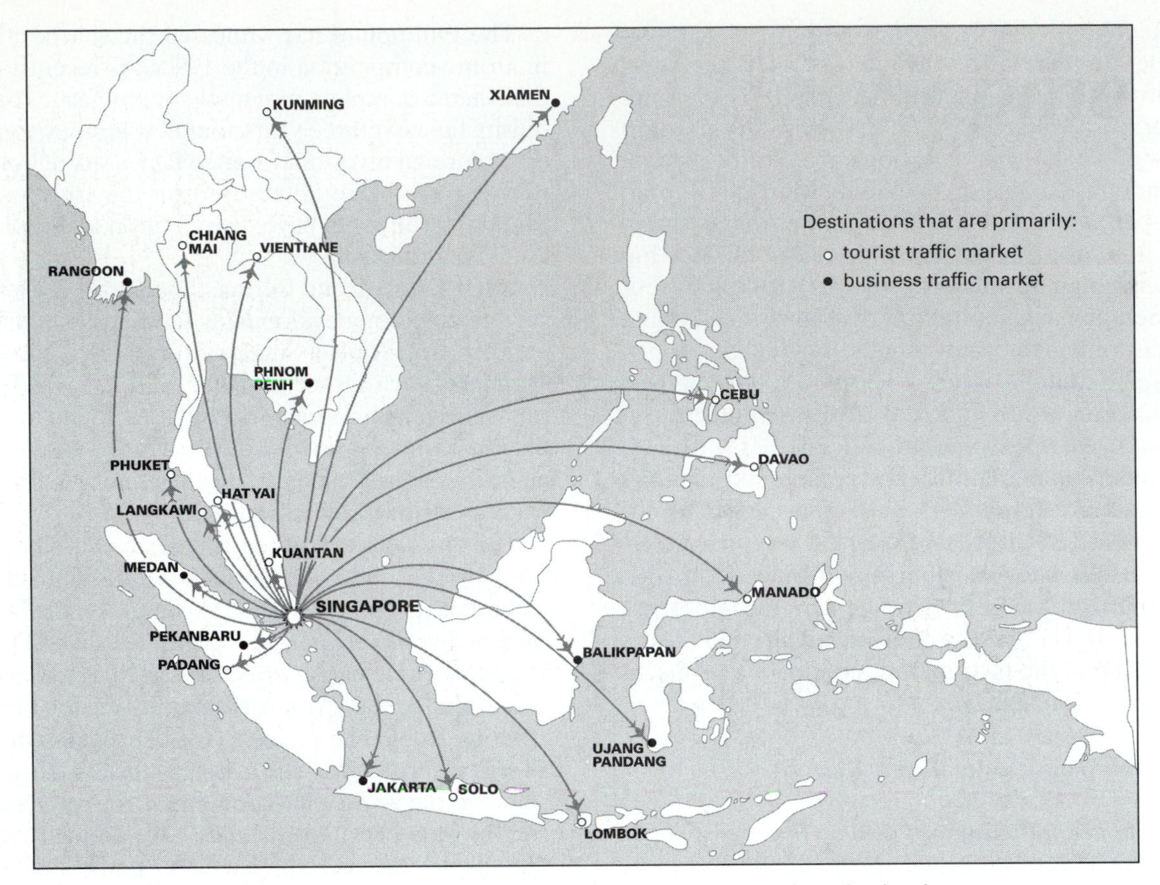

Map 9.7 Silkair, Singapore Airlines' regional subsidiary, serves secondary destinations across Southeast Asia, including a mix of tourist and primarily business traffic and primarily tourist traffic markets. *Source:* OAG Desktop Guide: Worldwide*, vol. 22, no. 9, November 1997.*

interests. Traffic on the route surged in the early 1990s as thousands of United Nations personnel were deployed to Cambodia. In 1997, two Cambodian carriers served the route: Royal Air Cambodge, which is backed by Malaysia Airlines, and Kampuchea Airlines, which is a Thai–Cambodian joint venture. Moreover, Thai involvement in Cambodia's air transport sector is not limited to airline investments: A Thai firm is upgrading the Pochentong International Airport near Phnom Penh. It will expand the passenger terminal, extend the runway, and build the new cargo terminal. The same firm has forged a joint venture with the Laotian civil aviation authorities to upgrade that country's national airline and its airports.

The country poised to record the largest growth in air traffic is Vietnam. The lifting of the United States trade embargo against Vietnam and signing of the trade agreement clears the way for Vietnam Airlines to rebuild its fleet. Previously, aircraft produced in the United States or aircraft with a significant proportion of U.S.-manufactured components (such as the engines) could not be sold to Vietnam. Now Vietnam is expected to emerge as an important market for Western aircraft manufacturers. In addition to major markets in Asia, Vietnam Airlines has also begun to serve points in Europe and Australia, some of which have large Vietnamese immigrant populations. Air France, which was the first Western airline to resume service to Vietnam, is helping the Vietnamese carrier acquire new aircraft and is also supplying maintenance support and staff training.

CHALLENGES

Although the quality and the availability of transportation infrastructure and services vary widely throughout the region, it is clear that the gradual expansion of these has played a key role in developmental growth. It is important to recognize that the need to maintain a competitive posture in the global economy means that infrastructure improvements and the increased efficiency of transport are key as both local and international investors search for opportunities in the regional economies.

In many ways it is in the largest cities, the so-called megaurban regions, where transport constraints and needs are most conspicuous. Mass transit projects have been required for many years; yet the outcomes in these efforts have often been disappointing. Governments must work more closely with the private sector to develop reasonable solutions to move these projects forward. In addition it is critical that a stronger rationalization of the transport system take place. This means essentially that authorities must devise better ways to utilize the existing transport situations. One example is to develop workable means of restraining traffic growth. Singapore's example is noteworthy in this regard.

The process of port modernization in the region must continue, but it is clear that here too rationalization must be employed as these nodes are critical interfaces with the global economy. Air facilities, which are now nonexistent or are unsafe, must also be constructed and upgraded, particularly in Indonesia, Burma, and Vietnam, to assist regional economies as well as to be able to meet a growing international demand for tourism. In this respect there is great potential in Southeast Asia (see Chapter 8). Finally, and in many ways most important, it is necessary that governments renew efforts to respond to the needs of the rural populations in providing the means of transport that will allow the more efficient delivery of social and economic services. A needed critical impact is improved access to employment for both men and women. The solution lies in the creation of stronger incentives and appropriate institutions that will allow transport to transform the lives of those at the economic and social margins.

10 The Role of the State

CAROLYN L. CARTIER

Across the Asian region, from the countries of Southeast Asia to China and Japan, the state plays a strong role in guiding economic and social development. The state is the set of institutions, policies, and worldviews that constitute a government, define national territory, maintain society, and engage in international relations. All countries in Southeast Asia became modern states in the post-World War II period. Yet the practices of the contemporary state demonstrate simultaneously the goals of postcolonial development as well as traditional values of society inherited from the monarchical era. All of the countries in the region, except the Philippines, trace their histories to kingdoms or sultanates, whose values were selectively folded into state ideology during the colonial and postcolonial eras. Just as Southeast Asia exhibits diversity of cultures and histories, no one model of state organization or state ideology prevails. The contemporary state in Southeast Asia is a complex institution that represents both the internationalized political economy of the global order and the cultural and political traditions of deeply rooted local practices.

In most Southeast Asian countries the state has played a direct role in planning social and economic development. The state also defines and defends national boundaries, inscribes values about nationalism in the human landscape, and plays a major role in urban and regional planning for both industrial activity and human settlement. At the regional scale, important interstate agreements, especially ASEAN have defined international diplomatic affairs and helped direct foreign investment and to develop trade relations among member countries. ASEAN has transformed since its post-World War II formation from five original countries Indonesia, Malaysia, the Philippines, Singapore, and Thailand to embrace the goal of encompassing all the countries of the region and is currently playing an important role in defining regional environmental policy. Southeast Asia as a political geographic region and its central multilateral association, ASEAN, have become well recognized in the world order. In other regions where countries do not share cultural and political systems, regional stability and cooperation has remained an elusive goal. In Southeast Asia regional stability has become a reality. How do these diverse nation-states remain distinctive and increasingly unified?

STATE IDEOLOGY

State ideology is the set of values the state uses to underwrite its own legitimacy and connect individual members of society, or citizens, to the national order. It "refers both to a psychological belief system, and also to the articulation of that belief system in the form of a program for political action" (Brown, 1994, p. 6). The origins of the concept of the *nation-state* concept are decidedly Western European and were introduced to Southeast Asia as a result of the colonial period and the process of postwar decolonization. The nation-state ideal is that a nation, a nationwide community with a cohesive national culture, is coextensive with state territory. Of course the reality is that contemporary nation-states are composed of diverse peoples or many nations; this compels the state to create common values and goals that will create unifying national themes and a sense of a shared territory. The successful working of state ideology depends on the state's ability to encompass both individual and community values and beliefs and to tie those values to class and ethnic affiliation, religion, language, dialect group association, and ultimately the ruling national order. The state seeks to achieve a shared national outlook among all its population groups and attempts to build nationalist values that are general enough for all people to embrace. With some variation among individual states, the most dependable elements of a shared national consciousness are generated and controlled through policies on official state history and ideology, language, education, religion, the press, political participation, population size and settlement, and human rights. The state transmits such ideology through government campaigns, in schools and the workplace, and through the media; it also often enforces its perspectives through subtle and sometimes not-so-subtle systems of rewards for compliance and punishments for noncompliance. The state also regularly seeks to suppress localized alternative views and belief systems that challenge the national order.

But how does a population of millions of people attain a common national outlook? Benedict Anderson (1983) has formulated the idea of the *imagined community* to conceptualize how nationalism is a thoroughly constructed concept that has depended on the diffusion of ideas about state ideology and national culture through the press and other media. This concept captures how people cannot *know* their relations to the large community of people that constitutes the nation; because the population is simply too large, they have to *imagine* that they share values and views with everybody else. When all people have read the same textbooks in school, have read the same national newspaper, or have watched the same national news programs on television, they begin to become bound together in a shared national consciousness. The technological mediums of transmission may be television, radio, and newspapers or books, but activating nationalisms for the popular level and transfer to citizen consciousness has often depended on the oratory skills of charismatic leaders, especially in the original ASEAN member countries. Heads of state for Indonesia, Malaysia, the Philippines, Singapore, and Thailand have been particularly effective at transforming state ideology into popular platforms, launching social and economic development goals, and sketching visions of the development path on both the domestic and national scenes. In this biographical approach to state leadership, the Southeast Asian state as an institution mirrors the worldviews of its foremost monarchs, prime ministers, and presidents. The next section focuses on the deployment of state ideology by major heads of state to highlight how the state creates national culture, which contributes to national and regional political stability.

Heads of State and the Deployment of State Ideology

Southeast Asia's hereditary rulers have continued to retain an important if not central presence on the political scene, especially in Thailand and Brunei, and also in Cambodia, Laos, and to a lesser degree Malaysia. The monarchs and sultans of the region's traditional kingdoms and sultanates regularly represent or deploy state ideology that melds powerful symbolisms of traditional cultural values with contemporary state goals. The nonhereditary politicians and elected leaders of the postcolonial order have ranged in leadership style from popularly elected officials serving out designated terms of office to apparent "presidents-for life." Long-ruling regional leaders have particularly contributed to the evolution of state ideology.

When the other countries of Southeast Asia were fighting for or defining independence from colonial rule, Thailand made the transition from absolute monarchy to constitutional monarchy as the independent country it had always been. Despite a history of tensions in Thailand between the military and the government, the country's political forces always proclaimed allegiance to Buddhism and the king. Through it all King Bhumipol Adulyadej has retained the aura of a god-king and has earned extraordinary *prabaramee* (respected charisma) with the Thai people for his moral integrity, nearly faultless leadership, and utter devotion to the country. Much more than a figurehead touring the periphery to implement rural development projects and the like, King Bhumipol has selectively intervened in Thai domestic affairs, especially at periods of crisis in government leadership, and has been credited for restoring order to the nation during political impasses. Elsewhere in the region, only Prince Norodom Sihanouk of Cambodia has come close to possessing so much symbolic royal power. In Cambodia the *deva-raja* or god-king tradition, based in the syncretic forces of Hindu and Buddhist belief, created a pyramidal social order where the king, at the apex, derived power from his position in the cosmos. Prince Sihanouk, though, never enjoyed the degree of popular support that in Thailand translates into real political power and the potential for societal transformation. Nevertheless, from the 1940s to the 1990s, through French colonial occupation, a period of populist rule, military rule, the reign of terror of the Khmer Rouge, Vietnamese invasion, protracted negotiations over a coalition government, occupation by United Nations peacekeeping forces to oversee a national election, and, in 1997, the failure of the coalition government after a coup by Second Prime Minister Hun Sen, Prince Sihanouk has remained the major personification of traditional Cambodian society as well as the country's singular representative on the international scene. In 1993, about to reascend the throne as king after the elections, he was the only figure who continued to hold respect from all the factional groups vying for leadership in Cambodia. In Laos the kingdom was maintained until 1975, when King Savang Vatthana abdicated in favor of a coalition government. But after the communist victory in Vietnam and the rise of the Khmer Rouge in Cambodia, the Laotian communists moved to center stage in Laos.

Sultan Sir Muda Hassanal Bolkiah, who also holds the title of prime minister, defense minister, national university dean, and chief religious leader, celebrated 25 years of rule over the Sultanate of Brunei in 1992. The sultan embodies both religious and royal titles and rules by decree. He is the central arbiter of the official state ideology *Malay Islam beraja*, or Malay Muslim monarchy, that ties Malay ethnicity to Islam as the two fundamental underpinnings of society. Because Brunei's major state activity is to manage the revenue derived from its largesse of petroleum resources, it is referred to in political economic terms as a *rentier state,* in which a country's sources of income derive from external sources and not primarily from domestic taxation (see Chapter 13). In this way the sultan and the royal family need not concern themselves with the range of issues common to most governments' agendas and are able to rule for all intents and purposes as they please. In Malaysia the hereditary system of nine ruling sultans continues to exist in the *Yang di-Pertuan Agong* (king of the nation) system. Every five years, one of the nine Malaysian states with a sultan yields its hereditary ruler to assume the titular and symbolic position as the traditional king of all Malay society. In reality, Malaysia's sultans have gradually lost power and some status over time. For example, in 1993 an act of parliament supported by the prime minister stripped them of their traditional legal immunity. The political force of state ideology in the Malaysian political order lies in the office of the country's prime minister.

Two prime ministers in Malaysia's postcolonial history, Prime Minister Tunku Abdul Rahman and Prime Minister Seri Dr. Mahathir bin Mohamad, have served as the guiding pillars of state ideology. Among the countries that were former colonies, Malaysia experienced a relatively peaceful transition to independence, and for a few short years it seemed that the Federation of Malaysia would encompass Singapore in a new nation-state. But differences over the role of the Singapore's political leadership and Singapore's 75 percent ethnic Chinese population in an increasingly Malay-oriented Malaysia led to a split, and Singapore left the Federation. Prime Minister Tunku Abdul Rahman, the new government's first prime minister, supported equal opportunities for all ethnic groups in Malaysia, but in 1969 a national election led to rioting, especially between Chinese and Malay political factions. Ever since, *1969* in Malaysia has become a national metonymy for potential social chaos should political economic inequities not be managed. In response, other powerful forces in government increasingly called for Tunku's policies of equal treatment to be replaced with new forms of pro-Malay nationalist ideology and economic policy. The events led to a new national ideology, called *rukunegara* (pillars of the nation), which emphasizes the supremacy of the constitution, the rule of law, mutual respect and good social behavior, loyalty to king and country, and belief in God, in which the monarchical and religious values point to Islam and the system of hereditary Malay sultans. The New Economic Policy (NEP) was designed as an explicit policy tool by which to reorder capital accumulation in favor of *bumiputra* (sons of the soil) or Malays. The guiding force of the NEP has been Prime Minister Seri Dr. Mahathir bin Mohamad, who assumed the leadership of Malaysian government in 1981 and has maintained his position ever since. Prime Minister Mahathir has gained legitimacy by a swift and clear-cut decision-making style, promotion of rapid economic development, and support not only of Malay national culture but also Asian culture more generally, partly via critiques of Western culture and Western international political hegemony. Concerning a number of international issues, Mahathir has actually asserted himself as the "spokesperson of the Third World." His proposal for an East Asia Economic Grouping and his role in articulating a platform for the industrializing countries at the United Nations Conference on Environment and Development, the Earth Summit, 1992, are discussed in the sections on regional integration and regional environmental governance, respectively. In the summer of 1997, at the onset of what has become the "Asian financial crisis," Prime Minister Mahathir lashed out at the world's currency traders, accusing them of driving down regional currency exchange rates and thereby undermining regional economic development. In this case, Prime Minister Mahathir's political rhetoric can be examined as anti-Western discourse that serves to mollify his domestic constituency and maintain domestic stability at the risk of some international criticism.

It is difficult to imagine how Singapore would be today without the leadership of Prime Minister Lee Kuan Yew. Prime minister and architect of Singaporean society and economy from independence in 1959 to 1990, now Senior Minister Lee continues to be widely influential in government decision making and international policy. As leader of the main Singapore political party, the People's Action Party (PAP), Prime Minister Lee is widely credited for leading an export-oriented industrialization drive that transformed a small island lacking natural resources into an industrialized country, based on OECD (Organization for Economic Cooperation and Development) standards, by 1996. In tandem with economic development goals, Prime Minister Lee also actively promoted strict social and moral values, based on a mandatory savings scheme and regulation of behavior in the public sphere against smoking, gum chewing, eating on subways, and the like. These smaller-scale regulations prompt Western views of Singapore as a "country of rules" that lacks freedom, but visitors encounter the results in Singapore's sparkling urban environment. As the country became tied to the world economy, Lee was also outspoken about guarding values of Singaporean society against globalizing Western cultural forms. In 1984 he invited the world's reigning neo-Confucian scholars to Singapore, as consultants to the state, to advise the government in school curriculum planning and how to best maintain and adapt Confucian values for modern Singaporean society (Tu, 1984). In these ways, Singapore state ideology depends on a "corporatist model" in which clearly defined expectations character-

Photo 10.1 Billboard advertisement for a "drug-free" Malaysia. (Leinbach)

ize the development path in both the economic and social spheres.

In postcolonial Indonesia, state ideology has been bound up in the leadership of the country's only two presidents, Sukarno and Suharto. Sukarno, by all accounts a great orator and mesmerizing leader, convinced people that the newly installed constitutional democracy was not right for the country and replaced it in 1955 with "Guided Democracy," based on concepts of Indonesian tradition. The cornerstone of his domestic state ideology was the *Pancasila*, or the five principles: monotheism, nationalism, humanitarianism, social justice, and democracy. These principles remained vaguely defined as ideals rather than becoming actual rights. Even though he was a founder of the Non-Aligned Movement, ASEAN, and led the move to join the United Nations and OPEC, Sukarno eschewed democratic ideas of citizen participation on the domestic scale. He turned instead to the traditional concept *gotong royong* (tradition of mutual aid) as the guiding ideology. From villages to the office of the president, decision making would proceed through *musjawarah* (consultation and discussion) and *mufakat* (consensus), rather than at the ballot box. *Gotong royong* invoked traditional patron–client relationships: subordinates accepted the rule of their superiors in harmonious consensus with the expectation that leaders ensured the basic needs of society. Patron-client relationships are the basis of the "patrimonial state," in which the client is loyal to the patron who offers protection. Sukarno projected himself as the *bapak* or father, and people treated him accordingly. In the neopatrimonial state that characterizes Indonesia, state promotion of development and democratic society means that the terms of the relationship

Photo 10.2 Social engineering through advertising, Singapore. (Leinbach)

have changed: Now people expect some material resources in exchange for political support (Brown, 1994). In other words, if the economy does not distribute rewards, the state loses legitimacy, as witnessed in the riots of May 1998. Perhaps Sukarno's greatest political strength was his ability to balance power between the military, the PKI (Communist Party of Indonesia), and conservative Muslim leaders. His greatest weakness was economic mismanagement of the country: *Gotong royong* could not turn a melange of agricultural societies into an industrialized state. By the 1960s his regime faltered and an abortive coup in 1965 attributed to the PKI led to widespread killing of communist party members. Chinese citizens, by simple race-based association with China as a communist state, were next. The bloodbath that brought down Sukarno has been likened to the reign of terror wrought by Pol Pot in Cambodia and the exodus of the "boat people," many of whom were also ethnic Chinese from Vietnam after 1975, as a baseline measure of harrowing social upheaval.

In 1967 Suharto came to power and proclaimed a "new order" for Indonesia. Suharto retained the *Pancasila* but restructured state ideology by banning communism, sanctioning *dwi fungsi* (dual function), which stands for allowing military participation in civilian politics, banning the "floating masses" from organized politics except during elections, and promoting economic development. In these ways he became the sole leading voice in Indonesian politics and more powerful than even Sukarno (Suryadinata, 1997b; Suryadinata, 1996). His relationship with the military has been especially direct and strong, partly as a result of appointing military officers to civilian administrative posts at provincial and national levels. At the international level, Suharto played a key role in formation of the regional association ASEAN and developed close links with Japan and the United States. Suharto has extended his domestic empire through his children, who have entered the business elite through their family connections and amassed a fortune. The Suharto regime appeared for a time to have brought enhanced stability and economic growth to Indonesia, which lent him considerable popular support. But the Asian financial crisis of 1997–1998 led to his political demise. Suharto stepped aside, and without popular elections installed as president his longtime confidant B.J. Habibie. In the midst of the political and economic turmoil, the Chinese were scapegoats for the economic problems even though Suharto's family had squandered an embarrassment of riches.

The Philippines has also witnessed wrenching leadership transitions in the postcolonial period. In the past half-century, two presidents, Ferdinand Marcos and Corazon Aquino, have symbolized the differences and possibilities in Philippine state ideology. President Marcos, first elected in 1965, led the country as both popularly elected official and as dictator during nearly a decade of martial law from 1972–1981. A series of political and economic threats, including challenges to his political authority by Liberal Party leader Senator Benigno S. Aquino, Jr., provoked Marcos to declare martial law in 1972, which was coincident with the end of his second and final term of office. Marcos articulated his clearest form of state ideology under martial law, the New Society Movement, which promised to address a range of economic, political, and social ills. During this era Marcos also sought to enhance his legitimacy by employing traditional forms of social organization. He reached back to the precolonial era to revive the indigenous system of community organization, the *barangay*, as the local level form of political representation to replace the barrio system created by the Spanish. But Marcos continued to rely on maintaining power through cronies and patron-client relationships; he allowed the economy to remain in the hands of a tiny economic elite, and he restricted freedoms on the famous Philippine press. When the regime appeared secure in 1981, Marcos lifted martial law, but two years later former political adversary Benigno Aquino delivered the regime its fundamental challenge by returning from self-imposed exile in the United States. Aquino must have posed the decisive threat to the Marcos dictatorship: He was assassinated on the tarmac as he deplaned at Manila airport. An international incident of dire proportions, few people anywhere believed the government claim that a lone gunman, who was immediately killed by security personnel, shot Aquino. The credibility of the Marcos regime was shattered; this was underscored when the commission investigating the assassination directly implicated high-ranking military officers. In the next national elections, multiparty support coalesced around Corazon Aquino, the martyred widow of the slain leader and ultimate antithesis of Marcos. After ballots were cast, Aquino emerged as the apparent winner as Marcos attempted to claim victory again, against charges of massive fraud and in the face of large-scale public demonstrations. His own supporters began to defect, and some of his key military leaders mutinied and occupied strategic military bases. The U.S. government read the situation for what it was and offered the Marcoses exile in Hawaii. Still struggling to hold power, Marcos attempted to put down the mutiny by his generals and sent tanks into the streets. But in a dramatic confrontation the armored vehicles turned back from thousands of ordinary unarmed citizens, led by nuns, who blocked the major road to the military base outside Manila. U.S. military helicopters landed to pluck Marcos, his wife Imelda, and an assorted coterie of cronies out of Malacañang palace and set in motion his exile from the country. In this "people power" revolution, history was made in the Philippines and simultaneously around the world in spectacular television coverage.

In 1986 Corazon Aquino, an internationally sophisticated and deeply religious Catholic housewife and mother from a wealthy sugar-plantation-owning family, became president of the Philippines. State ideology "Cory style" was people power, which centered at first on large scale public nonviolent demonstrations and more generally on the return of public accountability in government. Aquino institutionalized popular democracy by rewriting the constitution, which established a bicameral legislature and a single six-year presidential term. Her government also attempted land reform, but reforms that threatened powerful interest groups came to little long-term effect. The Aquino government also faced overwhelming international debt, inherited form the Marcos era, and seven military coups. Popularity did not easily translate into leadership, as Filipinos waited for stronger guidance and political economic change. Under the Aquino administration the Philippines became the "capital of NGOs" in Asia, which attests to both the weakness of the state in fulfilling particular societal functions and the openness of the state in tolerance of democratic practice and the evolution of civil society. In these ways, citizen politics in the country has preceded national economic achievements.

The role of the state in the economy began to shift under the presidency of Fidel Ramos, the first non-Catholic president of the Philippines. President Ramos, an army general, served as secretary of defense under the Aquino administration and withstood numerous attempted coups. His loyalty and popularity were important in the stabilization of the country, and these qualities allowed him to win the presidency in 1992. For Ramos, state ideology was commensurate with a solid Philippine plan for economic progress. He generally succeeded through his construction of the "*Philippines 2000*" plan. Although Ramos probably would have been elected to a second term if the constitution allowed, the 1998 national election offered less-experienced candidates and problematic choices: Imelda Marcos emerged to offer herself as a candidate for the presidency. The front runner emerged in a movie actor named Joseph Estrada, and so national leadership in the Philippines has shifted again to a relatively inexperienced populist candidate.

For two millennia state ideology in Vietnam has centered on the themes of nationalism and independence, which Vietnam has earned in victorious military confrontations over major world powers. Vietnam has beaten back China in both modern and historic times. In 1954 Vietnam's victory over its French colonizers brought independence to all of French Indochina and liberated Laos and Cambodia as well. In 1975, Vietnam drove out all U.S. troops after a prolonged war. In 1978–1979 when Vietnam invaded Cambodia to control Pol Pot, China, then allied with the Khmer Rouge, attacked Vietnam. Vietnam, true to traditional form, prevailed on both fronts and beat back both the Khmer Rouge and the Chinese. The source of twentieth-century inspiration for this extraordinary history of Vietnamese patriotism was Ho Chi Minh, the founder of the communist party in Vietnam. In 1945 after the Japanese retreat, Ho Chi Minh proclaimed Vietnamese independence, but the French were not willing to take leave after the war. Ho's army battled the French and then the U.S. forces. The U.S. effort to turn southern Vietnam into a democratic market economy could not prevail in a country where pan-Vietnamese nationalism has always won out over international influence. The ultimate proclamation of the strength of Ho Chi Minh's nationalist vision was fully recognized after the fall of Saigon in 1975 when the victorious communist government renamed the city after their heroic leader, who died in 1969. As the communist bloc continued to collapse through the 1980s, Vietnam lost its economic support and subsidies from the former Soviet Union. The country then turned to a new guiding policy, *doi moi* (renovation), formally established in 1986. *Doi moi* has become the new state ideology of economic and social development in Vietnam.

Government in the Lao People's Democratic Republic (LPDR) continues to be intertwined with the ruling order in Vietnam, and the idea of a truly independent postcolonial state in Laos is just beginning to emerge. Laos existed as a kingdom for some 700 years, centered on the historic settlement of Luang Prabang. The kingdom actually survived until 1975, but after the communist victory in Vietnam and the rise of the Khmer Rouge in Cambodia, the Laotian communists moved to center stage. Future state ideology in Laos continues to lie in the hands of the Lao People's Revolutionary Party (LPRP), the sole political organization, which is dominated by the lowland Lao Loum peoples. Unlike political bodies in Cambodia, the communist party in Laos has maintained relative stability during the 1980s. In 1986 at the Fourth Party Congress, leaders introduced economic reform policies through the New Economic Mechanism and have promoted the new development ideology as the "Lao way to development," which blends market economic principles with enduring Marxist values of the ruling party and traditional cultural ideas about Buddhism. It remains to be seen whether the new Lao way will become an acceptable nationalist ideology and development platform for the country's 47 ethnic groups. In tandem with Laotian reforms, ASEAN has brought Laos more directly into the regional and international arenas by admitting the country to the regional organization in 1997.

Burma's postcolonial history of political representation has repeatedly collapsed into a series of military dictatorships and brutally suppressed popular movements. The state model in Burma has been based on deeply acrimonious politics between ruling state nationalism and ethnic nationalism, in which the lowland Burmans, who inherited the state from colonial rule, have created an "ethnocratic state" model designed to assimilate border peoples, often through violent military forays. State goals of unification translated into "Burmanization," as ruling Burmans compelled other population groups to adopt their perceived more advanced cultural values. But real fears of loss of cultural identity, lack of self-determination, and economic marginalization led the country's minority groups, especially the Shan, Karen, Karenni (Kayah), and Kachin, to organize against Burmese "internal colonialism." For most of the postwar period up to 1995, the country has been at civil war. After 1974 army General Ne Win matched the political assimilationist policy with an autarkic economic policy called the Burmese Way to Socialism. Ne Win outlawed all political parties except his own Burma Socialist Program Party and nationalized the economy, which has led to near-economic collapse. The 1988 political crisis in Burma was provoked in part by a crazy scheme of currency devaluation whereby the state announced the demonetization of 25-, 35-, and 75-kyat banknotes based on the advice of Ne Win's numerologist. New banknotes were issued in alternative denominations, and more than 60 percent of the money in circulation became worthless. People in the capital took to the streets. In September 1988 another army general, Saw Maung, seized power and sought to suppress the popular movement. He set up the contemporary Burmese state with an acronym whose very sound connotes an authoritarian and repressive regime: SLORC. The State Law and Order Restoration Council violently put down the popular movement in 1989 and defined its state ideology as the dual goals of guarding the nation against foreign aggression and internal dissidents. In the case of the former goal, the state has faced no real foreign aggression; in the case of the latter, SLORC has faced a formidable challenge in domestic politics from 1991 Nobel Peace Prize winner Aung San Suu Kyi and her followers. Suu Kyi, daughter of Burma's great nationalist hero and founding father, Aung San, has voluntarily separated herself from her husband and son in England to promote domestic human rights and the removal of Burma's military regime. In 1990 General Saw Maung called for elections in order to confirm the SLORC regime. The country's previous election had been held in 1960. Undaunted by SLORC, 73 percent of Burma's eligible voters turned out for the election from 492 districts, each electing one representative to constitute a people's assembly; they cast ballots overwhelmingly in favor of the opposition (Lintner, 1994b). Opposition candidates received 80 percent of the assembly seats, but SLORC generals nullified the election and the assembly never convened.

ASEAN, with some internal disagreement among its members, has taken the position that embracing Burma through multilateral diplomacy and economic engagement may be better for the country and the region than an isolationist policy that would punish the SLORC regime for its mismanagement. Burma became a member of ASEAN in 1997. In November 1997, the regime changed its name to the State Peace and Development Council (SPDC) in an apparent attempt to highlight its declared aim to ensure the emergence of an orderly and democratic system and to establish a peaceful and modern state. But some of the same leaders remain in charge, and so it remains to be seen when Burma will be able to chart a truly transformative political course.

POLITICAL GEOGRAPHY AND TERRITORIAL PRACTICES

The carving out of Southeast Asian countries from mainland mountains and peninsulas and island archipelagoes has been a long process of territorial definition by kingdoms, colonial powers, and independent nation-states. Defining the political geography of the nation-state and the region has been a continuing enterprise in which colonial and then postcolonial state powers defined sovereignty over territory by fixing territorial boundaries. *Territoriality*, or the regulation and control strategies employed to defend the territorial state, is a critical element of state legitimacy. Although overall regional stability has prevailed in postcolonial Southeast Asia, localized political problems have threatened the stability of the territorial state in both historic and postcolonial times. How the state has cohered through such localized challenges to ruling authority can be examined through territorial practices, or how the government defines and defends its territorial boundaries.

"Mapmaking" and Territorialization

In Thailand, the one Southeast Asian state whose political boundaries should not have been the specific focus of colonial cartographers, the concept of the territorially bounded nation is also a modern one that was definitively shaped by French and British colonial powers (Winichakul, 1994). Before the late ninteenth century, overlapping or multiple sovereignties were common in the nation known as Siam. The making of a nation-state out of the kingdom of Siam required the imposition of a new political geography in which nationhood was defined, at least in part, by the practice of mapping territorial boundaries around the land between Burma and Cambodia. The British mapped Burma; the French mapped Cambodia. The Thais took their information and mapped the land in between. The map of Thailand that we know today first appeared as a byproduct of colonial cartographers at the end of the nineteenth century (Map 10.1). Mapped for the first time, the enduring Thai kingdom gained a visualizable geography

BOX 10.1 The Free Burma Coalition

In 1997 the Free Burma Coalition, a U.S.-based nongovernmental organization (NGO) promoting human rights in Burma, claimed a major victory in its campaign to stop Western corporations from investing in Burma: the U.S.-based soft-drink producer PepsiCo announced that it would discontinue all sales in Burma. Pressure on PepsiCo mounted after the Free Burma Coalition network influenced Stanford and Harvard Universities to decline contracts for PepsiCo products, and major U.S. municipalities, notably the university towns of Madison, Wisconsin, headquarters of the Free Burma Coalition, Ann Arbor, Michigan, and Berkeley, California, endorsed the Pepsi boycott. PepsiCo's decision to pull out of Burma followed similar actions by other major multinational manufacturers, including the Heineken and Carlsburg brewing companies, and the Eddie Bauer and Liz Claiborne clothing companies. How does the work of the Free Burma Coalition illuminate the role of the state in Burma? A military junta known by the acronym SLORC (State Law and Order Restoration Council) has headed the state in Burma and has resorted to oppressive and violent measures to neutralize challenges to its authority. The founder of the Free Burma Coalition is a native of Burma, a student who fled the country in 1988 after the government violently suppressed a student movement. In Burma the absence of state measures guaranteeing the functioning of civil society means that people have been forbidden to organize against the state. In contrast to the ASEAN policy of "constructive engagement" (discussed later in the chapter), which holds that increasing international economic ties with Burma will lead to political economic liberalization and democracy, the Free Burma Coalition takes the position that the state in Burma has used the presence of foreign investment and corresponding international brand names as means of legitimizing its totalitarian regime.

whose "shape," or boundaries, came to constitute defensible territory. Mapped, it was ever more clearly possessed. The colonial exercise of mapping was, of course, inherited by postcolonial governments across the region. Often, new leaders of the independent Southeast Asian states accepted unquestioningly, especially when it was in their favor, the power of the map.

In its historic condition, the kingdom of Cambodia extended to the lands of Laos and the Mekong delta in modern Southern Vietnam, but its more expansionist neighbors squeezed it into its current territorial configuration. Vietnam annexed Cambodia's southeast coast at the end of the seventeenth century, an act that fundamentally defined Cambodian perceptions of Vietnam's enduring imperialist stance. Thailand seized Cambodian territory as well. In the fourteenth and fifteenth centuries the Thai occupied Angkor four times, but these forays did not result in permanent occupation. From the ninth to the fourteenth centuries, the Khmer culture centered at the magnificent temple complex at Angkor and had no rival in the region (see Map 3.5). The Thai intrusions brought to an end the reign of Angkor as the center of the Cambodian kingdom. In the late eighteenth century, five Cambodian provinces came under Thai control. It was this history that France marshaled as justification for their incorporation of Cambodia into French Indochina: They "saved" the Khmer kingdom from the Thai and the Vietnamese (Gordon, 1966). The French actually retrieved lost provinces from Thailand and reincorporated them into Cambodia; on the other hand, they employed Vietnamese to administer their Cambodian protectorate, which only increased Vietnamese presence in Cambodia and set the basis for Vietnamese control of the economy. As a result of this history, Cambodia has lost both territory and territorial control, and its borderlands have come to represent zones of instability.

Territorial practices acted out on Borneo have brought Indonesia, Malaysia, Brunei, and the Philippines into direct and sometimes contentious dialogue over state territory and boundary definitions. Borneo's complex political geography owes to the colonial era when the "white rajahs," really mercantile emissaries of the British crown, prevailed in Sabah and Sarawak. In the North Borneo territories that became Sabah, complex and overlapping historic claims between the Sulu sultanate and the sultanate of Brunei, which antedated the British presence, actually led the Philippines, in 1962, to lodge a claim against the Federation of Malaysia that Sabah was its sovereign territory. This claim, which historians have assessed as unfounded, found support in Sukarno's Indonesia. In the international diplomatic issue known as the *Konfrontasi*, or Confrontation, Sukarno took an offensive position against colonial powers in the region, in part to gain the western half of New Guinea from the Dutch. He especially protested the proposed merger of the Malayan peninsular states, the British Borneo territories of Sabah and Sarawak, and Singapore. Britain defended Sabah against Indonesia and took the view that future stability for its colonies at the heart of the region hinged on a plan to amalgamate all its territories into one large state entity. Singapore of course left the federation, while Sukarno worked on turning the western half of the island of New Guinea into the Indonesian province of Irian Jaya. Brunei's unusual political geography is a product of both the historic sultanate and colonial forces. Sandwiched between Sabah and Sarawak, Brunei, is also partly bifurcated by a tongue of land in

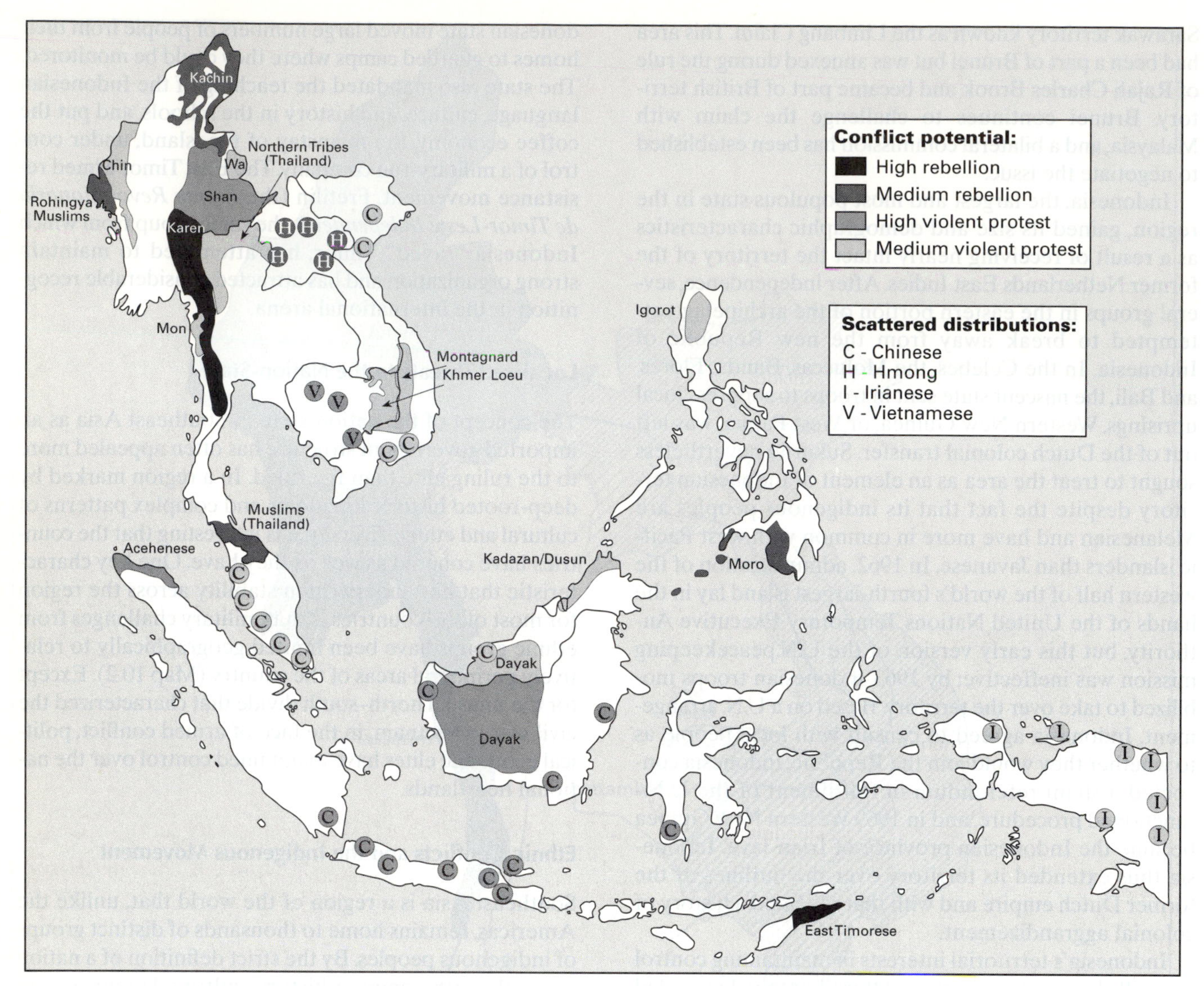

Map 10.2 Recent and potential ethnic conflicts, 1970 - 2000. *Source: Ulack, 1998, p. 26.*

of ideological threats to the "developmental state." Indigenous peoples, such as the Dayaks and Igorots, who desire to maintain traditional nonindustrial, nonurban lifeways, represent values that are antithetical to the goals of national economic development.

Ethnic separatist movements in Burma have presented the greatest challenge to the stability and legitimacy of the state in Southeast Asia. Beyond the Burman heartland of Burma, the indigenous ethnic minority peoples waged war against the state virtually since independence. From the state's perspective, the military conflict has been a civil war; from the perspective of the borderlands, incorporation of the Shan, Karen, Karenni, Kachin, and other nations has been a violation of national sovereignty. The problems originated in a set of anomalies that occurred in the transition from the colonial era. Under British rule, the territory of the Burman heartland was administered as "Ministerial Burma," while the smaller ethnic groups on the periphery were designated the "Frontier Area." This strict division left traditional local leaders in power in the Frontier Area; it was only *after* the country achieved independence that peoples of the frontier were no longer allowed to maintain autonomy. The new constitution of Burma, forged in 1947, gave the Shan and the Karenni peoples the voluntary right of secession after a 10 year trial period, whereas the traditional homelands of other groups, specifically the Mon and Rakhine, were not acknowledged. The Karen, whose goals for a free Karen state were entertained by the British but denied under the new Burman government, declared the Free State of Kawthoolei in 1949 and came the closest to winning sovereignty by military contest.

After seizing power in 1962, General Ne Win nationalized all land and set out to extend state control over the ethnic minority states. Decades of fighting on the frontiers came to a slow, deadly end in the mid-1990s when the state negotiated cease-fire agreements with ethnic group armies. In the case of the Karen, the state had to take the Kawthoolei capital, Mannerplaw, by military force first. In 1994, the notorious leader of the Shan State, Khun Sa, who was wanted in the United States on opium drug trade charges, actually volunteered to surrender himself if international powers would grant political recognition to the sovereignty of the Shan State. In Burma, most analysts view the state's position to gain peace beyond the heartland as motivated more by access to valuable natural resources than humanitarian concerns for peace and stability. Already Burma has negotiated with Thai forestry industries to log teak just inside the Burma border with Thailand, areas that were long under the de facto control of several indigenous groups. Thailand banned the export of raw logs in 1987, its own teak forests in many areas of the country clear-cut beyond predictable ecosystem recovery.

The Philippines and Thailand have faced separatist movements of another sort, where Muslim groups located in areas of historic sultanates have resisted incorporation into the state. In the southern frontier of the Philippines on Mindanao and in the Sulu archipelago, the Malay Muslim groups known as Moros have challenged state authority since independence. In 1972, with the imposition of martial law and the directive to lay down arms, long-standing Muslim opposition to the central government coalesced in the formation of the Moro National Liberation Front (MNLF), which brought together all Muslim groups opposed to the state, internationalized their cause, and effectively led to government acceptance of "separate development" for Muslims (Funnell, 1990).

In Indonesia, nationalist policies of ethnic accommodation rather than assimilation have depended on the acceptance of state authority by indigenous groups. But several groups have remained committed to self-determination, and Indonesia's territorial conflicts with localized peoples range over the vast archipelago. In northern Sumatra the people of the "special region" of Aceh have pursued a separatist movement since the colonial era. The Acehenese most recently mounted a military offensive in 1989–1991 that brought reaction from the Indonesian army. Peoples of East Timor, incorporated into the Indonesian state against their will, have continued to press for autonomy. In East Timor, the Catholic Church has become a major institution of resistance and remains under direct supervision of the Vatican. In Irian Jaya, the *Organisasi Papua Merdeka,* or Free Papua Movement (OPM), has fought Indonesian rule since 1969. The OPM has reasons to suspect national accommodation policies. About indigenous Irianese (Papuan) people, one Indonesian minister was known to have exclaimed that the government must "get them down out of the trees, even if we have to pull them down" (Osborne, 1985, p. 136). An education minister in the Suharto cabinet explained that the government would do its best to bring progress to the people but that their life ways "would take time to be abolished" (Osborne, 1985, p. 137). In response to suggestions from Western observers that perhaps the Papuans ought to be allowed cultural self-determination, another government official responded that Indonesia "had no intention of trying to create a 'human zoo.'" In this case of Irian Jaya, state practices toward the indigenous people have amounted to what Bernard Nietschmann (1985) has termed "fourth world colonialism."

The state often takes a management approach to indigenous peoples, particularly when they occupy primary natural resource areas. In Malaysia the Department of Aboriginal Affairs developed "regroupment schemes" for *orang asli*, or original people of the Malay Peninsula who have traditionally lived in forested slopes of the peninsula's central mountain range. The name *orang asli* is an umbrella term that encompasses diverse indigenous peoples on the peninsula, most of whom speak a language of the Mon-Khmer family and whose presence in the area antedates that of the Malays themselves, which represents an intellectual challenge to *bumiputra* policies. Under such schemes, the *orang asli*, who by state definition had no legal claim over the land, were expected to give up their partially peripatetic subsistence lifestyle in favor of growing traditional cash crops on designated aboriginal reserve sites. The longer-term goal of the Department of Aboriginal Affairs has been to bring the *orang asli* out of the forests and assimilate them into Malay society and to promote conversion to Islam. In response, some *orang asli* as well as indigenous groups in the Borneo states have organized to promote their constitutional rights to maintain their cultural identity. The United Nations Working Group on Indigenous Peoples, established in 1984, is the major international-level forum through which indigenous peoples can gain support and recognition and form alliances with other indigenous groups.

LANDS AND LANDSCAPES OF NATIONALIST SENTIMENT

Geographical manifestations of nation-building practices are widespread and visible in symbolic landscapes of national culture. Such places may be new landscapes created by the state to monumentalize the ruling order and government achievements, such as central squares and plazas or prominent high-rise buildings. The historic landscapes and sacred sites of the monarchical era often endure as important symbolic places of national culture where political leaders and citizens alike visit to demonstrate their cultural values and beliefs. National ideology is also often represented in urban and regional development plans that seek to organize patterns of settlement and in the landscapes of ethnic groups where the state promotes economic development projects that would overwrite the place-based histories of local peoples.

The historic cities of courts and the modern cities of capitals lie at the national heartlands of Southeast Asian states. Heartlands, in the sense of national landscapes, can be thought of as real and symbolic centers of national culture and power that all people of a nation-state understand as fundamentally place-based representations of national meaning. All over the Southeast Asian region traditional centers of rule continue to serve as important symbolic national heartlands. In Burma, Pagan and Mandalay are the centers of great historic capitals that mark a proud historic Burman nation and the role of Buddhism in society. In Rangoon, the capital of the colonial and postcolonial eras, Shwe Dagon pagoda endures as the religious symbolic center. Thailand's former capitals, Sukhothai and Ayutthaya, north of Bangkok, tell a story of landscape history that fundamentally anchors the central plain as the enduring heartland of the Thai nation. In Cambodia, Angkor Wat is the historic politico-religious center; around Tonle Sap, the great lake, people find the fecund heartland of sustenance and the ecological basis of a historic agricultural society. The lowland Lao Loum peoples of Laos dominate the heartland of the country, and their culture has emerged to control the modern political leadership of the state. In Malaysia the hearth of the Islamic nation is at Malacca, the country's first sultanate, but the heartland of the turn of the century belongs to the capital district where Prime Minister Mahathir has presided over monumentalizing a successful rapid industrialization drive in the new "multimedia supercorridor" that includes a new capital city, Putrajaya, the high technology city Cyberjaya, and in Kuala Lumpur itself, some of the tallest buildings in the world, the Petronas Towers (See Box 7.1). The heartland of Indonesia belongs to Java in Jogjakarta, another historic sultanate, while massive Jakarta is the primary political economic center of the contemporary state. Manila is Jakarta's equivalent in the Philippines. The Philippines, unlike the rest of the Southeast Asian states, had no precolonial urban tradition, and the meaning of local society in the *barangay* concept remains a strong national force.

Rather than just describing where and what they are, here we can analyze landscapes of national significance to assess how the state seeks to inscribe or control nationalist symbolism in the built environment, how local people experience such national landscapes, and ways they interpret and rework localized centers of national identity. These "nationscapes" serve as constant reminders of the rule of the state and national culture, and as places where people experience national identity on their own terms.

In Malaysia the state declared Malacca its singular historic city in 1989 to commemorate the settlement's remarkable past as the location of the first sultanate on the peninsula, the fount of Islam in Malaysia, and as the Southeast Asian center of international trade during the early colonial era. Malacca was the first European colony east of India and historic structures from waves of colonial rule, Portuguese, Dutch, and British, mark Malaccas central historic district (See Photo 3.3). The state saw Malacca's historic landscapes as resources for heritage tourism planning and has promoted Malacca as a prime tourist destination with the slogan "Where it all began . . . ." But in Malacca the state also favors interpretations of history that promote "Malayness" and, sanctioned by the goals of the NEP, has invested in the built environment to create landscapes that represent Malay culture; for example, no major elements of the built environment marked the era of the Malacca sultanate, so the state sponsored construction of a grand replica of a sultan's palace. Malacca is also the site of the world's largest traditional Chinese burial ground, a 500-year-old cemetery called *Bukit China* (Chinese Hill) (Photo 10.3). In the middle of the 1980s, the state eyed the site for tourist development and proposed to develop the land into the world's largest cultural theme park. This proposal spurred the formation of a preservation group, supported by the major national Chinese political parties, to thwart the plan. Preservationists documented how the site represented a half-millennium of Malaysian history in a series of historic events that occurred on the hill, and they contended how the presence of sacred Malay graves there symbolized the history of Chinese-Malay community harmony. The preservation campaign grew into a national movement that led to the conservation of the site as a historic open space and jogging park (Cartier, 1997). As a result of widespread dissemination of important information about the landscape history of *Bukit China*, state leaders could only support the conservation of a landscape whose history emerged as a geographical metaphor for the nation itself.

In Thailand, Bangkok is the ultimate center of state and society, and the Chao Pharya River is its artery, connecting city to country and symbolizing the lifeblood of the nation. The great royal palace at the center of Bangkok is a formidable landscape of the Thai nation, but the palace is not a place for personalized, regular connection to the national order. For the nationscape that is both available and inspiring, Bangkokians gather at the broad square on Rajadamnoen Avenue where a bronze statue of King Rama V stands as a monument to his reign. Rama V died in 1910, but his reign is revered for instituting in Thailand a range of reforms from the abolition of slavery to the introduction of the modern school system. "On some nights, thousands of people throng the broad square . . . . They lay out rush mats on which they erect small alters and portraits of the King . . . . Offerings of brandy, cigars, and pink roses are favored because it is believed that the King was something of a *bon-viveur*. The ritual bears all the hallmarks of a cult" (Vatikiotis, 1996b, p. 137). What draws thousands of young people and middle-class professionals to this

Photo 10.3 Aerial view of Bukit China, Malacca, Malaysia. (Cartier)

place? People obviously find in this nearly pilgrimmatic activity a touchstone to the past, a connection to a proud Thai nationalist identity and one that symbolizes a counterpoise to the hypermodern experiences of place otherwise found all over high-rise, fast-moving Bangkok. This nationscape in Bangkok is the literal inscription of tradition *in* modernity. Such places in the Southeast Asian region also demonstrate the degree of tolerance that the state is able to maintain for localized belief systems and cultural practices.

ROLES OF THE STATE IN SOCIETY

Regional government leaders know that achieving the goals of modernization and development depends fundamentally on the participation and stable functioning of society. The state works to build social and institutional contexts that allow diverse peoples to cohere, work, and prosper ultimately as members of a civil society, or a society in which the citizenry is able to participate in political activities with the state's guarantee of citizen safety in the exercise of participatory activities. Without social coherence and stability, the state could not guarantee economic development. The following sections focus on the developmental consequences of state-society relationships in the region.

State Policies on Language and Education

In countries of cultural and ethnic diversity, the state regularly inculcates values of national culture and promotes national unity by teaching a common language. The role

Photo 10.4 Joggers stretching at the Eng Choon Association monument, Bukit China summit. The Eng Choon monument is a commemorative site for the Eng Choon Association, the largest place-based association of the Chinese community of Malacca. (Cartier)

other publications. During the 1992 demonstrations in Thailand against the military regime, the government attempted to censor newspapers critical of the military leadership, but the offending publications simply printed editions with large blanked-out sections to throw support to the popular movement. Several countries in Southeast Asia have maintained control over the media and have focused on thwarting criticisms of the state and its leaders. In Malaysia newspapers and magazines are regulated by the Printing Presses and Publications Act of 1984, and the state threatens the press with closure should they intimidate the government. Political parties in Malaysia own or control major newspapers and radio stations, and so their coverage can be expected to reflect political platforms. The major English language daily, the *New Straits Times*, is controlled by the UMNO, while *The Star* has been published by the Malaysian Chinese Association (MCA). In Singapore, sales of major regional news weeklies, such as the *Far Eastern Economic Review*, have been banned for periods of time based on reporting the Singaporean government deemed incorrect or libelous. Brunei has also censored the international press, and the state owns the country's only television station. In Indonesia most newspapers are privately owned but are subject to state regulations as well as general surveillance. In 1994 the government shut down three popular publications in reaction to stories about tensions within the president's cabinet. In the least developed countries, controls over the flow of information are more extreme. In Burma freedoms of speech and of the press have been nonexistent since 1962. SLORC has monitored the television, radio stations, and newspapers and controls the telephone system. Television and radio stations in Vietnam are all owned by the state, and the 1993 Law of Publishing prohibits any material that would appear to be critical of the state. But by one estimate, under the new market economy activities of *doi moi,* half the books and newspapers published in Vietnam are not licensed (Porter, 1993), which is perhaps a consequence of market deregulation the government is willing to tolerate.

The Formation of Civil Society

The political participation by citizens of a nation-state and the state's guarantee of citizen safety in the exercise of participatory activities are measures of evolution of civil society and the foundation of national political culture. Participation takes diverse forms, from formal state-defined freedoms of speech and the press and rights to organize, protest, and vote, to informal measures such as patron–client relationships and social networks based on personalism. These informal aspects of social participation in social networks and patron–client relationships particularly characterize Southeast Asian society and in some countries are the most important elements of citizen interaction. Personalism, or simply relying on personal relationships, means that who you know can be vastly more important than law or government institutions or even personal abilities in advancing through society or gaining access to special opportunities.

Across Southeast Asia, the evolution of civil society varies considerably, from the more open democratic states of the Philippines and Thailand and the burgeoning meritocracy of Singapore, to countries whose citizens still have few opportunities to participate in the political process. In Thailand the constitution grants freedom of assembly, speech, and religion, which have been guaranteed except during periods of martial law. Thailand has numerous NGOs, and such citizens' grassroots groups have been important in evolving the country's environmental movement. The only major restriction on exercising one's opinions is lèse-majesté, the law against criticizing the royal family. Advocating communism is also against the law. But Thailand has also exercised human rights limitations, in denying citizenship to political refugees, especially indigenous peoples who have fled into Thailand from neighboring Burma, and in outlawing worker strikes. Policies like these are always justified on the ground that tolerating them would threaten national security and, by extension, the entire development process. The 1995 Thai Constitutional Amendment Bill gave women equal rights to men and required laws to stop forms of discrimination by sex.

In the Philippines, which might be referred to as the capital of Western-style democracy in the region, full voting rights were achieved at independence in 1946. The United States introduced political reforms, created a commonwealth, and promoted a democratic government without undertaking land reforms to dismantle the pattern of concentrated land ownership that has always defined the Filipino elite. Critics charge that despite the semblance of democratic practice, attaining power in Philippine society continues to be determined by personalistic ties and is dominated by a very few wealthy landowning families who have come to dominate political offices as well. Some go as far as to say that the worst legacy of colonialism in the region was the U.S. impact in the Philippines because "it saddled the country with American-style democracy" (Vatikiotis, 1996b, p. 6) to underscore that democracy without land reform has actually impeded economic development. Under the Aquino administration the government created the Philippine Commission on Human Rights, which gained the country a certain legitimacy in the international arena by supporting many United Nations proposals. The role of NGOs in the Philippines has become increasingly important, especially in support of environmental problems and women's rights, but this reflects both the freedoms and limitations of the Philippine system. The Philippines, like Thailand, has been a major center of "sex tourism" and is also the major source in the region of international women migrant workers in domestic service jobs. In response, several women's NGOs have worked with government to create more and better opportunities for women (see Chapter 11). Overall the opportunities of an "open society" in the Philippines have both allowed the perpetuation of elite political economic power and the empowerment of formerly more marginalized groups, and in this complex and contested political landscape, achieving rapid economic development has remained elusive.

Compared to the freedoms of speech and the press in the Philippines, it is common to find Western perceptions of Singapore that characterize the country as a soft authoritarian state because of the role of the state in regulating society and the severe restrictions on criticizing government. Technically Singapore has a multiparty system, but the ruling People's Action Party has never been seriously challenged by opposition parties. NGOs that represent environmental concerns and family and women's issues are under evolution, and Singapore is home to one of the region's major NGOs promoting the status of women and gender in development (see Chapter 11). Still, where civil society is concerned, Singapore is less a center of citizen activism than Malaysia and Indonesia. The achievements of economic development in Singapore provide a clear refutation of the Western idea that high levels of economic development go hand in hand with civil liberties.

Malaysia and Indonesia closely monitor citizen participation in society and regard citizen critiques of government policies as potential crimes against the state. But participation in politics is an active element of Malaysian society, and the Malaysian opposition party, the Democratic Action Party, has enjoyed some solid achievements in the postcolonial period. In Malaysia, state control over citizen participation is maintained in the name of preventing ethnic tension. In 1987, during a state-defined period of potential racial disunity, Malaysian Prime Minister Mahathir invoked the Internal Security Act, which allows incarceration without trial, and detained more than 100 people including the leader of the country's main opposition party. Prime Minister Mahathir claimed that ethnic strife was brewing, while critics charged that many of those arrested had not been involved in racial issues but rather had begun to present a threat to the security of his own position (Means, 1991). Mahathir is planning Malaysia's future on all fronts in the state's broad-scale development plan, Vision 2020, and does not suffer challenges to this larger set of goals.

In Indonesia, human rights abuses continued to occur during the New Order, especially in East Timor, and during the economic crisis of May 1998. There is no right to free association in Indonesia, the state maintains control over NGOs, and activism is forbidden on university campuses. The *Pancasila* has served as a broad-reaching and generally defined social platform that tolerates diverse religions and cultures, but critics charge that it has been used to stifle dissent. Suharto restructured state ideology by sanctioning *dwi fungsi* (dual function), which stands for allowing military participation in civilian politics, banning the "floating masses" from organized politics except during elections, and promoting economic development. Suharto created an official political organization, Golkar, a hybrid kind of superparty that encompasses the bureaucracy, the military, and the business elite. Golkar representatives have overwhelmingly prevailed in every national election. The importance of informal participation in civil society is so strong in Indonesia that patron–client ties have been institutionalized in state ideology, originally in Sukarno's policy of *gotong royong*, discussed above. Partly as a result, the NGO sector in Indonesia expanded significantly in number and diversity during the postcolonial period.

Brunei has experienced limited democratic politics. In the 1960s the British encouraged the contemporary sultan's father to institute a measure of elective government, and Bruneians held their first elections in 1962. But the sultan refused to convene the legislature, and Brunei has not held elections since. Sultan Hassanal Bolkiah's domestic and international state policies run parallel courses: He has deemed that Bruneians are disinterested in the democratic process and has requested that the international community not attempt to influence Brunei's state practices with Western values.

Doi moi in Vietnam has resulted in some substantial changes in the political sphere that should enhance the evolution of civil society. Leaders have allowed more debate in the National Assembly, released thousands of political prisoners under government amnesties in 1987 and 1988, and cracked down on corruption within the communist party and the government. In 1989 Laos moved forward with the first elections since 1975 and installed the members of the new National Assembly. In Cambodia, by contrast, politics have remained chaotic. The United Nations Transitional Authority in Cambodia (UNTAC) worked to install a democratically elected government following 20 years of factional infighting between at least four rival claimants to state authority: Prince Sihanouk and his party, Funcinpec; the Khmer Rouge; the government of Hun Sen, who was installed by Vietnam; and the rightists led by Son Sann, a former prime minister under Sihanouk. In the elections Hun Sen's Cambodian People's Party (CPP) narrowly lost to the Funcinpec Party, led by the Sihanouk's son, Prince Ranariddh. But the CPP held control over the military and police, and Hun Sen forced himself into sharing leadership of the country. Prince Ranariddh and Hun Sen emerged as first and second prime ministers, respectively, but in July 1997 the internationally sanctioned coalition government collapsed when Hun Sen staged a coup while the prince was out of the country. ASEAN leaders stepped back and postponed the planned admission of the country into the ASEAN group that was slated to occur only two weeks hence.

State Policies on Population Planning

Across Southeast Asia the state has taken a direct role in population planning. State promotion of birth control policies relates directly to development goals. In the developed countries, industrialization and urbanization have generally led to a decrease in the rate of population growth. Thus, in tandem, the promotion of family planning with industrialization and urbanization through economic development are viewed as the keys to managing population size. Most Southeast Asian states have articulated specific population policies. Both Thailand and Indonesia have promoted slowing population growth through carefully managed family-planning programs. Currently, the Philippines and Vietnam are facing the greatest challenges to manage high birthrates. By contrast, in Malaysia and Singapore the state has promoted population growth for very different reasons.

In the Philippines, the powerful Catholic Church has made the implementation of birth control technology especially difficult, a position which former President Aquino tacitly upheld. Under President Ramos, the state began to stress the health benefits of contraception to mothers and children as a way to make birth control seem more reasonable. Whereas in all the other major ASEAN countries choices in birth control technology are available and widely used, in the Philippines the majority of people have continued to rely on traditional methods and the birthrate has remained high. The other country in the region whose state is concerned about population size is Vietnam, where in the period 1968–1988 the number of births per woman dropped from six to four, and the state goal of the two-child family could not be met. In Vietnam under *doi moi* and the return to household-based farming, the largely rural population has continued to regard children as valuable assets for farm labor. The state officially promotes a policy of two-child families, and sanctions have been applied for couples who have more than two children.

The state in Malaysia has taken a pronatalist policy and promoted a five-child family with the goal of economic expansion, while Singapore has called for increasing its low birthrate to ensure against future labor shortages (Tyler, 1992). Prime Minister Mahathir initially tied the pronatalist policy to Malaysia's early focus on state-planned strategy of import-substitution industrialization, in which a large population would be required to consume the fruits of the country's planned industrial production. But this justification for the policy has not proved realistic. More to the point, the policy has generally been viewed as a less clear-cut way to increase the proportionate size of the Malay population. In Singapore, by contrast, the low rate of population growth has led to concerns over the size of the future labor force, and as a result, the state has promoted a planned increase in the population growth rate. State promotion of population growth has been especially aimed at highly educated women, which stirred controversy about a government eugenics campaign.

In Singapore the state also promotes the family household as the basic social unit of society and has institutionalized marriage in the system of state housing. The Singapore Housing and Development Board (HDB) houses 85 percent of the population, and the vast majority of HDB housing units are apartments in high-rise towers in residential estates. Partly because private housing is so expensive, demand for HDB housing exceeds supply and the HDB has had to maintain a five-year waiting list that accepts only married couples. This situation has engendered an interesting response from young adults. Couples who are not ready to get married but who, if they did get married, would not want to live with family members while waiting for their HDB apartment to be assigned have officially married in a government civil ceremony to acquire the documentation that allows them to sign up for the HDB housing wait list. They do not consider themselves, nor do their families and friends, "really married" but only "registered married." If they do marry, it will be the traditional formal ceremony with family and friends that is regarded as the solemn matrimonial pledge. Here we see an ultimately pragmatic response to restrictive state policies. In 1997 Singapore shifted on this policy to consider allowing single people older than 35 years of age to purchase their own HDB flats.

In Indonesia and Thailand the national birth-control campaigns have become models for global success stories in state fertility policy. Indonesia's National Family Planning Coordinating Agency, a central government unit that reports directly to the office of the president, has set target fertility-reduction goals and has promoted family planning under the slogan "two is enough." The state's family-planning drive began under the rule of President Suharto who secured cooperation from Islamic religious leaders. The campaign works through a system of community volunteers called *kaders* who provide local family-planning advice; it also offers couples who limit their family size to two children a program of incentives, including a trip to Mecca for those who have pledged successfully for 10 years. Indonesia looks forward to an average fertility rate of two children per couple by 2010. Thailand initiated a state population policy in 1970 with the national policy slogan "many children make you poor" as the basis for an economic argument to limit the number of children per family to two in both rural and urban areas. The Thai Ministry of Public Health established the National Family Planning Programme (NFPP) to manage the campaign and make contraceptives widely available. Between 1979 and 1982 headmen of all rural villages were called on to attend family-planning seminars; for the Islamic minority in southern Thailand, the government arranged field trips to Indonesia to demonstrate how Muslims were also actively participating in family planning. By 1990 the fertility rate in Thailand had dropped to an average of just above two children per couple; the country had practically achieved its targeted goal, from an average of six children to two, in just one generation.

State Policies on the Chinese in Southeast Asia

Ethnic Chinese overseas comprise more than 28 million people, of which roughly 20 million, or nearly 5 percent of the total population of Southeast Asia, reside in the countries of Southeast Asia (Suryadinata, 1997a; Poston and Yu, 1990). The contemporary population distribution of Chinese in the region owes to a long history of migration between the provinces of South China, especially Fujian and Guangdong, and historic trading centers in Southeast Asian ports. European colonial powers in the region encouraged Chinese and Indian migrant populations to serve as "merchant middlemen," which created a pattern of uneven capital accumulation in favor of the urban-based Chinese. The postcolonial state has met the social and economic conditions created by this historic Chinese diaspora in different ways.

In Thailand, the state sought to control the Chinese community through the early twentieth century, but, in essence, Thai government policy permitted Chinese business activity for the trade-off of accepting some forms of assimilation, such as adopting Thai names and attending government schools. Thailand-born children of ethnic Chinese became, on the principle of jus soli, Thai citizens, and Chinese have been able to apply for and attain naturalized citizenship. The Thai, compared to the other Southeast Asian states, have maintained a strong policy of assimilation toward the ethnic Chinese.

In Indonesia the failed coup of 1965, which the state associated with communist Chinese influences from mainland China, led to invoking the Internal Security Act and placed controls on the ethnic Chinese. The act banned dragon dances, firecrackers, bringing Chinese-language materials into the country, and the public display of Chinese characters. Prominent signs declaring the national ban on the import of print materials in the Chinese language greeted visitors on arrival at the international airport in Jakarta. After 1990, with the normalization of relations between Jakarta and Beijing, the ban was partially lifted, and now it is permissible, for example, to print tourist pamphlets in Chinese for distribution to Chinese-speaking tourists. But as a June 27, 1997, *Jakarta Post* front-page headline proclaimed, there is no end in sight for the ban on the Chinese language in Indonesia, based on continuing concerns about racial integration. Indonesia does not accept the concept of jus soli; ethnic Chinese born in Indonesia are not citizens of the country and must apply (and pay) for citizenship later in life. During the May 1998 riots in response to the political economic crisis threatening to engulf Indonesia, the Chinese were scapegoats again. Ethnic Chinese in Indonesia form only a small part of the country's total population at just under 3 percent, but they form a disproportionately powerful group in the economy. Reports surfaced widely that Chinese were specifically targeted for random acts of terrorism including specific acts of violence aimed at Chinese women in forcible gang rapes by organized forces associated with the military. A history of regional racism against the Chinese has been horrifically refreshed in late twentieth-century Indonesia, as the state has faltered disastrously in coming to grips with the extremes of uneven development created under the Suharto regime (see Map 10.2).

Malaysia has the largest Chinese minority population (28 percent) by percentage of total population of any country in the world. The country has an explicit ethnic policy written into its economic development planning through the NEP. The first page of the Second Malaysia Plan set forth the goal of restructuring "Malaysian society to correct economic imbalance, in order to reduce and eventually eliminate the identification of race with economic function" (Malaysia, 1971, p. 1). Viewed critically, however, policies aimed at poverty alleviation actually assisted ethnic groups unevenly and virtually ignored the fact that there are also poor rural Chinese (L. Lim, 1996). The NEP's Industrial Coordination Act created an industrial policy that actively discriminated in favor of *bumiputra* businessmen. Compared to Indonesia, Malaysia's Chinese population forms a large portion of the total population and is politically active, and so state policies have actually had to be more accommodating rather than coercively assimilating.

In the postcolonial Philippines, a first order of nationalistic business was the enactment in 1954 of the Retail Trade Nationalization Law, which restricted retail trade—the specialty of Chinese business practice—to Filipinos or Filipino-owned corporations after 1964. Chinese proprietors who engaged in trade before 1954 could maintain their businesses, but they could not entertain the traditional practice of passing on the business to their children. Most Chinese did not have Philippine citizenship, and, unlike the practice in Thailand, immigrant Chinese could not easily gain naturalized citizenship; their children were also considered Chinese nationals. The Chinese tried to get around this by attempting to get some family members naturalized citizenship, but the process was uncertain and expensive. The whole situation for the Chinese in the Philippines changed dramatically with one presidential decree in 1975, when Marcos, under martial law, issued Presidential Letter of Instruction No. 270 requesting quick naturalization for qualified aliens. In the Philippines too, where economic performance had been marginal, the ethnic Chinese were a strong economic asset.

After the civil war came to an end in Vietnam, state-led economic restructuring included policies aimed against Chinese business owners. In 1975, the Declaration concerning Industry and Trade included the stipulation that "all property of compradore capitalists . . . , whether they have fled abroad or remain at home, is . . . confiscated in whole or in part depending on the nature and extent of their offenses" (Vo, 1990, p. 65). The declaration primarily affected ethnic Chinese, who were blamed as collaborators of the South Vietnamese. In 1975 and 1976 the state ordered raids on the companies of "compradore capitalists" in Saigon–Cholon, the city home to one-half the ethnic Chinese population in Vietnam. Many of them sought to leave instead, in waves of "boat people" whose often unseaworthy craft poured into the South China Sea through the second half of the 1970s. Some of the boat people were luckier than others and found homes in alternative countries, but thousands were also forcibly repatriated to Vietnam, especially from British Hong Kong, based on the rationale that they were economic and not political refugees and therefore not internationally recognized as deserving international protection afforded under the United Nations High Commission on Refugees, which only makes provision for political refugees.

REGIONAL INTEGRATION AND MUTILATERAL AGREEMENTS

Major changes in global politics and the world economy after World War II have promoted new relationships between nation-states and enhanced regional integration. To address transboundary issues and the set of transnational concerns created by flows of goods, capital, politics, people, and ideas, leading members of the Southeast Asian states have established regional organizations and multilateral agreements to set common policy and facilitate political and economic exchange.

Early efforts at regional integration in Southeast Asia were political groupings designed to carve a zone of security out of what appeared to be an unstable Asia open to the threat of communist movements. In this light, the first major regional organization, the Southeast Asian Treaty Organization (SEATO), was formed in 1955. SEATO, however, was heavily advocated by the United States and promoted by European powers as a sort of Asian answer to NATO. Indonesia and Malaysia refused to join. A more indigenous strategy, the Association of Southeast Asia (ASA) was proposed instead by Malaysia's then Prime Minister Tunku Abdul Rahman. ASA, launched in 1961, formed a regional group with Malaysia, Thailand, and the Philippines. But its existence was dashed by the British formation of the Federation of Malaysia in 1963, which brought into play competing territorial claims on Borneo between the Philippines and Malaysia. Malaysia faced more serious problems with Indonesia, which peaked in the period 1963–1966 in Sukarno's Confrontation policies by making very real the threat of intraregional military expansionism. This event provided the fundamental impetus for a regionally defined regional association (Vatikiotis, 1996).

The Rise of ASEAN and the Formation of AFTA and APEC

In 1967, leaders of the democratic market economies in Southeast Asia, Singapore, Malaysia, Indonesia, the Philippines, and Thailand, formed the Association of Southeast Asian Nations (ASEAN) to promote regional peace and security. The organization was formed during the Vietnam War, at the height of the cold war and on the heels of Sukarno's Confrontation. In the formation of ASEAN, regional leaders seemed to have agreed that they could not afford to contend with both regional instability and the communist threat. In 1984 Brunei joined the organization after it gained its independence from Britain. Vietnam joined the group in 1995, followed by Burma and Laos in 1997, and Cambodia in 1999. Discussions have been underway to consider membership for Papua New Guinea that will make the "ASEAN 11."

By the late-1990s ASEAN had made its mark on the international scene as a significant political economic regional organization, but its current prominence belies more than two decades of relative inactivity. In its early years ASEAN was more of a symbolic political grouping than an actively functioning economic organization. ASEAN did not hold its first summit meeting until 1976. This move to action was brought on by the end of the Vietnam War, the U.S. departure from the political scene, and uncertainty over future political events in the region, including the "reign of terror" led by Pol Pot in Cambodia. During these years the Cambodian crisis was the major issue facing the ASEAN nations. A decade later in 1987 members met in Manila in response to the new role of Japan in the regional economy—during the 1980s Japan surpassed the United States as the region's number one trading partner and aid donor. As a result, the association began to formulate more-substantial economic policy. In the 1990s, ASEAN emerged as a fully functioning regional association with clear political economic goals and has taken a pragmatic approach to the less-developed countries of the region. Where Burma is concerned, ASEAN has taken the perspective that constructive engagement would lead to greater economic stability and some dilution of the extreme characteristics of the military regime by measures of democratic political practice. On this basis, Burma, which has been the focus of investment boycotts based on charges of human rights abuses, has been admitted to ASEAN. ASEAN is willing to take the long-term perspective that expanding interstate economic linkages will lead to greater overall regional security and prosperity.

At ASEAN's fourth summit meeting in 1992, the group concretized regional economic integration in the plan for the ASEAN Free Trade Area (AFTA), which calls for implementation of a common tariff scheme for regionally produced manufactured products with at least 40 percent ASEAN-made content. The impetus for AFTA was a partial response to similar regional economic integration schemes around the world, the widespread adoption of export-oriented industrialization policies within the region, and recognition of the combined role of regional governments and the private sector in forging greater ties for economic cooperation in light of competition coming from outside the region (Tan, 1996). The original plan called for implementing regional free trade in 15 years, in 2008, but the time frame was revised and shortened in 1994 to the year 2003, as a result of the formation of other regional free-trade policies. Assessments of the potential impacts of AFTA suggest that the agreement may be more of a response to international economy than a plan that will directly impact the region in the near term. For example, intra-ASEAN trade accounts for less than 20 percent of the regional total, and 40 percent of this owes to already tariff-free trade in Singapore (L. Lim, 1996). At the international scale, AFTA has a larger role to play for the ASEAN states as a regional policy response to the Asia–Pacific Economic Cooperation (APEC) forum.

The formation of the APEC forum in 1989 by Canada, Australia, the United States, New Zealand, Japan, South Korea, and the ASEAN states now also includes in its membership China, Taiwan, and Hong Kong. This is one of the few international organizations that includes in its membership both China and Taiwan, which represents fundamentally how the organization is primarily economic in its constitution and goals. Similarly, need for independent representation by Hong Kong also speaks to the important roles of the Hong Kong and Taiwan economies irrespective of their political status with regard to the Peoples Republic of China (PRC). APEC gained momentum quickly and has held annual summits in cities around the Pacific. Its membership has also agreed to pan-regional free trade by 2010 for developed country members, including Singapore, and by 2020 for developing country members. The APEC free-trade plan prompted ASEAN to shorten its deadline for intraregional free trade in Southeast Asia with the goal of furthering its own regional development before the larger Asian regional economy and the United States become more intertwined (Lim, 1996). The United States has been heavily in favor of APEC and has widely promoted the organization. But for ASEAN, APEC challenges ASEAN's regional identity and coherence, on three general fronts. First, although ASEAN must be a core group of APEC countries, it may be difficult for ASEAN to preserve a distinct voice within the larger APEC grouping and direct focus on Southeast Asian regional interests. Second, the membership of several major industrialized countries in APEC necessarily shifts the focus of issues from that of a developing-country stance, which has been ASEAN's perspective, to that of a "north–south" dialog. Growth in the ASEAN states has been dynamic, but by comparison to the larger population base and GNP of the APEC group, the ASEAN population is at most one-fifth of the APEC total, and its total GNP constitutes and even smaller proportion at less than 5 percent. Enlarging ASEAN by the addition of new member states is one way the Southeast Asian regional association can maintain its position in the face of larger populations and economies. ASEAN may also be able to gain position within APEC by careful management of FDI because the larger APEC economies are its primary sources of investment.

Prime Minister Mahathir of Malaysia has been especially concerned about relations between Asian states and the larger world economy. In the international political economic arena Prime Minister Mahathir has promoted an Asia-centric perspective in the idea of an East Asia Economic Grouping (EAEG), which explicitly excludes the United States and other non-Asian countries. Mahathir saw the EAEG as a potential counterweight to the regional integration arrangements proceeding in Europe and the Americas and proposed the EAEG in 1990 as an Asia-driven alternative to APEC. He has also been concerned to anchor Japan's overseas economy securely in the Asian region. But APEC, which has been promoted by the United States and includes both Asian and non-Asian countries, has prevailed over other bodies as the largest-scale economic organization in the region. Two Southeast Asian states, specifically Indonesia and Thailand, were not in favor of the EAEG idea because they viewed their export-oriented economies as too tied to U.S. and European markets to risk appearance of creating protectionist economic blocs. Mahathir has scaled back his original proposal to the idea of an East Asian Economic Caucus (EAEC), which may function as a subgroup of APEC.

Growth Triangles

Growth triangle plans, discussed and planned at ASEAN meetings, are one major way that Southeast Asian countries with different levels of economic development cooperate rather than compete across national boundaries. The concept of the growth triangle (GT) scheme originated at the center of the region in Singapore where the state was seeking lower cost locations and new development sites for manufacturing production. Across the causeway in Johor state of Malaysia and south into the Indonesian islands of the Riau Archipelago, which are closer to Singapore than they are to Indonesia, land and labor costs are much lower. In this GT arrangement three states made the best of comparative advantage and factor endowments (see Map 14.6). Singapore companies had actually been moving some operations to Johor in the early-1980s, and so the GT scheme announced in 1989 by then Deputy Prime Minister of Singapore Goh Chok Tong followed prevailing economic patterns. Now prime minister, Goh is responsible for coining the *growth triangle* phrase in a plan to tie Singaporean investment to Batam Island in Riau Province (Kumar and Lee, 1991). Successful development of the Johor–Singapore–Riau GT scheme has led to other ASEAN area GT plans. Malaysia, Thailand, and Indonesia have designated a second regional GT scheme, centered at the Thai–Malay border and straddling the Straits of Malacca to northern Sumatra. The Indonesia, Malaysia, Thailand GT is planned to diversify the industrial base in the area and will give some preferential investor status to citizens from the three countries. Another ASEAN GT has been proposed between the southern Philippines, in Davao, Manado in the northern Celebes, Indonesia, and Sandakan in Sabah, Malaysia.

Southeast Asia and its East Asian Neighbors

In the context of the larger Asian regional political economy, Southeast Asian states must regularly negotiate relations with the two largest economies in the region, Japan and China. Japan has become the number-one source of foreign aid and FDI in the region and a major trade partner for many of the Southeast Asian countries. By the next century China may become the largest economy in the world and simultaneously represents a considerable political force in the continuing leadership of the Chinese Communist Party and the maintenance of an authoritarian state. During the twentieth century the larger Asian regional geopolitical strategies of both Japan and China have entered directly into Southeast Asia. Japan's march over Southeast Asia in World War II engendered considerable local animosity in Southeast Asia, but the memory of the wartime legacy of the Japanese occupation continues to fade with Japanese aid and foreign assistance in the region. Regional communist movements in several Southeast Asian countries operating in sympathy with "Red China" during the first two decades of communist China's Maoist era (1949–1976) threatened the stability of the region. Fears sowed in the period of regional communist uprisings have remained alive in regional politics and in discriminatory policies against ethnic Chinese citizens in some Southeast Asian states.

Japanese foreign aid and assistance. The role of Japan in the Southeast Asian region has evolved dramatically since World War II. Japan has become the number-one source of foreign aid or overseas development assistance (ODA) for Southeast Asian countries and the number-one source of FDI in the region. Japan's aid and investment presence in the Asian region is greater than the comparable U.S. role in Latin America or the role of European states in Africa. Japan's enhanced ODA commitment to the ASEAN nations began with Thailand during the Cambodian conflict as a way of helping to maintain stability in the country in the face of the Khmer Rouge. Similarly, Japan directed considerable FDI to Thailand to help fund its EOI policies. In the 1980s Japan broadened its efforts in ASEAN and spread its aid disbursements between Thailand, Malaysia, Indonesia, and the Philippines. In the 1990s, Japan has become the number-one source of ODA for all Southeast Asian countries except Vietnam and Malaysia. Japan sends US$2–3 billion a year in total bilateral aid to the countries of the region, mostly for the development of major transportation infrastructure projects (JMFA, 1996). In 1993, Japan announced a 50 percent increase in ODA over the next five years. Based on cumulative disbursements, Indonesia is the overall top recipient of Japan's bilateral aid, now followed by China, which is currently receiving the greatest amount of Japanese ODA per year, the Philippines, and Thailand. In the Southeast Asian region, the United States is a major provider of assistance only to the Philippines.

The 1985 Plaza Accord, which realigned currency exchange rates among major industrialized nations, led to a major increase in the value of the yen and a consequent shift of Japanese investment offshore in search of lower-cost manufacturing operations. Simultaneously, a recession in the ASEAN countries in 1985–1986 led to relaxation of regulations on the utilization of FDI. In this period, Japanese investors substantially increased their FDI in Southeast Asia. In regard to balance of trade, while Japan provides a market for ASEAN exports, ASEAN's economic relationship with Japan is the opposite of that with the United States: ASEAN actually has a trade deficit with Japan but enjoys a trade surplus with the United States. Japan has maintained a trade surplus with the region because of the massive amounts of equipment and machinery that Japanese companies have exported to Southeast Asia to set up manufacturing industries, and because of the quantity of Japanese-made components that are shipped into the region for final assembly. Japan's regional trade surplus with ASEAN should shrink as the Southeast Asian industrial base matures.

Having noted the important role of Japanese aid and investment in the region since the 1970s, this has now changed as a result of a severe recession and bank failures in Japan (see Chapter 7, p. 177 and Sender, 1999).

The China factor in ASEAN geopolitical relations. In the regional geopolitics of the postwar period, China's political economy has shaped ASEAN policy and ASEAN–China relations at several key junctures. On the mainland, Thai and Malaysian concerns over the potential geopolitical moves of China and allied communist countries led in 1971 to ASEAN's first policy declaration, the ZOPFAN (zone of peace, freedom, and neutrality) plan, which called for Southeast Asia to remain free from interference by foreign powers. The Cambodian conflict in the 1970s and 1980s, exacerbated by the Vietnamese invasion of Cambodia in 1978, was a serious regional political problem that pushed military pressure on Thailand's Cambodian border. By this time, China aimed to establish a de facto regional coalition with Southeast Asian countries against Vietnam, which it effectively achieved in 1979 when Thailand allowed Chinese arms to be shipped through Thai territory to the Khmer Rouge in Cambodia. In this way China secured Thailand's position against the Vietnamese and provided an important repository of reassurance for Malaysia and Singapore as well, the other countries whose leaders had reason to interpret Vietnamese aims as expansionist (Buszynski, 1995). Malaysia's Prime Minister Mahathir had remained particularly untrusting of China's moves—he continued to regard China as maintaining links with regional communist parties in Southeast Asia. In fact, Mahathir had some reason for doubt because the (increasingly nominal) existence of the Chinese Communist Party of Malaysia remained an obstacle in Malaysian–Chinese relations until 1989, when it was finally formally disbanded. But by the mid-1980s, China began to shift its position on the Cambodian issue and began to reduce its support for the Khmer Rouge. Instead, China endorsed factions of the anti-Vietnamese coalition headed by Sihanouk and Son Sann and later went on to play an important role in the Paris Conference that led to the U.N.-sponsored peace settlement in Cambodia in 1990.

In China's immediate post-Mao era, the new Chinese leadership called for the normalization of diplomatic relations with Southeast Asian countries as early as 1977, especially with Indonesia. Thailand had established diplomatic relations with China earlier than other ASEAN countries in 1975. But the failed communist coup in Indonesia in 1965, which threatened the heart of the country as the Chinese Embassy staff in Jakarta proclaimed support for the communists, decidedly dashed any effort Indonesia might have made to reestablish diplomatic relations with China early in the post-Mao era. The seriousness of the impasse in the Jakarta–Beijing axis led other ASEAN countries, notably Singapore, not to proceed on normalization of relations with China until the Indonesians resolved the issue. But unlike the Western expectation that political and economic relations are regularly mutually constitutive, ASEAN nations and China developed economic ties in advance of diplomatic normalization. On the economic front, as China and ASEAN countries independently pursued EOI policies, trade between the countries began to increase. The year 1985 marked an economic downturn in the region that motivated Prime Minister Mahathir to visit Beijing for the first time to develop economic relations and to encourage China to increase imports from Malaysia even before the CPM threat had dissolved. Direct trade commenced between Indonesia and China in 1985, fully five years before political ties were resumed. Economic relations gave way to political normalization, and Indonesia restored diplomatic relations with China in 1990. Singapore quickly followed suit, and this action fostered Singapore's economic opportunities in China. Well before 1990, Singapore had consistently been China's number-one economic partner in ASEAN, in terms of both trade and foreign investment. Indeed, Singapore's economic relations with China have been so strong that Singapore acted prudently not to bring greater regional attention to its growing economic ties with China. In 1978, China's premier leader, Deng Xiaoping, visited Singapore and was so impressed with economic and social development in the city-state and with former Prime Minister Lee Kuan Yew's leadership of the country that the two leaders established a series of high-level contacts that led to the largest joint-venture development project in China, the Singapore–Suzhou Township, a Singapore-style residential industrial estate. Chinese leaders have praised the "Singapore model" as one whose set of political economic conditions, high level of economic development, and strong social policy controls China might consider worthy of emulation.

ASEAN countries have tended to share with China a view that emphasis on human rights and democratization are Western policies that are not fully engaged with the realities and social conditions of Asian societies. In a region where development is the prevailing state goal, the most basic human right is the joint state and societal goal of economic development. This definition of human rights does not encompass individual human rights and views the concept of universal human rights as culturally specific. Indonesian, Malaysian, and Singaporean leaders have been outspoken on the issue of individual human rights viewing it as one of many ways that the West seeks to undermine the governing traditions of Asian countries and assert moral superiority. Indonesian President Suharto claimed that any attempt to impose

BOX 10.2 *"Shadow Ecology" and the Southeast Asian Timber Industry*

Southeast Asian sources of tropical timber and timber products are concentrated in the Philippines, the Malaysian states of Sabah and Sarawak, and in the outer islands of Indonesia, where about one-half of the forests remaining in tropical Asia are located. From the 1960s to the 1990s these three countries logged primary forests heavily and unsustainably and have been Southeast Asia's largest timber exporters. The primary destination of Southeast Asia's tropical timber exports is the most highly industrialized country in the larger Asian region, Japan. Since the 1960s Japan has been the world's largest importer of tropical timber, and more than 90 percent of its tropical timber imports have come from the Philippines, Indonesia, and Malaysia. The Philippines led the regional export boom in tropical logs, beginning in the 1950s when U.S. companies such as Weyerhauser and Georgia-Pacific began to provide technical assistance to develop the Philippine logging industry. The country quickly became the number-one source of raw logs imported into Japan, which consumed nearly two-thirds of Philippine log exports during the log export boom (1964– 1973). As the quantity of easily accessible high-quality logs declined in the Philippines, the regional center of log export shifted to Sabah and Sarawak and to outer islands of Indonesia, Kalimantan, Sumatra, Sulawesi, and the province of Irian Jaya in western New Guinea, where logging has proceeded at a frenzied pace. As a result, today only Sarawak continues to export raw logs legally (Table 10.1).

The relationships between Japan's foreign aid and investment, state development plans in Southeast Asia, and the regional tropical timber economy are intertwined (Dauvergne, 1997). In the 1970s Japan's ODA (overseas development assistance) and FDI directly and indirectly financed the development of the Indonesian timber industry. These transnational political economic relations demonstrate the complex problems behind mitigating tropical deforestation in the region. In 1967 Indonesia promulgated the Foreign Capital Investment Law, which initiated a wave of foreign investment in the timber economy and intensified logging in the outer islands, especially Kalimantan. Japan's early role in the timber industry began, like that of the United States in the Philippines, with technical assistance in exchange for timber export agreements. After 1967, the number of Japanese companies investing in the timber industry increased, including almost all the major Japanese trading companies such as Mitsubishi, Mitsui, and Sumitomo, while the Japanese Overseas Economic Cooperation Fund loaned money for infrastructural development including access roads in forest areas. Japan became the primary investor in Indonesia after 1967. In 1971 Indonesia replaced the Philippines as Japan's number-one source of raw logs, which were largely processed into plywood in Japan. Under *Repelita I*, Indonesia's first five-year plan (1969/70–1973/74), the state targeted increased foreign exchange earnings from raw log exports. Suharto promoted the timber industry by granting forest concessions to the military and a handful of important business associates, based on traditional patron-client relationships. As a result, the Indonesian Department of Defense business group controlled 14 timber companies during the heyday of the raw-log export era in the late 1970s, and the military continues to be centrally involved in timber extraction in East Kalimantan. Under *Repelita II* (1974/75–1978/79) the state shifted to efforts to developing the forest products industry, especially plywood, and by 1980 began to implement a phased ban on the export of raw logs. After the ban on export of raw logs took effect in 1988, most Japanese companies withdrew from the Indonesian timber industry, but Japan remains the largest consumer of Indonesian plywood. The environmental impacts of international timber demand in Southeast Asia are part of Japan's "shadow ecology," or the set of environmental impacts that one country's political economic activities produce in the natural resource base of other countries (Dauvergne, 1997). By the 1990s, Japan and Southeast Asian tropical timber exporting countries started to take steps to manage production sustainably, but the deforestation of primary forest in Southeast Asia has already become the central event in the region's postwar environmental history (see Chapter 2).

"foreign" values on another society actually constitutes a violation of human rights (Robison, 1996). Prime Minister Mahathir of Malaysia has promoted his idea of EAEG as a multilateral alliance that could assume a political dimension on the international human rights issue. So far, Mahathir has not received much support for the EAEG among ASEAN nations. It is in regard to the geopolitics of the South China Sea islands, known generally in the international arena as the Spratlys, that China's influence in ASEAN affairs continues to be most far reaching and undefined.

The Southeast Asian Regional Environment

Regional environmental planning is becoming increasingly important in Southeast Asia, especially after the first "haze

Year	Philippines	Sabah	Sarawak	Indonesia	Other*	Total
1970	7.542	3.960	1.872	6.091	0.585	20.050
1971	5.701	4.130	1.472	8.181	0.627	20.111
1972	5.136	5.409	1.377	8.977	0.738	21.637
1973	5.896	7.309	1.251	11.213	1.008	26.695
1974	3.886	6.997	0.951	11.450	0.891	24.175
1975	2.853	5.598	0.702	7.298	0.522	16.973
1976	1.692	8.490	1.738	9.656	0.591	22.167
1977	1.501	8.138	1.487	9.272	0.546	20.944
1978	1.559	9.212	1.496	8.986	0.539	21.792
1979	1.264	8.200	2.268	9.769	0.582	22.083
1980	1.073	6.306	2.260	8.639	0.663	18.941
1981	1.418	5.471	2.917	4.138	0.784	14.728
1982	1.308	6.442	4.049	2.453	0.869	15.121
1983	0.648	6.238	4.075	2.111	0.807	13.879
1984	0.935	5.483	4.256	1.328	0.941	12.943
1985	0.510	5.892	5.395	0.137	1.061	12.995
1986	0.264	6.019	4.778	0.000	1.056	12.117
1987	0.027	6.980	5.494	0.000	1.154	13.655
1988	0.033	5.351	5.260	0.000	1.001	11.645
1989	0.052	4.641	6.683	0.000	1.184	12.560
1990	0.023	3.420	6.749	0.000	0.909	11.101
1991	0.002	2.577	6.468	0.000	1.067	10.114
1992	0.000	2.064	6.363	0.000	1.543	9.970
1993	0.002	0.293	4.293	0.000	2.222	7.440
1994	0.000	0.000	4.463	0.000	2.339	6.802
1995	0.000	0.000	3.902	0.000	2.022	5.924

**Other includes Papua New Guinea, Solomon Islands, Vietnam, Burma, Laos, Cambodia, and other minor exporters in the region.*

Table 10.1 Japanese tropical log imports from Southeast Asia, 1970-95 (million cubic meters). *Source: Dauvergne, 1997.*

crisis" in 1997 resulting from the combined effects of west Pacific El Niño forces and extensive forest burning in Borneo and Sumatra. Environmental policy making is difficult business, however, and more so in the developing world where rapid economic growth is a regular goal. Moreover, economic development theory has not historically accounted for impacts on the natural environment. Why? Because polluting the air, water, soil, and other environmental impacts have generally been, in monetary terms, free of cost. These uncosted variables, called externalities, do not appear in systems of formal economic accounting and, as a result, have not been prioritized in economic development planning. But the rise of an international environmental regime, most recently articulated at the global scale in Rio de Janeiro in 1982 at the United Nations Conference on Environment and Development,

BOX 10.3 Territorial Claims in the South China Sea

Territorial sovereignty over the hundreds of small reef islands in the South China Sea has been a serious regional dispute. The Spratly Islands at the center of the territorial dispute are claimed in part or in entirety by six governments: China, Taiwan, Vietnam, the Philippines, Malaysia, and Brunei. China, Taiwan, and Vietnam claim the whole group; the Philippines claims the central portion of the area, and Brunei and Malaysia claim parts of the southern margin of the island group (Map 10.3). The Spratlys consist of more that 400 coral reef outcrops and low islands scattered over 300,000 sq mi (800,000 sq km) at the surface of the sea—hardly habitable territory. But projections about oil and natural gas concentrations under the South China Sea bed have made them a much sought after natural resource area. China, the nation-state furthest from the center of the island group, has been the most assertive claimant (Valencia, 1995). A violent clash took place on Johnston Reef in 1988 between China and Vietnam, in which three Vietnamese ships were sunk. In 1992 China negotiated with the U.S. firm Crestone oil to explore on the western fringe of the Spratly group, within Vietnam's 200-mi Exclusive Economic Zone (EEZ), and deployed submarines to patrol the area. China's push into active exploration led the ASEAN ministers to promulgate the formal ASEAN Declaration on the South China Sea, which called for all claimants to settle disputes peacefully. Fundamentally undeterred, China pressed on the Philippines claim in a more provocative incident when in 1995 it sent a naval squadron to occupy Mischief Reef, well within the EEZ of the Philippines. EEZs are 200-mi (321.8-km) zones defined under the international Law of the Sea and are thus generally recognized as extensions of territory under state sovereignty. This incident at the appropriately named Mischief Reef catalyzed a united ASEAN protest against China, and a U.S.-government policy statement on the problem (USIP, 1996). China has since responded that it will pursue a solution on the Spratly claims consistent with the U.N. Convention on the Law of the Sea, but the range of issues dividing the claimants to the Spratlys remain unsettled.

the Earth Summit, has led to worldwide recognition of a new development paradigm, sustainable development. Sustainable development promotes the perspective that resources should be used at a rate that does not compromise the natural resource base for future generations. Although a laudable goal, the reality of implementing sustainable development fundamentally questions high growth-oriented economic development and economies based on intensive nonrenewable resource consumption. Further, the sustainable development paradigm originated in Western environmentalism not in development studies or economics and, as a result, has yet to become adequately conceptualized for implementation by development planners around the world (Adams, 1990).

International politics of sustainable development came into play at the Earth Summit, and Southeast Asia's representatives were at the center of debate. In response to calls from North American and European constituencies to curtail tropical deforestation, Prime Minister Mahathir of Malaysia articulated a platform for industrializing nations: If tropical developing states such as Malaysia decreased logging, industrialized countries should compensate them financially for lost foreign exchange. Mahathir justified this spokesman's role by hosting leaders of 55 developing countries in Kuala Lumpur for the Second Ministerial Conference of Developing Countries on Environment and Development in advance of the Earth Summit, where he hammered out a platform pinpointing "North–South" differences in environmental perspectives:

> Fear by the North of environmental degradation provides the South the leverage that did not exist before. It is fully justified for us to approach it in this way. The developed countries have no tropical forest but by involving environmental issues they wish to control the exploitation of forests in developing countries. But we are also acutely conscious that we are a developing country which needs the wealth afforded by our forests. We do not cut down our trees foolishly. We need living space, we need space for agriculture, and we need the money from the sale of our timber. If it is in the interest of the rich that we do not cut down our trees then they must compensate us for the loss of income. Well, if we have to service the world's need for oxygen, for ecological balance, then we must be fairly compensated. Or else allow us our right to our timber wealth.(*"Let's Green the World . . . ,"* 1992)

Critics view Prime Minister Mahathir's antiindustrialized country stance as more rhetoric than reality because the economic development policies he really promotes are in line with economic policies of the World Bank and other Western-based economic institutions (Vatikiotis, 1992). But Mahathir's Asia-centric politics have gained him the regional spotlight on international diplomacy, and he has also assumed a role as spokesperson for the Asian regional economy and its relations with the United States and Europe.

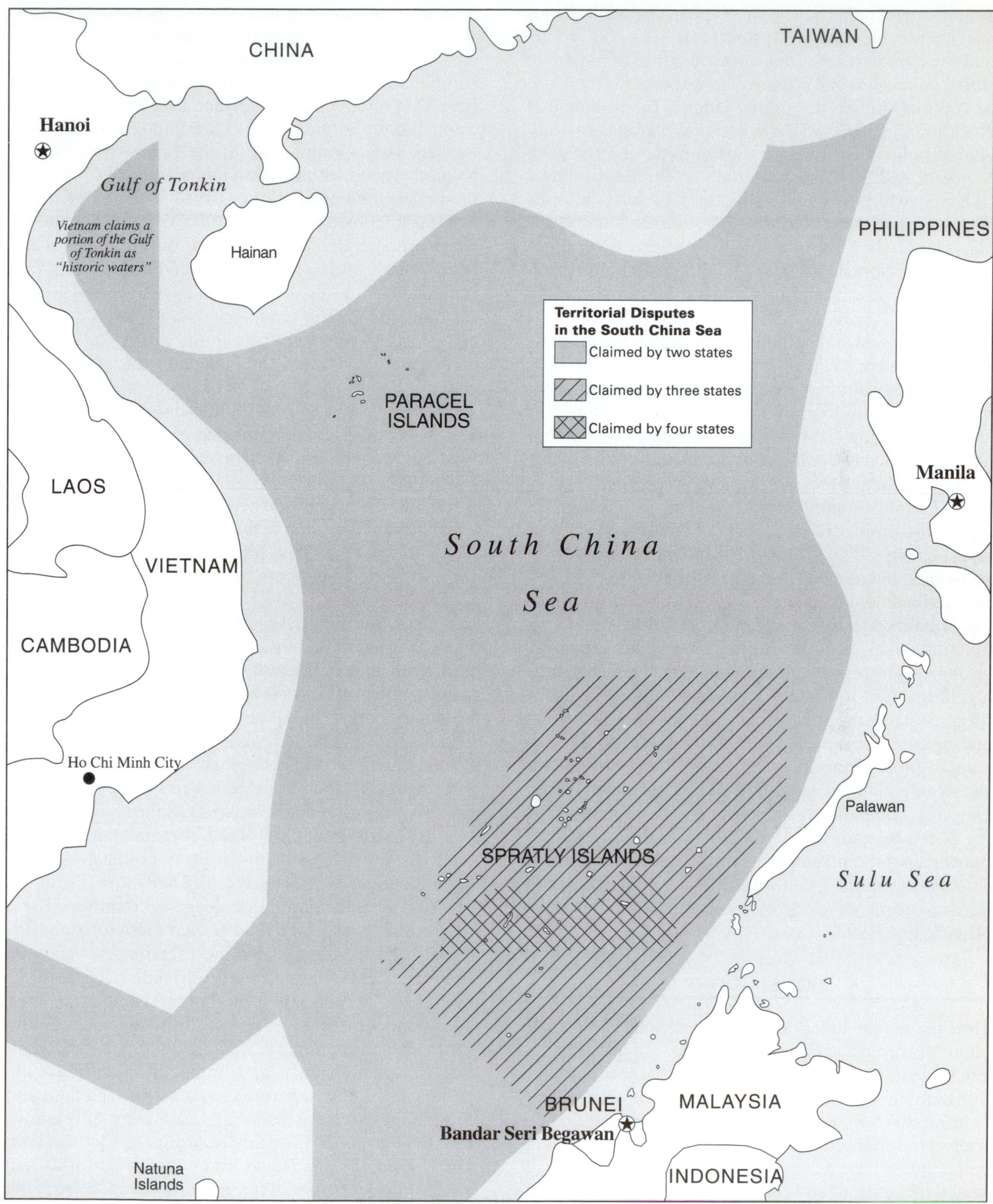

Map 10.3 The Spratly Islands. *Source: Various.*

Only a few countries around the world have devised a national sustainable development plan. The Netherlands plan is a notable example. In Southeast Asia, ASEAN actually responded earlier than most countries or regions to global environmental concerns. In response to the United Nations Conference on the Human Environment at Stockholm in 1972, ASEAN initiated regional environmental policy by forming a subcommittee on the environment, ASEAN Experts Group on the Environment (AEGE), which formulated the first of a series of three regional environmental planning measures under the name ASEAN Environment Programme. After 1992, ASEAN responded to Agenda 21, the major global policy document resulting from the Earth Summit, by upgrading the AEGE to ASEAN Senior Officials on the Environment (ASOEN), which formulated the ASEAN Strategic Plan of Action on the Environment, 1994–1998. This plan lays out five general objectives and ten more specific strategies to promote environmental protection and conservation (ASEAN, 1994). Highlights of the objectives include integrating environmental concerns into all development processes, achieving regional environmental quality standards, and studying implications of AFTA increasing free trade on the environment. To attain such objectives, ASOEN has recommended several strategies, including evolving a framework for integrating environmental concerns in regional decision making, enhanced state–private sector relations on environmental issues, improving the database on regional environmental information, strengthening institutional and legal capacities to implement multilateral environmental agreements, especially in regard to biological diversity and coastal zone protection, control over toxins and transboundary movements of hazardous waste, and the promotion of environmentally sound technology. However, as is commonly the case where environmental policy is concerned, there remains a considerable gap between policy and enforcement. It is also worth noting that among major specified strategies, none focuses on deforestation or conservation of tropical forests (Brookfield, 1993).

CONCLUSIONS

How the nation-states of Southeast Asia have remained culturally distinct and regionally unified is a question that can be answered only by careful country-by-country understanding and thoughtful interstate comparison. First, cultural diversity within the nation-state is traditionally rooted in inherited practices of historic kingdoms and sultanates, whose achievements and legitimacy are still selectively utilized and celebrated and whose actual descendants, like descendants of royalty in European countries that no longer are monarchies, still command presence on the national scene. Second, perspectives on traditional culture and cultural diversity are deeply rooted in national territory, in traditional heartlands and indigenous homelands. Indigenous cultures and indigenous peoples survived the colonial encounter in Southeast Asia. They have not necessarily fared well under the postcolonial state, but they have existed, endured, and transformed. The colonial state in the region transformed political economies more than it fundamentally reworked the underpinnings of culture. Third, a characteristic of cultural tolerance prevails in many countries of the region, demonstrated by syncretic cultural complexes and belief systems and by the existence of diverse forms of social organization, including less highly structured gender relations and greater tendencies toward matriarchal social structure than elsewhere in the world. Cultural tolerance has not become embedded in the ideology of all states in the region, but it has been an important, albeit sometimes elusive, element of state ideology in the region. Finally, what Western political theorists term patron–client relations in Southeast Asia means that the quality of an individual's relations with other people and the larger community is critically important. In simple terms, in spite of political and economic difficulties, people seek to get along. In these ways the state in Southeast Asia is, with some exception, one that respects the individual who respects the group.

Across Southeast Asia, regional coherence can be thought of as defined by sets of unique and overlapping transnational ties. Each of the states shares a different set of characteristics with each of its regional neighbors, whether a cultural base, common territorial outlook, political worldview or geopolitical problem, environmental conditions, or economic goals. In the cultural realm, three mainland Southeast Asian states, Burma, Laos, and Cambodia, share Theravadan Buddhist religion and politico-religious ideology; Laos also maintains a strong relationship with Vietnam based on political and economic goals and increasingly with Thailand as a result of Thailand's need for Laotian hydropower resources. These countries, as well as all the others, have also tolerated different belief systems within their territories. Malaysia, Indonesia, and Brunei share the Islamic religious base, and especially in Malaysia and Indonesia the state maintains a moderate association with the Islamic religion. On the Islamic periphery, both the Philippines and Thailand have significant Islamic populations and have experienced Islamic separatist movements. Thailand and the Philippines have also shared a history of relations with the U.S. military, especially during the Vietnam War. On Borneo, the one island in the region that is carved into political territory of three nation-states, Islam is the prevailing world religious force, although indigenous peoples constitute a pan-Bornean local cultural complex. At the geographical center of the region in Singapore, the state tolerates all cultural and religious groups and now ties itself to other countries in the region through diverse foreign investment projects. In the political and economic realms, as we have seen, most of the countries of the region share a similar outward-looking economic development perspective in export-oriented industrialization, and the region's central political organization, ASEAN, has been inclusive and has sought to encompass all regional states despite their level of economic development or political ideology. ASEAN, in its institutional tolerance, has steadfastly promoted regional stability and does not, as in the case of Cambodia in 1997, condone threats to the existing political order. ASEAN is trying to promote regional environmentalism as well; so perhaps shared environmental problems, especially deforestation, which has been a problem for all countries in the region save Brunei, can become new common ground for regional integration. These are just some of the more outstanding patterns of intraregional ties.

This geographical approach, an explanatory tour of the states in the Southeast Asian region, highlights the historically evolved practices and ideas that have shaped and guided each state. The region's diversity *and* increasing integration depend ultimately on the extraordinary ways that its communities have endured historically and cohere in the present. Recognition and understanding of distinct state origins, ideology, and territorial history is how the geographical approach avoids the "territorial trap" (Agnew, 1994), the pitfalls of viewing the state as just a piece of land on a map, an ahistorical and changeless "container" of political economic processes. Only when we understand the *located* evolution of the state can we comprehend its distinction and assess with precision how its leaders will represent the state, and further state goals, in the regional and international arenas.

11 Gender And Development

CAROLYN L. CARTIER JESSICA ROTHENBERG-AALAMI

Making senior partner at a major public accounting firm is a high level of professional achievement anywhere in the world. This is perhaps more so in Singapore in the heart of the Southeast Asian region where a first generation of highly educated women is beginning to reach the top ranks of their professions. Julia Soon is in her early 40s and has been with the same firm for nearly 20 years after obtaining both her bachelor's degree and MBA at the National University of Singapore, one of the best universities in the Asia–Pacific region. She describes her daily life routine before she and her husband arranged to hire live-in domestic help: "We didn't have a maid then. In the morning when I woke up I knew exactly what I needed to do. I would prepare the breakfast, feed my son, then dress for work. When my son was in primary school, I would teach him his homework, make sure he took his bath, tell him stories, play with him, and then send him to bed. Once he was asleep I would do what I needed to do for myself. My husband didn't help. He wouldn't. You see, before I became manager I had to wash his shirt, when I became manager I was still washing his shirt, when I became partner I still had to wash his shirt" (adapted from Low, 1993, p. 75).

How is development gendered? Analyzing gender and development means discovering how economic development produces differences in men's and women's economic activities and opportunities and how factors of gender characterize processes of development itself. But how are differences produced? Why have development planners increasingly realized that gender-aware development planning is critical to the success of the overall development agenda? This chapter will introduce the concepts of gender and the social construction of gender and will demonstrate how geographers, social scientists, and development planners use gender in development analysis in the Southeast Asian region.

In the public sphere of the media and popular culture, people regularly use the word *gender* in a simplistic way to refer to the categories men and women. The actual and more complex definition of *gender* is a process about how people take on male and female characteristics throughout their lives. Just as cultures vary around the world, so do processes of gendering. What is particularly interesting about the Southeast Asian region, especially before the colonial era, is that the status of women has been higher than in other world regions. Moreover, the status of women in Southeast Asia has been markedly higher than in neighboring regions of East Asia and South Asia. Much of this chapter will focus on how men's and women's opportunities and social and economic roles may be more subtle, flexible, and egalitarian in Southeast Asia than in other world regions. The chapter will also demonstrate how countervailing historic economic processes of the colonial era and the post-World War II development era have diminished women's relative equality in some areas. Colonialism and the impacts of the postwar economic development process have encouraged more traditional roles for women in some areas. How can development itself—whose legitimacy is based on economic growth and rising standards of living—make social conditions less desirable for women?

This geographical view of gender and development in Southeast Asia assesses the comparative status of men and women in the region from precolonial times to the contemporary era of rapid economic development. The evolving economic development trajectory, from economies based on primary production to diversified economies based on manufacturing and service sector industries, provides the framework for the chapter. From precolonial times to the present, women, like men, in Southeast Asia have been active participants in the economy. In the era of primary production economies based on agriculture and natural resource extraction, women regularly managed rice farming and vegetable growing and marketed crops. The transition from agricultural economies to manufacturing and service sector industries created new labor roles for both men and women, and in some cases children, especially in labor-intensive manufacturing production. Rapid industrialization in Southeast Asia has depended on manufacturing products for export, which in turn has depended on the availability of low-wage labor to produce low-cost goods. Characteristically, low-wage assembly work has become the work of women in Southeast Asia and around the world. How has export-oriented development come to rely on women's labor? How does rapid industrialization promote patterns of "gendered labor"?

Answers to these questions can be uncovered by examining the history of the post-World War II development process. In the 1950s and 1960s, development projects did not include gender as a component of analysis. By the 1970s, assessments of projects implemented in the 1950s and 1960s revealed that some development projects failed because planners did not find out exactly who was doing what work. Around the world, specific tasks of farming, manufacturing, and other sectors of work are gendered; in other words, some jobs are regularly the work of men, and other jobs are usually the work of women. For example, all over Asia, women traditionally have been planters of rice seedlings, while in the contemporary era of mechanized farming men have dominated machine broadcasting. In the electronics and clothing industries, labels on regionally assembled products read "Made in Thailand" and "Made in the Philippines" when, more precisely, the labels should probably read "Made by Women in the Philippines," and so on. Just after the first wave of postwar development projects had been implemented, development practitioners began to recognize how economic development affected men and women differently. Southeast Asia had become an important world region of low-wage manufacturing production, and women characteristically took jobs in electronics, garment, and footwear production. Which jobs were gendered male? In these assembly industries, the low-wage male-gendered jobs have been warehousing, packing, and stock management positions; at a higher wage level, men have been supervisors of women assembly line workers. In terms of sheer numbers, the construction industry has been the male equivalent of women's assembly work in the manufacturing sector. The role of men in literally building the infrastructure of the region's cities, especially in Singapore and Kuala Lumpur, has lead the construction industry to hire the lowest-cost male workers who migrate from outside the region, especially from Bangladesh, to take construction jobs. What this also means is that Southeast Asian men, especially in the more highly developed countries of the region, will not engage in construction work or will do so only as a last resort option because the pay is so low. Diane Elson (1995) has referred to this set of differences—the process of gendering jobs male or female, and the male-gendered values of the development process itself—as "male bias in the development process." *Male bias* means that the rewards of wealth, power, and status accrued through development have disproportionately favored men, and especially professional men. In the terms of gender analysis, this is how the development process is "gendered male."

The overall mission of gender-sensitive development work is to increase the security of livelihoods for both men and women and, as a result, the security of the household. The security of the family household directly contributes to strengthening local communities and thereby creates the basis for more fully functioning and stable societies and economies (Heyzer and Sen, 1994). At the global scale, the United Nations has taken the lead in promoting gender-sensitive development, especially through its subsidiary offices UNDP (United Nations Development Program) and UNIFEM (United Nations Fund for Women). At the regional and local levels, nongovernmental organizations (NGOs) devoted to issues of gendered development have become increasingly active since the 1980s. By contrast to the promotion of development at the highest scale such as through the offices of the World Bank and International Monetary Fund, or at the nation-state scale through government promotion of national development schemes, the focus on gender and development regularly calls attention to local scale realities. Local gender-based view of development differs from national and international scale perspectives in that it strives to promote the empowerment of ordinary men and women rather than the status and wealth of national elites. Such development work is barely a generation old and remains relatively uncharted territory.

ENGENDERING DEVELOPMENT

In the development field, the concepts of gender and gendered labor were virtually invisible in the early stages of post-World War II development work. Prior to 1970 development specialists actually thought that the development process affected men and women in the same ways. Now, of course, development specialists consider the opposite to be true (Box 11.1). We can identify three sets of important ideas that have contributed toward this shift. The first set of ideas concerns how researchers and development planners have changed their perspectives on gender issues and discovered the importance of gender in economic development planning. The second set of ideas concerns the recognition of the gendered division of labor. The third set of ideas concerns how *gender* is defined as a "process of social construction".

How do researchers see development issues in the world around them? How have development practitioners viewed issues in developing countries? First, the most important way of making sure development planners do their work without problematic biases is through the concept of "the gaze." The gaze, often termed the "patriarchal gaze" or the "male gaze," is based on understanding how the act of looking at a person is a powerful position for the viewer. When the object of the gaze is female, and when the person doing the gazing is male, the male

BOX 11.1 The Gender-Related Development Index (GDI)

The United Nations has formulated a gender-related development index (GDI) based on the major U.N. indicator of human development, the human development index (HDI). The GDI measures the same indicators as the HDI: income, life expectancy, adult literacy, and education, as well as factors in inequality between men and women in these categories. The U.N. has also developed a gender empowerment measure (GEM), an index that represents women's participation in political decision making, access to professional opportunities, and earning power. The Human Development Report compares the GDI for 130 countries and draws the interesting conclusion that gender equity does not correlate with income levels of a society: High economic development does not guarantee gender equity. The report also found that women work longer hours than men in nearly every country and that most of women's work around the world remains unpaid.

These are important discoveries, and yet the GDI and the GEM do not take into account other critical dimensions of gender equity. The GDI and GEM can only capture what is measurable in quantitative data. The GDI and GEM do not account for important qualitative factors contributing to gender equity, such as support and recognition in the family, at school, and in the workplace, access to and consumption of resources within the household, and participation in informal and community decision making. These realities of daily life are important factors that socially construct men's and women's experiences.

In the GDI ranking of 130 countries, the Southeast Asian countries hold places from position 96 to number 28. The position of each country in the list reflects the overall level of development of the country: Singapore (28) ranks the highest, followed by Brunei (31), Thailand (33), Malaysia (38), the Philippines (64), Indonesia (68), Vietnam (74), Burma (94), and the Lao People's Democratic Republic (96). The GEM ranking of Southeast Asian countries for which data exists is the Philippines (28), owing to a very high number of women in administrative and professional positions, followed by Singapore (35), Malaysia (45), Thailand (54), and Indonesia (56).

Source: United Nations Development Programme (UNDP), *Human Development Report,* 1995.

occupies the authoritative position and reports on the female. In past development work, this condition has meant that male researchers reported on their view of women's experience rather than women reporting themselves. Problems resulted. Women may have been especially reluctant to reveal details of their personal lives to male strangers, even when they are government researchers or official development officers. In some cultures, male heads of household regularly speak for their wives and other female members of the household so that a woman's perspective on household labor, for example, might not be revealed at all. Development planners can embody the male gaze and create plans that reflect male bias in development projects themselves. Only after household-level development projects did not achieve their stated goals did planners begin to realize that projects were inadequately conceived as a result of not having taken into account women's roles and points of view (Elson, 1992). In sum, researchers and planners never uncovered and never utilized much information on women's economic roles in the first decades of postcolonial development. Now development workers regularly solicit data from women about their own situations in their own terms. Second, and a closely related issue, people in developing countries have charged that some development researchers—men and women alike—imposed points of view that are ethnocentric or relative only in the context of the industrialized world. As a result, development practitioners now regularly report perspectives of local people, both men and women, in their own terms (see Momsen and Kinnaird, 1993; Marchand and Parpart, 1995). Third, legacies of First World–Third World political economic divides have left some development planners still unwilling to examine gender issues in development. Even though research has shown that many Western postwar macroeconomic development policies negatively impacted women, and both men and women especially in lower-income strata, some development specialists will not advocate policies that would enhance the status of lower-income people if such perspectives place them in conflict with U.S. national development strategies. This point is discussed further below in the section "Applied development strategies" and in the later section "Gender and structural readjustment."

How does gender characterize job roles? The first work to draw significant attention to the critical importance of gendered labor differences in economic activity was Ester Boserup's (1970) book *Women's Role in Economic Development.* Boserup focused on agricultural work and recognized how men's and women's labor participation varied depending on the type of agricultural system. She drew attention to the gendered labor processes that take place when wage labor is introduced through forms of agricultural modernization. When agricultural practice is transformed in favor of the use of mechanized production, high-yielding seed varieties, chemical fertilizers, and pesticides, the package of new technologies promotes cash-crop production to earn enough money to justify the new inputs. National governments regularly introduced these new technologies to heads of households—men—at centers of technology transfer run by men. Boserup's research confirmed these realities by showing that mostly men were engaged in the modernized agricultural sector and cash-crop production. Cash-crop production also diminishes the amount of land available for subsistence-crop production, which has typically been managed by women. As a result, household food consumption patterns are often impacted, which, if severe, can lead to problems such as inadequate nutrition and increased child mortality. These conclusions prompted the realization that development planning often reached the household through the person engaged in cash-crop production—the male—rather than the person engaged in foodstuff production for the household—the female—and the person whose labor activities were regularly the target of local-level development projects. As a result of Boserup's groundbreaking work, development planners began to look more closely at local conditions and a new concept: the gendered division of labor. The gendered division of labor refers to the patterns of concentration of men and women in particular job roles and occupations.

The concept used to explain how gendered divisions of labor arise and how gendered social roles emerge in society is called the social construction of gender. The social construction of gender is the set of processes through which a society promotes particular female and male norms of behavior. People act out behavioral norms in social contexts, in the household, at social functions, at school, and at work, but society accords lesser or greater importance to social behavioral norms based on female and male characteristics. For example, "being vulnerable" is usually considered a female norm and is socially constructed as "low power" or negative. In daily life this norm works out in school classrooms, where high school and university teachers may be concerned in some classes that women on average participate less in class discussion—and with less authoritative style—than the most active men. In such circumstances, what might be perceived as more vulnerable is a less assertive or less competitive style that is being crowded out by more active voices. Alternatively, conditions in schools and workplaces may encourage women to adopt male-gendered behavioral norms to hold certain jobs that are usually held by men. How men and women are socialized to adopt female and male behavioral norms and also social status based on such characteristics and roles can vary around the world and regionally within countries according to norms of a population group. Based on these ideas, the working definition of *gender* for scholars and development practitioners is the social construction of gendered characteristics—not biological sex. Theoretician Gerda Lerner (1986, p. 238) writes about gender that "unfortunately, the term is used both in academic discourse and in the media as interchangeable with 'sex.' In fact, its widespread public use probably is due to it sounding a bit more 'refined' than the plain word 'sex' with its 'nasty' connotations. Such usage is unfortunate, because it hides and mystifies the difference between the biological given—sex—and the socially constructed—gender." Social construction of gender recognizes that feminine-associated or masculine-associated behavior patterns are taught and learned at an early age and formed over the life course. This definition also recognizes how men can have female-gendered characteristics and women can have male-gendered characteristics. Most current work on development utilizes the social construction of gender to demonstrate how men's and women's experiences are different and are socially produced under varying circumstances of family, race or ethnicity, sexuality, age, class background, education level, and employment situation.

Even though gender is a complex concept, the reality is that social norms often reflect stereotypes about men's and women's roles and abilities, rather than men's and women's individual possibilities. Around the world, families and institutions often judge men's and women's contributions and opportunities based on characteristics of patriarchy. *Patriarchy* is social organization based on descent through the male line combined with societal values based on masculine norms and values and the dominance of men in the home and the workplace. But patriarchal social structures, by favoring men, can also yield gender discrimination and unequal access to opportunities and resources for women. The household based power of patriarchy has historically derived from the role of senior men in providing for family members. This historic "bargain," however, has broken down in many places of the world, especially where economic opportunities are limited. Further problems of patriarchy arise when men continue to appropriate patriarchal power without using that power to support the family, whether financially or morally. Historically, women have borne the brunt of gender-related discrimination and abuses of patriarchal power, but men, especially those who do not play by the rules of "the boys," also commonly experience discrimination when their gender characteristics are not "male enough."

Learning how to apply the concept of the social construction of gender—and how the social construction of gender is different and more complex than comparisons of the sex-based categories men and women—can be approached by assessing demographic data (Box 11.2). Population statistics from some developing world regions demonstrate that numbers of females are lower than expected based on the natural rate of births (Sen, 1990). Discovering these unexpected differences in the numbers of females and males in the developing world yields empirical data—information by sex. Analyses of data by sex can measure the proportion of men and women in a population, their comparative education and income levels,

BOX 11.2 Son Preference and Patriarchy in the Asian Region

In Asia, North Africa, and to a lesser extent in parts of Latin America, the numbers of women reported at all ages are lower than normal size for the human species. On a world scale, more than 100 million women are "missing," and approximately 60 million women are missing in the Asian region. (In general, the natural human birthrate is slightly lower for girls; the average ratio of female to male births around the world is 100 : 105/106, but females have higher survival rates than males.) In societies where biological sex is a factor in infant mortality and life expectancy, strong traditional social values about the worth of females and males in society—which are gendered characteristics—operate through reproductive choice, infanticide, health care, and quality of life. In South Asia and East Asia the especially high value placed on sons and the male line of descent favors the birth of boys and careful nurturing of male children. The generally lower value placed on a girl's or a woman's life is the main reason why the survival rate for females has been lower. Low status of women can be described in theoretical terms through patriarchy, which Lerner (1986, p. 239) defines as "the manifestation and institutionalization of male dominance over women and children in the family and the extension of male dominance over women in society in general." Strong patriarchal values characterize the Confucian cultures of the East Asian region and the Islamic and caste-based Hindu cultures of South Asia. Although the status of women has risen in some parts of the region with increased education and standards of living, improvements have been uneven. Indeed with such advances in medical practice as the sonogram and amniocentesis test, wich are designed to check for fetal birth defects, the abortion rate of female fetuses has actually risen in India, South Korea, and China. In China the one-child policy has also underscored son preference, especially in rural areas. As Table 11.1 shows, the number of women per 100 men is actually lower in urban areas of South Asia where people have greater access to medical technologies. In general, the ratios of women to men are higher in Southeast Asia than in South or East Asia (except Japan). For the Southeast Asian region the population ratio by sex demonstrates the existence of a more natural ratio of female to male births and survivability. The countries of mainland Southeast Asia which exhibit numbers of women per 100 men greater than 100 are also countries where Buddhism has prevailed as the major belief system. The ratio of women to men is higher than expected in Cambodia because of excessive deaths of men in military conflicts. In the wider Asian region only in Japan, the fully industrialized country, does the birth ratio demonstrate conditions which have not diminished the survivability of females.

Country	Rural	Urban
South Asia		
Bangladesh	99	79
India	96	88
Nepal	96	87
Pakistan	95	88
East Asia		
China	95	93
Japan	106	102
South Korea	98	98
Southeast Asia		
Brunei	94	95
Burma	101	100
Cambodia	109	104
Indonesia	101	100
Laos	103	103
Malaysia	99	98
Philippines	96	99
Vietnam	103	102

Table 11.1 Number of women per 100 men in the Asian region. *Source: United Nations, 1995a* (The World's Women: Trends and Statistics).

job roles, and so on. Statistical data, however, cannot answer why differences in numbers exist. Problematizing the data to ask why the differences exist is what yields analysis based on the social construction of gender. Why are women "missing" in the developing world? Why are women "missing" in the Asian region? In the Southeast Asian region specifically, the good news is that the disparity between female and male births is not as low as in South Asia or East Asia. This intriguing difference compels the question: Compared to the rest of the Asian region, why is it that in Southeast Asia the population imbalance by sex is less extreme?

At least part of the answer is that, more than any other world region, Southeast Asia has been an area of bilateral descent, matriarchy, and what has been termed by anthropologists as more "loosely structured" social organization (Evers, 1969). In societies that practice bilateral descent, both the mother's and the father's families play important roles in defining family organization and determining who inherits property. *Loose structure* refers to flexible social structures or flexible interpretations of social norms. In addition, a few cultures in Southeast Asia are matriarchies, in which the female line defines descent, women dominate, and female social norms lead to privilege and power. These conditions create in some areas of Southeast Asia more beneficial conditions for women and greater degrees of gender equality. Indeed outside observers have often viewed Southeast Asia as distinctly separate from South Asia and East Asia because of relatively high female status in a range of activities and arenas including the family and the household, matters of ritual and religion, and decision making in the economic sphere generally, especially in agriculture practice and marketing (Wolters, 1982; Reid, 1988a, 1988b; Atkinson and Errington, 1990; Andaya, 1995). Other contemporary analyses of gender assert that historic versions of gender equality have given way as a result of colonialism and industrialization (Mies, 1986; Stoler, 1991; P. Van Esterik, 1995) or that the state maintains traditional discourses of high status for women to suit its own goals (Sullivan, 1994). In some areas of Southeast Asia the position of women has been resilient enough to survive colonial impacts so that relatively high status for women, or at least the perception of high status, endures.

HISTORIC PATTERNS OF SOCIAL ORGANIZATION

In Java and southern Sumatra, local evidence from as far back as the tenth century records how rural women engaged in an array of economic affairs from entering into contracts, making and receiving loans and owning property, to village decision making (Andaya, 1995). In addition to female prominence in agriculture and marketing, women acted as ritual specialists in religious rites in their villages and communities. Traditional women's participation in indigenous religious practice predated the arrival of Buddhism, Hinduism, Islam, and Christianity to the region and was likely an ameliorating force that diluted the patriarchal structures associated with these world religions. The status of women in Southeast Asia also may be viewed through the lens of dowry custom. For example, in contrast to high dowry costs for women's families in much of India, marriage arrangements were reciprocal or could entail payment by the groom's family to the bride's. While evidence substantiates the idea that women's roles in the economic sphere in Southeast Asia have been generally more prominent than in other world regions, it is important to realize that the status of women in economic activity and in the domestic sphere has also varied widely over place and time.

Historic records about traditional life in Southeast Asia confirm the perspective that the status of women in the region differed from expected gender norms in Western and Confucian societies. Some early Chinese and Western explorers in Southeast Asia left accounts recording their surprise at women trading and marketing agricultural and handicraft products and holding major political positions. From the perspectives of most Western and Chinese male explorers and traders, women seemed to be out of place in such economic and political roles. The fifteenth-century Chinese Buddhist pilgrim Ma Huan found women's participation in the Thai economy so prevalent that he thought men must have all their affairs decided by their wives. Colonial writers marveled over the Burmese monarchy in Pegu, where queen Shin Saw Bu ruled over the Mon people in the fifteenth century. In a style characteristic of the biases of colonial-era writing, one observer wrote "when greater races bound the feet or veiled the face of their women, or doubted if she had a soul, the Burmese held her free and enthroned her as chieftainess and queen" (Harvey, 1925). Some colonialists saw relative equality of the sexes in Burma as a sign of backwardness. A political officer in the British colonial administration advised that "to overcome this backwardness, the Burmese men should learn to kill, to make war and to oppress their women" (Mies, 1986, p. 93). Apparently the appearance of relatively egalitarian gender roles in Southeast Asia challenged the worldviews of these European men. Twentieth-century Dutch colonial administrators on Java maintained such perspectives, couched in the language of "rational" market economics: "Women who, instead of taking up some honest business by which to earn a living decently, seek to make a little money by sitting all day by the road-side selling a few vegetables and other little things of small value, and do this in such multitudes that they jostle each other and create great disorder in the market place, beside depriving one another of profit and the possibility of obtaining a sufficient living from this trafficking" (Alexander, 1987, p. 55). Yet there were less ethnocentric colonial administrators such as Thomas Stamford Raffles who viewed local conditions in more local terms: "In the transaction of money concerns the women are universally considered superior to men, and from the common laborer to the chief of the province it is usual for the husband to entrust his pecuniary affairs entirely to his wife. The women alone attend the markets, and conduct all the business of buying and selling. It is proverbial to say the Javanese men are fools in money concerns" (Raffles, 1817, p. 353). Overall, these colonial worldviews set the stage for the transformation of women's roles in the economy.

The diffusion of major world religions to the region might be expected to have diminished the status of women, but examples from around the region demonstrate

Photo 11.1 Market vendor, Semarang, central Java. (Leinbach)

how women have negotiated oppressive tendencies of patriarchal religious beliefs. Influences of Islam in Malaysia, Indonesia, and in the southern Philippines have emphasized traditional roles for women, especially in regard to family honor, morality, and sexuality. But this is not to say that women themselves lack options. Wearing the veil, for example, which outside observers often view as a symbol of women's subordination under Islam, is regularly a matter of women's choice in Malaysia and Indonesia, where it is common for young women to wear the veil as a personal statement and against the preferences of their parents. In countries of mainland Southeast Asia where Theravada Buddhism prevails, the status of women may be expected to be somewhat lower because the religion defines women as being inherently lower in religious status than men, but this definition has not fundamentally limited the status of women in Buddhist society. In an example from Thailand, women Buddhist meditation teachers gained status through their work, in which they are not viewed simply as female but as persons en route to transcending the categorical distinctions "male" and "female" (Van Esterik, 1996). Other interpretations of social status and Buddhism explain that gender status is linked to hierarchies of age, seniority, and lifestyle, and thus only by considering these factors together can we begin to define the status of men and women (Karim, 1995a). In Vietnam the historic diffusion of the Confucian cultural complex from China introduced a strong patriarchal form of society, but local Vietnamese traditions venerating women and religious syncretism have diffused the potential power of Confucianism. Like the rest of the region, women in Vietnam have been associated with control over marketing and small-scale commercial activities.

Island Southeast Asia—from southern Sumatra and the Malay Peninsula, and from Indonesia west of the Torres Strait to the Philippines—has especially exemplified bilateral patterns of social grouping, flexible social organization, and centers of matriarchal culture (Karim, 1995b). This pattern created the circumstances for relatively equal status between men and women centered at the heart of the region. The major center of matriarchal social organization in the region is in central West Sumatra, the homeland of the Minangkabau people. Historic migration from Sumatra to West Malaysia brought matrilineal practice to the West Malaysian states, especially to Negeri Sembilan, but in an example from the Minangkabau society in contemporary Indonesia, the role of the state in contemporary development processes works to diminish the status of women (Box 11.3).

Impacts of the Colonial Era

Colonialism was the first economic force that influenced traditional regional forms of gender equality. Colonial policies promoted commercial agriculture and transformed gendered labor roles in agricultural production. In precolonial Southeast Asia, women tended to dominate economic transactions in the agricultural sector. When colonial policies promoted large-scale production of cash crops, this valuable new export sector of agricultural production came to be dominated by men. Women maintained their primary work roles in growing food for local domestic consumption, in padi field management and local village marketing of vegetable crops. Colonial policies often explicitly incorporated men into the export sector and marginalized women's participation. This pattern of male-gendered labor in external cash economies and of female-gendered subsistence production and local marketing has been a common gendered division of labor under colonial agricultural economies the world over. The following example from colonial Sumatra demonstrates how such conditions evolved.

In the sixteenth and seventeenth centuries, British colonial traders introduced a cash-crop economy to the ports of southeast Sumatra (Andaya, 1995). Southeast Sumatra became the Southeast Asian center of trade in *piper negrum*, or black pepper, which was the first cash crop introduced to the Indonesian Archipelago for widespread commercial production. When pepper first arrived from India, it was incorporated into the female-gendered realm of subsistence agricultural production in home garden plots. But as foreign traders at ports on Sumatra's southeast coast demanded ever larger quantities of pepper, local rulers responded by ordering pepper-growing families to plant more vines. The new pepper vines competed with other crops. Where once women grew and marketed diverse types of produce, the rise of the international pepper trade placed increased demand on the village gardens. As the demand for pepper grew, the centers of the trade shifted from local villages to ports at the distant coast. In this new production and trade regime, more men than women made the long journey to the shipping centers. Dutch and English merchants even recorded that they preferred men as agents of trade. Thus a women's agricultural and marketing economy became an international colonial export regime and a "man-to-man" enterprise.

Analysis of the pepper trade also reveals a darker side of the colonial transition for women. English colonial management of the pepper-plantation economy, following the norms of English law, recognized only males as primary cultivators and landowners. In 1766 colonial policy required all males older than 16 years to plant 500 pepper vines and doubled the regulation to 1,000 pepper vines for married men. But men without families of any sort could not hope to muster the labor required to tend that many plants—and young women, who traditionally tended the home gardens, remained single rather than becoming slave to 1,000 plants of pepper. A labor shortage developed. Local people traditionally solved labor shortages in the region by the custom of *ambil anak*, or adopting a child, or by alternative marriage patterns, such as the eldest daughter in a family marrying a poorer man who would live permanently with his wife's family and meet the dowry payment through his labor. The English colonialists, who wanted to increase the number of married households and thereby increase pepper production, prohibited this custom, citing "administrative difficulties." The English also sought to legislate traditional birth-control practices. The low population growth rate in the region at replacement levels of two children—which contributed to the "labor shortage"—was partly achieved through natural abortifacents, the use of which the English made punishable by death. Local resistance to colonial regulations frustrated colonial administrators who took oppressively excessive—and gendered—steps to force their production regime. The lack of control over the family-based pepper economy actually led to colonialists holding wives hostage for their husbands' prescribed pepper production. Some English officials punished and even enslaved females to compel a husband or father to produce more pepper.

Evidence from the Philippines substantiates the perspective that women enjoyed relatively higher status in the community before colonial rule (Feliciano, 1994). In indigenous Filipino society a woman could become chief of the *barangay*, but Spanish customs and law (as well as Catholicism) restricted women's opportunities in colonial society and economy. Spanish law defined a wife as fundamentally subordinate to her husband, and the colonial education system differentiated between boys and girls. Once again, though, colonial circumstances have not fundamentally overturned the status of women. As the section on contemporary gendered division of labor below demonstrates, Filipinas have maintained relatively high status in a variety of economic roles.

In addition to impacting local traditions of gendered labor and gendered household norms in parts of the region, the impact of colonial economies created a dualistic economic structure in formal and informal sectors of economic activity. The characteristics of the formal sector—in industrial development and wage labor in the manufacturing and service sectors—have contrasted sharply with the informal sector, which designates the economic environment of poorer income groups in low-technology, minimally capitalized, small-scale economic activities (see also Chapters 5 and 7). Informal-sector job activities include domestic labor, casual labor, street vending of food and consumer goods, scavenging and providing personal services such as shoe shining or letter writing, and strictly illegal activities such as drug dealing or prostitution.

The formal and informal sectors exhibit patterns of gendered geographies in the distribution of male and female laborers: in the world generally, the informal sector of the economy has been gendered female while the formal sector has been gendered male. But, again, as the discussion of gendered labor patterns in the region shows below, in Southeast Asia the proportion of women in the informal sector is not dramatically lower than that of men. The participation of women in the formal sector is higher than in most other developing-country regions, but still lower than that of men, especially in leadership positions in the public sphere. One way of explaining this difference is that as the formal sector evolved in major cities through the colonial and postcolonial periods, corporate businesses and the public roles of government became more regularly the realm of men, while village hinterlands remained more synonymous with women's activities. The trade-off for women is that they have yielded to men the public spheres of power in business, political, and religious life but continue to uphold important areas of decision-making in the informal sphere of the household, village, and town.

BOX 11.3 The Minangkabau: Matriarchy in West Sumatra

Minangkabau proverbs refer to women metaphorically as "center pillar of the big house" (the lineage house) and "holder of the key to the chest" (lineage property). In Minangkabau society, descent and inheritance are organized matrilineally from mothers to daughters. Social organization and landholding center on the lineage house or, more recently, several grouped houses where an extended family of three to four generations lives in proximity and holds a common land title. Men have regularly moved to the lineage home of the women on marriage, and senior women have held authority in kinship relations and economic matters in the household.

The Minangkabau people living in West Sumatra are the largest matrilineal group in the world. They number approximately 3.8 million people, and an equal number of Minangkabau, as a result of migration, live elsewhere in Indonesia and beyond. They are still predominantly agriculturalists, but surplus rice production, as a result of double-cropping since the 1970s, has created opportunities to engage in economies beyond the homeland area. In Minangkabau society, Islam and *adat* have existed side by side and so demonstrate how patriarchal structures of Islam were mediated by strong alternative traditions. Where daily decision making has been concerned, both elder men and women shared in the negotiation of village and family life and so created a considerable amount of gender reciprocity within the extended family. In Minangkabau society a male leadership role known as *penghulu*, or titled male, has existed as a means by which men represented their sublineage in matters outside the matriline. In this way, males occupied an important role in extrafamilial village affairs.

Analysis of impacts of Indonesian development planning on Minangkabau society demonstrates how the state can redefine society and, in doing so, rewrite the social construction of gender and challenge matriarchal practice. Blackwood (1995) reviews how Indonesian development plans have promoted particular visions of womanhood and the family that reflect normative Western models of modernization and development. Indonesian development officials themselves have promoted such transformations as a result of new views on modernization gained from their training at Western universities. The various Indonesian national development plans, *Repelita,* encouraged all citizens to take active part in national development but has promoted the primary roles of women as wives and mothers in the domestic sphere. This ideological and spatial contrast between the public and private spheres created a model that socially constructed more circumscribed roles for women than existed under traditional historic forms of economic and social organization. State policies prescribed that a woman had five basic duties: support her husband, manage the household, produce children, raise children properly, and conduct herself as a good citizen. The late Tien Suharto, wife of the president, articulated good citizenship for women in her own terms: "A harmonious and orderly household is a great contribution to the smooth running of development efforts. . . . It is the duty of the wife to see to it that her household is in order to so that when her husband comes home from a busy day he will find peace and harmony at home." These positions fostered the idea that women's economic activities were at best supplemental to their husbands' income.

How the Indonesian state has defined the household has also impacted the social construction of gender at a direct, place-based level. In a state-sponsored birth control program, Minangkabau villagers have been required to post on the front doors of their houses a sign naming the "household head" and the "wife," followed by the type of birth control practiced in the family. According to state law, all heads of household must be male, and all households must be registered with the village administrative office. Thus the state has constructed definitions about men's and women's roles that are different from Minangkabau practice and, by official public monitoring, altered the interpretation of who is household head in Minangkabau society. This state ideology constructs a "modern" household that upholds the ideology of national and Westernized development goals. This is a clear example of male bias in the development process.

In urban Malaysia, educated Minangkabau *penghulu* who hold professional positions have organized to maintain the cultural identity of their so-

Photo 11.2 Traditional transport in Minangkabau society. Bukit Tinggi, West Sumatra. (Leinbach)

ciety. The *Lembaga Kerpatan Adat Alam Minangkabau* or LKAAM (Association of Adat Councils of the Minangkabau World) formed in 1970 to preserve information about Minangkabau *adat*. This organization resembles the traditional role of male leadership in Minangkabau society in the extrafamilial sphere. But as Blackwood points out, because LKAAM is under state auspices, its publications promote national ideas about male and female-gendered leadership duties, and thus the new postcolonial state ideology of women's roles in the home. The prominence of this organization, combined with contemporary interpretations of women's roles in Islam and prompted by influences of the worldwide Islamic reform movement, is effectively deemphasizing the role of women's power in traditional Minangkabau culture. Blackwood (1995, p. 144) concludes, "Male adat experts, such as those in LKAAM, produce knowledge about adat that emphasizes male dominance, asserting the primacy of titled men in decision making and authority within the lineage. They thereby marginalize female elders and underscore the state ideology of domesticity."

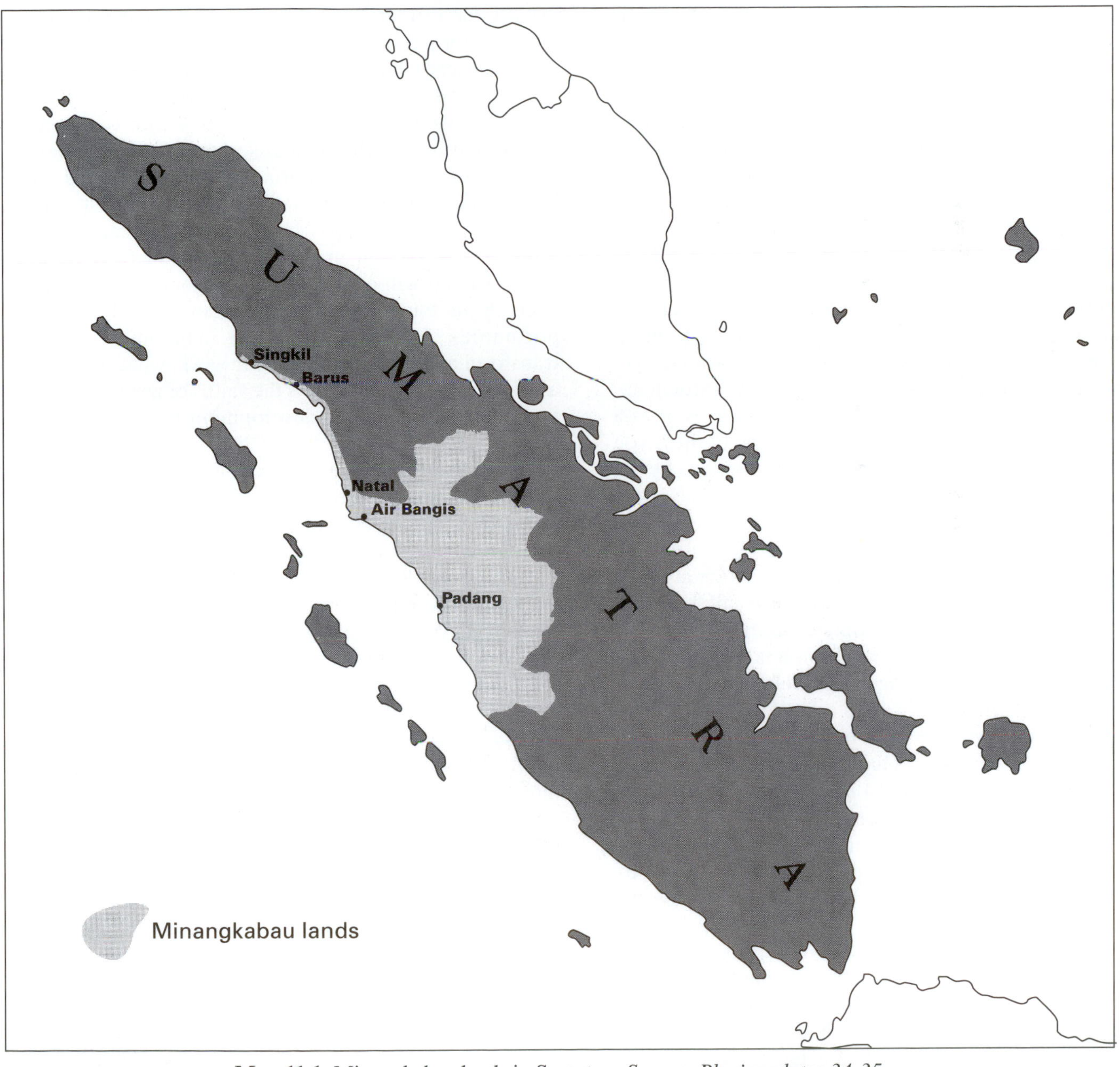

Map 11.1 Minangkabau lands in Sumatra. *Source: Pluvier, plates 34-35.*

THE GENDERED DIVISION OF LABOR (GDL)

Households and the gendered division of labor

After the colonial era, newly independent countries in Southeast Asia adopted development strategies to promote economic growth and raise standards of living. New industries created new jobs and people entered new types of work. This in turn created new distinctions between the place of the household and the workplace. For women the result has often been what development planners call the "triple workday." The triple workday refers to how women's labor particularly comprises three roles: the primary role in household domestic labor, child rearing, and care for the elderly; work in the community including gathering of communal resources such as water or fuel, or in industrialized societies such chores as shopping and running errands; and income-earning activities outside and beyond the household in the formal or informal sectors. In this way, the development process enlarged the scope of women's daily activities and often required them to juggle more responsibilities than ever before. Sometimes men experience the triple workday as well, but more often they are able to confine their labor to only one or two daily labor roles. A thirty–something-year-old female top bank executive in Singapore put it this way: "To be a success, a wife has got to be unchallengeable in the various spheres of home, garden, appearance, children, cooking, entertaining as well as a chosen career. If you fall down on any one of these you are open to attack. The successful man is judged by the position he reaches: a successful woman by her ability to keep a number of eggs in the air without smashes and yet to climb as high as she can" (Low 1993, p. 42).

In an actual example from an Indonesian household where both the husband and the wife have entered manufacturing work, only the woman has maintained the triple workday: "Soendani works in a large textile factory in the Indonesian Tangerang industrial area. She is 28 years old and a senior high school graduate. Soendani's husband, Mansul, used to work in the same company with her but now he has his own business making furniture. They live in a 4 × 4 meter rented room which functions as a sleeping room, guest room kitchen, and as Mansul's work space. Soendani is both primary income earner and housewife. Her typical work day is 15 hours while her husband works 5–8 hours depending on the number of customer orders he receives. Soendani's work schedule is also demanding. She regularly works the graveyard shift, from 11 P.M. to 7 A.M. (many of the manufacturing plants operate around the clock), after which she returns home to cook, clean, and do the washing. By about noon she sleeps, and then gets up to cook dinner by 6 P.M. By 9:30 she must get ready to leave for the factory again. She believes that domestic work like cooking, washing and cleaning is her duty and obligation as a wife. However, she expressed her tiredness by saying, 'I wish to be a single woman as before.' Mansul has more spare time than his wife but he never assists her with domestic work. Soendani's income has become the main one for the household, because of the regularity and continuity of her work. Mansul's income is irregular and cannot be relied upon" (adapted from CAW, 1995, p. 257).

In rapidly industrializing areas, the development process also involves shifts in the gendered division of labor (GDL). The GDL refers to the specialization of men and women in different occupations in both the formal and informal sectors. Formal sector workers are usually the primary income earners who earn reliable wages. Informal sector workers are often secondary income earners, whose wages are lower and serve to supplement the household income. Around the world, women have generally made up the majority of secondary earners in the formal sector and the primary group of earners in the informal sector. But during the last two decades of industrialization in Southeast Asia, women have increasingly become primary earners by holding new jobs in manufacturing industries (Ong, 1990). The growth of service-sector employment, especially in the tourism industry, has also created new opportunities. As Southeast Asian economies rapidly transform and people migrate for work, what new jobs are women taking and how is the status of women changing? How are economic development trends affecting gender equity?

Households are an important place to begin analysis of the GDL because the household is where families construct gendered norms and where people determine differences in power and status. A *household* is defined as a group of people who share the same residence and participate collectively, if not always cooperatively, in the basic tasks of producing food, earning money, raising children, and consuming goods. Household provisioning, food preparation, child care, education, health care, and laundry are the types of household activities that are usually gendered female. Activities that are gendered male regularly include wage labor, cash-crop production, and interacting as the head of household in institutional contexts such as with the government. These general patterns, though, are observations of trends and should not be used to maintain stereotypes or promote the notion that any given man or woman should maintain these roles. Within the household who has access to resources and who decides how to allocate those resources can affect the quality of family life in several critical ways. Decisions about allocating resources toward foodstuffs affect basic nutrition. Money spent on luxury goods such as tobacco or alcohol can impact whether children will be compelled to enter the labor force and forego education. Even when women spend much of their day at the workplace, what happens to the wages women earn is regularly defined in the household by who holds decision-making power over household finances.

Analyzing household structure as the basis of the formation of gendered norms and the gendered division of labor in Southeast Asia yields diverse results. Patterns of household based gendered roles do exist, but patterns are complex and differentiated. At the high end of the economic development spectrum in industrialized Singapore, Salaff (1992) found that even single women working outside the home regularly surrendered the bulk of their income to their families. Rather than providing these women the financial resources to set up independent households or pursue educational opportunities, women's income became family resources. The interests of the patriarchal family as a whole determined allocations of family income, and decisions to fund higher education often favored sons over daughters. Based on 1990 country-level statistics of unpaid women workers in family businesses, Singapore displayed a rate of 77 percent the highest in the region (United Nations, 1995, p. 149). Clearly in this case an overall high level of economic development neither guarantees women basic wages nor decision-making power over the fruits of their own labor. Even in contemporary Singapore, the state does not allow single people under the age of 35 to purchase government housing, thereby officially promoting marriage and compelling unmarried adult children to stay at home.

Women's traditional roles in marketing and trading in the region should equate to some control over household finances. This is certainly the case in Java, where women maintain key roles in marketing both agricultural and manufacturing products at the local scale. But does managing the household budget reliably construct high status for women? The power of local patriarchal values, not unlike Dutch colonial views, can socially construct small-scale economic activity as low status. "[I]n Java, money, however anxiously sought and coveted, is still perceived at another level as a grubby substance and money-handling an ignoble, crude practice. The business of managing household finances and making even the most crucial financial decisions for a family is valued no higher than washing dishes and minding babies. In general, it is quite simply considered beneath the dignity of the Javanese male" (Sullivan, 1994, p. 9). From this perspective, generated by contemporary research in a low-income urban area on Java, managing household finances has been gendered female and has become low status.

Gendered Divisions of Labor in Agriculture

Development is often equated with the transformation of an economy in which primary production goods, unprocessed natural resources, and agricultural products decline in their relative share of the national product. The share of jobs in the primary industrial sectors declines, and the proportion of jobs in manufacturing and service industries increases. Still, the agricultural sector in Southeast Asia remains critically important, and rice is the primary staple crop. Where the agricultural base has been transformed, the GDL is often impacted in the process. Within the household production system, men's and women's labor can be separated by crop types, field location, cash-crop or subsistence-crop production, specific farming tasks, and livestock rearing. While variations in gendered labor patterns are considerable across the

Photo 11.3 Sumatran chili traders. (Leinbach)

region, a few broad-scale conclusions maintain. First, as has been demonstrated already, subsistence-crop production is gendered female. Second, the gendering of cash-crop production varies, usually by the type of crop and its degree of integration into the world economy. Women maintain considerable control over local-scale marketing of vegetable crops. Industrial, large-scale, and high-value cash-crop production tends to be gendered male. Third, the introduction of Green Revolution technologies has been gendered male, and the associated reduction in labor requirements has freed women for either increased domestic activity or, in some cases, migration to centers of manufacturing production.

In Malaysia, shifts within the export-oriented primary production sector demonstrate how the GDL operates both within the country and at international scales. In Malaysia, oil palm has increasingly replaced rubber in the plantation economy, which resulted in an out-migration of male rubber tappers and a labor shortage. Local women, who had engaged in subsistence-agricultural production, were previously employed only as a reserve tapping workforce, were given the status of tappers with equal pay. Because of the labor shortage, plantations attempted to maintain this female workforce through the provision of maternity allowances, housing, basic child care, and nearby schooling. In the late 1980s, however, with the fluctuation in commodity prices, these women were replaced by illegal male migrant workers from lower-income countries such as Bangladesh, Sri Lanka, Burma, and lower economic growth areas of neighboring Indonesia (Ariffin, 1992). This shift encouraged rural-educated Malay women to migrate for work in manufacturing plants of the formal sector (Ariffin, 1992; Wee, 1995).

The implementation of Green Revolution technologies across the region has transformed gendered divisions of labor in agriculture. In Malaysia, the role of the state in introducing the Green Revolution package has affected the GDL in rice production by leading to a decline in female labor (Ng, 1991). Malaysian development planners did not consider women to be farmers in their own right but saw them as "farmers, wives," even though Malay women have always participated in all farming activities from land preparation to harvest (Ariffin 1992, p. 23). The state taught the new technologies of agricultural production to male heads of households, which gendered male the mechanization of agriculture. For example, the introduction of seed broadcasting by mechanical methods, undertaken by men, replaced seed preparation work in nursery plots that had typically been the work of women. In this process men participated more actively in market-oriented agricultural production, whereas the state encouraged women to focus on domestic production. Consequently the role of women farmers diminished. State programs through FELDA (Federal Land Development Authority) and other offices recast women's traditional roles by providing them courses on nutrition, tailoring, handicraft, and home economics. In the process of agricultural modernization, many women were relieved of agricultural labor but found their roles in society increasingly circumscribed within the domestic sphere.

Modernization of upland farming systems has impacted the GDL in more remote areas of the region as well. In the upland periphery of Sulawesi, Indonesia, the introduction of commercial tree crops for the export market transformed a swidden farming system (Li, 1997). The Tinombo region of Central Sulawesi is inhabited by about 30,000 Lauje people, 10,000 of whom are laborers, small entrepreneurs, and fishers living along the narrow coastal plain. The remaining 20,000 Lauje are swidden farmers who have been producing tobacco for sale in local markets since at least the 1920s and shallots and garlic since the 1950s. In the 1990s, as a result of the introduction of commercial tree crops such as cocoa, cloves, and cashew, land previously accessible to the community for swidden-farming production became a male-gendered privilege. Under the new tree-crop regime, a "pioneer" gained new permanent rights to land cleared from primary forest. Although women and men had equal access to land for their swidden-farming activities, their access to land in the tree-crop regime differed because women traditionally do not clear forest. In the context of this traditional GDL, some men have claimed that women cannot become sole owners of cocoa or trees based on the notion that they are not capable of doing the heavy work necessary for tree-crop planting. In this way men who stood to gain substantially from the new tree crops redefined legitimate work as heavy, men's work. Subsequently, women "only weeded the grass," which was a more important labor contribution under the mixed swidden system. Men also characterized women as lacking knowledge about tree crops, a view that local officials and agricultural extension workers—who had taught male heads of households about tree crops—strongly reinforced. Women, however, have resisted these limitations and some women have engaged in direct, independent action to plant trees and lay claim to tree crops. They have acquired land by various means, including outright purchase, in some cases by using their profit from shallots, and by paying for male labor to undertake heavy work while avoiding assistance from husbands, which could result in future claims on the trees. In this case from Sulawesi, women have sought to maintain their economic potential even as the introduction of new cash crops has not supported gender equity.

In Vietnam cultivated land remains the principle means of livelihood in more than 10 million rural households, and arable land per capita is low. In 1988 the state made rural households individual economic units, autonomous in selling farm produce and accounting for profits and losses. These factors frame the GDL because many agricultural tasks in rural households are gendered and valued differently in an increasingly market-oriented and monetized economy (Lich, 1994). In Vietnam women perform the majority of the agricultural work, while men's labor has become increasingly highly valued. A 1990 survey conducted by the Hanoi Centre of Women's Research on 200 households showed that women performed 80 percent of the work of rice cultivation and food processing; 60 percent of work in transplanting, weeding, fertilizing, and harvesting; and most of the work of breeding domestic animals. Each day, every rural woman worked between 12 and 14 hours; men, by contrast, worked an average of 8 to 10 hours. Although the women worked longer hours than their male counterparts, their actual labor has been valued as less significant because men have attended government training courses on agricultural technology and possess technical knowledge about how to increase crop production. Between 1992 and 1994, only 13 percent of participants at technology-transfer educational programs were women, and the majority of those were from the Red River Delta region where agricultural land per capita is the lowest.

As the region has industrialized, work in the agricultural sector has increasingly overlapped with other types of economic activities. In some areas, women's roles in agriculture have not been constrained to low-value products but instead have evolved into trading and marketing in higher-value manufactured goods. In Java, an important traditional center of women's dominance in marketing, many women have shifted their product emphasis from vegetable crops to other types of cash crops and manufactured products, especially cloth and clothing. Success in marketing higher-value goods also brings new risks, as the following vignette illustrates: In Kebumen, central Java, Bu Zainuddin began small-scale trading later than many women, only after her husband deserted the family (Alexander, 1987, p. 33). She based her initial transactions on a small amount of cash gained by selling household subsistence rice to buy village eggs, which she resold to a wholesaler in the Kebumen market. But she found the supply of eggs unreliable, so she switched to preparation of coconuts for copra production. At a small-scale level, trade in coconuts is a high-risk venture because the quality of coconuts varies substantially and it is difficult to judge the quality of unhusked nuts; coconut trade bankrupted Bu Zainuddin within a year. She then switched to selling manufactured clothing, which proved a more successful trade commodity for her. She financed her first purchase of clothing by selling her gold earrings,which had been bought at US$35 an ounce, when the price of gold reached US$600 an ounce. After overcoming this initial hurdle of the lack of finance capital, she has continued to sell clothing successfully. From the local scale of the village household, Bu Zainuddin's choices of wholesaling and marketing activities connected her increasingly to the world economy. Bu Zainuddin's experience also demonstrates why gender-sensitive development planning now regularly emphasizes the provision of small-scale loans to women who are heads of households.

Photo 11.4 Women are predominant participants in local vegetable production and agricultural marketing in Kejajar, central Java. (Cartier)

Although in many areas women have continued to succeed in the marketing sector, overall, the processes of agricultural modernization in Southeast Asia have taken women from the heart of the agricultural production process. Where women's work remains central to agricultural production, it is largely in the realm of subsistence-crop management and vegetable marketing. Cash-crop economies introduced a new dynamic production regime to households, and the effects have registered unevenly. In some places women have lost control over household standards of living, whereas in other areas women retain control over the household budget and decision making. Where women have entered the labor force in disproportionate numbers to men, different patterns of gendered labor and household maintenance emerge. Industrialization of the Southeast Asian region through the rise of manufacturing industries has been the most important new source of dependable paid work for women, and has challenged traditional household organization, particularly when young women leave rural areas for work in export processing zones (EPZs).

Internal and International Migration

Compared to other world regions, population patterns of men and women in Southeast Asia exhibit unusually even distribution, and neither sex migrates disproportionately in the region as a whole (see Chapter 4). Analyses of migration patterns typically focus on migration for work-related reasons and measure numbers of men and women leaving the rural, small-town, or small-city household for manufacturing or service sector employment in larger urban areas and sometimes abroad. But as data from Table 11.2 shows, among Malaysia, the Philippines, and Thailand, in the 1960s and 1970s rural-to-urban migration was the dominant pattern only in the Philippines. In Malaysia and Thailand, rural-to-rural patterns of relocation have actually been more important for both men and women. Before the era of export-oriented industrialization (EOI), migration for young unmarried women in Malaysia was unthinkable, and the majority of relocations in rural-to-rural movement reflected marriage patterns. But the data do not reflect increased rural-to-urban female migration to urban areas in the 1980s and 1990s; systematic research on gendered migration patterns is limited, and the comparative data only covers the period of the 1960s and 1970s (United Nations, 1993). In Thailand, the migration stigma for unmarried women never prevailed, and young women increasingly migrated in large numbers to the capital to work in a variety of industries. In the Philippines the lower level of development of the Philippine economy meant that more women migrants entered domestic service in urban areas, although the establishment of free-trade zones in the 1980s has provided new opportunities in manufacturing assembly jobs for women. The second major migration pattern in the Philippines is international migration for domestic work for women and construction labor for men, and this is discussed in the section on women domestic migrant workers below. Figure 11.1, comparing sex based population distribution in rural and urban areas for major world regions, underscores relative gender equity in migration for the Southeast Asian region.

Country, census year	Migration stream	Male (percentage)	Female (percentage)	% Female
Malaysia, 1970 residence in 1965	rural - urban	8.9	8.7	46
	urban - urban	20.6	19.3	45
	rural - rural	38.5	39.2	47
	urban - rural	32.1	32.7	47
	total # of migrants (000s)	484	421	46.5
Philippines, 1973 residence in 1965	rural - urban	35.0	42.6	60.6
	urban - urban	25.5	25.1	55.5
	rural - rural	22.4	17.5	49.8
	urban - rural	17.0	14.8	52.5
	total # of migrants (000s)	1,321	1,676	55.9
Thailand, 1980 residence in 1975	rural - urban	13.6	17.3	53.7
	urban - urban	17.1	20.0	51.6
	rural - rural	58.6	53.1	45.3
	urban - rural	10.6	9.7	45.5
	total # of migrants (000s)	1,430	1,308	47.8

Table 11.2 Migrants by sex, according to urban or rural origin and destination. *Source: United Nations, 1993a* (Internal Migration of Women in Developing Countries).

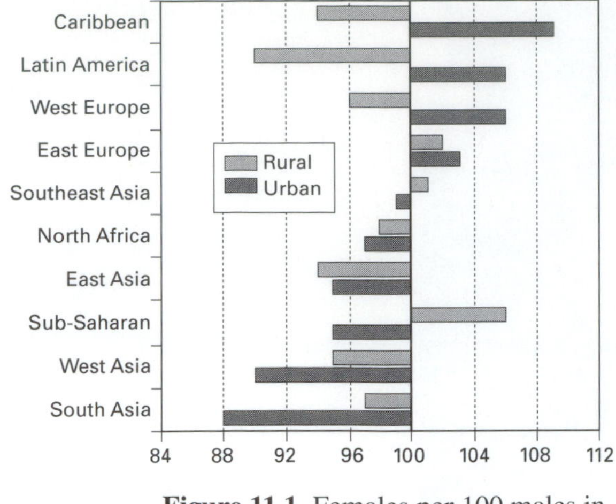

Figure 11.1 Females per 100 males in urban and rural areas, 1995.

Gendered Divisions of Labor in Manufacturing

The evolution of manufacturing industries for large-scale production of consumer goods creates a new range of job opportunities for men and women. Manufacturing takes place for both domestic and foreign consumption and at a variety of levels of capitalization. In some areas, the expansion of traditional domestic craft production has transformed formerly unpaid work into wage-labor jobs. In major tourist destinations such as Bali and Chiang Mai in northern Thailand, large-scale production of local craft products has become a mainstay of the economy in the second half of the twentieth century. In other areas, the arrival of multinational corporations seeking low wage labor to manufacture export goods for international consumption has introduced entirely new production regimes.

Regional emphasis on the economic development strategy of EOI has depended on a strong manufacturing sector to produce consumer goods for the world market. Across the region, low-wage manufacturing employment has brought a first generation of women factory workers into direct contact with the world economy. Most of the jobs available for women in manufacturing plants are assembly-line work in electronics, sewing in garment production, and mixed assembly and sewing in athletic-footwear production. Women have often migrated to work in manufacturing zones, leaving rural areas for industrial development zones to live in factory dormitories. For some the experience of dependable wage labor has offered upward mobility, new standards of livelihood, and personal options. For others, assembly work has offered marginal labor conditions and little hope for advancement into supervisory roles—jobs that regularly remain the domain of men.

Manufacturing companies prefer women workers based on notions that women are easier to employ than men. Chant and McIlwaine (1995, p. 166) describe these notions as a product of the social construction of gender, in which manufacturing industries base their perceptions of women workers on "a range of dubious physiological and psychological stereotypes, such as women being more manually dexterous, physically capable of sitting for long hours, being docile and malleable, and having a high threshold for boring, repetitive work." The consequences of this set of gendered expectations mean that employers mold women into fitting this range of characteristics, keep them in these job roles, and do not promote them

Photo 11.5 Women dominate both traditional small scale handicraft production piece work in garment production. (Cartier)

Photo 11.6 In Bor Sang, east of Chiang Mai in northern Thailand, women workers maintain the traditional umbrella industry and take jobs in the newer machine-based local garment industry. (Cartier)

into more highly skilled jobs. As a result, women can be paid low wages for a long period of time for doing the same thing. The notion of the docile Asian worker is problematic, imbued with cultural and gender bias, and capable of stripping women of their own possibilities. The following localized analyses, from Malaysia, the Philippines, Indonesia, and Vietnam attest to a range of experiences in the region.

In Malaysia export-oriented electronics manufacturing plants are concentrated in EPZs on the west coast of the peninsula, especially in Penang. In the decade between 1970 and 1980, the Malaysian government established 59 industrial estates throughout the peninsula and nine free trade zones (FTZs) (Ong, 1987, p. 145). From 1960 to 1990, the share of manufacturing in the Malaysian GDP grew from 9 percent to 27 percent, but perhaps its most startling characteristic was the massive and sudden employment of young, single Muslim Malay women from rural areas. Even in the first decade of production, the number of Malaysian women in the manufacturing sector increased 12 times while the proportion of male workers declined. Never before had Malaysian women left their traditional village occupations in such numbers. Most of these women came from families twice as large as the national average and with very low incomes. Many chose factory work to reduce economic dependency on their households, to supplement family income, and to gain some independence. Although factory wages could be just marginally higher than wages from agricultural work, factory jobs were more stable and offered benefits of formal sector employment such as subsidized meals, medical services, transport to work, uniforms, sports facilities, and some leisure activities. For manufacturers, these young unmarried employees are an attractive labor pool. They average in age between 16 and 24, and employers regularly view them as more easily "disciplined" than older women because they are accustomed to obeying their elders. The education levels of the women may be low but they are literate, and, because they are single, employers also view them as more available for overtime assignments because their household responsibilities are less demanding. Contemporary economic analyses tend to uphold these gendered stereotypes: "cheap and docile labor is important to maintain competitiveness in labor-intensive industries. . ." (Rokiah, 1996, p. 37). What is the real use of "docility" as a gendered social construction? On average, young women in developing-world manufacturing zones are paid only 5–10 percent of the wages of workers in manufacturing jobs in industrialized countries.

Living away from the protection of the family has raised concerns in Malaysian society that young women factory workers have been affected by influences that conflict with traditional Malay Muslim values. Aihwa Ong's (1987, 1990) work on Malay women factory workers in both U.S.- and Japanese-owned manufacturing plants analyzes the complexity of the feminization of manufacturing labor and how women's new labor roles threaten existing social norms. What is especially interesting are the ways in which the young women actively resist the highly structured labor regime on the factory floor. Ong's (1987, p. 152) work also confirms how factory managers socially construct assembly jobs as gendered female: "At the electronics and micro machinery plants, management's definition of the semiskilled operations as biologically suited to 'the oriental girl' in effect required Malay peasant women to adopt such 'feminine' traits," as "fast fingers, fine eyesight, and the passivity to withstand low-skilled, unstimulating work." Contrasts between influences in Japanese- and U.S.-owned factories points to some elements of the problem. Society views women who work at the U.S.-owned plants as morally corrupted by such plant activities as cosmetics and beauty shows and thus more "Western" than their counterparts in Japanese factories. Some Japanese firms, by contrast, stress traditional connections with the women's *kampung* homes by having monthly meetings with workers' parents. Nevertheless, some women do obtain new independence as a result of factory work and make plans for future education and alternative careers. These choices threaten the values of young women's roles in traditional Islamic Malay rural society. Slang expressions for the factory women connote them as "bad girls:" *minah letrik* is a name that translates as the local equivalent of "hot stuff." This does not have a positive connotation in Malay society. By the early 1980s the association between women factory workers and ideas of immorality had become a national issue and programs to help young women maintain traditional values were set up in some *kampung*. A major government official was even quoted as saying that rural women who work in the zones are said to become "less religious and have loose morals" (Ong, 1990, p. 408).

This combination of workplace and societal conditions—long hours of tedious work assembling semiconductors under microscopes, and the social surveillance of traditional society—has given rise to unexpected forms of resistance: spirit-possession incidents. During a spirit-possession incident, women claim that spirits, which often take the form of an ancient man, loom over their shoulders as they stare into the microscopes. Although reported in newspapers as "mass hysteria" in FTZs, spirit possession is based in Malay tradition and associated with experiences in which women face adverse circumstances. In traditional Malay society, women are socially-constructed as being prone to irrational and disruptive behavior, and young girls are held to attract spirits, especially when they venture out alone after dark. Seen in this context, spirit possession on the factory floor satisfies Malay tradition while serving as a form of resistance against oppressive factory managers and overall working conditions by disrupting the production process.

In the Philippines, three major FTZs are in Luzon, and a fourth is located on Mactan Island in the Visayas. These zones are centers of garment and electronics production, which together constitute 50 percent of Philippine exports by value (see Figure 15.8). In 1991, garments were still the top Philippine export, although the share of electronics had increased significantly. Women workers overwhelmingly dominate employment in garment factories at more than 80 percent of total employees. In electronics manufacturing, women workers comprise about 75 percent of the labor force (Chant and McIlwaine, 1995, p. 139). As in other countries, assembly jobs in electronics firms are at least 90 percent female. Nonassembly jobs in the factories, like warehousing maintenance, are more likely to be held by men. The social construction of gendered characteristics in the labor process also prevails in the Philippines, where one manager even stated that women were more suited to assembly work because "their anatomy and biology allows them to sit in one place for longer than men," while other managers reported that they did not want to hire men because they were "wont to gamble and drink on the premises" and were "too problematic to employ" (Chant and McIlwaine, 1995, p. 141–7). In these ways, managers construct gendered views of men and women that justify their patterns of hiring.

In the 1980s, Indonesia launched a comprehensive economic program to encourage foreign investment, and the country has since attracted hundreds of export-oriented manufacturing plants. Like elsewhere, female workers comprise approximately 75–85 percent of the labor force and a higher proportion of assembly-line workers. Many of the women assembly workers have migrated from rural areas to urban industrial zones to work in textile, garment, and footwear production industries. Some have migrated to more peripheral economic zones in the Indonesian portion of the Singapore–Malaysia–Indonesia growth triangle, that is, to the industrial parks on Batam island in the Riau Archipelago. The advantage Indonesia offers foreign multinationals is lower-cost wages than other countries in the region: in 1994 U.S. dollars, an Indonesian worker earned US$1–1.50 per day, a Thai worker earned US$4–5 per day, and a Malaysian worker earned US$10–20 per day. Low wages have particularly attracted labor-intensive sewing industries of clothing and athletic shoes assembly, and industry growth has been considerable (CAW, 1995). In 1980 Indonesia had 134 garment companies. By 1990, 1,766 garment companies had located in the country. Growth has also been strong in the footwear industry. In 1980, 57 footwear companies were operating in Indonesia; by 1990, there were 234 shoe-producing companies. Most of these companies produce brand name athletic shoes such as Nike, Avia, and Reebok. Multinational production in the footwear and garment industries, unlike the electronics industry, is usually subcontracted and thus offers less-secure job opportunities. Subcontracting production means that foreign companies contract local producers to manufacture for them, instead of building their own branch plants like most electronics firms. In this way the multinational producer minimizes its local commitment and does not invest in constructing its own factory. This type of arrangement is ultimately flexible for the multinational and less secure for local producers and local workers. Indonesia is especially known for its footwear industry, in which women are the predominant workers in assembly line and sewing jobs. In 1990, Nike, utilizing the strategy of subcontracting to take advantage of the lowest-cost wages in the region, located some of its production in Indonesia. As wages and union organizing in South Korea and Taiwan took off, Nike shifted production to Indonesia, China, Thailand, and Vietnam. Nike produces 99 percent of the 90 million shoes it sells every year in Asia by a contracted workforce of more than 75,000 (Wee and Heyzer, 1995, p. 101–2). Because the manufacturing process socially constructs the value of women's labor as low, for Nike shoes produced in Indonesia that retail for US$100 the cost of a woman's labor is only 12 cents! The advantages of production in Indonesia demonstrate the flexibility of multinational corporate location in creating a new international division of labor (NIDL) with gendered characteristics.

For workers who have entered manufacturing production and perform contract work for Nike in Vietnam, manufacturing jobs offer a completely new working environment and an alternative to rural farm labor. The following vignette is based on the perspective of a young woman whose experience at the manufacturing plant is better than average: "Hoa, 18 years old, lives with her extended family in Cu Chi, an apparently typical rural town in Vietnam. Cu Chi was pulled into the global economy in the summer of 1995 when South Korea-based Samyang Company built a factory there to make shoes exclusively for Nike. The company hired 5,300 people—half the local population. Hoa lives in a typical house: three rooms, woven bamboo walls, no running water with her parents, her brother and his family. 'I had the option to stay home and work in the fields,' Hoa said. 'I chose the factory. I can make money for the family.' Like her fellow workers, she was paid a $35-a-month training wage during her first three months, below the $45-a-month legal minimum wage in Cu Chi. She has since been promoted to supervisor and is earning about $53 a month. In keeping with Vietnamese tradition, she hands over her paycheck to her family. She would take home more money but for several payroll deductions. Among other costs, the factory deducts for drinking water and increased the deduction from 30 cents to 50 cents a month during the first year. The work is repetitive and dull. Some workers lace shoes or simply place finished products in boxes all day. 'It takes a lot of energy,' Hoa said, 'it's hard. But its OK. We can do it'" (Manning, 1997).

The Service Sector

Service-sector industries in Southeast Asia encompass a wide range of both formal- and informal-sector jobs, from street vending to the tourism industry and professional positions in health, education, law, international business, and government service. Around the world, low-end service-sector jobs have commonly been gendered female, whereas highly paid jobs in the professional arena have been gendered male. Comparison of statistics on men's and women's work in the larger Asian region points to some interesting trends that underscore how women of Southeast Asia are more integrated into this economic sector than women from other regions. This section examines geographical conditions of women's employment in the formal and informal economies of two major service-sector industries: the tourism and "hospital-

Photo 11.7 Young women migrant workers from Java work on two-year contracts at electronics factories in the Singapore-managed Batamindo Industrial Park on Batam Island in Riau Province, Indonesia. These women are assembling compact disk car stereos for the U.S. market. (Cartier)

Asian subregion	Professional, technical, and related		Administrative and managerial		Clerical and related; service		Sales		Production and transport	
	1980	1990	1980	1990	1980	1990	1980	1990	1980	1990
East Asia	35	43	7	11	41	48	40	42	32	30
Southeast Asia	42	48	13	17	40	48	45	53	25	21
South Asia	30	32	8	6	15	20	8	8	26	16
West Asia	30	37	4	7	19	29	6	12	4	7

Table 11.3 Women's share of employment in the major occupational groups, 1980 and 1990. *Source: United Nations 1995a* (The World's Women: Trends and Statistics).

ity" industry and the domestic-service industry. Table 11.3 provides comparative data for the Asian region to substantiate the claim that women are more actively employed in the formal sector in Southeast Asia than in neighboring Asian subregions. In the majority of sectors of the major occupational groups, the share of women workers in Southeast Asia are higher than in East, South, or West Asia. Notably, the percentage of women employed in professional and technical jobs is particularly high in Southeast Asia. At the university level, women's share of employment is higher than in other professional positions. Indeed, in the Philippines and Thailand women comprise more than one-half of all university teachers (Figure 11.2). This high share of university faculty employment is unsurpassed in the Asian region and is unequaled around the world for countries of this size. Unfortunately, salaries are usually not commensurate with these professional positions.

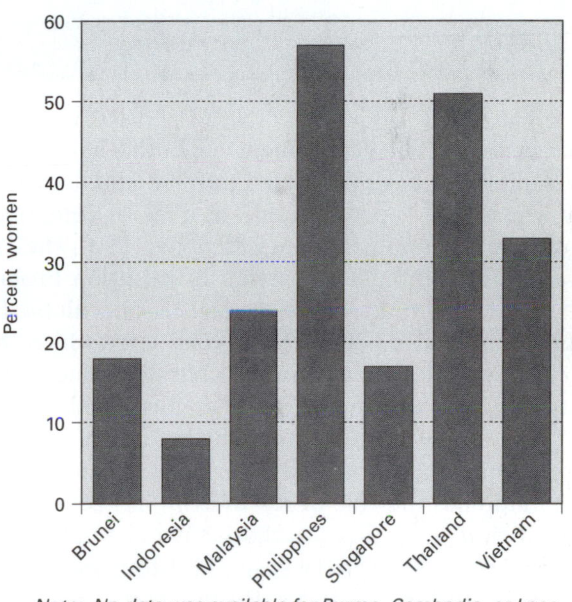

Figure 11.2 Female university teachers in Southeast Asian countries. *Source: United Nations, 1995a* (The World's Women: Trends and Statistics).

Gendered Development in the Tourism and "Entertainment" Industries

Tourism is the most visible service-sector industry employing large numbers of women in Southeast Asia in both formal and informal sectors. Tourism employment for women has been important in the formal sector of the tourism industry in urban areas, in the informal sector through ethnic tourism, and in the entertainment or hospitality industry, which often serves as a euphemism for sex tourism. Chapter 8 covers the postwar growth of the tourism industry, which has served as a major source of employment for women. The role of women in the tourism industry has varied substantially from the heavily promoted image of the "Singapore girl" in anchoring the region's number-one airline, Singapore Airlines, to the illicit side of the so-called entertainment industry in cases of child prostitution. Analysis of most of these roles yields the conclusion that the Asian woman serves as the "exotic other," the different, seductive symbol of Asian culture that is attractive to tourists embodying the "male gaze" or the "tourist gaze" (Urry, 1990). All across the region, ethnic tourism based on the encounter with "exotic" indigenous peoples, such as trekking in northern Thailand, has depended considerably on the participation of both men and women in traditional roles and clothing, even though they might prefer to don running shoes and international-style sportswear.

Some Southeast Asian governments have regularly promoted tourism development policy with gendered

Photo 11.8 Sign depicting "long-necked Karen women" directing trekkers to the Karen village homelands in northeastern Thailand. (Cartier)

characteristics. The governments of both Thailand and the Philippines have promoted tourism and associated "entertainment"—prostitution—as a major component of postwar development policy (Truong, 1990). The Thai government, although it passed a Prostitution Prohibition Act in 1960, undermined the legislation with the Entertainment Places Act passed six years later, which "had enough loopholes to encourage coffee shops and restaurants to add prostitution to their menus" (Enloe, 1990, p. 35). In 1986, Thailand earned more foreign exchange from tourism than from any other economic activity. In the Philippines the role of sex tourism has been so significant that several researchers have concluded that prostitution has formed the basis of the tourism industry, especially during the 1970s. By the 1980s, significant protests against sex tourism in the region rose in Japan where people organized to petition the Prime Minister to at least curtail advertisement of sex-oriented male holidays. Under the government of Corazon Aquino raids were made on Manila's entertainment district (Ermita) where police arrested hundreds of women but virtually no pimps or male clients (Enloe, 1990, p. 39). Women's NGOs in the Philippines—especially Gabriela, the major umbrella feminist organization, whose mission is to improve the status of women by providing alternative employment—were not consulted in the crackdown. Gabriela takes a comprehensive approach to specific women's issues and treats these apparent "women-specific" issues in the context of the state's focus on export-oriented production and high national economic growth at the expense of local-level services for women and families. Since 1989, the Philippine Department of Tourism has adopted a formal stand against sex tours. Although common sense might suggest the HIV/AIDS pandemic should curtail prostitution in the region, in the short run at least it appears to be fueling demand for young girls as men seek "safe" commercial sex. Bangkok, Manila, Phnom Penh (which has become a new center of child prostitution), and Ho Chi Minh City have all been major centers of sex tourism and prostitution.

The gendered division of labor in prostitution also produces gendered geographies. Women who work as prostitutes regularly leave their homes and natal communities for work in places where the stigma of their occupation will not reflect on their families. Consequently they have few family ties and ephemeral support systems. Men who work as pimps, by contrast, more commonly work in their natal communities and do not suffer the same social marginality. In a prostitution district that serves local demand in the city of Jogjakarta, central Java, local pimps controlled the prostitution economy and retained on average one-half the fee contracted with the customer. The female prostitutes, all from other areas of central Java, had so little local control over local lodgings and domestic space that they commonly had to rent beds in houses other than the ones they rented to live in, incurring further costs up to 50 percent of their earnings (Sullivan, 1994, p. 35). Prostitutes rented beds from the wives of pimps, and thus the households of the pimps regularly received 75 percent of the income from prostitution. If female sex workers are the focus of demand in prostitution, why do pimps earn more income? The role of the pimp is socially constructed as more powerful and the agent of control over female labor in this industry, and the low social status of the female prostitute leaves the women disempowered to demand more income.

Women Domestic Migrant Workers

High growth economies in the Asian region have created more opportunities for professional women in the formal sector and, as a result, have also created a niche in the labor market for paid domestic household work. This has resulted in an international division of female labor in Southeast Asia where in high-income countries women professionals work outside the home, while women from low income countries migrate to high-income countries to manage dual-income professional households. The Philippines and Indonesia are the primary sending countries of domestic workers, while Singapore and Malaysia and, to a lesser extent, Brunei receive domestic workers. For labor-exporting countries like the Philippines, the export of women's domestic labor has been a source of foreign exchange and a buffer to high levels of internal unemployment. In Indonesia the national five-year development plan *Repelita V* (1989/90–1993/94) projected a target number of migrant workers in terms of their foreign exchange-earning capabilities. Thus the national development explicitly plans on the overseas earning power of migrant workers to supplement national income. In 1998 after Indonesia began to experience economic problems, the state communicated to Hong Kong labor immigration that it would like to increase the numbers of domestic workers in Hong Kong from the current 24,700 to more than 70,000 to decrease domestic unemployment and increase remittances. Dubbed the "maid trade," the recruitment network for domestic workers spans sending and receiving countries and links villages to towns and ultimately to cities abroad. It is a multimillion-dollar transnational trade and is closely related to other businesses that facilitate the migration process, such as banks, money lenders, hotels, airlines, translation services, medical clinics, and training institutions. Placement companies organize jobs for women migrants; the state regulates migration and visas. Most women migrating legally for domestic-service employment work on two-year contracts, and they regularly remit money home. In 1988, the Indonesian government counted 100,000 overseas domestic workers, all assumed to be women, who were distributed among several countries but mainly located in Malaysia, Singapore, and Saudi Arabia (Heyzer and Wee, 1993, p. 41). The largest migrant stream is from the Philippines, where more than a half-million men and women leave for contract work each year, the majority for countries of the Middle East. Indeed the destinations of the largest number of Filipinas lies outside the boundaries of the Southeast Asian region in Hong Kong and countries of West Asia, especially Saudi Arabia. Women are almost all employed in domestic service; the largest number of men find employment in construction work. In Hong Kong alone there are currently about 150,000 Filipino women working in domestic service (Photo 11.9). Many of them are mature educated women with bachelor's degrees who are still able to earn more in Hong Kong than at home. Once they return to the Philippines, reintegration in the home country proves an especially difficult problem for women as their work experience in domestic service does not improve their job possibilities.

Gender analysis of this international labor market in domestic work reveals how the women workers experience the social construction of gender. In gendered perceptions of women domestic workers in Saudi Arabia and Indonesia especially, married overseas domestic workers are regarded as "good women" because they are helping their families; single domestic workers have been treated with suspicion, thereby discouraging single women from solo participation in society (Heyzer and Wee, 1993). Unlike the pattern of women's low-wage assembly work in manufacturing industries, about 50 percent of women migrating for domestic service are married and have two to three children at home. Researchers have also revealed how placement companies sometimes advertise Filipinas by appearance of their skin color and thus create racist stereotypes that lead to light-skinned women being offered more opportunities and higher wages (Tyner, 1994).

PLANNING ENGENDERED DEVELOPMENT

In applied project work on gender and development, two major approaches to integrating women and gender have evolved since Ester Boserup first challenged the professional development community to incorporate the concept of the gendered division of labor. These two approaches are "Women in Development" and "Gender and Development."

Applied Development Strategies: "Women in Development" vs. "Gender and Development"

Recognition that successful development projects often depend on incorporation of gender perspectives has led to the evolution of two widely accepted strategies for formulating development plans. The policy perspective known as Women in Development, or WID, evolved in the early 1970s and was promoted by the Washington, D.C. chapter of the Women's Committee of the Society for International Development as part of a conscious strategy to bring to the issues revealed by Boserup to the attention of U.S.-based development policy makers (Rathgeber, 1990). The formulation of WID strategies coincided with the onset of the United Nations Decade for Women, 1976–1985, and sought to add awareness of women's work to the development planning process. The WID strategy aimed to raise the status of women, enhance implementation of development goals, and contribute to the overall betterment of households and

Photo 11.9 Filipino women call home from this mobile international direct dial telephone station set up especially on Sundays in the central district of Hong Kong. Thousands of Filipino domestic workers traditionally meet in the parks and public squares of the Hong Kong central business district on their day off to renew ties, share news, and maintain social support networks. (Cartier)

communities. In some ways WID appeared to be a radical departure from 1950s and 1960s development truisms because WID projects regularly trained women for new income-generating activities. But after a decade of "adding women" to development projects, analyses of the status of women in the workforce continued to demonstrate that postwar modernization programs actually fueled the decline of the status of women in the developing world. In some areas, women came to hold the majority of the lowest-paying jobs in the industrial sector; in others, agricultural modernization increasingly circumscribed women's roles within the domestic sphere. In addition, although many women took wage labor jobs for the first time, they were continuing to meet the gender-based responsibilities of their households, and hence the "triple workday." The WID strategy assumed that the benefits of modernization would eventually "trickle down" to women, but except for highly educated women, the overall quality of life for women was not reliably improving. Limitations of the WID strategy led many people to reconsider how raising the status of women might alternatively be reached. Obvious to many policy analysts in the developing world was the fact that the WID strategy asserted the importance of equal opportunities for women, but the strategy did not examine how or why it was the case that women continued to hold the lowest-paying jobs in the new industrial sectors. In essence, the WID strategy did not examine the social construction of gender that limited women's opportunities: WID focused on women, not gender.

The policy known as Gender and Development, or GAD, has risen to supplant the WID perspective in many organizations and world regions. The GAD policy incorporates explicitly the perspective of the social construction of gender. The GAD perspective, unlike the WID platform, has also sought to identify connections and contradictions among gender, class, race, and development, and it encourages women to empower themselves. GAD, by acknowledging the demands of the triple workday, encourages government to step in to provide social services that women depend on. This has become a controversial issue in countries facing structural adjustment programs that call for restricting social services (see the section "The World Bank and structural adjustment programs" below). GAD also promotes the role of the state in supporting the legal status of women in property ownership and property inheritance. In its extreme, the GAD platform has been viewed by WID promoters as too interventionist for questioning the power and wealth of entrenched elites. As a result, the GAD approach was not quickly embraced by existing WID-oriented development programs in the industrialized West. In the developing world, by contrast, organizations representing a GAD platform have proliferated. The main international NGO promoting the goals of the GAD platform is DAWN (Development Alternatives with Women for a New Era), which was launched in 1985. DAWN's main policy orientation places the needs of people before the drive for profits (DAWN, 1995). DAWN proposes a threefold political economic strategy: reclaiming the state to act in the benefit of the majority, challenging market economies to meet goals of social responsibility, and building institutions that strengthen civil society, with priority attention to the role of the women's movement (Box 11.4). As a result of the efforts of DAWN and other organizations, by the 1990s this political economic ap-

BOX 11.4 Engender: Gender and Development in Southeast Asia

The NGO known as ENGENDER (Center for Environment, Gender, and Development) takes its name from this trilogy of interrelated concerns. ENGENDER was formed in 1992 in Singapore as an autonomous regional organization to promote a GAD platform through local and regional projects. ENGENDER's projects regularly take a sustainable-livelihoods approach that gives careful consideration to meeting both basic needs of households and environmen-tal sustainability. ENGENDER is also a major center of policy formulation on gender and development in the Southeast Asian region.

ENGENDER's projects in the Riau Archipelago exemplify the incorporation of both gendered labor and sustainable-development concerns. In the participatory marine-resources management project off the north coast of Bintan Island in Indonesia, women are the primary managers of a diverse range of resources for daily human sustenance, including water, agricultural products, and forest and coastal resources. The project focuses on the link between gender and environment by stressing women's knowledge of the marine environment as a basis for sustainable development; in this community, such resources as women's environmental knowledge are not valued according to market transactions but on the basis of their potential to support the local community (Wee, 1995).

ENGENDER has also initiated the Gaia Livelihoods Program, which promotes sustainable craft making based livelihoods for low-income and disadvantaged communities in the Riau Islands, Kalimanatan, Java, northern Thailand, and even Singapore. Gaia Livelihoods creates jobs and new income opportunities by identifying markets and products for traditional crafts. In a project based at the Danau Sentarum Wildlife Reserve in West Kalimantan, for example, a Gaia crafts project tapped women's traditional basket-making skills to make baskets to specification for the Body Shop. The Body Shop, a chain of personal-care products stores across the United States, has become a multinational chain and has multiple branches in major Asian cities, especially Singapore. In this example from Kalimantan, women used rattan canes, which are cut from various species of vine plants indigenous to the tropical forest, to make the baskets. ENGENDER singled out this location for a Gaia crafts project because the timber economy had transformed local conditions: men had increasingly migrated for work in the timber industry and women were left with fewer reliable income sources. In the process, rattan, many species of which are endangered, was increasingly being stripped to be sold at low cost to timber companies to bind large hardwood logs onto trucks. By making rattan baskets, women were able to decrease pressure on the collection of rattan and increase their local income base 100 times by selling the baskets directly to the Body Shop. The brilliance of this project is that ENGENDER "eliminated the middle man;" the profits from sales went directly to the basket makers and the community.

The director of ENGENDER, Vivienne Wee, coordinates the sustainable-development platform of the transnational board of DAWN. Dr. Wee is a key leader and organizer in NGO activity on gender and development in the Southeast Asian region. A former co-director of ENGENDER, Noeleen Heyzer, was also a member of DAWN and was appointed director of UNIFEM, in 1994. Thus ENGENDER brings GAD leadership from the Southeast Asian region to the world forum.

proach to restructuring gender relations had become so common around the world that most development practitioners, from NGOs to the World Bank, began to add gender-based analyses to their existing WID programs. As a result, most of the larger international development agencies now employ some combination of a WID/GAD approach (Karim, 1995).

State Planning for Gender and Development

The majority of Southeast Asian countries have government organizations that oversee the status of women, but government views vary on what constitutes women's problems and how to improve gender equity.

Singapore has a demographic problem that is unique in the Southeast Asia region: a rate of population growth so low that the government has projected a future labor shortage. As a result, the state promotes population growth to maintain a "youthful, dynamic workforce" (Karim, 1991: 144). Research on the problem has demonstrated low marriage and child birthrates among the highest-educated women. In response, in 1984 the state established the Social Development Unit (SDU) to promote marriage among highly educated professionals. Dr. Eileen Aw never imagined heading such an organization but knew that "the traditional Chinese male habit of preferring wives less educated than themselves" combined with the condition that Singapore women have been unwilling to "accept men of inferior education" made the problem ripe for social construction of new gender values (Aw cited in Momsen 1991, p. 33–35). In 1986 the government officially changed its family-planning program slogan from "stop at two" to "have three or more, if you can afford it" and has instituted a package of incentives for larger families that includes tax rebates, subsidies for day care, and priority in school enrollment (DeGlopper, 1991, p. 75). Marriage and birthrates among highly educated professionals have risen as a result.

Cambodia, by contrast, faces a population shortage of a very different sort. In Cambodia, after decades of war, the Royal Government supported a "baby boom" in the 1980s to help rejuvenate the population base of the country. Because of the impacts of war, women constitute 60–65 percent of the adult population and 30 percent of household heads. The highest rates of poverty also occur in these households. The current development plan of the state, however, includes neither poverty reduction nor economic stimulus packages for these heads of household, and women widowed by war are not compensated by the state. However, the 1993 constitution guaranteed women equality in political, economic, social, and cultural spheres and abolished all forms of discrimination against women. A Comprehensive Womens Code is being developed to deal with basic rights, labor and employment, marriage, and family and criminal law procedures.

In the Philippines the state has increasingly taken into consideration gender issues, in the revision of the family code, inauguration of a Development Plan for Women, outlawing of the mail-order bride business, and recommendations for community-based child care facilities (Chant and McIlwaine, 1995). Corazon Aquino wrote into the constitution of 1987, "the state recognizes the role of women in nation-building and shall ensure the fundamental equality before the law of women and men" (Chant and McIlwaine, 1995, p. 38). The Family Code of 1949 has been revised to enhance gender equity, including provisions to allow a woman to apply for credit without her husband's permission, make the family home common property of husband and wife, and define household management as the right and duty of both sexes. Yet the code remains gender biased: the infidelity of the wife "poses more danger" than the infidelity of the husband, based on the rationale that it "seriously injures the honor of the family, impairs the purity of the home, and may bring children not of the husband" (Pineda, 1991, p. 116). This effectively creates a double standard for men and women and socially constructs higher moral standards for women than for men. By contrast, the state Development Plan for Women sets out a series of proposals designed to "alter the traditional concept of a woman's self-worth as being subordinate to a man's" in individual, family, sociocultural, economic, political, and legal spheres. The National Commission on the Role of Filipino Women focuses on education of rural women and the development of livelihood projects for low-income groups. Finally, a state Republic Act has called for the "integration of women as full and equal partners in nation building" by creating a legal obligation on the part of the National Economic and Development Authority to transfer a "substantial portion of official development assistance funds from foreign government, multilateral agencies and organization" to support programs for women (Chant and McIlwaine, 1995, p. 40). Clearly in the Philippines the state has recognized that women do not have equal status in the household and have take steps to institutionalize a higher status for women.

In Vietnam the role of the state in supporting women is unique and is based in the country's history of socialist government. In 1930 the Vietnamese government founded the Vietnam Womens Union (VWU) as a national body to bring support to women and households at all levels from the village to the national capital (Lich, 1994). The VWU is a mass organization of more than 11 million members and functions to assist families in both daily problems, such as the settlement of family conflicts, and relations with the state by formulating legislation and demanding government policy changes. VWU programs focus on job creation and income enhancement and actually critique the WID "just add women" approach to economic development. During the past decade, the VWU has worked with the state bank to increase the availability of credit to poor women with revolving loan programs at low interest rates. These funds are used to stimulate income-generating activities for small businesses, purchase of pigs or other livestock, stocking ponds with fish, expansion for rice padi production, and supplies for handicrafts which can be sold in local and regional markets. In these ways the VWU continues to support the traditional regional role of women managing local-level economic activities.

The World Bank and Structural Adjustment Policies

In 1980 the major multilateral lenders, the International Monetary Fund (IMF) and the World Bank, made structural adjustment policies (SAPs) a requirement for loans granted to shore up faltering economies and service debts. SAPs require countries to promote economic growth by deflation (decreasing the amount of money in circulation), devaluing currency, privatizing industry, and increasing exports and reducing imports (Elson, 1992, p. 30). Although these adjustments are designed to enable long-term economic growth, the short-term costs bear disproportionately on the poor (Beneria and Feldman, 1992). SAPs also include ending state subsidies, public expenditures, and price controls. Eliminating such costs as state-subsidized education, child care, and health care benefits have had considerable gendered impacts. In Southeast Asia, Malaysia, the Philippines, Indonesia, Vietnam, and Thailand have undergone SAPs to make prescribed adjustments in national economic planning.

International critiques of SAPs by UNICEF (1987) and NGOs prompted the World Bank to begin to assess their impacts; for example, a World Bank review of its own programs for the urban poor revealed that SAPs targeted men as the "providers" in households even when a resident adult male was not present and thereby deemphasized the role of women (Sparr, 1994). SAPs had not acknowledged gendered divisions of labor in the household and in that way reflected 1950s and 1960s standards in development work. In response to the critiques, the World Bank began a WID initiative in 1987. But even the World Bank's 1990 *World Development Report* promoted labor-intensive employment for poverty alleviation, completely ignoring the fact that women already maintain labor-intensive roles, or the triple workday. In 1996 the World Bank began a series of reports on how to implement gender policies as part of a broader strategy to promote dialogue with the U.N. and NGOs. DAWN and affiliated NGOs have increasingly pressured the bank to implement gender policies through GAD initiatives.

In the Philippines, the first SAP was implemented in 1980 accompanied by loans from the World Bank and the IMF. By the end of 1983 the unemployment rate stood at 16 percent, nearly five full points higher than the 10.9 percent rate in 1980 (Floro, 1994). What went wrong came from complex interrelations between conditions in the world economy and local production in the Philippines. A global recession in the early 1980s negatively impacted the export trade, and certain commodity prices, especially that of sugar, which has been a mainstay of the Philippine economy, were at their lowest level in five decades. In rural areas, the SAP called for withdrawal of government subsidies in agriculture, reduction of agricultural extension services, and promotion of export crops. Although the state removed many subsidizes, it continued to pay subsidized prices to sugar planters, who are dominated by a landed elite. On the island of Mindanao, farmers followed government recommendations and switched production emphasis from staple food crops to sugarcane, bananas, and pineapple. When the price of sugar fell, rural household incomes further declined. Surveys on women's labor and household nutrition under the new export regime showed that women's domestic labor had increased, as had their hours of work in farming tasks. Women put in extra time to meet new shortfalls in the household food supply and the household budget due to elimination of government social services and increased prices on household goods. "Neither the Philippine government nor the IMF or World Bank had any major gender sensitive economic policy that recognized women's productive contribution. The SAP was formulated and implemented without any consideration of women's primary roles in maintaining the main economic unit: the household. Overall these policies were designed with no regard to the importance of unpaid housework or home production activities" (Floro, 1994, p. 123–4).

As a result of the critiques of SAPs, the bank now prepares Country Assistance Strategies (CAS) that include gender as an integral planning element (World Bank, 1996). On the basis of this new system, 28 percent of World Bank operations contained gender-specific actions, and a further 9 percent contained a discussion of gender issues in 1995 ; this compares with 11 percent and 5 percent, respectively, in 1988. Most of the gender focus, however, is concentrated in areas that represent traditionally female-gendered concerns: population, health, and nutrition (76 percent), education (65 percent), agriculture (60 percent), and social issues (33 percent). In other important sectors that impact both men and women (finance, industry, transportation, environment, urban, and water supply and sanitation), the proportion of projects with a gender component has been much lower. Further improvements in World Bank gender policies may result from a global campaign to transform the bank. Women's Eyes on the World Bank is a coalition of NGOs (from Africa, Latin America, Europe, ENGENDER from Singapore, and a unit of DAWN; see Box 11.4) and a global campaign to transform the World Bank to meet women's needs and address issues of gender and development.

The UN and Regional NGOs: Making New Linkages for Engendered Development

UNIFEM (the United Nations' Development Fund for Women), founded in 1976, is an autonomous unit of the UNDP (United Nations Development Program) and functions as the collective voice of women within the United Nations system. UNIFEM brings women's perspectives into decision-making at all levels, especially through a Women's Development Agenda to promote political empowerment by advancing women's rights. In 1994 UNIFEM made Noeleen Heyzer, a co-director of ENGENDER, director of the organization. Heyzer brings to UNIFEM and the UNDP a long tradition of gender-aware planning from the heart of the Southeast Asian region. Among a range of projects UNIFEM funds are two important programs that bring qualitative gender analysis to quantitative government statistical procedures. One program, for Indonesia's Central Bureau of Statistics, seeks to enhance gender responsiveness of the government statistical system, and another regional project designs gender-sensitive statistical procedures for the national statistical offices of the Philippines, Thailand, and Vietnam. These are important large-scale projects that could fundamentally transform how countries measure not only the status of women and standards of living but overall levels of development.

Each decade the United Nations holds an international conference on the status of women. The Fourth World Conference on Women took place in China in 1995 at Huairou, a small town outside Beijing. China originally planned to hold the conference in Beijing, but China's leaders switched the site location based on the realization that the conference would bring with it the activity of hundreds of NGOs whose representative members would speak out and demonstrate in Tiananmen Square. Because Tiananmen was the site of antigovernment protests that led to the Tiananmen massacre in 1989, Chinese leaders could not permit demonstrations at the site. The political geography of changing the site location, however, did not alter the outcome of the conference or NGO activity at Huairou. NGO members demonstrated at the site and produced a 12-point platform on a dozen critical areas of concern that were considered to be the main obstacles to women's advancement around the world. The platform was the highlight of the International NGO Forum. It underscored the role of human rights as women's rights and was adopted unanimously at the conference by representatives of 189 countries. Recognition of the importance of women's human rights reflects the fact that three-quarters of the United Nations member states have become parties to the Convention on the Elimination of All Forms of Discrimination Against Women (CEDAW) established in the late 1970s. Perhaps surprisingly, Singapore, the country at the top of the Gender Development Index within the Southeast Asian region, is not signatory to CEDAW.

The U.N. promotion of human rights as women's rights is a formula that is perhaps nowhere in greater need in the region than Burma and Cambodia. Conflict in Burma between the state and ethnic nationalities since the outbreak of civil war in 1949 has had considerable impacts on women, who have left strife-torn areas of their country for uncertain futures and marginal economic opportunities in Thailand. Many of the women come from nationality groups along the border where settlements have been unstable. For some women, crossing the border into an uncertain future in Thailand has appeared as a better choice than being killed in a border war, raped by Burmese soldiers, or enslaved by the Burmese army. Conservative estimates of girls and women from Burma working in brothels in Thailand now range between 20,000 and 30,000, with approximately 10,000 new recruits brought in each year (Human Right's Watch, 1993). This situation must be understood in terms of the political economic conditions in which they emerged: increased Thai-Burmese border trade, Thailand's image as a sex-tourism destination, and the role of Burma's ruling State Law and Order Restoration Council's (SLORC) in ending borderland conflicts with ethnic minorities by military force. The Asia Watch division of the international NGO Human Right's Watch has documented the trafficking of underage women from Burma into Thailand and has concluded that trafficking occurs with the direct complicity of the Thai police and other officials. This of course is in direct violation of CEDAW and other international laws. Many of the young women are forced to work in bonded labor conditions and seem to be infected with the HIV virus at a rate approximately three times higher than prostitutes in Thailand generally. Asia Watch has responded by making urgent recommendations to both the Thai and Burmese governments to control and eliminate this illegal trafficking and has called on the United Nations to assist in the implementation of human rights in Burma to curb the abuse.

Cambodia has been a subject of human rights focus as a result of atrocities committed during horrors of the reign of Pol Pot and even for abuses of local women committed by United Nations Peace keeping troops stationed in the country (LICADHO, 1995). The mission of LICADHO (Cambodian League for the Promotion and Defence of Human Rights), in conjunction with the state, is to raise gender awareness and women's rights by strengthening human rights. The organization works throughout the country by sending trainers who teach women's rights and human rights and distributes illustrated booklets, posters, and curricula that include information on basic nutrition and health. Their effort is widely appreciated in Cambodia since the civil war destroyed the education system and, in some areas, schools are still sparsely distributed, especially high schools. Only 20 percent of adult women are literate in Cambodia, as compared to 50 percent of men, and so the work of LICADHO is making a major contribution to education and literacy for women in Cambodian society.

CONCLUSIONS

From human rights to the household, gender analysis has ultimately transformed how development is planned and practiced. Organizations from the local level like LICADHO and ENGENDER to the national level (the Vietnamese Women's Union) and the international level (UNIFEM and even the World Bank) are now all working together more than ever before to ensure the promotion of gender equity in development. The transformation of the development agenda to adopt a gender perspective is a slow and incremental process that depends on the willingness of people and organizations to change many of their basic views about social roles and behavioral norms. In Southeast Asia existing values about gender equity have facilitated the role of gender analysis in development and allowed greater progress in combined gender and development planning goals.

Several important conclusions about gender and development in Southeast Asia emerge from this analysis. First, the status of women in many parts of Southeast Asia seems to have been historically higher than in other Asian regions. Centers of matriarchal social organization have contributed to patterns of high status for women. Second, the status of women in Southeast Asia has remained relatively high, especially by comparison to other parts of Asia. By statistical measures, Southeast Asia stands out regularly as a region characterized by relatively even male and female population sizes and labor-force participation rates. Third, it is equally important to recognize, though, that the status of women has not been uniformly high across the region and that patterns of gender equity vary within countries and cultural groups. Strong patriarchal conditions also prevail within the region. Fourth, the status of women in the region and gender equity more generally have been changing as a result of the development process. Postwar modernization and international economic development planning have often carried male bias, which has destabilized some of the existing patterns of gender equity. By diminishing relatively egalitarian gender relations between men and women, postwar economic development socially constructed new uneven gender relations from which the region may never fully recover. Fifth, and finally, it is important to recognize that despite problematic conditions for women in some parts of the region, including prostitution and marginal working conditions, women have not been passive victims of the development process but have engaged actively in new opportunities and worked to resist or transform problematic conditions brought with the development package.

Postwar economic development in Southeast Asia has dramatically transformed societies and economies and in many places has brought a first generation of men and women into the workforce. The role of the Southeast Asian region in the world economy as a center of low-wage manufacturing production has especially impacted young women and changed the stakes of their potential life paths. While factory work is not a long-term desirable option, the experience has allowed some women to postpone early marriage and household formation. Some women factory workers living in manufacturing zones have been able to gain independence by living at a distance from the family and by saving a portion of their income. When young women refuse to return home from manufacturing zones to marry, the stage is set for the traditional order to break down (Wolf, 1992). Changing conditions such as these challenge analyses that characterize women as victims of society and docile workers who serve the goals of economic development. Even as women may be oppressed by working conditions, families, and social norms, they have also evolved strategies for coping with difficult circumstances. Some of these strategies, such as spirit-possession incidents on the factory floor in the export processing zones of Malaysia, may seem unusual. Indeed male-gendered perspectives on such incidents would seek to construct them as irrational, but this is an attempt to discipline the young women workers, not take seriously the problems of oppressive working conditions, and to maintain their low salaries. In many places women are attempting to enhance their situations by engaging in forms of resistance to social and economic forces and using their new resources to strike out on alternative life paths. When they do, society often does not welcome them but instead socially constructs their activities as inappropriate or immoral, as in the Malaysian case of *minah letrik* (hot stuff), to keep them within the bounds of the existing social order. Ultimately, for many women, social control strategies do not constrain them but propel them to move on to new communities and new places where their decisions are affirmed and supported, especially to major cities and urban areas. Regularly and around the world, it is the role of women to initiate and carry out fundamental social and institutional change.

12 Indonesia

GRAEME HUGO

Indonesia is the world's fourth-largest state with a population in 1999 of 207 million and hence must loom large in any discussion of Southeast Asian countries. Stretching across some 40 degrees of longitude in a vast archipelagic realm as broad as the 48 contiguous U.S. states, Indonesia is a diverse collection of islands and peoples struggling to become a nation. Stitched together by two centuries of impermanent Dutch colonial rule, Indonesia since independence in 1945 and until 1998, had been ruled by just two leaders, both authoritarians with firm ideas of how to rule and develop the diverse nation. Divided by the watershed year of 1965 when leaders changed, Indonesia has formed an outward-looking, pro-Western posture that has brought rapid albeit uneven development to the huge population. Indonesia faces tremendous problems in part because of its rich diversity of peoples, resources, and physical environments and more recently the 1997–1998 Asian financial crisis, but it does have the potential to become a powerful state in the world community. As by far the largest country in Southeast Asia, the future of the region depends on Indonesia being successful in its struggle to develop.

Like the Southeast Asian region as a whole, Indonesia has experienced rapid economic growth, restructuring, and substantial social change during the last two decades. This has seen Indonesia make the transition from the World Bank's (1994) "Low Income" economies with GNP per capita in 1992 being US$670 to a "Low-Middle Income" economy in 1994 with US$880 per capita (World Bank, 1996). Nevertheless, Indonesia remains a poor country with the largest labor surplus in the Southeast Asian region. Some 40 percent of its workers are underemployed, infant mortality levels hover around 66 per 1,000 live births, and one-half of its workers are employed in agriculture. In the late 1990s its key characteristics are its physical and cultural diversity and the rapidity of the social and economic change it is experiencing. With 4 out of 10 Southeast Asians being Indonesians, its destiny is of critical importance to Southeast Asia.

In mid-1997 the now well-known Asian financial crisis began in Thailand and soon became a "contagion," spreading to other Asian economies, ultimately affecting most severely Thailand, Malaysia, South Korea, and Indonesia (see Boxes 7.3, 16.1, 13.1, and 12.3). Indonesia has the dubious distinction of having witnessed the greatest fall in the value of its currency (the rupiah) of any of the Asian economies. In part because it is such a large and diverse country—the world's fourth-most-populated—the crisis has perhaps been the worst here, one in which the repercussions will be felt long into the future. Among other impacts, the economic crisis in Indonesia has brought about political turmoil that included a fall from power of President Suharto in 1998. Another outcome of the economic turmoil has been that long-standing ethnic differences have been further aggravated, bringing about conflicts between various groups and the local Chinese population, between transmigrant settler groups such as the Madurese in Kalimantan and indigenous groups, and in Timor (Box 12.3).

UNITY IN DIVERSITY—INDONESIA'S GEOGRAPHY AND SOCIETY

Indonesia's national motto of "Unity in Diversity" is a very appropriate one, and an appreciation of its enormous diversity is fundamental to any understanding of the country's contemporary economic and social situation. There are more than 300 separate ethnolinguistic groups that are being welded together into a nation-state, and they are spread over more than 13,000 islands. There is a vast range of ecological situations from the richest agricultural lands in Java, one of the world's most densely settled agricultural areas, to the swampy eastern coastal areas of Sumatra, the mountainous interiors of the Outer Islands, and the dry, drought-prone areas of East Nusa Tenggara. The contrast between densely settled Inner Indonesia (Java–Madura–Bali) and the Outer Islands (Sumatra, Kalimantan, Sulawesi, and the other islands) is striking with Java–Madura having 60 percent of the nation's population on 7 percent of the land area.

The Inner/Outer Indonesia dichotomy in population density largely reflects disparities in resource endowments, and although there is considerable potential for development in the Outer Islands they are frequently incorrectly portrayed as being "empty" (Map 12.2). In fact,

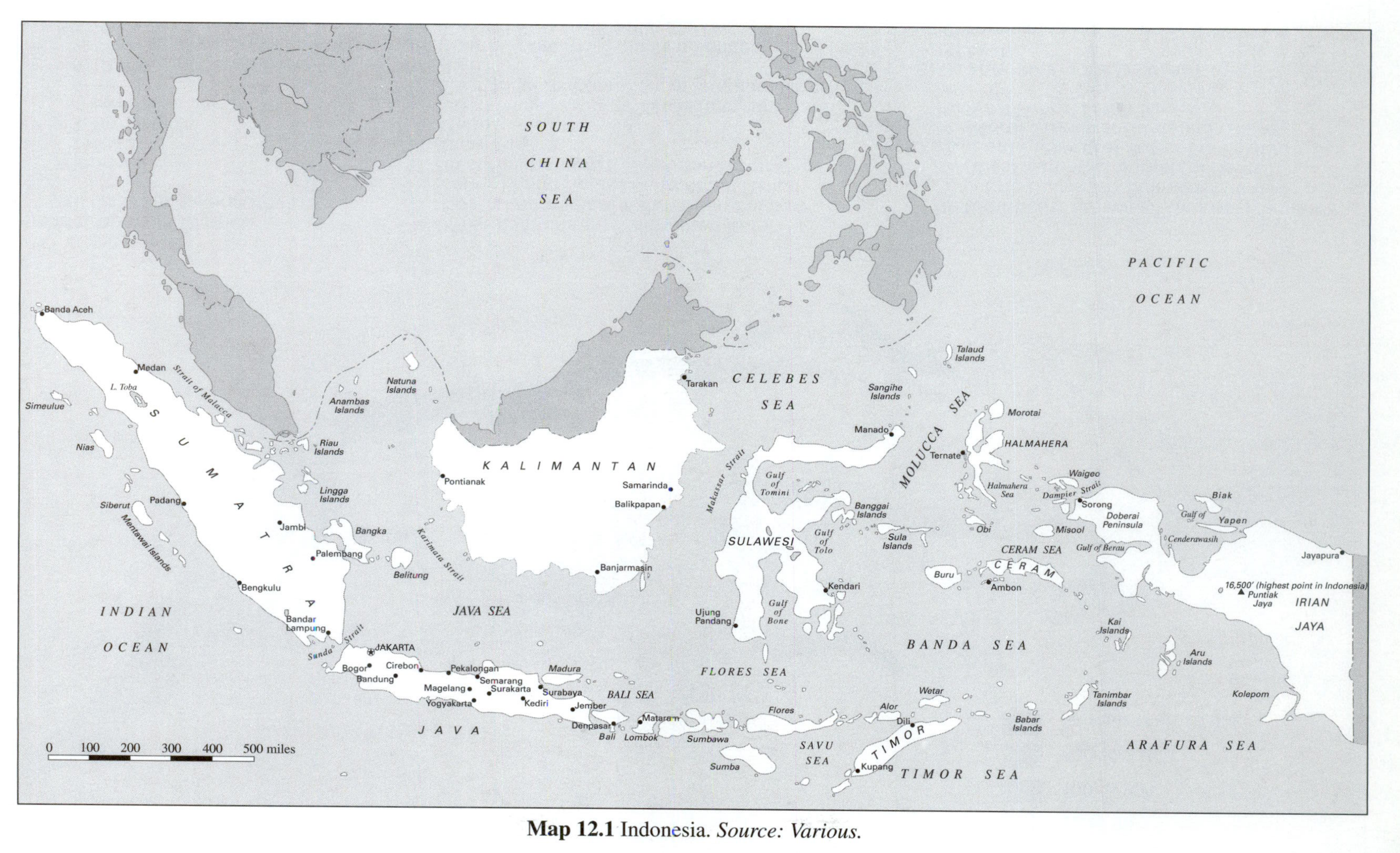

Map 12.1 Indonesia. *Source: Various.*

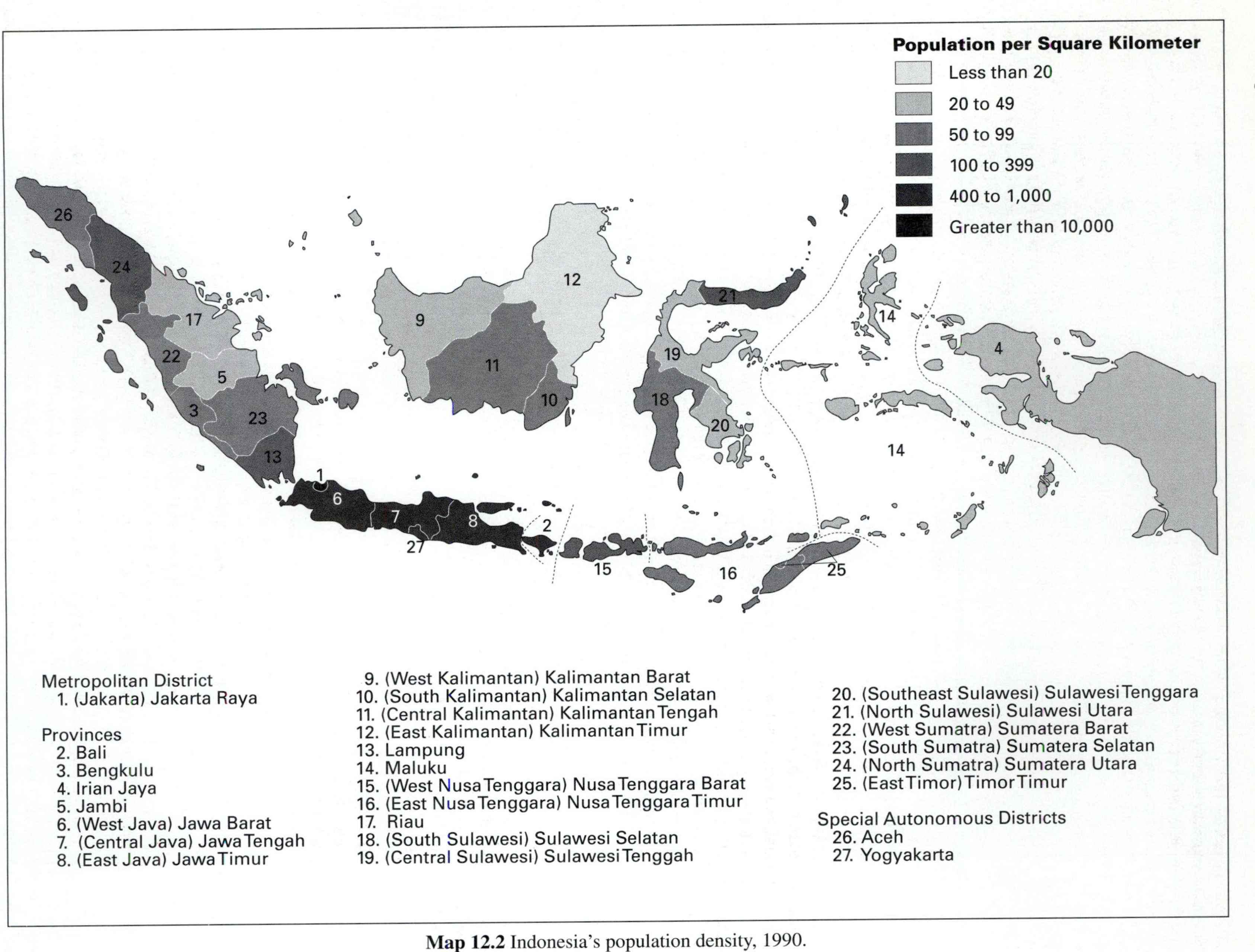

Map 12.2 Indonesia's population density, 1990.

Sumatra alone, with a population in 1990 of 36.5 million, would be the twenty-ninth-largest country in the world if it were a separate state.

Indonesia is the world's largest archipelago, and the geographical fragmentation of the state has had a great deal of influence on patterns of human activity. There is not only enormous interregional variation in physical landscapes and resource endowment but also in degree of development and access to services. The archipelagic nature of the country hampers national integration efforts, and although interisland and intra-island transport have been improved enormously, it remains a limiting factor on development in the more peripheral parts of the country. Most mining, oil exploration, and plantation activity occurs in the Outer Islands whereas manufacturing, especially large- and medium-scale industry, is disproportionately concentrated in Java.

Indonesia's strategic maritime location has opened it over the centuries to various waves of migrations and the spread of ideas, beliefs, goods and technology from both the north (China) and the west (India, the Arab world, and finally Europe). As a result, it is probably the most ethnically and culturally heterogeneous of the world's largest nations. Clifford Geertz (1963b, p. 24) summarizes this diversity as follows:

> There are over three hundred different ethnic groups in Indonesia, each with its own cultural identity, and more than two hundred and fifty distinct languages are spoken . . . nearly all the important world religions are represented, in addition to a wide range of indigenous ones.

Within ethnic groups, too, Indonesians have loyalties to kinship and regional and local groupings, and frequently their behavior is influenced by group norms, formalized into a body of customary law (*adat*).

The ethnic complexity of Indonesia is reflected in Map 12.3, which depicts the major language groups within the country. Many Indonesians today speak *Bahasa Indonesia,* a *lingua franca* that developed as a trading language around the archipelago, was adopted by the newly independent Indonesia in 1945 as the national language, and has subsequently been vigornously promoted as a major unifying element in the diverse nation. [In 1995, 41 percent of Indonesians reported being able to speak Indonesian compared with 19 percent in 1971.] Most Indonesians also speak the language of their own ethnolinguistic subgroup as their primary language. Two major cultural divides can be superimposed on the complexity in Map 12.3 (Fisher, 1965, p. 239). The first is a contrast between the coastal and interior groups and the second is between the western and eastern islands. The western two-thirds of Indonesia is dominated by two groups. The smaller, comprising the Proto-Malay (Nesiot) groups, were the earliest arrivals from northwest India or Burma and now are found mainly in the interior of the islands—the Gajo of Sumatra, Badui and Tenggarese of Java, Sasak of Lombok, Dayaks of Kalimantan, and Toradja of Sulawesi (Missen, 1972, p. 85).

The much larger group of Deutero-Malay (Pareoean) people dominate in Java and in the coastal areas of the islands of the western two-thirds of Indonesia. These groups evolved from a series of waves of migration over several thousand years from Southern China (see Map 3.1). The largest group here are the Javanese who made up 47 percent of the population in 1930, the last occasion an ethnicity question was included in an Indonesian Census. [The ethnicity question has not been included in post-Independence Censuses because of the imperative of both the Old Order (1945–1965) and New Order (1965–present) governments to unify the diverse nation and deemphasize ethnic differences.] The Javanese continued dominance is reflected in the fact that 39 percent of Indonesians reported speaking Javanese as their daily language in the home at the 1990 Census. Some 86.4 percent of Javanese speakers lived on Java. The second-largest group, the Sundanese (14.5 percent in 1930 and 16 percent in 1990) is predominantly found in West Java. Some 96 percent of Sundanese speakers lived in Java in 1990. The third-largest group are also predominantly based in Java. The Madurese made up more than 7 percent of the total indigenous population in 1930, and 4 percent of Indonesians spoke Madurese at home in 1990.

The largest groups outside Java are also of Malay origin. The Minangkabau of Western Sumatra were the largest group in 1930 (3.4 percent), and 2.4 percent reported speaking that language at home in 1990. The Batak groups of Northern Sumatra were also significant—2 percent in 1930 and 2.2 percent in 1990. The next-largest group are the Bugis of Southern Sulawesi who made up 2.6 percent of the indigenous population in 1930, and their language was spoken in 2.2 percent of homes in 1990. Other significant groups include the Banjarese of South Kalimantan (1.5 percent in 1930 and 1.8 percent in 1990) and the Balinese (1.7 percent in 1990). Despite attempts by the government to Indonesianize the total population, these groups each retain their identity and have separate languages, cultures, and characteristics. Although there has been substantial intermarriage and interregional migration that has blurred the patterns evident in Figure 12.2, the distinctiveness of the various groups and their dominance of their heartland regions remains a significant element in Indonesia.

The ethnic and cultural situation is even more complex in the eastern one-third of Indonesia, which is one of the world's greatest cultural divides, representing a transition zone between Malay peoples of Asia and the

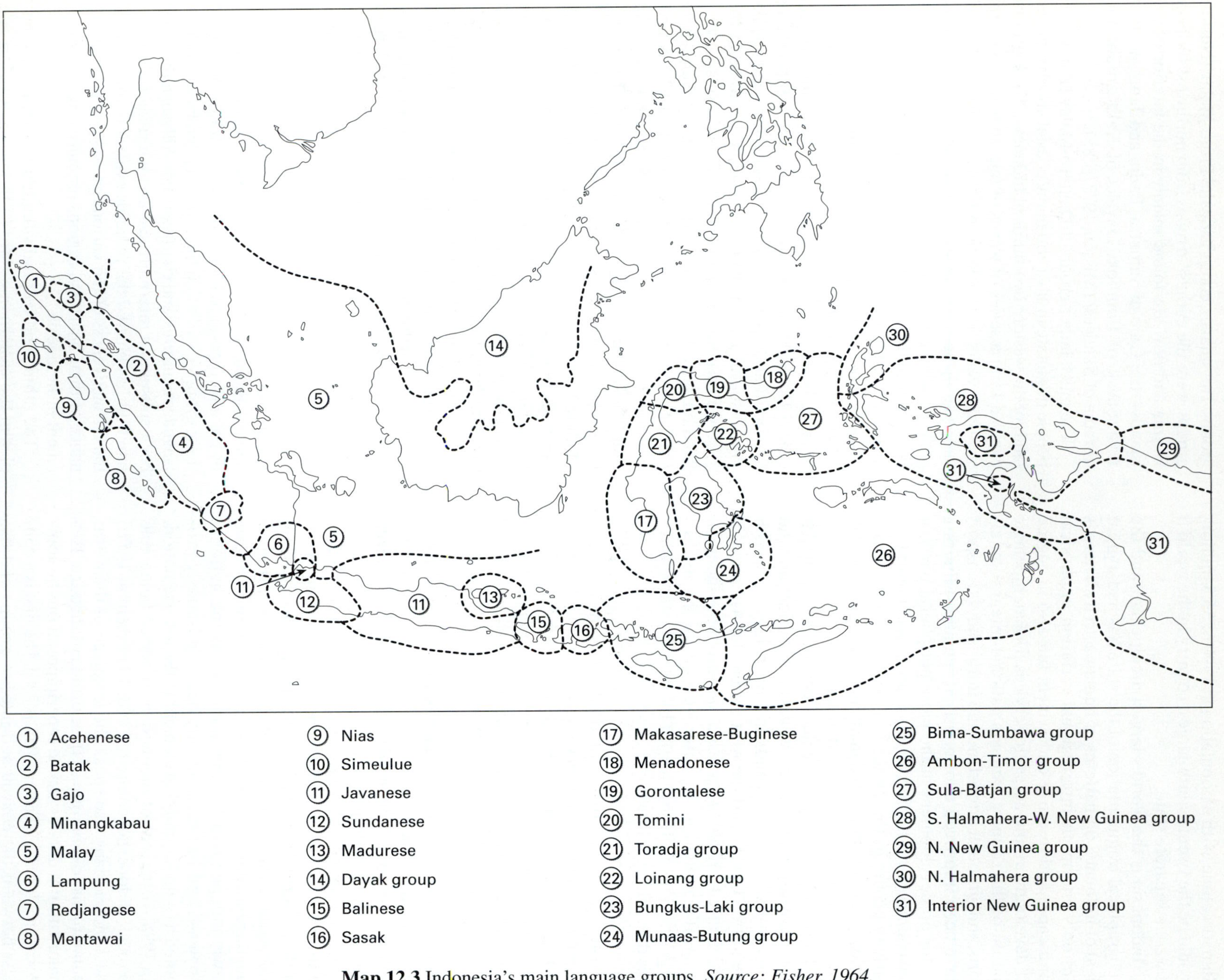

Map 12.3 Indonesia's main language groups. *Source: Fisher, 1964.*

Photo 12.1 Javanese women in the Johar market, Semarang, Central Java. (Leinbach)

Melanesians of the Western Pacific. Although people of Melanesian origin dominate in Irian Java and the Moluccas, there has been considerable mixing with Malay groups, especially in East Nusa Tenggara. In contemporary times the substantial migration of Malays from Java and Sulawesi into such predominantly Melanesian areas as Irian Java has attracted controversy (Manning and Rumbiak, 1989; Adicondro, 1986).

As in other Southeast Asian countries, the Chinese are the largest nonindigenous group in Indonesia. Ethnic Chinese make up about 3 percent of Indonesia's population (Coppel, 1980, p. 792; Suryadinata, 1997a, p. 21), although their dominance in some sectors of the economy is perceived by some to give them a significance out of proportion to their numbers. Chinese presence in Indonesia long predated European contact with Chinese traders, being established in coastal trading settlements and cities when the first Europeans arrived. However, during the last century of colonialism, heavy immigration of Chinese occurred, often under Dutch auspices, so their numbers increased from 221,438 in 1860 to 1,233,234 in 1930 (Hugo, 1981, p. 118). Other Asian groups like South Asians also migrated into Indonesia during those years increasing from 8,909 in 1860 to 115,535 in 1930. The formal and informal sensitivity regarding the Chinese is reflected in the fact that no separate language category was allocated to them in the 1990 Census. The Chinese are significantly represented in most areas, mainly in the urban areas of provinces. However, they are also especially conspicuous proportionately in areas where they had been brought in by the Dutch to work as coolies on plantations and in mines (northern and eastern Sumatra and West Kalimantan).

The ethnic diversity in Indonesia has occasionally erupted into violence. The 1950s was characterized by struggles for independence from Indonesia in West Java, South Sulawesi, West Sumatra, and the Moluccas (McNicoll, 1968). However in more recent times there have been attempts to gain independence at the western and easternmost extremities of Irian Jaya and the westernmost limit of Aceh. In 1975 when East Timor was decolonized by Portugal, it was taken over by Indonesia, and there are still claims for independence from East Timorese overseas and in the province. There are other conflicts too that arise from time to time. For example in early 1997 there were clashes between indigenous Dayaks in West Kalimantan and settlers from Madura off the island of Java (McBeth and Cohen, 1997). This is one of the conflicts that has arisen from government attempts to settle people from inner Indonesia to the outer islands, leading to conflict between very different ethnic groups. Similarly the onset of the economic crisis in 1997–1998 was marked by anti-Chinese riots in West Java in which the Chinese who dominate among merchants were blamed for rising prices.

Indonesia's cultural mosaic has been influenced over the centuries by migrations, trade, and influences not only from the north and east (China) but also from the west. Trade with India and the spread of Hinduism and Buddhism to Indonesia dates back to the second century A.D. (Missen, 1972, p. 98). During the next millennium a number of kingdoms arose and fell in different parts of the archipelago and Hindu (e.g., Mataram) and Buddhist (e.g., Borobodur) temples are all that remain to indicate the scale and significance of these indigenous kingdoms that were based around significant inland cities at a time when urbanization in Europe was limited (see Maps 3.3–3.6). In addition, a number of coastal trading ports arose and fell around the archipelago trading, especially with India and China (McGee, 1967).

Islam first spread to Indonesia in the late thirteenth century by Arab merchants and traders who settled in Aceh in North Sumatra. It then diffused via Arabs and Gujeratis slowly through Indonesia, especially in the sixteenth century as trade expanded. By the end of the seventeenth century the population in most coastal areas had been converted to Islam as had the interior of Java and Sumatra. At the 1990 Census, 87 percent of the population was reported to be Muslim, 10 percent Christian, 2 percent Hindu, and 1 percent Buddhist. These data do not give a true picture, however, because there is an enormous interregional variation in the strength with which Islam is embraced, which Muslim practices are followed, and the extent to which the religion is infused with local traditional religious ideas and practices.

There is a strong regional dimension to religion in Indonesia and this is apparent in Map 12.4 (see also Map 3.8). This shows that Bali and West Lombok are the only relics of the Hindu dominance of the archipelago. Islam is dominant in the west and devout Muslims are concentrated in northern and western Sumatra, West Java, and South Kalimantan. Traditional religions still retain a hold in the highland areas of many of the Outer Islands. The arrival of the Europeans in the fifteenth century saw the overlaying of another religion (Christianity), but it became dominant in only a few areas—North Sumatra, North Sulawesi, and over much of Eastern Indonesia.

The arrival of Europeans in the region resulted in trade similar to that which the Chinese and Indians had been doing for centuries. The Portuguese established themselves in several parts of the archipelago but especially in the Moluccas (now Maluku), which was a major source of spices for European markets. Indeed, this was the area that Christopher Columbus was seeking when he discovered the Americas. During the next two centuries the Dutch gradually supplanted the Portuguese and increased their presence and influence (working through local rulers) in the region. Accordingly, the VOC (Dutch East India Company) came to be a dominant force in Indonesia during the seventeenth and eighteenth centuries. The VOC was abolished in 1798, and the Dutch government assumed direct responsibility for administering the "East Indies." However, it was only after 1830 that the Dutch began to establish sovereignty over much of the area that is now known as Indonesia. The European exploitation of the East Indies had a profound impact. The raison d'être of colonialism was exploitation, and the Dutch aimed to extract raw materials like rubber, oil, sugar, and spices at minimum cost so that development of indigenous economic activity was suppressed. Moreover, education and health programs were not extended to the mass of people, industrial development shifted, and the rights of the local people greatly limited. The Dutch hold on the East Indies was not broken until World War II with the Japanese invasion in 1942. With the defeat of the Japanese by the allies, Indonesia declared its independence in 1945 but had to engage in a war of liberation with the Dutch until 1949. The post-Independence period has seen two eras. The Old Order under President Sukarno which lasted two decades and the New Order of President Suharto.

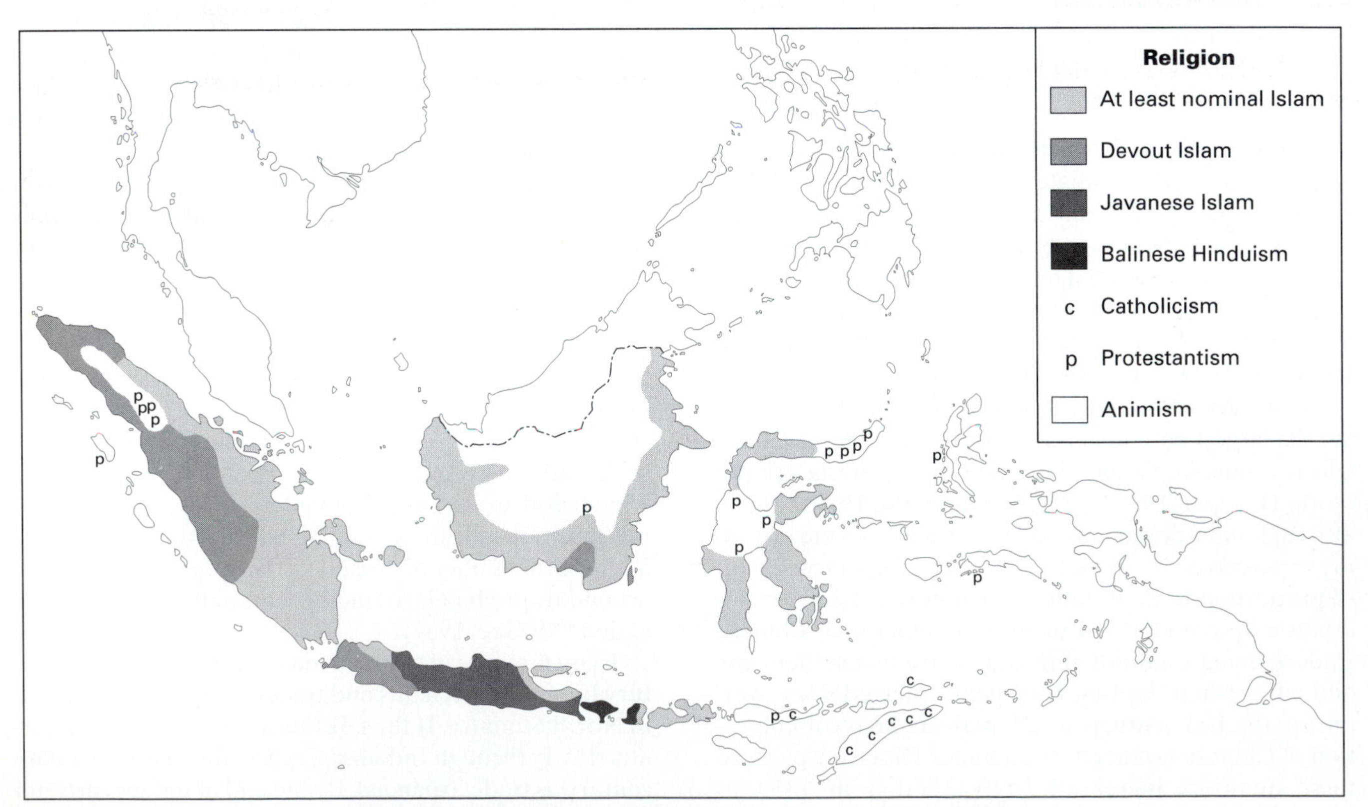

Map 12.4 Indonesia's major religions. *Source: Fisher, 1964.*

ENVIRONMENT

Indonesia's physical environment has a great deal of impact on human activity in Indonesia. The major features of Indonesia's physical environment are shown in Map 12.5. The alignment of the islands of Indonesia is dictated by a series of arcs of geological activity and in most cases their relief is dominated, virtually from end to end, by rugged mountain backbones, capped in the volcanic zone by numerous majestic cones of active volcanoes, many of which rise to well over 10,000 feet above sea level (Fisher, 1964, p. 17). Most settlement has been concentrated in the lowland areas, especially the well-drained, fertile alluvial plains of East and Central Java and, to a lesser extent, North Sumatra and parts of Nusa Tenggara. On the other hand, the low-lying coastal plains of eastern Sumatra, West and South Kalimantan, and southwestern Irian are swampy and less attractive to settlement.

Soil fertility varies enormously and is the major determinant of variations in rural population density. In common with most tropical areas, the bulk of Indonesia's soils are poor due to leaching. There are exceptions, however, and those of most significance are the extensive sections of East and Central Java, Bali and Lombok, and smaller areas of West Java, Sumatra, and Sulawesi with neutral or basic soils derived from recent volcanic ejecta, which are exceptionally rich and support some of the highest densities of agricultural population in the world. The other major exceptions lie in river valleys and well-drained coastal areas where recently deposited alluvium is conducive to intensive agriculture. However, in places where acidic volcanic or heavily leached lateritic soils predominate, agricultural potential is limited.

Indonesia has a tropical climate, with mean monthly temperatures invariably above 64.4°F (18°C) (Fisher, 1964, p. 21) with little seasonal or regional variation. There are, however, significant declines in average temperature with altitude, and these have a major influence on the types of agriculture practiced in particular areas. Rainfall is generally high and not as limiting an influence on agriculture as are soils, drainage, and temperature conditions. The west coasts of the main islands receive the heaviest rainfall (more than 100 in., or 254 cm, per year) under the influence of the southwest monsoon. There are, however, areas in which rainfall is not only lower (below 40 in., or 102 cm) but, more important, unreliable. These areas are predominantly in the eastern part of the archipelago (especially in the Nusa Tenggara islands stretching from Lombok to Timor); it is here where crop failures, food crises, and famines have been reported in recent years. Rainfall failure also occurs, albeit much less frequently, in the southern part of Sulawesi, the eastern part of Java, and the northwestern coast of Java.

We can summarize this brief consideration of Indonesia's geographical background with reference to what Buchanan (1967, p. 43) identifies as favorable ecological niches. These are areas in which the conjunction of the nature of the soils, the amount and distribution of rainfall, and the drainage conditions are such that the area lends itself most favorable to intensive agricultural activity. Such areas identified by Buchanan in Indonesia are shown in Map 12.6 and they include the following types:

1. coastal plains except on the eastern coasts where swampy conditions prevail
2. alluvial valleys and plains (often terraced) except where there are drainage problems
3. volcanic slopes of neutral or basic materials—over much of Java

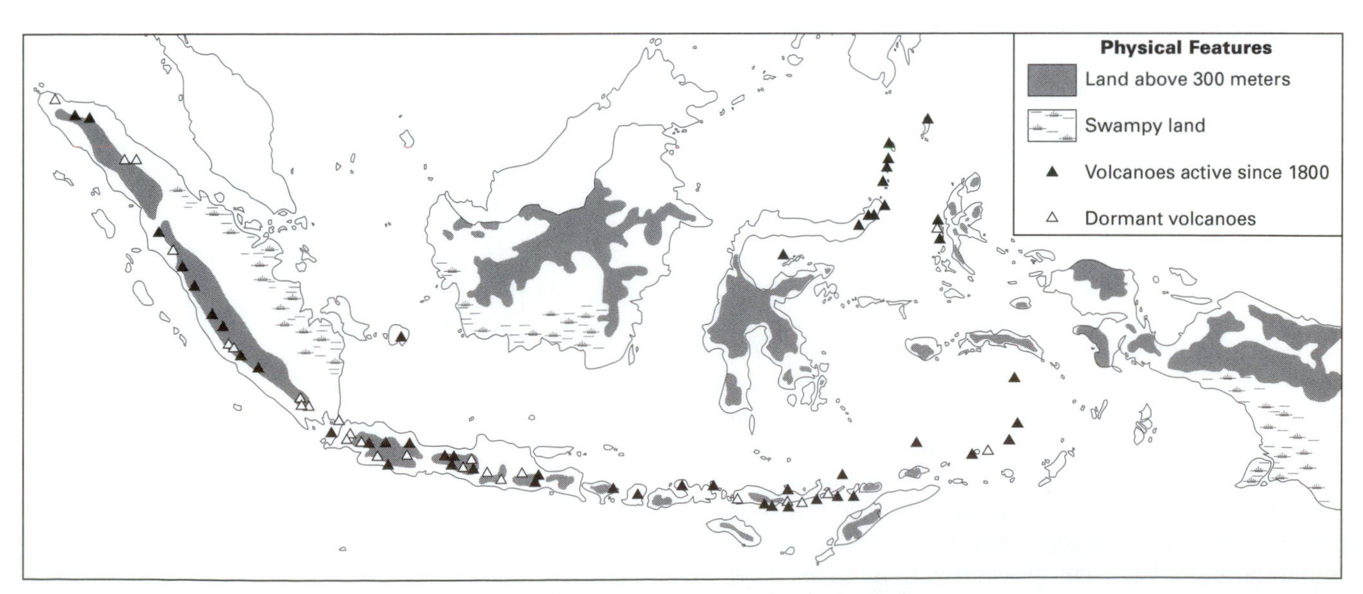

Map 12.5 Major features of physical relief.

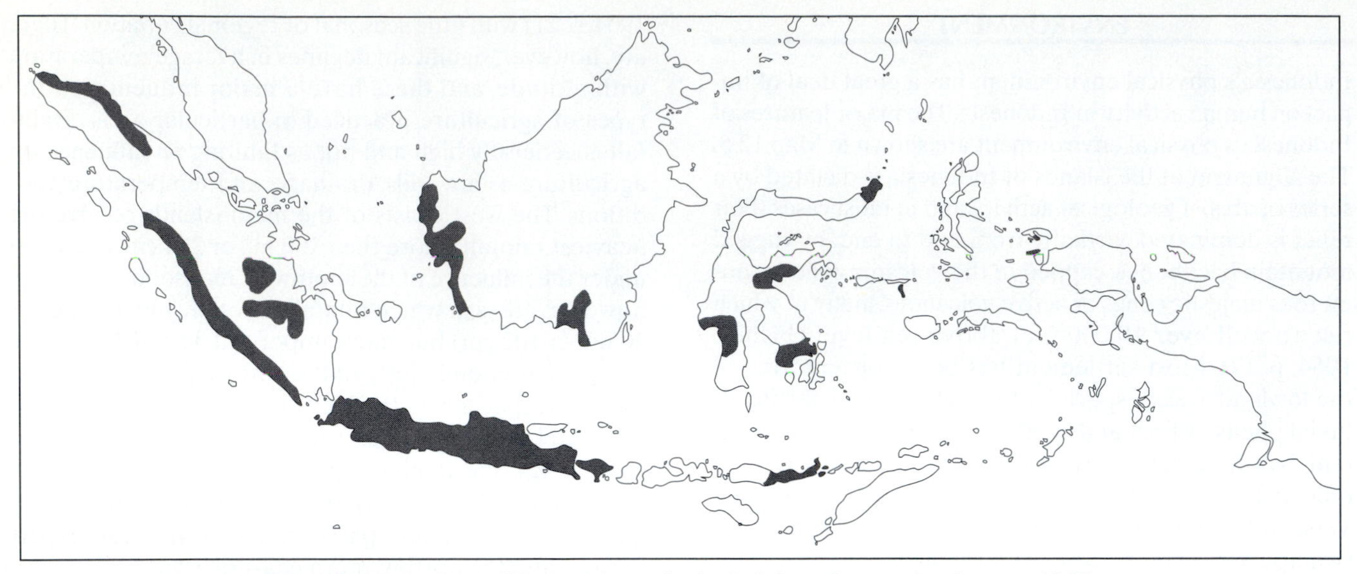

Map 12.6 Locations of favorable ecological niches. *Source: Buchanan, 1967.*

It is interesting to note that Buchanan classified virtually all of Java as a favorable ecological niche.

Also note the areas that are difficult environments for intensive agricultural production:

1. coastal swamp environments (for example, east coast of Sumatra)
2. forested steep inland hills
3. dry zones (for example, Nusa Tenggara; southern South Sulawesi)
4. key wet zones

Environmental issues loom very large in Indonesia with substantial population pressure on land, pressures to earn foreign exchange through export of natural resources, and lack of expertise and resources for environmental surveillance and control. These have been brought to a head in the global arena with the massive forest fires of 1997 (Rigg, 1999) (see Box 12.1). The environmental problems that face Indonesia are many and diverse and are of worldwide, rather than just Indonesian, significance; for example, Indonesia, with 279 million acres (113 million hectares) in 1994, has the second-largest area of tropical rain forest (*Jakarta Post,* May 31, 1994, p. 1) that has an important role as the lungs of the world. However this forest is being lost at 1.5 million acres (0.6–1 million hectares) annually and there is a thriving illegal forestry industry. As Figure 12.1 shows, Indonesia has a very large proportion of its total land area under forest, and it is crucial that this forest is used in a sustainable way (see also Maps 2.6a and 2.6b). However, it is estimated some 123.5 million acres (50 million hectares) of forest have been lost during the last 23 years. Land degradation is of significance all over Indonesia, and it has been shown that more than 20 percent of land is degraded. Thirty-six of Indonesia's 125 watersheds have been degraded, and erosion rates in Indonesia are up to 40 tons of soil per hectare. In Java, the home of 60 percent of Indonesians, there is a soil loss of about 50 tons per hectare per year. This is exacerbated with population pressure leading to clearing of upland areas in association with the high rainfall. Another major problem is the loss of biological diversity. Indonesia is one of the most biologically diverse countries in the world with 17 percent of the world's plant species (Cohen, 1996), but the increased clearing of land and replacement of natural environments with a monoculture is leading to a rapid extinction of species. This is most notable among larger animals

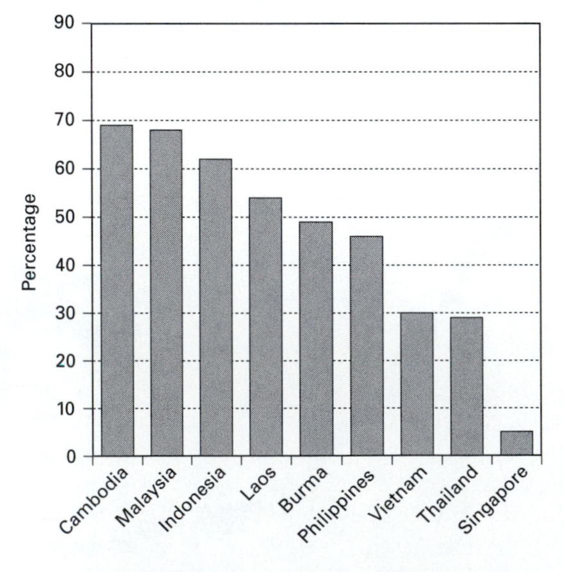

Figure 12.1 Percentage of land in forest and woodland, 1992–1994. *Source: World Resources, 1998, p. 299.*

BOX 12.1 Indonesia's Forest Fires

Any visitor to Indonesia, Malaysia, or Singapore in the second half of 1997 cannot fail to have noticed a gray-brown haze that blanketed much of Southeast Asia during this period. Each year the dry season in Indonesia has been one of substantial forest fires (for example, see Jayasankan and McBeth, 1994) but this was exacerbated in 1997 due to the El Niño effect and increased fires in Sumatra and Kalimantan, set especially by large forestry and companies clearing land for palm oil and other plantations. The dry season has traditionally been a period for clearing of forest by swidden agriculturalists but also by foresters and others wishing to clear land for commercial or subsistence agriculture. The smog comes largely from thousands of fires that were deliberately lit but subsequently raged out of control, often going underground and igniting coal seams that were extremely difficult to extinguish. While the delay of the monsoon by several months, which usually douses the annual fires in Sumatra and Kalimantan, by the El Niño effect is partly blamed for the massive fires, there has been a great deal of debate in the region about the causes of the haze. On the one hand, swidden agriculturists each dry season set fire to areas to establish agricultural plots for a few years. Later these are abandoned because the residual fertility of the soil had been used up as a result of leaching and the land needs to lie fallow for several years. This cyclic agricultural migration over established areas has been going on for centuries and represents a sustainable adaptation to the nature of the environment. On the other hand there have been massive clearings of the forests in the outer islands of Indonesia both by the government to establish transmigration settlements (see Box 12.2) and by the private sector as part of their logging operations. The establishment of large-scale palm oil and other export-oriented agricultural plantations has also been responsible. There can be no doubt that legal and illegal clearing of forestland by fire in Sumatra and Indonesia has increased. This has lead to substantial criticism of the Indonesian government from both within the country and internationally as undoubtedly large firms on government leased land must take part of the blame for the 1997 fires.

Hence the causes of the forest fires have been complex. On the one hand they are environmental, given the El Niño effect; the nature of tropical ecosystems and climate also have had a role to play. However, the causes are also economic in that there has been great pressure to increase production of export-oriented cash crops especially for palm oil. The operation of the swidden agricultural system is also a factor. There are political reasons because government's encouragement of land-clearing subsidies and development of outer islands areas results in fires.

The effect of the forest fires has been massive and has not been confined to Indonesia. Indeed their impact on Singapore, Malaysia, and to a lesser extent Thailand threatened the government relationships between countries within ASEAN. In October 1997, analysis of satellite photographs showed 4.25 million acres (1.7 million hectares) were burned. This, of course, represented a massive loss of biodiversity and other resources. For example, the fires destroyed village fishponds in the areas effected. The choking haze led to a huge increase in respiratory infections among Indonesians, Singaporeans, and Malaysians. In parts of Indonesia, Singapore, and Malaysia, the sun was not visible for months. A number of major plane and boat crashes occurred with massive loss of life; for example, a flight to Medan in North Sumatra crashed with a loss of more than 200 lives. Indonesian losses of revenue from tourism, crops, health costs, and legal compensation have been estimated to be more than US$20 billion. Falls in tourism and industrial output in Malaysia cost US$30 million and in Singapore US$80 million (Cohen, Dolven, and Hiebert, 1998). The tourism numbers in affected Southeast Asian countries fell for the first time after many years of massive growth. The economic effects of the fires undoubtedly were part of the causal factors influencing the currency crisis that gripped Southeast Asia in 1997–1998 (Rigg, 1999).

such as tigers, rhinoceros, elephants, and orangutan, but is also important for plants, fish, birds, and small animals.

The other type of environmental problems are those resulting from the use of the environment by people. Pollution of many kinds is increasing in significance year by year. The pollution of water resources in urban and rural areas is a particular problem. Use of pesticide increases exponentially, and the amount increased by 200 percent between 1981 and 1991. There are increased quantities of toxic liquid and solid wastes in the freshwater, seawater, and air of Indonesia. Indonesia's rapidly expanding vehicle fleet (Hugo, 1997a) uses a particularly polluting type of leaded gasoline. Nevertheless, as is the case in other LDCs, the per capita use of fuel in Indonesia remains quite low. Table 12.1 for example shows that Indonesia's energy use has more than doubled between 1980 and 1994, while that in MDCs increased by only 13.5 percent. However in 1994 energy use in Indonesia was only 7 percent the size of the average for MDCs.

These data put in perspective calls from people in MDCs for LDCs such as Indonesia to reduce their energy consumption. Indeed, energy use must increase if there is to be an improvement in levels of living in Indonesia. Similarly the carbon dioxide emissions in Indonesia, although substantial, are only 7 percent the size of their MDC counterparts.

A third major dimension to the environmental issue in Indonesia relates to use of resources (Hardjono, 1994). The ability of Indonesia to feed its growing population is addressed in the next section, although it should be noted

	Indonesia		High Income Countries		Low Income Countries	
	1980	1994	1980	1994	1980	1994
Energy use (kgm per capita)	169	366	4,464	5,066	248	369
Carbon dioxide emissions (metric tons)	95	185	9,837	10,266	2,063	3,880
Carbon dioxide emissions (metric tons per capita)	0.6	1.0	12.4	13.9	0.9	1.5

Table 12.1 Indonesia: Energy use and emissions, 1980 and 1994. *Source: World Bank, 1997.*

here that there are environmental problems that threaten food production and security. One of Indonesia's huge development advantages has been its substantial oil, gas, and coal resources. However, it must be recalled that these resources are finite; for example, in 1994 it was officially estimated that Indonesia's oil reserves were only sufficient to meet domestic demand for another decade (Thornton, 1995), although at the time it was exporting roughly 800,000 barrels of oil a day from its 1.5 million barrels per day daily output.

Environmental issues have not been given a high enough priority in Indonesia even though it was one of the first nations in the world to appoint an environment minister. Not only are the issues large but they are changing in type with massively increasing urbanization and industrialization (Douglass, 1989; Koestoer, 1992).

POPULATION PATTERNS

The growth of Indonesia's population in the nineteenth century has been a subject of considerable debate. Figure 12.2 shows that this was a period of rapid growth, although some commentators suggested that such growth was partly a function of considerable underestimation of the population in the early years of the nineteenth century (Breman, 1971). Focusing more on the contemporary period, and especially the early years of independence, relatively low rates of population growth occurred due to the disruption caused by economic depression, Japanese occupation, and the war of independence with the Dutch. Hence the population grew by only 1.5 percent annually during that period. With improvements in mortality in the 1960s and 1970s, the annual growth rate accelerated to 2.1 percent and 2.4 percent respectively.

The decline in fertility that began in the 1970s saw the annual population growth rate decline to 2.2 percent in the early 1980s and 1.8 percent in the latter half of that decade. That decline has continued such that the current annual rate of growth is estimated to be 1.6 percent per annum and over the Sixth Five-Year Plan (*Repelita VI* 1993–1998), the population increased at 1.5 percent annually. By 1999 the population exceeded 207 million.

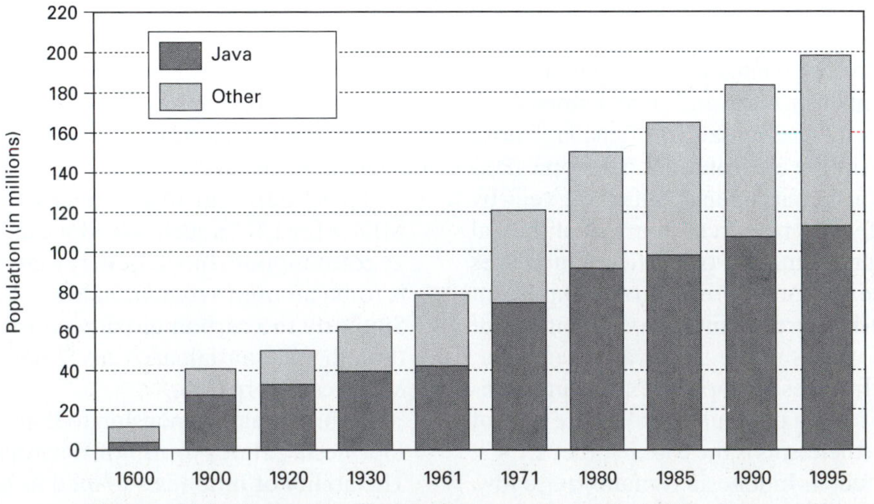

Figure 12.2 Indonesia: Population growth, 1600–1995. *Source: Hugo, et al., 1987, p. 31; Biro Pusat Statistik, 1991, 1997; World Bank, 1994.*

Fertility

The debate about the relationship between population growth on the one hand and economic and social change on the other has raged since Malthus's work of more than two centuries ago. In the contemporary Indonesian context, Figure 12.3 shows that economic growth has considerably outpaced population growth during the last three decades. Indeed, the average rate of growth in GDP has been three times the average rate of population growth. In examining population growth trends, it is important to examine the components of that growth separately. Of all the waves of social, economic, and demographic change that have transformed Indonesia during the last quarter century, none has been so striking or far reaching in its consequences as the decline in fertility. Contemporary women in Indonesia are having only one-half as many children as their counterparts in the 1960s. The TFR has declined steadily since the late 1960s when it stood at about 5.6 children per woman. This had declined to 5.2 in the early 1970s and 4.68 in the late 1970s. At the 1990 Census the TFR for the 1985–1990 period was measured at 3.3 and at the 1994 Indonesian Demographic and Health Survey a level of 2.9 was recorded for the previous two years.

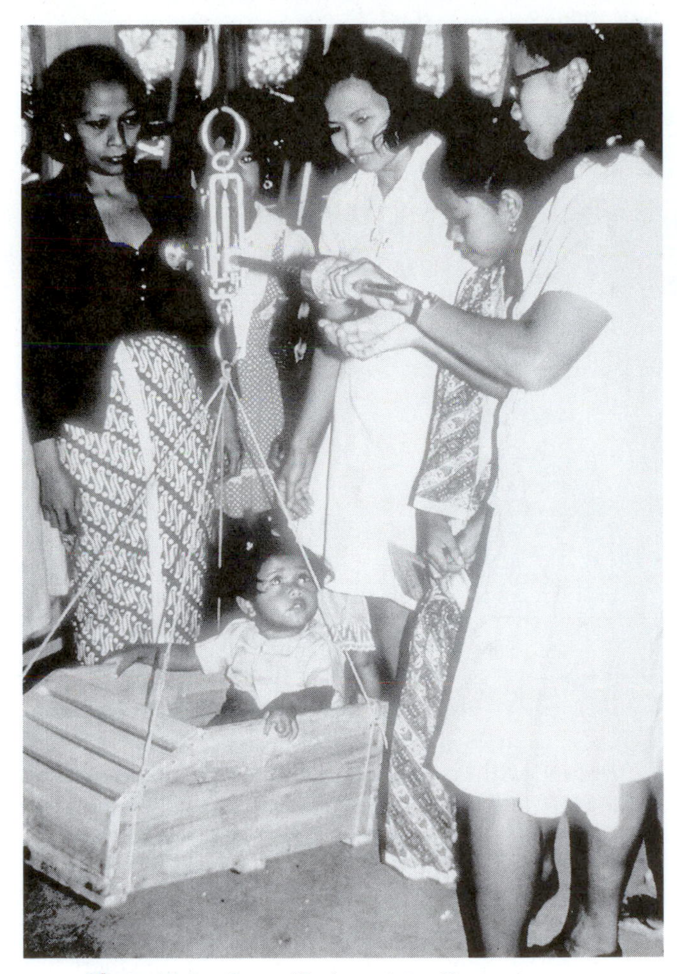

Photo 12.2 Part of Indonesia's effort to improve children's health focuses on family nutrition. A major activity is child growth monitoring. Monthly weighings take place at sub-village weighing posts by volunteers (*kaders*) and are intended to monitor weight in order to detect nutrition-related problems. (Leinbach)

The processes that shape the decline in fertility are complex but are definitely associated with significant changes both in the role and status of women and in the structure and functioning of the family. This involves a move away from emotionally extended to nuclear families, intergenerational relationships, selection of marriage partners and relationships between marriage partners. These social changes are associated with the introduction of universal education, the penetration of mass media, and the reduction in the significance of the family as a unit of production. The national family-planning program has undoubtedly played a significant role. In 1972, only 400,000 couples were practicing some form of family planning whereas in 1993, there were 21.3 million. In 1994, 96 percent of currently married women interviewed in the National Demographic and Health Survey indicated that they had knowledge of at least one method of family planning.

There can be no doubt that the high level of acceptance and use of contraception in Indonesia has been due in significant part to the success of the nation's Family Planning Program. With the installation of the New Order government, there was a government commitment to bring down rates of population growth. By 1969 the National Family Planning Coordinating Board (BKKBN) had been formed and the National Family Planning Program initiated. Initially the program concentrated its activity in Java–Bali and was progressively extended first to a group of larger Outer Island provinces and later to the remaining provinces. It gained strong support not only from government but also influential Muslim leaders.

Family-planning services are provided by the Ministry of Health through hospitals, township (*kecamatan*) health centers (*Puskesmas*), village health centers (*Puskesmas Pembantu*) and monthly neighborhood health posts (*Posyandu*). In addition BKKBN employs more than 25,000 field-workers to recruit acceptors and distribute pills and condoms through village-based volunteers. These efforts also involve key village leaders in achieving local support for the program. This has led to the program gaining a great deal of acceptance and support at grassroots level. Also at the community level approximately one-quarter of a million acceptor groups has been formed, and the BKKBN have been able to help them with resources for income-generating and other development activities. In this way and others, the program has focused not just on reduction of family size but improving the quality of life of families. Another key to the success of

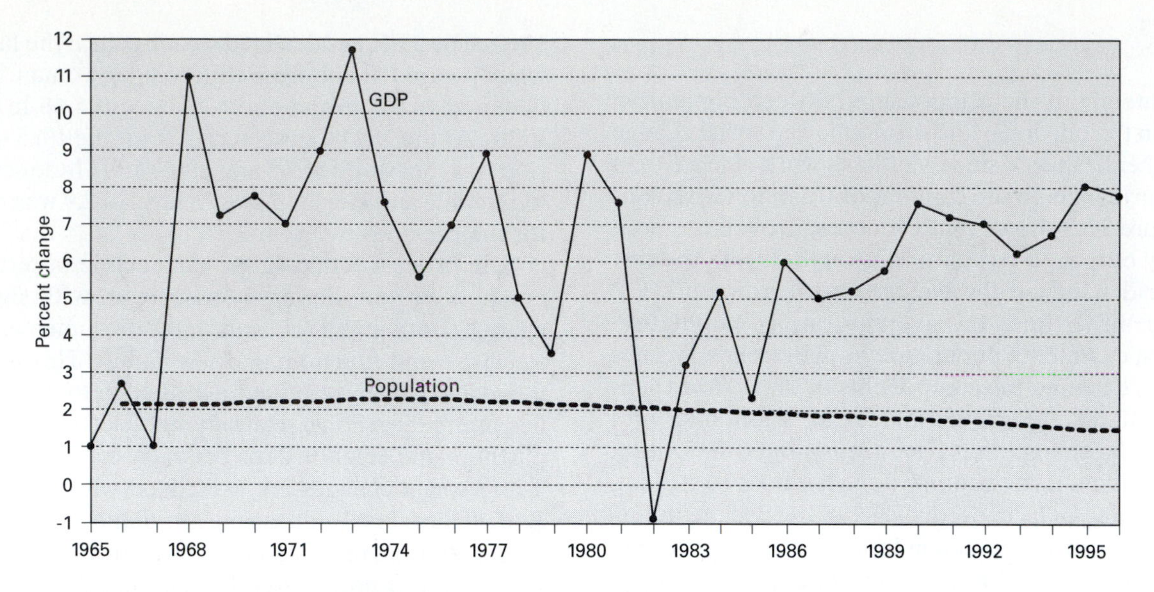

Figure 12.3 Indonesia: GDP and population growth rates, 1965–1996. *Sources: H. Hill, 1994; Indonesian Censuses of 1961, 1971, 1980, and 1990; Solomon, 1997.*

Photo 12.3 Indonesia's very successful family planning program is conspicuously identified on this billboard in central Jakarta. (Leinbach)

the program has been the highly developed logistics system that has generally ensured delivery of supplies in a timely way in diverse geographical circumstances.

Mortality and Health

The government's success in achieving a reduction in population growth has been achieved predominantly through a significant reduction in fertility during the last two decades. There has been a remarkable decline of 49 percent in fertility during the last quarter century. At the same time that fertility has declined, mortality levels have also been reduced, although Indonesia still has higher mortality than most of its ASEAN neighbors. Table 12.2 shows that as fertility has declined in recent decades, there has been a decline in infant mortality although data on fertility and mortality are of limited quality. The rapid reduction in infant mortality has also clearly been a function of the improved nutrition of children (Leinbach, 1988).

Figure 12.4 shows that during the last decade there has been a significant increase in the population of Indonesian children with good nutrition status (from 49 to 64 percent), although there has been little improvement in the population with poor nutritional status. This remains a significant problem in Indonesia. Other reasons for the decline in infant mortality include the decline in fertility

Year	Total Fertility Rate	Year	Infant Mortality Rate
1967–70	5.605	1967	145
1971–75	5.200	1971	142
1976–79	4.680	1980	112
1980–85	4.055	1987	75
1983–87	3.590	1991	74
1989–91	3.022	1994	67
1991–94	2.850		

Table 12.2 Estimates of fertility and infant mortality change, 1967–1994. *Source: Hugo, et al., 1987; IDHS, 1995.*

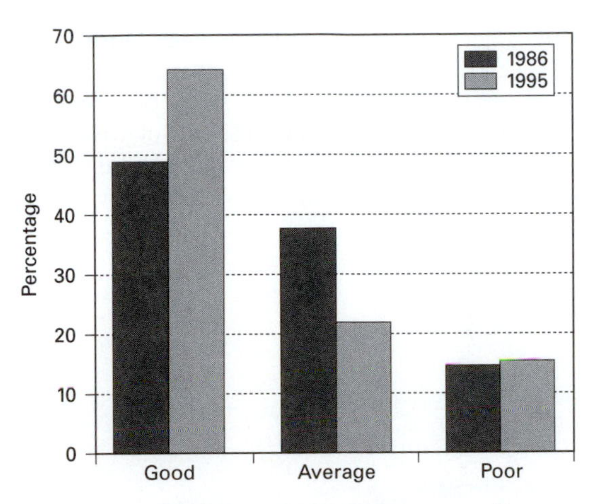

Figure 12.4 Indonesia: Percentage of infants according to nutrition status, 1986 and 1995. *Source: Biro Pusat Statistik and United Nations Development Programme, 1997, p. 67.*

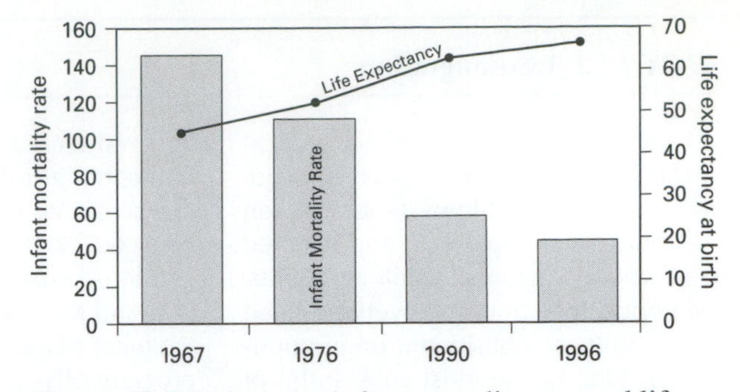

Figure 12.5 Indonesia: infant mortality rate and life expectancy at birth for select years, 1967–1996. *Source: Biro Pusat Statistik and United Nations Development Programme, 1997, p. 61.*

Indicator	Year	Percent of Population
Access to safe water	1994–95	63
Access to sanitation	1994–95	55
Percent under 5 malnourished	1994–95	39
Maternal mortality (mothers' deaths per 100,000 births)	1989–95	390

Table 12.3 Indonesia: Health indicators. *Source: World Bank, 1997.*

that has reduced the incidence of high-risk births, better training of midwives, greater availability of modern medicine, immunization, and the overall reduction in poverty. The close relationship between infant mortality and overall mortality is reflected in Figure 12.5 that shows the substantial increase in life expectancy during the last decade.

The expectation of life at birth has increased from about 47 in 1971 to 52 in 1980 and is presently about 62 (60 for males and 64 for females). The health of Indonesians has improved dramatically during the last three decades with improved nutrition, the spread of immunization, and development of a community-based health-care system. Nevertheless, life expectancy in Indonesia remains low compared with Malaysia, Singapore, and the Philippines. Maternal mortality remains a significant problem, and although many infectious diseases are declining in their effects, they still remain significant. Moreover, chronic and degenerative diseases are growing. This is particularly true of cancer and heart disease, with smoking being almost universal among men. Gaining access to adequate modern medical care remains a problem for many Indonesians, especially those outside of Java.

Improving the health of Indonesians remains a major challenge for the Indonesian government. Table 12.3 shows more than one-third of Indonesians still do not have access to safe water supplies, and almost one-half have no access to sanitation facilities. Despite the reduction in mortality, the death of women in childbirth, and child malnutrition remain large problems in Indonesia. Spending by government per person in the health area remains at an extremely low level of less than US$10 per person per year.

Internal Migration

Former President Sukarno frequently articulated Indonesia's population problem as being one of "unbalanced" distribution rather than excessive growth. Programs to "even out" Indonesia's population distribution go back to the Dutch colonial "*kolonisasi*" programs in the first decade of the twentieth century as part of the so-called ethical policy designed to improve the living conditions of people in Java. This program was also taken up by post-Independence governments and is referred to as Transmigration (see Box 12.2). This was given particular impetus in the late 1970s and early 1980s when some

BOX 12.2 Transmigration

Governments in many LDCs in tropical regions have identified one of their major development problems as an uneven distribution of population and initiated settlement programs to shift agricultural populations from the overpopulated areas to those considered underpopulated. One of the most substantial of these is Indonesia's Transmigration Program which was initiated in the first part of the twentieth century by the Dutch and continued through both independence regimes (Leinbach, 1989). In its original form this movement had both demographic and economic aims to reduce demographic pressure in Java–Bali and reduce the incidence of poverty. In modern day Indonesia the program has a number of objectives. Its main aim is developmental—to bring into production land that was hitherto unproductive and thereby assist regional development. It also has the aims to change the perceived demographic imbalance between Java–Bali and the Outer Islands and to populate sensitive areas in Irian Jaya and Kalimantan with people from the Inner Islands. The government has a program of selecting transmigrant families from Java–Bali–Madura and assisting the development of new agricultural settlements for them in the outer islands. Selection criteria has over the years changed but tends to be restricted to family groups with agricultural experience.

The program's level of success has been mixed. Up to the late 1970s, the program experienced a lack of success. It faced many problems (Leinbach, 1989; Leinbach, Watkins, and Bowen, 1992). These included lack of funds to properly prepare the area of settlement and assist settlers to establish themselves. Limited initial survey of the settlement areas resulted in unsuitable areas being settled. For example there were attempts to establish wet-rice agriculture on sandy soil. Some settlements were not provided with appropriate infrastructure. On some occasions settlers themselves did not have enough agricultural experience, as when the government briefly attempted to transmigrate urban vagrants from Jakarta. On other occasions, the land that was settled was not actually vacant, and there were local claimants to the land. In other cases there were cultural clashes between local inhabitants and the newcomers from Java, Bali, or Madura; for example pig-keeping Balinese settlers in South Sulawesi clashed with Muslim local inhabitants. Other Outer Island groups have expressed a

Photo 12.4 Transmigration has occurred in some very harsh environments. Settlers' homes shown here are in Sumber Jaya, located in the delta area of the Musi River in South Sumatra, where tidal swamp rice production is the economic base. No roads exist in this area and all travel occurs by boat through a system of canals. (Leinbach)

1.29 million families (about 5 million people) were moved under the auspices of the Transmigration Program. Although the goals of the program in recent years are now predominantly articulated in terms of regional development of the Outer Islands rather than in demographic redistribution terms, the uneven distribution of population in Indonesia is one of the most salient features of the nation's demography. For most of this century, Java's population has been growing more slowly than that of Outer Indonesia. Hence Table 12.4 shows that the proportion of Indonesians living in Java has declined from about 66 percent at the time of Independence to 60 percent at the 1990 Census and 58.9 percent at the 1995 Intercensal Survey.

However, the shift in government policy in the late 1980s to facilitate international and domestic private investment

fear of Javanization or new colonization of their areas and the resulting extinction of their ethnolinguistic group. These claims for example, have arisen in Irian Jaya (McBeth, 1994, p. 50). In some cases the problems have lead to significant return migration of the transmigrants and it has been shown that sometimes the transmigrants do not end up better off than they were in Java.

Many of these things changed in the late 1970s with the involvement of the World Bank, the massive funding of transmigration from loans and revenues earned from the increase in oil price. There was substantial investment in transport and land preparation and the introduction of greater quality control. In the 1980s as government funds became scarce, the idea of spontaneous transmigration was developed, whereby Java residents were encouraged to go to transmigration settlements but not receive government assistance to do so. There were also a number of semi-*swakarsa* schemes whereby the transmigrants received from the government only part of the transmigration package that was provided to full transmigrants. Figure 12.6 shows the large numbers of families who were moved from Java–Bali to the Outer Islands in *Repelita III* and *IV* when the World Bank became involved in transmigration differed greatly from those moved in the previous *Repelitas*.

However, it will be noted in Figure 12.6 that *Repelita IV* saw a substantial decline in the numbers of transmigrant families funded by the government and an increase in the number of spontaneous (*swakarsa*) transmigrants. This pattern was even more pronounced with the economic crunch of the mid-1980s—funding of transmigration was cut back so that numbers of families moved in *Repelita V* did not approach the ambitious target of 550,000. Indeed, sponsored families numbered about one-fifth of that target. Map 12.7 shows the provinces of origin and destination respectively of transmigrants during the first five *Repelitas*.

The dominance of Central and East Java as origins is clearly apparent. Sumatra remains the predominant destination of transmigrants, accounting for 58 percent during the first four *Repelitas* and 54 percent in *Repelita V*. Lampung's declining attraction is apparent in the fact that it was the second-most-important province of destination during *Repelita II-IV* accounting for 11 percent but it absorbed only 7 percent of transmigrants in *Repelita V*. Meanwhile, the proportions settling in Eastern Indonesia (mainly Irian Jaya) increased from 6.8 percent to 10.1 percent, those in Kalimantan from 22 to 23 percent and in Sulawesi from 8.5 to 12.5 percent (Oekan, 1993).

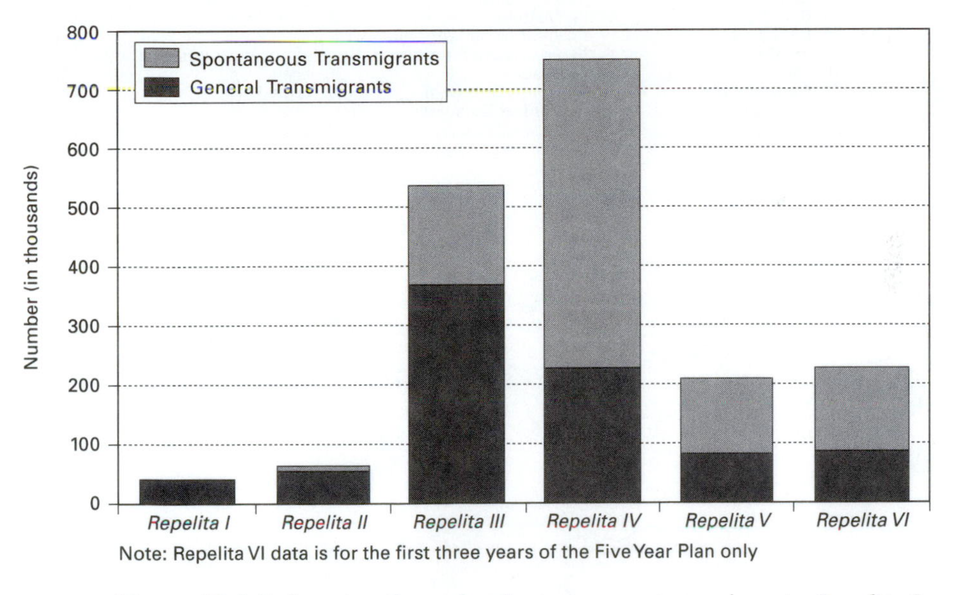

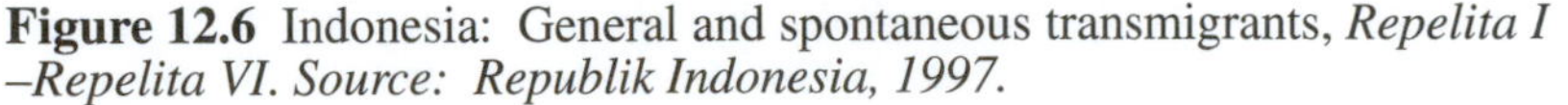

Figure 12.6 Indonesia: General and spontaneous transmigrants, *Repelita I –Repelita VI. Source: Republik Indonesia, 1997.*

and industrialization is tending to favor growth in Java. Between 1985 and 1990 the number of people moving into Java (773,789) was almost as great as the number moving in the opposite direction (973,340). Hence it would appear that there is a slowing down in the net redistribution of the population from Java to the Outer Islands, and it is the higher fertility and natural increase in the Outer Islands that is chiefly responsible for the more rapid population growth outside Java. The reasons for the slowing down in redistribution are several. One is undoubtedly a shift in government policy that has encouraged foreign and domestic investment in industry. This has greatly favored Java and hence attracted Outer Island migrants to Java and retained Java-born people who otherwise would have moved to the Outer Islands. Another factor has been a decline in government sponsored transmigration to Outer Islands.

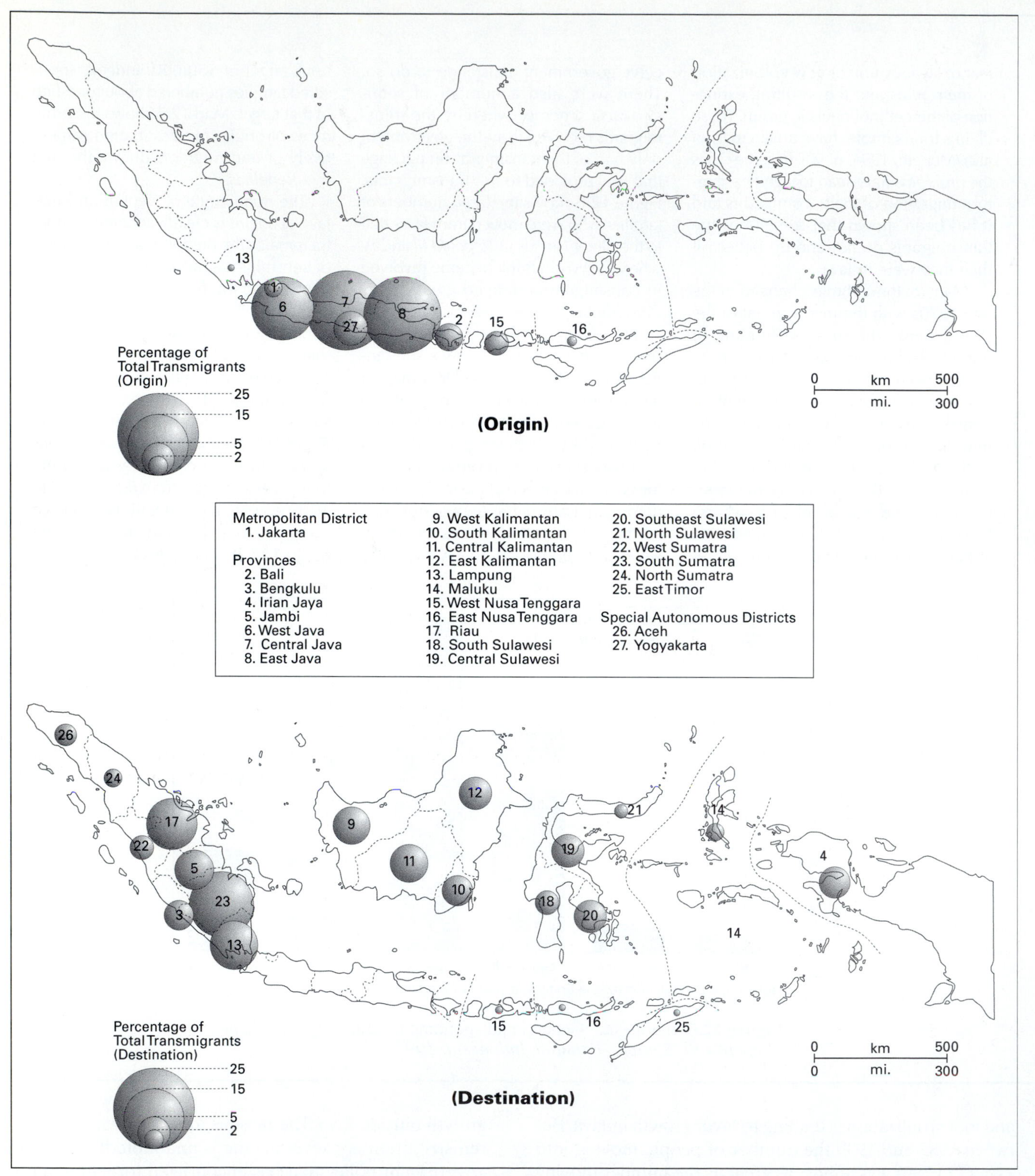

Map 12.7 Origin and destination of transmigrants among Indonesians 1968-1993. *Source: Republic of Indonesia, 1993.*

Year	Java		Other Islands		Total
	Number (in thousands)	Percent	Number (in thousands)	Percent	
1900	29,000	72.0	11,150	28.0	40,150
1920	34,984	71.0	14,171	29.0	49,155
1930	41,718	69.0	19,009	31.0	60,727
1961	62,993	64.9	34,026	35.1	97,019
1971	76,102	63.8	43,130	36.2	119,232
1980	91,270	61.9	56,220	38.1	147,490
1985	99,853	60.9	64,194	39.1	164,047
1990	107,574	60.0	71,748	40.0	179,322
1995	114,734	58.9	80,022	41.1	194,755

Table 12.4 Population of Java and other islands, 1900–1995. *Source: Hugo, et al., 1987, 31; Biro Pusat Statistik, 1991, 1997.*

Transmigration is only one element (and not the largest one) in a substantial migration from Java to the Outer Islands, with spontaneous moves being substantial. Table 12.5 shows that the number of people living in the Outer Islands but who had migrated there from Java increased by 73 percent between 1971 and 1980, while the number who had moved in the opposite direction increased by only 15 percent.

Hence there was a net migration loss to Java overall of 2.4 million people. However, during the 1980s there was a distinct change with a net increase of migrants from Java residing in the Outer Islands in the 1980s (1,576,910) (a similar magnitude to that recorded in the 1970s (1,510,354)), although the percentage increase of 44 percent was lower than that recorded in the 1970s. Thus, although there was a quite significant increase in outmigration from Java in the 1980s, it was somewhat lower than some projections had anticipated on the basis of the 1970s experience. The most striking change in Table 12.5 is in the number of migrants from the Outer Islands residing in Java—this doubled between 1980 and 1990. As a result there was only a comparatively small (16 percent) increase in the overall lifetime net migration loss from Java, from 2.35 million in 1980 to 2.71 million in 1990. Hence almost nine decades of transmigration and spontaneous migration from Java to the Outer Islands had provided a relatively small redistribution equivalent only to three years of population growth in Java.

The dominant population-mobility pattern in Indonesia in terms of its economic and social significance is that from rural to urban areas. The pattern of change in urban growth and urbanization in Indonesia during the last 35 years is shown in Table 12.6. The top row of the table shows the rate of urban growth in Indonesia. It can be seen that in each of the intercensal periods, the rate of urban growth far outstripped that of the rural population. It is especially notable, however, that in the 1980s, not only was the rate of urban growth more than six times greater

Java	1971	1980	1990	Percent Change	
				1971–80	1980–90
Total Outmigrants	2,062,206	3,572,560	5,149,470	+73	+44.1
Total Inmigrants	1,067,777	1,225,560	2,434,719	+15	+98.7
Net Migration	-994,429	-2,347,000	-2,714,751	+136	+15.7

Note: Based upon most recent migration data using census question on province of previous residence.

Table 12.5 Migration into and out of Java: 1971, 1980, and 1990. *Source: 1971, 1980, and 1990 Censuses of Indonesia.*

to urban areas in Indonesia. While there are no substantiating data collected in censuses or national surveys, it is clear that the tempo of nonpermanent movement has greatly increased during the last 25 years. There are a number of case studies that demonstrate this; in particular, those that resurveyed villages studied in the early 1970s discovered a substantial increase in nonpermanent moves (e.g., Kasai, 1988; Keyfitz, 1985; Singarimbun, 1986) and found that this change had been fundamental in improving the economic situation in those villages through a substantial inflow of remittances (see Chapter 4). Quotations from two such studies will suffice to document this pattern of an increasing tempo of temporary migration out of Javan villages during the last two decades. Edmundson and Edmundson (1983, p. 53), in their comparison of two East Java villages in 1969 and 1981, found that in one of their villages:

> The mid-1980 Census included 630 urban workers who retained village residence cards but they were only present on weekend visits. This intermediate category of part-time migrants forms an important new group whose freedom of movement and social adaptability represent a significant change in attitude from the traditional and highly territorial thought patterns of older villagers.

Manning's (1986, p. 28–31) report from six West Java villages from 1976 to 1983 concludes that:

> Despite substantial increases in rice production and incomes, there appear to have been relatively few jobs created in rural areas. Quite a substantial proportion of rural households seem to have benefited from the trickle down of urban income growth through entry largely into self-employed activities in transport and petty trade. This has been a major factor influencing agricultural income change in the survey villages over the seven year period studied . . . permanent movement to urban areas and movement out of agriculture was not a dominant pattern.

More than a decade ago the World Bank (1984, p. 20) estimated that at least 25 percent of rural households on Java have at least one family member working for part of the year in urban areas. This would imply that at least 3.75 million people are involved in this form of migration on Java, equivalent to slightly more than 50 percent of the measured 1980 urban employment in Java. Of course, because migrants are only working in the cities for part of the year, the average effect is less than this, but it is not unlikely that about one-sixth of the average daily urban workforce consists of temporary migrants who are not included in official employment figures. A series of studies sponsored by the Ministry of Population and Environment (Mantra and Molo, 1985) examined circular migration and commuting in six Indonesian cities. In addition to establishing how significant and widespread this mobility was in cities of varying sizes in both Inner and Outer Indonesia, it was found that the great majority of these movers had only been circulating to those cities since 1980. It would appear that this pattern of an increasing tempo of nonpermanent migration has continued during the last decade and has become of even greater significance with improvements in transport, advances in education, changes in the roles of women, and increased urban and industrial development.

Perhaps the strongest evidence of a pattern of continued increase in the scale and significance of nonpermanent migration is derived from a comprehensive longitudinal study of 37 villages in Java carried out over the period 1967–1991. In that study Collier et al. (p. 1) concluded that:

> Twenty five years ago many of the landless laborers on Java had very few sources of income. Now most of the landless rural families on Java have at least one person who is working outside of the village, and in a factory or service job.

In all of the villages in the 1992–1993 resurvey, massive migration out of the village to jobs in the larger cities and towns was recorded, and only 20 percent of households depended on agriculture for their total livelihood. The bulk of the movement recorded was on a temporary basis. The fact that those villages were deliberately selected to be representative of villages both in high and low accessibility areas suggest that the scale of nonpermanent rural–urban movement from Javanese villages has increased exponentially in recent years and that such movement still far outweighs permanent relocation from village to city. Moreover, with further labor-displacing developments in agriculture (Hugo, 1995b), it seems likely that this movement will continue to be of great significance in Indonesia.

Although there are many similarities in the contemporary situation with respect to patterns and processes of nonpermanent migration compared with the 1970s, there also have been some significant changes. Paramount among these is the increasing involvement of those directed from village to city. Another change is that the increasing size of the formal sector in Indonesian cities, especially Jakarta and other large cities in Java, has led to an increasing number of migrants having to be more or less permanently settled in the city and not free to come and go to their village as frequently and readily as was possible when they worked in the informal sector. In many cases, for example with many young women working in factories in and around cities such as Jakarta and Surabaya, there are intentions to eventually return to settle in their village, but the fixed time commitments of their work prevent them from circulating to and from the village on a weekly, biweekly, or monthly basis.

Indonesia's urban areas have not only recorded massive population gains during the 1980s but there also has been a huge increase in the lateral extent of urban areas. This lateral extension has tended to occur in corridors, along major transport routes radiating out from (and linking) major urban areas (McGee, 1991; Firman, 1992). This phenomenon, together with the rapid increase in rural to urban circular migration mentioned earlier, is producing a new form of diffuse urbanization in Indonesia, especially in densely settled Java. The overlapping of urban and rural populations and areas is producing a blurring of the distinctions between them and is most intense in the area around Jakarta (JABOTABEK), around Surabaya (GERANGKERTOSUSIDO), and along the transport corridors linking major cities (especially Jakarta–Bandung [already given the acronym of JABOPUNJUR—Jakarta, Bogor, Puncak, and Cianjur], Jakarta–Cirebon, Surabaya–Malang, Yogyakarta–Semarang). There may also be a situation emerging whereby such a pattern of diffuse urbanization surrounding a major metropolitan center is overlapping national boundaries with the development of the Southern Growth Triangle, or SIJORI (Singapore–Johor–Riau). This involves an overspilling of Singapore's industrial development into the adjoining Malaysian state of Johor and Indonesian province of Riau. In the latter case the rapid urban growth on the island of Batam is very much an extension of Singapore.

Indonesia's urban system is increasingly dominated by Jakarta, which now constitutes a giant megaurban region, or EMR, extending far beyond the official boundaries of the capital city district as Map 12.8 shows. In 1990 the official population of the capital city special district was 8.23 million. However, if the urban areas in the three surrounding *kabupaten* (provinces) shown in Map 12.8 are included, the population becomes 13.1 million; if the total population of the JABOTABEK region is included, the population of the EMR was 17.1 million (Hugo, 1996a). The population of the special district grew by 35.5 percent during the 1980–1990 period, but that of the JABOTABEK region grew by 49 percent. The expansion of foreign and domestic investment in industrial development since the mid-1980s has led to increased concentration of population growth and industrial development in the JABOTABEK region, putting increasing pressure upon housing, public transport, garbage disposal, utility provision, health, and education services. There are increasing signs of environmental stress in the region with air pollution levels in some areas being above the maximum WHO-recommended ceilings and aquifers becoming saline due to the encroachment of seawater (Hugo, 1996a). Airborne emissions in Jakarta exceed WHO Standards during one-third of the year (Park, 1997). The JABOTABEK share of national population increased from 7.1 percent in 1971 to 7.8 in 1980, 9.5 percent in 1990 and 10.4 percent in 1995.

Indonesia's age structure is undergoing substantial change as a result of the declines in fertility and mortality detailed earlier. Figure 12.7 shows that at the 1995 Intercensal Survey the numbers of Indonesians of less than 10 years of age was similar to that in 1980 as a result of the halving of fertility during the last quarter century.

However, the legacy of previous high fertility is evident in the large numbers in the teenage years who will enter

Photo 12.5 Jalan Thamrin, Jakarta. (Leinbach)

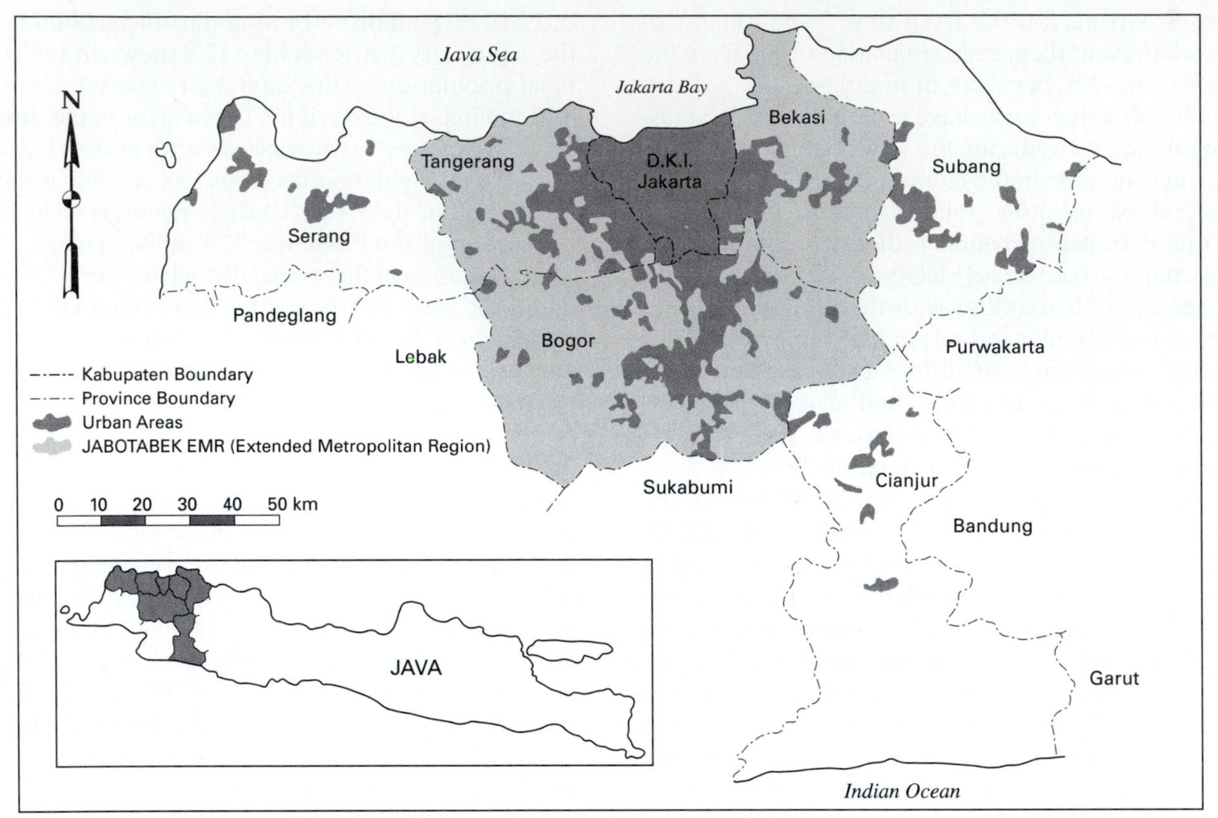

Map 12.8 The functioning urban region of Greater Jakarta.

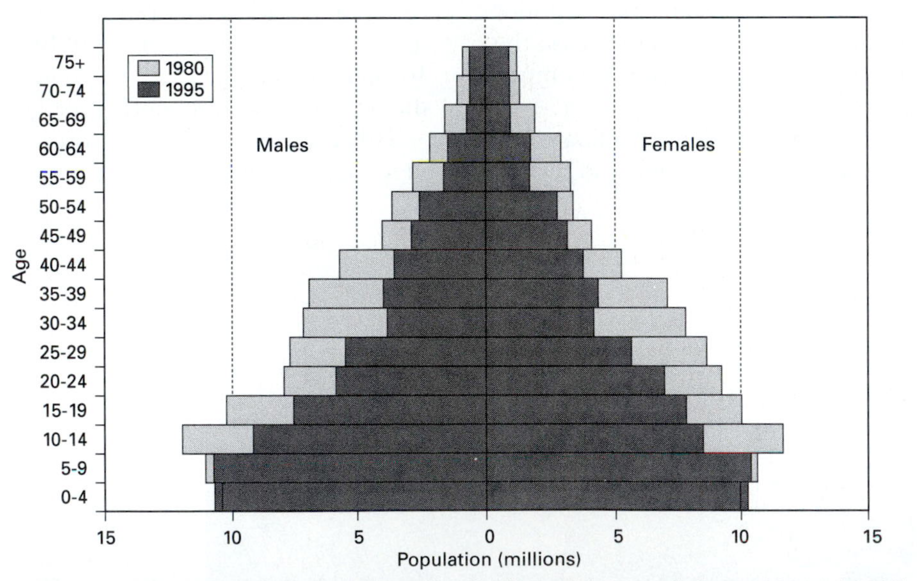

Figure 12.7 Indonesia age and sex distribution of the population, 1980 and 1995.
Source: Indonesian Census of 1980; Intercensal Survey, 1995.

the labor force over the next decade. Hence Indonesia's workforce is currently growing at almost twice the rate of the total population, with a net increment each year of around 2.5 million to the 80 million already in the workforce. It will also be noted in Figure 12.7 that Indonesia, like many countries in the Asian region, is poised for substantial growth of its elderly population. At present Indonesia has some 11.5 million people 60 years of age and over, but this will increase to 16 million by 2000 and 29 million in 2020. During this period they will increase from 6.4 percent of the total population to 11.4 percent in 2020. This represents a considerable challenge to policy makers because

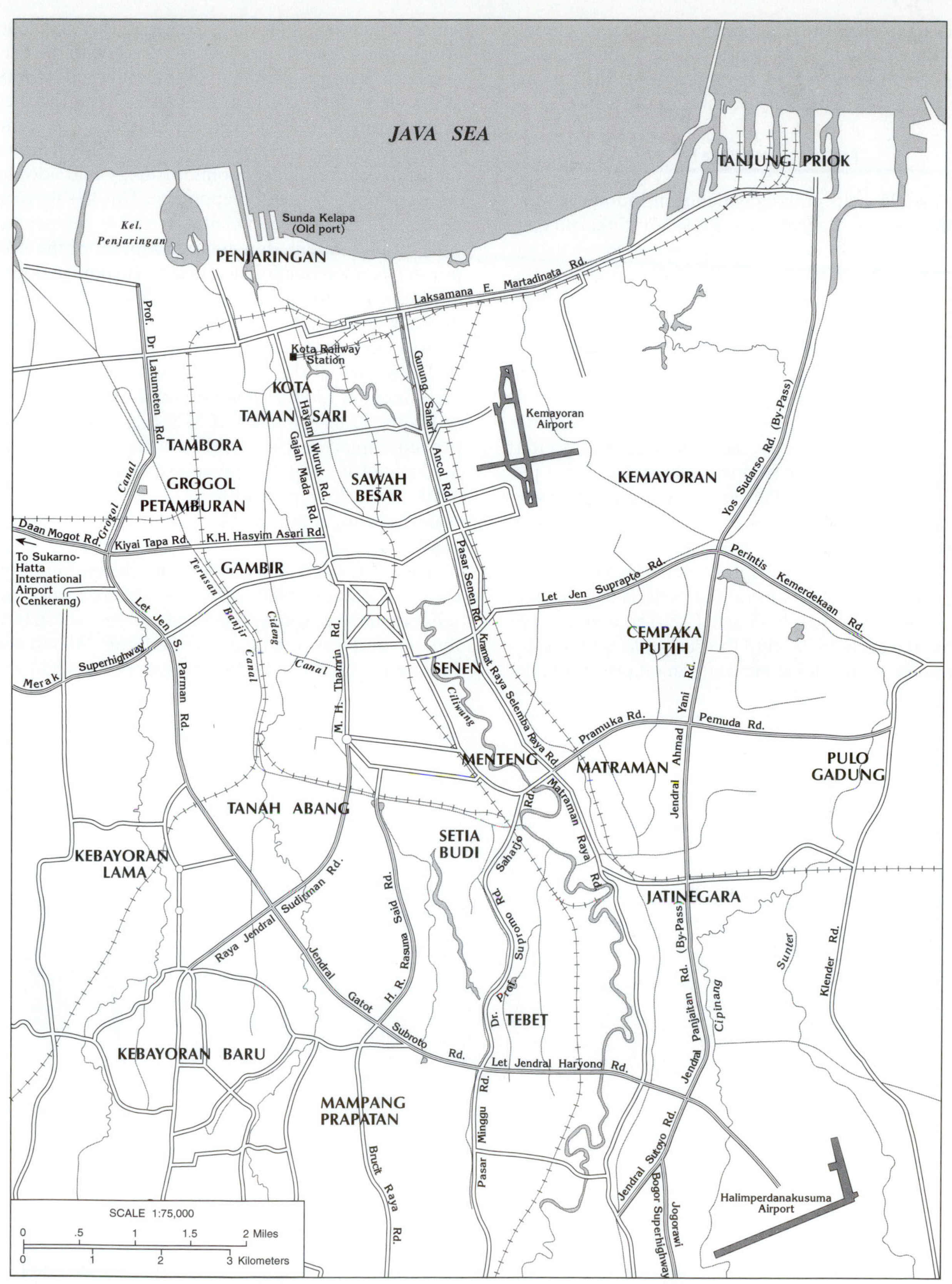

Map 12.9 Jakarta metropolitan area. *Source: Ulack and Pauer, 1989.*

the present availability of support for the dependent elderly is almost totally from family sources; contemporary changes in the Indonesian family may mean that this source will not be so readily available in the future.

ECONOMIC CHANGE

The first 25 years of independence in Indonesia were ones of economic stagnation as President Sukarno concentrated his efforts on wielding Indonesia into a nation and establishing its standing in Southeast Asia and the world. The New Order government of President Suharto placed greater emphasis on economic development and shaped national economic growth through a series of five-year development plans (*Repelita*) beginning in 1969. Under the New Order government, economic growth considerably outpaced population growth in Indonesia (Figure 12.3). Indeed, the average rate of growth in GDP during the 1969–1994 period was nearly 7 percent per year (Booth, 1994, p. 4). This was three times the average rate of population growth over the same period. However, it is apparent that there have been some fluctuations in the rate of growth of GDP. The Suharto era can be divided into two substantial periods of economic growth that are evident in Figure 12.3. The first, extending from the early 1970s until the early 1980s, was largely sustained by the global increases in oil prices and the windfall increases in foreign exchange earnings that Indonesia subsequently enjoyed. However, with the decline in oil prices and realignment of international currencies in the early 1980s, the heavy reliance of the Indonesian economy on oil and gas exports resulted in a precipitous decline in the rate of economic growth. The government, therefore, shifted its economic strategy and adopted a strongly market-oriented approach. This has involved a process of liberalization characterized by reduced controls on investment, deregulating aspects of the economy, and development of measures to encourage domestic and foreign investment. Accordingly, the economy became more diversified and high rates of economic growth were resumed through the mid-1990s.

The impact of the liberalization measures on foreign investment in Indonesia are dramatically demonstrated in Figure 12.8. The amount of foreign investment approved tripled in 1987–1988 and thereafter has been maintained at a high level despite downturns in the early 1990s. It is also important to note the dominance of the manufacturing sector in the upswing in investment during the last decade.

There has been, as a result of the changes initiated in the early and mid-1980s, a considerable reduction in the reliance of the economy on the oil and gas sector. In 1981 the oil and gas sector accounted for 24 percent of Indonesia's GDP, but by 1992 this proportion had almost

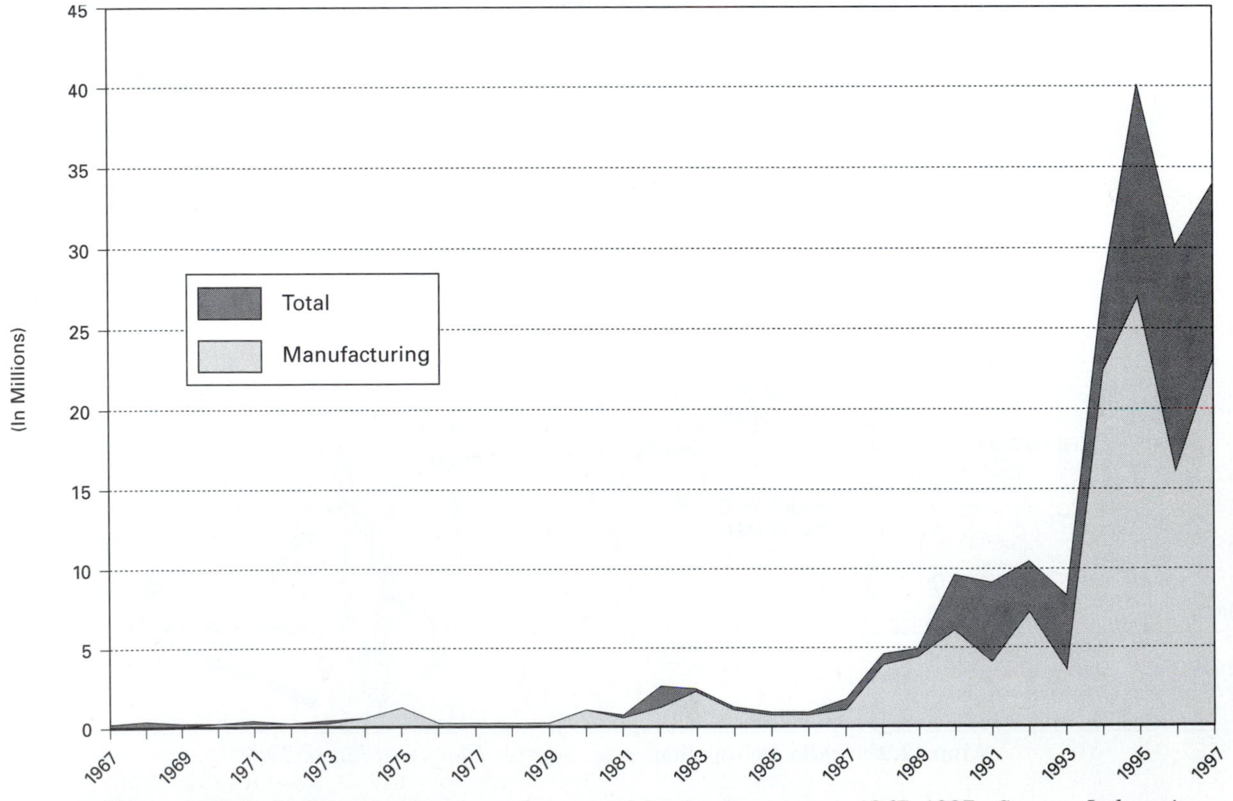

Figure 12.8 Indonesia: Employment of approved foreign investment, 1967–1997. *Source: Indonesian Capital Investment Board.*

BOX 12.3 Indonesia In The Financial Crisis by Thomas R. Leinbach

As recently as mid-1997, Indonesia's economy seemed to be performing well. Inflation was moderate, output had grown by more than 7 percent during the year and investment had grown by more than 16 percent. In addition, the balance of payments was so strong that international reserves had increased by 38 percent to US$28 billion in the 10 months prior to the crisis. Financial institutions underwent a remarkable transformation from 1983 to 1991 as a result of two major deregulatory reform packages (McLeod, 1998). Private banks grew dramatically and provided new life to a sector dominated previously by state banks.

Following the run on the Thai *baht* and the Malaysian *ringgit,* the Indonesian *rupiah* also came under duress. The latter development would have, many economic experts agree, taken place without the Thai pressure as it was a crisis waiting to happen. However the more severe impact in Indonesia came about because of counterproductive policy decisions by the government and the growing political instability associated with the issue of presidential succession. The failure to marshal an effective response and the ultimate severity of the crisis are explained as the consequences of a lengthy process of politically related economic and financial activity. In addition investment and financing decisions had become driven by euphoria brought on by years of steady expansion. This was fueled by the "Suharto connection" that became the guarantee or collateral underlying the viability of many enterprises. Although the *rupiah* depreciated, exports failed to take off as was expected. The strong shock to the private sector caused by the increased value of foreign debts forced firms to postpone or cancel investment spending. Individuals also cut back on discretionary spending. Unemployment soared into the millions as the crisis spread. The government also closed 38 private banks and took over and recapitalized still others. Essentially banks were allowed to borrow excessively outside of Indonesia and to engage in risky investment behavior with no central bank supervision. As in Malaysia and Thailand, the investment in the property market caused a supply that far exceeded demand.

Economists believe that a fundamental flaw in Indonesia's macroeconomic policy was largely responsible for igniting the crisis. Essentially this relates to the inability to control price levels, money supply, the exchange rate, and the interest rate simultaneously (McLeod 1998). The deeper implications of this shortcoming were that Indonesia succumbed to a crisis of confidence and raised doubts about the way the economy and state were being managed. The undermining of the Suharto connection caused massive uncertainty and exposed previously ignored weaknesses. The common perception was (and perhaps still is) that it was futile to hope for a more democratic system where there would be a lessening of abuses for economic gain. Compounding the already difficult situation was the lack of a coherent strategy in applying a series of International Monetary Fund packages amounting to more than $40 billion (Robison and Rosser, 1998). One interpretation is that the economic policy makers viewed the looming crisis as an opportunity to accelerate the lagging progress of reform in the domestic economy and to attack the privileged position of President Suharto's family and business associates (McLeod, 1998). The government's move to reduce fuel and electricity subsidies in early April 1998 provided a reason for students to protest. Student fatalities brought devastating riots that claimed still more lives and huge property damage. As previously it was the Chinese community that bore the brunt of this backlash.

In addition to the political situation, it is important to realize that the impact of the crisis was exacerbated by the both the forest fires (the Asian Development Bank revealed that last year's Indonesian fires destroyed $4.5 billion worth of forest) and a deepened drought brought on by an El Niño effect. Both of these natural disasters have produced great hardship especially for people in rural areas. In addition the country has been hit by low world petroleum prices. While the crisis produced perhaps only an increase of 1 million in unemployment, there have been huge impacts reflected in underemployment escalation, increases in poverty, decreases in earnings, and a suppression of employment growth in agriculture. Furthermore it is important to note that although the whole of the country has been affected, the severity of the crisis is "regionally patchy." Most significant in this regard is that the Outer Islands have not been as sharply affected as has Java.

By May 1999, the *rupiah*'s value had been stabilized somewhat, and President Habibie had responded to the crisis with a number of important measures and promises. The establishment of an Indonesian Debt Restructuring Agency was critical. Perhaps most appealing is his plan for regional autonomy, which, he hopes, will create a foundation for stability and new energy in development. This plan, an important response to *reformasi,* would allow outlying regions to arrange their own business deals with foreign investors, process their own commodities, and set their own development priorities (McBeth and Cohen, 1999). A recent development has been to offer East Timor the option of becoming an autonomous region or independence. Although cronyism has been blunted, it has not disappeared. Better insight as to the future of Indonesian recovery and growth must wait until after the presidential election process has been completed and the new leader has implemented a campaign platform.

halved to 13 percent (Booth, 1994, p. 4). In exports the change has been even more dramatic as Figure 12.9 shows. Each of the last two Five-Year Plans has seen a doubling of annual non-oil/gas export earnings. As Booth (1994, p. 4) points out: "In 1981 petroleum products, including natural gas, accounted for over 80 percent of total export earnings, by 1992 their export share had dropped to under 25 percent. Oil and gas revenues accounted for 70 percent of government revenues in 1981–1982; by 1993–1994 their contribution had fallen to under 30 percent."

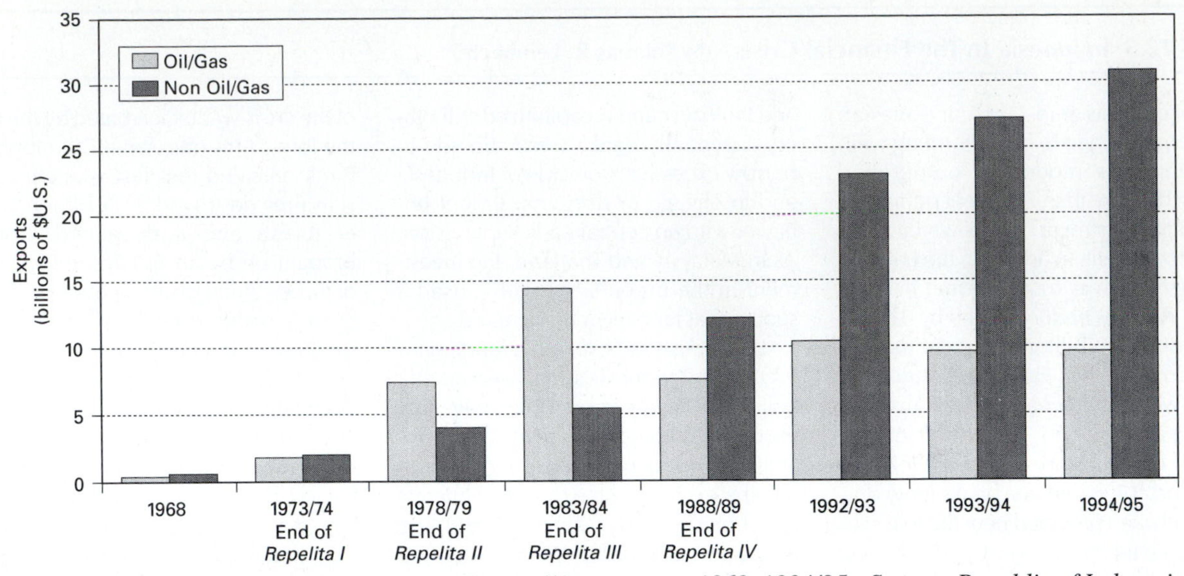

Figure 12.9 Indonesia: Oil/gas and non-oil/gas exports, 1968–1994/95. *Source: Republic of Indonesia 1993; Parker and Hutabarat, 1996.*

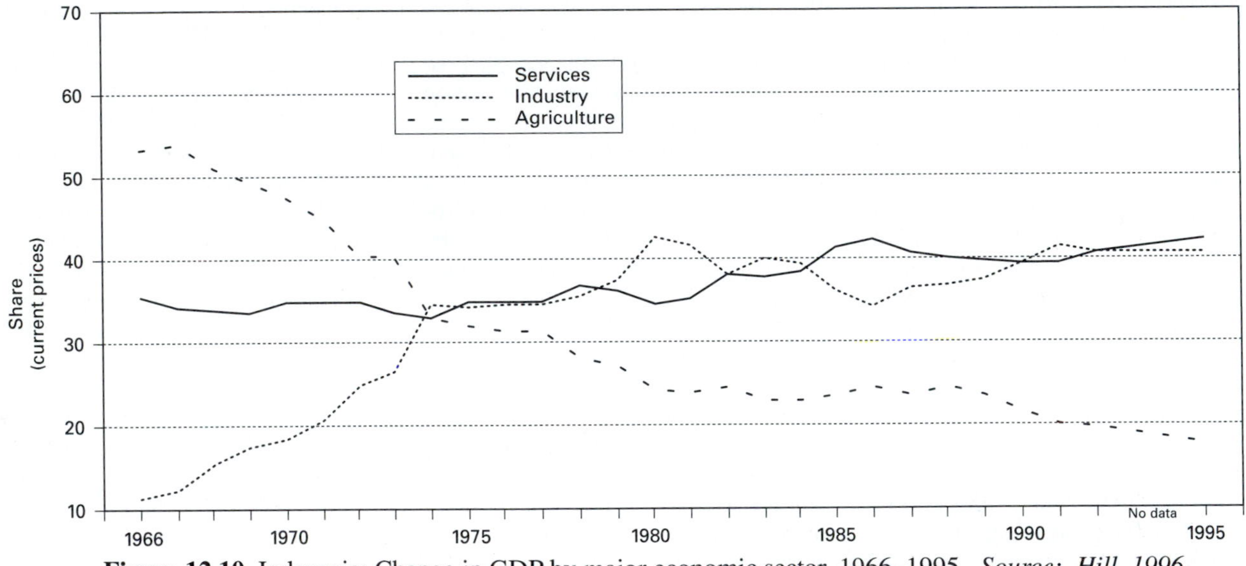

Figure 12.10 Indonesia: Change in GDP by major economic sector, 1966–1995. *Source: Hill, 1996.*

The rapid economic growth of the last three decades has been accompanied by substantial structural change in the economy. Figure 12.10 shows that whereas agriculture provided substantially more than half of GNP in 1966, by 1988 this had declined to 22.5 percent and by 1995 to 17 percent (Hill, 1996, p. 19; Manning and Jayasuriya, 1996, p. 5). Although the services sector has increased its share of the economy from 35 to 41 percent, the bulk of the structural change has been in the industrial sector (broadly defined to include mining, manufacturing, utilities, and construction). The industrial sector increased its share of the economy more than fourfold to account for 42 percent in 1995. The manufacturing sector alone increased from 8.5 percent of the economy in 1968 to more than 25 percent in 1996.

Despite this massive structural change in the economy, it would be incorrect to conclude that the primary sector is stagnating. Indeed, one of the most striking changes of recent decades has been Indonesia's transformation from the world's major rice-importing nation to a country making occasional rice exports. The output of rice during the 1968–1996 period has trebled from 17 to 51 million tons annually (Figure 12.11). However the crippling effect of the El Niño inspired drought in 1997–1998 meant that Indonesia in this period changed from being a rice exporter to the largest rice importer, importing more than 1 million tons of rice.

Nevertheless, the increase in production was largely achieved through an increase in productivity because the area under rice increased by only 35 percent over that pe-

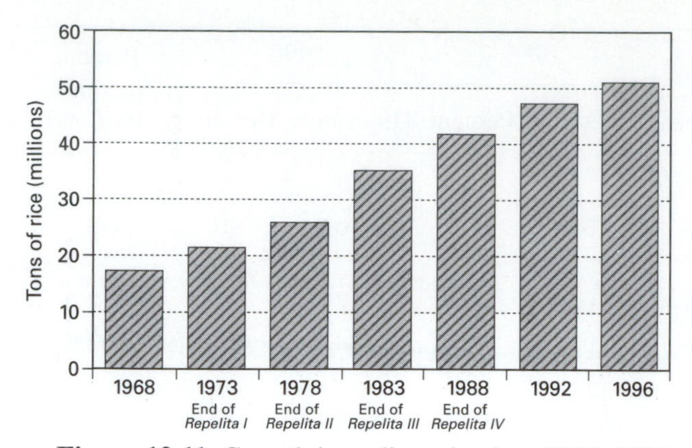

Figure 12.11 Growth in padi production, 1968–1996. *Source: Republic of Indonesia, 1993; Manning and Jayasuriya, 1996.*

riod while the output per hectare rose from 2.13 tons per hectare to 4.35 (Republic of Indonesia 1993). Hence availability of rice produced in Indonesia has increased more than three times faster than population during the last quarter-century. Nevertheless, some would question the sustainability of such a rapid increase in food production, given its reliance on substantial inputs of fertilizers and pesticides and the evidence of increasing environmental degradation.

Much of the expansion in rice production has been achieved through increasing inputs other than labor. The increasing productivity of Indonesian agriculture is demonstrated in Table 12.8; however, cumulative growth in real agricultural labor productivity was only 10 percent in the 1980s (Tomich, 1992, p. 8). The table also shows that the last decade has seen agriculture's share of Indonesia's workforce fall below one-half for the first time, while its share of GDP has fallen below one-fifth. It is anticipated that Indonesia's agricultural workforce will begin to decline in numbers in the late 1990s (Hugo, 1993, p. 75). Indeed, by the mid-1990s it was shown that the absolute numbers of workers in agriculture had begun to decline quite sharply in Java and to stabilize in the Outer Islands (Manning and Jayasuriya, 1996, p. 32).

Like several other Asian countries, Indonesia experienced a Green Revolution in the 1970s and 1980s. This saw the introduction of high-yielding varieties of rice, improved irrigation and agricultural techniques, as well as greater use of pesticides and fertilizers. This resulted in the increased commercialization of agriculture and involved higher levels of mechanization and commercial labor arrangements. Greatly increased agricultural productivity was achieved together with expanded overall production, but there was also significant associated labor displacement from agriculture and changes in the social relationships in Indonesian villages. In the traditional system all villagers could take part in agricultural activity for a share of the harvest. However, with commercialization all agricultural work was put on a pay basis so that the patron-client relationship between landowners and landless began to break down.

The manufacturing sector has been most dynamic in the second half of the 1980s and early 1990s, growing at more than 10 percent per annum in most years, well above the average for all sectors. This is particularly true of the non-oil and gas manufacturing subsector. In 1991 the share of the economy made up by manufacturing exceeded agriculture for the first time in the nation's history. As Hill (1992, p. 5) points out: "This is an historic turning point in Indonesia's economic development. Although the event was hastened by poor rice harvests, its fundamental origins lie in the impressive export-oriented industrial growth which has occurred since 1986." The growth in this sector is reflected in employment. In the 1980s employment in large- and medium-size manufacturing enterprises increased by 117 percent, but that in small enterprises also expanded by a very healthy 63 percent (Table 12.9). Indeed, in the late 1980s small-enterprise employment grew as fast as that in large- and medium-size enterprises. It is interesting to note that employment in large- and medium-size enterprises increased three times faster in the Outer Islands than in Java, although Java still retained 76 percent of all employment in this sector in 1990.

The proportion of small-enterprise employment in Java was even higher (77 percent). The fact that much investment in manufacturing since 1990 has been in and around Jakarta would suggest that manufacturing employment has

Year	Agriculture's Share of Labor Force (Billion rupiah at constant 1983 prices)	Agricultural GDP (Percent)	Agriculture's Share of GDP (Percent)	Agricultural Labor Force (Million)	Agricultural GDP/ Agricultural Labor (Thousands of rupiah per person at constant 1983 prices)
1980	16,399	25.0	28.8	55.9	569
1985	19,300	22.7			
1990	22,357	19.4	35.5	49.2	626

Table 12.8 GDP and labor force in agriculture. *Source: Tomich, 1992, p. 9.*

	1980		1985		1990		Percent Increase 1980–90
	Thousands	Percent	Thousands	Percent	Thousands	Percent	
Indonesia							
Total manufacturing employment	3,959	100	4,888	100	7,163	100	80.9
Employment in large- and medium-size enterprises	1,300	33	1,900	39	2,821	39	117.0
Other manufacturing employment	2,659	67	2,988	61	4,342	61	62.6
Java							
Total manufacturing employment	3,023	100	3,704	100	5,503	100	82.0
Employment in large- and medium-size enterprises	1,119	37	1,494	40	2,152	39	92.3
Other manufacturing employment	1,904	63	2,210	60	3,351	61	76.0
Off Java							
Total manufacturing employment	936	100	1,184	100	1,660	100	77.4
Employment in large- and medium-size enterprises	181	19	406	34	669	40	269.6
Other manufacturing employment	755	81	778	66	991	60	31.3

Table 12.9 Manufacturing employment by region and scale, 1980–1990. *Source: Hill, 1992, p. 38.*

	1989–90	1992–93	Percent Growth
Number of Enterprises			
Java	68,414	80,835	18.2
Other Islands	48,324	67,007	38.7
Total	116,738	147,842	26.6
Number of Employees			
Java	3,188,824	4,280,461	34.2
Other Islands	1,772,947	2,312,544	30.4
Total	4,961,771	6,593,005	32.8

Table 12.10 Change in registered manufacturing enterprises, 1989–90 to 1992–93. *Source: Indonesia, Department of Labor, 1995.*

become even more concentrated in Jakarta, especially that in large enterprises. In the first three years of the 1990s employment in large- and medium-scale manufacturing enterprises increased by almost one-third—with a higher rate of increase in Java than in the Outer Islands (Table 12.10). On the other hand, the increase in the *number* of enterprises was considerably faster in the Outer Islands than in Java, suggesting that in Java the growth is mainly in large concerns while that outside Java is in smaller enterprises.

Unemployment levels remain quite low in Indonesia. In fact it has been said that "unemployment is a concept that is of little relevance among Indonesia's vast rural labor force, and of only limited relevance among urban workers: low productivity and low earnings constitute a more important issue." (Hugo et al., 1987, p. 286). This assessment arose out of the observation that the poor in Indonesia could not afford to be unemployed and had to take on any work they could regardless of its low income, and low productivity. Only people from relatively well-off families could afford to be unemployed. Hence unemployment levels in Indonesia have traditionally been very low (1.7 percent in the 1980 Census, and 1.1 percent at the 1985 Intercensal Survey), but these figures cannot be interpreted in the same way as they are in Western economies. With the changes occurring in the Indonesian labor force however, unemployment is becoming of greater significance. These changes include the increased levels of education, increased participation in urban wage employment and associated formal-sector labor market structures, and the shift out of agricultural employment (Jones and Manning, 1992). Hence there were significant increases overall in unemployment (from 1.4 to 2.8 percent for males, and from 2.3 to 3.9 percent for females) between 1980 and 1990 and 5.6 and 10.1 in 1995. The overall rates are three times as high in urban as in rural areas. It would appear then that with the structural changes occurring in the national economy, open unemployment is becoming more common, especially in Indonesia's towns and cities. Nevertheless, unemployment remains concentrated in the young, educated groups.

Underemployment is very difficult to measure in Indonesia and a great deal of care needs to be exercised in the interpretation of the data on hours worked, which is used to measure underemployment. Some have estimated the current extent of underemployment in Indonesia to be as high as 44 percent of the workforce ("Labor Trends in Indonesia," 1990, p. 9) if the conventional less-than-35-hours-of-work definition is used. However, hours worked can give a misleading indication of the extent of underemployment because of:

- Multiple job holding—many Indonesians have more than one job and the number of hours reported may relate only to the main job.
- The seasonal nature of much work, especially in the agricultural sector.
- Also in some areas (e.g., agriculture) many workers working short hours (especially females) are part-time workers and do not want additional work.
- Many urban jobs in the informal sector involve long hours but large amounts of downtime in which little productive is accomplished, so hours worked is not a very good indicator of underemployment in such contexts.
- Jones and Manning (1992) also draw attention to the fact that there have been large discrepancies observed between hours worked recorded by village studies and national surveys. The former report low levels of underemployment and the latter high levels.

Therefore, a considerable degree of caution must be exercised in interpreting the hours-worked data that is collected in the Indonesian Census enumerations. Nevertheless, there can be no doubt that underemployment levels are very high in Indonesia.

One of the distinctive features of Indonesia's labor force is the significance of what has come to be called the informal sector (Photo 12.6). This is dealt with in more detail in Chapter 4, Box 4.1, but in recent years in Indonesia as well as elsewhere there has been a change in attitude of policy makers and researchers towards this sector. It was formerly frequently viewed negatively as involving low productivity and making little contribution to economic and social development. This view has greatly changed as our knowledge of the sector has increased. It has come to be appreciated not only because of its substantial labor absorptive capacity and income redistribution capacities but also because of its linkages to other sectors and its important roles in economic growth and change.

Although the informal sector is significant in most Asian nations, in few countries it is as large as it is in Indonesia. Hence Table 12.11 shows that compared with its ASEAN partners a greater proportion of Indonesia's workers are in this sector. The table also indicates the small representation in Indonesia of white-collar workers.

It was stressed earlier that one of the quintessential features of Indonesia is its great regional diversity, and this is certainly true of economic development. Hill and Weidemann (1989, p. 5) identify as one of the key features of regional development in Indonesia, the concentration of economic activity in Java and to a lesser extent Sumatra. In 1971 these two areas accounted for 55 and 29 percent of Indonesia's GDP and by 1983, 50 and 32 percent,

Photo 12.6 Toy vendor in the informal economy of Solo, Central Java. (Leinbach)

Country	Informal Workers as a Percent of Nonagricultural Employment		White Collar Workers as a Percent of Total Employment
	Male	Female	All
Indonesia	43	63	9
Malaysia	23	25	19
Philippines	23	36	11
South Korea	26	31	22
Thailand	31	42	5

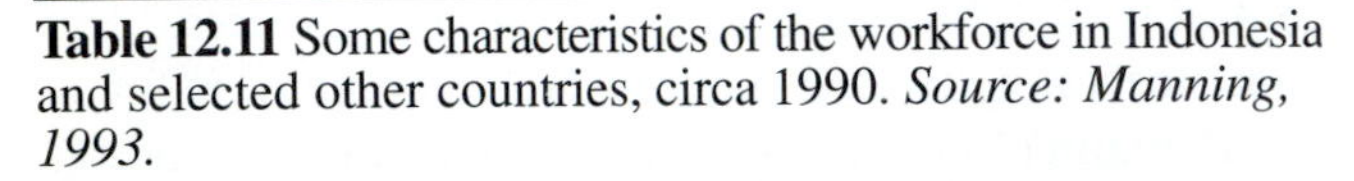

Table 12.11 Some characteristics of the workforce in Indonesia and selected other countries, circa 1990. *Source: Manning, 1993.*

respectively. By 1989, however, Java had once again increased it share to 54 percent, and Sumatra's share had fallen to 29 percent; these trends were maintained so that by 1993 some 59 percent of gross regional domestic product (GRDP) were accounted for by Java and 22.6 percent by Sumatra. Moreover, Java accounted for 69.1 percent of foreign investment during the 1967–1992 period and 62.2 percent of domestic investment between 1968 and 1992 (Hill, 1996, p. 218). Regional Gross Domestic Product figures for provinces in Indonesia such as those presented in Table 12.12 are somewhat misleading because as Hill and Weideman (1989, p. 5) point out, these figures are "distorted by the inclusion of oil and gas output in the regional accounts of provinces in which production occurs, even though almost all the revenue accrues to the central government, Pertamina (the state-owned oil company, in effect a common entity), and foreign petroleum companies." Hence the resource rich regional economies of Indonesia—Aceh, Riau, East Kalimantan and to a lesser extent Central Kalimantan, South Sumatra, and Irian Jaya display as having the highest regional GRDP per capita in 1990 (Table 12.12). It also is shown in the table that their GRDP per capita falls precipitously if oil and mining are excluded.

It is also interesting to note that the provinces that have an GRDP per capita below the national average (minus oil and mining production) are strongly spatially concentrated. All of the provinces of Java except Jakarta are below average, as are the four Sumatran provinces that do not have oil and gas exploitation activities. However, the heaviest concentration of low GRDP provinces are in Eastern Indonesia. Indeed, only Irian Jaya had an above-average figure, and the lowest three figures were recorded by West and East Nusa Tenggara and East Timor. Low levels of GRDP per capita also apply throughout Sulawesi and Lampung in Sumatra.

Table 12.12 also shows the extent of real growth in provincial GRDP per capita between 1973 and 1990. Note that the oil and gas provinces had the slowest growth. The major exception to this is East and West Nusa Tenggara which, besides having the lowest GRDP per capita in the nation, had some of the slowest growth rates.

Hill and Weideman (1989) identify a second major feature of Indonesia's regional economies as rapid structural change during the last two decades. Whereas in 1971 in more than half of the provinces more than one-half of GRDP was provided by agriculture, this now applies only to East and West Nusa Tenggara. Agriculture declined in significance in all provinces. It is especially interesting to consider the provinces of Java in this context. In the three very large provinces, the share of agriculture fell from almost one-half to much less than one-third.

It is interesting to note that the manufacturing sector by 1989 accounted for 34 percent of GRDP in Jakarta and 37 percent in West Java. This reflects the substantial industrialization of that region.

The measurement of poverty is a much discussed and controversial topic in Indonesia. Official estimates of poverty made by the Indonesian Central Bureau of Statistics are shown in Table 12.13. There is obviously a great deal of subjectivity in determining poverty lines and establishing proportions of the population living below that line, and this is apparent in the significant difference between the official poverty estimates and those made by the World Bank. Nevertheless, there is no argument that overall there has been a substantial improvement in overall living conditions in Indonesia during the last two decades.

The regional differences referred to earlier are reflected in substantial differences between provinces in the incidence of poverty. Again, in Table 12.14 (see also Map 4.1), although one can argue about the levels of poverty, the differences between the regions are a meaningful reflection of variations in patterns of well-being between Indonesia's regions.

The lowest incidence of poverty is clearly in the resource rich provinces of the Outer Islands—East Kalimantan, Riau, and Jambi as well as in Bengkulu and West Sumatra. Eastern Indonesia remains a problem area with a very high incidence of poverty, especially in West (19.5 percent) and East (21.8 percent) Nusa Tenggara, in Maluku (23.9 percent) and Irian Jaya (24.2 percent). Clearly the Eastern Indonesian region should be the subject of special attention in Indonesia's development efforts and there are signs that this is the case. It should be noted too that poverty remains high in Central (15.8 percent) and East (13.3 percent) Java in Inner Indonesia. However, the industrialization occurring in the western one-third of Java is reflected in the incidence of poverty being lower in West Java (12.2 percent) and Jakarta (5.7 percent).

Province	GRP 1990 (Billions of rupiah)	GRP per Capita, 1990 (Thousands of rupiah)	GRP per Capita, Real Growth 1973–90 (Percent)	Net Exports as Percent of GRP 1989	Percent of Exports		Percent of Foreign Investment, 1967–92	Percent of Domestic Investment, 1968–92
					1976	1991		
	(1)	(2)	(3)	(4)	(5)	(6)	(7)	(8)
Sumatra								
Aceh	8,290	2,448	12.5	66.8	0.2	10.7	1.9	1.6
(excluding oil)	2,897	737	6.4					
North Sumatra	10,833	1,063	5.3	10.3	7.1	6.1	7.4	3.1
West Sumatra	3,297	829	7.1	19.9	0.7	0.7	0.1	0.8
Riau	13,231	4,493	-3.2	48.8	43.2	17.3	5.0	7.1
(excluding oil)	2,672	907	3.4					
Jambi	1,414	709	3.3	20.0	0.8	0.9	--	1.5
South Sumatra	8,268	1,304	3.4	9.5	4.4	2.4	2.2	2.6
Bengkulu	795	684	7.2	-13.8	--	--	0.1	0.3
Lampung	3,217	540	2.5	19.8	1.5	1.2	1.1	1.6
Java-Bali								
Jakarta	22,855	2,481	5.9	71.2	11.2	26.1	28.5	8.9
West Java	31,358	917	5.5	18.8	1.8	0.9	30.3	35.8
Central Java	21,689	673	6.0	9.6	0.7	2.5	4.2	7.7
Yogyakarta	1,901	654	4.2	-28.1	--	--	-0.1	0.6
East Java	29,161	769	5.8	2.7	1.6	7.9	6.0	9.2
Bali	3,018	1,090	8.0	-1.1	0.1	0.6	4.0	2.6
Kalimantan								
West Kalimantan	2,743	860	6.0	-5.6	1.2	1.9	0.2	3.1
Central Kalimantan	1,376	998	5.8	-8.0	1.0	0.4	0.5	0.4
South Kalimantan	2,326	887	4.5	32.0	1.1	2.1	1.0	1.3
East Kalimantan	10,770	5,821	4.8	55.4	17.6	12.5	3.1	4.6
(excluding oil)	4,410	2,383	4.7					
Sulawesi								
North Sulawesi	1,507	593	4.8	-6.6	0.2	0.2	0.6	0.9
Central Sulawesi	982	581	4.7	9.2	0.2	0.1	0.1	0.6
South Sulawesi	4,241	610	5.2	3.9	0.3	1.6	2.3	0.9
Southeast Sulawesi	821	616	5.0	4.7	0.4	0.2	0.1	0.5
Eastern Indonesia								
West Nusa Tenggara	1,290	383	4.6	-5.8	--	--	0.1	0.6
East Nusa Tenggara	737	361	4.7	-16.5	0.1	--	0.1	0.3
East Timor	269	364	4.7	-14.4	--	--	0	--
Maluku	1,463	809	5.1	33.3	0.6	1.3	0.1	1.1
Irian Jaya	2,047	1,247	0.7	52.5	4.0	2.5	0.8	2.4
(excluding mining)	1,217	742	6.3					
Indonesia	196,919	1,098	3.6		100.0	100.0	100.0	100.0
(excluding mining)	171,471	956	5.0					

Table 12.12 Indonesian provinces: Indicators of regional development. *Source: Hill, 1996, p. 218–9.*

TRADE, TRANSPORT, AND COMMUNICATION

The growing significance of globalization trends in the Asian region are strongly reflected in Indonesia. Foreign investment has increased exponentially since the mid-1980s. Similar trends are evident in other international flows of goods and people. There has been a massive change in both the level and composition of international trade during the last decade. A major part of the "new" Indonesian economic policy has been a succession of

Year	Official Estimate (percent)	World Bank Estimate (percent)
1976	40.1	54.2
1978	33.3	47.2
1980	28.6	39.8
1984	21.6	33.0
1987	17.4	21.6
1990	15.1	n/a
1993	13.7	n/a

Table 12.13 Estimates of percent of total population living in poverty, 1976–1993. *Source: Pangestu and Aziz, 1994; World Bank, 1991.*

market-oriented deregulation packages since 1983 (Parker, 1991, p. 3). As Schwarz (1990, p. 40) points out: "promotion of exports and private investments, both local and foreign, have replaced a policy of import substitutions marked by high tariffs and heavy government intervention in distributing capital, licenses and other means of production. In a series of deregulation packages Indonesia's economic technocrats . . . have whittled away non-tariff barriers, relaxed investment restrictions, simplified industrial licensing, overhauled banking regulations, revitalized a listless stock market, cleaned up a corruption plagued customs service and streamlined the tax code." There have been substantial efforts to remove or reduce tariff and nontariff barriers to trade (Leinbach, 1995). For example, in 1986 a government ruling allowed exporters to apply for rebates on customs duties paid on imported raw materials. Although some improvement in the lifting of import controls has occurred, progress has been slow and uneven in this area.

The pattern of change in exports and imports during the last 15 years is striking (Figure 12.12). It can be seen that during this period the composition of exports has been transformed. The devastating impact of the mid-1980s oil-market collapse on Indonesia's export earnings is evident. The value of petroleum and gas exports fell from US$20.6 billion in 1981 to US$8.3 billion in 1986. During the same period, total export earnings fell from US$25.2 to US$14.8 billion. It was not until 1990 that Indonesia's export earnings again exceeded US$25 billion, but in 1990 petroleum and gas made up only 43 percent compared with 82 percent in 1981. By 1996, when export revenues totaled US$23.5 billion, oil and gas made up only 22 percent of the total. Clearly, there has been a massive change in the structure of Indonesia's exports that has seen increased diversification and reduced dependence upon oil and gas. This is of crucial importance to Indonesia's long-term economic development because heavy reliance on oil obviously exposes the economy to the mercy of international oil prices and because the finite nature of the oil and gas resources means that high levels of exports will not be able to be maintained in the long term.

The second half of the 1980s and early 1990s witnessed a recovery in export earnings they were matched with an increase in imports (Figure 12.12). Nevertheless, Indonesia has maintained a positive balance of export earnings over import costs during the last decade. The import growth is associated with the rapid industrialization (such as purchase of equipment) and the economic growth that has fueled greatly expanded consumer spending.

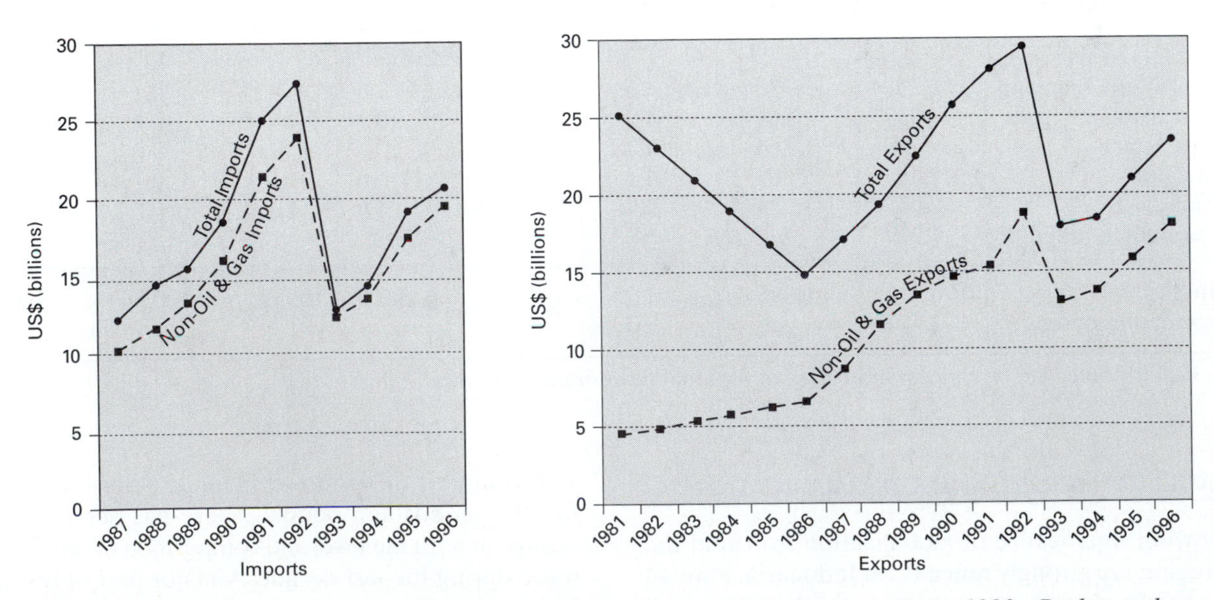

Figure 12.12 Indonesia: Imports 1987–1996 and exports 1981–1996. *Source: Hugo, 1993; Parker and Hutabarat, 1996.*

Province	Mean Income per Capita Rupiahs per Month 1990	Incidence of Poverty (Percent)		
		1987	1990*	1993
Aceh	19,702	12.50	15.9	13.5
North Sumatra	17,964	22.32	13.5	12.3
West Sumatra	22,155	6.98	15.0	13.5
Riau	20,657	7.85	13.7	11.2
Jambi	20,120	6.48	n/a	13.4
South Sumatra	20,929	15.02	16.8	14.9
Bengkulu	19,837	8.77	n/a	13.1
Lampung	17,428	34.41	13.1	11.6
Jakarta	n/a	n/a	7.8	5.7
West Java	18,365	22.97	13.9	12.2
Central Java	15,407	40.95	13.5	15.8
Yogyakarta	17,681	25.19	15.5	11.8
East Java	16,840	38.83	14.8	13.3
Bali	19,236	39.99	11.2	9.5
West Nusa Tenggara	15,214	47.05	23.2	19.5
North Nusa Tenggara	16,315	53.00	24.1	21.8
East Timor	18,697	45.27	n/a	36.2
West Kalimantan	17,385	27.64	27.6	25.1
Central Kalimantan	18,524	17.65	n/a	20.9
South Kalimantan	18,076	15.37	21.2	18.6
East Kalimantan	24,979	8.17	n/a	13.8
North Sulawesi	21,881	27.72	14.9	11.8
Central Sulawesi	21,766	29.77	n/a	10.5
South Sulawesi	16,564	42.01	10.8	9.0
Southeast Sulawesi	15,310	55.81	n/a	10.8
Maluku	18,093	24.95	n/a	23.9
Irian Jaya	17,402	40.96	n/a	24.2
Indonesia		17.42	15.1	13.7

** See also Map 4.1 for percentage of population living in poverty.*

Table 12.14 Provincial variation in selected measures of poverty, 1987, 1990, and 1993. *Source: World Bank, 1991, p. 113; Pangestu and Aziz, 1994.*

There are several other dimensions to the growing strength of international linkages between Indonesia and the rest of the world. Like most LDCs, Indonesia has a current account deficit. This presently is about US$8 billion, or about 3 percent of GDP, which is commonly accepted as manageable and in line with the needs of a dynamic and growing economy (Parker and Hutabarat, 1996, p. 21). Total foreign debt before the 1997–1998 financial crisis was about US$100 billion. The increasing flows of people to and from Indonesia are also notable. Tourism has expanded rapidly with the numbers of arrivals reaching 1 million for the first time in 1987, 2 million in 1990, and doubling to 4 million in 1994. As a major labor-surplus nation in the region, Indonesia also exports significant numbers of overseas contract workers with about a quarter of a million workers being overseas in the mid-1990s under the auspices of the nation's official program. More than four times that number are overseas working (mainly in Malaysia) but are undocumented (Hugo, 1996b).

One of the major changes in Indonesia during the last three decades has been the increase in personal mobility facilitated by a rapid growth in the nation's transport and communications networks, transforming the movement of goods and people around the archipelago (Leinbach, 1986). The fragmented nature of the country has constituted a significant barrier to the movement of people, and goods and improvements of sea, land, and air transport infrastructure have been an important priority in the national five-year plans (see Chapter 9: Transportation). Interisland shipping has improved greatly as

Chartered Company in 1881 that was given the right to administer and commercially exploit natural resources. In Sarawak, the British adventurer James Brooke was appointed the governor of Sarawak by the sultan of Brunei in 1841 and successive "white rajahs" remained in power until 1946 when Sarawak was ceded to the British Crown. The successive Brooke governments were paternalistic autocrats who played potential native groups against one another and prevented the spread of any large-scale commercial concerns that would threaten their political and economic autonomy. Both Sabah and Sarawak became British protectorates in 1888, and both territories as well as Labuan became Crown colonies in 1946. By this late date, Brunei had lost much of its tribute territories through land concessions to the British. This impoverished feudal state also became a British protectorate in 1888 and was assigned a British resident in 1905.

During the short colonial period following World War II whereby a number of apprentice governments ruled, Malaya became the parliamentary-based, independent state of Malaysia in 1963. Singapore was part of the federation until 1965 when it became an independent political entity. In the nine peninsular states, sultans continue to wield significant power over both cultural and land-resource-related matters. While both Sabah and Sarawak are equal members of the federation, their responsibility is somewhat different in that certain safeguards were built into the federal constitution to ease their entry into the federation. Negotiations for tiny Brunei to enter the Malayan Federation reached an impasse in 1963, primarily over questions regarding control of oil revenues. Brunei remained a British protectorate until 1984 when it became Southeast Asia's youngest independent state, governed by an Islamic monarchy.

Sociocultural Evolution of Ethnic Groups

Ethnicity permeates almost every aspect of national life in Malaysia. By the time the British arrived, Malays had already converted to Islam and were predominantly village or *kampung* folk practicing fishing, agriculture, and trade along coastal and riverine margins and the few small interior basins (Mahmud, 1970). Because of the uneven spread of colonialism and policies aimed at maintaining the traditional peasant economy, there developed two Malay socioeconomic classes. The Malay feudal aristocracy retained their privileged social status and were enlisted by the colonial government to fill middle and lower level civil service positions. The Malay peasant class, however, did not participate in the colonial economy anchored by tin and rubber exports. Economic life for the Malay peasant not only stagnated but grew worse during the colonial period; although Malay reservation lands were extensive, the introduction of private landownership created a large rural and landless class that was not connected to the expanding cash economy.

Enlisted by the colonial government to assume the function of laborers and middlemen wholesalers and retailers in a colonial economy, the Chinese became the urban-based entrepreneurial class who overtime controlled ever-increasing amounts of private wealth (Purcell, 1967). Also enlisted were Indians who provided the backbone of the plantation labor force but also occupied, beyond what their numbers might indicate, the ranks of the educated professional services class (Sandhu, 1969). Providing a wide array of forest products to coastal trading centers during the colonial period, relatively few aboriginal Semang, or *orang asli,* continued to practice a hunting-and-gathering economy in the isolated localities of the peninsula (Dunn, 1975). Because each culture was separated by residence, education, and workplace defined by colonial "ethnic management" policies, ethnic integration was practically nonexistent. In other Southeast Asian states, ethnic Chinese comprise a much smaller percentage of the total population, and assimilation into the dominant native culture is far more common.

When compared to Peninsular Malaysia, the ethnic mosaic of East Malaysia is substantially different because of the often dominant non-Muslim and non-Malay aboriginal peoples (Cleary and Eaton, 1992). The most numerous of these indigenous groups are the Ibans and Bidayuhs of Sarawak and the Kadazans and Muruts of Sabah. Chinese immigrants also arrived in East Malaysia in substantial numbers. With noninterference in indigenous affairs even more aggressively pursued in colonial East Malaysia, the dispersed and isolated indigenous groups continued the traditional practice of swidden agriculture in forested interior regions throughout the colonial period. There are few differences separating the traditional Malay culture of Brunei from Malaysia. Fishing and padi farming remained common pursuits during the colonial period as minor extraction of coal and cutch (a by-product of mangrove trees used in dyeing and tanning) did little to transform traditional life. The exploitation of oil in 1932, however, began the transformation of Malay life as many abandoned agriculture for government employment. Much like Peninsular Malaysia, a significant urban-based Chinese minority merchant class exists, but more like East Malaysia, Brunei is populated by a minority of aboriginal peoples tied to subsistence agricultural activity.

POPULATION PATTERNS, MIGRATION, AND URBANIZATION

During the past quarter-century, the pervasive effects of the NEP have transformed the national settlement system of Malaysia. Indeed, few development policies illustrate so clearly the impact of government on the interlinked

BOX 13.1 The NEP and Malaysia's Cultural Mosaic

The NEP was devised to elevate the economic status of Malays through the redistribution of economic activity under conditions of an expanding economic pie. In part through government-sponsored programs and economic institutions, the mainstreaming of Malays into the country's rapid economic growth trajectory has been substantial. No longer just the stereotypical government official and farmer, affirmative action policies have provided the psychological confidence that Malays required to engage in economic activity once deemed only appropriate for Chinese and westerners. There has emerged during the past 25 years then, a substantial number of Malay doctors, engineers, accountants, lawyers, university professors, and even entrepreneurs. The rise of a Malay middle class is evidenced by the proliferation of suburban tract homes and the resulting congestion of highways filled with family-owned automobiles engaged in the grind of daily commuting. Shopping malls are jammed with adolescents seeking out trendy fashions and recent music CDs and videos. Although single groups of females covered from head-to-toe in Arab dress is common, just as common are groups of mixed gender teenagers dressed in jeans and T-shirts. As the most urban-industrial Muslim majority country in the world, modern Malay society possesses little resemblance to the Muslim cultures of the Middle East.

The foundation of Malay affluence rested on access to educational opportunities at the university level. Although it is true that educational standards have eroded as a result of reserving 64 percent of university places for Malays, far greater economic opportunities are now open to one-half of the country's population. Two substantial barriers exist, however, to the continued healthy contribution of Malays to economic growth, particularly within the context of the global economy. First, the institutionalization of the Malay language as the only medium of instruction in public education, and the absence of the systematic use of English reduces Malaysia's opportunity to further plug into the global economy. With East Asian NIE economies booming, thousands of concerned Malay parents are paradoxically sending their children to Mandarin Chinese language schools. Second, education among Malays lacks a sufficient focus on sciences and practical skills training that in the long term is a barrier to attracting foreign investment and Malaysia's pronouncement of *Vision 2020* calling for a fully industrialized country in the next quarter-century.

Although no longer holding an economic monopoly on the more modern sectors of the economy, the economic status of ethnic Chinese has changed relatively little since the NEP. At the corporate level, Chinese continue to capture comparative advantage by systematically entering into growth areas of the economy. During the 1970s, for example, Chinese dominated the automobile trade, but with the emergence of the Proton national car, they dominate the car-parts business. At a smaller scale, ethnic Chinese control some 90 percent of the medium- and small-size firms on the Kuala Lumpur Stock Exchange. And as was true during the pre-NEP period, Chinese continue to dominate the small family-run businesses that line the streets of so many Malaysian urban places.

Ethnic Indians, however, have not fared as well since the NEP was instituted. Although there exists a small and well-educated Indian professional class, the vast majority of Indians remain poor and disenfranchised from the larger Malaysian economic growth boom. The conditions under which Indians arrived in colonial Malaya tell much about their present marginalized position. Most arrived as plantation laborers and a good many today remain on the estates. When leaving the estates for urban areas, most remain dependent on low wage and manual labor employment. Economic marginality is also attributable to the rigid caste system that accompanied them from India. As members of the socially marginal lower castes in India, deep-seated feelings of helplessness and inferiority persist to the present day. Indeed, those Indians that make up part of Malaysia's professional class are the children of higher-caste Indians brought from India by the British as plantation supervisors and civil servants.

Although Malaysia has made great strides in creating a more harmonious plural society, to claim that ethnic animosity has been eliminated is shortsighted. Ethnic tensions are not based on mutually exclusive cultural traditions but on economic opportunity. The reduction of ethnic tensions has materialized under conditions of economic expansion, and a prolonged economic recession and resulting increases in unemployment would only aggravate interethnic animosity. The source of this economic-induced ethnic tension would materialize among younger Malays who have become very comfortable with the existing culture of entitlement fashioned during the NEP years.

elements of population growth and distribution, migration, and urbanization. The greatest impact of the NEP has been on the relative population share and geographic distribution of the Malay community; Malays have become far more mobile and urbanized in response to economic opportunities provided by the NEP.

With a total 1997 population of 21 million and a population density of 166 persons per square mile, Malaysia does not suffer from land and natural resource shortages. The distribution of the national population is, however, geographically uneven. Peninsular Malaysia accounts for only 40 percent of national territory but supports 80 percent of the national population. Furthermore, the distribution of the peninsular population is spatially uneven: The five more urbanized west coast states of Negeri Sembilan, Malacca, Selangor (plus the Federal Territory

Photo 13.1 The Arab-influenced dress of these women first appeared in urban areas rather than in the traditionally more conservative countryside. Such apparel possesses two sometimes overlapping symbolic functions. One is to provide resistance to the Western cultural values pervasive in cities and the second, as protection against those who might question their Islamic values. (Ulack)

built-up areas" around gazetted areas of 10,000 or more people. Redefinition to more truly reflect levels of urbanization was needed because during the 1980s, for example, only 64 percent of the population of Kuala Lumpur resided within the gazetted urban area.

Some observations regarding levels of urbanization by state or region illustrate the impact of economic growth on the urbanization process during the NEP period. The west coast states of Johor, Selangor, Perak, Penang, and the Federal Territory of Kuala Lumpur accounted for approximately 59 percent of the country's urban population in 1991. Selangor, Penang, and Perak are characterized by urbanization levels of 75, 75, and 54 percent, respectively. Between 1970 and 1991, the number of urban places increased from 55 to 127, and these same three states increased their share of the total number of urban places from 50 to 60 percent. The historical inertia of existing colonial urban places functioning as transport, plantation, and mining centers and more recently as locations for foreign investment explains greater rates of urbanization. Well below the national average of 51 percent are the east and northwest coastal states of Kelantan (33), Kedah (33), Pahang (30), and Perlis (27) whose economies remained tied to the agricultural sector. Although Sarawak and Sabah, much like other less-developed states have experienced some of the highest rates of urbanization during the 1970–1991 period, the relative absence of an urban-based manufacturing economy and persistence of a resource-based regional economy explains the relatively low percentage of urban dwellers (37 percent in Sarawak, and 33 percent in Sabah).

Kuala Lumpur and the Klang Valley Conurbation. Primarily as a result of the discovery of alluvial tin and subsequent Chinese settlement in the mid-1800s, Kuala Lumpur began to grow and has continued to evolve over time as a center of political control and industrial development. During the colonial period, however, the city's economic power was localized because of the commercial dominance of Penang and Singapore. It was only in 1930 that the city's population surpassed 100,000 people. Rapid economic growth only truly commenced in the 1970s in response to NEP-induced growth strategies. After the Sultan of Selangor ceded the city to the federal government, Kuala Lumpur was designated as a federal capital in 1974. Anchoring the Klang Valley Conurbation (KVC), Kuala Lumpur has been transformed into the economic, political, and cultural core of the country (Map 13.3).

The KVC extends from the foothills of the Main Range bordering Pahang west to the Malacca Strait. Encompassing 1,108 sq mi (2,842 sq km), the KVC includes the federal district of Kuala Lumpur and a number of satellite cities in the state of Selangor, each supporting a mix of industrial, commercial, administrative, and residential land uses. The rapid emergence of the KVC with burgeoning satellite towns is a product of land-price increases associated with industrial growth and government employment during the 1970s and 1980s. The first evidence of an emerging conurbation took the form of an east–west urban corridor between Kuala Lumpur and the "outport" of Klang (Aiken and Leigh, 1975). Petaling Jaya was established in 1952 based on British "new town" planning designs and by 1966 supported 264 factories

Photo 13.2 Built in 1907, the Masjid Jame (or Friday Mosque) is located at the confluence of the Kelang and Gombak Rivers in the heart of Kuala Lumpur (which means "muddy confluence"). This was the location where the cities' founders, Chinese tin miners, first settled Kuala Lumpur in 1857. (Airriess)

producing light manufactured goods. In 1991, the two satellite cities of Petaling Jaya and Klang were the nation's fourth- and fifth-most-populous urban places. Adjacent Sungei Way was established in the 1970s as an industrial free trade zone. Further to the west is Shah Alam, where the newly planned Selangor state capital, MARA (Institute of Technology), and Southeast Asia's largest mosque are located.

More recently, the shape of the Klang Valley conurbation has been transformed with the spread of urbanization along the corridor of the north–south highway. Serdang supports the National Electricity Board Training Institute, the Palm Oil Research Institute, the Malayan Banking Training Institute, and the Agricultural University of Malaysia. The National University of Malaysia is located further south at Bandar Baru Bangi. Further residential, commercial, and industrial complexes are now appearing along the border with Negeri Sembilan. Although the hinterland of Kuala Lumpur has made less headway north of the city, in part because of a wide, hilly greenbelt, suburban residential and light industrial activity is now appearing in distant Rawang, approximately 18.6 mi (30 km) north of Kuala Lumpur.

Kuala Lumpur proper projects a Western if not modern urban landscape image. It is an architectural mix of government buildings both colonial and modern, commercial space of either the older Chinese shophouse or shopping mall variety, and high-rise office buildings. The prestigious Golden Triangle district of hotels, upscale shopping and condominium development is a landscape testimony to the adoption of western cultural values. Much like U.S. cities, the aggressive redevelopment of older buildings to suit modern consumer patterns has proceeded apace; the historic Central Market has been transformed into a shopping mall, the grand colonial period Majestic Hotel is now the National Art Gallery, and the architecturally magnificent, Moorish-style railway station will become a first-class hotel. Completed in 1997, the 88-story and 1,483 ft (452 m) Petronas "Twin Towers" anchors the future 100-acre (40.5-ha) Kuala Lumpur City Center complex (see Photo 7.1). The commercial and residential Twin Towers were the world's tallest buildings when completed. Because a handful of costly megaprojects are partially government funded, critics have claimed that finances would be better spent on basic needs development. The planned new federal capital of Putrajaya south of Kuala Lumpur is seen as another "growth at any cost" policy.

Nestled within the modern urban landscape of the KVC are pockets of squatter settlements, most of which are located on the outskirts of Kuala Lumpur. With approximately 17 percent of the metropolitan population living in squatter settlements in the mid-1980s, the greater Kuala Lumpur area possesses a lower percentage of squatters when compared to most other Southeast Asian primate cities. Most squatter settlements occupy abandoned tin mine lands or stream flood plains. The squatter population is multiethnic with Chinese comprising 52 percent and Malays and Indians comprising 33 and 15 percent, respectively. Individual squatter settlements are multiethnic as well, with each ethnic group occupying geographically discrete sections of the settlement. Squatter settlement origins are traced to the late 1960s and 1970s when rural to urban migrants arrived for the purpose of obtaining better paying jobs. Many

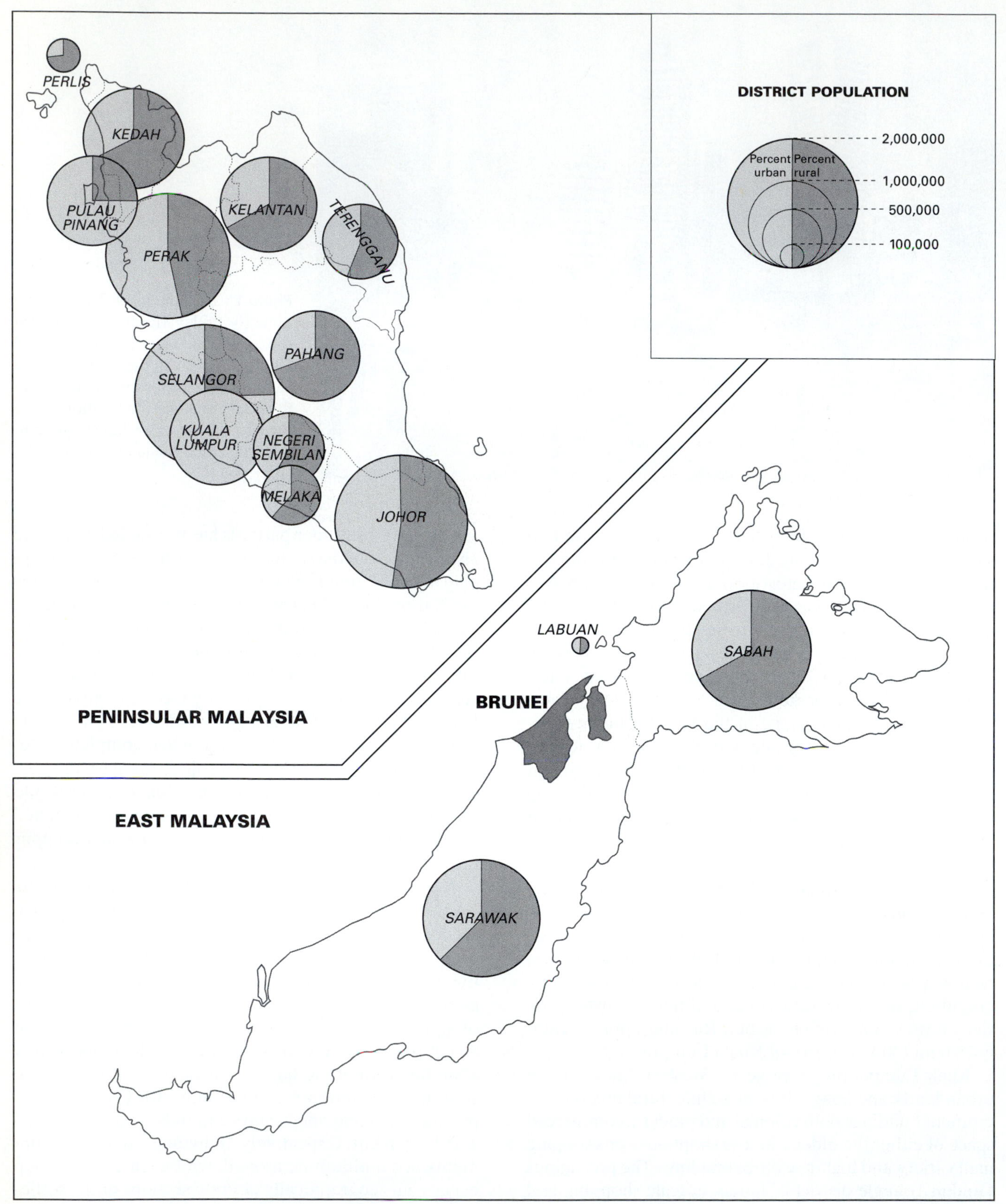

Map 13.2 Urban hierarchies in Malaysia.

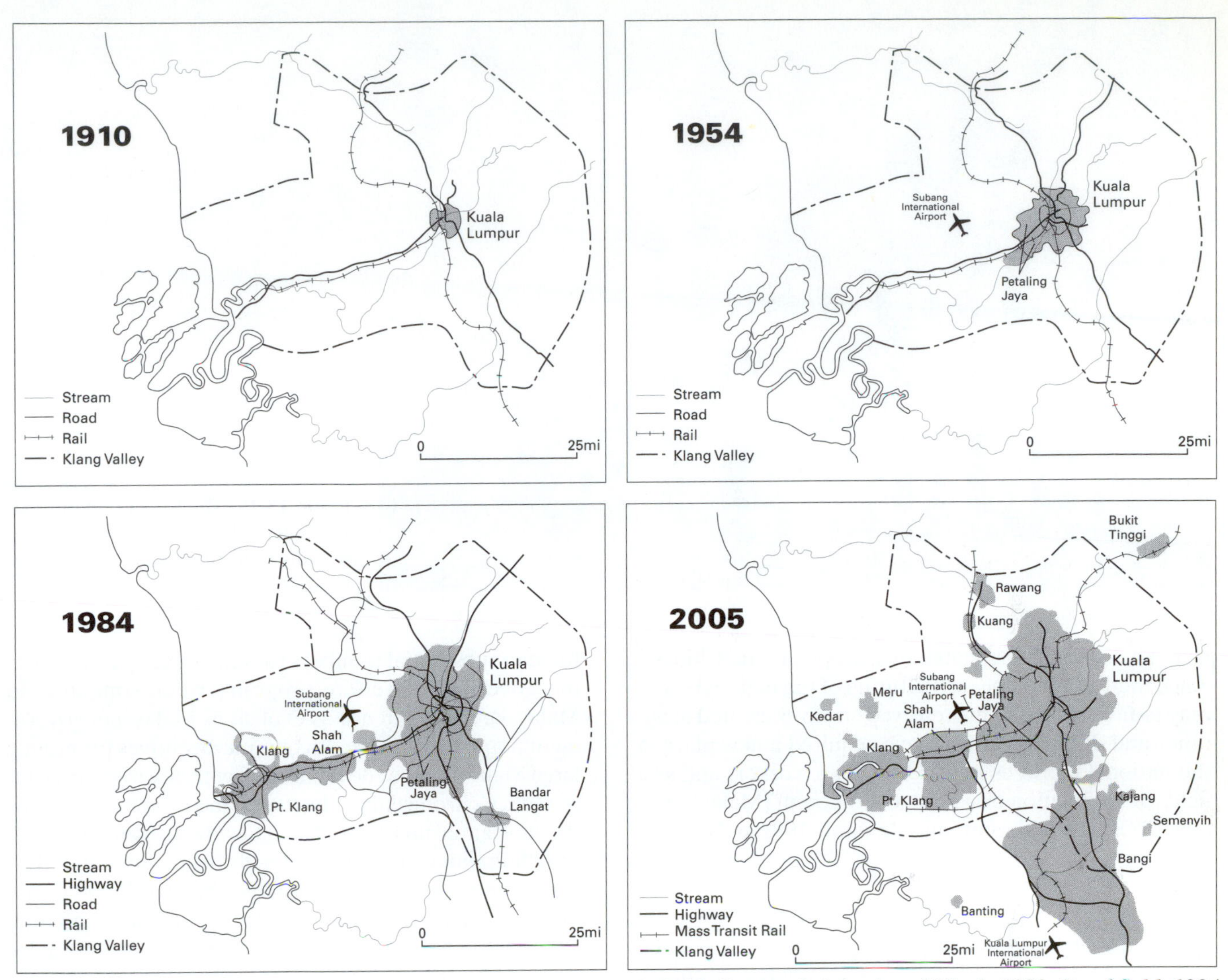

Map 13.3 Urban growth in the Klang Valley for select years, 1910–2005. *Source: Leinbach and Ulack, 1993; Brookfield, 1994 p. 261.*

present-day squatter inhabitants are descendants of these original migrants.

Because of the shortage of low-cost housing in the capital, the municipal government is in the continual process of demolishing squatter settlements to make way for multistory public housing. Aside from forfeiting what took so long to own, squatter settlers oppose redevelopment because not living at ground level deprives children of play space, gardening space, as well the ability to operate a home-based industry on which their survival developed. Although squatters do not currently hold land titles, many have problems qualifying for a bank loan to purchase flats. In addition, resettlement projects often are in locations distant from employment and services. Much like squatter redevelopment elsewhere in the developing world, governments are often not sensitive to the specific needs of squatters when formulating redevelopment plans.

Although not fitting the classic "desakota" model (McGee, 1991) because of the absence of a densely populated and extensive wet-rice hinterland, the KVC does share some attributes of this Asian megacity phenomenon. Between the transport spokes lined by residential development and industrial estates remains much interstitial green space dotted by villages and agricultural land that is being physically integrated into the urban economy. Villages have experienced in situ urbanization as lower and middle class commuters seek out affordable living accommodations made possible by the low-cost transport revolution of the bus and motorcycle. Many original inhabitants supplement subsistence rice production and cash-crop income with urban-based, off-farm employment. Much of this interstitial space experiencing "settlement transition" is Malay Reservation land originally established during the colonial period to protect traditional Malay rural life

Photo 13.3 The Kuala Lumpur railway station, built in 1911, is one of the world's most interesting railway stations. (Ulack)

from an economy dominated by Europeans and Chinese. Since the 1970s and the restructuring of the national economy to favor Malays, "the reserved enclaves created in another and different context have acquired a new place in the national fabric: as homes for urban workers and as a field for speculation by a new generation of [Malay] capitalists" (Brookfield, Hadi, and Zaharah, 1991, p. 41).

Brunei

Mirroring the settlement patterns of East Malaysia, Brunei's 1990 population of 253,400 is largely concentrated in coastal and estuarine regions. The smallest district of Brunei is Muara, where the capital of Bandar Seri Begawan is located. Muara supports approximately 65 percent of the national population on roughly 10 percent of the land with much of the balance of the population occupying the two western districts of Belait and Tutong with 21 percent and 11 percent, respectively. The eastern district of Temburong is sparsely populated with only 2.9 percent of the national population total. In 1990, Malays comprised approximately 67 percent of the total population, which includes a handful of indigenous groups long acculturated into the dominant Malay culture. The remaining ethnic groups are comprised of Chinese (16 percent), other indigenous people from East Malaysia (6 percent) and a high proportion of both Asian and Western expatriates (11 percent) connected to the oil and gas industry.

The 1990 annual rate of natural population increase of 2.8 percent is relatively high (2.3 percent in 1997), despite a substantial decline from 4.5 percent in 1971. High growth rates are explained in part by continued high fertility among Malay women who have yet to enter the labor force in substantial numbers as well as even more generous government welfare programs when compared to Malaysia. Although not as explicit as Malaysian government population policies, economic incentives promoting large Malay families dovetails well with government plans to reduce chronic labor shortages and ultimately replace the substantial number of foreign workers from Malaysia, the Philippines, and Thailand with Brunei nationals. Recent immigrants comprise a large share of Brunei's national population; approximately 31 percent of the total 1981 population were foreign born and among non-Malay groups, 50 percent were foreign born.

Common to rentier economies is a workforce dominated by government and oil and gas industry employment, much of which are found in urban areas. It is not surprising then that 59 percent of the national population resided in urban places in 1990. The four most populous urban centers in 1990 possess particular functions; Bandar Seri Begawan is the national capital; Muara functions as the national port; and Seria and Kuala Belait in the far west serve as centers for the oil and gas industry and the capital of this fossil-fuel-rich district, respectively. Bandar Seri Begawan is the overwhelming primate city of this micro-state; in 1992, approximately 21 percent of the national population resided here, and in 1990, the capital city accounted for 44 percent of all urban residents in the country.

Until the 1920s, the core of historic Brunei was Kampong Ayer or "water village," one of the most durable water-based urban settlements in the world (see Photo 13.4). Kampong Ayer is a mass of distinct but contiguous 40 urban wards perched on stilts over the Brunei River that in 1991 housed approximately 53 percent of the city's residents. Once the center of the sultanate engaged in trade,

Photo 13.4 From the vantage point of Kampung Ayer, the Omar Ali Saifuddin Mosque stands on the edge of downtown Bandar Seri Begawan. (Mark Cleary)

artisan, and fishing activity, Kampong Ayer became economically and locationally marginalized as Bandar Seri Begawan has expanded over land through nearby low density housing estates. It has, however, been integrated into the economy of the capital because Kampong Ayer now functions as a residential suburb, particularly for the burgeoning government workforce (Cleary and Eaton, 1992).

AGRICULTURE AND INDUSTRY

The transformation of Malaysia's national economy from one largely dependent upon the agricultural sector through the 1970s to an economy anchored by higher value manufacturing in the 1990s is testimony to Malaysia's status as an NIE (Brookfield, 1994). By the mid-1980s, manufacturing replaced agriculture as the most important economic sector contributing to GDP (Table 13.1). Manufacturing surpassed both agriculture and mining and quarrying combined by the early 1990s and would have earlier if not for the expansion of the timber and oil industry during the 1980s. The share of most other sectors remained stable with the exception of transport, storage and communications, finance, real estate, and business service, whose productive linkages with manufacturing explain their increased percentage shares of the GDP.

Agriculture

Despite the relative decline in the percentage of GDP accounted for by agriculture, this sector was a critical component of Malaysia's economic takeoff in the early 1970s and continues to be an important generator of national wealth today. Indeed, the agricultural sector in three of the four national development plans between 1970 and 1990 received the largest share of government-development spending. The NEP goal of raising incomes of the predominantly Malay rural workforce explains this concentration of investment. But the interrelated factors of stemming rural to urban migration and the promotion of cash-crop export production are important as well. Judged solely upon poverty eradication, agricultural policies have generally been successful as rates of rural poverty have decreased from 59 percent in 1970 to 15 percent in 1995. There exists, however, some present and future policy-induced problems in the agricultural sector. The focus here is upon the two most important subsectors of smallholder cash-crop production and rice or "padi" production with brief attention given to recent developments in other primary economy subsectors.

Smallholder land-development schemes. The greatest agricultural investment in the eradication of rural poverty has been the development of large-scale resettlement schemes in resource frontier areas; between 1970 and 1990 approximately 50 percent of government investment in agriculture was dedicated to land development. This rural-led "resource frontier" development strategy was the basis for Malaysia's dramatic economic growth beginning in the 1970s. These settlement schemes were established by The Federal Land Development Authority (FELDA), Regional Development Authorities (RDA) and other state and federal agencies (Aiken et al., 1982). With the exception of Penang, FELDA schemes have been established in all states, with the greatest concentration in Pahang and Johor (see Map 13.4). By 1985, approximately 2,172.5 sq mi (562,940 ha) had been developed

Industry	1970	1980	1990	1995
Agricultural, Fishing, and Forestry	30.8	22.2	18.7	15.5
Mining and Quarrying	6.3	4.6	9.7	7.3
Manufacturing	13.4	20.5	27.0	32.4
Transport, Storage and Communications	4.7	5.6	6.5	8.0
Wholesale and Retail Trade Hotel and Restaurants	13.3	12.6	11.0	11.8
Finance, Real Estate, Business Services	8.4	8.2	9.7	10.6
Construction	3.9	4.6	3.5	3.6
Government Services	11.1	13.0	10.7	9.2

Table 13.1 GDP - percent by industry of origin.

with FELDA accounting for 30 percent of the total (Sutton, 1989). By 1990, approximately 120,000 families had been resettled, involving about one-half million people. In the same year, 290 of the 422 individual schemes possessed settlers. Although the responsibility of each development authority varies, most provided a total development package: selecting settlers, clearing of land, planting crops, providing financial allowances to settlers until the crop matured, processing of harvests, and final marketing of produce. Each family was given a holding of approximately 10 acres (4 ha) that would be turned over to them once all loans were repaid. Each scheme, which averages several thousand acres, is anchored by a village providing a host of urbanlike facilities and services (Map 13.5).

The growth of the smallholder sector in both area and production has been dramatic. Cash crops such as cocoa and rubber are common, but because of lower prices and demand, particularly with rubber, oil palm is the most widespread crop cultivated; smallholders accounted for 45 percent of total area planted in 1990, almost equal to that of the plantation estate sector. With production efficiency almost equal to that of private estates, these two sectors combined allowed Malaysia to be the largest exporter of oil palm in the world in 1990. Whether the vantage point is from the air or ground, the vast carpet of oil palms spreading over miles of level ground and rolling hills speaks to the traveler of the importance of this single crop to Malaysia's export economy (see Photo 13.5).

FELDA activities have witnessed dramatic evolutionary changes through time. From 1956 to 1960, the responsibilities of providing funds and technical advice to state governments establishing land settlement schemes shifted to active involvement in establishing and operating schemes during the 1961–1967 period. The third or post-1967 stage not only witnessed the dramatic areal expansion of FELDA schemes but also included greater vertical integration into processing and marketing and the production of consumer products such as soaps and cooking oil. The fourth stage ushered in the establishment of large-scale and integrated regional development complexes comprised of many individual schemes. The Jengka Triangle in Pahang was the first of these large-scale development complexes, followed by the Johor Tenggara and Pahang Tenggara in the early 1970s. A fifth stage, which is currently materializing, entails the withdrawal of FELDA from direct administrative responsibilities and the parallel promotion of a more self-reliant attitude among both settlers and local scheme management.

As the only truly successful national rural land settlement program in the developing world, FELDA has received great praise from a wide circle of experts. Based upon a number of measures, such praise is well deserved. In every five-year planning period, FELDA has achieved or exceeded land development targets. Settlement schemes have increased semiurbanization in rural areas and stemmed the tide of rural to urban migration. Socioeconomic goals have been achieved through increased levels of education among settler children and dramatically raised rural income levels; in the Jengka Triangle, settlers earn four times the income of other rural poor and it is not uncommon for scheme settlers to own TV sets, motorcycles, and in some cases automobiles. Most importantly, though, FELDA aggressively pursues the goals of land ownership and through scheme development councils instills the importance of self-reliance and local decision-making power critical to the process of modernizing behavior. Despite the goals of landownership, however, only 5 percent of scheme settlers possessed land titles in 1987.

The overall success of FELDA in improving Malaysia's rural economy must, however, be qualified because of a

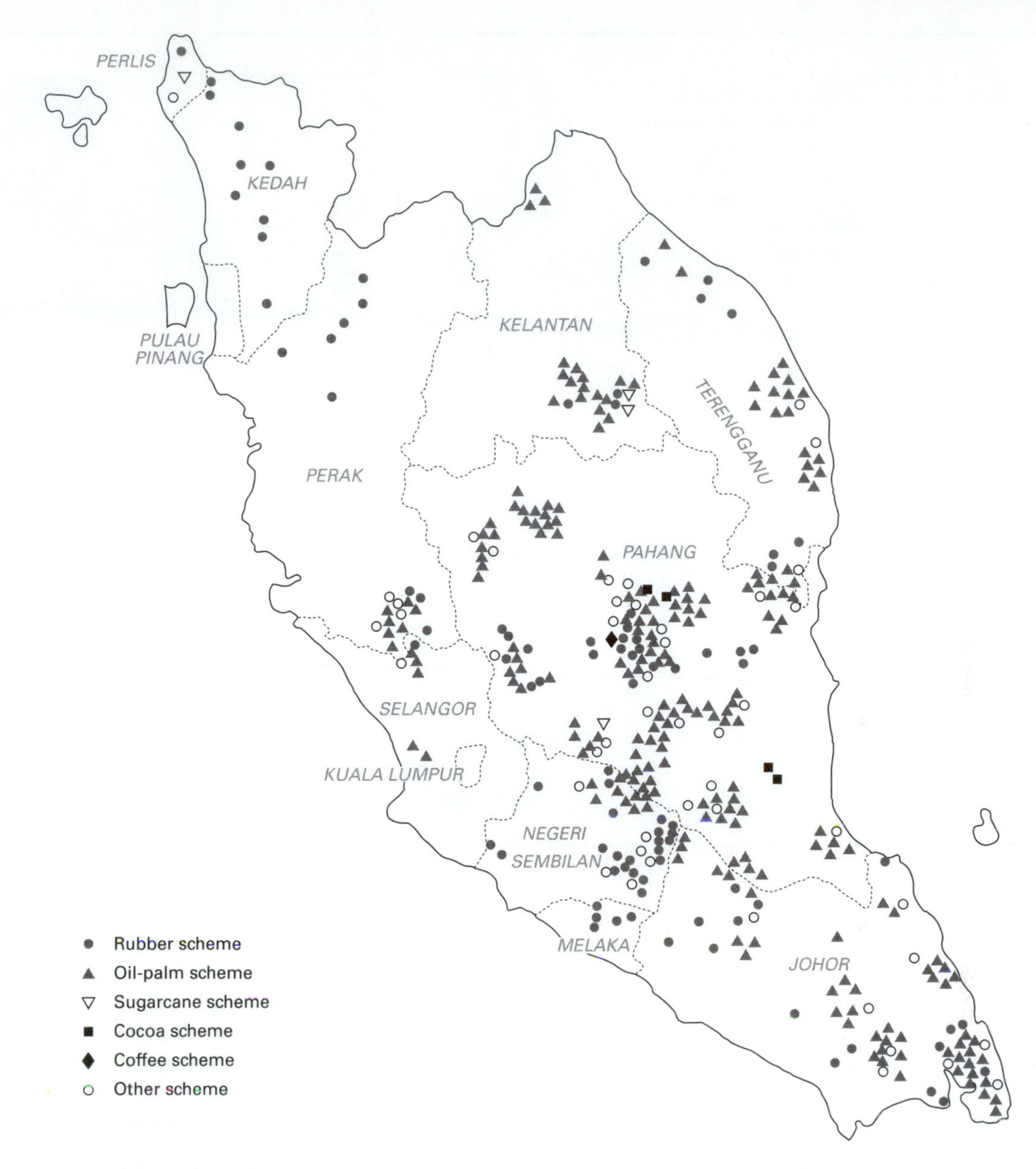

Map 13.4 FELDA's schemes and installations in Peninsular Malaysia, circa 1988.

host of current problems with the potential of aggravating future production. First, because resettlement schemes have only effected a small fraction of rural households living in poverty, there has emerged significant income differences in the rural sector. Second, the cost of resettlement has been very high because of administrative and infrastructure costs. Third, schemes and associated regional development programs have not resulted in the hoped for "trickle-up" effect because agro-based industries established by RDAs have not generated much off-farm employment opportunities. Fourth, because of a pervasive top-down planning philosophy, there has developed a dependency syndrome that reduces the participation of settlers in the economic development process; some critics claim that human development has been sacrificed to the interest of the state and or the international market. It is interesting to note, however, that the majority of settlers do not wish for FELDA to divest itself of management responsibilities and that anxiety over complete independence is in part explained by the similarity of the hierarchical nature of FELDA society to the structured patron-client relationships characteristic of feudal classical Malay society. Fifth, the children of scheme settlers have shown little interest in pursuing farmwork. On the oldest schemes, for example, only 33 percent of second generation settlers remain on the scheme. In a sense, FELDA is a victim of its own success; accompanying increased educational attainment is the desire of future scheme farmers to seek better employment opportunities elsewhere. This is why in 1995, approximately 432,000

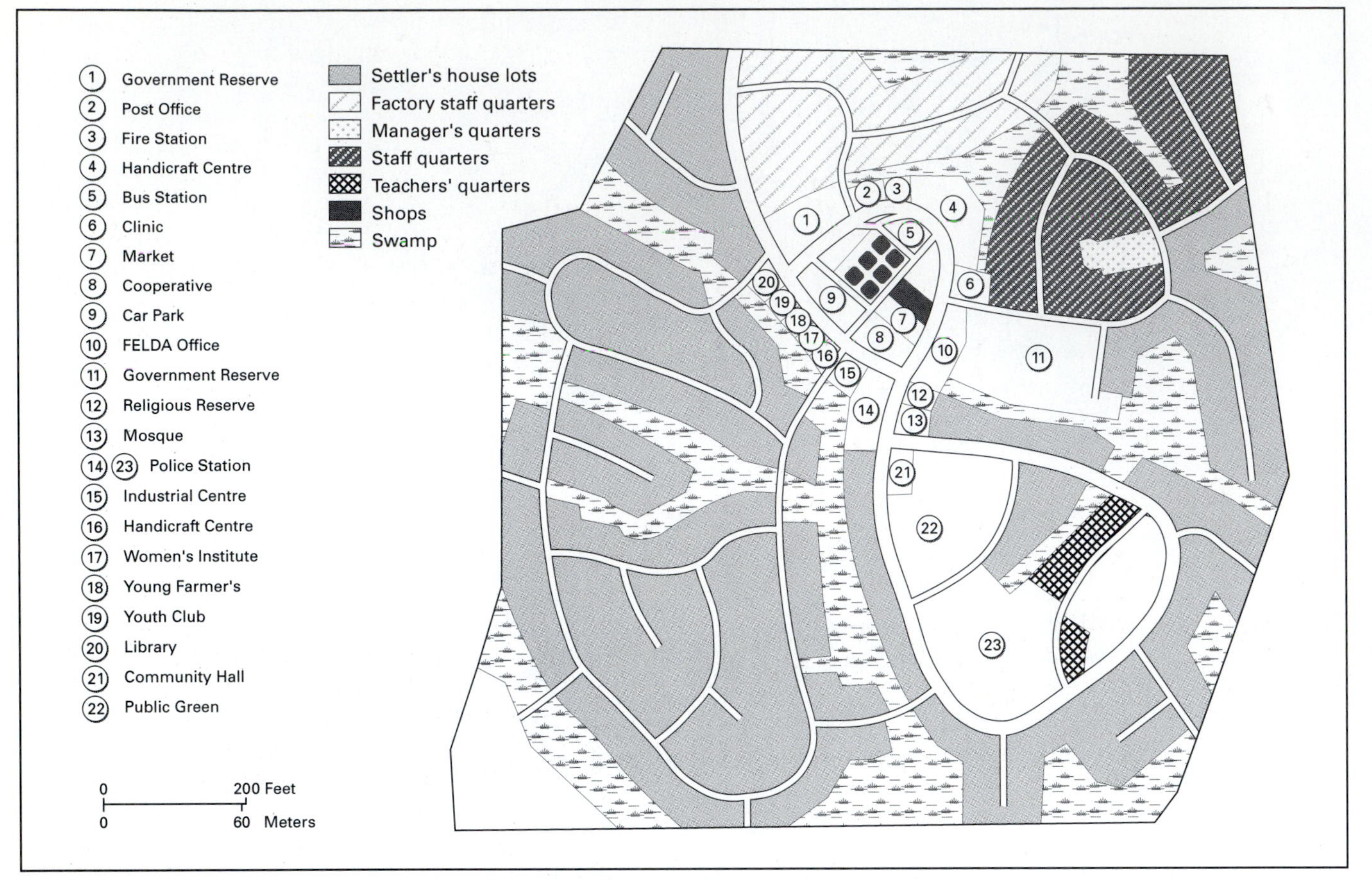

Map 13.5 FELDA newtown.

foreign laborers working on a wage or share-cropping basis were employed in the agricultural sector. Finally, it is imperative that schemes diversify away from oil palm and rubber because of fluctuating commodity prices and demand; the United States for example has substantially reduced palm oil imports because of the dietary and health problem of high saturated fat content. Since the early 1990s, FELDA halted new land development schemes in anticipation of injecting greater degrees of privatization because of substantial government debt. Privatization, however, introduces the possibility of losing critical political patronage because the top-down philosophy ensures the persistence of traditional power structures (Aiken et al., 1982).

Aside from the expected problems of isolation and lack of infrastructure, land development schemes in East Malaysia have met with little success as an economic development tool. In Sarawak as well as Sabah, Native Reserves and Customary Land hinders developing large tracts of land. Most important, however, is that the FELDA model is ill suited because native peoples possess little desire to engage in the regimented work of resettlement schemes. As a result, schemes in both Sabah and Sarawak have been forced to contract Indonesian workers from neighboring Kalimantan at a massive scale. This is especially true for the huge 246,290-acre (107,000-ha) Sahabat regional development complex in eastern Sabah. In situ improvement through technical advice, subsidies, and credit schemes appear to be the best option for raising incomes of many shifting agriculturalists.

Declining padi production. No single agricultural subsector has hindered rural development as much as padi (Courtenay, 1986). The introduction of Green Revolution hybrids, double cropping, irrigation of two-thirds of padi land, and substantial government subsides have raised productivity per hectare and worker, but total production, hectarage, and incomes of padi farmers have decreased over the past 20 years. Between 1985 and 1995 production decreased by 14 percent, and cultivated area decreased by 0.6 percent during the 1990–1995 planning period. As a result, rice self-sufficiency declined from 80 to 75 percent between 1990 and 1995. Although the percentage of padi farmers living in poverty declined from 88 to 58 percent between 1970 and 1985, poverty has increased during the 1980s to levels almost two and one-half times the national average for other agricultural sectors. The most basic productivity problem is that the majority of holdings are of inadequate size to allow individual households to achieve poverty line income. As

Photo 13.5 Oil palm in Pahang, Peninsular Malaysia. (Leinbach)

a result, the majority of households in the traditional northern rice bowl regions of Kedah and Kelantan must either engage in off-farm employment to cross the poverty line threshold or migrate to urban areas. Migration of young workers during the early 1980s meant that population growth rates in padi-growing areas were commonly 1 percent lower than the national average, resulting in the substantial aging of the padi farming population. When traveling through traditional padi regions, it is not uncommon to encounter numerous abandoned fields breaking up the once contiguous expanse of checkerboard green. The seemingly tranquil village is commonly inhabited by farm families composed of an aging father and mother and perhaps younger daughters, with the older sons and daughters having already joined the migration stream to urban areas.

The goal of reducing poverty among padi farmers while at the same time increasing relative productivity through technology inputs that require labor shedding and land aggregation has created the present paradox. The economic growth model based on capitalist agriculture, particularly in economically peripheral regions has resulted in the uneven spread of Green Revolution technology and the marginalization of many producers. A number of options to improve the padi economy are worthy of consideration. First is the free-market option to abandon the goal of self-sufficiency in rice production. Because Malaysia imports substantial amounts of rice, it is accepted at least at the national level that the government has already acquiesced to importing less-expensive rice. Second, despite subsidies already being too high, the government could raise subsidies to make padi farming more financially rewarding. Third, the government might continue to provide in situ alternative employment under the condition that farming remains the focus of household energies.

Other agricultural crops. Because of relatively high and increasing food import bills, the government has placed greater emphasis on financially more-rewarding domestic agricultural activities that in some cases also possess promising export potential. Much like value added processing of palm oil exports, the government has recently promoted the emergence of food-processing industries, particularly fish, vegetables, and fruit. After decades of dwindling nearshore fish resources and traditional problems of surplus labor, coupled with a high reliance on fish for animal protein, the government has targeted deep-sea fishing as a growth sector. Between 1985 and 1990, the proportion of deep-sea tonnage of the total marine catch increased from 3 to 12 percent despite a 24 percent decline in the number of fishers during the 1980s. Livestock, particularly poultry and eggs, is seen as a growth area as production between 1985 and 1995 increased 55 and 40 percent, respectively. Vegetable production increased 25 percent between 1985 and 1995, but was only able to meet 87 percent of domestic demand. One solution to the diversification problem on monocrop FELDA schemes includes the greater production of livestock and vegetables. The other agricultural commodity for which Malaysia is well known is timber (Kumar, 1986). Before the ban on log exports in the early 1990s, Malaysia was the largest exporter of logs in the world; in 1981, Malaysia's world share of hardwood log exports amounted to an amazing 48 percent. In 1990, East Malaysia accounted for 73.2 percent of national timber production, the majority of which entailed little value-added processing before export. Increased levels of timber processing, particularly furniture, is a primary government goal; during the 1985–1990 period, wood product exports almost quadrupled.

Industrialization

The dramatic transformation of Malaysia's national economy now anchored by higher-value manufacturing is a developing country success story. Malaysia's industrial profile includes electronics and textiles primarily for export and a host of heavy industries such as auto assembly and steel for domestic consumption. A discussion of

BOX 13.2 The Malaysian Financial Crisis by Thomas R. Leinbach

Initially, Malaysia did not appear vulnerable to the financial crisis because of full employment, continuing high inflow of long-term capital, and a seemingly strong currency rating. However, the crisis sent the country's economy reeling by January 1998. The country's vulnerable situation became serious because of excessive short-term foreign borrowing, weaknesses in the financial system, and a significant increase in the real exchange rate, suggesting deterioration in competitiveness. Especially conspicuous was a rapid expansion of bank loans to the private sector, many of which were invested in real estate. Similarities to the Thai situation are apparent here. The initial policy response of the government was one of denial because economists felt that Malaysia's economy was above susceptibility. Prone to xenophobia (i.e., fear of foreigners), Prime Minister Mahathir claimed that the country's economic downturn was a result of a foreign conspiracy to rob Malaysia of its fledgling developed country status. Eventually, however, a set of austerity measures was announced whereby government spending was cut by nearly 20 percent, new overseas investment by Malaysian firms was halted, costly and high-profile public-works projects were delayed, and salaries of government ministers were cut.

Domestic market-oriented firms have been struck by contraction in demand, weaker stock prices, and a property market slump. Increased competitiveness will result from the depreciated currency, and export-oriented manufacturing may benefit. A bright note here is that a new product cycle has emerged in the computer industry and demand has risen for low end PCs that Malaysia can supply. Fortunately, much of the electronics trade is U.S. and European dependent rather than intraregionally ASEAN dependent. Policy makers suggest that the recovery pace will depend on the extent to which multinational enterprises continue to view Malaysia as a prime investment location. More than 45 percent of the country's manufacturing production and more than 75 percent of total manufactured exports are accounted for by multinational enterprises (MNEs). On the surface, the risk of social and political instability is low compared to Indonesia. However, the criminal conviction of Deputy Prime Minister Anwar Ibrahim in 1999 (on dubious corruption charges) has increased calls for *reformasi* (reform). If problems of unemployment loom critical, the country can repatriate foreign workers, which account for 25 percent of the workforce. By mid-1999 the government had begun to implement reform of the fragmented domestic financial system. Perhaps more important was the stronger rationalization of the public investment program which the crisis forced upon government decision makers.

Malaysia's fledging NIE status must be qualified by pointing out that these export-oriented industries are overreliant on foreign investment and narrowly based, and domestic heavy industries are economically inefficient because they are government owned, or what is referred to as public sector state-owned enterprises. In addition, both industrial strategies have created industrial dualism through depressing growth in small- and medium-size industries. Lastly, the growth of the industrial sector favored western states and has created substantial geographic inequalities in levels of economic development.

Although Malaysia's pre-NEP industrial strategy was successful in terms of dramatically raising industrial production, this import substitution strategy (ISI) of domestically producing manufactured goods was disappointing. The focus of production was on lower-value consumer goods with little export potential. Moreover, ISI was dominated by foreign and non-Malay domestic investment; by the late 1960s Bumiputra interests controlled less than 5 percent of shares in commercial and industrial enterprises. Nor did ISI address the problem of geographic inequalities of industrialization; in 1971, the four states of Johor, Selangor, Perak, and Penang accounted for 75 percent of employment and output.

The failure of ISI to promote Malays successfully in commerce and industry, coupled with the 1969 race riots in many urban areas, prompted the government to implement the NEP in 1971. It was through the NEP that Malays were to participate in the modern urban–industrial economy through positive discrimination policies such as preferential treatment and quotas in tertiary education, government loans and licensing preferences for business creation, hiring quotas in medium- and large-scale industries, and ownership share in limited companies. The NEP philosophy was anchored in "growth redistribution" to replace the simple pre-NEP "growth" philosophy.

The industrial goals of the NEP were generally achieved. By 1995, foreign ownership in the corporate sector had fallen to 21 percent. Although falling short of the NEP goal of 30 percent, the Malay share increased from a tiny percentage to 21 percent. Chinese economic power continued as their share of the corporate sector remained high with 41 percent in 1995. Employment shares by ethnic group between 1970 and 1995 witnessed dramatic change as well. The Malay employment share in the manufacturing sector increased from 26 percent to 51 percent and in the service sector from 31 percent to 45 percent. Chinese and Indian employment shares experienced parallel declines respectively; from 58 to 31 percent and 16 to 12 percent in the manufacturing sector and 46 to 22 percent and 23 to 7 percent in the service sector. Achieving these NEP goals materialized through two forms of industrial growth; the promotion of foreign-owned, export-oriented industries (EOI) based upon the union between state, local, and foreign investment and the creation of government-owned public enterprises (PEs).

Foreign investment. Because of political stability and a relatively advanced transport infrastructure dating from the colonial period, Malaysia has always been a promising location for foreign investors. Since the early 1970s, foreign investment has dramatically increased, anchored by relatively cheap labor and a host of tax breaks and other financial incentives associated with the establishment of free trade zones (FTZs), export processing zones (EPZs), and licensed manufacturing warehouses (LMWs). Additional financial incentives were offered to foreign investors in the mid-1980s in response to the serious economic recession in the early 1980s. As a consequence, foreign direct investment (FDI) increased eighteenfold from 1985 to 1990 and accounted for 42 percent of total manufacturing investment value and 44 percent of new jobs created during the same period. Employment in this "assembly" economy greatly benefited Malays, particularly young females, who accounted for 60 percent of total new employment in the manufacturing sector.

This most recent flush of foreign investment in EOI is East Asian based, replacing Western-dominated investment of the 1970s. Rising currency evaluation and rising labor costs in Japan and Taiwan forced the establishment of offshore or branch manufacturing plants. During the 1985–1990 period, Taiwan and Japan accounted for 53 percent of total foreign investment value, up from only 31 percent in 1985. Primarily investing in electronic, and textile and clothing apparel industries, these two industrial subsectors alone accounted for approximately 50 percent of growth in manufacturing employment and 65 percent of gross manufactured export value from 1985 to 1990. By 1995, the electrical and electronics products subsector accounted for 66 percent of gross manufactured export value.

Although foreign investment has become a substantial employment generator allowing for a much higher standard of living, long-term economic benefits are still questionable. First, foreign investment has created overdependence on the volatile global economy. The situation is exacerbated because investment has narrowly focused on electronics and textile industries. Second, the location of FTZs only aggravates the problem of industrial concentration in a handful of choice regions; in 1992, half of the 12 FTZs were in Selangor and Penang; only one was located in a less-developed region (Sarawak). Most importantly, FTZ manufacturing linkages with the domestic economy have been poor, although there exists substantial potential. In the government's focus upon modern, export-oriented industry, it has neglected labor intensive and financially poor small-and medium-size industry (SMI) possessing the potential to create the necessary production linkages with FTZ located facilities. Since the implementation of the NEP, for example, relative employment and gross output among SMIs has gradually declined. By 1992, SMIs comprised 84 percent of the number of total firms but contributed approximately 28 percent of manufacturing added value. Perhaps one reason for such neglect is that SMI owners are predominantly ethnic Chinese and thus did not conform to the socioeconomic agenda of the NEP. Greater government attention to SMI growth to create productive upstream linkages as suppliers to large scale industry is required to reduce the serious dualism in Malaysian industry.

To illustrate some of these problems associated with foreign investment, we examine Japanese foreign investment because this Asian industrial giant possesses a special economic relationship with Malaysia (Tsuruoka, 1992). Being the single most-significant foreign investor, trading partner, and foreign aid provider during the NEP period, Malaysia has become overdependent on Japanese capital. In the early 1990s there were 250 Japanese companies in Malaysia, 40 of which were wholly Japanese owned, with the balance being joint ventures. Indeed, no other Southeast Asian government has courted Japanese investment capital as aggressively as Malaysia. The government's "Look East" policy promoting the Japanese work ethic among Malay workers as well as an important source of investment, business skills, and technology transfer has in a sense signaled Malaysia's acceptance of Japan's economic leadership. Critics of the policy have charged that Japanese corporations possess great influence over government economic decision making; the Japanese Chamber of Commerce president in Kuala Lumpur for example, is sometimes referred to as Malaysia's "deputy finance minister." Other critics claim that the Japanese are unwilling to share technology and that the technology that is shared is already outdated. The transfer of business skills at joint venture manufacturing plants is a particularly contentious issue. Malaysians are relegated to lower- and middle-level management positions with little chance for advancement, while important decisions are made by Japanese executives; the appointed Malaysian manager is often just a figurehead. In addition, Malaysians view Japanese managers as being too autocratic and formal. This stands in direct contrast to U.S.-owned manufacturing plants where Malaysians often assume upper management and supervisory roles (O'Brien, 1994), particularly in the important electronics industry.

The Penang electronics industry. Although not viewed as a microcosm of foreign investment-led industrial growth in Malaysia, the Penang electronics industry provides a suitable example to illustrate the problems associated with Malaysia's drive toward achieving NIE status (Rasiah, 1989). Immediately after losing its "free port" status in 1971, the state government established a number of FTZs and was able to attract foreign electronics firms

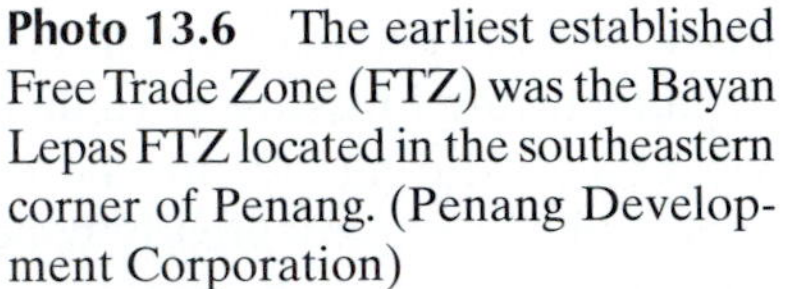

Photo 13.6 The earliest established Free Trade Zone (FTZ) was the Bayan Lepas FTZ located in the southeastern corner of Penang. (Penang Development Corporation)

through tax incentives and the provisioning of cheap labor engaged in lower-value and labor-intensive assembling and testing stages of production (see Photo 13.6). By 1981, U.S. firms such as National Semiconductor, Hewlett Packard, and Intel as well as numerous Japanese and Korean firms employed approximately 20,000 workers, most being young, single Malay women from northern labor-surplus states (see Chapter 11, Gender and Development). Despite cramped, dormitory-like accommodations and rapid labor turnover, these female "children of the NEP" sent sufficient amounts of income to become the primary income earners for their rural families. This has been loosely likened to the process of employing female labor characteristic of the emerging industrialized west a century ago (Armstrong and McGee 1985, p. 205).

Malaysia's vulnerability to a foreign-dominated export economy surfaced in the early 1980s when the global semiconductor market bottomed-out as a result of worldwide recession. Substantial reduction of FTZ labor ensued and Penang's economic base was threatened with the possibility of geographically mobile multinational corporations (MNCs) searching for cheaper sources of labor elsewhere. For a number of reasons, however, Penang benefited from changes in product and production technology during the mid-1980s. First, the government extended and improved upon previous tax incentives to retain foreign-owned production facilities. Second, there developed through the 1980s a growing regional Asian market for both components and final products for which Penang is well established to serve. In response to this growing market, MNCs have chosen to source, both locally and regionally, materials from the ASEAN region that further solidifies Penang's position. With increased automation, FTZ plants now require fewer unskilled workers and greater numbers of skilled workers with technical knowledge; at Intel's facility for example, one of six employees is an engineer compared to 1 in 40 in 1980. In 1993, unskilled workers accounted for only 32 percent of the electronics industry workforce. At the same time, employment in foreign-owned electronics firms increased to approximately 78,000 workers in 1992.

The success of Penang, now referred to as Silicon Island, has attracted other foreign and domestic industries engaged in the production of textiles, plastics, metalworking, and packaging. These are now spread across seven free industrial zones, industrial parks, or estates. An industrial conurbation has in fact emerged as many new firms have located to mainland Prai (Province Wellesley), which is connected to Penang by a recently constructed bridge. As a result, manufacturing contributed to 48 percent of Penang's GDP and employed approximately 30 percent of the local workforce in 1992. The electronics industry accounted for 21 percent of the number of factories, 54 percent of the total manufacturing workforce, and 25 percent of paid-up capital in manufacturing in the same year. Although still part of the dominant west coast industrial corridor, the dramatic growth of Penang-based industry does contribute to the geographic deconcentration of industrial development away from the KVC.

No longer considered just an offshore enclave producing low-end intermediate electronic inputs, Penang is beginning to capture higher value, front end production stages such as wafer manufacturing (wafers are a thin slice of semiconductor used as the base material on which transistors and integrated circuit components are formed). In addition, Malaysian engineers in U.S.-owned plants are independently engaged in innovative research and development that was once reserved for design engineers in the United States. Although still in its infancy stage, production linkages with the domestic economy are increasing as parts subcontracting becomes more common. Only if these linkages are developed will Malaysia escape from the economic vulnerability inherent in the boom-and-bust cycle nature of the global electronics industry.

Johor—An emerging industrial growth center. Although part of the more developed west coast peninsular states, the recent emergence of southern Johor as an industrial growth center provides for more spatial equity in regional development (Guinness, 1994). Beginning in the early 1970s, both the state and federal governments dedicated substantial investments to establish both the Pasir Gudang Industrial Area and Johor international port on land covered by mangrove swamp and aging rubber trees directly east of the state capital of Johor Baharu. Much of the early industrial investment was concentrated in public enterprise palm oil refineries and metal-processing plants. By the mid-1980s, Pasir Gudang supported a more diverse industrial base in response to the establishment of a Singapore-centered growth triangle. This more-recent industrial growth based on private investment is a result of industrial spillover from Singapore whereby foreign investment in chemicals, textiles, rubber products, and low value electronic components by East Asian and some western firms are commonly of the joint venture variety with Singaporean interests; in 1988, approximately 48 percent of foreign investors were Singaporeans. With cheaper land prices, lower labor costs, and access to the General System of Tariffs in Malaysia, the relocation of labor-intensive manufacturing to Johor dovetails well with Singapore's Second Industrial Revolution, which focuses upon higher-value manufacturing. This international cooperation, or "twinning," phenomenon, allows foreign investors to establish the labor intensive stages of manufacturing in Johor before the product gains further value through additional product enhancement in Singapore.

Although the Malaysian government welcomes an alternative site for foreign investment, this development introduces two potential problems of a political economy nature. First, the federal government is concerned that the Johor regional economy is once again becoming too dependent on Singapore and thus threatens the traditional Kuala Lumpur-based core-periphery political dynamics. Second, the coupling of foreign and domestic Chinese capital poses a serious test to the stated NEP goals of economic equity. These perceived problems are expected because Johor possesses strong historical, ethnic, and economic links with the dominant ethnic Chinese island city-state to the south.

Public enterprises. The rapid pace of industrialization in Malaysia during the past two decades is also evidenced by the proliferation of government-owned public enterprises (PEs). Indeed, the activities of PEs accounted for 29 percent of total government development expenditure during the 1985–1990 period for 24 percent of GDP in 1987, far greater than in Indonesia (7 percent in 1988) and the Philippines (3 percent in 1983). The most important function of PEs is NEP based: to promote Malay ownership in commerce and industry and to increase Malay participation in the modern sector through employment opportunities. Managed by Malays, PEs function as trust agencies for the people until such time Malays are financially able to buy shares and acquire capital through private investment. PEs also allow for economic diversification through the establishment of heavy industries. In this sense, the government's early 1980s "Look East" policy was to emulate the success of East Asian NIEs and their "state-directed" economic growth policies. Under the administrative umbrella of the Heavy Industries Corporation (HICOM) established in 1980, a number of heavily subsidized showcase projects have come on line; the Perwira steel plant in Terengganu, a shipyard in Labuhan, a petrochemical plant in Sarawak, a motorcycle engine plant in Penang, and the most ambitious of all, the "Proton" auto assembly plant in Shah Alam. These were the largest of some 700 PEs in 1987.

Although the economic objectives of PEs are arguably worthwhile, economic performance has been dismally poor. PE debt as a whole is substantial and would be far greater if not for the profitable state-owned petroleum company PETRONAS. Poor financial performance is based on two primary factors: First, government subsidies and price controls provide protection from international competition, and, second, the socioeconomic goals of the NEP oftentimes became more important than economic goals and thus severely affected efficiency because of widespread mismanagement and corruption.

Perhaps the most widely publicized government financed venture and one that symbolizes the difficulty of heavy industrial development is the Proton auto assembly project (Means, 1991). Showcased as Malaysia's national car replete with the Islamic star and crescent designer emblem, this pet project of Prime Minister Mahathir is a joint venture between HICOM and Mitsubishi, holding 70 and 30 percent shares, respectively. Since the first car rolled off the assembly line in 1985, the joint venture has experienced operational deficits and numerous other financial problems that only add to government debt. Although the Proton captured 61 percent of the domestic passenger auto market in 1992, targeting the middle and upper-middle classes in a low-volume market such as Malaysia's has resulted in the plant operating at less than full capacity. Export markets have been targeted, but problems of intense international competition, establishing dealer networks, and satisfying foreign emission and safety standards have severely limited export potential. In response, the government implemented subsidization policies to promote the Proton further at home. The two most important indirect subsidies that are part of the total subsidization package are the elimination of the hefty 40 percent duties on imported Proton parts while the 18 other assembly plants continue to pay duties and of government loans to civil servants to purchase only the Proton auto (see Photo 7.4). This government-subsidized competition has displaced labor and the market share of other assembly plants. Finally, less than 30 percent of parts were locally sourced in 1992, pointing to the absence of substantial domestic productive linkages with small- and medium-size firms.

Since the mid-1980s, the government has had some success in privatizing a handful of PEs, particularly in the transport and communications subsector. The Seventh Malaysia Plan (1996–2000) has also outlined an even more aggressive privitization program. These privatization measures serve the interests of the NEP by promoting entrepreneurship and corporate equity among Malays. The reality of privatization, however, is that a small group of ethnic Malay elites with government ties have taken advantage of the government divestiture because of the absence of qualified bumiputra investors and the inability of Chinese to purchase PEs. While these former PEs are now run more efficiently and the government has been freed of financial burden, an economic monopoly nonetheless continues.

Conclusions and outlook. Taking stock of NEP-directed economic growth is warranted at this juncture. Without a doubt, government policies to foster economic growth through a more productive agricultural and industrial sector have proven quite successful during the past 25 years. This growth, however, has been geographically uneven (see Figure 13.2). Of the 14 political units, only five are characterized by per capita GDP above the national average in 1995 and four of those five are west coast states. As the country's preeminent service center and industrialized state, respectively, Kuala Lumpur and Selangor will retain their positions primarily because these locations are preferred by the expanding private sector. With industrial spillover from Singapore, the state of Johor experienced the greatest gain in GDP share during the past decade. "The continued persistence of these development enclaves, which are characterized by strong colonial influence and ethnic Chinese settlement, is quite conspicuous." Trengganu remains a poor agricultural state, and only the enclave nature of the state's rich oil endowment inflates per capita GDP as a ratio of the national average. Indeed, mean monthly household income as a ratio of the national average in Trengganu ranks tenth. The middle income states of Negeri Sembilan and Malacca have already benefited and are expected to continue to benefit from industrial spillover from the congested Klang Valley. Although Sabah and Sarawak occupy the upper tier of middle-income states, these states remain dominantly rural and relatively poor. Their inflated GDP positions are traced to significant log exports as well as to substantial federal revenues dedicated to basic infrastructural improvements. Although substantial progress has been made to alleviate poverty in lower-income states, their economies remain tied to the agricultural sector. If it were not for industrial growth replacing the once flourishing tin mining sector around Ipoh, Perak's national share of GDP would be much reduced. Stretching across the northern tier of the peninsula are the four endemically poor and rural states. Indeed, the benefits of the NEP have yet to reach Kelantan—this state's mean monthly household income in 1995 was only 26 percent of that of Selangor.

Brunei. The very nature of oil-rich rentier economies such as Brunei's inhibits any serious efforts at economic diversification. Substantial oil and gas export revenues allows for a positive balance of trade, no personal income taxes, and the investment of these revenues into the public sector to create a "Shellfare" state satisfies the economic needs of the population. There exists in Brunei, then, a relatively affluent consumer class in the absence of a producer class (Gunn, 1993). It is this unidimensional nature of economic growth in Brunei that erects barriers to the successful development of a more diversified economic base; the oil and natural gas sector accounted for 53.7 percent of GDP in 1990, and manufacturing, coupled with mining and quarrying, only accounted for 2 percent of non-oil GDP in 1990.

As might be expected, the primary sector activities of agriculture, fisheries, and forestry are not well developed. It is not because of the absence of arable land because only approximately 4.5 percent of used land in the late 1980s was under agriculture, but rather it is primarily due to greater economic opportunities in urban-based civil service employment. In 1960, one-third of the national labor force was engaged in these three primary subsectors, but by 1986 it had dropped to 5 percent, and by 1990 the combined share of these primary subsectors to GDP was only slightly above 1 percent. Although government pronouncements in the 1991–1995 national plan called for greater food security because approximately 80 percent of food needs were imported during the 1980s, only 2 percent of public sector allocation was dedicated to these subsectors. Brunei still must import approximately 75 percent of its food needs. Substantial opportunity is seen in high-value vegetable production as well as livestock; the government imports a substantial amount of livestock from a government-owned ranch in Australia that

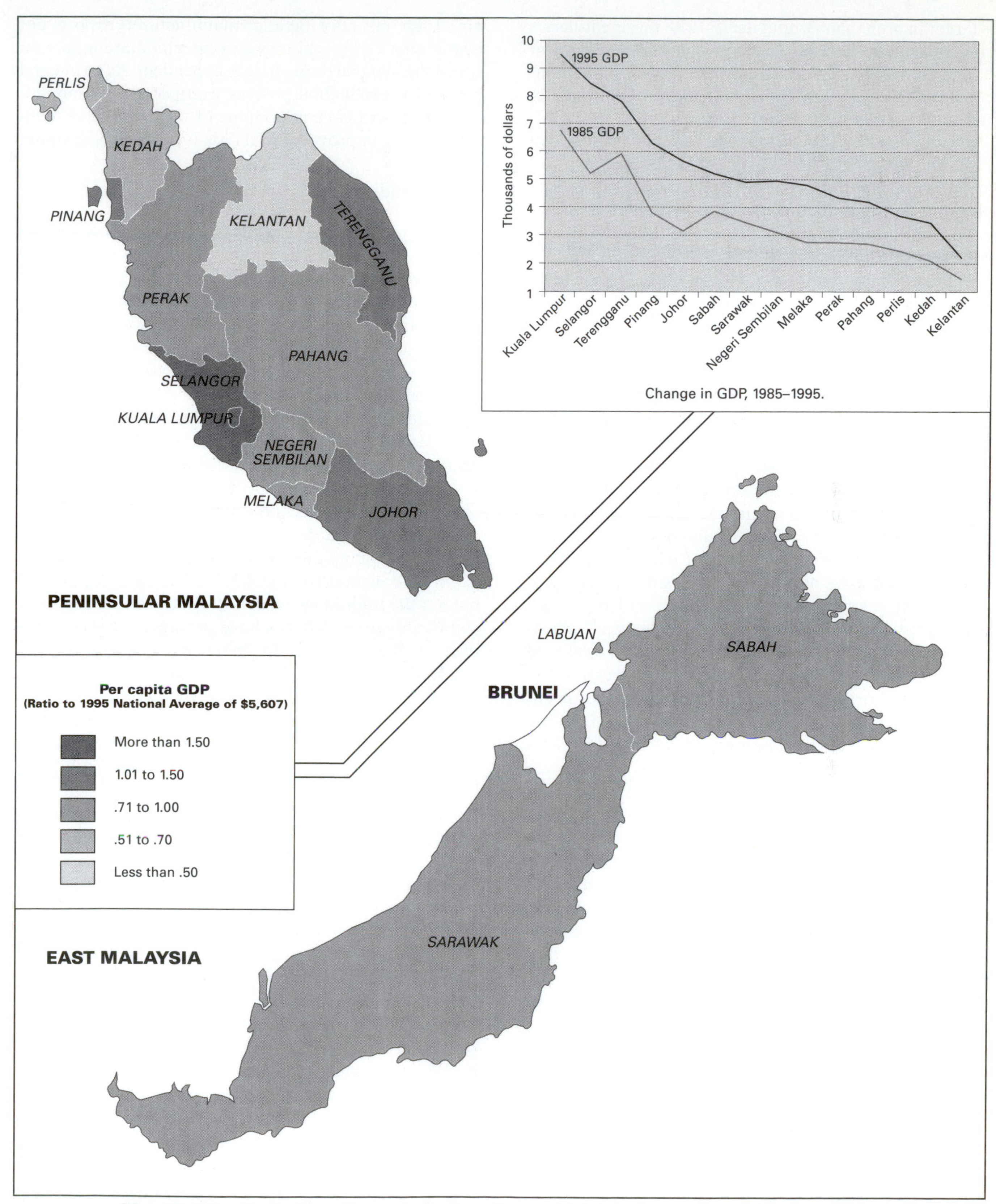

Map 13.6 Per Capita Gross Domestic Product. *Source: Malaysia, Department of Statistics, 1995.*

is larger in area than Brunei itself! Both marine fisheries and aquaculture have also been given a higher priority in budget allocations. The forestry industry is still in infancy stage by virtue of low domestic market demand and restrictions on forest product exports.

Production of oil began with the onshore Seria field in 1932, and by 1990, almost 90 percent of the country's oil and all of it's natural gas originated from seven offshore oil-producing fields. Brunei produced 150,000 barrels per day in 1990, down from 240,000 barrels per day in 1980 because of the oil conservation policy to limit output. At present production rates, reserves of both oil and natural gas are expected to last approximately 40 years. Four companies account for much of the oil and natural gas production and distribution; Royal Dutch Shell is dominant, and the Brunei government holds a 50 percent equity share in each of the companies. In 1990, the oil and natural gas sector accounted for 53.7 percent of GDP, down from the 84 percent share in 1980.

Although the obvious strategy toward economic diversification is value-added processing of oil and gas, little has yet to emerge. Downstream processing of oil takes place at facilities in Miri and Bintulu in Sarawak, and although the gas liquefaction plant at Lumut is important, the capital intensive nature of the investment limits substantial employment generation. Value-added production seems economically inconsequential when overseas investments derived from windfall oil profits now exceed oil revenues. The path to greater economic diversity then is private sector ISI, but despite numerous government financial incentives, a fledgling manufacturing sector has not emerged as a result of numerous obstacles. First, revenues from oil and gas exports appreciate the Bruneian currency causing high exchange rates which in turn makes imported goods relatively less expensive. Second, the emergence of a productive ISI is hindered by a small national population base. Most important, however, is that potential private sector manufacturing is unable to match the high wage rates of government employment, which include generous house loan and pension programs. It is not surprising then that the percentage of the labor force employed by the government sector between 1971 and 1986 increased from 39 percent to a staggering 51 percent, while in the non-oil private sector, employment declined from 54 to 45 percent during the same period. In the absence of a manufacturing sector and producer class, Brunei's economy has progressed from a primary to a government-based tertiary sector without ever developing an intermediary secondary sector.

TRADE, TRANSPORT, AND COMMUNICATIONS

Because Malaysia's economic growth rests on foreign investment and dependence on the global economy, trade becomes a critical ingredient in reaching NIE status. Foreign trade not only includes manufactured exports but also imports of machinery and intermediate inputs on which the manufacturing base is dependent. As an integral part of the production process, transport and communications are critical components of trade and must continuously be improved if Malaysia is to reach NIE status.

Foreign Trade

Illustrative of Malaysia's robust economic growth is the sixfold increase in the value of exports during the 1980–1995 period (see Figure 13.2). The dramatic increase in the value of exports that accompanied the diversification of trade products from primary to secondary sector exports is characteristic of an emerging industrial economy; the value of manufactured exports account for 80 percent of total export value in 1995. Accounting for approximately 66 percent of manufacturing export value was electrical and electronics products, much of which originate from foreign-owned facilities.

Much of the decrease in the value of primary sector exports during this decade is the result of the increased value share of manufactures, but it is also attributable to declines in commodity prices rather than reductions in export production volume. Such is the case of palm oil, but not the traditional exports of rubber and tin because commodity price declines have prompted reduced production. The significant decline in saw-log exports is the result of the depletion of timber resources and the government's desire for greater downstream processing, particularly in the furniture and wood product industries, which are projected to be leading export subsectors. The value of crude petroleum exports dropped significantly because of depressed global oil prices, but the volume did as well because of the rapidly developing domestic chemical industry. During the 1985–1990 period, exports of LNG from the Bintulu facility in Sarawak expanded rapidly, but they diminished by 1995 with increased domestic demand for energy.

As is common among rapidly industrializing countries, Malaysia's volume of imports is substantial. This high dependence on imports is exacerbated in economies such as Malaysia's that are relatively open and increasingly dedicated to free market trade. Although Malaysia possessed a healthy 1980 trade balance favoring exports, by 1995 the value of imports surpassed that of exports. Manufactures accounted for 73 percent of import value. In the absence of significant production linkages with SMIs, intermediate goods in the form of parts comprised a hefty share of imports in Malaysia's "assembly" economy; in 1995, intermediate goods comprised 44 percent of import value. Another variety of manufactured imports for which there exists no statistics to measure relative importance is consumer manufactures. Industrialization has created substantially higher personal in-

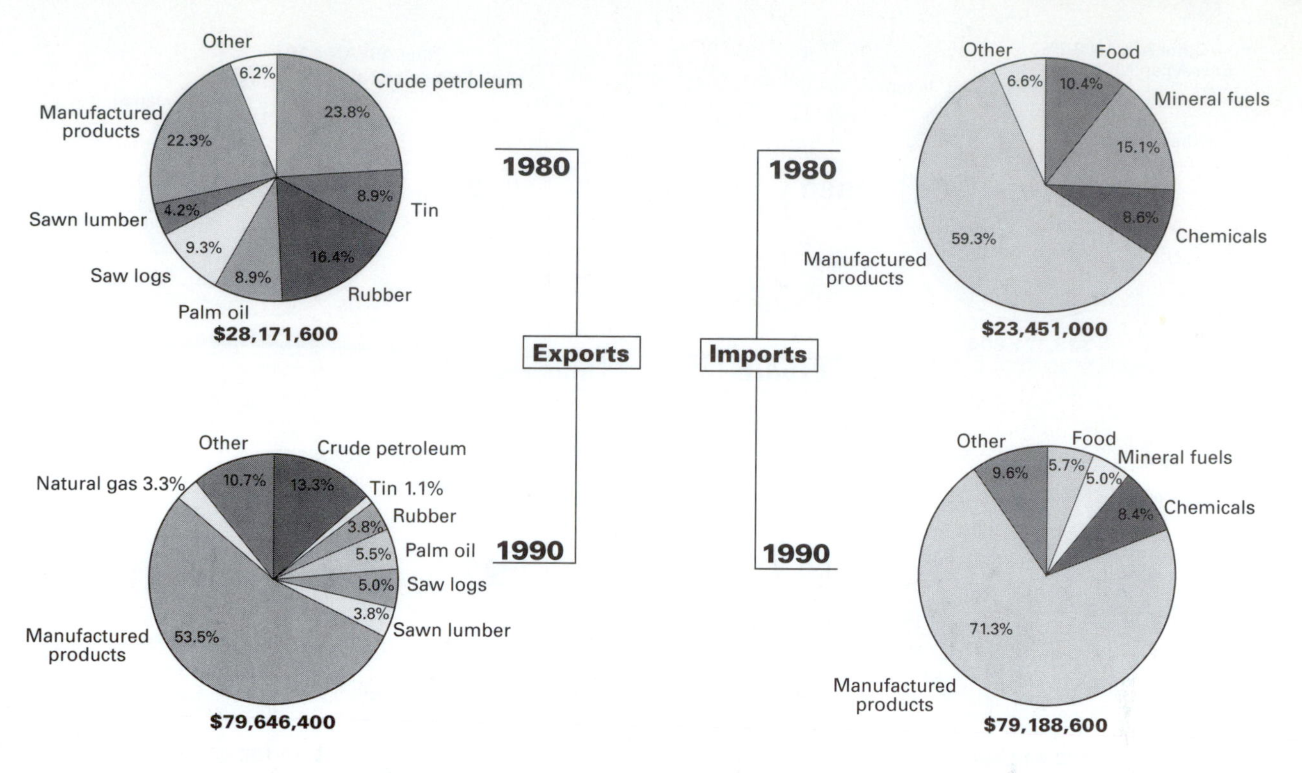

Figure 13.2 Value of exports and imports, 1980-1990.

comes, levels of urbanization, Westernizing lifestyles, and consumer habits. One only needs to witness the phenomenal growth of shopping malls in even the smaller cities to appreciate the diffusion of popular Western consumption habits in Malaysia that can only be satisfied by imports.

Despite the dramatic increase in Malaysia's trade volume between 1980 and 1995, the number of primary trading partners remained relatively narrow (see Figure 13.3). During the 1980–1995 period the direction of exports remained relatively stable with some few exceptions. The export value with ASEAN and East Asian NIEs more than doubled, while the export value to Japan dropped significantly. The explanation is that Japanese offshore manufacturing facilities now produce for the regional and U.S. markets rather than simply re-exporting to Japan. The increase in the value of Japanese imports during the 1980–1995 period testifies to this development. Although it is difficult to generalize trade structure, Malaysia imports a variety of manufactures from these partners and exports a narrow range of manufactures and primary commodities. For example, by import value in 1990 (the most recent trade figures available) Japan, Singapore, and the United States accounted for 47 percent of chemicals, 71 percent of machinery and transport equipment, and 41 percent of manufactured goods; in export value, these three countries accounted for a high share of Malaysia's exports—67 percent of semiconductors and 53 percent of crude petroleum. Japan alone accounts for virtually all LNG exports, and before policies promoting domestic value-added processing in Malaysia, (this resource hungry nation consumed half of all sawn log exports during the late 1980s).

The narrow range of Malaysia's trading partners is slowly expanding, however, particularly toward the East Asian NIEs of South Korea, Taiwan, and Hong Kong. With the second wave of East Asian foreign investment during the 1980s, the slice of intra-Asian trade has dramatically increased. In 1990, for example, 19 percent of imported manufactured goods and 8 percent of machinery and transport equipment originated from the East Asian NIEs. South Korea accounted for an increased share of Malaysia's commodity exports; 13 percent of rubber, 14 percent of crude petroleum, and 9 percent of sawn timber. Intra-ASEAN trade also provides opportunities for weaning Malaysia from its dependence on a handful of traditional export markets. It is Malaysia's comparative advantage in manufacturing exports that has led the government to promote the Asean Free Trade Area (AFTA) aggressively. Finally, Malaysia has also developed increasingly important bilateral trade ties with other developing countries. Broadened trade ties is a priority as government sponsored export promotion activities increased thirteenfold from the 1986–1990 to 1991–1995 planning periods.

Despite the rapid transformation of Malaysia's economy during the past decade, the current account of the balance of payments is of great concern. During the 1980s, the current account deficits rapidly increased from $620 million or 1 percent of GNP to $5,245 million or

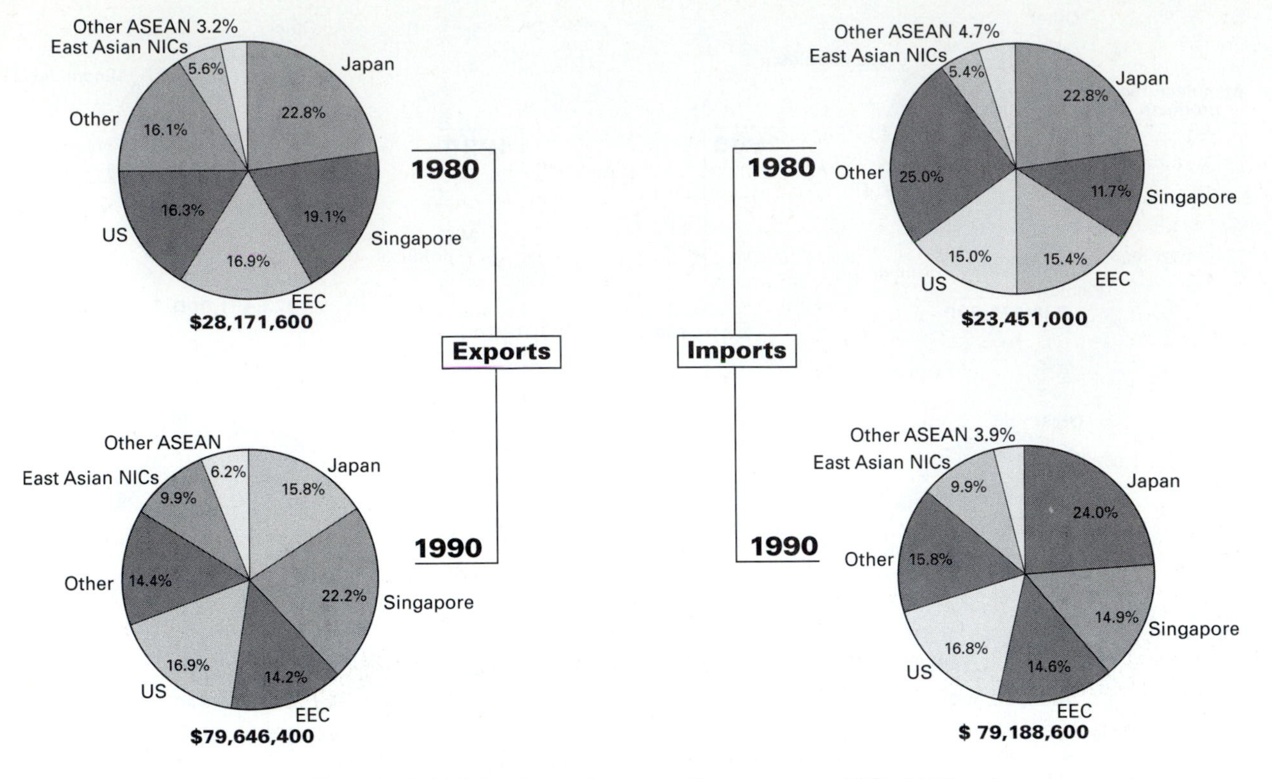

Figure 13.3 Malaysia's primary trading partners, 1980-1990.

5 percent of GNP. One reason, common among NIEs with open markets, is the higher growth of import merchandise over export merchandise. This situation highlights the importance of promoting domestically produced intermediate goods. Many of the current account-deficit problems, however, are due to a negative service account because of charges for freight transport and insurance as well as repatriation of investment. The deficit created by these two services alone accounted for 9.3 percent of GNP in 1995, up from 8.8 in 1990.

Transport

With the exception of Singapore, no other Southeast Asian state supports a more developed transport and communications network then Malaysia (see Map 13.7). The present road, rail, port, and other communication systems are an integral component of the entire developmental process and are being constantly improved so that they do not function as a barrier to economic growth. The government is well aware of the importance of transport in the development process because with 20 percent of the federal government development budget allocation, transport is second only to agriculture and rural development under the 1986–1990 and 1991–1995 national plans.

Roads, which accounted for three-quarters of the total allocation for transport and communications during the 1991–1995 planning period, are the most important transport subsector. Total length for both federal and state roads in 1995 was 39,883 mi (64,328 km), an increase of approximately 19 percent over 1990. Unlike most other development indicators, the spatial distribution and quality of roads in Peninsular Malaysia is rather uniform. The poorer states support an equal length of road infrastructure for their territorial size when compared to the more-developed west coast states, and equity in road infrastructure is matched by the quality of roads as well; with approximately 88 percent of peninsular roads paved in 1995, no state is characterized by less than 75 percent of roads paved. Such is not the case for frontier East Malaysia; with 60 percent of the Malaysian land area, Sarawak and Sabah accounted for only 21 percent of road length in 1990, and the rate of road construction during the 1985–1995 period was only one-half that on the peninsula. In addition, only 46 percent of roads are paved when compared to the 1995 national average of 75 percent.

During the past 25 years, three general areas of road improvements have been targeted. First is a system of rural roads associated with regional development and FELDA schemes. These are comprised of feeder roads linking settlements to the main road network and farm to market roads that are part of the total rural development package to improve agricultural productivity. Second is the construction or improvement of interurban highways to facilitate more efficient transport between centers of industrial concentration. The most important is the almost

BOX 13.3: Tourism Development in Malaysia

When compared to other Southeast Asian states, Malaysia is not commonly thought of as a popular international tourist destination. Perhaps the most basic explanation is that unlike Thailand and Indonesia, for example, the opportunities for sun and surf and cultural heritage tourism are not that substantial; much of the peninsular west coast and the coast of East Malaysia are clothed in mangrove, and Malaysia does not possess any precolonial architectural artifacts to which tourists might be attracted. Nor does Malaysia possess the shopping opportunities for which Singapore is so famous, although Kuala Lumpur is gradually developing a foreign tourist shopping infrastructure. This does not mean, however, that tourism development is not a government priority; government budgetary allocation during the 1991–1996 planning period quadrupled that of the 1985–1991 planning period.

It is the west coast states that have traditionally received the lions share of foreign tourists. The sun and surf destinations of Langkawi Island and Penang Island in the northwest are the country's most developed. The colonial cultural heritage landscapes of Kuala Lumpur, Penang, and Malacca as well as the former hill stations of Cameron Highlands and Fraser's Hill are the most important tourist destinations. Indeed, the western core states benefit from their deeper colonial past through the marketing of colonial landscapes and multiethnic festivals to the heritage tourists. The more isolated and Malay World east coast states where colonialism was not as pervasive remains an international tourist backwater. The bulk of east coast sun-and-surf destinations are found along the beaches of the southeast coast and offshore islands of Johor state, as well as north of Kuantan in Pahang state. These destinations, however, are few, and the economic multiplier effect is minimal because of their enclave nature. The potential heritage tourism of the east coast that comprises padi fields, fishing villages, and traditional handicraft cottage industries are slowly disappearing as macroeconomic policies of the national government draw Malays into more financially remunerative occupations. Spatial access, a critical factor in tourism development, is an obvious problem in these more isolated east coast states.

Malaysia's foreign-tourist source countries are quite narrow in number. In 1995, some 76 percent of foreign tourists originated from the ASEAN region, with Singapore accounting for some 60 percent of this total. Much like the international border crossing at Tijuana, Mexico, the vast majority of Singaporeans engaged in day trips for shopping in Johor Baharu or overnight package sun-and-surf trips to the beach resorts in Johor state. The throng of returning Singaporeans, waiting for hours to clear Malaysian customs in Johor Baharu on Sunday evenings, is testimony to this short-visit phenomenon. The rapid development of sun-and-surf destinations on the Indonesian islands of Batam and Bintan just south of Singapore threatens Malaysia's monopoly on this slice of the tourist trade. Domestic tourism has yet to develop fully because Malays continue to vacation in their home regions to visit relatives and friends rather than travel for the sake of leisure. Predicting that rising affluence among Malays will modernize their travel habits, the government is aggressively promoting greater domestic tourism activity.

Photo 13.7 Real estate development and other building projects have been slowed down by the Asian financial crisis. Here is pictured oceanfront development on the west coast of Peninsular Malaysia. (Ulack)

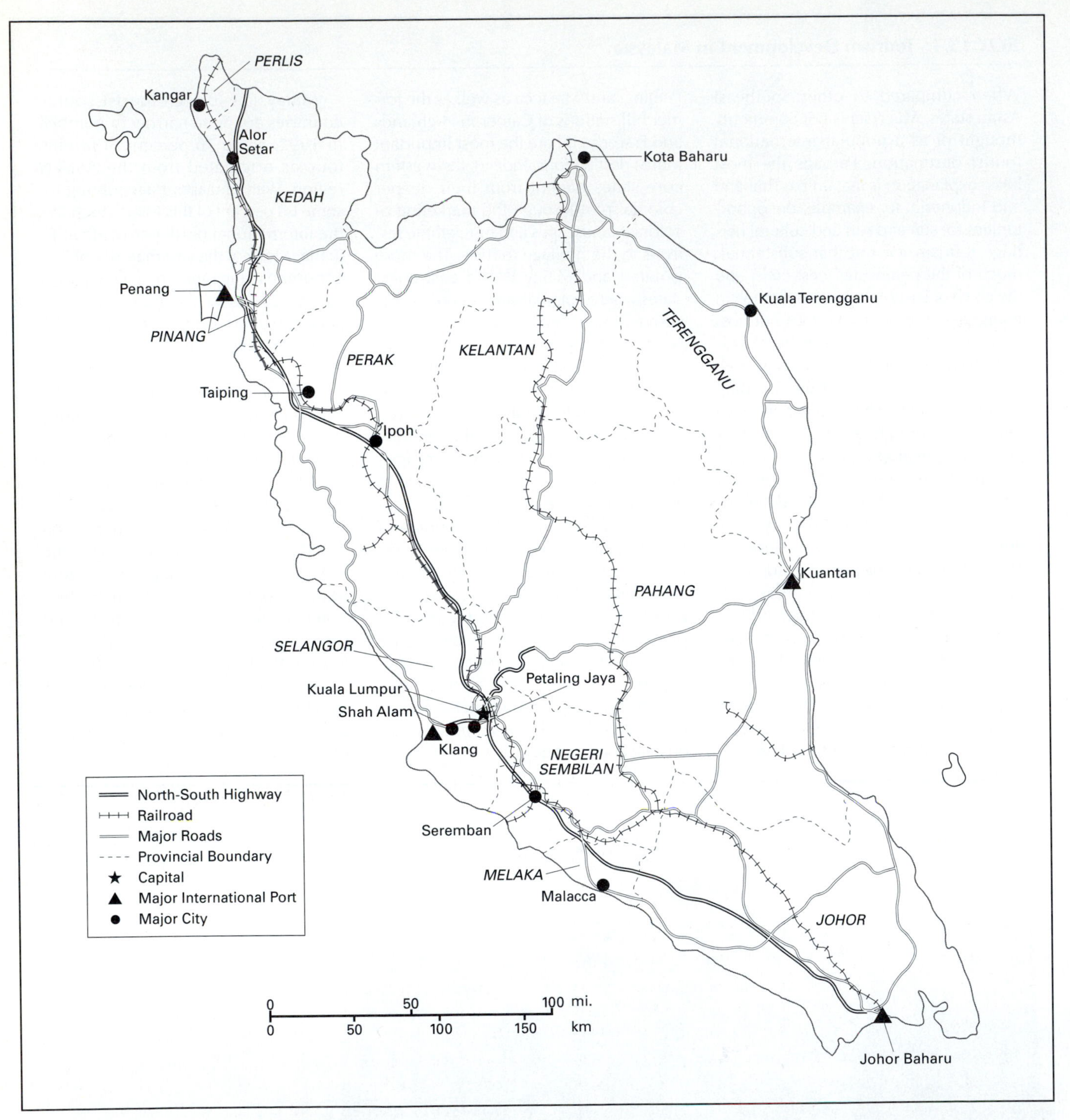

Map 13.7 Peninsular Malaysia's transportation network, c. 1990.

completed North–South Toll Expressway linking urban centers within the west coast states. The East–West Highway completed in 1983 is also crucial for it provides Kelantan access to west coast ports and stimulates resource-based industry in the northeast quadrant of the peninsula. East Malaysia has also been the target of road investment, as trunk highway construction to connect the primary coastal urban centers is now complete. In Sabah, it has only been since the 1960s that construction of an all-weather road to link the east and west coasts began.

The third focus of road investment is in the larger urban areas where congestion associated with industrial growth, population increase, and residential sprawl has become unmanageable. This is particularly true for the

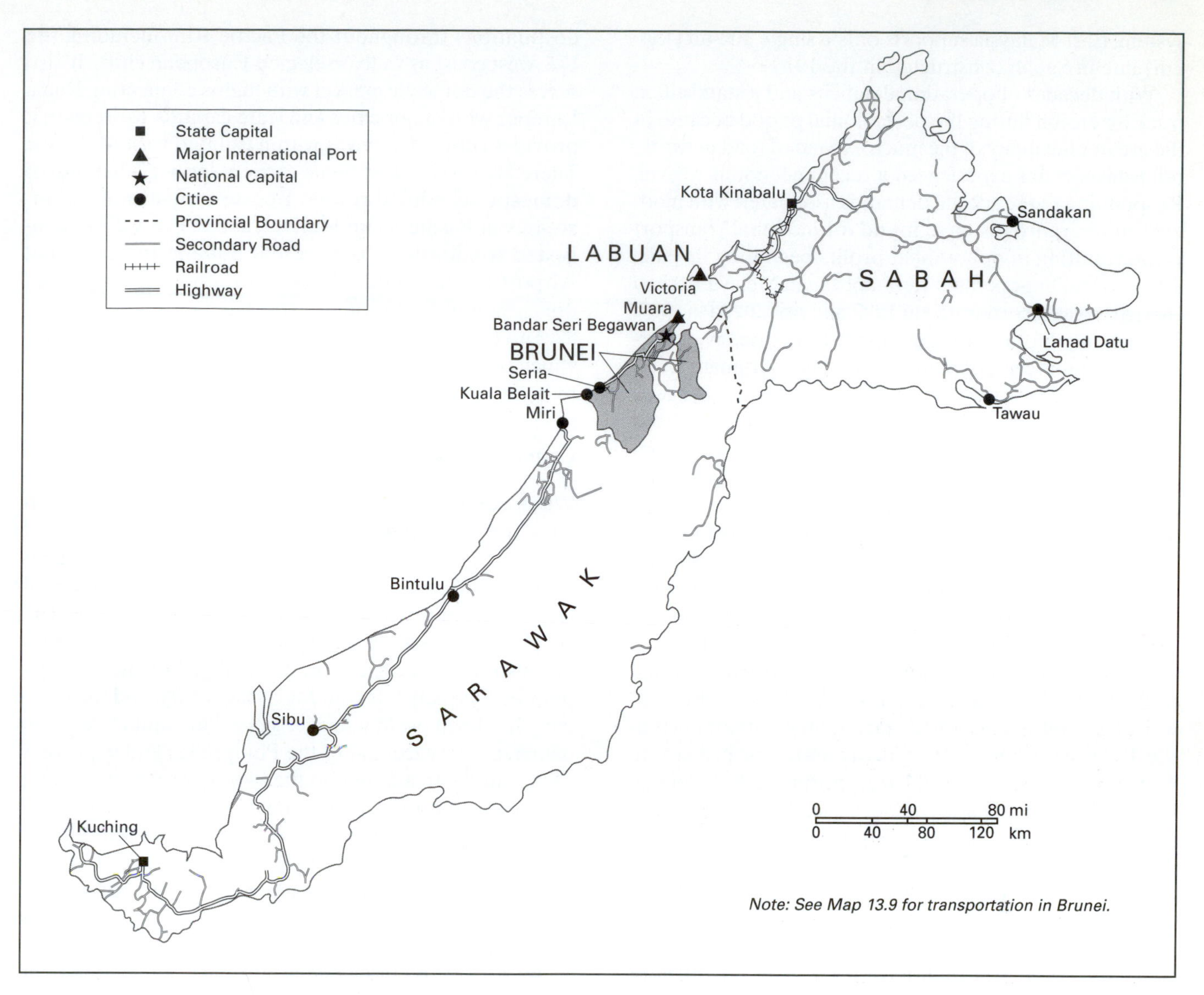

Map 13.8 East Malaysia's transportation network, c. 1990.

KVC where car sales in 1991 accounted for 43 percent of the national total. With rising affluence and status linked to consumption, it is projected that three of four households in the KVC will own cars by 2005. In response, a number of intraurban, limited access thoroughfares such as the New Klang Valley Expressway between Kuala Lumpur and Klang and the Shah Alam Expressway have been completed or are in the planning stages. Such improvements, however, will not significantly reduce road congestion because of the highly fragmented and inefficient system of public transport comprised of taxis and minibuses. Although certainly not as congested as Bangkok, Manila, or Jakarta, the long-term solution to congestion is not road- but rail-based transport. In response, the government has implemented the nation's first metropolitan rail commuter service in the form of a light rapid transit system or LRT (light rail system), whose many lines will eventually reach to the edge of the KVC and involve intermodal transport based upon a bus feeder system centered on LRT stations.

Although viewed as a relatively good system by Southeast Asian standards, rail transport in Malaysia has never played a dominant role in economic development. With a total of 1,017 mi (1,640 km) of track, the majority of freight and passengers moves along on the west coast line stretching from Johor to Perlis. An east coast line terminating in Kelantan connects the two coasts but only indirectly. A more-direct east coast rail line between Kuantan and the west coast was being considered during the late 1980s but has been shelved because of the huge financial investment involved. Unlike other Southeast Asian rail systems, Malaysia rail service interfaces with both Singapore's mass transit system as well as Thailand's state-run railroad. Despite only a rudimentary road system, East Malaysia supports only a single 102-mi (154-km) line in Sabah constructed in the 1910s.

With decades of operational deficits and a standstill in track extension during the postcolonial period because of the greater flexibility of the much expanded road network, rail transport has experienced a recent economic revival. Responding to the special demands associated with modernization, railroads have found a functional transport niche resulting in operational profits beginning in 1988. The government-operated Keratapi Tanah Melayu Berhad (KTMB) was incorporated in 1992 and privatized in 1997. Since its incorporation, the company has aggressively pursued a commercial orientation resulting in dramatic quality improvements as well as a tripling of revenues between 1992 and 1995. Passenger traffic for example, increased slightly between 1990 and 1995, and KTM's share of container traffic increased 27 percent during the same period. With the establishment of inland container depots and double-tracking, as well as investment in a Kuala Lumpur-centered commuter train system which includes double-tracking electric trains along the rapidly expanding urban corridor to Seremban to the south, KTM provides a model for other rail systems in Southeast Asia to emulate.

With so much national wealth derived from foreign trade, ports assume a critical link in Malaysia's drive toward NIE status. This is why the country's major ports in the 1990s have been either incorporated or privatized. The three busiest general cargo ports are Port Klang, Penang and Johor (Pasir Gudang) each serving the major industrial regions of the peninsula. As part of Malaysia's primate economic region, Port Klang accounted for 55 percent of cargo tonnage at these three ports in 1995. The port of Penang has witnessed steady growth as more offshore plants are established. Pasir Gudang has experienced dramatic growth based not only on offshore manufacturing facilities but also as an exit point for agricultural exports from the southern one-third of the peninsula. Commissioned in 1977, Pasir Gudang plays an important role in slowing the traditional transshipment function of Singapore; although much greater in past decades, by 1995 approximately one-fifth of Malaysia's foreign trade continued to pass through Singapore. The remaining ports are, for the most part, highly specialized. Although envisioned to function as an international port for the east coast, Kuantan port is underutilized because most agricultural products exit west coast ports. Port Dickson in Negeri Sembilan as well as Miri and Bintulu in Sarawak serve the petrochemical industry almost exclusively. In the absence of efficient long-distance land transport, the principal Sabah ports of Kota Kinabalu, Sandakan, and Tawau function as important gateways for primary product exports and manufactured imports.

Once a government-operated service, but now a privatized national airline, Malaysia Airlines serves major destinations throughout the Pacific Rim including the U.S. west coast as well as selected European cities. It also serves the domestic market with flights connecting Kuala Lumpur with major cities and state capitals. Moreover, it provides rural air services within isolated East Malaysia. There also exists two private carriers serving a handful of domestic scheduled routes. Because of substantial increases in business and tourist travelers, as well as congested conditions at the present Subang International Airport located within the Kuala Lumpur–Klang corridor, a new international airport is being constructed 37.2 mi (60 km) to the south at Sepang. The new airport, which opened in 1998, as well as existing ones are slated to be incorporated to increase operational efficiency.

Communications

When compared to most other Southeast Asian countries, the early 1990s communication infrastructure of Malaysia is well developed. As relatively affluent consumers, Malaysians possess four times the numbers of telephones per capita when compared to Filipinos. The geographic spread of telephone service into rural areas has also been dramatic; the federal government incorporated Syarikat Telekom Malaysia in 1986 and the company has been profitable ever since. Per capita television ownership is twice that of the Philippines and triple that of Indonesia. In addition to the two government stations, Malaysia supports a commercial private television station; it is through the medium of television that almost one-half of total advertising dollars were spent to beam messages to a growing consumer society. Foreign products are in high demand; eight of the top ten advertisers in the early-1990s were foreign companies.

As part of the basic needs approach to development policies, Malaysia's rural electricity infrastructure is well developed; approximately 80 percent of peninsular rural households were electrified. In East Malaysia, however, partly because of the isolated nature of rural settlement, only 50 percent of households were covered. The demand for energy by a rapidly expanding industrial sector and increasingly affluent consumer society is, however, putting a severe strain on the power grid. Despite the aggressive diversification of well-endowed domestic energy supplies whereby gas and hydropower are gradually replacing oil, energy demands have fallen short of supply. Power distribution is a problem as well; in September 1992, for example, 9 of the 11 peninsular states experienced a complete blackout for most of a single day because of distribution problems. Because manufacturing and transport accounted for 70 percent of total final energy consumed in 1990, shortcomings of the present distribution system undermine the ability of Malaysia to attract future foreign investment in manufacturing. A number of hydroelectric projects, including the huge Bakun Dam project in Sarawak that will flood an area the size of Singapore and transmit power to the peninsula via submarine cable, is expected to help alleviate power shortage problems.

Brunei

Despite its great economic wealth, Brunei is the exception among ASEAN states in that the value of external trade is overwhelmingly derived from the primary sector (Neville, 1985). In 1990 oil and gas exports comprised approximately 96 percent of the total value of exports (54 oil, 37 LNG, and 5 petroleum products) and approximately 67 percent of the total value of external trade. The volatility of overdependence on oil and gas exports is great because revenues decreased from $11,084 billion to $6,155 billion from 1980–1990 as a result of depressed global oil prices and voluntary conservation measures imposed by the government. As is common among rentier economies where the people are consumers rather than producers, imports during the 1980s continued to rise; the value of imports to the total value of external trade increased from 1 to 3 percent during the 1980–1990 period. In 1990, the imports of greatest value were manufactured goods (33 percent), machinery and transport equipment (29 percent), and food and beverage (25 percent).

The profile of Brunei's primary trading partners resembles that of Malaysia's. In 1989, Japan, Korea, and Singapore accounted for 58 percent of oil exports, and Japan takes 100 percent of LNG exports; much of that is based on long-term contracts to supply electric utility firms in Tokyo and Osaka. Despite negative annual percentage growth rates of exports at −9 percent and imports at −7 percent, as well as an annual balance of trade growth rate of −14.5 percent during 1986–1990, Brunei has not experienced severe financial problems; overseas investments and surplus oil and natural gas revenues allow the government to invest in blue chip shares of stock, primarily in Japan and the United States, yielding approximately $2 billion annually.

Brunei's transport infrastructure reflects the importance of the oil and natural gas industry (see Map 13.9). The only major highway is one that straddles the developed coastal region between the capital of Bandar Seri Begawan in the east and the oil center of Kuala Belait in the west. A handful of short-distance feeder roads are linked to the coastal highway, but the southern one-half is void of motorable highways and is only accessible by river or helicopter. Almost one-half of all motorable roads are concentrated in the capital district of Muara-Brunei. With one privately owned auto for every three persons and urban public transportation poorly developed, Bandar Seri Begawan experiences serious road congestion. Brunei supports three general cargo ports, the largest of which is Bandar Seri Begawan's outport of Muara, located 7 mi (11 km) northeast of the capital city. The port of Kuala Belait serves both small general cargo vessels primarily from Singapore, and most important, offshore oil field vessels. Crude oil is loaded onto tankers from a 6-mi- (9.7-km-) long offshore pipeline at Seria, and LNG is loaded in a similar fashion from a 3-mi- (4.8-km-) long jetty at Lumut. The only international airport is at Bandar Seri Begawan. The government-owned Royal Brunei Airlines commenced operations in 1975 and now supports service to all ASEAN capitals and selected European cities. But because of Brunei's small areal size, regularly scheduled domestic service is not economically justified. Unlike the oil industry, the number of foreign managers and technicians have slowly been reduced through "Bruneization."

Brunei's communications infrastructure is well developed. With 17 telephones per 100 persons, Brunei supports twice the number of phones per capita when compared to Malaysia. The government operates a single television station and two radio stations, one transmitting in English and one in Chinese. The government also publishes several daily and weekly newspapers and magazines in both Malay and English. The government is also able to provide basic public utility needs to all. Generating most of the country's electricity needs from the Lumut LNG plant, approximately 90 percent of the population was served by electricity.

POLITICAL ASPECTS OF DEVELOPMENT

Because the political dimension of Malaysia's post-NEP economic growth has been so pervasive, it is both necessary and realistic to examine the process of economic transformation from a political perspective. Perhaps the single most identifiable characteristic of Malaysia's political system is an overwhelming degree of postindependence stability to ensure continued foreign investment. It is this overdependence on foreign investment that has greatly influenced the persistence of ethnic or communally based political parties and suppressed the emergence of class-based political parties that normally accompany the kind of social and economic change experienced under the NEP. The emergence of multiethnic parties each united by the economic and political needs of the middle and lower class would threaten the stable political conditions to which foreign investment is attracted. Indeed, the bloody ethnic riots of 1969 between Malays and Chinese possessed an obvious racial dimension because two non-Malay political parties unexpectedly gained significant electoral power, but we must also point out that intraethnic Malay income differences were widening because of government economic policies during the 1960s.

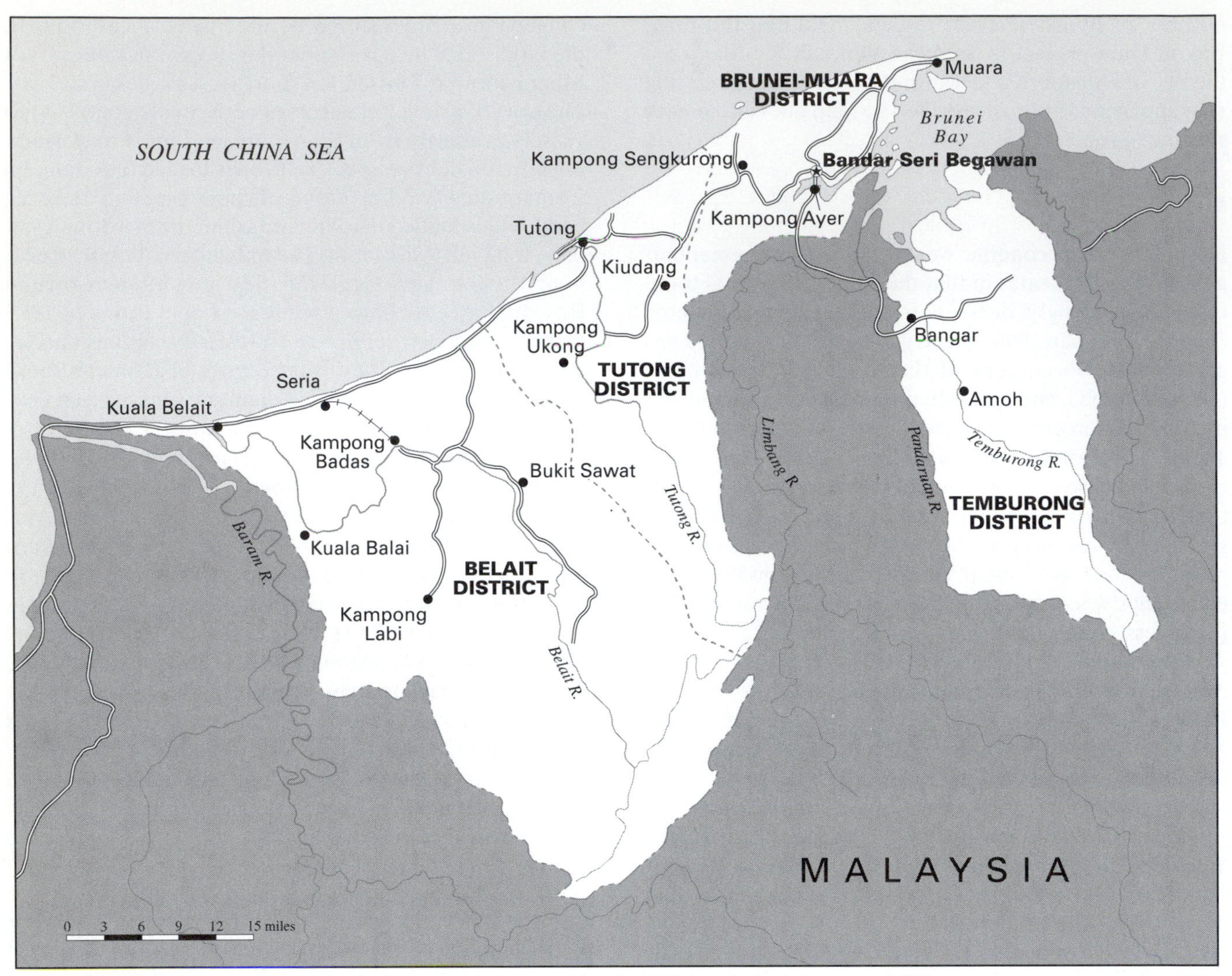

Map 13.9 Brunei. *Source: Wright, 1991; Ulack and Pauer, 1989.*

Class-Communal Politics

The ruling class is composed of numerous groups and is multiracial (Brennan, 1985). Political power was inherited by the former Malay bureaucrats in the preindependence colonial government. Once drawn exclusively from the aristocracy, today's second generation bureaucrats are more representative of Malay society. The balance of the ruling class is comprised of rural Malay landlords and the Chinese business class with interests in finance, industry, plantations, and mines. Although excessively wealthy from land resource earnings, the privileged state aristocracy, particularly the sultans, function as protectors of bumiputra cultural and religious traditions but possess no great political and economic power at the national level. This elite class, particularly higher level bureaucrats and Chinese corporate interests, require political stability because both are heavily reliant on foreign investment to maintain their respective political and economic power.

The balance of the class–race structure is comprised of the middle class and the "dominated" classes. The Malay middle class has expanded dramatically as rural to urban migration in response to educational opportunities and government employment that has secured political allegiance for the ruling bureaucratic elite. By 1990, the white-collar middle class increased to 33 percent of the workforce, and almost one-half of the Malay workforce was middle class. Although the Chinese and Indian middle class have expanded, their proportion of the total middle class has declined since 1970. Below the middle class are the dominated classes of the Malay peasantry, the rural mining and plantation workers comprised primarily of Chinese and Indians, and the multiracial urban working class which possesses little political power.

Although individual economic classes are multiracial, the structure of the political organizations remains ethnically based. Since independence, the political system has been controlled by a coalition of ethnically based parties within the "National Front," or *Barisan Nasional* (BN). The BN is primarily composed of the United Malay Nationalist Organization (UMNO), the Malaysian Chinese Association (MCA), and the Malaysian Indian Congress (MIC). UMNO possesses paramount decision-making power within the coalition and utilizes a number of political mechanisms to reduce the electoral strength of the MCA and MIC partners, as well as opposition parties. The most effective mechanism in diluting the non-Malay vote is through blatant gerrymandering; in the mid-1980s Malays barely comprised the majority of the national population but were the ethnic majority in 74 percent of federal election districts. Not only are rural Malay districts heavily weighted, but political patronage is doled out to landlords who deliver the peasant vote to ensure UMNO's continued strength in the coalition. Government employment and contracts to the fledgling Malay business community accomplish similar ends. UMNO's political power has been strengthened through aggressive privatization efforts as well (Milne, 1986). Supporters of UMNO purchase public enterprises at bargain prices and UMNO itself has extended its corporate investments through the privatization process; the country's only private television station, the North–South Toll Expressway and the second bridge span between Johor Baharu and Singapore are just a few of the ventures of UMNO-connected companies. Political patronage at both the village and corporate levels to further the ends of UMNO can thus be described as *politik wang*, or "money politics." Lastly, UMNO strength is guaranteed through the sedition laws of the loosely interpreted Internal Security Act that allow the government to imprison individuals without trial who are perceived as being a threat to the political status quo. This makes it difficult for both non-UMNO and multiethnic opposition parties to attract Malay voters.

Politics and Development in East Malaysia

The ability of Kuala Lumpur to link politics and development in East Malaysia is equally as strong. However, the form of the influence changes because of the numerical superiority of non-Malay ethnic groups. First, Kuala Lumpur has gradually withdrawn the special state's rights used to entice Sarawak and Sabah into the federation in 1963. Second, the ability to self-manage state economic affairs has been eroded through the implementation of national development policies that are often not suited to East Malaysian economic circumstances. Third, political patronage plays a critical role in solidifying Kuala Lumpur's control over the national periphery because, lacking state aristocracies, it is the government that determines the rights to land development and allocation of revenues derived from substantial resource exploitation crucial to export growth.

In Sarawak, UMNO has held power since the 1970s because of the early unification of various Muslim–Malay political parties, with the assistance of UMNO, and the continued political fragmentation of indigenous non-Muslim groups, particularly the Dayaks and Chinese. UMNO accomplished political control through patronage and thus has determined the present direction of economic development, as timber- and oil- and gas-derived wealth is held in the hands of the government bureaucracy and the growing urban middle class to the detriment of the peasant agricultural economy. In a penetrating interpretation of development landscapes in the capital of Kuching, G. Dixon (1991, p. 155–156) concludes that "opulent government buildings can be interpreted either way, but there can be no doubt the government is its own top priority." In Sabah, political stability has been far more unsettled (Kahin, 1992). While greater state control of economic revenues and policies is an important factor in perennially strained state–federal relations, cultural identity and preservation issues assume importance in the political agenda of the powerful Sabah Unity Party (PBS) comprised of Christian Kadazan and ethnic Chinese. Two of the many important cultural issues include the halting of the Kuala Lumpur-assisted flow of illegal Muslim Filipino in-migrants who are given the right to vote in state and federal elections and the greater "nativization" of employment in federal departments and agencies. From the Sabahan perspective, the strategy of UMNO to fragment the multiethnic PBS and create communally based political parties as found on the peninsula is part of a grand plan for greater Malay–Muslim dominance in East Malaysia, particularly in Sabah.

Islam and Development

The role of Islam in Malaysia's political community allows us to examine the role of religion and politics in economic development. Central to this relationship is the political party of Partai Islam Se-Malaysia (PAS), which possesses only a handful of members in the national parliament and has only intermittently controlled the Kelantan state government. It is however, UMNO's primary opposition party and has systematically pressured the government to adopt laws promoting a more Islamic economy and state. Increased Islamic programming on the government TV station and enforcement of stricter Islamic food taboos, particularly among the young and middle class, are two important examples. Much of its political activity has been peaceful, although PAS has been indirectly linked to a number of protests that have turned violent.

Perhaps the greatest threat of PAS to the legitimacy of Malaysia's UMNO-led government is to question the basic nature of capitalist economic development during the NEP period. Because Islam in Malaysia is associated with issues of economic justice, Islam has become a "channel of protest" against the uneven geographic development process inherent in capitalism. Although the legislative power of PAS is very limited, its effective power is substantial because only Islam "enjoys sufficient credibility in Malay society to allow it to perform this [political] function" (Muzaffar, 1986, p. 68). In Kelantan and Terengganu, for example, where PAS is strong, the effect of new technology and investment and the emergence of a new economic elite has been to increase prices for land, goods, and services. Large businesses connected to west coast UMNO elites have caused many small businesses to fail. PAS is seen as a champion of many Malays who remain poor as NEP policies have either passed them by or worsened their economic position. More importantly, PAS has attacked the culture of the West that has accompanied capitalism. The greater levels of urbanization and the parallel demise of social cohesion common to village life has resulted in the glorification of materialism, secularism, and obsession with the self. In a sense, "Islamic resurgence is perhaps a response to this culture that capitalism cultivates" (Muzaffar, 1986, p. 69). It is important to point out that as the government is pressured to adopt a more Islamic tone to its policies, the challenge of nation building in this ethnically fragmented country becomes greater.

Demise of Democracy?

The top-down form of economic development in Malaysia has engendered the domination of a single Malay-based political party within a multiracial coalition and the suppression of class-based politics that normally accompanies capitalist economic development (Crouch, 1993). Institutions such as labor unions, and an independent national media and judiciary critical to the formation of a class-based politics have also been suppressed. Although the dramatic growth of the Malay middle class accompanying capitalist economic development should in principle function as a check on the growth of an authoritarian government, most Malays view the present political system as necessary to protect Malay political dominance and to promote their economic ascendancy under the NEP. The present class-race configuration being determined by an export-based economic development strategy leads Mehmet (1986, preface) to conclude that current government trustees are "successors of the colonial elite." This class-based analysis of capitalist economic development in Malaysia is supported by the disturbing trend of increasing intraethnic wealth disparities during the period of the NEP.

Despite the apparent demise of a more democratic society paralleling NEP-induced economic growth, it is difficult to characterize the Malaysian government as authoritarian because politically severe measures such as arrests of political opponents under the Internal Security Act are not applied at a grand scale but are only selectively enforced when the power of the ruling elite is directly threatened. Nevertheless, at a time when democracy is advancing in many regions of the world, Malaysia's present record of civil rights is deteriorating. With great strides made in economic growth, there exits little reason for Malaysia's civil rights record during the 1970s and 1980s to be only marginally higher than the average for developing countries.

International Politics

Malaysia's increased participation in the global economy has been accompanied by a more aggressive role in international politics, particularly since Dr. Mohammad bin Mahathir became Prime Minister in 1981. In the broadest sense, Malaysia's present foreign policy priorities are ranked in order of importance as being ASEAN and regional issues first, Islamic countries second, and the nonaligned movement third. As a testament to ASEAN's success, Malaysia possesses no serious political problems with other ASEAN states. The conflicts of Philippine claims on Sabah during the 1960s and 1970s as well as the formal end of the Communist insurgency along Thailand's southern border in 1989 no longer causes friction between Malaysia and its two neighbors. In matters pertaining to regional security, Malaysia has taken an aggressive stance. The government for example took the initiative among ASEAN states to mediate the Cambodian civil war following the withdrawal of the United States from Vietnam in 1975. Despite the role of China in aiding Malaysia's domestic Communist insurgency during the 1950s and 1960s, Malaysia established diplomatic relations with the People's Republic of China in 1974, the first ASEAN state to do so. Malaysia is still wary, however, of China's present and future political intentions in Southeast Asia. This helps explain in part, Malaysia's occupation of three Spratly Islands, an archipelago in the southern South China Sea over which China and other countries claim complete sovereignty (see Map 10.3).

During the 1980s, Malaysia established much stronger ties with other Islamic states as the pan-Islamic movement moved to center stage of global international affairs (Hussin 1990). Malaysia's participation in this global political and cultural movement has earned it sufficient credentials for joining the Islamic family of nations. In a short period of time, Malaysia has become an activist member in the Organization of Islamic Conference and the Islamic Foreign Minister's Conference. The government has assisted in the establishment of an Islamic Bank, an International Islamic University, an Islamic Medical Center, in addition to sponsoring Islamic Conferences and international Koran reading contests. In backing the Palestinian homeland cause and consistently engaging in anti-Israeli political rhetoric, Malaysia appears to be more sympathetic to the Palestinian Liberation Organization than some Arab states. Malaysia has sent peacekeeping troops to Somalia and Bosnia and even sponsored the resettlement of Bosnian Muslims in Malaysia. Although Malaysia's increasingly activist role in pan-Islamic politics is obviously anchored by religion, economic and trade issues are important as well; foreign policy as "political entrepreneurship" has proved fruitful as Malaysia's trade with Islamic countries, particularly Iran, blossomed during the 1980s.

Malaysia's political association within the nonaligned movement has earned both the government and Prime Minister Mahathir the well-publicized role as maverick. During the 1980s Mahathir increasingly viewed himself as a forceful champion of developing country rights. Malaysia has established bilateral payment schemes to bypass what is perceived to be the exploitative financial institutions of the west. Mahathir has also been systematically critical of the International Labor Organization for questioning Malaysia's labor practices. Criticism of Malaysia's Internal Security Act by Western-based human rights organizations has also earned frequent tongue lashings from Mahathir. Perhaps the most publicized confrontation between Mahathir and Western critics has been on environmental issues; the prime minister has always defended rampant deforestation by questioning the past environmental practices of Western countries.

Brunei

The linkages between political and economic development in Brunei are far more clear-cut than in Malaysia in that the concentration of political and economic power rests with the absolutist monarchy headed by Sultan Hassanal Bolkiah as prime minister and minister of defense with full executive power, his son and his brother function as the minister of foreign affairs and minister of finance, respectively. Although the monarchy is assisted by constitutional councils and a legislature, both are powerless because the sultan has ruled by decree since 1962 when key provisions of the constitution were withdrawn, the electoral process was suspended, and a state of emergency was enacted following an unsuccessful rebellion. The most effective institutional influence on monarchy-centered government policy is the ministerial cabinet, divided into technocratic, pro-Western, and ideological Islamic camps.

Because rentier state monarchies often weigh political over economic considerations, the institutional barrier of power concentration by the monarchy partially explains the absence of a more diversified industrial base. Social and cultural modernization accompanying a diversified economic base and the greater influence of both domestic and foreign private investment would erode the absolute power of the royal family. Similar to Malaysia, but more so, pressure to democratize the political system is hindered by the fact that the consumer middle class is wholly dependent on government employment. In describing Brunei as a modern-day Malay *negara* state, Gunn (1993, p. 127) astutely observes that Brunei's membership in ASEAN has prompted the state to adopt "economic perestroika and the rhetoric of developmentalism [although] beyond the rhetoric lies the dead hand of tradition and history."

THE ENVIRONMENT AND DEVELOPMENT

The primary issue in regard to Malaysia's natural environment is not whether its present status is significantly better or worse than other Southeast Asian states but the rapid rate in which the natural environment has been negatively impacted as a result of NEP policies. Indifference to environmental quality throughout the 1970s and 1980s is reflected in national development plans that pay either scant attention to environmental issues or sacrifice the environment to NEP goals; the tone of the Fifth Malaysia Plan (1986–1990) was clear in that "environmental standards will be made consistent with the development goals of the country (Malaysia, Government of, 1986, p. 289). The realization of a more environmentally sustainable economic development policy, however, is possible on the horizon as the opening sentence of the 21-page environmental chapter of the Sixth Malaysia Plan (1991–1995) reads: "As efficient management of the environment and natural resources is an essential condition for ensuring balanced development, the government will develop better techniques for integrating environmental considerations in the formulation of programs and projects" (Malaysia, Government of, 1991, p. 389). Because such pronouncements are mere speculation, it would be wise to restrict this section on the environment to an examination of how economic development conflicts with the sustainability of Malaysia's rich and varied environmental endowment.

Forests

Perhaps the one aspect of the environment that has experienced the greatest transformation and international attention during the past two decades has been the forest cover. Although deforestation in the form of swidden-type plantation development for pepper, gambier, and cassava during the nineteenth century and rubber plantations during the early twentieth century resulted in significant loss of forest cover, planned land development schemes and logging over the last two decades have resulted in substantially greater environmental modification (Kumar, 1986; Brookfield et al., 1990). In Southeast Asia, Malaysia is third after Brunei and Indonesia in terms of the percentage of this forest cover. In 1991,

Malaysia maintained 56 percent of its forest cover. The spatial distribution of remaining forest cover is, however, uneven. Sarawak and Sabah support 69 and 60 percent, respectively; forest cover in Peninsular Malaysia is only 42 percent. Greater long-term agricultural development and urban sprawl on the peninsula has taken its toll because in the early 1950s, when systematic forest removal commenced, forest cover was approximately 74 percent.

The most significant source of peninsular forest conversions has been agricultural development schemes. During the early 1960s FELDA schemes were small and noncontiguous, but beginning in the early 1970s, increasingly large regional development schemes destroyed substantially large forested areas. By the early 1980s in fact, FELDA accounted for approximately two-thirds of annual forest conversions. Brookfield (1995, p. 269) refers to this quarter-century of environmental change as being similar to the "great ages of clearance" characteristic of Medieval Europe and early modern North America. Although FELDA is to be blamed for the rapid rate of forest removal, the focus on government-organized cash-crop production alleviated the common deforestation problems elsewhere in Southeast Asia caused by the uncontrolled pioneering of land-hungry peasants. With few or no new schemes being developed, it is projected that agricultural conversions will have a much reduced effect upon forest-cover loss. Nevertheless, if rubber and oil palm were classified as forest, Malaysia's tree cover would increase from 42 to 74 percent!

The second source of forest-cover loss, accounting for approximately one-quarter of existing deforestation, is logging. Once a log exporter, Peninsular Malaysia banned log exports of certain species in 1972 and a total ban was enacted in 1985. Not only was this required because annual harvests were exceeding by five times the resource recovery rate, but logs were needed to supply the rapidly expanding domestic wood industries. Another important source of deforestation includes hydroelectric dam construction, which by the early 1990s had consumed more than 247,000 acres (100,000 ha) of forest within the peninsular Main Range. Additional sources of forest loss include the reclamation of wetlands and coastal mangrove swamps for high-value aquaculture as well as urban growth; the spread of Kuala Lumpur for example will no doubt destroy the forests on the western face of the Main Range and the forests in the immediate interior of the mountain divide.

The geographic and ecological consequences of deforestation on the peninsula are wide ranging. Aside from scattered patches, undisturbed lowland dipterocarp is virtually absent, leaving remaining dipterocarp forests to clothe hill and mountainsides. Large, undisturbed tracts of forest are geographically restricted to the upland north but even this tract will be subdivided into two blocks as the settlement frontiers of southern Kelantan and northern Pahang meet. Of the many negative ecological consequences of forest clearance, soil erosion, river siltation, and accompanying changes in hydrologic systems possess the greatest impact. Chances for downstream urban flooding only aggravates the historic problem of river-mouth siltation, and forest clearance in upstream watersheds creates water shortages in downstream, water-dependent rice regions during periods of below-average precipitation.

In East Malaysia, logging accounts for the great majority of forest-cover loss. In Sarawak, logging concessions are given to private timber companies who then hire contractors, but in Sabah, the colonial legacy continues in that the industry is dominated by a handful of multinational corporations. Timber extraction in Sarawak has taken place at a massive scale; between 1963 and 1985, 30 percent of Sarawak's forests were logged. Much of the logged area is state land in the interior, over which indigenous people claim traditional rights. State intentions to convert much of this land to permanent forest estates have yet to materialize. It is this conflict between logging interests, the state, and indigenous peoples (particularly the once nomadic Penan and longhouse-dwelling Kenyah) that has provided much publicized scrutiny from both domestic and international environmental groups. The commodification of the environment through logging activities has badly damaged the natural landscape to which these people belong: Heavy machinery damages nontimber trees and other plants with subsistence and commercial value; once clear streams providing fish and drinking water are now polluted; logging debris makes river transport dangerous; and crop-destroying flash floods as a result of forest-cover loss are becoming more common (Brosius, 1992). Not only has logging transformed the natural environment in which indigenous peoples live but their economic circumstances as well. An increasing number of indigenous peoples now rely on wages earned as laborers during the initial extraction phase of logging (Brookfield et al., 1995, p. 132). Although government bans on the export of logs in the early 1990s and reforestation programs will no doubt reduce rates of deforestation, the growth of timber-processing facilities might in fact circumvent these environmental goals because of the substantial increase in value-added profits associated with an expanding market for finished wood products.

Other Environmental Problems

A handful of other environmental problems have emerged as by-products of Malaysia's accelerated development goals. Many of these problems are associated with water, particularly the pollution of rivers. In the late 1980s, approximately 48 percent of Malaysia's rivers were determined to be seriously polluted. Although tin mining activity has declined dramatically since the early 1980s, surface erosion from abandoned mines and tailings continue to contribute to river siltation. More recently, government leases have required operators to level off and resurface mining areas with grass, but enforcement has been lacking. In the Klang Valley, which was one of the more important tin mining regions, plans are to phase out tin mining activity altogether. Around Kuala Lumpur, a number of reclaimed mines have been transformed into city parks, market gardens, and squatter settlements.

With increased industrialization and urbanization, other sources of water pollution have negatively impacted river quality. Among industrial sources, the food and beverage industry (40 percent) and agro-processing industries (21 percent) account for the greatest number of industrial sources discharging into inland waters. Industrial sources of heavy metal pollution are low but are increasing as the chemical and manufacturing subsectors grow. In terms of the relative amount of organic pollutant load, however, domestic sewage accounted for approximately 76 percent of the total load in 1989. As expected, the geographic distribution of the number of stationary sources of industrial pollution were concentrated in west coast states; in 1989, Penang, Perak, Selangor, and Johor accounted for approximately 60 percent of the total.

An additional and far more visible source of environmental degradation is atmospheric pollution. Airborne pollution is common in rural areas because of quarries, wood, and agro-processing industries, but it is greatest in urban areas, particularly in the KVC, which supports high densities of industry and motor vehicle traffic. Although not as severe as Bangkok or Jakarta, the combination of landforms (the Main Range to the east) and human activity (the urban heat island and high rates of private automobile ownership) sometimes create intolerable atmospheric conditions. Severe smog and particularly haze during a two-week dry spell in August 1990, for example, forced the delay and even rerouting of shipping and air traffic and caused various forms of human discomfort. Because much of the atmospheric pollution load is directly caused by motor vehicles, the planned reduction of gasoline lead content will only partially alleviate the problem. The hyper-growth of the KVC has also engendered more-frequent water shortages because infrastructure has been unable to keep pace with the increased demand from industrial and residential users. In the summer of 1998, for example, residential water users were forced to ration water, and many were without water for days. Government plans call for a privatized water transfer project involving the construction of pipelines from neighboring Pahang state.

Economic Versus Environmental Goals

It is fair to conclude that Malaysia's dramatic economic growth rate has been accompanied by significant negative environmental impacts. A handful of examples illustrating the sacrifice of the environment to NEP goals make this relationship poignantly clear. A ministerial level post charged with environmental matters has yet to be created, and the enforcement power of the understaffed and underfunded Department of the Environment remains toothless; major development projects often commence without serious environmental impact assessments. It was not until 1980 that a National Parks Act was introduced, but by the late 1980s not a single national park had been created. However, by the early 1990s six national parks were established, but only Taman Negara in the north central peninsula is of significant areal size. Although active, environmental nongovernment organizations (NGO) are viewed with suspicion by the government and play a minimal role in environmental legislation. Similarly, the Malaysian Forest Service is the most professionally dedicated in Southeast Asia, but lacks legislative clout. Serious reduction in forest clearance has been limited by the single minded promotion of a cash-crop export sector and the creation of a politically loyal Malay peasant class. The goal of manufacturing the Malaysian Proton automobile and the privatization of toll roads benefiting the politically connected only discourages the emergence of a more environmentally sustainable public transport system. Finally, the 70-million-population policy to create a domestic economies-of-scale market and Malay voter base places even greater strain on the country's natural resources (Jomo, 1993).

It is also fair to describe the absence of a nationally integrated environmental policy as being the product of institutional weaknesses between federal, state, and municipal governments. The federal constitution provides jurisdiction over agriculture, forestry, and land resources to the individual states. In most cases then, the federal government is able to legislate environmental laws, but these laws are only enforceable if individual states pass similar legislation. The federal government, however, possesses substantial persuasive powers to influence state legislation and does so when national goals are paramount; national parks and hydroelectric projects are two cases in point. Federal-state fiscal relations also contribute to environmental degradation. Because the federal government controls the allocation of development funds, states often compensate for perceived financial constraints through exploiting natural resources in an unsustainable manner. At the municipal level, environmental authorities are poorly funded and staffed, and any costly infrastructural improvements are almost wholly dependent on grants from the federal government; throughout the 1980s, federal grants remained stagnant. More important, the responsibilities of municipal authorities are largely restricted to enforcing environmental laws enacted by the federal or state governments, rather than developing environmental management plans or strategies promoting a sustainable environment specific to communities under their jurisdiction.

Brunei

The nature of Brunei's national economy precludes both the variety and scale of environmental problems more characteristic of Malaysia (Chuan, 1991). With the majority of the population being relatively affluent urban civil service wage earners, and with the oil and gas industry dampening the emergence of a more diversified economic base, more common sources of environmentally degrading activity have yet to materialize. This is particularly true with reference to forest cover. In 1990, approximately 80 percent of the country's land area was forested. What little logging does take place is exclusively for domestic consumption, and any future growth of wood-based industries are to be supplied from already planned tree-crop plantations. A decline in agricultural activity during the last quarter-century has actually resulted in abandoned fields reverting to secondary forest, and unlike in neighboring Sarawak, shifting cultivation is not common.

Brunei does possess localized environmental problems, most involving various forms of water pollution. A serious form of environmental degradation originates from offshore oil exploration and production facilities. The industry experienced two serious blowouts in the 1980s that contaminated stretches of coastline; this form of pollution is an always present danger. Another serious source of marine pollution that threatens fish resources and human health were highly toxic red tides occurring in Brunei Bay during 1976, 1980, and 1989. The causes of these algae blooms are as yet unknown. Perhaps the most persistent and visible source of water pollution is from urban waste in the Brunei River adjacent to Bandar Seri Begawan. Much of the waste is from the rapidly expanding population of the capital, but especially from Kampung Ayer where the absence of sewage facilities necessitates disposing of organic wastes into the river. The situation is aggravated because tidal influence and the maze of stilts supporting the Kampung Ayer houses tends to delay the flushing of waste into downstream and eventually estuarine areas. Without serious attention to this problem, aquatic resources are seriously threatened. By 1991, the government had yet to create an autonomous, national level department addressing environmental problems.

CONCLUSIONS AND PROSPECTS

Few would have predicted, in the late1960s, the rapid pace of Malaysia's dramatic economic and social transformation during the past quarter-century. Despite a relatively high GNP, the replacement of agriculture by industry as a primary income earner, the material comforts of most of its citizens, and the growth of cities dotted by gleaming office towers and shopping malls, Malaysia has remained the same in numerous respects.

The ethnic divide between Malays and non-Malays persists as a result of ethnocentric NEP goals. Malaysia is thus no nearer attaining true nationhood status than it was just before the racial riots of 1969. Although NEP policies have significantly raised the equity of Malay corporate ownership, this has only been possible through massive government subsidies. Much like during the colonial and early independence period, private enterprise continues to be dominated by foreign investment, and the implicit anti-Chinese goals of the NEP have resulted in a substantial outflow of Chinese capital that could have been invested at home. Forgotten of course are the majority of ethnic Indians whose economic condition has worsened since the late 1960s. The opportunities for higher-value and technology-intensive economic growth are hindered by a poorly educated workforce. Because of the welfare largesse provided by the NEP, too few Malays attend university, and those who do traditionally choose more-secure government-sector employment. The corps of scientists and engineers employed by foreign electronics firms are overwhelmingly ethnic Chinese. The increased privatization of the public sector as planned by the government will hopefully alleviate the narrow curriculum focus of Malay university students. Raising levels of development in the peninsular northwest and east, as well as East Malaysia, must also become a primary concern of development planners. Only time will tell whether increased affluence will bring a more democratic government as witnessed in the NIEs of South Korea and Taiwan.

Problems of Malay-centered, ethnocentric government policies and government ownership of industry are quite similar in oil-rich Brunei. In a country in which the business interests of the royal family and the state are almost indistinguishable, the process of self-sustaining economic diversification away from a resource-based economy seems almost intractable. Post-Persian Gulf War developments in rentier Kuwait whereby more democratic institutions are being introduced do not seem to appear on Brunei's horizon for the foreseeable future.

14 Singapore

JOHN T. BOWEN, JR.

INTRODUCTION

Singapore is the smallest country in Southeast Asia and one of the smallest independent nations in the world, but the island republic packs a powerful punch in the global economy. Indeed, on a per capita basis, Singapore is one of the wealthiest countries in the world. In the *1995 World Bank Atlas*, the city-state ranked ninth in Gross Domestic Product (GDP) per capita after adjusting for differences in the cost of living (Table 14.1). Singapore is an important example of a country that, through an advantageous location and talented leadership, has moved rapidly into the first tier of developed countries.

Yet the outlook from Singapore was not always so bright. Singapore's racial composition (unlike its predominantly Malay neighbors, Singapore's population is mainly Chinese) and its small size created an acute sense of vulnerability in the nation when it achieved independence in 1965. The external security threats faced by the new nation were compounded by serious domestic problems. Unemployment was high, there was severe crowding in the central city, and communal violence among the nation's diverse ethnic groups had erupted on several occasions. In light of these treacherous beginnings, Singapore's achievements have been remarkable.

Today, the city-state's economy is increasingly based on sophisticated manufactured goods and knowledge-intensive services. In the mid-1990s, for example, Singapore was the world's leading producer of computer disk

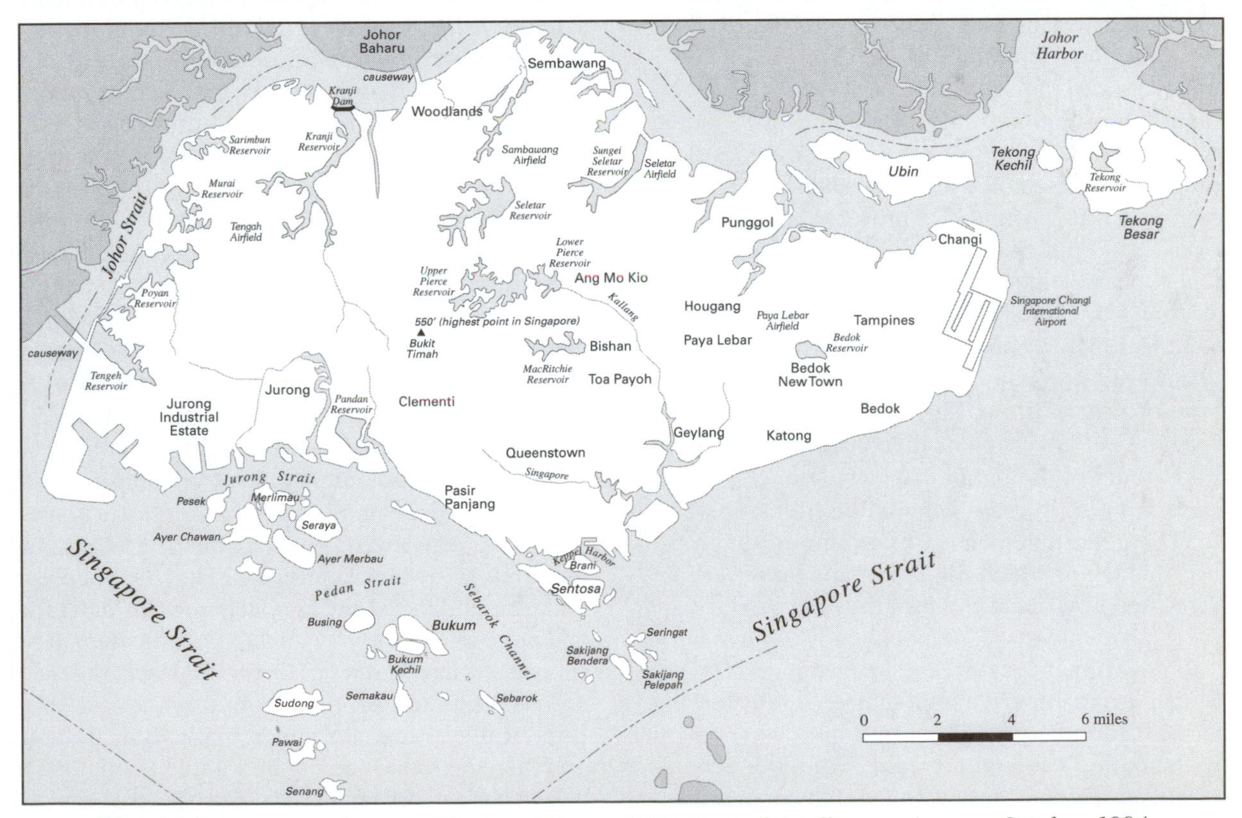

Map 14.1 Singapore. *Source: Ulack and Pauer, 1989; Central Intelligence Agency, October, 1994.*

Country	1994 Gross National Product (U.S. Dollars)	1994 Gross Domestic Product per Capita* (U.S. Dollars)
Luxembourg	14.2 billion	29,510
United States	6.3 trillion	24,750
Switzerland	254 billion	23,620
United Arab Emirates	38 billion	23,390
Qatar	7 billion	22,910
Hong Kong	104 billion	21,670
Japan	3.9 trillion	21,090
Germany	1.9 trillion	20,980
Singapore	55 billion	20,470
Canada	574 billion	20,410

**The figures have been adjusted to take into account differences in cost of living, creating a measure called GDP Purchasing Power Parity per capita. For instance, the U.S. ranks higher in terms of GDP (PPP) per capita than in terms of simple GDP per capita because the cost of living in the U.S. is relatively low compared to other wealthy countries.*

Table 14.1 The world's richest economies. *Source: "The World's Richest Economies," 1995.*

drives. Other high-technology goods manufactured in the city-state include pioneering software, pharmaceuticals, computer chips, and aircraft components. Similarly, the service sector is highly advanced by comparison with other Southeast Asian nations. Firms offering banking, insurance, finance, and similar services cater to multinational corporations—many of which have established regional headquarters in Singapore to coordinate their activities throughout Southeast Asia.

Singapore's striking economic performance attracts much favorable attention from around the world, but the country's successes go well beyond economic development. Singapore is also noted for its racial harmony, clean and green urban environment, innovative public housing program, and political stability. Singaporeans enjoy one of the highest living standards in Asia, not simply in terms of their per capita income but also in their ability to afford their own homes and to walk the city's relatively crime-free streets.

Yet the quality of life in Singapore has come at the cost of significant government intervention, not just in the broad affairs of the nation such as defense and development but also in more personal matters of everyday life. The government has, for example, sought to influence the age at which Singaporeans marry, whom they marry, and how many children they have once married. Even seemingly trivial personal choices have been subject to government regulation; for instance, it is illegal in Singapore to import, manufacture, or sell chewing gum because too many wads of used gum were found stuck to the nation's commuter trains. For older generations of Singaporeans—those who lived through the crisis years—an intrusive government has been well worth the dramatic improvements in living standards that have been achieved in just one generation. A key issue for the future, however, is how younger Singaporeans—those who have known only a cosmopolitan affluent (at least until the recent financial crisis of 1997) Singapore—will reshape the government's role in their lives.

HISTORICAL EVOLUTION AND CULTURE

Rapid economic development has transformed the face of Singapore. Sprawling housing estates have sprung up where only a few years before there was forest. The downtown cityscape changes almost yearly with emergence of yet another glass-and-steel office building. The boundary between sea and land has itself been shifted as a result of an intensive, decades-long land reclamation program. Today, Singapore has about 10 percent more land than at independence. Indeed, in many parts of the country, little physical evidence remains from the 140 years of British rule in Singapore. Yet many characteristics of contemporary Singapore reflect its colonial past. These features include the racial mix of Chinese, Malays, and Indians; the blend of Asian and Western institutions; and the wide use of English as the language of government, education, business, and everyday life.

The Founding of a British Settlement

When a small British squadron arrived in Singapore in January 1819, the island was home to no more than 1,000 people, most of whom earned their living at sea (Lim, 1991). The island and its residents were of no great significance to the British, but its location was increasingly strategic due to the rapid growth of trade among Europe, China, India, and the East Indies (Ken, 1991a). Much of this burgeoning trade flowed through the Strait of Malacca, the shortest passage between the South China Sea and the Indian Ocean. The northern approach to the strait is nearly 310 mi (500 km) wide, but it narrows to a width of just 40.3 mi (65 km) at the southern outlet. Singapore was one of many islands near this outlet (Map 14.2) from which access to the strait could be controlled.

Thomas Stamford Raffles, an agent with the British East India Company urgently wanted to establish a British settlement in this area as a counter to Dutch power (Ken, 1991a). Raffles viewed British commercial interests as being undermined by the expansion of Dutch influence in Java, Sumatra, and the Moluccas (all in present-day Indonesia) and more particularly by the Dutch settlement at Malacca (in present-day Malaysia). From Malacca, they dominated the Strait and profited from taxes levied on passing ships. Raffles's search brought him to Singapore in 1819, and there he found the site he wanted (Lim, 1991). It had a deep harbor, dry land along the coast, plentiful drinking water, and a commanding hill from which passing ships could be observed and the British settlement defended.

Map 14.2 Singapore's strategic location.

Suitably impressed, Raffles made an agreement with the local Malay chief for the establishment of a British "factory" on a narrow strip of coast adjacent to the Singapore River. The factory would not produce anything but would instead serve as a major gathering point for goods flowing into and out of the region (Chew, 1991).

Singapore soon emerged as the most important entrepôt in Southeast Asia (Ken, 1991b). In nineteenth-century Singapore, there were two main trading seasons, both tied to the arrival of the monsoon winds. The northeast monsoon in January, February, and March brought ships from China, Siam, and Cochin-China (southern Vietnam). They carried immigrant laborers, dried and salted foods, silk, tea, and porcelain. The cargo was sold in Singapore and, with the arrival of the southwest monsoon in May, the ships returned home with spices, tin, and gold from the Malay Archipelago, opium from India, and English manufactured products such as weapons and cotton fabric. The second major trading season, in September and October, brought traders laden with goods such as rice, spices, and shark fins (a prized food in Asia) from throughout the East Indies, and they in turn headed back for points across the archipelago carrying opium, salt, guns, silk, and cotton goods.

An Immigrant Society

As Singapore's importance as an entrepôt grew, its population grew too. By the late 1820s, the Chinese were the largest ethnic group (Turnbull, 1980). Initially, they came from Malacca, Penang, and other cities in the region that already had Chinese populations. Many more Chinese immigrants later came directly from China, especially from Southeastern China which was devastated by civil war in the 1850s. The newest migrants spoke a variety of Chinese dialects and relied on their connections to earlier migrants from the same dialect group to get jobs in Singapore. The result was that each dialect group became associated with particular occupations (Kong, 1990). For example, the Hokkienese, the largest dialect group, worked in trade, shipping, and banking (LePoer, 1991).

Singapore was also characterized by residential segregation (Eng and Savage, 1991). The earliest British plan called for each group to live in a separate part of the settlement. The British occupied the area adjacent to the government buildings on the eastern bank of the Singapore River; Chinatown was established next to the commercial quarter on the river's western bank. Farther from the core of the settlement were areas set aside for the colony's other major groups: Arabs, Malays, and Indians (Map 14.3).

Most of the immigrants to Singapore were drawn by the economic opportunities in the growing city. Apart from its advantageous location, several factors favored Singapore's growth during the nineteenth and early twentieth centuries (LePoer, 1991). First, the opening of the Suez Canal in 1869 stimulated the expansion of commerce between Asia and Europe, with much of the increased traffic flowing through the Strait of Malacca. Second, the displacement of sailing ships by steamships gave Singapore new importance as a coaling station. Most European shipping companies established coal depots in Singapore to service their fleets, adding to the level of activity in the harbor. Third, industrialization in Europe and North America created great demand for Malayan rubber and tin, which became important components in Singapore's entrepôt trade. Though Singapore prospered economically, living conditions were very difficult for most of its population. Many lived on the brink of poverty in crowded tenements. Malaria, cholera, and opium addiction gave Singapore a very high mortality rate—higher than India, Hong Kong, or Ceylon (present-day Sri Lanka) in 1896 (LePoer, 1991). Educational facilities were poorly developed, and crime, much of it attributable to Chinese secret societies, was rampant.

World War II

On the eve of World War II, the majority of Singapore's one-half million people had been born in other countries (LePoer, 1991). In particular, the China-born population was substantial. Not surprisingly therefore, anti-Japanese sentiment became increasingly fervent as Japan waged war in China during the 1930s. Nevertheless, there was relatively little fear in Singapore of a Japanese attack on the city. Beginning in 1923, the British had built a naval base in Singapore as a counter to growing Japanese sea power. By 1941, the base was one of the largest in the world and British officials in Singapore considered the

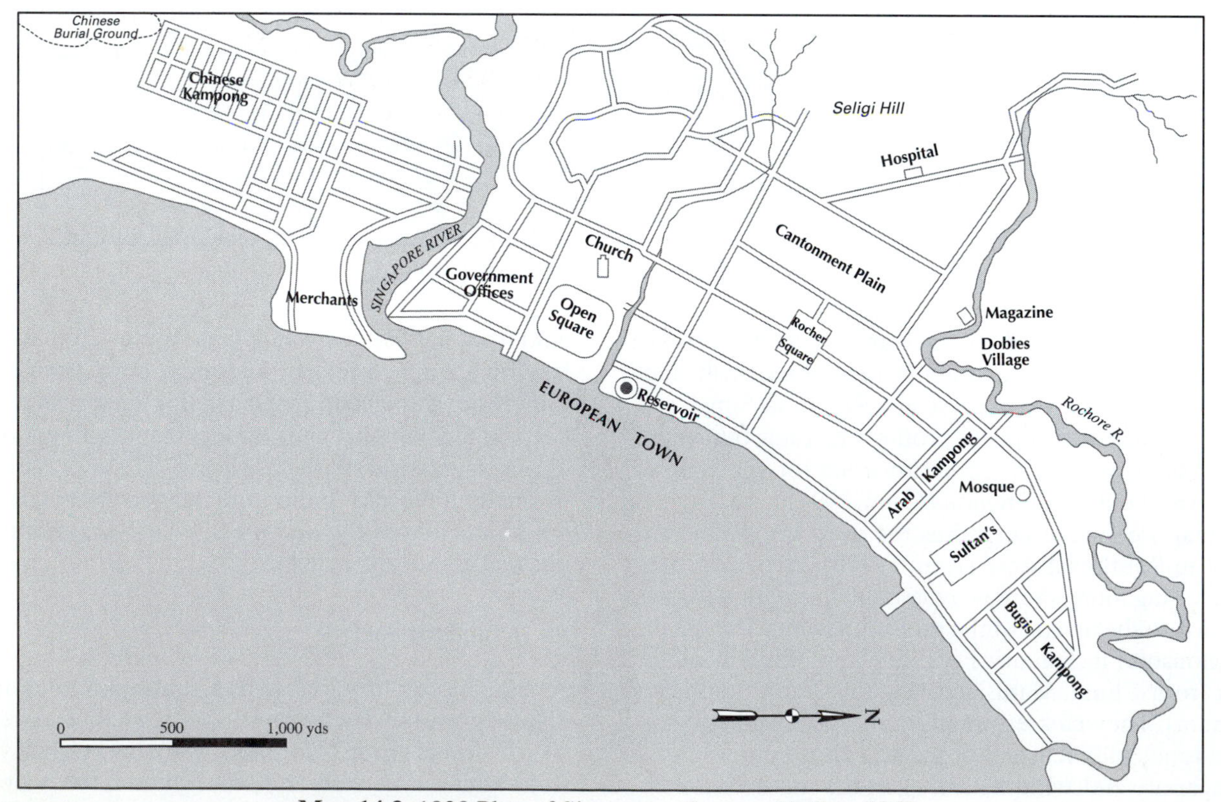

Map 14.3 1828 Plan of Singapore. *Source: McGee, 1967.*

city impregnable. In the end, however, Singapore fell quickly to the Japanese as British military resources, especially warships and aircraft, were deployed elsewhere in the global conflict (Chippington, 1992). Air raids began on December 8, 1941, and the island itself was invaded on February 8, 1942. The Japanese advanced from the causeway linking the Malay Peninsula to Singapore island, and during the next week they drove Singapore's population and defenders into a smaller and smaller area near the downtown. On February 15, 1942, with a million people crowded within a 3-mi (4.8-km) radius, the British finally surrendered (LePoer, 1991). The occupation of Singapore, renamed *Shonan* ("Light of the South") by the Japanese, lasted three and one-half years and was the most difficult period in the city's history. It was especially hard for the Chinese, who were punished for their support of China in its war with Japan. Thousands were executed, including teachers, civil servants, journalists, and intellectuals. Nearly all Singaporeans, regardless of race, suffered as a result of the chronic food shortages and other privations which lasted throughout the war.

The Road to Independence

The surrender of Japan to the Allies in 1945 signaled the beginning of a tumultuous period during which Singapore would make the transition from British colony to independent country. In the late 1940s and throughout the 1950s, the city was beset by frequent strikes and protests, many organized by the Malayan Communist Party. The colonial government in Singapore responded to the strikes and protests by giving Singaporeans greater representation in the Legislative Council, but the disturbances continued (Drysdale, 1984). Finally, in August 1958, an agreement was forged on self-government (Drysdale, 1984). Britain was left with responsibility for foreign affairs and defense, but all other matters were devolved to the local government. The first election under the new constitution, held in May 1959, was won by the People's Action Party (PAP), and Lee Kuan Yew became the first prime minister (Drysdale, 1984). Lee had studied law at England's Cambridge University and was one of the PAP's young middle-class leaders (Milne and Mauzy, 1990). He would become the most influential figure in Singapore's history. Lee held the post of prime minister, without interruption, until 1990, when he stepped down to let a younger generation PAP leader take his place. But even after stepping down, Lee continued to play a highly influential role in the government from his position as senior minister.

The most severe test of Lee's leadership came early in his tenure. In the early 1960s, the leader of Malaya proposed the formation of a new state called Malaysia that would merge Malaya, Singapore, and the British territories on the island of Borneo (Sabah and Sarawak). The merger proposal was very controversial in Singapore, but Lee considered it economically essential and also believed it would give the government more power to suppress communist elements on the island (Sesser, 1993). Against the cries of critics who denounced the plan as a sellout, Lee was able to gain approval for the merger in a September 1962 islandwide referendum, and one year later, Singapore became part of Malaysia. The merger did not last, however. From the beginning, it was under tremendous pressure both internally and externally. Indonesia strongly opposed the merger and adopted a policy of *Konfrontasi* ("Confrontation") to destroy it. Within Malaysia, meanwhile, tensions flared between the Malay-dominated government in Kuala Lumpur and predominantly Chinese Singapore. Fearing communal violence, the Malaysian prime minister decided that Singapore should separate from Malaysia. With the entire Singapore delegation abstaining, the Malaysian parliament voted unanimously in favor of separation on August 9, 1965, forcing complete sovereignty on an unwilling Singapore (Drysdale, 1984).

Survival and Prosperity

The new nation faced many problems. Economically, the country's prospects were grim. The conflict between Indonesia and Malaysia had severed the entrepôt trade with Indonesia. Moreover, further trade with Malaysia was dampened by high tariffs imposed by the Malaysian government. Singapore had virtually no natural resources; even its water supply was inadequate for the growing population. Militarily, the city-state was in a weak position, sandwiched between unfriendly neighbors. But perhaps the most profound challenge was to forge a new national identity among the diverse peoples of Singapore. The urgency of the country's situation was reflected in the survival-oriented approach of the PAP government. To accelerate economic growth, the government stepped up its efforts to attract foreign firms (Rodan, 1989). Singapore had adopted a strategy of enticing foreign, export-oriented manufacturing firms as early as 1961, but the number of jobs created was insufficient. Many firms were skeptical of Singapore's workforce, which was well-known for its frequent strikes, so the government clamped down on independent trade unions. Legislation was passed to extend the length of the work week, reduce the number of holidays, and give employers greater power over hiring and firing. In 1969, as a result of the new government policies, there were no strikes and economic growth rates surged (Figure 14.1).

The PAP government also took aggressive action to redress the country's military weakness (Katz, 1991). To counter the larger troop strengths of nearby countries, Singapore established mandatory national military service for male citizens. Every young man had to serve two years (later extended to two and one-half years) of full-time service in the military. The national service requirement was controversial, but the PAP's near monopoly on

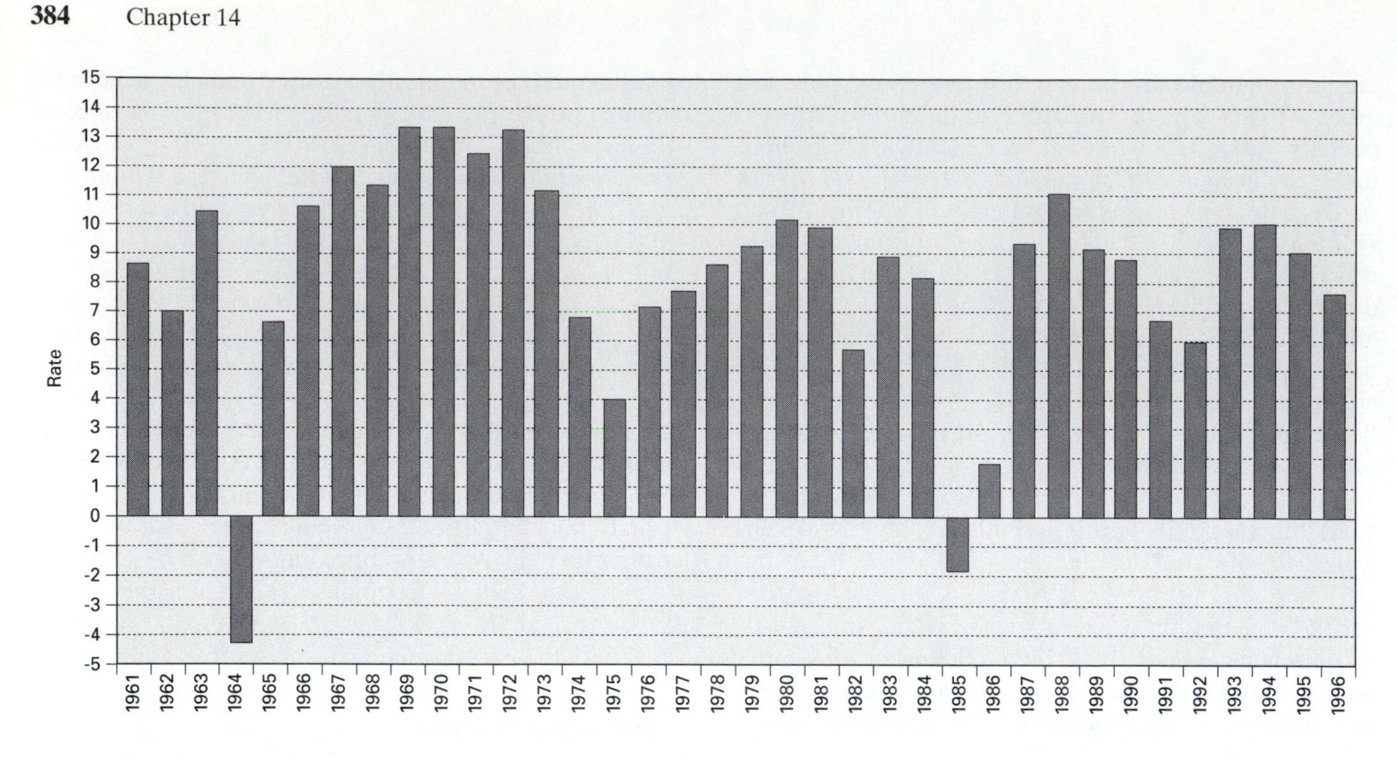

Figure 14.1 Economic growth: Annual rate of increase in Gross Domestic Product. *Source: Singapore Department of Statistics, 1997.*

political power (see Table 14.5—it won every seat in Singapore's Parliament in 1968) gave the government the power to push the measure through. Since then, the government has continued to pour resources into defense. In 1995, Singapore's defense spending was greater than that of any other Southeast Asian nation. (Figure 14.2)

Another critical goal for the new government was to forge a new national identity—to get Singaporeans to think about Singapore as a nation to which they belonged and to which they owed their allegiance. Nation building was a sensitive task. So that none of the major ethnic groups would feel slighted, each was allowed to keep a measure of its separate identity (Kong, 1991). Four official languages were designated: Chinese (Mandarin), Malay, Tamil (the language of most Indians in Singapore), and English. At the same time, however, national service, the national education system, and other institutions served to unite Singaporeans through their shared experiences (Quah, 1990).

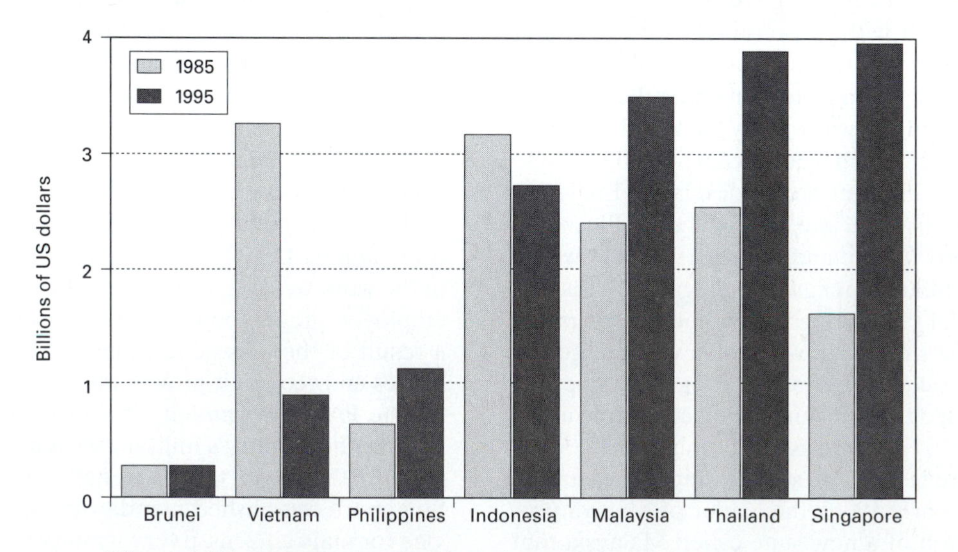

Figure 14.2 Defense spending by selected ASEAN countries. *Source: "Asian Defense Spending," 1997.*

A Cosmopolitan, Asian Culture

Singapore's economic success has fostered an exuberant pride in its Asian heritage, and the government has sought to reinforce and propagate the Asian values of Singapore's immigrant peoples. One way it has done so is to require young Singaporeans to become fluent in one of the main Asian languages of Singapore (Mandarin, Malay, or Tamil). The expectation is that, in learning one of these languages, students will be instilled with the values of the society where the language originated. "Asian" values are defined as hard work, respect for authority, willingness to subordinate one's own interests to the interests of the group, decision making through consensus rather than conflict, and devotion to family. These values are also formally taught in the nation's schools.

Perpetuating these values is important for two reasons. First, they enhance Singapore's global competitiveness by helping to create a hard-working, co-operative workforce. Second, they make it more likely that government leaders will be able to continue to direct Singapore's social, political, and economic development. The government's paternalistic role in the city-state would be difficult to sustain if more Singaporeans adopted Western-style individualism or advocated a more participatory democracy and greater civil liberties. So far, many Singaporeans seem to agree with the government's view that the Western approach to such issues is not appropriate to their country. This view of the West represents a significant change in Singaporeans' attitude. Decades ago, Western ideas and products were held in great esteem, especially by the elite. Lee Kuan Yew himself took a Western name, Harry, by which he was known until his political career began in earnest. In contrast, Lee in the 1990s became one of the staunchest critics of the Western foreign policy toward Asia (e.g., U.S. treatment of China and Burma) and Western values of liberal democracy.

Nevertheless, Western influence in Singapore remains strong. Indeed, it would be difficult to exclude such influence in a city-state that is so cosmopolitan. Popular culture, in particular, bears the imprint of the West. A wide variety of U.S. television programs, ranging from *The Simpsons* to *Wheel of Fortune* are shown on Singapore channels, and Western movies, novels, and pop music also enjoy great popularity in Singapore. A much different aspect of Western influence in Singapore is the increasing popularity of Christianity. In 1990, 13 percent of Singaporeans were Christians, up from 10 percent in 1980 (Figure 14.3). The increase came at the expense of Buddhism and Taoism, the traditional Asian religions of the majority Chinese population. Indeed, Christianity has spread especially rapidly among the growing population of young, English-educated Chinese Singaporeans. Two other important religions in Singapore are Islam, the religion of almost all Singaporean Malays and some Indians, and Hinduism, a major religion among the Indian population.

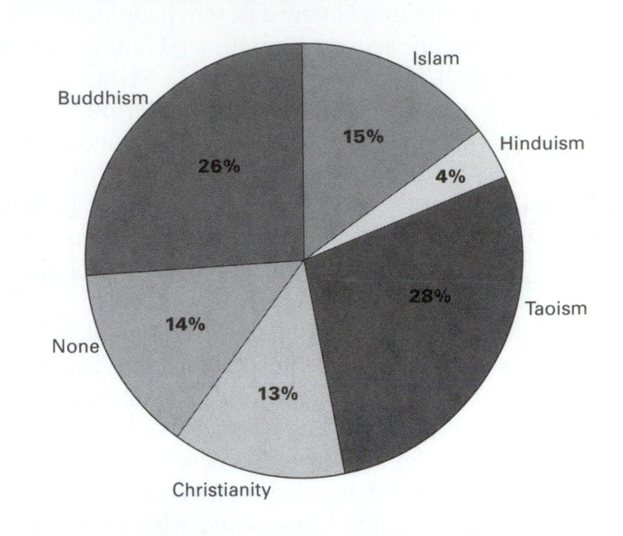

Figure 14.3 Religious preference in Singapore, 1990.

Photo 14.1 A restored building in Chinatown. Many of these architectural treasures now house professional offices and upscale shops. (Bowen)

Ultimately, the distinguishing feature of Singaporean culture is neither its Western nor its Eastern dimension but rather the mixture of many different peoples and their cultures. Singapore society is multiracial, multireligious, and multilingual. It is not simply Chinese, Malays, and Indians who comprise the Singapore population. Each of these groups includes people with a bewildering variety of origins. Consider the group "Indians." It includes anyone with ancestors from India, Pakistan, Bangladesh, or Sri Lanka. They may be Hindus, Buddhists, Muslims, Christians, or Sikhs. They speak English, Tamil, Hindi, Punjabi, Urdu, Malayalam, Gujurati, and other languages.

These myriad minority groups have, in general, retained some of their distinctiveness in contemporary Singapore. The U.S. American "melting pot" image of many different peoples being molded into a new homogeneous whole does not fit this immigrant nation. Still, a new Singaporean culture is slowly emerging. It incorporates elements from the city-state's different cultures into a unique blend. "Singlish," for example, is a slang form of spoken English. Its notable features include the use of the word *lah* as a kind of exclamation and the borrowing of words from other languages commonly spoken in Singapore. Standard English grammar is also relaxed in Singlish. A student might say "The test so *terok*, lah!," meaning "The test was very difficult!"

Photo 14.2 The Sri Mariamman Temple is in the heart of Chinatown and is the oldest Hindu temple in the city-state. A temple of the south Indian Dravidian style, it was originally built in 1827 and rebuilt in 1862. Pictured here is the *goporum* (tower) which is over the entrance gate. (Ulack)

The country's pressure-packed, highly competitive lifestyle has given rise to a distinctly Singaporean outlook on life. Singaporeans jokingly poke fun at each other for being too *kiasu*, a Chinese word meaning "afraid to die." In Singapore, *kiasu* refers to desire of Singaporeans to be first in everything and to get as much as possible from every opportunity; for example, many parents send their children, even preschoolers, for hours of private tutoring each week, in everything from gymnastics to mathematics. The seriousness with which Singaporeans take education is evident in the consistently high ranking of Singapore students in international comparisons of scholastic achievement. But *kiasu* behavior can also be unattractive: In an infamous incident in the mid-1990s, a crowd of parents created a near riot as they rushed for free textbooks at a school that had wanted to donate the books to needy families. The incident aroused widespread public indignation, especially when it was noted that many of the parents had driven to the giveaway in Mercedes-Benzes and that a community leader was among the rudest members of the crowd. The government's response to such displays of kiasuism has been to advocate a more gracious and courteous society. Each year in July, the government launches a high-profile Courtesy Campaign, complete with courtesy banners, catchy jingles, courtesy contests, TV advertisements, and courtesy variety shows featuring the nation's best-known TV entertainers (see Photo 10.2).

POPULATION

In 1996, the population of Singapore was 3 million with an annual growth rate of 1.9 percent (Figure 14.4). The majority of Singaporeans (77 percent) are of Chinese ancestry. The remainder include Malays (14 percent), Indians (7 percent), and a smaller number with other ethnic backgrounds.

The nature of the demographic problems faced by Singapore have changed markedly since Worth War II. In the early postwar period, Singapore faced problems common to many developing countries. Population growth was rapid, and the absolute size of the population was greater than local resources (especially the housing supply) could support. In 1957, the population was surging at an annual rate of 4.4 percent (at that rate, the population would double every 16 years), fueled by a high rate of natural increase and heavy migration from Malaya (Hock, 1991). The population's rapid growth exacerbated poor living conditions along the island's southeastern shore, where 75 percent of Singaporeans lived (Tan and Phang, 1991). Both of these problems were addressed by aggressive gov-

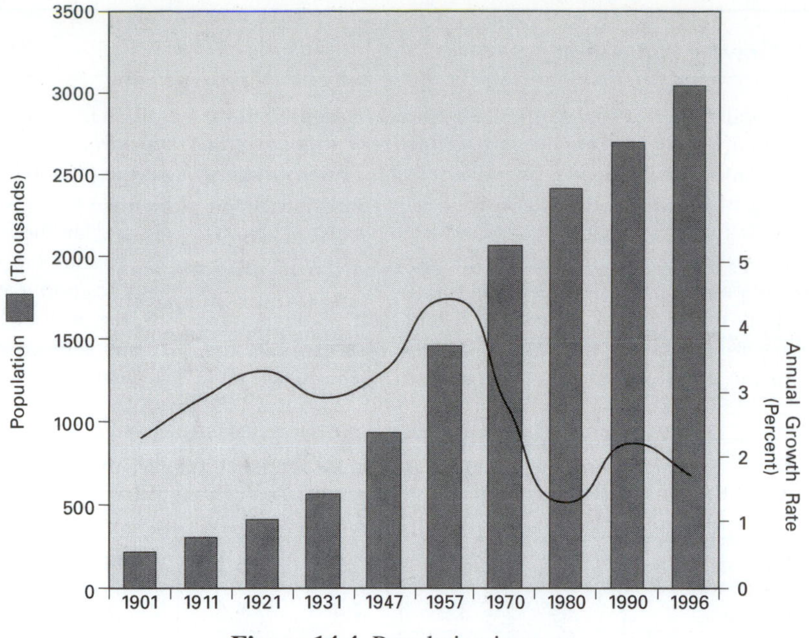

Figure 14.4 Population increase.

ernment action after independence. Singapore's public housing program resettled hundreds of thousands of people to previously undeveloped areas of the island, giving many families the opportunity to own their own home for the first time. Efforts to reduce Singapore's population growth rate also met with great success; indeed, since the mid-1980s, the most serious demographic issue in Singapore has been the country's *falling* birth rate. A related issue, the aging of Singapore's population, raises troubling issues for the city-state's future (Box 14.1).

Population Growth

In the mid-1960s, a government program was launched to reduce Singapore's population growth. The program sought to bring down the country's total fertility rate (TFR: the average number of children a woman will have in her lifetime) to 2.1, or the replacement level. At 2.1, the children of each generation will just replace their parents in number, and population growth will eventually stop. Singapore's plan was to achieve a TFR of 2.1 by 1980 and zero population growth by 2030 (Yap, 1992).

During the 1960s and 1970s, Singaporeans were bombarded with the message "Two is enough." Financial incentives and disincentives were also used to encourage couples to have small families. These policies reinforced and amplified the effects of rapid economic development on birth rates. As the city-state prospered, more women entered the workforce, more young people pursued secondary and tertiary education, and people attached more importance to their careers. These changes meant that people married later and, once married, had fewer children. By the early 1980s, Singapore's total fertility rate had fallen to 1.6, far below the desired level (Figure 14.5).

In an important 1983 speech, Prime Minister Lee Kuan Yew drew attention to this new problem. It was not simply the overall decline in Singapore's fertility that sparked the alarm of government leaders, but also the fact that the trend was considerably stronger for university-educated women than for those having little formal education. Government planners feared that Singapore's workforce would decline in average intelligence and would be unable to excel in the global economy if this pattern of reproduction continued. To counter this threat, the government launched a battery of measures in 1984 to increase the fertility rate of well-educated women and reduce that of their lesser-educated counterparts (Yap, 1992). The year these measures began is slightly ironic because Singapore's new population policy seemed to some like something from George Orwell's haunting novel *1984*. Under the policy, low-income couples with only primary school qualifications were given cash grants if they agreed to sterilization after their first or second child. Conversely, couples with secondary qualifications were eligible for large tax rebates if they had three or more children. The government also gave priority in primary school registration to the first three children of university graduate mothers, making it more likely that these children would get into the nation's best primary schools and gain an early head start in their all-important educational careers. A final important measure was the establishment of the Social Development Unit (SDU) (Social Development Unit, 1991) to encourage university graduates to marry and to provide opportunities for them to meet. The SDU operates a computerized matchmaking

BOX 14.1 Singapore's Graying Population

The graying of Singapore's population is a consequence of the sharp decline in fertility rates since the 1980s (Ju and Jones, 1989). As a result, the aging of Singapore is expected to be especially rapid. One moderate United Nations projection forecasts that in 2012, 10 percent of Singaporeans will be elderly (age 65 and over). By 2033, the figure will be 20 percent, a doubling of the elderly share in just 21 years. In contrast, Britain reached the 10 percent threshold in 1945 but will not reach 20 percent until 2031. The accelerated aging of Singapore's population will place tremendous pressure on the country's health and retirement systems.

The small size of today's young families means that the responsibility for supporting the aged in the future will fall on fewer working adults. Moreover, the structure of Singapore's families is shifting gradually toward more nuclear families and fewer intergenerational, extended families. The greater affluence of the country has meant that more young families find it affordable and desirable to live on their own. This trend could mean a more difficult future for the growing elderly population.

Country	Percent of Population Aged 65 and Over					
	1980	1990	2000	2010	2020	2025
Indonesia	3.3	3.8	4.9	6.1	7.4	8.7
Malaysia	3.7	3.9	4.4	5.6	7.8	9.1
Philippines	3.4	3.5	3.8	4.7	6.3	7.5
Singapore	4.7	5.6	7.1	9.2	14.5	17.0
Thailand	3.5	3.9	4.7	5.7	7.5	9.1
United States						17.2
Britain						18.7
Switzerland						23.8
Japan						23.8

Table 14.2 Aging in Singapore and other ASEAN members. *Source: Ju and Jones, 1989.*

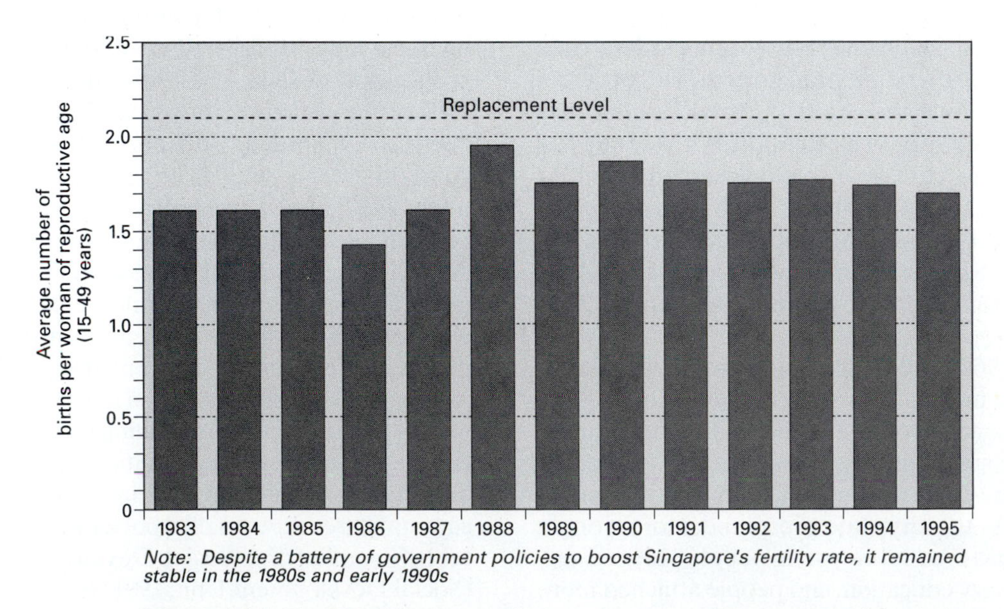

Figure 14.5 Total Fertility Rate. *Source: Singapore Department of Statistics.*

service, sponsors picnics and theme parties, and offers instruction in dancing and other social graces.

All of these policies were controversial. The public response to the primary school registration policy was so heated that the policy was withdrawn in 1985. The sterilization incentive was also withdrawn, and the elitism of the other policies was softened somewhat. For example, the eligibility requirements for tax rebates to well-educated couples were reduced so that more families could qualify, and sister organizations to the SDU were established to also encourage marriage among individuals who did not graduate from a university. Still, Singapore's efforts to engineer the future makeup of its population comprise one of the most comprehensive eugenics programs in the world. The government's strategy has had some success. The SDU, for instance, claimed credit for 1,557 marriages between university graduates in its first eight years of existence. Women at opposite ends of the educational attainment spectrum seem to have responded to the government's goals for family size: Mothers with at least some secondary education now account for a greater proportion of families with three or more children (Yap, 1992). However, the proportion of university graduate women age 35 to 44 who have not married continues to grow and reached nearly 25 percent in 1995 (Leow, 1997).

A more worrisome problem is that Singapore's TFR has not exceeded the replacement level of 2.1 in any year since Lee launched his pronatalist policies (Singapore Department of Statistics, 1997). It briefly jumped from 1.6 to 2.0 in 1988; but that was an isolated phenomenon. The Chinese Year of the Dragon, believed by many Singaporeans to be the luckiest of the 12 animal signs to be born under, coincided with 1988. The following year the TFR returned to 1.8 and has since fallen to 1.7. If sustained, this rate will eventually result in negative population growth in Singapore.

Negative growth would have serious consequences. It would worsen Singapore's chronic labor shortages, undermining the nation's competitiveness, and it would exaggerate the problems associated with the aging of Singapore's population. To some extent, the lack of Singaporean workers can be offset through increased immigration and through increased participation of women in the labor force. The latter option is of limited appeal, however, because raising female labor-force participation (51 percent of women worked in 1993, up from 46 percent in 1983) would probably exacerbate the nation's low total fertility rate.

Foreign workers are the most readily available means of easing the country's labor shortage. More than 10 percent of Singapore's workforce is composed of foreign citizens (Tat, 1992). They include both unskilled guest workers from such nearby countries as Bangladesh and the Philippines as well as technical and managerial personnel from North America, Europe, Japan, and other advanced economies. The government has liberalized permanent residency requirements for highly skilled foreigners. By 1995, the number of permanent residents had leaped to nearly 200,000, or about twice the level of 1990 (Leow, 1997). Special efforts were made after the Tiananmen incident in 1989 to recruit skilled Hong Kong citizens before the Chinese takeover of Hong Kong in 1997.

Public Housing and Population Patterns

One of the distinctive features of development in Singapore has been the key role played by the government's massive public housing program. Public housing was launched on a large scale after a huge 1961 fire left 14,000 families homeless in central Singapore. Today, approximately 87 percent of Singaporeans live in one of the high-density housing estates built by the Housing Development Board (HDB), making it one of the most comprehensive public housing programs among non-Communist countries (Housing Development Board, 1992).

Indeed, public housing must be regarded as one of Singapore's greatest achievements. The contrast with public housing in other countries is striking (Huat, 1988). Elsewhere, public housing projects are often drab and austere concentrations of poverty, crime, and social breakdown. Yet in Singapore, the HDB has built a very livable environment, with a mixture of low-income and middle-income families sharing each neighborhood. One of the defining features of Singapore's public housing program is its universality. The popularity of HDB apartments is attributable chiefly to their affordability. Public housing is heavily subsidized by the state, and the subsidies are passed on to those who purchase apartments from HDB. A three-bedroom HDB apartment sold for about US$170,000 in 1997, versus about US$400,000 for a comparable apartment in a private development. Even with the HDB subsidy, however, the cost of housing is extraordinarily high. How can Singaporeans afford such huge sums? The key for most families is the Central Provident Fund (CPF). The CPF is a mandatory social security savings program to which employees and employers contribute equally each month, with the combined contributions amounting to about one-third of a worker's salary. Singaporeans can withdraw the funds in their CPF accounts to invest in housing, education, and government-approved shares. The ability of individuals to control how their social security savings are invested marks an important difference between Singapore's system and that of most other countries.

By relying on their CPF savings, the great majority of people living in HDB apartments are able to own their homes (Huat, 1988). Home ownership has been stressed as a way of giving ordinary citizens a stake in the city-state. In fact, an HDB apartment is the most important financial investment that most Singaporean families make. The rapid escalation of land costs since the mid-1980s has caused home prices to skyrocket, enriching many lower-income and middle-income families who sell their HDB apartments for several times what they originally paid.

When the large-scale public housing program was launched in the 1960s, government leaders believed that ordinary people would feel more nationalistic if they owned a home in Singapore. In addition, home ownership is a key element of Singapore's political stability; people with a significant personal financial investment in the city-state are unlikely to favor radical parties. In fact, the decades-long effort to increase the number of home owners in Singapore is considered one of the keys to the PAP's hold on political power.

In addition to home ownership, racial desegregation is another important achievement of the public housing program. The sale of apartments is controlled to prevent any building or neighborhood from being disproportionately occupied by a single race (Quah, 1990). In the past for example, Malays were found to prefer developments along the eastern coast, resulting in racial enclaves. Now, no building may be more than 25 percent Malay and no neighborhood more than 22 percent Malay. Sales to other racial groups are also constrained by percentage limits (Table 14.3). As a result, a family wishing to sell its HDB apartment may be told that they can sell it only to buyers from a particular racial group. One effect of the desegregation of Singapore housing is that it is difficult for parties appealing to minority groups to win seats in Parliament.

Such controversies aside, it is clear that public housing has had a profound impact on Singapore's social environment. The massive public housing effort has also transformed the city-state's physical environment. The earlier housing projects were located close to the central business district (CBD). Toa Payoh New Town, for instance, ranging from 3.7 to 5 mi (6–8 km) north of the CBD, was begun in 1965. Later new towns were built further and further from the CBD. In 1973, for instance, Ang Mo Kio New Town was begun a few kilometers further north of Toa Payoh. And near the northern shore of the island, about 12.4 mi (20 km) from the CBD, Yishun New Town was established in 1976. New towns also spread to the east and west. By the 1980s, HDB developments were finished or underway across most of Singapore. In addition, spaces between earlier projects were developed. For example, the Bishan housing estate was begun in 1984, filling the gap between Toa Payoh and Ang Mo Kio New Towns. By the mid-1990s, the most important regions of new development were the far north and northwestern sections of the island (Map 14.4).

	Maximum Percentage Occupancy by Race	
	Neighborhood *(typically 4,000 to 6,000 families)*	Building *(typically 100 to 200 families)*
Chinese	84	87
Malay	22	25
Indian/Other	10	13

Table 14.3 Racial limits in public housing. *Source: Quah, 1990.*

Photo 14.3 Pasir Ris, one of the most recently developed new towns, is designed with distinctive features such as the small tower atop the building on the right. (Bowen)

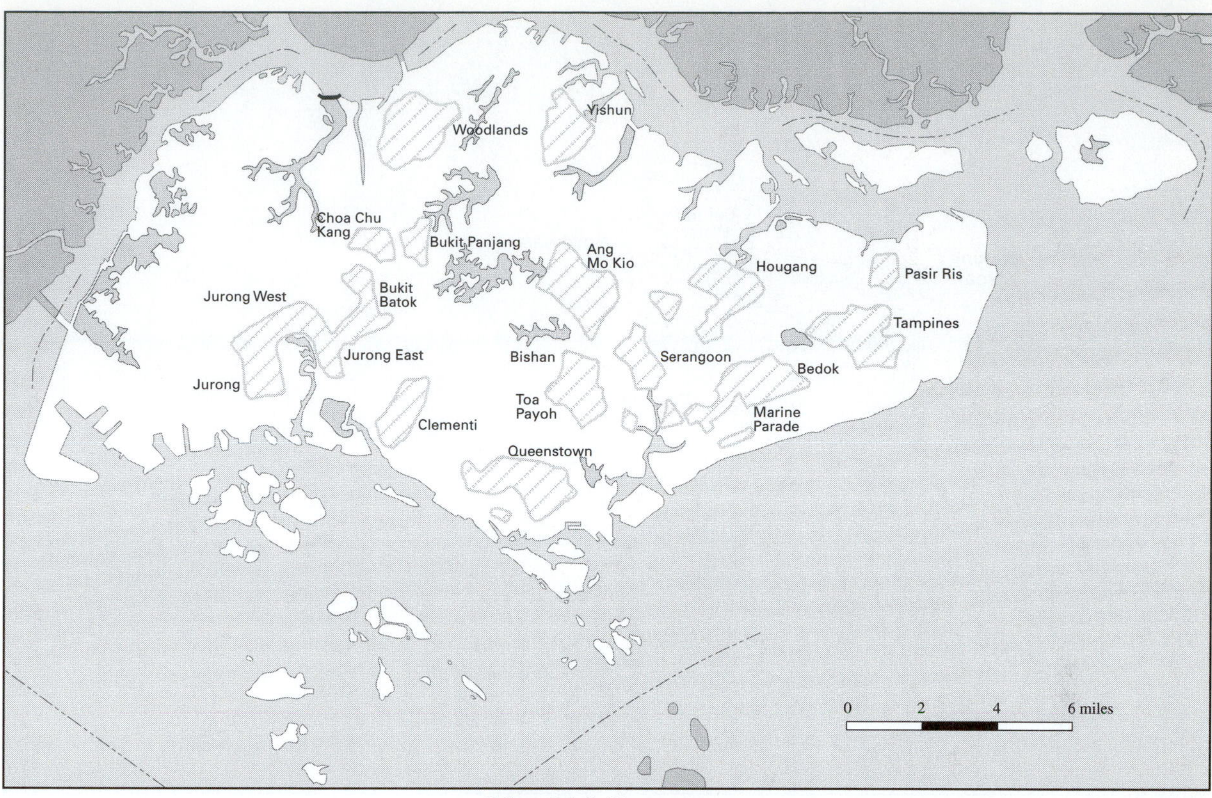

Map 14.4 New towns.

New towns are made up of hundreds of high-rise apartment buildings. Typically, each new town houses 150,000 to 300,000 people. Ang Mo Kio New Town, for example, has a population of slightly more than 200,000. Like other new towns, Ang Mo Kio is designed to be largely self-sufficient in many functions (Tan and Phang, 1991). Originally, planners hoped that residents would be able to work in their new towns, but employment opportunities have remained concentrated in the CBD and the heavily industrialized area of Jurong. But new towns do offer many services, reducing the need to commute. Ang Mo Kio, for example, features dozens of coffee shops and hawker centers (cheap, informal restaurants), hundreds of shops, a few supermarkets, 12 churches, schools, two swimming complexes, and a variety of other community amenities. The availability of so many facilities within walking distance is stressed as a means for new town residents to meet one another. The lack of pedestrian activities in new town developments in other countries has forced people to rely on their cars and has been a deterrent to the emergence of community cohesiveness. A newer strategy to forge a genuine community spirit among public housing residents is to give each development a distinct look. Previously, one housing estate looked very much like another, with an endless sea of off-white concrete buildings. Ang Mo Kio, Bedok, and Clementi looked like mirror images of one another despite their different locations in the city-state. Newer developments, however, incorporate different designs and building materials.

Improving living standards is politically important in Singapore. In a 1990 book called *The Next Lap*, the government promised that each person would have an average of 35 square meters of living space by mid-twentyfirst century up from the current level of 20 square meters (Urban Redevelopment Authority, 1991). The percentage of the population living in high-density housing will also decline as public housing types become more diverse. Yet there are limits imposed by Singapore's meager land resources. The government's own estimates call for a population of 4 million by the mid-twentyfirst century (Urban Redevelopment Authority, 1991). With large areas of land set aside for water catchment, military training, and nature preserves, high-density housing developments are sure to remain a prominent feature of Singapore's physical (and political) landscape. Further decentralization will help to alleviate crowding in the city-state. The HDB program has already brought about significant decentralization of Singapore's population (Map 14.5) (George, 1994). Gone are the days when most Singaporeans lived within a few kilometers of the CBD. Now the government wants to decentralize office employment away from the CBD, too. Four regional centers have been established along the periphery of Singapore (Urban Redevelopment Authority, 1991). Each will mix office buildings with hotels, entertainment complexes, and shopping malls.

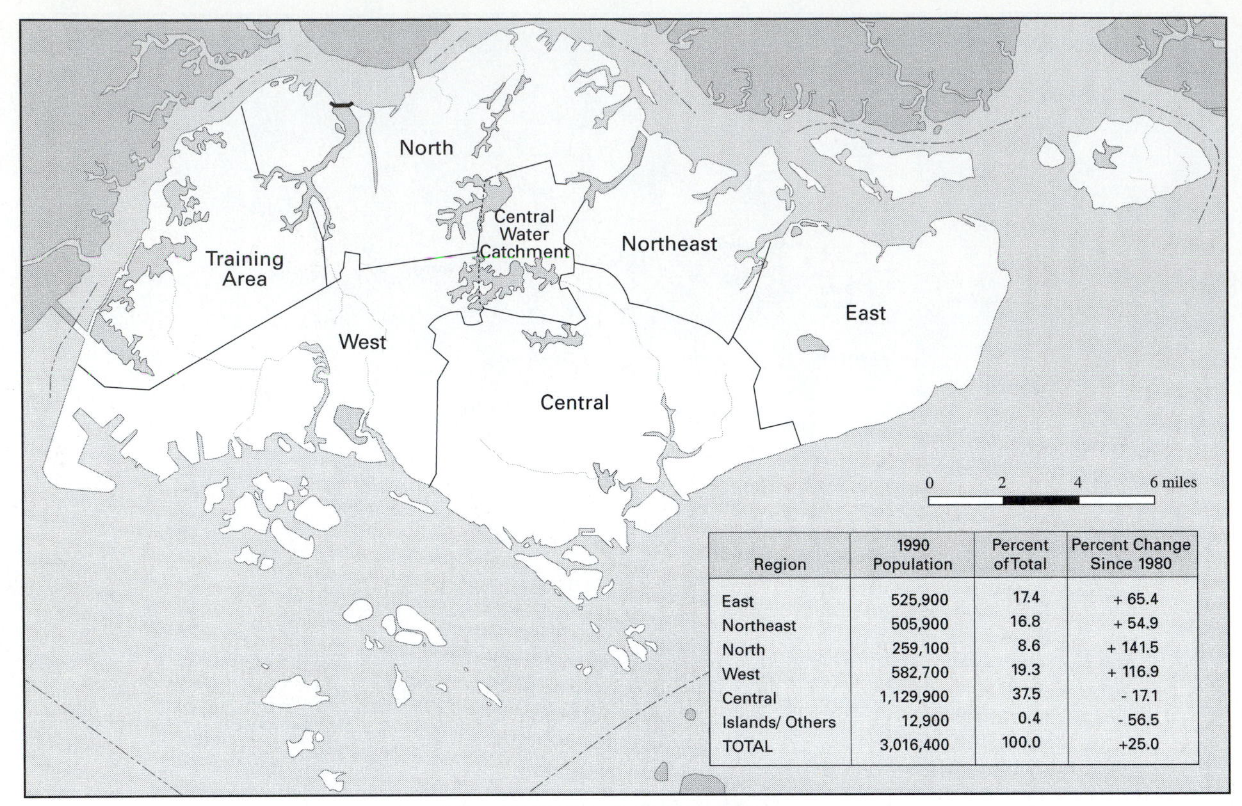

Region	1990 Population	Percent of Total	Percent Change Since 1980
East	525,900	17.4	+ 65.4
Northeast	505,900	16.8	+ 54.9
North	259,100	8.6	+ 141.5
West	582,700	19.3	+ 116.9
Central	1,129,900	37.5	- 17.1
Islands/ Others	12,900	0.4	- 56.5
TOTAL	3,016,400	100.0	+25.0

Map 14.5 Population resettlement.

ENVIRONMENT

Visitors to Singapore are often impressed by the country's orderly and clean appearance. The showpiece expressway from the airport to the city center is lined with carefully manicured tropical trees and shrubs. The city's streets are cleaned daily by a small army of rubbish collectors and street sweepers. The Singapore River shimmers in the bright sunlight as it passes through the CBD, its banks lined with posh outdoor restaurants and bars. Parks and gardens abound, and the air is generally free of the smoke and exhaust fumes which accompany clogged roads in other Southeast Asian cities.

The environment is yet one more arena of government planning in Singapore; and there is little doubt that decisive government action has improved environmental quality. In the early 1970s, rapid economic development created environmental problems typical of today's developing cities. Industrialization and a growing population of motor vehicles exacerbated air pollution, and public health was undermined by the lack of sanitary sewage facilities in some neighborhoods.

The strength of Singapore's government has allowed it to deal with these concerns decisively and economic development has enabled funds to be allocated to environmental problems that would be accorded little priority in a poorer country. The importance of both political ability and money is illustrated in the clean-up of the Singapore and Kallang Rivers (Ministry of the Environment, 1987). These two rivers flow through the CBD and were grossly polluted by the mid-1970s. In 1977, Prime Minister Lee Kuan Yew declared that within 10 years, people would again be able to fish in the largely lifeless rivers. His goal was met but not before tens of millions of dollars had been spent and more than 25,000 families were resettled, most to HDB apartments. Street hawkers were moved from their traditional places along the streets of Chinatown and other historic districts to permanent stalls in government-built food centers replete with proper sewage and freshwater systems. Pig farms and duck farms near the river were closed down. Muddy river banks were replaced with parks, tiled walkways, and clean sandy shores.

As a result of such efforts, Singaporeans enjoy better environmental conditions than many people in other Southeast Asian quarters. Ultimately, however, environmental quality in Singapore is dependent on what happens in those other nations. The city-state is physically so small that events in Indonesia and Malaysia can have a great environmental impact on the city-state. In 1994, and again in 1997, severe forest fires in Sumatra and Kalimantan caused weeks of suffocating, tear-inducing haze in Singapore. An especially important environmental link between Singapore and its neighbors is the city's water supply (Box 14.2); indeed the limited supply of freshwater

Photo 14.4 The Singapore River flows through the Central Business District (CBD). High-rise buildings are the backdrop to Boat Quay, a newly renovated area of shops, restaurants, and nightlife. (Ulack)

Photo 14.5 Another office building emerges in Singapore's skyline. (Bowen)

BOX 14.2 Water: Singapore's Achilles Heel?

Singapore draws only half of its water supply from the precipitation that falls on the country's territory. The other half comes from southern Malaysia through a pipeline buried in the causeway between Singapore and Johor. The 50 billion gallons that Malaysia pumps into Singapore each year are governed by two treaties between Singapore and Malaysia. When the first treaty expires in 2011, Singapore will face a potentially crippling water shortage unless the treaty is renewed or alternative water sources are found. The question of water electrified both sides of the causeway in 1997 when Malaysia–Singapore relations plummeted to their lowest level in decades. The tension between the two neighbors was sparked by a series of angry exchanges following some disparaging remarks made by Lee Kuan Yew about the crime rate in Johor. The controversy quickly escalated into a diplomatic crisis, and issues not related to crime in Johor, especially water, took center stage.

To encourage the Malaysians to provide the city-state with a longer-term water supply, Singapore has offered to help Malaysia realize its dream of a high-speed passenger train service between Singapore and Kunming, China, via Kuala Lumpur. Singapore is also looking to its southern neighbor for water. A treaty was signed in 1991 between Indonesia and Singapore to provide the city-state with up to 1 billion gallons per day, four times Singapore's current consumption. But this arrangement has not allayed concerns about the water supply, in part because the new water supply is to come from Indonesia's Riau Islands, an area of growing water pollution (Ibrahim, 1995). To increase its self-sufficiency further, Singapore will build a desalinization plant in the western part by 2003, and the price that consumers pay for water will be doubled between 1997 and 2000 to discourage waste (Leong, 1997).

has been identified as one of the most important constraints on future economic growth (Ibrahim, 1995).

Singapore economic success has resulted in a rapid increase in waste water and solid waste. Dealing with these problems is made more difficult by the shortage of land in the country; for example, faced with rapid escalation in land prices, the government plans to locate its next big landfill on an offshore island and to build one of the world's largest, most expensive incinerators near the heavily industrialized area in Jurong. An even more-expensive plan calls for a new 49.6-mi (80-km) network of tunnels buried 25 stories below ground to carry waste water to eight sewerage treatment plants ("ENV's priority," 1997).

ECONOMY

Since the mid-1960s, Singapore has proclaimed itself one of the fastest growing economies in the world. Between 1965 and 1996, real GDP in Singapore grew at an average annual rate of 8.7 percent per year (Lee, 1989; Singapore Department of Statistics, 1997). Few countries in the world could rival that statistic. Yet even more impressive than the sheer growth of Singapore's economy has been the remarkable evolution, some would say *revolution*, that has made the economy one of the most sophisticated in the world. In the 1960s, Singapore invited multinational corporations (MNCs) to come to the city-state and employ low-wage workers for *labor-intensive* manufacturing operations. By 1990, Singapore was a world leader in *knowledge-intensive* manufacturing. Singaporean workers are now involved in designing, engineering, and testing new high-technology products such as computer software and silicon memory chips. The rapid growth of manufacturing has been complemented by the flourishing international business services sector.

The main beneficiaries of this transition have been the people of Singapore. The competitive strength of their nation has brought Singaporeans ever higher living standards. Yet Singapore's competitiveness is by no means guaranteed. Other nations are trying to duplicate the city-state's strategy. Indeed, Singapore need look no further than its Southeast Asian neighbors to find eager rivals targeting the same industries in which the city-state is competing. Their more abundant land and labor resources and their larger domestic markets may ultimately enable them to overcome Singapore's head start. The challenge for Singapore then is to constantly forge into new industries in which less developed countries cannot yet compete.

The Development of Export-Oriented Industrialization

In the 1960s, a traditional pillar of the economy, entrepôt trade, was ravaged by tensions among Singapore, Malaysia, and Indonesia. Even after relations improved, entrepôt trade continued to be dampened by the efforts of Indonesia and Malaysia to develop their own trade links with other countries, effectively bypassing Singapore. In 1968, another economic calamity struck Singapore: The British announced that by 1971 they would close all of their bases east of the Suez Canal, including the sprawling facility in Singapore. For Singapore, this decision was a turning point. The British base employed, directly or indirectly, one-fifth of Singapore's workforce and accounted for one-quarter of the city-state's GDP (LePoer, 1991; Sullivan, 1991). Absorbing the shock of the withdrawal required more rapid growth in the other sectors of the economy.

Singapore's economic success became export-oriented industrialization (EOI). Three other Asian economies, South Korea, Taiwan, and Hong Kong, also pursued an aggressive strategy of EOI at this time, but Singapore's approach was somewhat different from the others for it relied much more heavily on investments from foreign multinational corporations (MNCs). Singapore's leaders felt that the city-state had too little experience with manufacturing and too few entrepreneurs to rely on domestic firms. For MNCs looking for an inexpensive place to do business, Singapore had several attractions (Rodan, 1989). The labor force was partly English speaking and relatively well educated, and after the PAP government clamped down on unions in 1968, strikes were rare. Infrastructure was excellent, especially the airport, seaport, and the massive industrial park the government established at Jurong in southwestern Singapore. Finally, special tax concessions were available to foreign investors. One of the industries to take advantage of these opportunities was the U.S. semiconductor industry (Henderson, 1986). Silicon wafers that had been manufactured in the United States were airfreighted to Singapore where the labor-intensive, tedious work of cutting the wafers into integrated circuits was carried out. Then the circuits were shipped by air cargo to the United States to be incorporated into a wide variety of electronic goods. The success of Singapore's EOI strategy was reflected in the rising importance of manufacturing in the national economy during the 1960s and 1970s (Chee, 1991). In 1957, 14.3 percent of Singapore's workforce was in the manufacturing sector. By 1970, the proportion was 22 percent, and 10 years later it reached 30 percent, making manufacturing the most important sector in Singapore's economy (Figure 14.6).

The "Second Industrial Revolution"

Singapore's export success has powered the thriving economy. Between 1966 and 1973, the annual rate of GDP growth never fell below 10 percent. But the city-state's dependence on exports also meant that it was sensitive to any downturn in major importing nations. The 1973 Arab Oil Embargo triggered a recession in Singapore's most important markets, leading to more feeble

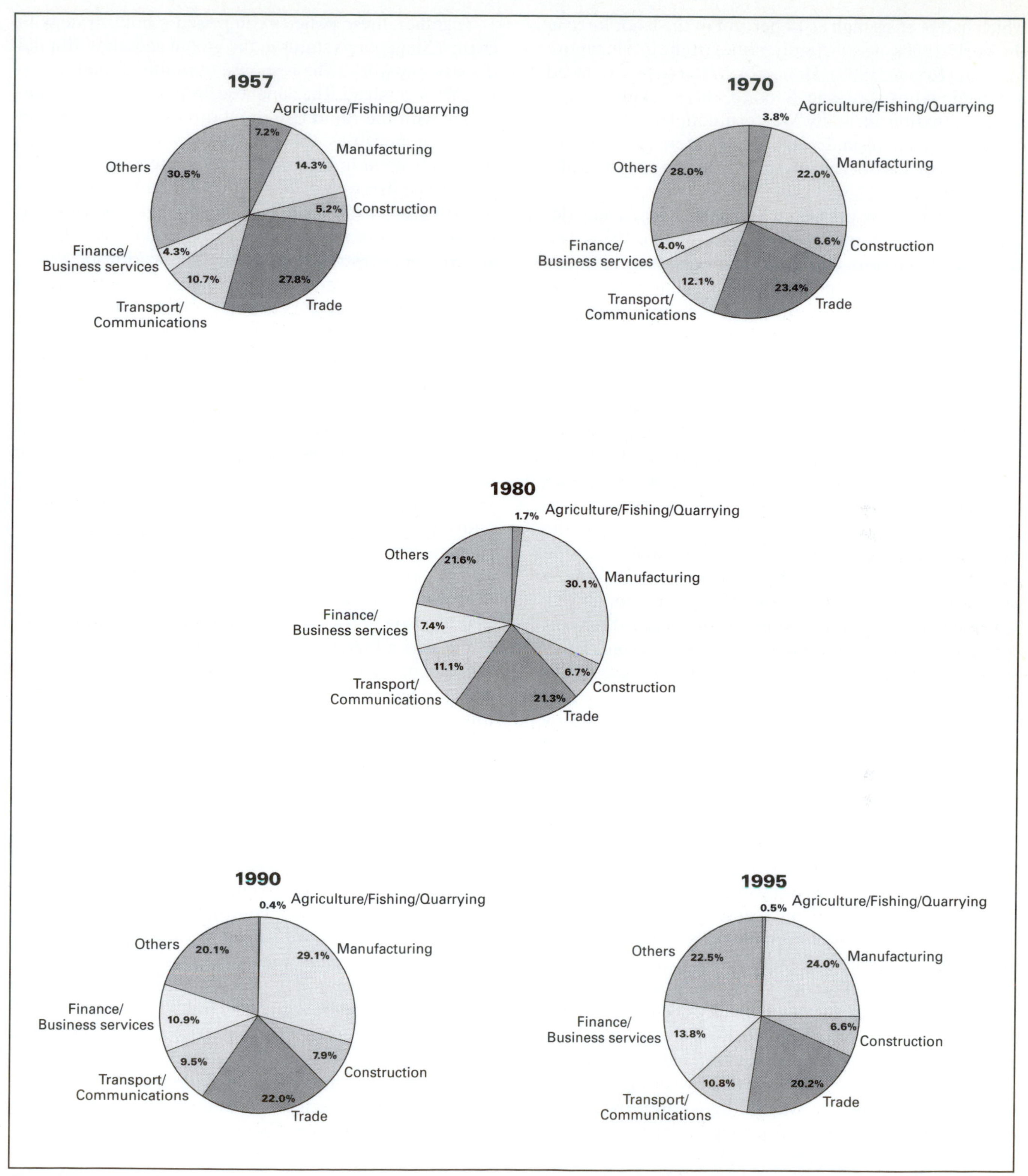

Figure 14.6 Employment by economic sector for selected years, 1973–1995. *Source: Singapore Department of Statistics, 1997.*

growth in Singapore too. A few years later, the economy grew more rapidly again, but the lesson of the mid-1970s remained: Singapore's economic health was closely linked to the overall health of the global economy.

There were two other threats to Singapore's export-oriented economy during the 1970s. First, the country's workforce became more expensive, making other countries potentially more attractive to MNCs. Unemployment, which had been as high as 14 percent of the labor force in the early 1960s, was virtually nonexistent in Singapore by 1976 (Rodan, 1989). Henceforth, the city-state faced a chronic labor shortage. Second, the weakening economic situation in many Western countries led to increased protectionism. To save jobs, Western governments erected trade barriers against imports from countries like Singapore.

Faced with these pressures, Singapore's leadership devised a new development strategy in the late 1970s. The economy was to move away from labor-intensive goods like textiles and cheap electronics equipment. In their stead, the new specialties of Singapore's manufacturing sector would be high-technology goods made by highly skilled, well-paid Singapore workers. Industries targeted for development included specialty chemicals, precision engineering products, automotive components, medical and surgical instruments, computer peripheral equipment, and computer software. The transition to an economy emphasizing more sophisticated goods was to be Singapore's "Second Industrial Revolution" (Rodan, 1989). If successful, this revolution was designed to solve some of the problems associated with exporting labor-intensive goods. First, it was recognized that these new industries are relatively insensitive to recession. Consumers might buy fewer shirts during a recession but the demand for, say, surgical instruments is not greatly affected. Second, it was unlikely that protectionism would impact these industries because they were still dominated by highly competitive firms from the West. Finally, countries offering lower wages than Singapore could not easily compete in these industries because they lacked the well-educated workforce and specialized infrastructure needed to manufacture such technically demanding products.

To launch the Second Industrial Revolution, in 1978 the government adopted a set of aggressive measures. Expenditures for tertiary education, especially in engineering, were increased radically, and a special board was established to promote computer skills at all levels of the labor force. The objective of these investments was to train workers in high technology skills. The government also built new industrial estates specially designed for high tech firms. In those industries which might not attract large amounts of foreign investment at first, such as aerospace, the government took the initiative by setting up state-owned firms to pave the way. Although the government adopted various measures to attract foreign firms in high-technology industries, it also sought to discourage new investment in labor-intensive activities. Its most daring measure in this regard was a government-mandated pay increase for all Singaporeans. The National Wages Council ordered large hikes in 1979, 1980, and 1981 that resulted in average wage increases of more than 50 percent (Rodan, 1989). Such a large increase made Singapore too costly for many types of labor-intensive manufacturing.

Together, these policies comprised a bold strategy to change Singapore's status in the global economy. But did the strategy work? The response of multinational corporations was mixed. The Japanese, in particular, were unhappy with the rapid rise in labor costs. Investment in high-tech industries did occur but was slow. Meanwhile, new investment in traditional, labor-intensive industries evaporated. The worst effect of the Second Industrial Revolution was to exacerbate the 1985–1986 recession in Singapore. The recession was not caused exclusively by the government's restructuring policy. The worldwide glut of oil, which devastated Singapore's refining industry, was also important. But the city-state's high wages did lead many manufacturers to look elsewhere for possible investment sites. In 1986, nearly 7 percent of Singapore's workforce was unemployed.

To accelerate the recovery from the recession, the government reversed its wage policy. Wages were brought down, and labor-intensive operations were welcome once more in Singapore. But the other policies of the Second Industrial Revolution were maintained and seem to have had some effect on Singapore's development. The comprehensiveness of the government's strategy is illustrated by its efforts to establish an aerospace industry in Singapore. The aerospace industry, which includes the manufacture and repair of aircraft and aircraft components, was a logical choice since the city-state already had one of the busiest airports in Asia as well as one of the most successful airlines in the world, Singapore Airlines (SIA). To make Singapore a "one-stop hub" for the aerospace industry, the government adopted a multipronged approach (Low, 1995). First, it established an industrial estate adjacent to the airport to cater specifically to aviation-related firms. Second, the government gave aerospace firms "pioneer industry" tax status, entitling them to large tax breaks for 10 years. Third, the government funded worker training in skills directly relevant to the aerospace industry. Fourth, the government set up a new state-owned firm, Singapore Aerospace (SAe), to foster the growth of the industry in Singapore. SAe repairs and refurbishes aircraft as well as manufacturing components like the landing gear doors for the Boeing B777 and passenger side doors for the Airbus A320. By 1995, there were more than 50 aerospace firms in Singapore (Low, 1995). Many are small MNCs that specialize in repairing and refurbishing specific aircraft components like brakes and electrical generators. These companies, along with SAe and Singapore Airlines Engineering Company (a subsidiary of SIA) have made Singapore the largest airplane repair center in Asia (Low, 1995). Could Singapore have achieved this distinction without the special government policies? Perhaps; but the government seems convinced that its policies have accelerated the development of targeted industries. Its efforts to orchestrate Singapore's Second Industrial Revolution will continue.

New Directions for the Economy

Since the 1985–1986 recession, the government has sought to diversify the city-state's economy away from its traditional reliance on EOI and investment by foreign MNCs (Lee, 1989). Economic diversification in Singapore has three dimensions:

1. Reinvigorating the Service Sector. Historically, the service sector in Singapore has consisted primarily of low wage workers providing domestic services. An excellent example is the food hawker who, working from a simple cart or stall, sells inexpensive meals. In fact, hawkers facilitated the industrialization of Singapore because they helped keep the cost of living, and therefore wages, low. Food hawkers and other domestic service providers still have this effect in Singapore.

Yet the potential of the service sector to contribute to economic development in Singapore goes well beyond its favorable impact on living costs. International business services have become an important dimension of the Singapore economy. Like those who work in the industries targeted for development under the Second Industrial Revolution, employees of firms providing international business services are highly skilled and well paid. They often have expertise in fields like banking, finance, advertising, engineering, or management.

This pool of talent has grown as Singapore has emerged as Asia's third-most-important financial center after Tokyo and Hong Kong (Sullivan, 1991). Banks from around the world established offices in Singapore during the 1970s and 1980s. From Singapore, they coordinate financial flows to and from Asia. These banks also encouraged the growth of ancillary services such as insurance.

The government has sought to encourage the growth of more complex international business services. After the 1985–1986 recession, the Operational Headquarters Programme was initiated. It offers a variety of incentives, including tax breaks, to MNCs that establish a major regional headquarters in Singapore (Ministry of Information and the Arts, 1993). By 1997, dozens of MNCs, mostly Fortune 500 companies such as the Dutch electronics giant Philips, had taken advantage of the government's offer and had broadened the scope of their activities in Singapore from just manufacturing to include research and development, product development, marketing, logistics management (coordinating the movement of resources in a firm's operations), financial management, and technical services. Hundreds more MNCs operate smaller regional headquarters operations in Singapore.

Another dimension of Singapore's service sector is tourism. Until the 1990s, much of Singapore's tourist industry was fueled by visitors who came to shop in the malls and boutiques along Orchard Road. However, the sharp appreciation of the Singapore dollar has made the city less attractive to foreign shoppers. To develop a more diverse tourist industry, new attractions have been developed. Sentosa, a small island off Singapore's southern coast, has been developed as a tourist haven and features everything from "Volcanoland" to a wax museum depicting Singapore's history. To attract business and conference tourists, a mammoth new convention center was opened in 1996.

2. Stimulating Local Entrepreneurship. The overwhelming importance of MNCs in Singapore's economy is illustrated by the contrast between the computer industry in Singapore and Taiwan. In the early 1990s, Singapore and Taiwan were the fourth- and fifth-most-important producers of computers and computer-related products, respectively (after the United States, Japan, and Britain); yet their involvement in the industry was very different. Foreign MNCs account for more than 70 percent of the output of Singapore's industry. Firms such as Apple and Hewlett-Packard have major facilities in the city-state. By contrast, foreign firms generate less than 20 percent of the Taiwan industry's output. Small, indigenous companies are much more important in Taiwan. Their aggressive innovation has enabled Taiwan to capture large global market shares for products such as monitors and computer mice.

With its advantages in infrastructure, its international transportation links, and its English-educated workforce, Singapore continues to attract big investments from MNCs. For example, in 1994, a joint venture among Singapore's Economic Development Board, Texas Instruments, Canon, and Hewlett-Packard established a US$330 million silicon wafer fabrication plant. Such capital-intensive facilities are a tribute to the success of Singapore's Second Industrial Revolution; yet the government would also like to see Singapore spawn more indigenous firms that will lead innovation in the computer industry as well as other industries. One example of a successful Singaporean innovator is Creative Technologies. In the mid-1990s it became a technology leader in the development of sound cards, products that enhance the capability of computers to make sounds. Creative Technologies is still the exception, however. To become a center for innovation in the burgeoning computer and information technology industries, Singapore will need many more homegrown entrepreneurial firms.

Cultivating more indigenous entrepreneurs will take time. Tax incentives, excellent infrastructure, and stable government—factors crucial to Singapore's attraction to MNCs—cannot stimulate entrepreneurship. Indeed, Singapore's very importance as a center for MNCs may have curtailed entrepreneurship. Many young people have opted for a safe job in one of the large MNC operations rather than make a risky venture in forming their own business.

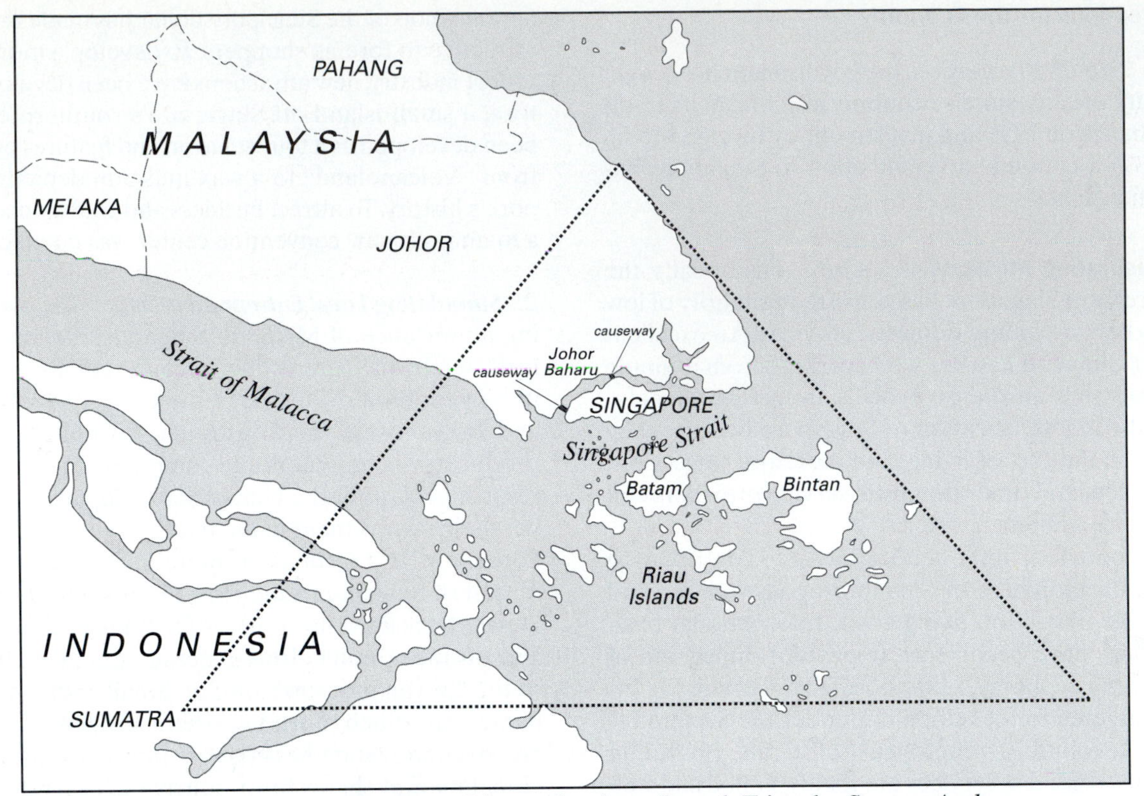

Map 14.6 Johor–Singapore–Riau or Southern Growth Triangle. *Source: Author.*

3. Singapore's Second Wing: Investment in the Region. The final prong of the government's economic diversification strategy has been to strengthen ties with other Asian nations. The most ambitious government program to achieve this aim is the Growth Triangle (Map 14.6). Singapore is one of three nodes in this project that was announced in 1989 (Kumar and Lee, 1991). The others are Johor state in Malaysia and the Riau Islands of Indonesia. Restrictions on the movement of people, goods, and capital are to be eased so that a firm can invest in the triangle as though it were a single extended metropolitan area. The resources of the three partners complement each other (Table 14.4). Singapore offers excellent communications and transportation infrastructure, a highly skilled workforce, and a wide range of international business services. Riau and Johor offer more ample land and labor resources. The motivation for Singapore's participation in this project is clear. Even as it loses its competitiveness for investments requiring abundant land or labor, it can still benefit if those investments flow instead to one of its partners in the Growth Triangle.

Yet Singapore also stands to benefit from investments further afield in Asia. Public sector institutions have been effective in marketing their expertise to governments eager to learn from Singapore's example. For instance, HDB set up a corporation to market its public housing expertise and has so far been involved in the planning and design of new towns in China and Indonesia. Singapore's government is also involved in a gigantic project to build a Jurong-style industrial estate in Suzhou, a city near Shanghai in China. Singapore's experience in providing inexpensive, efficient public goods is, some believe, the most attractive resource it can sell to newly emerging economies.

Although Singapore's public sector experience is particularly relevant in such countries, its private sector firms are also aggressively exploring opportunities in the region. It is the private sector which has led Singapore investment in Thailand, for example. Singapore was the third-most-important source nation for foreign direct investment in Thailand in 1993 (Schwarz, 1994). Only Japanese and U.S. firms invested more. Singaporean firms have also invested in the other countries of Southeast Asia as well as the potentially much bigger economies of China and India. Even in Vietnam, where

	Industrial Land Costs (US$/Sq.Mi.)	Labor Costs (US$/Month)		
		Unskilled	Semi-Skilled	Skilled
Johor	4.08	150	220	400
Singapore	4.25	350	420	600
Riau Islands	2.30	90	140	200

Table 14.4 Comparative costs in the Johor–Singapore–Riau growth triangle, 1989. *Source: Kumar and Lee, 1991.*

Singaporeans were forbidden from investing until the 1991 Cambodian peace accord, Singapore has become the eighth-most-important source of foreign capital ("Chinese Wave in Vietnam," 1994). Together, the overseas investments by Singaporean firms and institutions comprise what Lee Kuan Yew has called the second wing of the city-state's economy, complementing the first wing of investments in Singapore itself by foreign MNCs. By amplifying both of these linkages with the global economy, Singapore maximizes its competitive advantages; it gains access to resources and markets far greater than the small island could provide. But Singapore's ties to the region and to the world beyond also bring risks. Singapore's increasingly close relationships with other Southeast Asian economies made it inevitable that the financial crisis which struck other regional economies in 1997 would spill over into the city-state's borders. In 1998, Singapore experienced its first recession since the mid-1980s.

TRADE AND TRANSPORT

Trade is the lifeblood of the Singapore economy. That was the case when Singapore was chiefly an entrepôt, it remained so when the economy's driving force became export-oriented manufacturing, and it will also be true of the more diversified economy that is emerging in Singapore. In the future, trade in services will become more important, but manufactured goods will still be critical in the country's trade flows. To ensure that trade between Singapore and the rest of the world can continue to grow, the government has invested heavily in the country's seaport and its airport, making them among the among the largest and most efficient in the world. The domestic transportation system is also well developed, featuring an excellent network of public buses and trains. As a result, Singaporeans are less dependent on private automobiles than their counterparts in other wealthy countries. Public transportation is vital to Singapore because it reduces the need to build new highways or widen existing ones, both of which would use up more of Singapore's limited land resources.

Trade Flows

Though the importance of trade is a constant in the history of Singapore, the nature of trade has certainly changed. As an entrepôt, most of the city's exports were re-exports. Goods exported from a country were imported into Singapore, processed, and then re-exported to their final destination. Tin from the Malay Peninsula, for instance, was collected in Singapore and then loaded onto ships bound for the industrialized countries. Today, slightly more than half of the value of Singapore's exports are accounted for by domestic exports, products manufactured in the city-state.

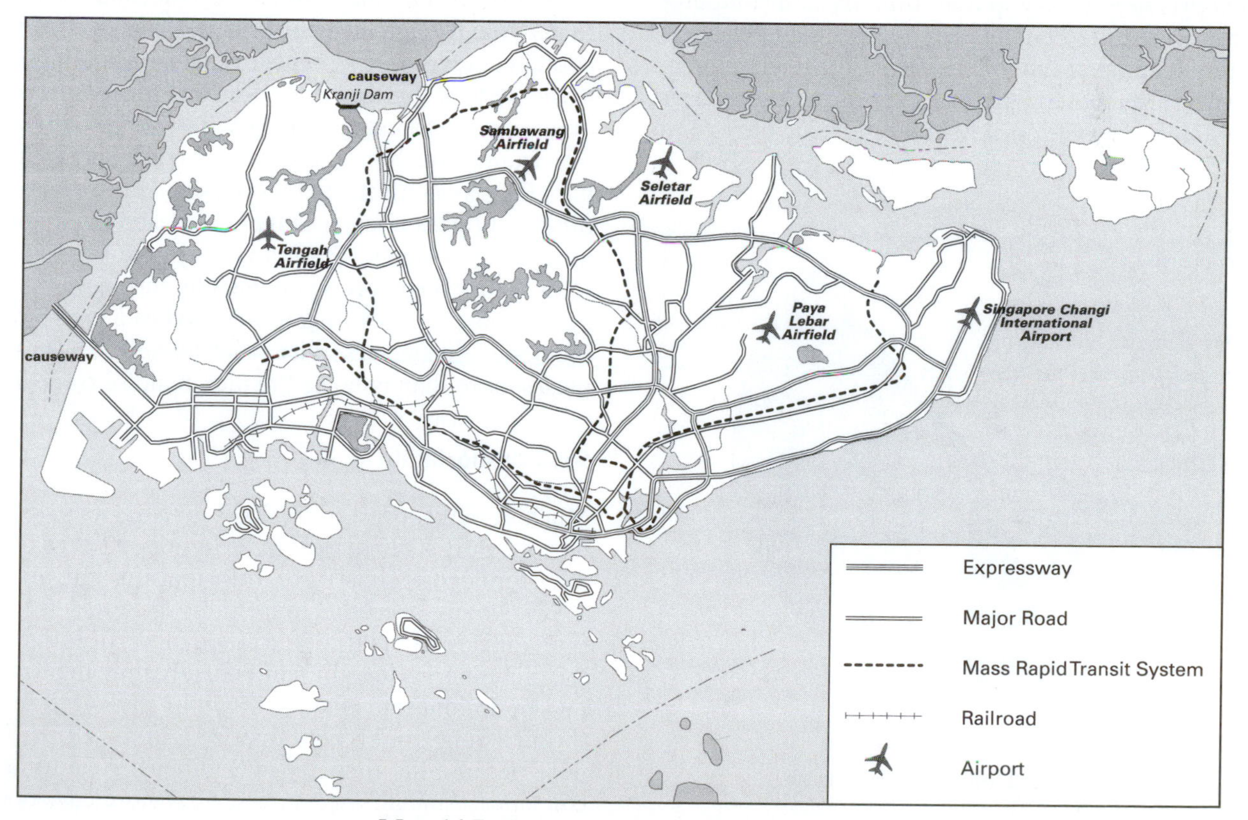

Map 14.7 Singapore's transportation network.

Refined petroleum products are among Singapore's most important exports (Figure 14.7). Indeed, Singapore is the world's third-most-important refining center after Houston and Rotterdam (Sullivan, 1991). The city-state has no oil resources of its own but has established a niche in the refining industry as a "balancing" center (Fesharaki, 1989). Singapore's refineries balance the weaknesses of the refining industry in the Asia–Pacific region: too little capacity to meet the region's growing energy demand and an inability to meet the product mix needed by Asian energy consumers. The refineries in Singapore have been able to get more high-value petroleum products out of crude oil than most other refineries in the region. Singapore's five major refiners (British Petroleum, Shell, Esso, Mobil, and the state-owned Singapore Petroleum Company) export to a changing mix of countries as some countries in the Asia–Pacific region expand their own domestic refining capacity. In the early 1990s, for instance, both South Korea and Japan became more self-sufficient in refining and imported less from Singapore. In 1993, the single most important export market for Singapore's refiners was Hong Kong (Singapore Department of Statistics, 1995). Thailand, China, and India are also likely to be growth markets for Singapore's petroleum products.

Singapore's other important exports include products from the computer industry. Disk drives were the single most important export item in the early 1990s. Singapore also exports desktop computers, printers, and computer monitors. Integrated circuits are another significant export item. The rapid rate of change in information technology suggests that Singapore's exports in this industry will continue to evolve quickly. Much of Singapore's computer industry output is exported to the United States as U.S. MNCs with operations in the city-state export their output back to the home market. More generally, the United States is the most important export market for Singapore (Singapore Department of Statistics, 1997). Americans consume roughly one-fifth of Singapore's exports. Singapore's second-most-important export market is Malaysia, and Japan ranks third. Exports to Asian markets are growing more rapidly than those to the United States, and Singaporean exporters are eagerly exploring the region's emerging economies; for instance, Singapore is Burma's most important trading partner.

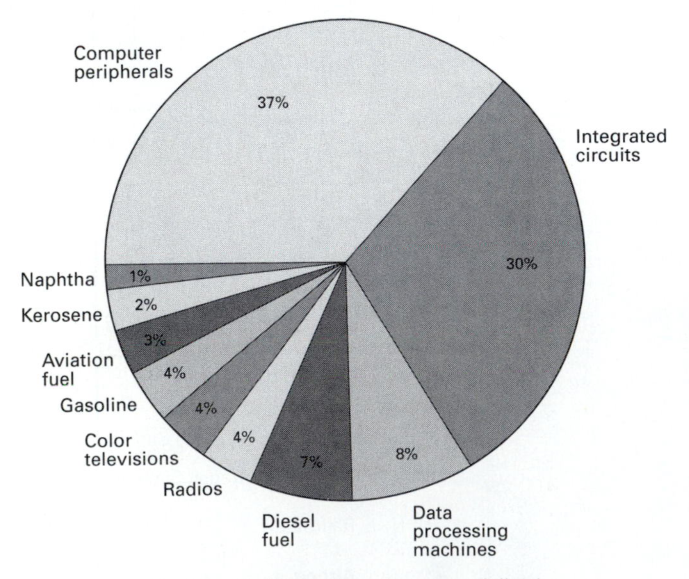

Figure 14.7 Leading export items, 1996.

Despite Singapore's success in tapping export markets, it has a trade deficit—more specifically, a merchandise trade deficit. Its trade deficit with Japan is especially large (Figure 14.8). In part, this is attributable to the practices of Japanese MNCs operating in Singapore (Sullivan, 1991). Whereas U.S. MNCs operating in Singapore produce mainly for the U.S. home market, Japanese MNCs produce mainly for third-country markets (including the United States). Relatively little of the Japanese MNCs' output is directed back to Japan. Consequently, the flow of goods to Japan is weak. At the same time, however, Singaporeans, like consumers around the world, have bought a multitude of Japanese cars, video cassette recorders, kitchen appliances, and a host of other items. The result is a big trade deficit with Japan.

Singapore's deficit in merchandise trade with the rest of the world is financed in part by the steady inflow of funds from foreign investors, especially major MNCs. In addition, the country benefits from its substantial surplus in the trade of services. For example, Singapore's superior air and sea transport services draw cargo and passengers from throughout the region who use Singapore as a hub. The payments made by shippers and travelers from other countries to such Singapore-based companies as Singapore Airlines (SIA) and Neptune Orient Lines (NOL) give Singapore a very healthy surplus in the trade of transport services. Singapore also has a big surplus in the trade of financial services such as banking. These surpluses more than balance the merchandise trade deficit.

Sea Transportation

Much of Singapore's trade flows through its seaport. In fact, the port is the busiest in the world in terms of shipping tonnage (Ministry of Information and the Arts, 1993). At any one time, more than 700 ships are in the port, ranging from ultralarge crude oil carriers that serve the oil refineries to light bumboats that ply the routes to nearby Indonesian islands. Much of the cargo moving through Singapore is containerized. In the 1970s, Singapore became the first city in Southeast Asia to open a container terminal; since then, several more have opened, and Singapore has become the second busiest container

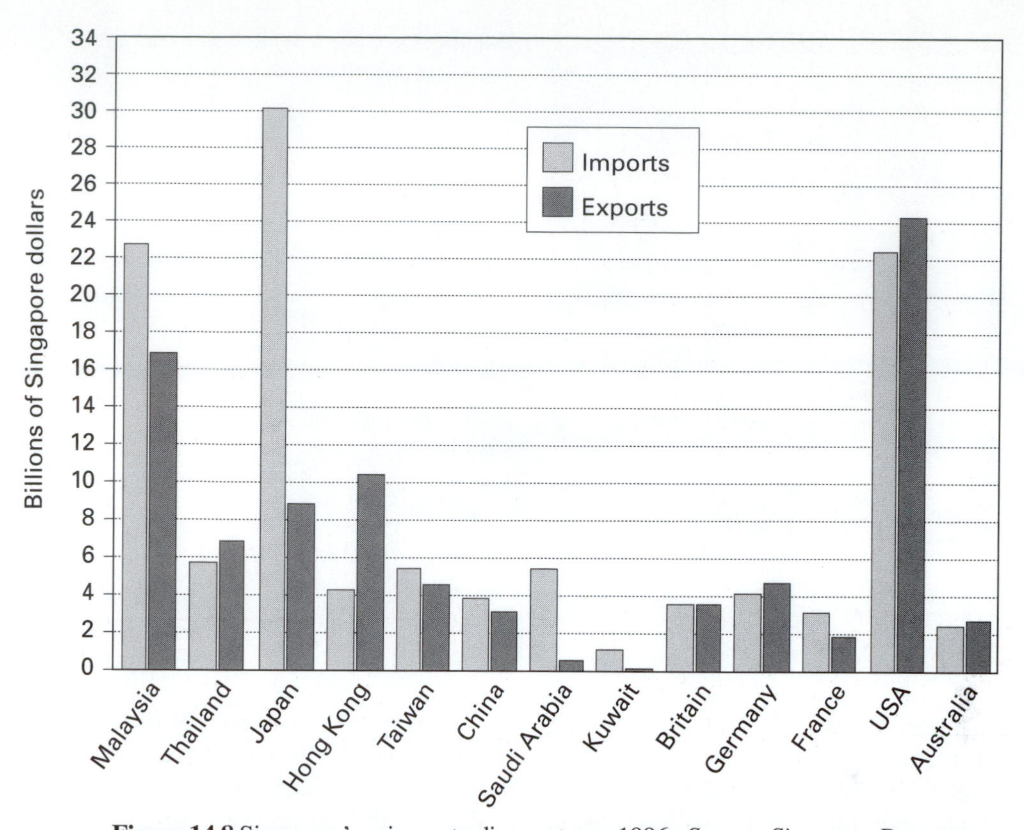

Figure 14.8 Singapore's primary trading partners, 1996. *Source: Singapore Department of Statistics.*

port in the world, handling 13.0 million TEUs in 1996, just behind the 13.2 million TEUs handled by Hong Kong (Ng, 1997). One TEU is equivalent to the capacity of a 20 ft (609.6 cm) shipping container.

The port extends all along the southern coast of Singapore island. In addition, terminals have been established on offshore islands to make room for growth. In the west, offshore islands have become peninsulas as the sea has been reclaimed to permit the growth of Jurong Terminal, which serves Singapore's most heavily industrialized area. The lack of land in the port area is a key constraint on further port development. To make more efficient use of the limited land and labor available to port activities in Singapore, the port has invested heavily in automation and computerization. As a result, the time that cargo takes to pass through the port is much less than for ports elsewhere in the region, reinforcing Singapore's historic advantage as the major hub port for the region. A substantial share of the cargo to and from Malaysia, Thailand, and Indonesia still passes through Singapore; for instance, Malaysian port officials estimate that about 55 percent of containers exported from Kuala Lumpur's Port Kelang pass through Singapore ("More M'sian-owned," 1996). To counteract this trend, the Malaysian government has launched a major initiative to get Malaysian importers and exporters to use Malaysian ports. Singapore's response to this threat is to continue its successful strategy of massive new infrastructure development and further investment in automation and computerization. A new container terminal in western Singapore will more than double the capacity of the port to 34 million TEUs by 2009 ("Futuristic Container Terminal," 1995).

Shipbuilding and ship repair are two important activities along Singapore's waterfront. After the British abandoned their naval base, the site was converted to a giant shipyard. The new shipyard and others built subsequently absorbed the workers who might otherwise have become jobless after the pullout (Sullivan, 1991). Today, Singapore's shipyards manufacture and refurbish merchant vessels and warships and even "Loveboats." In 1994, the two cruise ships that hosted the 1970s television series *Loveboat* docked at Singapore for midlife facelifts.

Air Transportation

Singapore's importance as a seaport is mirrored by its status as one of the world's major international airline hubs. Changi Airport is served by more than 60 airlines from around the world, including a growing number of all-freighter operators that cater to Asia's booming trading economies. These airlines, both passenger and cargo, are drawn to Singapore because the city serves as the regional air traffic hub for Southeast Asia. Indeed, the

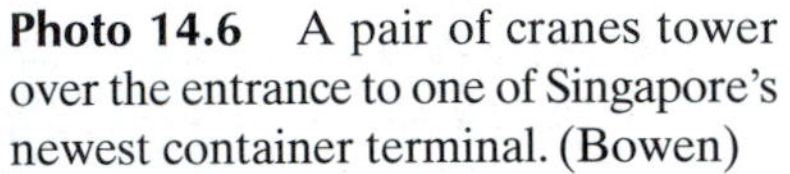

Photo 14.6 A pair of cranes tower over the entrance to one of Singapore's newest container terminal. (Bowen)

routes between Singapore and nearby capitals are among the most densely traveled in the world; for instance, the route between Singapore and Kuala Lumpur is one of the 10 busiest international routes in the world (Bowen, 1993). By serving Singapore, international airlines have access to the whole Southeast Asian market: A passenger on a United Airlines flight from Chicago, for example, can easily make a connection to one of the dozens of flights each day to Indonesia once he or she arrives in Singapore. Changi Airport is, therefore, much more attractive to international airlines, even small ones, than the city-state's population would suggest.

Singapore has been one of the leading advocates of "open skies." Open skies treaties eliminate most restrictions on which airlines can serve which routes with what type of aircraft and how often. Singapore was the first Asian country to sign an open skies treaty with the United States. Singapore's liberal approach to airlines that want to serve Changi Airport is part of its strategy to counter the threat posed by new airport developments in the region, especially Malaysia's new Kuala Lumpur International Airport (KLIA). KLIA is expected to present Changi Airport with a keen new rival for air traffic. Other competing airport developments are also planned

Photo 14.7 United Parcel Service (UPS), which uses Changi Airport as one of its major worldwide air freight hubs, makes deliveries in Singapore's Central Business District (CBD) with these motorcycles. (Bowen)

or underway in Bangkok and Manila. Even nearby Hang Nadim Airport on Indonesia's Batam Island has been upgraded to accommodate jumbo jets. Singapore's view is that with air traffic (both passenger and cargo) growing rapidly in the region, there is more than enough business for all of these new and upgraded facilities. Changi Airport is also expanding, with a third passenger terminal to be completed by 2003 and land reclamation underway for the airport's third and fourth runways.

SIA is Singapore's flag carrier and one of the largest airlines in the world, flying to 73 cities in 39 countries. In 1995, it was ranked twelfth among airlines worldwide in terms of passenger traffic and sixth in terms of air cargo traffic. SIA has exploited the strategic location of its hub to tap international traffic within Asia and between Asia, Europe, North America, and Australia. The airline is Singapore's single largest private employer; roughly 1 out of every 80 workers in the city-state is employed by SIA (Bowen, 1993).

In the 1990s, SIA has given greater emphasis to cargo traffic as a complement to passenger traffic. In fact, cargo traffic is expanding more rapidly in Singapore than passenger traffic. An increasing proportion of the city-state's manufactured goods are high-value, light, and quickly obsolete goods from the computer industry such as chips, disk drives, and printers—just the sort of products that are likely to be exported by air rather than by the cheaper but much slower seaborne alternative. But Singapore's exports by air also reflect the tiny remnant of the country's agricultural sector. Every day, shipments of expensive orchids and live tropical fish are airlifted from farms in Singapore to customers in Japan, Europe, and the United States.

Land Transportation

Although the government of Singapore has made massive investments in its seaport and airport to allow international traffic to expand as rapidly as possible, its investment in land transportation has a different objective. The government has sought to funnel traffic growth on the island into the public transportation system rather than private automobiles. The mainstay of the public transportation system are buses, with the two main operators deploying more than 3,000 buses (Ministry of Information and the Arts, 1993). To make buses more attractive to the increasingly affluent Singapore commuter, both of the companies operate mainly air-conditioned buses, and one began to introduce double-decker, air-conditioned tourist-style "SuperBuses" on heavily traveled routes in 1993.

The other main element of the public transportation system is the Mass Rapid Transit (MRT) system (Map 14.8). The first phase of the MRT was completed in 1990 and covered a 41.54-mi (67-km) network (Sullivan, 1991). The US$2 billion investment in the MRT, Singapore's largest ever infrastructure project, helped boost the economy out of the 1985–1986 recession (Sullivan, 1991). In 1996, a new branch was opened in northernmost Singapore, adding another 9.9 mi (16-km) of track. Two more lines, to northeastern Singapore and to the airport will be opened early in the twentyfirst century. Reflecting the high cost of land in Singapore, the two new lines will be built almost entirely underground.

Most of the country's public housing estates are already served by the MRT, and bus routes are integrated with the MRT so that commuters from across the island

Photo 14.8 Throughout much of the city-state, the Mass Rapid Transport (MRT) system operates on elevated tracks. (Bowen)

Map 14.8 Mass rapid transit system, 1996.

can use the train, especially for trips to work in the CBD. In several large estates, the government is also building smaller light rail transit (LRT) systems using driverless trains to feed traffic into MRT stations.

As a result of such investments to increase the ease and attraction of using buses and trains, about 51 percent of daily trips are now on public transport. The government's goal is to raise that figure to 75 percent (Land Transit Authority, 1996). Achieving this target will be difficult because public transport can never match the convenience of the private automobile. To deter people from choosing the latter alternative, the government has made it extraordinarily expensive to own and drive a car. A person who wants to buy a new car in Singapore must first bid for a certificate of entitlement (COE). The government restricts the number of COEs that are made available, and for popular classes of cars (especially for medium-size and large cars) a COE is a huge expense. A COE for a medium-size car such as the Toyota Corolla, a popular seller in Singapore, cost US$40,000 in June, 1997 (Leong, 1997). Nor do the financial disincentives end with the COE. One must also pay a 45 percent import duty on the car's open market value and an additional registration fee of 150 percent of the car's open market value, all of which raised the cost of a Toyota Corolla to approximately US$90,000 in June 1997. By contrast, the same car cost about US$15,000 in the United States.

As incomes in Singapore continue to grow, the demand for automobiles also increases. The government has promised that the ratio of cars to people will rise from 1:10 in 1996 to 1:7 by 2010 (Land Transit Authority, 1996). To allow more people to own cars without having to expand the road network greatly, the government has implemented tough measures to control demand (see box), for instance, motorists must pay special fees to drive in the CBD or to use the country's main expressways during the morning rush hour. The idea of such measures is to smooth out demand away from the most congestion-prone areas and time periods. If demand can be spread more evenly throughout the day and diverted to less popular roadways, Singapore's existing highway infrastructure will be able to support many more cars than it does now. That will allow the government to release more COEs, and the cost of owning an automobile should stop escalating so rapidly.

POLITICAL DEVELOPMENTS

The People's Action Party (PAP) has governed Singapore continuously since 1959. During that period, the party and especially its top leadership have enjoyed con-

BOX 14.3 Electronic Road Pricing

Since the 1970s, Singapore's government has used road pricing to discourage motorists from using congested roadways. The first measure adopted was the Area Licensing Scheme (ALS) that required motorists to buy a special license to enter the CBD during the peak morning and evening rush hours. In 1997, a Road Pricing Scheme (RPS) was extended to three most important expressways on the island with fees again being charged for use during the rush hours. The fees charged for the ALS and RPS amount to a few dollars per day, enough to compel many commuters to find alternate routes or other modes of transport.

The next step for Singapore is Electronic Road Pricing (ERP) (Land Transport Authority, 1996). Automobiles in the country will be fitted with a transmitter that will "communicate" with sensors on overhead gantries along the country's major roadways. The sensors will record the movements of every vehicle, and a certain amount will be deducted from a smart card built into each car's transmitter. The charge for driving along typically congested roads and at peak hours will be greater than for other roads and times.

Singapore will be the world's first country to try ERP on a large scale. Singapore is well suited to the ERP because its traffic network is largely self-contained. Except for the causeway and the new bridge to Johor completed in 1997, Singapore's traffic system is cut off from the outside world. This containment means that it is easier for the government to educate motorists about the ERP and to ensure that all cars using the affected highways are fitted with the proper device.

Singapore's political culture also facilitates ERP. In the 1980s, Hong Kong experimented with ERP but eventually abandoned the effort because it was politically unacceptable. One concern about ERP in Hong Kong was that the colonial government (and after 1997 the government in Beijing) would be able to track the movements of individual motorists, using the ERP as a kind of domestic spy system. This issue has not been raised in Singapore, partly because there is no tradition of aggressively challenging PAP government initiatives and partly because the success of the PAP's policies since the 1960s has creaited a political culture in which most people simply trust the government to solve the country's problems.

siderable political autonomy because few groups or individuals outside the PAP have had the political clout to challenge the government's policies seriously. The government has used its power to promote Singapore's economic development, often through policies that would be untenable in other countries; for example, it would be politically suicidal for most governments to levy taxes and fees that triple or quadruple the price of an automobile, even if those taxes and fees serve the worthwhile purpose of saving limited land resources for uses other than new streets and highways. Also, people in many other countries would be deeply suspicious of government efforts to promote marriage and reproduction among certain elite groups. Yet these and other potentially controversial measures have been justified as necessary to sustain the city-state's remarkable economic development.

Foreign MNCs, which account for much of the economic activity in Singapore, have been attracted by the government's pragmatic commitment to economic development. The PAP's policies toward labor, infrastructure, education, and other economic variables have created a very favorable business environment in Singapore. Foreign investors are also attracted by the longevity of the PAP government. Singapore has been politically stable for decades, and foreign investors have confidently made billions of dollars worth of commitments in the city-state with little fear of a radical change in government that might render those investments worthless.

The People's Action Party

The PAP government has certainly played a leading role in Singapore's development. It has set the goals for the economic development process, such as those of the Second Industrial Revolution, and has established the conditions necessary to meet those goals. In turn, Singapore's prosperity has bolstered the political power of the PAP. It is very difficult for any rival party to argue, "We can do better. Vote for us." Indeed, opposition parties have difficulty even attracting candidates to run for Parliament, in part because it is feared that running against the PAP will damage one's career. That does not mean that the PAP is invincible. On the contrary, its share of the popular vote declined continuously from 1980 to 1991 (Table 14.5). In the 1991 general election, the PAP received 59 percent of the vote—a solid majority, of course—but the lowest percentage for the PAP since 1963 (Ministry of Information and the Arts, 1993). What factors lay behind this trend? Today's electorate is dominated by voters who have little or no memory of Singapore's crisis years. Younger voters do not feel the same gratitude toward PAP leaders that is still common among those who witnessed the astonishing improvement in Singapore's fortunes during the 1960s and 1970s. Today, the country's prosperity is taken for granted. Recognizing that the voters in the 1990s are increasingly willing to cast their ballots for the opposition, the PAP approached the 1997 election with a more fiercesome political strategy.

Year	Percentage of Popular Vote Won by PAP	Seats in Parliament Out of Maximum Possible
1959	53	43 of 51
1963	47	37 of 51
1968	84	58 of 58
1972	69	65 of 65
1976	72	69 of 69
1980	76	75 of 75
1984	63	77 of 79
1988	62	80 of 81
1991	59	77 of 81
1997	65	81 of 83

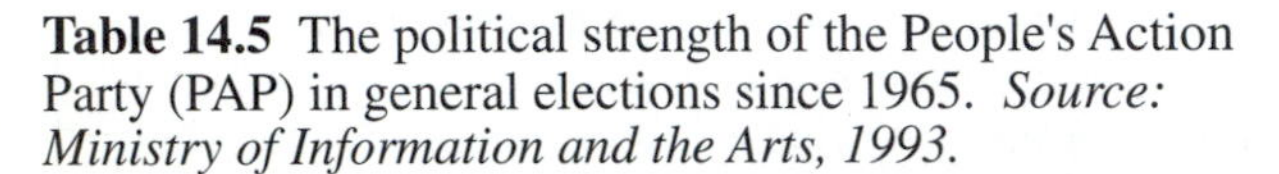

Table 14.5 The political strength of the People's Action Party (PAP) in general elections since 1965. *Source: Ministry of Information and the Arts, 1993.*

PAP leaders promised to accelerate the government-subsidized upgrading of HDB apartments (e.g., adding an extra room to each apartment) in districts that voted heavily for the PAP, while those that voted for the opposition would be placed last on the list for upgrading. Indeed, in the heat of battle, Prime Minister Goh Chok Tong even suggested that districts voting for the opposition would become slums. The election resulted in the highest PAP share of the vote since the 1970s, with 65 percent of the vote.

The PAP's political dominance seems sure to continue. In the wake of the 1997 election, the major opposition parties were in shambles. The Singapore Democratic Party, which had three of the four opposition members of Parliament in 1991, was left with none in 1997, and the Worker's Party position was undermined when one of its most popular candidates, Tang Liang Hong, was accused of being a pro-Chinese, anti-Christian, anti-Malay "racial chauvinist." Tang said that the PAP leaders were lying about his record; in retaliation a group of PAP heavyweights including Senior Minister Lee Kuan Yew, Prime Minister Goh Chok Tong, and the two deputy prime ministers sued the Worker's Party candidate for defamation. The PAP leaders' suit was successful and they were awarded nearly US$6 million in damages by the court. In the meantime, Tang fled the country.

The heavy-handed approach toward Tang was viewed negatively by many Singaporeans, reflecting the fact that the increasingly well-educated Singapore electorate approves of the PAP's economic policies but not necessarily its politics. Perhaps because the attitude of Singaporeans toward the PAP is changing, the PAP's attitude toward the people it governs has also changed to some extent under the leadership of Prime Minister Goh Chok Tong, who took over from Lee Kuan Yew in 1990. The scope of government intervention in everyday life has narrowed; for example, there are no longer government-imposed standards on the appropriate length of a man's hair. Once, a man with hair below a certain length could be denied government services and, if a foreigner, could be denied entry to Singapore. Another example concerns censorship of movies: Until the early 1990s, only movies deemed acceptable for a general audience, including children, were permitted to be shown in the country's movie theaters. Movies not meeting this standard were either banned or edited to remove the offending scenes and language. In 1991, a new rating was introduced to allow movies restricted to adult audiences. There is a limit, however, to how far the PAP government can change in response to popular demand. The PAP is not, after all, a populist party; rather, it reflects the meritocratic nature of Singapore more generally. In a meritocracy, leaders derive their right to rule, their legitimacy, from their superior knowledge and ability. The role of the populace is to defer to those with authority. For the PAP to defer instead to the governed—to change important policies in accordance with the changing views of Singaporeans—would be a fundamental contradiction of the party's philosophy.

Singapore's Innovative Parliamentary System

Though Singapore's form of government is based on the British parliamentary system, the city-state's Parliament has several unique features, two of which were introduced in the late 1980s and early 1990s. The first reform was the creation of group representation constituencies (GRCs) (Rodan, 1993). Each GRC is served by four to six members of Parliament (MPs) rather than the usual norm of one MP per constituency. In the 1997 election, there were 15 GRCs (which together elected 74 MPs) and 9 single member constituencies (which sent one elected MP each). The rationale for GRCs was to ensure that racial minorities would be represented in Parliament. At least one of the four MPs from each GRC must be a Malay, Indian, or member of another minority group.

In addition to serving this purpose, however, the GRCs also allow the PAP to make it even harder for the opposition parties to win seats in Parliament. Voters in GRCs must select a party's slate of four to six candidates as a whole rather than picking and choosing from among the candidates put forward by different parties. The PAP has taken advantage of this rule by teaming nationally known leaders with more obscure candidates in each GRC. For example, former Prime Minister Lee Kuan Yew is one of the six MPs from one GRC while Prime Minister Goh Chok Tong is one of the six MPs from another GRC. It would be very difficult for most Singaporeans to vote against these two political giants no matter how weak the other candidates on the PAP teams are or how strong the opposition teams are. The PAP can therefore use one unbeatable candidate to win up to six seats in Parliament.

The second recent innovation in Singapore's parliamentary system is the nominated MPs scheme. Responding to the criticisms of those who claimed that Parliament served little purpose beyond confirming the PAP leadership's decisions, the constitution was amended to allow between one and six nominated MPs (Rodan, 1993). The nominated MPs are not politicians but rather individuals deemed to have made significant contributions to Singapore; for example, those serving in the mid-1990s included a surgeon, a professor, and a businessman. Because they have no political links, the nominated MPs are supposed to be freer in voicing their opinions of government policies. To become a nominated MP, however, one must be chosen by a committee made up of elected MPs as well as other state officials. Not surprisingly, then, those serving in this special role are politically moderate. Unlike MPs elected from opposition parties, the nominated MPs pose no threat to the PAP government because they cannot vote on certain important issues and cannot form a government.

The State and Development in Singapore

Singapore's PAP leaders make no apologies for the rather half- hearted nature of democracy in the city-state nor for the myriad restrictions on political freedoms—political freedoms that are commonplace in other countries as economically developed as Singapore. Instances abound of tight restrictions on freedom of the press, freedom of speech, and freedom of assembly. For instance, a newspaper columnist who wrote a mildly critical editorial about Goh Chok Tong in the mid-1990s was publicly scolded by the prime minister and warned that if she wanted to make comments about politics she should join a political party; several political opponents of Lee Kuan Yew have been sued, like Tang Liang Hong, into bankruptcy after being charged with libel; a group of Filipino workers in Singapore who wanted to hold a public prayer vigil after a Filipino maid was hanged for murder in the city-state in 1995 were warned that such a vigil was illegal and they would be arrested if they went ahead with the plan. PAP spokespeople vigorously defend the prerogative of the government to establish a society that does not follow the Western model of liberal democracy; Singapore's more authoritarian style, they argue, reflects the Asian values of its population, its vulnerability to internal and external threats, and its desire to avoid the mistakes of the West. Indeed, having orchestrated the growth of an economy that now rivals those of many Western nations, Singapore's leaders are unlikely to discard what has proven to be a very effective political system.

But although the government seeks to control the political development of Singapore, the openness of the economy means that their control can never be complete. The limitations on government power are illustrated by the mixed reception that the Internet has received in the city-state. On the one hand, Singapore's leaders want businesses and households to be linked to the Internet, which is one element of the global information superhighway. But messages carried over the Internet have also included scathing criticisms of the PAP, incorrect information about government policy, and pornography. The government would like to prevent such transmissions from the anonymous, amorphous electronic network but can do relatively little to stop them.

Slower economic growth in the future, especially given the recent financial crises will also present the government with new challenges. In addition, keener competition from regional rivals and the maturation of Singapore's own economy (and the aging of the population) will make it difficult to achieve the average annual growth of almost 9 percent that was recorded between the mid-1960s and the mid-1990s. Slower economic growth will also mean a slower rate of improvement in living standards. This is to be expected as the quality of life in Singapore is already high by world standards. The transition to slower growth could threaten the PAP because it has based its legitimacy in the past on delivering rapid economic growth and better living standards.

The PAP has shown remarkable resilience in the past, and it is likely that it will survive these new challenges too. In part, the PAP's longevity is attributable to the institutions it has put in place that serve to perpetuate its power. The HDB public housing program, the system of Group Representation Schemes, the tradition of meritocracy are some of the institutions which help the PAP to keep its lock on political power. But the PAP also enjoys broad popularity, confidence, and trust. The challenge for the PAP's new generation of leaders is to build up the same level of public support among younger Singaporeans.

With a view toward this challenge, Prime Minister Goh Chok Tong set out a vision in the 1990s for a more gracious society that would value artistic achievement, diversity, and civic participation in addition to economic success. The new emphasis placed on these values reflects Singapore's growing affluence. From independence through to the 1980s, the country was very single minded in its focus on economic development. Now, with Singapore among the most economically developed nations on Earth, attention has shifted to creating a more livable Singapore, a place that is not just a supercharged economy but also the "best home" for Singaporeans.

15 The Philippines

RICHARD ULACK

The Philippines is somewhat of an anomaly compared to the other Southeast Asian nation-states. Partly because it is the only contemporary nation that had no major pre-European indigenous kingdom or state within its present territory, it was to become a colony that was quickly and almost totally subjugated by Spain, its colonial "master" from 1565–1898. There had been no large, well-established kingdoms, cities, or trading ports such as Angkor, Palembang, or Malacca; in short, the archipelago was, compared to much of the rest of Southeast Asia, an economic and political backwater, and with the exception of Islam, which first gained a foothold in the southern Philippines in the fifteenth century, religion and other major external cultural influences had not yet penetrated the islands in a major way. In short, Spain found a land and people rife for the taking. The Spanish, impelled by the twin urges of acquisition and proselytization, had a greater impact on its colony than perhaps any of the other colonizers. Filipinos became Roman Catholics and remained so, and Spanish landlords became rich from the *encomiendas* that were established and from the profits that were to be made from trading the primary commodities produced in the islands. Indeed, contemporary Filipino surnames and place names are of Spanish origin. Also, unlike other Southeast Asian states, the Philippines had a second Western colonizer, the United States from 1898–1946.

The impacts of these two colonial influences means that the Philippines is today the most Westernized of the Southeast Asian nations. "The least Oriental country in the Orient" (Wernstedt and Spencer, 1967, p. 115) and "Three centuries in a Catholic convent and fifty years in Hollywood" (Karnow, 1989, p. 9) are apt summaries of Philippine history during the 350 years of colonial rule. Although the people are overwhelmingly Mongoloid in racial characteristic and speak as their primary language one or another of about 75 Malay–Polynesian languages and dialects, their religion is Roman Catholicism, and the principal common language and the language of education and the national media is English.

Photo 15.1 Among the numerous legacies from the American colonial period is basketball, today among the Philippines' most popular participant and spectator sports. (Ulack)

Today, the Philippines is well integrated into the global economy, and although there are recent indications that the economic situation is improving (see, for example, "Back on the Road," 1996; Hutchcroft, 1996; Kulkarni and Tasker, 1996; Tiglao, 1995), it has not thus far made the social and economic progress that the other original ASEAN members have. President Ramos declared a *Philippines 2000* program whereby the country will join the newly industrializing economies (NIEs) by the year 2000. But the nation is not yet poised, as are Thailand and especially Malaysia, to join Singapore among the NIEs even though by some measures (for example, percentage of GDP from manufacturing and percent of labor force in manufacturing), the Philippines compares favorably with those nations. On the other hand and for a variety of reasons, the Philippines has not been as negatively affected by the Asian financial crisis of the late 1990s as have Indonesia, Thailand, and Malaysia. Nonetheless, the continuation of economic and political control by a small oligarchy created during the colonial period, a near total absence of fossil-fuel energy resources, and political issues including the continuing Moslem insurgency in the south, among other factors, have meant that the country cannot yet be counted among those capitalist nations that have begun to move out of the world's periphery. It will be interesting to follow the administration of newly elected (in 1998) former movie actor President Joseph Estrada, not part of the Philippine oligarchy, to see if political change will occur.

HISTORICAL EVOLUTION

Filipino Roots

Because the archipelago is the farthest area in the region from the Asian landmass, the islands were among the last areas in Southeast Asia to be settled by people. Whereas the earliest archeological evidence for human habitation is more than 20,000 years old, it is thought that the first people arrived on Palawan more than 30,000 years ago during the last great ice age. At that time, Palawan and other parts of archipelagic Southeast Asia that were surrounded by the Continental Shelf was connected by land to Borneo and the Indonesian archipelago and were thus accessible by land to the Asian mainland. These earliest peoples were the ancestors of today's Negritos, or *Aetas,* black-skinned people of short stature. Never very populous and always widely scattered in their distribution, very few people of this racial stock remain in the Philippines. By the 1960s, perhaps as many as 15,000 relatively pure blooded Negritos remained, and these were found in the remote, forested areas of the larger islands (Wernstedt and Spencer, p. 150).

A second, larger wave of immigrants to the archipelago were the proto-Malays. The first seafarers to the Philippines, these Mongoloid settlers probably arrived in the northern Philippines from southern China and Tonkin as many as 10,000 years ago. These were the forebears of the present-day Bontoc, Ifugao, Kalinga, and Apayao peoples of interior, upland northern Luzon, ethnic groups that are today collectively known as the Igorots, the Tagalog word for mountain-dwelling people (Wernstedt and Spencer, 1967, p. 349). Among other notable achievements, the mountain peoples of northern Luzon built the magnificent terraced rice padis into the steep mountain slopes of the Central Cordillera. These practices were probably learned from the Chinese, and today such terraces in the Philippines are found only in northern Luzon. Other proto-Malay groups came from the south, and their descendants include the Manobo and Bagobo ethnic groups of Mindanao, the Mangyan of Mindoro, and the Tagbanua of Palawan.

Before the dawn of the Christian era, the final and largest wave of immigrants, peoples also of Mongoloid racial stock called the Deutero-Malay, began to arrive from the Indonesian archipelago. They "pushed" the earlier immigrants to more interior and remote upland locations and quickly acquired control of the coastal districts. These were organized into small, independent communities made up of kinship groups called *barangays*, the basic unit of structure, each ruled by a chief, or *datu*. Unlike Java and the other major core areas of Southeast Asia, there were no great unified kingdoms that emerged in the Philippines before the arrival of the European colonizers.

About five centuries before the arrival of the Europeans, commercial relations with China, Indochina, the Malay Peninsula, India, and the Arab lands intensified. These seafarers brought their porcelain wares, silk, cotton, gold, and jewelry, as well as other manufactures and traded for shell and coral products, bird's nests (for soup), mattings and baskets, abaca, coconuts, rattans, and aromatic woods and other products of the sea and forest. During the fifteenth century another external impact, Islam, had gained a foothold in the Sulu Archipelago. By the 1570s the religion had diffused as far north as Manila where the Spanish established their colonial capital.

The Spanish and U.S. Colonial Period and After

The year 1521 marks the beginning of European influence in the archipelago as in that year Ferdinand Magellan, Portuguese by birth but a citizen of Spain, arrived in the archipelago. First landing in the central Philippines near the island of Leyte, Magellan's ships sailed on to the small island of Mactan, near present-day Cebu City. On Mactan, Magellan was killed by a local *datu* named Lapu-Lapu and his men. Magellan's ships sailed on and one of them

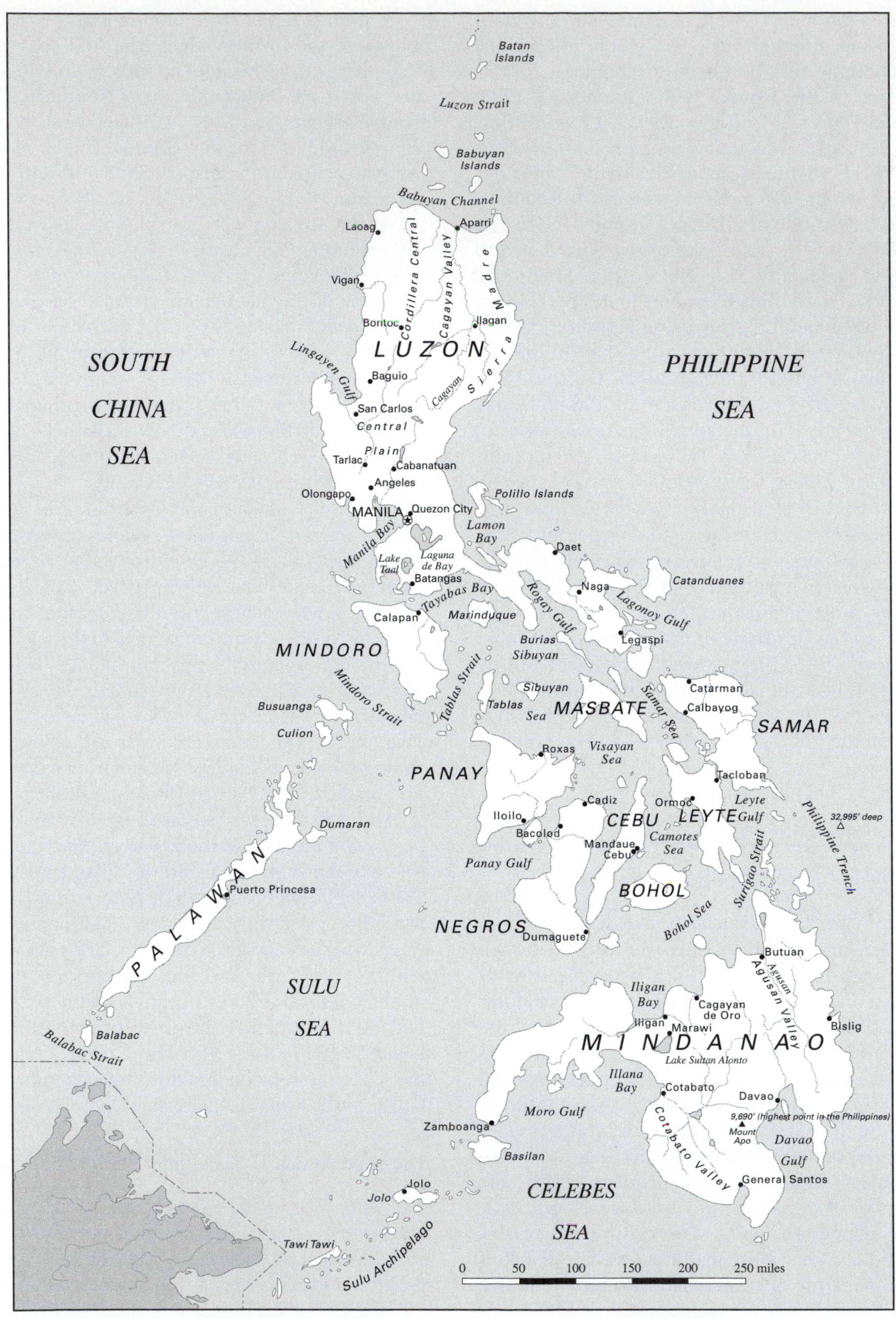

Map 15.1 The Philippines. *Source: Ulack and Pauer, 1989.*

survived, eventually to complete the first European circumnavigation of the world. Lapu-Lapu, who has become something of a national hero, has been credited with being the first Filipino leader to have repelled European aggression.

By the mid-sixteenth century other Spanish vessels had arrived and the colonization began in earnest. Miguel López de Legaspi, the first royal governor, established in 1565 the capital at Cebu, located on the eastern coast of Cebu Island. After defeating the Muslim ruler Rajah Soliman at Maynilad, Legaspi moved the capital to Manila in 1571. Manila's much better site and situation allowed Spain to realize more effectively its early objectives in the Philippines, most notably to provide a safe, intermediate harbor for the Manila galleon trade between Mexico and China which lasted from 1565 to 1815 (Lyon, 1990). By the early 1600s Spain's control of the Philippines was complete, and Roman Catholicism soon replaced the animistic beliefs of most Filipinos and, except in the Muslim south and in the most remote interior areas, it became the predominant religion.

Photo 15.2 Monument to Lapu-Lapu, Mactan Island, Cebu Province. (Ulack)

The nearly 350 years of Spanish rule greatly altered the Philippines. In addition to the introduction of Roman Catholicism, other important early legacies included new systems of landholding such as the *encomienda* system (landholdings covering several villages that were granted to Spaniards); payment of taxes and tribute to the *encomenderos*; the commercial production and export trade of tropical crops including the important crop of sugarcane but also abaca, coconuts, and tobacco; the introduction of new plants including maize, sweet potatoes, manioc, and the pineapple; and the introduction of education through Catholic schools, including the region's first college (in 1611), now the University of Santo Tomas, in Manila. Interestingly, the clergy did not use the Spanish language nor offer much in the way of European literature and thinking in schools as there was some concern that the "... indigenous people might get restive if they received sophisticated education. Consequently, the education was religious and anti-intellectual" (Sardesai, 1989, p. 68). During the period the Philippines was closely linked by the galleons that carried Chinese porcelain and silk from Manila to Mexico and carried priests and silver bullion on the return voyage. By the 1590s a number of Chinese had permanently settled in Manila and had become an economically important community, serving not only as traders but also as bookkeepers and craftsworkers. Many Chinese, aware of the political and social advantages enjoyed by the Roman Catholics in the colony, converted to Catholicism and married Filipinas. As was so all over Southeast Asia, distrust and envy of their economic successes often caused great difficulty for Chinese in the Philippines as they and their *mestizo* (mixed-blood) descendants became a powerful and dynamic force in Philippine society. Today, the Chinese and their descendants, together with Spanish *mestizos*, control a disproportionate share of the wealth and power in the Philippines. Although difficult to define because of assimilation through intermarriage, in 1990 the Philippine census reported there were some 600,000 ethnic Chinese, or less than 1 percent of the total population.

Whereas Spain firmly controlled the colony at the top, it governed the countryside through indirect rule by creating an indigenous Filipino upper class, referred to as the *principales.* This was the beginning of an oligarchic system of control that remains to the present day. During this time the original Filipino notion of communal land ownership was replaced with the concept of private, individual ownership of land, and it is the descendants of these *principales* that control most of the land today (Dolan, 1993, p. 6).

Gradually, during the latter half of the nineteenth century, discontent with Spanish policies brought about a growing nationalism in the Philippines. A number of incidents of Spanish repression enraged Filipino intellectuals, students, and clergy who were led by Dr. José Rizal y Mercado, physician, writer, and poet. Rizal espoused integration with Spain rather than independence; however, by the mid-1890s Rizal and his colleagues had become disillusioned and had abandoned hope for a peaceful solution. Revolution seemed the only answer. The execution of José Rizal in Manila on December 30, 1896, sealed his martyrdom and facilitated the revolt against Spain.

The United States acquisition of the Philippines came as a result of its victory in the Spanish-American War in 1898. Filipinos, led by General Emilio Aguinaldo, expected U.S. support for independence following the defeat of Spain. The United States, however, refused to give the Filipinos their independence, and bitter conflict between U.S. and Philippine forces lasted until 1901, when Aguinaldo was captured. In 1934 the U.S. Congress passed the Tydings-McDuffie Act whereby the Philippines would be granted complete independence after a 10-year period of internal self-government. In 1935 the Commonwealth of the Philippines was established, but World War II delayed independence for the Philippines; the Japanese occupied Manila in 1942 and soon afterward the entire country was under Japanese occupation. Philippine independence from the United States finally came on July 4, 1946, although the independence day celebrated as the national holiday is June 12, the day the Philippines became independent from Spain in 1898.

The United States's presence in the Philippines during the colonial period and since has left some lasting impacts. Among these are a public education system that uses the English language as its primary medium of instruction; political institutions, including the notion of democracy (until martial law in 1972) and the constitutional separation of powers between the presidency and Congress; and an economy that has remained closely linked and dependent on the former colonizer, although economic relations with other nations, most notably Japan, have become increasingly important in the last two decades.

After independence, Americans maintained some privileges in the newly independent Republic of the Philippines. For example, the Philippines' first president, Manuel Roxas, gave U.S. citizens equality with Filipinos in exploiting the country's natural resources, and in 1956 under the Laurel-Langley Agreement, this parity was extended to all economic activities until 1974. Ties between the Philippines and its former colonizer have remained strong through, for example, the presence of U.S. military bases until 1992 when they closed, most notably Clark Air Base and Subic Bay Naval Base, both on Luzon Island. But in all cases the economic agreements and relations that were promulgated after independence worked in favor of the former colonizer.

CULTURAL GEOGRAPHY

Contemporary Ethnic Identities

The vast majority of the Philippine population is comprised of Malay–Polynesian-speaking peoples of Mongoloid racial stock. Collectively, the population is called Filipino, but in reality it is comprised of about 75 distinct ethnic groups. According to the results of the 1990 national census, it was reported there are more than 950 dialects spoken in the country, but only a few of these were spoken by significant numbers, and the eight most important dialects spoken accounted for more than 85 percent of the population (Figure 15.1). Four of the languages, Tagalog and Ilocano spoken on Luzon, and Cebuano and Ilongo (or Hiligaynon) spoken in the Visayas, were the primary languages for nearly three-quarters of the population. Spoken by about 28 percent of the population, Tagalog—the language of the indigenous population of the Manila region and central Luzon—and Cebuano, spoken by nearly one-quarter of the population, are the largest single languages, but each has its own dialects. Filipino, the national language of the Philippines, is based on Tagalog. Filipino and English are the two official languages, and whereas both are used as languages of instruction in schools, English has remained the primary medium in education, business, government, and the national media. Although President Aquino in 1990 ordered that all government offices use Filipino as a language of communication, most Filipinos have not accepted it as the national language in the 1990s.

Given the diversity of languages and dialects in the Philippines, one would expect difficulty in communication between the different culture regions of the island nation. Creation of a national language, Filipino, has been difficult because several other languages with large populations exist. Today, English remains the *lingua franca* in most of the Philippines, and the country's major newspapers, magazines, and radio and television stations are in English. It is likely that English and Filipino will remain the "national" languages, while one's native language or dialect will continue to be that which is spoken at home.

Religion

Religion plays a central role in Philippine life. Whereas nearly all Filipinos practice some form of Christianity or Islam, these major religions have been superimposed on ancient traditions and belief systems, and the blend, especially when combined with the strong personal faith of Filipinos, that has resulted is a modification of the way in which the religion is practiced elsewhere and is thus uniquely Filipino. Supernaturalism, elaborate rituals such

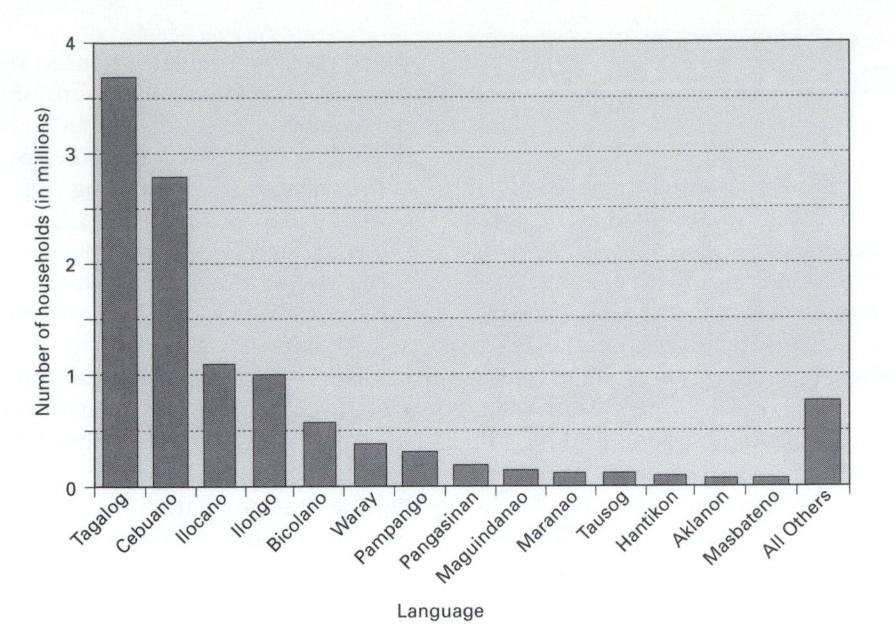

Figure 15.1 Major ethnic groups, 1990.

as the reenactment of the crucifixion, and ceremonies characterize religion in the country.

One of the main legacies of the Spanish period, of course, was the introduction of Roman Catholicism. Today it is by far the most important religion in the country with about 83 percent of the population professing adherence, making it the third-largest religion in the 10-nation Southeast Asian region after Islam and Buddhism, respectively. Other forms of Christianity include a number of imported Protestant beliefs that account for 5.4 percent of the total population and two important indigenous Christian beliefs, Aglipayan (2.6 percent) and *Iglesia ni Cristo* (2.3 percent), both of which emerged in the early 1900s. Authoritarian and evangelical, membership in *Iglesia ni Cristo* has grown rapidly in recent decades. About 4.6 percent of the population adhere to Islam, and the vast majority of Muslims reside in western and southern Mindanao and on the islands of the Sulu Archipelago. Most Chinese are Christians, but there are a small number of Buddhists and Taoists.

Photo 15.3 Iglesia ni Cristo is an indigenous Christian religion in the Philippines and today comprises over two percent of the population. Iglesia ni Cristo churches, like this one in provincial Cebu, are found in cities and towns throughout the archipelago. (Ulack)

POPULATION PATTERNS, MIGRATION, AND URBANIZATION

Population Growth and Policy

The Philippines' demographic transition began sometime around 1930. Prior to that, death rates fluctuated greatly, as for example during the period 1870–1903 when there were repeated "national and local epidemics of cholera, smallpox, and influenza, as well as direct and indirect military deaths" (Concepción and Smith, 1977, p. 14). Since 1930, and with the exception of the Japanese occupation and World War II during which death rates increased and birth rates declined, population has increased rapidly. The 1990 population was officially listed at 60.7 million, and by 1995, the population was estimated to be 68.4 million (Population Reference Bureau, 1995). As the accompanying diagram (Figure 15.2) illustrates, the Philippine population has been increasing at an annual rate of between 2.1 and 3 percent since 1972—during the period shown it has been one of the highest in Southeast Asia. This is the result of a high crude birthrate (CBR), which has ranged from 30 to 45 per 1,000 population between 1950 and 1995, and a low crude death rate (CDR). The CDR has declined from about 15 per 1,000 in the 1950s to less than 9 per 1,000 presently. Thus, whereas fertility levels have declined, so too have mortality levels, which means that the overall growth rate has remained at a high level. The current annual rate of growth, about 2.1 percent, implies that the population will double itself in about 33 years, to more than 140 million by the year 2030. According to the United Nations and other data-gathering sources, a "reasonable" population projection for the Philippines in 2025 is 113.5 million (Population Reference Bureau, 1996).

Whereas the official government view of the nation's population is that it is too high, it is also true that the strong support for family planning programs that existed during the Marcos years no longer exists. During these years, the government espoused that improvements in the nation's economy, and therefore the lives of individual families were closely related to correction of the population problem. The Philippine government agency that is responsible for controlling population growth, Popcom, in 1985 set a target of a 1 percent growth rate by the year 2000. To achieve that goal, the agency recommended a maximum of two children per couple spaced at three-year intervals, that women delay marriage to age 23, and that men wait to marry until they are 25 (Dolan, 1993, p. 74). Such goals have not been widely attained, and this is partly related to the changing view of the relationship between population growth and economic development. In short, some social scientists now believe that population pressures are not the cause of the nation's poverty. But rather it is a common belief that exploitation of the poor by the wealthy has caused and exacerbates poverty, and because widespread poverty persists, especially in the rural areas of the Philippines, children continue to be viewed by parents as economic assets and not as liabilities. In short, if significant progress is to be made in reducing fertility levels, the conditions that cause poverty must be addressed. Former President Ramos, a Protestant, is an advocate of population-growth reduction, but advocating birth control through the use of contraceptives is not popular in a culture where more than four-fifths of the population is Roman Catholic.

Health conditions in the Philippines are typical of most other Southeast Asian nations. Conditions in most areas have generally improved, as is suggested by the increase in life expectancy from 51 years in 1960 to 64 years in 1995 and by the decline in the infant mortality from 101 (per 1,000 children during their first 12 months) in 1950 to 40 per 1,000 in 1995. Ongoing health concerns include malnutrition among children, especially in rural areas, and a new area of concern is the increase of persons infected with acquired immune deficiency syndrome (AIDS). As of mid-1991, the Department of Health re-

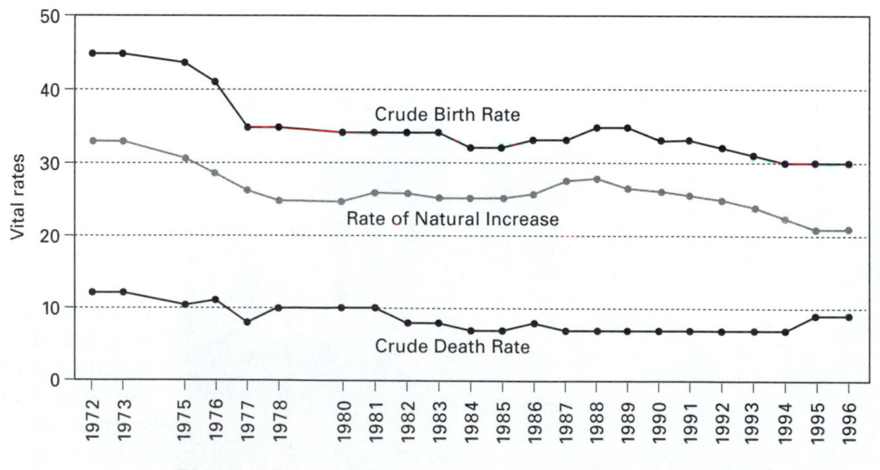

Figure 15.2 Philippine population growth, 1972–1996.

ported that 240 Filipinos were so infected (Dolan, 1993, p. 113). Although the number sounds relatively small and certainly underestimated, there is some concern that the country will follow the pattern of Thailand, "where AIDS has spread into the general population due to the practice among Thai men of having unprotected sex with prostitutes. These men then pass the virus on to their wives and girlfriends" ("AIDS and the Philippines," 1994, p. 2). The Philippines is a late-marrying society wherein cultural mores discourage the use of condoms and where there exists a sizable sex industry frequented by both Filipinos and tourists from abroad. Furthermore, a significant number of young Filipinas travel abroad to work as hostesses, singers, and maids; in some cases, prostitution results and with it, the spread of sexually transmitted diseases.

Population Density and Distribution

Except for the city-state of Singapore, the Philippines has the highest absolute density of population of any of the Southeast Asian countries. In 1994, the population density of the country was almost 600 persons per square mile, nearly double the average for the region (Figure 15.3). With extensive areas of mountainous and upland terrain, this means that the great majority of the population is concentrated in the much smaller area of lowland interior, river basins, and, especially, along coastal plains. Population pressures are especially serious in the lowland areas as a result of the uneven distribution.

The Philippines is one of the more urbanized countries of Southeast Asia (only Brunei and Malaysia, and Singapore of course, have higher percentages), and about one-half of all Filipinos reside in cities and towns. This

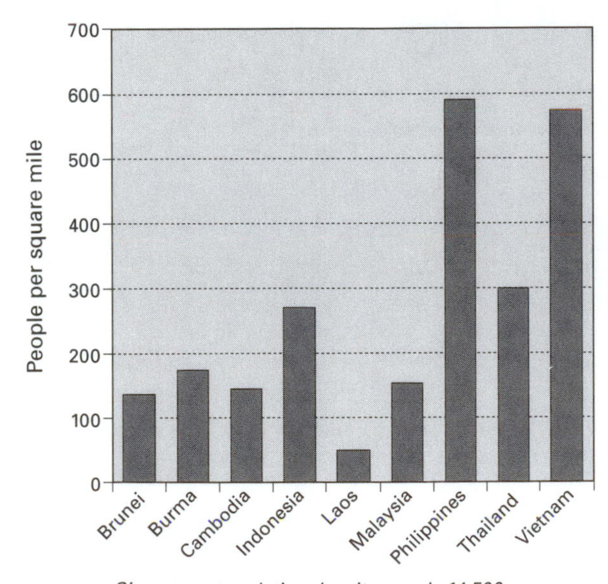

Singapore population density equals 14,500 people per square mile

Figure 15.3 Population density, 1994.

characteristic of urbanism in part derives from the country's Spanish heritage. Metro Manila, with a population of more than 9 million in the mid-1990s contains more than 13 percent of the Philippines' total population and more than one-quarter of all Philippine urban dwellers. Manila is a huge and sprawling megaurban region that overshadows all other cities in the Philippines; indeed, except for Thailand, the Philippines has a higher index of primacy than any of the larger Southeast Asian nations (about 9, which means that Manila's population is about nine times greater than that of the next largest metropolitan area).

Rural areas of high density in the Philippines include the nation's major rice-producing region, the extensive Central Plain north of Manila; the Ilocos coast of northwestern Luzon; the lowlands of the Bicol Peninsula; the coastal areas of the Visayan Islands, especially Cebu, Negros, Panay, Samar, and Leyte; and in the country's major river basins, notably those of the Cagayan in northeastern Luzon, and the Agusan and Cotabato river basins in Mindanao (Map 15.2). Generally fertile volcanic-derived soils on Luzon have made the Central Plain and the Bicol rich agricultural areas with high rural population densities. The Philippines' two best-known volcanoes, Pinatubo and Mayon, are located on the edge of the Central Plain and in Bicol, respectively. The Visayas are centrally located in the archipelago, and portions of them, notably Cebu, were less affected by malaria than other parts of the country (Vandermeer, 1967). These areas have a long history of high population densities, and today such rural areas as the Visayas can no longer support such high concentrations of population because there is little land left to farm. As a result, numerous individuals, mostly younger adults, have been "pushed" from these places to seek opportunities in other less densely populated rural "frontiers" such as Mindanao, to cities and towns, and even to foreign destinations.

Internal Migration and Urbanization

During the twentieth century the Philippine population has witnessed two major internal migration flows: (1) from densely populated rural lowland areas on Luzon and in the Visayas to sparsely populated rural frontier areas, most notably Mindanao and the Cagayan Valley of northern Luzon (see, for example, Ulack, 1977) and, (2) since World War II, heavy migration from rural areas to cities, especially to metropolitan Manila. The rural to rural migration, especially since the 1930s, has meant that over time the population has become more evenly spread throughout the nation than it was in the nineteenth century, with 54 percent of the total located on Luzon (includes Palawan, Mindoro, and other nearby islands), nearly one-quarter on Mindanao and adjacent small islands (the Sulu Archipelago), and 22 percent in the Visayan Islands (Figure 15.4). The data clearly show the increased growth in Mindanao through in-migration, mostly from the Visayas,

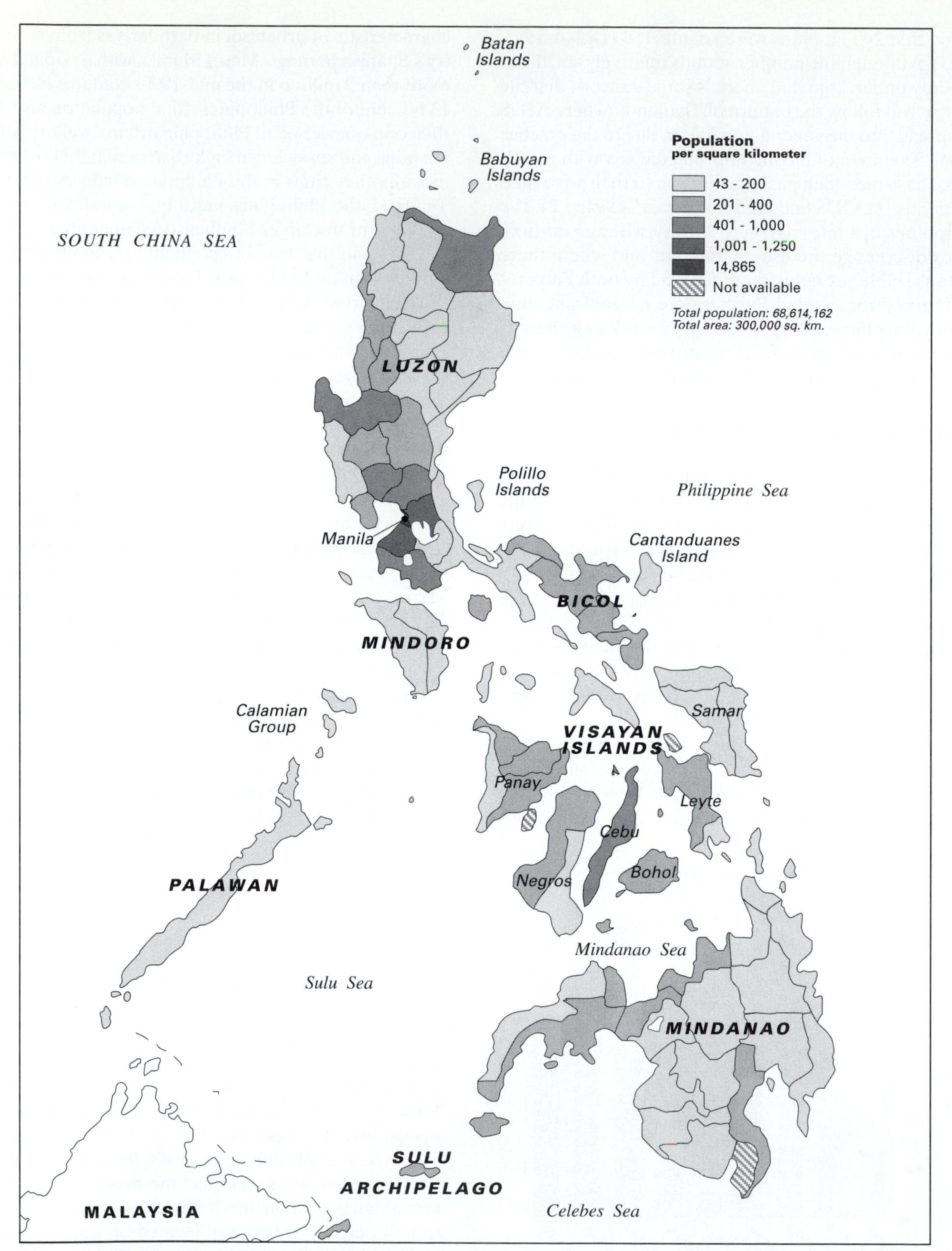

Map 15.2 Philippine population density, by province, 1995. *Sources: Ulack and Pauer, 1989, p. 67; National Statistics Office, 1998.*

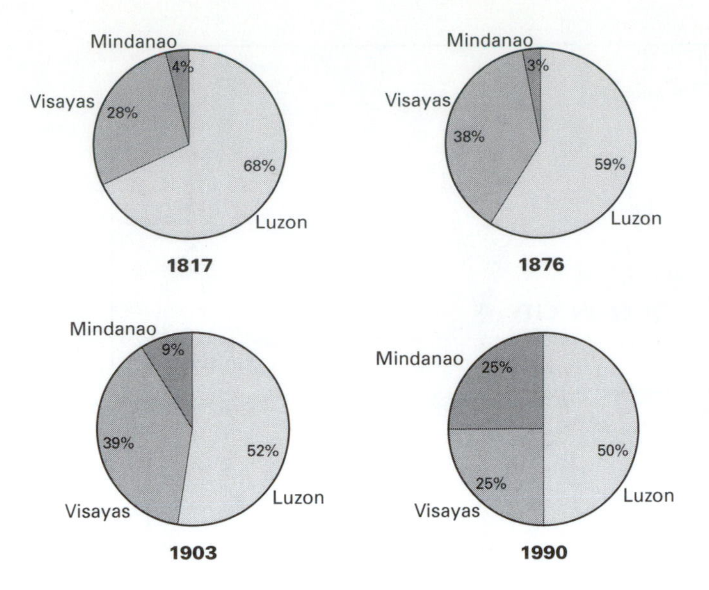

Figure 15.4 Population distribution.

whose population declined from 37 percent of the national total in 1903 to 22 percent in 1990.

Although the Philippines is still a rural nation, nearly one-half (49 percent) of the population resides in urban areas. The cities and towns, especially the largest ones, are growing most rapidly in terms of population, and that is because of the high rate of in-migration to these places. Besides the metropolitan Manila region, other large places include metropolitan Cebu, the country's second-largest urban area, with a population in excess of 1 million. Other cities of importance include Davao, Cagayan de Oro, and Zamboanga on Mindanao; Bacolod and Iloilo in the Visayas; and Baguio, Angeles, and Olongapo on Luzon. Baguio, called the City of Pines, is the chief mountain resort and the informal summer capital of the Philippines. Angeles and Olongapo have experienced great change in the past several years due to the closing of the two major U.S. military installations. The major dramatic and sizable eruption of Mount Pinatubo, a long-dormant volcano about 15 miles east of Angeles, in June 1991, forced the evacuation and closure of Clark Air Base and left the future of the city in doubt. However, by the mid-1990s the former U.S. Subic Bay Naval Base at Olongapo had been transformed into a Philippine free trade zone, and 200 companies, including a large number from Taiwan, had pledged investments of US$1.2 billion in the zone ("Back on the Road," 1996, p. 12). In 1995 the U.S. company Federal Express established Subic as its Asian air hub for overnight parcel deliveries in the region.

Metropolitan Manila

Manila is the nation's chief port for international trade as well as its leading industrial, financial, cultural, and educational center. The site of the early Spanish city is near the mouth of the Pasig River and includes the area known as Intramuros, the original walled fortress built during the early colonial period (Map 15.3). The "new" Manila is found in the rapidly expanding peripheral areas, or suburbs. For example, Makati is the modern center for finance, business, and tourism and a residential area where the Filipino elite and international businesspeople as well as government bureaucrats have concentrated. Quezon City, the site selected as the new capital of the Philippines in 1937 and which officially replaced Manila as capital from 1948 to 1976, is the area where the newer government buildings, medical centers, and the main campus of the University of the Philippines are located. Other administratively separate "cities" that are in reality part of the metropolitan Manila region include Caloocan, Pasay, and Pasig. In 1975 the four politically separate cities of Manila, Quezon City, Pasay, Caloocan, and 13 other urbanized municipalities (or "towns") were restructured and integrated into what is known today as Metro Manila (MMA) or the administrative region called the national capital region (NCR). Interspersed throughout these peripheral areas, as well as in some of the older areas of the city such as Tondo near the port area, are the poor of the city who live in the numerous slum and squatter areas. Some estimates put the city's share of the population residing in such "blighted" communities at as much as one-third.

Like other primate cities in Southeast Asia, Manila's population has grown very rapidly since 1950 when it was about 1.5 million. By 1996 the population was well over 9 million. Massive urban sprawl has intensified the problems of this megacity, the second largest in the Southeast Asian region after Jakarta. Manila has been the destination for large numbers of rural in-migrants from all over the Philippines, people seeking a better life through the employment that is perceived to exist in the city. Unfortunately, the better life is often elusive because employment, if it is found at all, is often in low-paying jobs in the informal economic sector—street vendors selling balut (a duck egg with a partially formed embryo), public market vendors, stevedores, drivers of a variety of vehicles including the colorful jeepneys, or workers in the city garbage dump salvaging whatever might be of value. In the mid-1980s, more 16 percent of the labor force was unemployed, and more than two-fifths was underemployed. When a national economic crisis hits the Philippines, it usually affects Manila especially hard; for example, job losses among industrial workers in the 1980s were heavily concentrated in Metro Manila, where nearly three-fifths of the Philippines' industrial activity is located. The necessary infrastructure for the thousands who arrive in the city each year is inadequate, and thus housing, educational and health facilities, and other social needs are often unmet. Housing continues to be a major problem; the Philippine government estimated that at

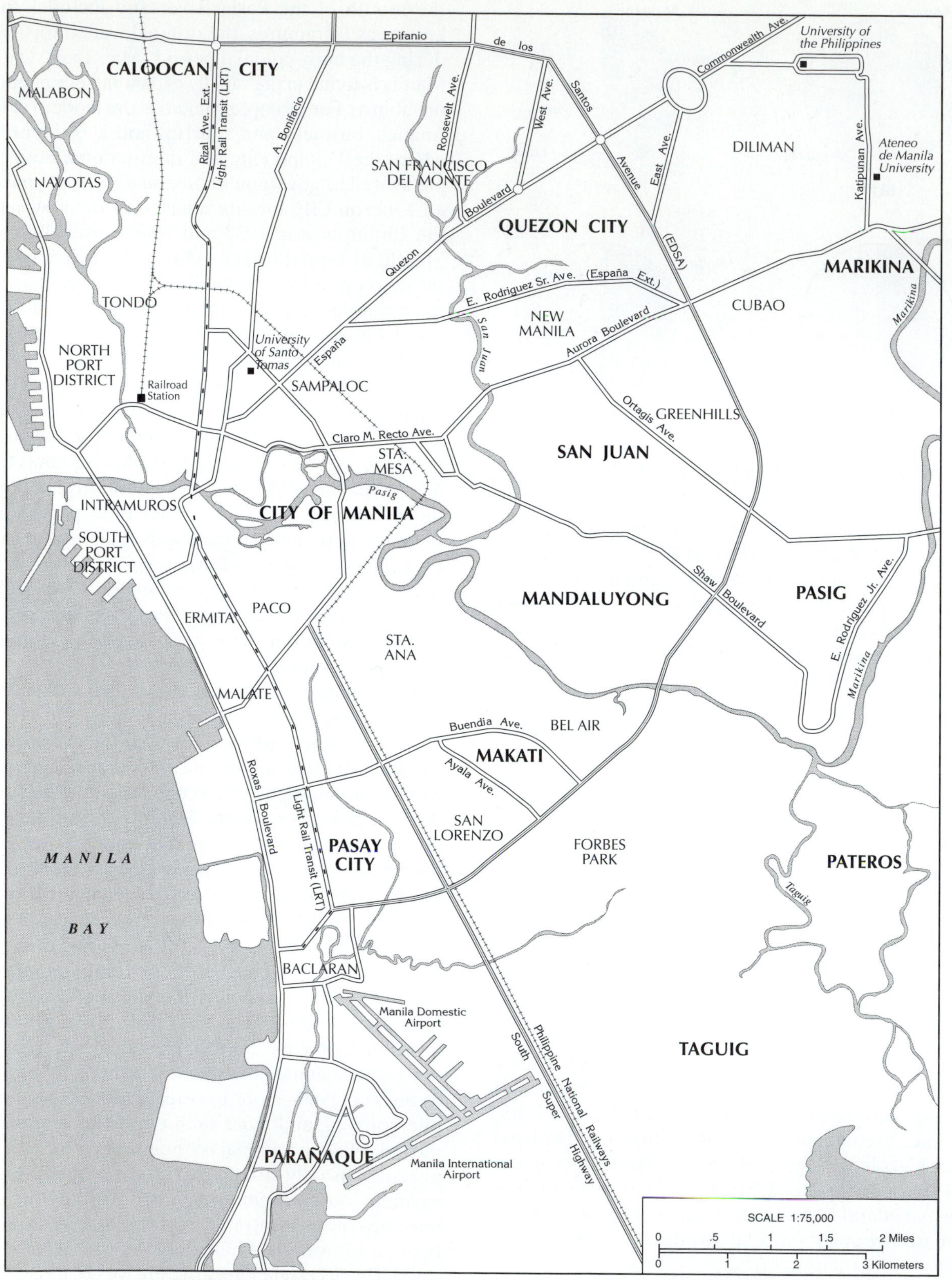

Map 15.3 Manila metropolitan area. *Source: Ulack and Pauer, 1989.*

least two-thirds of all new housing being constructed in the metropolitan area in the mid-1980s was illegal and uncontrolled. Such informal housing provides shelter for probably more than one-third of the city's population, sometimes under deplorable conditions. The government has provided some public housing for the poor, but a large share of this population is not able to afford even the low-cost public housing that has been constructed. Thus, such public housing is purchased by the wealthy and then either occupied or rented by persons in higher socioeconomic groups. Housing has become unattainable for many of the poor in Manila because the price of land has increased dramatically. Such high urban land values mean that informal-sector housing (squatter housing) and settling in the urban fringe areas (where land is less expensive but jobs are less accessible) will continue to dominate the settlement options of the poor. Unfortunately, because land prices have increased from 100 to 200 percent in the central areas of Manila, wealthy speculators and developers are competing with the poor in seeking less-expensive land on the fringe to build their shopping malls, condominiums, residential subdivisions, and public buildings.

During the last two decades much of the government planning and construction has been focused on alleviating the heavy traffic congestion found in Metro Manila, a problem typical of all the large Southeast Asian cities. New highways, the construction of "flyovers" (bridges) at major intersections, reconsideration of banning the ubiquitous jeepneys in Manila, and extensions to the light rail transport (LRT) system which opened in 1984 are strategies being employed or considered.

Since the 1960s the government has tried a number of measures to divert growth from Manila including the development of growth centers elsewhere in the country (for example, the industrial growth pole of Iligan City in northern Mindanao and, more recently, General Santos City in southern Mindanao), squatter relocation projects outside Manila, a ban against new industries locating in Manila, and the construction of export processing centers such as those at Baguio, Bataan, Mactan (Cebu), and, most recently, the Subic Freeport. Another recent plan, developed during the Aquino administration and known as the Calabarzon project, is a long-range-plan (1991–2000) that would bring about the spatial dispersal of industry and population to provinces near Manila through acceleration of agricultural and industrial growth in nearby areas. The plan was prepared by a Japanese agency under the auspices of the Philippine government and is likely to be supported financially by Japan. *Calabarzon* is the acronym for the five provinces surrounding Manila (Ocampo, 1995). Such spatial strategies, as elsewhere in the region, have been "mothballed" or assigned a much lower priority, given the financial instability that prevailed in the late 1990s.

THE ECONOMY

Overview of Resources

In 1990, primary commodities (agriculture, fishing, forestry, mining, and quarrying) contributed less than one-quarter of the Philippine gross domestic product (GDP), but nearly 11 million persons, or more than two-fifths of

Photo 15.4 Another example of informal sector activity is this sidewalk "band", performing in Manila for contributions. (Ulack)

Photo 15.5 Squatter housing, as is the case all over the Third World, is found in cities throughout the Philippines. The housing seen here is located near the bridge that connects Metro Cebu with Mactan Island. (Ulack)

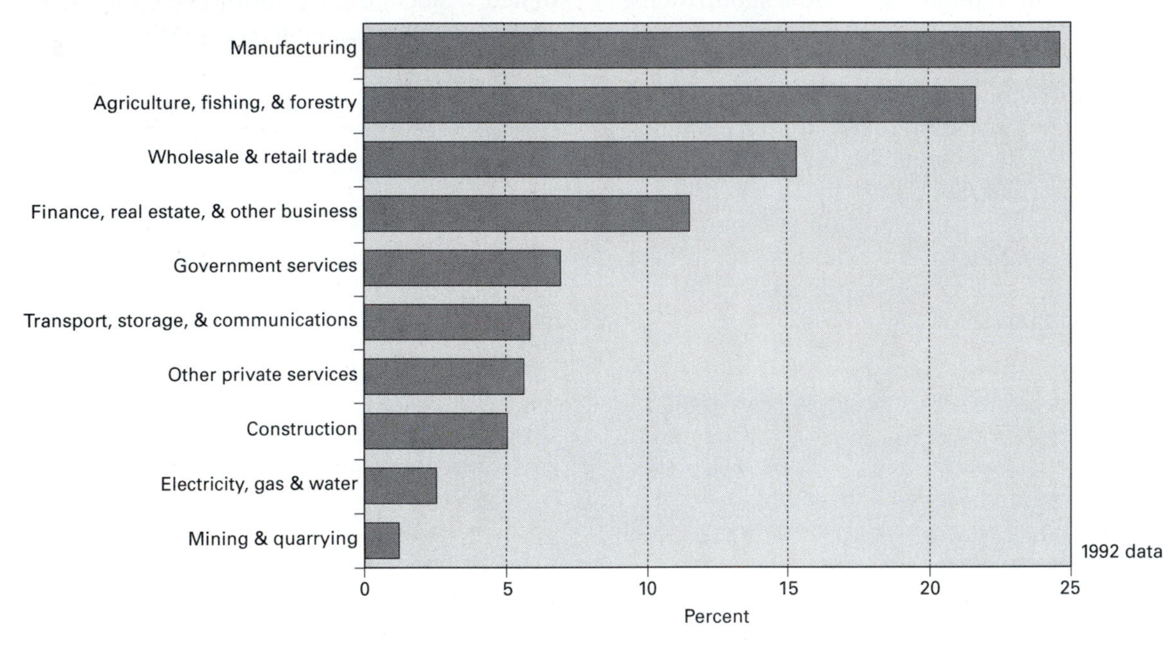

Figure 15.5 Gross domestic product, by economic activity.

the total civilian labor force, were engaged in the production of such activities (Figures 15.5 and 15.7). The Philippines is often described as a nation that is wealthy in mineral and natural resources, but in reality the country is rich in but a few natural resources; only marine, forest, and a few mineral resources are significant. Among the most important mineral resources are copper, gold, silver, chromite, and iron, but only 1 percent of the labor force was engaged in mining and quarrying. There are only small commercial deposits of a few other minerals and, thus far at least, no major deposits of fossil fuels have been found (the discoveries near the island of Palawan, while promising, have yielded only relatively small amounts of petroleum).

Three Southeast Asian nations, Thailand, Singapore, and the Philippines, depended heavily on imported energy in 1995 (Figure 15.6). In the early 1970s the country was dependent on imported oil for 95 percent of its energy supply, but by the 1990s that dependency had decreased to less than two-thirds, which reflected the development of domestic sources of energy notably hydroelectric, geothermal, and nonconventional sources in-

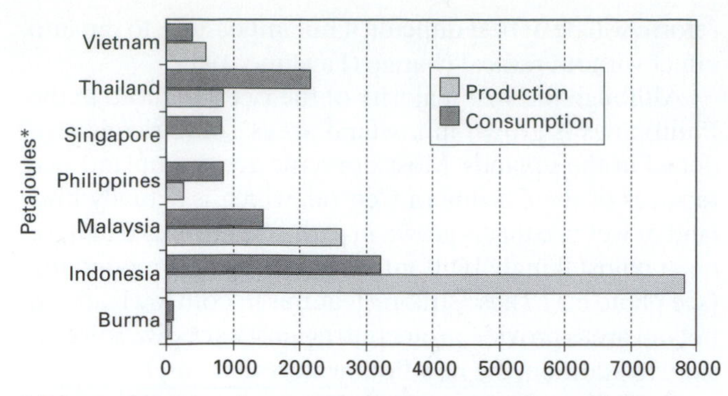

** Primary electricity values are calculated to equate them with the amount of coal or oil required to produce an equivalent unit of thermal electricity and are expressed in joules. The conversion from kilowatt-hours to joules, assuming a thermal efficiency of the primary source of 100 percent, is 0.0036 petajoules per million kilowatt-hours.*

Note: Cambodia and Laos are not shown because energy production and consumption totaled less than 10 petajoules.

Figure 15.6 Energy production and consumption, 1995. *Source: World Resources, 1998–1999, p. 333 and 340.*

cluding agricultural wastes (Demaine et. al, 1994, p. 851). The Philippines is the world's second-largest producer of geothermal energy after the United States and, in 1995, produced more than 1,200 megawatts compared to 2,800 megawatts for the United States. The ultimate potential for geothermal power generation in the Philippines is estimated to be between 3,000 and 4,000 megawatts (Huttrer, 1996). Power shortages, as evidenced by frequent blackouts and brownouts, especially in the Metro Manila region, were common into the 1990s. Newly-elected President Ramos immediately attended to the power-shortage problem by obtaining emergency powers and importing the generators needed. The power crisis was officially declared to be over in 1994. The Philippines hopes to benefit from potential oil and natural gas discoveries in the South China Sea. However, the resolution of the boundary dispute over the Spratlys will have to be settled first.

Manufacturing, the leading contributor to GDP, with 24 percent of the value, employed 10 percent of the labor force, or slightly more than 2.5 million workers in 1992. In terms of employment, community, social, and personal services ranked second with 4.2 million employees, and wholesale and retail trade with 3.3 million workers ranked third. The unemployed accounted for 9 percent of the labor force, or 2.2 million persons age 15 years and over. Significantly, unemployment in the national capital region is at least twice that of the nation as a whole. Such unemployment figures do not include those who are underemployed and in the informal economic sector.

That the country's economic situation had improved by the mid-1990s is indicated by the fact that in 1994 the growth of the gross national product was 4.3 percent, a significant increase from the 2.1 percent growth of 1993 and years of stagnation prior to that (Tiglao, 1995, p. 36). Indeed, forecasts for 1996 suggest that GNP growth will be more than 6 percent, causing one internationally known financial company to state that "the archipelago has moved from Latin American-style instability to become a newborn Asian tiger" (Tiglao, 1996b).

Agricultural Patterns

In the Philippines, rice is the staple food for three-quarters of the population. Rice and maize are the major food crops, and each comprises about one-quarter of the total cultivated land area. The major rice-producing areas include the Central Plain and the southern Luzon region.

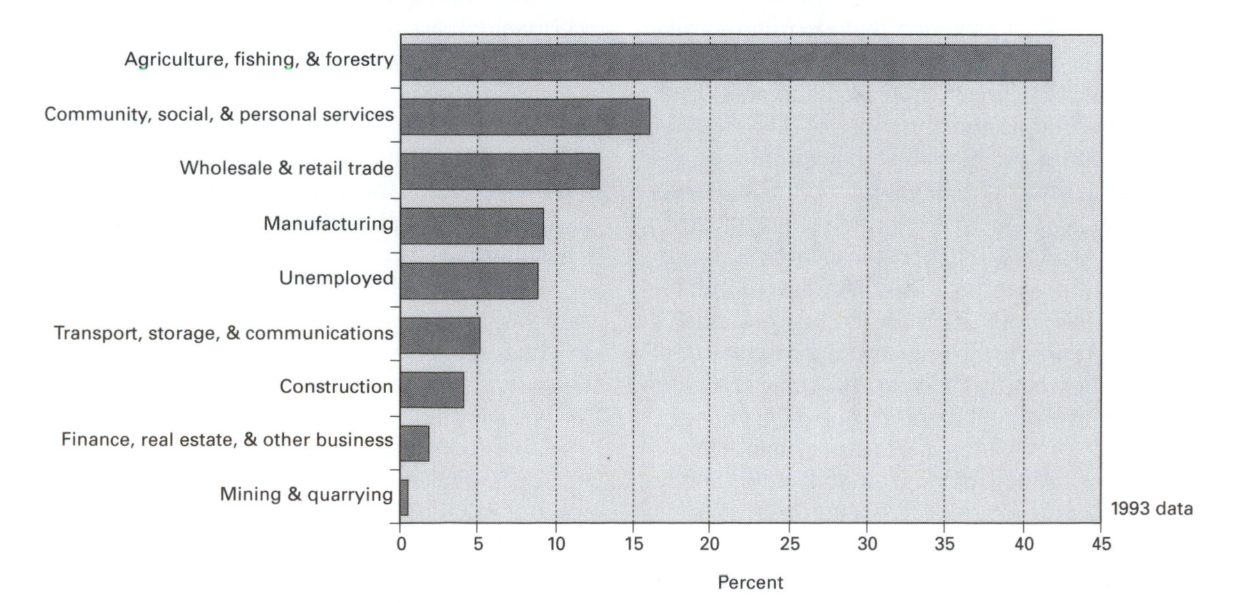

Figure 15.7 Civilian labor force, by economic activity.

It was on Luzon, at Los Baños in Laguna Province, that the International Rice Research Institute (IRRI) was established. It was through research begun here in the 1960s that many of the improvements in rice cultivation and production, commonly known as the Green Revolution, resulted. New high-yielding varieties (HYV) of rice were developed at Los Baños, and these varieties have been adopted and diffused throughout the developing world. Such varieties yield two to three times the production of traditional varieties. Where farmers have utilized these strains, production has increased, sometimes dramatically. But the new varieties require expensive inputs of fertilizer and pesticides. Consequently, the cost, as well as other reasons such as the different taste of the new varieties and new production techniques, have meant that not all farmers have been able or willing to adopt the new varieties. Nonetheless, the new technology was in large part responsible for enabling the Philippines to double its rice yields per hectare between 1965 and 1986 (Demaine et. al, 1994, p. 852). In the late 1970s the Philippines became self-sufficient in rice, although the rapidly increasing population and rise in demand has meant that imports were again needed in some years. Maize production has increased as well and by the mid-1980s was sufficient to fulfill human needs, but imports were still needed for domestic livestock.

Another ongoing problem that has an impact on Philippine rice production and agriculture more generally, is the issue of tenancy. Land reform, though an espoused policy of recent Philippine political leaders including Ferdinand Marcos, Corazon Aquino, and Fidel Ramos, is still not widespread and remains an obstacle to Philippine development. As early as 1972 President Marcos launched a limited land reform program, and again in 1988 President Aquino introduced a Comprehensive Agrarian Reform Program (CARP), which provided that eventually each landowner could own up to 12.4 acres (5 ha) [and each heir would be granted 7.4 acres (3 ha)]. Both programs have remained behind schedule and have not produced the desired results. The more recent plan will never be realized by many peasant farmers due to government conversion of agricultural land to industrial, commercial, residential, and tourism uses, which will dislocate, according to one estimate, perhaps one-quarter of the nation's 10 million farmers (Nieva, 1994, p. 19). Former President Ramos's *Philippines 2000* program, whose goal was to produce NIE status for the Philippines by the year 2000, was based on attracting foreign investment in industry, logging, mining, and the energy sector. Owners of the large sugar haciendas (including that of the Cojuangco family—the family of Mrs. Aquino) and large commercial plantations (e.g., banana, pineapple) found loopholes in CARP and have been able to maintain control of their estates (Banks, 1994). In short, given that the political elites are also the nation's principal landowners, it would seem that real land reform will be at best difficult, if not impossible, to put into effect without radical change (Faustino, 1993).

Although the vast majority of the rice produced in the Philippines is grown in lowland areas, some rice is produced in the uplands. Most dramatic are the upland rice terraces of the Cordillera Central, which is actually lowland or wet rice that is grown in padi-like terraces that have been painstakingly built into the sides of the mountains. (see photo 6.3). These striking features in both lowland and upland areas provide protection against excessive soil erosion. In addition, the new floodwaters that enter the padis each growing season mean that essential mineral matter is replaced. Such padis often produce rice for centuries without serious loss of soil fertility. Upland or dry rice is normally grown in higher areas where population densities are lower. It is one of a number of crops found in the upland areas farmed by shifting cultivators. Dry rice does not require large amounts of water; its yields, however, are lower than that of wet rice.

Because maize does not have the moisture requirements of wet rice, it is widely cultivated in areas like Cebu and other central Visayan Islands, where annual rainfall amounts are sometimes below 60 inches. Although rainfall amounts are higher in northern and western Mindanao than they are in Cebu, maize is the staple crop for many there as well. The principal reason for this is because northern and western Mindanao has received the rural to rural migrants from the Visayas who brought with them their agricultural practices, including maize production. The area planted to maize is actually slightly greater than that planted in rice; however, because rice yields so much more grain per unit area of land—rice yielded nearly twice as much grain per hectare—its importance as a food crop far surpasses that of maize.

Other significant Philippine food crops include cassava (manioc), sweet potatoes (camote), yams, peanuts, vegetables, and fruits. Of the root crops, cassava is the most important and accounts for more than one-half the area in such crops. Most fruits are grown for local consumption or sale, but a few, notably bananas and pineapples, have become important foreign exchange earners. In 1990 bananas and pineapples were the second- and fourth-leading agricultural exports by value; coconuts and coconut products (including copra and coconut oil) and sugar ranked first and third, respectively, among agricultural exports. Bananas are grown widely in the Philippines, but most of the commercial plantations producing both pineapples and bananas for export are found on Mindanao.

As is so of all former colonies, considerable emphasis has been placed on the production of plantation crops for export. The coconut palm is a prime example and is widespread everywhere south of Manila; the most important producing regions are Mindanao and southern Luzon. Most of the coconuts are used to produce copra, or dried coconut meat, a rich source of vegetable oil. The

Philippines produces one-quarter of the world's coconuts and is the world's major exporter of coconuts and coconut products, including copra. Unfortunately, evidence of the health risks associated with high cholesterol levels has meant that the export of tropical oils, including that derived from the coconut, has been adversely affected. In 1990, more than 6 percent of the total value of Philippine exports was derived from the coconut, making it the Philippines' third-leading export overall (after electronics and garments; Figure 15.8). Sugarcane, the Philippines' first plantation crop, now follows coconuts, bananas, and pineapples in terms of export value among agricultural commodities and is no longer the major export that it once was. Most of the sugar is grown in the traditional lowland areas of the western Visayas region.

Forests and Fisheries

Timber and manufactured wood products account for about 8 percent of the total value of Philippine exports. Most Philippine woods are classified as hardwoods and, although not true mahoganies, are usually marketed as "Philippine mahogany." These woods and several other forest species including rattan, are widely used in furniture, paneling, interior finishing, and for plywood and veneer. The forests of Mindanao provide most of the timber.

Fish and other seafoods are the principal source of protein in the average Filipino's diet. The annual per capita consumption of seafoods is about 70 pounds, which is twice the national average for the Southeast Asian region. About 80 percent of the total fish catch is consumed fresh; the rest is salted, dried, or smoked. Fishery resources are subdivided into two categories: marine and aquaculture. Marine fisheries consist of commercial and municipal fisheries whereas aquaculture includes brackish water and freshwater fisheries and sea farms. In recent years tuna, shrimp, and prawns have been among the top Philippine exports, and Japan is the leading customer. In 1991, shrimps and prawns and tuna ranked as the nation's fifth- and thirteenth-leading exports by value, respectively (see Figure 15.8).

As is so with the forests, exploitation of marine resources has caused severe environmental damage, including species extinction and biodiversity loss, especially through the destruction of coral reefs. In general, coral reefs are much less resistant to disturbance than other coastal habitats and degradation and destruction is occurring as a result of pollution, erosion of nearby land areas and subsequent smothering of corals by sediments, cyanide poisoning (used to stun fish for food and for the international aquarium trade), and damage from recreation and tourism. Although coral reefs cover less than 0.2 percent of the worlds ocean beds, they are the most species-diverse areas of all explored marine habitats, approaching tropical rain forests in their species richness. Up to one-quarter of all marine species and one-fifth of all known marine fish live in coral reef ecosystems. The Southeast Asian region accounts for 30 percent of the world's coral reefs and is "the most species diverse area of all explored marine habitats." (*World Resources Institute, 1996–97,* 1996, p. 254)

Manufacturing Industry

The traditional function of manufacturing in the Philippines was the processing of agricultural produce. Thus, for example, rice- and corn-processing mills and sugar-refining centrals were established to process those crops. Soon after independence in 1946, like most newly independent Third

BOX 15.1 Deforestation: A Continuing Environmental Problem

Deforestation, which can be described as severe in the Philippines, is related partly to rapid population increase. This is so because in rural areas especially, fuelwood is often the major source of energy for domestic consumption and therefore population increases mean there is an increased demand for fuelwood, as well as an increased demand for land for agricultural uses, including that in shifting cultivation. Additionally, the Philippines suffers from heavy legal and illegal logging and exportation, and inadequate reforestation. Among Southeast Asian nations, only Thailand has experienced a deforestation rate as great as the Philippines during the 1980s. In 1980, nearly 27.2 million acres (11 million hectares) of natural forest remained in the country; by 1990 the figure stood at 19.3 million (7.8 million). If only rain forests are considered (which comprised almost one-half of all Philippine forests in 1990), then of the 40 or so nations in the world that still had rain forests only four nations—Jamaica, Haiti, Bangladesh, and Thailand—had higher deforestation rates than the Philippines (*World Resources Institute, 1994–95,* 1994, p. 306–9). It has been estimated that about 90 percent of lowland forests in the Philippines have disappeared in the past 30 years, which among other problems, has caused massive losses in biodiversity.

In addition to the tropical forests, the environmentally important mangrove forests that line the coastlines are also being destroyed rapidly. The Philippines has lost a greater share of its mangrove forests than any other Southeast Asian nation. It is estimated the country lost 70 percent of its total mangrove area between 1920–1990 (*World Resources Institute, 1996–97,* 1996, p. 253). Human activities that contribute to this loss include overharvesting for fuelwood, agriculture, mining, pollution, damming of rivers which alters water salinity, and conversion to fishponds.

World nations, the Philippines embarked on an import substitution industrialization (ISI) strategy whereby factories were established to produce commonly used items that historically had been imported. Thus, factories that bottled, packaged, and reassembled goods manufactured overseas for the Philippine market were established along with steel mills, fertilizer plants, cement factories, and the like. The idea was to make the country less dependent on such imported goods; unfortunately, the strategy did not work very well because components still had to be imported and local markets were small. The Philippines' largest company, the San Miguel Corporation, was begun as a brewery in 1890 and in the 1920s diversified to include other soft drink and food items, as well as investments in agribusiness such as coconut oil and copra, banking, commodities, insurance, and real estate. It is an example of an import-substituting industry, albeit one that has diversified and is among the most successful companies in the nation.

Most recently many Third World nations like the Philippines have been placing much greater emphasis on export-oriented industrialization (EOI) policies whereby goods for export are emphasized in development strategies. Coupled with this strategy has been the establishment of government-run export processing zones (EPZ) where multinational corporations have situated plants because of the tax and export-fee exemptions and, of course, because of the availability of cheap labor. The benefits derived from the plants that have located in such zones, except for the limited employment and training they provide, have had a minimal impact on the national economy, and because companies in these zones have little stake there (e.g., they don't own the land or facility), it is not difficult to move where labor is cheaper and profits are greater. In the Philippines, EPZs are located in Cavite and Bataan near Manila, on Mactan Island near Cebu, and in Baguio. In addition, there are the recently established Clark Special Economic Zone (CSEZ) and the Subic Bay Freeport (SBF), both of which have been converted from the former U.S. military installations into industrial estates and business centers. Industries that are encouraged to locate in these latter two zones are non-polluting light to medium industries such as electronics and computers, machinery, textiles and garments, banks and other financial institutions, and tourism-related industries ("Investment Opportunities in the Philippines," 1995). Finally, there are several private industrial estates that have been established, mostly in the Metro Manila area, that are registered with the government Board of Investment. These private estates also provide incentives that are designed to attract industries. EOI plants produce or assemble a wide variety of products including textiles and clothing, electronic goods, and watches. Today the two leading exports of the Philippines in order of importance are garments and electronics (Figure 15.8). Additionally, local entrepreneurs (often with foreign financial partners) continue to process primary commodities for export including such items as plywood, refined sugar, canned pineapples, copra, and coconut oil. Other industries that have been operating in the Philippines (especially Manila) for a long period include numerous footwear (shoes, sandals) establishments and cigar and cigarette plants.

As in most Third World nations, manufacturing accounts for a relatively small share of total employment; as we have noted, in the Philippines the share is about 10 percent. On the other hand, manufactured goods account for nearly one-quarter of the country's gross domestic product and the major share of all Philippine exports—more than 70 percent in 1990. That is a tenfold increase from the 6 percent share of exports that manufactured goods accounted for in 1960. In short, whereas the percentage employed in manufacturing jobs has not increased greatly during the past three decades or so, the importance of manufacturing to the national economy has gained dramatically.

Philippine manufacturing is highly concentrated in or near Metro Manila. This is due to a number of factors including Manila's role as the principal port of entry for imported raw materials and other goods, the existence of a huge local market, a pool of skilled labor, economies of scale, and the presence of financial, governmental, and cultural institutions. The Manila metropolitan area accounts for more than one-half the total Philippine manufacturing employment, a number of large manufacturing establishments, and value added in manufacturing. As we have already seen, a variety of strategies have been employed in an attempt to decentralize this heavy concentration of economic activity.

The second- and third-largest metropolitan areas in the Philippines, Cebu (Box 15.2) and Davao, also have important concentrations of manufacturing including plywood and lumber mills, furniture firms, food-processing plants, and cement factories. Small, often family-operated firms known as cottage industries produce traditional handicrafts to meet tourist demands and local needs, as well as for export. Products include wood carvings, basketry, woven items, brassware, matting, and pottery.

The Service Economy

Frequently migrants, as well as local residents, are unable to find gainful employment in the cities. Thus individuals are either forced to return to their origins, join the long list of unemployed in the city, or secure work in what is called the urban informal sector. The informal sector includes many different types of service-related jobs for which one does not earn a regular wage or salary. People who work as vendors with stalls in public markets or roaming streets, beggars, drivers of certain types of vehicles like the pedicab in Indonesia and the jeepney in the Philippines, and itinerant stevedores are all examples of informal-sector workers. Informal-sector workers are often described as underemployed; that is, they do not work full-time but rather only when work is available. On the other hand, many informal-sector workers do work full-time and quite frequently earn more than those in the formal sector. Today in many Third World cities, urban informal jobs account for a greater share of urban employment than do those in the formal sector. The latter jobs include those in which employees receive a regular wage or salary, as for example government workers, clerks in large department stores and supermarkets, professionals such as lawyers and doctors and teachers, and factory workers. It is in the service sector, which includes both formal and informal workers, where most new employment has emerged in Southeast Asia.

Thus the urban informal sector plays a very important role in cities in that it provides employment for the many low-income, often poorly educated migrants who come from the nation's densely populated rural areas. It has been argued by some that such cities, like the rural areas, are also places of involution because the informal sector eventually absorbs many of those who are not successful in finding formal sector jobs but choose to remain in the city. Again, the question of how long such places can continue to absorb migrants without some kind of more radical transformation must be raised. Clearly, rapid decline of the rate of natural population increase coupled with rapid improvements in economic conditions will do much to forestall or eliminate involution in both rural and urban areas.

Transportation

Interisland ships are the principal means of transportation between the islands of the Philippines, and hundreds of small interisland freighters carry cargo and passengers between ports all over the nation. Cebu, with its central location in the Philippines, and Manila are the hubs of interisland (domestic) shipping. Cebu, in fact, has a greater number of interisland ships calling at its port than does Manila. The Port of Manila ranks first in the Philippines as an international port and both Manila and Cebu have modern, containerized ports. Manila's containerized port, built in 1979, was the nation's first modern container terminal complex.

The land transportation system of the Philippines includes road and railroad networks. Each island's road network focuses on its ports. As is so in other former colonial areas, these networks were established to transport primary commodities from the hinterland to the ports for marketing and export. Railroads are much less important than roads; only 500 mi (805 km) of railroads are located on Luzon and Panay Islands. Within Manila, the light rail transit (LRT) system has reportedly helped the traffic congestion problem. It is estimated that approximately one-quarter million passengers are served daily by the LRT, and new lines are being planned to serve more of Metro Manila. Other major modes of transportation for passengers in Manila and elsewhere in the Philippines include buses, taxis, pedicabs, horse-drawn carriages, and the jeepney.

Photo 15.7 The Philippines is famous for its jeepneys like these two in Manila. (Ulack)

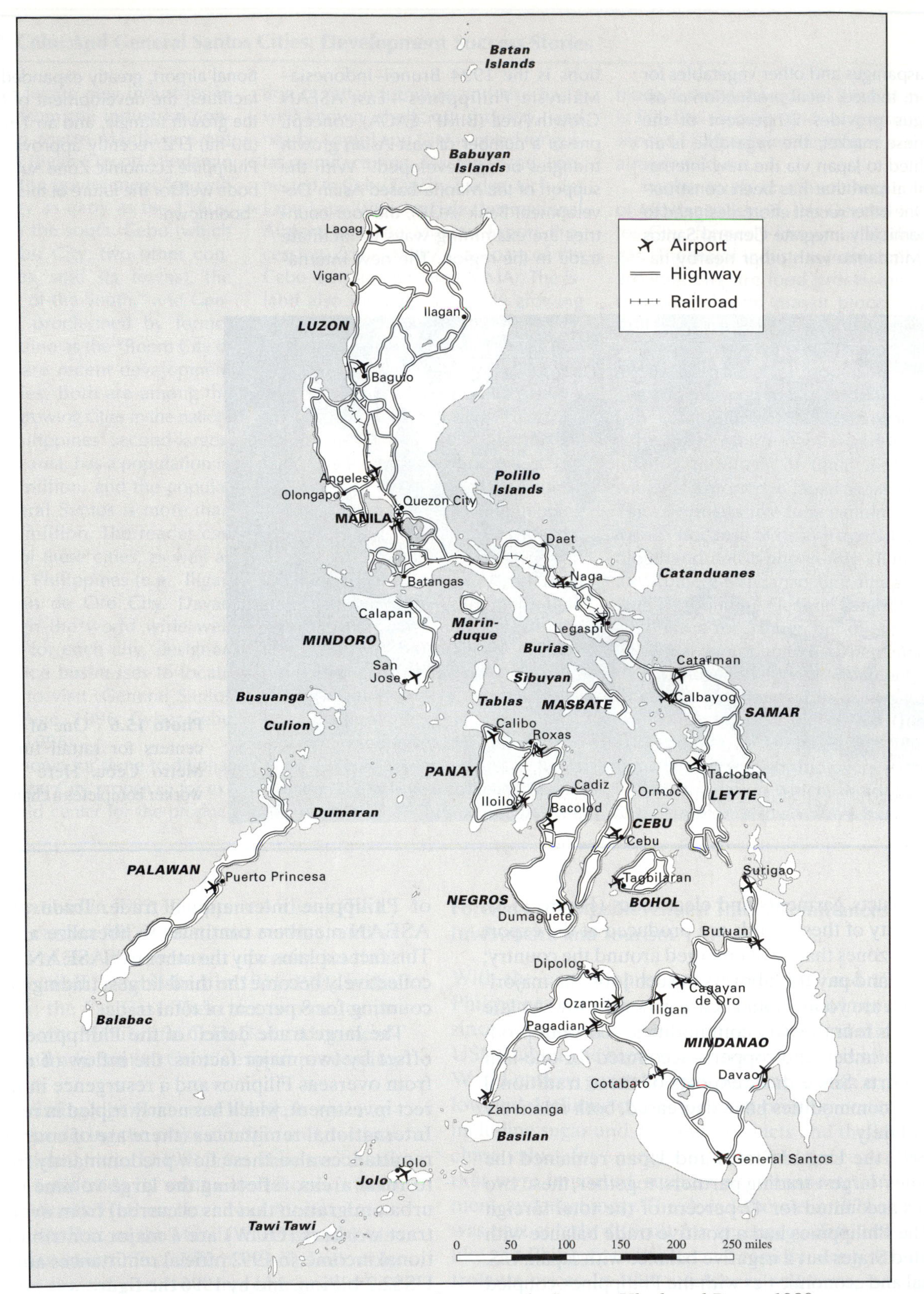

Map 15.4 The Philippines' transportation network. *Source: Ulack and Pauer, 1989.*

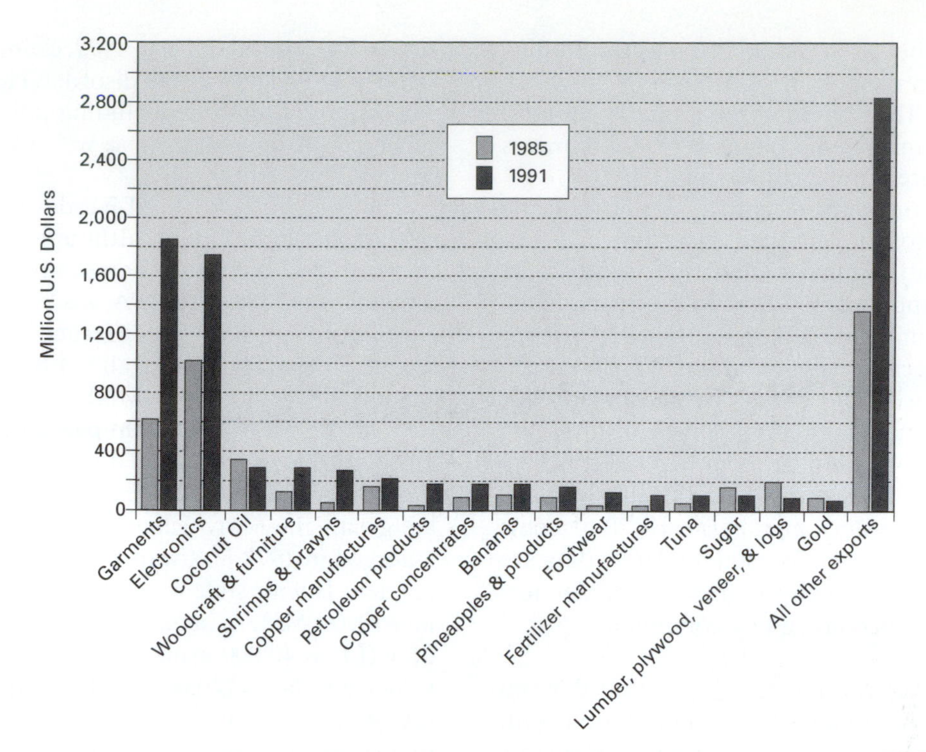

Figure 15.8 Principal exports: 1985 and 1991.

foreign exchange inflow. Overall, the Philippines' situation with regard to foreign exchange revenue inflows improved significantly during the Ramos administration (see Table 4.12).

The first large emigration of overseas Philippine workers occurred in the early 1900s when thousands of Filipinos went to Hawaii to work on plantations (today, persons of Filipino ancestry comprise more than 10 percent of Hawaii's population). Many of these workers, others who later went to work on U.S. defense and war-related projects in such places as Guam and Vietnam, or those who sought jobs that were opening in Canada eventually became residents and sought citizenship. These *balikbayans* (overseas Filipinos) are today an important source of revenue and gifts for relatives still in the Philippines. Recent censuses put the number of permanent or settler Filipino immigrants at 1.4 million, the vast majority in the United States (93 percent), Canada, and Australia (Martin, 1994, p. 642). Additionally, the Labor Department estimated there were 4.3 million Filipinos who had jobs outside the country in 1995 (Kulkarni and Tasker, 1996, p. 26); about 600 thousand Filipino OCWs travel overseas annually, and about two-thirds of the total work in the Middle East, especially the oil fields of Saudi Arabia, Kuwait, and the United Arab Emirates (National Statistics Office, 1992, p. 687–8). A major issue for the Philippines in regard to migrant workers is that nearby Asian nations, notably Japan, Hong Kong, Singapore, and Malaysia offer the best prospects for providing work for Filipinos, but such employment is in the form of mostly low-paid domestics and entertainers. Although these women earn on average two to four times what they would make in the Philippines, they must pay labor recruitment fees that can reduce earnings by as much as one-quarter. Furthermore, while Asian employers often complain of untrained workers, Filipinos, some of whom turn to prostitution, often endure abuse and homesickness (Martin, 1994, p. 642). One recent example of an issue involving a contract worker that

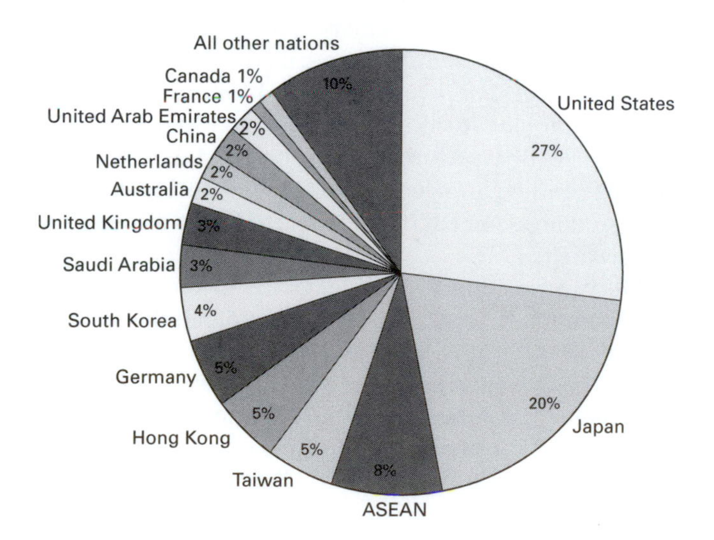

Figure 15.9 Total trade by country, 1991.

strained relations between Singapore and the Philippines was the 1995 execution of a Filipino domestic worker for murder. The Philippine government believed that she had been mistakenly blamed for the murders and had been tortured into a confession.

The second major factor offsetting the large trade deficit is the inflow of foreign direct investment. Beginning in the Aquino administration, the capital sent abroad by the Philippine elite began to return, and during the Ramos administration direct foreign investment and foreign loans became very important. What has been most significant during the Ramos administration has been foreign investment in the Philippine stock market, amounting to US$9.4 billion through September 1995. During the first half of 1996, a further $1.6 billion was invested, most of it in the stock market (nearly 70 percent) and the rest as direct foreign investment. These data are clear evidence that the economy is growing and that foreign investment has become a key component to economic growth.

Another source of foreign revenue is that derived from tourism. The Philippines has great potential with regard to tourism but due to recent political events (e.g., New People's Army and Muslim rebellions, the coup attempts against Mrs. Aquino in 1989, the Gulf War in 1991) and natural disasters (e.g., eruption of Mt. Pinatubo in 1991), there have been several recent off years for tourism. Excluding such events, tourism in the Philippines has experienced an upward trend, and in 1996 more than 2 million tourists visited the country.

THE PHILIPPINES: PROBLEMS AND PROSPECTS

The Political-Economic Situation

The mid-1990s has brought positive signs of change, although since independence in 1946 the political and economic situation of the Philippines can be described as one of instability. As we have seen, growing socioeconomic inequalities, rapid population growth, the continued entrenchment of the oligarchy, and a large and cumbersome bureaucracy are among the factors that help explain such instability. In 1965 then-popular Ferdinand E. Marcos was elected president, but very soon, declining social and economic conditions brought about a resumption of the Communist insurgency, led by the Communist Party of the Philippines (CPP) and its military arm, the New Peoples Army (NPA). In the south, violent outbreaks by the Moro National Liberation Front (MNLF), an organization of armed Muslim rebels in Mindanao and Sulu who sought their own autonomy, ensued. Ostensibly because of these dissident movements, Marcos declared martial law in 1972, claiming that it was the last defense against rising disorder. He jailed politicians who opposed his rule, including political foe Benigno ("Ninoy") Aquino, Jr., whose wife, Corazon was later to become president.

The NPA, CPP, and MNLF today remain thorns in the side of the government, although recent events have improved the situation. Though still a presence in a few outlying provinces, the NPA has been marginalized because of recent internal differences with the CPP and because the CPP was legalized by President Ramos. The government and the MNLF have been negotiating for decades for a permanent peace in Mindanao, and in 1995 the situation improved when the government and the MNLF came to terms. However, a much less moderate group, the Moro Islamic Liberation Front (MILF), emerged in 1978 and wants nothing less than an independent Islamic state in the southern Philippines. In the mid-1990s, Western intelligence estimates put the number of MILF at 40,000 armed members (though the MILF leadership claims 120,000 and the Philippine government says there are only 8,000), more than the 24,000 soldiers that comprised the NPA at their peak in 1985 (Tiglao, 1996a, p. 26).

Incomes of workers declined steadily during the 1970s, and general disillusionment grew because of martial law and the consolidation of control by the Marcos regime and his increasingly dictatorial posture. The economy continued to disintegrate amid charges of overwhelming corruption by Marcos, his wife Imelda, and other cronies. Increasingly, the population opposed the Marcos's rule. On August 21, 1983, Aquino was assassinated as he disembarked an aircraft in Manila on returning from a U.S. visit. The government blamed Communist rebels for the murder, but the evidence pointed toward the government itself. In 1986, Marcos was declared winner in an election that was marred by violence and fraud. Strong public outcry and the so-called EDSA revolt (the acronym for *Epifanio de los Santos Avenue* where an uprising of thousands of Filipinos blocked Marcos's military) finally drove Marcos from power. Marcos, his wife Imelda, and several companions fled to Hawaii and Mrs. Aquino assumed the presidency. Marcos died in exile in Hawaii in 1989.

During the Aquino presidency there were six unsuccessful coup attempts, but Mrs. Aquino had the support of the military (headed by her successor, General Fidel Ramos) and thus none of these attempts were successful; nonetheless, they do suggest the difficulties of her presidency and of the generally poor social and economic climate that continued to exist. Political turmoil and increasing discontent in rural and urban areas were the order of the day.

BOX 15.3 The Philippines And The Asian Financial Crisis

Unlike the situations in Thailand, Malaysia, Indonesia, and South Korea most experts agree that the Philippines, though seriously affected by the Asian financial crisis that began in Thailand in 1997, has weathered the storm better. Indeed, former President Ramos said of the Filipino people in 1997 that they should "count their blessings because the crisis is worse in Thailand, Malaysia, and Indonesia."

Why has the crisis been less severe in the Philippines? Although the reasons for this are numerous and complex, three explanations are essential. First, the Philippines' economic performance during the decade prior to the crisis onset was not as strong as the economies of Indonesia and Thailand and so it was not as affected by the "crisis of success." In other words, "it has weathered the crisis relatively well partly because it was not experiencing economic 'success' in comparison with the very high growth rates of the previously high-performing East Asian economies" (Intal et al., 1998, p. 161). The average annual increase in GDP in the Philippines, for example, was slightly more than 2 percent between 1990 and 1995, whereas in the "high-flying" economies, average annual increases in GDP were more than 8 percent during the same time period. Second, the Philippines experienced its own "crisis" in the early 1980s during the final years of the Marcos administration. Cronyism was conspicuous in that period and continued into the early years of the Aquino administration. Thus, the Philippine government began to resolve some of its problems earlier, through measures which tightened control of the banking system and imposed foreign exchange regulations. Third, whereas the current account deficit has been large (because of the trade deficit and interest payments on the debt) and is similar to other situations in the region, this has been offset in the Philippines by two important factors: the inflow of remittances from overseas Filipinos and a resurgence in foreign direct investment, which has nearly tripled in recent years. International remittances from overseas contract workers (OCW) is a major contributor to national income; in 1992 official remittances amounted to US$2.2 billion and by 1996 the figure was estimated to be over US$3.5 billion (see Table 4.12). In short, the Philippines' situation with regard to foreign exchange revenue inflows has improved significantly beginning in the Ramos administration. Further, President Estrada has promised to continue the economic policies of the Ramos administration, for example, through trade liberalization and privatization, and thus Filipinos are optimistic about their economic future.

In mid-1997 the now well-known Asian financial crisis began in Thailand and soon became a contagion, spreading to other Asian economies, and ultimately affecting most severely Thailand, Malaysia, South Korea, and Indonesia (see boxes in Chapters 7, 13, 15, and 16). Indonesia has the dubious distinction of having witnessed the greatest fall in the value of its currency (the rupiah) of any of the Asian economies. In part because it is such a large and diverse country (the world's fourth-most populated), the crisis has perhaps been the worst here, one in which the repercussions will be felt long into the future. Among other impacts, the economic crisis in Indonesia has brought about political turmoil that included a fall from power of President Suharto and his cronies in 1998. Another outcome of the economic turmoil has been that long-standing ethnic differences have been further aggravated, bringing about conflicts between various groups and the local Chinese population, between transmigrant settler groups such as the Madurese in Kalimantan and indigenous groups and in Timor.

In 1991 the United States began to withdraw from Clark Air Base, a process that was speeded up by the eruption of nearby Mt. Pinatubo in June. In September 1992, the United States began to vacate Subic Bay Naval Base. Also in 1992, President Aquino, who decided that she would not run for another term, endorsed her National Defense Secretary, General Fidel Ramos, who was elected and became president in 1992. Since 1992 there has been significant economic progress in the country. One of the first successes of the new administration was to end the regular electrical brownouts that occurred in Manila and elsewhere. The electric power industry was privatized (as was the central bank, oil refineries, and the national airline to name a few; in all a total of some 400 government activities have been privatized) and by 1994 electric power generation had been modernized and the brownouts ceased. The Ramos government cut import tariffs to make exports more competitive and ended some of the largest monopolies such as the vast telecommunications business. In addition, he is credited with the return of much of the Philippine capital that had fled the country during the Marcos years. Economic liberalization and infrastructure development will be among the major legacies of the Ramos administration.

But for the poor in urban and rural areas little has changed: The slum and squatter communities persist in the cities, inflation has been more than 10 percent per year (although in 1996 it fell to under 5 percent), land reform has not been successful, unemployment and underemployment are high, infrastructure problems persist, and the land-based oligarchic families remain in power. Relative to its neighbors, the Philippines has a long way to go: "According to the World Bank, the proportion of people living in poverty in the Philippines is 39%, compared with 22% in Thailand, 19% in Indonesia, 14% in Malaysia and 5% in South Korea.... Even more strikingly, income distribution has hardly changed in 30 years. Most Filipinos live just above or just below the poverty line. The middle class remains tiny. Yet the people at the top end of the scale are not just well off, they are immensely rich ... the average income of the richest fifth of Indonesians is about 4.5 times that of the poorest fifth, whereas in the Philippines the multiple is almost 11 times." ("Back on the Road," 1996, p. 5) Nevertheless, as we have seen there is some evidence that the economic situation is beginning to improve in the Philippines; President Ramos has attacked some of the Philippines' major problems, including the political and economic dominance of oligarchic groups, head on. Whether or not Ramos's plan for the Philippines to become a newly-industrializing economy (NIE) by the year 2000 (a goal of *Philippine 2000*) is realized remains to be seen; to say the least, there was much to be accomplished before Ramos's six-year term ended in 1998.

Regional Relations

In the international arena the Philippines has several issues that also must be resolved. There have been historic differences between the Philippines and Malaysia over boundaries with Sabah, and though this issue has been resolved for the present, Filipino Muslims in large numbers have sought and found refuge in nearby Sabah because of the long-term Muslim insurgency in the southern Philippines. The Philippines recently began bilateral negotiations with Malaysia to deal with the estimated 190,000 Filipinos in Sabah. Malaysia recognizes 80,000 Filipinos as refugees, and is negotiating a bilateral agreement with the Philippines over the fate of unauthorized Filipino migrants in Malaysia ("Philippines Proposes Labor Agreement," 1994).

As noted already, one other potential problem area is the treatment of Philippine OCWs. In addition, the Manila-based Institute for Strategic and Development Studies estimated there were 300,000 unauthorized Filipinos working in neighboring East Asian nations, so that illegal Filipino migrant workers in East Asia outnumber legal migrant workers. Recently, the Filipino National Security Adviser proposed that labor migration be formally included in the agenda of the Asian Free-Trade Area (AFTA) to help resolve problems associated with illegal Philippine workers in Brunei, Singapore, Malaysia, and elsewhere in the region. It was proposed that regionwide ground rules be established to deal with migrant workers in the region and that Japan, South Korea, and Taiwan be persuaded to join the regional labor agreement. Clearly, as the largest labor exporter in the region, the Philippines has a special interest in the problems of its migrant workers ("Philippines Proposes Labor Agreement," 1994).

A major regional dispute, which includes in varying degrees China, Taiwan, Malaysia, Brunei, Vietnam, Indonesia, and the Philippines, involves the tiny, unpopulated Spratly Islands in the South China Sea (see Map 10.3). The nations are vying for fishing rights and rights to the oil and natural gas fields that may exist in the area. The current dispute over the Spratlys can be dated to 1978 when China occupied six of the islands, taking them from Vietnam (in 1974 China occupied the Paracel Islands, also in the South China Sea; Vietnam also claims the Paracels, which are located about equidistant from the Chinese and Vietnamese coasts). In 1995, the Chinese built some concrete markers on several of the tiny islands of Mischief Reef, located about 120 mi (200 km) from the Philippines' Palawan Island where reserves of petroleum have been discovered. Because they were inside waters claimed by the Philippines, the Philippine government ordered its air force to destroy the structures. Additionally, 60 Chinese fishers were arrested by the Philippine navy for being within the waters of the Spratlys. One unexpected outcome of the Chinese incursion is that relations between the Philippines and the United States improved, as demonstrated by a joint U.S.–Philippine naval exercise near the Spratly Islands. The Philippines, and the ASEAN nations more generally, continue to regard the United States as a counterbalance to Chinese dominance in Southeast Asia. The area claimed by China extends to include a portion of the huge Indonesian gas field north of Natuna Island, a gas field that the U.S.'s Exxon Oil Company signed an agreement in 1994 to develop with Indonesia. In 1991, Indonesia organized the first of several meetings to examine the possibility of those nations involved agreeing to jointly develop the resources of the Spratlys ("The Spratly Island Dispute," 1995). Thus far, no agreement has been reached and the potential for military conflict over the Spratlys continues.

Prospects

The period since independence has been marked by repeated internal crises—political upheavals including peasant insurrections, student demonstrations, Communist insurgencies, Muslim rebellions in the south, and martial law, as well as natural disasters including typhoons, flooding, earthquakes, and volcanic activity. Continued high population growth rates combined with high population densities mean that there is also still a population problem in the country. Although there have been some recent advances in the economic arena, the political, environmental, and population problems that exist mean improvements in social conditions and the economy still have a long way to go. Furthermore, the Philippines is still heavily dependent on the production and export of primary commodities, including the traditional commodities of sugar, timber, copper, gold, and coconut products as well as new ones—fish and shellfish, bananas, and pineapples. Certainly such dependence has diminished in recent years, but many of the new economic activities that have emerged such as garment manufacturing, electronic assembly production, and the tourism industry are low-wage and low-skill level types of activities often located in special processing zones where benefits (e.g., taxes) are minimal to the host country. In short, whereas some macrolevel statistics suggest that economic growth has been significant in the 1990s (for example, see Table 4.12), the question of *who* is benefiting must be asked. Simply put, the poor of the country (i.e., the majority of people) have not yet witnessed the benefits of this growth. Perhaps recent events in places such as the Subic Freeport, the Clark Special Economic Zone, and Cebu and General Santos Cities are symbols of change and of the potential that the Philippines is said to hold. Notwithstanding this positive note, it is doubtful that the country will be counted among the NIEs at the beginning of the new millennium.

Map 16.1 Thailand. *Source: Ulack and Pauer, 1989.*

Photo 16.1 Wat Benchamabophit, the Marble Temple, Bangkok. (Ulack)

a period in Thai history that has seen nine kings ascend to the throne and marked the beginning of the dynasty that has held the throne to the present day.

Modern Thailand: A New Political Order

The mid-nineteenth century marks the beginning of modern Thai history. In the last 100 years Thai society has been increasingly transformed by both internal and external political and economic forces. The economy became more closely tied to the world economic system, the contemporary boundaries of the nation-state were established, a national bureaucracy similar to most modern nation-states was created, and the political system changed from an absolute monarchy to a constitutional form of government. Successive governments under increasing military influence have also reformulated the central elements of national culture—the monarch, the Buddhist religion, and the notion of "Thai" national identity.

By 1850 the expansion of British and French colonial dominions in Burma and Indochina forced Siam to think in terms of territorial boundaries rather than of vassals and subjects. Western colonial expansion in Burma and Cambodia had forced Siam to relinquish its old claims to portions of these areas and confront demands for equal trade rights. A critical watershed in this process was the Bowring Treaty signed with Britain in 1855. This treaty opened the country to international trade, fixed low import and export duties, and established the concept of extraterritorial rights for foreign nationals. In response, King Mongkut (Rama IV) (1851–1868) and his successors began a series of internal reforms that were to transform Siam from a traditional to a modern nation-state. Members of the nobility were encouraged to acquire formal education, especially English and Western science and technology. Various administrative reforms reorganized the system of provincial administration, the state bureaucracy was redefined as a series of functional ministries, and the taxation system was changed to increase central government revenues. Rice soon became the major export product of the country followed by teak, rubber, and tin. Padi cultivation expanded significantly in the lower Chao Phraya delta and in the Northeast as well after 1900, rapidly changing a subsistence economy into a commercial one. Commercialization brought with it the growth of a cash economy, rising consumption of imported Western goods, and the decline of traditional handicrafts.

The most significant transformations took place in Bangkok which emerged as a major port, a trading and commercial center, and the hub of a growing transportation network. The new jobs created by this growth were filled predominantly by Chinese migrants, mainly from southeastern China, who soon controlled much of the internal trade of the kingdom and dominated the population of Bangkok by the late nineteenth century. This led to an increasing ethnic division of labor similar to the plural societies that developed in colonial dependencies elsewhere in Southeast Asia. From the mid-nineteenth century onward then, Siam became increasingly integrated into an international order dominated by Western Powers that led to Siam's transformation into a modern nation-state.

An important step in this process has been the emergence of a new political order. In 1932 the absolute monarchy was toppled by a revolution that retained the monarchy but within the framework of a constitutional system or constitutional monarchy. The coup also resulted in the replacement of *Siam* with *Thailand* as the official name of the country, a move directed toward uniting all Tai-speaking people and defining *Thai* as only those speaking the Tai-language and who were born of Buddhist Thai parents. With the abdication of the king in 1935, the monarchy was effectively removed from Thai political life for the next 30 years. In its place emerged a new generation of leaders in which the military has played an increasingly powerful role in defining the character of political leadership, national interests, and national development policy. Beginning with the administration of Prime Minister Phibun Songkhram in 1938, the modern political history of Thailand has been described by Wyatt as "... a dreary succession of military coups and attempted coups ..." (1984, p. 243), interspersed with popularly elected but often weak and corruption-prone democratic governments.

This pattern became evident in 1947 when the military engineered the first of 17 coups or attempted coups that have taken place in the last 40 years. This step has been seen as a clear indication of the intent of the military to determine the character of the national governments that rule Thailand. A decade later the longest period of authoritarian rule, 1957–1973, began under the successive leadership of Field Marshal Sarit Thanarat, Gen. Thanom Kittikachorn, and Gen. Prapas Charusathien. This period also saw the government make "development" as both a national goal and an ideology on which the legitimacy of the government was based. United States involvement in the war in Vietnam, Laos, and Cambodia since the 1960s contributed to these goals through increased foreign economic investment, military aid, and the growth of the business sector. However, this legitimizing ideology had far less popular appeal than the nationalism associated with the monarchy. The resurgent role of Thailand's reigning monarchs, King Bhumiphol and Queen Sirikit, helped to redefine the monarchy-nationalism relationship and check abuses by a succession of military-dominated governments. The prominence of the military in Thai political and domestic affairs has also been reinforced by events in Indochina and the potential threats they have posed to the security and stability of the Thai state. Since 1975 for example, Thailand has given sanctuary to millions of refugees fleeing the war in Vietnam, Laos, and Cambodia. The largest numbers of Cambodian refugees were concentrated in temporary camps along the Thai-Cambodian border; other refugees from Laos and Vietnam were housed in camps in the north, northeast, and in the southern peninsula, respectively. Since the early 1990s when these camps were "officially" closed, refugees who had not been granted third-country resettlement have been increasingly repatriated back to their home country. Their long presence on Thai soil has served to justify and reinforce the involvement of the military in the affairs of the state.

As new generations of the Thai middle class became exposed to democratic ideals through secondary and higher education, agitation grew for greater democracy. Since the student-led revolt of 1973, there have been brief periods of civilian-led democratic governments followed by returns to more conservative military-backed governments. The ousting of a military-dominated government in 1992 by democratic forces and its replacement by the popularly elected government of Prime Minister Chuan Leekpai was hailed as the beginning of a new democratic order. This initial optimism has waned somewhat as indecision, political maneuvering, and corruption following a scandal over a new land reform program for the poor forced a call for new elections in July 1995. This new administration has been a five-party coalition government that remains untested. Despite this unsettled political situation, there is a general feeling that the era of authoritarian military governments has ended and the country has reached the level of political maturity needed to further its continued economic and social progress.

THE HUMAN DIMENSION

Compared with other developing countries, Thailand has experienced particularly rapid social and economic change since 1960. Its per capita Gross National Product (GNP) grew faster than all but 11 other developing nations through 1990, school-age enrollments have grown, land under cultivation has increased 2.5 times, and an expanding urban industrial core region has emerged in the lower Central Plain around the capital city, Bangkok. Many of these developments have been accompanied by rapid population growth, increases in the size of the urban population and in levels of urbanization, and the continuing flow of rural-to-urban migrants to the metropolitan Bangkok area. Indeed, demographic trends in the last two decades suggest that the country has entered a period of demographic transition that will see a drop in growth rates, a marked aging of the population, and changes in family size and composition (Campbell, Mason, Pernia, 1993). These dimensions of Thailand's human landscape and some of their implications for the future direction are considered in the following section.

Population Patterns

The contemporary pattern of population distribution in Thailand is the result of the long-term interaction with the natural environment coupled with the effects of social and cultural institutions. These dynamics are suggested in the historical shift of the dominant Tai-speaking population southward from Sukhothai into the lowland wet-padi landscapes of Central Thailand. Population growth and associated technical and organizational developments made movement into these more extensive and potentially productive landscapes both possible and necessary. Yet, as recently as the early nineteenth century the geographic limits of the emerging Siamese state encompassed only a portion of the area which today comprises modern Thailand. For example, the northeast region was only lightly settled by Tai populations and

	1960	1970	1980	1990	2000	2010
Population (millions)	26.2	34.4	44.3	56.3	64.4	70.9
Distribution by Region (millions)						
Bangkok	2.1	2.9	4.6	5.5		
Central	7.3	7.7	9.6	12.7	-	-
North	4.4	7.5	9.0	11.1	-	-
Northeast	8.8	12.0	15.5	19.8	-	-
South	3.6	4.3	5.5	7.2	-	-
Growth Rate (percent)	3.1	3.2	2.5	1.9	1.4	1.0
Total Fertility Rate	6.6	5.6	3.6	2.6	2.0	1.7
Life Expectancy						
Male	56.0	58.0	60.0	62.6	66.0	68.5
Female	62.0	64.0	66.0	68.1	70.2	72.2
Age Distribution (percent)						
Under 15 years	43.2	45.1	40.0	33.4	27.4	23.0
15 to 59 years	52.2	50.0	54.6	60.6	65.0	67.7
60 and older	4.6	4.9	5.4	6.1	7.5	9.3

Table 16.1 Population statistics and projections, 1960–2010. *Source: Campbell, Mason, and Pernia, 1993.*

remained under the titular control of Laos until the early decades of the last century. By 1800 the population of Siam was only 4 million and had increased to perhaps 8 million by 1900 (United Nations *Statistical Yearbook*, 1976). Since the beginning of this century, the population has grown at a much faster rate.

Between 1911, when the first Thai census was completed, and 1947, the population grew from 8.3 million to 17 million under rather low growth rates of 1.4 to 2.2 percent per year. This pattern changed rapidly in the next two decades when annual growth rates rose to more than 3.2 percent and the total population had increased to 34.4 million by 1970. Since then however, the impacts of family planning technology and socioeconomic change have contributed to a lowering of the growth rate to 1.9 percent, and this declining trend is projected to continue until the year 2010 when the rate of growth drops to 1.0 percent per year. Despite regional differences in growth rates and shifting patterns of internal migration during the past 80 years, the distribution of the current population has changed very little in this century. One-third is concentrated in the central region where the highest densities are in the Bangkok Metropolitan Region (BMR) and the adjoining coastal provinces along the upper gulf (Map 16.2). Another 35 percent is located in the Northeast, the most populous of the Thailand's four main regions but with less than 5 percent of the population being urban. The two least densely settled regions of the country, the North and the South, accounted for 20 percent and 13 percent of the population in 1990, respectively (Figure 16.1).

Changes in fertility rates, increases in life expectancy, and the aging of the population indicated in Table 16.1 have led to suggestions that Thailand is in the midst of its demographic transition. This shift from high to low levels of mortality and fertility resulting in a declining population growth is generally associated with the transition from an agrarian to industrialized economy. Since 1960 the percentage of the population below 15 years of age has declined from 43 to 34 percent; the working age population (15–59) rose to more than 64 percent in 1994. Although the elderly population is still small, their numbers are growing faster than any other age group and will account for nearly 1 in 10 people by the year 2010. These changes are expected to have important social and economic implications in the future.

The number of households will continue to increase, but their average size is expected to decline to 3.5 members and the percentage of elderly households will grow to 11 percent by early in the next century. A slowing rate of population increase will also create a demand for labor that will exceed available supplies by almost 12 percent in 2000 and by more than 50 percent in the year 2015 . With an increase in real wages during the next 20 years, the increasing scarcity of labor may begin to change Thailand's comparative advantage and the relative profitability of many of its labor intensive economic enterprises. These changes will also have implications for the amount and composition of goods and services such as education, housing, health care, employment, and wages, and their related effects on the

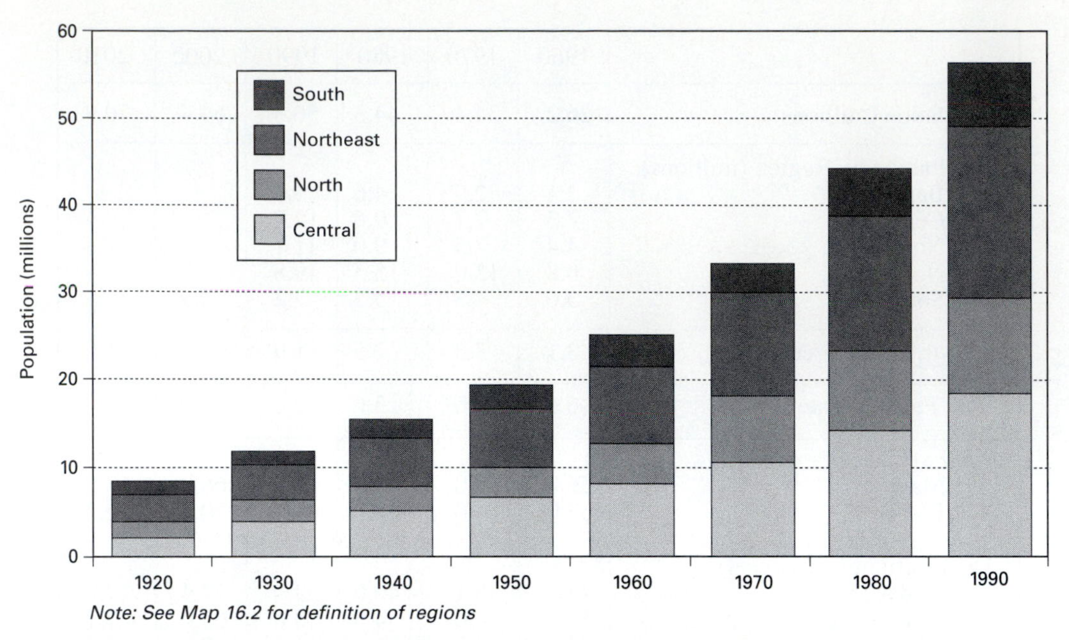

Figure 16.1 Thailand's regional population distribution, 1920–1990.

	Bangkok	Central	Northeast	North	South	Kingdom
Population, 1991						
Number	5,620,591	13,012,489	20,044,480	11,075,736	7,207,732	56,961,030
Percent	15.3	17.4	35.2	19.4	12.6	100.0
Density per km^2	2,965	94	92	53	78	86
Growth Rate (percent)						
1960–70	4.1	2.9	3.3	3.1	3.1	3.1
1970–80	4.3	2.4	2.5	1.8	2.7	2.5
1980–90						
Percent Urban, 1980	100	9.9	4.0	7.5	12.6	17.0
Percent Below Poverty Line, 1990	3.4	16.0	37.5	23.2	21.5	23.7
Mean Household Income, 1988	257	138	100	111	129	134
Per Capita Gross Regional Product (GRP)	769	270	100	160	190	258

Table 16.2 Selected regional demographic and economic indicators. *Source: Thailand, National Statistical Office, 1988, 1992; Hutaserani and Pornchai, 1990.*

future direction of the country's social and economic development.

Mobility and Socioeconomic Dynamics

One feature of modern Thailand that has attracted increasing attention is the rate of geographical mobility of its people. These movements involve temporary (circulation) and permanent migration within and between regions of the country as well as a significant amount of international labor migration. Domestically, these movements are reflected in both lifetime and five-year migration patterns. The former refers to those living in a province different from that of their birth, and the latter to people who changed their province of residence within the past five years. Between 1960 and 1990 the levels of lifetime

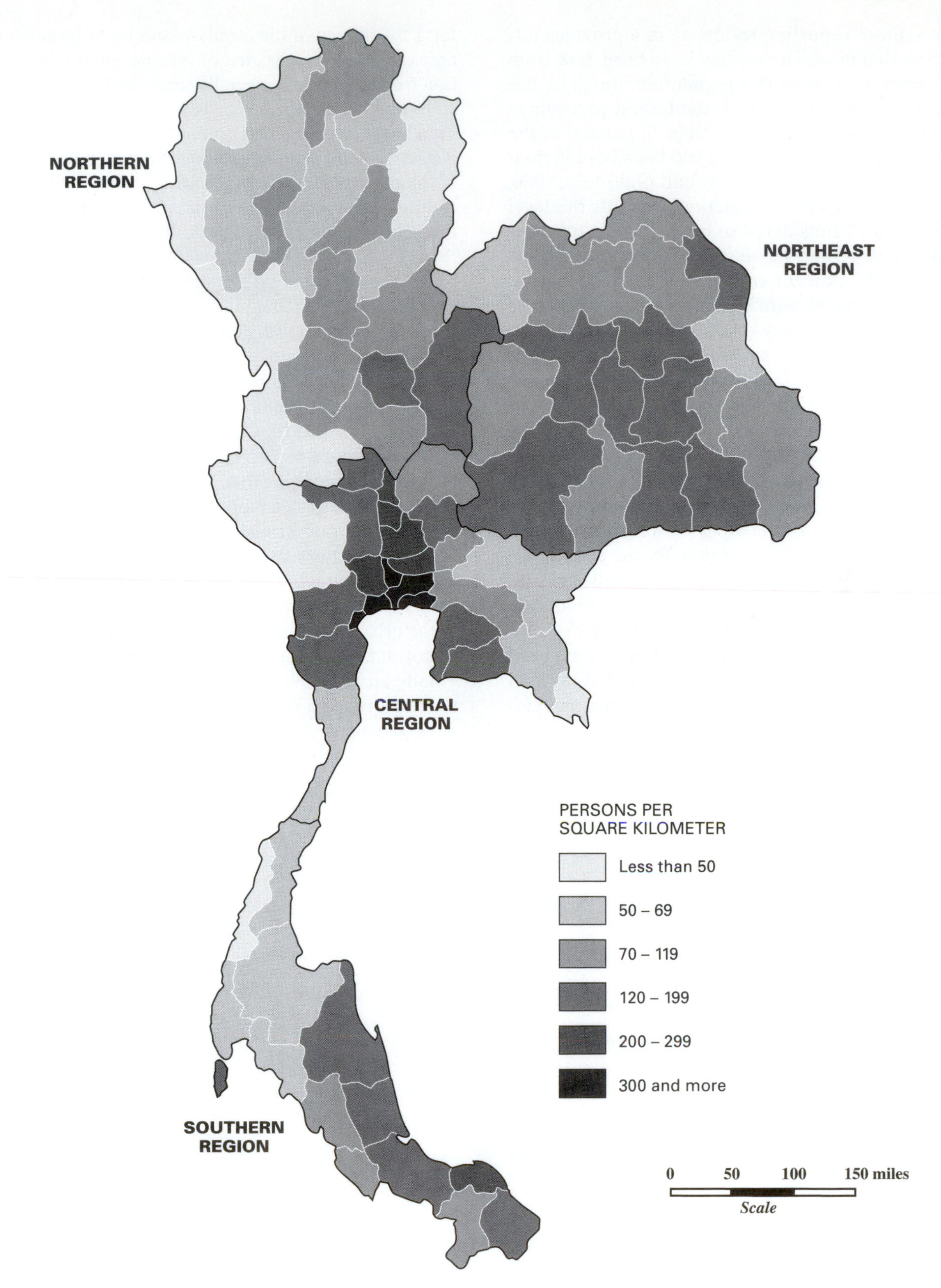

Map 16.2 Thailand's population density, 1990s.

migration, those reporting residence in a province different from that in which they had been born, rose from 11 to 17 percent. Among this population, Bangkok has consistently been the area with the largest percentage of lifetime migrants, while more than 80 percent of the population in each region in 1992 had been born in their province of residence (Goldstein and Goldstein, 1986, p. 17). Despite the increase in lifetime migrants, this level of mobility is still considered to be rather low for a developing country. The 1960 and 1990 population census reported that 3.2 percent and 4.3 percent, respectively, of the total population were living in a province different from the one in which they resided five years earlier.

During the last 30 years rural to rural migrants have accounted for both the largest number and percentage of all migrants in Thailand (Figure 16.2). Until the mid-1970s, this population accounted for more than one-half of all migrations but has declined consistently since then. Most of the moves were over relatively short distances, within provinces or between contiguous provinces, and to more lightly settled "frontier areas" in the upper Central Region and upper Northeast region. By the early 1990s intraregional movements had declined significantly although a sharp increase in urban to rural migration from Bangkok was recorded (Map 16.3). Among the explanations for the decline in rural to rural movements has been more limited opportunities due to the closing of the land frontier, some success in localized rural development efforts, the impact of the family-planning program in rural areas, and the importance of international labor migration from Thailand, especially since the late 1970s (Goldstein and Goldstein, 1986, 40). In 1980 almost 100,000 Thai labor migrants had gone abroad, most to the Middle East. Given recent conditions affecting the economies of the oil-exporting nations in the Persian Gulf, this flow of international labor migrants from Thailand may already have declined (Ashakul, 1989). On the other hand, the rural-to-urban migration stream, which has been the focus of much attention throughout the developing world (see discussion in Chapter 4), has risen in Thailand, with Bangkok being the single most important destination.

Between 1990 and 1992, Bangkok showed a net gain from migration of more than 302,000 migrants, of which 51 percent came from the Northeast and another 22 percent from the North. Considering the rapid growth of Bangkok and the perceived opportunities available there, it is also not surprising that Thai census reports have consistently noted that economic considerations such as seeking work or job transfers accounted for more than 60 percent of the reasons given for such movements. Although this rural to urban stream is small relative to the rural to rural flow, the numbers are significant in relation to the urban population. Moreover, such migrants play an important role in the economic aspects of urban growth. Finally, an often overlooked migration pattern is that involving urban to rural migrants. These flows increased

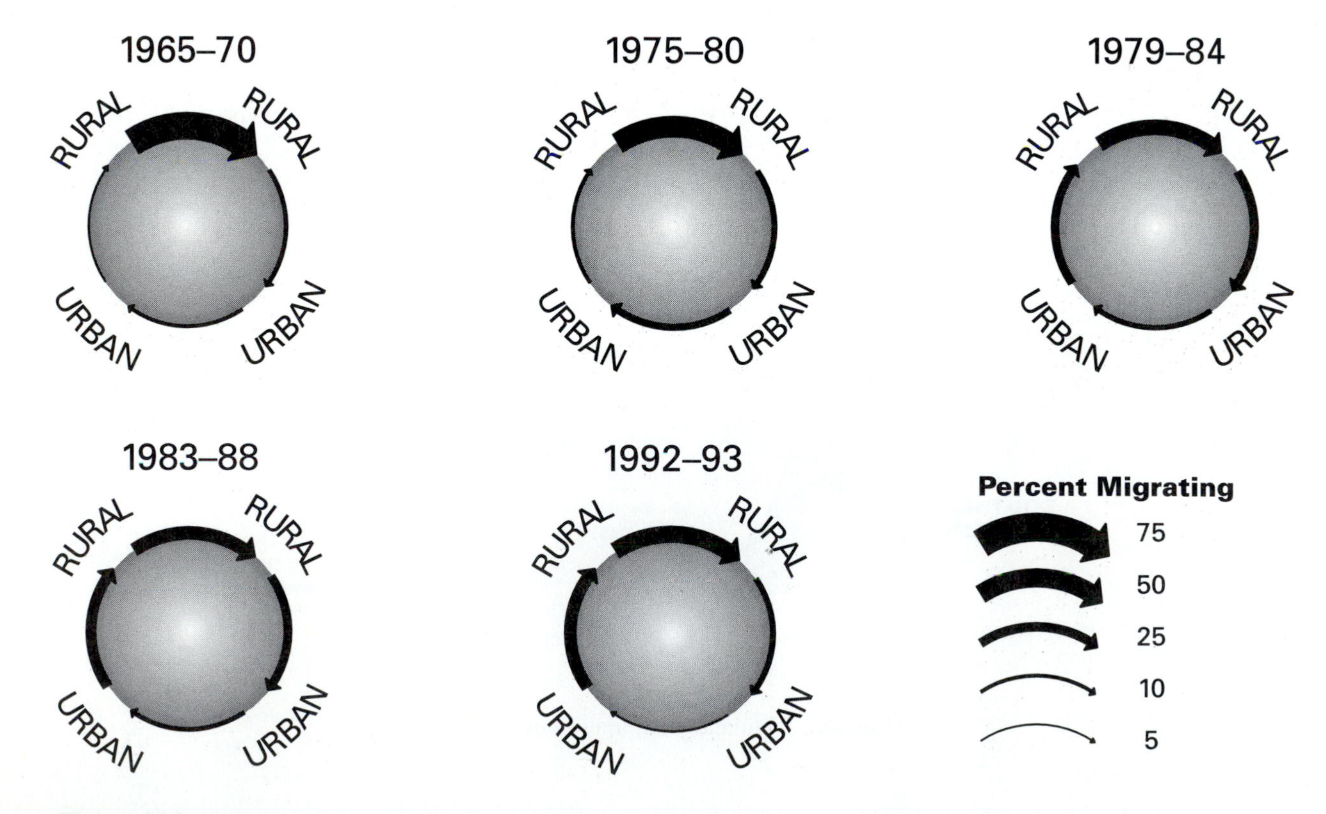

Figure 16.2 Migration streams in Thailand. *Source: Ashakul, 1989, p. 8; Thailand, 1994, Table D, p. 38.*

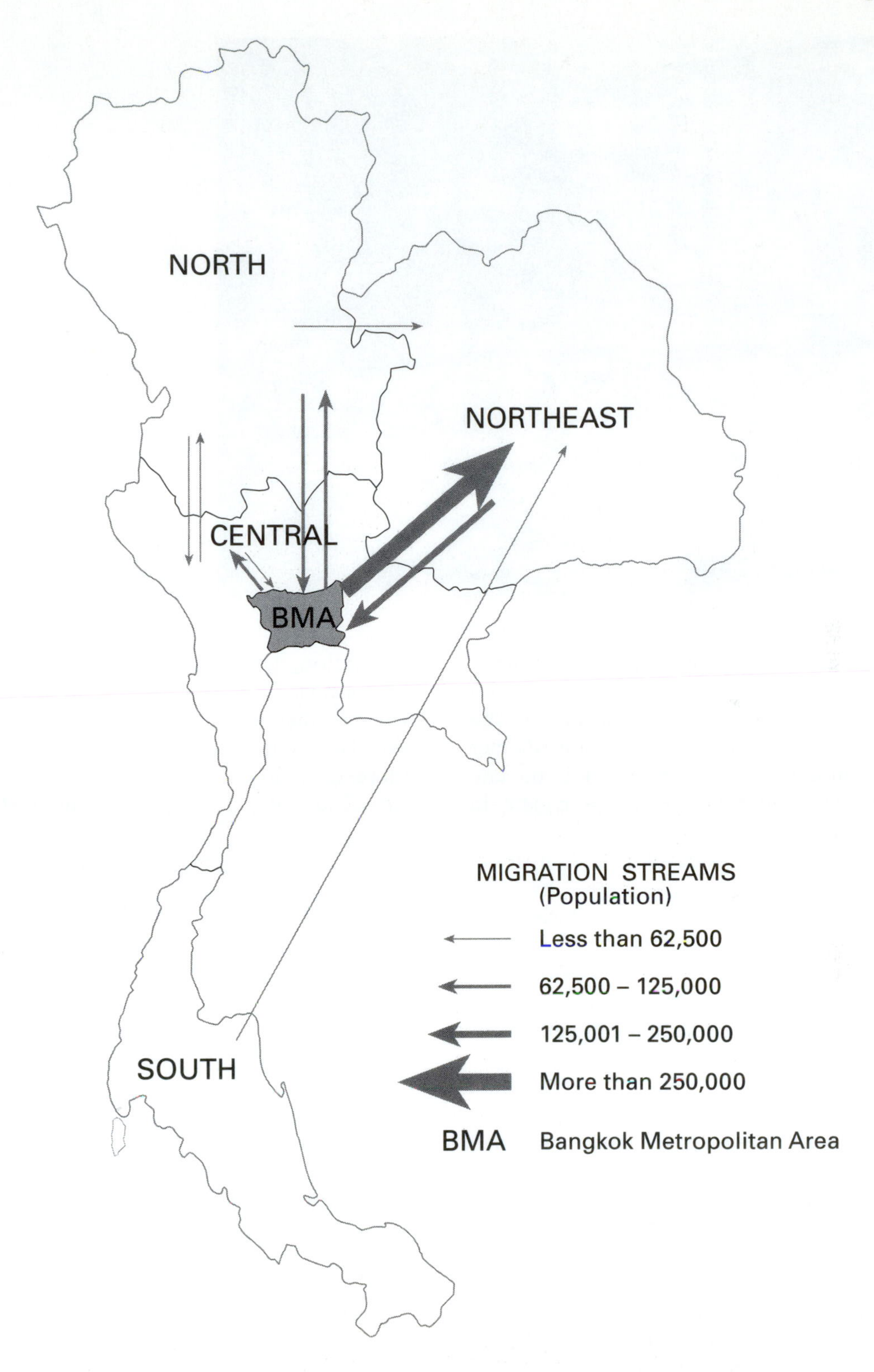

Map 16.3 Thailand's major interregional migration streams, 1990–1992.

particularly rapidly into the 1990s and have accounted for almost one-third of all migrations. For example, between 1990 and 1992 more than 1.1 million people moved from an urban to a rural place, a fourfold increase over the 1975–1980 period. Of this number, 545,000 or 47 percent involved migration from Bangkok to the Northeast Region, and another 188,000 to provinces in the Central Region adjoining the BMR. One explanation for the increase in urban to rural movements may be related to the recovery of the agriculture sector, especially increases

Photo 16.2 Floating market west of Bangkok Metropolitan Area. (Leinbach)

in crop prices, better transport, off-farm employment, and, more important, the "push factor" associated with the economic recession of the urban economy during the 1980s. Although it may be too early to determine whether the small decline in movements to Bangkok in the late 1980s is a temporary or long-term trend, developments in the early 1990s suggest that it has continued.

Thailand is one of many less-developed countries in which population growth and redistribution have assumed increased importance both demographically and economically. Given the economic and social changes that it is undergoing and the generally held assumption that migration will increase as economic development progresses, Thailand's first four national development plans included policies to reduce migration to the BMR. In addition these plans promoted regional "growth poles" such as Khon Kaen and Khorat and encouraged movement to other provinces. But the government concluded in the early 1980s that these policies had been unsuccessful. Rural to urban migration had increased, urban-rural disparities in income and poverty levels had risen, and significant gaps in economic opportunity between regions remained. Cited as reasons for this failure were economic pressures pushing migrants from less- to more-developed regions, unsuccessful efforts to decentralize authority to regional and local levels that may have impeded regional and rural development programs, and the likelihood that the BMR may have been made more attractive by efforts to cope with continued rural to urban migration. Although all subsequent development plans have given these problems high priority by focusing larger resource allocations to the poorest sectors of the rural population, their results have been widely questioned (Suthasupa, 1987). Because no other regional NIE has achieved that status with such wide urban-rural and regional economic disparities, these issues will have to be addressed before the country can claim to display a pattern of equitable development (Figure 16.3).

Primacy and Urban Development

Although Thailand is still a predominantly rural society, there has been an increasing urbanization trend since 1947. The share of population residing in urban areas in-

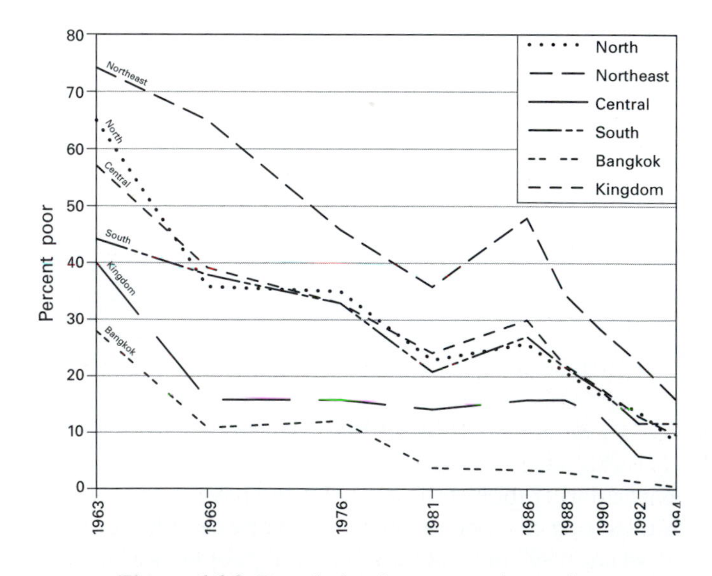

Figure 16.3 Population in poverty, by region.

creased from about 10 percent in 1947 to 23 percent in 1991, and this trend is expected to continue. High rates of natural increase among the urban population have also contributed to this trend: Between 1965 and 1980, Thailand's urban population grew at an average annual rate of 4.6 percent, and this rate has risen to almost 5 percent since 1980 . These rates have consistently been above the rate for the country as a whole and higher than the regional urban averages for other ASEAN nations during the same period. Despite this growth, Thailand's level of urbanization is still below that for other ASEAN countries and for developing countries in general as well. More important, however, is the fact that most of this growth has been concentrated in a single city, Bangkok.

Of all the cities in Southeast Asia, Bangkok is without question the best example of a primate city (see Box 5.1). Bangkok ranks fifteenth among the world's largest cities, and the 1991 population of 8.7 million for the Bangkok Metropolitan Area (unofficial estimates place it closer to 10 million) accounts for more than four-fifths of Thailand's total urban population. This number is projected to increase to 11.5 million by the year 2000. Bangkok, therefore has a population almost 34 times larger than the next largest city, Chiang Mai with a population of less than 250,000. The country's other large cities, Khorat and Khon Kaen in the northeast and Hat Yai in the south, respectively, all have populations of approximately 200,000. Among the three components of Bangkok's growth—natural increase, migration, and expansion of city boundaries—it is estimated that migration has accounted for nearly 40 percent of Bangkok's growth during the past 20 years (Angel and Ponchokebai, 1989). In light of the previous discussion of internal population mobility in Thailand, this fact underscores the importance of migration in development and its implications for urban growth.

The increasing concentration of industrial, administrative, and trade activities in Bangkok and the five peripheral provinces that comprise the BMR have made this the economic core of the country. In 1988 the BMR accounted for 51 percent of the Gross Domestic Product (GDP), including more than one-half of all registered factories and industrial estates in the country as well as a disproportionate share of the country's essential medical, telecommunications, and educational resources (Mekvichai et al., 1990; Rigg, 1994). This concentration of economic activity has also increased the primacy of Bangkok because the powerful economies of scale that exist in a large urban area serve as locational advantages that attract industry. Historically, the capital city area has had strong cost and locational advantages for most firms, especially smaller ones. Although industry in the BMR is diversified, rough zones of industry by type and size have evolved. Approximately 50 percent of the industries in Bangkok are labor-intensive export industries with fewer than 20 employees. In the adjoining provinces of the BMR where land and other costs are relatively cheap, larger firms with more than 65 employees and more heavy, capital intensive textile industries are located. In 1987 the BMR contained more than 26,000 factories and its share of value added from manufacturing had risen to 78 percent (Biggs et al., 1990). Although this suggests a trend toward decentralization of industry away from Bangkok, most such firms have only shifted to the outer periphery of the BMR. Such figures lend support to the arguments that the pattern of urban primacy and centralization of economic activity in Bangkok have contributed to increases in urban-rural and regional inequities in income distribution and poverty (Jitsuchon, 1988; Ikemoto, 1992). For example, in 1988 the average per capita income in the BMR was US$3,596, or ten times greater than the US$365 for the Northeast, the poorest region of the country. In many regards then, Thailand is not really an NIE, but rather Bangkok is a newly industrializing city whose primacy has been increased by this centralization of the country's economic growth.

This degree of centralization in economic activity, investment, and urban population growth in Bangkok has not developed without increasingly high social and environmental costs. Congestion, traffic chaos, expanding urban slums, environmental pollution, mounting piles of uncollected refuse, and inadequate and steadily deteriorating infrastructure are some of these unwanted costs. This has resulted in the steady deterioration of the urban environment and the quality of life for most urban residents. Urban Thailand now exhibits levels of vehicle-based congestion and air pollution that are at least equal to if not worse than in many industrialized areas of the world. In 1991 the number of motor vehicles in the BMR totaled 2.1 million; motorcycles accounted for 42 percent, private passenger cars 36 percent, and buses only 1.2 percent of the total. Not only does this mass of vehicles confound efforts to alleviate the cities persistent gridlock but World Bank estimates suggest that this congestion causes the loss of more than US$7.6 million a day for wasted fuel and time lost in traffic jams (*Bangkok Bank Monthly Review,* 1993). Congestion has also given rise to worsening environmental conditions. Lead, carbon monoxide and dust levels in Bangkok's atmosphere have risen 4.5 times in the last five years, one in eight city residents seeks medical treatment for respiratory problems, and noise levels in the main streets are considered health hazards. Only 2 percent of the city's population is connected to Bangkok's limited sewerage system, most solid waste is discharged into waterways, and more than 15 percent of the garbage disposed of daily is left uncollected. These conditions are perhaps worst in the city's more than 1,000 slum

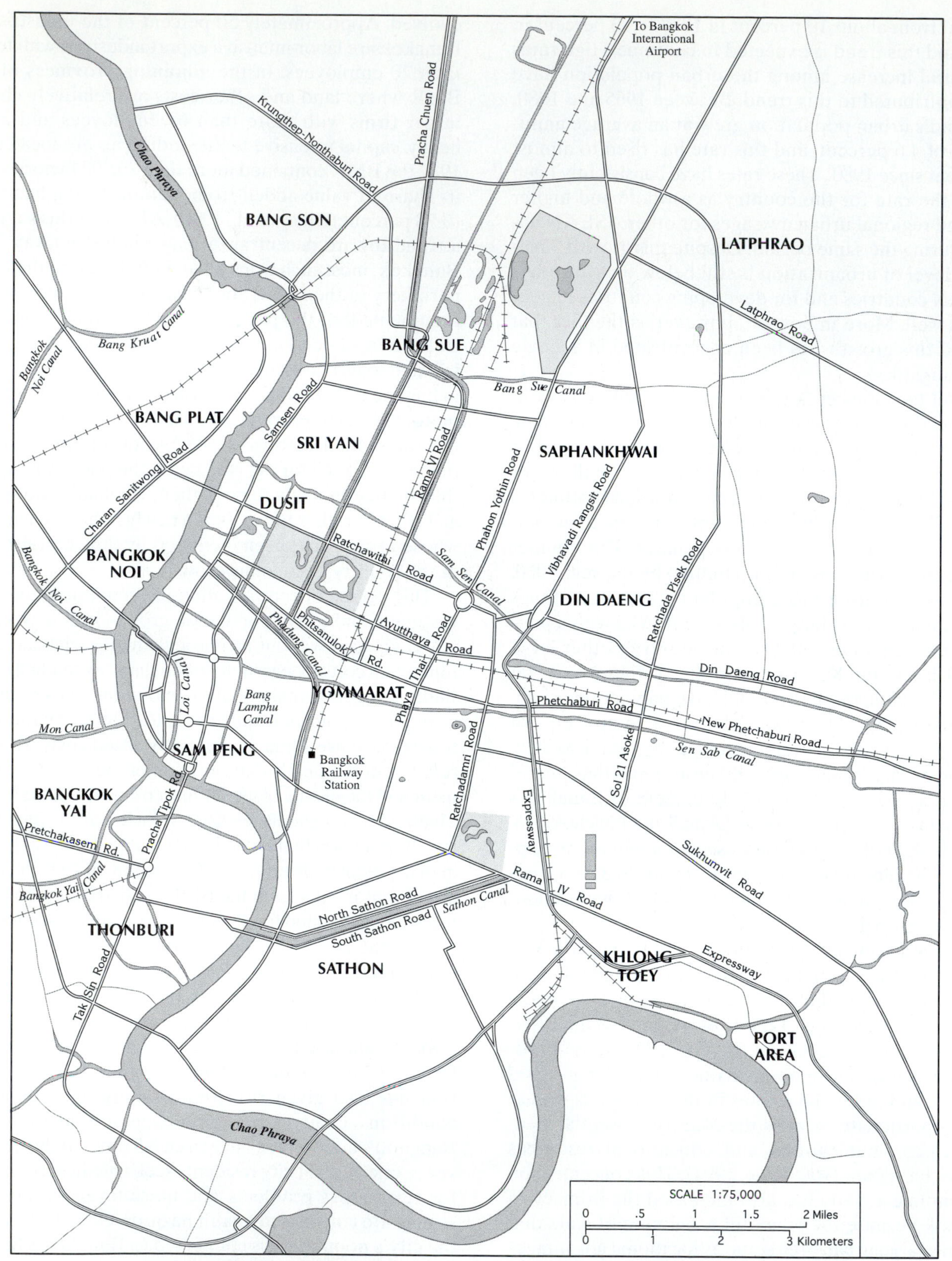

Map 16.4 Bangkok Metropolitan Area. *Source: Ulack and Pauer, 1989.*

Photo 16.3 New residential and industrial estates on the expanding periphery of the Bangkok Metropolitan Area (BMA). (Hafner)

Photo 16.4 The changing cityscape of Metropolitan Bangkok. (Hafner)

neighborhood areas that contain an estimated 2 million people (Angel and Pounchokchai, 1989). Unlike the situation in many cities in the developing world, these squatter settlements are rarely illegal. Perhaps two-thirds of these slum residents live in houses built on private land whose owners obtain a land rent from the resident squatters; one-third live on public land (Rigg, 1994, p. 143). For more than 20 years the National Housing Authority has made periodic and generally unsuccessful efforts to provide low-cost housing to meet this need.

This evidence suggests that the capital does dominate the country to a remarkable degree. In fact the Thai state is in many respects a Bangkok-based state. This has led to rising criticisms that the city is too large, that it is parasitic, and that its growth must be controlled. Furthermore, it is argued that the size of the city is resulting in diseconomies of scale with excessive congestion, high land prices, a deterioration of the urban environment, lagging urban and rural development in the periphery, and growing urban-rural and regional inequalities.

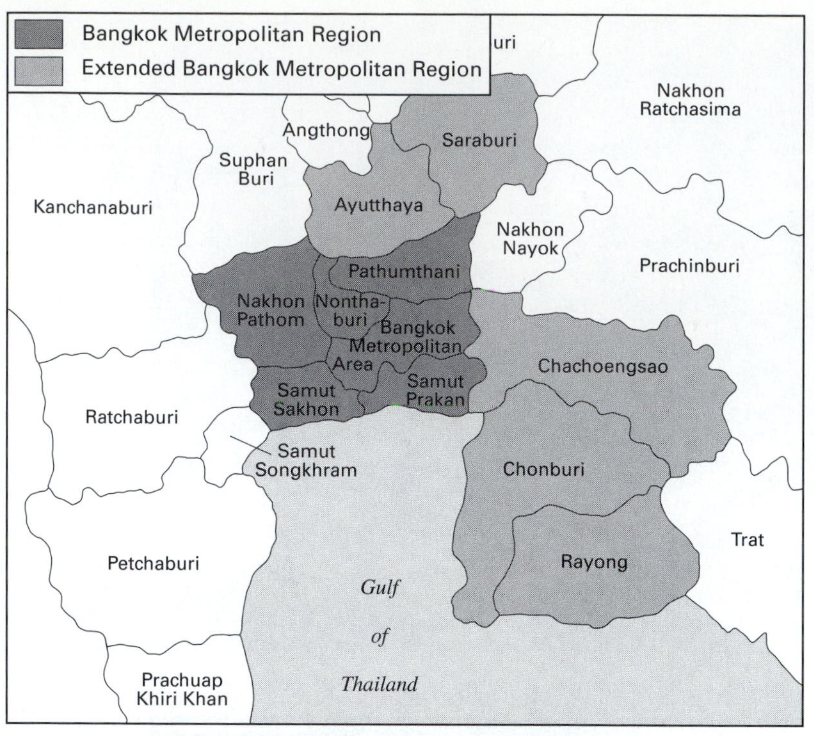

Map 16.5 Bangkok Metropolitan Region.

Photo 16.5 Squatter housing and clogged canals are symptoms of Bangkok's poorly regulated growth. (Hafner)

The rapid changes occurring in the Thai economy are also expected to have a significant impact on urbanization patterns and urban development issues for the next two decades. Dramatic increases in urban growth in both the Extended Bangkok Metropolitan Region (EBMR)—the Bangkok Metropolitan Region, the Eastern Seaboard, Chachoengsao, Ayutthaya, and Saraburi—and regional cities are anticipated to occur during the next two decades. By the year 2010, almost 50 percent of the population will be urbanized, with the greatest population gains expected in the south and the EBMR. Indeed, even with the declining rate of population growth, the BMR is expected to account for one-third of the country's total population increase between 1990 and 2010 (Thailand Development Research Institute, 1987, p. 199).

The government, recognizing these problems and trends in urbanization, has recently begun to develop new policies for the long-term urban growth and development of the country. In addition to addressing existing problems, it proposed to (1) expand highway and communication links between urban centers in the BMR; (2) emphasize different roles for urban centers throughout the country; (3) broaden urban planning practices to include region-serving development functions such as agroprocessing services, regional marketing, and communications functions; and (4) formulate a regional network or cluster strategy of urban and regional development (Thailand Development Research Institute, 1991, p. 27). Rather than dividing the existing urban system into a hierarchy of regional urban growth centers, a strategy that has been ineffectively pursued since the 1960s, the network approach views all cities and towns in a region as members of a cluster, the whole being greater than the sum of the parts. This strategy is expected to generate agglomeration economies while allowing the government to target corridors between major urban centers as sites for industrial parks and provisioning higher-order services—hospitals, universities, and recreation—to nearby towns. The development of a high-technology industrial park in Ayutthaya based on low land prices, proximity to the Don Muang International Airport, and access to the services and amenities of Bangkok is one example of this approach. Whether all of these recommendations are adopted and implemented during the period of the Seventh National Development Plan is still uncertain. What is increasingly clear, however, is that without an integrated package of programs and policies to resolve the social, economic, and environmental problems in the BMR, it will become an increasing liability and drag on national development.

AGRICULTURE AND THE ECONOMY

The image of Thailand as a productive agrarian society with abundant land and forest resources has changed dramatically in recent decades. For centuries agriculture has been the main productive activity and way of life for a majority of Thailand's population. Before the 1950s, almost 90 percent of the population was dependent on farming for most or all of their income. Agricultural products together with farmland were the main sources of government revenue and foreign exchange, and rice was the single most important crop. While agriculture remains an important component of the economy, its position relative to manufacturing and services has been steadily declining for almost three decades. During this same period the basis of the economy has shifted from primary product exports driven by the expansion of the agricultural sector to export-led growth based on industrial production. In this relatively brief period of time then, Thailand has begun the transition from an agrarian economy into a rapidly industrializing country .

Despite these changes, land under cultivation and padi production have more than doubled, a marked trend toward crop diversification has taken place, and field crops now rank with rice among Thailand's top 10 exports by value (Figure 16.4). This suggested persistence of a padi-based agricultural economy disguises some profound changes in the agricultural sector. In this section we examine this structural change in the economy, focusing on the issues of change and stagnation in the agricultural sector, changing regional land use patterns, the shift from import-substitution to export-led economic growth, and how these dynamics are reflected in income distribution and poverty.

Change and Stagnation in Thai Agriculture

Since 1950 the agricultural sector's share of the GDP has declined from 50 to 12 percent while the industrial sector's share has grown to more than 38 percent (Bangkok Bank Monthly Review, 1993). This structural shift in the economy has also been accompanied by an increase in the share of the total labor force employed in the industrial and service sectors while the share in the agricultural sector share has shrunk from 80 to 61 percent (Asian Development Bank, 1993b). Despite these changes, agriculture has remained an important component of the Thai economy. Cropland has more than doubled, the production of most major crops has grown by almost 62 percent since 1960, and agriculture still accounts for two-thirds of the labor force. There has also been a marked trend toward crop diversification. The area used for field crops

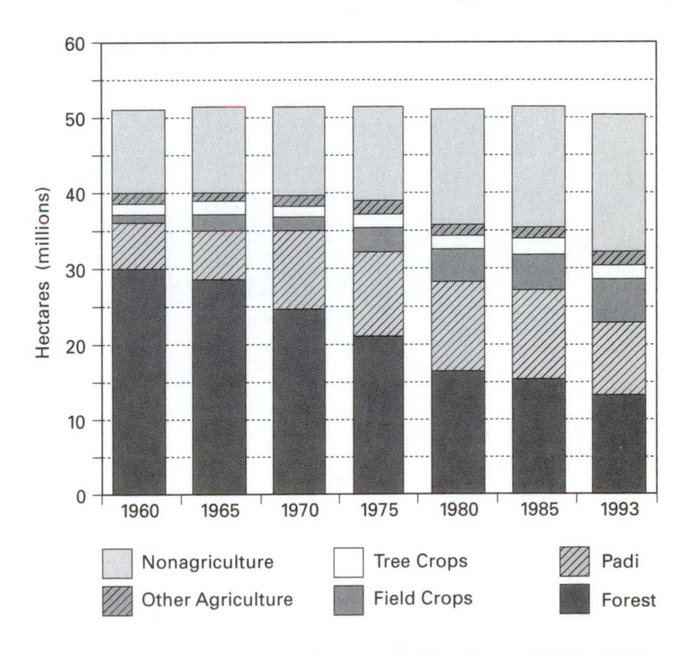

Figure 16.4 Land utilization in Thailand, 1960–1993.

such as cassava, maize, and sugar has risen from 2.7 to 11.6 million acres (1.1–4.7 million hectares) and their share of total land under cultivation has increased from 12 to 23 percent. The extent of this trend toward crop diversification is suggested in the increased production of cassava (tapioca), sugarcane, and maize (Table 16.3). Tapioca emerged as one of Thailand's major agricultural exports in the 1980s as cultivated area rose from 1.1 to 3.2 million acres (440,000–1.3 million hectares) between 1975 and 1984. Total cassava output is now second only to Brazil, the world's leading producer, and 85 percent of this output during the 1980s was purchased by the European Economic Community for cattle feed. Rubber has also become an increasingly important crop, moving ahead of padi in value among the top agricultural exports in 1994. This growth in rubber production also moved Thailand ahead of Malaysia in the early 1990s as the world's leading producer of natural rubber. Thailand is also Asia's biggest net exporter of processed foods, primarily canned tuna, frozen chicken, and frozen shrimp.

In general, three factors account for this growth in agricultural output. Farmer response to higher crop prices, especially between 1966 and 1977 have contributed to the growth in cultivated area and crop production. Lower fertilizer prices after 1975 also facilitated a rise in farm productivity, especially in the double-cropped padi areas of the Central Region. The expansion of land under cultivation has unquestionably been the most important factor in this growth. Between 1960 and 1975 cultivated area increased from 23.7 to 44.2 million acres, (9.6–17.9 million hectares), with much of this growth taking place in the northeast. Since then, the shrinking supply of arable land has slowed the rate of increase in cultivated area and diminished land as a factor in crop output. The significant expansion in cultivated area that has sustained this growth in crop production has for the most part been made possible by a dramatic reduction in forest area (Figure 16.4). Between 1960 and 1988 national forest area declined from 79 million acres (32 million hectares) to less than 33.1 million acres (13.4 million hectares) and almost 80 percent of the area lost to forest has been converted to cropland. The environmental, economic, and social implications of deforestation in Thailand has become one of the country's most challenging issues and will be examined in greater detail later in this chapter.

These trends in production and cultivated area have tended to obscure the general stagnation of what is arguably the most important sector of the agricultural economy, padi production. Between 1960 and 1986, the harvested area, total production, and yields of padi showed consistent improvement as land under cultivation expanded, investments in irrigation were made, and the Green Revolution technology began to be adopted. Most of the benefits from these factors had been exhausted by the mid-1980s, and since then harvested area and total production have declined from 4 to 6 percent and yields have changed little from the levels of the mid-1980s. This situation has resulted from a combination of factors, including the small area of all padi land under irrigation, the low level of adoption of high-yielding vari-

	1960	1970	1980	1990
Land Use (percent)				
Crop Land	16.9	29.6	37.1	43.3
Pasture	-	3.4	2.1	1.5
Forest and Woodland	58.4	45.1	32.2	27.6
Cropland per Capita (hectares)	0.28	0.39	0.54	0.40
Irrigated Land as Percent of Cropland	14.6	15.8	18.1	19.0
Agricultural Labor as Percent of Total Labor	80.0	79.3	70.8	66.1
Crop Production (thousand metric tons)				
Sugarcane	3,382	6,620	12,460	40,660
Cassava	203	3,341	17,744	19,705
Rice	7,789	13,270	16,800	17,024
Maize	44	1,950	3,150	3,800
Coconut	281	744	537	1,141
Fiber Crops	640	510	380	253

Table 16.3 Agricultural statistics for Thailand. *Source: Thailand, Ministry of Agriculture and Cooperatives.*

eties (HYVs) of padi, and the low padi yields that resulted from expanding land under cultivation to increase production rather than intensifying the use of existing land. For example, in 1986 the average padi yield was 1,769 lb per acre (1,983 kg/ha), one-third that of Japan, the United States, and Taiwan. Similarly, only 13 percent of the nation's total padi area was planted with HYVs, most of which was on irrigated land in the Central Region. This level is comparable to Brazil, Burma, and India but considerably lower than Vietnam and Indonesia. This adoption pattern has had important implications for the widening differences in regional agricultural productivity and income distribution that have emerged in recent decades. Thailand also has one of the lowest levels of irrigated cropland in Asia. Only one-fifth of all cropland is irrigated, more than 60 percent of all padi land is nonirrigated, and 80 percent of that area is in the more commercialized padi production areas of the Central Region. Drought, water shortages, and the growth of nonagricultural water use in this region has also meant a shortage of water for dry-season crop production, causing many farmers to shift from paddy to field crops, which require less water. Finally, it should be noted that Thailand's average fertilizer use of 16 lbs per acre (18 kg/ha) is lower than in most other Asian nations due to high prices, a reliance on imports, and high costs of institutional credit.

This situation is in part due to the government's tepid response to the technology of the Green Revolution, because the environment of large areas of the country especially in the northeast are unsuited to the new HYVs, and because until recently land was abundant and padi surpluses have been available (Rigg, 1994). Historically only 10 percent of government investment has been in agriculture, more than one-half of which has been spent on irrigation, especially in the Central Region which has three times as much land under irrigation as in any other region. Government support for diversifying agriculture away from padi toward export crops due to high world-market price, its use of price support policies, and encouragement of agribusiness have also weakened the position of the padi sector. This neglect has been accompanied by a historical reluctance for "big capital" to be invested in the direct production of padi despite the importance of the padi industry during the early years of this century. However, one of the major beneficiaries of government agricultural and padi policies has been agribusiness, the fastest growing agricultural sector since 1975.

Finally, the land frontier, which has been a source of agricultural growth and poverty alleviation in the past, is now all but exhausted. Of the country's total land area of 126.7 million acres (51.3 million hectares) 60 million acres (24.3 million hectares) were under cultivation in 1985. This represents almost 95 percent of the land considered suitable for upland, padi, and perennial crops under current production conditions (Thailand Development Research Institute, 1987, 25). Because the 5 percent historical growth rate for the agricultural sector has come primarily through an expansion in land under cultivation, the closing of the land frontier has posed a number of critical economic and social issues for policy makers, resource managers, and much of the rural farm population. Some of these issues are suggested in the regional patterns of land-use change that have taken place in the last several decades.

Regional Agrarian Landscapes

Land use is a dynamic process, changing over time and in response to population growth, new cropping systems and technology, commercialization, the influence of the state, and other factors. In the last several decades these forces have increasingly affected the patterns of agricultural land use and the social and economic fabric of rural society in most regions of Thailand. These changes can be more clearly understood if we examine them at a regional level. For this purpose we have divided the country into four subregions; the North, Northeast, Central, and South. The boundaries of these divisions are consistent with those shown in Map 16.2. Although there are other regionalization schemes, the four divisions used here are the most consistent with available statistical data.

The economic transformation taking place in Thailand has had the greatest impacts in the Central and Northern Regions of the country. From many perspectives, these changes have been the most profound in the Central Region. As both the cultural and economic center of Thailand, this region is actually a conglomerate of three geographic subregions: the western mountains, the flat alluvial Bangkok Plain cut by the Chao Phraya River, and the southeast coast. Between 1850 and 1934 the Central Region emerged as the principal padi-growing area of the country, and during that period it accounted for most of the estimated twenty-five-fold increase in padi production and the near quadrupling of the area planted to padi. Even as recently as the 1950s, the monocultural economy of the Central Region produced more padi than all other regions of the country. Since then however, agricultural land use has become more diversified, padi acreage has declined, and more land has been devoted to vegetables, fruit, and field crops. Padi still accounts for more than one-half of all land under cultivation, but total output was less than one-quarter of Thailand's production. Much of the region produces at least two padi crops per year, a rainy season (May–October) or main crop and a second dry season (October–January) crop. While the main season padi crop accounts for 81 percent of all planted area and 60 percent of the Central Region's total production, yields are generally only one-half those of the dry season crop. One reason for this disparity between wet and dry season padi production have been water shortages and increasing competition for water between agricultural and non-agricultural users.

Even the casual observer will be able to identify some of the ways these changes have affected rural communities. Fields are now plowed with gasoline or diesel powered tractors replacing water buffalo as the major source of power for pulling plows, harrows, and other farm equipment. There is widespread use of fertilizers and pesticides, mechanical pumps have made water use more efficient, and the new HYVs have allowed farmers to grow two and three crops of padi per year. This increased productivity of land and commercialization of farming has also contributed to a rise in land values, tenancy, and landlessness. The number of farmers who rent rather than own the land they cultivate is as high as 75–85 percent in some areas close to Bangkok and as much as 50 percent in other areas of the lower Central region. As control of land has transferred to the wealthy urban classes and agribusiness, the traditional social and economic solidarity of villages has been weakened and class divisions have begun to emerge. Despite these changes, most rural people still value their shared traditions in contrast to the power and wealth of urban residents.

Another economically important change in land use has been the expansion in area devoted to vegetable and fruit production for the urban market. These forms of intensive market gardening, traditionally occupying the low lying coastal tracts south and west of Bangkok, have increased by 630,000 acres (255,000 ha) in the last 30 years, or a gain that accounts for almost three-quarters of the area in other crops. Much of this has been in response to growth in export and urban market demand, rising land prices, and the greater economic returns that fruit and vegetables provide over padi. These factors have also caused more-intensive forms of agricultural land use to be displaced further from the city while padi fields in the coastal tracts have been converted to commercial fishponds and freshwater shrimp farms. More diversified farming of corn, soybeans, mungbeans, cotton, and sugarcane occupies large areas on the less fertile soils around the periphery of the Central Region. The trend toward more-diversified agricultural land use has also extended into the eastern coastal provinces from Chonburi to Trat. Export market crops such as cassava, sugarcane, and corn including 398,000 acres (161,000 ha) of rubber, now cover more than 3 million acres (1.2 million hectares) of formerly forested land in this subregion. Indeed, in a single generation the Central Region's monocultural padi landscape has become one of the most diversified and productive agricultural regions of the country.

Northern Thailand has also experienced some significant changes in agricultural land use and production conditions. Spread across an area slightly larger than the state of Michigan, its 17 administrative provinces (or *changwat*) encompass an upper subregion of forested mountains and valleys, and a more undulating and less-densely forested lower region bordering the northern margins of the Central Plain. But with a population density of only 151 people per sq mi (58/sq km), which is the lowest among Thailand's four regions (see Map 16.2), and the concentration of agricultural activity in the many small intermountain basins and narrow river valleys, farm sizes are only 8.9 acres (3.6 ha), significantly below the national average of 10.4 acres (4.2 ha). However, improved irrigation has allowed for increased double-cropping of padi, higher yields, and a greater diversity in cropping patterns. As a consequence, land values have risen and commercial farmers and agribusinesses control an increasing share of the land.

Since 1975 the area under cultivation has increased by 11 percent per year, Northern Thailand has the second-fastest growth rate among all regions of the country. Much of this growth has come from increases in field and tree crops rather than padi that still occupies 61.5 percent (8.4 million acres, or 3.4 million hectares) of all cultivated land. Although garden crops, tobacco, and soybeans are also grown as an irrigated second crop in the dry season, the most significant change in land use since 1975 has been the 40 percent increase in area devoted to field crops such as corn, peanuts, soybeans and sugarcane. These crops now cover 4.9 million acres (2 mil-

Region	Total Area	Forest	Farm Holdings						Unclassified Land
			Total	Padi	Field	Tree Crops	Other Crops		
North	16,964	8,652	5,568	3,427	2,019	60	62		2,744
Northeast	16,885	2,532	9,960	6,226	3,722	0	12		4,363
Central	10,390	2,607	5,667	2,675	2,433	162	388		2,116
South	7,072	1,612	3,131	1,152	51	1,586	342		2,329
Kingdom	51,311	15,403	24,326	13,480	8,179	1,808	804		11,552

Table 16.4 Land utilization (thousands of hectares) by region, 1985. *Source: Thailand Development Research Institute, 1987.*

lion hectares), the second-largest land-use activity in the region. A smaller but economically significant amount of land is also devoted to tree crops and vegetables, which are included under "Other Crops" in Table 16.4. Hot and temperate climate fruit and vegetables, including strawberries, are also grown in the North. Their output has supported a growing agribusiness, particularly canning and frozen-fruit processing factories that emphasize processing for export markets. Despite the nationwide loss of forestland, the North with more than 21.1 million acres (8.6 million hectares) in forest, still retains its traditional position as the most heavily forested region of the country.

The second-largest, most populous, and poorest region is the Northeast. Unlike the North and Central Regions, the effects of an expanding market economy have had a more limited impact in the Northeast. Commonly identified as *Isarn*, a reference to its strong ethnic and cultural ties to Laos, this is a region where the population has struggled for decades with poor soils, periodic drought, and highly irregular rainfall (Box 16.2). Only the narrow flood plains along the Mun and Chi Rivers provide alluvial soils and water supplies able to support irrigated padi cultivation. Elsewhere, environmental conditions have made agriculture heavily dependent on the seasonal monsoon rains. Consequently, agricultural productivity is low, poverty has been a persistent fact of life for most rural people, and permanent or seasonal out-migration to find employment has been a chronic pattern for decades.

The Northeast's share of agricultural land is more than 40 percent of the national total; yet agricultural output accounts for only one-quarter of national production. The most significant changes in land use during the past several decades have been the relative decline in padi area, the growth in field crops, and the dramatic depletion of forested area (Table 16.4). Since 1967 the cultivation of both glutinous and nonglutinous padi has increased from 7.7 to 15.6 million acres (3.1–6.3 million hectares), or 83 percent of the cultivated area. Yields for both wet season 1,483 lb/acre (or 1,662 kg/ha) and irrigated dry season (2,599 lb/acre or 2,912 kg/hectare) padi have remained below the national and regional averages, and second-crop or dry season padi output accounts for only 3 percent of the annual total. The glutinous or "sticky" padi variety is the basic food staple in the Northeast, and its cultivation accounts for almost two-thirds of all padi acreage. However, the relative share of all agricultural land planted to padi has declined, primarily due to the growth in field crops.

The 9.1 million acres (3.7 million hectares) currently devoted to field crops represents an increase of more than 3.5 million acres (1.5 million hectares) since 1965. Corn, kenaf, and cassava are the main cash crops planted on converted forestland and account for 95 percent of the upland crop acreage. Much of this growth has been due to farmer response to market and price incentives, government efforts to promote crop diversification, and the conversion of large tracts of forest to agricultural uses. Unquestionably, the most significant change in land use in the Northeast has been the depletion of its forest resources. Since 1961 forests have declined from 42 to 12 percent of the regional land area, an aggregate loss of almost 7.4 million acres (3 million hectares). Perhaps, more than in any other region, this loss of forested area illustrates the collapse of the country's land frontier. Of the 28.2 million acres (11.4 million hectares) of land suitable for agriculture in the Northeast, almost 27.4 million acres (11.1 million hectares) are already under cultivation. Solving this problem so that the region's agricultural resources remain sustainable, forestland is preserved, and the related problems of poverty and low farm productivity are alleviated will remain one of the main challenges facing the Thai government into the twenty-first century.

Perhaps the most distinctive agricultural landscapes are found in peninsular or Southern Thailand, the smallest of the country's four main regions. The peninsula includes

BOX 16.2 An Isarn Family

The lives of rural farm families in the Northeast have changed significantly in the last half-century. When Ban Lat village in Roi Et Province was settled 80 years ago, land was plentiful, forests were full of wild animals (including deer, tigers, and wild boar), cash was seldom needed, and families lived reasonably comfortable self-sufficient lives. In time, population growth reduced land availability, soils became less fertile, crop yields declined, and farmers were urged to take loans to pay for chemical fertilizers, pesticides, tractors, and new varieties of rice and other crops. Instead of growing rice for their own consumption, it is now increasingly produced to meet the needs of a market economy and to pay for the new goods that roads and electricity have brought to the village.

These changes are reflected in how Ban Lat households now spend their annual farm income of US$795. In an average year, US$139 is spent to buy food in the market, US$46 for clothing and accessories, US$160 for firewood and building materials, US$53 for medicines, US$43 for educational costs, US$16 for entertainment, and US$476 for agricultural production costs such as hired labor for crop planting and harvesting, field plowing, fertilizer, and seeds. These expenses mean an average household spends almost US$140 more than they earn, 8 of 10 households are in debt, and loans plus seasonal nonfarm employment in Bangkok have become part of their way of life.

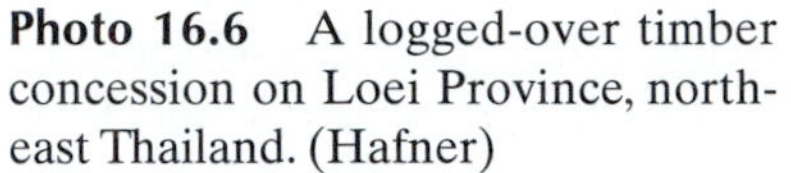

Photo 16.6 A logged-over timber concession on Loei Province, northeast Thailand. (Hafner)

14 provinces extending from Chumphon southward to the border with neighboring Malaysia. Three distinct subregions can be identified here; the densely forested western mountains that form a natural boundary with Burma; a series of inland basins, river flood plains, and coastal plains opening to the east; and clusters of offshore islands that have become popular tourist attractions in recent years. The peculiar linear shape of the peninsula, its length of 367 mi (592 km), and its position between two oceans contribute to a distinct tropical rain forest climate with heavy annual rainfall and a short dry season. Unlike the other regions of Thailand, tree crops, particularly rubber, coconuts, fruit trees, cashews, and oil palm rather than padi are the dominant form of agricultural land use. Rubber is by far the most important both economically and in area under cultivation. The South contains more than 90 percent of the country's total rubber area, the vast majority of which is cultivated by 800,000 smallholders with average landholdings of only 7.9 acres (3.2 ha). Padi occupies slightly more than one-third of the agricultural land, much of it being on the east coast of the peninsula, and there is more limited growth or field crops because of the heavy annual rainfall. The region's extensive tidal flats are used for mangrove charcoal production and thatch from nipa palms. The peninsula is also the least heavily forested of Thailand's major regions with slightly less than 3.7 million acres (1.5 million hectares), a decline of 42 percent since 1961.

Export-Led Growth and Industrialization

The growth of Thailand's economy in the last three decades has taken place in two phases, an initial one of import-substitution industrialization (ISI) followed by its current phase of export-oriented industrialization (EOI). These phases generally encompass the period between 1960–1980 and 1980–1990, respectively. The ISI strategy pursued into the 1970s was intended to promote growth in the domestic industrial sector by protecting it from foreign competition while encouraging production of goods that would substitute for imports. This was accomplished by imposing tariff barriers and import quotas to nurture the growth of domestic industry, the expanding agricultural sector, and exporting primary products such as wood, tapioca products, tobacco, and sugar. During these years international development assistance from the United States and the International Bank for Reconstruction and Development (IBRD) enabled the government to expand its essential infrastructure, and rising international market prices stimulated agricultural expansion in export crop production, especially corn, kenaf, and cassava. Much of this increased production came through extending agriculture into forested areas around the margins of the Central Region and into the North and Northeast. Although the success of this ISI strategy is reflected in GDP growth rates of 6.9 to 7.9 percent in the 1960s and 1970s, only 26 percent of Thailand's total exports and less than 8 percent of the labor force was employed in manufacturing in 1976 (Hussey, 1993).

This period of economic expansion began to slow by the mid-1970s as a number of factors adversely affected the performance and growth of Thailand's economy. U.S. military troop withdrawals from Vietnam and the closure of its bases in Thailand resulted in the loss of important sources of foreign exchange and aid. Rising petroleum prices after the global oil crisis in the 1970s increased

Thailand's trade deficit and declining world market prices for many of its primary agricultural exports, especially in the early 1980s, caused an abrupt economic slowdown, rising debt, slower growth in exports, and a growth rate that slumped to 3.2 percent by 1985. These developments combined with a growing awareness of the limitations of the ISI strategy led the government and private sectors to embark on an aggressive policy of EOI to sustain levels of growth.

Thailand's economic performance during the last decade was unparalleled. Improved commodity prices and a 40 percent annual growth in manufacturing exports led to a 30 percent annual growth in total exports. Primary agricultural commodities have been increasingly replaced by manufactured goods among the country's top 10 exports by value and GDP growth rates had reached double digit levels in the late 1980s. Although these hot rates of growth had cooled to only 6.7 percent in 1996, the success of this EOI strategy has been facilitated by internal reforms and external events. Internal fiscal, monetary, and trade reforms were introduced in the early 1980s that increasingly favored foreign and local industrial enterprises producing for export. During the 1980s the Board of Investment (BOI), the main government agency responsible for industrial policy, also shifted toward an export promotion strategy. Promotional privileges and incentives were granted only to firms that exported a certain percentage of their output. Between 1986 and 1995, foreign direct investment (FDI) increased from approximately US$405 million to more than US$2.7 billion. Most of these investments were in projects that exported 80–90 percent of output, and the majority of this capital came from Japan, Taiwan, Hong Kong, Europe, and the United States. By 1991 FDI had increased to US$2.4 billion, and GDP growth rates rose from 4.5 to 13 percent. Thailand has also provided a number of other advantages that have attracted foreign investment and relocation of industrial facilities. It has a cheap, unskilled, but easily trainable labor supply. Infrastructure such as roads, airlines, port facilities, and water and power supplies were adequate for manufacturing and industrial needs. Tax and investment policies were restructured to be more favorable to foreign investors, and the government's BOI simplified procedures for entrepreneurs to gain investment promotion privileges. Several external factors have also affected the country's entry into EOI. Since 1989 when the United States removed Hong Kong, Taiwan, and South Korea from the list of countries with trade preferences under the General System of Preferences, Thailand has gained a comparative advantage in textile and garment manufacturing. Rising production and labor costs in the region's more advanced economies have also made Thailand a more competitive location for industrial investment due to its lower wages, educated labor force, and weak environmental and zoning controls. In the face of currency appreciation in South Korea, Taiwan, Hong Kong, and Japan, industrialists have been encouraged to relocate manufacturing plants to Thailand to remain competitive. Finally, Thailand's reputation for the lack of racial conflict and tight government controls on labor union organizing enhanced its favorable investment climate. Another aspect of its reputation—a prime destination for tourists—has had a more adverse social and environmental impact on the country.

Like many other developing countries, Thailand has seen the promotion of tourism as one way to put itself on the fast track to economic growth. Investing in the tourist industry is assumed to be less costly than other industries as well as an effective method of creating jobs and increasing local income. This general view propelled the marketing of this rapidly growing sector of the economy as Thailand was changing from a resource-based to a labor-based economy, an adjustment well suited for tourism development. Its success is reflected in the country's emergence as one of the best-known and -desired international tourist destinations. Since 1987 the number of tourist arrivals have doubled, and tourism income has more than tripled to US$5.4 billion, a level higher than any other ASEAN country except Singapore. However, as the country looks toward the future, the image of the Thai tourist industry has been increasingly tarnished.

Despite its success, the tourist industry has been confronted with a widening range of problems and controversy. Critics charge that tourism promotion has given

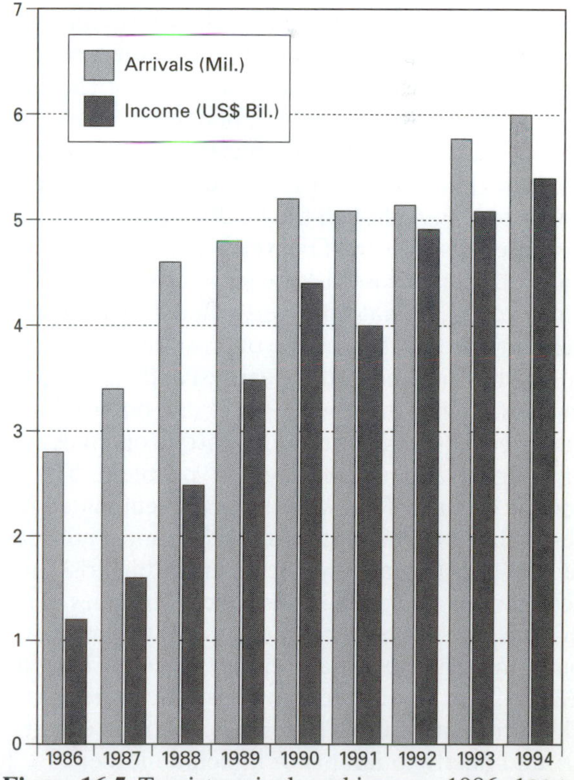

Figure 16.5 Tourist arrivals and income, 1986–1994.

BOX 16.3 Tourism And The Environment

The rapid pace of industrialization has increasingly eroded the images of pristine rice fields, unspoiled white sandy beaches, and even Bangkok's famous floating market that have lured tourists to Thailand for decades. The transformation of Pattaya, a small fishing village southeast of Bangkok, into an internationally known seaside resort in 20 years is an example of how unmanaged and uncontrolled tourism has damaged both the environment and some of the attractions that bring tourists to Thailand. One in three tourists to Thailand travels to Pattaya, which ranks second to Bangkok, in the number of tourists it receives. In 1991 the city's earnings from tourism were US$600 million, or 15 percent of the national tourist income. However, uncontrolled and unregulated growth has made most of Pattaya's beaches too polluted for swimming, poor rubbish and waste-water management has destroyed fish and other marine life, and beachfront construction has diminished the nature-based amenity value of the resort and contributed to a dramatic decrease in tourist visits since 1989. Similar conditions have affected many of the country's other resorts, including Khao Yao National Park and the famous Pa Tong Beach in Phuket.

priority to the quantity rather than quality of tourists, that financial gains favor large tourism enterprises while costs are disproportionately born by local people, and that tourism has exacted a rising toll on the environment and led to an excessive commercialization and decline of Thai culture (Box 16.3). The country's HIV/AIDS epidemic, political instability, and major problems with infrastructure, pollution, and deterioration of tourist attractions for short-term economic gain are other factors adding to this tarnished image.

Most distressing has been the country's rising HIV/AIDS epidemic that has had a decided impact on the tourist industry and the impact that tourism has had on the spread of this disease. Tourist-oriented prostitution first developed during the 1960s when it became popularized by the R&R activities of U. S. soldiers serving in Vietnam. After the war, the attractiveness of Thailand as a destination for sex tourism expanded because of low costs, open and free access, unofficial protection of the industry, and the emergence of a new clientele—white, male foreign tourists. Although the health risks from sexually transmitted diseases (STDs) were substantial, the industry grew because those risks were relatively low in comparison to those that emerged when AIDS became a significant factor in the practice of prostitution in the 1980s. Thailand has documented the earliest and most explosive growth of HIV in the region, a fact due both to international press coverage and the country's openness in acknowledging and addressing the problem. The Thai epidemic began in 1984 with the first documented AIDS cases in Bangkok, but until early 1988 almost no measurable HIV prevalence was detected in any high-risk groups. Since then the epidemic has advanced in waves of infection. The first wave occurred among the highest-risk groups, male homosexuals and intravenous drug users (IVDUs), then through female prostitutes, through those seeking treatment in STD centers, partners of infected men, and finally to infants of infected mothers. By July 1988, HIV prevalence among IVDUs in Bangkok was estimated at 40 percent, and in 1991 official estimates among this group ranged from 200,000 to 400,000 individuals. Since then the AIDS epidemic has advanced in waves of infection, starting with the highest-risk groups—male homosexuals and IVDUs—then spreading into the general population. All five waves of the epidemic are now evident in Thailand, and in some northern provinces infection levels among 21-year-old males in the general population are already at the 20 percent level. This suggests that the epidemic has not peaked, except perhaps among the highest risk IVDU groups, because widespread transmission only began in the late 1980s. Although the future infection rate is difficult to predict, a conservative projection estimated more than 400,000 HIV infections in 1993, rising to almost 800,000 by the year 2000, with an increase in cumulative AIDS deaths to more than 200,000 by the end of this century (Figure 16.6). Even with this more conservative estimate, the impacts of the epidemic will be to increase mortality, decrease the quantity of the labor supply, and diminish government investment as more resources will be required to meet the social and health costs resulting from this epidemic. All of these forces may slow or reverse economic development, diminish levels of foreign investment, and seriously damage the tourist industry.

During the past quarter century, Thailand has maintained an average annual growth rate of about 7 percent, with significant developments in many areas. Its dramatic economic growth of the late 1980s improved its chances for joining the ranks of upper-middle-income countries by early in the next century, becoming the region's next NIE, and laying the foundations for self-sustained growth. Despite the tarnishing of this performance by the political disturbances of May 1992, weak government leadership, and the rising specter of the HIV/AIDS epidemic, the economy was expected to continue to grow although at a slower rate. If this trend continues the country will have succeeded in lifting itself from its present state of underdevelopment and reduced persistent poverty and inequities in income distribution. The most fundamental issue facing

BOX 16.4 The Golf Business: Handicapping The Environment

Thailand's rapid economic growth and the contribution of the golf business to its thriving tourist industry has resulted in a serious environmental handicap. Golf, a fashionable sport among the business community, the growing middle and upper class, and foreign tourists, has seen the number of Thai golf enthusiasts increase to more than 450,000 since 1984, three-quarters of whom reside in Bangkok. The additional 30,000 to 50,000 foreign golfers, particularly from Japan, Korea, and Taiwan, who visit the country every year has spurred a dramatic growth in golf course development. In 1993 there were 116 courses in operation or under construction, occupying an estimated 45,843 acres (18,560 ha): 14 in Bangkok, 50 in other provinces, and 52 under construction. Although the golf boom has generated significant income and created jobs, it has also imposed a high cost on the environment. More than 40 of the new courses have encroached on national parks and forest reserves, their high water demands (6,535 cu yd, or 5,000 cu m per day) have reduced water availability for crop irrigation, and continuous use of toxic pesticides and fertilizers has adversely affected soils, contributed to water pollution, and created public health hazards from windblown pesticide residues. Despite these environmental and health costs, tourist authority promotions hope to attract as many as 100,000 foreign golfers by mid-1995.

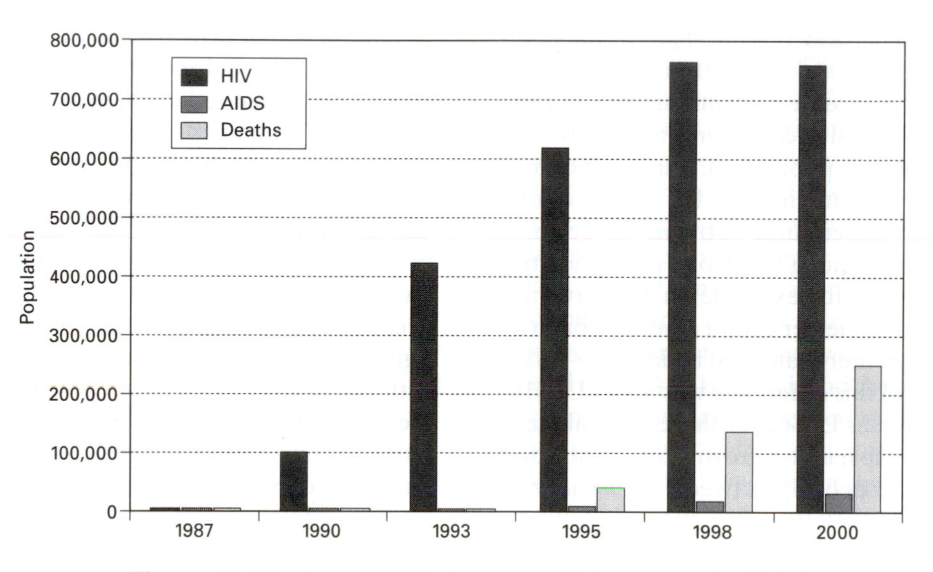

Figure 16.6 Projected HIV infections, AIDS cases, and AIDS deaths in Thailand, 1987–2000.

the country is the question of sustainability—can the country maintain its historical growth rates and build a sound economic base for self-sustained growth by the year 2000?

The answer to this question depends on two interrelated dimensions, one concerned with the external or global environment and the second with internal domestic conditions. Thailand's increasing integration into the global economic system has made it more vulnerable to developments confronting the world economy: rising international debt, volatile currency markets, energy prices, trends toward protectionism, and regional market integration (see Box 16.1). Domestically it faces a number of problems, among which are the growing inequity in income distribution; unacceptable levels of poverty; the depletion of its natural resources; the social, economic, and health costs related to the HIV/AIDS epidemic; and lagging levels of educational attainment. As technological change transforms the manufacturing sector, the demand for technically skilled secondary school graduates will increase. In 1991 only 52 percent of those finishing primary school went on to secondary school, a level well below that in Taiwan, Indonesia, and the Philippines, although efforts of the Ministry of Education had increased this level to 80 percent in 1994. This improvement will not quickly affect the educational levels of Thailand's labor force where almost 80 percent still have only a primary school education. University educated engineers for example, are in such short supply that private sector firms have begun to "pirate" engineers aggressively from their competitors. If Thailand hopes to participate fully in the expected "Asian Century," the resolution of these issues will have to be among the top priorities for the country's leadership and its citizens.

Development, Poverty and Income Inequality

One of the most perplexing problems encountered in Thailand's high rate of growth has been maintaining this performance while reducing absolute and relative poverty.

which now handles more than 50 percent of all cargo shipped in the country. Domestic Thai Airways service has also expanded, many provincial airports have been upgraded, and frequent daily flights link the capital with every region of the country and to most major tourist destinations. These developments have put the State Railway of Thailand (SRT) in an increasingly difficult position but not unlike the situation of many railroads in Southeast Asia—the Indonesian system is an example (see Chapter 9). The SRT's relative share of total cargo volume has declined, its financial position has continued to weaken, its rail facilities are deteriorating, and it has experienced persistent net operating losses (Thailand Development Research Institute, 1993).

Only two decades ago, Thailand's public service and transport infrastructure provided adequate facilities for commerce, trade, and communications throughout the country. Arguably, there were emerging concerns about the quality and adequacy of these facilities, but they were obscured by the resurgent growth of the economy in the last decade. However, this prosperity has seen an intensification of several long-standing problems for Thailand's transport and communications infrastructure that pose serious threats to future economic and social development. These include the need for expanded port and cargo-handling facilities, an inadequate national telecommunications system, urban highway congestion and inadequate mass transport, and insufficient provincial transport facilities. In 1993 a number of major projects were being planned or were under construction to attract increased investment to rural areas, relieve the growing gridlock in the BMR, and meet the government's top priority of narrowing the urban-rural income gap. These included 3 million new telephone lines in the country, the construction of more roads and mass transit systems in the BMR, the alleviation of traffic congestion through improved city and regional planning, and the integration of the regional networks of towns and cities to enhance service delivery and attract investment.

The initial focus of the projects most likely to gain approval and funding has been in the BMR and the EBMR. These will involve constructing a 2-million-line telephone network for Bangkok, building 34 new roads including a second-stage expressway, completing the elevated mass transit railway, or "Sky Train," an elevated tollway linking the Don Muang airport with the city center, and launching of the second phase of the Eastern Seaboard Development Program begun in 1981. This latter project will include the new Nong Ngu Hao International Airport east of Bangkok, a high-speed train on the Rayong–Bangkok route, and new highways linking Bangkok to several towns in the EBMR. These megaprojects were scheduled for completion between 1993 and 1998 at an estimated initial cost of US$3.8 billion or almost 18 percent of the entire 1993–1994 national budget. Only the elevated Don Muang tollway and the "Sky Train" were under construction in 1994 and their date of completion remains in doubt. A related and considerably more ambitious project of the Bangkok Metropolitan Administration (BMA) is the proposed 43-mi (70-km) underground subway. With a central "circuit" ringing the city and four lines radiating out to the city's "suburbs," the addition of this mass transit system is expected to enable the public transport network to handle up to 14 million people, or 70 percent of the daily commuter trips to Bangkok. Scheduled for completion in 2002, the estimated costs for this underground system are almost US$8 billion.

Apart from the obvious need to resolve the city's serious gridlock are the larger arguments about the effects of these conditions on the country's economic growth. These concerns include the deteriorating quality of the BMA urban environment, potential loss of foreign investment, the impact on the country's important tourist industry, and the loss of opportunities to exploit new markets expected to open in the neighboring Mekong Basin countries. If serious efforts are made to implement the proposed decentralization of the urban system and encourage investment outside of the BMR, transport and communications improvements will be critical to those goals.

Trade and Expanding Regional Markets

The increasing integration of Thailand's economy into the global economic system has been one factor in its rapid development in the last several decades. This has been fueled by the accelerated pace of world economic recovery in the 1990s, rising prices for padi and other key agricultural commodities, increased foreign investment, and a healthy growth in exports. During the last decade, Thailand's major foreign markets have been North America, Japan, and the European Union (EU), which together accounted for 55 percent of total manufactured and agricultural exports in 1994 (Figure 16.7). Although the share of total exports to these markets remained stable or showed modest declines, rising exports to its regional trading partners in ASEAN had moved these markets into first place by 1996.

Anticipated changes in global trading conditions related to the North American Free Trade Agreement (NAFTA) and conclusion of the General Agreement on Tariffs and Trade (GATT) negotiations have fostered greater caution in Thailand about its prospects for expanding its global trade. The implications of NAFTA, especially for Thailand and other ASEAN nations, are potential losses in both trade and investment as the trend toward global trade protectionism grows, although this still remains debatable. In a move to counter the potential negative impact of this trend, Thailand has begun to

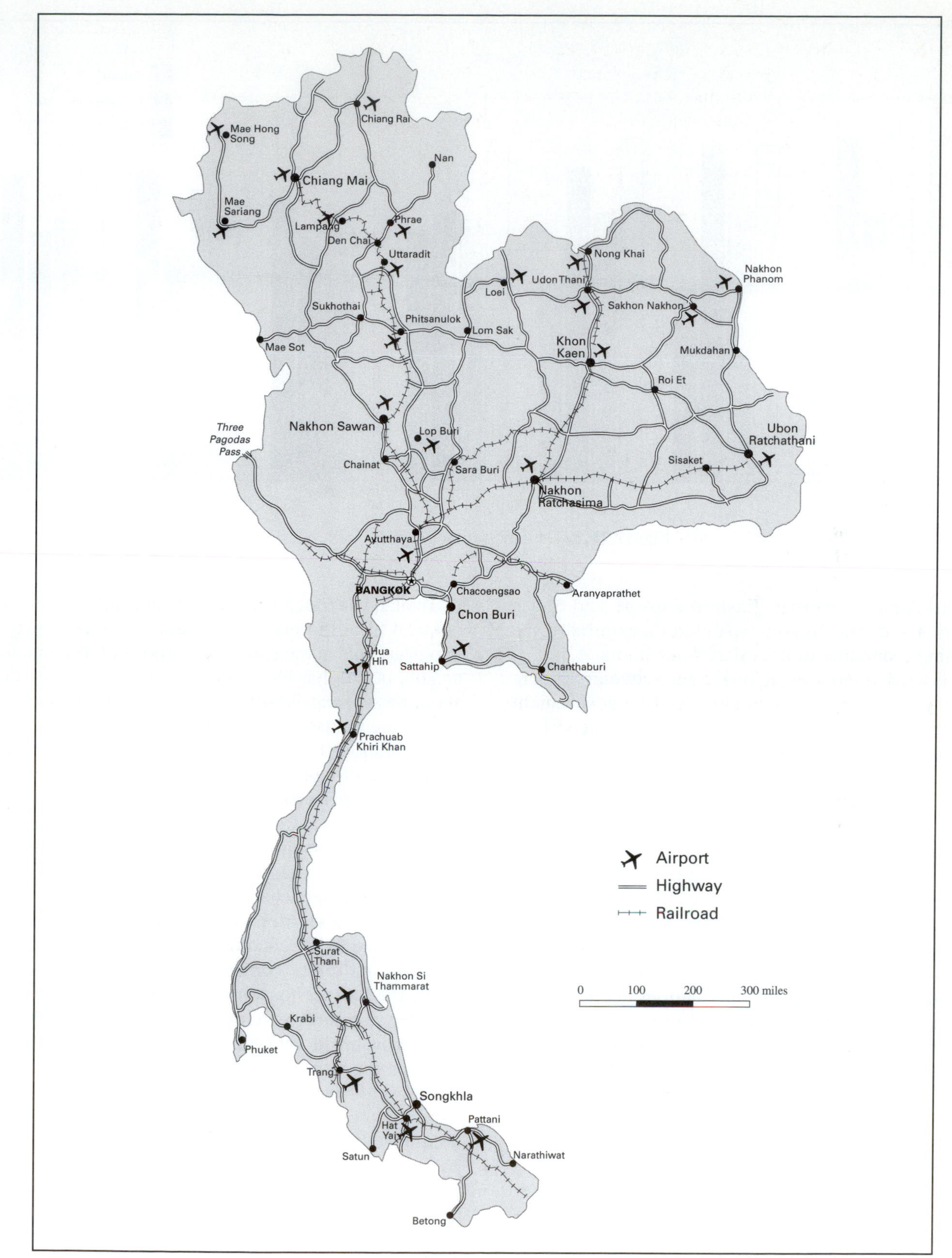

Map 16.6 Thailand's transportation network.

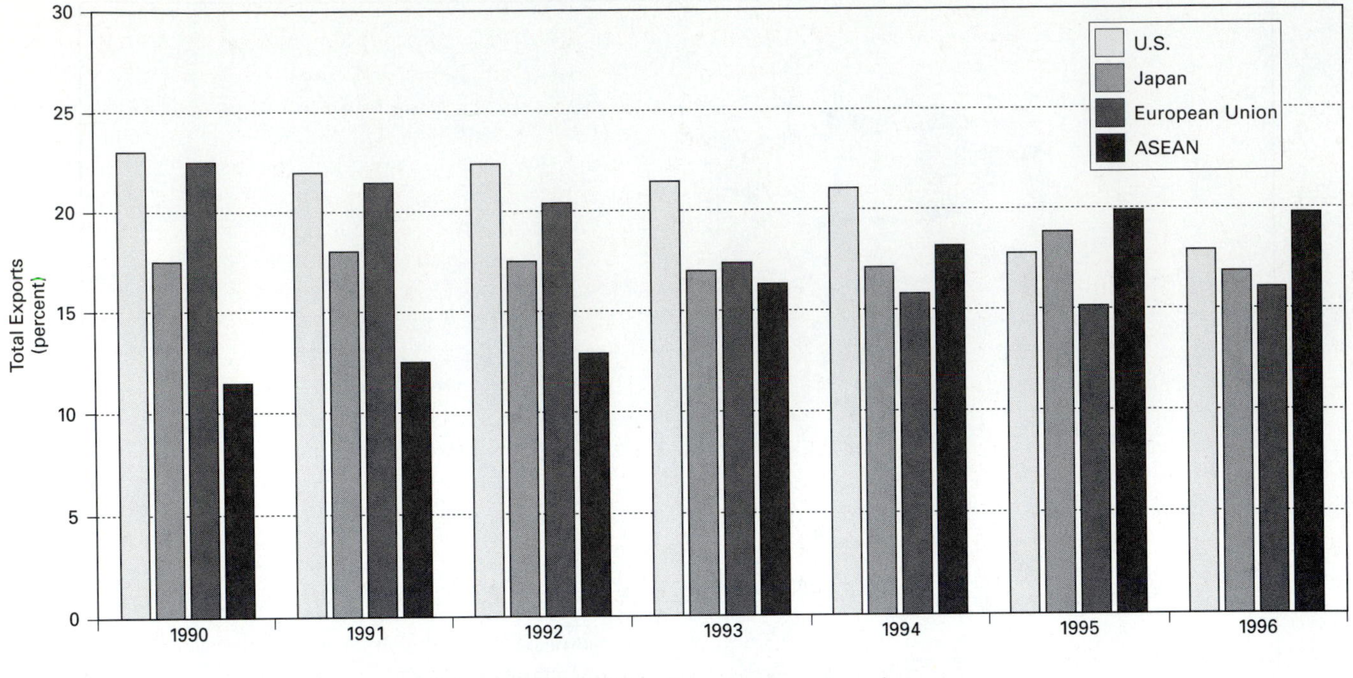

Figure 16.7 Thailand's major export markets.

look for new markets in Eastern Europe and South America and has already expanded its exports to the booming economies of its ASEAN neighbors.

This need for greater regional trade activities has generated increasing interest among the Thai government and exporters in the ASEAN Free Trade Area (AFTA). Although still in the initial stages of formation, this projected regional free trade zone is seen as a way to strengthen economic cooperation among ASEAN countries so that they can cope with changes in the global trading environment. In general, AFTA would stimulate expanded trade by removing regional trade barriers and tariffs, contribute to lower consumer prices due to cheaper imports, enhance the competitive edge in exports of ASEAN countries because of lower raw-material costs, stimulate employment in export-oriented industries, and increase trade bargaining power of the member nations in international trade. By reducing regional tariffs to as low as 5 percent on imported goods it has been argued that AFTA will create a robust free market within ASEAN. The benefits of this for Thailand may be an increase of US$1.8 billion in the volume of Thai trade and a 27 percent reduction in import price levels. With greater emphasis on market-oriented economic policies, the renewal of subregional cooperation on Mekong Basin Development also holds out the prospects for expanded economic cooperation with the other Mekong Basin countries, Burma, and Yunnan Province in southern China. Ultimately, what is hoped for is a region that is increasingly attractive to foreign investment, that is internationally competitive, and that is capable of realizing its full economic potential.

Thailand has been positioned by recent economic and political trends in mainland Southeast Asia to play an important role in the emerging markets of its immediate neighbors. This has also stimulated recent consultations about ways to expand subregional economic cooperation among the six nations that comprise the Mekong Basin region. Although official exports to its Indochinese neighbors account for only 2 percent of Thailand's total export trade, more open economic policies in Vietnam and Laos, the fragile peace in Cambodia, and longer-term prospects for their membership in ASEAN have focused increasing Thai attention on these countries. In Laos, relaxed central government control over the economy since 1986 has provided a greater role for market mechanisms in economic activity, encouraged private investment in business, and lifted controls on domestic and foreign trade. The Lao economy has expanded at a rate of 9 percent per year, and the number of new private industries has mushroomed. By 1992, the number of approved foreign investment ventures totaled 225, of which Thailand had 93 ventures totaling more than US$138 million, the most of any country. Traditional Laotian exports to Thailand have been wood products and electricity that account for one-third of total Lao national export earnings. In 1994 sawn timber made up more than 90 percent of Thai imports from Laos, and considering the expected tripling of Thailand's demand for pulp and paper and its 1986 ban on domestic logging, increased imports of Laotian timber will be of considerable importance for the immediate future. New hydroelectric projects are expected to increase electric power exports to Thailand, especially the Northeast, because of lower rates than from sources within Thailand. Thai trading opportunities with Laos are also extensive, and Thai bicycles, textiles, consumer goods, and many basic food commodities are popular and readily available. There are presently six trading points along the Lao–Thai border including the new Mittraphab or "Freedom" bridge linking Nong Khai with Vientiane, over which 63 percent of all Thai–Lao cross-border trade moved in 1994. However, several factors may affect any aggressive Thai involvement in the Laotian economy in the short term: the environmental effects resulting from unregulated foreign investment, the lack of a skilled labor force and technical personnel, and the small size of the domestic Laotian market.

Thailand's present trade with Vietnam is small but increasing. The composition of this trade provides one indication of the relative needs of each country. Thailand's primary exports have been sugar, motorcycles, chemicals, and fabrics; its imports are dominated by raw materials, paper, pulp, timber, crops, iron, fish, and minerals. Despite a relaxing of state control over the Vietnamese economy since 1986 and economic reforms included in the 1992 constitution, a rapid expansion in Thai trade and economic investment is unlikely in the near future. More immediate economic opportunities have been aggressively pursued by Thailand in Cambodia and, to a lesser degree, Burma. Trade between Thailand and Cambodia grew rapidly between 1989 and 1991 despite a persistent trade deficit. This deficit is based on increasing Thai imports of raw materials such as wood pulp, metal ores, livestock, and crops and more-limited exports of rice, sugar, scientific and medical equipment, plastic products, and pharmaceuticals. New foreign investment regulations in Cambodia have placed Thailand at the top among all foreign investors with almost 20 percent of all approved investment applications. The primary focus of Thai commercial interests have been in the bordering provinces of Sri Sophon, Battambang, Siem Reap, Pursat, and Koh Kong and have focused on gems and logging. A similar pattern of raw material imports and basic consumer goods and agricultural machinery exports is also typical of the more limited Thai–Burma trade. However, its close relationship with China and increasing foreign investment in Burma's extensive oil, gas, and mineral resources by Singapore, the United States, United Kingdom, and Japan have to some extent coopted Thai opportunities to invest heavily in these areas. This cautiously optimistic appraisal should not, however, disguise the serious constraints facing future Thai economic expansion into these markets. Uncertainties about the political direction and economic policies in Cambodia, an even more murky political situation in Burma, poor infrastructure and shortages of electric power, shortages of foreign exchange, and strong consumer goods competition from China are but a few of the obstacles to be overcome.

In the current global economic situation, Thailand is beginning to lose its comparative advantages from manufacturing labor-intensive goods with low wage rates and products using large amounts of natural resources. Indeed, these factors may already have begun to affect the economy. In 1996, export growth slowed to less than 1 percent, and increased competition from China, Indonesia, and Vietnam in the production of labor-intensive export goods was one major reason. Finding external sources for these raw materials has become an increasingly important challenge for the Thai public and private sectors. How well Thailand is able to seize opportunities to export its considerable expertise in manufacturing, services, business management, and the skills of an experienced educational and scientific community may well determine the direction of future economic and trade relations with ASEAN and its former Indochinese neighbors.

THE ENVIRONMENT AND DEVELOPMENT

Are economic development and the environment fundamentally at odds, or can they coexist in a more or less harmonious relationship? International concern over the environment in recent years has done much to stimulate debate over these conflicts between environmental quality and economic growth. Two general arguments tend to frame the opposing views on this debate. On the one hand are claims that technological and economic development will lead to enhanced resource conservation and a cleaner environment, once the transition has been made from natural-resource-based development to high-tech industrialization. Increased industrial efficiency will then create more opportunities to conserve natural resources and promote sustainable development. In this context the challenge lies in assuring that the technology is made available for all countries to follow a path of environmentally sustainable growth. A contrary view is that uncontrolled development is detrimental to the quality and sustainability of the environment, creating physical, economic, and social burdens that are disproportionately born by the poor. Although environmental quality and development are increasingly seen as interrelated global issues, perhaps nowhere in Southeast Asia is the conflict between these two clearer than in Thailand.

Natural resources—forest, land, water—have made indisputable contributions to Thailand's rapid economic growth. However, the social, economic, and environmental costs of that development have been increasingly defined in terms of degraded forests, polluted water supplies, increased soil erosion, and growing conflicts between competing users of scarce resources. These conditions have also added fuel to the debate about whether the present course of development is sustainable. This section briefly summarizes the status of Thailand's natural resources and profiles government efforts to improve the management of its most seriously threatened natural resource—its forest.

A Natural-Resource Profile

Since the 1970s it has become increasingly apparent that the current path of development in Thailand has contributed to an unsustainable trend in natural-resource consumption. This is evident in the state of the country's natural resources and the kinds of issues that must be addressed if this trend is to be reversed.

The contributions of the agricultural sector have grown significantly in the past quarter-century. However, under past conditions of rapid population growth, the once seemingly unlimited land frontier has now been all but exhausted, and almost 90 percent of the land suited to agriculture is now under cultivation. This will inevitably require more efficient use of existing land and more-equitable distribution of that land. Unfortunately, government progress in issuing basic land title documents has been slow and often ineffective. Only 12 percent of the agricultural land has legal title documents while 40 percent or more than one-third of land holdings have no form of documentation, especially in encroached areas of national forest. Mean farm size has also declined throughout the country since 1975, and almost 60 percent of households in some areas of the Central Region and Upper North are landless. Nationwide, at least 500,000 households have no land, a condition that has aggravated the problems of deforestation, lagging rural development, and rural poverty. Consequently, the shortage of land available for agriculture, the increasing conversion of more productive land to urban and industrial uses, and the urgency of accelerating land titling and land redistribution programs are the main issues requiring attention. Critical to the success of these efforts is also simplification of the responsibilities of the more than 24 government agencies and 10 committees that share responsibility for land resources.

Resource	Conditions	Key Issues
Land	• 90% land suited to agriculture under cultivation • 0.5 ha arable land per capita • 38% of agricultural land without legal titles.	• Shortage of arable land • Conversion of arable land to urban and industrial uses. • Land classification, titling, and redistribution
Soils	• 33% (171,000 ha) land with medium to high levels of soil erosion • Estimated loss of 47.5 mil. tons of topsoil annually • Saline soils a growing problem, especially in the Northeast	• Limits on sustained or increased agricultural productivity • Declining crop yields • Soil conservation
Forests	• Loss of 50% of forest area since 1960 • Degradation of watersheds • Deforestation rate exceeds rate of reforestation • 8.7 million people live and farm in national reserved forests	• Implementing National Forest Policy to reach 40% forest cover. • Improved watershed management • Promoting social forestry • Improved forest management • Growing deficit in timber
Water	• Sirikit and Bhumiphol reservoirs at 50% normal water levels • Growing use and competition for water resources • Daily groundwater use in BMA is 60% above safe extraction rates. • Water shortage for dry season crops	• Irrigation systems maintenance • Inefficient water useage • Flood and drought protection • Rate of urban subsidence in BMA is increasing • No national water management master plan or agency.
Fisheries	• Marine fishery production accounts for 90% of total fish production. • Eighth among world's fish producers and top exporter of shrimp • Marine production in the Gulf exceeds maximum sustainable yield	• Industrial pollution and untreated waste water threaten industry • Disease epidemics in fresh water fishing areas in all provinces. • Better marine fishery management in Thailand's Exclusive Economic Zone in the Gulf and Andaman Sea

Table 16.6 The status of Thailand's natural resources.

The increasing pressure on land resources is also manifest in serious problems with soil erosion. Almost 65,640 sq mi (170,000 sq km) or one-third of Thailand's land area suffers from medium or high degrees of soil erosion, particularly upland areas where slopes are greater than 5 percent. In some areas, the average annual soil loss per acre exceeds 65 tons (160 tons per ha), a situation that has both immediate and long-term effects on crop productivity and yields. The effects of siltation from soil erosion on dams and sedimentation along waterways has also reduced their storage capacity, raised the annual sediment load deposited by the Chao Phraya River in the Gulf of Thailand to 11 million tons, and contributed to current problems of water shortages and flooding in Bangkok. Salinization, another type of soil degradation, has also become an increasingly serious problem that affects agricultural activities in some portions of the Northeast.

Perhaps most symptomatic of the cost of development on natural resources has been the dramatic depletion of Thailand's forest resources. Since 1960 forest area has declined by 50 percent or more than 48,260 sq mi (125,000 sq km). The main causes of this problem have been commercial logging, the expansion of agriculture in forest areas, forest encroachment by the growing rural population, especially the poor, and unsustainable forest management practices by timber concessionaires and government forestry agencies. This has left many critical upper watersheds degraded, reduced water flows to some of the major dams, contributed to soil erosion, and by recent estimates, left more than 8 million people illegally occupying national reserved forestland. Although rates of deforestation have slowed in the last decade, deforestation continues at a pace 10 times the rate at which government and private interests can reforest degraded forested areas. Deforestation is thus not only an environmental issue but also an increasingly serious social and political problem. In this context, the key issues confronting the government include implementing the National Forestry Policy, rehabilitating degraded watersheds, and promoting programs such as social forestry to improve forest conservation and ensure their sustainability and productivity for future generations.

The quality, availability, and distribution of water resources has become an increasingly serious problem, especially in the BMR and the Central Region. Rising demand from the industrial and service sectors, increased domestic use in urban areas, weak regulation of consumption rates, and low rainfall during the last decade have contributed to this situation. Upstream growth in water consumption and changes in land use due to watershed destruction reduced water levels in the two largest dams, the Bhumiphol and the Sirikit, to a 40-year low in 1993 (Thailand Development Research Institute, 1993). Only the combined capacity of 809 billion cubic feet (22.9 billion cubic meters) of water in these facilities serve the 8 million people and 1,778,400 acres (720,000 ha) of irrigated padi land in the Central Plain, and one-third of this capacity is needed for electric power generation. This has reduced the availability of dry season water for irrigation to one-third normal levels and has caused farmers to shift from producing a second dry season padi crop to field crops that require less water.

Like many countries, Thailand has long used its rivers for waste-water disposal, and rapid industrialization and urbanization have seriously aggravated this problem. The

Photo 16.7 Mechanization has accelerated the pace of timber extraction. (Hafner)

development of public utilities—sewerage, waste-water treatment facilities, and municipal water supplies—have lagged far behind the rising demands of the urban population, and domestic sewerage is regularly discharged into waterways at such high rates that the Chao Phraya River has become unfit for domestic use. As the Thai economy has shifted toward more manufacturing activities, the number of industries that produce hazardous waste has continued to increase. By 1991, 351 factories used the Chao Phraya River for waste disposal, and 376 other plants used the nearby Tha Chin River in Samut Prakan for this purpose. Contamination of coastal waters, depleted marine resources, and the destruction of large areas of important mangrove ecosystems are but a few of the effects of this problem. The solution to these problems will depend on addressing both supply and demand side issues; improving the maintenance of irrigation systems, affecting changes in water consumption behavior, increasing water storage capacity, developing urban flood control systems, and creating a national water management plan and agency to coordinate water use and conservation.

Thailand's marine fishery industry has grown significantly in the last quarter-century and is now ranked eighth in the world in total fish production. The introduction of trawl-net fishing in 1963 resulted in a spectacular increase in marine fish catch, and within the last two years shrimp production has also moved Thailand into first place in world production. However, overfishing and the use of inappropriate technology has pushed fish capture rates in the Gulf of Thailand beyond maximum sustainable yields, and prospects are poor for further development of fishing in the gulf. Claims by neighboring countries to areas of Thailand's Exclusive Economic Zone in the gulf and the Andaman Sea have also reduced its access to traditional fishing grounds. The destruction of coastal mangrove systems by industrial pollution and their conversion to fishponds has seriously impacted coastal marine ecology, especially in the South. Furthermore, pollution, untreated waste-water discharge, and a variety of diseases have affected freshwater fisheries in almost every province in the country.

Although the government has slowly begun to recognize the high costs of environmental degradation, its efforts to protect the environment have yielded mixed results. National bans on logging, watershed protection schemes, energy-demand management programs, subsidies for unleaded gasoline, and investment in more waste treatment facilities are some of the government responses to environmental degradation. The Seventh National Development Plan (1992–1996) and the National Environmental Quality Act of 1992 have also added new policy and legislative tools to help improve environmental protection. Yet, most current efforts have still neglected the important area of prevention. Until a firmer government commitment is made to enforce both "end of pipeline" sanctions and increase emphasis on prevention, natural-resource degradation will continue to impose a high social, economic, and political cost on the country's development.

The Quagmire of Deforestation

No single issue illustrates the trade-offs that Thailand has made between economic and environmental priorities more than deforestation. Once recognized as a resource-rich country with abundant forests that yielded important wood exports, Thailand has become a net importer of wood and wood products and is increasingly harvesting timber from Burma, Cambodia, and Laos to meet rising domestic demand. It has lost 338,000 sq mi (130,000 sq km) of forest, or one-half of the country's estimated forest area in 1960, and only 347,155 sq mi (133, 521 sq km) of forest including reforested areas remained in 1983. Although the depletion of these resources has taken place nationwide, the greatest absolute and percentage declines since 1961 have been in the Northeast and South (Figure 16.8). Like many other developing nations, the exploitation of natural resources such as forests has added immeasurably to development and economic growth. However, deforestation in Thailand has also been driven by population growth, the uneven distribution of arable land, poor forest management and illegal logging, and rising poverty. If one weighs the exceptional growth in Thailand's GDP and economic progress against the use of its natural resources, the balance between exploiting natural wealth for economic gain would appear to be sound. Unfortunately, the environmental and social costs of this path are not adequately reflected in conventional economic indicators.

All forestlands and resources in Thailand are the property of the state, which grants limited rights to individuals for their use to meet domestic household needs. Between 1850 and 1950 the estimated rate of forest depletion was less than 1 percent per year, but in the last 30 years it has more than doubled. Various legislation and laws were passed to protect forest resources by creating national parks, setting aside areas as national forest reserves, limiting public exploitation of forest resources, and increasing penalties for illegal use of these resources. Since 1960, all of the country's seven national development plans have set targets for the land area to remain as protected national forest using reforestation, conservation, and tree plantation schemes to achieve these targets. Yet, each successive plan was forced to lower these levels as deforestation increasingly reduced the existing forested area. Having failed to meet these targets, and with the area of natural forest down to 29 percent at the

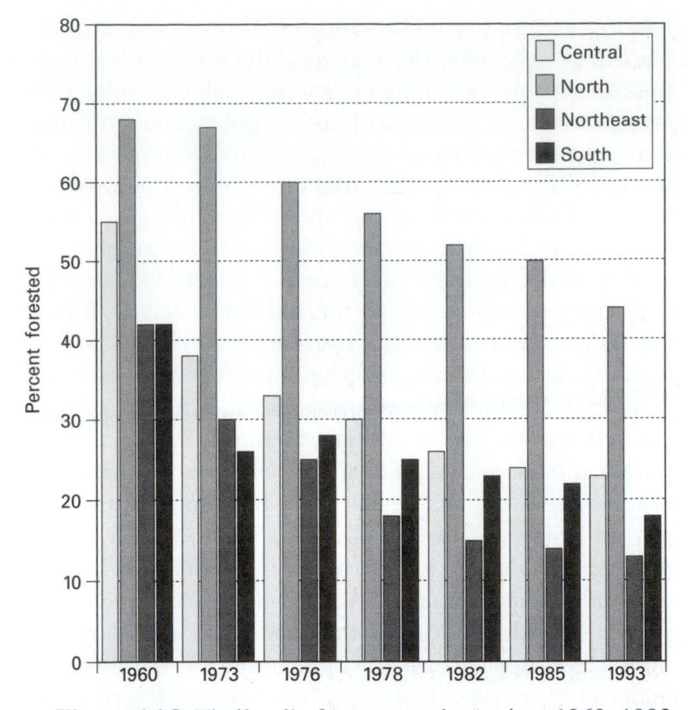

Figure 16.8 Thailand's forest area, by region, 1960–1993.

end of the Fifth Plan in 1985, a ban on logging was imposed and the government created a National Forest Plan to increase forest area to 40 percent: 15 percent to be set aside as protected conservation forest and 25 percent as economic forest for timber production. Reaching this target would require reforestation of 14.8 million acres (6 million hectares), but because the rate of deforestation is estimated to be 10 times the rate of reforestation, the current goal of creating an economic forest appears unattainable. Even assuming a complete halt in deforestation and a tripling of the rate of reforestation, it would take at least 35 years to reach the 40 percent target set in the 1985 National Forest Plan (Thailand Development Research Institute, 1987).

Evidence from both past efforts to improve management of these resources indicates the complexity of this issue. Since the 1970s the government has promoted the development of community woodlots, resettled forest populations from degraded forests to discourage encroachment and promote reforestation and, most recently, involved the private sector in creating tree plantations on denuded public forestland. Although many of these strategies are consistent with national reforestation goals, the design and implementation of programs to achieve those goals have too often been inadequate (Hafner, 1992). The nature of these problems are perhaps best illustrated by two recent government programs. Between 1975 and 1980 the Royal Forest Department created the Forest Village Program (FVP) and the Land Rights for Residents of Degraded Forests (STK) Programs to control illegal encroachment on reserved forests and promote reforestation. The FVP organized dispersed forest populations into villages where they were provided with basic services, allocated 5.9 acres (2.4 ha) of land for farming, and promised wage-labor employment in reforesting degraded forestland. Between 1975 and 1983 this program organized 144 "Forest Villages" throughout the country and allocated more than 69,160 acres (28,000 ha) of degraded forest to project participants, most in the Northeast where rates of deforestation were highest. Due in part to the high costs incurred in implementing the Forest Village Program, inadequate budget allocations, and problems surrounding land rights for program participants, a new project called the STK Program was implemented in 1980. This was essentially similar to the FVP except that no basic village services were provided and more-flexible limits on individual land allotments were allowed.

Although both programs were consistent with national policies on reforestation, there were serious flaws in their design and implementation. Under the FVP, degraded forestland allotted to villagers was often poorly suited to farming, the size of these allotments were often too small to meet basic household subsistence needs, and villagers were not given legal title to the land but only "user rights" or "certificates of use." These "certificates" are not acceptable as collateral for bank loans, and their user rights may not be sold or transferred except through inheritance. In many cases, the area set aside for land allotments was already claimed by other farmers, and because of massive migrations to FVP project sites, land set aside for agricultural use was insufficient for the number of qualified households. This has contributed to intervillage conflicts, a large landless population, destruction of reforestation areas, and further encroachment on the forest (Hafner and Apichatvullup, 1990). As demonstrated in these programs, one of the fundamental issues which complicates reforestation, aside from poor program design and implementation, is land tenancy and ownership.

Recognizing these difficulties, the government shifted its focus in the late 1980s to involving the private sector in reforestation by making denuded public forestland available for private tree plantations. These lands are offered at a nominal rent to private plantation operators along with generous tax concessions and exemptions from import duties. The experience since private reforestation began and logging was banned in 1989 has not been encouraging. Much of the denuded forestland rented to private concerns for tree plantations was already occupied by farmers, and evicting them was impossible; where attempted, this has created serious social problems. In some instances, offers to buy off squatter claims to the land have been partially successful, but this has produced problems of illegal transfers of untitled land and the suspicion that those same farmers have simply moved to clear new land in other reserved forests (Panayoutou and Phantumvanit, 1991). This suggests that privatization of denuded forests may simply result in replacement of natural forest with plantations while the remaining forests continued to be encroached on. A further problem has arisen from the preference of private firms to plant fast-growth tree species such as eucalyptus, which many argue are detrimental to soils, water supplies, and adjoining cropland. Strong opposition to this privatization program has also been mounted by NGOs, rural communities, and environmental groups that prefer a more ecologically sustainable community forestry approach to reforestation. Under these conditions, government efforts at reforestation have been a stalemate, at least for the moment.

Thailand is no longer a predominantly agrarian and natural resource-based society but has a rapidly industrializing and service-based economy. Primary resource sectors such as forestry and fisheries have lost their status as leading growth and foreign exchange earners. Even agriculture, which made Thailand one of the few major food-exporting countries in the developing world, has declined as a source of income and export earnings. As more value is generated from human resources and capital, natural resources are slowly losing their quantitative significance. Obviously, the traditional approach to natural resource management that originated during times of relative resource abundance is inappropriate to the general problems of resource scarcity that now confront Thailand. A new approach that will balance demands and supplies for all natural resources at sustainable levels of use and that makes the best use of scarce resources is needed, but for this to succeed, key policy reforms are needed in education, land titling, access to credit among the rural sector, increased agricultural productivity, and an expansion in nonagricultural employment opportunities.

THE STATE AND DEVELOPMENT

The success of Thailand's transition from import-substitution to export-led industrial growth is regularly cited as evidence for the country's impending emergence as Southeast Asia's next newly industrialized economy. This ongoing process is transforming not only the economy but many aspects of Thai society as well. These changes raise the question about the role that the state has played in development, particularly the extent and consequences of state intervention in the economy. One view holds that the state should exercise a minimalist laissez faire role by allowing growth to result from the operation of free market forces while the government maintains low barriers to trade and investment, provides basic infrastructure, and ensures stability of law and order. On the other hand, there are arguments that rapid economic growth has occurred because the state has intervened extensively in the economy by managing foreign exchange rates and credit markets, structuring industrial policy, and requiring that industries selected for government promotion meet the objectives of state development policy. The answers to this question for Thailand are at best ambiguous and can perhaps be illustrated with some brief examples.

Where the state has intervened to promote economic growth, most notably in the industrial sector, it has often done so in a passive, even ad hoc, manner. Government macroeconomic policies have fostered a climate conducive to trade, investment, and the growth of private enterprise. Stable foreign exchange rates and low inflation have encouraged increased foreign direct investment and the shift away from the raw material export industries of the 1960s and 1970s to higher-value manufactures such as electronics. However, this transition has resulted primarily from the investment and trade strategies of the multinational firms on which the country has become increasingly dependent for markets and technology, rather than from a coherent government industrial-sector policy. Furthermore, even the BOI established in 1960 to create policies that provide incentives for industrial investment such as tax holidays and exemptions of import duties on machinery has done little more than make investment in Thailand "at least as attractive" as investing in other ASEAN countries (Christiensen and Siamwalla, 1994).

In the agricultural sector, arguably the most important foundation of Thai economic growth until recently, the impacts of government intervention have been uneven at best. Historically, only about 10 percent of government investment has been in agriculture and more than one-half of that in irrigation facilities, primarily in the Central Region. Through the 1970s agriculture absorbed a large share of the expanding labor supply through the extension of land under cultivation, increases in cultivated area per farmer, and diversification of crop production. This was made possible by a land surplus that resulted in increased government revenue through expanded production and export of both padi and field crops. However, because Thailand had been a food surplus country there was little emphasis on promoting staple food production to achieve domestic self-sufficiency. Thus, the government taxed rather than subsidized farm output by raising domestic export taxes on rice, the "rice premium" initiated in 1955, above world market price to ensure adequate local supplies of low cost rice. At the same time, industrial and trade policies taxed agriculture inputs, such as fertilizer and machinery, that could have increased output. These contradictory and short-term policies have had little effect on the long-term issues of regional and rural-urban income disparities, the persistence of high rates of rural poverty, or the need to buffer the rural sector from declining agricultural prices and the declining land surplus. Similar arguments can also be raised concerning government failures to extend and upgrade irrigation facilities, address growing inadequacies in internal communications facilities, and advance levels of education beyond the primary level.

The influence of the Thai state on development has therefore been a mixture of effective macroeconomic management and benign neglect of sectoral level needs. This performance does not augur well for the demands created by a rapidly changing economy nor for the country's preparation for the next stage of industrialization. The growing educational deficits in the output of technicians and engineers, and the lack of state institutions capable of managing the social welfare functions of an advanced industrial economy are critical symptoms of the obstacles to attaining that goal. Any return to an era of strong state intervention typical of Thailand's recent military-dominated governments is unlikely to lead to an equitable and effective resolution of the complex problems that will undoubtedly occur in the future.

17 Burma

ROBERT E. HUKE

When Burma gained independence in 1948, the future looked bright. At that time several important observations could be made concerning Burma's likely future. First, the country was rich in natural resources, mineral as well as timber, and the rational utilization of these seemed to auger well for sustained economic development far into the future. Second, Burma had a population density well below that of its neighbors on all sides. There appeared to be ample land of good quality for future settlement and production in a country that already controlled two-thirds of the entire world's exportable rice—and that in a world with an ever-tightening supply of grains. Third, the British had left a well-trained civil service that appeared capable of ensuring smooth operation of the government. Fourth, Burma had become a sovereign parliamentary, democratic republic with a strong cultural unity among the native speakers of Burmese, this bequeathed from the old Burma. In addition there were the beginnings of representative institutions developed under the colonial administration.

Unfortunately, very little has progressed as well as was forecast. During much of the period since 1948, the country has been torn asunder by a series of armed insurrections, some originating from diverse political philosophies among the Burmans and others developing out of ill will between the Burmans and the Karens, Mons, Kachins, Shans, Was, and other nonnative speakers of Burmese. Insurgent activity climaxed in the mid-1950s when rebel forces controlled major portions of the Irrawaddy delta, most of the Dry Belt, large areas of the Shan State, the Kachin State, and Kayah State, all of the Karen State, and smaller portions of the Chin and Rakhine States. It was unsafe to travel outside of Rangoon except along the 40-mi (64.4-km) road to Pegu—and even that trip was not to be undertaken after dark. Exploding shells could often be heard from the campus of Rangoon University, and the capital city's water supply was interrupted on a regular basis (Lintner, 1994a).

Government forces have worked relentlessly to regain lost ground and to vanquish all insurgent forces. The military has often been overzealous in its efforts, using a far broader definition of *insurgent* than would be accepted by the world outside of the military. Civilians, especially students and religious leaders, have perished in significant numbers, and others have been incarcerated and held without charge. Major portions of the budget have been swallowed by the military operations, and civilian services have deteriorated seriously. With the government long in the hands of the military, it was inevitable that civil rights would be lost and that Burma would be transformed from what was once a land of happiness and relative prosperity to a land of frustration and discontent. Burma faces harrowing problems at the approach of the twenty-first century (Carey, 1997; Taylor, 1987).

THE HUMAN LANDSCAPE

The peopling of Burma has been strongly influenced by the basic environmental conditions of the country. All around the inland margins of Burma, the country has been shut off from the rest of Asia by rugged mountains or deeply dissected uplands, all of which are clothed with heavy forest cover. This topography does not encourage surface transportation, especially in the border areas, and today's map shows very few road and no rail connections between Burma and her neighbors (Map 17.12). Geography, political as well as physical, has precluded such developments. Mountains, gorges, torrential rains, endemic malaria, and thick forests however were not sufficient to stop tribal migrations from the north and northeast. During more than a millennium, peoples dislodged from their ancestral homes by flood, drought, pestilence, or war have drifted south and west to occupy new territories in what has become Burma, nor is there any reason to believe that the movement has ceased. Formal borders and the occasional border guard can do little in the face of unorganized tribal migration through the dense forests of the north.

Early Migrants

The population of present-day Burma owes its origin chiefly to three or perhaps four historic movements of people from the north and east . The earliest was that of the Tibeto-Burmans who entered from the mountain

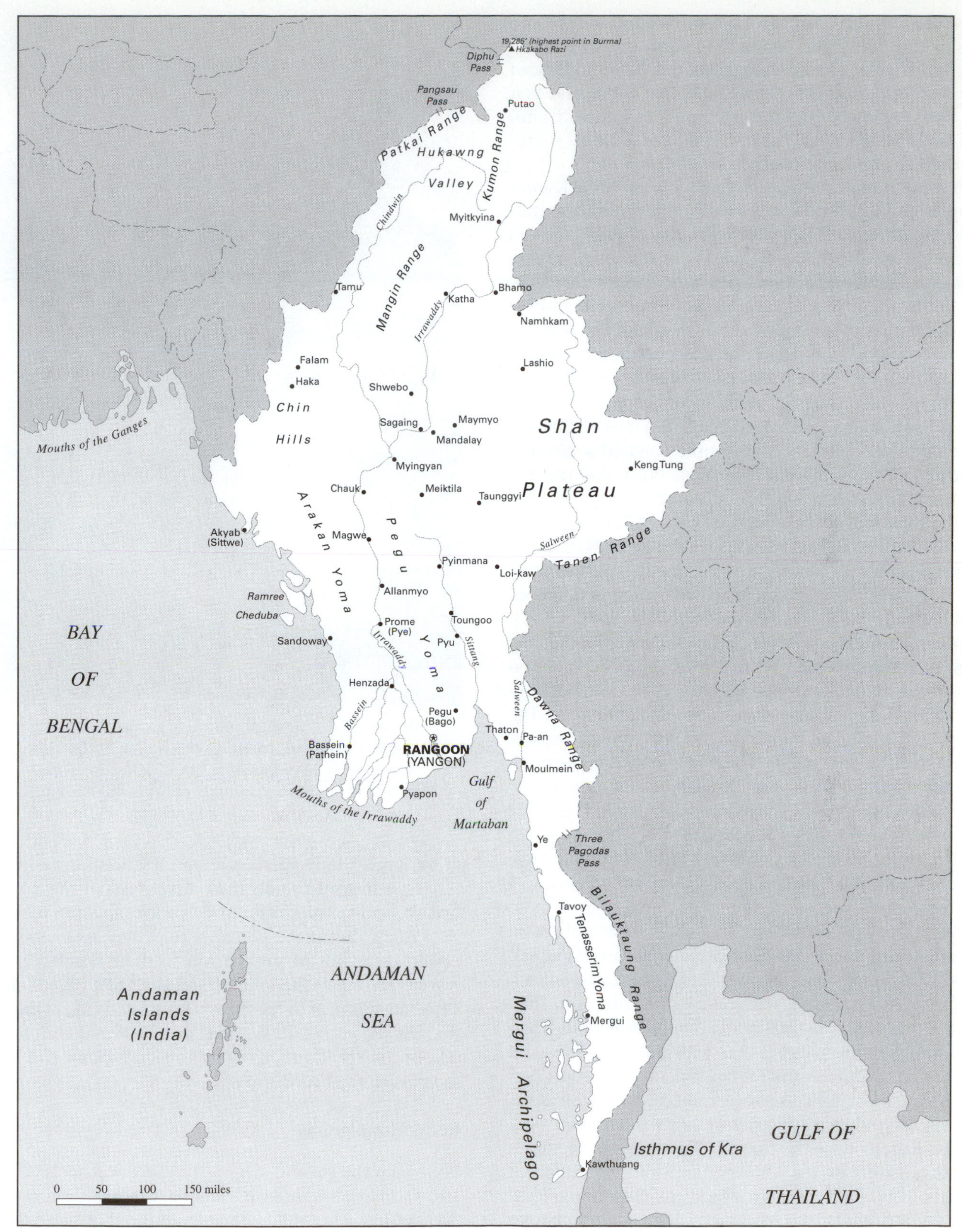

Map 17.1 Burma. *Source: Ulack and Pauer, 1989.*

passes to the north and moved southward along the valleys of the rivers that combine to form the Irrawaddy. Archaeological evidence suggests that this movement was well underway by 200 B.C. and that it brought first the Pyus, followed by the Burmans, Arakanese, Chins, and lastly the Kachins. The movement of Kachins southward along this route continues to this day.

A second major migration included the Mon-Khmer peoples and followed a southward path generally parallel to the Mekong River, well to the east of that followed by the Tibeto-Burmans. Segments of this flow, largely Mons, broke off and drifted to the west, eventually reaching the lower Salween Valley. This group settled along the coastal strip between the Salween and the Sittang, occupied portions of the Irrawaddy delta, and moved south along the Tenasserim coast. These people were known to the Burmans as Talaings and established capitals at Pyay, Thaton, and Martaban. The Mons or Talaings were perhaps a bit darker in complexion and somewhat shorter in stature than the Burmans but today the distinctions, if ever real, have disappeared.

The third wave was composed of Tai-Chinese people who moved into Burma from the high tablelands of southwestern Yunnan, the location of the kingdom of Nan Chao (Keyes, 1987). This movement, which overlapped temporally although lagging somewhat behind the earlier flows, began in about the seventh century, increased gradually from then through the twelfth century, and became even more significant following Kublai Khan's conquest of Yunnan in 1234. It continues as a trickle to the present. The Shan, the Kayah, and numerous smaller tribes were associated with this movement. The Shans settled in the often isolated basins and valleys of the Shan Upland, where they established numerous small independent states. Here they share the region with several lesser tribes who have been pushed onto the higher slopes in a form of human vertical zonation. Shans also occupy the deep broad gorges of the Salween River where they enjoy a warmer climate and a far heavier forest cover than do their neighbors on the adjacent plateau. The Shan State area east of the Salween River is remote from Burmese influence and difficult to access from Rangoon. It shares a long border with Thailand, China, and Laos. In this area, connections with Thailand have long been easier than with Burma proper, and it is here that Shan insurgent movements were most persistent and most difficult to overcome. Here, too, present-day opium trade has proven impossible to contain.

The Karen chose the rugged hills marking the southern extensions of the Shan upland and the adjacent plains of the Sittang and Irrawaddy as their home location. In portions of the Irrawaddy delta, Karen peoples outnumber Burmese speakers. In the hills they have become the chief occupants of the traditional routes between Burma and Thailand. In the plains they have mixed widely with peoples of Mon and Burmese cultural heritage. Many Karens adopted Christianity in the nineteenth century and took advantage of the educational opportunities offered by various mission centers. As a result, during the British period, they assumed leadership roles out of proportion to their numbers. The Karens have had the longest and strongest objection to what they think of as the army's iron-fisted rule of the nation. By the end of 1996, the Karen National Union was one of four holdouts refusing to sign a cease-fire with the government of Burma.

Photo 17.1 Several minority groups, including the Padaung, are found in the Kayah State. This young woman is the daughter of a headman and at age 14 has almost a full set of brass rings around her neck. (Huke)

Recent Immigrants

Following the second Anglo-Burmese war in 1852, the Irrawaddy delta became a major target for migration. Agriculture, which had been of a shifting subsistence nature supporting but a few people, was transformed to a commercial mono-cropping of rice. The potential for accommodating increased population numbers was enormous. Today's Irrawaddy, Prome, and Rangoon divisions were equally attractive to land-hungry migrants from several areas. Large numbers came from within Burma. The largest single source was the already crowded Burmese-speaking area of the Dry Belt, traditional Burma. Significant numbers also originated from the uplands along the Thailand–Burma border, especially from among the Karen speakers, thus sowing the seed for later problems.

A second major source of immigrants was the overpopulated lands of Madras and Bengal Provinces of British India. Landless peasants from Madras, laborers from Bengal, and Chettiar money lenders set sail across the Bay of Bengal in the hope of making a future in Burma. Many came without their families expecting to remit sufficient funds to make life easier at home; others saved in the expectation of bringing wives and children at some date in the future. Still others, almost all male, came unattached hoping for a more rewarding life in the new country. During the last half of the nineteenth century and the first four decades of the twentieth century, the migration provided an important part of the human power required to settle and work the delta rice system.

Chinese migrants also came to Burma, although in lesser numbers, and found jobs in mining operations that were frequently shunned by the locals and avoided by the Indians. The Chinese also became merchants and traders and, throughout the delta regions, developed the rice trade, including milling operations. In village after village a Chinese businessman, often married to a local woman, operated the only shop available.

By 1940 the population of Indians totaled well over 1 million and accounted for somewhat in excess of 6 percent of the total population of Burma. By this date they had dispersed widely over the country, but the greatest concentration remained in the delta where they served as laborers, money lenders, traders, and mill operators. Many Indians fled Burma in the face of the Japanese invasion early in World War II and never returned. Of those who fled with the retreating British army, at least one-half died of hunger, cholera, malaria, or pure exhaustion during the terrible retreat across the Arakan Yoma mountain range (Tinker, 1959).

In 1962 the military, led by General Ne Win, replaced the civilian government for the second time and established a set of national guidelines known as the Burmese Way to Socialism. This Marxist program proved to have a devastating impact on the national economy. It also caused important demographic changes and placed a continuing freeze on democratic traditions. The impact was especially felt by people of Indian or Chinese heritage. Nationalization of land, retail and wholesale trade, banking, and mining severely weakened the economic base of these communities. Punitive laws deprived even those who were second-generation Burma-born of their rights of citizenship. Finally, demonetization of nearly all currency wiped out the life savings of several hundred thousand families and left the "foreigners" (Indians and Chinese) destitute in a country where, at least officially, they were no longer welcome.

Large numbers of Indians, with considerable humanitarian aid from the government in New Delhi, emigrated to a homeland some of them had never seen. For the Chinese, emigration was more difficult, and many remained in Burma to start life anew. Unfortunately for them this was not the end of their problems. In 1967 economic mismanagement by the Ne Win government and a serious rice shortage led to widespread civil unrest. The legacy of diverting antigovernment feelings toward foreigners resulted in renewed anti-Chinese riots with considerable loss of life.

Present Population and Its Distribution

To people on the Indian subcontinent or in China, Burma must indeed seem to be the promised land. This is especially true in regard to the ratio between numbers of people and living space (Table 17.1). The overall density in 1994 was 65 persons per sq km. This compares with 810 in Bangladesh, 257 in India, 113 in China, and 111 in Thailand. By contrast, the numbers for Burma-size areas in North America are California combined with Oregon 48, or Texas 25. In comparison to neighbors in the Asian setting, Burma is lightly populated, but compared to North America it is far from empty. It should be noted that the states, actually minority homelands, together comprise more than one-half the total area of Burma, yet hold only 28 percent of the population. Population density in the divisions, or Burman homelands, is triple that of the states. It is also true that environmental conditions are more favorable for human settlement in the divisions than in the states.

Map 17.2 is based on population estimates for 1994 and indicates population distribution by state and division. The map shows Rangoon division to be the dominating focus of population and, as the eye moves away from this division in any direction, the population density is seen to decrease. The four political areas next after Rangoon in density are Irrawaddy Division, Pegu Division, Mandalay Division, and the Mon State. All four of these areas have extensive portions of their areas in agriculturally productive river or coastal plains. Four of these five most densely peopled units are homelands to Burmans, and only one, the Mon State, is a minority homeland. If the entire population of Greater Rangoon is subtracted from that of Rangoon Division, the area still falls well within the density range of the other four.

At the other end of the scale, the units comprising the two categories of lowest population density are six in number and include five states or minority homelands and only one division, Tenasserim. From the geographic point of view, it is interesting to note the location of these

	Area* sq km	1983 population	1941 population	% increase 1941–1983	1983 density (pop/sq km)
STATE					
Kachin	89,042	903,982	427,625	111	10
Kayah	11,733	168,355	70,493	139	14
Kayin	30,383	1,057,505	353,411	199	35
Chin	36,019	368,985	220,410	67	10
Shan	155,801	3,718,706	1,699,612	119	24
Mon	12,297	1,682,041	889,596	89	137
Rakhine	36,778	2,045,891	1,152,733	77	56
Subtotal	372,054	9,945,465	4,813,880	107	27
DIVISION					
Sagaing	94,603	3,855,991	1,811,360	113	41
Mandalay	37,024	4,580,923	1,907,703	140	124
Magway	44,812	3,241,103	1,719,404	89	72
Irrawaddy	35,172	4,991,057	2,659,126	88	142
Rangoon	10,140	3,973,782	1,346,867	195	392
Tenasserim	43,328	917,628	392,582	134	21
Pegu	39,404	3,800,240	2,088,441	82	96
Subtotal	304,483	25,360,724	11,925,483	113	83
TOTAL	676,537	35,306,189	16,739,363	111	52

** Area as of 1983.*

Table 17.1 Population and area of Burma's states and divisions. *Source: Census of Burma, 1941 and 1983.*

six. All are on the periphery of the country, all are distant from the Irrawaddy–Sittang lowland permanent field rice areas, and all are located in the mountain ring that surrounds the heartland.

The rate of population growth during the 42-year period illustrated is roughly equal in states and divisions. Current annual increase is at about 1.9 percent. This modest increase results from a birthrate of roughly 31 per thousand per year and a death rate of about 12 per thousand per year. The birth rate is 70 percent higher than in neighboring Thailand, perhaps due in large part to the fact that Burma has no family-planning program, while Thailand has a strong program. The death rate in Burma is 70 percent higher than in Thailand, reflecting the limited health facilities, the high cost and low availability of medical drugs, and the scarcity of public health programs. Fortunately, today's numbers show marked improvement since 1983 when the Census reported even higher rates: births 38, deaths 14, and annual growth rate 2.4 percent. Conditions in Burma improved a good deal between 1983 and 1997 but leave much to be desired.

For Burma as a whole, excluding land used for shifting cultivation, there are 425 persons for every 100 ha of farmed land (a bit over 1/2 acre per capita). This number contrasts sharply with corresponding figures for Bangladesh, 1,300, and for China, 1,275 (United Nations Food and Agricultural Organization, 1993). This contrast, more than any other single bit of evidence, provides the key to an understanding of the unique character of Burma's geography. Farmland is not yet considered to be in short supply. Both farmers and government officials have a very relaxed attitude toward increasing the intensity of farm operations. Food is not in short supply except in isolated pockets of urban poverty, and increasing the rice surplus available for export appears to be the prime objective of agricultural planning.

Sixty percent of Burma's population is 25 years of age or younger, and 36 percent have yet to reach the age of 15. Persons older than 25 are generally married and have children of their own; only a small proportion of those younger than 25 have formed families. The population as a whole is young, and children are quite conspicuous. Thirty-six percent of the country's population is composed of married couples. These couples have an average of somewhat more than three children each, and perhaps one-half of the families have one grandparent at home. Such groupings indicate a family size approximating five and one-half individuals. A young population with an average of more than three living children per family indicates rapid population growth and suggests that the rate may increase as the children reach marrying age (Huke and Huke, 1990b; United Nations, 1998).

Urbanization

Burma is largely a rural nation that boasts few major population centers. In fact less than 25 percent of the nation's people can today be classed as urban, and this number

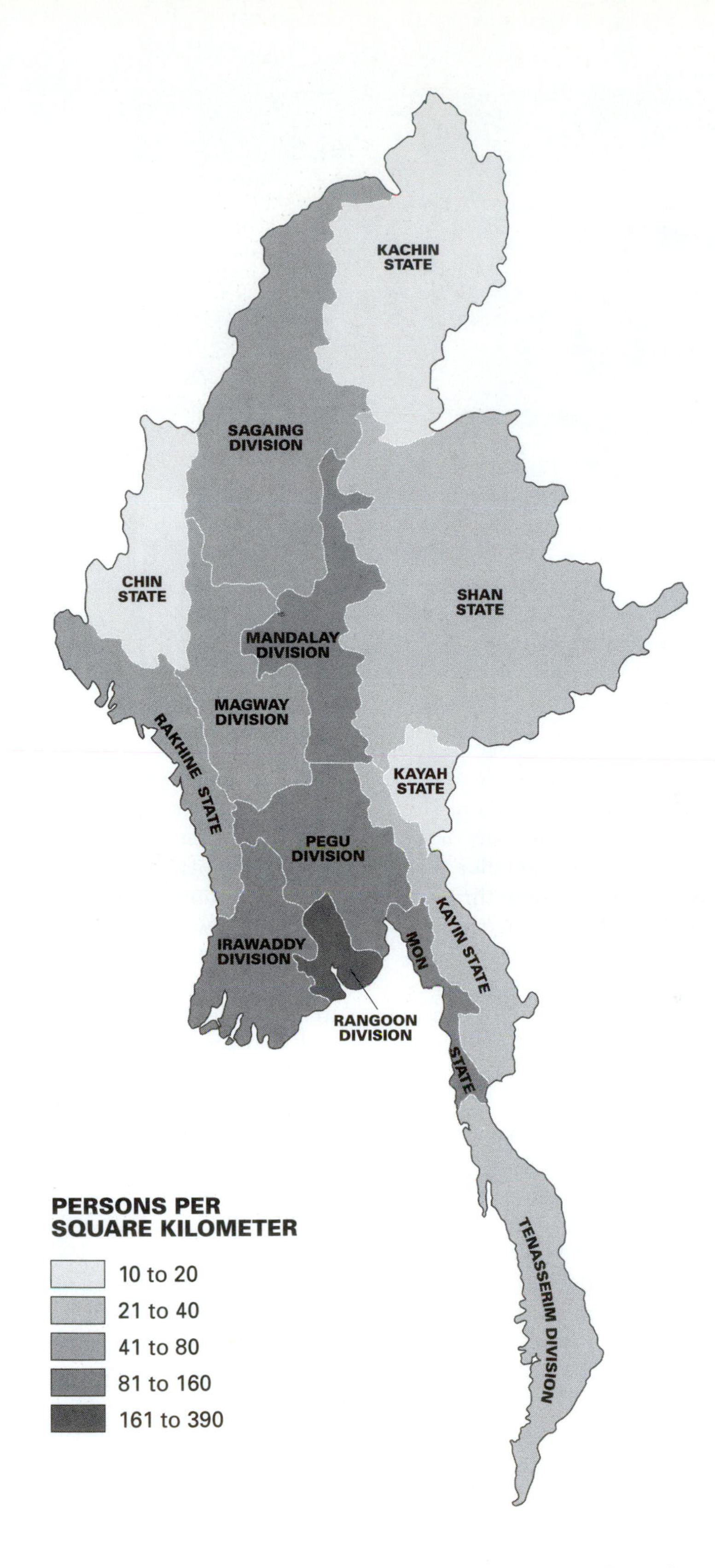

Map 17.2 Population density in Burma, 1994. *Source: Central Intelligence Agency, The World Factbook, 1994.*

Photo 17.2 Rangoon is a dowdy city although it does have several magnificent pagodas. Here is a view of the main intersection in the Central Business District (CBD), two blocks from the city hall and three blocks from the U. S. embassy. (Huke)

has not changed for more than a decade. Rangoon, the capital and primate city, is a sprawling, largely pre-World War II city with a population of about 2.5 million. This number seems too large for a city that lacks many of the modern urban amenities, but it is much smaller than Bangkok to the east or Dacca and Calcutta to the west. The city houses most government offices, Rangoon University, as well as technical schools and teachers colleges, port and warehouse facilities, and dozens of light manufacturing facilities. Many pagodas, most notably Shwe Dagon, and monasteries are spread throughout the city. Private cars are few, and people move about by bus, by trishaw, or by foot. One blessing is that traffic jams are nonexistent. Rangoon in many ways seems like a city of the 1930s: Several decades of isolation and long periods of internal fighting have left the city and the country with little progress toward modernization.

Rangoon and the other cities of Burma have not yet experienced the dramatic growth seen in urban centers over much of Southeast Asia. Rangoon's most dramatic growth and most chaotic period since independence took place during the years of 1958 through 1961 at which time insurgent armies controlled large areas of the countryside, security was at an all time low, and tens of thousands of peasants flooded into the city in the hope of finding protection. In the early 1960s downtown Rangoon was an unbelievable jumble of makeshift shanties with little in the way of sanitation, inadequate water and fire protection, and transport facilities so overloaded as to be useless. Early on the morning of March 2, 1962, army tanks rolled into the city from the cantonment near the airport, captured the city, and imprisoned the civilian government headed by Prime Minister U Nu. The army, under the leadership of General Ne Win, had completed the first of two separate coups with little opposition.

Shortly after the army came to power, refugees were forcibly removed from the city proper to a series of new army-built suburbs to the north and east of the city proper. These new towns were provided with minimal electric service and with barely adequate running water. Army detractors pointed out that the new towns flooded terribly in the rains and had poor sanitation, and that each had access to the city by only one road thus providing a ready-made control point.

During the 1970s and 1980s Rangoon grew constantly but slowly (although this observation is made suspect by government-issued estimates of the city's size). In 1963 the official number was 1,758,731; by 1973 it had grown to 3,662,312; and in 1983 the Census reported 2,458,712 (*Statesman's Yearbook*, 1965, 1975, 1985). Despite government attempts to prevent migration, small numbers of rural folk drifted to the capital city, which appeared to offer a more rewarding life than did the farms. Farm gate prices (value of the crop when sold at the farm) were very low, and even though employment opportunities in the city were limited, there was a well-established sawmilling operation, a small but growing pharmaceutical industry, several modest textile mills, many small printing establishments, a number of kilns and other manufacturers of building materials, boatyards, food processing facilities, opportunities for wood or stone carving, and many three- to six-person shops where employment might be found polishing gems, natural or synthetic, on foot-powered wheels.

In the mid-1990s Rangoon, a city without high-rise buildings where the tallest building is still a pagoda, and a city completely lacking in the hectic, congested traffic conditions so common in Southeast Asia, appeared on the verge of a growth spurt. One can only hope that it will not choke Rangoon as it has Bangkok. Around Rangoon's Royal Lake and Inya Lake, several very large four- and five-star hotels have been constructed by firms from Singapore and Thailand. These are desperately needed to supplement the three "luxury" (actually three-star) facilities and 20 or so guest houses that were in operation in 1997. Roads throughout the city are being improved and widened. Traffic lights have been installed at critical intersections. Bus terminals have been built, the main railway station modernized, and, perhaps most important, communications networks, including a functioning telephone system, are in place. Once the airport is brought up to an approximation of international standards and the port facilities improved, Rangoon may once again take its place as a destination of significance for the tourist industry.

Mandalay was established as a new city in 1857 by King Mindon who then transferred his court to the city in 1861. This was seen as a campaign to unify all Buddhists in the country to better oppose the British. As part of this effort, Mindon, in 1872, hosted the Fifth Great Synod of Buddhism in Mandalay—but all to no avail as the city and all the rest of the country was captured by the British in 1885. The city occupies a narrow portion of the Irrawaddy flood plain between the river and the escarpment of the Shan upland. Immediately south of the city, the Irrawaddy makes an unlikely right angle bend to the west to follow the path of the stream that captured it long ago. At the northeast corner of the city is the 775-ft (236-m) high outlier of the Shan upland, Mandalay Hill, the site of furious fighting between the Japanese and the British and Indian troops holding the stronghold at the summit. Just south of the city is the magnificent Arakan Pagoda built in 1784 by King Bodawpaya to house a Buddha image captured in Arakan and carried across the Arakan Yoma. In the years following independence, little was done in Mandalay to upgrade facilities. The damage of World War II was magnified by devastating fires in both the late 1950s and the 1960s. The city did, however, remain a transportation focus for the Dry Belt, a center for the processing of timber from northern forests, and a focus of the arts, especially weaving and silversmithing.

Mandalay is Burma's second-leading city and as Kipling might have put it, "a city furiously 'abuilding'." Location has proven to be everything. The city of more than 650,000 is one-quarter the size of Rangoon but in the late 1990s is rapidly increasing in importance precisely because of its location astride the trade routes linking Burma and China. In 1993 the Chinese government completed a highway bridge across the Shweli River, on the Burma–China border, allowing direct truck traffic to flow between Mandalay and China's Yunnan Province and simultaneously giving new importance to the rail line from Mandalay to Lashio. The connection has provided an impetus for growth in Mandalay. In 1994 hotels, banks, improved roads, and dozens of trading facilities were being built on the road to Mandalay. Chinese goods of all kinds are available in the bazaars, Chinese is spoken everywhere in the city, and many signs are expressed in both Burmese and Chinese.

Burma's third-largest city, Moulmein, at 250,000 is less than one-half the size of Mandalay. Moulmein lies at the mouth of the Salween River and controls the rail and road connections between the Tenasserim coast and the rest of Burma. The site was one terminus of the World War II Japanese-built rail line through Three Pagodas Pass, the subject of a classic movie, *The Bridge on the River Kwai.* Moulmein remains a trade center and port of modest importance. A number of other cities, including Pegu, Bassein, Akyab, Taunggyi, Monywa, Toungoo, Meiktila, and Lashio, have populations between 75,000 and 150,000. Without exception these are trading towns that serve extensive hinterlands. Each of them has at least one large and well-known pagoda along with several monasteries. Residential high schools and industrial or agricultural training schools are found in most, and these serve a large surrounding area. Bus and truck service along poorly maintained roads provide connections with neighboring centers. Many of these smaller cities have important rice milling or other food processing functions; most are rail or steamer hubs, but none have been modernized. Electric and water service are primitive; sewage, garbage, and trash are disposed of on an ad hoc basis.

The Village and Rural Character

With 75 percent of the population classified as rural, Burma has one of the highest rural percentages of any large nation. Throughout much of the country, villages have little or no obvious form of organization. They are inevitably nucleated and each is often characterized by a conspicuous pagoda. In the delta areas, villages often extend shoestringlike along the natural levee. Such a village is seldom more than one home deep, but it may be a quarter of a mile long and there are often two or even three small pagodas scattered throughout its length. Village size ranges from 15 or 20 homes to as many as 200. Each building is home to a nuclear family that may be as small as two persons but may also be as large as eight. In many homes one or two grandparents may live with the mother, father, and children.

In those upland areas where taungya (slash-and-burn farming) is the rule, homes are often of the longhouse variety individually providing shelter to an extended family

Photo 17.3 *Taungya,* or shifting or slash-and-burn agriculture, is the dominant form of agriculture in the uplands of much of Burma. Here in the Kachin state is a village of several longhouses, an altar in the foreground and several active *taungya* fields on the distant hills. (Huke)

of 40 to 60 individuals. A longhouse village seldom has more than a half-dozen homes. Each of these homes will support one or two outbuildings, much smaller than the main building but built on the same plan. One outbuilding serves as the granary where rice and other grains are stored. In front of the storage area is an unwalled but roofed portion of the building where each morning just before dawn the family's children handpound the day's supply of rice. The second outbuilding often encloses the brewing and distilling facility where rice and other grains are converted to alcoholic beverages. In a longhouse village, the entire settlement is often enclosed by a simple stockade or a fence of thorns. Individual homes are almost never fenced.

In lowland areas villages are seldom fenced, but individual dwellings often are. Homes are raised 3 to 6 feet (1–2 m) above the ground to improve ventilation and to prevent flooding of the living space during the heavy rains. Chickens and sometimes a pig or two are kept underneath, where they help to dispose of household waste and later contribute to the food supply. Fencing acts to contain the pigs. Villages include both families to whom the state has granted permission to farm a particular area, as well as other families who are without that right and thus are landless. For the country as a whole, the ratio between families with land and those without is roughly 4 to 1 (Huke and Huke, 1990b).The landless families provide agricultural labor, often on a shared basis, and engage in a wide variety of other tasks. There may be one or two homes serving as rudimentary general stores offering candles, plastic slippers, a few canned items, betel nut supplies, matches, and perhaps a few bottled soft drinks. Many but not all villages have a family or two who provide rice milling service, usually by handpounding. Almost always there are several families engaged in the manufacture of bamboo matting for walls or roofing. Commonly several of the women in the village make sweets for sale locally; others produce hats or rain capes from local grasses or bamboo. A few men find employment in the gathering of sap from bamboo or nipa palm and its processing into a modestly alcoholic beverage. Beside the pathway leading into almost every village, there is a small shrine with food, flower, and drink offerings for the local guardian spirits (*nats*).

Most homes are of a similar design, and there is little to distinguish economic or social class. Floors are made of split bamboo or, in more modern homes, of sawn lumber. Raised floors provide vastly improved ventilation and keep the waters of the heavy monsoon rains away from the living space. Poles that support the home are usually equipped with a rat guard to keep the pests at a distance. A granary, joined to the home much as is an attached garage in the United States, contains a large woven basket that is reinforced with a mixture of dried cowdung and clay. This provides a semiprotected location for storing that portion of last season's rice crop that will be consumed at home. Cattle and water buffalo are normally penned beside the home. Fruit trees and a small vegetable or herb garden are found in the immediate vicinity. Flower gardens are not found in such villages.

The Role of Religion in Burma

In Burma, pagodas and shrines dominate much of the landscape but none more than Shwe Dagon dominates Rangoon (Photo 17.5). "It is as I suppose the fairest place in all

Photo 17.4 Many delta village homes are raised well above the ground for ventilation and to avoid the wetness of the rainy season. Many homes have their own rice storage or granary. The young son is holding a bow used to shoot clay pellets at the birds in the rice fields. (Huke)

the world" wrote the English traveler Ralph Fitch concerning his visit to that pagoda in 1586. A common reaction for many on first exposure to Burma is that this must be the most Buddhist of all lands in the world. The traveler arriving at Rangoon by air will long remember the gold of Shwe Dagon, glistening in the sun by day or glowing in the sodium vapor spotlights by night. In an approach to Rangoon by ship, Shwe Dagon will be seen piercing the horizon long before other buildings of the capital are visible.

The religious doctrine practiced in Burma is Theravada Buddhism. This is also known as the lesser vehicle and as the way of the elders. These names derive from the third Synod of Buddhism held in 235 B.C. at Pataliputra, India. At that meeting there was a sharp break between Mahayana and Theravada (or Hinayana) doctrines. The Theravada elders were more conservative and orthodox: They placed very great emphasis on individual achievement and allowed few options (thus the lesser vehicle) toward the achieving of nirvana, the Buddhist equivalent of utopia, and they taught that only through service to others does one enhance ones' own status.

Buddhism in Burma shows many influences from the spirit worship that preceded it. Spirits or *nats* abound and must be appeased because it is believed that they have the potential for causing harm. Thus there is a *nat* shrine on a large banyon tree at the entrance to almost every village. Here rice, tea, candles, and flowers are offered every day. Individual homes also have a shelf devoted to both Buddha and the local spirits. The belief in spirits also suggests a faith in divination. Indeed Burmese of all persuasions interpret as auspicious or inauspicious all kinds of omens, especially those related to astrology.

The numerical dominance and special role of Buddhism (Table 17.2) in Burma was acknowledged by the constitution of 1947, which recognized it as "the faith professed by the great majority of the citizens." In 1961 the constitution was amended to make Buddhism the state religion, but in 1963 the Revolutionary Government, led by General Ne Win, amended it yet again; today the constitution "recognizes the right of everyone to freely profess and practice his religion."

Shwe Dagon Pagoda may be the "fairest place in all the world," but the real center of Burmese Buddhism and the treasury of national art is at Pagan, in the dry zone of Central Burma. The city may have been founded as early

Professed Affiliation by Percent of Population		
Religion	1931	1991
Buddhist	84.3	89
Animist	5.2	1
Muslim	4	4
Hindu	3.9	<1
Christian	2.3	
Protestant		3
Roman Catholic		1
Others	0.3	2

Note: roughly 2/3 of the Christian population are Karen, half of whom live in the Karen State.

Table 17.2 Religion in Burma. *Source: Census of Burma, 1931; Central Intelligence Agency, The World Factbook, 1994.*

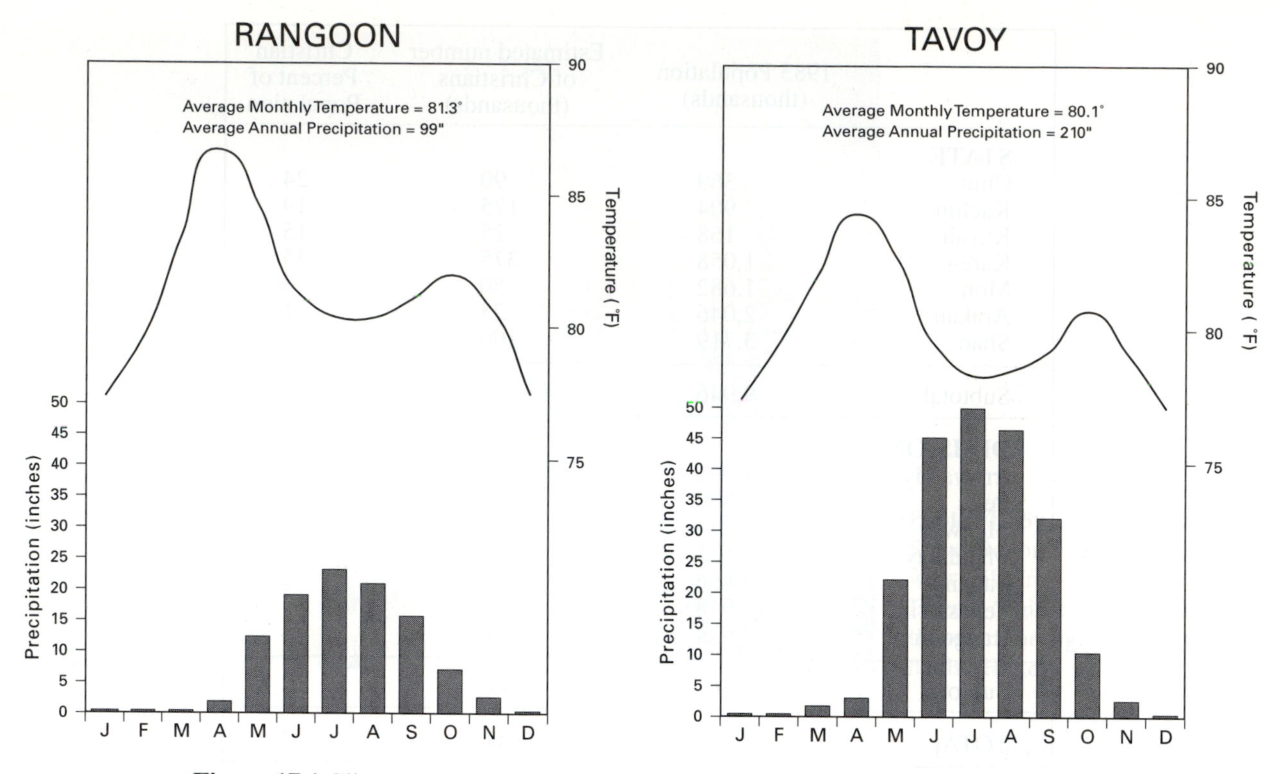

Figure 17.1 Climographs for Rangoon and Tavoy. *Source: Ulack and Pauer, 1989.*

A particularly striking characteristic of Burma's climate is the Dry Belt in the central portion of the country, south and west of Mandalay. Much of the air flowing into central Burma during the southwest monsoon season travels from the Bay of Bengal across the rugged Arakan Yoma. Very heavy precipitation falls on the windward side. The town of Akyab for example holds the world record for rainfall in one hour, 17.8 in. (406 mm). Once the air masses cross the crest line at between 6,900 ft and 9,200 ft (2,100–2,800 m), they flow downslope toward the Dry Belt at an elevation of 325 ft to 500 ft (100–150 m). During the descent, the air is compressed and heated at the rate of 1°F for every 328-ft (1°C for every 100-m) loss in elevation. It becomes so dry that it acts to pick up rather than deposit moisture. Such flows create a powerful rain shadow in the middle course of the Irrawaddy. This Dry Belt, one of the most distinctive features of Burma's climate, stands out sharply on Map 17.3 and is one of the most profound of such belts anywhere in the world.

Burma's surface configuration contributes mightily to the rainfall pattern and also provides relief from the oppressive lowland temperatures that are high all year and where even the cool season is cool only by comparison with the hot season. In Rangoon, for example, the coolest month is January, with a mean temperature of 77°F (25°C), roughly the same as Washington, D.C.; in August the warmest month is April which, at 87°F (30.5°C), is somewhat cooler than Phoenix, Arizona. Despite these high monthly averages, the hot season in the Irrawaddy delta is rarely as uncomfortable as are many summer days in such U.S. cities as St. Louis, Cincinnati, or Louisville. By far the most oppressive characteristic of weather in the lowlands is the uniformity from day to day and even day to night. During the wet season, humidity is constantly high and light physical exertion brings out rivers of perspiration.

To the east the elevations of the broad upland provide considerable relief. Taunggyi, at 4,430 ft (1,350 m), reports July as the warmest month averaging 72.9°F (22.7°C), identical to Boston, Massachusetts. The coolest month is January which at 56.8°F (13.8°C) is identical to January readings in Jacksonville, Florida. To many this temperature range is ideal and when combined with the modest 68 in. (1,750 mm) of rainfall each year is a weather regime difficult to improve on. Is it any wonder that the British established their "hot season capital" at the hill station of Maymyo, a mile above sea level in the Shan uplands?

Surface Features

The basic outline of Burma appears to be much like a huge kite, with mountain and upland regions comprising the four sides of the diamond and with a long tail extending southward toward Singapore. The central portion of the country is a vast lowland created in large part by the relentless work of monsoon-fed rivers operating over the ages. This lowland is itself broken into eastern

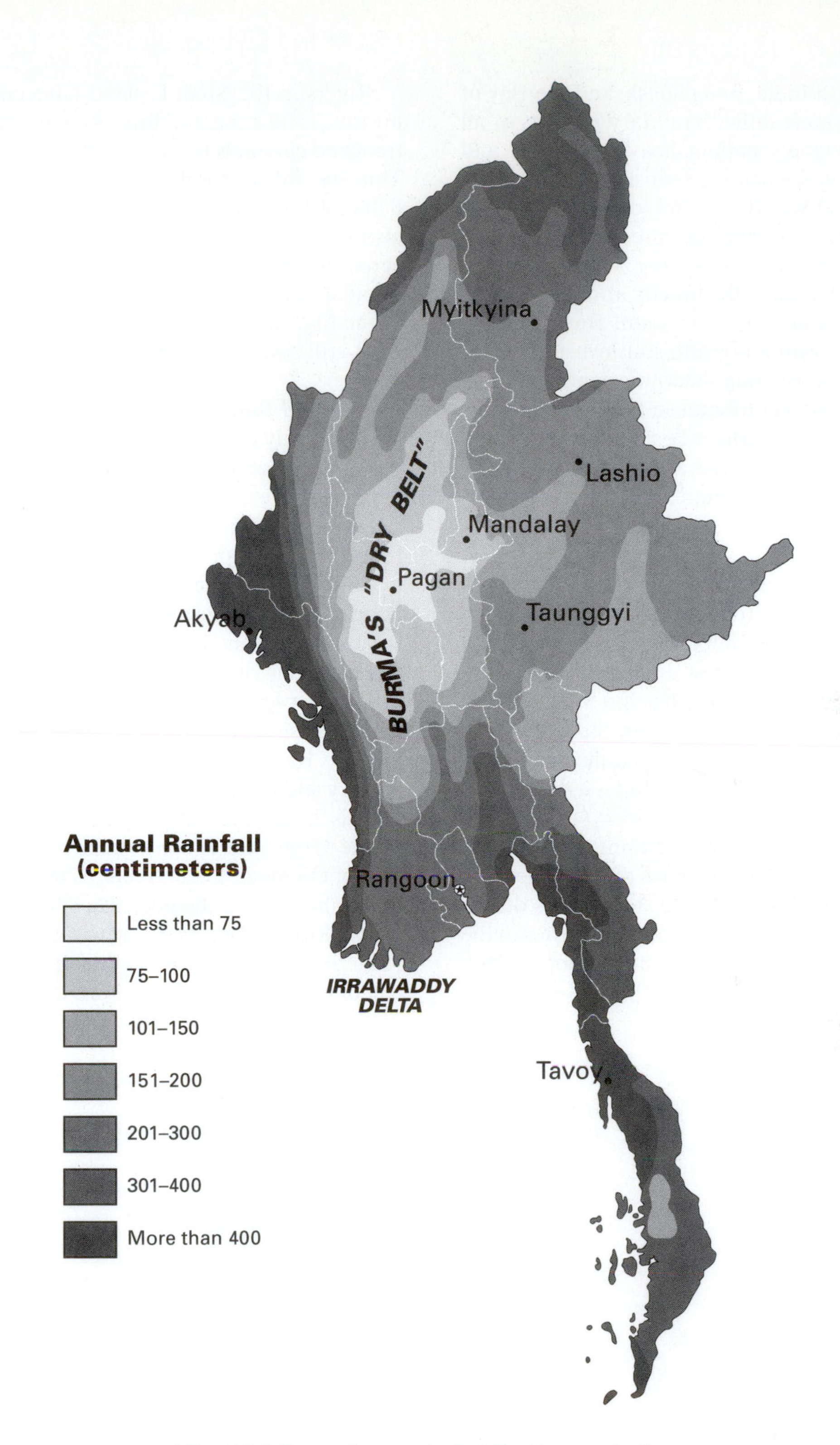

Map 17.3 Burma's annual rainfall. *Source: Author.*

and western components by a series of north–south trending hills that have resisted the work of the rivers and by isolated intrusive volcanic peaks.

The Western Mountains. At the northernmost point of Burma along the borders of Tibet and the Yunnan Provinces is a great snow-covered mountain mass known to the Burmese as Hkakabo Razi and to European map-makers as the Putao Knot. This is Southeast Asia's highest peak, the 19,309-ft (5,885-m) "Great Snowy Mountain." From the Putao Knot an arc of tightly faulted and heavily folded mountain ranges extend southwest then southeast along the border with India, Bangladesh, and the Bay of Bengal. One of the most difficult military retreats of all time crossed these steeply sloping, heavily forested, and malaria-infested ranges when the British withdrew from Burma early in World War II. Parallel to the Bay of Bengal, a narrow coastal plain separates the mountains from the sea. A few miles inland, peaks—the Arakan Yoma—reach 11,000 ft (3,350 m) and lie directly athwart the prevailing winds of the southwest monsoon. This gives rise to heavy orographic rainfall on the southwest or windward slopes and a powerful rain shadow in central Burma. The coastal plain is subject to intense rains during June, July, August, and September. This is best typified by Akyab which for 90 years has recorded an average of 200 in. (5,151 mm) a year with a maximum of 320 in. (8,219 mm) and a minimum of 119 in. (3,063 mm.)

The Shan Upland. The eastern one-third of Burma lies on the Shan Upland. This too can be traced from the Putao Knot and is a more complex and older series of mountains than those on the west. Included on this upland are the eastern portion of the Kachin State, the entire area of the Shan, Kayah, and Kayin States, eastern portions of the Mon State, as well as a significant portion of Tenasserim Division. The Shan Upland is a deeply dissected plateau, averaging roughly 4,265 ft (1,300 m) in elevation. Throughout much of its extent, the western edge is marked by a well-defined series of fault lines that often rise 1,970 ft (600 m) in a single step. This is one of the sharpest and most obvious physical boundaries in the country. The line marks changes in rainfall, temperature, vegetative cover, and many aspects of human use.

Rivers in the Shan Upland have cut deeply into the massive sandstone and limestone to create dramatic entrenched channels marked by numerous rapids and falls. Thus the Salween is navigable for only a few miles at its delta and the Myitnge not at all. The entrenched rivers also created difficult conditions for surface transport across the region. For the rail line to reach Lashio from Mandalay, it must cross the 1,000-ft (305-m) deep gorge of the Myitnge, which it does on a viaduct completed in 1903, still considered an engineering wonder.

The Central Burma Lowland. The Central Burma lowland is largely the gift of major rivers. The two great flood plains, the Irrawaddy–Chindwin to the west and the lesser Sittang to the east, are, however, separated by a south–north range of hills. These hills, known as the Pegu Yoma, extend from the rise supporting Rangoon's Shwe Dagon pagoda to Mount Popa, a few kilometers east of Pagan. Mount Popa is the most conspicuous landmark in the country and is a textbook example of an explosive volcano (Photo 17.7). The mountain rises abruptly from a base at 1,500 ft (460 m) to a peak at 4,981 ft (1,518 m). It has a circular crater whose northwest side was blown out in 442 B.C. according to local folklore. Rich soils and favorable climate have resulted in a heavy and varied floral cover that accounts for its name. Popa is Sanskrit for "flower." It is also known as the home of the Gods and is sometimes referred to as Burma's Mount Olympus. Legend tells us that here is the home of the Mahagiri Nat, the guardian spirit; ancient history describes Mount Popa as the hearth of Burmese culture, and modern history tells us that General Aung San, the architect of modern Burma, was born in its shadow.

Photo 17.6 The Shan Uplands are lightly forested rolling uplands, settled by a wide range of nationalities. Farming is not highly sophisticated as seen by this Shan farmer, his water buffalo, and a light wooden toothed harrow. Beans and vegetables are being planted. (Huke)

Photo 17.7 Mount Popa, near Pagan, is a sacred landform that is considered the hearth of Burmese culture. (Reed)

From Mandalay, today's Irrawaddy River leaves its ancient eastern valley adjacent to the Shan Upland and moves sharply west to a valley originally cut by the Chindwin. Geologically speaking, the ancestors of today's Irrawaddy, Chindwin, and Sittang Rivers were the proto-Chindwin to the west and the proto-Irrawaddy to the east. An east bank tributary of the proto-Chindwin cut more rapidly than did the proto-Irrawaddy and diverted its enormous flow to the west. What is known as the Irrawaddy today was formerly the proto-Irrawaddy in the north and the proto-Chindwin in the south. This is one of the world's best examples of a phenomenon known as river capture.

North of Mandalay the zone is marked by a series of dormant volcanoes, in the midst of which are located the world-renowned jade mines of the Kachin State. This is a well-forested and lightly settled part of the country with continuously cultivated areas both to the east and to the west.

RESOURCES

During the years preceding and immediately following Burma's emergence from colonial rule as an independent nation, most writers were sanguine concerning her potential for rapid economic development. This confidence was based on the twin concepts of a relatively low population pressure and a very rich resource base. Unfortunately the development has not yet taken place, and the population pressure has begun to increase. On the other hand, the resource base remains one of the strongest possessed by any country of Southeast Asia.

One of Burma's great resources are immense forests, which have provided an almost inexhaustible supply of construction material, charcoal for domestic cooking, and raw material for export. Today forests of one type or another cover almost one-half of the nation's total area (see Table 17.5). They are concentrated chiefly in the horseshoe of hill and mountain areas and in the central Pegu Yoma. Some of the forests are very heavy and highly productive; others are light and scattered. Most of the timber is of evergreen varieties, but a liberal mixture of semideciduous and deciduous stands are located in the drier regions. The forests are rich in highly prized cabinet woods and are especially known for holding much of the world's remaining teak. Of the total forest cover, about two-thirds is classed as "closed forests," meaning it is open to exploitation only with government license, but this stricture has been much violated in recent decades. The remaining one-third has also been seriously degraded by overcutting or by excessive practice of *taungya.*

Of the six forest types (Map 17.4) the tropical broadleaf deciduous is the most valuable in terms of its direct contribution to the economy. This is the natural home of teak which, here, is often found in close association with pyinkado, Burma's second-leading timber resource. Pyinkado is used principally for heavy construction and for railway ties. It is tough, hard, heavy, and resistant to termites and to rotting. Unfortunately it is so heavy as to have a specific gravity higher than 1 even after drying. Thus it cannot be floated down rivers to milling and collecting centers; instead it must be suspended under rafts or moved overland. In either case the cost is great. The

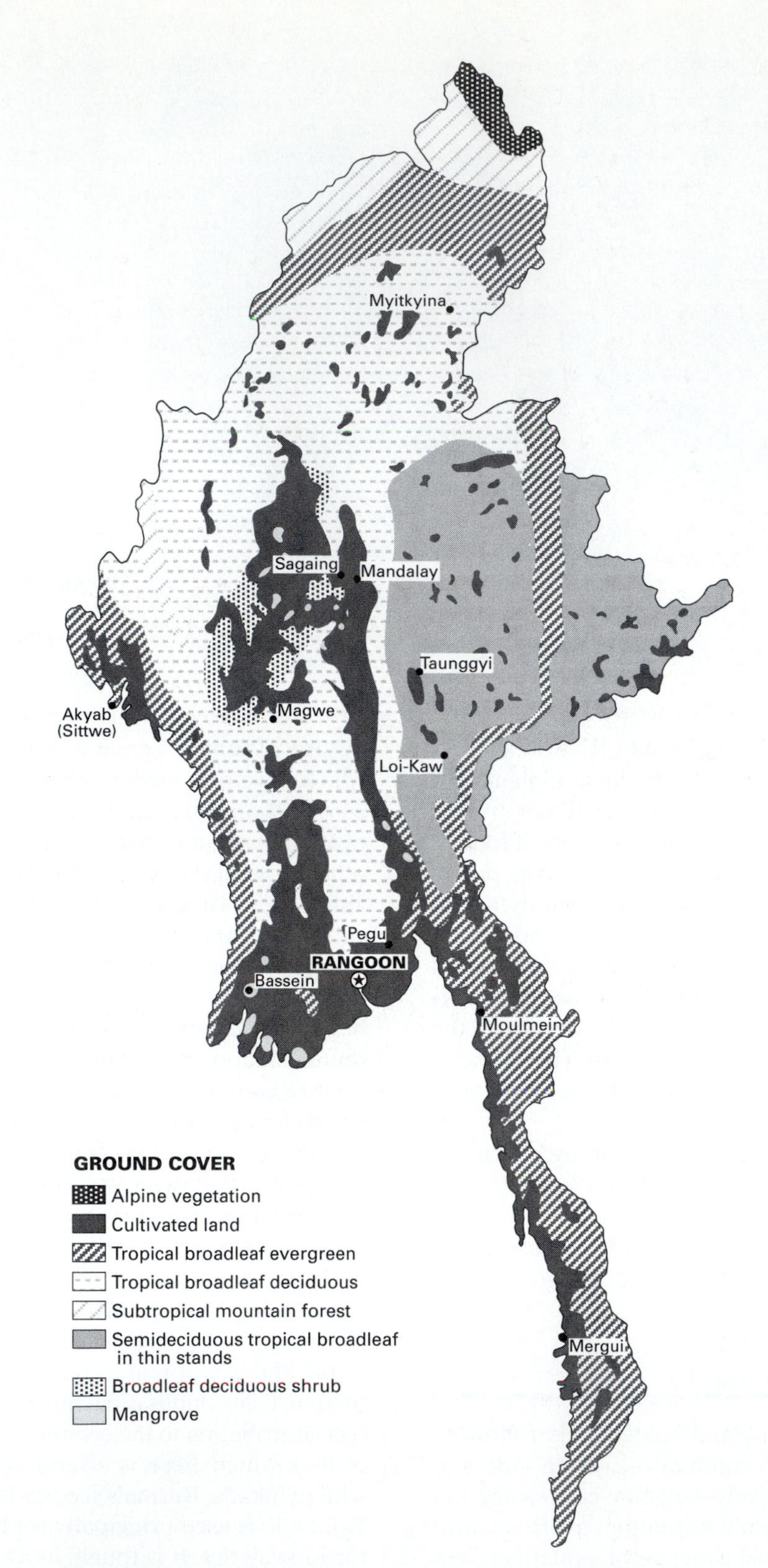

Map 17.4 Forests and cultivated land in Burma, 1995. *Source: Author.*

wood is of fine cabinet quality, takes a beautiful polish, and has a muted grain with a warm red color.

The special place of teak within the forest has long been recognized. In 1752, during the Alaungpaya dynasty, teak was proclaimed a "royal" tree and all such trees were reserved for the king. Shortly after the first Anglo-Burmese war, the colonial government established a strong Forest Department. A policy based on the already established principle that all teak, standing or felled, living or dead, belonged to the state was instituted. Forests were surveyed and classified; teak, pyinkado, and some other woods were inventoried, and a system of licensing for timber removal was instituted. Measurements of the growth rates of different species in various settings were used to determine the annual allowable cut. The Forest Department attempted to control *taungya* cultivation, which was seen to destroy large areas of high-quality forest. Shifting cultivation was believed to have been responsible for accelerating erosion and permanently degrading the quality of the ecosystem. Extensive planting of teak seedlings was undertaken in favorable areas where *taungya* could not be prevented (Bryant, 1994).

Teak does not occur in pure stands but rather is scattered widely through the forest. At best it constitutes about 12 percent of the stock. Trees have a rounded crown and grow to large size with tall cylindrical trunks. The tree is easily distinguished by its huge leaves that individually measure from 10 to 20 in. (25–50 cm) in length. Girths of 20 ft (6 m) and a clear trunk of 100 ft (30.5 m) to the first branch–with a total tree height of 150 ft (45 m) were common at the turn of the century but today are seldom found and then only in remote areas difficult to access.

Timber exports, chiefly teak, have long been one of the major exports of the country. For most of the twentieth century, rice has been far and away the number-one export and teak a poor second or third. However, since the mid-1980s when Burma's rice exports came on hard times, teak has taken over as the leading earner of foreign exchange. Prior to World War II the annual exportable surplus was a sustainable 500,000 cubic tons. Today the records show only 250,000 cubic tons, but much of the logging in frontier areas, especially along the Thai and the China borders, results in unreported exports.

Teak is in great demand for high-quality furniture and especially for shipbuilding. It is strong, durable, free of movement (does not shrink or expand with changing humidity), and usually straight and even-grained. It works well by hand or by machine and peels easily to thicknesses of 1/32 of an inch without breaking. It is practically impervious to fungus and resists the attack of most insects. Teak lasts very well in contact with water and needs no artificial preservative. For shipbuilding, teak has the added advantage of containing a wood tar that helps prevent corrosion of iron plates and fastenings with which it is in contact.

Burma still has one of the world's finest reserves of tropical forest. If these can be preserved and managed wisely, the forests could be a key factor in the development and modernization of the country. It augers well for planned development in the years ahead and owes much to several generations of conservation-minded leaders (Bryant, 1997).

During the latter years of the colonial period, the production and export of a wide variety of minerals provided about 40 percent of Burma's export earnings; these are the same resources that are productive today. Economically the most important of these are petroleum, tin, tungsten, and lead-silver. The most romantic are jade, rubies, sapphires, and gold.

Petroleum has been produced for at least 80 years from a series of fields at Chauk (the major area), Yenangyayng, and Myingyan, all in Central Burma. Production has never been important on a world scale, but in the decades prior to World War II, Burma Oil did export kerosene and paraffin to India. Modest refineries presently operate at Syriam, across the river from Rangoon, and at Chauk, but total output in 1994 was less than 7.3 million barrels, a figure insufficient for domestic needs and lower than the output achieved in 1970. Exploration rights have been granted to several foreign concerns, but few new finds have been made. In 1994 gasoline was rationed at between 2 to 4 gallons per week. Price per gallon at the government pump was Kyat (Kt) 25 (officially US$4), but on the black market the same gasoline was sold for Kt 200. Supplies beyond the ration came largely through resale by privileged government officials and military personnel.

Major natural gas reserves have long been suspected both in the Dry Belt and in the offshore areas of the Arakan and Tenasserim coasts. Reserves have been estimated at several trillion cubic feet, and exploration rights both in the traditional oil areas and in offshore areas have been awarded to Australian, Japanese, French, British, and U.S. companies. Production is already sufficient to provide roughly 45 percent of Burma's electric generation and is finding increasing demand among manufacturing plants. In March 1995 the Petroleum Authority of Thailand joined with France's Total and U.S.'s Unocal in a project to build a gas pipeline from the Yadana field, in the Gulf of Martaban, along the route of the former Japanese-built rail line between Moulmein and Bangkok. Completion of the project will require settlement with the Mon and the Karens who presently occupy portions of the proposed route.

Ores of tin and tungsten occur together in a belt running from the west-central part of the Shan State south through the Kayah State, the Karen State, and into Tenasserim. The largest production has come from hydraulic mining of eluvial and alluvial deposits at Tavoy, but the most consistent output has come from underground mining of quartz veins at the Mawchi Mines.

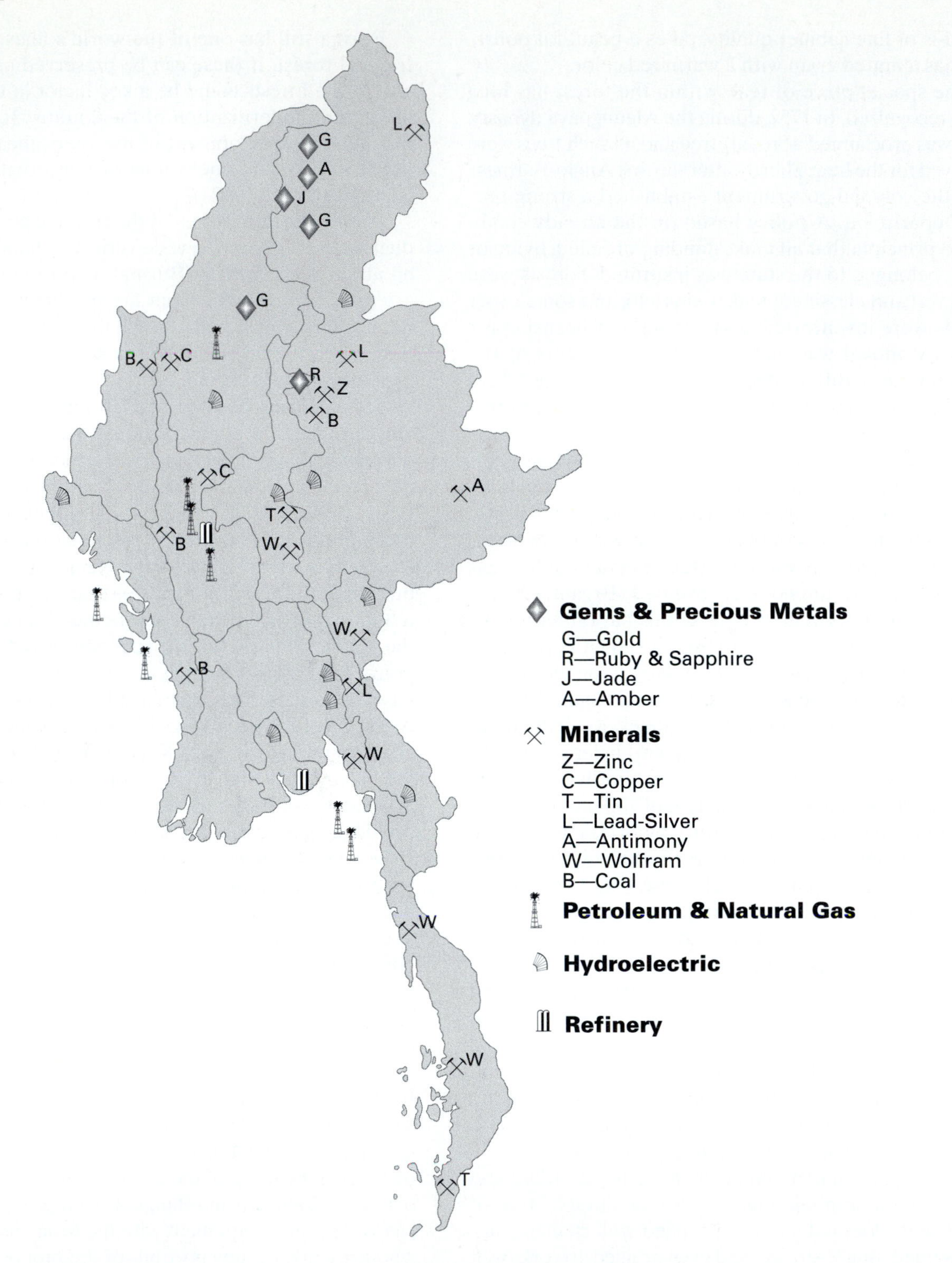

Map 17.5 Gems, metals, minerals, petroleum and natural gas, hydroelectric, and refineries in Burma.

Annual production totals for the two metals together had once averaged 8,000–9,000 tons, but by 1994 tin concentrates totaled only 386 tons with another 275 tons of refined metal and tungsten concentrates amounted to only 97 tons (Economist Intelligence Unit, 1995b).

Lead and silver are produced at the Bawdwin Mines, 30 mi (48.3 km) northwest of Lashio, one of the largest high-grade silver–lead–zinc ore bodies in the world. Here too, problems associated with nationalization, lack of capital, energy shortages, and political turmoil have seriously curtailed output. In 1994 production totals were 2,500 tons of lead and 180,000 ounces of refined silver, both roughly 15 percent of the outputs achieved several decades ago.

Jade has been mined since time immemorial in a rough mountain area about 60 mi (96.6 km) west of Myitkyina. A majority of the output has found its way to Hong Kong where it enjoys the reputation of being the finest quality stone in the world. Moderate quantities of ruby, sapphire, garnet, and aquamarine have come from near Mogok. Gold is known in several places along the banks of the Chindwin and Irrawaddy. For the past 32 years jade, precious stones, and pearls, both natural and cultured, have been auctioned at the annual gems emporium held by the state-owned Myanmar Gems Enterprise. Representatives from 15 countries attended in 1995 and sales totaled about $7.2 million, a modest decrease from previous years (Economist Intelligence Unit, 1995b).

HISTORICAL EVOLUTION OF BURMA

Nation building can be dated from A.D. 1044 when Anawrahta ascended the royal throne in Pagan. He conquered the lands of central Burma, subdued the Mons to the south and the east, overran the Kachins to the north, and sent his army into the Chinese domain of Nan Chao. Finally he conquered the Arakan Coast on the shores of the Bay of Bengal and managed in so doing to establish lasting enmity between the Arakanese and the Burmans. The Shan States remained apart during this period. Anawratha strengthened the place of Buddhism and built much of what can be seen today of the magnificent ruins of Pagan. For almost 1,000 years Burma thrived as an independent nation, although the locus of power shifted away from Pagan to Ava, Toungoo, Pegu, and back to Ava as various dynasties rose to power and were then supplanted by others. Burma finally lost her independence to Britain through a sequence of three wars fought at roughly 25-year intervals (Map 17.6).

Conquest by Britain and Independence

The First Anglo-Burma War was fought in 1824–1825 and grew out of unsettled conditions along the Arakan coast. Superior British forces overwhelmed the Burmese, and the peace treaty resulted in Burma's loss of both the Arakan and Tenasserim coastal regions to Britain. Burma also renounced all claims to Assam.

A second conflict arose 27 years later. That British administrators saw control of Burma's ports as providing a backdoor entry to the growing China trade was sufficient incentive to add those ports to the British Empire. Superior arms quickly carried the day, and Britain, without even the benefit of a peace treaty, annexed the Pegu and Martaban Provinces. This action, the Second Anglo-Burmese War, was imperialism at its ugliest and gave control of the delta areas of both the Irrawaddy and the Sittang to England. Burma was now cut off from access to the sea and had become economically and militarily subject to Britain's wishes.

In 1885 after a brief and almost bloodless campaign, British forces captured Burmese King Thibaw, sent him into exile in India, and took over administration and control of all that remained of Burma. This was the Third Anglo-Burmese War. When it was over, so-called upper Burma had been added to the British Empire. The underlying motivation for this conquest was also geopolitical.

Photo 17.8 In some portions of the Dry Belt, especially near Mandalay, pagodas built before independence by well-to-do landowners or by the king stand in the midst of padi areas. (Huke)

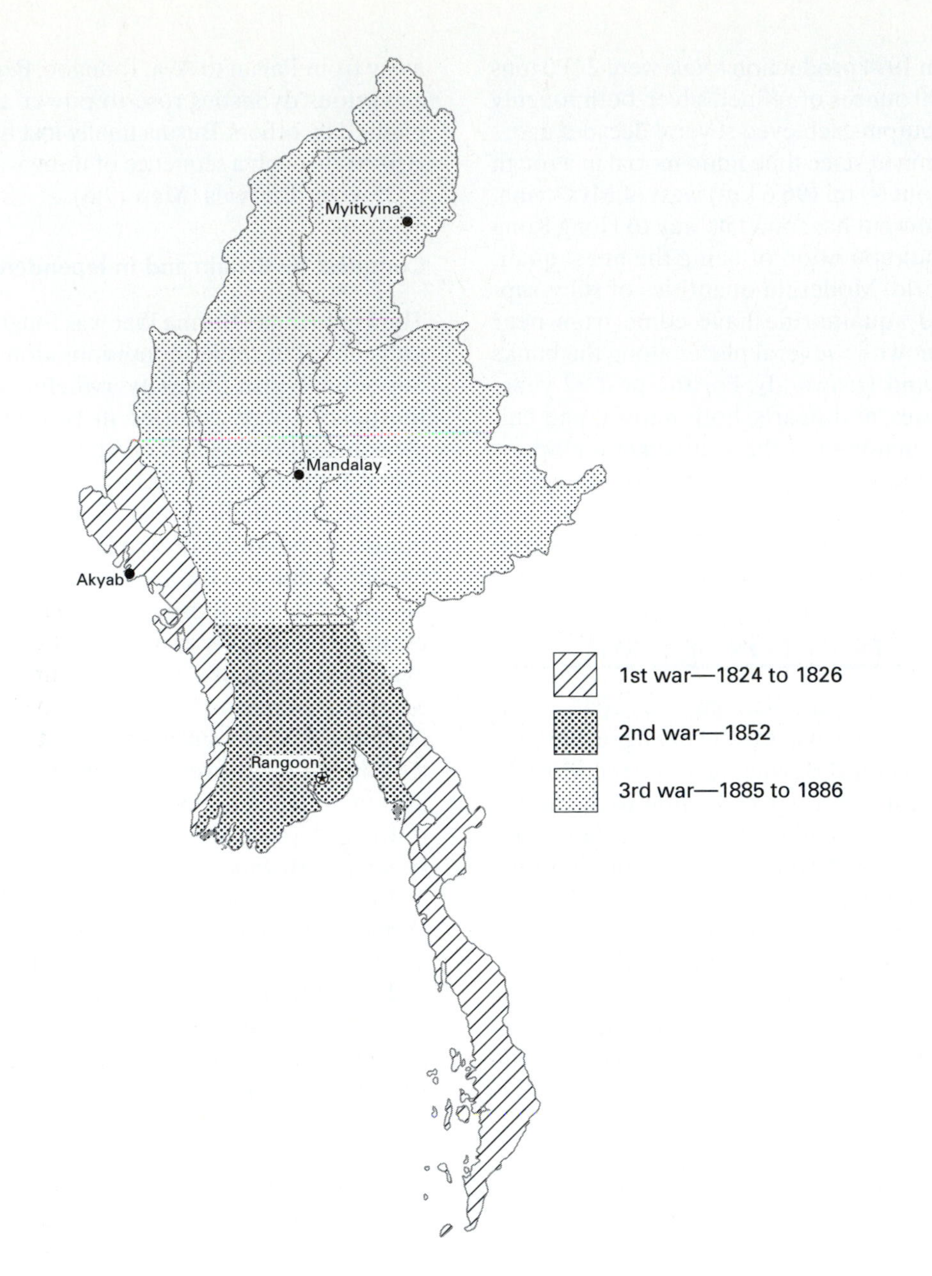

Map 17.6 Annexation by Britain followed three Anglo-Burmese wars.
Source: Author.

The British believed that Burmese authorities were allowing expansion of French interests in north Burma, an area critical for access to China.

Under British rule, Burma was administered as a province of India and remained such until 1937. The majority British attitude at the time was that Burma was an uncivilized country whose people would be only too grateful to exchange native despotism for the blessings of British rule. There were, however, some British who recognized that colonial rule was not always designed to benefit the native peoples but rather was meant to aid social and economic exploitation of the country for the benefit of the European masters. George Orwell's novel *Burmese Days* speaks magnificently for this point of view.

On April 1, 1937, the administration of Burma was divided from that of India, and Burma became a separate entity within the British Empire. Military, police, civil service, and foreign affairs continued to be controlled by the British, but internal affairs were left to the Burmese. This was not real freedom in Burmese eyes, but it was a move in the right direction. From 1937 until the Japanese intervention early in 1942, Burma had a limited experience with responsible parliamentary government. Burmans—as opposed to Karens, Shans, or Kachins—came to head many government departments, to hold most seats in the lower house of parliament, and to share the upper house with appointed members, some of whom were Europeans. The British continued to hold full control of the highland border areas covering 43 percent of today's area of Burma. The Burmans were still not masters in their own land, and ill feelings between them and the British intensified.

In the years preceding World War II, many members of the "hill tribes" had converted to Christianity and thus accentuated the contrasts between themselves and the Burmans who had remained largely Buddhist. The converts attended mission schools and learned a number of trades and skills in addition to the English language. As a result, Shans, Kachins, and Karens in particular found their way in disproportionate numbers into the British-controlled army, police, and civil service of pre-war Burma.

Full independence came to Burma on January 4, 1948, when Sir Hubert Rance, the last British governor, handed over authority to Sao Shwe Thaik, the first president of the Union of Burma. Burma was free again—but not all was well.

Insurgency

The constitution of newly independent Burma provided for a nominal form of federalism under which the Shan, Kayah, Karen, and Kachin States were to be created and given considerable freedom to govern themselves. The Chin Special Division was to enjoy somewhat less independence but with the clear understanding that full statehood would be forthcoming. The Shan, Kayah, Kachin, and Chin units were duly formed, but the Karen situation posed problems that could not be resolved. Major portions of the Karen population were settled in the Irrawaddy and Sittang delta regions. Here they lived in their own villages but were widely interspersed with Burmese settlements. No agreement could be reached on a state boundary.

To make matters worse, the Karen and other minority group representation in police, army, and civil service was lessened as Burmans took over government operations. The lack of state recognition and the dispossession from previously held posts was too much for the Karen minority. Less than a year following Burma's independence, they took up arms and open rebellion began.

Minority peoples soon came to believe that promises from Rangoon were honored more in word than deed. Within another year both the Kachins and the Shans joined the Karens in open rebellion, and for the next 45 years the country was constantly in turmoil (Map 17.7). The Burma army battled various ethnic insurgent groups, three different internal communist movements, and remnant Chinese KMT (the military arm of the Chinese Nationalist Party or Kuomintang) army groups along the Yunnan border. By the beginning of the 1990s the KMT and the communist organizations were no longer a threat, so government attention was directed toward military solutions to ethnic fighting. One-by-one cease-fire agreements were reached, most recently on January 29, 1994, involving the government of Burma and the Kachin Independence Army. The Karen insurgents have been contained in a group of fortified villages on the Thai border but remain defiant.

On the western edge of Burma, Arakanese dissidents keep the border area adjacent to Bangladesh in constant turmoil. The basis for disillusionment with the Burmese authorities is the charge that persons of the Muslim faith have been forcibly removed from their land in favor of Buddhist claimants, that Moslem shrines have been desecrated, and that individuals have been tortured. Many of the dissidents, who refer to themselves as Rohinga, have chosen to flee before the Burma army. In 1992 nearly 300,000 of them struggled across the Naaf River, which forms the boundary between Burma and Bangladesh. One must conclude that in times of particularly difficult economic or political problems at home, Burma's military regime has used the Rohinga as a scapegoat to divert attention from its own difficulties.

Today's Political Pattern

Burma is comprised of seven states, designed as homeland to seven of the largest minorities, and seven divisions that are designed as manageable units of the Burmese homeland (Map 17.8). States and divisions were seen as having equal status in the original constitution and were united in a federal structure in which the individual units were to have had considerable scope in the management of their own affairs. Unfortunately, this federal status was more nominal than real, and decisive power always remained with the central government.

In September 1972 the name of the country was changed to *Socialist Republic of the Union of Burma* and a new constitution was written. The new order created a unified state and removed all specific federal arrangements. The Burma Socialist Programme party, another name for the army, gained monopolistic power. In September 1988 the original name was reestablished, but no new initiatives toward federalism were instituted. Exactly one year later the country's name was again changed, this time by the generals comprising the State Law and Order Restoration Council (SLORC), the actual governing body of the state. The name used by SLORC is *Myanmar,* a literary term of self-reference that is thought to be derived from the Chinese word *mien.* This name is used by the United Nations and by a majority of the world's countries; it has not been adopted by the United States or Britain.

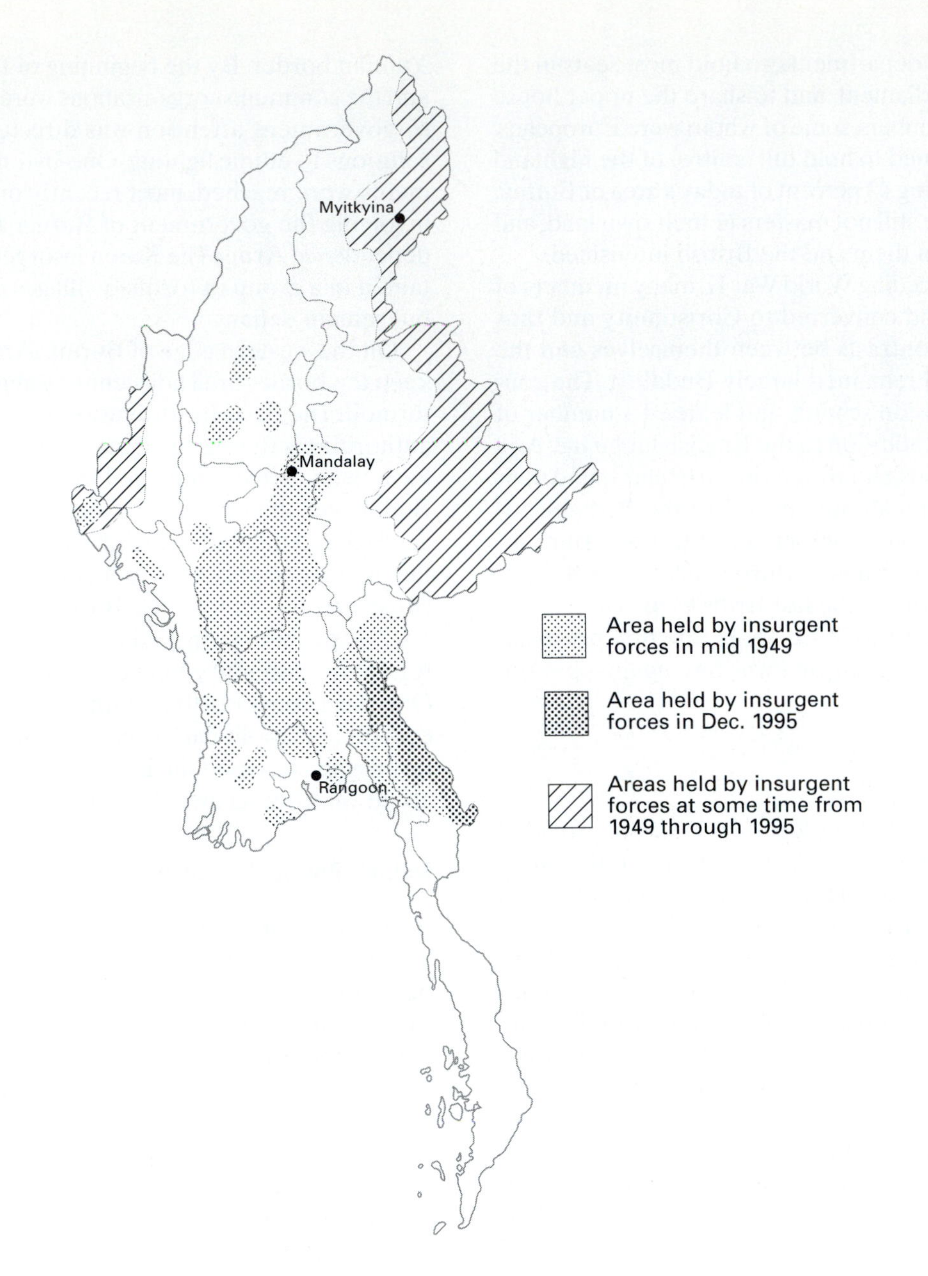

Map 17.7 Burma and the insurgents, 1948 to 1995. *Source: Author.*

In November 1997 SLORC was dissolved by its own order, and a new organization called the State Peace and Development Council (SPDC) was created. The change suggested a shift in emphasis from the establishment of internal security to matters of economic development. However, Senior General Than Shwe remained in charge and the new council looked much the same as SLORC. It does have a somewhat broader base—military leaders draw from several of the minority regions have been appointed to the council. The SPDC currently consists of 19 members. A 40-member cabinet consisting of 27 military officers and 13 civilians was appointed at the same time.

The organization to challenge the government most seriously by generals has been the National League for Democracy (NLD) formed in September 1988 by Aung San Suu Kyi and two imprisoned military officers. The objective of the NLD was to bring about social and political changes to guarantee a peaceful, stable, and progressive society where human rights are protected by the rule of law. Aung San Suu Kyi was particularly believable as leader of the political opposition to the military as she is the daughter of General Aung San, the assassinated hero of Burma's fight for independence. In anticipation of general elections, Suu Kyi campaigned vigorously for 10 months before being placed under house arrest, on July 20, 1989.

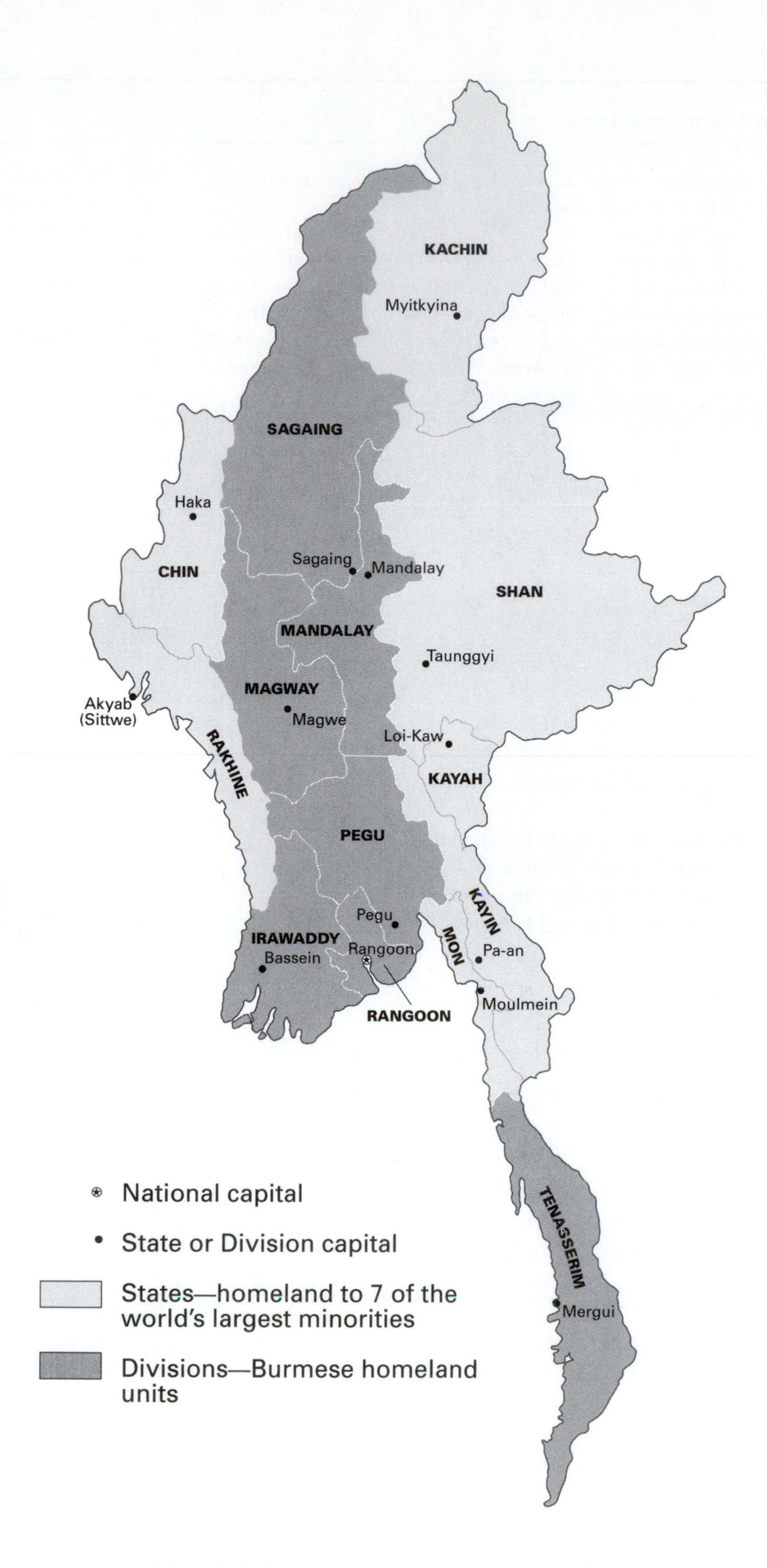

Map 17.8 Burma's states and divisions, 1999.

BOX 17.1 A Note on Names in Burma

In September 1989 the generals comprising the State Law and Order Restoration Council (SLORC) adopted the name Myanmar as the official designation for the country. *Myanmar* is a literary term of self-reference that is thought to be derived from the Chinese word *mien.* Many other place name spellings have been changed as well; for example, *Rangoon* has become *Yangon, Pegu* is now *Bago, Akyab* is *Sittwe,* the *Irrawaddy River* is the *Ayeyarwady,* the *Salween* is known as the *Thanlwin,* and *Arakan State* has become *Rakhine State.*

The change in country name has never been recognized by the United States or by Great Britain. In addition, the United States maintains no ambassador in Burma as a protest to a continuing lack of respect for human rights on the part of the government of Burma.

Throughout this chapter the term *Burman* is used to refer to a native speaker of Burmese. *Burman* is also used in the name of the broad language group from which modern Burmese evolved, thus Tibeto-Burman. The term *Burmese* is used to refer to a citizen of the country without reference to native language or to a nationwide organization such as the Army. Media reports sometimes use the term *Burma* (i.e., Burma army) in the same sense that *Burmese* is used above in reference to an organization.

Party	Percent of Vote	Number of Seats
National League for Democracy (Suu Kyi's party)	59.9	392
Shan Nationalities League for Democracy	1.7	23
Rakhine Democracy League	1.2	11
National Unity Party (the Government party)	21.2	10
Mon National Democratic Front	1.0	5
National Democratic Party for Human Rights	0.9	4
Chin National League for Democracy	0.4	3
Party for National Democracy	0.5	3
Union Paoh National Organization	0.3	3
Kachin State National Congress for Democracy	0.1	3
Others	12.8	19
Total	100	476

Note: 485 constituencies were contested but results were announced for only 476.

Table 17.4 General election results, May 1990. *Source: "A Tale of Two Systems," 1994.*

Despite her house arrest, the NLD party won an overwhelming victory (Table 17.4). Results of the election were canceled and the now embarrassed military government remained in power. (Suu Kyi, 1991; Guyot, 1991)

The release of Suu Kyi in July 1995, possibly as a result of U.S. economic and political pressures, has had less immediate effect on the people of Rangoon than had been expected by Western observers. The crowds in the streets outside of Suu Kyi's home have been modest, and even students have yet to be heard from. Possibly this is because stringent laws controlling public gatherings remain in place, or maybe it is because so many student leaders have been incarcerated. On the other hand, it may be related to the fact that economic liberalization has begun to have an impact. Goods are flowing into Burma more freely today when compared to earlier years. Also, the GNP rose 6 percent in 1994 and was expected to exceed that growth in 1995. At the end of 1997 the road past Suu Kyi's home remained blocked, and weekly gatherings of the faithful to talk to their leader were strictly supervised.

Burma's generals have often stated that their ultimate goal is the establishment of tranquillity in the country, to be followed by the return of government to the people. In fact it was a step toward this goal that caused the military much embarrassment in 1990. In May of that year the National League for Democracy, the party headed by Aung San Suu Kyi, carried 392 of the 485 contested seats in the National Assembly. Ten months earlier Suu Kyi had been placed under house arrest for continual violation of the government regulations against the holding of public gatherings. Clearly the election humiliated the military and the results were voided.

Some promises have been made and some steps taken to achieve a less rigid government eventually. In January 1993 a constitutional convention was convened to develop a constitution suitable for Burma. As of July 1995 its work "was half done" but members were sent home for an extended break. A paranoia concerning Aung San Suu Kyi was clearly evident in the work of the convention as the proposed constitution included a provision barring from the presidency "those who have lived abroad or have foreign spouses." Suu Kyi has lived abroad and is married to an Englishman. It is clear that the generals that comprise SLORC admire the dual role of the military in Indonesia and intend that in Burma, as well, the military should have a role in both the defense and the government of the country, an expectation they appear able to enforce.

LAND USE

When Burma achieved independence in 1948 the land-use picture held great promise for a bright future. The country had large reserves of arable land, a surprisingly small population, and an enormous resource of what arguably might be classed as the richest forestland in the world. By the mid-1990s the picture had changed profoundly (Table 17.5). Between 1947 and 1992 the area under cultivation had grown by only a modest 9 percent—this despite the fact than in the early years of independence, both domestic and foreign analysts spoke of vast areas for agricultural development and resettlement especially in the delta areas and the extensive middle and upper portions of the Chindwin Basin. During this same 45-year period, population grew by roughly 178 percent, with the result that the land farmed per capita decreased from more than 1.2 acres (1/2 ha) per person to less than 0.62 acres (1/4 ha) per capita. In a nation of only modest yields per unit area, the impact was devastating. In addition, the once magnificent forest cover area had decreased by 17 percent, but qualitatively it had been seriously impoverished by exploitation of the richest stands for short-term gains and for military expediency.

Agriculture: Burma's Life Blood

In Burma, the state is the owner of all lands, and the right to work farmland is exercised only with permission of the state. Land rents have been abolished and land sales are illegal. Production is concentrated on small family-operated holdings. Tenancy, which historically had been a very serious problem, especially in the rice lands of the Irrawaddy–Sittang deltas, is no longer of concern—or even recognized. Land transfer between individuals is not permitted, and reallocation of operating units is handled by the state (Huke and Huke, 1990b).

Farm units throughout the country average almost 5 acres (2 ha), and those in the major nonirrigated rice areas are somewhat larger than are those in any other portion of the country. Farms in the Dry Belt are slightly smaller than those elsewhere and are almost always planted to a variety of crops rather than being mono-cropped as is common in rice areas (Map 17.9). All areas outside of the Central Lowlands have farms of smaller average size. In part this is due to the fact that swidden is more prevalent in the hills and that mean farm size reflects only the land in actual use during a given year (Huke and Huke, 1990b).

A farm size of about 5 acres (2 ha) does not seem very large or terribly demanding of time based on the idea Americans have of a 350-acre (140-ha) Iowa corn-hog farm, an 800-acre (320-ha) Oklahoma wheat farm, or a 4,000-acre (1,600-ha) Texas ranch. But what does such a farm mean to a family in the delta of the Irrawaddy? First, such a family is far more representative of Burma's population than is a farm family in the United States. Here only about 4 percent of the population is engaged directly in agriculture; in Burma about 65 percent are farmers. Also, in Burma about 59 percent of all land used for agriculture is planted to rice. Thus our Irrawaddy delta farm family may be considered as representing a very large proportion of the country's population. Such a family has little mechanical help, certainly no tractor. Work is done

	1947		1992		1947–1992
	1,000 ha	% of area	1,000 ha	% of area	% change
Total area	67,754	100	67,658	100	
Land area			65,755		
Arable land	8,754	13	9,534	14	9
Permanent crops	?		505	1	
Permanent pasture	?		359	1	
Forest and woodland	39,094	58	32,387	48	-17
Other land	20,906	31	22,970	34	10
Population (000)	16,033		44,613		178
Arable land/capita (ha)	0.55		0.21		-62

Note: "Other land" includes unused but potentially productive, built on, parks, barren land, mangrove, roads.

Table 17.5 Land use change in Burma, 1947 – 1992. *Source: UN FAO Production Yearbook, 1993; UN FAO, Yearbook of Food and Agricultural Statistics, 1950, Part 1.*

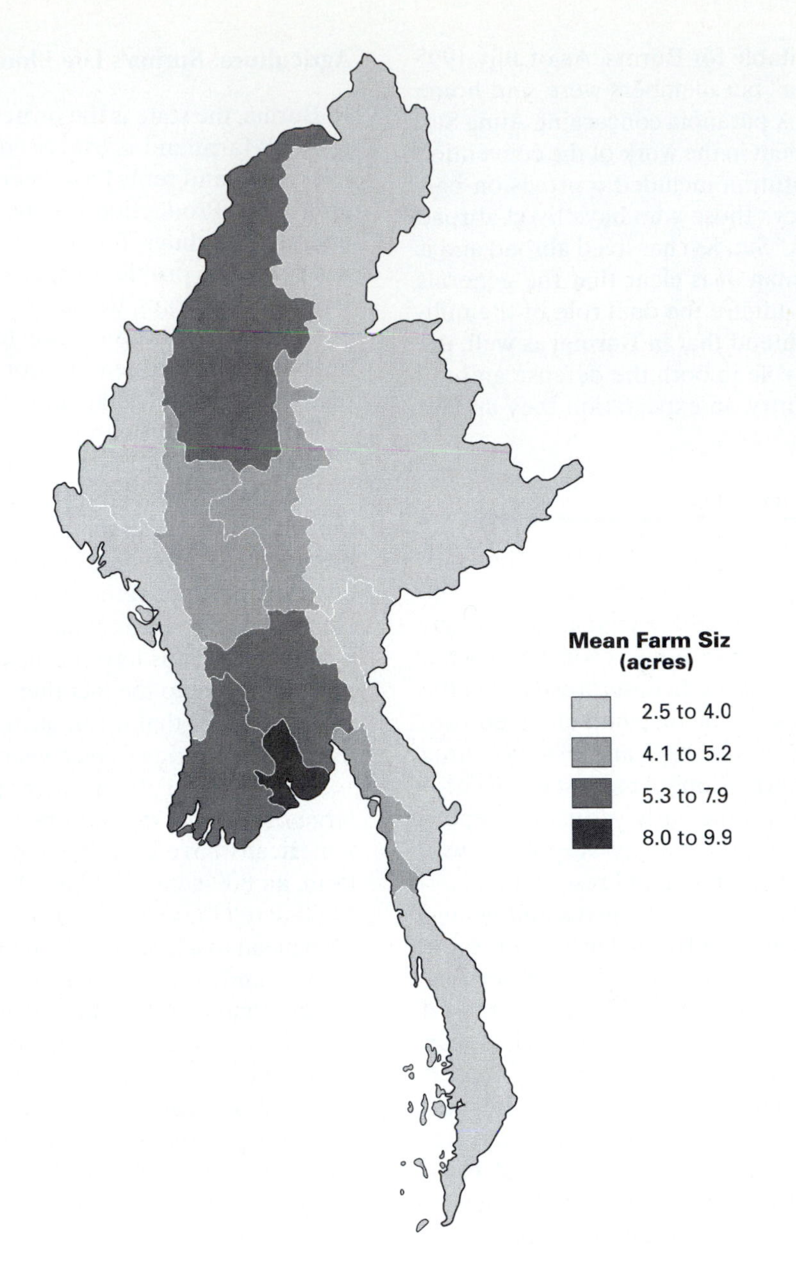

Map 17.9 Mean farm size. *Source: Huke, 1995.*

by hand or with the help of a water buffalo or two. Of course, during the dry season the soil is baked hard, too hard to be worked by a light plow pulled by water buffalo or oxen. Work can not begin until the fields have been thoroughly saturated by the early monsoon rains. Except for the small number of delta farms with irrigation, the ebb and flow of agricultural work is dependent on the vagaries of the monsoon season. The rainy season provides a window of opportunity which, in the delta, opens sometime in late June and exhausts itself by late October.

The average holding in the Irrawaddy District, within the third category on Map 17.9, is 5.8 acres (2.36 ha). Of this, about 5.6 acres (2.25 ha) are planted to rice with the remainder being paths and field boundaries. The Burmese family, as we have learned, consists of a couple with three children and perhaps one grandparent. Children attend school, care for the farm animals, stand bird watch over the rice fields, and collect firewood, and the oldest cares for the younger siblings. The farmer and his or her spouse are responsible for almost all of the work with rice. There are 122 days from June 20 through October 20, and the number of days of labor required for minimum preharvest care of 5.6 acres (2.25 ha) of rice is about 127 days. Labor requirements are not spread uniformly over the season but peak very strongly at transplanting time. Forty days are required to transplant 5.6 acres (2.25 ha) and the entire job must be done within two weeks. Think for a minute about the task of transplanting. One acre covers an area

209 ft by 209 ft. Hills of rice are transplanted at 10 in. (25 cm) on center. Thus there are 62,500 hills of rice for each acre of farm. With the mean area planted to rice being 5.6 acres (2.3 ha) in the Irrawaddy District, each housewife is responsible for bending over from the waist to plant 350,000 hills—a daunting task indeed and one impossible without help. Here is an opportunity to employ the landless, to share work with neighbors, or to involve children, old folks and—more and more commonly—the males of the household (Huke and Huke, 1990a).

When the rains let up and the fields dry out, the grain ripens and a second labor peak is imminent. Harvesting, hauling, drying, and threshing require about 34 additional days and must be completed as quickly as possible to minimize loss to birds, insects, and rats. Thus a second peak in the demand for labor in the cycle of rice farming. In Burma there is often insufficient labor available to meet the needs of these two peak periods. One of the reasons often given for the relatively low padi yield in Burma is the fact that transplanting, weeding, and harvesting are often a bit haphazard.

Rice is the premier crop in Burma. It occupies 60 percent of all land planted to crops and utilizes 56 percent of all farm labor. It is omnipresent in the landscape beyond the Dry Belt, and even there the brilliant green of the irrigated padi fields are locally dominant. Rice has maintained its leading position, although slightly reduced percentagewise from its position in 1960 (Table 17.6). Among the next leading crops, there have been several important shifts in emphasis during the past 35 years. Most dramatic is the case of sunflower, which in 1959–1960 was planted to only 24,500 acres (10,000 ha) but by 1993 was found on 329,000 acres (133,000 ha), mostly in Magway and Mandalay Divisions. In the Dry Belt, too, both maize and wheat have shown modest increases, but cotton, chickpeas, and especially tobacco have fallen off. Around the margins of the Dry Belt, sugarcane has increased its area well above the rate for all agricultural land, but even that is not sufficient to satisfy local demands. In the delta regions, jute has found modest acceptance as a cash crop, largely in areas where floodwater levels have proven to be too deep for success with the shorter varieties of rice being raised today.

Rice farming is intimately involved in the culture, food ways, and economy of almost every group in Burma. For example folklore tells us that when the Kachin people were sent forth from the center of the earth, they were given the seeds of rice and were directed to a wondrous country where everything was perfect and where rice grew well (Armstrong, 1997). Rice is an integral part of their creation myth and remains today as their leading crop and the most preferred food. Among the Kachins, rice means rice grown on the steeply sloping *taungya* fields, not the grain of diked and flooded fields, so common in much of Burma, that the Kachins believe is hardly fit for human consumption. Fermented rice from which

Crop	1993 area (000 ha)	Percent of crop area 1993	Percent of total 1993 arable area	1959-60 area (000 ha)	Percent of 12 crop area 1960	Percent change in area 1960–93
Rice	5,794	65	61	4,007	69	45
Sesame Seeds	992	11	10	592	10	68
Beans	660	7	7	378	6	75
Groundnuts	485	5	5	403	7	20
Maize	220	2	2	59	1	273
Seed Cotton	158	2	2	143	2	10
Chickpeas	149	2	2	117	2	27
Wheat	145	2	2	29	0	400
Sunflower Seed	133	1	1	10	0	1,230
Sugar Cane	76	1	1	26	0	192
Jute	48	1	1	8	0	500
Tobacco	38	0	0	46	1	-17
Total	8,898	100	93	5818	100	53
	1993			1960		% increase 1960–1993
Total Population	43,500,000			21,300,000		104

Table 17.6 Burma's 12 leading crops, 1993 and 1959–60. *Source: UN FAO Production Yearbook, 1993.*

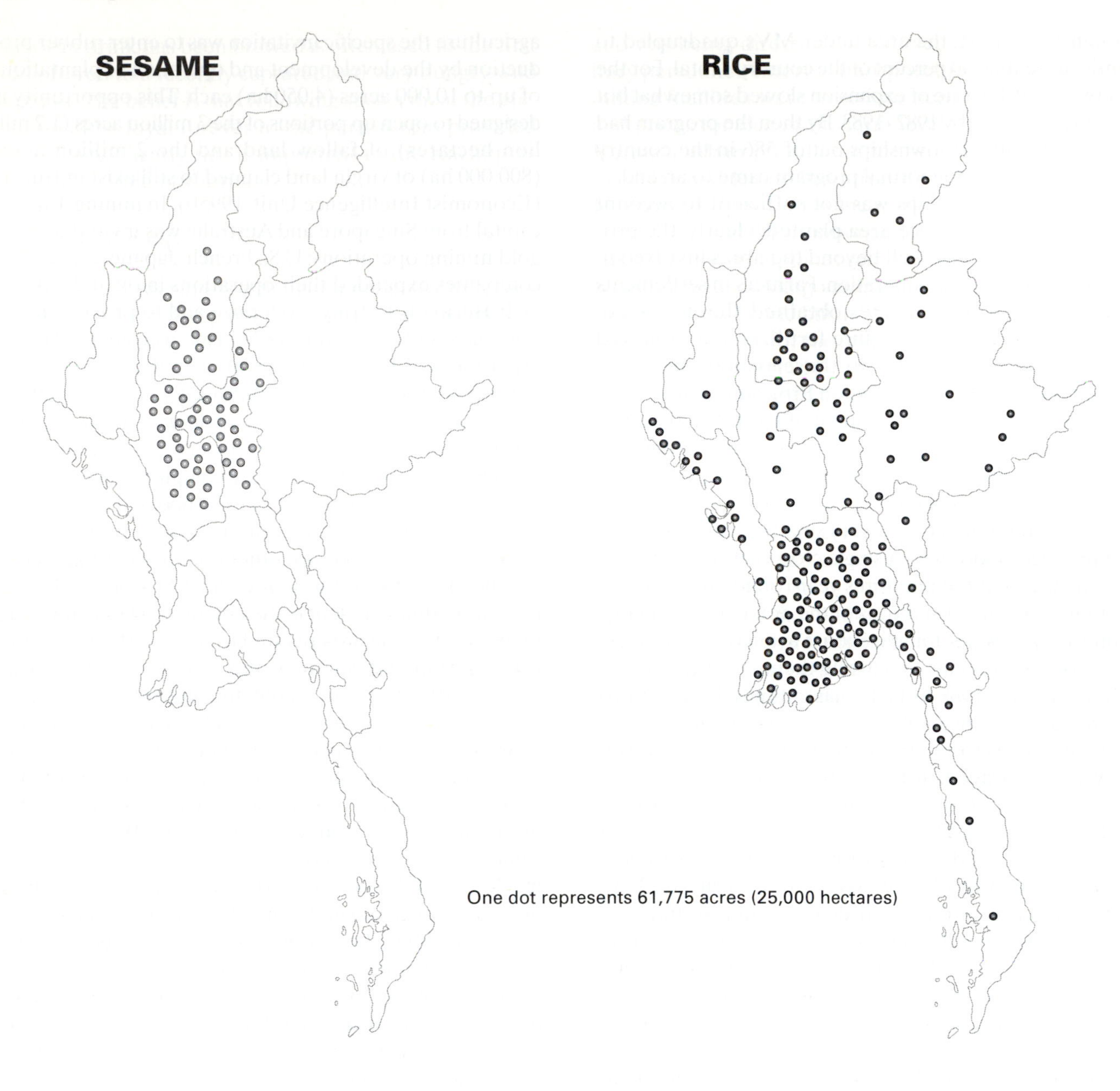

Map 17.10 Sesame and rice crops planted in Burma in 1993.

reached its zenith. The Dry Belt is distinctive for its low rainfall combined with high temperatures and for the fact that its agriculture contrasts sharply with that found in other portions of the country. The Dry Belt includes the central Irrawaddy and lower Chindwin valleys and the northern extension of the Sittang Valley. The rainfall map of Burma shows clearly that the Dry Belt area has less than 40 in. (100 cm) of rainfall per year (see Map 17.3). The core area around the towns of Pagan, Pakokku, and Myingyan is easy to identify. The outer limit of the Dry Belt is more difficult to demarcate because it is a climate boundary and as such shifts location from year to year. The area holds about one-fourth of the nation's people and one-third of its cultivated land. With its low average rainfall occurring mostly between May and October, water deficiency is a problem. Tanks, wells, canals, and streams provide irrigation water for about one-third of the agricultural land, and these fields are devoted largely to rice. Nevertheless, the Dry Belt does not normally grow enough rice for local needs and must depend on imports from the delta region.

In contrast to the delta, the Dry Belt produces a great variety of crops. Food crops include maize (corn), millet, sesamum (sesame), peanuts (groundnuts), chili, pulses (edible seeds of plants such as beans and peas), and a variety of vegetables. These, together with cotton and tobacco, occupy twice the acreage of rice. Although dry-farming techniques are used, crops frequently fail because of lower-than-normal rainfall. At one time such failures

BOX 17.2 Foods in Burma

Nowhere in the world does a national population consume as much rice per capita or derive as high a percentage of its food energy from rice as the people in Burma. In Burma, each person consumes 409 lb (186 kg) of cleaned rice every 12 months. That provides almost three-quarters of their annual caloric intake. By comparison, the average U.S. citizen eats about 21 lb (4 kg) of rice per year, accounting for less than 2 percent of her/his calories.

In a Burmese household, a measurement of roughly one and one-quarter cups of uncooked white rice is prepared for each person for each meal. This enormous mound of rice is eaten with the addition of vegetable, fish, shrimp, or chicken, often curried and invariably with a spoonful or more of *ngapi,* fermented fish or shrimp paste. Beef is rarely consumed. Goat meat finds its way into curry occasionally, and pork is common in rice or noodle dishes based on Chinese traditions.

Mohinga, a soup with fish, shrimp, or vegetables and hot chilies is commonly served at breakfast or lunch. *Kyaukswe,* a delicious souplike food rich enough to serve as an entire meal, is made with chicken, noodles, coconut milk, split peas, onions, ginger, garlic, lime, *ngapi,* and chilies.

An enormous variety of fruit and nuts are prized in season, and food stalls dispense fried banana chips, pancakes, dried shrimp, Shan (Chinese) sausage, pickled eggs, sweets, and sugarcane juice.

Tea is a common drink. In the hill areas especially, pickled tea leaves are also consumed. Coffee is served at roadside shops and restaurants but is seldom found at the village level. Alcoholic beverages are uncommon.

caused famines, but in recent decades rice surpluses from the Irrawaddy delta have been available. Sesame, sunflowers (note the enormous increase in the cultivation of this crop between 1960 and 1993 as shown on Table 17.6), and peanuts from the Dry Belt supply a large part of Burma's requirements of cooking oil; the cotton, even though of a rather short-staple length, finds a ready market in the nation's textile mills and increasingly in exports along the Burma Road to China. The Dry Belt is also home to several million cattle and water buffalo, kept as work animals rather than as a food source. Very few find their way to market as Burma's Buddhist population eats little meat of any kind, especially beef.

Broad river terraces that are currently dry farmed and parallel both the Irrawaddy and the Chindwin appear to be well suited to irrigation. Once funds are made available for the construction of dams, storage reservoirs, and additional canals, at least 3.7 million acres (1.5 million hectares) of now low-yielding farmland could be converted into high-yielding irrigated land capable of growing two or perhaps three crops a year. In 1994 the government announced a project for the "greening of the Dry Belt" that is eventually to irrigate some 617,500 acres (250,000 ha). A series of small dams, extensive embankment improvement, and much canalization is to be done. All of this must be accomplished to better utilize the waters of about 150 small rivers and streams. No use is yet planned for the waters of the Irrawaddy.

On the southern margin of the Dry Belt, a very intensive and highly sophisticated cropping system is already finding increasing acceptance in the 1990s. This rice–wheat system involves the production of a harvest of each of these grains sequentially on a given plot within a single calendar year. The soil must be puddled, that is, have all internal structure broken down, to allow the ponding of water for the rice crop. Following rice harvest the soil must be dried sufficiently to allow it to be plowed and aerated to support the wheat crop. Temperatures during the winter months in the Dry Belt are fine for wheat, and with only modest irrigation this system allows a significant increase in productivity. The potential appears to be great for major agricultural advances in the Dry Belt.

Shifting Cultivation

In the mountainous areas of Burma, swidden or shifting (slash-and-burn) cultivation, known locally as *taungya,* supports as many as 7 million persons. *Taungya* involves cutting and burning the forest cover and then planting and harvesting the crops. Rice is almost always the dominant crop, occupying about two-thirds of the surface. Maize, beans, tobacco, cotton, sesame, potatoes, and numerous vegetables are interplanted with the rice in most fields. Leaching of the soil is rapid under the conditions of excessive rainfall and constantly high temperatures of the area. A cultivated field loses its fertility after one or two years; it is then idled in favor of a new site. Where soils have not been severely eroded, the vegetation cover gradually reestablishes itself on the idled plots, although even in the best of cases the new cover will be missing several of the most sensitive members of the original ecosystem. Ideally the field is not planted again until it has regained a substantial part of its fertility 12 or 15 years hence. In this respect, *taungya* may be thought of as field rotation as opposed to crop rotation. This technique of land use often results in a moderately well-balanced diet. In part this is so because folks who practice *taungya* continue the traditions of hunting and gathering as well. This effort makes a small contribution of animal protein to the food intake.

The magnitude of *taungya's* impact on the environment of Burma may be judged by knowing that to provide the food and fiber required for one person for a

Photo 17.9 This Kachin woman will spend the day planting crops in her *taungya.* On the ground is her iron-tipped planting stick, several kinds of seed, and the ash from the burning that took place recently. To the left is her head basket. (Huke)

single year requires the yield from about three-quarters of an acre (0.3 ha). Even under ideal conditions a given plot is used for only two successive crops. Thus, every year a bit more than 2.5 million acres (1 million hectares) of Burma's land must be cut, cleared, burned, and readied for use. This is an area almost the size of the entire state of Connecticut. This is done just to support the population currently depending on swidden—no provisions are made for population increase.

In the southern portions of the Kachin State, northern parts of the Shan State, much of the Kayin State, and large parts of Tenasserim, population pressures have increased to the point that *taungya* farmers are returning to the original plot after only eight, six, or even four years of fallow. In such situations delicate soils are further degraded, anticipated yields plummet, and soil erosion is hastened. Hillsides once covered with magnificent forests are soon reduced to scrub or wasteland. Extensive areas have thus been converted to a cover of tall tough grasses that are too high in silica content for even water buffalo to graze and too tough to be worked by simple tools.

Opium

The hill areas that harbor insurgent activities on Burma's eastern borders are areas where *taungya* is the rule rather than the exception and are also areas where the opium poppy is produced in abundance. The greatest concentration is in the far northeastern portion of the Shan State, that area inhabited by the Wa people. Remaining portions of the Shan State, especially that portion east of the Salween River, and the eastern and southern sectors of the Kachin State also have long histories of production. The poppy has been raised for its yield of opium, and the drug used locally for many generations. Addiction among Burma's lowland population has not been widespread; the same can not be said for the producers.

During the 1960s and 1970s world events conspired to encourage the rapid expansion of production in northeastern Burma. An eager and wealthy market existed in the West, driven by social turmoil among the youth and greatly abetted by the easy access to drugs by service men and women during the Vietnam War. Northeastern Burma and adjacent areas had ideal environmental conditions, had experienced opium producers, resident entrepreneurs, and access to established trade routes.

The border areas provided a safe haven for a wide variety of insurgent operations all of which were in desperate need of money. The drug trade seemed an ideal source. Areas of Burma, together with adjacent portions of northern Thailand, northern Laos, and small areas of China's Yunnan Province became known as the Golden Triangle (Map 17.11) and soon became the world's chief producer of opium. So great was the production that convoys of 200 or 300 mules guarded by 1,000 or so heavily armed troops were sometimes used to carry the product from production and collection areas toward the point of sale (Renard, 1995).

Production of raw opium in Burma is on the order of 2,250 tons per year, a dramatic increase from the 800 to 900 tons produced annually in the 1980s. Two and one-quarter thousand tons is converted into about 225 tons of morphine, which is moved clandestinely to ports ranging from Chittagong to Shanghai and shipped mainly to North America. Late in 1995 an important battle was fought on the outskirts of Doi Lang on the Burma border north of the Thai city of Chiang Rai. The Mong Tai army of drug warlord Khun Sa, reinforced by Wa, Kachin, Palaung, and Lahu troops, was defeated by the Burmese army. Heroin refineries disappeared from the area and the flow of drugs across the Thai border dried up, only to be replaced by more numerous refineries to the north

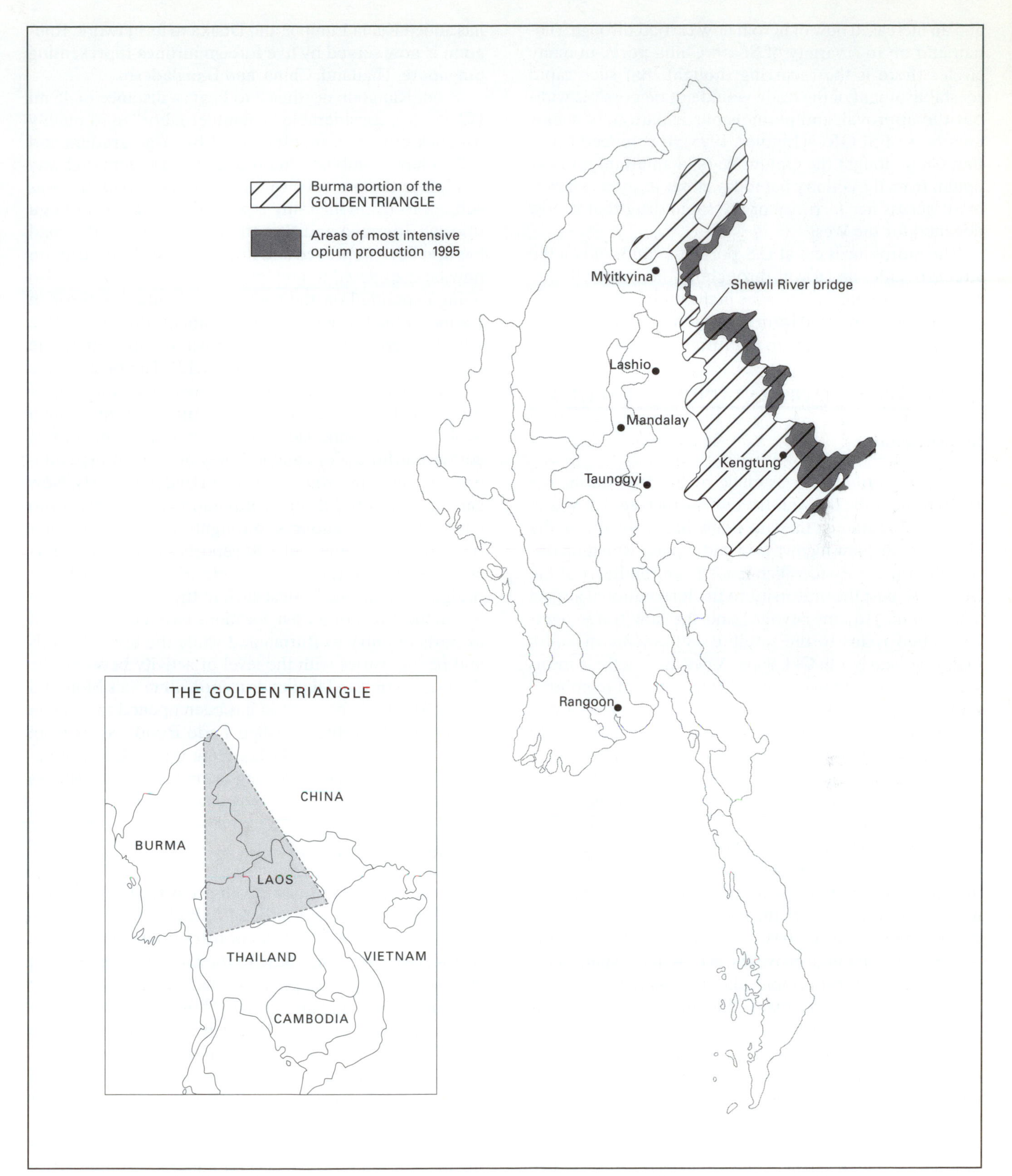

Map 17.11 Chief poppy producing areas of Burma. *Source: Author; Ulack and Pauer, 1989, p. 13.*

and an increased flow of heroin now carried through Yunnan and on to a variety of South China ports. In many circles there is the recurring thought that such rapid reestablishment of the trade would not be possible without the approval, and perhaps participation, of senior members of SLORC (Lintner, 1995). It is indeed ironic that China fought the Opium Wars in an attempt to bar opium from the country but today closes its eyes to movement across her territory of the same drug that is now destined for the West.

The morphine is cut at U.S. port cities and sold on the streets of cities and towns throughout the United States. Morphine arriving in U.S. cities in the mid-1990s is in such pure and concentrated form that it is now snorted rather than shot. At what price to the future of U. S. society?

ECONOMIC DEVELOPMENT UNDER THE MILITARY

Infrastructure

A great deal of effort was made by SLORC during the 1990s to upgrade the country's infrastructure that was in a deplorable state in the late 1980s. Most obvious to the visitor are the new highways in and around Rangoon (especially Prome Road which leads from the heart of the city, north past the university, to the international airport some 10 mi [16 km] beyond) and the new traffic light-controlled roads to the satellite settlements of South Okkalapa and North Okkalapa. A decade ago these towns were little more than organized collections of temporary shelters with unpaved roads, inadequate drinking water, very poor sanitation, and no reliable electric supplies. In the late 1990s the situation is considerably improved.

The Rangoon and Hlaing Rivers, which mark the western margin of Rangoon, have been deepened and docking facilities have been improved. The rivers are bordered by low-lying flood plains; the major urban development occupies slightly elevated land to the east. At the break in slope is located Insein Road, a major avenue to the north and west as well as the route serving much of Rangoon's manufacturing base. This crowded but recently modernized avenue provides access to a small steel rolling mill, several cement plants, glass and tile factories, a number of textile plants, and numerous small pharmaceutical industries.

Two new container facilities are being constructed at outports several miles seaward of the capital on the Rangoon River. Here, too, the age-old problem of silting is being attacked by newly purchased dredges. Five Star Lines, the government ocean carrier, has started to upgrade its aging 21-vessel fleet with the acquisition of two new ships from China (Economist Intelligence Unit, 1995a). Myanmar International Airline (MIA) has expanded its service with a weekly flight to Kunming and has added Kuala Lumpur and Dhaka to its network. Rangoon is now served by five foreign airlines representing Singapore, Thailand, China, and Bangladesh.

From Rangoon northeast to Pegu, a distance of 45 mi (72.45 km), considerable "volunteer labor" used mainly for rock crushing, supplemented by large graders and rollers have combined to produce an excellent highway to replace the previous winding two-lane road that was subject to frequent rainy season flooding. From Pegu through Toungoo and Yamethin to Mandalay, the road has also been improved and the 380-mi (612-km) trip can now be completed in one long day. Much work has also being expended on the route from Mandalay through Lashio to the border of China's Yunnan Province. Here a bridge across the Shweli River was completed by the Chinese in February 1993 (Map 17.12). The bridge gave an entirely new importance to the road. Today manufactured goods of all kinds flood into Burma from Yunnan while trucks, automobiles, and carriers of cooking oil, raw cotton, hardwood logs, and a variety of concentrated mineral ores clog the route to China (Badgley, 1994). New cars are imported through Rangoon and are driven to China. The flow amounts to roughly 1,200 automobiles per month, and unconfirmed reports say that the Mercedes dealer on the Chinese side of the Shweli River bridge is one of the busiest such in the world.

On the Thai border five locations have been approved as ports of entry to Burma and while the flow of goods and people varies with the level of activity between the Burmese Army and that of drug lord Khun Sa's Mong Tai Army. The India border too has been opened to trade at the point where the so-called Ledo Road passes from India into Burma (Map 17.12). Here the trickle of smuggled goods and people crossing the border since 1948 has been recognized.

Industrial Development

Burma has one of the least developed manufacturing sectors of any country in Southeast Asia (Hill, 1984). An extremely low portion of the labor force is involved and the sector contributes very little to the GNP. During the first 40 years of independence, there was little change, and it is only from the beginning of the 1990s that hopes for development were renewed. In fact, between the mid-1960s and the reforms of 1988, no foreign investment was allowed in Burma and foreign aid was strictly limited to "humanitarian aid." Even earlier, the British did little to foster industry other than that related to the processing of locally grown agricultural or forestry products and the mining and concentration of several mineral resources. Rice milling was far and away the leading industrial employer and, if adequate statistics were available, there is little doubt but that it is only slightly less dominant in 1995.

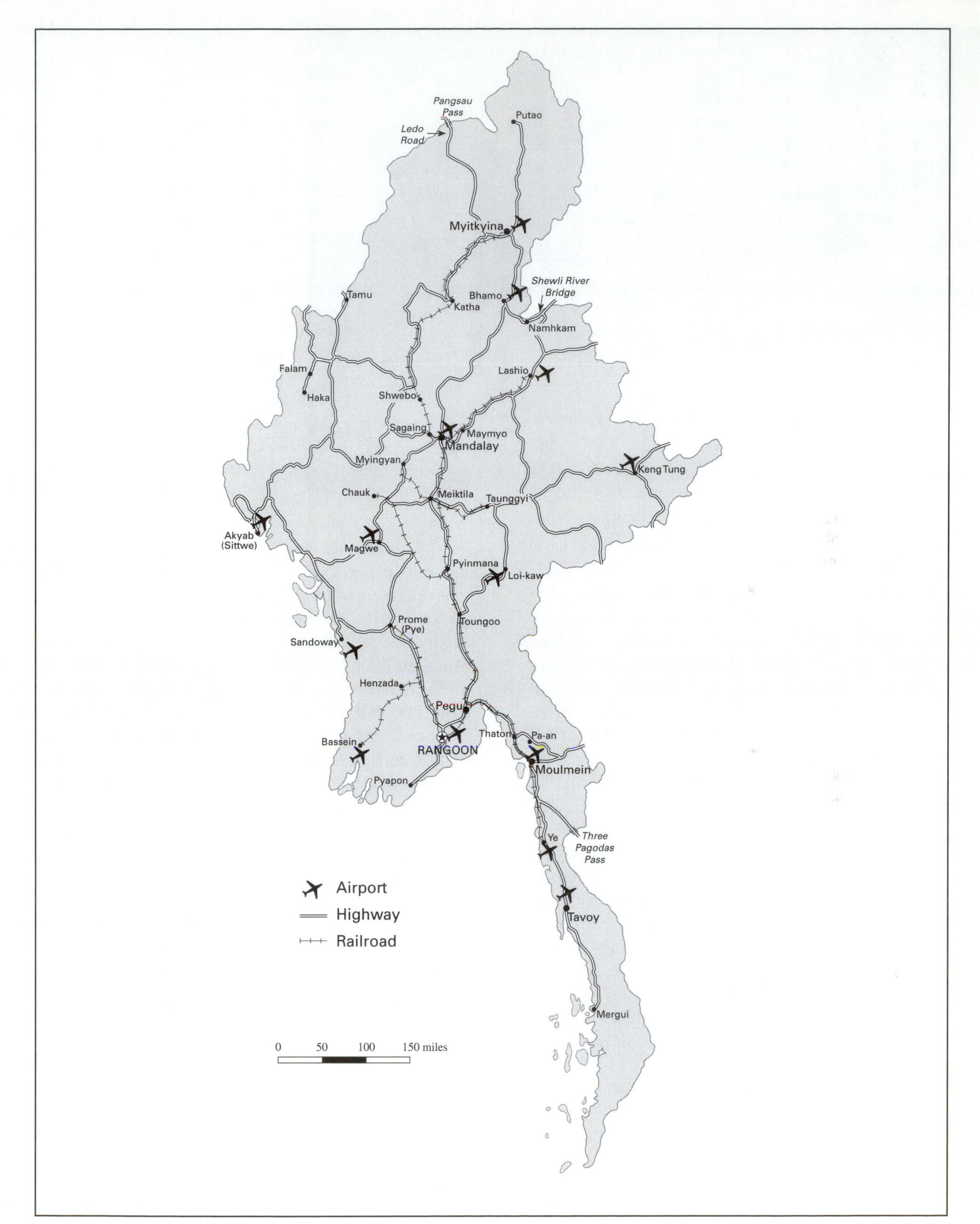

Map 17.12 Burma's transportation network. *Source: Ulack and Pauer, 1989.*

Photo 17.10 Roads are now being improved throughout Burma. The methods would be foreign to most Americans. Here, outside Rangoon in 1994, men and women are making gravel for the road by pounding large stones with heavy hammers. (Huke)

The major facets of recent economic policy have varied from year to year and from spokesman to spokesman but have been remarkably consistent in four items: (1) to achieve all-around development of the economy based upon the primacy of agriculture; (2) to develop a "properly functioning" market economy; (3) to attract technical know-how and investment capital; and (4) to retain control of the economy in the hands of the state and its citizens. Clearly, the military retains a morbid fear that the old colonialism will return to Burma in a new form. This fear is so strong that leadership trusts few citizens other than those in the army to control the nation's economic destiny.

Even so, in the wake of the euphoria engendered by the withering away of the "Burmese Road to Socialism" and the replacement of General Ne Win as the nation's leader, a number of manufacturing operations from around the world were attracted to Burma (Jacobs, 1996). Among the U.S. companies were Eddie Bauer, Liz Claiborne, Levi Strauss, and Federated Department Stores (Macy's). By late 1995 all four U.S. companies had found that they could not operate at a profit, and all four closed operations in Burma placing the blame on poor infrastructure and the inability to meet quotas. On the other hand, in April 1995, Caterpillar opened an office in Rangoon and shortly thereafter shipped 10 bulldozers to Burma for work in the border areas (Economist Intelligence Unit, 1995b).

In the area of food processing the picture has been a bit more encouraging. During 1994 and 1995 firms from Thailand and Japan have entered the economy in a small way—the Japanese in prawn farming and exporting, and the Thais in prawn farming as well as traditional fishing. The Thais are also involved in ice making, canning and freezing of fish, the manufacturing of fish powder, and the export of fish products (Economist Intelligence Unit, 1995b).

An especially vexing problem has been the inability to secure adequate and reliable electric power. New generating facilities put on line during the 1993–1995 period are both hydro and diesel and have vastly improved the situation in Rangoon and Mandalay. Beyond these two centers, supplies do not approach the needs of mining or manufacturing enterprises. In 1994 the state-owned capacity nationwide was about 950,000 kW of which 45 percent was from natural gas, 35 percent from hydro, 10 percent from diesel, and 10 percent from other sources. In that year the actual generated output was reported as 3,061 m kwh or sufficient only for each person in the country to burn a 100-w bulb for two hours each day (Hunter, 1995). Clearly, many village folk were without any access to electric power, and more than half of the total production went to industrial establishments in and around Rangoon and Mandalay.

Tourism

In 1996 approximately $70 million flowed into Burma from the tourist industry, and this represented the nation's largest singe source of foreign exchange. During that year roughly 160,000 visitors arrived by land through six approved border crossing points between Yunnan Province and the Shan or Kachin States. Most of these folks traveled south to Lashio where a magnificent Chinese temple has become a tourist destination. From here many continue on to Mandalay and a few to Rangoon.

Another 90,000 tourists, largely European, Japanese, and American, arrived by air at Rangoon, and most undertook travel in the Rangoon–Mandalay–Pagan triangle.

Political

As noted earlier, the 1991 Nobel Peace Prize winner, Daw Aung San Suu Kyi, was released unconditionally in 1995 from house arrest that had kept her silent for almost six full years. She was given freedom to travel wherever she wished and to meet with whomever she pleased. However, she did not attend the mid-September 1995 United Nations Fourth World Conference on Women in Beijing, reportedly because she was not certain that she would be readmitted to Burma.

Cease-fire agreements have been achieved with most of the rebel groups, and foreign investments, at least from neighboring states, are increasing (Jacobs, 1996). Trade with Thailand and Malaysia has improved markedly in the 1990s. Burmese businesspeople are finding it easier to travel abroad, and increasing numbers of students are being trained overseas. In July 1997 Burma was admitted to full membership in ASEAN (Guyot, 1998). In recent months, several ministers and other high-ranking officials of ASEAN nations have visited Burma in line with that organization's policy of "constructive engagement."

There is much more to be done, even if democracy returns to the country, but a tentative start has been made. By mid-1997, several four- and five-star hotels had been completed in Rangoon and in Mandalay by firms based in Singapore, Malaysia, and Thailand. Less-pretentious but highly serviceable facilities had been completed in Pagan, Heho, Taunggyi, Mamyo, Moulmein, and several other sites largely by local entrepreneurs. Major improvements were underway to public transportation and to the roadway system in the capital, and highways as far north as Lashio were being upgraded. Rail service between Rangoon and Mandalay had improved dramatically. Unfortunately, Rangoon's Mingladon airport, the nation's only field offering international connections, still lacks modern electronic equipment and has grossly outmoded passenger and baggage handling operations. It cannot serve the needs of the country. Port facilities may be sufficient for the shipping of rice and teak but are far below international standards for anything else—especially for passengers. Even the political situation is not yet fully acceptable in the eyes of the rest of the world, as was demonstrated in June 1995 when the International Committee of the Red Cross closed its office in Rangoon because SLORC continued to refuse to allow Red Cross officials to visit political prisoners on a regular basis (Economist Intelligence Unit, 1995a).

Of all the difficulties facing Burma in the development of international trade, foreign investment, and the development of the tourist industry, the most bothersome is the currency situation. The official rate of exchange hovers around Kt 6 for the U.S. dollar, while the unofficial rate that is available in the grey market and to certain favored business folk and military officers was 40 times that in November 1997. Despite currency problems, continuing military engagement with the Karen revolutionaries, the alleged suppression of human rights, and a U.S. embargo on trade with Burma (for other than "humanitarian purposes"), Burma has shown remarkable economic progress. During 1995 Burma's real G.D.P. increased by 9.8 percent and in 1996 by 5.8 percent.

If the oft-promised return to democracy actually takes place, Burma could become one of the "tigers" of Southeast Asia. Given its favorable geography, a return of democracy could easily be followed by a new influx of investment. In such an optimistic future, there is every likelihood that Burma could surpass both Thailand and Malaysia in the race to provide a better life for its citizens. This will not occur soon, unfortunately.

SUMMARY

When Burma achieved independence from the British in early 1948, the world was sanguine concerning the future of the country. Population pressure was considerably lower than in many neighboring countries; the rice economy was healthy and the world had need for Burma's surplus. There was considerable area of potential agricultural land, some in the delta areas and more in the basin of the Chindwin. But little effort had been made to utilize the vast water resources of the country, either for the generation of hydropower power or for irrigation. Petroleum, timber, tin, lead, silver, jade, rubies, and sapphires all offered the prospect of enhancing development.

Burma remains a country of impressive natural beauty, rich archaeological treasures, high tourist potential, and a rich culture. The production of sufficient food has seldom been problematic; yet over the past 30 years Burma has failed to achieve the progress expected and anticipated by an array of observers both inside and outside of the country. Clearly the explanation for this lies with the lethargy and stagnating policies of the state. The junta's interest in eliminating political opposition has overwhelmed the real need to develop an outward-looking development policy. The isolationist posture of the state must be altered and the best hope for this remains with international pressures and especially the influences of ASEAN members.

18 Vietnam, Laos, and Cambodia

DEAN FORBES CECILE CUTLER

Vietnam, Laos, and Cambodia are at a crossroads. Along with Burma they are the poorest countries in Southeast Asia, and they remain burdened by the inheritance of a turbulent past, unwieldy political institutions, and uncertainty about how to deal with the global capitalist economy. The Vietnamese government waivers over its commitment to economic reform, while Cambodia remains troubled by persistent political instability and domestic insurrection. Laos is the most committed to maintaining its journey along the socialist path, yet is also slowly opening to foreign investment.

Dynasties and empires have overlapped the present state boundaries of the three countries, changing the political geography of the region. These empires were formally, but loosely, brought together under the colonial influence of the French, and the three former colonies of Vietnam, Cambodia, and Laos were collectively known as French Indochina during the colonial period. Unlike other parts of Southeast Asia, the colonial union fragmented at independence into three separate states. Decolonization plunged each of them into civil war, economic stagnation, and sometimes chaos. Vietnam was divided into two, while Cambodia and Laos both experienced long periods of local insurrection before socialist governments were installed in the 1970s. During the 1970s and again during the 1980s, fraternal bonds strengthened between the socialist states, only to disintegrate in the fighting between Cambodia and Vietnam. As is so for the entire region, the three states of former French Indochina are culturally quite diverse, but one ethnic group predominates in each of the countries: Vietnamese, or Kinh, in Vietnam; Khmer in Cambodia; and Lao in Laos. Of the three, Laos is the most culturally diverse.

With the adoption of new economic policies aimed at building market-based economies and the return of instability in Cambodia, the links between the countries have remained frayed. Each now looks outward for an increased share of global trade, aid, and investment. Yet to develop economically, all three countries are in desperate need of improved infrastructure (including better roads, railroads, and bridges), and investment in human resources. The ruling regimes have yet to demonstrate that they have the skills to cope with the new demands, and they fear the political consequences of accelerating economic progress. As the last countries in Southeast Asia to embark upon the long path to true economic development as part of the global economy, Vietnam, Laos, and Cambodia will each need to signal clearly the direction they will choose at the crossroads.

HISTORICAL BACKGROUND

The origins of the ethnic Vietnamese are rooted in the coastal regions of southern China. As noted in Chapter 3, thousands of years ago the ancestors of the present population drifted southwards, fleeing the expansionist Chinese to the north, while searching for arable land for settlement. The Red River delta was the first area occupied by ethnic Vietnamese, and gradually the Vietnamese diffused southward, first into central Vietnam and soon after into the Mekong delta region. During this migration the Vietnamese conquered various groups, notably the Cham and the Khmer, who were either absorbed or pushed into new lands further south and west into present-day Cambodia.

The Vietnamese have for most of the last 2,000 years maintained an uneasy relationship with China; what is today northern and central Vietnam was for more than 1,000 years between 111 B.C. and 939 A.D. part of the Chinese empire. A product of this long-term connection is that Chinese culture has had a most important influence on the Vietnamese, although contemporary leaders continue to hold deep-seated concerns about Chinese ambitions regarding Vietnamese territory. Recall that most of the rest of the Southeast Asian region can trace its pre-Islamic and pre-European external cultural influences not to China but rather to India (see Map 3.2).

In the area of present-day Cambodia, the foundation for the Khmer state began in the third century A.D. and

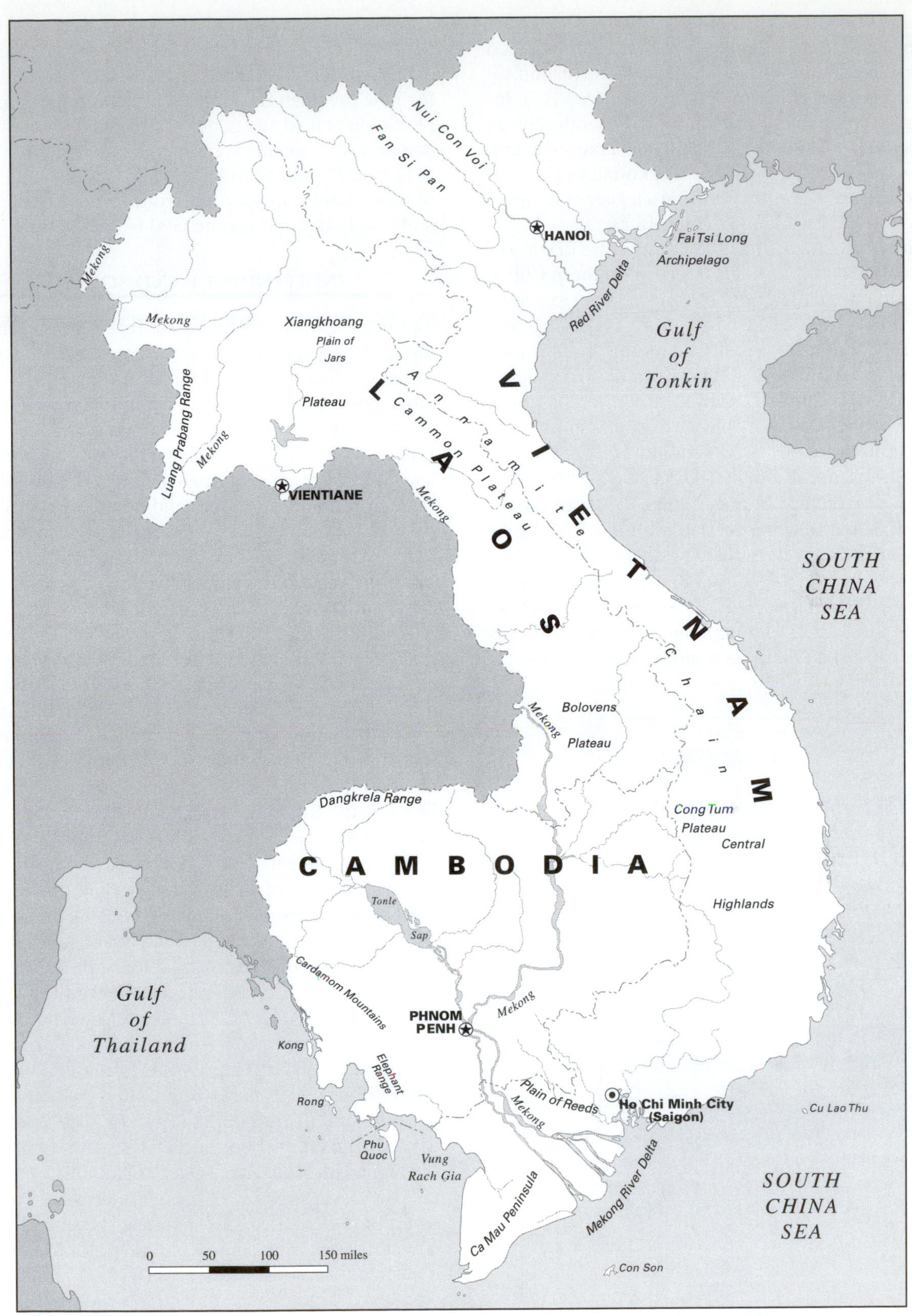

Map 18.1 Vietnam, Laos, and Cambodia. *Source: Ulack and Pauer, 1989.*

evolved from Funan, the kingdom that was located in what is today coastal southern Vietnam. Several states emerged culminating in one of the region's most powerful and sophisticated kingdoms, that of Angkor (see Map 3.5). Indeed, Angkor attained its zenith around the tenth century A.D. and could be included among the most powerful empires in the world at that time. The Angkorian empire collapsed in the fifteenth century leaving a legacy of temples, or *wats,* and irrigation channels (Map 18.2).

Like the Vietnamese and the Khmer, the origins of the various Thai and Lao groups were in southern China. The contemporary nation of Laos can be traced to a fourteenth century principality called Lan Xang which was linked with Cambodia. When Cambodia's importance declined, so too did the original state, only to reemerge in the nineteenth century as three separate kingdoms. These were under constant threat of sovereignty from their large and more powerful neighbors in present-day Vietnam, Burma, and Thailand (Fisher, 1964, p. 102–125).

The French influence in Indochina began with the establishment of Catholic missions in the 1600s. Gradually, France gained control over all of what was to be called French Indochina: Cochin China in the south became a colony in 1862, Cambodia in 1863, Annam and Tonkin in 1885, and France seized control of the three Lao kingdoms in 1893 and absorbed them, as a single entity, into the Indochinese Union (Map 18.3). The initial colonial thrust by the French was part of the complex process of trade and expansion that took place as European nations searched for new raw materials to fuel industrial growth. The exploitation of land and goods encouraged the view that the colonies were mainly suppliers of materials and revenue to France. The main products included the plantation crops of rubber and tea, especially from Annam, Cochin China, and Cambodia, and tin and coal from Tonkin.

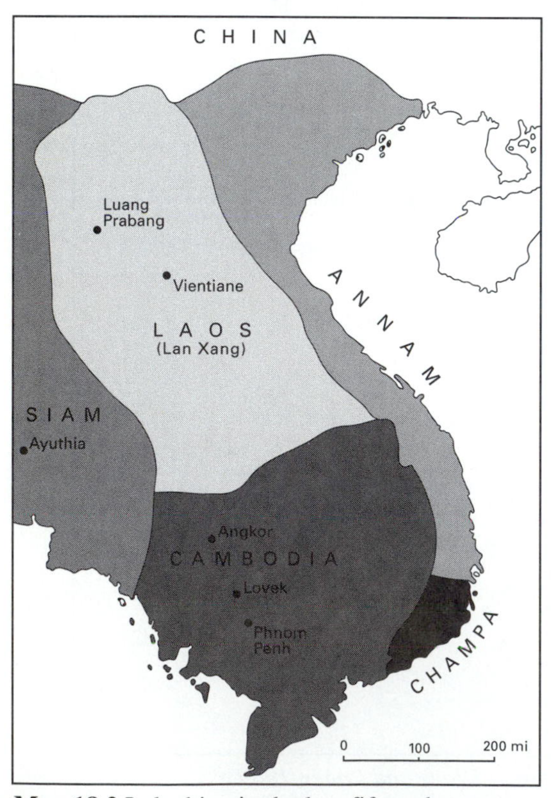

Map 18.2 Indochina in the late fifteenth century.
Source: Fisher, 1964, p. 118.

INDEPENDENCE AND SOCIALISM

The decade of the 1940s was a turning point. Independence struggles encompassed a series of internal and external conflicts, influenced at different times by larger, more powerful nations that became involved for political, ideological, and economic reasons. The influence of France on the internal politics of the Indochinese colonies had different results, although their locations exposed each to encroachment by China and communist ideology. At the same time, the various democracy movements attempted to unite the individual countries, rebuff communism, and rebuild countries ravaged by incursion and the aftermath of World War II (Fryer, 1979).

Vietnam declared its independence in 1945, but it did not achieve this goal fully for a number of years. French colonial forces fought back across all the states of Indochina, a reaction known as the First Indochina Conflict. Ultimately, French forces were defeated by the forces of Ho Chi Minh at Dien Bien Phu, which culminated in a decision to split Vietnam into two states: the Democratic Republic of Vietnam (North Vietnam) and the Republic of Vietnam (South Vietnam) in 1954.

The government of the North had as a priority the eventual reunification of the two parts of the country. Aided by the National Liberation Front (NLF) in the south, their persistence led to the long, 20-year war against the armies of the Republic of Vietnam and their overseas supporters, led by the United States. Known in the West as the Vietnam War, in Vietnam it is often called the American War. In this chapter it is referred to as the Second Indochina Conflict.

Following the withdrawal of the remaining U.S. troops from Vietnam in 1973, resistance in the South crumbled steadily until the fall of Saigon in April 1975 effectively marked the end of separate states in Vietnam. In 1976 the unified Socialist Republic of Vietnam was established, and the new government immediately set about trying to unite the country and socialize the economy of the south. It soon became apparent the task was greater than expected; the Vietnamese economy stagnated. Repression, especially of Vietnam's Chinese population, and other inappropriate domestic political actions provoked a flood of Vietnamese refugees, known as boat people because they departed Vietnam in small fishing boats

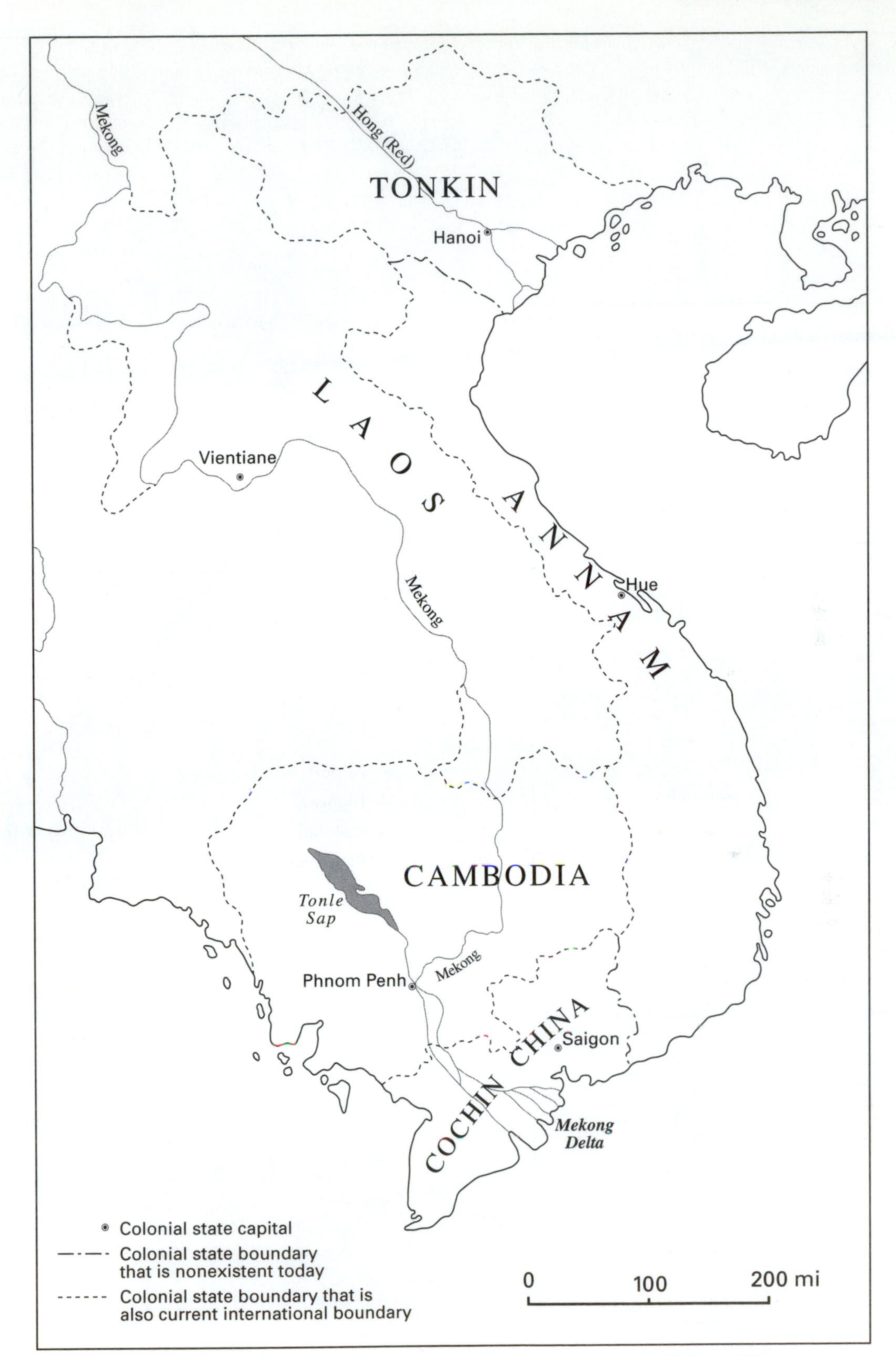

Map 18.3 The colonial states of Indochina. *Source: Fisher, 1964, p.150.*

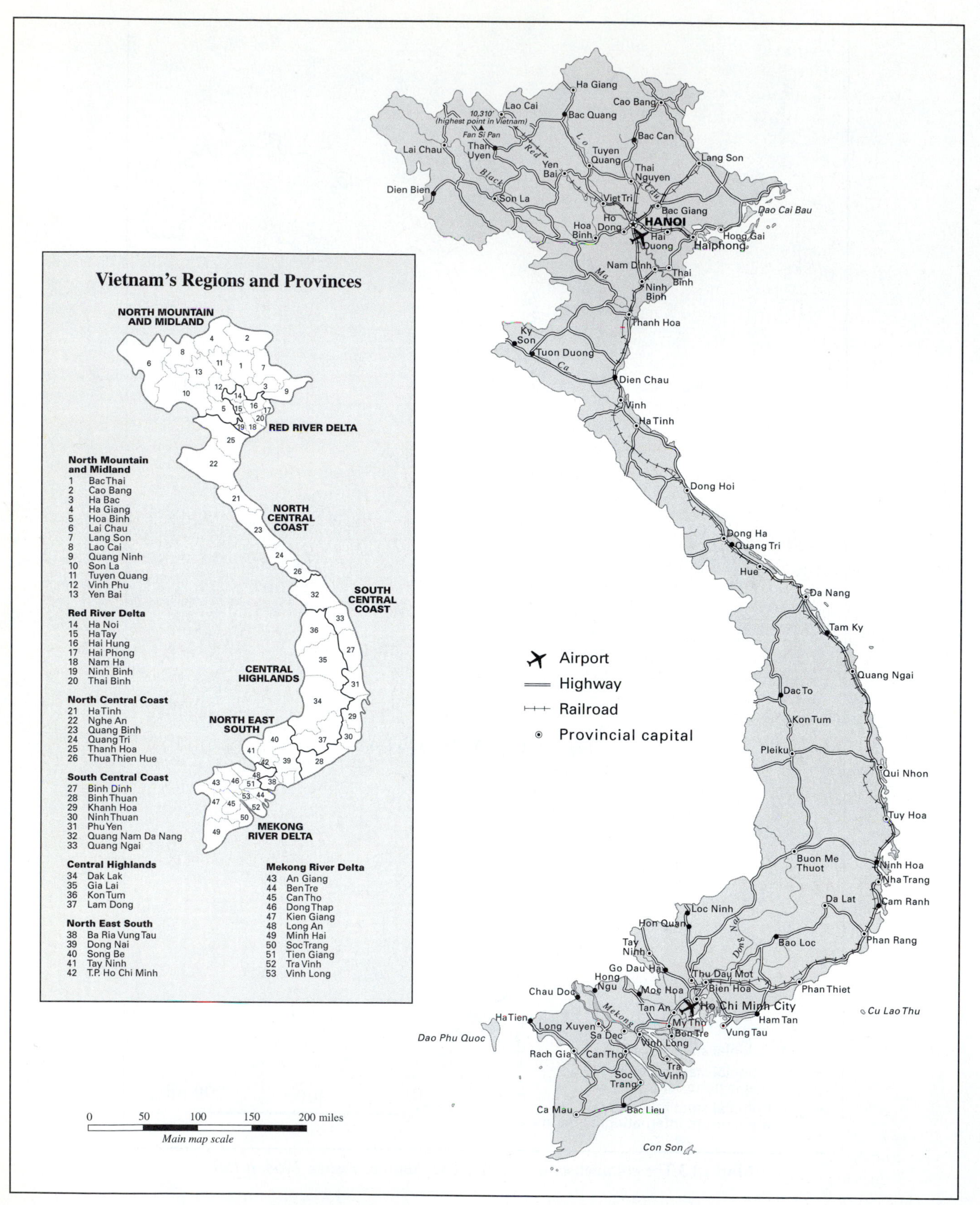

Map 18.4 Vietnam. *Source: Ulack and Pauer, 1989.*

(Thrift, 1987; Thrift and Forbes, 1986, p. 133–7). The Communist Party of Vietnam has remained in power since reunification. Fundamental political reform has not been part of its agenda, although a new Constitution was released in 1992 and the party has eagerly pushed a platform of economic reform or *doi moi,* since the mid-1980s.

In Cambodia Prince Sihanouk was the nominal leader following World War II. Cambodian society was hierarchically structured and Sihanouk attempted to reinforce traditional roles for the people, which had minimized the importance of formal education. Young people who were educated frequently traveled to France to obtain higher degrees, and this experience changed the way in which their political ideals developed. Although political parties began to emerge before independence was declared in 1950 by the People's Liberation Central Committee, the French remained in the country supporting the rule of Prince Sihanouk. The character of the prince has been a crucial part of the story of Cambodia's fight for independence and peace. Thought infallible by many of his subjects, others both inside and outside the country considered him irrational and unreliable. The mid-1950s saw the establishment of the Khmer type of socialism whereby some state ownership developed but much of the agriculture and commerce remained in private hands. Prince Sihanouk urged Cambodian socialists to avoid foreign models of economic development and to base their strategies on more traditional methods, which he believed had achieved the past glories of the country.

The role of Vietnam and of the Second Indochina Conflict was critical in the political development of Cambodia. A steady build-up of guerrilla opposition to the U.S.-supported government climaxed in 1975 when the Khmer Rouge (the military forces of the Communist Party of Kampuchea) led by Pol Pot, seized the capital of Phnom Penh and overthrew the U.S.-backed government of Lan Vol. The Khmer Rouge emptied the cities and turned life upside down for the Cambodian population. Democratic Kampuchea came into existence and the date was proclaimed Year Zero! Between one and two million Cambodians perished as the Khmer Rouge murdered its "class enemies," notably the educated and professionals, during the next four years. Others fled to the countryside; still others died as a result of inadequate food and medicine, hard labor, and the appalling way in which they were treated. The atrocities committed by the Khmer Rouge led to Democratic Kampuchea becoming popularly known as the "killing field" (Kiernan, 1985; Chandler, 1991).

Vietnam's invasion of Cambodia in late 1978 saw the retreat of the Khmer Rouge to highland areas in the west of the country. The Vietnamese supported the formation of the Hun Sen government, which ruled Cambodia (then called the People's Republic of Kampuchea) for the next decade, while simultaneously fighting a running battle with Khmer Rouge guerrillas and forces loyal to Prince Sihanouk. The continuing conflict between government and Khmer Rouge forces ranged from open combat between the two sides in some areas to guerrilla activities in others. One tactic used by the Khmer Rouge was to ambush trains, capture supplies, and take hostages. Some of the hostages have been from Western countries, but ethnic cleansing of Vietnamese and non-Khmer Cambodians also occurred.

International pressure finally resulted in the withdrawal of Vietnamese forces from Cambodia in September 1989, and the country was renamed the State of Cambodia. The signing of the Peace Accord in Paris in 1991 paved the way for the United Nations to attempt to disarm and demobilize the various fighting forces, repatriate 365,000 refugees from camps in Thailand, create a neutral political environment, and hold elections to establish a democratic government.

UN-sponsored elections, in which nearly 90 percent of enrolled voters participated, were held in Cambodia in May 1993. However, the outcome was not as many had hoped, although perhaps it was not surprising. Funcinpec, the French acronym for the National United Front for an Independent, Neutral, and Cooperative Cambodia, secured the largest number of seats; however, it failed to achieve a two-thirds majority in the Constituent Assembly. With the Khmer Rouge not participating and no party achieving a decisive electoral win, it gradually became apparent that a peaceful sharing of power was unlikely. Nevertheless, a coalition government was formed consisting primarily of two main political parties. Funcinpec leader Prince Norodom Ranariddh, son of Prince Norodom Sihanouk, was designated First Prime Minister. The Cambodian People's Party (CPP) representative, Hun Sen, was designated Second Prime Minister. The Khmer Rouge, still a major political and military force in the country, refused to participate in the election and therefore took no part in the political settlement. They retreated once again to provincial strongholds and resumed guerrilla war against the elected government of Cambodia (Frost, 1994). Although Pol Pot died in 1998, factionalism still prevailed in the country.

Laos became an independent country in 1953 when the Kingdom of Laos was created and a government installed in Vientiane. However, the Pathet Lao (the Lao Patriotic Front), which controlled a number of northern provinces, was not included in the administration. For the next two decades one unstable government after another was installed in Vientiane. Conservative politicians used links with the United States to bolster their stance against the Pathet Lao, which formed the Lao People's Revolutionary Party (LPRP) in 1955. Pathet Lao forces, which had sanctuaries in the north and enjoyed the support of the Vietnamese, were pitted against CIA-funded Hmong tribes.

distanced themselves from Laos, instead opting to support the right-wing guerrilla groups opposed to the LPRP (Gunn, 1990; Zasloff and Unger, 1991).

Although political reform is moving slowly in Laos, new leadership has edged the country toward economic reform. The death of key revolutionaries such as former President Kaysone Phomvihane in 1992 and Prince Souphanouvong in early 1995 highlighted the passing of leaders with strong links to Vietnam. In their place have appeared technocrats impressed by the rapid economic progress made in Thailand and China. Nevertheless, conservative, former revolutionaries retain power in the LPRP. They oppose reform, pointing especially to the social problems: prostitution, gambling, and drugs, for instance, that accompany increased openness.

POPULATION GEOGRAPHY AND URBANIZATION TRENDS

Population Size and Growth

Population growth has been uneven in Vietnam, Cambodia, and Laos, and this is related to the particular political and social events within each country. Since independence Vietnam and Laos have held systematic censuses, but Cambodia has conducted only one population census (in 1962). With these data and other more recent surveys, it is possible to give some idea of numbers, spatial patterns, and ethnicity.

A national census was held in Vietnam in 1989, when the population was enumerated at 64.4 million, making it the thirteenth-largest country in the world (Vu and Taillard, 1994, p. 41). By 1997 Vietnam's population was estimated to have topped 75 million. In 1995 there were 4.6 million Laotians enumerated in that country's census. There is no precise population count for the Cambodians, though a census was planned for 1998. Their previous official census of population was in 1962, but estimates put the approximate number at 11.2 million in 1997 (see Table 4.6).

Age structures. Population age structures of the three nations are similar with large proportions under age 15, despite the high infant mortality (see Figure 4.10). In Cambodia and Laos the estimated proportion in this category is 46 and 45 percent, respectively; in Vietnam it is slightly lower at 40 percent. Very low proportions of the populations are over 65 years—only 3 percent in Laos and Cambodia, and 5 percent in Vietnam. In short, the overall populations are very young, and the number of individuals in need of education and health services and approaching working age is very high.

Growth rates. Vietnam has had a target of 1.7 percent annual population growth for a number of years, and in 1997 its annual rate of natural population increase was 1.6 percent, below the goal (see Table 4.6). In Laos, estimates of current rates of natural increase range from 2.6 to 2.9 percent per annum. As a result of the lack of accurate data it is very difficult to calculate precisely the real growth rates of the Cambodian population. This is so because prolonged friction and war has had immeasurable effects. War not only caused death but reduced the birthrate due to separation of the sexes. During the period of Khmer Rouge rule there were bans on relationships between males and females, separation of the sexes, and restrictions on marriage. All effort was focused on the production of food and the development of the commune. Poverty and poor health facilities have resulted in high infant mortality rates. Nevertheless it is clear from the United Nations Transitional Authority in Cambodia's (UNTAC) data that there have been continuing highbirth

	Vietnam	Cambodia	Laos
Population Indicators:			
Population, mid-1992 (millions)	69.3	9.1	4.4
Annual population growth rate, 1960–1992 (percent)	2.2	1.5	2.2
Life expectancy at birth, 1992 (years)	67	51	51
Education:			
Adult illiteracy, 1990 (percent)	12	65	nd
Female illiteracy, 1990 (percent)	16	78	nd
Urbanization:			
Urban population, 1992 (proportion of total population)	20	12	19
Urban population growth rate, 1960–92 (percent per annum)	3.6	1.8	5.1
Area (thousands of square kilometers)	332	181	237

Table 18.1 The populations of Vietnam, Cambodia, and Laos. *Sources: World Bank, 1994 Tables 1 and 1a; United Nations Development Programme, 1993, p. 179, 181.*

rates (Economist Intelligence Unit, 1993, p. 96). Overall it is estimated that population growth rates averaged nearly 3 percent per annum for the period 1990–1995 (World Bank, 1994).

Fertility and mortality. Fertility levels throughout the three countries have been affected by the traumatic events of the last few decades. Separation of families, wartime restrictions, and the effects of famine have all had significant effects on the number of children born. In addition infant mortality has been high. In Vietnam the crude birthrate fell from 47 per 1,000 in 1960 to 31 per 1,000 in 1989 (Economist Intelligence Unit, 1994, p. 17). The pattern in Cambodia reflects the instability of society with a high rate of 50 per 1,000 in the early 1980s dropping to about 26–28 per 1,000 between 1985 and 1992, but increasing to 48 in 1993. Earlier data for Laos are not available but it is estimated that between 1990 and 1995 it was about 45 per 1,000 (UN Department for Economic and Social Information and Policy Analysis, 1994) and the rate is decreasing.

The total fertility rate (TFR) is a measure of the average number of children a woman would have if the present age specific fertility rate occurred during her whole lifetime. The 1997 figure for Laos was an unusually high 6.1 children per women. This TFR is the highest in Southeast Asia and Cambodia has the second-highest rate at 5.8 children per woman. Among the three countries Vietnam's TFR was lowest at 3.1, comparable to the regional average of 3.2 (see Table 4.6).

Infant mortality is also a significant factor that impinges on the fertility level. The tendency of people in the developing world to have larger families is well documented, and one of the accepted causes is high infant and early childhood mortality. Knowing the risk of losing children, parents plan on having more than they would if they were sure of their survival. Poor sanitation and nutrition levels in rural Laos have kept infant mortality at a substantial level of 102 deaths per 1,000 live births, whereas in Cambodia it is significantly higher at 111 per 1,000. As a comparison, the figures are 66, 32, and 4 in Indonesia, Thailand, and Singapore respectively (see Table 4.6). There has been a notable improvement in Laos and Cambodia since 1960 when the equivalent figure was 155 per 1,000 live births. Infant mortality in Vietnam has been reduced dramatically from 157 per 1,000 in 1960 to 38 in 1997, perhaps reflecting the better access to medical services that had been established and also the availability of potable water. There are doubts, however, about whether this improvement can be sustained.

One further measure that is useful when considering the overall pattern of population growth and family formation is the mortality rate for those under five years of age. An average for all developing countries is 100 per 1,000 live births, and when measured against this, Vietnam is in a relatively good position with about 50. However Laos is well above average with 145, and in Cambodia the number is even higher at 185. Factors affecting this are not only access to adequate food and water but also the poverty level, which means that children are forced to work from a very young age and that medical services are inadequate. In Cambodia, for example, there were 25,000 people per doctor in 1990, which means that basic health care was minimal.

Changes in the populations in all three countries are not only related to high birthrates but also to increased life expectancy. Even though there has been a major outflow of refugees during the past two decades and the deaths of young people due to many years of war, life expectancy has increased considerably since 1960. In Vietnam, for example, life expectancy rose from 43 to 67 years between 1960 and 1997. However, life expectancy remains fairly low in both Cambodia and Laos at about 50, though both have improved from about 40 years three decades earlier (United Nations Development Programme, 1994, p. 137).

The reasons for the differences between countries are the variations in social development and the higher poverty levels that exist throughout Cambodia and Laos. Khmer Rouge disruption in Cambodia led to many deaths when forced collectivization and the egress from the urban areas resulted in the closure of medical facilities and the dispersal of the educated elite. The Khmer Rouge were also opposed to Western medicine and instead insisted on reliance on traditional methods. Due to the lack of basic medical knowledge, many people did not survive the forced move to rural areas where endemic diseases such as malaria and dysentery are a problem. Their position was made worse by the fact that those who had come from the cities were not always welcomed or supported by the rural dwellers with whom they lived.

Tropical and other diseases are also a serious problem in Laos and appear to be increasing as a result of the pressure on available freshwater supplies. In Vietnam this unfortunate situation is exacerbated by funding problems. Health care has been made a responsibility of the provincial governments, and the lack of available finance has led to cutbacks in the quality of health care. As a consequence the poorer areas in particular are showing a deterioration in health facilities and hence an increase in infectious diseases including malaria, dengue fever, typhoid, and cholera (Economist Intelligence Unit, 1994, p. 18).

The threat of HIV and AIDS in the three countries has not yet been fully researched. Nevertheless with increasing urbanization and the opening up of the countries to tourism through such routes as the Friendship Bridge, which spans the Mekong River linking Laos and Thailand, the escalation of HIV/AIDS is an issue of concern (see Photo 18.7). The growth of the disease in neighboring Thailand, for example, has been alarming, with the United

Nations Development Program (UNDP) estimating that in 1994 there were at least 500,000 cases. The global spread of HIV is changing, and it is predicted that most new cases will be in Asia. Laos and Cambodia are especially likely to be affected by the spread of the disease along the main transport routes connecting them to Thailand.

The World Health Organization (WHO) claimed that at the end of 1994, 3.5 percent of blood donors in Phnom Penh were already infected; HIV had been introduced into the general Cambodian population. The link with the sex trade, which has been developing throughout the area, was also stressed by the WHO. Although fewer than 400 cases of HIV had been positively identified, WHO claimed that a conservative estimate of between 2,000 and 4,000 was probably more accurate and that because it was first identified in Cambodia, the disease had been increasing at a rate twice that of Thailand.

In Vietnam the disease was first discovered in 1990, and HIV carriers had been identified in 34 provinces by 1997. It is estimated that although about 8,000 cases had been identified in mid-1997, by the year 2000 the number could increase to about 300,000. Again, the role of blood donors is significant with at least 100 people becoming infected as a result of donated blood. The situation in Laos is similar. Relatively small numbers who carry the disease have so far been located, but the possibility of increase is of considerable concern. All three countries have HIV/AIDS control schemes, but they are limited by the inadequate health-care services and the difficulty faced in changing social activities, especially prostitution. The influence of the growing number of visitors, including overseas tourists, the UN peacekeeping force (in Cambodia), and the various relief and development personnel have compounded the overall situation (Chin, 1995).

Population distribution. The pattern of population distribution in Vietnam has been likened to two rice baskets suspended from a pole carried by a peasant. The largest geographic concentrations of people are in the Red River delta in the north and the Mekong River delta in the south. The Red River delta contained 13.6 million people in 1989, or 21.4 percent of Vietnam's total. The Mekong delta had a population of 14.3 million, 22.6 percent of Vietnam's total. The remainder of Vietnam's population is distributed in the highlands that ring the Red River delta and along the thin north–south spine of the country, analogous to the pole balanced on the shoulders of the peasant.

Population densities in the Red River delta in the north of Vietnam are higher than in the south and among the highest in the entire region (Map 18.7). Hai Hung province in the Red River delta had the highest rural densities in the north, estimated to be more than 4,500 per sq mi (1,747 per sq km) in 1994. Provinces in the Mekong delta have lower densities, the highest being for Tien Giang at 1,834 per sq mi (708 per sq km). The population densities in the two deltas are comparable with the most densely settled river basins on Java, which range from 1,000 (400 per sq km) to more than 2,500 persons per sq mi (1,000 per sq km) (Vietnam, *Statistics Yearbook of 1994*, internet).

In Cambodia and Laos, the provinces with the largest populations and the highest population densities are concentrated along the Mekong River (see Map 4.3). Cambodia's largest provinces include Kompong Cham, which had a population in 1994 estimated at 1.5 million, Prey Veng, Phnom Penh, and Takeo. Around the Tonle Sap lake is another cluster of well-populated provinces including Siem Reap, Kompong Thom, and Battambang.

The Lao population is concentrated in the three lowland provinces along the Mekong River. According to preliminary results from the 1995 Census, Vientiane, which includes the city and its immediate hinterland, had a population of 528,109 and the highest density of population (349 person per sq mi or 135 per sq km) in the country. The next-largest provinces are Savannakhet, which is located in the midsouth of the country; along the Mekong River; and Champasak, which straddles the Mekong further to the south. These three provinces account for almost one-third of the total population of the country (Laos National Statistical Centre, 1995, p. 18).

Internal migration and urbanization. Low levels of urbanization occur in Vietnam, Laos, and Cambodia by comparison with neighboring Southeast Asian countries. This is because socialist governments have generally exercised effective control over rural to urban and other forms of migration. In Vietnam three particular forms of direct government controls have been especially important in slowing rural to urban migration. The first has been government control of a significant proportion of jobs in the absence of a sizable private sector. The second point is that socialist governments have generally controlled a large segment of urban housing. Third, socialist governments were able to provide access to subsidized food for those with official permission to reside in particular areas (Forbes, 1996).

Together, these factors have made it very difficult for rural residents to migrate spontaneously to Vietnamese cities, as they have in most other Southeast Asian countries. Although rural migrants are generally able to flout controls on migration, such as the need for a resident's card to live in cities, being unable to secure jobs or housing and needing to purchase expensive rice on the private market are serious disincentives to migration. However, as the impact of economic reforms has spread, the ability of governments to control migration has eroded. This has made it easier for people to migrate, legally or illegally, to the larger urban areas. Thus urbanization rates

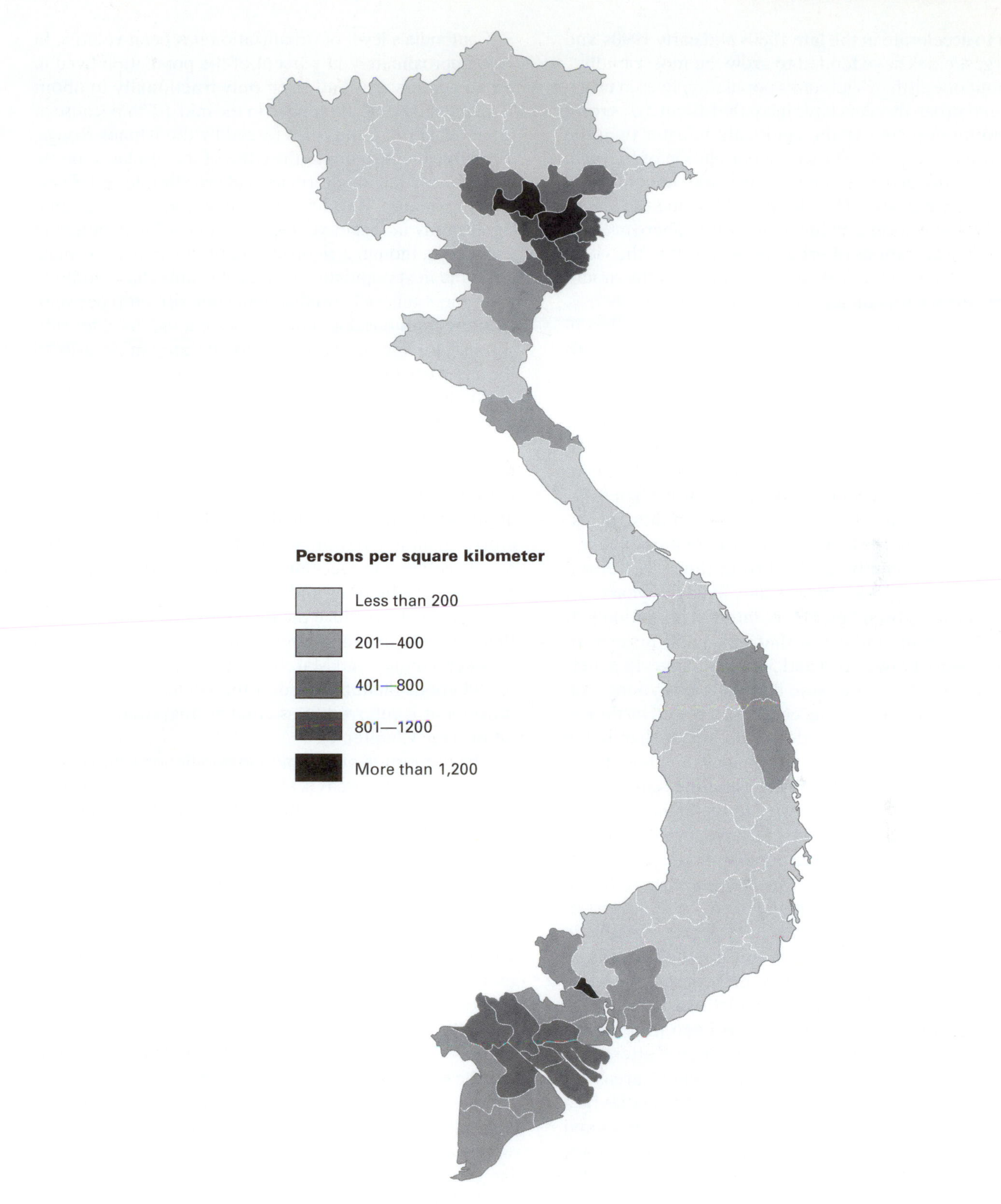

Map 18.7 Population densities in Vietnam, 1994. Vietnam, *Statistics Yearbook of 1994.*

began to accelerate in the late 1980s and early 1990s, and the largest cities have tended to grow the most rapidly.

About one-fifth of Vietnam's population lives in urban areas, and urban dwellers have increased about 3.6 percent per annum in recent years, significantly faster than the population as a whole. The urban population of Vietnam is unevenly divided between the north and the south with the largest urban area, Ho Chi Minh City, and the higher level of urbanization existing in the south. There are also greater concentrations of urban populations in the vicinity of the ports and industrial areas, whereas the mountainous regions remain least urbanized (Forbes, 1995).

The income gap between rural and urban dwellers in Vietnam is of some concern, not least because it fuels rapid urbanization. Foreign investors favor urban locations, especially the largest cities, exacerbating the problem. The government plans to improve the economic environment in remote regions, hoping to build new towns populated by shifting cultivators, and to upgrade rural infrastructure, but it recognizes the need for foreign assistance to help achieve the most ambitious of these plans.

An insight into the broader pattern of internal migration in Vietnam can be derived from the 1989 Census. Respondents were asked where they lived five years earlier, producing a broad picture of internal migration patterns. The results indicated that nearly 60 percent of migrants were between 15 and 34 years of age. In general the migration streams were dominated by young couples with no children or one child. As a result, birthrates in the destination areas tended to be higher than in the origin areas. More migrants were males than females.

Provinces in the Red River delta and the central coast (such as Quang Nam-Da Nang) had the highest rates of out-migration. Both these regions have high population densities and problems due to population pressure on resources. The highest rates of in-migration were to the provinces in the central highlands and the southeast of the country, where agricultural land was still available in New Economic Zones (NEZs). In sum, the majority of moves were from the north and central coast to the south and southern central highlands of Vietnam.

A program developing NEZs has been in place for several years (Desbarats 1987). In the period after unification, the NEZs were used as resettlement areas for temporary urban residents ushered out of the cities by a government hoping to reduce the size of urban areas and improve their productivity. Vietnamese authorities have persisted with the resettlement program because it distributes population along hitherto sparsely settled border regions and is intended to improve agricultural productivity in key highland regions. However, the program has achieved limited successes. Settlers arbitrarily chosen to move to NEZs have often returned home when government subsidies ended.

Cambodia's level of urbanization has been volatile. In 1960 approximately 11 percent of the population lived in urban centers; this had risen only fractionally to about 12 percent by 1992. It plunged in the mid-1970s because of the evacuation of the cities forced by the Khmer Rouge. The subsequent return to the cities of the displaced population fueled the long-term annual growth rate of 1.8 percent per annum. Rural-to-urban migration has escalated significantly in recent years and is a growing problem in Cambodia, though a rigorous quantitative analysis must await the next population census in Cambodia. Cambodian farmers generally produce only one rice crop per year. In many cases this does not produce enough food for subsistence, forcing rural dwellers to migrate temporarily to the city, particularly during the dry season.

An illustration from the Srok Phnom Penh area adjacent to the capital illustrates the point (Paul, 1995, p. 7–10). The New Year is celebrated in April, with relatives returning to the village. The padi fields are prepared in May and seedlings emerge in June and July. The fields are plowed and the seedlings transplanted in August; the crop is maintained from September until November, before harvesting in December. Drought can seriously affect rice production, and many families do not have enough land to produce the food they need. Farmers and their families are then able to migrate to Phnom Penh between January and March to take up jobs in the city or to sell goods manufactured in the village. This seasonal movement is referred to as circular migration, or circulation. (see Chapter 4).

The proportion of the Laotian population estimated to be living in urban areas is about 19 percent, a little higher than Cambodia but lower than in Vietnam. The urban population had a rapid average annual growth rate, 5.1 percent between 1960 and 1992.

Major cities. Ho Chi Minh City, formerly known as Saigon (and still called Saigon colloquially), is the largest city in the three countries and the fourth-most-populous metropolitan area in the region. In 1989 it had a population of 3.2 million (Figure 18.1). In contrast, Hanoi is the capital of Vietnam and contained 1.1 million urban residents, significantly fewer than Ho Chi Minh City, but is still the second-largest city in Vietnam, Cambodia, or Laos (Forbes and Ke, 1996). The major secondary cities of Vietnam include the northern port of Haiphong (456,000), which serves as Hanoi's "outport," and Danang, the largest city in central Vietnam (371,000).

Ho Chi Minh City is the economic center of Vietnam. Its entrepreneurial dynamism is evident from the new buildings growing up all around the city, and its commercial role is demonstrated by the billboards that dot the main thoroughfares advertising western consumer goods. Fears in early 1996 of creeping "cultural pollution"

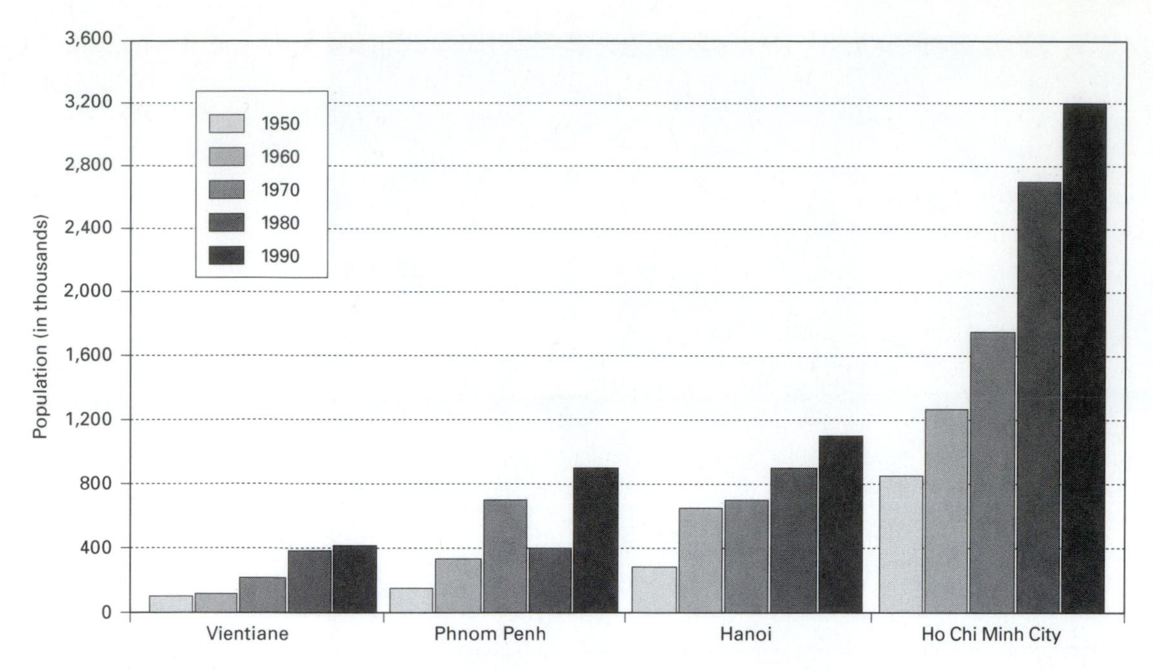

Figure 18.1 Population of major cities, 1950–1990. *Source: Forbes, 1996; Thrift and Forbes, 1986, p. 147.*

Photo 18.1 An aerial view of contemporary Ho Chi Minh City with recently constructed office and residential structures. Taken from the Saigon Trade Center, the Saigon River is in background. (Leinbach)

resulted in the government insisting that English language signs be painted over, leaving only Vietnamese language messages. The economic dynamism of Ho Chi Minh City is proving attractive to rural migrants who are anxious to improve their cash incomes. Poor peasants travel to the city during the off-season and take up informal sector jobs or beg for cash in the city streets before returning to the rural areas to work on the planting or harvesting of crops.

Although it lacks the entrepreneurial dynamism of Ho Chi Minh City, Hanoi is increasingly competitive as a destination for foreign investment. As the capital city it contains many of the symbols of Vietnam's nationhood, such as the Ho Chi Minh Mausoleum and the more recently constructed Ho Chi Minh Museum (Logan, 1994). The city is also the home of the National Assembly and, more important, the headquarters of the Vietnam Communist Party (VCP), where much policy is formulated

Photo 18.2 A typical street scene in "old" Hanoi where similar competing enterprises are concentrated together. This shop depicts "tin smithing" or "metal working" goods for sale. (Leinbach)

and power still resides, despite economic reform. Dispersed around the city are the main offices of the key government departments, such as the Ministry of Planning and Investment, which are the essential contacts for foreign businesses interested in operating in Vietnam.

Since the mid-1980s the rate of destruction of existing buildings has aroused strong concerns about threats to Hanoi's architectural heritage. Having been isolated from the depredations of the global economy for so long, the city possesses some unique architectural features. Concentrated in the old CBD, to the northwest of Lake Hoan Kiem, is an area known as the 36 streets, which is dominated by small Vietnamese shop-houses (Photo 18.2). Ringed around the other side of Hoan Kiem is an extensive area of villas, offices, and cultural buildings, such as the opera house, built in French provincial style and dating back to the colonial era (Photo 18.12). Though many buildings remain in sad disrepair, concerns that these areas could quickly lose many of their unique qualities have stimulated attempts to preserve parts of the inner city. Although some buildings have been painstakingly restored, the rapid growth of Vietnam's economy has triggered the need for extra commercial and office space. Consequently, the appearance of the inner city will almost certainly change beyond recognition.

Phnom Penh is the capital and primate city of Cambodia. Despite its recent turbulent history, including the evacuation that followed the Khmer Rouge occupation of the city in April 1975, Phnom Penh had an estimated population of 900,000 in 1990 (see Figure 18.1). Located at the junction of the Mekong, Tonle, and Bassac rivers,

Photo 18.3 Colonial French architectural style in Vientiane. (Forbes)

the topography of Phnom Penh is low and flat so that when the Mekong River rises between July and October it threatens to flood the city. The infrastructure remains poor. Water and electricity supplies are still erratic, though being improved.

Phnom Penh once rivaled Hanoi as the most charming of the formerly French colonial administrative centers in the region. The monuments within the city, the pagodas, and the royal palace combine with tree-lined avenues to enhance the urban ambience. Phnom Penh is a low-density, "green" city, with villas surrounded by trees dominating the landscape. However, the city shows more and more signs of the influx of rural populations. Many of the villas are overcrowded with families crammed inside, while outside spaces are filled by makeshift dwellings. Squatter settlements are thought to provide homes for up to 15 percent of the city's population (Paul, 1995, p. 12). An influx of foreigners in search of business opportunities is accelerating the construction of new buildings and the renovation of the old, often in an incoherent way. Increasing urban densities and the loss of trees threaten to jeopardize the city's architectural heritage, in a disturbing parallel with the transformation of Hanoi (Blancot, 1994).

It is estimated that Phnom Penh contains from 5,000 to 10,000 street children. Many are orphans or the product of dysfunctional families that are traumatized by the harsh experiences of Cambodian life since the 1970s. The children have migrated to Phnom Penh to escape physical violence or poverty. Once in the city they scrape a living through the informal sector, begging, recycling garbage, working on construction sites, or selling sex. Up to 15 percent of prostitutes in 1994 were believed to be under 16 years of age (Paul, 1995, p. 18).

Battambang (about 100,000) is Cambodia's second-largest city, while other major towns include the ocean port of Kompong Som, formerly Sihanoukville (16,000), and Siem Reap (10,000), which is the town nearest the famous temples of Angkor Wat and the walled city of Angkor Thom, constructed in the late twelfth Century.

Vientiane ("city of the moon") is the capital and largest city in Laos. The municipality had a population of 528,105 in 1995, although the city itself is less than a third that size, making it small compared to most other Southeast Asian capitals. Features remain that stem from its role as a center of colonial administration, including a handful of French colonial buildings clustered in the older parts of the city adjacent to the Mekong. Recent Soviet influence is manifest in some of the drab "socialist-realist" style of architecture. During the 1950s more than half of Vientiane's population was Vietnamese, brought in by the French to run the administration. They have subsequently returned home. Ethnic Chinese, who constitute up to about 5 percent of the population of Laos, are concentrated in Vientiane and Savannakhet, and there are a small number of Pakistanis and north Indians involved in retail trade. The next largest cities in Laos are Pakxe, which was established as an administrative post by the French early this century, Savannakhet, and the old royal capital of Louangphrabang.

Refugee emigration. Vietnam, Cambodia, and Laos have all endured conflict within their borders. One of the results of wars and political upheaval has been large numbers of refugees fleeing from their home countries. For some it has been a forced migration within the country and then escape overseas; for others, direct action to remove themselves and their families.

At the close of the Second Indochina Conflict, many South Vietnamese supporters were admitted into the United States; others remained in Vietnam unable to escape. However, the late 1970s saw an exodus of boat people who fled the country in poorly equipped craft. Many did not survive the harrowing journey; pirates, sharks, and ill health caused untold numbers of deaths. Those that eventually reached land were given temporary refuge in camps in Hong Kong, Thailand, Malaysia, Indonesia, the Philippines, and in some cases, Australia (Foley, forthcoming). For some these camps became long-term internment centers as their fate was decided by various overseas governments. Many were eventually granted refugee status and settlement elsewhere, for example North America, Europe, and Australia. After 1989 large numbers who failed to gain refugee status were repatriated to Vietnam.

Conflict in Laos has also led to refugee emigration. The United States has been the main destination of those who have escaped overseas (118,000 mostly Hmong; 124,500 lowland Lao), though Australia has also taken some Laotians (1,120 hill tribe Lao; 9,000 lowland Lao) (Map 18.8). Large numbers attempting to avoid the conflict have been unable to travel further than the border with Thailand, and although up to 14,000 have been repatriated as a result of an agreement between the United Nations High Commission for Refugees (UNHCR) and Bangkok, as recently as mid-1995 it is estimated that there were still about 10,000 Lao in the camps, most of whom are from the Hmong ethnic minority.

Within Cambodia the refugee problem is complex. Many thousands of people were removed from the cities in 1975 and became refugees in their own country. Some later escaped overland to Thailand and ultimately overseas to resettle. The actual numbers displaced are still not known, although some became boat people. The situation is further complicated by the exodus of ethnic Vietnamese from Cambodia as a result of increased tension with the Khmer Rouge during the early 1990s. While the Vietnamese were fleeing to Vietnam, ethnic Cambodians were leaving for Thailand with estimates of as many

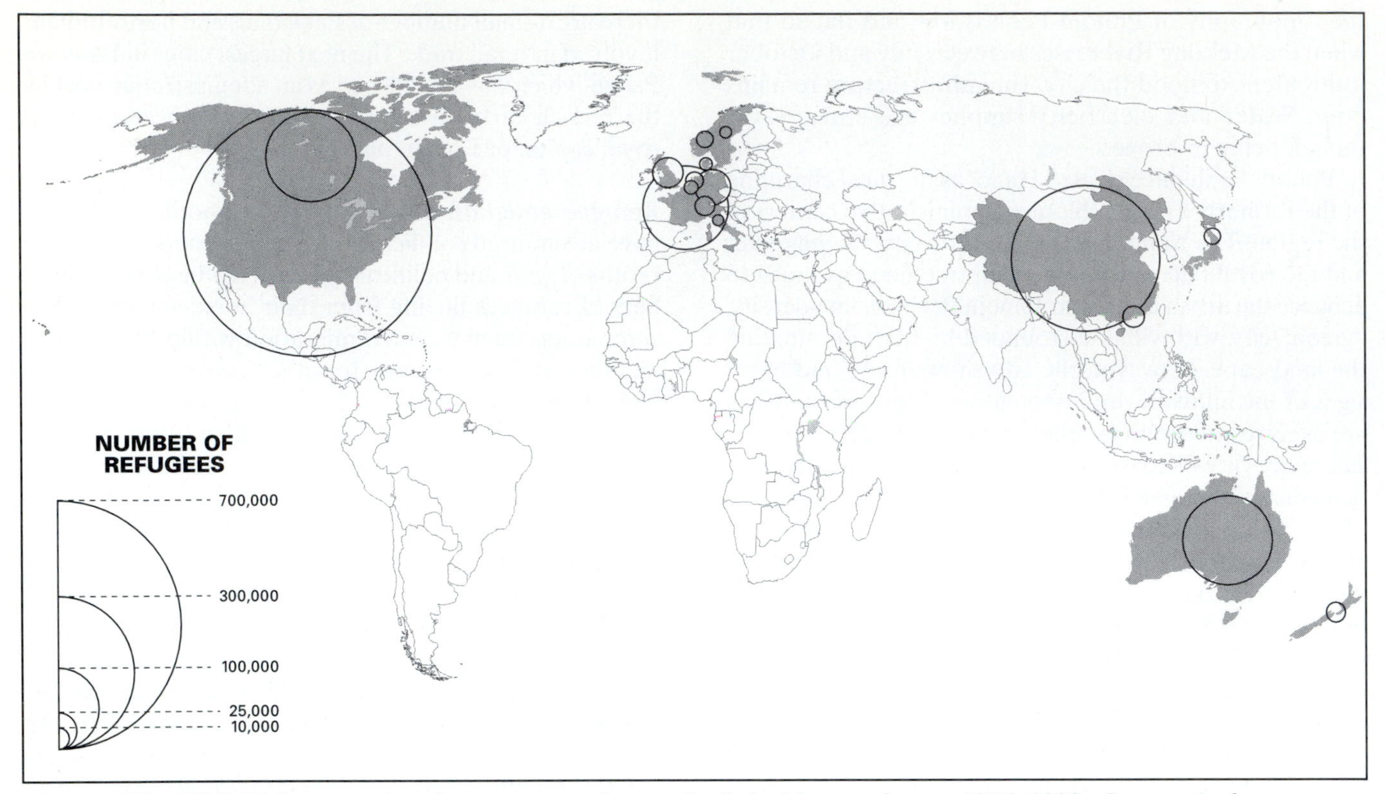

Map 18.8 Major countries of permanent settlement for Indochinese refugees, 1975–1988. *Source: Author.*

as 10,000 in camps on the Thai border in mid-1995. It is hoped that this refugee problem will be temporary, but without strong government action there is little hope for the permanent settlement of the displaced people. In late 1997 there were still many thousands of Cambodians crossing and recrossing to and from Thailand as political tensions and conflict between government forces under Premier Hun Sen and those who support the ousted co-premier Prince Norodom Ranariddh, continued.

One important outcome of the settlement of refugees overseas has been their later return to their home country, either permanently or temporarily. Many have taken advantage of the changing circumstances to invest in their home countries; others have returned to assist in the rebuilding of democratic structures. For example, in March 1995, 10 ministers of the Cambodian government and 30 members of parliament had foreign passports as a result of their time spent as refugees.

Cultural and Ethnic Diversity

Societies in the three countries are changing as a result of political and economic developments, as well as the upheavals within each society. Traditional village belief systems were swept away by political actions, but in many areas the fundamental village structure remains one of the layers of society. As mentioned earlier in this chapter, there has been a movement of people into the cities, but particularly in the case of Cambodia, this move to the city was interrupted by the forced exodus under the Khmer Rouge. At the present time the agricultural systems are changing as populations increase, and cash cropping supersedes subsistence farming and slash-and-burn activities. Despite these changes a strong sense of family and a high regard for ancestors remains. Whereas the whole area is generally perceived as being dominated by the Buddhist religion, Christianity is also important particularly among some Vietnamese.

Each country is ethnically diverse. This is the result of the complex settlement patterns that characterized the area's history and the more recent cross-border incursions and invasions that have inevitably led to further mixing. Nevertheless there are dominant populations within each country. In Vietnam approximately 88 percent of the population are ethnic Vietnamese (or Kinh). In Cambodia almost as high a proportion are Khmer in origin, but Laos is far more heterogeneous. Up to 60 percent of Laotians are ethnic Lao, but there are as many as 65 other groups. The three main subgroups whose settlement pattern is related to their time of arrival in the area are the midlanders known as Lao Theung, the lowland Lao or Lao Loum (the dominant group), and the highland Lao Sung (Ireson and

Ireson, 1991). However there are also large numbers of ethnic Vietnamese in both Cambodia and Laos.

There are still many ethnic or overseas Chinese scattered throughout Vietnam, Cambodia, and Laos, and their role, as elsewhere in Southeast Asia, has been critical in the development of the area. Although overseas Chinese settled in Vietnam as a result of the ebb and flow of links with China, the numbers increased in the nineteenth century during the French colonial period when they were important in the administration of the territory. Communities developed in Cholon, which was later absorbed by Ho Chi Minh City, and in port towns such as Haiphong. Although presently comprising just 1.5 percent of the population, they have played a key role especially in commercial activities, retail, and finance.

In more recent times the settlement of Vietnamese in both Cambodia and Laos occurred when troops were stationed in these countries. Some 50,000–60,000 Vietnamese troops were stationed in Laos, but were withdrawn by 1988. The Vietnamese are a high profile subset in Cambodia, but the number of illegal immigrants from Vietnam who live in Cambodia appears to be increasing and has caused alarm in parts of the country (Amer, 1994).

Religion and culture. Village life in Vietnam has traditionally been complex and diverse. The main religion is Buddhism. Most villages have at least one temple, and there may be both monks and nuns attached to the temple. Clearly, Vietnamese traditional society has changed due to the imposition of communism and the changes related to war and invasion. Buddhism is also tempered by a wide range of worship of cults or spirits. These exist in all manner of living and nonliving entities. The worship extends from that of spirits of dead people to trees and inanimate objects such as stones. Complex rites are associated with particular actions or times of life, including fishing, plowing, harvesting, childbirth, sickness, and death. It is the role of the sorcerer to identify the spirit associated with sickness, and this is then dealt with by rites of gift giving and appeasement to remove the bad spirit that has brought the sickness. Acknowledgement of spirits in all aspects of life is extremely important in villages and takes place alongside the more structured Buddhist worship. It remains important even where Christianity has taken hold (Hickey, 1964; Mabbett, 1989).

One of the most significant aspects of cultural life in Laos is the great respect which the young have for their parents and elders. This has resulted in close-knit Lao families living in relatively large, mixed generation households. Although there has been a move toward a wider range of crops being grown, in the main part glutinous rice is the staple food for most people, with the hill tribes supplementing their diet with forest foods when these are available. Many people use a spoon and a fork to eat, rather than the chopsticks used in other nearby countries such as Vietnam. Within the home people sit on low seats or cushions. Traditional dress for women is varied, but the basic style is a blouse with colorful sarong skirt. The style of dress for men is more Western in appearance.

Peasant society has persisted in Laos despite collectivization. There are fewer landless peasants than formerly (Evans, 1995, p. 67), but many farm very small areas. Cattle trading is important and buffalo are used as draught animals. Women and girls may supplement their family income with weaving silk and cotton. There has been an upsurge in this activity since 1975, and the reemergence of traditional dress among the population is partly a result of the ideological nationalism stimulated by the postrevolution government. In addition, income is derived through the sale of woven materials via Thai and Lao traders, some of whom also supply raw materials. Consequently a traditional skill has been developed for commercial purposes.

Village life and culture in Cambodia has been subject to so much disruption in many parts of the country that it is difficult to generalize about the present village pattern. However there are differences between the structure of society in Cambodia and that in Laos and Vietnam. Although most of the population follow Theravada Buddhism, Islam has a large number of adherents in Cambodia especially among the Chams and Christianity, particularly Roman Catholicism, is significant as a result of the French influence.

Cambodian village life is less structured than in Vietnam or Laos. Unlike the Vietnamese, Cambodians do not live in large villages but in small scattered hamlets, often occupied by single families. The most important acts in life for a man are building a house, buying a buffalo, and taking a wife. Each involves communication with the local Buddhist achar (priest) to set the propitious dates. Women share the work in the fields with the men. As there are many widows who are heads of households, they are often responsible for everything including plowing, sowing, transplanting, and harvesting. The irrigation work may also be their responsibility. Within the household women take care of their homes and tend vegetable gardens. Many lack education. The house, which is likely to be sparsely furnished and built on stilts, faces the rising sun. Fish, which are often traded for rice, vegetables and fruit, are the main elements in the diet. A major activity is raising irrigated rice. The strong role played by religion means that little meat is eaten because Theravada Buddhists are forbidden to kill living creatures. The Khmer dress is similar to that of the other groups described here. The sampot, a piece of cotton or silk wrapped around the waist and between the legs to appear as a pair of trousers, is now generally reserved for ceremonial wear, and men and women are more likely to wear simple trousers and tops.

ECONOMIC STRUCTURE AND SPATIAL ORGANIZATION

Economic Reform

Since the mid-1980s Vietnam, Laos, and Cambodia have veered from the socialist path, adopting economic reforms that have propelled them toward mixed-market economies (Ljunggren, 1993). They are engaging in the global economy long after most of their neighbors have already seized the opportunities open to labor-intensive manufacturers. The world economy is much more competitive than it was a decade ago, and new entrants are finding it hard to compete. Moreover, the countries are handicapped by rigid one-party systems (Vietnam and Laos) or chronic political instability (Cambodia). Added to this are the problems of undeveloped social and economic infrastructure.

Economic reforms were first introduced in Vietnam in 1979 but quickly stalled when the government became concerned by what it perceived as negative effects. A flourishing of coffee shops and other small service-sector enterprises, especially in Ho Chi Minh City, caused leading party members great concern. New reform measures were introduced into agricultural production in the early 1980s. The measures permitted farmers to sell on the open market any surplus production not required for their quota to the state. It was not until 1986 that the government and ruling Communist Party gave their strong support to the idea of economic renovation, or *doi moi*, as the Vietnamese termed the process. The sixth Party Congress in 1986 was the venue for the party's official endorsement of *doi moi*. The strategy was subsequently reaffirmed, though not without qualification, at the seventh Congress in 1991, and the eighth Congress in 1996.

Notable measures in the economic reform process have included the dismantling of the communes and the reallocation of land for the use of family farms, an opening of the country to foreign investment, reform of the banking system, and a gradual withdrawal of government financial support for state-owned enterprises. Official Vietnamese documents detail the end of the state-subsidy system and the emergence of market socialism. The government still claims to be committed to a socialist path but recognizes the value of market mechanisms and a greater level of integration into the global capitalist economy.

Like Vietnam, Laos recognized as early as 1979 that the transition to socialism would require a more protracted coexistence of socialist and capitalist forms of production than had originally been expected. Existing restrictions on private trade were therefore dropped, and a number of other reforms were implemented. Agricultural cooperatives were to be consolidated and no new cooperatives formed. Since 1986 Laos has been adopting reforms designed to shift away from a centrally managed economy and to rely more on market forces (Rigg, 1995). The new economic mechanism, as it is termed, continues to be broadened and refined. Measures parallel those introduced in Vietnam, including a shift in agriculture from cooperatives and state farms to household-based agriculture and a withdrawal of subsidies from state-owned industrial enterprises and banks. A liberal foreign investment law was introduced in 1989 with the result that Laos has successfully lured a small number of foreign investors to the country. The legal infrastructure continues to develop with the creation of new commercial, property, and tax laws and the establishment of a central bank (Pham, 1994).

While political stability has been a hallmark of Vietnam and Laos for over two decades, allowing them to concentrate on economic reform, Cambodia remains blighted by instability. In part due to the political turmoil, Cambodia's progress toward economic reform has been halting (St. John, 1995). Initiatives to reform the economy by the Vietnam-supported government commenced after the postponement of agricultural collectivization in 1986. The following year saw the beginnings of official encouragement of the private sector and the family economy. In 1988 national industries began to be freed from direct control through government plans, and in 1989 rights to inherit land and traditional usufruct land titles were restored.

The establishment of the coalition government after the UN-sponsored elections in 1993 has seen Cambodia inch further along the path of economic reform. By the mid-1990s the majority of state-owned enterprises had been privatized, price setting largely removed from direct state control, and agricultural lands reverted to family ownership. A new law in early 1994 brought taxation under the control of the Ministry of Finance, whereas previously ministries collected and spent their own taxes without centralized control (Asian Development Bank, 1994, p. 94). However the government has been deeply troubled by persistent accusations of corruption and internal discord. A key economic minister, Sam Rainsy, was forced out of office in 1994, expelled from the ruling party Funcinpec, and dismissed from his elected position in the National Assembly during 1995. He had antagonized key interests by drawing international attention to corruption in government.

Structure of the Economies

It is only in the 1990s that collection of economic data using Soviet Union-style categories has been abandoned and systems of conventional UN national accounts compiled. This limits the ability of geographers to compare shifts in these economies over time, as the two systems of measurement highlight different aspects of the economy. The major source of economic data on developing countries, the World Bank, publishes very little information on Laos and even less on Vietnam and Cambodia. An additional complication stems from the political instability in Cambodia, which makes the collection of accurate statistics problematic.

Vietnam's economy is about five times larger than the economies of Cambodia and Laos combined. This is not because the Vietnamese are more productive or their economy more developed but is rather a reflection of their much greater population size. Per capita GNP measures the approximate amount of economic output for each person in the nation. The World Bank (1999) calculations of GNP for the three countries are the lowest in Southeast Asia. Vietnam's 1996 per capita GNP was about US$290, Cambodia's US$300, and Laos US$400. Many would quibble about the accuracy of these figures but not about the picture they present. Vietnam, Cambodia, and Laos are among the least-developed countries in Asia economically.

Economic indicators. Since the mid-1980s the economies have achieved steady and impressive rates of growth. Vietnam's economic growth rate of 6.7 percent per annum is the highest, followed by Cambodia's 5.8 percent, and 5.1 percent in Laos (Table 18.2). Measuring GDP growth in per capita terms eliminates the influence of population growth on the economy. Not surprisingly, the resulting figures are less impressive. In per capita terms the Vietnamese economy grew at 4.5 percent per annum, a significantly better rate than Laos's economy, which increased by 2.7 percent, and Cambodia's 2.2 percent growth.

As the impact of economic reform has spread and improvements have been made to infrastructure and the skill levels of the labor force, the pace of economic growth has accelerated. Vietnam's economy has achieved the most spectacular gains, leading to exuberant claims that it is the next Asian "tiger" economy. Although such optimism may be premature, it was anticipated Vietnam would achieve growth rates of about 9–10 percent per annum in the second half of the 1990s. Reliable forecasts for the Cambodian and Laos economy could be made, given the lack of statistical information. Moreover, the long-term impact of the economic crisis that hit countries throughout the region in 1997, that of greater severity of impact in Thailand, Indonesia, and Malaysia, has slowed down overall economic growth in Vietnam, Laos, and Cambodia.

Rapid inflation of the currency has bedeviled all three governments. Under the socialist style of economic management, currency exchange rates were controlled for many years, but economic reform has resulted in deregulation. Laos adopted multiple exchange rates in an effort to maintain trade during the 1980s but abandoned them in 1987. The result was a surge in inflation. In 1986 US$1.00 would buy 95,000 kip; by 1990 it would purchase 714,000 kip (Asian Development Bank, 1991, p. 302). However in 1993 inflation had dropped to about 6 percent (Handley, 1993).

Agriculture is the dominant sector of the three economies. In 1993 it accounted for 36 percent of the GDP of Vietnam, 60 percent in Laos, and 48 percent in Cambodia. Despite the emphasis that socialist governments have given to industry, it contributed 25 percent of Vietnam's GDP and just 18 percent and 17 percent in Laos and Cambodia, respectively. The service sector of each of these countries is small and underdeveloped. Its share was 39 percent of Vietnam's GDP, 35 percent in Cambodia and 26 percent in Laos (Asian Development Bank, 1994, p. 236). By comparison with their Southeast Asian neighbors such as Indonesia and Thailand, agriculture is more than twice as important to the economies of Indochina, and the services sector about half as important.

Labor force. The sectoral structure of the economies is mirrored in the composition of the labor force. Between two-thirds and three-quarters of the economically active population are employed in agriculture (Table 18.2). Industry provides jobs for only a small fraction of the workforce—12 percent in Vietnam, and 7 percent in Laos and Cambodia. According to Laos's State Planning Commission, workers in industry in 1986 totaled just 27,000 (1.6 percent), while another 39,200 (2.3 percent) were employed in trade.

	Vietnam	Cambodia	Laos
Economic indicators:			
GNP per capita ($US)	220	170	250
Annual growth of GDP, 1988–93 (percent)	6.7	5.8	5.1
Annual growth of GDP per capita, 1988–93 (percent)	4.5	2.2	2.7
Structure of labor force, 1990–92:			
Agriculture (percent)	67	74	76
Industry (percent)	12	7	7
Services (percent)	21	19	17

Table 18.2 Important economic indicators. *Source: Asian Development Bank, 1994, p. 231, 232; United Nations Development Programme, 1994, p. 163.*

Photo 18.4 An expanding route network, purchase of Airbus aircraft, and the badly needed upgrading of Noi Bai Airport in Hanoi have come about as a result of economic reform and a more "outward-looking economy." (Leinbach)

The services sector is correspondingly small, as noted above. However, it is growing quickly, as a walk along streets in the center of the large towns will demonstrate. The informal-services sector, in particular, has become a feature of city streets in all three countries. Human-powered trishaws provide local passenger services. Food sellers congregate on street corners, selling staple indigenous foods such as bowls of soup (called *pho* in Vietnam) or exotic concoctions such as bread rolls spread with ham, paté, and local delicacies, revealing the culinary influence of the French (Photo 18.5). In addition sellers of books, motorbike parts, clothes, and cosmetics crowd the pavements, especially in the inner city areas.

In comparison with most developing countries, females are well represented in the region's labor forces. Women constitute 45 percent of the Lao labor force, 47 percent in Vietnam, and 56 percent in Cambodia (United Nations Development Programme, 1994, p. 147; Ngaosyvathn, 1993). The female majority in Cambodia's labor force is explained by the deaths of large numbers of males of working age during the Second Indochina Conflict and as a result of the brutal Khmer Rouge regime.

Unemployment has been a major problem for all three countries. Unemployment in Laos was officially estimated at 17 percent in 1990 (Economist Intelligence Unit, 1993, p. 75); in Vietnam the figure hovers around 20 percent. An accurate estimate is almost impossible. Many people may give the appearance of being unemployed but earn a small, irregular income from various informal-sector activities. Technically they may or may not be unemployed, depending on the statistical criteria adopted. Most will certainly be underemployed, meaning that they would like more work and a better income.

Photo 18.5 Food vendor in Hanoi's informal service economy. The informal sector is especially conspicuous in Hanoi. (Leinbach)

households were poor. This contrasted dramatically with farmers, of whom 60 percent were classified as living in poverty (World Bank, 1995, p. i–iii).

By comparison, information on incomes and poverty in Laos is harder to come by. Some insight can be gleaned from a survey of the urban workforce in the four main urban areas—Vientiane, Savannakhet, Pakse, and Louangphrabang—in mid-1992. It found that 46 percent of workers were self-employed in farming or retail trade; 32 percent earned wages in manufacturing, farming, and government jobs; and 21 percent worked in family businesses without earning wages. Average annual earnings of urban employees were about 360,000 kip (US$500), with men earning slightly more than women. Public-sector wages continue to drop further behind wages in the private sector (Dommen, 1994, p. 172). Income distribution across the whole country is more difficult to characterize accurately. In general, as in Vietnam the urban populations are better off than those in the rural areas, but the rural areas are not an undifferentiated lump—people living along the relatively rich plains adjacent to the Mekong River and in the south of the country are considerably wealthier than the farmers in the mountainous regions, especially in the north (Handley, 1993).

The Space Economy and Transport

The space economy, or geographic distribution of economic activities, of each country is biased toward the large urban areas that form the core political and economic centers. Thus Phnom Penh is situated in the middle of Cambodia's principal industrial region, and Vientiane provides a focus for the Lao economy.

Being significantly larger in population and different in shape, with more major cities, Vietnam's space economy is of greater complexity. Vietnam has a tripolar regional structure which is not dissimilar to the colonial territories into which the country was once divided. The northern Red River delta and its surrounding fringe of highlands constitutes one major region of contemporary Vietnam (Vu and Taillard, 1994, p. 398–406). The central coastal regions and adjacent plateaus form a second major region. The southern region includes the toe of the nation and consists of the Mekong delta and its surrounding plains. Each of these regions centers on an emerging urban corridor. The Hanoi–Haiphong corridor links the capital city with a major urban center and port in the north. The corridor connecting the third-largest city and port of Danang with the nineteenth century royal capital of Hue provides the urban locus of the central region. Vietnam's largest urban area, Ho Chi Minh City and the corridor that connects it to the industrial center at Bien Hoa and the port city of Vung Tau, is the economic center of the southern region.

BOX 18.1 Child Labor

The children of Vietnam, Laos, and Cambodia have a traumatic history, and their lives are to a great extent shaped by that history. Unlike most U.S. children, access to education is not automatic. Childhood may be very short as earning or unpaid work is required to contribute to the household well-being. Children work in a wide variety of occupations, and unfortunately many are drawn into the sex industry.

Lack of education and opportunity are fundamental elements in the use of child labor. Poverty can only be alleviated by children's wages, and child labor may be seen as a normal part of village life and culture. Where cultural systems are linked with a regime of exploitation of the poor, some children may be sold into bondage to pay debts accumulated by the family. These children are seen as the property of their owners and can be used as they see fit. Other parents, especially those in rural poverty, have been forced to sell their children, believing that these children, usually girls, are being sent to jobs in the city in restaurants, as maids, or in factories. However, unfortunately, this is rarely so.

Employers see young workers as cheap and less likely to cause industrial unrest; consequently employment may be available to them where there is no work offered for adults. The work hours are long and tedious, and jobs repetitive and dangerous. With little job security and lack of comprehension, the children can be forced into physical exhaustion and situations that are damaging to their health and well-being.

In Cambodia the issue of child labor has been complicated by the dislocation caused by war. Many children were left homeless—possibly as many as 10,000 children live on the streets of Phnom Penh. They earn money as sex workers and may even be bought or sold as sex slaves; some live by begging; others make bricks, sort garbage, work in the construction industry, sell food and drinks, and generally live by their wits. In the formal sector, factory employment is available to some young people, but their poor level of education assures that they will be locked into a low standard of living.

In nonurban areas, children work in the fields often as unpaid family workers. Traditional farming systems may require long hours of work, and young people working outside can suffer from exposure due to the extremes of climate and may be exposed to the danger of agricultural diseases, chemicals, and unsuitable tools and machinery. Their lack of education and illiteracy leads to misuse of modern agricultural technology and further danger. Young boys in particular have responsibility for herds of animals for long periods of time. The isolation they endure is not providential for mental development and maturity.

HOW CAN THE CHILDREN BE HELPED?

Alleviation of poverty and economic development are of major importance for the future of children. Smaller families and economic and political security will assist the successful development of the country; these factors are part of the strategy being undertaken by many countries giving aid to Cambodia.

A range of factors, including economic reform, have contributed to high levels of unemployment. The dismantling of rural communes, the decline of state-owned industry and reductions in the size of the civil service have resulted in fewer job opportunities. Inflows of foreign investment have generated new jobs but not sufficiently to compensate for the large numbers unable to find government-sponsored employment. In Vietnam the reduction in the size of the military following the withdrawal of troops from Cambodia released soldiers into the labor market, as has the repatriation of refugees from the camps in neighboring countries.

The Geography of Poverty and Income Distribution

Accurate information on poverty can be hard to find, especially for Vietnam, Cambodia, and Laos, which do not publish data that might put the country in a bad light. This situation has been partly rectified by a large survey carried out by Vietnam's State Planning Committee and the General Statistics Office in 1992–93. Titled the *Vietnam Living Standards Survey*, some 23,000 people living in 4,800 households were questioned about their incomes and patterns of consumption. Results revealed that the average annual per capita income in Vietnam's urban areas was 1.8 million dong (US$150) compared with the rural average of 0.9 million dong (US$75) (State Planning Committee, 1994, p. 217).

The story told by the survey about poverty was startling, though not unexpected. A poverty benchmark was defined as the consumption of 2,100 calories per day, which was equivalent to an annual income of 1.3 million dong (US$108) in urban areas, and 1.1 million dong (US$92) in rural areas. Some 51 percent of the Vietnamese population was below the poverty line, and about half that number (one-quarter of the total population) do not, on average, have sufficient food to eat on a daily basis.

The geographical distribution of poverty is important (Map 18.9). About 90 percent of Vietnam's poor live in the rural areas. Some 57 percent of the rural population was in poverty compared to 27 percent of the urban population. In the North Central region of the country, 71 percent of the people live in poverty, as do 59 percent in the Northern Uplands. By contrast, in the Southeastern region, which includes Ho Chi Minh City, about 29 percent of the population were poor. Reflecting the better incomes in the city, only 2 percent of white-collar workers and 2 percent of government workers'

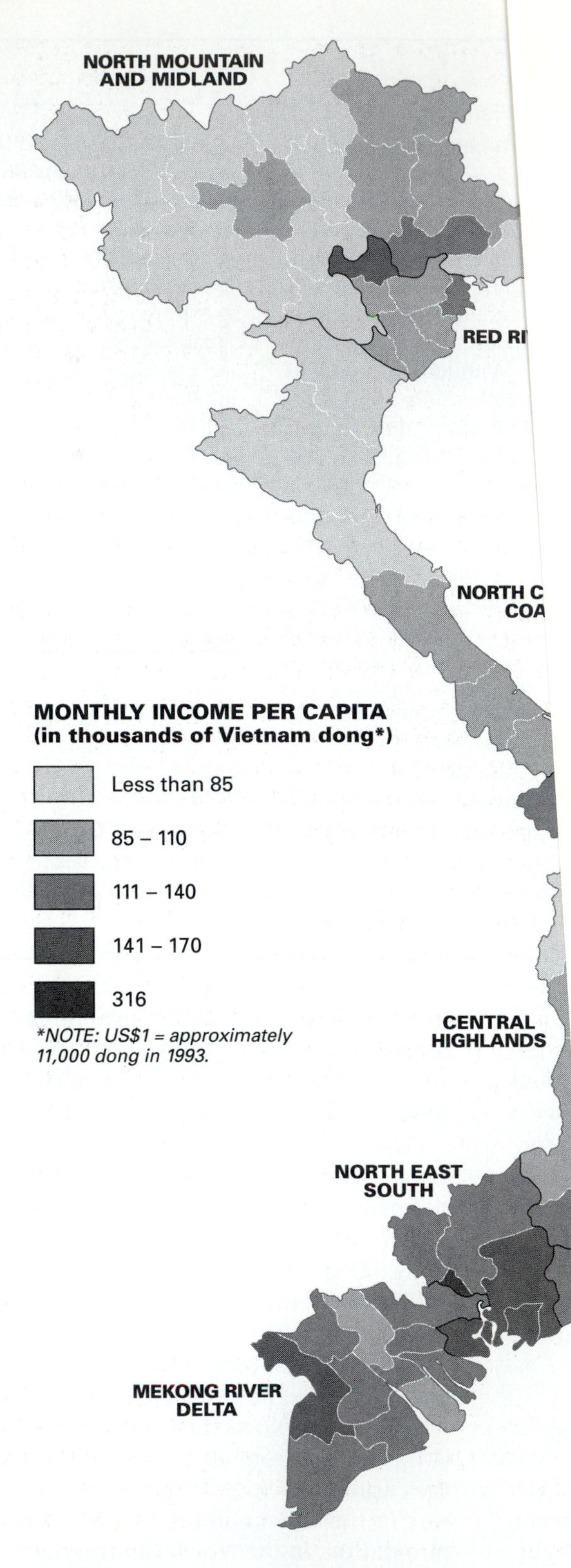

Map 18.9 Income distribution in Vietnam, 1993. *Sourc*
of 1994.

Photo 18.6 Highway 1 runs the length of Vietnam and as a secondary function allows farmers space and a sun-exposed surface for the drying of newly harvested rice before it is milled. (Forbes)

the road network during the Second Indochina Conflict, though it led to a general lack of investment and the existing poor rail services. There is a national rail link from Phnom Penh to Battembang and to Kampong Som, but other lines are unserviceable or in disrepair. The railways have also been subject to ambush from the Khmer Rouge, and the subsequent kidnapping of western and local hostages has affected their use. Air services have been improved as a result of the work done in Cambodia by the UN. It is hoped that better air services will promote tourism.

Laos has similar transport problems to those of Cambodia. Its roads are in poor condition, with less than one-quarter of the 8,675 mi (13,971 km) actually asphalted, though a further quarter are gravel. The main north–south road link—Highway 13—is still being completed, while other parts are in need of repair. This is a particular problem in Laos as it has no railway and is landlocked. Although Laos depends to a degree on river transport for internal trade, the Mekong contains too many rapids for it to be of great use. Most goods are carried by road, though passenger transport between such cities as Vientiane and Savannakhet is by river boat. Laos sends the bulk of its exports overland to the seaports in Thailand, and less frequently along Highway 9 to Danang in Vietnam.

The Mekong, which forms the western boundary of the country, has been a barrier to land transport between Thailand and Laos. In 1991 construction commenced on a three-quarter-mile (1.2-km) long bridge over the Mekong River, the first, connecting Nong Khai in northeast Thailand with Tha Naleng in Laos. It was designed to carry two lanes of traffic and pedestrian paths, with provision for a future railway running down the center. Known as the Friendship Bridge (Mittaphab Bridge in Lao), it was officially opened in 1994. Construction was funded by the Australian government, and the maintenance costs are being recouped from a toll on bridge users.

The impact of the bridge, which replaces a very slow, unreliable ferry service, is bound to be significant. It opens a road connection between Bangkok, 385 mi (620 km) away, and Vientiane, about 12 mi (20 km) northwest of the bridge. The Thais hope the bridge will speed up development of Thailand's relatively poor northeastern region, which has close cultural and ethnic links to Laos. In the early years it is anticipated that freight flows will be dominated by consumer goods being imported into Laos, but in the longer term the bridge will facilitate the movement of Lao exports to the main seaports. On the Thai side the bridge is connected to a well-developed road system and a rail terminal. To be effective, the Friendship Bridge needs to connect into a far more developed Lao road network than presently exists.

However, it will take some time for the full significance of the bridge to be assessed. The Lao government fears the potentially negative social side effects of close links to Thailand, including increased prostitution and drug use and the spread of HIV/AIDS. The government therefore insists that private vehicles from Thailand are left at the border and their occupants transported to Vientiane by taxi or bus. This has not stopped either government from discussing the possibility of another bridge over the Mekong, further to the west, linking Thailand's Chiang Khong with the Lao town of Ban Huey Xai (Lintner, 1995a), and yet another in the south of the country in the vicinity of Savannakhet (Handley, 1993).

Photo 18.7 The Friendship Bridge allows road movement over the Mekong between Thailand and Laos. (Forbes)

Improving the transport network is undoubtedly one of Laos's main priorities. This means upgrading the links with neighboring countries and, even more important, improving the domestic infrastructure. As an example, roads need to be developed between the provinces of Xiangkhoung and Khammouan. Along this corridor are located a number of the agricultural and industrial developments on which Laos's economic future depends.

Mekong River Commission

With its headwaters deep inside southwest China, the Mekong meanders south along the border between Laos and Burma, and Laos and Thailand, cutting across Cambodia before emptying into the sea in southern Vietnam. It provides a unifying thread for the affected countries. As far back as 1957 a Mekong Committee had been established by the four countries that share the Mekong lower basin: Cambodia, Laos, Thailand, and Vietnam, to secure cooperation among the riparian states (Fryer, 1979) (see Map 2.3). The withdrawal of the Cambodian government when it was under Khmer Rouge control in the mid-1970s resulted in the replacement of the original body by an Interim Mekong Committee. Cambodia agreed to participate in the discussions in 1991, although progress was slow because of tensions between Thailand and Vietnam.

In 1994 Cambodia, Laos, Thailand, and Vietnam agreed to establish a new Mekong River Commission intended to support joint development activities. The commission will take responsibility for compiling a river-basin development plan that will manage the resources needed for hydropower generation, navigation, flood control, fishing, tourism, and the transport of timber. It affords the countries involved an opportunity to exploit the economic advantages of the river for the region, at the same time protecting the Mekong environment. In November, 1995 Cambodia, Laos, and Vietnam met as the Mekong River Commission to adopt a formal work program aimed at transforming the underdeveloped river system. The countries put together a Mekong Work Program that aimed at obtaining funding for 96 projects to develop the basin. Feasibility studies are concentrated on hydropower generation, dam building, navigation, and agriculture (*Cambodia Times,* November 19, 1995).

The Greater Mekong Subregion

Cross-border economic interaction among the Mekong countries often occurs through smuggling and other forms of unrecorded, unregulated exchanges. Map 18.10 gives an indication of some of these flows of commodities. The Asian Development Bank (ADB), a major regional financial institution with its headquarters in Manila, is eager to see the rehabilitation and expansion of the transport network within the zone it calls the Greater Mekong Subregion. By supporting a comprehensive program of infrastructure improvement, it aims to link the economies of Cambodia, Laos, Vietnam, Thailand, Burma, and southwest China into a regional market approaching 220 million people.

At the core of the infrastructure needs of the region is an integrated road network (Map 18.11) that would connect the major inland centers such as Kunming (China), Mandalay (Burma), and Chiang Rai (Thailand) with Vientiane and Phnom Penh and the major ports adjacent to Bangkok, Ho Chi Minh City, Hanoi, Quang Tri, Danang, and Rangoon. Plans to develop the transport

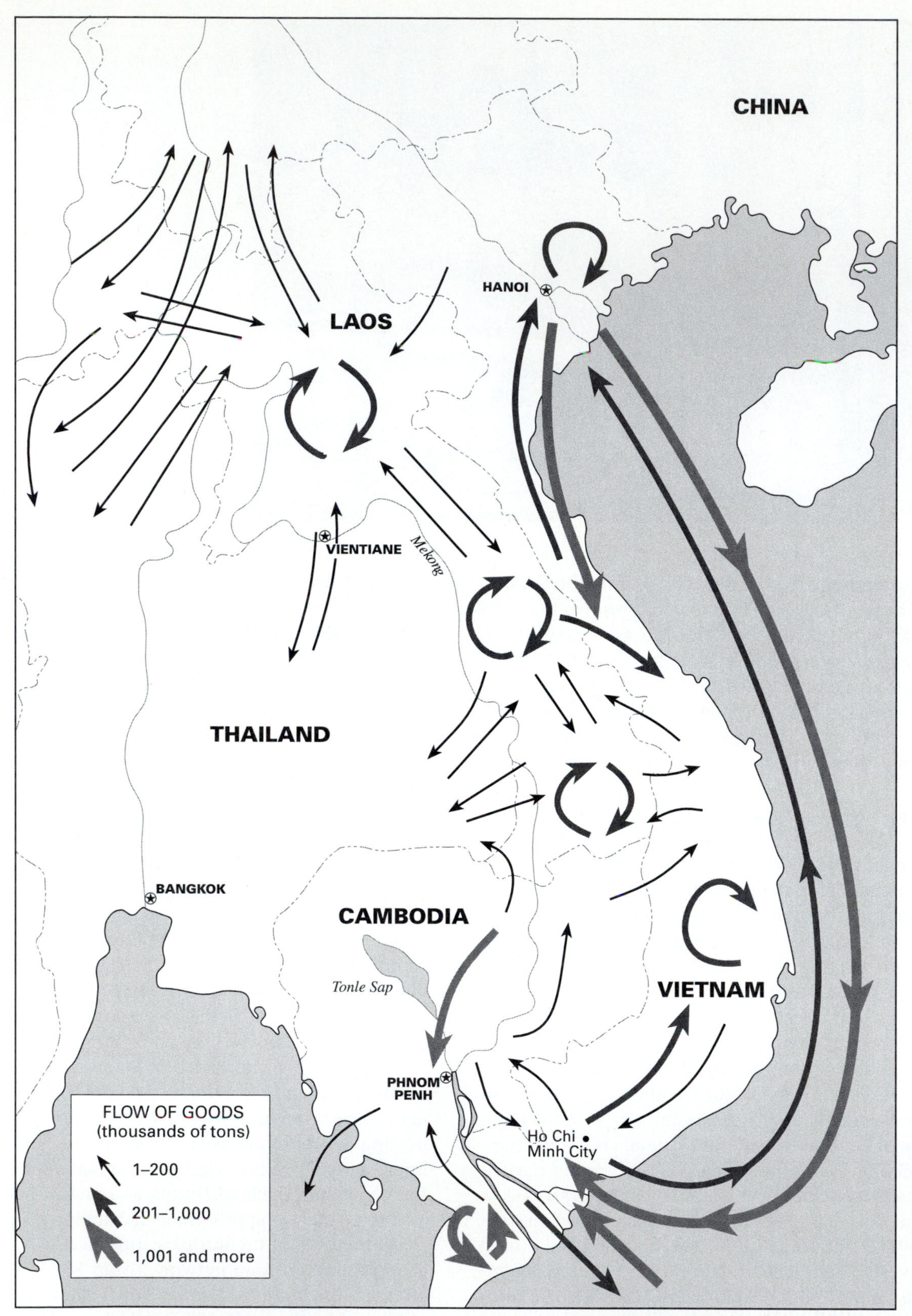

Map 18.10 Flow of goods. *Source: Asian Development Bank, 1993b, p. 40.*

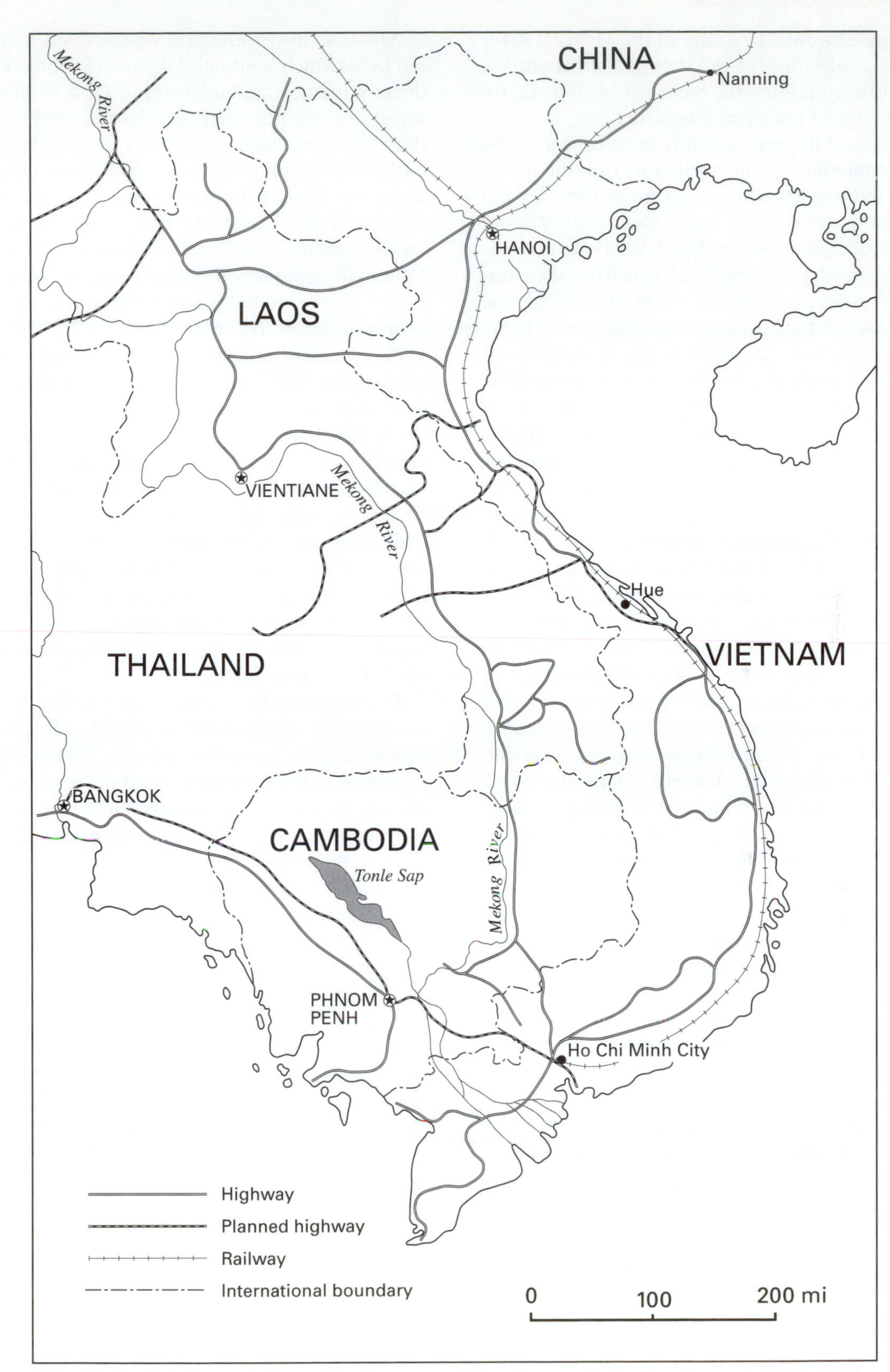

Map 18.11 Existing and planned land transport links. *Source: Asian Development Bank, 1993a, p. 54.*

and communications infrastructure of the Mekong River zone have received support from the relevant countries, although no firm commitments have yet been made toward the financing of the overall scheme.

Development of the area has not been welcomed by everyone. The building of dams will interrupt the natural river flow and may cause destruction of fisheries and agriculture and initiate costly water treatment (*Phnom Penh Post,* October 20—November 2, 1995). As many as 60,000 people may have to be displaced from their traditional home areas as the water regime is altered. The Mekong is the second-most-important fishing river in the world (after the Amazon), and there are believed to be more than 400 varieties of fish in the river. As some of the species are migratory, any interruption to their natural habitat would have disastrous results on the livelihood of those dependent on the river. Fish provide the major protein source for as many as 40 million people living in the basin area, and fishing or fish products are the mainstay of the peoples' economic livelihood.

Organizing the Mekong River's development is complicated by the number of countries involved and their varying objectives. Thailand is at a different stage of development from Vietnam, Laos, and Cambodia and as an upstream user has attempted to alter the objectives of the management of the river to allow it to capture water for its dry northeastern provinces. Despite Vietnam's great concern about this and the subsequent intervention of the UN to keep the Mekong Committee functioning, the role of the new commission will be difficult. The Mekong is also affected by actions taking place in China. China is not part of the commission and has continued to treat the river as if it were sole owner and is oblivious to the downstream consequences of its behavior. This has reduced water flows in Laos at the same time as many new dams are under construction.

RESOURCES AND ECONOMIC REFORM

The Resource Base

Vietnam has the most-abundant natural resources of the three countries. Its offshore oil and natural gas reserves are regarded as particularly important. Oil production in 1994 totaled 51 million barrels (148,000 barrels per day), and oil exports were worth $976 million, or 27 percent of Vietnam's total export revenue. Since the mid-1980s crude oil production in Vietnam has increased steadily (Figure 18.2). In the late 1990s all of Vietnam's crude oil was exported to Japan, Singapore, the United States, and South Korea. Oil reserves are estimated to be from 3 to 5 billion barrels. About 90 percent of Vietnam's oil output came from a joint venture between Vietnam and Russia through a company with the awkward title of VietSovPetro (Schwarz, 1995b).

Most of this production was derived from the Bach Ho field, which is situated 93 mi (150 km) out to sea on the continental shelf adjacent to Ho Chi Minh City and Vung Tau, the port that services the field. The Bach Ho field also provides gas, which is sent by pipeline in unliquified form to Vung Tau. Vietnam's other major oil fields are Rong ("Dragon") which is next to Bach Ho ("White Tiger"), and Dai Hung ("Big Bear"), which is located about 93 mi (150 km) further to the southeast. Other oil exploration concessions are located around these existing oil fields and further north in the coastal waters adjacent to Danang.

A significant concern for the Vietnamese is that China claims ownership of a large swath of the South China Sea. Chinese claims include the Paracel and Spratly Islands, but even more important for Vietnam, most of the seas that contain the country's offshore resources (see Box 10.3). Not only Vietnam is affected: The Philippines, Indonesia, and Malaysia also lay claim to areas within China's grasp, causing significant tensions between countries in the region. Anthracite coal, phosphates, copper, tin, zinc, iron, antimony, and chromium are all found in the northwest of Vietnam, which contains most of the country's exploitable minerals. Vietnam is a net exporter of coal and gemstones.

Fishing has been an important contributor to the subsistence diet of Vietnam's coastal population, as well as an increasingly valuable export. In 1994 marine products represented the third-largest export revenue after crude oil and textiles and garments. However, to meet long-term export targets, the capacity of the deep-sea fleet needs to be considerably expanded.

Cambodia has a modest resource base, with few mineral resources of significance. In the northwest the Khmer

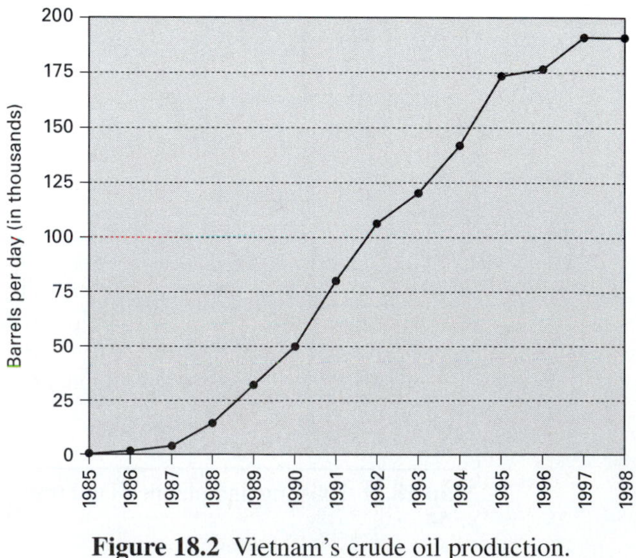

Figure 18.2 Vietnam's crude oil production. *Source: United States Energy Information Administration, 1998.*

Photo 18.8 Fish vendor on a beach in Vung Tau, a popular tourist destination for international and especially Vietnamese tourists. (Forbes)

Rouge have leased substantial areas of land to Thai companies to enable them to mine sapphires and rubies. Needless to say, neither the government nor the people of Cambodia benefit from this arrangement. Phosphates are abundant, and the country has hydropower potential (see Table 2.1). For most Cambodians, including the citizens of Phnom Penh, firewood remains the main source of energy, leading to concerns about the depletion of the forests. Electricity production is very limited, with most being consumed by households or government, leaving little available for industry. Offshore and onshore oil and gas exploration licenses have been issued by the government in an attempt to investigate previously unlocated energy potential.

In Cambodia timber production is of concern as logging has taken place in an uncontrolled way for many years, causing deforestation and consequent drought and flooding (Collins et al., 1991, p. 111–115). Heavy logging has continued, even though the dangers are known, as it is an important source of income for the government. Khmer Rouge guerrillas have often used illegal logging activities to finance their activities. Cambodian and Lao companies provide fronts for the export of timber to Thailand. In a surprising move in the mid-1990s in the eastern province of Stoeng Treng, the Khmer Rouge took to campaigning against logging activities. This represented a move to attract local support by demonstrating their support for the environment. The guerrillas have previously demonstrated their environmental awareness by supporting the protection of endangered species. However, cynics argue it was a crude attempt to establish a monopoly of the local timber trade (*Phnom Penh Post*, April 21–May 4, 1995).

Laos's location in between the rapidly growing, energy hungry economies of Thailand and southern China has focused attention on the energy resources of the country, particularly the potential hydroelectricity production. Estimates suggest the hydropower potential of Laos could be more than 22,000 megawatts, although current production is closer to merely 200 megawatts (see Table 2.1). Thailand is presently the largest market. The Electricity Generating Authority of Thailand purchases about 100 megawatts per annum from the Nam Ngum and Xeset hydropower stations. This makes electricity one of the top exports from Laos (Gill, 1994; Lintner, 1994). The Lao government, through its agency Electricité du Laos (EDL), has ambitious plans to increase electricity output to 1,500 megawatts by the year 2000, and to 4,000 megawatts by the year 2010. Some 58 possible sites for generating hydroelectricity have been identified. This would require the construction of many huge dams, although not all of the proposed sites are feasible.

The largest project is a dam on the Nam Theun River, a tributary of the Mekong. Plans provide for a 600-megawatt hydroelectricity plant, called Nam Theun II, to be built at a cost in excess of US$1 billion and completed by the year 2000. The project is a joint undertaking between Electricité du Laos and companies from France, Australia, and an Italian–Thai consortium. It will be a Build–Operate–Transfer (BOT) arrangement; that is, these companies will finance, build and operate the plant until they have achieved an agreed profit on their investment. Then the facility will be transferred to the Lao government. Thailand has agreed to purchase the total electricity output of the installation.

The physical resources of Laos provide it with a regional comparative advantage in the production of hydroelectricity. However, the current strategy has been questioned. Only a handful of the 58 identified sites are likely to prove economically feasible, and even then there are concerns about the availability of finance. Nam Theum II is the first Laos power station to be developed by private investment. It is unlikely that the major international development agencies will provide funding for further commercial development of Lao energy, and the projects required are likely to be too expensive for the private sector alone. Critics point to the development of power supplies in Vietnam, Burma, and Yunnan and the limited demand for power in Cambodia, leaving Thailand as the major purchaser of Lao hydroelectricity. Dependence on a single market creates a major economic risk, reducing the attractiveness of the huge investments required.

Although power stations are unlikely to be developed at most of the identified dam sites, the timber at many of these is already being harvested. Environmentalists complain that the ambitious hydroelectricity plans are providing a cover that allows loggers to expand activities that have otherwise been restricted by the government (Lintner, 1994).

As with Cambodia, Laos has in the past been a significant exporter of timber, but recent conservation pressure has reduced the export of valued woods. The expansion of the road network has allowed loggers to penetrate and exploit lowland stretches of tropical rain forest (Collins et al., 1991, p. 166–173). Despite attempts to control deforestation, illegal exports have continued. Members of the Lao and Cambodian military have been involved in illegal logging operations in southern Laos (*Phnom Penh Post*, April 21–May 4, 1995). A combination of legal and illegal cutting, traditional shifting agriculture, and forest fires has led to the destruction of considerable areas of forested land and concerns about environmental degradation.

Forests are earmarked by the Lao government as an important source of future income. Following the Indonesian model, the government intends restricting the export of whole logs and increasing exports of processed timber products such as chipboard. At the same time forests will be planted, with the assistance of the hill tribes, for later commercial exploitation. Laos has few known mineral resources, with the exception of some tin deposits.

Agricultural Production and Reform

As discussed earlier, agriculture is the mainstay of the economies of the three countries, especially Laos. Two major changes have affected agriculture since the early 1980s. First, and most significantly, agriculture was the focus of most of the initial economic reforms introduced in Vietnam, Cambodia, and Laos and by the mid-1980s were a key component of economic reform, notably through the redistribution of land from state to private control. Second, the introduction of Green Revolution innovations, such as better-managed irrigation systems and improved pesticides, have helped to increase agricultural production in selected regions, such as in the Mekong delta in Vietnam.

A key aim of the socialist government of Vietnam, beginning in the 1950s, was to collectivize agriculture, replacing individual holdings with communes. Not surprisingly the process of collectivization was often resisted, sometimes resulting in bloodshed. Nevertheless the creation of communes went ahead in the north of the country, eventually incorporating a significant proportion of the arable land. After unification in 1976 the collectivization process was extended to the southern part of the country. Once again it was resisted. As a result, the program was well short of completion, especially in the Mekong delta, before broader economic concerns began to overtake the desire to collectivize agriculture.

Following the disappointing agricultural output in Vietnam in the late 1970s, in 1981 the government introduced the product contract system into agriculture. Households were provided with land and agricultural inputs; in return they were required to provide a quota of crops to the state and could sell the surplus to the government or on the free market. Rural output and incomes rose as a result of these measures, though the shortage of consumer goods due to the stagnant industrial sector led to much frustration for successful farmers.

It was not until 1988 that the collective system of agriculture was totally abandoned and the land and equipment that previously belonged to the communes was transferred to individuals and families. The process has not always been straightforward, with many objecting to the corrupt way in which former state resources have been reallocated. Party officials, so the complaints go, have often ended up with the prize assets of the communes. In 1989 official prices for agricultural products were abandoned, meaning that all agricultural production had to be sold at market prices (Beresford, 1993).

About 22 percent of Vietnam's land area is cultivated. The Red River and Mekong River deltas are the centerpieces of Vietnamese agriculture, accounting for more than half the total cultivated land. Rice is the cornerstone of Vietnam's agricultural production, with some 65 percent of Vietnam's agricultural land being under rice cultivation. High-yielding varieties (HYV) were first introduced in the north in 1965 and in the Mekong delta before reunification, and again in the 1980s. As new lands are brought into cultivation they are generally planted with cash crops. Increased rice production will therefore depend on intensification through HYV, as well as improved technical inputs such as fertilizers, irrigation systems, and greater levels of mechanization for land preparation.

Rice production has been one of the success stories of Vietnam's economic reform. After many years in deficit, Vietnam started to export rice in 1989, quickly establishing itself as one of the world's major rice exporters. It remains one of the country's key exports. Other key agricultural crops are sugarcane, cassava, and sweet potato. Principal cash crops include coffee, tea, rubber, and soybeans.

In Cambodia collective organization of land and farming, especially of rice, was imposed during the Pol Pot era from 1975 to 1979. Since the mid-1980s the collective system has given way to private control of land, as in Vietnam (Frings, 1994). Commune leaders distributed land

Photo 18.9 A wide range of rice types are grown in the Mekong delta where high-yielding varieties produce incomes that are double those in the North's Red River region. Prices for a kilo (2.2 pounds) of rice, shown here in a Ho Chi Minh City market in Vietnamese dong, range from 32¢ (4,500 VND) to 69¢ (9,500 VND) where US$1 = VND 13,850. (Leinbach)

to rural families based on the number of children in each family. However, the land was insufficient in numerous cases; besides many families have since increased in size, outgrowing the plots available. Landlessness is increasing as key landholders acquire larger plots, and young married adults find the land of their parents insufficient to sustain their new families.

Most agriculture in Cambodia is inefficiently organized, and the poor soils and unreliable rainfall have contributed to keeping the population in a subsistence lifestyle. Although rice remains the main crop, there are secondary crops being grown, including maize, root crops, mung and soybeans, sesame, and groundnuts. Significant exports come from rubber, which is grown on plantations, and timber.

Although some progress is being made in the rural areas, barriers remain to the improvement of agricultural production through Green Revolution inputs. Khmer Rouge activity in rural areas has made the delivery of additional farm inputs problematic. The control of crucial inputs such as fertilizers and seed multiplication by inefficient government departments is another concern. The lack of sufficiently trained rural extension workers and the poor communications infrastructure slows the diffusion of innovations. Finally, the overall lack of rural infrastructure makes the transport of inputs and outputs costly and slow. Overseas assistance with agricultural development up until the mid-1990s centered largely on nongovernment organizations and humanitarian agencies, but this is changing as Cambodia is drawn into the global economy.

Laos covers a larger area than many realize. Its area of more than 91,000 sq mi (237,000 sq km) is significantly larger than countries such as Cambodia, South Korea, and Taiwan. However, it has been estimated that only 16 percent of land suitable for cultivation is under rice or secondary crops, with another 16 percent devoted to pastureland and ponds in which freshwater fish are raised. Total output of rice annually exceeds one million tons, although this is significantly below the government's production target (Economist Intelligence Unit, 1993, p. 76–77).

Following the communist takeover in 1975 the new government recognized the need to restructure the economic system. The dominance of subsistence agriculture and the lack of natural resources had produced an economy that depended on overseas aid, but this was cut off when the communists took power. To organize all aspects of economic activity, cooperatives were established. There was to be no coercion of the population but persuasion by example and encouragement. The collective movement gathered momentum until the late 1970s but then slowed down: It proved impractical for nonpadi farmers using slash-and-burn techniques to form collectives, and those farmers who normally farmed in an informal collaborative way could see little value in reorganizing themselves more formally. The collective movement remained at the center of government plans even though production targets were rarely achieved. Problems included the overcapitalization of farms and poorly managed and underutilized irrigation systems (Evans, 1995).

At its peak the state controlled more than half of Laos's rice fields and employed a similar proportion of all rural families. Economic reforms meant that from 1988 there was a shift in the organization of agriculture away from cooperatives and state-owned farms into family-based farming. By 1990 the majority of cooperatives and state farms had been disbanded and land and farm equipment transferred to families (Bourdet, 1995). Laotian agriculture is centered on rain-fed rice, though the upland Lao Theung people practice swidden agriculture. It was anticipated that economic reforms would result in increased rice production, as in Vietnam. Instead farmers have begun to produce such secondary crops as sugarcane and tuber vegetables and cash crops including tobacco, coffee, and even *cannabis.* These crops provide peasants with a more substantial cash income, but the change has meant that Laos has become dependent on food imports and aid. Lao farmers are largely unaffected by the Green Revolution because they do not have the required inputs for modern rice production. The result is that productivity levels of agriculture are among the lowest in the world. Parts of Laos are included within the so-called Golden Triangle (see Map 17.11) where farmers illegally produce significant quantities of raw opium, heroin, and marijuana for export. In an attempt to curb this production, aid programs have been put in place to encourage poor farmers to substitute legal crops.

Industry and the Decline of State Enterprises

Economic reform has had a significant impact on the industrial structure of all three countries. It has resulted in unprofitable state-owned industrial enterprises being cut loose from state control. Companies are no longer tied to government-set prices for inputs or quotas and fixed prices for their outputs. In addition the government will no longer cover losses incurred by unprofitable industries. The result is that managers must reduce costs and increase production to become profitable. Inevitably, many former state enterprises have become bankrupt and closed down; other companies have joined forces with foreign investors and their production has been revitalized.

There was little development of industry throughout Indochina during the colonial period as the French discouraged any activities that could compete with their own economy. In the 1950s and 1960s the strategy pursued in the north of Vietnam favored the establishment of heavy industries, following the model of the USSR. A significant amount of this was subsequently destroyed by U.S. bombing that targeted the industrial regions in the 1960s and early 1970s. Vietnam's major industries are concentrated in a number of large centers, including Hanoi, Ho Chi Minh City, and Bien Hoa, along with the key ports at Haiphong and Danang.

From the mid-1980s Vietnam has, in tandem with economic reform, given more emphasis to light industry and the expansion of manufacturing exports. After a slow start, value-added in industry grew 12 percent in 1992 and 11 percent in 1993. This was a faster annual rate of industrial growth than any other Southeast Asian country, including rapidly industrializing Thailand. It also was expected to be sustained through the 1990s (Asian Development Bank, 1994, p. 234). Merchandise exports grew at even faster rates, averaging in excess of 22 percent each year between 1988 and 1992 (Asian Development Bank, 1994, p. 241). This impressive rate of increase was on a par with neighboring countries such as Malaysia and Thailand and well above the rates achieved by Indonesia and Singapore. An influx of foreign investment has been drawn by Vietnam's cheap and moderately skilled labor force, as well as the perception that Vietnam represents "the last frontier" in Asia.

Photo 18.10 Although foreign direct investment (FDI) is producing modern industries, labor intensive operations such as this ceramics producer in Ho Chi Minh City are quite conspicuous. (Leinbach)

Just under one-half of Vietnam's industrial output in 1992 was from heavy industry, with light industries accounting for the rest. State-owned enterprises were responsible for 71 percent of production, and nonstate enterprises 29 percent. About two-thirds of state enterprises were owned and run by central government ministries, with the remainder in the ownership of local people's committees at the provincial, municipal, district, and ward levels. With allegiances to different levels of government, state -owned enterprises are far from homogeneous entities (Economist Intelligence Unit, 1994, p. 52). An additional concern has been a blurred distinction between state-owned firms and private firms owned and operated by the management of state-owned firms. This causes confusion for foreign business investors and has resulted in the often corrupt transfer of resources from the state to private enterprises.

Industry in Cambodia is small in scale and in total number of enterprises. Manufacturing industries in the towns emphasize production of light consumer goods including household items, soft drinks, alcohol, and pharmaceuticals. The country's largest factories, which are located in Phnom Penh, specialize in textiles, cigarettes, and tires. Small-scale rice mills are the most frequent type of rural industry.

Most industry performs poorly because of erratic supplies of electricity, raw materials, and spare parts, together with the poor quality of management. The stagnation of the industrial sector has been compounded by corruption as the ownership of state enterprises has been transferred to the private sector. The poor quality of physical infrastructure throughout the country has also affected industrial performance. When former state industries in Cambodia became "self-managing" in 1988, some fell under the control of local party committees, while others disappeared altogether—private investors shunned manufacturing industries, preferring to invest in urban real estate (Economist Intelligence Unit, 1994, p. 108).

As in Cambodia, the industrial sector in Laos is very small. Totaling an estimated 600 factories, few of which employ more than about 300 workers, it is concentrated in the area around Vientiane. Most factories produce for local consumption, and cigarettes, beer, soft drinks, household plastics, and clothing are common manufactures. Since 1992 increases in the export of garments and assembled motorcycles have been important. Textile and garment makers find Laos attractive because these goods have ready access to European markets. In the long term, manufacturing industry development is likely to be limited, given the small size of the labor force and the intense competition from export-oriented manufacturing countries such as Thailand and China.

The impact of economic reform in Laos has resulted in the state's share in industry falling from 82 percent in 1985 to 44 percent in 1992. At the same time, industrial production has been growing very quickly. In 1991 alone industrial output increased by 18 percent over the previous year (Economist Intelligence Unit, 1993, p. 80–81).

THE ENVIRONMENT

The environment is emerging as an important concern in Vietnam, Cambodia, and Laos (Kaosa-ard et al., 1995). Except in parts of the Red River valley and delta, none of these countries face intense population pressures comparable to Java, nor are the major cities as large and unwieldy as Jakarta, Manila, and Bangkok. Nevertheless, the environment of each was heavily damaged during the Second Indochina Conflict, and environmental degradation as a result of conflict continues in Cambodia. Although the respective governments demonstrate a modest interest in environmental matters, the absence of a civil society or a pluralistic political structure means that there are few opportunities for the community to add environmental matters to the political agenda (Beresford and Fraser, 1992).

War and the environment. The extraordinary post-1945 history of former French Indochina has left its imprint on the landscape. Each country has suffered and continues to suffer as a result of the impact of war on the environment. In Laos war has left large scars in the forests along the eastern edge of the country. Cambodians are at risk because of the very high numbers of live land mines and unexploded bombs.

Vietnam's environment suffered extensive damage during the Second Indochina Conflict. It is estimated that about 101,000 tons (72 million liters, or 92 million kilograms) of herbicides and defoliants were used by U.S. forces in Vietnam. Almost two-thirds contained the extremely toxic ingredient dioxin. About 4.2 million acres (1.7 million hectares) of rural south Vietnam, or 1 hectare in 10, were targeted. Nearly 50 percent of the forests and arable land in the south were sprayed more than once with toxic chemicals. As recently as 1979–80, buffer forests in border provinces in the north and southwest of the country were deliberately destroyed as a result of war.

The massive damage done to Vietnam's environment during warfare has had severe consequences. Many of the people who were exposed to the kinds of herbicides and defoliants used in Vietnam have sustained genetic damage and suffer from a greater incidence of birth defects and a higher rate of cancers. Vast amounts of vegetation were destroyed, along with the animals that were dependent on these inland, coastal, and aquatic habitats. The damage has been so severe that many areas have been slow to recover. The destruction of microorganisms in soils has often resulted in reduced soil fertility, whereas areas previously covered by woods and trees have been reduced to low productivity savannas.

The Second Indochina Conflict ended two decades ago; yet the damage done to the environment continues to the present and is likely to extend into the future. Dioxins and other toxins take years to decompose. In the meantime they are absorbed into plant species and then ingested by animals, becoming part of the food chain that leads to human consumption, poisoning people only remotely connected to the war. At the same time, toxic chemicals have been carried by rivers into the lowlands of Vietnam and into neighboring countries, as well as into the coastal seas.

Deforestation

Although deforestation is a major concern for each of the three countries, the Lao government identifies forest loss as its greatest environmental problem. Forest covered 70 percent of the land of Laos in the 1950s, but has declined to 47 percent in the 1990s (see Map 2.6). About 750,000 acres (300,000 ha) of forest are lost each year. Laos had a faster rate of deforestation than anywhere else in Southeast Asia except Thailand in the early 1980s.

The causes are varied and include unsustainable legal and illegal logging operations, the practices of swidden cultivators, the expansion of cleared land for farming, and the loss of forests due to city growth and the construction of new roads. The low incomes of the Lao cause them to rely heavily on fuelwood. On average individuals use one cubic meter of fuelwood per person per year. This represents one of the highest rates of fuelwood use in the region and is another cause of rapid deforestation.

The absence of forest management systems is exacerbating the rapid loss of forest in Laos. The reasons for the poor quality of forest management include the lack of funds for surveying and monitoring forest resources, extreme shortages of qualified staff, and the absence of any environmental legislation. However, a Tropical Forestry Action Plan and a National Conservation Strategy have been drafted and the Ministry of Science and Technology given primary responsibility for environmental issues.

All three governments are committed to the settlement of shifting cultivators and their adoption of permanent (i.e., sedentary) forms of agriculture. Socialist governments, especially, have typically seen shifting cultivation as a primitive and inefficient form of agriculture. As population densities increase and the rotation cycle of shifting cultivators is compressed, long-term damage is inflicted upon the environment. Timber is felled and burned, smoke causes air pollution, the erosion of soil from denuded slopes accelerates land degradation, flash flooding occurs due to faster rates of water runoff, and rivers become thick with soil and less useful as sources of hydroelectric power. But in many circumstances, shifting cultivators at low population densities do less damage to the environment than is generally thought and far less than illegal loggers and others using modern technology. The shifting cultivators are often unfairly targeted because they appear primitive and the technical sophistication of their agricultural practices are simply not well understood.

Environmental policy. Perhaps because of the severity of the damage done to Vietnam's environment during war, the government has made some effort to address the country's environmental problems. Environmental issues received comprehensive treatment in the *National Conservation Strategy* which was drafted in 1984. The document sought to take stock of the nation's environment and natural resources and devised a strategy to promote restoration of the environment and minimize future damage. A review of environmental legislation in Vietnam in 1990 concluded that existing laws provided some support for the conservation of land, vegetation, and wildlife.

The *National Conservation Policy* was followed by a *National Plan for Environment and Sustainable Development 1991–2000* which was formally adopted by the Council of Ministers in June 1991. The document was completed on the eve of the United Nations Conference on Environment and Development in Rio de Janeiro. A centerpiece of Vietnam's growing concern about environmental issues was the *Law on Environmental Protection* which was passed by the National Assembly in December 1993. Included in the national strategies is recognition of the importance of social and economic processes to Vietnam's hopes of improving environmental conditions. The importance of slowing and then stabilizing population growth, especially in the midlands, is a critical factor; another is the important role that environmental education will have in ensuring that Vietnamese are fully aware of the significance of the environment to sustainable development.

INTEGRATION WITH THE GLOBAL ECONOMY

Relations with the USSR and China

During the 1960s and especially in the 1970s, Vietnam and Laos (and to a much lesser extent Cambodia) were drawn into the orbit of the Union of Soviet Socialist Republics (USSR). However the breakup of the Soviet Union in the late 1980s encouraged the three countries to extend their links to the global capitalist economy, a process that has accelerated further in the 1990s.

The Soviet Union was an important political and economic influence on Vietnam and Laos. The Council for Mutual Economic Assistance (CMEA) was an international trade group, run by the Soviet Union, to which Vietnam and Laos belonged through the 1970s and 1980s. The Soviet Union financed trade with CMEA member countries, granting long-term concessional credit and expecting repayment in the form of commodities, a prac-

BOX 18.2 Ecological Effects of The War in Vietnam

The Second Indochina War left its mark on the landscape of the peninsula in many ways. Ecological damage was widespread, varied, and continuing. The weapons of war were far more wide ranging than had been used in any previous conflict and the effects more complex. The ecological changes were to some extent the result of the U.S. strategy to push South Vietnamese people into urban centers by making the rural areas impossible for them to live in (Lewallen, 1971, p. 34). Destruction took place on a massive scale in a number of areas of South Vietnam and was aimed at controlling the populace and making the environment inhospitable.

Methods used included:

Carpet bombing of the landscape
Extensive use of land mines
Defoliation of forests
Land clearance using "Roman ploughs"

Some of the effects:

Bomb craters covering extensive areas
Soil displacement and exposure of the subsoil
Compaction around the craters
Changes in the water table
Disruption of local drainage patterns
Accelerated erosion
Increased runoff; further soil erosion
Localized flooding
Destruction of flora
Loss of food supplies
Death and loss of animal, bird species
Loss of biodiversity
Chemical residues
Loss of soil fertility
Polluted water supplies
Possible birth defects
Mangrove destruction
Loss of fish breeding areas
Loss of food source for humans and birds
Major effect on the riverine regime

This brief list of the possible effects illustrates the range of problems besetting the people of Vietnam in their attempt to reestablish a degree of normality in their environment. Currently there are many schemes underway to clear mines, fill in bomb craters, and replace vegetation. It is however a very long process, and in many areas the environment cannot be restored to its prewar status. New and different ecological successions have developed.

Photo 18.11 Bombing in the Red River delta has left behind craters in the padi fields. (Forbes)

tice known as countertrade. In fact both Vietnam and Laos incurred huge debts to the Soviet Union. Some two-thirds of Laos's long term external debt of US$2.1 billion is owed to former CMEA countries. As a result of an agreement with the USSR in 1991, Laos's debts were to be repaid in so-called freely convertible currencies (rather than the currencies of the socialist bloc, such as the ruble) and through countertrade, but the latter has not materialized. A large proportion of Laos's trade is now with Western countries. The USSR also provided concessional aid to Laos, as it did to Vietnam, but this also ceased in the early 1990s.

Russia, which inherited the links to Vietnam and Laos that the former USSR had, has seen its standing in the two countries deteriorate significantly in recent years. It no longer has any significant political influence. However, there are still some important economic linkages; one is the joint venture oil company VietSovPetro. The opening of the economies of the three countries has shifted the dominant political orientation away from their main allies in the 1970s and 1980s, Russia and China, and toward neighboring Asian countries and the capital-exporting countries of the West.

Relations between Vietnam and China, historically poor, deteriorated further during the 1970s. Motivated by concerns about Vietnam's invasion of Cambodia, Chinese military pushed into the northern highlands of Vietnam in the late 1970s. The subsequent battles along the border between the two countries became known as the Third Indochina Conflict. Relations have since improved, and the road across the border between the two countries now carries considerable legal and illegal trade.

Opening Economic Relations

None of these countries have been easily integrated into the Western capitalist economy, though Laos's progress has been fractionally smoother than either Vietnam or Cambodia. Vietnam's stumbling block has been the U.S. economic embargo that was imposed initially on North Vietnam and subsequently extended to the whole country after unification. The *Trading with the Enemy Act* made it illegal for U.S. citizens to have any economic transactions with Vietnam. The embargo was eventually lifted in early 1994. Laos was subject to a similar order from the United States, and this was rescinded in 1995.

The impact of ending the embargoes has been relatively minor. Practically all the Western countries that had at one time given some support to the economic embargoes had openly ignored them by the late 1980s. In addition, Asian investors from South Korea, Taiwan, Singapore, Thailand, and Malaysia were undeterred and as a result were at the forefront of economic activity in Vietnam and Laos as those economies opened to the world.

In 1988 both Vietnam and Laos passed laws facilitating and promoting overseas investment. The result has been that foreign investment in Vietnam has accelerated in recent years. In the period from 1989 to 1994, more than 1,000 foreign investment licenses worth US$11 billion were issued by the Vietnamese government. Manufacturing industries are the most important foreign investment in Vietnam, followed by hotels and tourism, oil and gas, and services. Not all of these licenses will result in projects. Many companies will not take up the licenses because of problems with Vietnamese counterparts, because of difficulties with the foreign investment rules, or because the company is unable to fulfill its part of the arrangement. Vietnam is sometimes seen, quite falsely, as a frontier region where quick profits can be made. This reputation has attracted disreputable firms that frequently have not delivered the guarantees promised as part of their investment license. The result has been the inability to sustain a presence in the country.

The so-called Overseas Chinese investors have played an important role in the recent development of Vietnam. Three of the top four countries investing in Vietnam

Photo 18.12 Growing external investment through joint ventures and other forms of participation has begun to strengthen the economy and has produced the development of new tourist infrastructure that Vietnam badly needs if it is to launch a tourism promotion effort. The newly completed and refurbished Hanoi Hilton and Opera House are shown here. (Leinbach)

through the mid-1990s were Taiwan, Hong Kong, and Singapore. In addition Chinese entrepreneurs in neighboring countries such as Thailand and Malaysia have also been prominent. Whereas investors from these regional economic powers have dominated the initial phase of Vietnam's development expansion, it remains to be seen whether they will continue in this role as the United States, Japan, and the larger European economies consolidate their postembargo interests in Vietnam.

The main countries investing in Laos in 1993 were Thailand (US$61 million), China (US$15 million), Australia (US$12 million), Singapore (US$7 million), and Hong Kong (US$7 million) (Economist Intelligence Unit, 1994). As is evident from these figures, Laos's closest economic links are with Thailand. The relations, however, have not always been smooth. This is evidenced by Thai closures of the border between the two countries in the mid-1980s, and the occurrence of border skirmishes in 1988. Currently, Laos sells a broad range of goods and services in Thai markets. These range from electricity to timber and cattle. In addition to investment, the Thais have commenced a modest aid program in Laos (Handley, 1993). The Thai and Lao languages are very similar, and Lao watch Thai television with alacrity. Not surprisingly, the Lao struggle to establish an identity separate from the Thai (Ngaosyvathn and Ngaosyvathn, 1994).

The geographical location of Laos has helped it become a land bridge between China and Thailand. Chinese traders bring goods to Laos through the northeast, selling them to Lao consumers and Thai merchants in the northern Laohan town of Luang Namtha (see Map 18.6). Likewise, Thai traders bring manufactured goods, such as secondhand motor vehicles, for export across the border into southern China. Chinese traders have settled throughout the northeast provinces, raising Lao fears that increased local trade will result in more Chinese migrants to the region (Lintner, 1995a, p. 19). Politically Laos is in the process of forming closer links with China. Laos and China established full diplomatic relations in 1987. Former LPRP chief Kaysone Phomvihane visited Beijing in 1989, and Chinese Premier Li Peng responded with a visit to Vientiane the following year.

Aid to Vietnam

Most Western countries suspended their overseas development assistance programs to Vietnam in the late 1970s in protest over the invasion of Cambodia. Through most of the 1980s Vietnam received small amounts of support from countries such as Sweden; other countries, including Australia, funneled resources through the United Nations Development Program. The indirect flow of funds was maintained presumably in recognition of Vietnam's desperate need for foreign assistance. At the same time, donors were concerned that such flows not be contrary to the economic embargo on Vietnam.

A number of countries began resumption of aid to Vietnam in the early 1990s, and in 1993 the International Monetary Fund cleared the way for involvement by the World Bank and the Asian Development Bank. This aid ultimately will be of considerable significance to Vietnam's development. Annual meetings of multilateral and bilateral providers of development assistance to Vietnam are intended to ensure that aid is coordinated and in the highest-priority fields.

War in Cambodia

Cambodia represents a special problem for the international community. The capture, ransom, and subsequent murder of hostages by the Khmer Rouge has created a dilemma for those Western nations that have assisted Cambodia with nonlethal aid, in particular the United States, Australia, and France. In an effort to clean up the many thousands of mines remaining from the years of conflict and the new ones being planted to terrorize villagers into providing the Khmer Rouge with supplies, Western nations have been supplying aid such as engineers and equipment to remove land mines. The slaying of nonmilitary nationals puts political pressure on such countries to curtail their aid and possibly refocus on military solutions. The role of Western aid and trade in Cambodia's future is difficult to assess. The current government is precarious, lacking basic authority and political will. The Khmer Rouge have been losing ground during the 1990s, but few would confidently predict the permanent demise of the guerrilla force. As long as life for the Cambodian people remains fraught with difficulties, they are always at risk of further uprisings.

Toward Formal Regional Integration

The economic geography of the region is changing very quickly. While political borders remain intact, inevitably the complex and growing cross-border connections diminish their importance in the everyday lives of local residents. Yet paralleling the growth of local economic interaction has been the emergence of institutions designed to facilitate the links between countries within the region.

The Indochinese Union of the colonial days is long past. The fraternal relations that existed between the Communist parties of the three states were exploded in the 1970s by the bitter antagonism between the Khmer Rouge and the new rulers of Vietnam. The incursion of Democratic Kampuchea soldiers into Vietnam saw the latter retaliate and invade Cambodia in the late 1970s. Yet the withdrawal of Vietnamese troops in the late 1980s only partially managed to heal the wounds. Violent attacks on ethnic Vietnamese residents in Cambodia have

BOX 18.3 Land Mines in Cambodia

A tragic reminder of the recent turmoil in Cambodia is the constant danger faced by many people especially those in the countryside. During the Second Indochina War and its aftermath, many thousands of mines were laid that are still active and likely to detonate. The damage done to the people continues, many are maimed, some are killed. As a consequence 1 in every 230 Cambodians has suffered some degree of mine injury. The result of this continuing onslaught is devastating to the injured and their families.

As many as 10 million mines may remain live in Cambodia, covering an estimated 1,235 sq mi (3,200 sq km). The cost of removal is very high, up to US$1,000 per mine, or US$7 million per annum for 4 sq mi (10 sq km). Moreover, it is a dangerous task and requires well-trained operators. Removal of mines is therefore time consuming and requires meticulous work; in one instance it took up to 4 years to clear 10.4 sq mi (27 sq km).

THE EFFECT ON THE PEOPLE

Death or injury to people by land mines is only the beginning of the problems that effect Cambodian families. The inability of people to work their fields for fear of buried mines results in many areas being excluded for farming. The World Food Programme claimed that 45,000 Cambodians were unable to support their household's food needs fully as a result of mines and that as many as 150,000 others could not be adequately resettled due to the prevalence of mines. The damage and danger are not restricted to fields; forests, waterways, canals, and roads are also mined.

The effects are significant. People who are unable to farm need food supplies. If the parents in a family are injured, children will be required to remain at home and look after their family rather than attend school. This in turn reduces their chance of education and improved prospects. The medical costs to the country and individuals are significant. Aid must be spent on support for amputees; other aid money is diverted into mine clearance rather than redevelopment support; the family and village structure can be totally destroyed by resettlement and increased poverty.

WHO SUFFERS?

There is no doubt that Cambodia's children are the major victims of mines. Children's activities and their lack of literacy skills make them more vulnerable; their natural inquisitiveness and belief in themselves leads them into situations that adults avoid. Maimed children suffer physical and mental trauma, and their chances of a full education and the ability to develop their potential are greatly reduced.

Children suffer further if they are orphaned or have injured parents (see Box 18.1). They are left to cope with the burden of care for their families. Their childhood is destroyed.

Economically the family, village, and nation all suffer. People unable to work need support, land excluded from production puts pressure on usable land. Livestock cannot be grazed, and forest products become too dangerous to collect if mines are not removed.

pointed up the tense relationship between the Vietnamese and the Khmer Rouge in particular. Despite Lao involvement with China and Thailand, the link between Laos and Vietnam remains reasonably strong.

The Association of Southeast Asian Nations (ASEAN) is the main grouping of countries in the region. Set up in 1967, few would argue it has fulfilled all its original promise; nevertheless, despite its failure to improve the economic links between member countries significantly, it has established itself as an effective forum. Vietnam became the seventh member of ASEAN in mid-1995, Laos was admitted in 1997, and Cambodia was admitted in 1998. This represented a symbolic gesture of recognition of the fundamental shift that has occurred in the economic and political orientation of Vietnam and the real political importance of closer, more formal links between Vietnam and its neighbors (Tuan, 1994).

CONCLUSION

The second half of the twentieth century has produced much turmoil and change for Vietnam, Cambodia, and Laos. It began with a protracted period of decolonization and political struggles, interspersed with war. This provided an unstable foundation for each nation to make the transition to a socialist economy. Since the mid-1980s the region's transformation has been determined by an abandonment of key features of socialist economics and their replacement by a mixed-market economic system. Inevitably, Vietnam, Cambodia, and Laos are rapidly being integrated into the global economy. Yet both Vietnam and Laos are trying to preserve the existing system of government which is dominated by Communist parties. A palpable tension has been created by the juxtaposition of rapidly changing economies and inert political structures.

The impact on economic, social, and spatial structures and the environment has been profound. High levels of economic growth have been achieved, especially in Vietnam. The larger cities have been able to capitalize on this growth, with manufacturing industries emerging to take advantage of low labor costs. Escalating rural-to-urban migration is one consequence of this pattern of regionally uneven development. Meanwhile in the rural areas the dismantling of the government health-care system and the erosion of other social services, such as education, has left many people worse off than before.

On the positive side, after discussions over a five-year period, Vietnam agreed on July 29, 1999, to the terms of a trade agreement with the United States. It published its first state budget and has earmarked $13 million to set up a stock exchange. Although foreign investors remain on the sidelines, exports in 1999 were up 12 percent from the previous year. Inflation is under control at 4.5 percent per annum. Yet many problems remain, and among the highest priorities is reform of the public or state sector (FEER, 1999; *Economic Monitor,* Vietnam 1999).

Laos and Cambodia have been less affected by economic growth. In the case of Laos this is due to the government's ambivalence about opening up the economy, whereas Cambodia's political instability continues to scare off foreign investors. Both countries face an enormous task to find a niche in the global economy and at the same time raise domestic standards of living.

Predictions about the future of the region are often polarized. Glossing over the difficulties, some see no end to the economic blossoming of the region and predict that Vietnam will eventually emerge as an economy to rival the Four Tigers. An alternative interpretation points to the negative consequences of recent economic reform on a uniformly poverty-stricken population. On balance, the structural problems confronting the three countries, such as shaky political institutions and hopelessly inadequate physical infrastructure, will curb the pace of economic "take-off" of the region. However, now that the process of economic reform has begun, it is unlikely to be reversed. For years to come, progress in Vietnam, Laos, and Cambodia will be determined by the vicissitudes of its engagement with the global economy.

19 A Closing View: Development Dynamics-Prospects In The New Millennium

THOMAS R. LEINBACH RICHARD ULACK

As we noted at the outset of this volume, our objective has been to examine the character and great diversities in the Southeast Asian region and its individual countries by focusing upon the theme of past and present patterns of development. We have attempted to emphasize the key forces that affect and produce current developmental change as the individual countries seek to gain footholds in the international economy. In our introduction we called attention to several critical themes that we felt were seminal in assessing development in the contemporary environment. In this final chapter, rather than review and summarize the basic threads of progress that have been attained and the forces and constraints that have impeded this, we have chosen to focus on the major issues that we feel will be critical to further development in the coming millennium. We do this both to inform our readers and to use this as a charge to students, academics, and other researchers who we hope will use this volume as a stimulus for further inquiry on development in the region.

SETTING THE STAGE: THE ECONOMIC DIMENSION

The recent past in Southeast Asia has produced, as we have noted, rapidly growing economies. According to the World Bank there has been an increased equality of income distribution coupled with a reduction in poverty; at the same time, high growth rates have been sustained. Yet in a number of the Southeast Asian countries, there is a "wealth gap" that has become an important political preoccupation. Although some have shared in the fruits of the rapid growth, such expansion breeds high expectations and allows the majority to witness the wealth of an "elite" few. In Indonesia, the wealth gap is in fact one major reason for the toppling of the Suharto "New Order" in May 1998. Similarly, in the Philippines, Thailand, Malaysia, and even Vietnam, the "unevenness" of income has led to social unrest. Certainly the financial crisis that began in mid-1997 has dramatically changed the immediate future prospects for Southeast Asia. Whereas there is considerable variation among the countries, the financial crisis by mid-1998 had affected Indonesia, Thailand, and Malaysia the most seriously (see Boxes 12.3, 13.1, and 16.1; see also Box 15.3). Among the immediate effects were dramatic changes in the exchange rates, a decrease in GDP within the regional economy of at least 15 percent, sharp inflation, and tripling of unemployment to 10–12 percent. Moreover, the number of people living in poverty doubled from 26 to 52 million between 1997 and 1998, and some 30 million people are experiencing food insecurity.

The crisis has affected Indonesia most deeply. The failure to put together an effective initial response and the ultimate severity of the crisis must be understood and explained as the consequence of a lengthy process of politicization of economic and financial activity within the country. Indonesia's agreement with the IMF has been revised (as of September 1998) four times in response to deteriorating macroeconomic conditions. The result is a complex, multifaceted program to address macroeconomic imbalances, financial weaknesses, real sector inefficiencies, and the loss of private-sector confidence. The program covers fiscal and monetary policies, structural reform and deregulation, corporate debt and bankruptcy proceedings, banking sector reform and restructuring, and restoration of trade financing to promote exports.

Finally an important aspect of the recovery involves the Indonesian Chinese community. Ethnic Chinese form about 3.5 percent of Indonesia's 200 million population but hold the vast majority of the country's corporate wealth. Chinese businesses were specific targets of the politically orchestrated riots. As a result, some Chinese fled the country and may not return. The entrepreneurial skills the Indonesian-Chinese have developed will now be crucial to an economy in shambles. Of great significance is the fracturing of the retail and distribution network that was largely controlled by ethnic Chinese. Foreign investors, including Chinese businesspeople from Taiwan, Hong Kong, and elsewhere in Asia, will not likely step in until Indonesia guarantees the way that it does business. Yet within the country all groups may have to agree that a more equitable distribution of wealth is crucial to keep Indonesian society from unraveling.

Although the income inequality gap shrank between 1970 and 1990, the boom years of the 1990s suggest another trend. The share of national wealth held by the lowest 40 percent of the population has been contracting. One possibility, and indeed many ethnic Chinese and pribumis agree, is that an orderly redistribution of wealth such as occurred in Malaysia under the New Economic Policy may be possible in Indonesia. This is a longer run goal, and in the meantime there are other short term fixes to be implemented. Of critical importance is the outcome of the national election held in early June 1999 and thereafter the selection of a president. These decisions will have a powerful influence on the direction and progress of recovery during the next seven years.

THE ROLE OF THE STATE

The state will continue to be a strong force in Southeast Asia through the next century, particularly as governments come to grips with the unknown future of changed economic conditions. New issues will emerge on state policy agendas, perhaps particularly environmental and natural resource related issues. For example, Singapore and Thailand have both been searching for more water from their neighboring countries. Malaysia has promised to continue to supply Singapore through at least the first half of the century, but already Singapore wants more water than Malaysia is willing to channel to the island. Laos is constructing dams that will store water resources for Thailand. In the forest-resources sector, complete deforestation of primary forest for timber resources is possible early in the next century. In addition, the results of extensive damage to forests from fires on Sumatra and Borneo during the El Niño phases will not be fully understood for some years. The region will finally likely be compelled to restructure its forestry industries and review the environmental damage of widespread and rapid deforestation practices that began in the 1960s.

The region's new economic problems may compel ASEAN to evolve further and to strengthen its leadership role. The role of the state in the region will experience a new round of assessment as people look widely for causes and contributions to the recent Asian financial crisis. One particularly geographical contribution to the regional economic downturn is overextension in the real estate sector. In the later phases of rapid industrialization, the more economically developed of the states plowed foreign exchange earned from export-oriented manufacturing into "megadevelopment" projects. The "Twin Towers" in Kuala Lumpur is the most obvious of these, but the state widely supported many less glamorous but more land-hungry housing estate and golf course developments during the same period. As property values fall with the onset of recessionary conditions, both private developers and the state will not be able to recoup investments, thus delaying recovery.

How effective the state will be in mediating social concerns remains to be seen. Will Indonesia adequately address the disastrous social problems resulting from the riots that ultimately brought down former President Suharto? Citizens groups around the world have protested the distinct lack of a considered role of the state in seriously investigating the scores of documented crimes against women that occurred during the May 1998 riots. Untangling the complexity of this issue will be a long process, and only active participation by the state will bring resolution to the Indonesian polity.

The financial crisis, coupled with the region's colonial history, placement in the world economy, and the role of the state will, of course, have major implications for future prospects in a variety of areas including population and social prospects, cities, the cultural arena, and the environment. It is to these topics we now turn.

POPULATION AND SOCIAL PROSPECTS

The devastating impacts of the crisis will undoubtedly hamper social progress in Indonesia for at least a decade; its effects elsewhere will also be evident, though less so. Although some aspects of future social change must be speculative, some appear almost certain to occur; for example, we can be sure that the regional population will age. The latest population projections from the United Nations indicate that the number of Southeast Asians age 65 years and older will increase by 118 percent between 1995 and 2020, while the population as a whole will increase by only 37 percent. The proportion of the population age 65 years and older will increase from 4.3 to 6.8 percent. On the other hand, the growth of the population age younger than 15 years will slow, and their proportion will decline from 34.4 to 24.3 percent (see Figure 4.9). This will reduce the pressure on education systems and reduce the growth rate of the workforce, allowing a greater emphasis on quality education. Of course the wide variations in the countries with respect to age structure will continue to be significant, and the pressure for population to move, at least on a temporary basis, from such labor-surplus countries as Indonesia to such labor-shortage states as Singapore, will increase.

Population mobility undoubtedly will continue to increase both within and between countries. This is partly a function of increased education, improved and lower-cost transport, and the greater and more rapid spread of information about alternative places. However, it is also due to the expansion of the social networks created by the migrations of the 1980s and 1990s, which have greatly increased the number of contacts that Southeast Asians have with family members and friends in locations distant from their origins. Although part of the response to the financial crisis in Indonesia and Thailand has been a return of former rural-to-urban migrants to their village homes, it is certain that the urbanization of the population is likely to resume, and this will continue across the region, leading to one-half the region's population living in urban places by 2020, compared with one-third at present. Absolute declines of rural populations began in some areas in the 1990s, and these will become more widespread so that there will be absolute reductions in the number of rural dwellers in several countries by the early decades of the twenty-first century.

Regional declines in fertility and mortality have been dramatic in the last two decades, and they can be expected to continue, although countries such as Singapore and Malaysia have initiated pronatalist policies. The significance of declines will be much greater in some countries than others, but further reductions in both fertility and mortality can be expected in all countries, although much will depend on the extent to which the region remains peaceful.

The region's most important resource for development is its people, and the next two decades should witness substantial improvements in the quality of human resources. Almost all countries in the region have given educational development a high priority, and whereas there is substantial variation, there is a need to improve both the coverage of education and its quality. It is also necessary in some countries to overcome mismatches between the needs of restructuring modern economies and the current formal education system. Such mismatches, as have occurred in Indonesia, have sometimes created high levels of unemployed educated youth that will become an even more significant problem during the next decade. Indeed, the youth population in the region will be of crucial significance to political, social, and economic development.

The family will remain the fundamental social unit in the region in coming decades, but it will change significantly in its size, structure, and functioning. The transition from an extended family to a nuclear family will continue, families will become smaller, and age at marriage will continue to increase. In the long run, the role of women should also experience a change. With increased female participation in both education and the workforce outside the home, empowerment of women will continue, and hopefully the gap between men and women in the economy, society, and family will narrow.

In the shorter term, the economic downturn in the region will widely affect women and women's work disproportionately by comparison to men. Across the region, especially in countries implementing structural adjustment programs (SAPs), women will, on average, be the first to lose jobs. In addition, the demands of their "triple workday" will increase, as women look to informal ties and community networks to obtain household provisions that they are no longer able to buy or procure in the market. Similarly, families with girls who are old enough to work will often encourage the girls to leave school for paid work of any type to help with the household income. Although it can be generally expected that the employment status of women will deteriorate during a serious economic downturn, it may be possible that the relatively high status of women in Southeast Asia will mediate the extremes of these trends.

Many Asian students studying abroad, both men and women, have been called home by their families as Asian currencies have depreciated. For families who have been able to send their sons and daughters abroad for education, there does not seem to be a clear trend impacting females more than males. Only research on the next generation will tell just how the changing political economy of the region will have affected men and women's lifepaths.

CHANGING URBAN STRUCTURE AND FORM

The most obvious urban trend is the continued growth of the very largest cities, with the exception of Singapore, due to rural-to-urban migration. Although the recent economic crisis has caused a partial slowdown of the pace of economic development, industrialization will continue, drawing rural populations into low-wage employment and causing expansion of the cities into surrounding areas, creating what have been called extended metropolitan regions (EMRs). The phenomenon of the Southeast Asian megalopolis has arrived, witnessed by the sprawling urban regions of Kuala Lumpur in the Klang Valley (KVC), JABOTABEK, the Bangkok Metropolitan Region (BMR), the Manila Metropolitan Region, and on a smaller scale, Hanoi–Haiphong and Saigon–Cholon.

In such cases there are multiple urban centers within a complex mix of land uses and ways of life, effectively combining and confusing the former boundaries between the rural and urban.

Growth rates of 3 to 4 percent in the largest cities place considerable burden on already stressed urban services, leading to potentially serious urban environment and social problems. Industrialization and rapidly expanding ownership of private vehicles cause land degradation and pollution of water and air resources, as increasing population density puts stress on inadequate drainage, water supply, and waste management services. Diseconomies of scale in the urban centers—most obvious but not limited to the infamous traffic congestion and air pollution of Bangkok, Jakarta, and Manila—contribute to the expansion of the city into the surrounding periurban region. Most of these cities are on or close to the coast—in fact, 50–60 percent of the urban population of Southeast Asian countries lives in coastal areas—so to the problems of industrialization and population growth we must add the potential effects of local and global environmental change and natural hazards for which governments are inadequately prepared.

The diseconomies of scale in the large cities combined with government incentives could conceivably lead to investment in secondary cities. Certainly, smaller cities will grow, often at faster rates than the very large cities, but absolute numbers are much smaller. No matter how well intentioned and conceived the planning intervention, therefore, it is unlikely to slow growth effectively in the largest, "million" cities. Moreover, national governments committed to economic growth have tended to underplay the environmental and social costs of rapid urbanization and have adopted remedial rather than proactive planning strategies. It is unlikely, however, that this can continue, first as the problems accumulate and come to hinder economic growth itself, and second as the beneficiaries of economic development begin to demand improved quality of life. To some extent the urban middle classes can escape the negative effects of urban growth, and one witnesses miles of bumper-to-bumper traffic heading out of Jakarta, Bangkok, and Manila for the countryside on weekends, but as the city expands this is obviously a less viable option. Meanwhile, community-based groups among the lower classes and urban poor are combining with NGOs to meet their own needs for urban services or to make demands on government. Clearly the future of urbanization in Southeast Asia will depend on how governments meet the challenge of civil society and are able to acknowledge and incorporate local initiatives into a creative urban-planning process that alone they lack the economic and political capacity to effect.

It is clear that forces of social and economic change are strongly influencing rural areas, and concomitantly urban areas, in a variety of ways. One is seen in the process of deagrarianization, which is becoming more and more conspicuous. Nonagricultural rural employment (NARE) for households is not only desirable but is indeed a necessary alternative to farming in many communities. This is so because of constraints on land availability, contraction of urban employment as a result of economic crises, and an increasingly educated population that is much less satisfied with the agrarian lifestyle. Agriculture is clearly being squeezed by nonagricultural pursuits. Part of the rural change also involves a "restructuring of the household as genders and generations contest and renegotiate their respective roles" (Rigg, 1998). The diversification of the household economy and the increasingly blurred boundary between urban and rural has created unique microeconomic family strategies as individuals and households move between agriculture and industry and urban and rural spheres. The harsh economic climate in Southeast Asia will, at least for the near future, force individuals and families to transit increasingly to quite different livelihood schemes. Yet despite these ongoing changes we see within the region a clear resilience in its leaders and people that will endure and succeed in the years ahead.

HUMAN RIGHTS AND CULTURAL–ETHNIC IDENTITIES

Southeast Asia has been described as a cultural shatter belt, a region that has experienced conflict based on cultural differences for centuries. The story of how the region's fragmented physical nature coupled with its excellent relative location influenced its cultural geography and immigration patterns is well known. It was, however, the eventual colonizers from Europe and the United States that have perhaps had the greatest impact on the contemporary cultural conflict and human rights problems that have persisted in the region since independence. As Hirschman (1995, p. 29) notes, it was these colonizers who brought to the region racist beliefs and, also, the processes by which Chinese immigrated into the region and were accommodated. The colonizers also created the state political boundaries and many of the political institutions that exist today; boundaries that were superimposed on the region, sometimes including many diverse groups within a single colony and sometimes dividing groups between colonies. These boundaries and institutions of course form, and to a large extent, guide the 10 states that comprise the region today. Any analysis of the potential for ethnic conflict in the region must recognize the importance of these boundaries and the policies and politics of the states within and among which the various groups are located. National governments in Southeast Asia vary greatly in regard to their relations with minority groups. In short, ethnic consciousness and the potential for ethnic conflict, at least in Southeast Asia, are closely related to state policies and politics. What is crucial to human rights and cultural minorities throughout the region is the way in which the state addresses these issues. In Southeast Asia today, policies of state governments (which are controlled by the majority or the largest ethnic group) play the key role in the status of minority groups. Overall, as David Brown argues, "in each of the Southeast Asian states, . . . ethnic interests have been deemed democratically legitimate by the state only when they have been supportive of the incumbent regime. Ethnicity thus inhabits a shadow world—liable to be designated as subversive communalism at one moment, but applauded as the legitimate articulation of cultural values and interests at another" (Brown, 1996, p. 310).

The current financial crisis has exacerbated the conflicts that have occurred in the past between classes as well as between ethnic groups in several parts of the region. Perhaps the most serious conflicts will arise in Indonesia as the social and economic situation there deteriorates. Indeed, this has already negatively affected Indonesia's Chinese minority and has perhaps brought out long simmering tensions among Dayaks, Malays, and transmigrant Madurese in Kalimantan for example. Other areas of tension remain in Aceh, East Timor, and Irian Jaya. In addition other regional groups likely to be involved in conflict include the Muslims of the southern Philippines, southern Thailand, and western Burma, as well as such other Burmese groups as the Karen, located in the periphery of that state (see Map 10.2).

ENVIRONMENTAL IMPACTS AND CONCERNS

The recent extensive forest fires in Indonesia (see Box 12.1) especially and the El Niño triggered drought in the region are only two strong reminders of the importance of the environment in development and its often precarious situation and impact. Drought cannot be prevented, but its impact upon society can be mitigated. This is certainly a responsibility of the state. Similarly the role of the state and its policies must be questioned in connection with widespread burning. In addition to the major question of preserving forest cover and/or harvesting it in a sound manner is the focus on the sources and consequences of human activity in landscapes already cleared of natural forest cover. Recognizing that much of the forest has been removed, attention must now be directed to the sustainability of cleared areas. Because forest conservation areas that exclude people are often more susceptible to species degradation, sustaining biodiversity is more promising in areas where people—especially in lightly populated areas—possess opportunities to manage their own resources. Providing a sustainable future must involve small-scale swidden farmers who have gradually shifted to highly productive forms of padi and complex agroforestry. The key to the sustainability of logged areas once timber-based employment declines is the promotion of evolving agroecosystems, as opposed to centrally controlled tree cropping that culturally discriminated against forest people. Only then may the present stage of endangerment not evolve into a stage of criticality. Despite the concerns of both the government and environmentalists, grassland agriculture based upon extensive forms of production, such as grazing, padi, and cash crops, has proven to be quite successful. The ability of modified environments to become sustainable, however, requires local decision-making power rather than the continued concentration of political and economic power among the few.

For more than two decades policy makers and administrators in the region have been abundantly aware of the economic and social implications of deforestation, pollution, haphazard land-use decisions, and a variety of other unwanted by-products of development. Yet given national self-interest, the preoccupation of governments with short-term priorities, and especially the vested interests of so many individuals and groups in growth and capital accumulation, the future sustainability of common property resources remains as fragile and uncontrollable as ever. Market processes contain powerful mechanisms to combat both the physical scarcities of "developmental" resources, with examples ranging from the timber of Kalimantan to the hydroelectric potential of the Mekong River basin and any geopolitical threats to remove resource supplies. But these processes have not acted to protect the renewable and environmental resources, nor can they do so unless there is the political will to change management institutions and to alter the conventional values and perceptions that have been placed on the environment.

CLOSING THOUGHTS

The last two decades have seen unprecedented changes in the economic, political, and social fabric of most of Southeast Asia. Traditional ways of livelihood have been challenged as never before, and it is difficult to see how this will change in the near future. The sum of the influences has gradually during an even longer span of time had a deep impact. The mutually reinforcing dynamisms of commercialism and demography especially have had a tremendous impact on the traditional social fabric of the region–the peasantry (Elson, 1997, p, 239). Development has in some ways ushered in a cultural crisis where local identities have been lost. There are indeed strong arguments suggesting that the end of the peasantry has come to Southeast Asia even though there is an enduring existence of rural production and of the social and economic circuits of life and livelihood that accompany it. "But the rural producers of today are productive and enduring only insofar as they have moved from being peasants into new and different worlds of production, and consequently, of social and economic life. In fact the peasantry has been radically reconstituted through its own agency and that of the broader forces of change" (Elson, 1997, p. 240). The nature, scale, and pace of development will of course continue to vary substantially over the region. But Southeast Asia has in many ways developed a place in the global economy. Although we are confident that the countries most severely ravaged by the financial crisis of 1997–1998 will rebound and indeed are doing so now, the path ahead will be difficult and progress slow. Infrastructure must be upgraded, and a financially sound and globally competitive environment must be redeveloped. Both tourism and manufacturing are keys in the new progress. The construction of these favorable factors will be made all the more difficult because of the now ensconced spirit of democratization and reformation as well as the creation of new personal and community identities.

Glossary

abaca—a Philippine plant related to the banana whose stalks yield a fiber that are the source for Manila hemp

adat—indigenous system of custom, belief, and laws in Indonesia

Association of Southeast Asian Nations (ASEAN)—founded in 1967 for the purpose of promoting regional stability, economic development, and cultural exchange. ASEAN's founding members were Indonesia, Malaysia, the Philippines, Singapore, and Thailand; Brunei joined ASEAN in 1984; Vietnam in 1995; Laos and Burma in 1997; and Cambodia in 1999

Bahasa Indonesia—the Indonesian national language, also known as Indonesian; an Austronesian language reported to be modelled on Riau Malay and 80 percent cognate with Standard Malay

barangay—indigenous system of community organization in precolonial Philippines; recently replaced Spanish barrio system as the local level form of political representation

becak—trishaw in Indonesia

bemos—3-wheeled vehicles in Indonesia

Brahman—highest (priestly) Hindu caste

bumiputra—"sons of the soil" in Malaysia; basically, the indigenous Malay population

betel nut—the seed of the fruit of the betel palm; when chewed with betel leaves and lime it is a mild drug

break-of-bulk point—a location along a transport route where cargo and/or passengers must be transferred from one carrier to another, which is sometimes another mode of transport (e.g., in a seaport from an ocean-going vessel to a train)

cassava—a starch derived from the root of the cassava plant, used in tapioca and as a staple food in some tropical areas

citizen politics—a state in which there is toleration for democratic practices and the evolution of a civil society; in Southeast Asia best epitomized by the Philippines

chain migration—a process whereby initial successful cityward migrants impress upon their rural friends and relatives their urban lifestyle, and provide contacts and resources in the city, both reinforcing the desire and providing the necessary means for other migrants to follow. Such chain migration is a cumulative process, initially selective of the most informed and ambitious, but expanding to incorporate elaborate migrant networks reaching down the urban hierarchy into remote rural areas, drawing the populations of towns and villages to the city and into the urban economy

changwat—province in Thailand

cinchona—a tree whose bark yields quinine; native to South America

circulation (sometimes referred to as **circular migration**)—form of human mobility characterized by temporary movement where the intent of the circulator is to eventually return to his or her origin (the intent of migration is to permanently reside at the destination). Usually defined as involving at least a one-night stayover at the destination but usually is much longer. "Migrant" farm laborers, college students, and overseas workers are examples of circulators

closed economy—an economy in which there are not foreign transactions or any other form of economic contact with the outside world

copra—the dried meat of the coconut from which coconut oil is extracted

current account deficit—exports and imports of goods and services, net factor income from abroad, and unilateral transfers (remittances, gifts, and foreign aid)

datu—local chief, in the Philippines

desakota—the term derives from the Bahasa Indonesian words for village (desa) and town (kota), and refers to extended metropolitan regions (*see EMR*)

deva-raja—god-king

dipterocarp (*Dipterocarpaceae*)—family of tall tropical hardwood trees dominant in Southeast Asian forests. Species range in height from 125 to 200 feet (38 to 61 meters) and have a full canopy of broad leaves that are usually evergreen. Among the more common species are red and white lauan, narra, and ipil

division of labor (international or **spatial)**—in the most basic sense allocation of tasks among laborers such that each one engages in tasks that he or she performs most efficiently and this promotes worker specialization and thereby increases overall labor productivity (from Adam Smith's *The Wealth of Nations*). In the context of globalization increasingly the reference is to countries or regions that specialize in tasks for which they have a comparative advantage based upon the cost of factors of production and especially labor

doi moi—means "renovation" of the Vietnamese economy. Formally established in 1986, it has become the new state ideology for economic and social development

economic infrastructure—underlying amount of capital accumulation embodied in roads, railways, and other forms of transport and communications as well as water supplies, financial institutions, electricity, and public services such as health and education

encomienda (ecomenderos)—system of land holding whereby several villages were granted to Spanish colonizers in the Philippines; payment of taxes and tribute was made to the encomenderos

entrepôt—a seaport that serves as a transit point for imports and exports moving into and out of a region

exclusive economic zone (EEZ)—a wide belt of sea and seabed adjacent to the national boundaries where the state claims preferential fishing rights and control over the exploitation of mineral and other natural resources. Boundary disagreements with neighboring states sometimes prevent the extension of the EEZ to the full limits claimed. For example, the Philippines claims a 200-nautical mile EEZ, now considered the international standard

extended metropolitan region (EMR)—term derived from the work of geographers T.G. McGee and N. Ginsburg which refers to the emergence of large and distinct urban regions wherein the difference between urban and rural is disappearing. Jabotabek is such a region

footloose industries—manufacturing industries whose location is not dependent upon the location of natural resources (such as tin, coal, or timber) or agricultural products, but only to local labor

free trade area—a form of economic integration in which there exists free internal trade among member countries but each member is free to levy different external tariffs among nonmember countries (e.g., AFTA)

gambier—an astringent extract obtained from leaves and shoots of a tropical Asian shrub which is used in dyeing, tanning, and medicine

gendered division of labor (GDL)—refers to the patterns of concentration of men and women in particular job roles and occupations

GERBANGKERTOSUSILA—an acronym refering to the large conurbation that incorporates Surabaya and the adjoining *kabupaten* of Bangkalan, Mojokerto, and Sidoarjo

Green Revolution—agricultural revolution which began about 1950s whereby high-yielding varieties (HYVs) of rice and other cereal grains were developed and produced in certain developing countries. The world center for research on HYVs of rice is located in Los Baños, the Philippines

infrastructure (*see economic infrastructure*)

International Monetary Fund (IMF)—an autonomous international institution whose main purpose is to regulate the international monetary exchange system and in particular to control fluctuations in exchange rates of world currencies in a bid to alleviate severe balance of payments problems

JABOTABEK—an acronym made up of Jakarta and the three adjoining West Java *kabupaten* of Bogor, Tanggerang, and Bekasi

jute—tropical Asian plant that yields a strong, coarse fiber used for sacking and cordage

kabupaten—Indonesia is made up of 27 provincial-level administrative units. The provinces in turn are subdivided into districts called kabupaten. According to 1991 statistics, Indonesia had 241 districts (kabupaten)

kampung—the general term for village in Indonesia (also *desa*) and Malaysia

kenaf—(*Hibiscus cannabinus*) an annual herbaceous member of the mallow family. The kenaf plant is considered one of the most promising alternatives to virgin soft and hardwoods and is used in industrial applications such as paper production. Related to cotton and okra, kenaf is indigenous to West Africa. In 1960, the U.S. Department of Agriculture chose kenaf from among 500 candidates as the most promising non-wood fiber for pulp and paper production

kotadesasi—term used by the geographer T.G. McGee to describe a uniquely Asian form of urban development, whereby non-agricultural activities develop in rural areas on the urban periphery and in corridors between cities. Derived from the words town/village in the Bahasa Indonesian language

ksatriya—warrior caste in Hinduism

liberalization—policies of economic reform highlighted by deregulation and privatization

matriarchy—female line defines descent, women dominate, and female norms lead to privilege and power

modern variety—refers to the short-statured, stiff-strawed, fertilizer responsive, nonphotoperiod-sensitive Indica padi (rice) varities typified by the IR8 variety which was developed at IRRI between 1962–1966

nagara—capital city

nats—spirits in Burma from the pre-Buddhist period which are still revered

nipa (palm)—a palm that grows in coastal areas and has long leaves which are used for thatching

orang asli—"original people" of the Malay Peninsula

outport—port that serves an interior, larger city. Examples include Muara (Bandar Seri Begawan), Belawan (Medan), and Port Klang (Kuala Lumpur)

padi—flooded rice field

Pancasila—introduced by Sukarno, it is the cornerstone of Indonesia's domestic state ideology. It is based on the five principles of montheism, nationalism, humanitarianism, social justice, and democracy

pandanus—any of a variety of palm-like trees and shrubs whose leaves yield a fiber used in woven articles such as mats

patriarchy—social organization based on descent through the male line combined with societal values based on masculine norms and values and the dominance of men in the home and workplace

patron-client relationship—whereby subordinates accept the rule of their superiors with the expectation that leaders will supply the basic needs of society; the client is loyal to the patron who offers protection

political economy—the attempt to merge economic analysis with practical politics, i.e., to view economic activity in its political context

polity—an organized society, such as a nation or state, having a specific form of government

pribumi—literally, an indigene, or native of Indonesia. In the colonial era, the great majority of the population of the archipelago came to regard themselves as indigenous, in contrast to the non-indigenous Dutch and Chinese (and, to a degree, Arab) communities. After independence the distinction persisted, expressed as a dichotomy between elements that were pribumi and those that were not. The distinction has had significant implications for economic development policy

privatization—process of growth of private ownership as a means of production and in contrast to state owned and operated enterprises (SOEs)

rentier economy or **rentier state**—where a country's sources of income are largely externally derived and are not primarily from domestic taxation

Repelita—Indonesia's five-year national development plans

sawah—Indonesian word meaning "flooded padi fields"

sericulture—the raising of silkworms for the production of raw silk

shadow ecology—set of environmental impacts one country's political-economic activities produce in the natural resource base of other countries

social construction of gender—the set of processes through which a society promotes particular female and male norms of behavior

Southeast Asia Treaty Organization (SEATO)—established in September, 1954 as a result of the 1954 Geneva Agreements to halt the spread of communism in Southeast Asia. SEATO never had an active military role and was ultimately disbanded in 1977 following the success of the communist movements in Cambodia, Laos, and Vietnam in 1975. The original signatories to SEATO were Australia, Britain, France, New Zealand, Pakistan, the Philippines, Thailand, and the United States

structural transformation—the process of transforming the basic industrial structure of an economy so that the contribution to national income by manufacturing increasingly becomes higher than that of the agricultural sector

subduction zone—area wherein one crustal plate descends under an adjoining crustal plate; the subducted material melts and it is an area where vulcanism and other tectonic activity is intense

sunrise and sunset industries—refers to industries with differing emphases of labor, capital, raw materials, and technology, which are dominant at different stages in the industrialization process. In part the changing structure of industry is influenced by available technology which redefines the nature of jobs performed, skills required and the training and qualifications needed. "Sunset" industries such as paper and pulp, textiles, footwear, and food products are industries which have high raw material inputs, little value added, and are normally associated with the beginning stages of the industrial process where ISI prevails. "Sunrise" industries such as computer and electronics appear as the industrial base undergoes further restructuring, are also often labor intensive but demand somewhat more skilled labor, rely on imported parts, and reflect demand for new technology driven products.

Suvarnadvipa—Hindi for Peninsula, or "Island of Gold"

swidden—slash-and-burn, shifting agriculture

Tai—used here to refer to any people speaking a language belonging to the Tai family of languages; examples include the Shan of Burma and southern China, the Siamese or Central Thai of Thailand, and the Lao of Laos (*see also Thai*)

tapioca—a starch obtained from the root of the cassava plant, used for puddings and a thickening agent in cooking

taro—plant cultivated in tropics which is valuable for its large, starchy, edible root

taungya—slash-and-burn farming in Burma

territoriality—regulation and control strategies employed to defend the territorial state

Thai—refers here to Tai-speaking people (and also some non Tai-speaking peoples) who are also citizens of modern Thailand (*see also Tai*)

total fertility rate (TFR)—the average number of children women would bear if they went through their child-bearing years conforming to the age-specific fertility rates of a given year

trishaw—three-wheeled motorized vehicle used for public transportation

urban growth—simply, the growth of cities and urban population

urban involution—the capacity of a city's service, and especially the informal, sector to absorb more and more laborers and to split, share, or fractionalize employment opportunities so that most are maintained at least at the subsistence level

urbanization—a process involving the movement of people from rural to urban places and their subsequent change from a rural to an urban life-style, with its associated values, attitudes, and behaviors

wat—Buddhist temple

References

Adams, W.A. *Sustainable Development: Environment and Sustainability in the Third World.* London and New York: Routledge, 1990.

Adas, Michael. *The Burma Delta: Economic Development and Social Change on an Asian Rice Frontier, 1852–1941.* Madison: University of Wisconsin Press, 1974.

Adicondro, G.J. *Datang Dengan Kapal, Tidur di Pasar, Buang Air di Kali, Pulang Naik Pesawat (Arrive by Ship, Sleep in the Market, Urinate in the River, Leave by Plane).* Jayapura: Yayasan Pengembangan Masyarakat Desa Irian Jaya, 1986.

Agnew, John. "The Territorial Trap: The Geographical Assumptions of International Relations Theory." *Review of International Political Economy*, vol. 1, no. 1, 1994, p. 53–80.

Ahlburg, D.A., ed. *Independent Inquiry Report into Population and Development.* Australian Government, April 1994.

"Aids and the Philippines." *Asia Pacific Observer*, vol. 1, no. 1, April–June 1994, p. 2.

Aiken, R., C. Leigh, T. Leinbach, and M. Moss. *Development and Environment in Peninsular Malaysia.* Singapore: McGraw-Hill, 1982.

Aiken, S. Robert. "Early Penang Hill Station." *The Geographical Review,* vol. 77, no. 4, October 1987, p. 421–39.

Aiken, S. Robert. *Key Economic Indicators of Developing Asian and Pacific Countries, 1993, vol. XXIV.* Manila: Economics and Development Resource Center, 1993.

Aiken, S. Robert, and C.H. Leigh. "Malaysia's Emerging Conurbation." *Annals of the Association of American Geographers,* vol. 65, 1975, p. 546–63.

Airriess, C.A. "Export-Oriented Manufacturing and Container Transport in ASEAN." *Geography,* vol. 78, 1993, p. 31–42.

Airriess, C.A. "Port-Centered Transport Development in Colonial North Sumatra." *Indonesia,* vol. 59, 1995, p. 65–91.

Airriess, C.A. "The Spatial Spread of Container Transport in a Developing Regional Economy: North Sumatra, Indonesia." *Transportation Research A,* vol. 23, 1989, p. 453–61.

Airriess, C.A., and T. Kohno. "Interdependent Industrialization: Japan, Indonesia and The Asahan Project." *Tijdschrift voor Economishe en Sociale Geografie,* vol. 84, 1993, p. 207–19.

Alavi, Rokiah. *Industrialisation in Malaysia: Import Substitution and Infant Industry Performance.* London and New York: Routledge, 1996.

Alburo, F.A., C.C. Bautista, and M.S.H. Gochoco. "Pacific Direct Investment Flows Into ASEAN." *ASEAN Economic Bulletin,* vol. 8, no. 3, 1992, p. 284–308.

Alexander, Jennifer. *Trade, Traders, and Trading in Rural Java.* Singapore: Oxford University Press, 1987.

Ali, Anuwar. *Malaysia's Industrialization: The Quest for Technology.* Oxford: Oxford University Press, 1992.

Ali, Anuwar, and Wong Poh Kam. "Direct Foreign Investment in the Malaysian Industrial Sector." *Industrialising Malaysia: Policy, Performance, Prospects*. Ed. K.S. Jomo. London: Routledge, 1993, p. 77–117.

Amer, R. "The Ethnic Vietnamese in Cambodia: A Minority at Risk?" *Contemporary Southeast Asia*, vol. 16, no. 2, 1994, p. 210–38.

Andaya, Barbara Watson. "Women and Economic Change: The Pepper Trade in Pre-Modern Southeast Asia." *Journal of the Economic and Social History of the Orient*, vol. 38, no. 2, 1995, p. 165–90.

Andaya, Barbara Watson, and Leonard Y. Andaya. *A History of Malaysia.* Hong Kong: Macmillan Publishers, 1982.

Anderson, Benedict R. O'G. "East Timor and Indonesia: Some Implications." *East Timor at the Crossroads: The Forging of a Nation*. Ed. P. Carey and G. Carter Bentley. London: Cassell; New York: The Social Science Research Council, 1995, p. 137–47.

Anderson, Benedict R. O'G. *Imagined Communities: Reflections on the Origin and Spread of Nationalism*. London and New York: Verso, 1983.

Angel, S., and Sopon Pornchokchai. "Bangkok Slum Lands: Policy Implications of Recent Findings." *Cities,* vol. 6, no. 2, 1989, p. 136–46.

Antonio, J.J. (revised by W.W. Fegen). *Guide to Bangkok and Siam*. Bangkok: Siam Observer, 1904.

Ariffin, Jamilah. *Women and Development in Malaysia*. Selangor Darul Ehsan, Malaysia: Pelanduk Publications, 1992.

Armstrong, Ruth. *The Kachins of Burma*. Bloomington, Indiana: Eastern Press, 1997.

Armstrong, W., and T.G. McGee. "Revolutionary Change and the Third World City: A Theory of Urban Involution. *Civilizations*, vol. 18, no. 3, 1968, p. 353–78.

Armstrong, W., and T.G. McGee. *Theatres of Accumulation: Studies in Asian and Latin American Urbanization.* New York: Methuen, 1985.

Armstrong, W., and T.G. McGee. "Women Workers or Working Women? A Case Study of Female Workers in Malaysia." *Theatres of Accumulation: Studies in Asian and Latin American Urbanization*. London: Methuen, 1987, p. 202–17.

Arnold, E.P. *Earthquake Hazard Mitigation Programme in Southeast Asia: Technical Evaluation, Executive Summary, and Final Report*. Denver: U.S. Geological Survey, 1986.

ASEAN. *ASEAN Statistical Indicators.* Singapore: Institute of Southeast Asian Studies, ASEAN Secretariat, 1997.

ASEAN. *ASEAN Strategic Plan of Action on the Environment.* Jakarta: ASEAN Secretariat, 1994.

Ashakul, Teera. *Migration: Trends and Determinants.* Bangkok: Thailand Development Research Institute, 1989.

Asia Watch Women's Rights Project. *A Modern Form of Slavery: Trafficking of Burmese Women and Girls into Brothels in Thailand.* New York: Human Rights Watch, 1993.

"Asian Defence Spending." *The Straits Times,* May 6, 1997, p. 25.

Asian Development Bank. *Asian Development Outlook 1994.* Hong Kong: Oxford University Press, 1994.

Asian Development Bank. *Economic Cooperation in the Greater Mekong Subregion.* Manila: Asian Development Bank, 1993a.

Asian Development Bank. *Key Indicators of Developing Asian and Pacific Countries, 1993,* vol. XXIV. Manila: Economics and Development Resource Center, Asian Development Bank, 1993b.

Asian Development Bank. *Subregional Economic Cooperation.* Manila: Asian Development Bank, 1993c.

Asian Development Bank and Winrock International. *Proceedings of the Regional Workshop on Sustainable Agricultural Development in Asia and the Pacific Region.* Manila: Asian Development Bank, 1992.

Athukorala, Prema-chandra. "Malaysia." *East Asia in Crisis.* Ed. R. McLeod and Ross Garnaut. Routledge, 1998.

Atkinson, Jane M., and Shelly Errington, eds. *Power and Difference: Gender in Island Southeast Asia.* Stanford: Stanford University Press, 1990.

Aung-Thwin, Michael. *Pagan: The Origins of Modern Burma.* Honolulu: University of Hawaii Press, 1985.

"Back on the Road: A Survey of the Philippines." *The Economist,* May 11, 1996, p. 1–20.

Badgley, John. "Myanmar in 1993: A Watershed Year." *Asian Survey*, vol. 34, Feb. 1994, p. 153–59.

Bangkok Bank Monthly Review, "Key Economic Indicators of Thailand," vol. 34, no. 5, 1993, p. 32–5.

Bangkok Post Economic Review Mid-Year, 1994. Bangkok: Post Publishing Co. Ltd., 1994.

Banks, Tony. "The Politics of Agrarian Reform Implementation: The Case of the Multinational Corporate Pineapple Plantations Under the Comprehensive Agrarian Reform Programme." *Pilipinas: A Journal of Philippine Studies,* no. 23, Fall 1994, p. 39–60.

Barff, R., and J. Austen. "It's Gotta Be da Shoes: Domestic Manufacturing, International Subcontracting, and the Production of Athletic Footwear." *Environmental and Planning A,* vol. 25, 1993, p. 1103–14.

Barwell, I., and J.D. Howe. *Appropriate Transport Facilities for the Rural Sector in Developing Countries.* Geneva: International Labor Organization, 1979.

Barwell, I., G.A. Edmonds, J.D.G.F. Howe, and J. de Veen. *Rural Transport in Developing Countries.* London: Intermediate Technology Publications, 1985.

Bellwood, Peter. *Man's Conquest of the Pacific: The Prehistory of Southeast Asia and Oceania.* New York: Oxford University Press, 1979.

Bellwood, Peter. *Prehistory of the Indo-Malaysian Archipelago.* Sydney: Academic Press, 1985.

Bellwood, Peter. "Southeast Asia Before History." *The Cambridge History of Southeast Asia. Volume One: From Early Times to c. 1800.* Ed. Nicholas Tarling. Cambridge: Cambridge University Press, 1994, p. 55–136.

Beneria, Lourdes, and Amy Lind. "Engendering International Trade: Concepts, Policy, and Action." *A Commitment to the World's Women: Perspectives on Development for Beijing and Beyond.* Ed. Noeleen Heyzer. New York: United Nations, 1995, p. 69–85.

Beneria, Lourdes, and Shelley Feldman, eds. *Unequal Burden: Economic Crises, Persistent Poverty, and Women's Work.* Boulder, San Francisco, Oxford: Westview Press, 1992.

Bentley, G. Carter. "Indigenous States of Southeast Asia." *Annual Review of Anthropology*, vol. 15, 1986, p. 275–305.

Beresford, M. "The Political Economy of Dismantling the 'Bureaucratic Centralism and Subsidy System' in Vietnam." *Southeast Asia in the 1990s: Authoritarianism, Democracy and Capitalism.* Ed. K. Hewison, R. Robison and G. Rodan. Sydney: Allen and Unwin, 1993.

Beresford, M., and S. Fraser. "Political Economy of the Environment in Vietnam." *Journal of Contemporary Asia*, vol. 22, no. 1, 1992, p. 3–19.

Bernard, Stephane, and Rodolphe De Koninck. "The Retreat of the Forest in Southeast Asia: A Cartographic Assessment." *Singapore Journal of Tropical Geography*, vol. 17, no. 1, 1996, p. 3–4.

Bhattacharya, Amar, and Johannes Linn. *Trade and Industrial Policies in the Developing Countries of East Asia.* Washington, D.C.: World Bank, 1988.

Bidani, B., and M. Ravallion. "A Regional Poverty Profile for Indonesia." *Bulletin of Indonesian Economic Studies*, vol. 29, no. 3, 1993, p. 37–68.

Biers, Dan. "The Thai Economy in 1993 and Trends for 1994." *Far East Economic Review,* vol. 35, no. 1, 1994, p. 9–18.

Biggs, Tyler, Peter Brimble, Donald Snodgrass, and Michael Murray. *Rural Industry and Employment Study: A Synthesis Report* vol. 58. Bangkok: Thailand Development Research Institute Foundation, 1990.

Billard, A. "The Largest Concentration of Kampuchean Refugees in Thailand." *Refugees*, vol. 3, 1983, p. 12–3.

Bilsborrow, R.E., G. Hugo, A.S. Oberai, and H. Zlotnik. *International Migration Statistics.* Geneva: International Labour Office, 1997.

Biro Pusat Statistik (BPS). *Penduduk Indonesia: Hasil Survei Penduduk Antar Sensus 1995 (Population of Indonesia: Results of the 1995 Intercensal Population Survey)*. Jakarta: Biro Pusat Statistik, 1996.

Biro Pusat Statistik (BPS). *Penduduk Indonesia: Tabel Pendahulian Hasil Sub Sampel Sensus Penduduk 1990.* Jakarta: Biro Pusat Statistik, 1991.

Biro Pusat Statistik and United Nations Development Programme (BPS and UNDP). *Laporan Pembanguanan Manusia (Report on Human Development) 1986*; *Laporan Pembanguanan Manusia (Report on Human Development) 1996.* Jakarta: United Nations, 1987 and 1997.

Blackwood, Evelyn. "Senior Women, Model Mothers, and Dutiful Wives: Managing Gender Contradictions in a Minangkabau Village." *Bewitching Women, Pious Men: Gender and Body Politics in Southeast Asia.* Ed. A. Ong and

M.G. Peletz. Berkeley: University of California Press, 1995, p. 124–58.

Blaikie, Piers, and Harold Brookfield. *Land Degradation and Society.* London and New York: Metheun, 1987.

Blancot, C. "Phnom Penh: Defying Man and Nature." *Cultural Identity and Urban Change in Southeast Asia: Interpretative Essays*. Ed. M. Askew and W.S. Logan. Geelong: Deakin University Press, 1994, p. 71–84.

Blomqvist, Hans C. "Intraregional Foreign Investment in East Asia." *ASEAN Economic Bulletin,* vol. 11, no. 3, 1995, p. 280–97.

Boonyabancha, S. "Causes and Effects of Slum Eviction in Bangkok." *Land for Housing the Poor*. Ed. S. Angel, et al. Singapore: Select Books, 1983, p. 254–80.

Booth, A. "Repelita VI and the Second Long-Term Development Plan." *Bulletin of Indonesian Economic Studies*, vol. 30, no. 3, 1994, p. 3–40.

Booth, A. "The Tourism Boom in Indonesia." *Bulletin of Indonesian Economic Studies*, vol. 26, no. 3, December, 1990, p. 45–73.

Borthwick, Mark. *Pacific Century: The Emergence of Pacific Asia.* Boulder, Colorado: Westview Press, 1992.

Boserup, E. *The Conditions of Agricultural Growth: The Economics of Agrarian Change Under Population Pressure*. London: Allen & Unwin, 1965.

Boserup, E. *Women's Role in Economic Development.* New York: St. Martin's Press, 1970.

Bourdet, Y. "Rural Reforms and Agricultural Productivity in Laos." *The Journal of Developing Areas*, vol. 29, 1995, p. 161–82.

Bowen, J.T., Jr. *The Airline Industry and Economic Development in Singapore.* Ann Arbor, Mich.: UMI, 1993.

Bowen, J.T., Jr. "Global Economy, Global Airlines: Singapore and the Geography of International Air Transport." *Singapore Transport and Logistics,* vol. 1, 1992, p. 20–32.

Bowen, J.T., Jr., and T.R. Leinbach. "The State and Liberalization: The Airline Industry in the East Asian NICs." *Annals of the Association of American Geographers,* vol. 85, 1995, p. 468–93.

Boxer, Charles R. *The Dutch Seaborne Empire, 1600–1800.* London: Hutchinson, 1965.

Boxer, Charles R. *The Portuguese Seaborne Empire, 1415–1825.* London: Hutchinson, 1969.

Boyce, James K. *The Philippines: The Political Economy of Growth and Impoverishment in the Marcos Era.* London: Macmillan Press, 1993.

"A Brazilian Tale." *Economist,* vol. 310, no. 7590, February 18, 1989, p. 31–2.

Breman, J.C. *Djawa: Pertumbuhan Penduduk dan Struktur Demografis (Java: Population Growth and Demographic Structure)*. Transl. Sugarda Purbakawatja. Jakarta: Bhratara, 1971.

Brennan, M. "Class, Politics and Race in Modern Malaysia." *Southeast Asia: Essays in the Political Economy of Structural Change*. Ed. R. Higgot and R. Robison. London: Routledge, 1985, p. 93-127.

Britton, Stephen G. "The Political Economy of Tourism in the Third World." *Annals of Tourism Research,* vol. 9, no. 3, 1982, p. 331–58.

Britton, Stephen G. "Tourism, Capital, and Place: Towards a Critical Geography of Tourism." *Environment and Planning D: Society and Space*, vol. 9, 1991, p. 451–78.

Broad, R., and J. Cavanagh. "Of Rainforests and Robber Barons." *Amicus Journal,* vol. 16, no. 2, 1994, p. 18–26.

Broad, R., with J. Cavanagh. *Plundering Paradise: The Struggle for the Environment in the Philippines.* Berkeley: University of California Press, 1993.

Broek, Jan O.M. "Diversity and Unity in Southeast Asia." *Geographical Review,* vol. 34, 1944, p. 175–95.

Brookfield, H. "The End of the 'Resource Frontier.'" *Transformation with Industrialization in Peninsular Malaysia*. Ed. H. Brookfield. Kuala Lumpur: Oxford University Press, 1994, p. 82–94.

Brookfield, H., and Y. Byron, eds. *Southeast Asia's Environmental Future: The Search for Sustainability*. Singapore: Oxford University Press; Tokyo: United Nations University Press, 1993.

Brookfield, H., A.S. Hadi, and Zaharah Mahmud. *The Village in the City: The In-Situ Urbanization of Villages, Villagers and Their Land Around Kuala Lumpur, Malaysia*. Singapore: Oxford University Press, 1991.

Brookfield, H., F.J. Lian, Kwai-Sim Low, and L. Potter. "Borneo and the Malay Peninsula." *The Earth as Transformed by Human Action*. Ed. B.L. Turner et al. New York: Cambridge University Press, 1990, vol. 2, p. 495–512.

Brookfield, H., L. Potter, and Y. Byron. *In Place of the Forest: Environmental and Socio-Economic Transformation in Borneo and the Eastern Malay Peninsula.* New York: United Nations University Press, 1995.

Brosius, J.P. "Perspectives on Penan Development in Sarawak." *Alam Sekitar,* vol. 17, no. 1, 1992, p. 6–17.

Brown, David. "Ethnicity, Nationalism and Democracy in South-East Asia." *Ethnicity*. Ed. John Hutchinson and Anthony D. Smith. Oxford: Oxford University Press, 1996, p. 305–10.

Brown, David. *The State and Ethnic Politics in Southeast Asia.* London and New York: Routledge, 1994.

Brown, K., and D. Pearce, eds. *The Causes of Tropical Deforestation.* London: UCL Press, 1994.

Brown, R.P.C., and J. Foster. "Some Common Fallacies About Migrants Remittances in the South Pacific: Lessons From Tongan and Western Samoan Research." *Pacific Viewpoint,* 1995.

Brown, T., and P. Xenos. "AIDS in Asia: The Gathering Storm." *Asia–Pacific Issues*, no. 16. Honolulu: East–West Center, August 1994.

Bryant, Raymond. *The Political Ecology of Forestry in Burma, 1824–1994.* London: Hurst, 1997.

Bryant, Raymond. "Shifting the Cultivator: the Politics of Teak Regeneration in Colonial Burma." *Modern Asian Studies,* vol. 28, May 1994, p. 225–50.

Brydon, Lynne, and Sylvia Chant. *Women in the Third World: Gender Issues in Rural and Urban Areas.* New Brunswick: Rutgers University Press, 1989.

Buchanan, K. *The Southeast Asian World: An Introductory Essay.* London: G. Bell and Sons, 1967.

"Bumper to Bumper." *The Economist,* August 17, 1996 p. 51–2.

Bury, J.B. *The Idea of Progress: An Inquiry Into Its Origin and Growth.* London: Macmillan and Co., 1920.

Buszynski, Leszek. "China and the ASEAN Region." *China as a Great Power: Myths, Realities and Challenges in the Asia–Pacific*

Region. Ed. S. Harris and G. Klintworth. New York: St. Martin's Press, 1995, p. 161–84.

Butler, R.W. "The Concept of a Tourist Area Cycle of Evolution: Implications for Management of Resources." *Canadian Geographer*, vol. 24, no. 1, 1980, p. 5–12.

Buttinger, Joseph. *Vietnam: A Political History.* New York: Frederick A. Praeger Publishers, 1968.

CAW (Committee for Asian Women). *Silk and Steel: Asian Women Workers Confront Challenges of Industrial Restructuring.* Hong Kong: Committee for Asian Women, 1995.

Callaham, R., and R. Buckman. *Some Perspectives of Forestry in the Philippines, Indonesia, Malaysia, and Thailand.* Washington, D.C.: U.S. Department of Agriculture, Forest Service, 1981.

Cambodia Times. www.cambodiatimes.com

Campbell, Burnham O., Andrew Mason, and Ernesto M. Pernia. *The Economic Impact of Demographic Change in Thailand, 1980–2015.* Honolulu: University of Hawaii Press, 1993.

"Car Industry Told to Cut Reliance on Japanese Parts." *Business Times,* Singapore, February 16, 1996, p. 5.

Carey, P.B.R. *Burma: The Challenge of Change in a Divided Society.* New York: St. Martin's Press, 1997.

Cartier, Carolyn L. "The Dead, Place/Space, and Social Activism: Constructing the Nationscape in Historic Melaka." *Environment and Planning D: Society and Space*, vol. 15, 1997, p. 555–86.

Castells, M., L. Goh, and Y-W Kwok. *The Shep Kip Mei Syndrome: Economic Development and Public Housing in Hong Kong and Singapore.* London: Pion, 1990.

Central Intelligence Agency. *The World Factbook.* Washington, D.C.: U.S. Government Printing Office, 1994.

Central Intelligence Agency. 1:150,000 scale map of Singapore, October 1994.

Chalmers, Ian. "International and Regional Integration: The Political Economy of the Electronics Industry in ASEAN." *ASEAN Economic Bulletin,* vol. 8, no. 2, 1991, p. 194–209.

Chambers, R. *Rural Development: Putting the Last First.* London: Longman, 1983.

Chandler, D. *A History of Cambodia.* 2nd ed. Boulder, Colo: Westview Press, 1991.

Changrien, P., and R.J. Stimson. "Bangkok: Jewel in Thailand's Crown. *New Cities of the Pacific Rim.* Ed. E.J. Blakely and R.J. Stimson. Berkeley: University of California, Institute of Urban and Regional Development, Monograph 43, 1992, p. 15.1–15.39.

Chant, Sylvia, and Cathy McIlwaine. *Women of a Lesser Cost: Female Labour, Foreign Exchange & Philippine Development.* Manila: Ateneo de Manila University Press, 1995.

Chapman, Graham P., and Kathleen M. Baker, eds. *The Changing Geography of Asia.* London: Routledge, 1992.

Chee, Chan Heng. "Political Developments." *A History of Singapore.* Ed. Ernest C.T. Chew and Edwin Lee. Singapore: Oxford University Press, 1991, p. 157–81.

Cheema, G. Shabrir. "Rural Development in Asia: Case Studies on Programme Implementation." New Delhi: Sterling Publishers Private Ltd., 1985.

Chew, Ernest C.T. "The Founding of a British Settlement." *A History of Singapore.* Ed. Ernest C.T. Chew and Edwin Lee. Singapore: Oxford University Press, 1991, p. 36–40.

Chew, Ernest C.T., and Edwin Lee, eds. *A History of Singapore.* Singapore: Oxford University Press, 1991.

Chia, L.S. "The Port of Singapore." *The Management of Success: The Moulding of Modern Singapore.* Ed. K.S. Sandhu and P. Wheatley. Singapore: Institute of Southeast Asian Studies, 1989, p. 314–36.

Chiew, Seen Kong. *see* Kong, Chiew Seen.

Chin, J. "Scenario for the AIDS Epidemic in Asia." *Asia-Pacific Population Research Reports*, no. 2. Honolulu: East–West Center Program on Population, 1995.

"China, Myanmar Draw Trade Ties Closer." *Beijing Review*, vol. 38, no. 2, January 9–15, 1995, p. 5.

"The Chinese Wave in Vietnam." *Singapore Business,* vol. 18, no. 6, July 1994, p. 32–8.

Chippington, George. *Singapore: The Inexcusable Betrayal.* Worcester, United Kingdom: Self Publishing Assoc., 1992.

Cho, G. *The Malaysian Economy: Spatial Perspectives.* London: Routledge, 1990.

Christensen, Scott R., and Ammar Siamwalla. "Muddling Toward a Miracle: Thailand and East Asian Growth." *TDRI Quarterly Review*, vol. 9, no. 2, June 1994, p. 13–19.

Chuan, G.K. "The State of the Physical Environment of Brunei Darussalam." *Malaysian Journal of Tropical Geography,* vol. 22, 1991, p. 19–27.

Clad, James. *Behind the Myth: Business, Money, and Power in Southeast Asia.* London: Unwin Hyman, 1989.

Cleary, M., and P. Eaton. *Borneo: Change and Development.* Singapore: Oxford University Press, 1992.

Cœdès, Georges. *Angkor: An Introduction.* Transl. and ed. E.F. Gardiner. Hong Kong: Oxford University Press, 1963.

Cœdès, Georges. *Histoire Ancienne des Etats Hindouisés d'Extrême-Orient.* Hanoi: Imprimerie d'Extrême-Orient, 1944.

Cœdès, Georges. *The Indianized States of Southeast Asia.* Ed. Walter F. Vella and transl. Susan Brown Cowing. Honolulu: East–West Center Press, 1968.

Cohen, M. "Forest Fire—The Biodiversity Debate Heats Up in Asia." *Far Eastern Economic Review*, January 11, 1996, p. 66–9.

Cohen, M. "High Anxiety: Government Proposal Could Crimp NGO Activities." *Far Eastern Economic Review,* September 29, 1994, p. 32.

Cohen, M., B. Dolven, and M. Hiebert, M. "Yes, Again—And This Time Indonesia's Fires are Harder to Stop." *Far Eastern Economic Review*, March 19, 1998, p. 22–3.

Cohen, M. "Where There's Smoke... Spread of Indonesian Oil-Palm Plantations Fuels the Haze." *Far Eastern Economic Review,* October 2, 1997 (see also www.feer.com).

Colchester, M. *Pirates, Squatters and Poachers.* London: Survival International, 1989.

Collier, W., L. Santoso, K. Soentoro, and R. Wibowo. "New Approach to Rural Development in Java: Twenty Five Years of Village Studies in Java." Mimeo.

Collins, N.T., J.A. Sayer, and T.C. Whitmore. *The Conservation Atlas of Tropical Forests: Asia and the Pacific.* London: Macmillan Press, 1991.

Concepcion, Mercedes B. "Population Growth in Southeast Asia: Pushing the Limits." *South-East Asia's Environmental Future: The Search for Sustainability.* Ed. Harold Brookfield and Yvonne Burton. Tokyo: United Nations University Press, 1993, p. 33–46.

Concepcion, Mercedes B., and Peter C. Smith. *The Demographic Situation in the Philippines: An Assessment in 1977.*

Papers of the East–West Population Institute, no. 44. Honolulu: East–West Center, June 1977.

Containerisation International Yearbook, 1990–1995.

Cook, Paul, and Martin Minogue. "Economic Reform and Political Change in Myanmar (Burma)." *World Development*, vol. 21, no.7, 1993, p. 1151–61.

Coppel, C.A. "China and the Ethnic Chinese in Indonesia." *Indonesia: Australian Perspectives*. Ed. J.J. Fox, R.G. Garnaut, P.T. McCawley, and J.A.C. Mackie. Canberra: Research School of Pacific Studies, ANU, 1980, p. 729–34.

Corey, K.E., R.G. Fletcher, and B.J. Moscove. "Singapore: The Planned New City of the Pacific Rim." *New Cities of the Pacific Rim.* Ed. E.J. Blakely and R.J. Stimson. Berkeley: University of California, Institute of Urban and Regional Development, Monograph 43, 1992, p. 14.1–14.31.

Courtenay, P.P. "The Dilemma of Malaysian Padi Policy." *Australian Geographer,* vol. 17, 1986, p. 178–85.

Cowan, C.D. *Nineteenth Century Malaya: the Origins of British Political Control.* London: Oxford University Press, 1961.

Crossette, Barbara. *The Great Hill Stations of Asia*. Boulder, Colo.: Westview Press, 1998.

Crouch, H. "Malaysia: Neither Authoritarian Nor Democratic." *Southeast Asia in the 1990s: Authoritarianism, Democracy and Capitalism*. Ed. K. Hewison, R. Robison, and G. Rodan. St. Leonards, New South Wales, Australia: Allen and Unwin, 1993, p. 133–57.

CyberCebu Home Page, 1996, http://www.gsilink.com/

DAWN (Development Alternatives with Women for a New Era). "Rethinking Social Development: DAWN's Vision." *World Development*, vol. 23, no. 11, 1995, p. 2001–4.

Daniels, Peter. "Services in a Shrinking World." *Geography,* vol. 80, no. 2, 1995, p. 97–110.

Dauvergne, Peter. *Shadows in the Forest: Japan and the Politics of Timber in Southeast Asia*. Cambridge: MIT Press, 1997.

Davan, Janadas, ed. *Southeast Asia: Challenges of the 21st Century.* Singapore: Institute of Southeast Asian Studies, 1994.

David, Cristina C., and Keijiro Otsuka, eds. *Modern Rice Technology and Income Distribution in Asia.* Boulder, Colo.: Lynne Rienner Publishers, 1994.

Dearden, Philip. "Tourism and Sustainable Development in Northern Thailand." *The Geographical Review*, vol. 81, no. 4, October 1991, p. 400–13.

DeGlopper, Donald R. "The Society and Its Environment." *Singapore: A Country Study*, 2nd ed. Ed. B.L. LePoer. Washington, D.C.: Headquarters, Dept. of the Army, 1991, p. 65–117.

Del Casino, Vincent. "Creating 'Tourism Space': The Social Construction of Sex Tourism in Thailand." Unpublished M.A. thesis, University of Wisconsin-Madison, 1995.

Del Casino, Vincent. "Mapping the Tourist Gaze or Creating It: Revivifying Notions of Other." *Paper/Scissors/Rock: Proceedings of an Interdisciplinary Graduate Student Conference on Nationalism, Empire and Post-Colonialism.* Ed. Jacques Critchley, David Feeney, Neil Hanlon, and Mike Ripmeester. Kingston, Ontario: Queen's University, 1996, p. 63–86.

Demaine, Harvey, Ian Brown, and Edith Hodgkinson. "The Philippines." *The Far East and Australasia 1994.* London: Europa Publications Limited, 1994, p. 837–77.

Demaine, Harvey, Ian Brown, and Edith Hodgkinson. "The Philippines." *The Far East and Australasia 1995.* London: Europa Publications Limited, 1995, p. 870–912.

Desbarats, J. "Population Redistribution in the Socialist Republic of Vietnam." *Population and Development Review*, vol. 13, no. 1, 1987, p. 43–76.

Dick, H.W. *The Indonesian Interisland Shipping Industry: An Analysis of Competition and Regulation*. Singapore: Institute of Southeast Asian Studies, 1985.

Dicken, Peter. *Global Shift: Industrial Change in a Turbulent World.* London: Paul Chapman Publishing, 1986.

Dicken, Peter. *Global Shift: Transforming the World Economy.* 3rd ed. New York: Guilford, 1998.

Dicken, Peter. "Mining and Manufacturing." *Southeast Asian Development: Geographical Perspectives.* Ed. Denis Dwyer. London: Longman, 1990, p. 193–224.

Dillon, R. "Indian Squatters in Kuala Lumpur." *Transformation with Industrialization in Peninsular Malaysia*. Ed. H. Brookfield. Kuala Lumpur: Oxford University Press, 1994, p. 210–33.

Din, Kadir H. "Dialogue With the Hosts: An Educational Strategy Towards Sustainable Tourism." *Tourism in South-East Asia.* Ed. Michael Hitchcock, Victor T. King, and Michael J.G. Parnwell. London: Routledge, 1993, p. 327–36.

Dixon, Chris. *Southeast Asia in the World Economy*. Cambridge: Cambridge University Press, 1991.

Dixon, Chris, and David W. Smith, eds. *Uneven Development in Southeast Asia.* Ashgate, 1997.

Dixon, G. "The Landscape of Development in East Malaysia: A Memoir." *Borneo Research Bulletin,* vol. 23, 1991, p. 152–60.

Doeppers, Daniel F. "The Development of Philippine Cities Before 1900." *The Journal of Asian Studies*, vol. 31, no. 4, August 1972, p. 769–92.

Dolan, Ronald E., ed. *Philippines: A Country Study*. Washington, D.C.: U.S. Government Printing Office, Area Handbook Series, 1993.

Dommen, A.J. "Laos: Consolidating the Economy." *Southeast Asian Affairs 1994*. Singapore: Institute of Southeast Asian Studies, 1994, p. 167–78.

Douglass, M. "The Environmental Sustainability of Development: Coordination, Incentives and Political Will In Land-Use Planning for the Jakarta Metropolis." *Third World Planning Review,* vol. 11, no. 2, 1989, p. 211–38.

Douglass, M. "The Future of Cities on the Pacific Rim." *Pacific Rim Cities in the World Economy*. Ed. M.P. Smith. New Brunswick: Transaction Books, 1992, p. 9–63.

"Dr M Launches Sale of Proton Wira in Vietnam." *Business Times,* Singapore, March 11, 1996, p. 3.

Drakakis-Smith, D.W. "Urban Food Distribution in Asia and Africa." *Geographical Journal,* vol. 157, no. 1, 1991, p. 51–60.

Drakakis-Smith, D.W., and P.J. Rimmer. "Taming the "Wild City": Managing Southeast Asia's Primate Cities Since the 1960s." *Asian Geographer,* vol. 1, no. 1, 1982, p. 17–34.

Drysdale, John. *Singapore: Struggle for Success*. Singapore: Times Books International, 1984.

Dunn, F.L. *Rainforest Collectors and Traders: A Study of Resource Utilization in Modern and Ancient Malaya*. Kuala Lumpur: Royal Asiatic Society, Malaysian Branch, Monograph no. 5, 1975.

Dunn, James. "The Timor Affair in International Perspective." *East Timor at the Crossroads: The Forging of a Nation*. Ed. P. Carey and G. Carter Bentley. London: Cassell; New York: The Social Science Research Council, 1995, p. 59–72.

Dunne, Michael. "GM's Thai Gambit" *Asian Wall Street Journal,* June 21–22, 1996, p. 6.

Dwyer, D. "The City in the Developing World and the Example of Southeast Asia." *Geography,* vol. 35, 1968, p. 353–69.

Dwyer, D. "Urbanization." *Southeast Asia Development: Geographical Perspectives.* Ed. D. Dwyer. Harlow, Essex, U.K.: Longman, 1990.

"ENV's Priority: Tackling Waste Problems." *The Straits Times,* Singapore, March 1, 1997, p. 25.

"Economic Monitor: Vietnam." *Far Eastern Economic Review.* August 5, 1999, p. 51.

"Economies: Yunnan Aims to Be Asia's New Entrepot; Links With Burma and Laos; Coping With the Tourist Influx." *Far Eastern Economic Review*, vol. 160, no. 37, 1997.

Economist Intelligence Unit (EIU). *Country Profile: Burma, Cambodia, Laos 1995–96.* London: Economist Intelligence Unit, 1995a.

Economist Intelligence Unit. *Country Report: Cambodia, Laos, Myanmar: 3rd quarter 1995.* London: Economist Intelligence Unit, 1995a.

Economist Intelligence Unit. *Country Report: Cambodia, Laos, Myanmar: 2nd Quarter 1995.* London: Economist Intelligence Unit, 1995b.

Economist Intelligence Unit. *Country Profile: Philippines 1995–96,* London: Economist Intelligence Unit, 1995b.

Economist Intelligence Unit. *Country Profile: Thailand—Myanmar.* London: The Economist Intelligence Unit, 1995c.

Economist Intelligence Unit. *Indochina: Vietnam, Laos, Cambodia 1993–94*. London: Economist Intelligence Unit, 1993.

Economist Intelligence Unit. *Indochina: Vietnam, Laos, Cambodia 1994–95*. London: Economist Intelligence Unit, 1994.

Edmundson, W., and S. Edmundson. "A Decade of Village Development in East Java." *Bulletin of Indonesian Economic Studies*, vol. 19, no. 2, 1983, p. 46–59.

Elson, Diane. "From Survival Strategies to Transformation Strategies: Women's Needs & Structural Adjustment." *Unequal Burden: Economic Crises, Persistent Poverty, and Women's Work*. Ed. Lourdes Beneria and Shelley Feldman. Boulder, Colo.: Westview Press, 1992.

Elson, Diane. "Male Bias in Macroeconomics: The Case of Structural Adjustment." *Male Bias in the Development Process*. Ed. Diane Elson. Manchester: Manchester University Press, 1990.

Elson, Diane, ed. *Male Bias in the Development Process*, 2nd ed. Manchester, England: Manchester University Press, 1995.

Elson, R.E. *The End of the Peasantry in Southeast Asia: A Social and Economic History of Peasant Livelihood, 1800–1900s.* London: Macmillan Press, 1997.

Emmerson, Donald K. "Southeast Asia: What's in a Name?" *Journal of Southeast Asian Studies*, vol. 15, no. 1, March 1984, p. 1–21.

Eng, Peter. "The Media and Democratization in Southeast Asia." *Current History*, vol. 96, no. 614, 1997, p. 437.

Eng, Teo Siew, and Victor R. Savage. "Singapore Landscape: A Historical Overview of Housing Image." *A History of Singapore.* Ed. Ernest C.T. Chew and Edwin Lee. Singapore: Oxford University Press: 1991, p. 312–38.

Enloe, Cynthia. *Bananas, Beaches, and Bases: Making Feminist Sense of International Politics.* Berkeley, California: University of California Press, 1989.

Enloe, Cynthia. *Making Sense of International Politics: Bananas, Beaches, and Bases*. Berkeley: University of California Press, 1990.

"Environment in Asia." *Far Eastern Economic Review*, vol. 156, no. 43, October 28, 1993, p. 48–62.

Etherington, K., and D. Simon. "Para-Transit and Employment in Phnom Penh: the Dynamics and Development Potential of *Cyclo* Riding." *Journal of Transport Geography,* vol. 4, 1996, p. 37–53.

Europa Publications Limited (1991, 1993, & 1995 editions). *The Far East and Australasia.* London: Europa Publications Limited.

Evans, G. *Lao Peasants Under Socialism and Post-Socialism*. Chiang Mai, Thailand: Silkworm Books, 1995.

Evers, H-D. *Subsistence Production and Wage Labor in Jakarta*. Sociology of Development Research Centre working paper no. 29. Bielefeld: University of Bielefeld, 1981.

Evers, H-D. "Urban Landownership, Ethnicity and Class in Southeast Asian Cities." *International Journal of Urban and Regional Research,* vol. 8, 1983, p. 481–96.

Evers, H-D., ed. *Loosely Structured Social Systems: Thailand in Comparative Perspective*. New Haven: Yale University, Southeast Asia Studies, 1969.

Evers, H-D., and R. Korff. "Subsistence Production in Bangkok. *Development,* vol. 4, 1986, p. 50–5.

Fairclough, Gordon. "Child Labour: It Isn't Black and White." *Far Eastern Economic Review,* vol. 159, no. 10, March 7, 1996, p. 54–8.

Falkus, Malcolm. "Thai Industrialization: An Overview." *Thailand's Industrialization and Its Consequences.* Ed. Medhi Krongkaew. New York: St. Martin's Press, 1995, p. 13–32.

Faustino, Jaime. "Mining the State: Dominant Classes, Marcos and Aquino." *Pilipinas: A Journal of Philippine Studies,* no. 21, Fall 1993, p. 55–67.

Fauveau, V., K. Phimmasone, M. Oudom, I. Godin, and P. Pholsena. "Socio-Cultural and Economic Determinants of Contraceptive Use of the Lao PDR." Paper prepared for the XXII General Population Conference of the IUSSP, Montreal, Canada, August 24 to September 1, 1993.

Feeney, G., and P. Xenos. "The Demographic Situation in Vietnam: Past, Present, Prospect." *The Challenges of Vietnam's Reconstruction*. Ed. by N.L. Jamieson, M.H. Nguyen, and A.T. Rambo. Vietnam: Indochinese Institute, 1992.

Feliciano, Myrna S. "Law, Gender, and the Family in the Philippines." *Law and Society Review*, vol. 28, no. 3, 1994, p. 547–60.

Fesharaki, Fereidun. "Singapore as an Oil Centre." *The Management of Success: The Moulding of Modern Singapore.* Ed. Kernial Singh Sandhu and Paul Wheatley. Singapore: Institute of Southeast Asian Studies, 1989, p. 300–13.

Firman, T. "The Spatial Pattern of Urban Population Growth in Java, 1980–1990." *Bulletin of Indonesian Economic Studies*, vol. 28, no. 2, 1992, p. 95–109.

Fisher, Charles. "Indonesia." *Education and Political Development*. Ed. J.S. Coleman. Princeton: Princeton University, 1965, p. 92–122.

Fisher, Charles. *South-East Asia: A Social, Economic and Political Geography*, London: Methuen, 1964.

Fisher, Charles. "Southeast Asia: the Balkans of the Orient?" *Geography,* vol. 47, 1962, p. 347–67.

Fisher, Charles. "A View of Southeast Asia." *Southeast Asia: An International Quarterly*, vol. 1, nos. 1–2 (Winter–Spring, 1971), p. 5–40.

Floro, Maria Sagrario. "The Dynamics of Economic Change and Gender Roles: Export Cropping in the Philippines." *Mortgaging Women's Lives: Feminist Critiques of Strucutural Adjustment.* Ed. Pamela Sparr. London and New Jersey: Zed Books, 1994. p. 116–33.

Foley, P. "From Hell to Paradise: The Stages of Vietnamese Refugee Migration Under the Comprehensive Plan of Action." Unpublished Ph.D. thesis, Australian National University, Canberra, forthcoming.

Forbes, D. "The Urban Network and Economic Reform in Vietnam." *Environment and Planning A,* vol. 27, no. 5, 1995, p. 793–808.

Forbes, D. "Urbanization, Migration and Vietnam's Spatial Structure." *Sojourn. Journal of Social Issues in Southeast Asia*, vol. 11, no. 1, 1996, p. 24–51.

Forbes, D., and Le Hong Ke. "A City in Transition: Socialist Reform and the Management of Hanoi." *The Dynamics of Metropolitan Management in Southeast Asia*. Ed. J. Ruland. Singapore: Institute of Southeast Asian Studies, 1996, p. 71–98.

Forbes, D., and N. Thrift. "International Impacts on the Urbanization Process in the Asian Region: A Review." *Urbanization and Urban Policies in Pacific Asia.* Ed. R.J. Fuchs, G.W. Jones, and E.M. Pernia. Boulder, Colo.: Westview Press, 1987, p. 67–87.

"Ford to Invest U.S. $53 Million in Two Components Plants in Thailand." *Business Times,* Singapore, January 26, 1996, p. 20.

Ford, L. "A Model of Indonesian City Structure." *Geographical Review,* vol. 83, no. 4, 1993, p. 374–96.

Fox, James J. "Ecological Policies for Sustaining High Production in Rice: Observations on Rice Intensification in Indonesia." *South-East Asia's Environmental Future: The Search for Sustainability.* Ed. Harold Brookfield and Yvonne Burton. Tokyo: United Nations University Press, 1993, p. 211–24.

Freedman, R. "Asia's Recent Fertility Decline and Prospects for Future Demographic Change." *Asia–Pacific Population Research Reports*, no. 1. Honolulu: East–West Center, 1995.

Friedland, J. "High-Rise Greed." *Far Eastern Economic Review,* November 20, 1990, p. 54.

Friedmann, J. "The World City Hypothesis." *Development and Change,* vol. 17, 1986, p. 69–83.

Frings, V. "Cambodia After Decollectivization." *Journal of Contemporary Asia*, vol. 24, no. 1, 1994, p. 49–66.

Frost, F. "Cambodia: From UNTAC to Royal Government." *Southeast Asian Affairs 1994.* Singapore: Institute of Southeast Asian Studies (ISEAS), 1994, p. 79–101.

Fryer, Donald W. *Emerging Southeast Asia: A Study in Growth and Stagnation*, 2nd ed. London: George Philip and Son; New York: Halsted, 1979.

Fryer, Donald W. "Agriculture and Fisheries." *South East Asian Development*. Ed. D.J. Dwyer. New York: Longman, 1990, p. 168–82.

Fui, Lim Hin. *Poverty and Household Strategies in Malaysian New Villages*. Petaling Jaya: Pelanduk Publications, 1994.

Funnell, Victor. "The Philippines: In Search of the Nebulous State." *The Post-Colonial State in Asia: Dialectics of Politics and Culture*. Ed. S.K. Mitra. London: Harvester Wheatsheaf, 1990, p. 131–53.

"Futuristic Container Terminal in Four Phases." *The Straits Times (Life!),* Singapore, December 21, 1995, p. 2.

Gardiner, P. "Urban Population in Indonesia: Future Trends." *Spatial Development in Indonesia: Review and Prospects.* Ed. T.J. Kim, G. Knaap, and I.J. Azis. Brookfield, Vt.: Avebury, 1992, p. 267–91.

Gargan, Edward A. "Subic Bay Rises as an Industrial Hotbed." *The New York Times*, May 30, 1995, p. C8.

Garnaut, Ross. *Asian Market Economies: Challenges of a Changing International Environment.* Singapore: Institute of Southeast Asian Studies, 1994.

Geertz, C. *Agricultural Involution: The Process of Ecological Change in Indonesia.* Berkeley: University of California Press, 1963a.

Geertz, C. *Peddlers and Princes, Social Change and Economic Modernization in Two Indonesian Towns.* Chicago: University of Chicago Press, 1963b.

General Santos City Home Page, 1996, http://www.teleport.com/~rmartin/gensan.shtml

General Statistical Office. *Vietnam Population Census 1989: Detailed Analysis of Sample Results*. Hanoi: General Statistical Office, 1991.

George, Cherian. "More People Living in Outlying Areas." *The Straits Times,* Singapore, July 21, 1994, p. 1, 24–5.

Gibney, Frank Jr. "Detroit Gets Into Gear." *Time,* July 15, 1996, p. 20–6.

Gill, I. "Lao PDR's Dilemma: Is It Just Hydropower Versus Environment?" *Focus*, December 1994, p. 10–12.

Gillis, M. "The Logging Industry in Tropical Asia." *People of the Rainforest.* Ed. J. Denslow and C. Padoch. Berkeley: University of California Press, 1988, p. 177–84.

Ginsburg, N., B. Koppel, and T.G. McGee, eds. *The Extended Metropolis: Settlement Transition in Asia.* Honolulu: University of Hawaii Press, 1991.

Goldberg, M.A. *The Chinese Connection: Getting Plugged In to Pacific Rim Real Estate, Trade and Capital Markets*. Vancouver: University of British Columbia Press, 1985.

Goldstein, S. "Circulation in the Context of Total Mobility in Southeast Asia." *Papers of the East–West Population Institute,* no. 53, 1978.

Goldstein, Sidney, and Alice Goldstein. "Migration in Thailand: A twenty-five year review." Papers of the East–West Population Institute, no. 100. Honolulu: East–West Center, July 1986.

Gordon, Bernard K. *The Dimensions of Conflict in Southeast Asia*. Englewood Cliffs, N.J.: Prentice-Hall, 1966.

Goss, J., and B. Lindquist. "Conceptualizing International Labor Migration: A Structuration Perspective." *International Migration Review*, vol. 29, no. 2, 1995, p. 317–51.

Goss, J.D. *Production and Reproduction Among the Urban Poor in Metro Manila, Philippines: Relations of Exploitation and Conditions of Existence.* Ph.D. Dissertation, University of Kentucky, 1990.

Goss, J.D. "Right to the City: Forms of Land Allocation and the Struggle for Living Place and Working Space in Manila." *Philippine Sociological Review*, forthcoming.

Gould, W.T.S. "Urban Development and the World Bank." *Third World Planning Review,* vol. 14, no. 2, 1992, p. ii–vi.

"Government Will Remain Involved in Fostering Economic Growth." *The Straits Times,* Singapore, June 6, 1996, p. 50.

Govindasami, P., and J. DaVanzo. "Ethnicity and Fertility Differentials in Peninsular Malaysia: Do Policies Matter?" *Population and Development Review,* vol. 18, 1992, p. 243–67.

Greenhouse, Steven. "Nike Shoe Plant in Vietnam is Called Unsafe for Workers." *The New York Times,* November 8, 1997, p. A1, B2.

Grice, Kevin, and David Drakakis-Smith. "The Role of the State in Shaping Development: Two Decades of Growth in Singapore." *Transactions, Institute of British Geographers,* vol. 10, 1985, p. 347–59.

Groslier, Bernard, and Jacques Arthaud. *Angkor: Art and Civilization.* New York: Frederick A. Praeger Publishers, 1966.

Guinness, P. "The State and Industrial Development: Johore Port and the Pasir Gudang Industrial Area." *Transformation with Industrialization in Peninsular Malaysia.* Ed. H. Brookfield. Kuala Lumpur: Oxford University Press, 1994, p. 189–209.

Gullick, J.M. "Indigenous Political Systems of Western Malaya." London: University of London Press, 1958.

Gunn, G.C. *Rebellion in Laos: Peasant and Politics in a Colonial Backwater.* Boulder, Colo.: Westview Press, 1990.

Gunn, G.C. "Rentier Capitalism in Negara Brunei Darussalam." *Southeast Asia in the 1990s: Authoritarianism, Democracy and Capitalism.* Ed. K. Hewison, R. Robison, and G. Rodan. St. Leonards, Australia: Allen and Unwin, 1993, p. 109–32.

Guyot, James. "Burma in 1997: From Empire to ASEAN." *Asian Survey*, vol. 38, no. 2, 1998, p. 190.

Guyot, James F. "Myanmar in 1990: The Unconsummated Election." *Asian Survey,* vol. 31, Feb. 1991, p. 205–11.

Gwatkin, D.R. "Indications of Change in Developing Country Mortality Trends: The End of an Era." *Population and Development Review*, vol. 6, no. 4, 1980, p. 615–44.

Hadiwinoto, S., and J. Leitmann. "Jakarta." *Cities,* vol. 11, no. 3, 1994, p. 153–7.

Hafner, James A. "Greening the Village: The PDA Community Woodlot Program in the Northeast." *Regional Development Dialogue*, 1992.

Hafner, James A. "Highway Network Expansion in Central Thailand, 1917–1967." *International Geography.* Ed. W. Adams and Helleiner. Toronto: University of Toronto Press, 1972, p. 1200–4.

Hafner, James A., and Yaowalak Apichatvullop. "Farming the Forest: Managing People and Trees in Reserved Forests in Thailand." *Geoforum,* vol. 21, no. 3, 1990a, p. 331–46.

Hafner, James A., and Yaowalak Apichatvullop. "Forces and Policy Issues Affecting Forest Use in Northeast Thailand 1900–1985." *Keepers of the Forest: Land Management Alternatives in Southeast Asia.* Ed. Mark Poffenberger. West Hartford, Conn.: Kumarian Press, 1990b, p. 69–94.

Hafner, James A., and Yaowalak Apichatvullop. "Managing Forests and People in Reserved Forests in Thailand." *GEOFORUM* 2, 1990, p. 1–14.

Hall, Colin Michael. *Tourism in the Pacific Rim: Development, Impacts, and Markets.* Melbourne, Australia: Longman Cheshire Pty Limited, 1994.

Hall, D.G.E. *Burma.* London: Hutchinsons University Library, 1950.

Hall, D.G.E. *Atlas of South-East Asia*, London: Macmillan & Co., 1964.

Hall, D.G.E. *A History of South-East Asia*. London: Macmillan & Co., 1955.

Hall, Kenneth R. "Economic History of Early Southeast Asia." *The Cambridge History of Southeast Asia. Volume One: From Early Times to c. 1800.* Ed. Nicholas Tarling. Cambridge: Cambridge University Press, 1994, p. 183–275.

Hall, Kenneth R. "The Indianization of Funan: An Economic History of Southeast Asia's First State." *Journal of Southeast Asian Studies*, vol. 13, no. 1, March 1982, p. 81–106.

Hall, Kenneth R. *Maritime Trade and State Development in Early Southeast Asia.* Honolulu: University of Hawaii Press, 1985.

Handley, P. "River of Promise." *Far Eastern Economic Review*, vol. 156, no. 7, September 16, 1993, p. 68–72.

Handley, P. "Victims of Success." *Far Eastern Economic Review,* September 19, 1990, p. 44.

Hardjono, J. "Resource Utilization and the Environment." *Indonesia's New Order—The Dynamics of Socio-Economic Transformation*. Ed. H. Hill. Sydney: Allen and Unwin, 1994.

Harvey, D. *Social Justice and the City.* Oxford: Blackwell, 1973.

Harvey, Godfrey E. *History of Burma: From the Earliest Times to 10 March, 1824, The Beginning of the English Conquest*. London and New York: Longmans, Green and Co., 1925.

Havener, Robert D. "Food Production After the Green Revolution: Addressing Sustainability Issues." Proceedings: *Sustainable Agricultural Development in Asia and the Pacific x Region.* Manila: The Asian Development Bank and Winrock International, 1992, p. 5–14

Heine-Geldern, Robert von. "Conceptions of State and Kingship in Southeast Asia." *The Far Eastern Quarterly*, vol. 2, no. 1, November 1942, p. 15–30.

Henderson, J.W. "The New International Division of Labor and American Semiconductor Production in Southeast Asia." *Multinational Corporations and the Third World.* Ed. C.J. Dixon, D. Drakakis-Smith, and H.D. Watts. Boulder, Colo.: Westview Press, 1986.

Hermalin, A.I. "Aging in Asia: Setting the Research Foundation." *Asia–Pacific Research Reports*, no. 4. Honolulu: East–West Center, 1995.

Heyzer, Noeleen, ed. *Women Farmers and Rural Change in Asia.* Kuala Lumpur, Malaysia: Asian and Pacific Development Centre, 1987.

Heyzer, Noeleen, G. Nijehold, and N. Weerakoon. *The Trade in Domestic Workers: Causes, Mechanisms and Consequences of International Migration,* vol. 1. Kuala Lumpur: Asian and Pacific Development Centre; London and New Jersey: Zed Books, 1994.

Heyzer, Noeleen, and Gita Sen, eds. *Gender, Economic Growth and Poverty: Market Growth and State Planning in Asia and the Pacific*. New Delhi: Kali; Utrecht, Netherlands: International Books, 1994.

Heyzer, Noeleen, and Vivienne Wee. "Domestic Workers in Transient Overseas Employment: Who Benefits, Who Profits." *The Trade in Domestic Workers: Causes, Mechanisms, and Consequences of International Migration*, vol. 1. Ed. N. Heyzer, G. Lycklama Nijeholt, and N. Weerakoon. Kuala Lumpur:

Asian and Pacific Develoment Centre; London: Zed, 1993, p. 31–101.

Hickey, G. *Village in Vietnam*. New Haven: Yale University Press, 1964.

Hidayat. "The Urban Informal Sector of Indonesia: An Empirical Search for a Solution." Department of Manpower and Transmigration working paper. December 1978.

Hiebert, J. "On the Offensive." *Far Eastern Economic Review,* February 6, 1997, p. 38–9.

Hiebert, M. "Flowers in the Dirt." *Far Eastern Economic Review,* June 20, 1991, p. 84–5.

Hiebert, M. "A Fortune in Waste: Scarcity Forces Vietnam to Reuse Its Resources." *Far Eastern Economic Review,* December 23, 1993a, p. 36.

Hiebert, M. "In the Family Way." *Far Eastern Economic Review*, April 22, 1993b, p. 72.

Hiebert, M. "Single Mothers: Women in Men-Short Vietnam Are Having Children Out of Wedlock." *Far Eastern Economic Review*, February 24, 1994a, p. 60–1.

Hiebert, M. "Stuck at the Bottom: Despite Vietnam's Reforms Many Still Live in Poverty." *Far Eastern Economic Review,* January 13, 1994b, p. 70–1.

Hiebert, M. "Taking to the Hills." *Far Eastern Economic Review,* May 25, 1989, p. 42–4.

Hiebert, M. "Trial by Fire: Smog Crisis Tests ASEAN's Vaunted Cooperation." *Far Eastern Economic Review,* October 16, 1997.

Hill, Hal. "The Economy." *Indonesia's New Order: The Dynamics of Socio-Economic Transformation*. Ed. H. Hill. Sydney: Allen and Unwin, 1994.

Hill, Hal. *The Indonesian Economy Since 1996*: Cambridge: Cambridge University Press, 1996.

Hill, Hal. "Industrialization in Burma in Historical Perspective." *Journal of Southeast Asian Studies,* vol. XI, no.1, March 1984, p. 134–49.

Hill, Hal. "Survey of Recent Developments." *Bulletin of Indonesian Economic Studies*, vol. 28, no. 2, 1992, p. 3–41.

Hill, Hal. "Southeast Asian Economic Development: An Analytical Survey." *Southeast Asia.* Economics Division working papers, 93/4. Research School of Pacific Studies, Australian National University, 1993.

Hill, Hal, and A. Weidemann. "Regional Development in Indonesia: Patterns and Issues." *Unity and Diversity: Regional Economic Development in Indonesia Since 1970*. Ed. H. Hill. Singapore: Oxford University Press, 1989, p. 3–54.

Hill, Hal, and Pang Eng Fong. "The State and Industrial Restructuring: A Comparison of the Aerospace Industry in Indonesia and Singapore." *ASEAN Economic Bulletin,* November 1988, p. 152–68.

Hill, Richard Child. "Global Factory and Company Town: The Changing Division of Labour in the International Automobile Industry." *Global Restructuring and Territorial Development.* Ed. Jeffrey Henderson and Manuel Castells. London: Sage Publications, 1987, p. 18–37.

Hilling, D. *Transport and Developing Countries*. London: Routledge, 1996.

Hirsch, Philip. *Development Dilemmas in Rural Thailand*. Singapore; New York: Oxford University Press, 1990.

Hirsch, Philip, and Larry Lohman. "Contemporary Politics of Environment in Thailand." *Asian Survey*, vol. XXIX, no. 4, April 1989, p. 439–51.

Hirschman, Charles. "Ethnic Diversity and Change in Southeast Asia." *Population, Ethnicity, and Nation-Building.* Ed. Calvin Goldscheider. Boulder, Colo.: Westview Press, 1995, p. 19–36.

Hitchcock, Michael, Victor T. King, and Michael J.G. Parnwell. *Tourism in South-East Asia.* London: Routledge, 1993.

Ho, D.K.H. *The Seaport Economy: A Study of the Singapore Experience.* Singapore: Singapore University Press, 1996.

Hock, Saw Swee. "Population Growth and Control." *A History of Singapore.* Ed. Ernest C.T. Chew and Edwin Lee. Singapore: Oxford University Press: 1991, p. 219–41.

Hoffman, M.L. "Unregistered Land, Informal Housing, and the Spatial Development of Jakarta." *Spatial Development in Indonesia: Review and Prospects.* Ed. T.J. Kim, G. Knaap, and I.J. Azis. Brookfield, Vt: Avebury, 1992, p. 329–49.

Holloway, R. *Doing Development-Governments, NGO's and the Rural Poor in Asia.* London: Earthscan Publications Ltd., 1989.

Hong, E. *Natives of Sarawak: Survival in Borneo's Vanishing Forests.* Kuching: Institut Masyarakat, 1987.

Hori, H. "Development of the Mekong River Basin, Its Problems and Future Prospects." *Water International,* vol. 18, no. 2, 1993, p. 110–15.

Housing Development Board. *HDB Annual Report 1991–1992*. Singapore, 1992.

How, Tan Tarn. "Meeting the Computer Challenge: How Singapore and Taiwan Plan to Stay in the Big League." *The Sunday Times: Sunday Review,* Singapore, June 19, 1994, p. 1, 6.

Howe, J., and P. Richards, eds. *Rural Roads and Poverty Alleviation*. London: Intermediate Technology Publications, 1984.

Htin, Aung Maung. *A History of Burma.* New York and London: Columbia University Press, 1967.

Huat, Chua Beng. "Public Housing Policies Compared: U.S., Socialist Countries and Singapore." Department of Sociology working paper no. 94. Singapore: National University of Singapore, 1988.

Hughes, Helen, and Bernahu Woldekidan. "The Emergence of the Middle Class in ASEAN Countries." *ASEAN Economic Bulletin,* vol. 11, no. 7, 1994, p. 139–49.

Hughes, W. "Social Benefits Through Improved Transport." *Transport and National Goals.* Ed. E.T. Haefele. Washington, D.C.: Brooking Institution, 1969, p. 105–21.

Hugo, G.J. "Asia on the Move: Research Challenges for Population Geography." *International Journal of Population Geography*, June 1996a.

Hugo, G.J. "Brain Drain and Student Movements." Paper prepared for Conference on International Trade and Migration in the APEC Region, University of Melbourne, July 10–11, 1995.

Hugo, G.J. "Changing Patterns of Population Mobility." *Indonesia Assessment: Population and Human Resources.* Ed. T.H. Hull and G.W. Jones. Canberra: Australian National University, 1997a, p. 68–100.

Hugo, G.J. "Circular Migration in Indonesia." *Population and Development Review*, vol. 8, no. 1, 1982, p. 59–84.

Hugo, G.J. "The Demographic Impact of Famine." *Famine as a Geographical Phenomenon*. Ed. B. Currey and G.J. Hugo. Dordrecht: D. Reidel, 1984, p. 7–31.

Hugo, G.J. "Indonesian Labor Migration to Malaysia: Trends and Policy Implications. *Southeast Asian Journal of Social Science*, vol. 21, no. 1, 1993a, p. 36–70.

Hugo, G.J. "Indonesia's Migration Transition." *Journal fur Entwicklungspolitik*, vol. 11, no. 3, 1995b, p. 285–309.

Hugo, G.J. "Intergenerational Wealth Flows and the Elderly in Indonesia." *The Continuing Demographic Transition*. Ed. G.W. Jones, R.M. Douglas, J.C. Caldwell, and R.M. D'Souza. Oxford: Clarendon Press, 1997b, p. 111–34.

Hugo, G.J. "Labor Export from Indonesia: An Overview." *ASEAN Economic Bulletin* (Special Issue on Labor Migration in Asia), vol. 12, no. 2, 1996b, p. 275–98.

Hugo, G.J. "Levels, Trends and Patterns of Urbanization." *Migration, Urbanization and Development in Indonesia*. New York: United Nations, ESCAP, 1981.

Hugo, G.J. *Manpower and Employment Situation in Indonesia, 1993*. Jakarta: Ministry of Manpower, 1993b.

Hugo, G.J. "Population Change and Development in Indonesia." *Asia Pacific: New Geographies of the Pacific Rim*. Ed. R.F. Watters and T.G. McGee. London: Hurst and Company, 1997c, p. 223–49.

Hugo, G.J. *Population Mobility in West Java*. Yogyakarta: Gadjah Mada University Press, 1978.

Hugo, G.J. "Urbanisasi: Indonesia in Transition." *Development Bulletin*, vol. 27, 1993c, p. 46–50.

Hugo, G.J. "Urbanization in Indonesia: City and Countryside Linked." *The Urban Transformation of the Developing World*. Ed. J. Gugler. New York: Oxford University Press, 1996c, p. 132–83.

Hugo, G.J. "Women on the Move: Changing Patterns of Population Movement of Women in Indonesia." *Gender and Migration in Developing Countries*. Ed. S. Chant. London: Belhaven Press, 1992, p. 174–96.

Hugo, G.J., L.L. Lim, and S. Narayan. *Malaysian Human Resources Development Planning Project Module II: Labor Supply and Processes, Study no. 4: Labor Mobility*. Discipline of Geography, School of Social Sciences, Flinders University of South Australia, First Draft of Final Report, January 1989.

Hugo, G.J., T.H. Hull, V.J. Hull, and G.W. Jones. *The Demographic Dimension in Indonesian Development*. Singapore: Oxford University Press, 1987.

Huke, Robert E., and Eleanor H. Huke. "Human Geography of Rice in Southeast Asia." Manila, Philippines: International Rice Research Institute, 1990b.

Huke, Robert E., and Eleanor H. Huke. "Rice, Then & Now." Manila, Philippines: International Rice Research Institute, 1990a.

Hull, T.H. "First Results of Vietnam's 1989 Census." *Population Today*, October 1990, p. 6–8.

Human Rights Watch. *A Modern Form of Slavery: Trafficking of Burmese Women and Girls into Brothels in Thailand*. New York, Washington, Los Angeles, London: Human Rights Watch, 1993.

Hunter, Brian, ed. *The Statesman's Yearbook: 1995–96*. Bedfordshire, England: The Macmillan Press Ltd., 1995.

Hurst, P. *Rainforest Politics: Ecological Destruction in South-East Asia*. London: Zed Books, 1990.

Hussey, Antonia. "Rapid Industrialization in Thailand, 1986–1991." *Geographical Review*, vol. 83, no. 1, 1993, p. 14–27.

Hussey, Antonia. "Tourism in a Balinese Village." *The Geographical Review*, vol. 79, 1989, p. 311–25.

Hussin Mutalib. "Islamic Revivalism in ASEAN States: Political Implications." *Asian Survey*, vol. 30, 1990, p. 877–91.

Hutaserani, Suganya, and Pornchai Tawong. *Urban Poor Upgrading: Analyses of Poverty Trends and Profile of the Urban Poor in Thailand*. Background report no. 6–2. Bangkok: Thailand Development Research Institute, 1990.

Hutchcroft, Paul D. "The Philippines at the Crossroads: Sustaining Economic and Political Reform, Asia Society." (22-page Asian Update pamphlet), New York, November 1996.

Hutterer, Karl L. "Prehistoric Trade and the Evolution of Philippine Societies." *Economic Exchange and Social Interaction in Southeast Asia: Perspectives From Prehistory, History and Ethnography*. Ed. K.L. Hutterer. Ann Arbor: University of Michigan Center for South and Southeast Asian Studies, 1977, p. 177–96.

Huttrer, Gerald W. "The Status of World Geothermal Power Production 1990–1994" (and a subsequent update), May 16, 1996. International Geothermal Association. http://www.demon.co.uk/geosci/wrphilip.html

Ibrahim, Zuraidah. "Singapore's Water: History, Politics and Future Options." *The Straits Times*, March 4, 1995, p. 15.

IDHS. *Indonesia Demographic and Health Survey 1991; Demographic and Health Survey 1994*. Jakarta: Central Bureau of Statistics, Ministry of Health, Ministry of Population/National Family Planning Coordinating Board, Macro International Inc., 1992, 1995.

ILO. *World Labor Report 1993*. Geneva: ILO, 1993.

"IPTN Targets Design Recognition in Second Decade of Operations." *Indonesia Development News*, vol. 12, no. 4, March/April 1989, p. 6–7.

Ikemoto, Yukio. "Income Inequality in Thailand in the 1980s." *Southeast Asian Studies*, vol. 30, no. 2, 1992, p. 213–35.

Indonesia, A Quarter Century of Progress (1969–1993). Jakarta: Republic of Indonesia, 1993.

"Indonesia: Okay, They Say." *The Economist*, July 13, 1996, p. 28.

Indonesia Development News, vol. 5.

Intal, Ponciano, Jr., et al., "The Philippines." *East Asia in Crisis: From Being A Miracle to Needing One?*" Ed. Ross McLeod and Ross Garnaut. New York: Routledge, 1998, p. 145–161.

"International Monetary Fund (IMF)." *Far Eastern Economic Review*, 1991.

Investment Opportunities in the Philippines, 1995, http://www.webquest.com/phil2000/oppty.htm

Ireson C.J. and W.R. Ireson. "Ethnicity and Development in Laos." *Asian Survey*, vol. 31, no. 10, 1991, p. 920–37.

Iyer, Pico. *Video Night in Katmandu*. New York: Vintage Books, 1989.

JMFA (Japanese Ministry of Foreign Affairs). *Japan's ODA: Annual Report, 1996*. Tokyo: Association for Promotion of International Cooperation, 1996.

Jackson, R.F. "What Follows the Frontier's Last Obsequies?—Interregional Migration in the Philippines, 1985–1990." *Asian and Pacific Migration Journal*, forthcoming.

Jacobs, Sheldon A. *Myanmar: Trade and Investment Potential in Asia*. New York: United Nations, 1996.

Jakarta Post, May 31, 1994.

Jayasankaran, S. "Balancing Act." *Far Eastern Economic Review*, December 21, 1995, p. 24–31.

Jayasankaran, S. "Made-In-Malaysia: The Proton Project." *Industrialising Malaysia: Policy, Performance, Prospects.* Ed. K.S. Jomo. London: Routledge, 1993, p. 272–85.

Jayasankaran, S., and J. McBeth. "Hazy Days-Forest Fires in Indonesia Irritate its Neighbours." *Far Eastern Economic Review*, October 20, 1994, p. 66–7.

Jayasuriya, Sisira, and Hal Hill. "The Economy." *The Philippines After Marcos.* Ed. R.J. May and Francisco Nemenzo. New York: St. Martin's Press, 1985, p. 130–51.

Jefferson, Mark. "The Law of the Primate City." *The Geographical Review*, vol. 29, no. 2, April 1939, p. 226–32.

Jellinek, L. "Circular Migration and the Pondok Dwelling System." *Food, Shelter and Transportation in South-East Asia and the Pacific.* Ed. R.J. Rimmer et al. Canberra: Australian National University, Department of Human Geography Publication HG/12, 1978, p. 135–54.

Jellinek, L. *The Wheel of Fortune: The History of a Poor Community in Jakarta.* Honolulu: University of Hawaii Press, 1991.

Jitsuchon, Somchai. "Thailand's Changing Income Distribution Pattern: 1975/76 to 1985/86." *TDRI Quarterly Newsletter,* vol. 3, no. 3, September 1988, p. 14–26.

Jocano, F.L. *Slum As A Way of Life*. Quezon City: University of the Philippines Press, 1984.

Johansen, F. "Toll Road Characteristics and Toll Road Experience in Selected South East Asia Countries." *Transportation Research* 23 A, 1989, p. 463–66.

Johnston, Bruce F., and William C. Clark. *Redesigning Rural Development: A Strategic Perspective.* Baltimore, Md.: The Johns Hopkins University Press, 1982.

Johnston, R.J. "The Challenge of a Rapidly Changing World Political Map." *Fennia*, vol. 172, no. 2, 1994, p. 87–96.

Jomo, K.S. "Financial Liberalization, Crises, and Malaysian Policy Responses." *World Development,* vol. 26, no. 8, August 1998, p. 1563–74.

Jomo, K.S. "Malaysia's Politicized Environment." *Aliran Monthly,* vol. 13, no. 3, 1993, p. 25–7.

Jomo K.S., and Chris Edwards. "Malaysian Industrialisation in Historical Perspective." *Industrialising Malaysia: Policy, Performance, Prospects.* Ed. K.S. Jomo. London: Routledge, 1993, p. 14–39.

Jones, G.W. *Marriage and Divorce in Islamic South-East Asia*. Melbourne: Oxford University Press, 1994.

Jones, G.W., and C. Manning. "Labor Force and Employment During the 1980s." *The Oil Boom and After: Indonesian Economic Policy and Performance in the Soeharto Era*. Ed. A. Booth. Kuala Lumpur: Oxford University Press, 1992.

Jones, J.H. "Rural Roads and Poverty Alleviation in Thailand." *Rural Roads and Poverty Alleviation*. Ed. J. Howe and P. Richards. Boulder, Colo.: Westview, 1984, p. 142–62.

Ju, Chen Ai, and Gavin Jones. *The Ageing of ASEAN: Its Socioeconomic Consequences.* Singapore: Institute of Southeast Asian Studies, 1989.

Judson, Ann H. *An Account of the American Baptist Mission to the Burman Empire in a Series of Letters Addressed to a Gentleman in London.* London: Joseph Butterworth and Son, 1828.

"KL Confident of Turning Round Perwaja Steel." *Business Times,* Singapore, March 22, 1996, p. 8.

Kahin, A.R. "Crisis on the Periphery: The Rift Between Kuala Lumpur and Sabah." *Pacific Affairs,* vol. 65, 1992, p. 30–67.

Kaosa-ard, M., S.S. Pednekar, S.R. Christensen, K. Aksornwong, and A.B. Rala. *Natural Resources Management in Mainland Southeast Asia*. Bangkok: Thailand Development Research Institute, 1995.

Karamoy, A., and G. Dias. "Delivery of Urban Services in Kampungs in Jakarta and Ujung Pandang." *Community Participation in Delivering Urban Services in Asia.* Ed. T.M. Yeung and T.G. McGee. Ottawa: IDRC, 1986, p. 191–210.

Karan, P.P., and W.A. Bladen. "The Geopolitical Base." *Southeast Asia: Realm of Contrasts*. Ed. A. Dutt. Boulder, Colo.: Westview, 1985, p. 20–35.

Karim, Wazir Jahan. "Bilateralism and Gender in Southeast Asia." *'Male' and 'Female' in Developing Southeast Asia*. Ed. W.J. Karim. Oxford and Washington, D.C.: Berg Publishers, 1995b, p. 35–74.

Karim, Wazir Jahan. "Introduction: Genderising Anthropology in Southeast Asia." *'Male' and 'Female' in Developing Southeast Asia*. Ed. W.J. Karim. Oxford and Washington, D.C.: Berg Publishers, 1995a, p. 3–34.

Karim, Wazir Jahan. "Research on Women in Southeast Asia: Current and Future Directions." *Women's Studies International: Nairobi and Beyond*. Ed. Aruna Rao. New York: The Feminist Press, 1991, p. 142–55

Karnow, Stanley. *In Our Image: America's Empire in the Philippines.* New York: Random House, 1989.

Kasai, S. "Remittances of Out-Migrants to Their Original Families: Evidence From Two Indonesian Villages." *Journal of Population Studies*, no. 11, 1988, p. 15–30

Katz, Rodney P. "National Security." *Singapore: A Country Study.* Ed. Barbara Leitch LePoer. Washington, D.C.: Federal Research Division, Library of Congress, 1991, p. 217–72.

Kaufman, J., and N. Gunawan. "Family Planning Knowledge and Practice in Vietnam: Results from a Microlevel Survey." Paper prepared for Population Association of America Annual Meeting, Miami, May 5–9, 1994.

Kaur, Amarjit. *Bridge and Barrier: Transport and Communications in Colonial Malaya, 1870–1957*. Singapore: Oxford University Press, 1985.

Ken, Wong Lin. "Commercial Growth Before the Second World War." *A History of Singapore.* Ed. Ernest C.T. Chew and Edwin Lee. Singapore: Oxford University Press, 1991b. p. 41–65.

Ken, Wong Lin. "The Strategic Significance of Singapore in Modern History." *A History of Singapore.* Ed. Ernest C.T. Chew and Edwin Lee. Singapore: Oxford University Press, 1991, p. 18–35.

"Key Economic Indicators of Thailand." *Bangkok Bank Monthly Review*. vol. 34, no. 5, p. 32–5, 1993.

Keyes, Charles F. *The Golden Peninsula.* New York: Macmillan, 1977.

Keyes, Charles F. "Introduction." *Reshaping Local Worlds: Formal Education and Cultural Change in Rural Southeast Asia*. Ed. C.F. Keyes. New Haven, Conn.: Yale University Southeast Asia Studies, 1991.

Keyes, Charles F. "Thailand: Buddhist Kingdom as Modern Nation-State." Boulder, Colo.: Westview, 1987.

Keyes, Charles F., Laurel Kendall, and Helen Hardacre. "Contested Visions of Community in East and Southeast Asia." *Asian Visions of Authority: Religion and the Moderns States of East and Southeast Asia*. Ed. C.F. Keyes, L. Kendall, and H. Hardacre. Honolulu: University of Hawaii Press, 1994, p. 1–18.

Keyes, W.J. *Metro Manila: A Case Study of Policies Towards Urban Slums.* Makati: Ministry of Human Settlements, 1983.

Keyfitz, N. "Development in an East Javanese Village, 1953 and 1985." *Population and Development Review*, vol. 11, no. 4, 1985, p. 695–719

Keyfitz, N. "The Ecology of Indonesian Cities." *American Journal of Sociology,* vol. 66, 1961, p. 384.

Khan, Habibullah, Sock-yong Phang, and Rex S. Toh. "Research Note: Tourism Growth in Singapore: An Optimal Target." *Annals of Tourism Research*, vol. 23, no. 1, 1996, p. 222–3.

Khin Win, U. "A Century of Rice Improvement in Burma." Los Baños, Laguna, Philippines: International Rice Research Institute, 1991.

Khin Win, U., U Nyi Nyi, and E.C. Price. "The Impact of a Special High-Yielding-Rice Program in Burma." Los Baños, Laguna, Philippines: International Rice Research Institute. IRRI Research Paper Series, no. 58, 1981.

Kiernan, B. *How Pol Pot Came to Power*. Verso, London: New Left Books, 1985.

King, A.D. *Urbanism, Colonialism and the World Economy*. New York: Routledge, 1990.

Kinnaird, Vivian, Uma Kothari, and Derek Hall. "Tourism: Gender Perspectives." *Tourism: A Gender Analysis*. Ed. Vivian Kinnaird and Derek Hall. Chichester: John Wiley, 1994, p. 1–34.

Koestoer, R.H. "Environmental Perspective on Land/Economic Processes in Rural, Urban and the Desakota Regions." *Masyarakat Indonesia*, vol. 19, no. 2, 1992, p. 161–86.

Kong, Chiew Seen. "Nation-Building in Singapore: An Historical Perspective." *In Search of Singapore's Values.* Ed. Jon S.T. Quah. Singapore: Times Academic Press, 1990, p. 6–23.

Kraiyudh Dhiratayakinant. "Public-Private Sector Partnership in Industrialization." *Thailand's Industrialization and Its Consequences.* Ed. Medhi Krongkaew. New York: St. Martin's Press, 1995, p. 99–115.

Krongkaew Medhi. "Development Lessons from Southeast Asian Economies: Can the Present Success Be Sustained?" *Development Strategies for the 21st Century.* Ed. Teruyuki Iwasaki, Takoshi Mori, and Hiroichi Yamaguchi. Tokyo: Institute of Developing Economies, 1992, p. 90–116.

Krongkaew, Medhi, Pranee Tinakorn, and Suphat Suphachalasai. "Rural Poverty in Thailand: Policy Issues and Responses." *Asian Development Review*, vol. 10, no. 1, 1992, 199–225.

Kulkarni, V.G. "Designer Genes." *Far Eastern Economic Review*, September 8, 1983, p. 23–4.

Kulkarni, V.G., and Rodney Tasker. "Promises to Keep." *Far Eastern Economic Review,* vol. 159, no. 9, February 29, 1996, p. 22–8.

Kulke, Hermann. *The Devaraja Cult.* Cornell Data Paper no. 108. Ithaca: Southeast Asia Program, Department of Asian Studies, Cornell University, 1978.

Kulke, Hermann. "The Early and the Imperial Kingdom in Southeast Asian History." *Southeast Asia in the 9th to 14th Centuries.* Ed. David G. Marr and A.C. Milner. Singapore: Institute of Southeast Asian Studies, 1986, p. 1–22.

Kumar, R. *The Forest Resources of Malaysia: Their Economics and Development*. Singapore: Oxford University Press, 1986.

Kumar, Sree, and Lee Tsao Yuan. "A Singapore Perspective." *Growth Triangle: The Johor-Singapore–Riau Experience*. Ed. Lee Tsao Yuan. Singapore: Institute of Southeast Asian Studies, 1991, p. 1–36.

Kummer, D.M. *Deforestation in the Postwar Philippines.* Chicago: University of Chicago Press, 1992.

Kummer, D.M., and B.L. Turner II. "The Human Causes of Deforestation in Southeast Asia." *BioScience,* vol. 44, no. 5, 1994, p. 323–8.

Kundstadter, Peter, et al., eds. *Farmers in the Forest: Economic Development and Marginal Agriculture in Northern Thailand*. Honolulu: University Press of Hawaii, 1978.

LICADHO (Cambodian League for the Promotion and Defence of Human Rights). *Women's Rights as Human Rights, NGO Forum on Women—Beijing 1995*. Phnom Penh, Cambodia: LICADHO, 1995.

Labita, Al. "Manila Approves Car Assembly by Proton." *The Business Times,* Singapore, March 15, 1996, p. 20.

"Labor Trends in Indonesia, 1989–1990." Jakarta: American Embassy, 1990, Mimeo.

Land Transport Authority. *A World Class Land Transport System.* Singapore: Land Transport Authority, 1996.

Laos, National Statistical Centre. *Lao Census 1995: Preliminary Report 2.* Vientiane: Committee for Planning and Cooperation, 1995.

Laquian, A.A. *Slums and Squatters in Six Philippine Cities.* Ottawa: International Development Research Center, 1972.

Lauridsen, L.S. "The Financial Crisis in Thailand: Causes, Conduct, and Consequences?" *World Development,* vol. 26, no. 8, August 1998.

Lea, David A.M., and D.P. Chaudhri, eds. *Rural Development and The State.* London: Methuen, 1983.

Leake, D., Jr. *Brunei: The Modern Southeast-Asian Islamic Sultanate*. Jefferson, N.C.: McFarland, 1989.

Lee, Soo Ann. "Expansion of the Services Sector." *The Management of Success: The Moulding of Modern Singapore.* Ed. Kernial Singh Sandhu and Paul Wheatley. Singapore: Institute of Southeast Asian Studies, 1989, p. 280–99.

Leete, R., and Alam, I. eds. *The Revolution in Asian Fertility: Dimensions, Causes and Implications*. Oxford: Clarendon Press, 1993.

Leete, R., and Tan Boon Ann. "Contrasting Fertility Trends Among Ethnic Groups in Malaysia." *The Revolution in Asian Fertility: Dimensions, Causes and Implications.* Ed. R. Leete and I. Alam. Oxford: Clarendon Press, 1993.

Legge, J.D. "The Writing of Southeast Asian History." *The Cambridge History of Southeast Asia. Volume One: From Early Times to c. 1800.* Ed. Nicholas Tarling. Cambridge: Cambridge University Press, 1994, p. 1–50.

Leheny, David. "A Political Economy of Asian Sex Tourism." *Annals of Tourism Research*, vol. 22, no. 2, 1995, p. 367–84.

Leinbach, Thomas R. "Child Survival in Indonesia." *Third World Planning Review*, vol. 10, no. 3, 1988, p. 255–69.

Leinbach, Thomas R. "Emerging Spatial Patterns of Development." *Development and Environment in Peninsular Malaysia*. Ed. R. Aiken, M. Moss, T.R. Leinbach, and C. Leigh. Singapore: McGraw Hill, 1982a, p. 81–100.

Leinbach, Thomas R. "Industrial Strategy in Malaysia: The Role of Export Processing Zones." *GeoJournal,* vol. 6, no. 5, 1982b, p. 459–68.

Leinbach, Thomas R. "Rural Transport and Population Mobility in Indonesia." *The Journal of Developing Areas,* vol. 17, 1983a, p. 349–64.

Leinbach, Thomas R. "Trade Regimes, Foreign Investment, and the New Liberalization Environment in Indonesia." *The Location of Foreign Direct Investment*. Ed. M. Green and R. McNaughton. London: Avebury, 1995, p. 151–74.

Leinbach, Thomas R. "The Transmigration Program in Indonesian National Development Strategy: Current Status and Future Requirements." *Habitat International,* vol. 13, no. 3, 1989, p. 81–95.

Leinbach, Thomas R. "Transport and Third World Development: Review, Issues and Prescription." *Transportation Research,* vol. 29A, no. 5 (1995), p. 337–44.

Leinbach, Thomas R. "Transport Evaluation in Rural Development: An Indonesian Case Study." *Third World Planning Review,* vol. 5, 1983b, p. 23–35.

Leinbach, Thomas R. "Transport Policies in Conflict: Deregulation, Subsidies and Regional Development in Indonesia." *Transportation Research,* vol 23A, no. 6, November 1989, p. 467-475

Leinbach, Thomas R. "Transportation and Development in Malaya." *Annals of the Association of American Geographers,* vol. 65, 1975, p. 459–68.

Leinbach, Thomas R. "Travel Characteristics and Mobility Behavior: Aspects of Rural Transport Impact in Indonesia." *Geografiska Annaler*, ser. B, vol. 63, 1981, p. 119–29.

Leinbach, Thomas R., and Bambang Suwarno. "Commuting and Circulation Characteristics in the Intermediate Sized City: The Example of Medan, Indonesia." *Singapore Journal of Tropical Geography,* vol. 6, no. 1, 1985, p. 35–47.

Leinbach, Thomas R., and Chia Lin Sien. *Southeast Asian Transport: Issues in Development.* Singapore: Oxford University Press, 1989.

Leinbach, Thomas R., and John Bowen. "Diversity in Peasant Economic Behaviors: Transmigrant Households in South Sumatra, Indonesia." *Geographical Analysis,* vol. 24, no. 4, October 1992, p. 335–351

Leinbach, Thomas R., and Richard Ulack. "Cities of Southeast Asia." *Cities of the World: World Regional Urban Development.* 2nd ed. Ed. Stanley D. Brunn and Jack F. Williams. New York: HarperCollins College Publishers, 1993, p. 388–429.

Leinbach, Thomas R., and John Watkins. "Remittances and Circulation Behavior in the Livelihood Process: Transmigrant Families in South Sumatara." *Economic Geography,* vol. 74, no. 1, April 1998, p. 45–63.

Leinbach, Thomas R., John Watkins, and John Bowen, Jr. "Employment Behavior and the Family in Indonesian Transmigration." *Annals of the Association of American Geographers,* vol. 82, no. 1, 1992, p. 23–47.

Leipziger, D.M. *Awakening the Market: Viet Nam's Economic Transition.* World Bank Discussion Papers, no. 157, Washington, D.C., 1992.

Lenz, Ralph. "On Resurrecting Tourism in Vietnam." *Focus*, vol. 43, no. 3, 1993, p. 1–6.

Leong, Chan Teik. "COE Prices Rise Sharply." *The Straits Times,* Singapore, May 10, 1997, p. 1.

Leow, Jason. "Profile of Singapore (1990–1995): Bigger Homes and Better Pay." *The Straits Times*, Singapore, June 2, 1997, p. 28.

LePoer, Barbara Leitch. "Historical Setting." *Singapore: A Country Study.* Ed. Barbara Leitch LePoer. Washington, D.C.: Federal Research Division, Library of Congress, 1991, p. 1–64.

Lerner, Gerda. *The Creation of Patriarchy*. New York: Oxford University Press, 1986.

"Let's Green the World by Year 2000, Says PM." *New Straits Times,* Kuala Lumpur, April 28, 1992, p. 3–4.

Leung, Y.W. *Development Land and Development Charge in Singapore.* St. Paul, Minn.: Butterworth, 1987.

Leur, J.C. van. "On Early Asian Trade." *Indonesian Trade and Society*. Transl. by James S. Holmes and A. van Marle. The Hague: W. van Hoeve Ltd., 1955, p. 1–144.

Lewallen J. *Ecology of Devastation: Indochina*. Baltimore: Penguin, 1971.

Li, Tania Murray. "Producing Agrarian Transformation at the Indonesian Periphery." *Economic Analysis Beyond the Local System*. Ed. Richard E. Blanton, Peter N. Peregrine, Deborah Winslow, and Thomas D. Hall. Lanham, New York, London: University Press of America, 1997, p. 125–46.

Lich, Hoang Thi. "The Development of Household Economies and Market Systems in Improving the Gender and Poverty Situation in Vietnam." *Gender, Economic Growth and Poverty: Market Growth and State Planning in Asia and the Pacific*. Ed. N. Heyzer and G. Sen. New Delhi: Kali; Utrecht: International Books, 1994, p. 318–33.

Lim, Arthur Joo-Hock. "Geographical Setting." *A History of Singapore.* Ed. Ernest C.T. Chew and Edwin Lee. Singapore: Oxford University Press, 1991, p. 3–14.

Lim, Linda Y.C. "ASEAN: New Modes of Economic Cooperation." *Southeast Asian in the New World Order*. Ed. D. Wurfel and B. Burton. New York: St. Martin's Press, 1996, p. 19–35.

Lim, Teck Ghee, and M. Valencia, eds. *Conflict over Natural Resources in South-East Asia and the Pacific*. Singapore: Oxford University Press, 1990.

Lim, Vivian. "Women Electronics Workers in Southeast Asia: the Emergence of the Working Class." *Global Restructuring and Territorial Development.* Ed. Jeffrey Henderson and Manue Castells. London: Sage Publications, 1987, p. 112–35.

Lindauer, D.L. "The Tondo Project: Whom Have We Served?" *Philippine Journal of Public Administration,* vol. 25, no. 3/4, 1981, p. 280–7.

Lintner, B. "Add Water: Laos' Hydroelectric Plans Seem Overambitious." *Far Eastern Economic Review*, vol. 157, no. 41, 1994a, p. 70–1.

Lintner, B. *Burma in Revolt: Opium and Insurgency Since 1948.* Boulder, Colo.: Westview Press, 1994b.

Lintner, B. "Laos: Ties That Bind." *Far Eastern Economic Review*, vol. 158, no 6, February 9, 1995a, p 18–23.

Lintner, B. "Lost for Words." *Far Eastern Economic Review*, vol. 160, no. 49, 1997, p. 90–1.

Lintner, B. "The Noose Tightens: Khun Sa Faces a Day of Reckoning." *Far Eastern Economic Review,* October 19, 1995b, p. 28–9.

Ljunggren, B. "Market Economies Under Communist Regimes—Reform in Vietnam, Laos and Cambodia." *The*

Challenge of Reform in Indochina. Ed. B. Ljunggren. Cambridge, Mass.: Harvard University Press, 1993.

Lloyd's List Maritime Asia, various 1996 issues.

Logan, W.S. "Hanoi Townscape: Symbolic Imagery in Vietnam's Capital." *Cultural Identity and Urban Change in Southeast Asia: Interpretative Essays*. Ed. M. Askew and W.S. Logan. Geelong: Deakin University Press, 1994, p. 43–70.

Logan, W.S. "Heritage Planning in Post-*doi moi* Hanoi." *American Planning Association Journal*, Summer 1995, p. 328–43.

"Logging On." *Economist,* vol. 339, no. 7909, April 8, 1995, p. 34–5.

Long, Ngo Vinh. "Agrarian Differentiation in the Southern Region of Vietnam." *Sociology of "Developing Societies" Southeast Asia.* Ed. John G. Taylor and Andrew Turton. New York: Monthly Review Press, 1988, p. 133–45.

Low, Guat Tin. *Successful Women in Singapore: Issues, Problems and Challenges*. Singapore: EPB Publishers, 1993.

Low, Rowena. *Singapore's Aerospace Industry.* ASEAN/Singapore Briefing, no. 9, Economic Research Department, DBS Bank, 1995.

Lyon, Eugene. "Track of the Manila Galleons." *National Geographic*, vol. 178, no. 3, September 1990, p. 3–37.

"MSC Plan Gets High Priority." *The Straits Times,* Singapore, August 3, 1996, p. 25.

Mabbett, I.W. "Devaraja." *Journal of Southeast Asian History*, vol. 10, no. 2, September 1969, p. 202–23.

Mabbett, I.W. "The 'Indianization' of Southeast Asia: Reflections on the Historical Sources." *Journal of Southeast Asian Studies*, vol. 8, no. 2, September 1977, p. 143–61.

Mabbett, I.W. "Religious Beliefs and Practices of the Vietnamese." Centre for Southeast Asian Studies working paper 60. Melbourne: Monash University, 1989.

MacCannell, Dean. *The Tourist: A New Theory of the Leisure Class.* New York: Schocken Books, 1989.

Machado, Kit G. "ASEAN State Industrial Policies and Japanese Regional Production Strategies: The Case of Malaysia's Motor Vehicle Industry." *The Evolving Pacific Basin in the Global Political Economy: Domestic and International Linkages.* Ed. Cal Clark and Steve Chan. Boulder, Colo.: Lynne Rienner Publishers, 1992, p. 169–202.

Mahmud, Zaharah. "The Period and the Nature of 'Traditional' Settlements in the Malay Peninsula." *Journal of the Royal Asiatic Society,* Malaysian Branch, vol. 43, 1970, p. 81–113.

Majumdar, R.C. *Hindu Colonies in the Far East,* 2nd ed. revised Calcutta: N.K. Gossain & Co., 1963.

Malaysia, Department of Statistics. *Laporan Am Banci Pendudukan (General Report of the Population Census),* vol. 1. Malaysia: Department of Statistics, 1995.

Malaysia, Government of. *Second Malaysia Plan, 1971–1975; Fifth Malaysia Plan 1986–1990; Sixth Malaysia Plan 1991–1995; Seventh Malaysia Plan 1996–2000.* Kuala Lumpur: National Printing Department, 1971, 1986, 1991, 1996.

"Malaysia's Proton Wira is Best-Selling Model Here." *The Straits Times,* Singapore, January 16, 1996, p. 19.

Mangin, W. "Introduction." *Peasants in Cities: Readings in the Anthropology of Urbanization.* Ed. W. Mangin. Boston: Houghton Mifflin, 1970, p. xiii–xxxix.

Manning, C. "Examining Both Sides of the Ledger: Economic Growth and Labor Welfare Under Soeharto." *Indonesia Assessment 1993.* Ed. C. Manning and J. Hardjono. Canberra: Australian National University. Political and Social Change Monograph no. 20, 1993, p. 61–87.

Manning, C. "The Green Revolution, Labor Displacement, Incomes and Wealth in Rural Java." Adelaide, Australia: Flinders University, 1986. Mimeograph.

Manning, C., and M. Rumbiak. "Irian Jaya." *Unity and Diveristy: Regional Economic Development in Indonesia Since 1970.* Ed. Hal Hill. Singapore: Oxford University Press, 1989.

Manning, C., and S. Jayasuriya. "Survey of Recent Developments." *Bulletin of Indonesian Economic Studies*, vol. 32, no. 2, 1996, p. 3–43.

Manning, Jeff. "Nike: Each Worker's Life is One Piece of the Puzzle." *The Oregonian*, Portland, November 10, 1997.

Mantra, I. and M. Molo. *Studi Mobilitas Sirkuler Penduduk ke Enam Kota Besar di Indonesia*, Kerjasama Kantor Menteri Negara Kependudukan den Lingkungan Hidup dengan Pusat Penelitian Kependudukan UGM, Laporan akhir Yogyakarta, Pusat Penelitian Kependudukan, 1985.

Marchand, Marianne H., and Jane L. Parpart, eds. *Feminism, Postmodernism, Development*. London and New York: Routledge, 1995.

Martin, Philip L. "Migration and Trade: The Case of the Philippines" (Conference Report). *International Migration Review*, vol. 27, no. 3, p. 639–45.

Mason, K. Oppenheim. "Family Change and Support of the Elderly in Asia: What Do We Know?" *Asia–Pacific Population Journal*, vol. 7, no. 3, 1992, p. 13–32.

Mason, K. Oppenheim. "Is the Situation of Women in Asia Improving or Deteriorating?" *Asia–Pacific Population Research Report*, no. 6, Honolulu: East–West Center, 1995.

Matsui, Yayori. *Women's Asia.* London and New Jersey: Zed Books, 1987.

Maxton, Graeme P. *World Car Forecasts: The Outlook for Sales, Production and Vehicles in Use to 2000.* London: EIU, 1995 ed.

McBeth, J. "At Loggerheads-Irianese Feel Threatened by Influx of Migrants." *Far Eastern Economic Review*, March 10, 1994, p. 50–2.

McBeth, J., and D. Goertzen. "Down in the Dumps." *Far Eastern Economic Review,* October 17, 1991, p. 30.

McBeth, J., and M. Cohen. "Loosening the Bonds." *Far Eastern Economic Review,* January 21, 1999.

McBeth, J., and M. Cohen. "Murder and Mayhem." *Far Eastern Ecnomic Review*, February 20, 1997, p. 26–7.

McCall, M.K. "Political Economy and Rural Transport: A Reappraisal of Transport Impacts." *Antipode,* vol. 9, 1977, p. 56–67.

McGee, T.G. "The Emergence of Desa Kota Regions in Asia: Expanding an Hypothesis." *The Extended Metropolis: Settlement Transition in Asia.* Ed. N. Ginsburg, B. Koppel, and T.G. McGee. Honolulu: University of Hawaii Press, 1991, p. 3–25.

McGee, T.G. *The Southeast Asian City: A Social Geography of the Primate Cities of Southeast Asia.* London: Bell; New York: Frederick A. Praeger, Publishers, 1967.

McGee, T.G. "Urbanisasi or Kotadesasi? Evolving Patterns of Urbanization in Asia." *Urbanization in Asia: Spatial Dimensions and Policy Issues.* Ed. F.J. Costa, A.K. Dutt, L.J.C. Ma, and A.G. Noble. Honolulu: University of Hawaii Press, 1989, p. 93–108.

McGee, T.G., and Ira M. Robinson, eds. *The Mega-Urban Regions of Southeast Asia.* Vancouver: University of British Columbia Press, 1995.

McGee, T.G., and Y. Yeung. *Hawkers in South East Asian Cities: Planning for the Bazaar Economy.* Ottawa: IDRC, 1977.

McKendrick, David. "Obstacles to Catch-Up: The Case of the Indonesian Aircraft Industry." *Bulletin of Indonesian Economic Studies,* vol. 28, no. 1, 1992, p. 39–66.

McLeod, Ross H. "Indonesia." *East Asia in Crisis.* Ed. Ross McLeod and Ross Garnaut. London and New York: Routledge, 1998.

McNicoll, G. "Internal Migration in Indonesia: Descriptive Notes." *Indonesia*, vol. 5, 1968, p. 29–92.

Means, G.P. *Malaysian Politics: The Second Generation.* New York: Oxford University Press, 1991.

Meesook, Oey Astra. *Income, Consumption and Poverty in Thailand, 1962/63 to 1975/76.* World Bank Staff working paper no. 364. Washington. DC: The World Bank, 1979.

Mehmet, O. *Development in Malaysia: Poverty, Wealth and Trusteeship*. London: Croom Helm, 1986.

Meilink-Roelofsz, M.A.P. *Asian Trade and European Influence in the Indonesian Archipelago Between 1500 and About 1630.* The Hague: Martinus Nijhoff, 1962.

Mekvichai, B., D. Foster, S. Chomchan, and P. Kritiporn. *Urbanization and the Environment-Managing the Conflict.* Research Report no. 6. Bangkok: Thailand Development and Research Institute, 1990.

Meng, Try Ea. "Kampuchea: A Country Adrift." *Population and Development Review*, vol. 7, no. 2, 1981, p. 209–28.

Menon, N. "Moving KL." *Malaysian Industry*, December 1995, p. 6–13.

Meyer, Walter. *Beyond the Mask: Toward a Transdiscplinary Approach of Selected Social Problems Related to the Evolution and Context of International Tourism in Thailand.* Saarbrucken, Germany: Verlag Breitenbach Publishers, 1988.

Mies, Maria. *Patriarchy and Accumulation on a World Scale: Women in the International Division of Labor.* London: Zed, 1986.

Milne, R.S. "Ethnic Aspects of Privatization in Malaysia." *Ethnic Preference and Public Policy in Developing States.* Ed. N. Nevitte and C.H. Kennedy. Boulder, Colo.: Lynne Rienner, 1986, p. 119–34.

Milne, R.S., and Diane K. Mauzy. *Singapore: The Legacy of Lee Kuan Yew*. Boulder, Colo.: Westview Press, 1990.

Ministry of Information and the Arts. *Singapore 1993.* Singapore, 1993.

Ministry of the Environment. *Clean Rivers.* Singapore, 1987.

Ministry of Trade and Industry. *The Strategic Economic Plan: Towards a Developed Nation.* Singapore, 1991.

Mirkinson, Judith. "Red Light, Green Light: The Global Trafficking of Women." June 14, 1994. Internet address: pfoc@igc.apc.org in igc:hr.women.

Missen, G.J. *Viewpoint on Indonesia: A Geographical Study*. Melbourne: Nelson, 1972.

Mohanty, Chandra T., Ann Russo, and Lourdes Torees, eds. *Third World Women and the Politics of Feminism.* Bloomington: Indiana University Press, 1991.

Moir, H. "The Informal Sector of Jakarta." Jakarta: LEKNAS-LIPI Monograph Series, 1978.

Momsen, Janet H. *Women and Development in the Third World.* London and New York: Routledge, 1991.

Momsen, Janet H., and Vivian Kinnaird, eds. *Different Places, Different Voices: Gender and Development in Africa, Asia and Latin America.* London and New York: Routledge, 1993.

"More M'sian [Malaysian]-Owned Ships Needed: Officials." *The Shipping Times,* Singapore, February 29, 1996, p. 1.

Murray, D. "From Battlefield to Market Place: Regional Economic Cooperation in the Mekong Zone." *Geography*, vol. 79, no. 345, 1994, p. 350–53.

Muzaffar, C. "Malaysia: Islamic Resurgence and the Question of Development." *Sojourn,* vol. 1, 1986, p. 57–75.

Mydans, Seth. "As Turmoil Builds, Thai Leader Shuffles Cabinet." *The New York Times,* October 25, 1997a, p. A3.

Mydans, Seth. "Depositors in Indonesia Count Losses as Banks Shut." *The New York Times,* November 14, 1997b, p. A3.

Myint, Hla. "Inward and Outward-Looking Countries Revisited: the Case of Indonesia." *Bulletin of Indonesian Economic Studies,* vol. 20, no. 2, August 1984, p. 39–52.

Nash, Allen. *Investment Opportunities in the Vietnamese Textile Industry.* A Report to the Asia Research Centre on Social, Political and Economic Change and the Australia-Taiwan Business Council, Asia Research Centre on Social, Political and Economic Change, Murdoch University, Western Australia, 1993.

National Research Council. *Sustainable Agriculture and the Environment in the Humid Tropics*. Washington, D.C.: National Academy Press, 1993.

Nelson, J. *Migrants, Urban Poverty and Instability in Developing Countries.* Occasional paper 22, Harvard University: Center for International Affairs, 1969.

Neville, W. "Economy and Employment in Brunei." *Geographical Review,* vol. 75, 1985, p. 451–61.

Ng, Cecilia. "Malay Women and Rice Production in Western Malaysia." *Women, Development, and Survival in the Third World*. Ed. Haleh Afshar. London and New York: Longman, 1991, p. 188–210.

Ng, E-Ching. "Economic Race: It's Getting Hotter in the Region." *The Straits Times,* Singapore, May 12, 1997, p. 30.

Ngaosyvathn, Mayoury. *Lao Women: Yesterday and Today*. Vientiane: State Publishing Enterprise, 1993.

Ngaosyvathn, Mayoury, and Pheuiphanh Ngaosyvathn. "Kith and Kin Politics: The Relationship Between Laos and Thailand." *Journal of Contemporary Asia Publishers.* Manila, 1994.

Nietschmann, Bernard. "Indonesia, Bangladesh: Disguised Invasion of Indigenous Nations." *Fourth World Journal*, vol. 1, no. 2, 1985, p. 89–126.

Nieva, Antonio Ma. "Land Scam: Agrarian Reform, Ramos Style." *Multinational Monitor,* vol. 15, no. 1, January/February 1994, p. 19–24.

OAG Desktop Guide: Worldwide, vol. 22, no. 9, November 1997.

O'Brien, L. "Some Characteristics of Work in the Manufacturing Sector." *Transformation With Industrialization in Peninsular Malaysia*. Ed. H. Brookfield. Kuala Lumpur: Oxford University Press, 1994, p. 169–87.

Ocampo, Romeo B. "The Metro Manila Mega-Region." *The Mega-Urban Regions of Southeast Asia.* Ed. T.G. McGee and Ira M. Robinson. Vancouver, British Columbia: UBC Press, 1995, p. 282–95.

O'Connor, David. "Electronics and Industrialisation: Approaching the 21st Century." *Industrialising Malaysia: Policy, Performance, Prospects.* Ed. K.S. Jomo. London, Routledge, 1993b, p. 210–33.

O'Connor, David. "Textiles and Clothing: Sunrise or Sunset Industry?" *Industrialising Malaysia: Policy, Performance, Prospects.* Ed. K.S. Jomo. London, Routledge, 1993a, p. 234–61.

O'Connor, K. "Airport Development in Southeast Asia." *Journal of Transport Geography,* vol. 3, 1995, p. 269–79.

O'Connor, R.A. *A Theory of Indigenous Southeast Asian Urbanization.* Research Notes and Discussion Paper 38, Institute of Southeast Asian Studies. Singapore: ISEAS, 1983.

Oekan Soekotjo Abdoellah. *Indonesian Transmigrants and Adaptation.* Berkeley: University of California, 1993.

Ogawa, N. "Urbanization and Internal Migration in Selected ASEAN Countries: Trends and Prospects." *Urbanization and Migration in ASEAN Development.* Ed. P.M. Hauser, D.S. Suits, and N. Ogawa. Honolulu: University of Hawaii Press, 1985, p. 83–107.

Ogawa, N., G.W. Jones, and J.G. Williamson, eds. *Human Resources in Development Along the Asia–Pacific Rim.* Singapore: Oxford University Press, 1993.

Ong, Aihwa. "Japanese Factories, Malay Workers: Class and Sexual Metaphors in West Malaysia." *Power and Difference: Gender in Island Southeast Asia.* Ed. J.M. Atkinson and S. Errington. Stanford: Stanford University Press, 1990.

Ong, Aihwa. *Spirits of Resistance and Capitalist Discipline: Factory Women in Malaysia.* Albany: State University of New York Press, 1987.

Ooi, J. *Depletion of the Forest Resources in the Philippines.* Field Report Series no. 18. Singapore: Institute of Southeast Asian Studies, 1987.

Oppermann, Martin. "Intranational Tourist Flows in Malaysia." *Annals of Tourism Research*, vol. 19, 1992, p. 482–500.

Orwell, George. "Burmese Days." New York: Harper & Brothers, 1934.

Osborne, Milton E. *The French Presence in Cochinchina and Cambodia: Rule and Response (1859–1905).* Ithaca and London: Cornell University Press, 1969.

Osborne, Robin. *Indonesia's Secret War: The Guerrilla Struggle in Irian Jaya.* Sydney: Allen & Unwin, 1985.

Owen, W. *Transportation and World Development.* London: Hutchinson, 1987.

Pakkasem, P. "Decentralization is Not the Answer." *The Nation,* Bangkok, July 5, 1987.

Panayoutou, Theodore, and Dhira Phantumvanit. "Rural Natural Resources Management: Lessons from Thailand." *TDRI Quarterly Review.* vol. 6, no. 1, 1991, p 17–23.

Pangestu, M., and I.J. Azis. "Survey of Recent Developments." *Bulletin of Indonesian and Economic Studies*, vol. 30, no. 2, 1994, p. 3–48.

Park, J. "Clearing the Air on the Haze." *Far Eastern Economic Review*, November 13, 1997, p. 34.

Parker, S. "Survey of Recent Developments." *Bulletin of Indonesian Economic Studies*, vol. 27, no. 1, 1991, p. 3–38.

Parker, S., and P. Hutabarat. "Survey of Recent Developments." *Bulletin of Indonesian Economic Studies*, vol. 32, no. 3, 1996, p. 3–32.

Parnwell, Michael J.G. "Environmental Issues and Tourism in Thailand." *Tourism in South-East Asia.* Ed. Michael Hitchcock, Victor T. King, and Michael J.G. Parnwell. London: Routledge, 1993, p. 286–302.

Paul, D. "Street Survival: Children, Work and Urban Drift in Cambodia." *Issues in Global Development*, no 3. Melbourne: World Vision Australia, 1995.

Pearce, Douglas. *Tourist Development.* New York: Longman Scientific & Technical, 1989.

Peleggi, Maurizio. "National Heritage and Global Tourism in Thailand." *Annals of Tourism Research*, vol. 23, no. 2, 1996, p. 432–48.

Peletz, Michael G. *Reason and Passion: Representations of Gender in Malay Society.* Berkeley: University of California Press, 1996.

Peluso, N.L. *Rich Forests, Poor People: Resource Control and Resistance In Java.* Berkeley: University of California Press, 1992.

Pelzer, K.J. "Man's Role in Changing the Landscape of Southeast Asia." *Journal of Asian Studies,* vol. 27, no. 2, 1968, 269–79.

Pham, Chi Do, ed. *Economic Development in Lao P.D.R.: Horizon 2000.* Vientiane: Bank of the Lao People's Democratic Republic, 1994.

Phelan, John L. *The Hispanization of the Philippines: Spanish Aims and Filipino Responses, 1565–1700.* Madison: The University of Wisconsin Press, 1959.

Philippines, National Statistics Office. *1992 Philippine Yearbook.* Manila: Philippines National Statistics Office, 1992.

Philippines, National Statistics Office. 1995 Philippines Population Census. See January 25, 1998 web page at http://www.census.gov.ph/data/sectordata/pop0.html.

"Philippines Proposes Labor Agreement." *Migration News,* vol. 1, no. 2, March 1994. (Newsletter produced at the University of California. See also http://128.120.36.171/By-Month/MN-Vol-1-94/MN_March_1994)

Phnom Penh Post, www.vais.net/wtapang/PPP/

Phongpaichit, Pasuk. "Bangkok Masseuses: Holding Up the Family Sky." *Southeast Asia Chronicle,* 1978, p. 15–23.

Phongpaichit, Pasuk, and Chris Baker. *Thailand: Economy and Politics.* Kuala Lumpur: Oxford University Press, 1995.

Picard, Michel. "Cultural Tourism in Bali: National Integration and Regional Differentiation." *Tourism in South-East Asia.* Ed. Michael L. Hitchcock, Victor T. King, and Michael J.G. Parnwell. London: Routledge, 1993, p. 71–98.

Pineda, Ernesto L. *The Family Code of the Philippinese Annotated*, 2nd ed. Quezon City: Central Lawbook Publishing Co., 1991.

Pluvier, Jan M. *Historical Atlas of South-East Asia.* Leiden, Netherlands: E.J. Brill, 1995.

Poffenberger, M., ed. *Keepers of the Forest: Land Management Alternatives in Southeast Asia.* West Hartford, Conn.: Kumarian Press, 1990.

Population Reference Bureau. *World Population Data Sheet*, various years. Washington, D.C.: Population Reference Bureau, Inc., 1983, 1994, 1996, 1997.

Porter, Gareth. *Vietnam: The Politics of Bureaucratic Socialism.* Ithaca: Cornell University Press, 1993.

Poston, Dudley L., Jr., and Mei-yu Yu. "The Distribution of the Overseas Chinese in the Contemporary World." *International Migration Review*, vol. 24, no. 3, 1990, p. 480–508.

Pranee, Tinakorn. "Industrialization and Welfare: How Poverty and Income Distribution Are Affected." *Thailands Industrialization and Its Consequences.* Ed. Medhi Krongkaew. New York: St. Martin's Press, 1995, p. 218–31.

Pressat, R. *Population*. Harmondsworth, England: Penguin, 1970.

Punpuing, S. "Bangkok and Its Environment As A Context of Commuting." Proceedings, *IGU Commission of Population and the Environment, Symposium on Population, Health and the Environment, Chiang Mai, January 7–11, 1997, Thailand.* Chiang Mai University: Department of Geography, 1997, p. 200–25.

Purcell, V. *The Chinese in Modern Malaya*. Kuala Lumpur: Oxford University Press, 1967.

Putzel, James. A *Captive Land: The Politics of Agrarian Reform in the Philippines.* New York: Monthly Review Press, 1992.

Quah, Jon S.T. "Government Policies and Nation-Building." *In Search of Singapore's Values.* Ed. Jon S.T. Quah. Singapore: Times Academic Press, 1990, p. 45–65.

Quibria, M.G., ed. *Rural Poverty in Asia:, Priority Issues and Policy Options*. Hong Kong: Oxford University Press, 1993.

Rafferty, Milton. *A Geography of World Tourism.* Englewood Cliffs, N.J.: Prentice-Hall, 1992.

Raffles, Thomas Stanford. *The History of Java, vol. 1*. London: John Murray, 1817.

Rasiah, Rasiah H. "Competition and Restructuring in the Semiconductor Industry: Implications for Technology Transfer and Its Absorption in Penang." *Southeast Asian Journal of Social Sciences,* vol. 17, 1989, p. 41–57.

Rajah Rasiah H. "Free Trade Zones and Industrial Development in Malaysia." *Industrialising Malaysia: Policy, Performance, Prospects.* Ed. K.S. Jomo. London: Routledge, 1993, p. 118–46.

Rathgeber, Eva M. "WID, WAD, GAD: Trends in Research and Practice." *Journal of Developing Areas*, vol. 24, no. 4, 1990, p. 489–502.

Redfield, Robert, and Milton B. Singer. "The Cultural Role of Cities." *Economic Development and Cultural Change*, vol. 3, no. 1, October 1954, p. 53–73.

Reed, Robert R. "The Colonial Genesis of Hill Stations: The Genting Exception." *The Geographical Review,* vol. 69, no. 4, October 1979, p. 463–68.

Reed, Robert R. *Colonial Manila: The Context of Hispanic Urbanism and Process of Morphogenesis.* University of California Publications in Geography, vol. 22. Berkeley and Los Angeles: University of California Press, 1978.

Reed, Robert R. "From Highland Hamlet to Regional Capital: Reflections on the Colonial Origins, Urban Transformation, and Environmental Impact of Dalat." *The Challenges of Highland Development in Vietnam.* Ed. A. Terry Rambo, Robert R. Reed, Le Trong Cuc, and Michael R. DiGregorio. Honolulu: East–West Center, Program on Environment, 1995.

Reed, Robert R. "Indigenous Urban Traditions in South-East Asia." *Changing South-East Asian Cities: Readings on Urbanization.* Ed. Y.M. Yeung and C.P. Lo. Singapore: Oxford University Press, 1976a, p. 14–27.

Reed, Robert R. "Remarks on the Colonial Genesis of the Hill Station in Southeast Asia with Particular Reference to the Cities of Buitenzorg (Bogor) and Baguio." *Asian Profile,* vol. 4, no. 6, December 1976b, p. 545–91.

Reid, Anthony. "Female Roles in Pre-Colonial Southeast Asia." *Modern Asian Studies,* vol. 22, no. 3, 1988a, p. 629–45.

Reid, Anthony. *Southeast Asia in the Age of Commerce 1450–1680, Volume 1, The Lands Below the Winds*. New Haven: Yale University Press, 1988b.

Reid, Anthony. *Southeast Asia in the Age of Commerce, 1450–1680. Volume 2, Expansion and Crisis.* New Haven and London: Yale University Press, 1993.

Reid, Anthony. "The Structure of Cities in Southeast Asia, Fifteenth to Seventeenth Centuries." *Journal of Southeast Asian Studies,* vol. 11, no. 2, 1983, p. 235–50.

Renard, Ronald Duane. *The Burmese Connection: Illegal Drugs and the Making of the Golden Triangle.* Boulder, Colo.: L. Rienner Publ. 1995.

Replogle, M. *Non-Motorized Vehicles in Asian Cities*. Washington, D.C.: World Bank, 1992.

Republic of Indonesia. *Indonesia: A Quarter Century of Progress (1968–1993) Tables and Graphs.* Jakarta, Indonesia, 1993.

Republik Indonesia. *Pelaksanaan Tahun Ketiga Repelita VI.* Jakarta, Indonesia, 1997.

Reynolds, Craig J. "A New Look at Old Southeast Asia." *The Journal of Asian Studies*, vol. 54, no. 2, May 1995, p. 419–46.

Rice, R. *The Informal Sector Employment in Indonesia: Some Issues and Suggestions for Improvement,* Report Series A, no. 11, Information System for Employment Development and Manpower Planning, Indonesian Department of Manpower and International Labor Office, 1992.

Richardson, H.W. "The Big, Bad City: Mega-City Myth?" *Third World Planning Review,* vol. 11, no. 4, 1989, p. 355–72.

Richter, Linda K. "Changing Directions in Philippine Policy Formation and Implementation: Land Reform and Tourism Development Under Marcos, Aquino, and Ramos." *Crossroads: An Interdisciplinary Journal of Southeast Asian Studies*, vol. 9, no. 1, 1994, p. 33–62.

Ricklefs, M.C. *A History of Modern Indonesia, c. 1300 to the Present.* London: Macmillan, 1981.

Rigg, Jonathan. "Grass-Roots Development in Rural Thailand: A Lost Cause?" *World Development,* vol. 19, no. 2/3, 1991a, 199–211.

Rigg, Johnathan, "Indonesia's Forest Fires." *Geography Review,* vol. 12, no. 5, May 1999, p. 20–21.

Rigg, Jonathan. "Managing Dependency in a Reforming Economy: The Lao PDR." *Contemporary Southeast Asia*, vol. 17, no. 2, September 1995, p. 147–72.

Rigg, Jonathan. "Rural-Urban Interactions, Agriculture and Wealth: A Southeast Asian Perspective. *Progress in Human Geography,* vol. 22, no. 4, December 1998, p. 497–522.

Rigg, Jonathan. *Southeast Asia: A Region in Transition. A Thematic Human Geography of the ASEAN Region*. Boston: Unwin Hyman, 1991b.

Rigg, Jonathan. *Southeast Asia: A Region in Transition.* London and New York: Routledge, 1994.

Rigg, Jonathan. *Southeast Asia: The Human Landscape of Modernization and Development.* London: Routledge, 1997.

Rimban, L. "Tales of Woe Don't Deter Pinay's Quest for Jobs Abroad." *The Manila Chronicle*, vol. 3, November 1995.

Rimmer, P.J. "Paratransit: A Commentary." *Environment and Planning A* vol. 12, 1980, p. 937–44.

Rimmer, P.J. "Regional Economic Integration in Pacific Asia." *Environment and Planning A,* vol. 26, 1994, p. 1731–59.

Rimmer, P.J. "The Role of Paratransit in South East Asian Cities." *Singapore Journal of Tropical Geography,* vol. 5, 1984, p. 45–62.

Rimmer, P.J., and H.W. Dick. "Improving Urban Public Transport in Southeast Asia Cities." *Transport Policy and Decision-Making,* vol. 1, 1980, p. 97–120.

"River of Promise." *Far Eastern Economic Review*, vol. 156, no. 37, September 16, 1993, p. 68–72.

Robinson, Geoffrey. "Human Rights in Southeast Asia: Rhetoric and Reality." *Southeast Asian in the New World Order*. Ed. D. Wurfel and B. Burton. New York: St. Martin's Press, 1995, p. 74–99.

Robinson, R. "Industrial Strategies and Port Development in Developing Countries." *Tidjschrift voor Economische en Sociale Geografie,* vol. 76, 1985, p. 133–43.

Robinson, R. "Regional Ports: Development and Change Since the 1970s." *Southeast Asian Transport: Issues in Development.* Ed. T.R. Leinbach and Chia Lin Sien. Singapore: Oxford University Press, 1989, p. 133–69.

Robinson, W.C. "Implicit Policies and the Urban Bias as Factors Affecting Urbanization. *Urbanization and Urban Policies in Pacific Asia.* Ed. R.J. Fuchs, G.W. Jones, and E.M. Pernia. Boulder, Colo.: Westview Press, 1987, p. 169–82.

Robison, Richard, and Andrew Rosser. "Contesting Reform: Indonesia's New Order and the IMF." *World Development,* vol. 26, no. 8, August 1998, p. 1593–1609.

Robison, Richard, and David S.G. Goodman, eds. *The New Rich in Asia: Mobile Phones, McDonalds, and Middle Class Revolution*. London: Routledge, 1996.

Robison, Richard, Kevin Hewison, Richard Higgott, eds. *Southeast Asia in the 1980s: The Politics of Economic Crisis.* Sydney: Allen and Unwin, 1987.

Rodan, Gerry. *The Political Economy of Singapore's Industrialization.* New York: St. Martin's Press, 1989.

Rodan, Garry. "Preserving the One-Party State in Contemporary Singapore." *Southeast Asia in the 1990s: Authoritarianism, Democracy and Capitalism*. Ed. Kevin Hewison, Richard Robinson, and Garry Rodan. St. Leonards, New South Wales, Australia: Allen and Unwin: 1993, p. 75–108.

Rogge, J.R. "Refugee Migration: Changing Characteristics and Prospects." Paper presented at Expert Group Meeting on Population Distribution and Migration, Santa Cruz, Bolivia, January 18–22, 1993.

Rohwer, Jim. *Asia Rising*. New York: Touchstone, 1996.

Rokiah, Alavi. *Industrialisation in Malaysia: Import Substitution and Infant Industry Performance,* 1996.

Rola, Agnes, and Prabhu Pingali. "Pesticides, Rice Productivity and Health Impacts in the Philippines." *Agricultural Policy and Sustainability: Case Studies from India, Chile, the Philippines and the United States.* Ed. Paul Faeth. Washington, D.C.: World Resources Institute, 1993, p. 47–62.

Rondinelli, D.A. "Decentralization, Territorial Power and the State: A Critical Response." *Development and Change,* vol. 21, 1990, p. 491–500.

Ruland, J. *Urban Development in Southeast Asia: Regional Cities and Local Government.* Boulder, Colo.: Westview, 1992.

Runciman, Steven. *The White Rajahs: A History of Sarawak from 1841 to 1946.* Cambridge: Cambridge University Press, 1960.

Rurakdee, Narissara. "The Golf Business." *Bangkok Bank Monthly Review*, vol. 33, no. 6, 1992, p. 14–17.

Russell, S.S. "Migrant Remittances and Development." *International Migration Quarterly Review*, vol. 30, no. 3/4, 1992, p. 267–87.

Russell, S.S. *Population and Development in the Philippines: An Update.* The World Bank Asia Country Department II, Population and Human Resources Division, Washington, D.C., 1991.

Russell, S.S., and M.S. Teitelbaum. *International Migration and International Trade*. World Bank Discussion Paper 160. Washington, D.C.: The World Bank, 1992.

Ruzicka, L. "Implications of Mortality Trends and Differentials in the ESCAP Region." Mimeograph, 1983.

St. John, R.B. "The Political Economy of the Royal Government of Cambodia." *Contemporary Southeast Asia*, vol. 17, no. 3, 1995, p. 265–81.

Saith, Ashwani. *The Rural Non-Farm Economy: Processes and Policies*. Geneva: International Labor Organization, 1992.

Salaff, Janet. "Women, The Family, and The State in Hong Kong, Taiwan, and Singapore." Ed. R.P. Appelbaum and J. Henderson. Newbury Park, Calif.: Sage, 1992, p. 267–88.

Salih, Kamal. "Urban Dilemmas in Southeast Asia." *Singapore Journal of Tropical Geography,* vol. 3, no. 2, 1982, p. 147–61.

Salih, Kamal, and Mei Ling Young. "Social Forces, the State and the International Division of Labour: the Case of Malaysia." *Global Restructuring and Territorial Development.* Ed. Jeffrey Henderson and Manuel Castells. London: Sage Publications, 1987, p. 168–202.

Samudavanija, Chai-Anan. "Economic Development and Democracy." *Thailand's Industrialization and Its Consequences.* Ed. Medhi Krongkaew. New York: St. Martin's Press, 1995, p. 235–50.

Sandhu, K.S. *Indians in Malaysia*. Cambridge: Cambridge University Press, 1969.

Sardesai, D.R. *Southeast Asia: Past and Present.* Boulder, Colo.: Westview Press; London: Macmillan, 1989.

Sauer, Carl O. *Agricultural Origins and Dispersals*. New York: American Geographical Society, 1952.

Savage, Victor. *Western Impressions of Nature and Landscape in Southeast Asia.* Singapore: Singapore University Press, 1984.

Saywell, T. "Workers' Offensive." *Far Eastern Economic Review,* May 29, 1997, p. 50–2.

Schendel, Willem Van. *Three Deltas: Accumulation and Poverty in Rural Burma, Bengal and South India.* New Delhi; Newbury Park, Calif.: Sage Publication. Series: Indo-Dutch studies on development alternatives, 1991.

Schwarz, Adam. "Just How Liberal: Thailand Tinkers with Foreign Investment Laws." *Far Eastern Economic Review,* vol. 147, no. 24, June 16, 1994, p. 70.

Schwarz, Adam. "A Miracle Comes Home." *Far Eastern Economic Review*, April 19, 1990, p. 40–4.

Schwarz, Adam. *A Nation in Waiting: Indonesia in the 1990s.* St. Leonards, New South Wales, Australia: Allen & Unwin, 1995a.

Schwarz, Adam. "Where Oil and Water Mix." *Far Eastern Economic Review*, vol. 158, no. 11, March 16, 1995b, p 54–8.

Scott, James. *Weapons of the Weak*. New Haven, Conn.: Yale University Press, 1985.

Season and Crop Report for the Year Ending June 30, 1960. Rangoon: Government Printing Office (in Burmese).

Seaward, N. "A New Pragmatism: Foreign Capital Dominates Malaysian Manufacturing Investment." *Far Eastern Economic Review,* January 21, 1988, p. 49–50.

Secrett, C. "The Environmental Impact of Transmigration." *Ecologist,* vol. 16, no. 2–3, 1986, p. 77–88.

Sen, Amartya. "More Than 100 Million Women Are Missing." *The New York Review*, December 20, 1990, p. 61–6.

Sen, Gita, and Caren Grown. *Development, Crises, and Alternative Visions: Third World Women's Perspectives*. New York: Monthly Review Press, 1987.

Sender, Henry. "Stemming the Flood." *Far Eastern Economic Review,* vol. 162, no. 30, July 29, 1999, p. 52–54.

Sesser, Stan. *Lands of Charm and Cruelty.* New York: Vintage, 1993.

Sethuraman, S.V. *Jakarta: Urban Development and Employment.* Geneva: International Labor Office, 1976.

Shamshul, A.B. "Religion and Ethnic Politics in Malaysia: The Significance of the Islamic Resurgence Phenomenon." *Asian Visions of Authority: Religion and the Moderns States of East and Southeast Asia*. Ed. C.F. Keyes, L. Kendall, and H. Hardacre. Honolulu: University of Hawaii Press, 1994, p. 99–116.

Sharma, M.L. "Role of Groundwater in Urban Water Supplies in Bangkok, Thailand and Jakarta, Indonesia." Environment and Policy Institute working paper. Honolulu: East-West Center, 1986.

Shaw, Gareth, and Allan Williams. *Critical Issues in Tourism: A Geographical Perspective.* Oxford: Blackwell Publishing, 1994.

Shields, Robert. *Places on the Margin: Alternative Geographies of Modernity.* London: Routledge, 1991.

Sicular, D.T. *Scavengers, Recyclers and Solutions for Solid Waste Management in Indonesia.* Berkeley: Center for Southeast Asian Studies, University of California, 1992.

Silverstein, Josef. *Burmese Politics: The Dilemma of National Unity*. New Brunswick: Rutgers University Press, 1980.

Simmons, A.B. "Slowing Metropolitan Growth in Asia: Policies, Programs and Results. *Population and Development Review,* vol. 5, 1988, p. 57–104.

Singapore, Department of Statistics. *Monthly Digest of Statistics*. Singapore, various issues.

Singarimbun, M. "Sriharjo Revisited." *Bulletin of Indonesian Economic Studies*, vol. 12, no. 2, 1986, p. 117–25.

Smith, D.A., and R.J. Nemeth. "Urban Development in Southeast Asia: An Historical Structural Analysis." *Urbanization in the Developing World*. Ed. D. Drakakis-Smith. London: Croom Helm, 1986, p. 121–39.

Social Development Unit. *The Need for Social Development in Singapore.* Singapore, 1991.

"Social Indicators: The Burden on Asia's Coastline." *Far Eastern Economic Review,* August 2, 1990, p. 50.

Solomon, J. "What Political Risk?" *Far Eastern Economic Review*, February 20, 1997.

Solon, O. "An Essay in the Theory of Urban Squatting and Land Development." Ph.D. Dissertation, University of the Philippines, Dilman, School of Economics, 1987.

"Southeast Asia's Wealth Gap." *The Economist.* April 13, 1996, p. 29–31.

Sparr, Pamela, ed. *Mortgaging Women's Lives: Feminist Critiques of Structural Adjustment.* London: Zed Books, 1994.

Spencer, A.H. "Urban Transport." *Southeast Asian Transport: Issues in Development*. Ed. T.R. Leinbach and Chia Lin Sien. Singapore: Oxford University Press, 1989, p. 190–231.

"S'pore Chip Industry Receives $21M Boost." *The Straits Times,* Singapore, August 29, 1996, p. 50.

"The Spratly Island Dispute." 1995. http://www.reedbooks.com.au/heinemann/hot/sprat.html

Sricharatchanya, P. "Jungle Warfare." *Far Eastern Economic Review,* vol. 137, no. 38, September 17, 1987, p. 86–8.

Srimonkol, Katin, and Gerald G. Marten. "Traditional Agriculture in Northern Thailand." *Traditional Agriculture in Southeast Asia: A Human Ecology Perspective.* Ed. Gerald G. Marten. Boulder, Colo.: Westview Press, 1986, p. 85–102.

Stahl, C.W. and PECC-HRD Task Force. "International Labor Migration and the East Asian APEC / PECC Economies: Trends, Issues and Policies." Paper presented at PECC Human Resource Development Task Force Meeting, Brunei, June 7–8 , 1996.

Standing, Guy. "Labour Flexibility in the Malaysian Manufacturing Sector." *Industrialising Malaysia: Policy, Performance, Prospects.* Ed. K.S. Jomo. London: Routledge, 1993, p. 14–39.

Stanley, Peter W. *A Nation in the Making: The Philippines and the United States, 1899–1921.* Cambridge: Harvard University Press, 1974.

State Planning Committee. *Vietnam Living Standards Survey*. Hanoi, 1994.

The Statesman's Yearbook: Statistical and Historical Annual of the States of the World. New York: St. Martin's Press, annual.

Steinberg, D.J., ed. *In Search of Southeast Asia: A Modern History*. New York: Praeger Publishers, Inc., 1971.

Steinberg, D.J., et al., *In Search of Southeast Asia: A Modern History.* Honolulu: University of Hawaii Press, 1985.

Steinberg, David I. "Burma, A Socialist Nation of Southeast Asia." *Westview profiles. Nations of Contemporary Asia.* Boulder, Colo.: Westview Press, 1982.

Sternstein, L. "The Growth of the Population of the World's Pre-eminent "Primate City": Bangkok at Its Bicentenary." *Journal of Southeast Asian Studies,* vol. 15, 1984, p. 43–68.

Stoler, Ann, L. "Carnal Knowledge and Imperial Power: Gender, Race, and Morality in Colonial Asia." *Gender at the Crossroads of Knowledge: Feminist Anthropology in the Postmodern Era*. Ed. M. di Leonardo. Berkeley: University of California Press, 1991, p. 51–101.

Stuart-Fox, Martin. *Buddhist Kingdom, Marxist State: The Making of Modern Laos*. Bangkok: White Lotus, 1996.

Sullivan, Margaret. "The Economy." *Singapore: A Country Study.* Ed. Barbara Leitch LePoer. Washington, D.C.: Federal Research Division, Library of Congress: 1991, p. 119–74.

Sullivan, Norma. *Masters and Managers: A Study of Gender Relations in Urban Java.* St. Leonards New South Wales, Australia: Allen & Unwin, 1994.

Suphachalasai, Suphat. "Export Led Industrialization." Thailand's Industrialization and Its Consequences. Ed. Medhi Krongkaew. New York: St Martin's Press, 1995, p. 66–84.

Surarerks, Vanpen. *Thai Governmental Rural Development Programs: An Analysis and Evaluation of the Rural Job Creation Programs in Thailand, 1980–1985.* Chiang Mai, Thailand: Department of Geography, Chiang Mai University, 1986.

Suryadinata, Leo. "Democratization and Political Succession in Suharto's Indonesia." *Asian Survey*, vol. 37, no. 3, 1997b, p. 269.

Suryadinata, Leo, ed. *Ethnic Chinese as Southeast Asians*. Singapore: Institute of Southeast Asian Studies; New York: St. Martin's Press, 1997a.

Suryadinata, Leo, and Michael Malley. "Indonesia's Foreign Policy Under Suharto." *Journal of Asian Studies*, vol. 55, no. 4, 1996, p. 1090.

Suthasupa, Paiboon. "Rural Development in Thailand." *Community Development Journal,* vol. 22, no. 2, April 1987, p. 81–6.

Sutton, K. "Malaysia's FELDA Land Settlement Model in Time and Space." *Geoforum,* vol. 20, no. 3, 1989, p. 339–54.

Suu Kyi, Aung San. *Freedom From Fear and Other Writings.* Ed. Michael Aris. New York: Penguin Books, 1991.

Svendsen, Mark, and Mark W. Rosegrant. "Irrigation Development in Southeast Asia Beyond 2000: Will the Future Be Like the Past?" *Water International,* vol. 19, 1994, p. 23–35

Swain, Margaret Byrne. "Gender in Tourism." *Annals of Tourism Research*, vol. 22, 1995, p. 247–66.

"A Tale of Two Systems." *The Economist,* July 9, 1994, p. 33.

Tambiah, Stanley J. "The Galactic Polity: The Structure of Traditional Kingdoms in Southeast Asia." *Annals of the New York Academy of Sciences*, vol. 293, 1977, p. 69–97.

Tan, Augustine H.H., and Phang Sock Yang. *The Singapore Experience in Public Housing.* Singapore: Centre for Advanced Studies, occasional paper no. 9, 1991.

Tan, Joseph. L.H. *AFTA in the Changing International Economy.* Ed. M. Ariff et al. Singapore: Institute of Southeast Asian Studies, 1996, p. 1–18.

Tarling, Nicholas, ed. *The Cambridge History of Southeast Asia. Volume One: From Early Times to c. 1800.* Cambridge: Cambridge University Press, 1994a.

Tarling, Nicholas, ed. *The Cambridge History of Southeast Asia. Volume Two: The Nineteenth and Twentieth Centuries.* Cambridge: Cambridge University Press, 1994b.

Tasker, R. "Bangkok on the Brink." *Far Eastern Economic Review,* November 29, 1990, p. 52–3.

Tat, Hui Weng. "Singapore's Immigration Policy: An Economic Perspective." *Public Policies in Singapore: Changes in the 1980s and Future Signposts.* Ed. Linda Low and Toh Shun Heng. Singapore: Times Academic Press, 1992, p. 170–93.

Taylor, Keith W. *The Birth of Vietnam.* Berkeley: The University of California Press, 1983.

Taylor, Keith W. "The Early Kingdoms." *The Cambridge History of Southeast Asia. Volume One: From Early Times to c. 1800.* Ed. Nicholas Tarling. Cambridge: Cambridge University Press, 1994, p. 137–82.

Taylor, Robert H. "Disaster or Relief?: J.S. Furnivall and the Bankruptcy of Burma." *Modern Asian Studies,* vol. 29, February 1995, p. 45–63.

Taylor, Robert H. *The State in Burma.* London: C. Hurst & Co., 1987, p. 373–86.

Thailand Ministry of Agriculture and Cooperatives. *Agricultural Statistics of Thailand.* Bangkok: Center for Agricultural Statistics, various years.

Thailand Development Research Institute. *Thailand Natural Resource Profile.* Bangkok: TDRI, 1987.

Thailand Development Research Institute. "National Urban Development Policy Framework." *TDRI Quarterly Review,* vol. 6, no. 4, 1991, p. 25–30.

Thailand, National Statistics Office. *Statistical Yearbook of Thailand.* Bangkok: National Statistical Office, various years.

Thailand, Office of the Prime Minister. *Report of the 1988 Socioeconomic Survey.* Bangkok: National Statistical Office, 1988.

Thailand, Office of the Prime Minister. *Report of the 1992 Migration Survey*. Bangkok: National Statistical Office, 1994.

"Thailand's Drought Crisis." *TDRI Quarterly Review*, vol. 9, no. 12, March 1994, p. 28–9.

Thanh-Dam, Truong. *Sex, Money, and Morality: Prostitution and Tourism in Southeast Asia.* London: Zed Books, 1990.

Thee, Kian Wie. *see* Wie, Thee Kian.

Thomaz, Luis Felipe Ferreira Reis. "The Malay Sultanate of Melaka." *Southeast Asia in the Early Modern Era: Trade, Power, and Belief.* Ed. Anthony Reid. Ithaca and London: Cornell University Press, 1993, p. 69–90.

Thorbek, S. *Voices from the City: Women of Bangkok.* London: Zed Books, 1987.

Thornton, E. "Day of Reckoning." *Far Eastern Economic Review*, May 11, 1995, p. 62–4.

Thrift, N. "Vietnam: Geography of a Socialist Siege Economy." *Geography*, vol. 72, no. 317, 1987, p. 340–44.

Thrift, N., and D. Forbes. "Cities, Socialism and War: Hanoi, Saigon, and the Vietnamese Experience of Urbanisation." *Society and Space*, vol. 3, 1985, p. 279–308.

Thrift, N., and D. Forbes. *The Price Of War: Urbanization in Vietnam 1954–1985*. London: Allen and Unwin, Croom Helm, 1986.

Tiglao, Rigoberto. "The Fire Next Time: Has Manila Let the Moro Movement Go Too Far?" *Far Eastern Economic Review*, vol. 159, no. 13, March 28, 1996a, p. 26–9.

Tiglao, Rigoberto. "Newborn Tiger." *Far Eastern Economic Review,* vol. 159, no. 43, October 24, 1996b, p. 44.

Tiglao, Rigoberto. "The Other Philippines: Cebu's Can-Do Business Climate is Alternative Magnet to Manila." *Far Eastern Economic Review,* November 28, 1991, p. 60–2.

Tiglao, Rigoberto. "Progress on Parade: Ramos Takes a Military Approach to Problem-Solving." *Far Eastern Economic Review*, July 6, 1995, p. 36–8.

Tiglao, Rigoberto. "Welcome Exchange." *Far Eastern Economic Review*, February 29, 1996c, p. 24–5.

Tinker, Hugh. *The Union of Burma, A Study of the First Years of Independence,* 2nd ed. London, New York: Oxford University Press, 1959.

Tomich, T.P. "Survey of Recent Developments." *Bulletin of Indonesia Economic Studies*, vol. 28, no. 3, 1992, p. 3–39.

Tregonning, K.G. *A History of Modern Malaya.* London: University of London Press, 1964.

Truong, David H.D., and Carolyn L. Gates. *Effects of Government Policies on the Incentive to Invest, Enterprise Behaviour and Employment: The Case Study of Vietnam.* working paper no. 57, World Employment Programme Research, International Labour Organization, 1992.

Truong, Thanh-dam. *Sex, Money and Morality: Prostitution and Tourism in Southeast Asia*. London: Zed, 1990.

Tsuruoka, D. "Malaysia: Look East—And Up." *Japan in Asia.* Hong Kong: Review Publishing Company, 1992, p. 119–35.

Tu, Wei-ming. *Confucian Ethics Today: The Singapore Challenge*. Singapore: Curriculum Development Institute of Singapore, Federal Publications, 1984.

Tuan, Hoang Anh. "Vietnam's Membership in ASEAN: Economic, Political and Security Implications." *Contemporary Southeast Asia,* vol. 16, no. 3, 1994, p. 259–73.

Turnbull, C. Mary. *A Short History of Malaysia, Singapore and Brunei.* Singapore: Graham Brash, 1980.

Tyler, Charles. "Baby Power: Population Policy in Malaysia and Singapore." *Geographical Magazine,* vol. 64, no. 1, 1992, p. 16–21.

Tyner, James A. "The Social Construction of Gendered Migration From the Philippines." *Asian and Pacific Migration Review,* vol. 3, no. 4, 1994, p. 589–617.

Ulack, R. "Final Summary Report: Minority Groups in Southeast Asia: Past, Present, and Potential Conflicts." Unpublished report for Research Corporation of the University of Hawaii, May 1998.

Ulack, R. "Migration to Mindanao: Population Growth in the Final Stage of a Pioneer Frontier." *Tijdschrift voor Economische en Social Geografie,* vol. 68, no. 3, 1977, p. 133–44.

Ulack, R. "The Role of Urban Squatter Settlements," *Annals of the Association of American Geographers,* Vol. 68, No. 4 (December 1978), p. 535–50.

Ulack, R., and Gyula Pauer. *Atlas of Southeast Asia.* New York: Macmillan, 1989.

Ulack, R., and T.R. Leinbach. "Migration and Employment in Urban Southeast Asia: Examples from Indonesia and the Philippines." *National Geographic Research,* vol. 1, no. 3, 1985, p. 310–31.

United Nations. *A Demographic Perspective on Women in Development in Cambodia, Lao People's Democratic Republic, Myanmar, and Vietnam.* New York, 1998.

United Nations. *Internal Migration of Women in Developing Countries.* New York: United Nations, 1993a.

United Nations. *Patterns of Urban and Rural Population Growth.* New York: United Nations, 1980.

United Nations. *Population Growth and Policies in Mega-Cities: Bangkok.* Department of International Economic and Social Affairs, Population Policy Paper no. 10. New York: United Nations, 1987.

United Nations. *Population Growth and Policies in Mega-Cities: Jakarta.* Department of International Economic and Social Affairs, Population Policy Paper no. 18. New York: United Nations, 1989.

United Nations. *Population Growth and Policies in Mega-Cities: Metro Manila.* Department of International Economic and Social Affairs, Population Policy Paper no. 5. New York: United Nations: 1986.

United Nations. *Production Yearbook.* vol. 46. Rome: FAO, 1993b.

United Nations. *The Sex and Age Distribution of the World Populations—The 1994 Revision.* New York: United Nations, 1994a.

United Nations. *Statistical Yearbook.* New York: United Nations, 1985, 1993, 1994, and 1995 editions.

United Nations. *Statistical Yearbook for Asia and the Pacific 1995.* Bangkok: United Nations Economic and Social Commission for Asia and the Pacific, Statistics Division, 1996.

United Nations. *World Investment Directory. Volume I: Asia and the Pacific.* New York: United Nations, 1992.

United Nations. *World Population Trends and Policies.* New York: United Nations, 1979.

United Nations. *World Urbanization Prospects: The 1994 Revision Annex Tables.* New York: United Nations, 1994b.

United Nations. *The World's Women: Trends and Statistics.* New York: United Nations, 1995a.

United Nations Department for Economic and Social Information and Policy Analysis. *World Population 1994.* New York: United Nations Department for Economic and Social Information and Policy Analysis, Population Division, 1994.

United Nations Development Programme. *Human Development Report.* New York: Oxford University Press, 1993, 1994, and 1995 editions.

United Nations Economic and Social Commission for Asia and the Pacific (UNESCAP). *ESCAP Population Data Sheet,* various years. Bangkok: ESCAP, Population Division, 1983, 1993a, 1997.

United Nations Economic and Social Commission for Asia and the Pacific (UNESCAP). *Population of Thailand.* Country Monograph Series no. 3. Bangkok: UN ESCAP, 1970.

United Nations Economic and Social Commission for Asia and the Pacific (UNESCAP). "Research on the Status of Women: Implications for Integrated Population Policies and Programs." *Population Headliners,* no. 106 Supplement. ESCAP: Population Division, 1984.

United Nations Economic and Social Commission for Asia and the Pacific (UNESCAP). *State of Urbanization in Asia and the Pacific 1993.* New York: United Nations, 1993b.

United Nations Economic and Social Commission for Asia and the Pacific (UNESCAP). *Urbanization and Socio-Economic Development in Asia and the Pacific.* New York: UNESCAP, Asia Population Studies Series no. 122, 1993c.

United Nations Food and Agricultural Organization (UNFAO). *Production Yearbook.* Rome: Food and Agricultural Organization of the United Nations. See also UN FAO homepage, http://www.fao.org

United States Energy Information Administration. December 1998 web page for Vietnam at www.eia.doe.gov

Urban Redevelopment Authority. *Living the Next Lap.* Singapore, 1991.

Urry, John. *The Tourist Gaze: Leisure and Travel in Contemporary Society.* London: Sage Publications, 1990.

USIP (United States Institute of Peace). *The South China Sea Dispute: Prospects for Preventive Diplomacy.* Washington, D.C.: USIP, 1996.

Valencia, Mark J. "The Spratly Imbroglio in the Post-Cold War Era." *Southeast Asia in the New World Order.* Ed. D. Wurfel and B. Burton. New York: St. Martin's Press, 1995, p. 244–69.

Van Esterik, John. "Women Meditation Teachers in Thailand." *Women of Southeast Asia.* Ed. Penny Van Esterik. DeKalb: Northern Illinois University, Center for Southeast Asian Studies, 1996, p. 33–41.

Van Esterick, Penny. "Beauty and the Beast: The Cultural Context of the May Massacre Bangkok, Thailand." Thai Studies Project, Women in Develoment Consortium in Thailand. Toronto: York University, Paper Number 10, working paper Series, 1994.

Van Esterik, Penny. "Rewriting Gender and Development Anthropology in Southeast Asia." *'Male' and 'Female' in Developing Southeast Asia.* Ed. W.J. Karim. Oxford and Washington, D.C.: Berg Publishers, 1995, p. 247–59.

Vandermeer, Canute. Population Patterns on the Island of Cebu, the Philippines: 1500 to 1900. *Annals of the Association of American Geographers,* vol. 57, 1967, p. 315–37.

Vatikiotis, Michael. "Going Global, Thai Style: Dusit Thani Hotel Group makes Big Overseas Push." *Far Eastern Economic Review*, vol. 159, no. 16, April 18, 1996a, p. 77–8.

Vatikiotis, Michael. "Making of a Maverick." *Far Eastern Economic Review*, vol. 155, no. 33, 1992, p. 18–20.

Vatikiotis, Michael. "Off the Streets." *Far Eastern Economic Review,* March 29, 1990, p. 34–5.

Vatikiotis, Michael. *Political Change in Southeast Asia: Trimming the Banyan Tree.* London and New York: Routledge, 1996b.

Vatikiotis, Michael, M. Clifford, and J. McBeth. "The Lure of Asia." *Far Eastern Economic Review*, February 3, 1994, p. 32–4.

Vietnam, *Statistics Yearbook of 1994.* "Area, Population, Administrative Unit of Provinces and Cities in 1994." (Table) Vietnam: Batin Ltd. (www.batin.com.vn/niengiam/Chapter 2/21 table.htm); *Statistics Yearbook of 1996.* "Area, Population, Administrative Unit of Provinces and Cities in 1996." (www.batin.com.vn/niengiam/Chapter 11/c112table.htm)

Vitug, M.D. *The Politics of Logging: Power From the Forest.* Manila: Philippine Center for Investigative Journalism, 1993.

Vo, Nhan Tri. *Vietnam's Economic Policy Since 1975*. Singapore: Institute of Southeast Asian Studies, 1990.

Vu, Tu Lap, and C. Taillard. *An Atlas of Vietnam*. Collection Dynamiques du Territoire Series no. 13, RECLUS-La Documentation Francaise, Paris, 1994.

Wall, Geoffrey. "International Collaboration in the Search for Sustainable Tourism in Bali, Indonesia." *Journal of Sustainable Tourism,* vol. 1, no. 1, 1993, p. 38–47.

Wall, Geoffrey. "Perspectives on Tourism in Selected Balinese Villages." *Annals of Tourism Research,* vol. 23, no. 1, 1996, p. 123–37.

Ward, M. "Port Swettenham and Its Hinterland, 1900–1960." *Journal of Tropical Geography,* vol. 19, 1964, p. 69–78.

Warr, Peter. "Thailand." *East Asia in Crisis.* Ed. Ross McLeod and Ross Garnaut. London: Routledge, 1998, p. 49–65.

Wee, Vivienne. "Women and Sustainable Livelihoods." *A Commitment to the World's Women: Perspectives on Development and Beyond.* Ed. Noeleen Heyzer. New York: UNIFEM, 1995, p. 208–12.

Wee, Vivienne, and Noeleen Heyzer. *Gender, Poverty and Sustainable Development: Towards a Holistic Framework of Understanding and Action*. Singapore: ENGENDER, 1995.

Weissman, Robert. "The Politics of Economic Chaos in the Philippines." *Multinational Monitor,* vol. 15, no. 1, January/February 1994, p. 13–18.

Wernstedt, Frederick L., and Joseph E. Spencer. *The Philippine Island World: A Physical, Cultural, and Regional Geography.* Berkeley: University of California Press, 1967.

Westfall, M. *On Borrowed Land*. Videorecording. Culver City, Calif.: Amphion Productions, 1990.

Westlake, M. "Mean Streets." *Far Eastern Economic Review,* November 29, 1990, p. 56–8.

Wheatley, Paul. *City as Symbol.* Innaugural Lecture delivered at University College London, November 20, 1967. London: H.K. Lewis & Co., Ltd., 1969.

Wheatley, Paul. *The Golden Khersonese*. Kuala Lumpur: University of Malaya Press, 1961.

Wheatley, Paul. *Impressions of the Malay Peninsula in Ancient Times.* Singapore: Eastern Universities Press, Ltd., 1965.

Wheatley, Paul. *Nagara and Commandery: Origins of the Southeast Asian Urban Traditions*. Research Paper Nos. 207–208. Chicago: The University of Chicago, Department of Geography, 1983.

Wheatley, Paul. "Satyanrta in Suvarnadvipa: From Reciprocity to Redistribution in Ancient Southeast Asia." *Ancient Civilization and Trade.* Ed. Jeremy A. Sabloff and C.C. Lamberg-Karlovsky. Albuquerque: University of New Mexico Press, 1975, p. 227–83.

Wheeler, Tony. *South-East Asia on a Shoestring.* Berkeley, Calif.: Lonely Planet Publications, 1985.

White, Christine Pelzer. "Socialist Transformation of Agriculture and Gender Relations: The Vietnamese Case." *Sociology of "Developing Societies" Southeast Asia.* Ed. John G. Taylor and Andrew Turton. New York: Monthly Review Press, 1988, p. 165–77.

Wiboonchutikula, Paitoon. *ASEAN and EC: Technological Trends and Their Impact on ASEAN Industries.* Singapore: Institute of Southeast Asian Studies, ASEAN Economic Research Unit, 1990.

Wickberg, Edgar. *The Chinese in Philippine Life, 1850–1898.* New Haven and London: Yale University Press, 1965.

Wie, Thee Kian. "The Surge of Asian NIC Investment into Indonesia." *Bulletin of Indonesian Economic Studies,* vol. 27, no. 3, 1991, p. 55–88.

Wilbanks, T.J. "Sustainable Development in Geographic Perspective." *Annals of the Association of American Geographers,* vol. 84, no. 4, 1994, p. 541–56.

Wilkinson, Paul, and Wiwik Pratiwi. "Gender and Tourism in an Indonesian Village." *Annals of Tourism Research*, vol. 22, 1995, p. 283–99.

Williamson, Peter A.T. "Tourist Developers on Koh Samui, Thailand." *Journal of Cultural Geography,* vol. 12, no. 2, Spring/Summer 1992, p. 53–64.

Win, K., and K. Win. *Myanmar's Experience in Rice Improvement, 1830–1985.* IRRI Research Paper Series no. 141. Los Banos, Philippines: International Rice Research Institute, 1990.

Winichakul, Thongchai. *Siam Mapped: A History of the Geo-body of a Nation*. Honolulu: University of Hawaii Press, 1994.

Withington, W.A. "Rural-Urban Change." *Southeast Asia: Realm of Contrasts*. Ed. A. Dutt. Boulder, Colo.: Westview, 1985, p. 91–101.

Wolf, Diane. *Factory Daughters: Gender, Household Dynamics, and Rural Industrialization in Java*. Berkeley, Los Angeles, Oxford: University of California Press, 1992.

Wolters, O.W. *Early Indonesian Commerce: A Study of the Origins of Srivijaya.* Ithaca: Cornell University Press, 1967.

Wolters, O.W. *The Fall of Srivijaya in Malay History.* Ithaca: Cornell University Press, 1970.

Wolters, O.W. *History, Culture and Region in Southeast Asian Perspectives*. Singapore: Institute of Southeast Asian Studies, 1982.

Wolters, O.W. "Khmer 'Hinduism' in the Seventh Century." *Early South East Asia: Essays in Archaeology, History and Historical Geography.* Ed. R.B. Smith and W. Watson. London: Oxford University Press, 1979, p. 427–42.

Wong, Jill. *Thailand: Prospects for the Export Sector,* ASEAN Briefing no. 25, DBS Bank, Economic Research Department, November 1993.

Wong, John. *ASEAN Economies in Perspective: A Comparative Study of Indonesia, Malaysia, the Philippines, and Thailand.* London: Macmillan, 1980.

Wong, K.C., et al. *Report of the Tourism Task Force.* Singapore: Ministry of Trade and Tourism, 1984.

Wong, Lin Ken. *see* Ken, Wong Lin.

Wood, Robert E. "Ethnic Tourism, the State, and Cultural Change in Southeast Asia." *Annals of Tourism Research,* vol. 11, 1984, p. 353–74.

Wood, Robert E. "International Tourism and Cultural Change in S.E. Asia." *Economic Development and Cultural Change*, vol. 3, 1980, p. 561–81.

Wood, Robert E. "Tourism, Culture, and the Sociology of Development." *Tourism in South-East Asia.* Ed. Michael L. Hitchcock, Victor T. King, and Michael J.G. Parnwell. London: Routledge, 1993, p. 48–70.

World Bank. "Development and the Environment." *World Development Report.* Washington, D.C.: World Bank, 1992.

World Bank. *The East Asian Miracle: Economic Growth and Public Policy*. New York: Oxford University Press, 1993.

World Bank. "Economic and Social Development: An Overview of Regional Differentials and Related Processes, Main Report." *Indonesia: Selected Aspects of Spatial Development*, Report no. 4776-IND. World Bank: East Asia and Pacific Regional Office, Country Programs Department, 1984.

World Bank. *Implementing the World Bank's Gender Policies*. Progress Report no. 1. Washington, D.C.: World Bank, March 1996. Also see http://www.worldbank.org/html/hcovp.gender.contents.html

World Bank. *Indonesian Poverty Assessment and Strategy Report.* Report no. 8034-IND. Washington, D.C.: World Bank, 1991a.

World Bank. *A Public Development Program for Thailand*. Baltimore: Johns Hopkins University Press, 1959.

World Bank. *Rural Sector Policy Paper.* Washington, D.C.: 1975.

World Bank. *Trends in Developing Economies*. Washington, D.C.: World Bank, 1995a.

World Bank. *Urban Policy and Economic Development: An Agenda for the 1990s.* Washington, D.C.: World Bank, 1991b.

World Bank. *Vietnam: Poverty Assessment and Strategy*. Report no. 13442-VN. Washington, 1995b.

World Bank. *World Development Report,* various years. New York: Oxford University Press, 1983, 1989, 1994, 1995, 1996, 1997, 1998, and 1999.

World Bank. *World Tables, 1988–89 Edition.* Baltimore: Johns Hopkins University Press, 1989.

World Resources Institute. *World Resources 1994–1995: A Guide to the Global Environment; World Resources 1996–1997: A Guide to the Global Environment;* and *World Resources 1998–1999: A Guide to the Global Environment.* New York: Oxford University Press, 1994, 1996, and 1998.

World Tourism Organization. *Current Travel and Tourism Indicators.* Madrid: WTO, 1991.

World Tourism Organization. *Tourism Development Report,* 1st ed. Madrid: WTO.

World Tourism Organization. *Yearbook of Tourism Statistics,* vols. 1& 2, 46th ed., Madrid: WTO, 1986; 1994.

"The World's Richest Economies." *The Straits Times Weekly Edition,* March 4, 1995, p. 14.

Wright, David K. *Brunei.* Chicago: Children's Press Chicago, 1991, p. 85 and 111.

Wyatt, David K. *Thailand: A Short History*. New Haven: Yale University Press, 1984.

Yap, Mui Teng. "Population Policy." *Public Policies in Singapore: Changes in the 1980s and Future Signposts.* Ed. Linda Low and Toh Shun Heng. Singapore: Times Academic Press, 1992, p. 127–43.

Yap, Mui Teng. Singapore's Three or More Policy: The First Five Years. *Asia Pacific Population Journal,* vol. 10, no. 4, 1995, p. 39–52.

Yeung, Y.M. "Great Cities of Eastern Asia." *The Metropolis Era, Volume 1: A World of Giant Cities.* Ed. M. Dogan and J.D. Kasarda. Newbury Park, Calif.: Sage, 1988, p. 154–86.

Yeung, Y.M. "The Housing Problem in Urbanizing Southeast Asia." *Urban Society in Southeast Asia*. Ed. G.H. Krausse. Hong Kong: Asian Research Service, 1985, p. 43–66.

Yoshihara, K. *The Rise of Ersatz Capitalism in Southeast Asia*. Oxford: Oxford University Press, 1988.

"Yuman Aims to Be Asia's New Entrepot: Links with Burma and Laos." *Far Eastern Economic Review*, vol. 160, no. 37, 1997, p. 54.

Zablan, A. *Vital Issues of the Urban Poor*. Quezon City: Ateneo de Manila University, 1990.

Zasloff, J.J., and L. Unger, eds. *Laos: Beyond the Revolution*, Macmillan, London, 1991.

Index

Graphic materials in this index are indicated as follows: page numbers for maps are followed by an **m**; page numbers for tables are followed by a **t**; page numbers for photographs and illustrations are followed by an **i**.